THE GEOLOGY OF EGYPT

THE GEOLOGY OF EGYPT

Edited by

RUSHDI SAID

*Consulting Geologist, Ph.D. (Harvard), Dr. rer. nat. h.c. (Technical University, Berlin),
Membre Institut d'Egypte, Honorary Fellow Geological Society of America, Honorary
Member Geological Society of Africa*

*Published for the Egyptian General Petroleum Corporation, Conoco Hurghada Inc.
and Repsol Exploracion, S.A. by*

A.A.BALKEMA / ROTTERDAM / BROOKFIELD / 1990

Dedicated to
Coy H. Squyres
who initiated this work

Published by
A.A.Balkema, P.O.Box 1675, 3000 BR Rotterdam, Netherlands
A.A.Balkema Publishers, Old Post Road, Brookfield, VT 05036, USA
ISBN 90 6191 856 1
© 1990 A.A.Balkema, Rotterdam
Printed in the Netherlands

Contents

PART 6: PALEONTOLOGICAL NOTES

PART 7: ANNEXES

Contents

Introduction

This book is an attempt to review and synthesize the large amount of data which has accumulated since the publication of my book *The Geology of Egypt* (Elsevier, 1962). It consists of 32 chapters organised in six parts and appended by a list of references and the indexes. The chapters are written by scholars whose names and addresses appear in the list of contributors p. 739. The past twenty-five years have been active years of geological research in Egypt during which time the desert became more easily accessible and the application of new methods to probe the earth and its constituents and to store and retrieve data became commonplace.

A book written by various authors will include varying views regarding the same subject. A reconcilliation of the different interpretations advanced by the authors has been attempted. However, where such reconciliation was not possible, the differences have been allowed to remain in preference to forcing the varying interpretations into a single overview.

Part 1 gives a broad outline of the geomorphology and tectonics of Egypt. It also includes chapters on the application of remote sensing in the preparation of the 1:500,000 new geological map, an explanatory note on the new gravity map and notes on the seismicity and geophysical regime of Egypt. The tectonic evolution is treated in chapters dealing with tectonic framework, vulcanicity and the position of Egypt in the frame of global tectonics. A plate tectonic model is developed where the evolution of Egypt is tied to global events.

Part 2 deals with the basement complex and includes Chapters 10 to 12. Chapters 10 and 11 deal with the Precambrian rocks of the Eastern Desert and Sinai and with the little-known basement outcrops of the south Western Desert respectively. Chapter 12 reviews the literature and describes in detail the rock units of the Precambrian of Egypt.

Part 3 deals with the geology of selected areas chosen because they had been the subject of intensive studies during the past years. Part 4 reconstructs the geological history of Egypt during the Paleozoic, Mesozoic, Cenozoic and Quaternary. A series of paleogeographic maps illustrates this part.

Part 5 examines the economic mineral deposits, oil potential and the groundwater reservoir of the Western Desert. Part 6 concludes the book with paleontological notes on fossil groups some of which had not received attention in previous studies. References used in the different chapters are assembled in one list in Part 7.

I owe special thanks to a large number of people who helped bring about the completion of this book. Foremost among these is my colleague Coy H. Squyres, Conoco, who initiated this project and maintained a keen interest in it throughout. To him this book is fondly dedicated. I am also grateful to Bob Handley and P. Close, Conoco, for the support they have given to this project. To E. Klitzsch, I owe a special debt for his generous cooperation and willingness to share the results of his work and the work of his teams in Egypt. To my colleague Maurice Hermina, I am indebted for his tireless effort in overseeing the seemingly endless work needed to prepare this book for the printer. Finally I am grateful to all the authors who have contributed to the book.

Acknowledgement is due to the following publishers and organisations for permission to reprint some of the figures appearing in this book: Allen and Unwin, London for Figures 2/5 & 2/6, The American Association of Petroleum Geologists for Figure 7/17, Blackwell Scientific Publishers for Figure 7/13, Editios Technip, Paris for Figures 2/8, 2/9 & 19/13, Journal Geodynamics for Figures 6/12, 6/13, 6/21 & 6/22, The Geological Society of America for Figures 8/12 & 22/3 and The Geological Survey of Israel for Figure 19/4. Acknowledgement is also due to AGIP, Conoco and Marathon Oil Companies for permission to use some of the material and figures included in Chapter 17 and to the Egyptian General Petroleum Corporation for allowing the use of some of the materials and figures included in the report *Petroleum Potential Evaluation, Western Desert, The Arab*

Republic of Egypt (1982) prepared for it by Robertson Research Limited in association with Associated
Resources Limited and ERC Energy Resource Consultants Limited in Chapter 15.

Rushdi Said
Washington, DC and Cairo, Egypt
January 1989

PART 1

General

CHAPTER 1

History of geological research

RUSHDI SAID
Consultant, Annandale, Virginia, USA and Cairo, Egypt

The published and unpublished literature on the geology of Egypt is extensive (for a bibliography, see Keldani 1939, Said & El-Shazly 1957, Said et al. 1975, Glenn & Denman 1980 and El-Baz 1984).

The history of geological research in Egypt may be divided into three episodes.

FIRST EPISODE

The first episode, extending from the French Expedition (1798-1801) to the establishment of the Geological Survey of Egypt in 1896, was characterized by sporadic research conducted by individual naturalists and travellers, by commissioned scientists and, in rare instances, by expeditions organised by institutes of learning in Europe. Lured by the mineral wealth of the country or its ancient civilization, these authors collected a great variety of information on the natural history of the country, and wrote several travel books which recorded their findings.

In this episode Napoleon's and Rholf's expeditions are of special importance. The first was responsible for bringing Egypt to the attention of the scientists of Europe by the publication of the memorable *Description de l'Egypte*. This work includes in many of its volumes, and especially the second volume (published in 1813), several chapters that are of interest to geologists and mineralogists and contains the first reliable map of the Eastern Desert of Egypt. The collections of the expedition, which are housed in Paris, were the subject of further study up to the end of the last century. Savigny's collection of Recent shells formed the basis of the magnificent work of Issel (1870) which still remains valid in its essence particularly in its discussion on the differentiation of the faunas of the Mediterranean and the Gulf of Suez.

Rholf's expedition (1874-1875) must be singled out for its significant contribution to the geology of Egypt. Led by Rholf and joined by such able men as Jordon and Zittel and accompanied on occasions by Ascherson, this expedition laid down the foundation of modern Egyptian geology, in fact, the modern scientific survey of the country. Ascherson published the results of his botanical observations; Jordon wrote on the geography and meteorology of the Western Desert; and Zittel wrote his notable geological contributions which were published in 1883 together with paleontological studies by such authorities as Schenck, Fuchs, Mayer-Eymr, Schwager, de la Harpe, Quaas, Wanner, de Loriol, Pratz and others. This expedition also published the first reliable geological map of the extensive deserts of Egypt to the south of the latitude of Fayum (scale 1:300,000). In his work Zittel benefitted from earlier literature, particularly from the results of Schweinfurth's studies which were used to complete the mapping of large tracts of the desert not surveyed by the expedition. This map remained the standard geological map of the country up to the publication of the Survey map of 1910. The divisions which Zittel made for the upper Cretaceous and lower Tertiary rocks of the Western Desert influenced subsequent authors. His conclusions on the distribution of Cenomanian strata in the northern part of the Eastern desert, which he had based on Schweinfurth's studies, were accepted for a long time and, in fact, formed the basis of the Survey geological map of 1910 and that of 1928.

The most prominent scientist of this episode was Georg Schweinfurth (1836-1925) who started his brilliant career in Egypt with the publication of the work describing his visit to the remote Gebel Elba region (1864) and who continued until the outbreak of World War I to produce excellent papers on the geology, geography, archeology, cartography and botany of Egypt. He became the central figure in natural history research in Egypt for more than fifty years. He supplied European scholars with large collections of rocks and fossils and kept up a fruitful correspondence with many outstanding scientists such as Beyrich, Martens, Heer and Eck. He wrote the section on the geography of Egypt in Baedecker's *Aegypten*. He was responsible for the discovery of the Wadi Araba

Carboniferous (1885), the Sad El-Kafara dam in the desert of Helwan (1885), the Cretaceous region of Abu Roash (1889) and the fossil vertebrates of Fayum (1886). His pioneer visit to Gebel Elba resulted in the monograph on the geography and botany of this remote region. Schweinfurth's wanderings off the beaten tracks of the Eastern Desert resulted in the publication of a detailed account of the central Eastern Desert with a map which was publishued in ten sheets between 1899 and 1910. Schweinfurth's capacities as organiser were notable; he founded the Société géographique d'Egypte and was the president of the Institut d'Egypte for many years.

Among the scientists whose interest was aroused by Schweinfurth was Walther, the famous German geologist. Walther's description of the Wadi Araba region (1890), the coral reefs of the Red Sea (1888), and the geomorphology of desert lands embodied in his classic *Das Gesetz der Wuestenbildung* (1900) are among the notable works of this episode.

During this episode practically every eminent European scientist of the age contributed to the studies of Egyptian geology in some form or another. Fraas crossed the desert between Qift and Quseir (1867) and published a map and a 'geognostic' profile of this area. Ehrenberg, who accompanied Minutoli's expedition to the oasis Jupiter Ammon in the twenties of the last century, reported in his classic *Mikrogeologie* on some rock collections he had made. D'Orbigny reported in his *Paleontologie franqaise* upon Lefebvre's collections in Paris. D'Archiac found Lefebvre's, Gaillardot's and Delanoue's collections in Paris of value when he wrote his famous *Histoire de Géologie* and his and Haime's *Description des Animaux fossiles du Group Nummulitique de l'Inde* (1853). Greco and Stefanini studied in detail the collections of Figari which were housed in the museum of Florence.

SECOND EPISODE

The episode which extends from the establishment of the Geological Survey of Egypt in 1896 until the revolution of 1952 firmly laid down the foundations of the regional geology of Egypt. This episode culminated in the publication of the geological map of Egypt in 1928 and Hume's *Geology of Egypt* (1925-1937).

The Geological Survey of Egypt was founded by H.G. Lyons and was staffed by competent men whose published memoirs still remain among the best and most authentic records on the geology of the surveyed regions. The history of the Geological Survey of Egypt from the date of its foundation until 1925 is given in Hume's *Geology of Egypt* and from that date until 1971 in Said (1971). Among the early members of the Survey were Beadnell, Barron, Ball, Hume and Blanckenhorn. Ferrar and Stewart joined the Survey in its early days. Fourtau, who had been active in Egyptian paleontology for years, finally joined the Survey as a paleontologist in 1911 and remained on its staff until his untimely death in 1920. Sadek joined the Survey in 1917 and gave his efforts mainly to the newly established Petroleum Research Board which was strengthened by the appointment of Madgwick and Moon in the same year. Sadek did not only distinguish himself as a brilliant scientist but was also instrumental in promoting geological studies in his executive capacities as Director of the Mines and Quarries department and later as a Cabinet Minister. Although primarily a geographer and surveyor, G.W. Murray's numerous publications attest to his thorough familiarity with the physical and human aspects of the Egyptian deserts.

Several workers outside the survey contributed to the geological study of Egypt during this period. Among these mention must be made of Barthoux, the geologist of the Suez Canal Company. He discovered the Jurassic exposures at Maghara, north Sinai, and sent fossil material to Paris where it was studied by Douvillé (1916). Barthoux's work on the *Roches ignées du Desert arabique* (1922) is a classic.

During this episode several eminent geologists visited the country. Of these mention must be made of Osborn, the famous vertebrate paleontologist, who collected with the assistance of the American Museum of Natural History, a large amount of the Fayum fossil vertebrates. Sandford and Arkell, aided by funds from the Oriental Institute, Chicago University, made their valuable study on the Egyptian Pleistocene. Stromer von Reichenbach's expeditions to the fossil vertebrate localities in the Western Desert and the upper Nile Valley resulted in a large number of classical publications on the subject.

From time to time the Geological Survey sent materials to be examined abroad. These made the subject of valuable papers. Andrews gave the descriptions of the Fayum fossil vertebrates; Bullen Newton identified many of the fossil invertebrates. Seward examined numerous fossil plants, and Chapman identified many foraminiferal species.

Although this episode of geological research resulted in the discovery of a large number of economically viable mineral deposits, there occurred a lull in geological activity in the latter part of this episode. The Geological Survey was relegated to a subsidiary position. Most of the contributions of these later years came from the universities which had been established during these years. Other contributions came from the oil sector which had seen an expansion in the thirties as a result of the promulgation of a new mining law encouraging oil exploration work. Almost the whole of the sedimentary area to the north of latitude 28°N was mapped to scales varying from 1:500,000 to 1:25,000 and occasionally to smaller scales. The area was completely covered by gravity survey and large parts were covered by magnetic and seismic surveys. A total footage of about 540,000 was drilled during this period. Although much of the information gathered by these oil companies was never synthesized in one major work, as had happened with the material of the earlier episode by Bowman (1925, 1926 and 1931), a great deal of the information filtered through the literature either in the form of papers or was abridged and included in relevant contributions.

THIRD EPISODE

The episode which extends from 1952 to the present was one of great expansion. Nineteen sixty two saw the publication of Said's treatise on *The Geology of Egypt* (1962) in which an attempt was made to find some order in the large amount of information that had accumulated over the previous years by fitting it into a conceptual framework.

Between 1954 and 1976, the Geological Survey of Egypt conducted an aggressive program of exploration for economic mineral deposits which were sought for the fulfillment of the five-year industrialization plan of the country. The survey underwent great changes during this period. The scale and urgency of the operation forced an enormous expansion. The number of scientists engaged was unprecedented, and new methods were introduced including geophysical and geochemical surveying as well as drilling and mining techniques. After the period of expansion, the Survey was reorganised in 1968 to become the central institute for applied geological research. The library was enlarged to file and keep all documents related to the applied earth sciences in Egypt.

The program of mineral research comprised three successive phases. The first was the re-examination of older mineral records. The second was the systematic study of the mineral deposits which could be used as raw materials for the industries that were to make use of the hydro-electric power expected to be generated from the Aswan High Dam which was then being built. The third was the carrying out of a detailed program to study the potential of part of the basement complex of the Eastern Desert of Egypt. The three phases spanned the years from 1954 to 1974.

The *first phase* resulted in the development of the Aswan iron ores which were to be used by a small iron and steel plant that was built in Helwan to the south of Cairo. Further work brought to light the potential of the Bahariya iron ores. These and the then recently discovered coal deposits at Gebel Maghara in Sinai were thoroughly studied during the second phase. The Bahariya ore was then developed and became the main source for the expanded iron and steel industry in Helwan.

The *second phase* comprised a survey of the raw materials which were to be used in the various metallurgical industries planned: the nepheline syenites of the ring complexes of the south Eastern Desert as a possible raw material for the aluminum industry which was later built in Naga Hammadi along the Nile in upper Egypt but which did not make use of this source, the phosphate deposits of the Nile Valley in south Egypt as a raw material for the phosphorous complex which never materialized, and the limestones along the Nile Valley as a flux material. The results of these and other studies are embodied in a report (Moharram et al. 1970). The phosphate project resulted also in the preparation of a geological map of the Idfu-Qena region (scale 1:200,000, 1968).

During the *third phase*, systematic work on the basement complex of the central Eastern Desert between latitudes 24° 30' and 25° 30' N was carried out. The area was mapped to scale 1:40,000 and close to 56,000 geochemical and panning samples were analysed. The work led to the discovery of the niobium-tantalum deposit of Nuweibi and Abu Dabbab and brought to light the great potential of the Barramiya gold field. A summary of the results of this phase is published in the Annals of the Geological Survey of Egypt (1976).

The period between 1974 and present was characterized by the diversification and increase of activity of foreign scientists and institutions and by an upsurge of oil exploration work by international companies. The oil sector, following its long established tradition, conducted more than 100 agreements with more than forty companies. Exploration drilling reached about five million feet, almost double the footage drilled in all of the previous episodes. In 1982 alone ninety wells were drilled totalling about 300,000 m. Gravity and magnetic surveys continued vigorously during this period. In 1974 alone a total of 17,000 km gravity survey in the Mediterranean, Gulf of Suez and Red Sea areas was carried out. Marine and aeromagnetic surveys covered large areas of the land of Egypt and its offshore areas (see Meshref, Chapter 10, this book). Seismic reflection methods were used extensively during this period. More than fifty oil and gas discoveries were made in the Gulf of Suez, Nile delta and the Western Desert. Reserves increased to 2.5 billion barrels of oil and a tremendous amount of gas. For a review of the history and activities of oil exploration work in Egypt, the reader is referred to the two books published by the Egyptian General Petroleum Corporation (1986) on the occasion of the hundredth anniversary of the drilling of the first oil well in Gemsa in 1886.

The mining and geological research sectors had been, since 1952, a domain reserved for national companies and institutions which sought outside help only within the framework of national policies and upon the initiative of these institutions. However, since 1976, many academic and applied institutions seeking joint or independent work in Egypt have overwhelmed the two sectors, and although it is too early to evaluate the results of these endeavours, it is clear that a great deal has been published. Of the foreign programs of this period Special Project Arid Areas conducted by various German institutes may be singled out for the systematic work it has carried out in the south Western Desert of Egypt resulting in the mapping of large tracts of this desert and the subdivision of the seemingly monotonous 'Nubian Sandstone' into mappable units (Klitzsch et al. 1979). Part of the results of this work are published in the Berliner Geowissenschaftliche Abhandlungen (1984, 1987) and numerous other publications. With the support of Conoco Oil Company and its partners, the German team published a series of eleven maps (scale 1:500,000) for the south Western Desert (Klitzsch & List 1980) and fifteen sheets of the Gulf of Suez at the scale of 1:10,000 (Klitzsch & Linke 1983). These were later revised and used in the raising of a new geological map covering the entirety of Egypt (see List et al., Chapter 3, this book).

Also worthy of mention is the joint expedition of the Geological Survey of Egypt, Southern Methodist University and Polish Academy of Science which pioneered in Quaternary and prehistoric research. Part of the results of this expedition is published in Wendorf & Schild (1976, 1980) and Said (1981). Scattered and occasionally meaningful papers, touching on various aspects of Egyptian geology, were published as a result of the work of many other expeditions. Interest in reclamation projects and mineral development brought into Egypt a flow of experts whose reports and feasibility studies touched upon the geology of certain areas. In most cases the material drew upon previous information and usually lacked substance.

The Geological Survey continued its efforts to publish the geological map of Egypt, a preliminary edition of which appeared in 1971 (scale 1:2,000,000) using no colors but lines and symbols only (Said 1971). This was followed by the publication of the basement complex (scale 1:100,000) of the Eastern and Western Deserts (El Ramly 1972). Both

maps followed a period of an active mapping program of which the following works are singled out for their significant contribution: Abdallah & Adindany (1963) on the west coast of the Gulf of Suez and Issawi (1969, 1971) on the Lybian Plateau of the south Western Desert. A greatly improved, updated and colored version of the 1971 1:2,000,000 map was published in 1981. Geological maps of the Aswan and Qena quadrangles (scale 1:500,000) were published in 1978. Three sheets of the 1:250,000 series, the Dakhla, Wadi Qena and Gebel El 'Urf quadrangles were issued in 1983.

From 1967 until 1984, Sinai fell under Israeli occupation and made the subject of numerous studies by Israeli scientists especially after the 1973 war. Close to 100 publications were made during the period of occupation (a collection of the reprints concerning Sinai was published by the Geological Survey of Israel, 1980-1984). A photo-geological interpretation map to the scale of 1:500,000 was published (Eyal et al. 1980) and close to sixteen wells were drilled after the search for oil in north Sinai. Most of the works endeavoured to integrate the geology of Sinai with that of Israel; the formational names of Israel were liberally used to designate the Sinai rocks.

The seas around Egypt have been the subject of intensive studies during recent years. Although the history of the investigation of the Red Sea and the East Mediterranean goes back to the late 19th century, the mid-sixties of this century witnessed the true systematic exploration work in both seas. The discovery of the basaltic nature of the axial rift of the Red Sea led to the conclusion that the sea opened by a process of oceanic accretion. The equally important discovery of metalliferous sediments covered by hot brines at the bottom of depressions or deeps usually situated along the rift axis of the sea aroused great interest and led to active oceanographic work by the Saudi-Sudanese Red Sea Commission (six expeditions between 1965 and 1971). International interest in the sea led to the launching of numerous expeditions: the USSR (two expeditions in 1976 and 1980), the USA (six expeditions between 1962 and 1972), Germany (two expeditions in 1965 and 1971), England (four expeditions between 1963 and 1967) and France (two expeditions in 1979 and 1981). A short summary of the results of these expeditions is found in Degens & Ross (1969), Stoffers & Ross (1974) and Thisse et al. (1983). Intensive oceanographic work in the East Mediterranean led to many interesting results. The discovery of a salt layer beneath the bottom of the sea led to the conclusion that this sea must have dried up during the Messinian. Geophysical work indicated that the African continent was being subducted underneath Europe. The results of the Deep

Sea Drilling Project in the Eastern Mediterranean are given in Ryan & Hsu (eds) (1973) and Hsu & Montadert (eds) (1977). The results of these oceanographic studies have had a great impact on our understanding of the evolution of Egypt's landscape and the development of the Nile.

CHAPTER 2

Geomorphology

RUSHDI SAID

Consultant, Annandale, Virginia, USA and Cairo, Egypt

Egypt forms the northeast corner of Africa and occupies nearly one-thirtieth of the total area of that continent. Bounded to the north by the Mediterranean Sea, to the south by the Republic of the Sudan, to the west by the Republic of Libya and to the east by Palestine, Israel, the Gulf of Aqaba and the Red Sea, it measures 1,073 km in greatest length from north to south, 1,226 km in greatest breadth from west to east and embraces a total area of almost one million square kilometers.

Situated between latitudes 22° and 32° N, Egypt lies for the most part in the temperate zone, with less than a quarter of its area south of the tropic of Cancer. The whole country forms part of the great desert belt that stretches eastward from the Atlantic across the whole of north Africa onward through Arabia; and like all other lands lying within this belt, it is characterized by a warm and almost rainless climate. The air temperature in Egypt frequently rises to over 40°C in the daytime during the summer and seldom falls as low as 0°C even during the coldest nights of winter. The average rainfall over the country as a whole is only about 1 cm a year. Even along the Mediterranean littoral, where most of the rain occurs, the average yearly precipitation is less than 20 cm and the amount decreases very rapidly as one proceeds inland.

The scanty rainfall of Egypt accounts for the fact that the greater part of Egypt consists of barren and desolate desert. It is only through the River Nile that a regular and voluminous supply of water, coming from the highlands lying far to the south, is secured. This water is channelled by artificial canals over the narrow strip of alluvial land on both sides of the river, the Fayum depression and the delta expanse. These tracts of fertile land, covering less than 3% of the total area of Egypt, support a dense population. The average density of population in the habitable part of Egypt is more than 1,500 persons/km^2 while there is only one inhabitant/6 km^2 in the vast desert areas.

The River Nile, therefore, represents a salient geographical feature that has shaped not only the physical traits of the country but also its history and the nature of its human settlements. The River Nile has given Egypt a strip of fertile land which has made possible not only the development of its famed ancient agricultural civilizatrion but also the growth of this civilization in peace and stability. Flowing midway through the country, the Nile so divides the land that there is no place in Egypt that is further than 300 km from the river. This natural occurrence made it possible for a sedentary society and a central authority to develop along the banks of that river.

The Nile, a perennial stream which makes its way over thousands of kilometers of desert, divides Egypt into two distinct morphological regions. The region to the east consists of a dissected plateau draining to the river, while the region to the west consists of a series of unconnected depressions. Different conditions affected the land on either side of the Nile, though the river is not everywhere the boundary between the two regions. The table-land between Kharga Oasis and the Nile is in continuation with the Maaza plateau of the Eastern Desert, and Wadi Gabgaba usurped the function of the Nile by intercepting the wadis draining westward from the Red Sea hills. Although the land to the east of the Nile forms one geomorphological region, it is divided geographically into the Eastern Desert and the Peninsula of Sinai separated by the Gulf of Suez. From the earliest of times, these geographical divisions were recognized not only because of the physical differences of the two areas but also because they were inhabited by different races. Herodotus may have intended to make this racial distinction when he named the deserts east and west of the Nile 'Arabian' and 'Libyan' respectively.

It has been the custom, therefore, to deal with Egypt as divisible geographically into the following major provinces:
1. Nile Valley and the Delta
2. Western Desert
3. Eastern Desert
4. Sinai Peninsula.
In addition to these geographical provinces of the

land of Egypt, the bordering Mediterranean and Red Seas are of special importance in understanding the geomorphological development of Egypt.

In the following paragraphs, some of the main geographical features of these provinces and the seas are given.

NILE VALLEY AND DELTA

The Nile Valley and the Delta occupy the alluvial tract along the terminal 1,350 km of the River Nile. These lie within the borders of Egypt. Along this course no tributary joins the Nile. After entering Egypt at Wadi Halfa it passes for more than 300 km through a narrow valley surrounded by cliffs of sandstone and granite on both its east and west sides until it reaches the First Cataract which commences about 7 km south of Aswan. The construction of the Aswan Dam at the beginning of this century inundated the strips of cultivable land along this stretch, while the construction of the Aswan High Dam rendered large tracts of the Nubian desert into a vast reservoir of water. This reservoir forms one of the greatest man-made lakes. It extends (at its 180 m asl) for almost 4½° of latitude from Aswan to the Dal Cataract in the Sudan. The total reservoir at this level has a surface area of 6,216 km^2, a mean width of 12.5 km and the length of its dendritic shorelines is 9,250 km. The volume of water held by this reservoir is 157 km^3.

The natural gradient of the river in Nubia (1 m/11 km) only slightly exceeds the slope of the river in the remaining 1,100 km of its course to the sea. North of Aswan, the Nile Valley broadens and the flat strips of cultivable land, extending between the river and the cliffs that bound its valley on either side, gradually increase in width northward. Near Esna, about 160 km north of Aswan, the sandstone of the bounding cliffs gives place to limestones; and at Qena, about 120 km north of Esna, the river makes a great bend bounded by limestone cliffs rising to heights of more than 300 m. Near Assiut, about 260 km north of Qena, the cliffs of the western side of the valley become much lower than those on the eastern side and continue so for more than 400 km to Cairo where the valley opens out to the delta. The average width of the alluvial floor of the Nile valley between Aswan and Cairo is about 10 km and that of the river itself about three-quarters of a kilometer. Throughout its entire course, the Nile tends to occupy the eastern side of its valley so that the cultivable lands to the west of the river are generally much wider than those to the east. In fact, in some places, the stream almost washes the eastern boundary cliffs.

After passing Cairo, the Nile pursues a northwest-erly direction for about 20 km and then divides into two branches, each of which meanders separately through the delta to the sea. The western branch (239 km in length) debouches into the Mediterranean at Rosetta, and the eastern branch, which is about 6 km longer, at Damietta.

Closely connected with the River Nile is the Fayum depression which lies at a little distance to the west of the Nile Valley and with which it is connected by a narrow channel through the desert hills. The lowest part of the depression is occupied by a shallow brackish lake (Birket Qarun) which is about 45 m below sea level and about 200 km^2 in area. The depression has a total area of about 1,700 km^2. Its floor slopes downward to the lake in a northwesterly direction from a level of about 32 m asl. The depression is a rich alluvial land irrigated by a canal that enters it from the Nile by way of the above-mentioned channel.

WESTERN DESERT

The Western Desert stretches westward from the Nile Valley to the borders of Libya and embraces an area, exclusive of Fayum, of about 681,000 km^2, that is, more than two-thirds of the whole area of Egypt. The Western Desert (Fig.2/1) is essentially a plateau desert with vast expanses of rocky ground and numerous extensive and closed-in depressions. It attains its greatest altitude in the extreme southwestern corner of the country where its general plateau character is disturbed by the great mountain mass of Gebel Uweinat lying just outside Egypt; but the northeastern flanks of the mountain are within the borders of Egypt.

Three subprovinces can be distinguished in this vast desert:

1 *Southern subprovince (Arba'in Desert)*

This subprovince lies in the shadow of the northern embayed escarpment which borders the Dakhla and Kharga depressions and extends southward into northern Sudan. This subprovince is named the Arba-'in desert by Haynes (1982) after the well-known Darb El-Arba'in caravan route which traverses this subprovince in its middle. Rainfall over this subprovince is negligible and individual storms at any particular place may be decades apart (Bagnold 1954). In the areas of Dakhla and Kharga oases, daily average wind velocities range between 4.3 and 18.5 km/h throughout the year, and the daily average relative humidity is between 28 and 56% (Ezzat 1974). In addition to having unique climatic conditions, it is singled out from other Egyptian geomorphic prov-

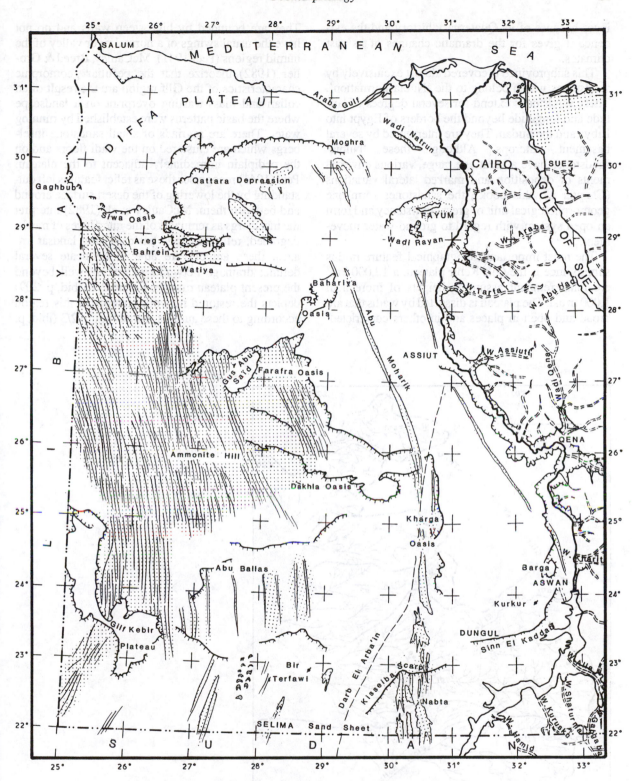

Figure 2.1 Location map of Western Desert showing major topographic features, scarps and dune belts.

inces because of its Quaternary history and the evidence it gives for the dramatic changes of its past climates.

This subprovince is covered almost exclusively by sandstones which belong to the 'Nubia Formation'. These sandstones extend for several degrees of latitude and longitude beyond the borders of Egypt into Libya and the Sudan. They are interrupted by several basement outcrops. Although these 'Nubia' sandstones have a great age range, various environments of deposition and marked lateral variations (Chapter 12, this book), the sandstones form one geomorphological unit of great consistency and form an open system with regard to ground water movement.

The most important topographical feature in this subprovince is the Gilf Kebir plateau, a 12,000 km^2 erosional remnant assuming heights of more than 1,000 m asl. The plateau is dissected by wadis that are broad and open in places and in others constricted.

They are bounded by high steep walls and do not have the usual swings of a normal river valley of the humid regions (Peel 1941). McCauley, Breed & Grolier (1982) theorize that the peculiar geomorphic characteristics of the Gilf region are the result of an eolian and mass-wasting overprint on a landscape where the basic patterns were established by running water. There are myriads of small sandstone inselbergs which are clustered on the wadi floors and on the pediplain immediately adjacent to the plateau. Peel (1941) recognises these as relict features left outstanding by the lowering of the desert surface around and between them. McCauley et al. (1982) interpret the inselbergs as remnants of the interfluves of much-degraded, relict drainage basins. Using landsat images, these authors are able to delineate several defunct drainage networks extending well beyond the present plateau margin. Figure 2/2 (ibid, p. 209) depicts the restored network patterns of this region according to these authors, while Figure 2/3 (ibid, p.

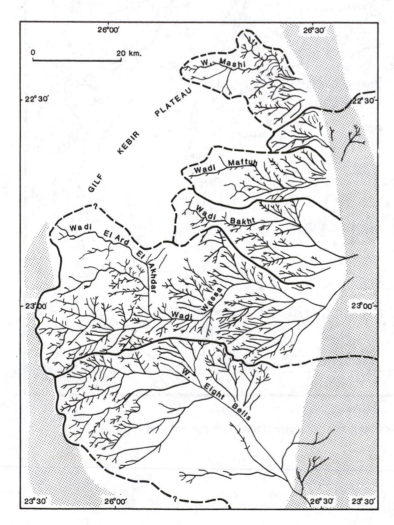

Figure 2.2 Restored drainage patterns of Gilf Kebir (after McCauey et al. 1982).

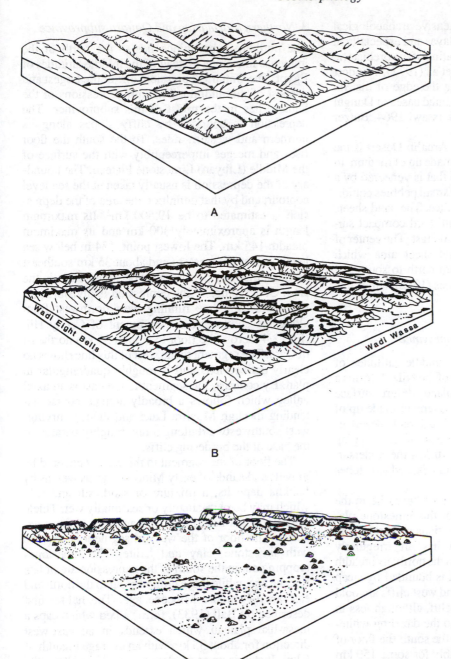

A

B

Wadi Eight Bells

Wadi Wassa

C

Figure 2.3 Block diagrams illustrating main stages in a suggested sequence of development of the southern Gilf Kebir landscape, view from southern margin of Eight Bells drainage system (after McCauley et al. 1982).

216) is a block diagram prepared by these authors to illustrate the main stages in a suggested sequence of development of the southern Gilf landscape.

The Kharga Oasis represents the largest depression of the Arba'in Desert. It extends from north to south for a distance of 220 km along the western edge of the limestone plateau and is up to 40 km wide. A fault runs along the depression (Beadnell 1909), and artesian springs are known to have occurred there as

far back as Acheulean times (Caton-Thompson 1952). As with all other depressions of the Arba'in Desert, the Kharga had playa lakes in Neolithic times. Another topographically depressed area is that which lies between the Kiseiba and the Sin El-Kaddab scarps which form the southern edge of the limestone plateau. The area assumes importance because it contains numerous subdepressions, many of which are filled with playa deposits carrying terminal Paleolithic to Neolithic artifacts. Extensive archaeological excavations in many of these playas and especially in Nabta have permitted precise dating of the prehistoric events of this period (Wendorf et al. (1977), Wendorf & Schild 1980). Situated along the edge of the Sin El-Kaddab scarp are the uninhabited oases of Dungul and Kurkur (Ball 1902, Said & Issawi 1964, Butzer 1964, Said 1969).

A remarkable feature of the Arba'in Desert is the presence of large flat stretches made up of medium to coarse-grained sand with a top that is veneered by a thin layer of lag, coarse sand or small pebbles equidistantly distributed over the surface. The sand sheets form remarkably flat plains with hard compact surfaces that make car travel easy and fast. The center of the Arba'in Desert has a sand sheet area which stretches for about 400 km from north to south and 300 km from east to west. Haynes (1982) names this sheet the Selima sand sheet.

2 *Middle limestone plateau subprovince*

The limestone plateau of the middle latitudes of Egypt extends along both sides of the Nile. It forms a rough-going, nearly level upland desert surface thinly veneered by an erosion pavement made up of alluvium and gravel. It is a virtual rock desert or Hamada. The pavement is the result of the chipping of the hard limestone bed that makes the undersurface. Rill stones, dreikanters and other edged stones are common on these surfaces.

The great oases of Dakhla and Kharga lie in the shadow of the southern edge of this limestone plateau. To the north lie the depressions of Farafra and Bahariya oases. The Farafra has an irregularly triangular shape with the apex to the north; its breadth increases as one goes south. It is bounded by steep cliffs on three sides. The east and west cliffs are bold and of great height; the north cliff, although less in height, is conspicuous owing to the dazzling whiteness of its precipitous face. To the south the floor of the depression rises imperceptibly for some 150 km until the Dakhla escarpment is reached.

The Bahariya is a natural excavation, about 1,800 km² in area. It differs from other oases, which are open on one or more sides, in being entirely surrounded by escarpments and in having a large number of hills within the depression. It is highly irregular in outline, more particularly on its western side, but the general shape of the depression is oval with its major axis running northeast and with a narrow blunt-pointed extension at each end.

3 *Northern Marmarica and Qattara subprovince*

The northern subprovince lies to the north of the immense limestone plateau of the middle latitudes of Egypt. The Qattara depression, one of the largest and deepest of the undrained natural depressions of the Sahara, lies in the midst of this subprovince. The depression is bounded by cliffy slopes along its northern and western sides. To the south the floor rises and merges imperceptibly with the surface of the Middle (Libyan) Limestone Plateau. The boundary of the depression is usually taken at the sea level contour; and by that definition the area of the depression is estimated to be 19,500 km². Its maximum length is approximately 300 km and its maximum breadth 145 km. The lowest point, 134 m below sea level, is near the western end about 35 km southeast of Qara, the only permanently occupied oasis in the area. Geographically the depression is divisible into two parts by a line running west northwest from along the southwestern escarpment of Cecily Hill (Fig. 2/4) across the floor to a point 35 km northeast of Qara. The segment to the west of this line shows no clearly defined axis and is roughly quadrangular in plan. By contrast, the segment to the east is an axial valley which follows a broadly arcuate course extending through Moghra Lake and thence curving-west southwestward along a line roughly parallel to the base of the bordering cliffs.

The floor of the segment to the west is covered by gravels and sands of early Miocene age as well as by Sabkha deposits, a mixture of sand, silt and salt, which may be permanently or seasonally wet. Thiele (1970) gives a description and analysis of these deposits. The floor of the western segment is covered with limestone, clay and halite. The limestones cropping out in the south of the depression are of late Eocene age, while those cropping out to the north and east are of middle Miocene age. The halite, first described by Ball (1933), forms a bed which caps a mesa-like plateau which extends in an east-west direction for about 35 km with an average breadth of 6 km. It stands up at an elevation of 30 m above the surrounding plain which lies below the –100 m contour. The halite is massive and translucent and is underlain by a bed of dark brown clay. Said (1979) relates this mass of halite to the Messinian crisis and concludes that it was probably deposited in a satellite basin of the receding Mediterranean.

In addition to the great Qattara depression, there are many other depressions extending from Jaghbub on the Libyan-Egyptian frontier to Faiyum and Wadi Rayan to the east. The greatest of these small depressions is that of Siwa Oasis. Other depressions include Bahrein, Wattiya, Areg, Numeisa and Sitra which fringe the southern margin of the Qattara depression.

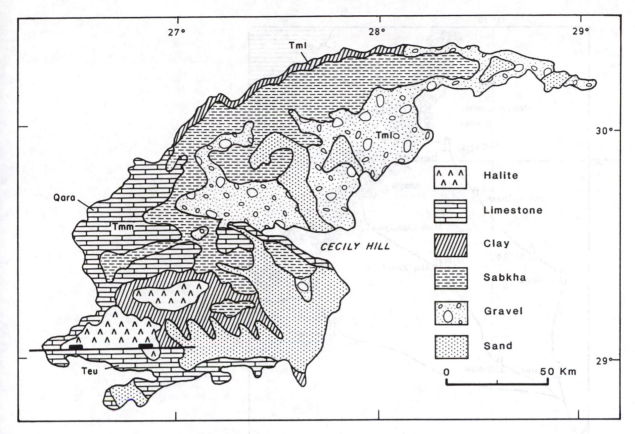

Figure 2.4 Surficial deposits of Qattara depression (after Said 1979).

To the east lies the Wadi Natrun depression.

The plateau to the north of Qattara slopes gently to the Mediterranean Sea and forms a rolling tableland covered with gravel and indented by shallow basins of sand surface material. The larger part of this plateau is made up of the white limestones of the middle Miocene Marmarica Formation. The plateau is named Diffa on the topographic maps and forms part of the Marmarica plateau, the desert country of North Africa along the Mediterranean Sea between ancient Cyrenaica and Egypt.

The coastline to the west of Alexandria is made up of a large number of open bays the most pronounced of which are the Gulf of Arabs in the east and the Gulf of Saloum in the west. In between these two gulfs there are a number of smaller bays separated by protruding points which seem to have been structurally controlled.

EASTERN DESERT

Extending from the Nile Valley eastward to the Gulf of Suez and the Red Sea, the Eastern Desert consists essentially of a backbone of high rugged mountains running parallel to and at a relatively short distance from the coast. These mountains are flanked to the north and west by intensively dissected sedimentary plateaux.

The igneous mountains commence in the neighborhood of Gebel Um Tenassib (lat. 28° 30' N) and extend southeastward beyond the Sudan border. The Red Sea hills do not form a continuous range, but rather a series of mountain groups, more or less coherently disposed in linear direction approximating to that of the coast, with some detached masses and peaks. The highest peak of Gebel Shayeb (near lat. 27° N) attains a height of 2,184 m asl, and hundreds of other peaks of considerable height occur throughout this range.

Northwest of Gebel Shayeb lie the red granite towers of Gebel Gattar separated by a gap from the dark rounded crags of Gebel Dokhan (Fig. 2/5). High on the mountain side of Gebel Dokhan lie the quarries of Imperial porphyry cherished by three centuries of Roman Emperors. A ruined fortress, three lifeless villages, temples and shrines, dry wells and other fossil whims of these Emperors lie at the footslopes of the mountain closed in by the hills from all sides. The quarried blocks of the Imperial porphyry were

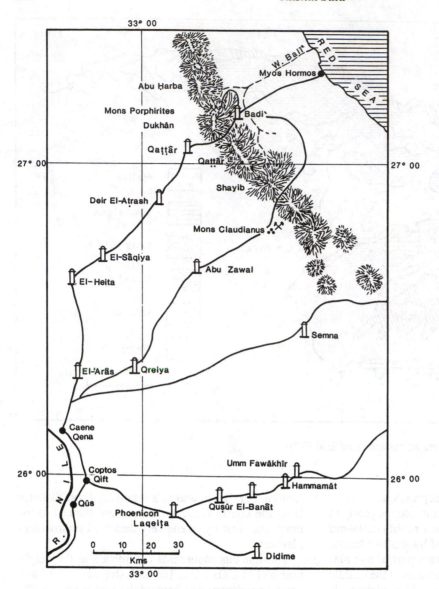

Figure 2.5 Map of central Eastern
Desert showing Roman roads
(after Murray 1967).

slid or lowered from the mountainside down a carefully prepared causeway or chute for more than a
kilometer to the wadi underneath, where they were
worked out and then shipped along a road to Qena.
Remains of six watering stations built along this road
at 25 km intervals can still be seen.

To the south of Gebel Shayeb lies the dead town of
Um El-Omeiyd (in Arabic, mother of pillars) which
the Romans knew as Mons Claudianus. This was the
site of a granite quarry and a garrison town. Mons
Claudianus granite is a sugary white variety spangled
with tawdry mica. The entire town is made up of
granite and is perhaps the best preserved of all Roman
remains in the desert.

Murray (1933) gives a picturesque description of a
bird's eye view of the Eastern Desert: a few glances

from these high pinnacles cover all that is worth
seeing in the Eastern Desert and Sinai. From the
highest pinnacle of wooded Elba (1,437 m) at the
extreme southeastern corner of Egypt, the eye travels
over the desert north of the audacious peaks of Berenice's Bodkin and the Mons Pentadactylus, now
Gebel Faraid (1,234 m) on the tropic of Cancer. On
their northern horizon lies the rose-red whaleback of
Gebel Hamata (1,978 m) which stares in turn at the
fin on the back of his brother leviathan Nugrus (1,505
m), only 80 km away. Nugrus sees the small jagged
peak of Abu Tiyur (1,099 m) near Quseir, once the
only town between Suez and Suakin. Little Abu
Tiyur looks humbly toward the vast Shayeb (2,187
m), the highest hill in mainland Egypt, whose glance
ranges over 320 km from the Nile at Qena to the

chapel crowned Katherina (2,641 m) above the convent of Mount Sinai, across the Gulf of Suez. Katherina sees the flat-topped Gebel Egma (1,620 m). Egma looks north to Halal (890 m) which stares over the frontier to the green hills of Hebron.

To the west of the northern portion of the Red Sea hills and partially separated from them by a wide valley (Wadi Qena) is an extensive limestone plateau, at places attaining a height of more than 500 m asl and extending southward to near Qena. This is the Maaza plateau which is in continuation with the Middle Limestone Plateau subprovince of the Western Desert across the Nile. Further south the mountains are flanked by a lower ever-broadening sandstone plateau extending further south beyond the Sudan border.

The Eastern Desert differs markedly from the Western Desert in that it is intensely dissected and all its drainage is external. While the eastward drainage to the Red Sea is by numerous independent wadis, the westward drainage to the Nile Valley mostly coalesces into a relatively small number of great trunk channels such as wadi Tarfa, Assiuti, Qena, Abad, Shait, Kharit and Allaqi. Wadi Qena is remarkable; its main channel, about 200 km in length, follows a direction almost exactly opposite to that of the Nile into which it debouches.

Belonging to the same geogrpahic province of the Eastern Desert are some islands in the Red Sea of which the principal are Ashrafi, Jubal, Gaysum, Tawila, Gefatin, Shadwan, Safaga, the Brothers, Daedalus and St John's (Zabargad). Most of the islands are situated in the Strait of Jubal at the entrance of the Gulf of Suez. The largest is the hilly Shadwan. A few islands are characterized by central peaks while many are flat and low and consist of sedimentaries; of these latter some may be only elevated portions of coral reefs.

SINAI PENINSULA

The Sinai Peninsula covers an area of 61,000 km². It is triangular in shape and is separated from mainland Egypt by the Suez Canal and the Gulf of Suez. It is continuous with the Asiatic continent for a distance of over 200 km between Rafa on the Mediterranean and Taba at the head of the Gulf of Aqaba. The core of the peninsula, situated near its southern end, consists of an intricate complex of high and very rugged igneous and metamorphic mountains. These rise to greater heights than any in the African part of Egypt. The highest peak, Gebel Katherina, attains an altitude of 2,641 m asl. Many other peaks and crests rise above the 2,000 m contour conspicuous among which are Gebel Um Shomer (2,586 m) and Gebel

Serbal (2,070 m). The core of the peninsula has a pseudo-Appalachian relief and shows all the signs of youthful physiography. It is dissected by numerous incised wadis that are everywhere showing signs of dowcutting. Drainage in the horst block of southern Sinai is toward the Gulfs of Suez and Aqaba. The water courses issuing from the mountains excavate deep ravines with deep sides and steep gradients. Their floors consist of bare rock, and their path is frequently obstructed by falls or cataracts. The main drainage channels of Wadi Feiran and El-Sheikh (Fig. 2/6), however, are filled with a series of yellowish silts and clays interbedded with minor sands and gravels (Barron 1907, Awad 1953). The beds seem to have originated in lakes which were formed as a result of the damming of these water courses by resistant porphyry dikes. The dikes acted as barriers to the groundwater flow and helped form the groundwater body of the famous oasis of Feiran. The lacustrine sediments were probably formed during early Neogene time and later intermittently disrupted by violent floods (Issar & Eckstein 1969).

Overlapping the nucleus of basement complex are the sedimentaries which form the tableland of Badiet El-Tih (Desert of the Wanderings) and the still higher plateau known as Gebel Egma. These two immense plateaux consist of almost horizontal strata which constitute a distinct geomorphological unit to which Hassan Awad gave the term 'Sinai tabulaire'. The Tih plateau is bounded on its east, south and west sides by vertical scarps whose descent can be made by certain routes or *Naqbs*. On the east, the scarp overlooks the depression of Wadi Araba of which the Gulf of Aqaba graben forms part. On the west the scarps are also fault-determined and terminate the Gulf of Suez depression.

In contrast to these extensive plateaux, northern Sinai is characterized by a large number of northeast-southwest trending elliptical anticlines and intervening synclinal depressions. These anticlines and synclines are breached by erosion and are fractured along lines that are more or less parallel to the axes of the anticlines. Geomorphologically this part of Sinai forms what Hassan Awad termed 'la region des domes'. The principal anticlinal mountains of northern Sinai (Fig. 2/7) are Gebels Yelleg (1,090 m), Halal (890 m) and Maghara (735 m). Beyond these and extending to the Mediterranean coast is a broad tract of sand dunes, some of which attain heights of over 100 m asl.

The greater part of Sinai is drained by Wadi El-Arish, a 310 km-long system which reaches the Mediterranean at the town of El-Arish. The wadi is the largest ephemeral stream in Sinai with a catchment area covering almost one-third of the entire area

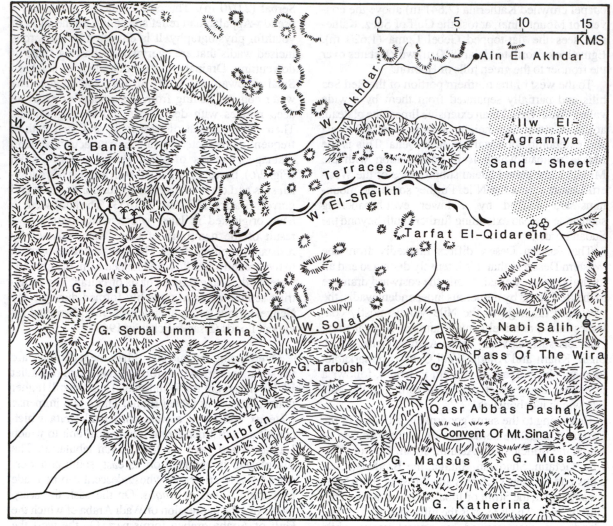

Figure 2.6 Basin of Wadi Feiran, Sinai (after Murray 1967).

of Sinai. Along its way, Wadi El-Arish crosses a
number of mountain blocks cutting deep narrow
passages called *daika*; otherwise it cuts its channel
through Pleistocene sediments which are of fluviatile
origin. According to Sneh (1982) the sedimentary
features indicate a long history of aridity extending
into the Pleistocene.

MEDITERRANEAN SEA

Located between the Eurasian and African blocks,
the present-day Mediterranean covers an area of
about 2,510,000 km². It has a length three times its
width. Its maximum length is 3,540 km between the
Strait of Gibraltar and Iskenderun, Turkey. The widest part, between Libya and Yugoslavia, is about

970 km. An underwater sill between Sicily and Tunisia divides the Mediterranean into two basins. The
eastern basin is deeper than the western one. The sea
has an average depth of 1,500 m. It reaches its
greatest depth (5,093 m) in the Hellenic trough. The
bathymetry of the eastern Mediterranean has recently
been compiled by Hall (1980) to the scale of
1:625,000.

The Mediterranean is all that remains of a great
ocean which, in early geological times, encircled half
the globe along a line of latitude. The modern sea is
made up of several small oceanic basins with broad
continental margins and numerous small dislocated
continental blocks inserted among the basins (Fig.
2/8). The oceanic basins are the remnants of a large
crust which disappeared by subduction as Europe and
Africa collided.

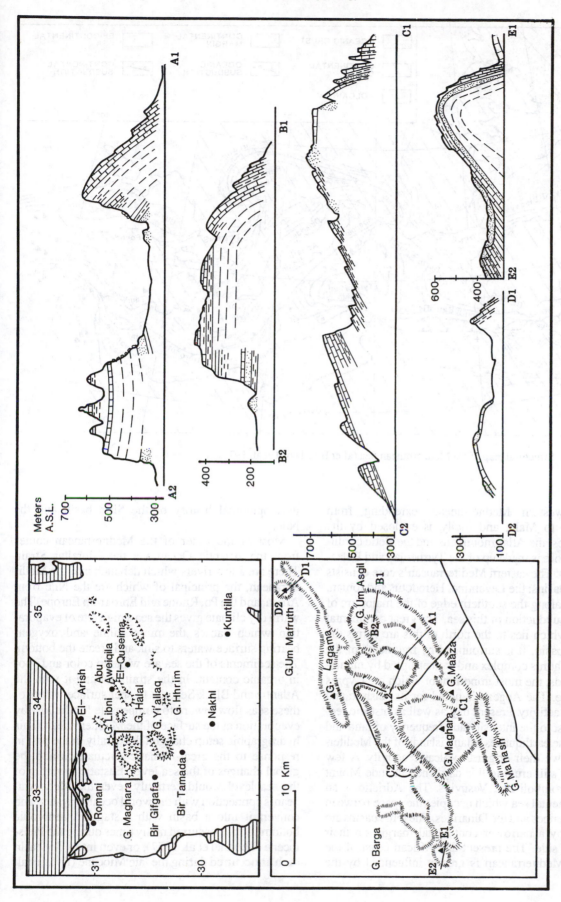

Figure 2.7 Cross-sections across Gebel Maghara, a principal anticlinal mountain of northern Sinai (after Bar Yosef & Phillips 1977).

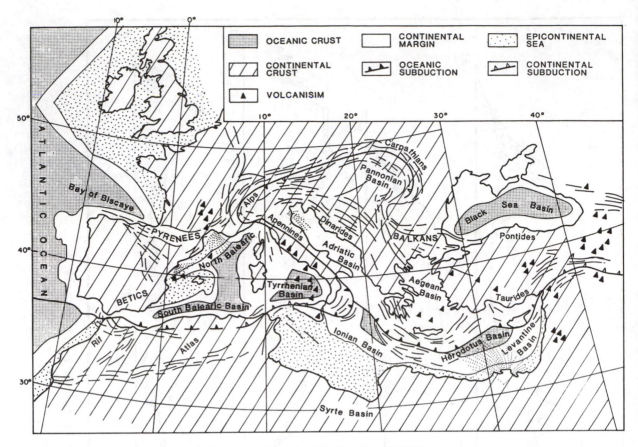

Figure 2.8 Structural map of the Mediterranean Sea (after Biju-Duval et al. 1977).

The western Mediterranean, extending from Gibraltar to Malta and Sicily, is enclosed by the Apennines, the Atlas mountains and the Betic cordillera. Its core is made up of the Tyrrhenian and Balearic basins. The eastern Mediterranean's core consists of three basins: the Levantine, Herodotus and Ionian. They lie along the southern edge of the major arc of oceanic subduction of this sea. The great continental margin which lies to the north of this arc forms the Aegean basin. It is surrounded by fragments of the Dinaro-Taurus complex and is interrupted by numerous islands, the most important of which are Cyprus and Crete. The Aegean is noted for its seismic and volcanic activity. Present day as well as historic records attest to the intensity and frequency of land and submarine earthquakes. Many islands of the Mediterranean owe their origin to volcanic activity. A few volcanos still erupt to this day. They include Mount Etna, Stromboli and Vesuvius. The Adriatic is an epicontinental sea which occupies the space between the Apennines and the Dinarides. The three basins are fringed by a narrower continental margin on their southern side. The present-day African coast of the eastern Mediterranean is greatly influenced by the developmental history of the Sirte basin and the Nile.

Most of the water of the Mediterranean comes from the Atlantic Ocean via the Gibraltar Strait. There are a few rivers which debouch into the Mediterranean, the principal of which are the Nile from Africa, and the Po, Rhone and Ebro from Europe. The warm dry climate gives the sea a high rate of evaporation which causes the more saline and oxygen-bearing surface waters to sink and areate the bottom. The sediments of the sea are white in color and poor in organic content. In the straits joining it with the Atlantic and Black Sea, the fresher surface waters of these seas flow inward to compensate for the loss by evaporation at the surface of the Mediterranean. This hydrographic setup changed drastically in the past in response to the great climatic fluctuations and the global changes of the sea level. Eustatic lowering of the sea level would certainly sever the Mediterranean's connection with the world oceanic system and convert it into a basin with a stagnant aenorobic bottom such as occurred many times during the Pleistocene (Thunnel et al. 1977), or even into a dry basin such as occurred during the late Miocene (Ryan et al.

1973). The Deep Sea Drilling Project (Leg 13) revealed the widespread occurrence of an evaporitic suite beneath the bottom of the sea which proved to be of late Miocene (Messinian) age. The evaporitic suite comprises halite, gypsum, anhydrite and dolomite. It assumes a great thickness and has a strong reflecting surface on seismic profiles thus hindering the unravelling of the nature of the older sediments and the earlier history of the sea.

The sediments of the beaches, coastline and the continental shelf off the coast of Egypt are the subject of an intensive study by an Egyptian-UNESCO joint project team (Nielsen 1974) and by a large number of authors (for a review, see Nir 1982, Said 1981 and El Ashry 1985). The Nile delta coastline evolved from arcuate to bird's foot as many of the distributaries of the Nile delta silted up in historic time, and as the sediments debouched by the river into the sea were reduced due to recent Nile control projects (Said 1958). The longshore currents which transport the Nile sediments along the shores of the Mediterranean to the east as far as Gaza seem to have been in effect since the late Pleistocene.

The evolution of the Mediterranean Sea has attracted the attention of a large number of authors. The land areas surrounding this sea, including the Alpine belt, are the subject of extensive and classical studies. Their folded belts have inspired some of the fundamental concepts in geology such as geosynclines and subduction. The idea that the modern Mediterranean is a remnant of a central ocean which separated Laurasia and Gondwana continents dates back to the last century. The evolution of this central ocean has recently been deciphered; and it has been shown that the Paleozoic marine sediments of the Himalaya-Alphs region were deposited in an ancestral ocean, the Paleotethys, which occupied a central but a more northerly position (Fig. 2/9). This Paleozoic ocean was consumed and a central but more southerly ocean, the Neotethys, developed. The concensus is that these two seas which succeeded one another formed vast expanses of ocean between Europe and Asia on one side and Afro-Arabia and India on the other.

During the Paleozoic, a Pleotethys covered west central Europe and extended to the Pamirs. It was separated from the lands to the south by a suture (Sengor 1985) and a ridge of minor uplift traceable, at least since the Ordovician, from northwest Afghanistan through northern Iran, southeast Turkey and eatward to possibly northern Italy (Sonnenfeld 1981). Numerous transgressions of this Sea covered northern Africa, the Levant, Arabia and the lands to the east during the Paleozoic. The sediments are mostly clastic and are of deltaic, tidal and/or marginal marine origin. They interfinger marine carbonates in the north. During many stages, evaporites were deposited in large basins along the southern edge of these marginal seas. During the Cambrian, for example, an evaporitic basin extended from the Salt Range of Pakistan to Oman, Hadramout and Arabia.

Several interpretations have been proposed to depict the evolution of the Neotethys and the modern Mediterranean (for a review of these and other aspects of the Eastern Mediterranean, see Dixon & Robertson 1984). The precursor of this sea seems to have come into existence during the late Triassic. It succeeded the Paleotethys which was consumed during the Mesozoic by a left lateral motion of Africa with respect to Europe. This motion was induced by the spreading of the Atlantic Ocean. It caused the rifting and breaking away of the Apulian and Anatolian plates and their accretion to Europe. The motion seems also to have been responsible for the attentuation of the crust of the Eastern Mediterranean and the formation of the African continental passive margin. The fact that the middle Triassic faunas of Egypt, like their predecessors in the late Paleozoic, are different from the Alpine or Germanic faunas and are related to those of Greece and Turkey is taken as evidence that the Anatolian and Apulian plates formed part of the southern Tethys until that time.

Figure 2/10 shows the relative position of Africa with respect to Europe (arbitrarily kept fixed in its present position) since the beginning of the opening of the Atlantic Ocean. From that time and until the late Cretaceous, Africa was moving left laterally at a relatively high rate (2-4 cm/year). During the late Cretaceous the motion slowed most probably because of the first major intracontinental collision between Arabia, Anatolia and Eurasia. A relatively modest right lateral motion of Africa followed between the late Cretaceous and early Eocene (49 my) in relation to the initiation of the spreading between Europe and North America. From then on, a state of intracontinental collision prevailed in which Africa was rotated around a pole situated in the vicinity of Morocco. This rotation was felt mostly in the east where it was essentially north-south.

RED SEA

The Red Sea occupies an elongate escarpment-bounded depression which extends in a south-southeast direction from Suez to the Strait of Bab-El Mandeb in a nearly straight line, separating the coasts of Arabia from those of Egypt, the Sudan and Ethiopia. Its total length is about 2,000 km, and its breadth varies from about 400 km in the southern half to 210 km at latitude 27° 45' N where it divides into two parts, the Gulf of Suez and the Gulf of Aqaba, sepa-

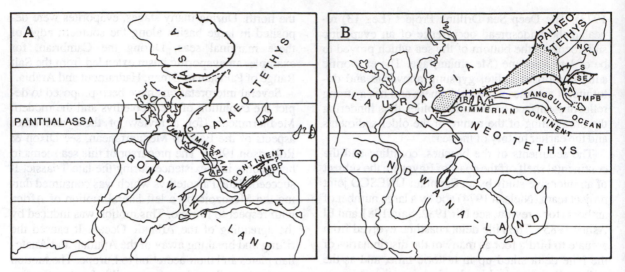

Figure 2.9 Paleotethys and Neotethys in: A. Late Permian time: geometry showing a gaping Paleotethys fringed to the south by the Turkish (T), Iranian (IR), Afghani (AF) and Thai-Malaysian peninsula (TMPB) blocks; B. Latest Triassic time: geometry showing the Neotethys replacing the closing Paleotethys (after Sengor 1985 with minor modifications).

rated from each other by the peninsula of Sinai.

Gulf of Suez

The Gulf of Suez is different from the Gulf of Aqaba and the Red Sea proper with regard to bottom topography, chemistry of water and type of sediment. The Gulf of Suez is flat-bottomed with a depth of 55-75 m. At its mouth, the Gulf descends to a depth five times its own and is thus a shallow shelf filled

with the surface water of the Red Sea. The exchange of water between the Gulf of Suez and the Mediterranean, which takes place through the Suez Canal, is of no importance to the water and salt budget of the Red Sea. The current regime of the Suez Canal and the changes it has undergone since the building of the Aswan High Dam are dealt with by Morcos & Messieh (1973), Hassan & El Sabh (1974) and Meshal (1975). Little water seems to be exchanged between the Mediterranean and the Red Sea via the Canal.

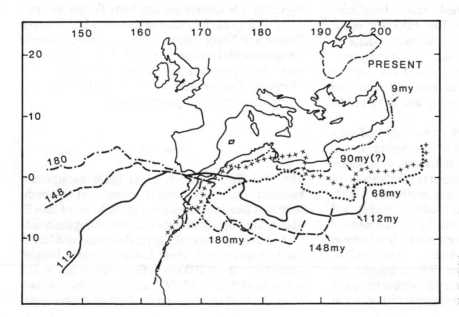

Figure 2.10 Plate kinematics of the Africa-Europe plate system (after Biju-Duval et al. 1977 with minor modifications).

This may explain the almost complete absence of any intermingling of faunas of the two seas since the cutting of the Canal in 1869 (Said 1950). The sediments of the Canal are examined by Stanley et al. (1982) while those of the Gulf of Suez by Shukri & Higazy (1944) and M.M. Mohamed (1979). The sediments of the Gulf of Suez are, like the sediments of the Red Sea, rich in carbonates although they include a larger precentage of detrital materials.

The exchange of water between the Gulf of Suez and the Red Sea is investigated by A. Mohamed (1940). Temperature and salinity variations in the Gulf suggest the inflow of a warm, less saline surface current from the Red Sea proper into the Gulf of Suez and the outflow of a bottom current, more saline and less warm. The latter mixes at the edge of the Gulf with the water in the upper strata of the Red Sea proper, the resulting admixture going down and contributing to the formation of the Red Sea bottom water.

Gulf of Aqaba

The Gulf of Aqaba is different from the Gulf of Suez. It has an irregular bottom topography of an average depth of 1,250 m. It is separated from the Red Sea proper by the sill of Tiran, about 250-260 m deep, outside of which the bottom is at a depth of more than 1,000 m. Near the eastern side of the Gulf troughs as deep as 1,850 m occur; and the contrast is great when we compare these depths with the heights of the mountains bordering the eastern side of the Gulf. The depth of the Gulf of Aqaba is nearly equal to the depth of the Red Sea which is about ten times as broad, although the Red Sea itself is remarkable for its great depth in proportion to its breadth. Along the beaches of the Gulf of Aqaba the mountains rise abruptly. The continental shelf which characterizes the Red Sea does not occur in the Gulf of Aqaba except in the south. A new bathymetric map of the Gulf to the scale of 1:250,000 is compiled by Hall & Ben-Avraham (1978). The map shows that the center of the Gulf is occupied by three deep and elongated basins separated by low sills (Fig. 2/11).

The mode of exchange of the water masses between the Gulf of Aqaba and the Red Sea proper across the shallow sill of Tiran is dealt with by Mohamed (1940). During the winter a surface current from the Red Sea to the Gulf is observed, while a bottom current moves from the Gulf to the Red Sea proper and contributes to the bottom waters of the Red Sea. This pattern is similar to the water exchange between the Red Sea and the Gulf of Aden in the same season.

The Gulf forms the southern segment of the Dead Sea rift. The structure is dominated by an-echelon faults which delimit three elongated basins that strike northeast (Ben Avraham et al. 1979). Seismic profiling indicates a composite thickness of 2-3 km which seems to belong to the late Neogene and the Quaternary. The sediment fill is made up of turbidites and pelagic segments.

Red Sea proper

The Red Sea possesses an irregular bottom. It has a characteristic continental shelf about 30 km in breadth. The center of the sea is made up of a wide 'main trough' extending from about 15° to 24° N, usually less than 50 km wide and characterized by steep walls and irregular bottom topography (Cochran 1983). South of 18° N, the continental shelves broaden considerably and are underlain by carbonate banks and reefs which effectively fill the main trough.

The Red Sea (Fig. 2/12) is located in a region which is characterized by such an arid climate that evaporation from the water surface greatly exceeds the small precipitation. The adjacent lands have no runoff and no rivers enter the Red Sea. Therefore, the surface salinity of the water of the Red Sea is high and reaches values between 40 and 41‰. Along the entire length the prevailing winds blow consistently from the north-northwest during the summer season; during the winter the north-northwest winds reach as far south as 21° ro 22° N, but south of this the wind direction is reversed and south-southeast winds predominate.

The climatic conditions and the prevailing winds determine the character of the exchange of waters of the Red Sea and the Gulf of Aden across the shallow sill which separates them. The exchange is subject to annual variation which is related to the change in the direction of the winds in winter and summer. In winter when south-southeast winds blow in through the strait of Bab El Mandeb, the surface layers are carried from the Gulf of Aden to the Red Sea, and at greater depths highly saline waters of the Red Sea flow across the sill. In summer when north-northwest winds prevail, the surface flow is directed out of the Red Sea by the wind. At shallower depths water from the Gulf of Aden flows in, having a lower salinity and lower temperature than the outflowing surface water. At still greater depths, highly saline Red Sea water appears to flow out over the sill but it is probable that this outflow is smaller than in winter.

The sediments of the northern Red Sea are studied by Shukri & Higazy (1944). They are richer in calcium carbonate than the sediments of the Gulf of Aqaba or the Gulf of Suez.

The origin of the Red Sea has attracted the attention of workers for a long time. For a review and

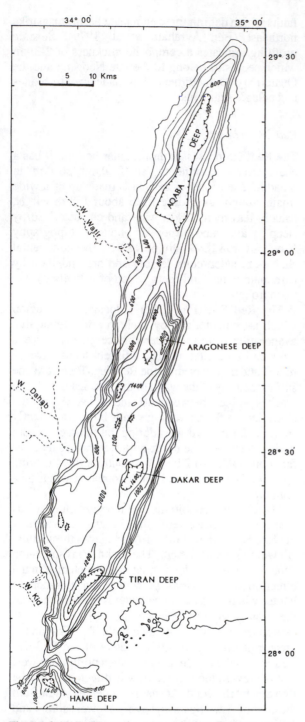

outline of the most recent results, the reader is referred to Degens & Ross (1969), Stoffers & Ross (1974), Cochran (1983), Thisse et al. (1983), Morgan Chapter 7 and Meshref Chapter 8 (this book). The fact that the axial trough is associated with large-amplitude magnetic anomalies and is filled with tholeiitic basalt has led to a concensus that the axial trough was generated by sea floor spreading. The magnetic lineation seems to indicate that the spreading occurred between five and six million years ago in the southern part and seems to be advancing from the south to the north. This must have followed a pre-rifting phase during the Oligocene and the formation of the main trough during the early if not the middle Miocene. The substratum underlying the main trough and shelves is difficult to ascertain because it is covered by a thick Miocene evapointe series which measures 3,000 m in thickness and even more in places. The top of this formation is marked by a strong seismic reflector (reflector S). The axial zone of the Red Sea has small (14 × 5 km) deeps which are characterized by high salinities and are thermally charged. Atlantis II deep (21° N) has two layers of brine each nearly 200 m thick covering the bottom sediments and screening them from the waters of the Red Sea. Under these brines, the sediments are metalliferous and seem to be hydrothermally generated.

Figure 2.11 Bathymetric map of the Gulf of Aqaba simplified after Hall & Avraham (1978).

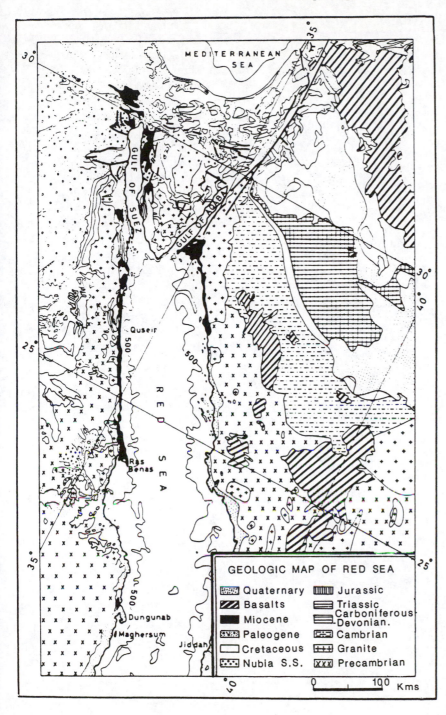

Figure 2.12 Generalized geological map of the northern portion of
the Red Sea (after Said 1969).

CHAPTER 3

Application of remote sensing and satellite cartography in preparing new geologic map 1:500,000

FRANZ K. LIST, BERND MEISSNER & GERHARD PÖHLMANN

Free University of Berlin and College of Engineering (TFH), Berlin, Germany

1 CONCEPT AND DEVELOPMENT OF REMOTE SENSING

The term 'remote sensing' has become increasingly familiar to the international scientific community during the last decade. It was coined in the early sixties by scientists of the Willow Run Laboratories in Ann Arbor, Michigan, that later became the Environmental Research Institute of Michigan (ERIM). However, even if the term itself is new, the methodology of remote sensing has been in use for a much longer time, if under different names.

Remote sensing – which means collecting information about an object without the use of physical contact – can be said to have started with the invention of photography more than 100 years ago. During 1814-1838, Joseph Nicéphore Niepce and Luis Jaques Mandé Daguerre in France developed the so-called 'daguerreotype', a photographic process using a polished silver plate as light-sensitive material. Despite the widespread acclaim and public acceptance of the daguerreotypes, the process used proved to be a dead end since it produced only positive unictes. The actual breakthrough in the development of photography was accomplished by the British physicist William Henry Fox Talbot who in 1831 used a silver halogenide coating on paper for the production of photographic images. Somewhat later, he produced the first negatives from which any number of positive copies could be made. So, for the first time, geometrically accurate and detailed images of natural objects became available for documentation, evaluation and, soon, also for measurements.

While the potential use of the photographic process for the production of maps and for the precise measurement of objects in the photographic image itself was realized and put to practical use almost instantly, it took about ninety years until geologists became aware and made use of the new tool. Shortly after the advent of the airplane, the first observation of geologic features from the air was described and discussed by Johnson (1921), Willis (1921), and Lee (1922). Soon it was realized that instead of the short-lived direct observations by a human observer in an aircraft, the study of aerial photographs, especially of pairs of photographs taken stereoscopically, provided a much higher degree of information on the earth's surface.

In 1938, a paper was published in the journal *World Petroleum* (Anonymous 1938) that described the operational use of aerial photographs for oil exploration in the nearly inaccessible rain forests of New Guinea by the 'Nederlandse Nieuw Guinea Petroleum Maatschappij'. This publication is generally regarded as marking the birth of geologic photo interpretation, or 'photogeology'. At the same time, an article was published in Switzerland by Helbling (1938) describing in great detail the application of photogrammetry (a science already well-established at that time) for geologic mapping. In contrast to the oil exploration project in New Guinea in which small-scale aerial photographs were used and a large area had to be surveyed looking mainly for structural anomalies, Helbling (1938) mostly made use of large-scale terrestrial photographs for detailed studies of the highly complicated alpine geology. In addition, he applied precision measurement and data transfer techniques as available to photogrammetrists.

These early and somewhat divergent trends continued after World War II. In the United states, photogeology was employed predominantly for small-scale regional mapping, often in connection with petroleum or mineral exploration. In Europe, on the other hand, the tendency toward large-scale mapping by means of aerial photography and toward the steps 'from interpretation to measurement' (Völger 1953) continued. In this context, emphasis was laid upon quantitative determination of structural data, like strike and dip, and of tectonic and morphometric data.

At that time, new developments in the United States were taking advantage of the progress made in the fields of semi-conductors, data processing and space technology; remote sensing, as it is understood now, came into being. Using these technological accomplishments, the 'classic' concept of aerial photographic interpretation was expanded in several ways, especially by:

– the development of *new sensors* that allow data acquisition in regions of the electromagnetic spectrum far outside the range of the visible light, like the near and far infrared, or the microwave region
– the development of *powerful computer systems* for electronic data processing, especially for digital image processing
– the development of new *carrier platforms* in addition to the airplane, like manned and unmanned spacecraft, e.g. Skylab, Spacelab, Landsat, Seasat, and Meteosat.

This enormous expansion of the potential of data acquisition and data interpretation required the development of a new concept of remote sensing that was termed 'multi'-concept by Colwell (1975). It is based on the idea that in order to get as complete information as possible on any given part of the earth's surface – which is the main objective of remote sensing – a single one-time image taken in one limited spectral region will normally not be sufficient. There is need for:

– multistationary imagery like stereoscopic aerial photography allowing spatial representation and measurement of the earth's surface
– multispectral imagery providing data recorded in different wave lengths like radar or thermal infrared
– multitemporal imagery monitoring changes of the earth's surface as necessary for the analysis of land-use patterns or hydrogeologic phenomena
– multiscale imagery, e.g. a combination of satellite and aerial photographic imagery, permitting the interpretation of otherwise unclassifiable features
– multiprocessing of image data like the creation of color composites for geologic and structural interpretation, the use of digital classification for land-use mapping, and the construction of digital mosaics for image maps and map bases
– multidisciplinary analysis of image data by scientists from different disciplines, and finally
– multithematic representation of different information levels for the user.

It is obvious that the potential of remote sensing methods exceeds that of a mere 'technique'. What is so aptly said by Estes, Jensen & Simonett (1980: 72) on remote sensing in the field of geography is equally valid for the realm of geologic mapping: 'Remote sensing is a reality within geography whose time has come. It is too powerful a tool to be ignored in terms of both its information potential and the logic implicit

in the reasoning process employed to analyze data. When allied with the traditional cornerstone of geography, i.e. cartography, in its new digital raiment, the two techniques can go far beyond being mere technologies'.

1.1 *Applications to geologic mapping*

Due to the large area covered by a single Landsat frame (roughly 34,000 km^2), the use of Landsat imagery for regional geologic studies is an obvious choice. If one considers that Egypt is covered by about 60 Landsat frames, and that one Landsat frame shows about the same area as 1,300 aerial photographs at a scale of 1:40,000, there can be little doubt that to map all of Egypt with the use of aerial photographs would be an extremely difficult undertaking at best. Of course, the aerial photographs would show much more detail than the Landsat-MSS imagery; on the other hand, this detail cannot be represented on a map at a scale of 1:500,000 anyway. Thus, for a regional mapping project like the one we are dealing with here, Landsat-MSS imagery is the logical choice, with aerial photography used sparingly as a source of ancillary information whenever needed.

The prime requirements for a geologic map are the representation of the *lithology* and *stratigraphy* of a given area, together with the relevant *structural elements*. The delimitation of the *boundaries* between the separate geologic units should be as precise as possible for the scale given, and enough *topographic detail* should be included to allow easy and reliable determination of one's location in the field. In more specialized geologic maps, information on the *mineral or hydrocarbon potential* may also be required.

It is a well-known fact and the basis for many decades of successful photogeologic work that *lithology* can be interpreted with high accuracy on conventional black-and-white aerial photographs. Interpretation, that is visual interpretation, of lithology from Landsat-MSS imagery is not too different from interpretation of small-scale aerial photography (Abrams & Siegal 1980, Baker 1975, Siegal & Gillespie 1980, Tator 1960). What is lost in information by the lower spatial resolution of the satellite images is gained by the higher spectral information contained in the four spectral bands of Landsat-MSS. Unfortunately, the spectral bands of Landsat-MSS are not optimally selected for discrimination of lithology but rather for classifiction of vegetation. This has changed with the introduction of the new Landsat-TM system which has seven spectral bands that are better suited for rock discrimination. At the start of the project, this type of data was not yet available, so only MSS data was utilized.

Today, visual interpretation of optimized color composite imagery (see 3.5.2) is still the best way for the extraction of geologic information from multispectral satellite imagery. Even though some promising results in digital classification for lithologic mapping have been reported from a mapping project in Saudi Arabia (Blodget & Brown 1982) or of a part of the Eastern Desert of Egypt (Jacobberger et al. 1983), this approach works only within limited areas with reasonably different lithologic units when using Landsat-MSS data. The introduction of algorithms for texture analysis and expert systems into the classification process may bring about a change in this situation sometime in the future. Until then, geologists will have to stick to visual interpretation for best results.

Where sedimentary rocks are concerned, lithologic information will only be meaningful when coupled with *stratigraphy*. Now it seems absurd at first glance to expect stratigraphic information from satellite imagery or from any remote sensing data, for that matter. No biostratigraphic age can be extracted from Landsat-MSS data; but relative stratigraphic information, e.g. the subdivision of seemingly uniform units in badly exposed or inaccessible areas, is another matter.

Calcareous sequences in particular show marked differences on the MSS image. Grootenboer et al. (1973) reported on a subdivision of the well-fieldmapped Transvaal dolomite in South Africa by means of Landsat-MSS data interpretation, a thing that would have been impossible otherwise. In Egypt, the vast carbonate exposures of the Western Desert are also a good example of the transfer of known stratigraphy into less known areas by means of Landsat-MSS interpretation, as used for the purpose of the present geologic map.

Structural geology has always found photogeology to be a valuable tool. This is equally true for the interpretation of Landsat-MSS imagery. Among the first meaningful results of Landsat-MSS interpretation were papers on structural problems, where the 'delineation of hundreds of hitherto unknown linear features' was found to be of great advantage (Isachsen et al. 1973: 603); also for the first time the structural pattern of large fracture systems like the African rift valley was revealed as a whole (Mohr 1973). This last point, the synoptic overview, was also important for the production of the present 1:500,000 map of Egypt, because it made it possible to map the fracture pattern of the entire country in a uniform and consistent way that would have been impossible otherwise.

The mapping and analysis of fracture patterns is closely related to the search for *mineral or hydrocarbon potential* by means of Landsat imagery (Bentz &

Gutman 1977, Collins et al. 1973, Halbouty 1976, 1980, Rowan & Lathram 1980, Smith 1977). The present geologic map originated from a petroleum exploration project, and it is also meant to serve as a base for future activities in that field.

In a treatise on geological remote sensing in the 1980s, Goetz (1980: 679) stated that 'the next decade will see remote sensing take its logical place in the roster of tools and techniques that the geologist will use in his everyday work'. While this is true, it is still interesting to note that while the scientific literature is prolific in papers on the practical application of remote sensing in geology, so far most of this has been of a local nature in spite of the fact that Landsat-MSS, by its very nature, should lend itself most successfully to regional projects. It seems that up to now nobody has used Landsat data for the geologic mapping of an entire country the size of Egypt.

1.2 *Applications to cartography*

Data collected by remote sensing methods and recorded, generally, on magnetic tape or photographically, can always be transformed into images on paper or film. These images, either in the form of color composites or of color-coded thematic classifications produced by digital computers, are similar to maps and often confused with maps. However, they are images and the 'often long and arduous' process of converting such images into maps has still to be performed (Doyle 1975: 1077). Long and arduous as this process may be, in many instances it is the only possible way to obtain a map base at all.

At present, if one speaks of remotely sensed data for cartographic applications, one generally means satellite images recorded by one of the satellites of the Landsat family. The use of space-borne imagery for mapping is only a continuation of a development that started in the 1920s with the growing importance of aerial photography and photogrammetry.

In photogrammetry, the main objective was to substitute time-consuming terrestrial methods, like plane-table mapping, with airborne techniques. This change in data acquisition did not result in a change in the subsequent steps of cartographic processing, because the elements of the cartographic content did not change. The fact that aerial photographs contained additional information apart from what was used for topographic purposes led to the development of topographic image maps and thus to new cartographic solutions. However, thematic maps were still based on classical topographic maps, not on image maps.

The increase in natural resources exploration activities in practically unmapped remote areas during the 1950s resulted in an increased use of aerial photographs as map substitutes. This led to an increased use of photogeologic methods since the basic data, the aerial photographs, were already available and commonly used for field orientation. Large areas were geologically mapped in this way, but the cartographic transformation of these interpretations was difficult due to the absence of adequate base maps.

When the transformation of the interpreted data is done by means of even a simple photogrammetric plotting instrument, sceletal topography can be plotted at the same time, allowing some orientation on the thematic map for the user. In maps of this kind, there is normally a gross imbalance between the detailed representation of geologic features and the scattered topographic information. Using a semi-controlled mosaic of aerial photographs as a base for a geologic map allows the spatial correlation of geologic data to image elements but the overall accuracy will be rather low. Areas with pronounced relief demand the production of expensive ortho-photo maps because of the relief-induced radial distortions in the aerial photographs.

The launch of the Landsat satellites opened a new dimension for mapping the earth's surface especially at small scales. Available maps of many third world countries are rarely suitable as base maps for thematic maps because of their low information content and their partial inaccuracy in the representation of topographic features. By classical photogrammetric methods, not even the demand in medium-scale maps (1:25,000–1:100,000) can be satisfied. Thus new ways of map production had to be found. A geometrically controlled digital mosaic of Landsat-MSS imagery provides an acceptable solution for a small-scale map base since relief-induced image distortions are minimal due to the high orbital altitude of the satellite. In a satellite image map, the mosaic is supplemented by additional topographic information like settlements, roads, names and so on.

For the present geologic map of Egypt, a satellite image mosaic served as a base for the accurate positioning of the geologic information and, in a differently processed form, as a means for topographic orientation within the thematic map. The Landsat data were first used in the form of conventional photomosaics ('Working sheet' series; see 3.3). In addition to the considerable manual labor involved in the preparation of these mosaics, the positioning errors inherent in the only system-corrected Landsat-MSS frames were too high to be tolerated. For the actual map, all frames were digitally corrected, mosaicked, and radiometrically matched in the full format of the 1:500,000 map (3.5.3).

Apart from the geologic interpretation of the Landsat scenes, topographic features were also interpreted in the imagery. First of all it was necessary to correlate known map elements with image elements since, as already mentioned, the positional accuracy of the available maps was, in part, rather low. Settlements, roads, wells, drainage networks, and other terrain features had to be allocated. Recent natural changes, e.g. of the coastline, or anthropogenic changes, like newly constructed roads, could be seen in the images and were used for updating the available topographic information.

Even though the production of the present map would not have been possible without the massive use of remotely sensed data, one still has to partly agree, for the time being, with the remark of Doyle (1975: 1103) that 'accurate topographic and planimetric mapping is not a truly operational procedure for any remote sensor'. In the present project, too, a good deal of experimentation was necessary to reach the goals set, and often operating procedures had to be modified to overcome unforeseen problems.

2 PREVIOUS GEOLOGIC MAPPING ACTIVITIES

In the preface to the second volume of Hume's *Geology of Egypt* (Hume 1934), H.H. Thomas stated in 1931 that 'the literature dealing with the geology of Egypt is most extensive'. The truth of that statement which includes the geologic mapping activities, is reflected in the first bibliography on the geologic literature of Egypt which appeared in 1910 and already comprised 155 pages (Sherborn 1910). It is not surprising tht the latest annotated bibliography on the geology of Egypt (El-Baz 1984) has grown to an impressive 778 pages.

Great strides have been made in the field of geologic mapping since the publication of the first map by Russegger (1842), following a report on the results of an expedition to Egypt (Russegger 1837). For a review of the history of geologic research and geologic mapping activities in Egypt, see Chapter 1.

Beginning with the establishment of the Geological Survey of Egypt in 1896, systematic mapping programs produced many maps at different scales of many parts of Egypt. However, up to the publication of the present geologic map at 1:500,000, the only map covering the entire country was the 1:2,000,000 geologic map of 1981 which succeeded those of 1971 and 1928.

In preparing the new geologic map use was made of all existing maps. Due acknowledgement is made to these in the legends of the different sheets.

2.1 *Use of remote sensing methods for geologic mapping in Egypt*

As outlined above, the geological interpretation of

aerial photographs, photogeology, is the oldest of remote sensing methods that have been used since the mid-thirties. Since a main application of photogeology is geological mapping of remote and inaccessible areas, it is not surprising that geologic fieldwork in Egypt has been supplemented for many years by aerial photographic work.

The wide-spread use of photogeologic methods in Egypt is documented by most of the geologic publications that deal with mapping: practically all geologic sketch maps that are published are based on the interpretation of aerial photographs. This is helped by the fact that the major part of Egypt – with the exception of the far southwest – is covered by aerial photographs of good quality. Due to the lack of vegetation cover in practically 97% of the country's surface, outcrops are generally well exposed and not too difficult to interpret.

On the other hand, desert varnish, iron crusts, sand and gravel sheets mask or obliterate a lot of geologic information (Greenwood 1962). In the basement complex, the effects of metamorphism mask the lithologic differences, making photographic interpretation difficult to impossible in some cases. Satellite image interpretation offers even more pitfalls; a lot of fieldwork is always needed to provide a reliable framework for the interpretation process.

One of the obvious applications of remote sensing methods in Egypt is the study of sand dunes, their morphology, their movement, and their cartographic representation. Many studies used aerial photography and satellite imagery for this kind of research, like Ashri (1973), Breed & Grow (1979), El-Baz et al. (1979), McKee (1979), Steffan (1983a, b) and others. Photogeologic methods were also used for medium-scale mapping, especially of basement rocks in the Eastern Desert (Sabet 1962, 1969). Structural geology has always been a domain of photointerpretation; the publications of Barakat & Ashri (1972), El Etr (1971), El Etr et al. (1973, 1974, 1979), Embabi & El-Kayali (1979), and El Khawaga (1979) are examples of the application of photogeology to fracture analysis. Moussa (1979) used aerial photographic interpretation for mineral exploration. Satellite imagery, especially Landsat data interpretation, was also applied in the description of more local features like craters (El-Baz 1981), or in studies of desert features in general (El-Baz 1978, 1980) or of regional geology (El Shazly et al. 1976, 1977).

2.2 *Existing topographic maps*

The 'Egyptian General Survey Authority' is responsible today for the execution of land surveying in Egypt, and for the production of topographic maps in general. This organization was established in 1898 as 'Survey of Egypt' which, at that time, was in charge of topographic as well as of geologic mapping. Today, the Geological Survey and the General Survey Authority are two separate entities. Since 1954, the 'Military Survey Department' has taken over part of the topographic surveying tasks.

Until 1950, a geodetic network was established throughout the entire Nile valley from Cairo to Aswan, and along the coast of the Mediterranean Sea to the eastern and western borders of the country.

The cultivated areas of Egypt have been surveyed at a scale of 1:25,000 and 1:100,000. A small part of the desert area, especially of the Eastern Desert, has been mapped at 1:100,000. The largest map scale at which the entire country is covered at the moment is 1:500,000. During the Second World War, and also for some time afterwards, some of the Survey of Egypt topographic maps were updated and reprinted by the 'British Survey Directorate, General Headquarters, Middle East Forces'. Many of the topographic maps now in use as the most recent issues come from that production. The newer US maps (AMS) at 1:250,000, 1:100,000 and 1:50,000 are not publically available. For the early history of topographic mapping in Egypt, see Said (1971: 8-10).

For the preparation of the new geological map at 1:500,000, sheets of the following map series were used as sources of topographic information:
– Egypt 1:100,000, normal series
– Egypt 1:25,000 for cultivated areas, desert areas appended at 1:100,000, Survey of Egypt, since 1927
– Egypt 1:100,000, topographical map, Survey of Egypt. The edition of this new series has been in progress since 1949; it is supposed to replace the existing maps by a consistent and uniform series. Sheets of the following old series were used: Western Desert, 1932 ff; North Coast, 1937; Northern Sinai, 1935; Southern Sinai, 1935; Eastern Desert, 1935.

The 1:500,000 map of Egypt, published by the Survey of Egypt from 1941 onward is not strictly a topographic map but a classic geographic map of the chorographic type. It does not provide sufficient detail to accurately locate geologic features. Of special value for the editing of the topographic base used for the preparation of the geologic map was the 1:250,000 Military Survey Map covering a large part of the country.

Recently the triangulation network in the Eastern Desert and along the Red Sea coast has been enlarged by the Egyptian authorities. It is planned to publish a new edition of the 1:500,000 topographic map following the international format and to extend the first order geodetic network of the entire Western Desert, and to supplement it by a second order network.

Along the 3,000 km long newly built roads elevations will be recorded.

For the southern part of the Western Desert super-wide angle aerial photography was acquired in 1982 for the production of 1:250,000 topographic maps. At the same time, 1:60,000 aerial photographs were obtained for 1:50,000 scale maps.

The lack of a uniform and up-to-date topographic map for the entirety of Egypt at present was taken into consideration when the program for the preparation of the new geologic map was planned (see 3.2).

3 NEW GEOLOGIC MAP 1:500,000

Normally, the publication of a geologic map of an entire country at a scale of 1:500,000 follows a long process of large-scale mapping of at least a large part of the country. If, as in the present project, the large scale mapping has not been completed, the process of producing the map has to be reversed. This calls for a 'top-down' approach, meaning that basically small-scale data have to be used and the research (field and laboratory work) that goes into the map has to be just as detailed as is required by the scale used.

Still, a map produced using the top-down approach is a most useful tool for the geologist confronted with any kind of practical work: the map documents the present level of knowledge, especially as far as regional geologic and stratigraphic problems are concerned and it can serve as a base and stepping-stone for further research. In former times, such an approach to small-scale mapping would have been problematic. With the development of remote sensing techniques, especially with the availability of a regional coverage of Egypt by Landsat-MSS image data, this method of mapping has become feasible.

3.1 'Preliminary Edition' series of the 1:500,000 map

In 1977, a cooperation program was initiated between the Academy of Scientific Reserach in Cairo and the German Research Foundation (DFG) under which geologic field work was conducted by both the Technical and the Free Universities of Berlin in southwestern Egypt. As a result of that work, the formerly undifferentiated 'Nubia' sandstone strata were subdivided into five mappable units (Klitzsch 1978, Klitzsch et al. 1979). Oil companies engaged in exploration work in the Western Desert of Egypt at that time, namely Conoco in partnership with Marathon and Pecten, became interested in this work. Consequently, Conoco supported additional field work and the production and printing of four sheets

of a preliminary geological interpretation map of southwestern Egypt at a scale of 1:500,000.

Since these sheets were principally intended as a basis for petroleum exploration activities in the Western Desert, the format and grid of the maps were chosen in such a way as to cover Conoco's concession acreage as economically as possible. At that time, no further extension of this map series was intended. However, since these first four map sheets – 2523 Gilf Kebir, 2525 Ammonite Hills, 2823 Baris, 2825 El Kharga (Klitzsch & List 1978) – proved to be helpful for exploration work as well as for orientation in the field (Beall & Squyres 1980), it was decided to map additional areas of Egypt in the same manner to keep up with the increasing exploration activities of Conoco and Marathon.

Thus four additional sheets were produced extending to the north and the south of the original area (2521 Gebel Uweinat, 2527 Farafra, 2821 Selima, 2827 El Minya), and finally three more covering the Nile basin in the east (3123 Aswan, 3125 Luxor, 3127 Asyut). The entire project, including the printing of the map was finished in 1980 (Klitzsch & List 1978, 1979, 1980).

The methods used in the preparation of this preliminary map relied heavily on the experiences gathered during a first satellite map project in the Tibesti Mountains following the launch of the first Landsat satellite (at that time still named ERTS-1) in 1972 (List 1976, List et al. 1977, 1978, List, Helmcke & Roland 1975, List & Pöhlmann 1976, List, Roland & Helmcke 1974).

At the time of the launch of Landsat-1, several studies had been conducted on the use of satellite imagery for geological mapping, mostly referring to Gemini or Apollo photographs (Lowman 1969, Lowman, McDivitt & White 1966, Lowman & Tiedemann 1971) or to simulated ERTS imagery (Short & McLeod 1972). But, not surprisingly, practical experience with space photography or satellite scanner imagery was still limited. In addition, few scientists had access to early versions of analog or even digital image processing systems.

Thus, at that early stage, only film products produced by NASA from the digital satellite data of the Tibesti Mountains were available for interpretation. These photographic products were treated very much like any other small scale aerial photographs, enlarged optically, and interpreted visually. From the same system-corrected film transparencies as provided by NASA a photographic mosaic was constructed and used as an image base for the satellite map. In the case of the Tibesti map, second order equidensities were produced by photographic means using Agfacontour film (Ranz & Schneider 1970, 1972), and printed over the original image mosaic for enhancement of structural detail (List & Pöhlmann 1976).

As mentioned, the production of the preliminary geological interpretation map of Egypt basically followed these lines. Avoiding some of the problems that were encountered during the Tibesti map project, the following stepwise procedure was set up:

- 'quick-look' field trips before the actual interpretation to be able to conduct a 'pre-controlled' interpretation (Gérards & Ladmirant 1962)
- visual interpretation of black-and-white Landsat-MSS band 6 film transparencies photographically enlarged to 1:500,000 from the system-corrected film products as available from EROS Data Center in Sioux Falls
- production and printing of a conventional semi-controlled image mosaic base map from the same MSS data at the final map scale of 1:500,000, using all available topographic information from existing maps
- compilation of the pre-controlled interpretation on the printed base maps, making use, in addition, of other sources of information like published reports, maps, etc.
- field check and correction of this first stage interpretation map, including field work and sampling for further laboratory investigations
- re-evaluation and re-interpretation of the first stage map, taking into account the field data and subsequent laboratory results; corrections and additions
- cartographic processing of the geologic 'thematic overlay'; production of the printing plates for the three base colors used to generate all the colors needed for the rendition of the lithologic units (List & Pöhlmann 1976).
- printing of the thematic overlay on top of the satellite image base maps
- depending on the geologic situation and the information available, in some cases the steps of image interpretation, field verification and re-evaluation of the map had to be repeated several times until a satisfactory result had been achieved.

By 1980, eleven sheets of the preliminary *Geological Interpretation Map of Southwest Egypt* had been published, covering about 60% of the national territory of Egypt.

3.2 *New 1:500,000 map*

While the preliminary map had served the purpose for which it was meant well, that is to provide a first and provisional overview of the geology of an area widely unknown and unmapped until then, it had never been intended as a final product conforming to international map standards. Also, enormous progress had been made in the field of digital processing of Landsat data between the years 1977 and 1980 when the preliminary map was being prepared. Thus it was decided that a new mapping project would be undertaken which would take advantage of this progress in satellite image processing and of the results of additional geologic field and laboratory work.

Consequently, steps were taken in 1981 to start the production of the map covering the entire area of Egypt in twenty sheets at a scale of 1:500,000. This map was to meet the internationally accepted cartographic standards and to comply with the following demands:

- while the preliminary map had been produced in an arbitrary format selected for purely practical reasons, the new map would conform to the format and the grid of the international 1:500,000 World Map series (see Fig.3/1)
- the preliminary map had used a semi-controlled image mosaic base constructed from system-corrected Landsat frames, containing therefore all the positional errors inherent in that type of imagery (up to several kilometers in the worst cases). Additional errors had been introduced by conventional mosaicking of the photographic products. The new map, on the other hand, would be based on a geometrically corrected image mosaic conforming to the topographic accuracy required for the corresponding map scale
- the imagery used for this mosaic would be preprocessed for superior image quality, applying digital filtering and contrast-enhancement techniques
- in order to avoid tonal changes in the mosaic from one Landsat frame to the other, radiometric corrections would be applied to all Landsat scenes
- in all cases where the geologic information given on the preliminary map had been based on insufficient field data or on interpretation only, additional field and laboratory work would be carried out to verify the geologic information content of the new map
- in this context, special attention would be given to a proper correlation of the many geologic formations existing in Egyptian geology that had never been represented and tied together in one single map
- since the new map, in view of the absence of reliable topographic maps for a major part of Egypt, would certainly be used for orientation in the field as well as for purely geologic information, the topography shown on the map would be updated to include all major new roads, settlements, etc.

These stringent demands on both the geologic content and the geometric accuracy of the new map could be met, given the short time of about six years for the execution of the entire project, only by proceeding

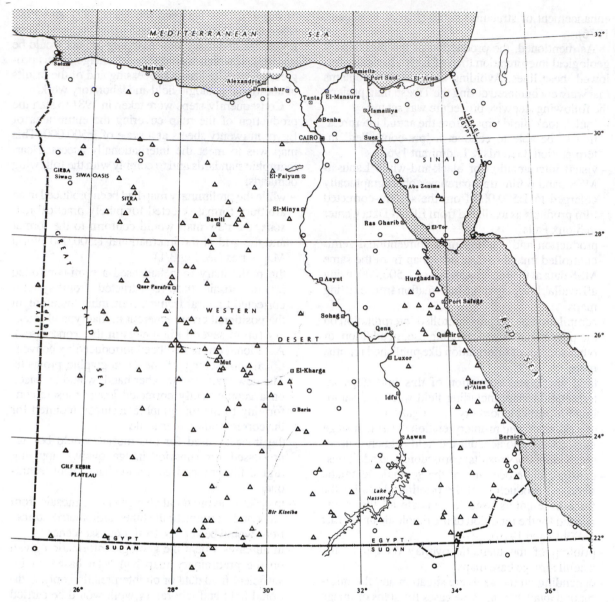

Figure 3.1 Names of map sheets and distribution of ground control points used for preparation of the 1:500,000 geological map. Circles = control points taken from existing maps; triangles = control points measured in the field.

along several parallel lines of work, making use of state of the art equipment and methodology together with a highly flexible and economic approach. Basically, two conflicting demands had to be satisfied:

On the one hand, the geologists needed an *appropriate base map from the very beginning of the project* on which all relevant geologic information as it became available from field and laboratory work, image interpretation, and literature reviews could be entered and compiled. On the other hand, in view of

the requirements on the positional accuracy of the map, it was obvious that only a geometrically and radiometrically *corrected digital mosaic* produced from the Landsat-MSS data could provide the kind of quality in the image base map that was called for. This meant, of course, that reliable control points identifiable in the satellite images had to be used for the geometric correction. Such control points were not available at that time. It followed that an intermediary stage had to be used in the mapping process to reconcile these conflicting objectives.

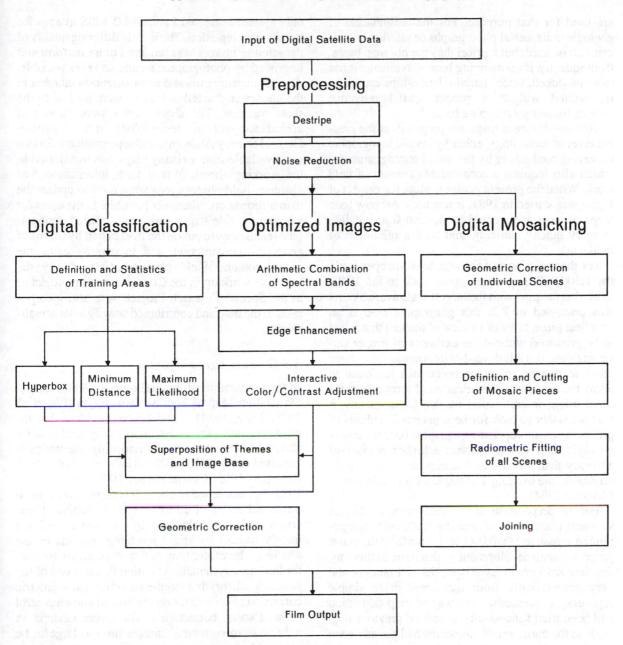

Figure 3.2 Flow diagram of digital image processing procedures.

This situation led to the development of the *Working Sheet* series at a scale of 1:250,000 (see 3.3). By proceeding in this way, the base map needed immediately for the beginning of the geologic part of the project became available on time; the first years of the project could also be used for the required field determination of control points (see 3.4). In addition, by careful planning of the logistically very demanding field trips, geologic field work and control point determination could be integrated, saving time and

expenses. Precise scheduling was a major concern in order to be able to fit together the different and divergent phases of the project at the right time, as shown in Figure 3/2.

3.3 *Working Sheet series 1:250,000*

An indispensable requirement for any geologic mapping activity is the availability of an appropriate mapping base. Normally, existing topographic maps are used for that purpose. For the delimitation of geologic units, aerial photographs or satellite images can also be used; but a priori they are not map bases. Consequently, if no mapping base is available, it has to be produced, since spatial relationships cannot be represented without a proper spatial reference system, meaning a mapping base.

Topographic base maps are prepared, at the present level of technology, either by classic topographic surveying methods or by the use of photogrammetry which also requires a considerable amount of field work. When the present geologic mapping project of Egypt was started in 1981, it was not clear how base maps of large areas could be derived from satellite imagery quickly, reliably, and with a minimum of field work.

For the major part of the area to be mapped, only the 1:500,000 map series going back to the 1940s existed as the map with the largest scale available. As was discussed in 2.2, this geographic map is an excellent piece of work in view of the fact that it had to be produced without the existence of proper surveying data. But it is devoid of information over large areas so that no reference points can be found to allow the necessary identification of terrain features or of image interpretation results. Consequently, it was necessary to look for new, untested methods of producing an appropriate topographic base if the new geologic map was to possess a higher amount of accuracy than would have been possibly by simply relying on the existing 1:500,000 map (Pöhlmann & Meissner 1984).

For the preparation of this new mapping base it was first planned to enlarge the 1:500,000 topographic base map to 1:250,000, and to perform the entire process of map development on that base. In this way, the final map could have been derived quickly and very economically from that base, using simple reprographic methods. This way of map derivation had been tried successfully in several previous map studies; the entire set of signatures had already been developed. However, this simple way could not be followed because it would have required geometrically corrected satellite image mosaics to be available from the very beginning of the project. As will be discussed in 3.4 and 3.5, the process of determining the needed field control points and of performing the geometric corrections in the computer took quite some time so that independent parallel routes had to be chosen for the execution of the project. The 1:250,000 scale Working Sheet Series (Pöhlmann, Meissner & List 1981, 1982, 1983, 1984) originated as one of these parallel routes.

Based on a grid construction calculated by G. Kramer (TFH Berlin), the individual working sheets were constructed using a semi-controlled mosaic of only system-corrected Landsat-1/3 MSS images for terrain representation. The widely differing quality of the satellite images was rendered more uniform and improved by photographic means, as far as possible. Any geometric errors and inconsistencies inherent in the uncorrected satellite images were pushed to the sheet margins. The sheet names were chosen in accordance with the terminology of the Egyptian General Survey Authority, and topographic information available from existing maps was transferred to the working sheets. Where such information was obsolete, field observations were used to update the map information, whenever possible. In the areas for which no usable larger-scale maps existed, topographic features were put on the map in part by means of geometric construction, and in part by using the image content. Field observations made by the scientists working in the Conoco mapping project or in the Special Research Project were also incorporated in the map and contributed heavily to its actualization.

3.4 *Geometric ground control*

In the preparation of the Preliminary Geological Interpretation Map of Southwestern Egypt (List et al. 1978, 1982, 1984) it had become evident that the positional accuracy of landmarks in typical system-corrected MSS images was not in the theoretically expected range of 150-200 m (Bähr & Schuhr 1974, Bernstein 1983, Colvocoresses 1972, Zobrist et al. 1983) but amounted to about 500 m, with occasional positional errors of up to several kilometers. These errors in images obtained by Landsat-1 to 3 are mainly caused by slight tumbling motions of the satellite. Therefore, they are random errors (unlike, for instance, systematic variation in the speed of the scanning mirror) that require an individual geometric correction of each scene on the base of known control points ('scene correction'). The errors inherent in only 'system-corrected' images are too large to be acceptable for the new map that was to conform to international map standards.

The theoretical positional accuracy of any map depends on the maximum drawing accuracy during production, that is 0.1 mm. At a map scale of 1:500,000 this value corresponds to ±50 m. In practical work, however, 0.2 mm is normally accepted as drawing accuracy which means ±100 m at our map scale. A precision of 0.1% in the representation of elevations as called for by Fritze et al. (1985) for orthophotos is not required in this case since relief-induced distortions in the Landsat images are minimal at the scale of 1:500,000, and since a reliable elevation network for the whole of Egypt is not available anyway.

A main objective was the production of Landsat image mosaics in the grid and format of the International World Map 1:500,000. The minimum number of 35 control points per Landsat scene (Kraus 1975, Baumgart & Quiel 1985) as deemed necessary for geometric corrections of Landsat-MSS data for middle Europe and the USA, or 135 control points for medium-scale map applications (Davison 1984) was out of the question for this project for practical reasons.

The usual way of obtaining ground control points from existing topographic maps at scales of 1:25,000 or 1:50,000 could not be followed with few exceptions since maps at these scales are only available for a few areas, e.g. the densely populated Nile delta. Generally, only topographic maps at 1:100,000 from different, partly obsolete, series could be used for point determinations. The areas covered by these maps, Nile valley, Mediterranean and Red Sea coasts, and Sinai amount to only 20% of the Egyptian territory. Consequently, for four-fifths of the country another solution had to be sought, that is control point determination in the field by means of satellite Doppler receivers.

Under the circumstances, the generation of a control point network according to European standards was out of the question. Only the absolute minimum of four control points per Landsat frame was aimed at. These control points had to be situated in the corners of the individual frames and thus in the overlapping parts of the images so that, ideally, one control point could be used in four adjoining image frames. The positioning and numbering of the point network was devised in such a way as to allow a possible later expansion of the network from four to nine points per scene. This would be of interest for further work, for instance for the production of selected sheets at a scale of 1:250,000 requiring more stringent geometric control. Whenever possible, more than four ground control points per frame were determined.

Orbital variations in the paths of the Landsat satellites make it impossible in most cases to find suitable overlap areas of four scenes that contain prominent landmarks that can be used for control points identifiable in several Landsat images. Normally, at least two out of four corner points have to be divided into two substituting points; sometimes, even three substitute points have to be selected. This procedure can only be followed if in the close neighborhood of the area selected for a control 'point' there are enough well-defined image features making it possible to pinpoint one 80 × 80 m sized picture element ('pixel') in the image as well as in the field. An additional prerequisite for the utilization of a theoretically suitable control point is accessibility by car. Since this can often be judged only on the spot, the point selection has to be handled in a highly flexible way, and often substitute points have to be found in the field. This requires field interpretation of the satellite images, preferably with the help of aerial photographs.

A special problem is posed by those Landsat frames that extend beyond the borders of Egypt where control point determination in the field is not possible. If no reliable maps are available for the areas in question, the distribution of control points will not be ideal, and 'floating' control points have to be used. In this procedure, image elements situated in the overlap areas of scenes with enough 'true' control points are selected and treated as auxiliary points.

The classic method for field determination of points in inaccessible areas by astronomic measurements has been substituted almost totally during the last decade by the development of Doppler satellite receivers for geodetic purposes. These receivers make use of the 'Transit' navigational satellite network that consists of four to six satellites operated by the US Navy. The polar-orbiting satellites transmit 150 and 400-MHz phase modulated signals that are received and processed by a portable antenna/receiver field unit. The Magnavox model MX-1502 used in the project records the phase-modulated data and Doppler frequency shift in both signals as a function of time. From the Doppler shift and digital data, it calculates the position of the reference marks on the antenna.

The calculations give the antenna position in terms of latitude and longitude, or of latitude, longitude and elevation above the reference ellipsoid and the geoid WGS 72. Each satellite pass within certain limits of orbital data and transmitting quality is used for improving the calculated values. As the theoretical positioning accuracy of better than 1 m is unnecessary for locating picture elements with a size of 80 × 80 m, and a drawing accuracy of 100 m in the map, a few good satellite passes are sufficient to give results with an accuracy of better than 10 m in horizontal position.

In practice this means that in many cases the determination of control points can take place during or together with the geologic field work, since generally flying camps are used (see 3.6) the position of which can often be adapted to the needs of control point measuring. This combination of geologic field work, control point determination, and survey of new topographic elements requires some compromises but results in a much increased efficiency of the field operations.

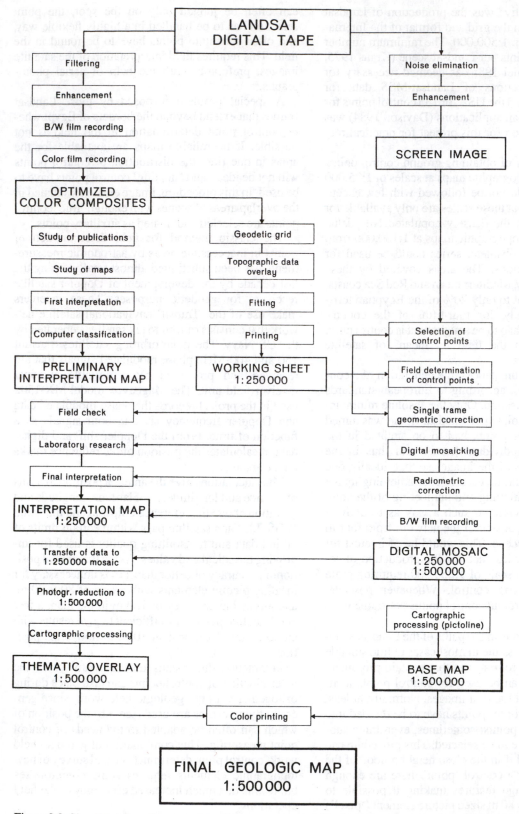

Figure 3.3   Simplified flow diagram of the different steps used for the preparation of the new 1:500,000 geological map.

3.5 *Digital image processing*

When the first satellite of the Landsat family, Landsat-1 (at that time still called ERTS-1), was successfully launched in 1972, digital image processing was a highly sophisticated and expensive field of research accessible only to a handful of selected scientists working in the very few research institutes around the world that had the necessary know-how and massive funding required at that time. Consequently, the first applications of Landsat imagery had to make use of whatever photographic products were obtainable from NASA and EROS Data Center that could be treated and interpreted like conventional photographic products, though they were of markedly lower image quality. With the rapid development of digital computers, computer memories, and mass storage devices, digital image processing is now within the reach of well equipped university institutes or private companies. As a result, the last few years have seen a dramatic increase in digital image processing capability and a marked improvement in the image quality obtainable and, as a direct consequence, an equally impressive rise in applications of remotely sensed image data. Today the success of any remote sensing project is directly linked to the digital image processing capability that can be brought to bear on the project. In this context, a good part of the improvement of the new geologic map over the preceding Preliminary Map (Klitzsch & List 1978, 1979, 1980) was possible only by the application of the full range of improved digital image-processing techniques (Fig.3/3).

All image processing activities necessary for the preparation of the new geologic map were performed on the *Geo*scientific *M*ulti-image *P*rocessing and *A*nalyzing *S*ystem, GEOMAPS, at the Free University of Berlin, Institute of Applied Geology, Remote Sensing Group. This system is built around an ISI/LogEtronics Earthview-IV image analyzer that is used for interactive image manipulation to determine the parameters necessary for the image transformation operations required. The Earthview-IV system can only handle screen-sized subimages of 512 by 512 pixels, so the actual image processing, mosaicking, and film plotting has to be done on a DEC 11/44 minicomputer, for which the necessary software had to be developed. Since the amount of data to be handled was rather large for that type of machine, a VAX computer was added to the system to improve the data throughput. A detailed description of the system used for this project is given by Schoele (1983), List, Meissner & Pöhlmann (1986), and List, Richter & Schoele (1978).

3.5.1 *Preprocessing*

The majority of the more than sixty Landsat frames required for a complete coverage of Egypt were purchased as computer-compatible tapes (CCT) from EROS Data Center, Sioux Falls. Most of these CCTs contain data recorded by Landsat-1, a few also by Landsat-2. Astonishingly enough, it is rather difficult to get cloud-free Landsat coverage over Egypt, especially over the Mediterranean coast. Therefore, an additional fifteen CCTs of Landsat-3 and Landsat-4 data were obtained from the European Landsat station ESRIN at Frascati, Italy. All these data are delivered in different digital formats and with a varying amount of corrections applied to the raw data.

Digital image preprocessing has to be applied to the raw data for the sake of image quality improvements that facilitate interpretation by the human interpreter and to provide a higher image quality where the data are used for corrected mosaics for the map base. Generally, there are three fields to which preprocessing has to be applied:

- contrast and edge enhancement, to produce a 'sharper' and 'clearer' picture
- removal of periodic noise, e.g. of the scan lines in the image
- removal of random noise that tends to 'blur' the picture.

The importance of preprocessing of digital Landsat data as a prerequisite for visual interpretation of geologic features is discussed by many authors. Since this is not the place to go into the details of image processing, it should suffice to point out that further information on this subject is to be found elsewhere.

3.5.2 *Computer-enhanced color composites*

The Landsat multispectral scanner, MSS, that is used on board Landsat-1 to 5, records the spectral reflectance of the earth's surface from a wavelength of 0.5 to 1.1 μm in four spectral bands. These spectral bands or 'channels' are numbered 4 to 7, denoting the green, red and two infrared sections of the electromagnetic spectrum. This multispectral subdivision of the electromagnetic radiation provides more information on a given surface than a 'pan-chromatic' recording in which the different wavelengths are lumped together.

Unfortunately, for rock surfaces this is true only to a limited degree. The information contained in the separate Landsat-MSS bands is highly redundant, the correlation coefficient between all four bands is always higher than 0.95 for different lithologic units under arid climatic conditions (Burger 1981). For vegetation, on the other hand, the situation is very different; the spectral differences allow digital classifications that are impossible to perform in the case of rock surfaces.

Still, the few percentage points of differing information recorded by the four spectral bands of

Landsat-MSS can provide quite valuable geologic information when properly enhanced and processed. The classic and still most widely used way of making the hidden spectral information in the four bands of Landsat visible to the human interpreter is the creation of so-called color composite images. To that end, the images recorded in the spectral bands 4, 5 and 7 (sometimes band 6 is used instead of 7) are color-coded and combined in such a way that a 'false color' image is generated which resembles a photograph taken with infrared color film. The main difference from the usual 'true color' image is that vegetation is shown in shades of red instead of green, due to the high and varying reflection of live vegetation in the near infrared. Since natural rock surfaces are characterized by highly similar spectral reflectance properties in the visible and the near infrared parts of the spectrum, for the purpose of geologic interpretation there is actually very little difference between true color and false color renditions of barren rock.

For the purpose of the map, color composites of the Landsat-MSS frames were generated on the GEO-MAPS system in such a way that the interpreter, who would normally have a good knowledge of the terrain shown on the image, participates in the process of the color composite generation on the screen. Thus by using the knowledge of the interpreter the image is manipulated interactively until the interpreter feels that the details important for geologic interpretation are emphasized in the best way possible. Subsequently the color composite is plotted on color film using the image transformation parameters that have been found to produce the best result for this given image and interpretation problem.

In the course of the present mapping project, new and better ways to create color images containing more useful information for the interpreter were tested. For instance, if a color composite is generated from band ratios instead of the bands themselves and then combined with a black-and-white 'intensity' image, the results were found to be clearly superior to a simple color composite (Salahchourian 1986). However, there are newer developments that also require a considerably higher amount of computer and operator time; for the interpretation requirements of the mapping project, 'normal' interactively optimized color composites were employed.

3.5.3 *Geometric correction and digital mosaicking*

As set forth in 3.4, the ground control points used for geometric correction had to be determined mostly in the field by means of satellite Doppler receivers, a rather time-consuming and expensive method. In order to keep within the time schedule set for the mapping project, the number of ground control points used for the correction of the more than 60 Landsat frames needed for a complete coverage of Egypt was reduced to the barest minimum of around 200 points. However, a controlled mosaic base relying on comparatively few ground control points is still quite an advance over what was available up to now, for instance, for the Western Desert of Egypt.

Because of the few control points available per frame for the geometric correction procedure, a linear interpolation algorithm was used. Cubic convolution would have given a better fit at the ground control points themselves but lower accuracy in between.

At the beginning of the project, the plan was to use individual geometrically corrected Landsat frames and to assemble these into a classic image mosaic. When during the course of the map project the need for a topographic base with the best accuracy obtainable became obvious, the decision was taken to also produce digitally matched mosaics for higher overall geometric accuracy and better tonal uniformity (Carlé 1982, Zobrist et al. 1983). This meant a considerably higher deployment of computer time that was partly offset by savings in the cartographic process.

The assembling of the Landsat-MSS frames for digital mosaics was done in the following way:

Subscenes of 512×512 pixels of the areas containing the landmarks used for ground control points were read from the computer tape and displayed on the image processing system. Since the ground control points were selected in such a way as to be mostly situated in areas where ideally four Landsat frames overlap, this generally meant that this process had to be repeated up to four times with different Landsat-MSS tapes. Using the data collected during the field determination of the control points, including photographs of the landmarks, relevant notes, and computer screen enlargements (up to 16 times), the pixels containing the measured ground control points were electronically marked in the image and stored.

This procedure is clearly the most critical part of the whole process for it is very easy to commit errors by misidentification of image elements. This is especially true when the same control point has to be identified in several satellite images taken under varying sensor angles, atmospheric conditions and during different seasons. In addition, the pixel raster of 80×80 m superimposed on the ground features is by its very nature different in the four images, making clean point identification even more problematic. The entire process is very time-consuming, and requires a great deal of experience. Preferably the point identification on the screen should be made by or together with the person who had selected and identified in the field the landmarks used for control points.

The input format of the ground control points is in

Universal Transverse Mercator (UTM) coordinates. Using the image coordinates of the marked pixels and the corresponding UTM values, the entire pre-processed Landsat frame is then recalculated to the UTM grid on the GEOMAPS system using, as mentioned earlier, a linear interpolation algorithm. In this process, the image is rotated so that the scanlines as they appear on the digital film plotter are oriented in an east-west direction. In addition, a remap of the Landsat-MSS image is performed that reformats the individual pixels (originally 79×56 m) to a size of 50×50 m, corresponding to the quadratic aperture of the raster plotter light-source. The result is a 'geocoded', geometrically correct image.

The residual errors in the variance/covariance matrix for each frame are then calculated, showing the accuracy of fit of the geometric correction, thereby immediately pointing out possible misidentifications of control points or input errors in one or more image subscenes by gross deviations in the variance of the calculated values. If the residual errors are considered satisfactory (typically 1 to 3 pixels), the actual geometric correction process is performed on a mini-computer (DEC PDP 11/44, with a 300 MByte hard disk storage capacity). Then the corrected scene is plotted on film by a raster plotter.

After all the frames needed for a single 1:500,000 map sheet (typically nine Landsat-MSS scenes) are corrected and plotted, they are manually assembled to determine the areas of overlap. Staying as close as possible to the ground control points, the corner points of the cuts for the mosaicking are determined, and input into the computer. The corrected Landsat frames are then digitally cut and joined together along the pre-determined lines, thus assuring a pixel-accurate assembly of the mosaic. This done, the mosiac is plotted on film, photographically enlarged to a scale of exactly 1:250,000, and checked for accuracy against the control points plotted in the UTM grid.

3.5.4 *Radiometric and contrast control*
An important feature of any good image mosaic is that it does not look like a mosaic composed of several individual images but rather like a single, uniform image. With respect to image geometry, this is achieved by the digital mosaicking process just described: terrain features continue uninterrupted – or nearly so – across the cut edge from one image into the next. Still, since in an east-west direction all, and in a north-south direction many satellite images are exposed on different days, and often in different seasons, under widely varying atmospheric conditions, the overall contrast and the range of gray levels in the individual images is generally rather dissimilar.

To produce the desired uniform appearance of the image mosaic that is essential when the mosaic is to be used as an image base for a map, the frequency distribution of gray levels in the image (representing the radiometric properties of the ground) and the overall range of gray levels (the contrast) have to be equalized from one image to the next. In this way the cuts between adjacent image frames become invisible to the eye.

In image processing theory, this is a fairly easy task: one has to select appropriate areas in the overlapping region of two images, and to calculate the frequency distribution in these areas with the help of the computer. Subsequently, the gray-value distribution of the second image frame is adjusted to that of the first one by 'histogram equalization', until a perfect radiometric match is achieved. In practice, unfortunately, there are several problems.

When two images that were taken at different seasons are combined, the sun elevations are different. If the spectral reflectance properties of the areas selected for radiometric adjustment vary with different illumination and sensor angles – and practically all natural surfaces show such a 'non-Lambertian' behavior to a certain degree – identical surfaces will produce different reflectance values, that is gray levels, in two neighboring images. That means that for purely physical reasons, no perfect fit can be achieved by this method. The problem becomes worse if the 'identical' areas used for radiometric adjustment are really different in their reflective properties. This happens whenever vegetation is present, and when the images used were exposed at different times. Then the same area may show lush green in one image and bare soil in another. In the desert, blown sand may significantly change the reflectivity of the land surface and make radiometric adjustment very difficult.

These examples show that the process of radiometric adjustment cannot be totally automated, and that the human operator is still needed to strike a compromise, when necessary. In some cases, like strong differences in vegetation, a complete adjustment may be impossible. Of course, a smoothing algorithm can be applied to such cuts but the basic problem still remains.

After the radiometric matching of the different image frames – which, as explained above, may sometimes not be quite perfect – the overall contrast of the mosaic has to be normalized. This standardization depends on the requirements of the reprographer and the printer, and is performed digitally. Since the contrast must not be too high for the ensuing printing process and the image, therefore, has a tendency to look 'flat', a mild edge enhancement algorithm is run over the final mosaic to give it a 'crisper' appearance.

3.5.5 *Plotting and photographic processing*

For the optimized color composites to be used for visual interpretation as well as for the digital mosaics to serve as image bases for the 1:500,000 map sheets, film outputs meeting the required accuracy standards had to be produced as the final stage in the image processing chain. For both products, an Optronics Colorwrite 4300 digital raster plotter was employed. The color composites were plotted on color reversal film at a scale of 1:1.18 million. This produced a color transparency with a format of 156.8 × 156.8 mm (skewed) when plotted with a pixel size of 50 microns.

There are basically two ways of producing a color composite image: in the first method, three black-and-white film plots are produced corresponding to the three spectral bands of the Landsat image used for the color rendition (generally bands 4, 5 and 7). These three images are carefully superimposed, matched, and perforated at the edge with a precision punch. In a photographic enlarger equipped with fitting studs the three images are consecutively projected onto color material, using yellow, cyan and magenta filters.

In the second method, the plotting is done directly on color film, resulting in a superior match of the three color images which are plotted with the inherent high precision of the instrument. The plots are repeated three times using appropriate color filters. The OPTRONICS instrument provides for both possibilities. For the map project, the second method of direct color plotting was used since it produced superior results for the requirements of the interpretation.

The mosaics were plotted on the same digital raster plotter employing the black-and-white mode. One 1:500,000 map sheet covers an area of 3 × 2°, corresponding to roughly 330 × 220 km. Since with regard to the accuracy of the map it was essential that the base mosaics were plotted in one piece each, the mosaics were recorded on film at a scale of 1:2 million. Using a pixel size of 25 microns, one map sheet covered an area of about 166 × 111 mm. With the use of fine-grained high-resolution film, the eight-times respectively four-times photographic enlargement of the master plot needed for the production of the 1:250,000 and the 1:500,000 image bases did not constitute a problem.

The color films as well as the black-and-white films were developed in an automatic COLENTA developing machine to guarantee the required quality and uniformity of the film products (Munier 1983). The fact that the plotting and film processing were performed by the same personnel helped considerably in making the necessary adjustments in the digital and photographic parts of the imaging process.

3.6 *Image interpretation and geologic data collection*

With the launch of the first earth observation satellite, ERTS/Landsat-1, several misconceptions about the potential of satellite data for geologic research arose; for example, that geologic field work, and with it geologic laboratory work, would shortly become close to obsolete because most of it would soon be taken over by automated computer classification of satellite data. Today these exaggerated hopes have come up against reality; and the reality is that remote sensing, while it cannot be expected to work miracles, has become an indispensible and extremely useful tool for geologic research (Baker 1975, Estes & Simonett 1975, Gold et al. 1973, Siegal & Gillespie 1980, Viljoen et al. 1975, Williams 1983). Without remote sensing, specifically without access to satellite image data and advanced image processing facilities, the present map could not have been produced within the existing constraints in timing and funding of the project. There is, however, no way to make sense out of remotely sensed data without considerable ground-based research.

The interpretation of the Landsat-MSS data for the map project was done visually by experienced interpreters, preferably the same scientists that also took part in the corresponding field work. Transparencies of optimized color-composite images (see 3.5.2) enlarged to the working sheet scale of 1:250,000 were used. For the actual image interpretation, the color composite film transparencies were put on a light table, and all pertinent data were annotated on clear transparent overlays with special ink.

A multi-scale approach is needed to obtain reliable geologic results (Sander 1948, 1950, Colwell 1975); satellite interpretation is but one source of information that has to be combined with field and laboratory work. Between 150 and 200 man-months of field work went into the preparation of the new map. An equal number of man-months was devoted to laboratory work involving biostratigraphic, sedimentologic and petrographic research.

The field work was planned and executed according to the progress of the satellite image interpretation. Typically, a preliminary, 'pre-controlled' image interpretation (Gérards & Ladmirant 1962) was performed first, based on whatever knowledge of the area was available from previous short field trips as well as from existing publications and maps. The results of this first-level interpretation were then transferred to the corresponding working sheets. Most important, all uncertainties and problem areas were marked on these maps for subsequent field checks. These maps, together with the interpreted original satellite images, were then verified in the field against the field evidence and corrected wherever necessary. If the need arose, an on-the-spot interpretation of the satellite imagery was also performed that could serve as a base for the following proper and careful re-interpretation of the images in the laboratory.

In this context, aerial photography was also used in order to clear up details not identifiable on the smaller-scale satellite images. However, the sheer number of aerial photographs covering the entire area of Egypt forbade a more extensive use of the classical photogeologic methods. Where necessary, the process of interpretation – field check, laboratory research, re-interpretation of the satellite imagery – was repeated iteratively until the results were considered satisfactory.

4 MAP PRODUCTION

The production of the present geologic map followed closely the methods of the 'Modular Map Production (MMP)' system as described by Pöhlmann & Meissner (1984). The basic idea of the MMP approach is to break up the cartographic processing into functional units, called modules (compilation, construction, reproduction, and printing) that can be individually and flexibly fitted to the exact requirements of the final product. In general, the sequence of operations is as follows:

0. Planning of the map production
1. Construction of the geometric base
2. Construction of the topographic base
3. Definition of thematic lines and symbols
4. Lettering
5. Coloring and screening
6. Image combination
7. Color proofing
8. Production of the printing plates
9. Printing.

The adaptability of the system was substantiated during the many changes and modifications that became necessary in the course of the project.

The geodetic calculations for the grid system were performed on the TFH computer in Berlin using programs developed by G. Kramer. The plots were output on a coordinatograph of the Survey Department of the Senate of Berlin. For the grid construction, the Universal Transverse Meracator (UTM) and the traditional Egyptian grid system were used. The accuracy of the calculation and the construction of the grid base are better than the drafting accuracy and the stability of the drafting base.

The coordinates of the topographic control points taken from field measurements and reliable maps were transformed in the same way. They were calculated with the same precision as the grid but their inherent accuracy is, of course, not as good since the basic data is of lower and varying reliability.

For the editorial preparation of the map, control points and, therefore, geometrically controlled mosaics were not yet available. For that reason, all preparatory work had to be done on the 'Working Sheet' series (see 3.3) with an inherently lesser accuracy. The definitive map content was defined and constructed on the base of the geometrically controlled image mosaic at the final map scale of 1:500,000.

An exception was made for the boundaries of the geologic units: their construction would not have been accurate enough had it been based on the geologic compilation on the (uncorrected) Working Sheets. In order to obtain the highest precision possible for these most important map elements, quarters of the final corrected image mosaics were enlarged to a scale of 1:250,000, the scale at which the interpretation of the Landsat images was performed. The geologic boundaries were then transferred from the original interpretations on the Landsat color composite transparencies to the corrected image base and transformed accordingly. The boundaries were then reduced to the final 1:500,000 map scale. After fitting together and mounting, a colored Cromaline copy would show the geologic information in its final form and position.

For field orientation and localization of the map content in relation to landscape and landmarks, a facsimile reproduction of the entire content of the image mosaic is, in principle, the best way. However, the presence of large rather dark-colored areas would result in a severe color shift of the hues used for the representation of the lithologic units. In order to avoid this color shift, as far as possible, and to still represent the major land structures, a pictoline process was used to get the desired results.

The lettering in the map set in type according to the preceding map design was then placed on the map taking special care not to interfere with geologic boundaries or structural lines like fractures or faults. Where necessary the lettering was set free. The geologic symbols, indispensable for unambiguous identification of the geologic units, were mounted into the remaining space.

The palette of colors for the geologic units was combined from the three base colors yellow, cyan and magenta. The portions of base colors for each unit was determined within a frame valid for the entire map, and photographic masks for each unit were prepared. Using these masks, the different levels of saturation for each color hue were then produced by screening. In some cases, especially within the Quaternary units, additional line patterns had to be introduced to make further differentiation possible.

The intermediate products that only existed as black-and-white screen masks up to this point were then combined in a first color proof copy to allow an evaluation of the effect of the color combinations used. After checking the color proof copies and making eventualy corrections, the maps were ready to go to the printer.

The map frame showing all ancillary information, geologic and otherwise, necessary for the proper use of the maps was defined in a separate layout. For reasons of economy of production, only in the very last phase was the transition from the smaller format of the colored map itself to the full format of the combined map and frame made. The maps were printed without a preceding proof at the facilities of the German Federal Institute of Applied Geodesy (Ifag), Berlin, using an eight-color offset printing process.

CHAPTER 4

Gravity map

HUSSEIN KAMEL

General Petroleum Company, Nasr City, Cairo, Egypt

The new Bouguer Gravity Anomaly Map of Egypt described in this chapter is compiled from available gravity surveys which were conducted by oil companies and from new measurements which were carried out for the unsurveyed parts of Egypt. In compiling this map, a National Gravity Standard Base Net (NGSBN,77) was established and tied to the International Base Net (IGSBN,71). The standard deviation of the observed gravity values of the national net ranges from +0.02 to 0.1 mgal. All previous gravity measurements were tied to this new net.

NATIONAL GRAVITY STANDARD BASE NET

The earliest absolute gravity measurements made in Egypt were carried out by the Anglo-Egyptian Oilfields Company using Holweck-Lejay pendulum in the years 1937-1940. The measurements covered a few stations (about 41 points) along a small area on both the eastern and western coasts of the Gulf of Suez.

In 1950-1951, Professor G. Woolard made 21 gravimetric measurements in Egypt as part of the world-wide gravity base net. This net was referred to the Potsdam absolute value which was then considered the world gravity datum. In 1974, a new world-wide gravity net was established (IGSBN,71) which introduced a new gravity reference datum replacing the Potsdam system. The enclosed gravity map of Egypt is based on a new National Gravity Standard Base Net (NGSBN,77) which was established on the basis of the IGSBN,71.

From January 1975 until June 1978, the General Petroleum Company carried out 624 gravity measurements throughout Egypt. These measurements were combined with the IGSBN,71 values to form 65 gravity base points of the NGSBN,77.

The gravity measurements were conducted using two Worden gravimeters. These instruments are temperature compensated and have a sensitivity of +0.01 scale units. The loop method was adopted and re-

peated observations were carried out at each point. Calibration runs to check the scale factor of each instrument were carried out occasionally. Instrument drift was recorded daily. Discrepancies in the performance of the two gravimeters used were checked by the repeated readings of the two instruments of one and the same station. The readings showed a correlation coefficient of about +0.02, which indicated that the recording of the two instruments throughout the whole net were consistent.

Gravity readings at stations located in the delta, the Nile Valley and along roads were made using suitable types of vehicles, while stations located in the remote areas of the south Western Desert were made using aircraft.

Measurements started mostly from IGSBN,71 base points and the computed gravity differences for each point were obtained from repeated back and forth measurements between consecutive stations. The mean square error for the obervations at the NGSBN,77 varies between +0.04 and +0.47.

Surveying and positioning of net

Geographic coordinates of the NGSBN,77 and their elevations above sea level were determined mainly by tacheometry. Most of the stations were tied to the available triangulation net. Stations located in the northern part of the Western Desert, the delta and the Nile Valley were tied to the second order triangulation net.

Points in the Gulf of Suez and the southern parts of the Eastern and Western Deserts were tied to the third and fourth order nets. The geographic coordinates of about twenty stations in the remote parts of the south Western Desert and in Sinai were interpolated from the available topographic maps of these areas.

Each station in the base net (except those located in airports) was monumented with a concrete incomplete cone about 80 cm high and 60×60 cm at the base. The notation of the station was engraved on an aluminum plate fixed on one side of the cone.

45

The location and distribution of the stations of the base net were governed by the extent of the measuring range and drift behaviour of the gravimeters used in the operation.

Adjustment of the NGSBN,77

The reduced gravity values, partially adjusted for drift, were analysed for final adjustments using a mathematical model based on U.V. Popov's method of successive iteration. Two sets of equations were formulated for both the polygons observed using vehicles and those observed using aircraft as the means of transportation. These equations were solved by the mathematical model through a prepared computer program. It is worth mentioning that the stations which were observed by both vehicles and aircraft were found to have consistent adjusted values after distributing the correction of the loops of observations measured by the vehicle or aircraft.

The adjustment is supposed to reduce the total gravity differences in all polygons of the net to zero, i.e. it would lead to sufficient accuracy relative to the IGSBN,71. The standard deviation of the NGSBN,77 varies between +0.02 and +0.1 compared to the IG-SBN,71 which ranges from +0.01 to 0.03 in Egypt. The final results of computations are to be found in the General Petroleum Company files.

Compilation of gravity measurements

Until the early forties, several gravity surveys were carried out in Egypt by the Anglo-Egyptian Oilfields, the Standard Oil, the South Mediterranean and the Socony Vacuum Oil companies. Instruments used were Thyssen, Carter and Mott-Smith gravimeters which have a sensitivity range between +0.5 and 1.0 mgal.

Gravity surveys using sensitive instruments such as Worden gravimeters were conducted by the Sahara Petroleum Company (SAPETCO) between 1954 and 1958. This company surveyed about 87,000 km² of the Western Desert to the north of latitude 28° N. Since 1964, several surveys have been conducted by Philips Petroleum, General Petroleum, AMOCO and GEOFISICA using sensitive instruments (Worden, North American and Lacoste). These surveys covered a large area of the Gulf of Suez and the Western Desert. All the data of the previous surveys were compiled, classified and ranked according to instrument sensitivity and measurement density (Table 1). The obervations amount to about 118,900, 60% of which was made by sensitive instruments. Tie-lines interconnecting these observations and the newly conducted survey were designed to take repeated observations along the interlocking nets of these surveys. Table 2 gives the statistics of these retied surveys, while Table 3 gives the statistics of the new surveys conducted in many areas chosen to complement the available data, and totalling about 100,000 km².

INTERPRETATION OF GRAVITY MAP OF EGYPT

The enclosed map is a Bouguer Gravity Map raised on the same scale (1:2,000,000) of the current Geological Map of Egypt (1981). It aims at depicting the main regional structural trends of Egypt. The data base of this map exceeds that required by international standards for maps of this scale.

Table 1. Previous gravity surveys (1922-1964).

Instrument	Sensitivity mgal	No. of stations	Area
North American	+0.01-0.05	20,700	ED, S, WD
Worden	+0.01	61,500	WD, D, NV, ED
Mott-Smith	+0.01	15,500	ED, S
Askania	+0.01	1,600	S
GAK-3M	+0.01	5,300	S, WD
Thyssen	+0.05	2,100	ED
Humble	+0.05	12,200	ED

ED = Eastern Desert; WD = Western Desert; S = Sinai; NV = Nile Valley; D = Nile Delta.

Table 2. Retied gravity surveys.

Area	No. of stations	Area (km²)	Retie
Fayum	6,500	12,200	15 base stations
ED	2,244	3,740	311 km + 2 base stations
NV	10,000	40,000	6 base stations
West Cairo	2,141	3,500	40 km
D	6,402	20,900	197 km
Sahara	26,000	100,000	89 km
South Siwa	1,259	6,250	89 km

ED = Eastern Desert; NV = Nile Valley; D = Nile Delta.

Table 3. Recent gravity surveys (1975-1978).

Area	Period mo/year	No. of stations	Area (km²)
Wadi Rayan	2/75–8/75	2,460	15,000
E. Qattara	8/75–3/76	3,733	16,000
NV	4/76–2/77	4,808	29,725
Wadi Natrun and N. coast	3/77–1/78	4,449	17,750
Gemsa and Hurgada	8/77–4/78	7,122	

The area to the north of latitude 28° N, which represents about 40% of the total area of Egypt, has a large density of gravity measurements (4-10 km²/station of survey) allowing for the construction of a gravity map on the scale of 1:100,000 with a contour interval of 1 mgal. One hundred and four sheets were prepared for this part of Egypt.

The area to the south of latitude 28° N has a smaller density (25-40 km²/station of survey) allowing for the construction of a gravity map on the scale of 1:500,000 or 1:1,000,000 with a contour interval of 5 mgal.

1 Northern Egypt

This part is characterized by a thick sedimentary section where most of the sedimentary basins are located. This part of Egypt is characterized by having several anomalous zones which are described as follows:

1. Anomalies of the Mediterranean coastal Hinge Zone: This zone is represented by a belt of intensive Bouguer anomaly variations; it extends along the coast from Salum to Port Said. Between Burg El-Arab and Alexandria, the belt is interrupted by a local increase of gravity values. In northern Sinai, the belt turns toward the sea, and the coast is outlined by local gravity highs. The regional gravity highs cover the Mediterranean coastal strip from Damietta to Rosetta and from Sidi Barrani to Mersa Matruh (Bouguer values 30-50 mgal) for a distance of 50 km inland. The latitudinal Nile Delta anomaly corresponds to the North Delta Embayment as outlined by Said (1981).

2. Anomalies of Syrian Arcing System: Local gravity anomalies form belts of northeastern-southwestern trend in north Sinai. The most prominent of these is the Gebel Maghara belt which is characterized by a narrow zone of high horizontal gradients on the southeast. This belt includes the local highs of Risan Aneiza, Maghara, Mafruth and Um Makhasa and seems to extend into mainland Egypt to include Genefe (Suez Canal zone), Helwan, El-Ayat (Nile Basin) and Bahrein (Western Desert) anomalies. A southeast shift in the anomalies of this belt would bring the local highs of Wasta and Abyad into this belt. To the north of Gebel Maghara belt lies a series of local highs extending in a southwest direction from west El-Arish to Katib Habashi.

A belt of local gravity highs crosses the north Sinai uplifts of Gebels Halal and Yelleg.

In the Western Desert, the Khatatba-Kattaniya High forms a distinctive belt of anomalies. Local highs occur along this belt: Camel Pass, Shoab Kasban, Rammak and Bahrein. Southeast of Khatatba, the belt bifurcates to form the Rammak and Shoab Kasban lines.

The gravity high anomalies of the Syrian arcing system are crossed at almost right angles by the lows of Khabra, Bardawil, Hassana (Sinai), Suez, Nile valley and Wadi Khadish-Beni Mazar (Western Desert).

3. Anomalies of Latitudinal Trend: The belts of anomalies with latitudinal trend reflect the Cretaceous structures of the offshore Mediterranean basin which forms part of the Unstable Shelf. They are well manifested along a belt which covers the local highs extending from Abu Sultan (Suez Canal zone) to Abu Za'abal (northwest Cairo) and then to Khatatba and onward to the Qattara ridge and Bir Fuad (Matruh-Siwa road).

A regional gravity low covers the east Mubarak-Gharadig-west Qattara basins. Lows of smaller size and lower intensity occur in Tanta, Bardawil (north Sinai) and Kanayis (Western Desert). The gravity low of Tanta (which coincides with the South Delta Block as outlined by Said 1981) is located between Alexandria, Damietta, Ismailiya and Wadi Natrun covering an area of about 25,000 km². It is outlined by the iso-anomalies −40 to −10 mgal. Its control portion is located in the Nile delta. In the area of Wadi Natrun and Burg El-Arab, this low is complicated by second magnitude local highs. In Kantara (north Sinai), the low has a gravity nose of latitudinal trend.

The low of Bardawil is located mainly under the Mediterranean waters. This low, as many other offshore lows, is lined to the south by a zone of local anomalies which extends all the way from the meridional line of Mamura-Gib Afia to the Great Bitter Lake. The latitudinal trend anomalies are also revealed in north Sinai, but against stronger anomalies of the Syrian arcing system.

4. Anomalies of Clysmic (Gulf of Suez) Trend: A belt of first, second and third order magnitude gravity anomalies extends in a northwest-southeast (Clysmic) direction in the Hurgada-Shadwan-Ras Mohamed areas. This belt is characterized by intensive gravity variations. It may be subdivided into the following three units:

a) Northern Zone of Gulf of Suez: This zone is characterized by first order gravity anomalies both in magnitude and size. The minimum value is about −50 mgal. It is bordered by a zone of intensive gravity variation within which the local gravity high of Ayun Musa and the gravity nose of Sudr-Lagia are located.

b) Central Zone of Gulf of Suez: This zone is also associated with first order gravity low anomalies of about −50 mgal. It is distinctly bordered by the strips

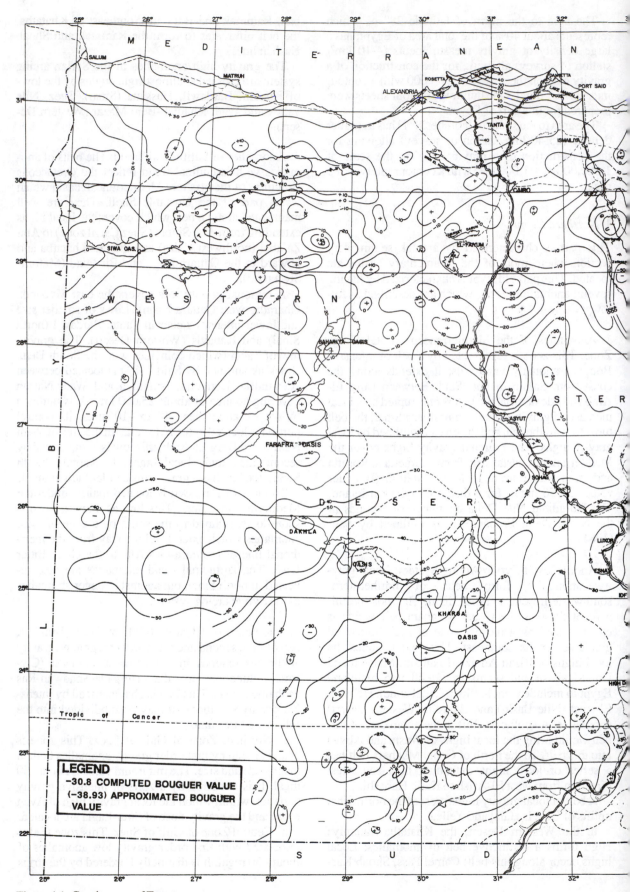

Figure 4.1 Gravity map of Egypt.

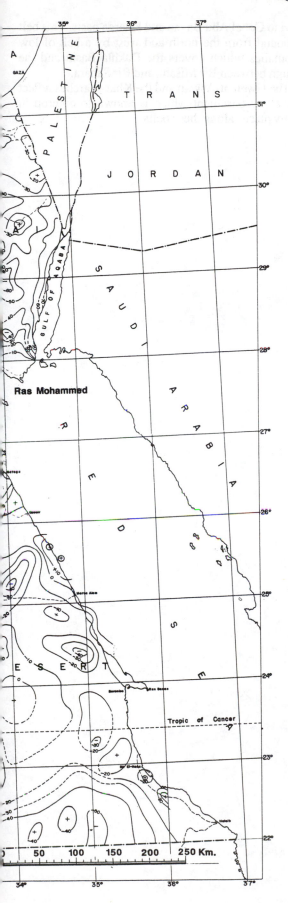

of intensive gravity variation of the meridional trend of Zafarana, Wadi Um Arta, Abu Rudeis-Belayim-Shukheir-Gharib and Lagia-Abu Zeneima areas. Within the axial part of the regional gravity low, local highs of clysmic trend occur: Ras Gharib, Bakr-Kareem and Gharib North. These are associated with the Kareem-west Ruahmi low which may form the Khashaba basin located to the west of Ruahmi and west Bakr.

c) Southern Zone of Gulf of Suez: Gravity anomalies in this zone differ from those of the other zones of the Gulf in being elongate in shape. They, like all others in the area between the Gulf and the Red Sea, follow the clysmic trend.

The first-order local low of Abu Rudeis–El-Tor lies in the Abu Durba area between Gebel Araba and El-Qaa plain. The minimum values are –50 mgal. A local high in the vicinity of Abu Durba is noticed in the Gebel Araba ridge. On the western side of the Gulf lies the Gebel Zeit local high which is followed to the west by a pronounced gravity low of about –40 mgal. This low may be attributed to the thick salt section of the subsurface Miocene in this part of the area.

Further to the south, at Esh El-Mellaha range a local gravity of 5 mgal is noticed. East of this positive gravity, there is a gravity low with a value of –40 mgal.

2 *Southern Egypt*

The Bouguer Gravity Map of Egypt to the south of latitude 28° N is compiled from maps raised to scales 1:500,000, 1:1,000,000 and 1:2,000,000. The most important anomalies shown on the map are summarized as follows:

1. A regional gravity low is observed in the Eastern Desert between the eastern bank of the Nile and the basement outcrops of the Red Sea hills. This low covers an area of about 80,000 km^2 and extends in a north-south direction between Qena in the south and Wadi Araba in the north. This low encounters four local lows of –50 mgal each. This low may reflect an old basin which could be older than the Gulf of Suez basin. The presence of this low was recently confirmed by the aeromagnetic survey conducted on this area in 1984.

2. A pronounced gravity high is seen north of Assiut. This anomaly is characterized by a steep horizontal gradient at its eastern flank. The amplitude of this high is about 10 mgal (between –20 and –10 mgal).

3. Several regional gravity anomalies are noted in the south Western Desert. They are in the form of two major belts:

a) The Uweinat-Aswan High which extends from Gebel Uweinat in a west-northwest direction to the Nile reach between Aswan and Abu Simbil. This broad belt is best manifest between Bir Dibis and Tushka, or from the intersecting point of longitude 30° E with the Egyptian Sudanese border and the Nile.

b) The Kharga High which extends from Bir Tar-fawi to Gebel Abu Bayan and Kharga oasis. This belt is bound from the north and west by a belt of low anomalies which covers the Dakhla Basin and the trough between Bir Misaha and Bir Sahara.

The Uweinat-Aswan and the Kharga arches reflect a shallow basement which is known to outcrop in many places along these belts.

CHAPTER 5

Seismicity

RASHAD M. KEBEASY

Helwan Institute of Astronomy and Geophysics, Egypt

Egypt is considered one of the few regions of the world where evidence of historical earthquake activity has been documented during the past 4,800 years. Instrumental recording of earthquakes started as early as 1899. Data of earthquakes which occurred in and around this country during the period from 2800 BC to 1984 have been gathered from local and international sources.

The aim of this work is to review critically the information of both the historical and instrumentally recorded principal earthquakes in Egypt and to investigate space and time distribution of the earthquake activity.

Data reliability and heterogeneity are important factors in defining the level of seismicity and determining its future recurrency. Although information of historical earthquakes is documented it cannot be regarded as complete as much of the old Egyptian literature was lost creating gaps in the earthquake record. Moreover, earthquake dating was the subject of differences among different authors (Poirier & Taher 1980). The location of these historical earthquake centers is almost exclusively along the Nile valley. The linear distribution of population causes little control on the east-west epicenter location.

Concerning recent activity (1900-1984), earthquake recording in Egypt started since 1899 with Helwan station. This station's instruments have been renewed several times and in late 1975 three other stations were added at Aswan, Abu-Simbil and Mersa Matrouh. As from July 1982 a radio-telemetery network of nine vertical short period stations was operated for monitoring microearthquake activity around the northern part of Lake Nasser. Configuration of this array covers an area of about 40 × 70 km. This network was expanded in 1983 to thirteen stations and some of them were equipped with horizontal and low gain components (Simpson et al. 1984). The signals from the individual stations are sent via radio-link in analogue form to a centre in Aswan where they are discriminated in real time and recorded on both magnetic tapes and drum recorders.

In case of instrumentally recorded data errors in the epicenter location of moderate and large earthquakes are within 30 km. This is due to the non-uniform distribution of teleseismic stations as well as local heterogeneity. Location of local and microearthquakes using only Helwan Station is precise in terms of distance but in some cases the azimuth can be erroneous by nearly 180°. The coverage is incomplete up to 1951. However with instrumentation improvements epicenter location has been improved considerably.

Maamoun et al. (1984) investigated the homogeneity of earthquake data and found that information of all earthquakes of magnitude equal to or larger than 5.0 is complete during the period from 1906 to 1981. Moreover, as from 1962, information of earthquakes of magnitude greater than 3.6 is also complete.

REVIEW OF HISTORICAL EARTHQUAKE ACTIVITY
(2800 BC TO 1900 AD)

Information on historical earthquakes is documented in the annals of ancient Egyptian history and Arabic literature. According to Sieberg (1932), Ambraseys (1961), Karnik (1969), Maamoun (1979), Ibrahim and Marzouk (1979), Poirier & Taher (1980) and Savage (1984), about 83 events were reported to have occurred in and around Egypt and to have caused damage of variable degrees in different localities. In the following paragraphs, the description of a few major historical earthquakes will be given.

2800 BC Sharquia Province earthquake
This unknown location earthquake was a severe one and caused deep fissures and soil cracks in Tell Basta, Sharquia province. The estimated maximum intensity is VII in a confined area near this village.

1210 BC near Abu-Simbil event
This event caused cracks in the temple of Ramses II

in Abu-Simbil, upper Egypt with an estimated intensity of VI. It is not certain, however, that the cracks are due to an earthquake.

221 BC earthquake

This earthquake had an intensity of VII at Siwa Oasis (Maamoun 1979). However, it caused destruction in about 100 locations in Libya. It is possible that this earthquake is the large one which took place in central Italy with an intensity of X and caused landfalls and diversion of rivers there.

27 BC Thebes, Upper Egypt, earthquake

This earthquake was a severe one and caused great damage leaving only four villages undestroyed in Thebes, upper Egypt.

1068 March 18, Aqaba earthquake

This is the first historical earthquake known to have strongly affected the Gulf of Suez area. It was located near Aqaba at the north end of the Gulf of Aqaba (Melville 1984, Ben-Menahem 1981). This event was felt strongly in Cairo where a mosque was damaged.

1303 August 8, offshore Mediterranean earthquake

This earthquake was placed by Sieberg south of Cairo because of the severe damage to many mosques and churches among which is the famous Amr-Ibn-El Aas mosque. Damage was also considerable in the Nile valley up to Qus in the south and Alexandria in the north where most of the town walls and the 120 m high beacon collapsed. It was also reported that this earthquake caused large scale damage in Rhodes and in Heraklion in Greece. Ambraseys (1961) placed its epicenter in the Mediterranean sea offshore of Egypt as As-Soyuti mentioned that the advance of sea submerged half of Alexandria. Kebeasy (1971) found that seismic waves from earthquakes occurring south of Greece and offshore Egypt travel to the south with abnormally low attentuation. Due to this phenomenon, effects of earthquakes from that region are usually found to be abnormally large in the south.

1847 August 7, Fayum earthquake

This earthquake was a remarkable event and had an intensity VIII in the Fayum region (Maamoun 1979). Eighty-five people were killed, 62 were injured, 3,000 houses and many mosques were destroyed. It was also felt over the whole area of Egypt. Heavy damage was reported as far as Assiut. In Cairo 100 people were killed and thousands of houses were destroyed. Moreover, thousands of people were injured and thousands of houses were damaged in different parts of the country. Intensity distribution of this earthquake is given in Figure 5/1(A).

1870 June 24, offshore earthquake

This earthquake was widely felt in Egypt and in different localities in Greece, southern Turkey and Palestine. This event, according to Maamoun (1979), had an intensity of VII in Alexandria and VI in many parts of the Nile Delta and Cairo, see Figure 5/1(B).

Recent earthquake activity (1900-1984)

Instrumental information of earthquakes during the period from 1900 to 1984 was collected from Sieberg (1932); Gutenberg & Richter (1954), Ismail (1960), Gergawi & El-Khashab (1968), Kebeasy et al. (1981), Maamoun et al. (1984), International Seismological Summary (ISS), National Oceanic and Atmospheric Administration (NOAA) as well as the Helwan station bulletin and Aswan radio-telemetry network. During this period, several significant earthquakes occurred in Egypt. Among these events are the following:

1955 September 12, offshore Alexandria earthquake.

This earthquake had a magnitude of 6.1 and was felt in the entire east Mediterranean basin and in Palestine, Cyprus, Dodecanise islands and as far as Athens (Rothe 1970). In Egypt, it was felt strongly and led to the loss of 22 lives and damage in the Nile Delta between Alexandria and Cairo. Destruction of more than 300 buildings of old brick construction were reported in Rosetta, Idku, Damanhour, Mahmoudya and Abu-Hommes. A maximum intensity of VII was assigned to a limited area in Bihira province where five persons were killed and 41 more were injured. Also an intensity of V to VII was reported in 15 or more localities, see Figure 5/1(C). Based on the isoseismal contours shown in Figure 5/1(C), Maamoun (1979) concluded that the epicenter of this event is located along the northern prolongation of the Red Sea-Gulf of Suez tectonic trend suggested by Sieberg (1932). According to Mckenzie (1972), the local mechanism of this event seems to suggest an oblique strikeslip motion, although the fit of the mechanism is poor due to conflicting data.

1955 November 12, Abu-Dabbab earthquake.

This earthquake had a magnitude of 5.5 and was felt in upper Egypt at Aswan and Qena and as far as Cairo but no damage was reported. Sykes & Landsman (1964) revised the location given in the International Seismological Summary using 95 stations and found little change. The depth is not well constrained but is clearly within the crust. The focal mechanism has normal and strike-slip faulting components produced by NNW minimum compressive stress and NE maximum compressive stress. Fault planes tend to strike roughly E-W or N-S to NE-SW.

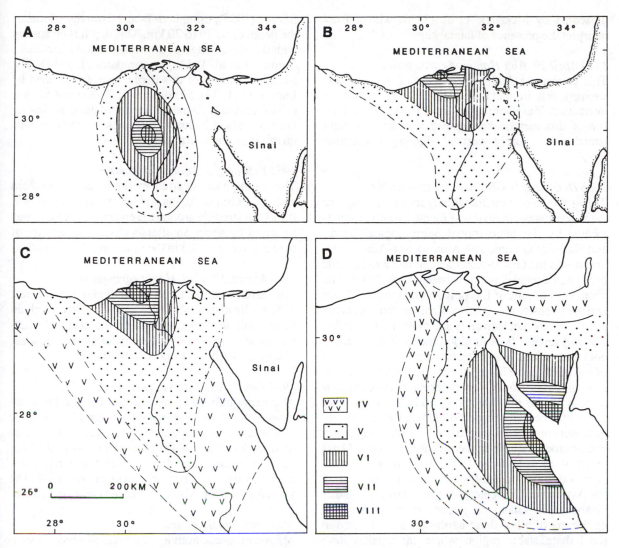

Figure 5.1 Intensity distribution of earthquakes of: A. August 1847; B. 24 June 1870; C. 12 September 1955 and D.31 March 1969 (A, B and C after Maamoun 1979, and D after Maamoun & El-Khashab 1978). IV-VIII = Earthquake intensity.

1969 March 31, Shadwan Island earthquake.
The magnitude of this earthquake is 6.3. Maamoun & El-Khashab (1978) studied the intensity distribution of this earthquake based on MSK scale and assigned a maximum intensity of IX to a small area in Shadwan Islands, Tawila and Jubal (Fig. 5/1(D)). On Shadwan Island landslides, earth slumps and rock falls were common. Fissures and cracks in soil were found with a main direction nearly parallel to the Red Sea-Gulf of Suez direction. At a distance less than 10 km west of the fractured area in the sea, one of the submarine coral reefs was raised by a few meters above the sea level after this event. A rumbling sound similar to thunder and noise caused by the sound of falling rocks on the land or the sea was accompanied by something similar to a large explosion and was asso-

ciated with sea wave disturbances. This earthquake was preceded by 35 large foreshocks during the last half of the month of March 1969. It was also followed by a large sequence of aftershocks. The epicenter of the main shock is located northwest of Shadwan Island. A rupture appeared to propagate to the south-east (Ben-Menahem & Aboodi 1971). The computed fault rupture length is 30 km. The depth of the focus was constrained by Ben-Menahem & Aboodi (1971) to be about 5 to 20 km using surface wave amplitude and spectral data. Savage (1984) using the data of Ben-Menahem & Aboodi calculated the depth to be 13 km. The focal mechanism as computed by Ben-Menahem & Aboodi (1971) point to normal faulting striking northwest with a component of left slip. This mechanism is consistent with the first motion data presented by Mckenzie et al. (1970) who did not interpret the presence of lateral slip.

1974 April 29, Abu-Hammad earthquake.

This earthquake had a magnitude of 4.9 and was strongly felt in the Nile Delta region. It had an intensity of V at Sharqia Province. The focal mechanism of this earthquake suggests a strike slip, right lateral with purely vertical dipping (Maamoun 1979).

1978 December 9, Gilf El-Kebir earthquake.

This is the largest instrumentally located earthquake in the southwestern region of Egypt. The epicenter as located by the International Seismological Center and National Oceanic and Atmospheric Administration lies in the Gilf el-Kebir area. It had a magnitude of 5.3 and a focal depth of between 7 and 10 km. The southwestern part of Egypt is an unpopulated desert and the intensity distribution of this earthquake is not estimated. Its mechanism is a combination of strike-slip and reverse fault motion. The fault plane could not be identified. The nearest station to the epicenter of this earthquake is Helwan at 850 km away and no fore- and aftershocks were detected. Microearthquake activity was not detected by Helwan. Many other non-instrumentally recorded earthquakes in this region were reported. Daggett et al. (1980, 1986) report one microearthquake in the vicinity of Dakhla Oasis about 400 km northwest of Aswan. Kebeasy et al. (1984) through the analysis of the data obtained by the Aswan network detect several small earthquakes at about 150 km to the northwest of Aswan. The epicenters of all of these earthquakes are located in the Sahara stable region where earthquakes have never been instrumentally recorded and was considered an aseismic zone (Gutenberg & Richter 1954).

1981 November 14, Kalabsha earthquake.

This earthquake had a magnitude of 5.5 (Helwan station) or 5.1 (NOAA) and it is of significance because of its possible association with Lake Nasser. Although its epicenter is located in Kalabsha about 60 km southwest of Aswan, it was strongly felt in Aswan, and in areas to the north up to Assiut and to the south up to Khartoum (Fig. 5/2). The intensity near the epicenter is between VII and VIII. Several cracks on the west bank of the lake and several rock-falls and minor cracks on the east bank were reported. The largest of these cracks is about 1 m in width and 20 km in length (Kebeasy et al. 1982). This earthquake was preceded by three main foreshocks and followed by a large number of aftershocks. The focal depth of this earthquake seems to be very shallow (Kebeasy et al. 1982). The ISC and NOAA estimate the depth to be 0 and 10 km respectively.

Savage (1984), using both P- and S-waves, estimates the depth to be 19 to 20 km. This depth is consistent with the depth range of the well-located aftershocks (Simpson et al. 1984, Toppozada et al. 1984). The focal mechanism of this earthquake was studied by Maamoun (1982), Simpson et al (1982) and Savage (1984) and was found to be right-lateral strike-slip fault (see also Chapter 6, this book, for more details on this important earthquake).

1983 February 3, Aqaba earthquake.

This earthquake occurred in Aqaba at the northern end of the Gulf of Aqaba and had a magnitude of 4.9. It was felt strongly around the epicentral area and was followed by about 56 aftershocks of magnitudes of between 1.7 and 4.85 in the following three weeks.

1984 March 19, Wadi Hagul earthquake.

This earthquake occurred in Wadi Hagul southwest of Suez. Its magnitude is 4.7 and was felt strongly in Suez, Ismailia and Cairo. Large numbers of aftershocks were recorded by nearby temporary stations. The focal depth is estimated to be 10 km.

1984 July 2, Abu-Dabbab earthquake.

This earthquake had a magnitude of 5.1 and was felt strongly in Aswan, Qena and Quseir. Five portable field stations were operated in the Abu-Dabbab area as from 19 June 1984 till the end of August of the same year. Large numbers of foreshocks and a tremendous sequence of aftershocks are recorded. The focal depth of the whole sequence is less than 12 km.

Microearthquake activity

Microearthquake activity has been observed in various regions in Egypt. The Helwan Station Observations point out such activity around Cairo, the Nile Delta and around the Gulf of Suez. Using these observations, Ismail (1960) located a number of microearthquakes around Cairo in the period from 1903 to 1950. Also Gergawi and El-Khashab (1968) have located a large number of microearthquakes around Cairo, the Gulf of Suez and the Nile Delta region and defined an active trend that runs along the Gulf of Suez and passes through the Nile Delta to the Mediterranean Sea. Kebeasy & Maamoun (1981) using Helwan observations found a microearthquake active trend that starts from Cairo and runs to the north along the west side of the Nile Delta. They attribute this activity to a probably active fault along this trend. Daggett et al. (1980, 1982) found two areas of intense microearthquake activity in the Abu-Dabbab area (25° 28' N, 34° 52' E) and the southern end of the Gulf of Suez (for more information on microearthquake activity, see Chapter 6, this book).

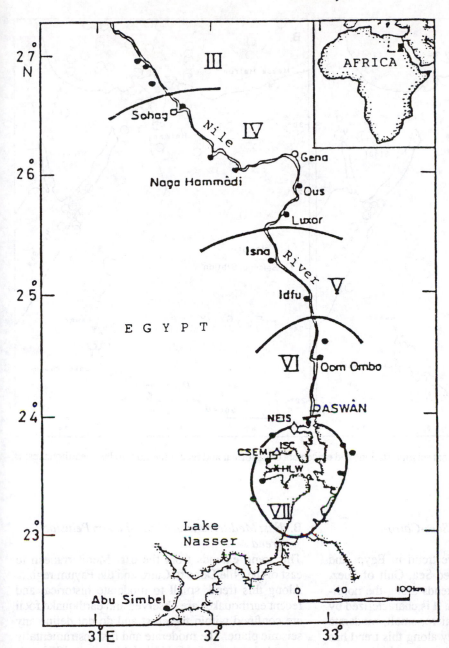

Figure 5.2 Intensity distribution of the 14 November 1981 earthquake as located by HLW = Helwan; NEIS = National Earthquake Information Service and ISC = International Seismological Center.

Tsunamis

Tsunamis or sea waves accompanying earthquakes are reported to have affected the Mediterranean coast of Egypt several times with varying intensities. The largest of these is the one due to the earthquake of 8 August 1303. As-soyuti (after Ambraseys 1961) stated that the advance of the sea due to that earthquake submerged half of the town of Alexandria. Other tsunamis of smaller intensity were reported in 24-20 BC, 1202, 1262 and 1908 AD.

SPACE DISTRIBUTION

Distribution of epicenters of both historical and instrumentally recorded earthquakes is given in Figure 5/3 for moderate to large and for small earthquakes. Figure 5/4 shows the epicenters of all of these earthquakes and the microearthquakes, the focal mechanisms of some principal earthquakes and the location of the seismographic stations. The distribution of earthquake epicenters suggests that the activity tends to occur along three main seismic active trends as follows.

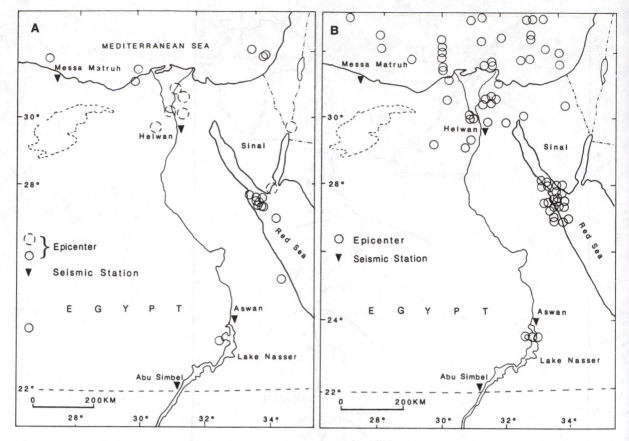

Figure 5.3 A. Location of permanent seismic stations and epicenters of historical and recent medium to large earthquakes; B. epicenters of small earthquakes.

A. Northern Red Sea–Gulf of Suez-Cairo-Alexandria Clysmic-Trend

This trend is the major active trend in Egypt and extends along the northern Red Sea, Gulf of Suez, Cairo and Alexandria and extends along the northwest in the Mediterranean Sea. It is characterized by the occurrences of shallow, micro, small, moderate and large earthquakes. Activity along this trend has increased in recent years. All earthquake focii are limited within the crust. Accuracy of depth determination does not allow definition of any seismic plane. The activity along this trend is mainly attributed to the Red Sea rifting as well as several active faults. A seismic gap exists in the Gulf of Suez where no earthquakes have been detected. Another gap is found between Cairo and Alexandria where only microearthquakes are frequently observed. If the activity in the Abu-Dabbab area is linked to this trend there will be a third gap between Abu-Dabbab in the south and latitude 27° N. However, the author believes that the Abu-Dabbab activity is of a local nature and can be attributed to intrusion and not to the Red Sea rifting.

B. East Mediterranean–Cairo-Fayum Pelusiac Trend

This trend extends from the east Mediterranean to east of the Nile Delta to Cairo and the Fayum region. Along this trend, small to moderate historical and recent earthquakes are observed and earthquake focii are confined within the crust and do not define any seismic plane. The moderate and first instrumentally recorded event in the Gilf El-Kebir area in 1978 may form the extension of this trend into the southwestern parts of the Western Desert although there are few earthquakes between the Fayum and Gilf El-Kebir areas. Aswan radio-telemetry network has detected several small events from the area to the northwest of Aswan.

C. The Levant-Aqaba Trend

This trend is a continuation of the Levant active fault and extends along the Gulf of Aqaba and southwest in the Red Sea and bisects the clysmic trend at about 27° N and 34.6° E. Earthquake occurrences are found mainly in both ends of the Gulf of Aqaba. Only

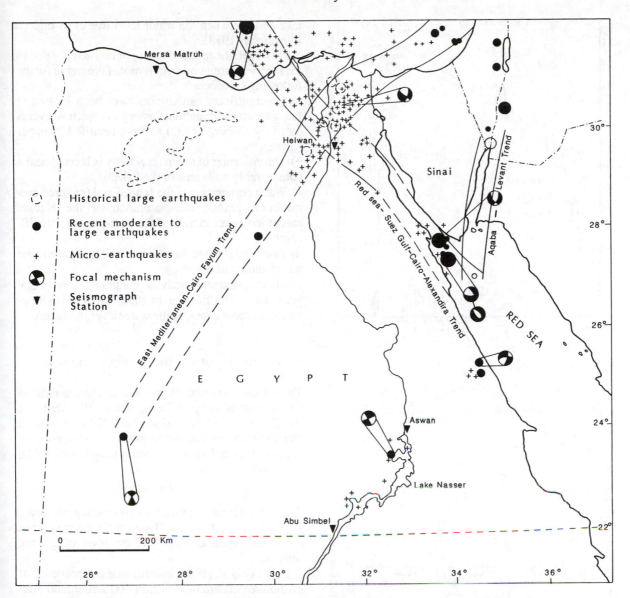

Figure 5.4 Epicentral distribution of all earthquakes, focal mechanisms of principal earthquakes and active seismic trends.

shallow small-size earthquakes are observed along this trend.

In addition to these trends there are several areas known to be active such as southwest of Aswan, Abu-Dabbab, Gilf El-Kebir and Wadi Hagul west of the Gulf of Suez. The activity in these areas is of a very local nature as previously discussed. Aswan activity is of particular interest and its nature is discussed here in detail.

Figures 5/5 and 5/6 show the sequence of earthquake activity which preceded and followed the earthquake of 14 November 1981 which took place in the Kalabsha area south of Aswan, together with the fluctuation of water level in Lake Nasser. It is likely

that the cause of this activity is tectonic rather than due to the filling of the lake for the following reasons:

a) The existence of sets of faults in an east-west direction in the Kalabsha area where the main shock took place;

b) The poor permeability of the Precambrian granite bedrock and the negligible number of microearthquakes recorded at least since late 1975 when the Aswan and Abu-Simbil stations were operated;

c) The occurrence of the earthquake of 14 November 1981 some 17 years after the filling of the

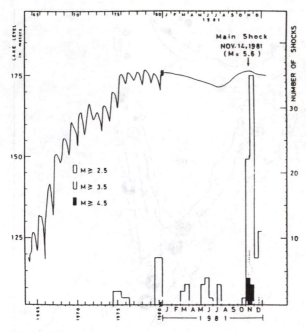

Figure 5.5

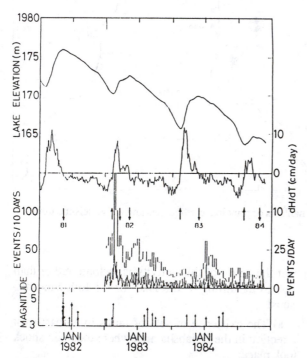

Figure 5.6 Water level fluctuations in Lake Nasser and number of earthquakes per 10 day intervals during the period July 1982 to December 1984 (after Simpson et al. 1986).

lake and not when the water level was at its highest (November 1978).

However, the induced nature of this activity due to the reservoir impoundment is more favourable for the following evidences:

a) No significant earthquakes have been located in the lake area throughout history except few events listed by Sieberg (1932) to have been felt in upper Egypt.

b) The epicenter of the main activity is located near a considerably wide area of the Lake.

c) Water penetration in the faulted and fractured precambrian granite and saturation of the Nubian sandstone cover increased the pore-pressure significantly.

d) The Lake is the second largest artificial Lake in the world and relatively deep.

e) Microearthquake activity continued for several years later. The number of these earthquakes fluctuates, in some cases, as the water level fluctuates.

FREQUENCY OF EARTHQUAKE OCCURRENCES

The number of earthquakes is a basic characteristic of the seismic activity of any given locality during a specific period of time. Gutenberg (1958) observed that the logarithm of the number of earthquakes of magnitude M or larger is inversely proportional to the magnitude M or

$$Log N = a - bM$$

where a and b are constant for a specific region during a specific period of time. However, they vary from region to region according to the level of seismic activity.

Kebeasy et al. (1981) investigated the frequency of earthquake occurrences within 200 km around Alexandria and determined a and b as 2.85 and 0.45 respectively and estimated that the largest event expected to occur in the region should have a magnitude of between 6.3 and 6.7 with a recurrency of eight times every 1,000 years. Kebeasy & Maamoun (1981) studied the seismic activity within an area of 200 km around Cairo and estimated a and b values as 2.31 and 0.37 respectively. They also estimated that the largest expected event should have a magnitude of between 6.0 and 6.5 with a recurrency of three times every 100 years. Both results were obtained on the basis of 25 year periods of time (1953-1978). Data used during this time period is considered to be homogeneous. Savage (1984) investigated the activity around the Aswan during the interval from 27 BC to 1984 and estimated the recurrence of earthquakes of a magnitude of 5.5 or more occurs once in approximately 300 years.

Kijko et al. (1985) used the aftershocks of the 14 November 1981 earthquake recorded by the Aswan radiotelemetry network during the period from July 1982 to December 1983 and estimated the b value for this homogeneous set of data as 0.7.

Based on this estimate, Savage defined a frequency-magnitude relationship for the Aswan region as

$$\text{Log } N = 4.7 - 0.7 \, M$$

for a period of 2011 years.

The frequency of shallow earthquake occurrences in the Gulf of Suez region during a period of 29 years (1953-1981) was studied by Kebeasy et al. (1984). The values of a and b are 2.46 and 0.39 respectively.

These studies indicate that, in general, the level of earthquake activity in Egypt is low.

CHAPTER 6

Some aspects of the geophysical regime of Egypt in relation to heat flow, groundwater and microearthquakes

FOUAD K. BOULOS

Geological Survey of Egypt, Abbassia, Cairo, Egypt

1 HEAT FLOW

Egypt lies in the northeast corner of the African continent close to the northeastern margins of the African plate (Le Pichon et al. 1973: 83). The northern boundary of the African plate in the eastern Mediterranean is poorly defined (McKenzie 1972) and the nature of the intermediate thickness of the eastern Mediterranean crust (20-25 km) is uncertain (Ryan et al. 1970, Lort 1971). To the east there is convincing evidence that the Red Sea is a young proto-ocean (see Girdler & Styles 1974) and, in the northeast, the Red Sea bifurcates into the Gulf of Suez and Aqaba around the small Sinai subplate (Le Pichon et al. 1973: 95-103). Heat flow determinations in the eastern Mediterranean have all resulted in similar but relatively low values (Ryan et al. 1970, Morgan 1979). In contrast, heat flow determinations from the Red Sea yield erratic, but generally high values, probably related to a median spreading center (Girdler 1970, Haenel 1972, Evans & Tammemagi 1974). Heat flow data from northern Egypt, therefore, provide a link between these two contrasting heat flow provinces. Different attempts have been made to estimate the geothermal gradient and heat flow in different regions of Egypt.

Geothermal gradients in the Gulf of Suez

Tewfic (1975) was the first to study the geothermal gradients in the Gulf of Suez from oil well temperature data, from the point of view of its relevance to hydrocarbon exploration. He calculated an average geothermal gradient in the Gulf of Suez (Fig. 6/1) of 26.1 mK/m (°C/km) from 321 oil wells in which Bottom Hole Temperature (BHT) data were available. The geothermal gradients were calculated from the equation

$$\text{Geothermal gradient} = \frac{\text{BHT} - \text{MST}}{D}$$

where MST is the mean annual surface temperature assumed to be 26.7°C and D is the depth at which the corresponding temperature is recorded. Gradients calculated in this manner are in error of less than 10% (Harper 1971, Evans & Coleman 1974) due to transient temperature disturbances encountered during conducting routine logging runs in newly drilled holes. Gradients ranged from 11 to 57 mK/m. Tewfic (1975) distinguishes three main regional hot trends for gradients higher than 27 mK/m in the Gulf of Suez running NW-SE subparallel to the shoreline: the first between Ras Shukeir and Amer field, the second between Amal field and the eastern part of Morgan field, and the third between the Nebwi area and Sudr (Fig. 6/1). No obvious direct relationship between thermal trends and oil accumulation has been found. For example, Ras Gharib and Bakr oilfields occur in rather hot areas, while Belayim Marine and Land oilfields represent cold areas of gradients lower than 24 mK/m. The distribution of geothermal gradient hot and cold areas in the Gulf of Suez, however, appeared in his opinion to be governed by major structural elements where the hot areas correlate well with gravity highs.

He also studied in detail the geothermal gradients at three oilfields and the data are summarized in Table 6/1. Studying a geothermal gradient profile passing by Bakr, Ras Gharib and July oilfields and its relationship with the Miocene evaporites and pre-Miocene relief in a diagrammatic cross section constructed between these oilfields, he noticed an excellent matching between the pre-Miocene surface and the geothermal profile. In his opinion, a direct relation appears to exist between the geothermal trends in the Gulf of Suez and the pre-Miocene relief. The hot trends are indicators of high pre-Miocene blocks while the cold trends indicate deeper pre-Miocene blocks. Besides, Miocene salt distribution accounts for most of the cold trends present in the Gulf of Suez.

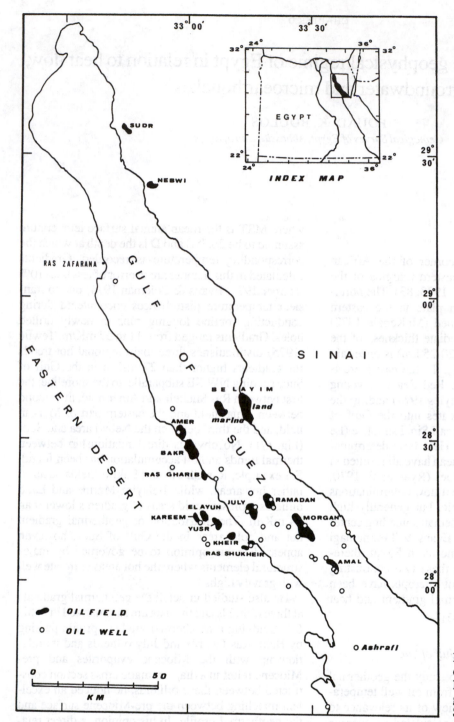

Figure 6.1 Location map of oilfields and oil wells from which BHT data were collected by Morgan et al. (1977) in the Gulf of Suez.

Geothermal gradients and heat flow in northern Egypt

To provide geothermal reconnaissance data for northern Egypt, geothermal gradients have been estimated by Morgan et al. (1977) using BHT data from 128 oil wells in the northern Western Desert of Egypt

Table 6/1. Geothermal gradient values in the Gulf of Suez.

Oilfield	No. of oil wells	Geothermal gradient (mK/m)	
		Average	Maximum
Ras Gharib	54	31.4	42.1
Belayim Marine	8	21.5	23.7
Morgan	80	23.9	32.5

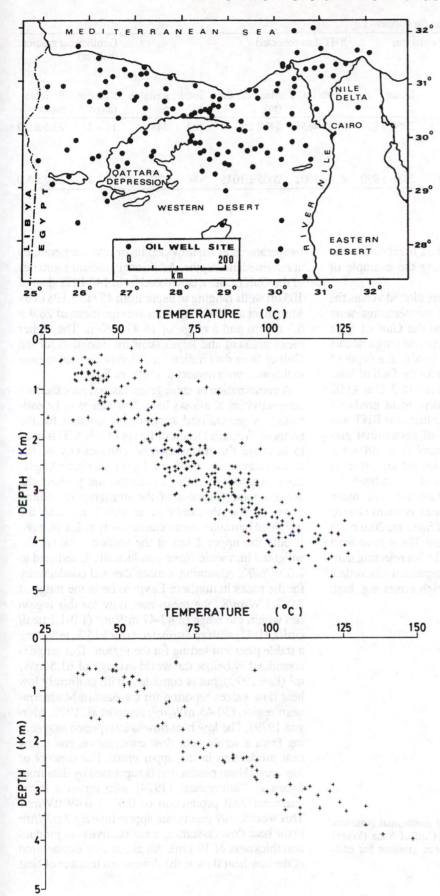

Figure 6.2 Locations of temperature data oil wells, northern Western Desert and Nile delta (after Morgan et al. 1977).

Figure 6.3 Data of 289 oil well BHT versus depth at northern Egypt. Calculated temperature gradient by least squares linear regression analysis is 18.98 mK/m (after Morgan et al. 1977).

Figure 6.4 Data of 76 oil well BHT versus depth at the Gulf of Suez. Calculated temperature gradient by least square linear regression analysis is 24.62 mK/m (after Morgan et al. 1977).

Fouad K. Boulos

Table 6/2. Geothermal gradient values for northern Egypt.

Region	No. of oil wells used for collecting data	Depth of wells (m)		BHT data collected						Geothermal gradient (mK/m)
		Range	Mean	Depth (m)	Mean depth (m)	Total no.	Reliable no.	Reliable range	Reliable mean	
North Western Desert of Egypt and the Nile delta	128	292 – 4656	2958 ± 937	114 – 4656	2408 ± 1112	289	248	13 – 27	20.6 ± 2.9	
Gulf of Suez	38	457 – 4002	2436 ± 970	457 – 4002	2005 ± 1045	76	70	12 – 40	26.7 ± 5.0	

and the Nile delta (Fig. 6/2) and 38 boreholes in the Gulf of Suez (Fig. 6/1) following the example of Tewfic (1975).

The collected data of BHT were plotted versus the corresponding depths at which temperatures were recorded for northern Egypt and the Gulf of Suez (Fig 6/3 and 6/4 respectively). Reported temperatures for northern Egypt ranged from 30.6°C at a depth of 372 m to 126.7°C at 4,197 m and for the Gulf of Suez from 36.1°C at a depth of 670 m to 148.3°C at 4,002 m. From a single borehole, independent gradients have been calculated where more than one BHT was available. The mean and range of geothermal gradients and the corresponding number of oil wells from which BHT data were collected are given in Table 6/2. Frequency histograms of all gradients for both regions are shown in Figure 6/5. The mean gradients estimated for the northern Western Desert/ Nile delta region and the Gulf of Suez are 20.6 ± 2.9 and 26.7 ± 5.0 mK/m respectively. These have been calculated from reliable gradients after rejecting temperature values considered to represent discordant measurements with unusually high errors e.g. high

temperatures at shallow depths or low temperatures at intermediate depths. Additional gradient estimates of the Gulf of Suez calculated from 143 BHT data of 105 oil wells ranging in depth from 457 to 5,198 m by Morgan et al. (1983) give an average mean of 26.9 ± 5.7 mK/m and a range of 14-47 mK/m. The higher mean gradient and larger standard deviation for the Gulf of Suez data indicate a different thermal regime in this area with respect to northern Egypt.

A combination of these gradients and rock thermal conductivities K allows heat flow values to be estimated. A generalized stratigraphic column for the northern Western Desert of Egypt (Table 6/3) is used to estimate the mean thermal conductivity of the rocks relevant to northern Egypt geothermal gradient. Assuming all the formations are present, the average total thickness of the stratigraphic column down to the Precambrian is 6,460 m, and the weighted harmonic mean conductivity is 2.3 W/m/K. If only the upper 2 km of the section is taken, the weighted harmonic mean conductivity is reduced to 2.0 W/m/K. Assuming a mean thermal conductivity for the rocks in northern Egypt to be in the range of 2.0-2.3 W/m/K, the mean heat flow for this region lies within the range of 42-47 mW/m^2 (1.0-1.1 μcal/ cm^2/s, HFU) with an estimated error ±15%, reflecting a stable platform setting for the region. This range is considerably below the world average of 61.5 mW/ m^2 (Lee 1970), but is consistent with uniformly low heat flow values reported for the eastern Mediterranean region (30-45 mW/m^2: Ryan et al. 1970, Morgan 1979). The low heat flow is interpreted as resulting from a stable heat flow contribution and a low heat production in the upper crust. The concept of low crustal heat production is supported by data from Evans & Tammemagi (1974) who report a mean basement heat production of 0.67 ± 0.49 μW/m^3. This would only contribute approximately 7 mW/m^2 to the heat flow (assuming a radioactive heat production thickness of 10 km). An alternative explanation of the low heat flow is the downward transfer of heat

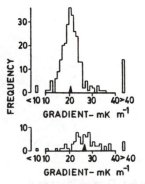

Figure 6.5 Histograms of calculated geothermal gradients for northern Egypt (upper) and the Gulf of Suez (lower) (after Morgan et al. 1977). The mean gradient for each region is indicated by the arrow head.

Table 6/3. Northern Egypt generalized stratigraphic column and assumed thermal conductivities.

Age formation	Aver. thickness (m)	Basic rock type	K (W/m/K)
Pliocene	58	Calcarenite and Sdy, Lst	2.1 (1)
M. Miocene Marmarica	183	Limestone	2.2 (2)
L. Miocene Moghra	213	Sand and gravel	2.1 (1)
Oligocene + U. Eocene Dabaa	503	Shale	1.5 (3)
M. + L. Eocene Apollonia	122	Limestone	2.2 (2)
U. Cretaceous Khoman	244	Chalk	1.8 (4)
U. Cretaceous Abu Roash	457	Limestone and shale	2.0 (3)
U.-L. Cretaceous Bahariya Kharita	488	Sandstone with shale	2.6 (5)
L. Cretaceous Alamein	219	Dol., Sdy, Dol., Lst.	3.8 (8)
L. Cretaceous Alam el-Buib/ Matruh Shale Sidi Barrani	1219	Sandstone and shale	2.6 (5)
U. Jurassic Masajid	305	Limestone and shale	2.0 (3)
M. Jurassic Khatatba	610	Shale	1.8 (3)
L. Jurassic Wadi Natrun	207	Limestone	3.4 (6)
Permian-Cambrian	1829	Shale and limestone	2.6 (5)
Precambrian		Basement rocks	2.9 (6)

The bracketed numbers indicate the reference from which the conductivity was assumed: (1) estimated from Clark (1966); (2) Coster (1947); (3) Clark (1966); (4) Evans & Coleman (1974); (5) Girdler (1970); (6) Evans & Tammemagi (1974).

Table 6/4. Thermal conductivity of drill cutting samples from southern Gulf of Suez oil well.

Description	Conductivity (W/m/K)	No. of samples
Siltstone	2.1	1
Siltstone/sandstone	2.7	1
Carbonate/shale/sandstone	2.4	1
Evaporite/shale	2.3	1
Evaporite	3.3	1
Granitic basement	3.1	2

N and longitudes 25 and 27° E, and other gradients appear to decrease in all directions away from this section. In the Nile delta region, it is thought that the gradients may be slightly reduced by the effects of recent sedimentation (see e.g. Benfield 1949). Other gradient variations are thought to result from bulk conductivity variations in the sections penetrated by the boreholes, notably a reduction in the thickness of the post-Eocene sediments to the south.

Estimation of the thermal conductivity in the Gulf of Suez is considerably more complex. Lithologic logs in the Gulf of Suez show large thicknesses of Miocene evaporites in the section that may reach several thousand meters in thickness (Said 1962). The thermal conductivities of evaporites, notably rock salt, are typically two to three times the conductivity of the majority of the sediments listed in Table 6/3 (Clark 1966: 464). Uncertainty of the thickness of evaporites in the section, therefore, leads to a large uncertainty in the bulk thermal conductivity. Sediment conductivities of a few drill cutting samples collected from one of the oil wells in the southern part of the Gulf of Suez were measured (Morgan et al. 1983) using the technique described by Sass et al. (1971) and their values are shown in Table 6/4.

A minimum estimate of the thermal conductivity of 2.3 W/m/K for the Gulf of Suez results in a minimum estimate of the heat flow of 60 mW/m², i.e. more than 30% higher than the north Egypt heat flow. With large thicknesses of evaporites in the Gulf of Suez section, however, the heat flow could be as high as 80 or even 100 mW/m² (Morgan et al. 1977). This order of heat flow is more consistent with the reported mean heat flow in the Red Sea of 116 mW/m² (Haenel 1972). There is also the possibility of some structural control on the geothermal gradients and the higher mean heat flow is thought to be related to shear motion taking place along a plate boundary passing along the Gulf (Picard 1970). This hypothesis is supported by the occurrence of hot springs on both sides of the Gulf (Swanberg et al. 1977).

by descending mantle material (Ryan et al. 1970). It is thought unlikely, however, that such a mechanism would produce such an extensive area of low heat flow, or that it would affect the intermediate eastern Mediterranean crust, and the continental north African crust in the same way.

There appears to be a slight systematic trend to the distribution of geothermal gradient values in northern Egypt indicatecd in a plot of mean geothermal gradients in 1 × 1 and 1 × 2 degree squares, as shown in Figure 6/6. The lowest mean gradients are found in the Nile delta region and south 29° N. The highest mean gradient is found between latitudes 30 and 31°

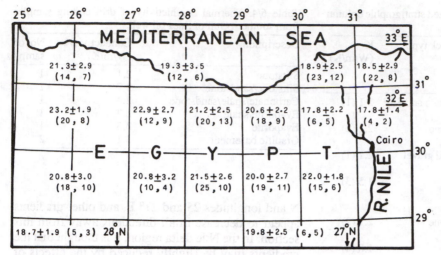

Figure 6.6 Mean geothermal gradients for northern Egypt in 1° × 1° and 1° × 2° areas (after Morgan et al. 1977). The mean gradient in each area is given in mK/m with the standard deviation. Given in brackets are the number of temperature data used and the number of boreholes in each area.

Geothermal data from measurements inside boreholes

The results obtained from the heat flow estimates of oil wells were supplemented by additional and more conventional heat flow measurements in existing mineral exploration and water boreholes (Morgan et al. 1980, 1983). A total of 52 mineral exploration drill holes and four water wells were selected at 13 sites in Egypt (Fig. 6/7 and Table 6/5), where temperatures were measured at 5 m intervals in the boreholes to a precision of 0.01°C using an electrical resistance thermistor. Some of these temperature data are shown in Figure 6/8. Temperatures were computed for linear sections of the temperature-versus-depth plots by least squares linear regression analysis and the mean gradients at each site are given in Table 6/5.

Figure 6/8 shows that the Nuweibi, Abu Ghalaga and Abu Tartur data have a large scatter at shallow depths due to surface temperature variations at different locations and annual surface temperature fluctuations. Little scatter is evident in the Abu Dabbab data as these boreholes were drilled from exploratory mine adits with relatively constant temperature. Little scatter is also evident in the data from the two boreholes at Sukkari, indicating similar surface temperatures for both locations at this site.

Data from western Egypt (Abu Tartur and West Kharga) indicate low geothermal gradients between 15 and 19 mK/m. These boreholes are in sediments of low thermal conductivity and low heat flow, thus extending the low heat flow province of the eastern Mediterranean as far south as latitude 26° N.

In eight boreholes at Abu Tartur (AT, Fig. 6/7), however, the calculated gradient was high averaging 74 mK/m ('phosphate' in Table 6/5). The high gradients were measured above a proven phosphate deposit at 80-150 m depth and the spatial coincidence of the thermal high and the phosphate suggests that the two are related. Natural gamma logging in the boreholes indicates that there are small levels of radioactivity that can account for such a dramatic increase in temperature gradients. The plausible explanation is that the heat may have been produced by oxidation, probably of pyrite within the phosphate beds. This would explain the rapid decrease in geothermal gradient as the phosphate horizon is penetrated and does not indicate a geothermal resource. The thermal anomaly over the phosphate is of interest as a possible exploration tool for the phosphate deposit. Shallow (1-2 m) temperature measurements could be used to map the lateral extent of the deposit (see e.g. Kappelmeyer 1957).

In the Eastern Desert, the geothermal gradients measured at eleven sites range from very low (8.2 mK/m at Gabbro Akarem) to high (50 mK/m at Abu Shegeila) (Table 6/5). At Abu Shegeila borehole, two distinct linear gradients of 30 and 50 mK/m have been observed at respective depths of 65-160 and 175-235 m probably due to a conductivity contrast. Taking into account the low thermal conductivity of the sediments at this site, the high gradient of 50 mK/m indicates a high heat flow of 105 mW/m² similar to the mean of 111 mW/m² reported for the Red Sea margin (Girdler & Evans 1977).

The other nine sites in eastern Egypt with the exception of the Shagar water well (SH, Fig. 6/7) are all in Precambrian basement outcrops which generally have a higher thermal conductivity than the sediments of western Egypt. The gradients from these sites are generally higher toward the east and when the thermal conductivities are taken into account, they define a band of high heat flow 30 km

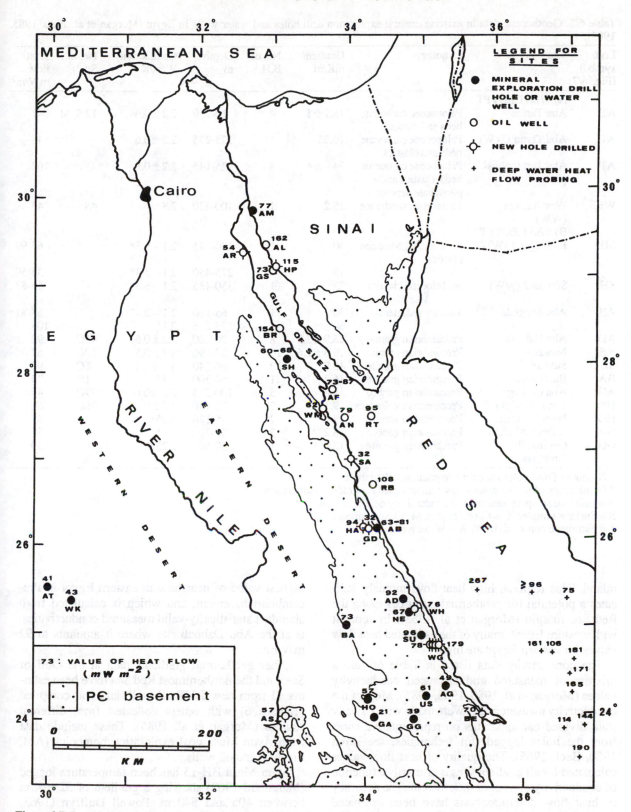

Figure 6.7 Location map of heat flow sites in Egypt and northern Red Sea (data in part after Morgan et al. 1985, and Girdler & Evans 1977). See Tables 6.5, 6.6 and 6.7 for site codes.

Table 6/5. Geothermal data in existing mineral exploration drill holes and water wells in Egypt (Morgan et al. 1980, 1983, 1985).

Loc. symbol (Fig. 6/7)	Site	Lithology	Gradient mK/m	No. of B.H.[1]	Depth[2] m	Conductivity W/m/K	No. of Sp.[3]	Heat flow mW/m^2
		A) WEST EGYPT						
AT	Abu Tartur	Palaeocene carbonate beds and clastics	18.7 ± 1	4	10-110	2.2 ± 0.6	12 S, M	41
AT	Abu Tartur (WW)	Palaeocene carbonate beds and clastics	20.25	1	125-235	2.2 ± 0.6		45
AT	Abu Tartur phosphate	Palaeocene carbonate beds, clastics and phosphate deposit	74 ± 6	8	25-145	2.2 ± 0.6		163
WK	West Kharga (WW)	Cretaceous sandstone	15.2	1	100-430	2.8	4S	43
		B) EAST EGYPT						
SH	E. Shagar-1 (WW)	Pliocene/M. Miocene clastics	30	1	100-275	2.1 – 3.3*		63-99
			18		275-450	2.1 – 3.3*		38-59
SH	Shagar-2 (WW)	M. Miocene clastics	25	1	350-485	2.1 – 3.3*		53-83
AB	Abu Shegeila	Tertiary sediments	30	1	65-160	2.1 – 2.7*		63-81
			50		175-235	2.1*		105
AD	Abu Dabbab	Precambrian granite	28.9 ± 2.9	8	20-200	3.2 ± 0.5	45C	92
NE	Nuweibi	Precambrian granite	20.3 ± 2.6	10	30-190	3.3 ± 0.5	60C	67
SU	Sukkari	Precambrian granite	18.9	3	60-240	5.1 ± 2.1	6C	96
BA	Barramiya	Precambrian granite	16.7	1	50-300	4.4	1S	73
AG	Abu Ghalaga	Precambrian gabbro	18.8	5	100-225	2.6 ± 0.6	39C	49
US	Umm Samiuki	Precambrian volcanics	19.1	2	25-140	3.2 ± 0.8	36C	61
HO	Homr Akarem	Precambrian granite	17.6	2	45-120	3.25*		57
GA	Gabbro Akarem	Precambrian gabbro	8.2	3	30-70	2.6*		21
GG	Genina El-Gharbiya	Precambrian granite	12.0	1	30-60	3.25*		39

1 Number of boreholes used for site gradient calculations.
2 Depth range of temperature measurements used for gradient calculations.
3 Number of samples measured in conductivity determinations.
S = Surface samples, C = Core samples, M = Mine samples.
* = Estimated conductivity, WW = Water well.

inland. This regional high heat flow anomaly indicates a potential for geothermal resources along the Red Sea margin (Morgan et al. 1983). In contrast with western Egypt, many of the estimated heat flow values for eastern Egypt are high.

The conductivity data listed in Table 6/5 are a mixture of measured and estimated conductivity values (Morgan et al. 1980, 1983, 1985). Most of the conductivity measurements were made with a steady state divided-bar apparatus on representative cores from the holes logged (for techniques, see Birch 1950, Beck 1965). The quality of heat flow values calculated for all gradient sites is primarily a function of conductivity control or lack thereof. The accuracy of heat flow determinations have been estimated (Morgan et al. 1985) to range from about ±10% for sites with good conductivity control (e.g. Abu Dabbab, Nuweibi) to ±30-50% for sites with poorly defined gradients and estimated conductivities. The

highest value of heat flow in eastern Egypt in Precambrian basement, and which is calculated from abundant statistically-valid measured conductivities, is at the Abu Dabbab site where it amounts to 92 mW/m^2.

Other geothermal gradients data at the Gulf of Suez and the northernmost Red Sea have been estimated from new BHT oil well data and compiled (Table 6/6) with others collected from different sources (Morgan et al. 1985). These include data from Ayun Musa coal exploration boreholes (AM, Fig. 6/7) and oil wells.

Ayun Musa BH-15 has been temperature logged (Mahmoud 1962) showing a gradient of 32 mK/m between 405 and 540 m. Powell Duffryn (1963) estimated the gradient at a neighbouring borehole BH-18 to be 75 mK/m where a single temperature of 50.5°C was measured at a depth of 400 m. Assuming a mean surface temperature of 29°C the mean gra-

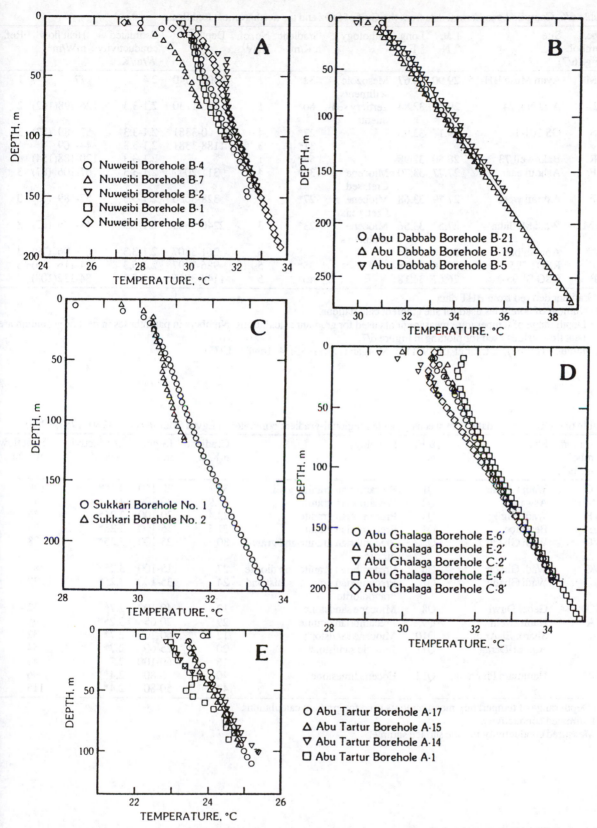

Figure 6.8 Examples of subsurface temperature data from: A. Nuweibi, B. Abu Dabbab, C. Sukkari, D. Abu Ghalaga and E. Abu Tartur boreholes (after Morgan et al. 1983).

Table 6/6. Compiled geothermal data from the Gulf of Suez and the northernmost Red Sea.

Loc. symbol (Fig. 6/7)	Site	Lat. °N	Long. °E	Lithology	Gradient mK/m	No. of BH[1]	Depth[2] m	Estimated conductivity W/m/K	Heat flow mW/m²	Ref.
AM	Ayun Musa BH-15	29.90	32.67	Mesozoic sediments	32	1	405- 540	2.4	77	1
AL	Asal No. 24	29.46	32.84	Tertiary sediments	60	1	130-1130	2.1-3.3	126-198(162)	2
GS	GS 101-1	29.17	32.95		27*	4	0-3381	2.1-3.3	57- 89 (73)	3
					31*	3	1188-3381	2.1-3.3	44- 69	
BR	Bakr well 73	28.50	32.98		57*	1		2.1-3.3	120-188(154)	4
AF	Ashrafi east	27.77	33.70	Miocene Cret. sed.	32*	4	311-2047	2.1-3.3	67-106 (87)	3
AF	Ashrafi west	27.79	33.68	Miocene Cret. sed.	27*	7	328-3656	2.1-3.3	57- 89 (73)	3
WM	Wadi Mellaha	27.57	33.56	Miocene sediments	23*	3	2240-3505	2.1-3.3	48- 76 (62)	3
AN	Abu Shiban	27.46	33.84		29*	6	261-3807	2.1-3.3	61- 96 (79)	3
RT	RSOT "95-1	27.46	34.18		35*	5	986-1897	2.1-3.3	74-116 (95)	3
RB	RSO B" 95-1	26.67	34.18		40*	5	1148-3016	2.1-3.3	84-132(108)	3

* Gradient derived from BHT data.
(1) Number of boreholes used for site gradient calculations.
(2) Depth range of temperature measurements used for gradient calculations. Numbers in parentheses in heat flow column are heat flow values used for plotting in Figure 6/7.
1. Mahmoud (1962), 2. Issar et al. (1971), 3. Morgan et al. (1985), 4. Tewfic (1975).

Table 6/7. Geothermal data from specially-drilled regional-gradient boreholes in Egypt (Morgan et al. 1980, 1983, 1985).

Location symbol (Fig. 6/7)	Site	B.H. no.	Lithology	Gradient mK/m	Depth[1] m	Conductivity W/m/K	Heat flow mW/m²
WG	Wadi Ghadir	Q1	Precambrian granitic gneiss	55	20-150	3.25*	179
AS	Aswan	Q2	Precambrian granite	13.9	20-100	4.1†	57
WH	Wadi Higlig	Q3	Precambrian granite	23.4	20-100	3.25*	76
BE	Berenice	Q4	Precambrian granite	21.5	35-100	3.25*	70
WG	Wadi Ghadir	Q5	Precambrian granite and granodiorite	30	25-100	3.25*	98
WG	Wadi Ghadir	Q6	Precambrian granite and diorite	27	15-100	3.25*	88
WG	Wadi Ghadir	Q7	Precambrian granite, schist and granodiorite	24	35-100	3.25*	78
GD	Gebel Duwi	Q8	Miocene sandstone	12	40-100	2.7*	32
HA	Hamrawein	Q9	Precambrian granite	29	20-65	3.25*	94
SA	Sharm El-Arab	Q10	Miocene sandstone	12	60-100	2.7*	32
AR	Abu El-Darag	Q11	Jurassic sandstone	20	25-60	2.7*	54
				16	60-100	2.7*	43
HP	Hammam Pharaoun	Q12	Eocene limestone	40	5-80	2.4*	96
				47.7	50-80	2.4*	115

1) Depth range of temperature measurements used for gradient calculations.
* Estimated conductivity.
† Measured conductivity for one surface sample.

dients at Ayun Musa were found to range between 13 and 54 mK/m. Higher gradients amounting to 35-40 mK/m are indicated from the data from two wells in the northernmost Red Sea (RT and RB, Fig. 6/7).

A significantly high gradient of about 60 mK/m was reported from temperature logs at Asal well No. 24 (Issar et al. 1971).

To complement the study of heat flow pattern in eastern Egypt and the Red Sea, data of heat flow probes from survey ships (Girdler & Evans 1977) at eleven sites in the northern Red Sea adjacent to the Egyptian territorial waters at a water depth of 1.1-2.2 km are given in Figure 6/7. Girdler & Evans (1977) notice that the average heat flow values decrease from 467 ± 116 mW/m² (N = 38) over the axial trough of the Red Sea to 111 ± 5 mW/m² (N = 13), over the region 50 to 170 km from the present spreading center. This decrease is about twice the world mean. Morgan et al. (1985) notice that five heat flow determinations in the coastal plain and offshore shelf of northern Sudan are very similar to the eastern Egypt coastal values. These values amount to 93 mW/m² (21.133° N, 37.083° E), 95 mW/m² (21.050° N, 37.283° E), 101 mW/m² (20.817° N, 37.283° E), 126 mW/m² (18.801° N, 37.644° E) and 100 mW/m² (18.317° N, 37.900° E) (Girdler 1970, Evans & Tammemagi 1974) thus reflecting a characteristic heat flow pattern of twice the world mean on the Red Sea margin.

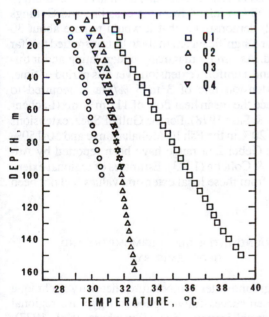

Figure 6.9 Subsurface temperature data from the four specially-drilled thermal-gradient boreholes: Q1. Wadi Ghadir; Q2. Aswan; Q3. Wadi Higlig and Q4. Berenice (after Morgan et al. 1983).

To test the conclusions arrived at from the study of the gradient data of the existing mineral and groundwater exploration boreholes, twelve heat flow shallow boreholes (≤100-150 m) were drilled in eastern Egypt (Morgan et al. 1980, 1983, 1985). The locations of these boreholes are shown in Figure 6/7 and their gradients are tabulated (Table 6/7). Some of the temperature measurements in the first four boreholes are shown in Figure 6/9. Temperature measurements inside these boreholes (Morgan et al. 1983, 1985) show that borehole Q2 close to Aswan (AS, Fig. 6/7) show a low gradient of 13.9 mK/m. Gradients become higher to the east amounting to 23.4 and 21.5 mK/m at boreholes Q3 (WH, Fig. 6/7) and Q4 (BE, Fig. 6/7) respectively, confirming the increase of heat flow toward the Red Sea axis.

At Wadi Ghadir borehole Q1 (located approximately 2 km from the Red Sea coast, WG, Fig. 6/7) a very high gradient of 55 mK/m was measured revealing a local geothermal anomaly. To investigate the lateral extent of this anomaly, three additional boreholes Q5, Q6 and Q7 (Fig. 6/10) were drilled to the west of Q1. Temperature measurements inside these boreholes (Fig. 6/11) show a rapid decrease of gradient to the west from 55 to 30 mK/m at a distance of 5 km to 24 mK/m at a distance of 18.5 km from the coast, thus reflecting a shallow source for the anomaly. Gravity data (Morgan et al. 1981) indicate a steeply dipping fault downthrowing to the east, approximately 0.5 km east of Q1, and the proximity of this structure to the Q1 anomaly suggests that the two may be related. Assuming a value of 3.25 W/m/K for the thermal conductivity at Wadi Ghadir (estimated from the mean conductivities of Abu Dabbab and Nuweibi), the heat flow values of the four sites Q1, Q5, Q6 and Q7 (Fig. 6/10) will be 179, 98, 88 and 78 mW/m² (Table 6/7). The very high heat flow at the first site Q1 may be in part due to thermal refraction caused by the juxtaposition of lower conductivity sediments and higher conductivity crystalline basement by the fault to the east of Q1 site (see e.g. Blackwell & Chapman 1977). An additional mechanism, such as the ascent of hot water along the fault, is required to explain the full magnitude of this anomaly (Morgan et al. 1981).

Temperature measurements inside other shallow gradient boreholes (Morgan et al. 1985), e.g. Gebel Duwi (GD), Hamrawein (HA), Sharm El-Arab (SA), and Abu El-Darag (AR) on the Red Sea coast (Fig. 6/7) are given in Figure 6/12. In general, low gradients in shallow sedimentary sites do not appear to be representative of the deeper thermal regime.

Red Sea coastal zone thermal anomaly

It has been shown that the observed heat flow values

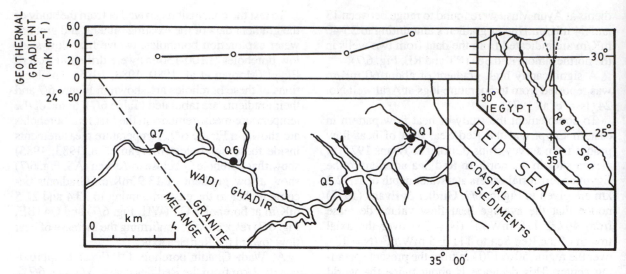

Figure 6.10 Locations of wells Q1, Q5, Q6 and Q7 across Wadi Ghadir; upper inset geothermal gradient profile (after Morgan et al. 1981).

indicate a general increase toward the Red Sea (Morgan et al. 1980) and this trend is shown more clearly in an updated plot (Morgan et al. 1985) of heat flow as a function of distance from the Red Sea margin (Fig. 6/13). Heat flow away from the Red Sea is probably in the range of 35-55 mW/m², a range consistent with a stable tectonic setting for these sites (e.g. Morgan 1984). Heat flow appears to increase relatively rapidly 30-40 km off the coast to 75-100 mW/m² (the very high value at Wadi Ghadir Q1 site is not thought to be regionally significant; see Morgan et al. 1981). The close spatial association of this thermal anomaly with the Red Sea suggests that it is related to the Red Sea opening.

In an attempt to interpret the Red Sea coastal zone thermal anomaly, Morgan et al. (1985) studied the basic mechanisms by which the high observed heat flow in the Red Sea coastal zone could be produced. The crustal heat production data of the coastal zone sites (Abu Dabbab, Wadi Higlig and Wadi Ghadir) is significantly below that required to produce the observed high heat flow at these sites with no additional component(s) of high heat flow.

Thermal effects of erosion may contribute to the coastal thermal anomaly. Preliminary apatite fission-track closing ages of samples collected from the Precambrian basement of the Eastern Desert of Egypt are in the range of 30 to 40 Ma indicating that these samples, now at the surface, have cooled from the closing temperature of apatite fission-tracks of 110°C (Gleadow & Duddy 1981) during the last 30 to 40 Ma. This indicates that 3 to 6 km of erosion took place during the last 30 to 40 Ma, based on the range of geothermal normal gradients observed in eastern

Egypt at present. Thus an erosion rate of around 0.1 to 0.2 km/Ma may account for some but not all of the coastal thermal anomaly.

Other mechanisms which could produce the high observed heat flow values in the Red Sea coastal zone are deformation within the lithosphere and heat input from below the lithosphere. It is possible that lateral conduction from the subsided Red Sea lithosphere contributed to this coastal anomaly. Heat flow data from eastern Egypt trace this anomaly a minimum of 25 to 30 km inland from the coastline and Gettings (1982) demonstrates that it would require about 30 Ma for a significant anomaly to be transmitted this far inland. Assuming magmatic underplating and intrusion and a uniform extension over this period of time, an extension rate of 3 to 5 %/Ma is required to produce the mean heat flow of 111 mW/m² (Lachenbruch & Sass 1978). For the Gulf of Suez, extensions of 10-20% in the Esh El-Mellaha range and 50-150% in the Gebel Zeit range have been reported by Angelier & Colletta (1983). Estimates of regional extensions from these local extension values had not been made.

2 GROUNDWATER: GEOTHERMOMETRY AND GEOCHEMISTRY

The groundwater temperature/chemistry technique has been successfully applied in Egypt for regional geothermal investigations (Swanberg et al. 1977). The water chemistry of thermal waters is a rapid and inexpensive exploration method for geothermal resources appraisal. This method provides information

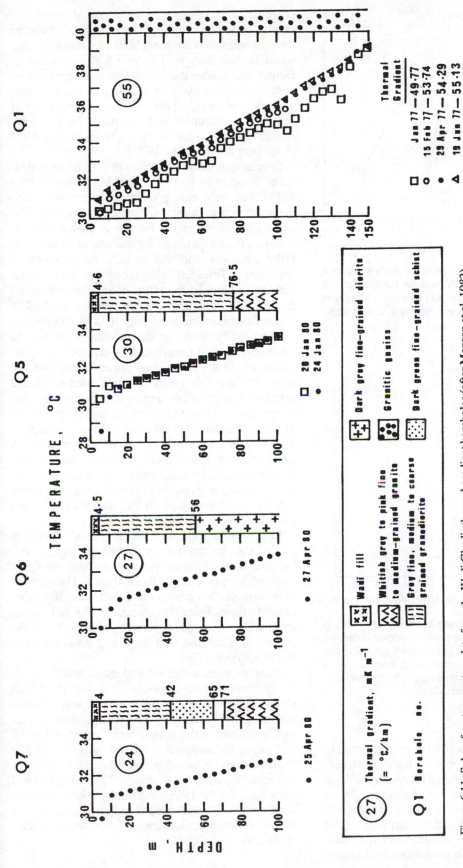

Figure 6.11 Subsurface temperature data from the Wadi Ghadir thermal gradient boreholes (after Morgan et al. 1983).

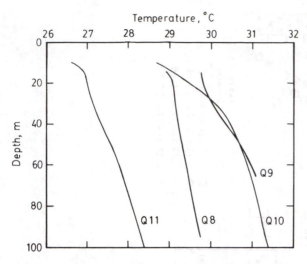

Figure 6.12 Temperature-depth plots for shallow gradient data from Gebel Duwi (Q8), Hamrawein (Q9), Sharm El Arab (Q10), and Abu El Darag (Q11) sites (after Morgan et al. 1985). Temperature data were recorded at 5 m depth intervals.

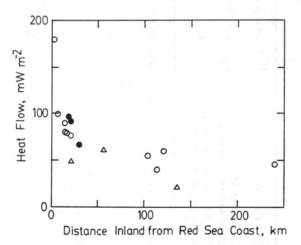

Figure 6.13 Heat flow in eastern Egypt south of 26° N plotted as a function of distance from the Red Sea coastline (after Morgan et al. 1985). Closed circles = granitic basement sites with measured conductivities; open circles = granitic basement sites with estimated conductivities; open triangles = non-granitic basement sites.

regarding the type of geothermal reservoir (liquid or vapour dominated) and its possible reservoir temperature. It also expands the general body of hydrologic knowledge of a given area by providing an indication of the water's origin, subsurface flow patterns and chemical quality. The study of non-thermal waters is also an important factor in geochemical exploration for geothermal resources. It establishes background

chemistry for comparison with thermal water chemistry which is required for the application of thermal water mixing models. Background geochemical studies also tend to reveal the presence of factors that render the use of chemical geothermometers invalid. Finally, it is also possible to utilize groundwater chemical data to detect the presence of geothermal resources that are not represented by surface features such as hot springs or hot wells (Swanberg & Alexander 1979).

Several qualitative indicators of subsurface temperature have been proposed (see Mariner & Willey 1976), but only two geothermometers have been demonstrated to have widespread application. The silica geothermometer (Fournier & Rowe 1966) is based on the temperature dependence of quartz solubility in water and the NaKCa geothermometer (Fournier & Truesdell 1973) is based on the temperature dependence of the ratios of sodium, potassium and calcium. A magnesium correction to the NaKCa geothermometer has been published by Fournier & Potter (1979). Both geothermometers attempt to determine the last temperature of water-rock equilibrium within the geothermal reservoir and both are subject to possible errors resulting from continued water-rock interactions as the water migrates from the geothermal reservoir to the sample point, mixing of waters that have equilibrated at different temperatures and the precipitation of the ions involved. Both geothermometers also require that the water chemistry be controlled by temperature dependent reactions. The basic assumptions of chemical geothermometry and the equations are given by Truesdell (1975) and Fournier & Potter (1979).

In a study carried by Swanberg et al. (1983) to assess the geothermal potential of Egypt, 160 samples were collected from many parts of Egypt (Fig. 6/14) and chemically analyzed. The data of these samples have been combined with fifty other samples taken from the literature. The only three areas which were not covered by this study are the interior of Sinai, the Nile delta, and the Great Sand Sea of southwest Egypt.

Samples were collected and made ready for investigations on two stages in 1976 and 1979 with almost half their total number in each stage (Swanberg et al. 1977 and 1983). At each site, the temperature and depth of the sample were recorded. One of the duplicate samples collected was chemically treated in situ by the addition of definite amounts of deionized distilled water (for the 1976 samples) and nitric acid (for the 1979 samples) in order to stabilize such constituents as silica and Fe. All treated samples were chemically analyzed within three weeks of their collection.

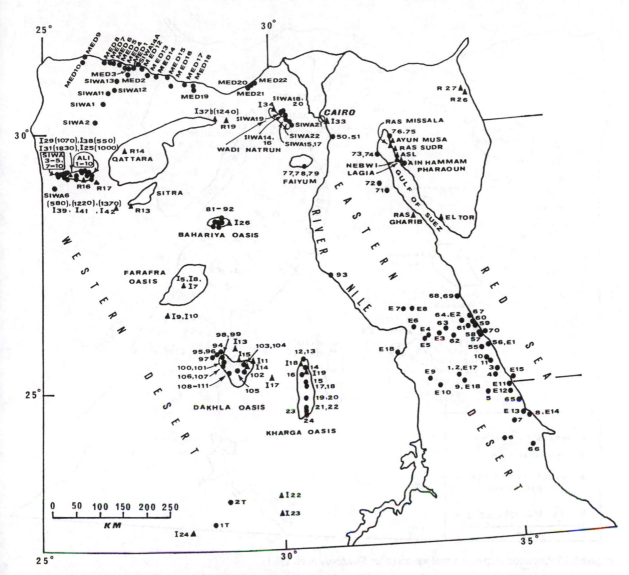

Figure 6.14 Locations from which water samples were collected. Solid dots represent locations of samples (Swanberg et al. 1983). Solid triangles indicate locations of samples which were taken from the literature: Prefix I (Ezzat 1974), R (El Ramly 1969), Gulf of Suez data (Issar et al. 1971). Well depth (m) for literature samples are shown in brackets.

Thermal water distribution

Swanberg et al. (1983) found that the mean air temperature in Cairo is 22°C and the mean ground temperature of Egypt from the temperature-depth data of Western Desert oases is 26°C. Taking into account Waring's definition of a hot spring as one being 8.3°C above mean air temperature, then temperatures of Egyptian springs need to exceed 30-35°C in order to be classified as thermal. Using this definition, many of the thermal springs reported by El Ramly (1969) cannot be strictly classified as thermal, even though their temperatures (25-35°C) may be sufficient for some geothermal applications.

Figure 6/15 shows the distribution of thermal springs (T > 30°C) and wells (T > 35°C) (Swanberg et al. 1983). All the thermal springs in Egypt are located along the shores of the Gulf of Suez: Ayun Musa, Ain Hammam Pharaoun and El Tor on the eastern shore of the Gulf, and Ain Sukhna on the western shore. These springs owe their existence to tectonic (or volcanic) heating associated with the opening of the Red Sea/Gulf of Suez rift. Helwan sulphur spring (Sample 51, Fig. 6/15) located just south of Cairo and reported as having a temperature of 36.1°C (El Ramly 1969) and 28.9°C (Swanberg et al. 1983) probably represents a deeply circulating groundwater

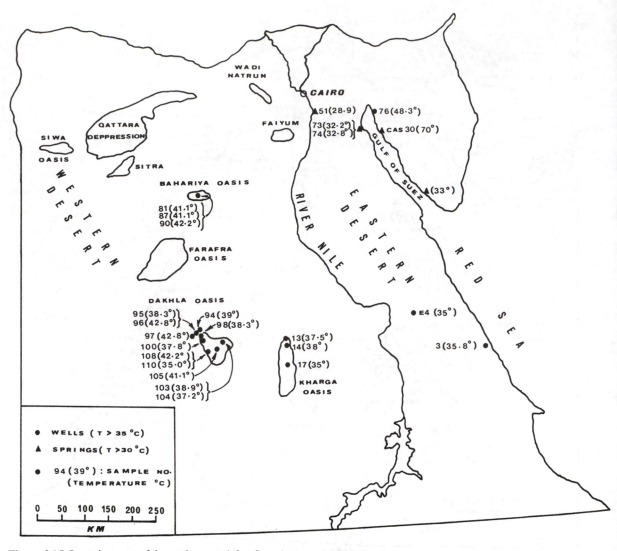

Figure 6.15 Location map of thermal waters (after Swanberg et al. 1983).

which has ascended to the surface along a fault zone.

In the Western Desert, there are no springs that can be strictly classified as 'thermal'. All the occurrences of thermal water are from deep wells.

Surface water temperature data have been plotted versus well depth (Swanberg et al. 1983) for wells from Kharga and Bahariya oases (Fig. 6/16). Since all these wells are either artesian or pumped continuously for agricultural purposes, the surface temperature should adequately reflect BHT and can thus be used to estimate the geothermal gradient. A least squares fit to these data yields a slope and intercept of 16.5 mK/m and 26.0°C respectively. The latter value is consistent on the basis of the temperatures observed at the top of the water table for the hand-dug wells of the Eastern Desert. Thus, it appears that the hot wells of these oases owe their thermal nature to

heating by a normal to low geothermal gradient and not to the presence of exploitable geothermal reservoirs.

A least squares fit to the temperature-depth data from wells at the Dakhla oasis (Fig. 6/17) yields a slope and intercept of 11.9 mK/m and 29.4°C respectively (Swanberg et al. 1983). It has been noted that wells that show anomalously high temperatures (average 40°C) are concentrated to the north (Fig. 6/18) near the escarpment forming the north boundary of the oasis. These data are most easily reconciled by assuming that water, heated by a normal to low geothermal gradient, is ascending along conduits at the north end of the oasis and migrating south through the principal aquifers. A reconnaissance microearthquake survey in this area subsequent to the discovery of the anomalous water temperatures rec-

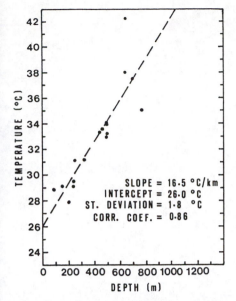

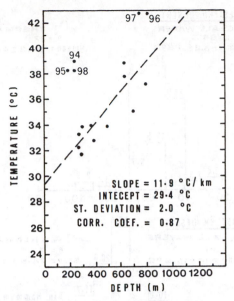

Figure 6.16 Plot of BHT versus depth for the deep artesian wells from the Kharga and Bahariya oases (after Swanberg et al. 1983).

Figure 6.17 Plot of BHT versus depth for the deep artesian wells from the Dakhla oasis (after Swanberg et al. 1983). Data of samples 94, 95 and 98 were omitted from regression analysis.

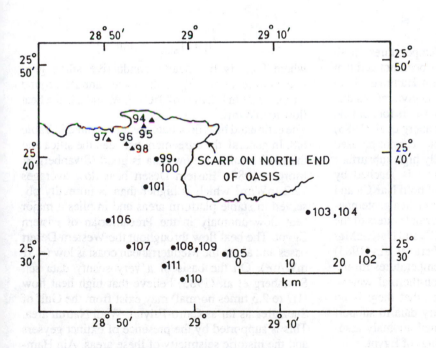

Figure 6.18 Dakhla oasis showing the locations of the hottest wells. Solid triangles: sites with temperatures ranging from 38.8 to 42.8° C (average 40.2° C); solid circles: sites with temperatures ranging from 31.7 to 42.2° C (average 35.7° C).

orded a single event with an epicenter on the escarpment (Morgan et al. 1980).

It is worth noting that two regions of Egypt have shown thermal activity in the recent geological past. These are the extinct geysers on both sides of the Cairo–Suez road and Gebel Uweinat area of the southwest corner of Egypt (El Ramly 1969).

Subsurface temperature estimates

The silica, NaKCa and NaKCaMg geothermometers have been applied to all the samples collected (Swanberg et al. 1983). A quick scan of these data fails to reveal any samples with abnormally high geotemperatures. Silica geotemperatures for the thermal waters have been plotted as a subset histogram beneath a

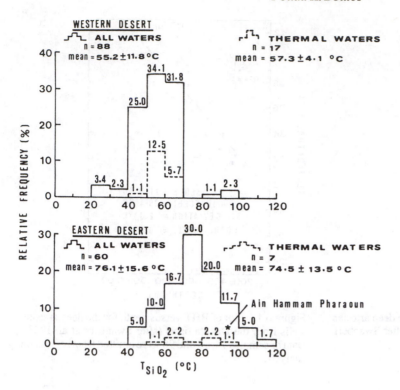

Figure 6.19 Histograms of silica geotemperatures for all groundwaters from the Western (top) and Eastern (bottom) Deserts (after Swanberg et al. 1983).

histogram showing the silica geotemperatures for all waters collected (Fig. 6/19). It has been noticed that with the possible exception of Ain Hammam Pharaoun (Fig. 6/14) the thermal waters give results that are comparable to the non-thermal waters for both the Eastern and Western Deserts. Swanberg et al. (1983) concluded that this geothermometer cannot be used to infer the presence of abnormally high subsurface temperatures. A similar conclusion is reached by least squares regression analysis of both NaKCa and NaKCaMg geotemperatures against silica geotemperatures for the thermal and non-thermal waters of the Eastern Desert. Plots of NaKCa/SiO$_2$ and NaKCaMg/SiO$_2$ geotemperatures by Swanberg et al. (1983) failed to show any elevated geotemperatures for the thermal waters relative to the non-thermal waters. Swanberg et al. (1983) concluded that there is no evidence from the geothermometry data to support the existence of a major geothermal anomaly associated with any of the thermal springs of Egypt.

Silica heat flow

Using the technique by Swanberg & Morgan (1979, 1980) to estimate the regional heat flow from the silica content of groundwater, Swanberg et al. (1983) calculated the silica heat flow for various parts of Egypt using the equation

$$q = (T_{SiO_2} - T_0)/m$$

where T_{SiO_2} is the quartz conductive silica geotemperature in °C, T_0 is the mean annual ground temperature in °C, m is 670°C m^2/W, and q is the heat flow in mW/m^2.

The estimated heat flow data are represented in Table 6/8. In general, the agreement between the silica and the traditional heat flow data is good (Swanberg & Morgan 1980). Eastern Desert heat flow averages 72.2 mW/m^2 which is higher than is normally observed in stable platform areas and implies a major heat flow anomaly in the Precambrian of eastern Egypt. The heat flow throughout the Western Desert oases and along the Mediterranean coast is low (< 51 mW/m^2). On the basis of a very scanty data set, Swanberg et al. (1983) believe that high heat flow (1.7 to 2.3 times normal) may exist from the Gulf of Suez area as far as Cairo-Faiyum-Wadi Natrun area. This is supported by the presence of extinct geysers and the historic seismicity of these areas. Ain Hammam Pharaoun, the hottest spring in Egypt at 70°C, lies on the border of the Gulf of Suez. This zone is the most favourable for geothermal exploration and development.

Geochemistry of groundwaters

The chemical analysis data of water samples were utilized to study the groundwater types in Egypt

Table 6/8. Heat flow estimates of Egypt based on the silica heat flow technique.

Location	Number of samples	T_{SiO_2} (°C)	T_0 (°C)	q (mW/m²)	Traditional q (mW/m²)
Eastern Desert	44	75.4 ± 15.3	27.9	72.2	77.6[1]
Kharga Oasis	13	47.5 ± 2.4	26.0	32.1	40-45[1]
Bahariya Oasis	12	54.8 ± 2.8	26.0	43.0	
Dakhla Oasis	18	55.7 ± 4.2	24.4	46.7	
Mediterranean coast	21	55.4 ± 17.3	21.2	51.0	
Siwa Oasis	22	60.3 ± 13.7	26.4	50.6	
Wadi Natrun	7	74.7 ± 19.4	26.0	72.7	
Cairo area	4	89.2 ± 13.4	24.9	96.0	
Sinai (west coast)	4	73.8 ± 14.6	25.0	72.8	80-100[2]

1. Morgan et al. (1980), 2. Morgan et al. (1977).

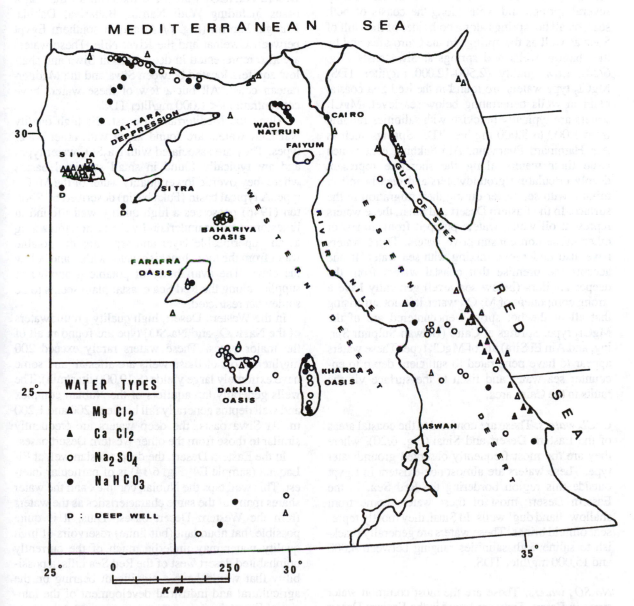

Figure 6.20 Geographical distribution of different groundwater types (reproduced by the author from individual distribution maps after Swanberg et al. 1988).

(Swanberg et al. 1988). For each major study area of Egypt, the chemical constituents of common anions and cations of water samples were represented on Sullins diagrams (Ezzat 1974, Swanberg et al. 1988). Figure 6/20 shows the geographic distribution of the four groundwater types as depicted by the Sullins technique. Samples where the ratio (total anions–total cations)/total anions exceeds 20% have not been used.

In this study, four types of groundwater in Egypt have been recognized: $MgCl_2$, $CaCl_2$, Na_2SO_4 and $NaHCO_3$ (Swanberg et al. 1988).

$MgCl_2$ waters (typical of sea water). These are represented by the Red Sea and Mediterranean Sea, by several springs and wells along the coasts of both seas, by all hot springs along both sides of the Gulf of Suez as well as the springs in the Cairo area and by the shallow wells and springs at Siwa oasis (Fig. 6/20). Low quality (2,500-12,000 mg/liter TDS) $MgCl_2$ type waters are found in the Red Sea coastal plain in wells penetrating below sea level. $MgCl_2$ waters are typically brackish with salinities ranging from 1,000 to 8,000 mg/liter TDS. Springs such as Ain Hammam Pharaoun, Ain Sukhna, etc., which issue their waters along the shoreline represent deeply circulating groundwaters and clearly reflect mixing with sea water during their migration to the surface. In the Eastern Desert and Sinai, these waters represent oil wells, waters pumped from mines, or other saline non-consumptive waters. These waters have also undergone mixing with sea water. If one accepts the premise that coastal waters from the deeper aquifers (below sea level) generally have a strong component of $MgCl_2$ water, it is not surprising that all of the hot springs encountered are of the $MgCl_2$ type. Springs at Cairo (Helwan Sulphur Spring, and Ain El Sira) are of $MgCl_2$ type. These waters appear to have penetrated to sufficient depth to encounter sea water and risen to the surface via the faults in the Cairo area.

$CaCl_2$ waters. These are confined to the coastal areas of the Eastern Desert and Sinai (Fig. 6/20), where they are the most frequently observed groundwater type. These waters are almost non-existent in Egypt outside this region bordering the Red Sea. In the Eastern Desert, most of these waters come from shallow 'hand dug' wells. In Sinai, they mostly represent oilfield brines. These waters are generally brackish to saline with salinities ranging between 4,000 and 15,000 mg/liter TDS.

Na_2SO_4 waters. These are the most common water types in Egypt. They are found in the Eastern Desert from the Precambrian complex of the Red Sea hills

and west to the River Nile (Fig. 6/20). They are also found throughout the major oases of the Western Desert and along the Mediterranean coast. They are generally fresh with typical concentrations of less than 500 mg/liter TDS. At the Dakhla oasis, most waters have a concentration of less than 200 mg/liter TDS.

$NaHCO_3$ waters. These are represented by the River Nile and by shallow groundwaters located in the major wadis adjacent to the River Nile (Fig. 6/20). Isolated occurrences of $NaHCO_3$ type waters are found in the Red Sea hills where they probably represent a local perched water table. In the Western Desert, $NaHCO_3$ waters are found in all the major oases including Wadi Natrun, Bahariya, Dakhla, Kharga, Farafra (Fig. 6/14) and in southern Egypt between Uweinat and the River Nile. These waters are also represented in deep wells at Siwa and shallow aquifers located between Siwa and the Mediterranean coast. All but a few of these waters have concentrations < 1,000 mg/liter TDS.

Along the Mediterranean coast, the high quality $NaHCO_3$ waters are found along with other water types. They are associated with Na_2SO_4 water types and are typically found in small synclinal basins where they overlie lower quality water of the $MgCl_2$ type. A typical basin (Fuka basin) described by Shotton (1946) possesses a high quality water found in limestone aquifer underlain by a clay horizon acting as an impermeable layer and separates the potable water from the more brackish water within and below the clays. The availability of potable groundwater supplies along the Red Sea coastal plain seems to be somewhat restricted.

In the Western Desert, high quality groundwaters of the $NaHCO_3$ and Na_2SO_4 type are found in all of the major oases. These waters rarely exceed 200 mg/liter. Most of these wells are artesian and some have extremely large yields (> 10,000 liter/min). The wells generally tap aquifers of the Nubian complex and well depths generally fall between 200 and 1,200 m. At Siwa oasis, the deep waters are frequently similar to those from the other Western Desert oases.

In the Eastern Desert, the deep artesian well at El-Laqeita (sample E4, Fig. 6/14) is of particular interest. This well taps the Nubian complex and the water shares many of the same characteristics as the waters from the Western Desert oases. Thus, it is quite possible that abundant (but finite) reservoirs of high quality water may underlie much of the currently uninhabited desert west of the Red Sea hills, a possibility that would have a significant bearing on the agricultural and industrial development of the interior of Egypt.

Recommended targets for future geothermal exploration

An ideal target geothermal reservoir would be a high porosity and permeability sedimentary rock near the base of the sedimentary section such as the 'Nubian sandstone' (Morgan et al. 1983) a series of well-sorted quartzose sandstones recorded as resting unconformably over peneplaned basement rocks of the Red Sea margin and attaining a thickness of up to 500 m in the Gulf of Suez (Said 1962).

The Nubian formation varies in depth from surface outcrops at a few localities along the margins of the Red Sea and Gulf of Suez to depths of several kilometers in the downfaulted blocks offshore (see Said 1962: Fig. 26). Suggested targets for further geothermal exploration would be areas where the Nubian formation lies at a depth of the order of 4 km. If this good reservoir can be found with a geothermal gradient of at least 32 mK/m (1 standard deviation above the mean for the Gulf of Suez oil wells), it should be possible to produce large volumes of geothermal fluid at approximately 150°C. All the data indicate that such a reservoir may exist beneath the Red Sea coastal plain and along the margins of the Gulf of Suez. There are indications that higher gradients, and thus higher temperatures, may be found at the same depth which makes the eastern Egypt geothermal prospects even more attractive.

3 MICROEARTHQUAKE STUDIES

Microearthquake studies were initiated in May 1976 as a result of a joint US-Egyptian Geological Survey project. The occurrence of an earthquake of 5.6 M_b magnitude in the Aswan High Dam area on 14 November 1981 brought to the fore the importance of monitoring microearthquakes which have, since then, become part of current activities of the Helwan Institute of Astronomy and Geophysics (Kebeasy et al. 1987). The US-Egyptian Geological Survey project aimed at defining the active tectonics of Egypt and the study of the microseismicity at different parts of the country. In this project, a program of microearthquakes recording has been conducted in Egypt (Daggett et al. 1980) with an emphasis on the region of the Eastern desert and the Red Sea margin. The collection of data continued until March 1981 using high sensitivity portable seismographs with gains up to a few orders of magnitude higher than is typical for the World Wide Standardized Seismograph Net (WWSSN) stations.

Until the initiation of this study, the northern Red Sea seemed essentially aseismic as very few earthquakes have been reported from this region by the WWSSN. Fairhead & Girdler (1970) reported an earthquake ($M_b = 6$) which occurred on 12 November 1955 on the western marginal scarp of the Red Sea (25.29° N, 34.58° E) in the vicinity of the Abu Dabbab area. They also reported the occurrence of a large earthquake ($M_b = 6$) near the mouth of the Gulf of Suez at 27.62° N, 33.91° E on 31 March 1969 preceded by three foreshocks and followed by 17 aftershocks ($M_b = 4.5$-5.2) in the neighborhood of Shadwan Island. Maamoun & El Khashab (1978) reported 35 foreshocks during the last half of March 1969 preceding the main shock. Ben Menahem & Aboodi (1971) located its epicenter at the northwest of Shadwan Island and found that the ruptured zone appeared to extend 30 km to the southeast. McKenzie et al. (1970) and McKenzie (1972) also reported a swarm of earthquakes at the southern end of the Gulf of Suez starting 1969. More detailed seismicity data for Egypt covering the period 1903 to 1966 have been reported by Ismail (1960) and Gergawi & El Khashab (1968) from the work at the Helwan Observatory, and these data indicate a higher level of seismicity than is recorded by the WWSSN. These data, however, do not indicate a coherent pattern of the seismicity of the northern Red Sea and report no additional epicenters in this area. It is possible that microearthquakes occur frequently in the northern Red Sea but are too small to be recorded by WWSSN.

Microearthquakes are defined as events with a Richter magnitude of less than 3.0. In the study made by Daggett et al. (1980), an attempt has been made to define more clearly the seismicity of Egypt and the Red Sea margin by deploying microearthquake seismograph stations. Sprengnether MEQ-800 high-gain single-channel portable seismographs have been used in conjunction with Teledyne-Geotech S-13 vertical 1 Hz seismometers. Five to six instruments were in most cases deployed in an array at a time. Two or three instruments were sometimes deployed for short periods for reconnaissance recording. Locations of microearthquake recording sites are shown in Figure 6/21. Most of the recording was concentrated in eastern Egypt and took place within several recording periods from 1967 to 1981.

At this stage, the determination of the hypocenter location of the events recorded on the arrays was made by using HYPO71, a standard iterative hypocenter location program (Lee & Lahr 1972). The P-wave crustal velocity model which was found (Daggett et al. 1986) to minimize the travel-time residuals at the array station locations was as follows:

0 to 10 km 6.0 km/s
10 to 25 km 6.5 km/s
> 75 km 8.0 km/s

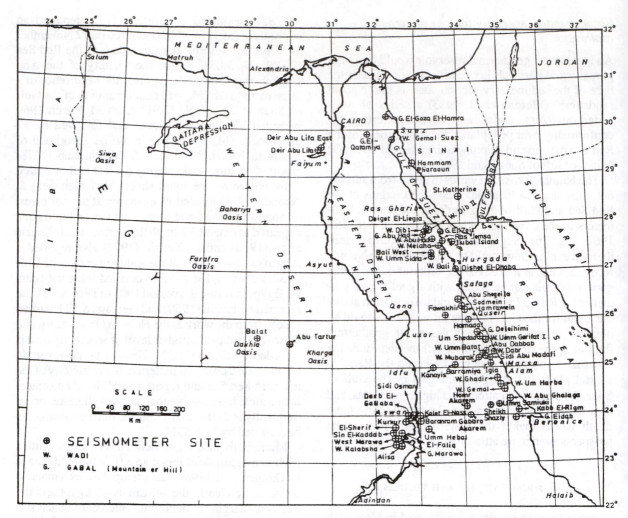

Figure 6.21 Recording sites for microearthquake studies in Egypt from May 1976 to 1982 (after Boulos et al. 1987).

This model is consistent with the regional geology (Precambrian crystalline basement), and the results of seismic refraction experiments performed on the Saudi Arabian Peninsula (Niazi 1968, Mooney et al. 1985) and along the Jordan-Dead Sea rift (Ginzburg et al. 1979). Poisson's ratio of 0.25 has been used to minimize S-wave travel-time residuals.

The magnitude (m) of locatable microearthquakes during this period was determined from the empirical relation

$$m = 2.0 \log (D) - 0.87$$

where D is the signal duration in seconds (Adams 1977).

Continuous recording of microearthquakes at the Aswan area (Fig. 6/21) has been conducted (Toppozada et al. 1988, Boulos et al. 1986) in a next phase (1981-1982) for about seven months after the occurrence of the Aswan earthquake (M = 5.5) on 14

November 1981 to locate the epicenters of the aftershocks.

Reconnaissance microearthquake recording

Reconnaissance recording was carried out in the following areas: 1) Abu Dabbab, 2) Idfu-Marsa Alam road, 3) south Eastern Desert, 4) Quseir-Qena road, 5) Hurgada-Ras Jemsa, 6) Faiyum, 7) Kharga-Dakhla oases, 8) Suez, 9) Wadi Ghadir, and 10) south Sinai. Of the above areas, only two show intensive microearthquake activity at the recording sites (Daggett et al. 1980). These are: a) Abu Dabbab and b) Hurgada areas.

With the exception of the two above-mentioned recording areas, only a few events have been recorded elsewhere. During recording at the Suez area, a few events have been observed probably originating from Ayun Musa hot springs (29.88° N, 32.65° E) at

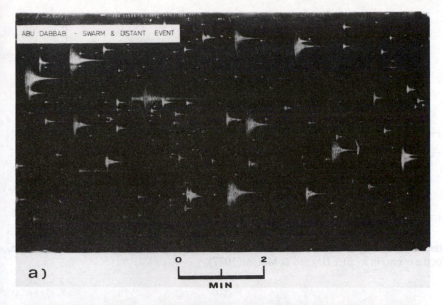

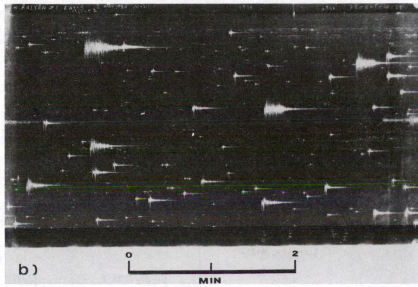

Figure 6.22 Abu Dabbab microearthquake swarm shown on two-day seismograms from a. Abu Dabbab and b. Wadi Raiyan recording sites started on 22 October 1976 and 2 November 1980 respectively (after Boulos et al. 1987).

the northern end of the Gulf of Suez. A single event was recorded during the Kharga-Dakhla oases reconnaissance survey (Daggett et al. 1980) and has its epicenter to the north of Dakhla oasis (25.6° N, 29.0° E) indicating an intraplate stress field and is thought to be related to an active fault controlling the flow of artesian hot water in this area reported by Swanberg et al. (1977). On a two-day record at Hammam Pharaoun, a local microearthquake was recorded having its epicenter near the Hammam Pharaoun hot spring on the Gulf of Suez shore.

Abu Dabbab seismicity

This area has been monitored on three occasions:

May/June 1976, October/November 1976 and April/May 1977, and on each occasion a high level of seismicity was recorded. The rate of activity ranged from 10 to 15 events/day to more than 60/day and sometimes reached 100 identifiable microearthquakes per day (Fig. 6/22). The rate of activity for a 12 day period in May/June 1977 is shown in Figure 6/23 together with a frequency versus magnitude plot. It has been calculated that this rate of activity is equivalent in energy release to one magnitude IV earthquake per year (Daggett & Morgan 1977), a very significant energy release from such a small area.

The reconnaissance array recording carried out in the area using a relatively large diameter array of 55 km for a period of 16 days during October/November

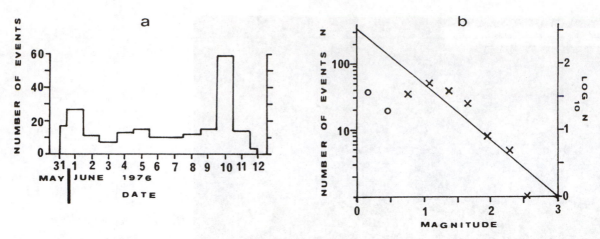

Figure 6.23 a. Histogram of number of events/day detected at Abu Dabbab seismic station (31 May to 12 June 1976). b. Frequency versus magnitude plot for the events in a (after Daggett & Morgan 1977).

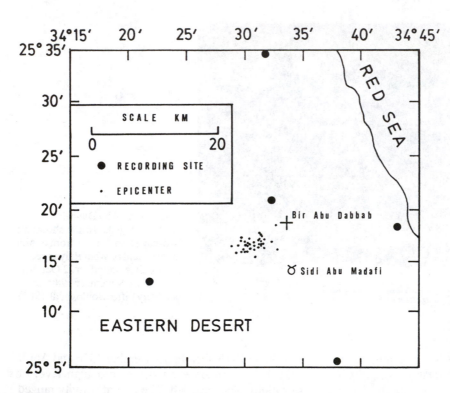

Figure 6.24 Location map for fall 1976 Abu Dabbab microearthquake array and recorded epicenters (after Daggett et al. 1986). Relative errors in hypocenter locations are less than 3 km.

1976 (Daggett et al. 1980, 1986) shows a clustering of microearthquakes in an area of about 50 km² (Fig. 6/24). In this period, 47 locatable events were recorded, 42 of which originated from a small area in the immediate vicinity of Abu Dabbab. Of the remaining five events, one event was located approximately 20 km south-southeast of Abu Dabbab in the Precambrian crust and the other four events, although poorly located, had epicenters in the Red Sea. The magnitudes of these events ranged from 1.3 to 3.6.

Detailed array recording at Abu Dabbab has been carried out after six months of the first reconnaissance recording with a much smaller diameter array of 12 km, centered on the active zone and operated for 15 days during Spring 1977 (Fig. 6/25). During this period, 140 locatable microearthquakes were recorded and the epicenters of all but one of these events originated from the immediate vicinity of Abu Dabbab and were restricted to an area of approximately 7 km east-west by 5 km north-south (Daggett et al. 1986). The epicenter of the lone event was located approximately 15 km northwest of the

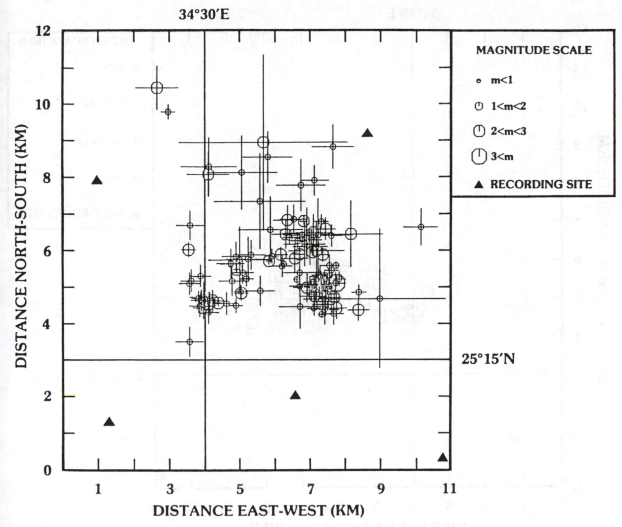

Figure 6.25 Location map for spring 1977 Abu Dabbab microearthquake array and recorded epicenters (after Daggett et al. 1986). Cross-lines show errors in epicenter locations.

center of the Abu Dabbab activity. The focal depth for all events was upper crustal ranging from 5 to 16 km, with a mode of 9 km. Magnitudes ranging from 0.2 to 2.5.

A plot showing the focal depth of the events on the epicenter map for the second recording period is shown in Figure 6/26. It is clear from this figure that the events do not follow linear trends or any other clearly defined spatial trend. Attempts to select events for composite fault plane solution have yielded internally inconsistent plots. Both the spatial distribution of the epicenters and the attempted fault plane solution determinations indicate that a complex stress field is responsible for the Abu Dabbab seismicity.

Earthquakes at Abu Dabbab have been reported by Morgan et al. (1981) to be accompanied by a sound of

distinct rumbling similar to the sound of a distant quarry blast. The sound is known to be heard by Bedouins for several generations.

The extremely tight clustering of the Abu Dabbab hypocenters and its high level of activity for long periods as indicated by Bedouin reports (Morgan et al. 1981) suggest that seismicity in this area is not directly related to regional tectonics and there is no obviously related structural feature (Daggett et al. 1986). One possible explanation is that the activity is related to the intrusion of a pluton into the Precambrian crust (Daggett et al. 1986). The rifted margin of the Red Sea is a likely environment for plutonic activity, but at present there is no direct evidence to support this hypothesis.

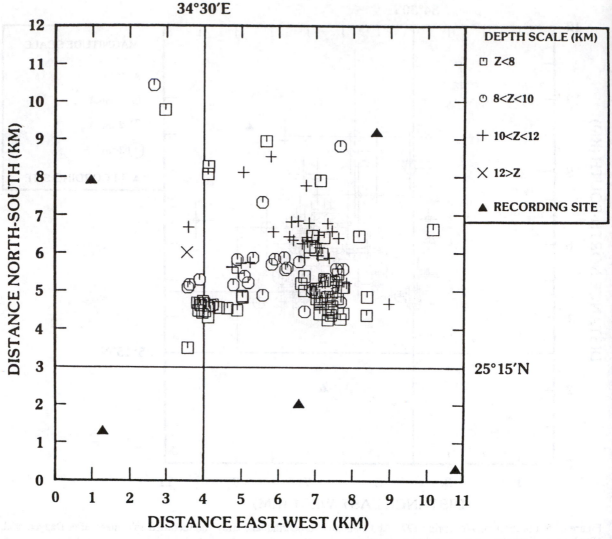

Figure 6.26 Plot of spring 1977 Abu Dabbab epicenters showing focal depths of events (after Daggett et al. 1986).

Seismicity of the Gulf of Suez

In February/March 1977, five seismic stations were deployed for 32 days at nine recording sites near the mouth of the Gulf of Suez (Fig. 6/27). One station of the array was placed on Jubal Island (27.63° N, 33.80° E) near the entrance of the Gulf, whereas the remaining instruments were deployed in the adjacent coastal region to the west.

The distribution of 93 locatable microearthquakes at the southern end of the Gulf of Suez was shown by Daggett et al. (1986) on the epicenter map (Fig. 6/27) north latitude 27° N with a clustering of the events beneath Jubal Island and a scattering of the events beneath the southern coasts of the Gulf. The average rate of activity in this area is 25 events/day, with swarms of as many as 200 events/day occurring once

every seven days. Magnitudes ranged from 0.7 to 3.1 and focal depth from 5 to 22 km.

Two clear northwest-southeast trends appear in the Gulf of Suez region on the epicenter map (Fig. 6/27). The first extends for 125 km in length and 25 km wide passing through the center of the mouth of the Gulf with a concentration of activity around Jubal Island to which the epicenter of the 31 March 1969 earthquake was very close. The second trend is defined along the southwestern coast of the Sinai Peninsula. These events appear to have originated on the major bounding fault(s) between the Precambrian outcrops of southern Sinai and the Gulf of Suez depression.

A linear regression fault plane analysis made of the projected hypocenters onto a vertical plane indicated fault planes striking parallel or subparallel to the axis

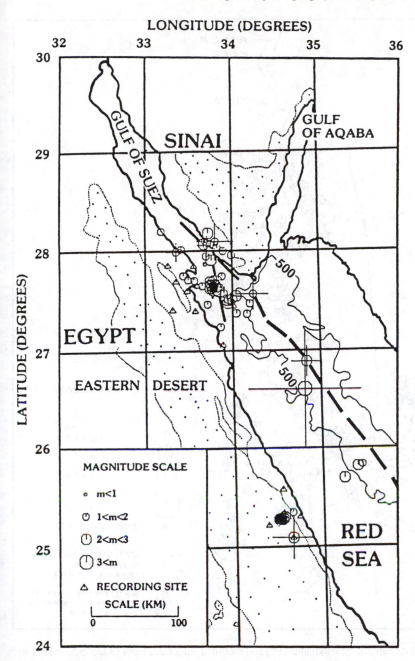

LONGITUDE (DEGREES)

Figure 6.27 Epicenters of all recorded microearthquakes in the northern Red Sea region (after Daggett et al. 1986). Cross-lines show errors in epicenter locations. Solid circles indicate numerous epicenters with overlapping symbols. 500 fathom depth contour and axis of deepest water are shown in the Red Sea. Oblique heavy lines indicate best fit strikes for the central Gulf of Suez and western Sinai epicenters. Strippled areas indicate outcropping Precambrian crystalline basement.

of the Gulf (Daggett et al. 1986). Strikes and dips of the fault planes are 349 and 46° for the central Gulf of Suez and 317 and 61° for the southwestern Sinai events. The fault plane analysis of the microearthquake hypocenters therefore indicates that the 1969 fault plane continues to be active and is consistent with the source parameters of Ben-Menahem & Aboodi (1971).

Daggett et al. (1986) relates the high rate of seismicity at the southern end of the Gulf of Suez to crustal movements among the Arabian and African plates and Sinai subplate as a result of the opening of the Red Sea extension in the Gulf of Suez and the left-lateral strike-slip motion in the Gulf of Aqaba.

Seismicity in the northern Red Sea region

Figure 6/27 compiles all the events recorded during the microearthquake study in eastern Egypt. North of 25.75° N, five poorly located events (Daggett et al. 1986) defines an active tectonic zone extending from the southern end of the Gulf of Suez into the median zone of the Red Sea. The report of Melville about two earthquakes felt at Brothers Island lighthouse (34.86°

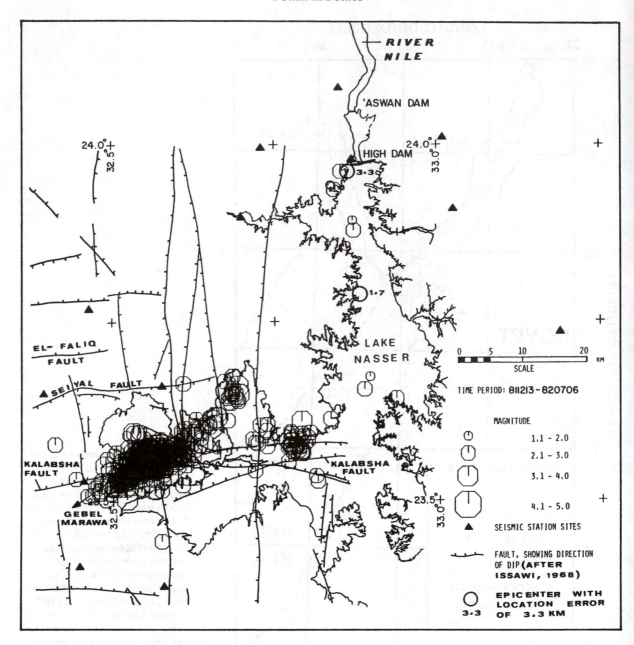

Figure 6.28 Location map of (1981/1982) Aswan microearthquake array and recorded epicenters from 13 December 1981 to 6 July 1982 (after Toppozada et al. 1988). Location error < 5 km.

E, 26.31° N) on 4 February 1908 and 6 January 1910 supports the presence of an area within the Red Sea near 26.25° N that has had continuing activity during this century (Woodward Clyde 1985). It is interesting to note that this activity follows the deep water axis (Laughton 1970). Some of these events were found (Boulos et al. 1987) to occur in the vicinity of hot brine deeps (Degens & Ross 1969), two of which namely Barbeque Deep (27.18° N) and Conrad Deep (27.05° N) have been discovered recently in the

northernmost Red Sea on the CONRAD 1984 Red Sea cruise (Cochran et al. 1985). The trend of these deeps and the epicenters of the northern Red Sea earthquakes possibly defines an active tectonic zone related to the opening of the Red Sea, extending from the median zone of the Red Sea to the Gulf of Suez.

Aswan seismicity

On 14 November 1981, a magnitude 5.6 earthquake

occurred 60 km southwest of the Aswan High Dam, under a large embayment of the lake. This embayment is a structural depression that has been submerged only since 1976. It marks the intersection of two major sets of faults trending approximately east-west and north-south.

A month after the main shock, the Geological Survey of Egypt installed a network of six MEQ-800 seismographs for about seven months from December 1981 to July 1982 for continuous recording of the aftershocks. In July 1982, the temporary network was replaced by a network of radio telemetered seismographs installed and operated by the Helwan Institute of Astronomy and Geophysics in cooperation with the Lamont and Doherty Geological Observatory of Columbia University (Simpson et al. 1987). The rate of microearthquake activity during the first recording period ranged from 10 to 20 events/day to swarms of 30 to 40 events/day and decreased according to Simpson et al. (1982) to 5 to 10 events/day in July-September 1982. In the first period, the magnitude of the located earthquakes (estimated using the duration method of Lee et al. 1972) ranged from 1.1 to 4 with the majority having magnitudes of about 2 to 3. The largest felt aftershocks observed during this period were between magnitude 4.0 and 4.4.

Initially five stations were installed, three east of the High Dam Lake (Lake Nasser) and two to the west (Fig. 6/28). The geometry of the network was varied along successive periods to achieve azimuthal coverage of the aftershocks as the epicentral distribution became known. In total, 14 sites (Fig. 6/28) were occupied during the seven months of monitoring.

More than 700 earthquakes of magnitude 1 or greater were located using HYPOELLIPSE computer program (Lahr 1979) using a three-layer crustal velocity model with P-velocity of 5.47 km/s in the top 4.4 km, 6.25 km/s from 4.4 to 20 km and 6.57 km/s below 20 km. The constructed model (Toppozada et al. 1988) was based on the seismic data recorded during blasting near Aswan Dam (Fig. 6/28) for the construction of the Aswan II power plant. More than 500 epicenters with a horizontal standard error of less than 5 km for the entire recording period were plotted (Fig. 6/28).

Analysis of Aswan microearthquakes (Toppozada et al. 1988) has shown that the majority of the epicenters are concentrated in the vicinity of Gebel Marawa about 65 km upstream southwest of the Aswan High Dam, along the east-west Kalabsha fault at the intersection with a northerly trending fault. Seismicity also clusters about 10 km north of the Kalabsha fault and also about 15 km further to the east near the intersection of the Kalabsha fault with another northerly trending fault. The focal mechanism of the main seismic zone is consistent with right-lateral strike-slip on the Kalabsha fault. A few well-located epicenters fall in the main stream of the Nile, approaching to within 5 km of the High Dam. Focal depth ranges from 10 to 25 km shallowing to the east.

It is worth mentioning that in addition to the local Aswan events, 30 regional earthquakes were located from Aswan seismograms (Toppozada et al. 1988, Boulos et al. 1987). Twenty of these events originated from the Eastern Desert around Abu Dabbab between latitudes 25 and 25.75° N and longitudes 34 and 35° E. Two events originated from the south Eastern Desert, and one from the Red Sea. Seven events originated from the Nile Valley north of Aswan between latitudes 24.25 and 24.78° N and longitudes 32.5 and 33.5° E.

With respect to Aswan seismicity, it is possible that the Kalabsha fault along which most of the earthquakes were concentrated was actively seismic before the filling of the lake and that unfelt microearthquakes occurred in the area. The few seismograms collected during the reconnaissance recording periods near Aswan in November 1977 and February 1978 unfortunately did not detect any of this activity (Boulos et al. 1987).

CHAPTER 7

Egypt in the framework of global tectonics

PAUL MORGAN

Geology Department, Box 6030, Northern Arizona University, Flagstaff, Arizona 86011, USA

During the decade following Said's (1962) synthesis of the geology of Egypt, a revolution has occurred in the earth sciences. The tantalizing questions surrounding the concepts of continental drift, which had remained controversial for more than half a century, were resolved with the recognition of the youth of ocean basins and the implications of sea-floor spreading. These concepts were linked with the theory of plate tectonics and many geologic events, thought once to be of only local significance, were recognized as global in origin. To a reader familiar with the concepts of plate tectonics, many of the elements of plate interactions are implicit in the framework of Egyptian geology outlined by Said (1962). The purpose of the present contribution is to build upon the excellent starting block provided by Said to explicitly describe the role of global tectonics in the tectonic and geologic history of Egypt.

The main global tectonic events that have directly or indirectly affected the geology of Egypt are summarized in Figure 7/1. These events are described chronologically below, emphasizing the regional geological implications of each event and leaving the details and local significance of each event to the authors of other chapters in this book.

PRECAMBRIAN – THE BASEMENT COMPLEX

The crystalline basement of Egypt is primarily Precambrian in age and is exposed extensively in the mountains of eastern Egypt and the southern part of Sinai (Said 1962). Smaller, but significant isolated exposures also occur across southern Egypt extending to the Precambrian complex or massif of Uweinat which straddles the borders of Egypt, Sudan and Libya (Schandelmeier et al. 1983, Schandelmeier & Darbyshire 1984). The Basement complex has been encountered in several wells in northern Egypt (e.g. Said 1962), and there is no reason to believe that it is not continuous throughout Egypt below the sedimentary cover from the exposures to the south and east.

The oldest dated rocks in the Basement complex are Late Archean (2673 ± 21 Ma) in age (Klerkx 1980) and are found in the Uweinat mountains of southwesternmost Egypt. These rocks, the Karkur Murr series, consist of charnockitic gneisses metamorphosed to granulite facies. The effects of retrograde metamorphism on the isotopic age-dating systematics are difficult to evaluate, but the age of 2673 Ma is considered to be a minimum age for the deformation accompanying the granulite facies metamorphism (Klerkx op cit). This metamorphism was associated with isoclinal folding with fold axes oriented north-south to north-northeast-south-southwest. Model neodymium ages for these charnockites are middle Archean (Harris et al. 1984). A second metamorphic event at about 1840 Ma is recorded in amphibolite facies migmatization and anatexis in the overlying 'Ayn Daw series, with east-west to northeast-southwest folds overturned to the north (Klerkx op cit). Newly obtained model ages from the Gebel Kamil and Gebel El Asr (southwest Egypt) areas range from 2300 to 1900 Ma (see Chapter 11, this book). These indicate the formation of a new crust in the Early to Middle Proterozoic and adds new evidence to the proposal of Harris et al. (1984) that the Late Archean nucleus of Gebel Uweinat is rimmed by a Middle Proterozoic fold belt. Finally a late (undated) stage of deformation formed large domes along north-south trending axes in both series.

Isotopic ages for other areas of the Basement complex indicate an upper age limit of 1200 Ma (Hashad 1980), consistent with model neodymium ages for these rocks (Harris et al. 1984). A wide range of younger ages have also been determined, indicating continual magmatic and tectonic activity in the Basement complex from 1200 to about 500 Ma. Hashad (1980) interprets a compilation of isotopic age data to indicate two major episodes of igneous intrusion in this period. The older episode from 1000 to 850 Ma

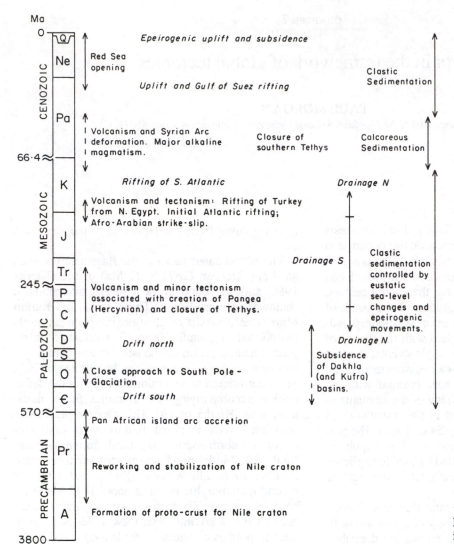

Figure 7.1 Major tectonic events in the geologic history of Egypt.

includes calc-alkalic diorites, tonalites, trondjhemites and granodiorites, the petrochemistry and initial strontium isotope ratios of which suggest a mantle source with little crustal contamination. The younger episode from 675 to 500 Ma is represented mainly by K-feldspar-rich granites, the petrochemistry of which suggests a crustal origin, while initial strontium isotope ratios suggest a mixed crust-mantle origin.

Thus, three distinct ages are indicated for the Basement complex in Egypt, a Middle or Late Archean age with Early Proterozoic reactivation in the Uweinat massif, an Early Proterozoic event of major crustal formation and reactivation, and a Middle to Late Proterozoic age with Eocambrian to Cambrian reactivation and possible addition. These three ages

represent a growth of the Egyptian basement complex from an Archean continental nucleus of undefined extent in southwestern Egypt to the modern extent of the continent by the addition of Late Precambrian juvenile crust. Little can be said at present regarding the formation of the Archean continental nucleus, except that the event responsible for the Late Archean granulite facies metamorphism, which was the stabilizing event of this nucleus, was probably analogous to a modern continent-continent collision (such as the Himalayan belt), as this appears to be the only setting in which granulite facies metamorphism can be formed at what finally becomes the top of a normal thickness continental crust (Morgan & Burke 1985).

In contrast to our poor understanding of the forma-

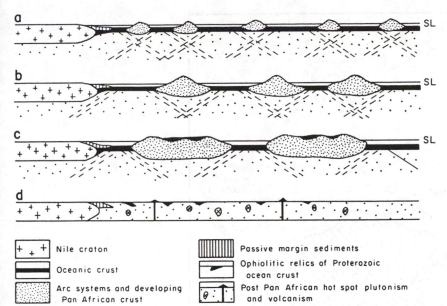

Nile craton

Oceanic crust

Arc systems and developing
Pan African crust

Passive margin sediments

Ophiolitic relics of Proterozoic
ocean crust

Post Pan African hot spot plutonism
and volcanism

Figure 7.2 Pan African arc accretion in eastern Egypt (after Gass
1982).

tion of the Archean continental nucleus, much progress has recently been made in understanding the
genesis of the Late Precambrian crust. The division
of these Precambrian rocks into plutonic, metavolcanic and meta-sedimentary sequences by the pioneering early geologists in this area, their careful
classification by Schurmann (e.g. 1957) and description by Said (1962) in terms of geosynclinal sequences strongly suggests, in a modern tectonic
framework, an accreting margin environment. This
association was first recognized in corresponding
Precambrian rocks of the Saudi Arabian shield in the
mid-1970s and shortly thereafter in eastern Egypt
(e.g. Bakor et al. 1976, Garson & Shalaby 1976,
Greenwood et al. 1976, 1980, Gass 1977, 1979,
Shackleton 1977, 1980). It is now generally accepted
that the continent grew during the Late Precambrian
by the accretion of island arcs onto the continental
nucleus as illustrated schematically in Figure 7/2
(from Gass 1982, see also Kroener 1979, Engel et al.
1980, Fleck et al. 1980, Shackleton et al. 1980,
Shimron 1980, 1984, Greenberg 1981, Stern 1981,
Reymer 1983, Ries et al. 1983 and others). Rapid
accretion of island arcs onto the continental nucleus
culminated in the Late Precambrian and late granitic
intrusions (Younger Granites) were probably formed
by crustal remelting of the newly formed and thickened crust, possibly in locally extensional environments (e.g. Greenwood et al. 1980, Stern et al.
1984).

Two basic factors remain unresolved in the Late
Precambrian growth of the Egyptian Basement: the
size of the Archean continental nucleus and the role
of strike-slip motion during island arc accretion. It
has been common in the past to include all of central
and eastern Saharan and northern Africa as non-
cratonic areas regionally affected by the 1100 to 500
Ma 'Pan African tectono-thermal event' as shown in
Figured 7/3 (e.g. Clifford 1970, Gass 1977, Kroener
1979, Windley 1984: 137-146). However, in Egypt
there is convincing evidence outlined above for the
existence of an Archean continental nucleus, represented in Egypt by the Uweinat massif, onto which
juvenile crust was accreted during the Pan African
event. Thus, it is probably more useful to limit the
mapped extent of the Pan African to those areas
where continental growth or major regional metamorphism of Pan African age can be demonstrated. A
revised interpretative map of Pan African orogenesis
and its implications are shown in Figure 7/4 (adapted
from Morgan & Burke 1985, and Burke & Sengor
1986) in which the Uweinat massif is shown to be
near the northeastern margin of a pre-1000 Ma craton
continuous with the Congo craton. The division of
two pre-1000 Ma cratons in Africa as shown in
Figure 7/4 is consistent with the definition of Pan
African events affecting only those areas in which
regional metamorphism of Pan African age can be
demonstrated (e.g. in Nigeria; Matheis & Caen-
Vachette 1983). A widely accepted explanation of the
tectonic significance of this division has yet to be
formulated, however, and as this area is peripheral to
the present discussion, it is not considered in detail
here.

The extent of the pre-1100 Ma craton (or cratons if
it was more than one craton) in northeast Africa is

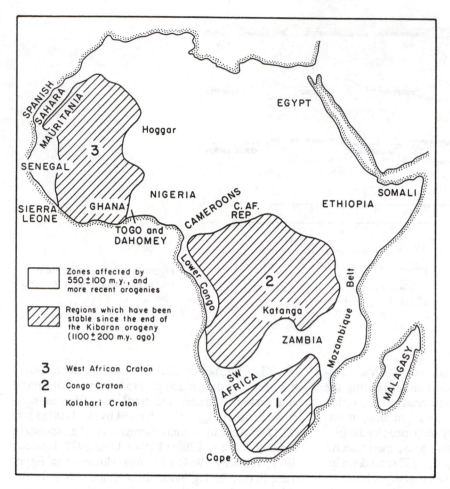

Figure 7.3 Extent of Pan African orogenesis as interpreted by Clifford (1970).

poorly constrained, but Archean to Middle Protero- zoic ages in this area support the concept of a nucleus onto which the Pan African of northeast Africa was accreted (see also Vail 1976). Arguments put forward by Church (1979, 1983) and Nasseef & Gass (1977) support the concept of a continental nucleus in Egypt onto which the Pan African was accreted, and Gass (1982) calls this nucleus the Nile craton, as shown in Figure 7/2, with the reasonable implication that it may have extended as far east as the Nile. Lead isotope data from late Precambrian igneous rocks are interpreted by Gillespie & Dixon (1983) to indicate that the boundary between the late Precambrian accreted terrain and the older craton to the west is in the Aswan area, providing additional support for the existence of a major basement boundary in the Nile area in southern Egypt. Further south in eastern Afri- ca the Pan African terrain includes granulite facies metamorphism, which as before is interpreted to indicate the site of a continent-continent collision (Morgan & Burke 1985). In northeast Africa the lack of Pan African granulite metamorphism indicates a

less intense accretion zone. The specific environment for the formation of the crust that was accreted to the continental nucleus in eastern Egypt, i.e. mid-ocean spreading center, marginal basin or island arc, cannot be unambiguously resolved from available data, and it is possible that more than one environment may be represented in the accreted crust. For example, Church (1979, 1983) and Hashad & Hassan (1979) present sedimentological and other evidence to sug- gest a marginal basin environment, whereas Stern (1979) presents chemical data which support the model of an island arc for the Younger Metavolcanics in eastern Egypt. Clearly, however, a significant vo- lume of new crust was accreted to eastern Egypt during the Pan African.

Reymer & Schubert (1984) have argued, based on studies of modern arcs, that the volume of the Pan African Arabian-Nubian shield is too large to be explained by simple island arc accretion. Their estimates of arc volume are minimum estimates based upon their study of relatively immature arc systems, however, and as the region from which arcs

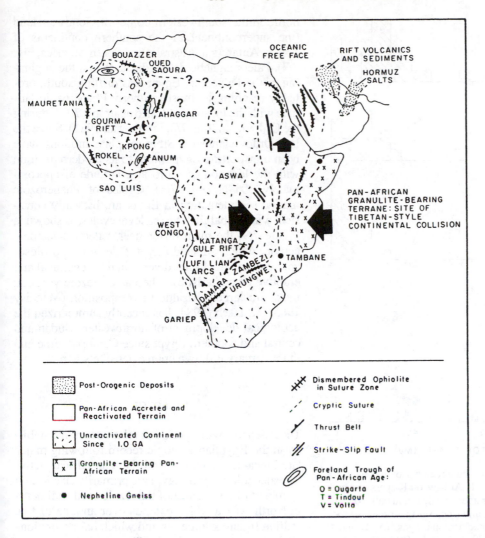

Figure 7.4 Revised interpretation of late Proterozoic African tectonics.

can be 'swept' into nucleus is essentially unconstrained over the 700 Ma period of arc accretion, the arc model can explain rapid continental growth in this area (K. Burke pers. comm., 1985). Furthermore, the evidence for a continent-continent collision to the south of the Arabian-Nubian shield suggests that arcs could have been 'swept' north into the unconstrained free-face zone of 'continental escape', giving further concentration of the arcs onto the continental nucleus. This concept implies major strike-slip motion approximately northward (modern orientation) during the arc-accretion episode, evidence for which is found in the northeast-southwest trending Najd fault system in Saudi Arabia (Moore 1979, Schmidt et al. 1979). This fault pattern has not been explicitly recognized in Egypt, but a parallel or complementary (northwest-southeast) system is predicted. Thus, major motion parallel to the margin of the continental nucleus is expected in addition to the component of

motion towards the nucleus during arc accretion. The main structural components of Africa relevant to the Precambrian Basement of Egypt are schematically summarized in Figure 7/4.

PHANEROZOIC – GENERAL OBSERVATIONS

Following rapid continental growth in the late Precambrian-early Paleozoic, Egypt and the continental mass of which northeast Africa was a part entered a period of relative tectonic quiescence. In fact, none of the later Phaerozoic tectonic and magmatic activity in Egypt matched the intensity of Pan African arc accretion. Phanerozoic rocks are dominated by relatively undeformed and unmetamorphosed sedimentary strata, with the most intense tectonic activity concentrating and, in fact, responsible for the modern continental margins of northeast

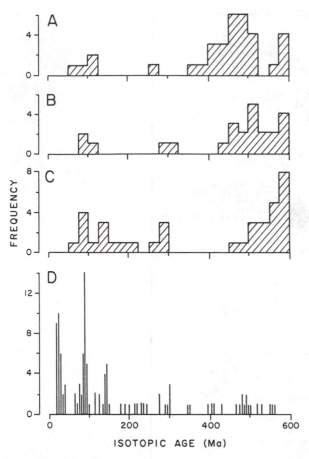

Figure 7.5 Histograms of Phanerozoic isotopic ages for Egyptian igneous rocks: A. K-Ar ages on K-spars and bulk granitoid samples; B. K-Ar ages on micas and bulk volcanic samples; C. Rb-Sr isotopic and isochron ages on whole rock samples; D. Undifferentiated isotopic ages. (A, B and C modified from Hashad 1980, D from R. Said pers. comm. 1984).

north on the southwestern corner of Gondwanaland (the supercontinent of the southern continents of Africa, Antarctica, Australia and South America, Fig. 7/7). Paleomagnetic data indicate that at the beginning of the Paleozoic, Egypt was drifting south, probably reaching its most southerly Phanerozoic latitude with the location of Cairo at almost 70° S during the Ordovician (Fig. 7/8, Smith et al. 1981). Since the late Ordovician/early Silurian, plate motions have been transporting Egypt north to its modern position with the location of Cairo at a latitude of approximately 30° N (Fig. 7/6). Cycles of Phanerozoic marine sedimentation in Egypt are basically consistent with global eustatic sea level cycles, as shown in Figure 7/9, the main differences reflecting tectonic activity which affected Egypt. Paleozoic and Mesozoic volcanic rocks are diverse in their chemical and eruptive characteristics, whereas Cenozoic volcanic rocks are largely basaltic in composition (Meneisy 1985). Klitzsch (1984) has recently summarized the geological development of northwestern Sudan and central and southern Egypt since Cambrian time and this summary is drawn upon extensively here.

PALEOZOIC

The Paleozoic was a period of relative tectonic stability in the Egyptian geologic record following major late Precambrian/early Cambrian Pan African activity. Paleozoic strata in Egypt are primarily clastic, and form part of a sequence of Paleozoic shelf sediments in North Africa which extends over an area of five million square kilometers and which has no obvious analogue anywhere else in the world (Burke pers. comm. 1984). Other shelf sequences, such as the interior of North America and the Russian Platform, contain much more carbonate than the Paleozoic North Africa strata. The general stratigraphic section for Upper (southern) Egypt deduced from recent hydrocarbon exploration in this region is shown in Figure 7/10. Four factors may have been important in the development of this shelf sequence: 1) erosion of Pan African mountains; 2) epeirogenic and structural movements in the craton; 3) glaciation due to the southerly latitude of North Africa for much of the Paleozoic (Figs 7/6 and 7/7); and 4) high global sea level in the late Cambrian/early Ordovician (Fig. 7/9).

Erosion of topography generated during Pan African tectonic activity is clearly the source for Paleozoic clastic sediments, but the long period of deposition of these sediments and the paucity of carbonates require some explanation. Glaciation of northeast Africa during the first half of the Paleozoic is suggested by its southerly paleolatitude, and is

Africa. Compilations of igneous age dates indicate relatively continual but minor igneous activity in eastern Egypt through the Phanerozoic (Fig. 7/5, Hashad 1980, Said pers. comm. 1985), but the major tectonic events recorded in the Phanerozoic sedimentary record indicate a relatively passive response to global eustatic sea level changes and plate interactions on the northern and eastern Egyptian margins. Using paleocontinental world maps (Smith et al. 1981), the approximate latitude of Egypt is plotted as a function of time through the Phanerozoic in Figure 7/6.

At the end of Pan African continental growth, Egypt was at mid-latitudes in the southern hemisphere. The Red Sea and eastern Mediterranean had not formed and Egypt was a continuous continental mass, with Saudi Arabia to the east and Turkey to the

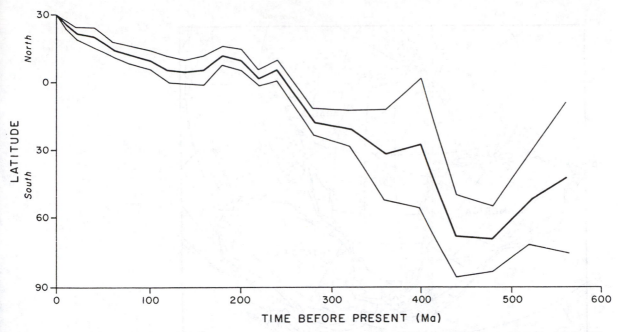

Figure 7.6 Phanerozoic paleolatitudes for Egypt. Heavy line shows approximate paleolatitude of the location of Cairo. Fine lines show the uncertainty in terms of the 95% confidence limits for the pole position (data taken from Smith et al. 1981).

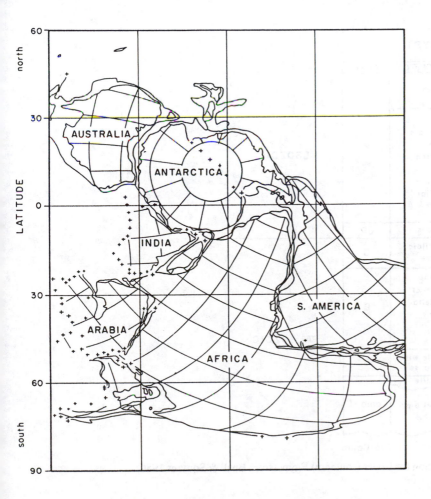

Figure 7.7 Reconstruction of the southern continents at 520 Ma. Latitude and longitude grid in 30° spacing (from Smith et al. 1981).

98   *Paul Morgan*

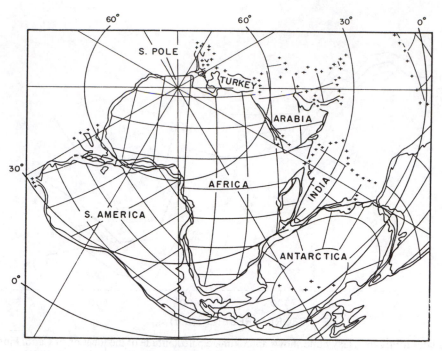

Figure 7.8 Reconstruction of the southern continents at 480 Ma, the time of the most southerly Phanerozoic latitude of Egypt (from Smith et al. 1981).

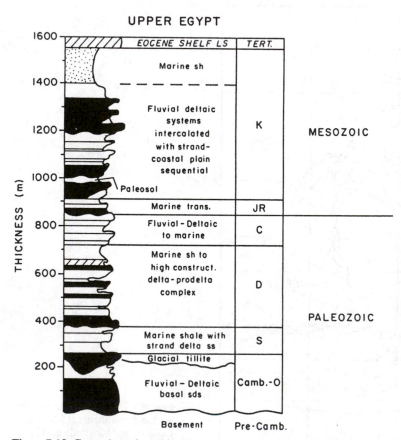

Figure 7.10 General stratigraphic section for Upper (southern) Egypt (from Beall & Squyres 1980).

SEA LEVEL CYCLES

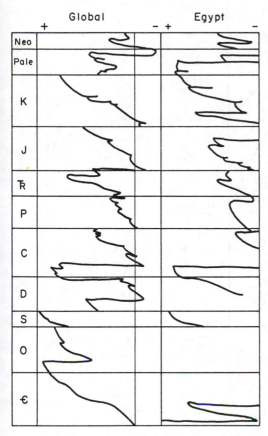

Figure 7.9 Comparison of global and Egyptian Phanerozoic eustatic sea-level cycles (from R. Said pers. comm. 1984).

confirmed by glacial tillite at the top of the Cambro-Ordovician section in Upper Egypt (Fig. 7/10, Beall & Squyres 1980), glacial deposits probably of Late Carboniferous age in southern Egypt and northern Sudan (Klitzsch 1983, 1984) and glacial deposits of late Ordovician age in the Ahaggar (K. Burke pers. comm. 1984). Glaciation as late as Permian-Carboniferous has been suggested for parts of the Arabian Peninsula (McClure 1978, 1980). If parts of the Pan African mountains were preserved under the ice-cap until at least the end of the Ordovician, the relatively long period of clastic sedimentation from the erosion of these mountains may be explained. High global sea level in the late Cambrian/early Ordovician and reworking of sediments associated with structural movements in the craton may have assisted in the distribution and redistribution of clastic sediments across North Africa at this time.

Paleozoic sediments in Egypt, as in other parts of North Africa, are predominantly clastic with interfingering calcareous sediments. Lack of fossils

in the Lower Paleozoic portion of this division prevents conclusive dating of most units of this age, and occurrences of pre-Carboniferous sediments in Egypt are spotty and widely separated. Said (1962) concluded, however, that northern Egypt was covered by an extensive marine transgression (or transgressions) during this period that probably did not overlap the southern portions of the country. Later studies have confirmed that during the Paleozoic sediments of shallow marine transgressions derived from the northwestern quadrant interfingered with fluvial deposits from the southeastern quadrant with frequent breaks in sedimentation (Klitzsch 1984). This distribution is consistent with pre-Pan African continental nucleus and the main Pan African mountains being to the south and east. The first well-dated and widespread Paleozoic sediments in Egypt are Carboniferous in age and record another major marine transgression. Again the thickest development of these sediments is to the north, suggesting a lingering control of sedimentation by the Pan African highlands.

Two areas of northeast Africa showed strong subsidence during the Paleozoic, the Kufra Basin to the southeast of Egypt and the Dakhla Basin in east-central Egypt (Fig. 7/11). More than 2500 m of Paleozoic sediments were deposited in the Kufra Basin, and Early Paleozoic to Carboniferous strata in the Dakhla Basin exceed 3000 m.

The tectonic origin of both of these basins is difficult to deduce from the available data. Possibly subsidence was initiated in both basins in response to stresses or the relaxation of thermal anomalies generated during the closing stages of Pan African crustal accretion. However, although the approximately 150 Ma period of slow subsidence in these basins is consistent with this hypothesis, corroborating evidence is lacking at present. Neev (1975, 1977) has suggested that the Kufra Basin is related to a major transcontinental shear, the 'Pelusium Line', which stretches from Anatolia down to the eastern limit of the Nile Delta, southwest across Africa to the Niger Delta, and into the Atlantic Ocean, but the range in ages of the features along this proposed trend make it unlikely that it is genetically related to Paleozoic subsidence in the Kufra Basin. Much of Egypt and adjacent portions of Sudan and Libya were marginal shelf to continental foreland to the Kufra and Dakhla Basins during most of the Paleozoic, on which a thin cover of very shallow marine sediments interbedded with fluvio-continental to deltaic clastics was deposited (Klitzsch 1984), probably primarily controlled by eustatic sea-level fluctuations.

Most of the land emerged from the sea at the end of the Carboniferous and Permian occurrences are minor (Said 1962). South Egypt formed a regional

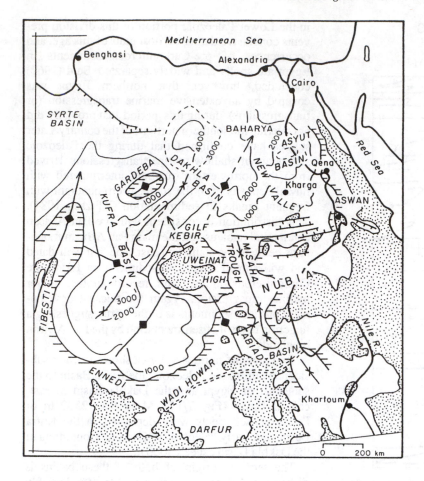

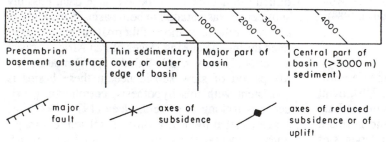

Precambrian basement at surface	Thin sedimentary cover or outer edge of basin	Major part of basin	Central part of basin (>3000 m) sediment)

major fault   axes of subsidence   axes of reduced subsidence or of uplift

Figure 7.11 Structural interpretation of part of the Eastern Sahara (from Klitzsch 1984).

structural high towards the end of the Paleozoic and eroded material (Paleozoic strata and Precambrian basement) was transported south into the Sudan, a reversal of the dominantly northerly transport direction earlier in the Paleozoic. The uplift was accompanied by faulting with an essentially east-west strike, volcanism and intrusions (Klitzsch 1984).

Although the post-Pan African Paleozoic was a time of relatively passive (in a tectonic sense) sedimentation in Egypt, isotopic age data indicate a wide range and distribution of Paleozoic magmatism (Fig. 7/5). Early Paleozoic magmatic activity is primarily restricted to late to post-tectonic granitoids, thought

to be crustal melts associated with thermal readjustment of the thickened crust towards the end of Pan African crustal growth (see above). Rb-Sr ages of these granites lie within the range 675 to 450 Ma, with a peak about 600 Ma (Hashad 1980). Younger K-Ar ages probably reflect minimum ages due to argon loss, and there is no evidence at present to dissociate any of these intrusions with the final stages of Pan African tectonism.

Later Paleozoic magmatism is represented by diverse igneous products including alkaline volcanics and ring complexes which range in age from 554 to 83 Ma (Serencsits et al. 1979, El-Shazly et al.

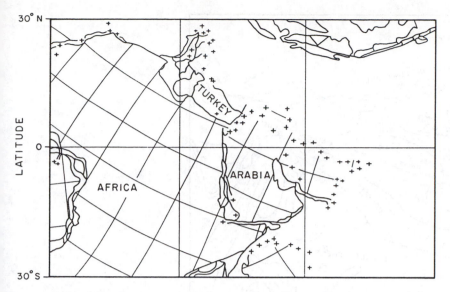

Figure 7.12 Reconstruction of the continents around Africa at 240 Ma, just prior to the assembly of Pangaea (from Smith et al. 1981).

1975, Hashad 1980, de Gruyter & Vogel 1981). Initial strontium isotope ratios for the alkaline volcanics are very low, indicating derivation from mantle source material (Hashad 1980). Geochronological data have been used to place the rocks into different age groups, but a universal recognition of discrete alkaline magmatic events has not been achieved (e.g. cf. Hashad 1980, de Gruyter & Vogel 1981, Meneisy 1985). De Gruyter & Vogel (1981), Meneisy (1986) and others have suggested that the alkaline complexes and other plutonic and volcanic assemblages were emplaced along reactivated Pan African fractures or pre-existing zones of weakness. De Gruyter & Vogel (1981) also suggest that the alkaline melts were generated in the asthenosphere by shear heating associated with rapid changes in plate motion.

Hashad (1980) suggests that there is a grouping of isotopic ages at 300-245 Ma and Meneisy (1986) identifies three volcanic events at 350 ± 10, 290 ± 10 and 230 ± 10 Ma roughly coincident with this group. This Carboniferous to Triassic magmatism was roughly coincident with the Hercinian (Variscian) orogeny in Europe, and the creation of the single supercontinent of Pangaea from the northern and southern supercontinents of Laurasia and Gondwanaland, respectively. Figure 7/12 shows the Smith et al. (1981) reconstruction of the continents around Africa at 240 Ma (Late Permian): Laurasia was moving westward relative to Gondwanaland at this time, resulting in orogeny in West Africa and some closure of the Tethys to the north of the African-Arabian land mass. There was some relationship between magmatism in Egypt and global tectonic events, but with the

limited precision of the current isotopic age data and our limited understanding of the genesis of mid-plate magmatism, the genetic association of the magmatism and tectonism remains to be unequivocally delineated. Tectonic movements in Egypt at this time controlled sedimentation, as outlined above, and include uplift of the Aswan-Uweinat block and the Um Baraka block in northeastern Egypt (R. Said pers. comm. 1984).

In summary, the Paleozoic of Egypt started with the final stages of Pan African arc accretion to the continental nucleus. Sedimentation for most of the Paleozoic was controlled by global eustatic sea-level changes, with clastic sediments probably derived primarily from Pan African highlands which were probably glaciated for at least some of the first half of the Paleozoic. Sporadic magmatic activity continued throughout the Paleozoic, but towards the end of this era there was a resurgence of magmatic and minor tectonic activity, coincident with and almost certainly related to a major global reorganization of the continents.

MESOZOIC

Magmatic and tectonic activity from the end of the Paleozoic continued into the Triassic in Egypt. Predominantly clastic sedimentation continued, primarily in northern Egypt, with cycles consistent with global eustatic sea-level changes at least into the Late Jurassic (Fig. 7/9). Differential vertical movements continued throughout Nubia in the Mesozoic, and the

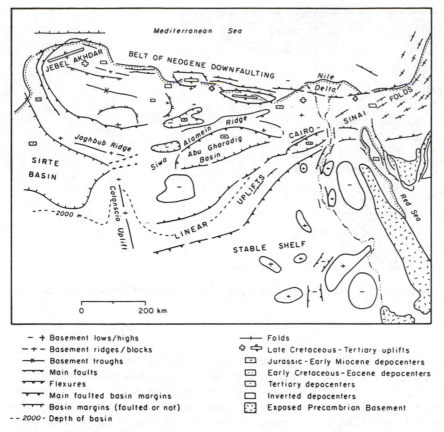

− + Basement lows/highs	⊢—— Folds
− + − Basement ridges/blocks	⬦ ⬦ Late Cretaceous-Tertiary uplifts
⊢—×— Basement troughs	▱ Jurassic-Early Miocene depocenters
⊤— Main faults	▭ Early Cretaceous-Eocene depocenters
⊤⊤⊤ Flexures	▱ Tertiary depocenters
⊥⊥⊥ Main faulted basin margins	▢ Inverted depocenters
⊤⊤⊤ Basin margins (faulted or not)	▨ Exposed Precambrian Basement
− − 2000 − Depth of basin	

Figure 7.13 Tectonic elements of northern Egypt (from Sestini 1984).

drainage regime which had switched from northward to southward late in the Paleozoic, reverted to northward once again in the middle of the Mesozoic era, probably in the Middle Jurassic (Klitzsch 1984). From the Jurassic to Late Cretaceous or Early Cenozoic time over a kilometer of Nubian strata were deposited in southwestern Egypt controlled primarily by marine transgressions and regressions (Klitzsch et al. 1979) with alluvial plain sand with interbedded channel and soil zone deposits interleaved with marine clay and silt and shoreline sand deposits. These cycles constitute the typical Nubian Sandstone and represent continuously changing environments from fluvial and deltaic deposition with local erosion and paleosoil formation to beach and swamp environments and shallow marine conditions (Klitzsch 1984). Extensive floral remains in the Nubia strata suggest a warm and humid to semihumid climate at the time of deposition (Klitzsch et al. 1979) which is consistent with the paleolatitude of Egypt just north of the equator during the Mesozoic (Fig. 7/6). To the east in the central Eastern Desert similar sequences are found but during the Middle Cretaceous sediments were transported in a dominantly westerly

direction from highlands in the area of the present Red Sea (Ward & McDonald 1979).

In northern Egypt sedimentation was similar during this period of time, but was controlled in addition by tectonic depocenters and positive structures (Fig. 7/13, Sestini 1984). These tectonic elements include features that continued to be active from the Late Paleozoic, new Mesozoic features, and features that were inverted in Late Mesozoic or Early Cenozoic times (Sestini op cit). Most of these features trend roughly east-west or east-northeast and fault systems with a similar trend affecting southern Egypt during the Cretaceous and possibly later (Klitzsch 1984). Many of these tectonic movements appear to be roughly coincident with initial Atlantic rifting and rifting of the South Atlantic (Fig. 7/1), but details of the regional Mesozoic tectonics in northeast Africa are not yet fully understood.

Hashad (1980) and Meneisy (1985) both identify an alkaline magmatic event in the Late Jurassic-Early Cretaceous (150 to 130 and 140 ± 15, respectively) which was coincident with initial rifting in the Atlantic (Figs 7/1 and 7/6). Rifting was probably also occurring along the north coast of Egypt at this time

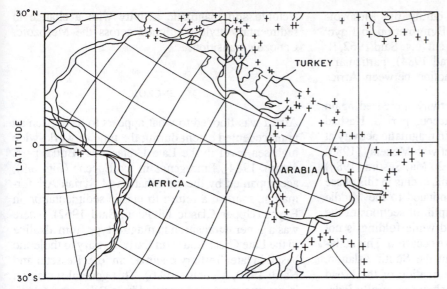

Figure 7.14 Reconstruction of the continents around Africa at 160 Ma during early Atlantic opening and rifting from northern Egypt (from Smith et al. 1981).

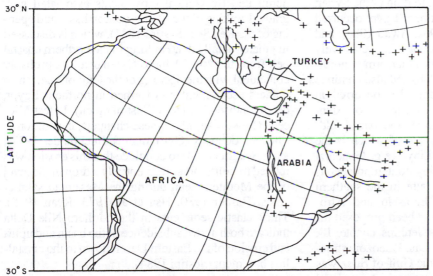

Figure 7.15 Reconstruction of the continents around Africa at 120 Ma after rifting from northern Egypt and prior to Syrian Arc deformation (from Smith et al. 1981).

(Figs 7/14 and 7/15, Burke & Dewey 1974, see also Morgan 1982, 1983, Duncan 1981, Van Houten 1983). Most reconstructions of Pangaea for the Late Permian to Early Jurassic suggest that the eastern Mediterranean was closed during this period with southern Turkey continuous with northern Egypt (e.g. Smith & Briden 1977, Smith & Woodcock 1982, Smith et al. 1981). If these reconstructions are basically correct, then it seems likely that the eastern Mediterranean basin originated from Early Mesozoic rifting (Monod et al. 1974, Hsu 1977, Bein & Gvirtz-

mann 1977, Biju-Duval et al. 1977, Biju-Duval & Dercourt 1980, Garfunkel & Derin 1984). The Mesozoic sea north of Egypt has been interpreted as part of a southern Neotethyan oceanic strand (e.g. Sengor et al. 1984), but Sestini (1984) can find no trace of a rifted continental margin close to the northern Egyptian coast. The nature of the crust beneath the eastern Mediterranean is still controversial (e.g. see Giese et al. 1982) and it could be either thinned continental crust or oceanic crust with a thick sediment blanket. However, all interpretations of this area indicate major extensional tectonics during the Mesozoic (Robertson & Dixon 1984). Deformation in Egypt was widespread during this event (e.g. Said 1962, R. Said pers. comm. 1984, Awad 1984), particularly associated with strike-slip motion between Africa and Arabia (Neev 1975).

Tectonic and magmatic activity increased again towards the end of the Cretaceous period. Hashad (1980) identifies a maximum in magmatic activity at 100 to 80 Ma in full agreement with Meneisy (1986) who identifies a peak at 90 ± 20 Ma. Sestini (1984) identifies three principal tectonic events in this phase of tectonic activity which continued to the Middle Eocene. The first event was uplift of sections of the northern Egyptian margin and while folding is not evident, block tilting could have occurred. This activity shortly followed rifting in the South Atlantic. Uplift was followed by the formation of the broad folds of the Syrian Arcs, which are generally independent of the trends of earlier structures. Growth of these folds occurred in various stages up to the Middle Eocene, but at different times in different places (Said 1962). Northeast trending folding of the Syrian Arcs in northern Sinai and northern Egypt occurred during this phase (e.g. Said 1962, Awad 1984), and these compressional features can be tentatively traced as far east as the Cyrenaica Platform in northeastern Libya (see Goudarzi 1980) and also continue offshore (Sestini 1984). Rifting also occurred in northeastern Africa at this time with an axis of extension approximately perpendicular to the axis of folding, the best documented example of which is the Sirte embayment in northern Libya (e.g. Goudarzi 1980). These events are roughly contemporaneous with the stages of oceanic closure in the southern Tethys and collision of the Eurasian and Afro-Arabian plates. Two models have been proposed for the evolution of the Syrian Arcs (Jenkins, chapter 18, this volume): 1) reactivation of Late Paleozoic structures formed when an embryonic Gulf of Suez was formed (Agah 1981) and, 2) compression associated with Late Cretaceous closing of the Tethys (e.g. Coleman 1981). The last event in this phase of activity was accentuation of uplift of the Mediterranean coast in the Early to Middle Eocene with the emergence and erosion of several positive structures and increased subsidence in the internal depocenters.

A widespread marine transgression accompanied the lateral compression of the Syrian Arcs and calcareous sediments became abundant (Middle Calcareous Division, Said 1962). This is the first indication of a widespread major change in the sedimentary environment from the clastic sedimentation that followed the Pan African. Thus, as the start of the Mesozoic in Egypt was accompanied by tectonic and magmatic activity continuous from the previous era,

tectonic and magmatic activity with a change in sedimentary style continued across the Mesozoic-Cenozoic transition.

CENOZOIC

Africa was flooded to what appears to have been an unprecedented extent during the time of world-wide high sea level in the Late Cretaceous (Burke pers. comm. 1984). Emergence in the Eocene (Fig. 6/9), accompanied by the culmination of Syrian Arc formation, marked a return to clastic sedimentation in Egypt (Upper Clastic Division, Said 1962). There was a general transition in magma type from alkaline in the Late Cretaceous and Early Tertiary to tholeiitic basalt in later Tertiary magmatism (e.g. Ressetar and Nairn 1980, Meneisy 1986). This general transition in magma type accompanied by a Paleogene transition in tectonic style from compression in the Syrian Arcs to extension in the Red Sea and Gulf of Suez. Since the Eocene the margins of Egypt have been dominated by vertical movements associated with gradual sinking of the Mediterranean basin and opening of the Red Sea (Sestini 1984), which is discussed in greater detail later. Faulting in the northern coastal belt from the Nile Delta to Cyrenaica was probably coincident with the opening of the Gulf of Suez and a marked escarpment was formed in northern Egypt, deeply incised in places (Sestini op cit). Latest Miocene (Messinian) fill in these channels, in addition to Pliocene deposits (Gvirtzman & Buchbinder 1977), leads Sestini (1984) to conclude that this erosion was related to uplift rather than entirely a drop in the level of the Mediterranean during the Messinian salinity crisis (Barber 1980, Hsu et al. 1973, Ryan 1978). Thick clastic sediments in the northern Nile Delta indicate both active subsidence of the basin and uplift and erosion of the Eastern Desert down to the crystalline basement (Sestini 1984).

RED SEA–GULF OF SUEZ RIFTING

The Red Sea-Gulf of Aden rift system has long been recognized as having been formed by the separation of Arabia from Africa (Lartet 1869) and is perhaps the best modern example of continental fragmentation and incipient ocean formation. These depressions were formed by the anticlockwise rotation of Saudi Arabia away from Africa about a pole of rotation in the central or south-central Mediterranean Sea (e.g. McKenzie et al. 1970, Freund 1970, Le Pichon & Francheteau 1978). Extension decreases westward along the Gulf of Aden and northward along the Red Sea, consistent with, and constraining

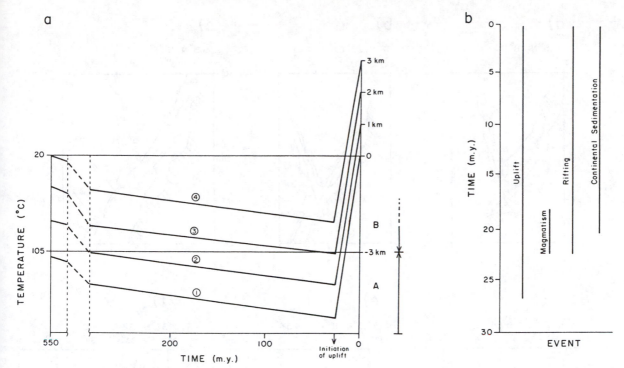

Figure 7.16 a. Thermal and uplift history of apatite samples from Sinai. b. Tentative time-table for rift-related events in Sinai (from Kohn & Eyal 1981).

the pole of opening. At the northern end of the Red Sea the opening is split between the opening of the Gulf of Suez and predominantly sinistral shear along the Gulf of Aqaba-Dead Sea rift system (e.g. Ben-Menahem et al. 1976). Despite the relative youth of this system, there are several major problems still remaining in interpreting its tectonic evolution. These problems include the absolute and relative timings of vertical and horizontal movements associated with rifting, the nature of the crust in the northern Red Sea (extended and thinned continental crust, a mixture of continental and oceanic crust, or oceanic crust), the relationship between opening and shear in the Gulfs of Suez and Aqaba to Red Sea opening and the mechanism of rifting.

Unequivocal dating of the first magmatic and/or tectonic events associated with Cenozoic rifting in eastern Egypt is difficult, a problem shared with most rifts (Ramberg & Morgan 1984). It has been suggested that the Late Cretaceous alkaline magmatism was associated with regional uplift (doming) of the Red Sea prior to extension (Ressetar and Nairn 1980), and Cretaceous doming in the area now occupied by the Red Sea is indicated by sediment transport directions (Ward & McDonald 1979). Most workers, however, do not consider the initiation of rifting to be this early. Meneisy (1986) identifies a late Eocene-early Oligocene magmatic event (40 ±

10 Ma) which he associates with Red Sea doming and extension. Kohn & Eyal (1981) used fission-track data to determine the uplift history of Sinai which are shown in Figure 7/16a. They deduced that uplift preceded extension by a few Ma, as shown in Figure 7/16b, but Steckler (1985) maintains that uplift and extension were synchronous, on the basis of a stratigraphic analysis of the Gulf of Suez. Meneisy (1985) also identifies a late Oligocene-early Miocene magmatic phase (24 ± 2 Ma) which he associates with Red Sea opening together with additional phases of magmatism to about 15 Ma. There is general agreement that major faulting associated with Gulf of Suez rifting began in the Latest Oligocene to Early Miocene (e.g. Robson 1971, Chenet & Letouzey 1983, Steckler 1985), at the same time as initial Red Sea rifting (Coleman 1974, Gass 1977).

Details of the exact direction and amount of extension in the Gulf of Suez and Red Sea are also currently controversial. In general, it is accepted that these depressions were formed by the anticlockwise rotation of Arabia away from Africa. As noted by Cochran (1983), however, and as shown in Figure 7/17, not all suggested poles of rotation are obviously consistent with coastline reconstructions. These problems arise because it is difficult to track the exact relative movements of Africa and Arabia during the early stages of Red Sea opening, due to the lack of

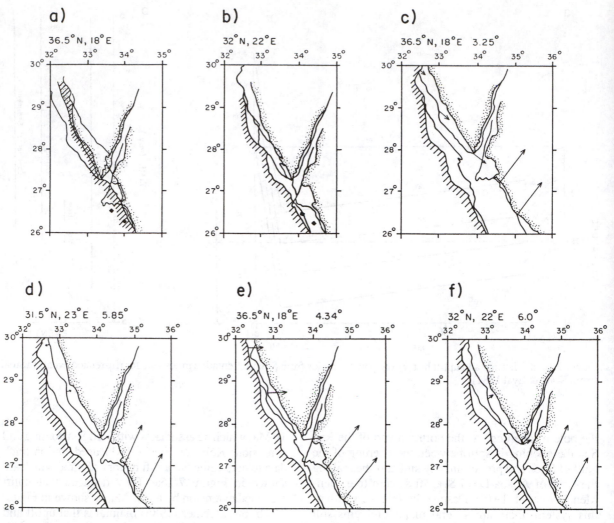

Figure 7.17 Reconstructions of Gulf of Suez and northern Red Sea and deduced plate motions for Sinai and Arabia (from Cochran 1983).

unambiguous marine magnetic anomalies recording these movements and the non-rigid behavior of the rift zones. Workers in this area generally agree that at least the central portions of the Gulf of Aden and southern Red Sea were formed by sea-floor spreading, although the extent and age of oceanic crust in these areas is less certain (e.g. Cochran 1981, 1982, 1983, Girdler & Styles 1982). Magnetic anomalies attributed to sea-floor spreading have been reported for the northern Red Sea (e.g. Hall 1979), but these interpretations have been questioned (Cochran 1983), and definitive data constraining the mode of extension in this region are lacking. Thick sediments, including evaporites reduce the resolution of the data (e.g. Cochran op cit). In addition, the Sinai subplate has moved relative to both Africa and Arabia, introducing further uncertainties in relative motions.

Other methods must thus be used to constrain plate movements in the northern Red Sea.

One possible constraint upon extension in the northern Red Sea and Gulf of Suez is provided by the complementary shear movements along the Gulf of Aqaba and the Dead Sea rift system (Lartet 1869, Dubertret 1932, Quennell 1958, 1959, 1984, Freund 1970, Freund et al. 1970). Offset of Miocene sedimentary and volcanic rocks was taken to indicate about 45 km of post-Miocene movement with an additional 62 km of movement thought to have occurred in the Early Miocene, possibly related to the initial faulting in the Gulf of Suez and Red Sea rifting. The amount of slightly more than 100 km of left-lateral offset has subsequently been corroborated by a corresponding offset of aeromagnetic anomalies, which unfortunately provide no information on

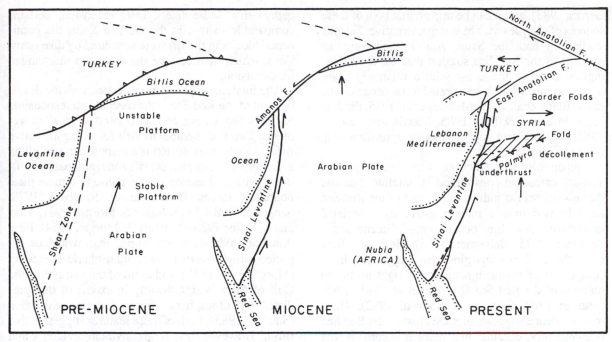

Figure 7.18 Plate tectonics of the west Arabian rift system related to the eastern Mediterranean region (from Quennell 1984, after Dewey & Sengor 1979).

the timing of the offset (Hatcher et al. 1981). Several studies have suggested two or more phases of opening of the Red Sea, possibly correlating with polyphase movement on the Dead Sea system (e.g. Girdler & Styles 1974, 1976, Richardson & Harrison 1976, Girdler 1985), although these studies are based primarily upon magnetic anomaly data which have alternative interpretations (Cochran 1983, Labrecque & Zitellini 1985). Even the timing of the 107 km shear along the Dead Sea-Gulf of Aqaba system has been questioned and suggested to be all Middle Miocene or later (Bartov et al. 1980). Thus, it seems reasonable to use the amount of shear on the Dead Sea system as a constraint on opening of the Red Sea and Gulf of Suez, but caution must be applied in interpreting the timing of these events.

From the evidence that initial rifting in the Gulf of Suez and Red Sea was approximately synchronous (Late Oligocene-Early Miocene), but that much, if not all, shear on the Dead Sea system occurred later (Pliocene), it appears that perhaps extension in the northern Red Sea was initially accommodated to the north by extension in the Gulf of Suez. Subsequent to the development of the Gulf of Suez, extension was accommodated by shear along the Dead Sea-Gulf of Aqaba system (Fig. 7/18, Dewey & Sengor 1979, Quennell 1984). This two-stage development for the northern termination of the Red Sea is almost certainly an oversimplification, and probably an overlap

in the Suez opening and Dead Sea shear occurred. However, it is thought to represent a reasonable first approximation to the timing of plate motions in this area. Opening of the Gulf of Suez may have had a component of strike-slip movement and did not progress far enough for complete separation of Sinai from Africa. The Gulf is underlain by thinned continental crust, with extension in the upper crust by brittle block faulting and rotation (e.g. Chenet & Letouzey 1983, Steckler 1985, Angelier 1985). This extension of the continental crust in this region must be taken into account in the evaluation of the plate reconstructions of the northern Red Sea (Fig. 7/17).

Attempts to identify the nature of the crust in the northern Red Sea rely upon geological evidence, geophysical data, and reconstructions of the extended crust. As noted above, magnetic anomalies are not sufficiently well defined to allow unequivocal interpretation of spreading patterns in the northern Red Sea as may be expected if the area was underlain by typical oceanic crust. However, the thick sedimentary section and possibly slow spreading rate for the northern Red Sea (half spreading rate of the order of 5 mm/a assuming 200 km of opening in 20 Ma) make it unlikely that linear magnetic anomalies would be distinct even if the crust was oceanic. Samples of the crust are limited to a couple of basement samples from oil tests (Cochran 1983) and Zabargad (St John's) Island (Bonatti et al. 1981, 1983, Styles &

Gerdes 1983), and it can be argued that both of these sources of samples may be unrepresentative. Seismic studies on both the Saudi Arabian and Egyptian margins of the Red Sea suggest that the crust thins rapidly close to the coast with a relatively sharp transition into what is interpreted to be oceanic-type crust (Blank et al. 1979, Mooney et al. 1985, Prodehl 1985, Milkereit & Fluh 1985, Makris pers. comm. 1983). Gravity data support this interpretation (e.g. Brown & Girdler 1982).

Palinspastic reconstructions of the Red Sea prior to Tertiary extension constrained by satellite data and the seismic results indicate that most of the Red Sea crust formed from a mantle source by rift-related igneous processes (i.e. 'oceanic crust', Greenwood & Anderson 1975, Bohannon 1986). High heat flow from the Red Sea margins also suggests a large component of young igneous crust right up to the margins of the Red Sea (Morgan et al. 1981, 1985, Gettings 1982, Gettings & Showail 1982). Thus, most evidence suggests that the crust of the Red Sea is dominantly 'oceanic' in affinity. It is unlikely that this crust was formed at a well-organized spreading ridge as such a feature is not evident in the modern

bathymetry. More likely, after extension, perhaps comparable to that in the Gulf of Suez, the continental blocks in the rift were separated by dike intrusions, which then became the dominant mechanism for extension.

The final major problem associated with the development of the Red Sea is the mechanism responsible for Red Sea rifting and associated vertical movements. Two end-member models for rifting are active rifting in which extension is a response to upwelling asthenosphere beneath the rift zone, and passive rifting in which extension is a response to remote plate boundary forces (Neumann & Ramberg 1978, Sengor & Burke 1978, Baker & Morgan 1981, Morgan & Baker 1983, Ramberg & Morgan 1984). Heat flow data indicate a hotter crust than would be expected simply from passive lithospheric stretching (Morgan et al. 1985), and uplift of the margins of the Gulf of Suez is significantly in excess of that predicted by lithospheric thinning during extension (Steckler 1985). Both of these features suggest active rifting. However, there is no convincing evidence that rifting was immediately preceded by major doming, which is the main characteristic associated with

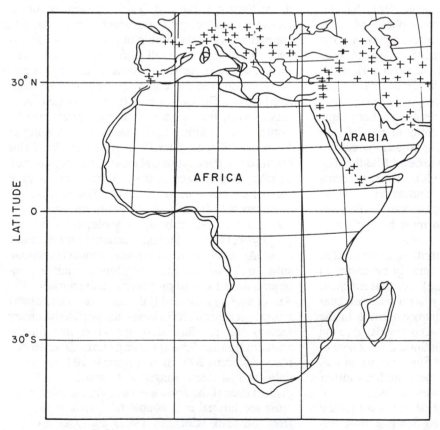

Figure 7.19 Reconstruction of continents surrounding Africa at 10 Ma, during the Red Sea opening (from Smith et al. 1981).

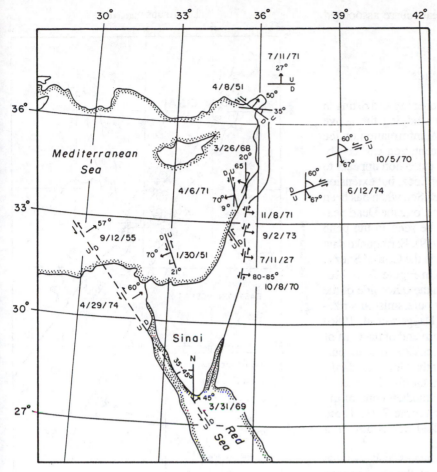

Figure 7.20 Earthquake slip vectors for Sinai (from Ben-Menahem et al. 1976).

models of active rifting. Steckler (1985) has suggested that initial extension created an instability in the lithosphere-asthenosphere system which caused additional heat input to the lithosphere and uplift after the initiation of rifting. Cretaceous magmatism and doming in the areas now occupied by the Red Sea suggest that the lithosphere in this region was preheated and thinned approximately 50 Ma prior to extension, implying that the rifting may be some hybrid of the active and passive mechanisms. This problem will only be resolved by detailed studies of the relative timing of uplift and extension in the Red Sea system.

RESPONSE TO RIFTING

Early sediments in the rifted zones include marine marls and shales, but as the rising rift margins isolated the Gulf of Suez and Red Sea from the Mediterranean, evaporites were deposited (Garfunkel & Bartov 1977). These basins were reconnected to normal

marine conditions at the end of the Miocene with an opening to the Indian Ocean through the Gulf of Aden. Elsewhere in Egypt the rising rift margins acted as a sediment source. Details of the stratigraphic sequence on the Red Sea margin correlate with relative changes in sea level on a regional and global scale, as well as phases of Red Sea rifting (Khedr 1984).

The ancestral rivers of the Nile appear to have their origin about the same time as the initiation of Red Sea rifting (Said 1976, 1981), and it is likely that topography associated with the uplift of the Red Sea margins controlled the development of this great river. Subsurface data indicate that the river appears to run in grabens in places with faults parallel to and possibly associated with the Red Sea (Said 1976). There is strong evidence also for Cenozoic uplift of the Uweinat massif in southwestern Egypt coincident with Red Sea rifting, and this uplift is associated with magmatic activity (Meneisy 1986). No significant Cenozoic extension has been recognized in the Uweinat massif, and the uplift is thought to be thermal in

origin, perhaps related to the heat source associated
with Red Sea rifling.

MODERN TECTONICS

Events associated with Red Sea opening and rifting in
the Gulf of Suez continue; by studying the active
tectonics of the region additional information on the
details of the continental fragmentation can be obtained. For the last 10 Ma little extension appears to
have occurred across the Gulf of Suez, but extension
has continued in the northern Red Sea whch has been
translated into strike-slip motion along the Dead Sea-
Gulf of Aqaba system, as can be seen in the plate
reconstruction for 10 Ma (Fig. 7/19). Earthquake data
indicate that the modern motion in the Gulf of Suez is
sinistral oblique-slip, as shown in Figure 7/20. Motion along the Gulf of Aqaba on the other side of the
Sinai subplate is approximately pure sinistral strike-
slip. Microearthquake studies (Daggett et al. 1986)
indicate that although the southern end of the Gulf of
Suez is active seismically, this activity does not currently extend along the length of the Gulf. The earthquake epicenters in the southern Gulf of Suez trend
towards active seismicity in the median zone of the
northern Red Sea, as shown in Figure 7/21. These
data suggest that the northern Red Sea is now in a
stage of median-ridge spreading.

Archaeological evidence indicates considerable
earthquake activity and shear along the Dead Sea rift
in historical times (Karcz et al. 1977), and folding
and uplift indicate late Holocene tectonic activity on
the margins of the Sinai subplate (Neev & Friedman
1978). However, there is little evidence for major
historical earthquakes having occurred with epicenters in Egypt (Poirier & Taher 1980, Poirier et al.
1980). A magnitude of 6.0 event in 1955 had its
epicenter in the Precambrian of the Red Sea Hills
(Fairhead & Girdler 1970) and seismic activity in this
area is thought to be common (Morgan et al. 1981,
Daggett et al. 1986). The source of this activity is
unknown, but it has been suggested that it may be
related with the intrusion of a pluton into the Precambrian crust associated with rifting (Daggett et al.
1986). In November of 1981 a magnitude 5.5 earthquake occurred near Aswan in southern Egypt
(Adams 1983). This event and its aftershocks occurred under a pronounced embayment in the Aswan
Lake which marks the intersection of two major sets
of faults which trend approximately east-west and
north-south. Much of the surface displacement occurred near the surface trace of the easterly trending
Kalabsha fault, and aftershock hypocenters clustered
beneath the surface trace of this fault at depths of
about 20 km, suggesting a near vertical attitude for

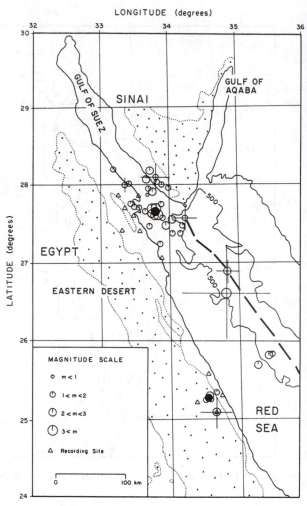

Figure 7.21 Microearthquake activity in eastern Egypt and
the Red Sea. Solid symbols indicate multiple coincident
epicenters (from Daggett et al. 1986).

this fault (T.R. Toppozada, pers. comm. 1984). Focal
mechanisms are consistent with dextral strike-slip
faulting along the east-west Kalabsha trend, approximately perpendicular to the Red Sea trend. Faulting
in this area appears to be a reactivation of older faults,
perhaps triggered by the filling of Lake Nasser.
Strike-slip motion perpedicular to the Red Sea trend
may be a reponse to a decrease in extension from
south to north in the northern Red Sea.

CONCLUSIONS

The analysis of the geologic history of Egypt in the
framework of global tectonics suggests that most of
the major geological features of Egypt can be explained in terms of the interaction of global tectonics.
The exact mechanisms linking remote global tectonic

events and local magmatic and/or tectonic events are, however, poorly understood at present. Egypt entered the Phanerozoic during a period of rapid continental growth through Pan African arc accretion. The end of the Mesozoic and the end of the Cenozoic appear to reverse this continental growth with rifting first from the north coast, then from the east coast of Egypt. Egypt therefore appears to be entering a new phase of the Wilson cycle of opening and closing of oceans.

ACKNOWLEDGEMENTS

Numerous people provided invaluable assistance at various stages in the preparation of this review. Kevin Burke is thanked for his major assistance in planning the review and Eberhard Klitzsch provided suggestions for major revisions. Rushdi Said provided ideas, data and encouragement at all stages in this study. H. Schandelmeier, C. Barnes and M.K. Morgan made thoughtful reviews of a draft of this manuscript. This work was carried out in part at the Lunar and Planetary Institute (LPI), which is operated by the Universities Space Research Association under contract no. NASW-3389 from the National Aeronautics and Space Administration, and in part at Purdue University. This is LPI contribution no. 625.

CHAPTER 8

Tectonic framework

WAFIK M. MESHREF

Gulf of Suez Petroleum Company, Cairo, Egypt

This chapter consists of two parts. Part I deals with the sequence of tectonic events which shaped the structure of Egypt, and Part II deals with the tectonics of the northern Red Sea-Gulf of Suez rift.

1 TECTONIC EVENTS

Basement rocks crop out in southern Sinai, the Eastern Desert and as isolated inliers in southern Egypt such as near Aswan, and at Gebel Uweinat at the extreme southwestern corner of Egypt – that is, within the craton area of Egypt associated with the Nubian-Arabian shield. The remainder of the surface of Egypt is covered by sedimentary rocks of both the stable and unstable shelf areas (Said 1962, Fig. 8/1).

Evolution of Egyptian basement
The Precambrian basement rocks of Saudi Arabia, Eastern Sudan and Egypt are considered to be the northeastern extension of the Mozambique belt of the African Shield (Fig. 8/2, Kroener 1977). Shackleton (1980) believes that these basement rocks are characterized by abundant granites and granodiorites, arc-type volcanics and well-authenticated ophiolites of Proterozoic age. He described these rocks as characterized by low grade metamorphism in contrast to the high grade metamorphic rocks of the southern Mozambique belt in Kenya and Tanzania which are of Archean age. Moreover, he considers the basement rocks in Egypt and Sudan to have been formed by an eastward younging succession of island arc/back arc complexes.

Gass (1977) estimates that more than 60% of the exposed basement rocks of the Nubian-Arabian Shield consists of calc-alkaline granitoid plutons emplaced into eruptive rocks and volcano-derived sediments of similar age and composition. Based on conclusions of Greenwood et al. (1976) and Neary et al. (1976), Gass suggests an intraoceanic island arc environment for the origin of these rocks.

Three ophiolite belts have been defined in Saudi Arabia (Brown & Coleman 1972) and are believed to represent cryptic sutures between orogenic belts. In Egypt, there are probably three ophiolite belts, the more easterly two of which are probably the crustally separated continuation of two Saudi Arabia ophiolite belts. Three or four orogenic episodes were, threfore, probably involved, following the initiation of sedimentation in the Red Sea geosyncline. These ophiolitic rocks occur in the form of predominantly northwest trending belts of folded thrusts (Krs et al. 1973).

Field, petrographic and geochemical data are consistent with the interpretation of lateral accretion from east to west of successive island arc systems during a period of almost 600 Ma, between 1100 and 500 Ma (Gass 1977, Nasseef & Gass 1977). Moreover, Garson & Shalaby (1976) suggested a plate tectonic model for the Red Sea geosyncline during the period from the Upper Proterozoic to the Cambrian. The proposed model suggests four cycles of marginal basins, flanked to the east and northeast by crustal cratonic portions, that were produced during continued subduction and steepening of the Benioff zone. These four cycles seem to have taken place at 1300, 1200, 1000-900 and 660 my BP.

El Shazly et al. (1973) described the basement rocks at Abu Swayel, Eastern Desert, as being of low grade and the oldest in Egypt, giving radiometric ages ranging from 1150 to 1300 my.

Compilation of 168 isotopic ages for rocks of the Egyptian basement complex, in the Eastern Desert and in Sinai, indicate an upper age limit of 1200 ma for the geosynclinal metasediments and metavolcanics, which are stratigraphically the oldest rocks in the basement (Hashad 1980).

The basement rocks atGebel Uweinat microcraton consist of medium to high grade gneisses which yielded radiometric ages between 2000 and 1200 ma. These ages, in some places, certainly represent a thermal event superimposed upon an even earlier amphibolite facies metamorphism dated at approximately 2600 ma. This is further emphasized by the fact that these metamorphic assemblages are intruded by the Late Pan-African syenitic ring complex.

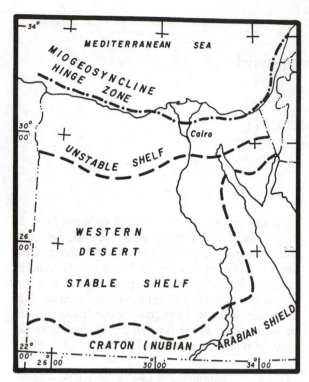

Figure 8.1 Sketch of the structural aspects of the Nubian-Arabian shield margin in northern Egypt.

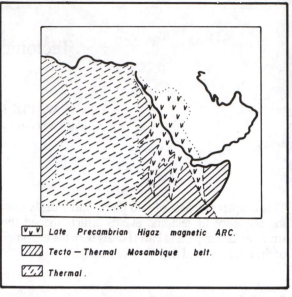

V V V	Late Precambrian Higaz magnetic ARC.
//////	Tecto – Thermal Mosambique belt.
/////	Thermal .

Figure 8.2 Areas affected by late Precambrian/early Paleozoic tectogenesis (modified after Kröner 1977b).

Accordingly, Precambrian basement rocks in Egypt can be subdivided into two main types based on composition, grade of metamorphism and radiometric age:

1. Includes the oldest rocks, of Archaean age, which outcrop at Gebel Uweinat at the extreme southwestern corner of Egypt. They correlate with the Tibesti massif of southern Libya.

2. Includes the relatively younger rocks of Proterozoic age, which outcrop in the Eastern Desert, Aswan, and as inliers south of Dakhla and Kharga Oases. They are considered to be the northern extension of the Nubian Shield, itself a continuation of the Mozambique belt.

Since basement rocks are the oldest rocks in Egypt, they are expected to record all tectonic events affecting Egypt since the early Precambrian to the present. Figure 8/3 shows the surface tectonic trends as traced from the 1981 geologic map of Egypt. This figure shows a large number of tectonic elements of varying lengths and trends. Moreover, it shows greater density of tectonic trends in areas where basement rocks crop out or in areas of relatively thin sedimentary cover. The tectonic elements show a much lesser distribution where the basement rocks are buried under a relatively thick column of sediments, as in central and northern Western Desert, the Delta, north Sinai, Gulf of Suez and the Red Sea.

The best geophysical tool for mapping tectonic trends of deeply buried basement rocks is magnetics. Usually aeromagnetic mapping reflects the horizontal variation in magnetic properties of underlying rocks. In this regard sedimentary rocks, excluding iron ores and basic extrusions intruded within them, are characterized by very low magnetic susceptibility, while igneous and metamorphic rocks, the main constituents of basement rocks, are strongly magnetized. The more heterogeneous and deformed the basement rocks, the sharper and stronger will their magnetic signatures be on the aeromagnetic map.

Magnetic trend analysis

Magnetic anomalies do not occur at random, but are generally aligned along definite and preferred axes forming trends. Magnetic trend patterns can thus be used to define magnetic provinces (Affleck 1963, Hall 1979).

A magnetic trend statistical analysis using results of all aeromagnetic surveys flown in Egypt was carried out. The results of this study are plotted in the form of rosetta diagrams (Fig. 4). Close examination of this figure yields the following observtions:

1. There are several tectonic trends that have affected Egypt throughout its geologic history: north-south (Nubian or East African trend); north-northeast (Aqaba trend); east-west (Tethyan trend); west-northwest (Darag trend); east-northeast (Syrian Arc trend); northwest (Red Sea or Suez trend), and northeast (Aualitic or Tibesti trend).

2. some tectonic trends are more characteristic of some parts of the country than others.

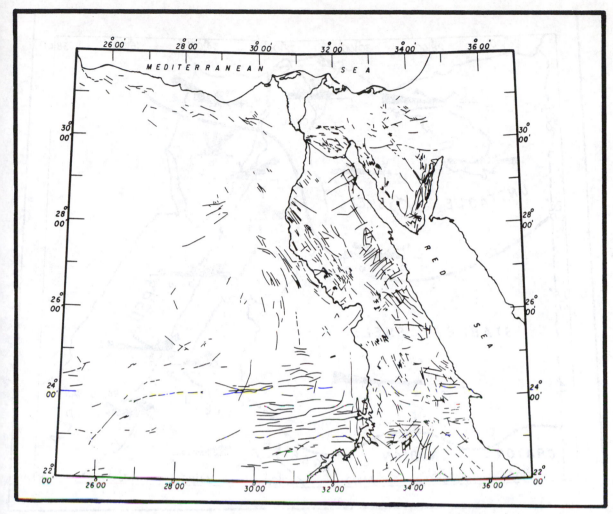

Figure 8.3 Surface tectonic elements (1981 geological map of Egypt, Geological Survey of Egypt).

3. Some trends show more strength in some parts of the country than in others.

4. Some tectonic lines show variations of 10-20° in the trend. This range over which a discernible linear occurs is considered as representing a spectrum of this linear, rather than a separate trend.

The following is a chronological discussion of the various tectonic trends that affected Egypt, shown in Figure 8/4, and their evolution. Only the Precambrian tectonic trends that affected the basement rocks will be considered when discussing the evolution of the Egyptian basement. Precambrian trends are those trends inherited during the cratonization process, during the formation of basement rocks. They should not be confused with basement trends, as the latter term refers to any tectonic trend that has affected the basement rocks throughout geological time.

The tectonic trends affecting the Phanerozoic section, from Paleozoic to Quaternary, will be discussed with reference to a series of isopach maps of various stratigraphic rock units in selected areas such as the north Western Desert and the Gulf of Suez. These isopach maps were constructed using subsurface geologic information from hundreds of wells drilled that encountered only complete sections, so that the trends of tectonic events affecting the deposition of the various rock units could be delineated.

The inliers of crystalline basement in southern Egypt are characterized by a dominant northerly trend which is the East African or Nubian trend of folds and metamorphic foliation. Acidic intrusives in the Eastern Desert can be related to Pan-African orogeny during which north-trending fold belts were formed (Brown & Coleman 1972, Fig. 8/5).

The Pan-African structures of the Tibesti area have a significantly regular, northeasterly trend extending northward to Gebel Uweinat microcraton and further beneath the Western Desert (Fig. 8/5). The Nubian trend and the Tibesti trend seem to intersect along a major structural discontinuity across Egypt. This is

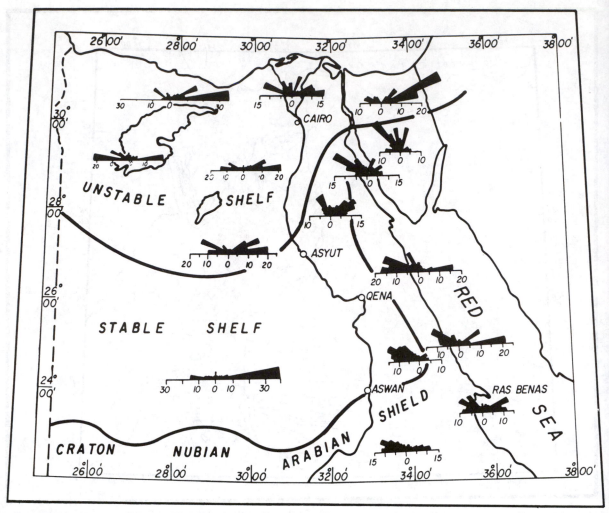

Figure 8.4 Rosetta diagram showing the major tectonic trends of Egypt.

probably the major transcurrent shear zone that belongs to the Pelusium Megashear system across Africa (Neev 1975, 1977, Fig. 8/6).

Neev postulates a megashear which extends from Turkey to the South Atlantic, running subparallel to the eastern margin of the Mediterranean Sea (where it was first identified and studied), then curving northeast-southwest across central Africa from the Nile delta to the delta of the Niger in the Gulf of Guinea (Fig. 8/6). It is presumed that this has functioned as a system of an echelon left-lateral megashears since Precambrian times. At least four other geostructures also cut across Africa, paralleling the Pelusium megashear system (Neev et al. 1982).

Precambrian tectonic trends
It is evident that the successive accretions of eastward younging successions of island arc/back arc complexes during the Hercynian Precambrian subduction

would result in northwest to north-northwest trending structures. The strength and density of this trend would be expected to fade away westward across Egypt. Yallouze & Knetsch (1959) recognized a northern basin that extends along Egypt in a northwest direction, parallel to the Red Sea (Fig. 8/7).

Aeromagnetic maps generally show broad, low frequency anomalies resulting from relatively deeper and older structures, as well as high frequency local anomalies that reflect shallow tectonic features. Thus regional magnetic maps are of vital importance in studying older regional basement tectonics, while residual magnetic maps are important in minerals exploration and for delineating shallow tectonics associated with them.

The regional magnetic map of the north Western Desert (Fig. 8/8) shows a series of strong positive magnetic anomalies of north-south to north-northwest-south-southeast trends that cut across

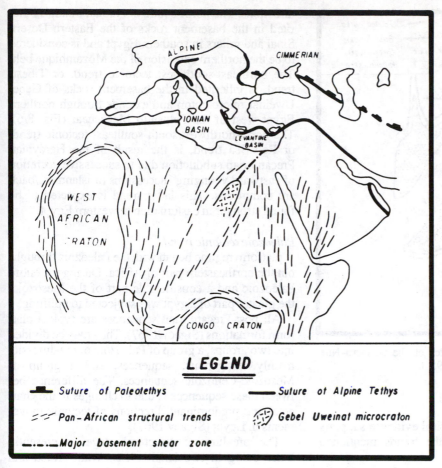

Figure 8.5 Plate and basement tectonic framework of north Africa (after Hsu & Bernoulli 1978).

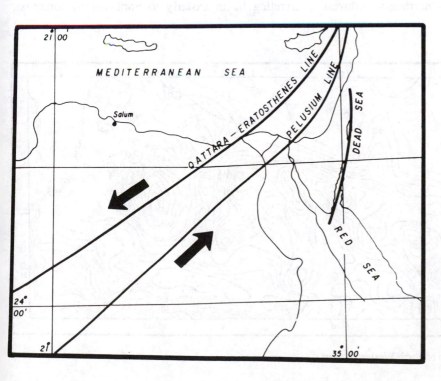

Figure 8.6 Photogeologic lineament interpretation showing the Pelusium megashear system (after Neev et al. 1982).

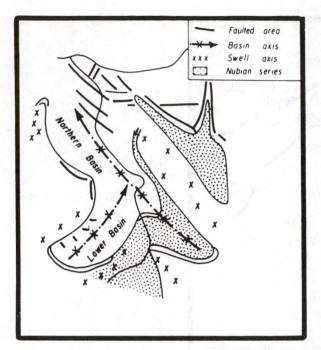

Figure 8.7 Structural block sketch of the northern Nile basin (after Yallouze & Knetsch 1954).

directions. The north-south, or Nubian trend, is inherited in the basement rocks of the Eastern Desert, Sinai and inliers in southern Egypt and is considered to be the northern extension of the Mozambique belt. The northeast-southwest tectonic trend, or Tibesti trend, is inherited in the basement rocks of Gebel Uweinat microcraton and extends through northern Egypt west of the Pelusium Megashear (Fig. 8/6). The north-northwest/south-southeast tectonic trend, or Red Sea trend, is the result of the Hercynian Precambrian subduction due to successive accretion of eastward younging successions of island arc/back arc complexes. This latter trend is expected to be found stronger in eastern than in western Egypt.

Paleozoic tectonic trends

A platform regime began with the Paleozoic throughout the northeastern part of Africa. During the entire Paleozoic and a considerable part of the Mesozoic, the greater part of Egypt was subjected to uplifting.

All post-Precambrian sequences are typical platform formations (Sigaev 1967). They may be divided into two groups, a group of Paleozoic or pre-Jurassic, mainly terrigenous, sequences, and a group of Mesozoic-Cenozoic sequences. The difference between these sequences is due to changes in the character and predominant directions of tectonic movements in Egypt (Sigaev 1967).

The pre-Jurassic structural stage is presently outcropping in two small areas, Wadi Araba and southeastern Sinai. In Wadi Araba, Carboniferous sediments occur in the core of a gentle anticline trending in an easterly to northeasterly direction,

northern Egypt. This geophysical evidence supports the assumed older age of the trends mentioned above.

In summary, the Precambrian tectonic trends in the Egyptian basement are of north-south, north-northwest/south-southeast and northeast-southwest

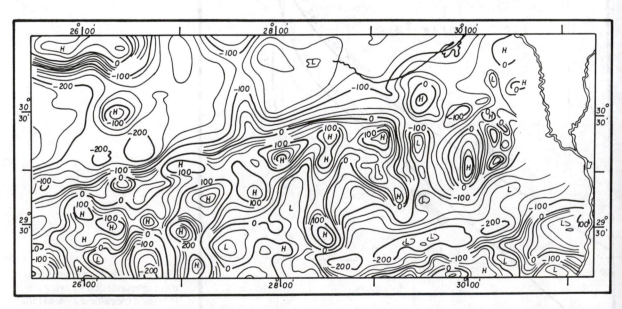

Figure 8.8 Regional magnetic map of north Western Desert.

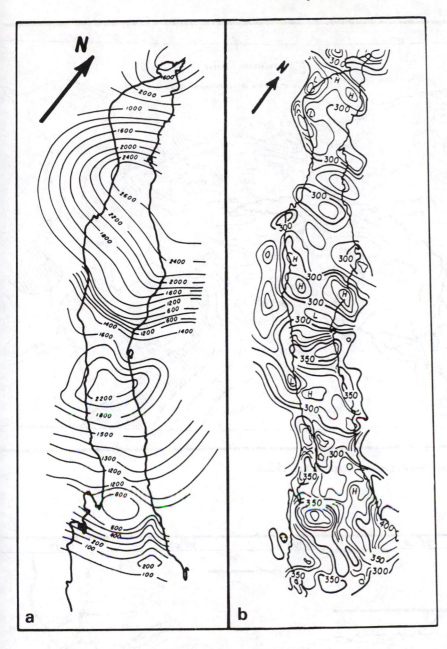

Figure 8.9 a. Total Paleozoic isopach map for Gulf of Suez (Beleity et al. 1986). b. Regional
magnetic map for the Gulf of
Suez.

interrupted by the Gulf of Suez Rift. In Sinai, this structural stage is also represented by Carboniferous sediments resting monoclinally on the northern flank of the south Sinai Precambrian domed uplift.

In Paleozoic time, the thrusting or uplifting of the African continent against the main block of the earth's crust might have been the cause of the formation of east-west trending swells and thrusts across the northern part of the African continent, including Egypt. The formation of swells took place at the edge of the platform. Although Paleozoic exposures are scarce, the Paleozoic is widely distributed, as the numerous subsurface occurrences in the Gulf of Suez

and north Western Desert show. The Paleozoic section, as mapped from data from 200 wells in the Gulf of Suez region, is classified from bottom to top into the Araba Formation and Naqus Formation of early Paleozoic age, the Um Bogma, equivalent to the Abu Durba Formation, and the Ataqa Formation, equivalent to the Rod El Hamal Formation of late Paleozoic age (Beleity et al. 1986). The isopach maps of the four Paleozoic formations show alternating thickening and thinning of Paleozoic rocks in the Gulf of Suez regime in an east-west direction. However, the isopach map of the total Paleozoic section in the Gulf of Suez shows a strong east-west tectonic trend that

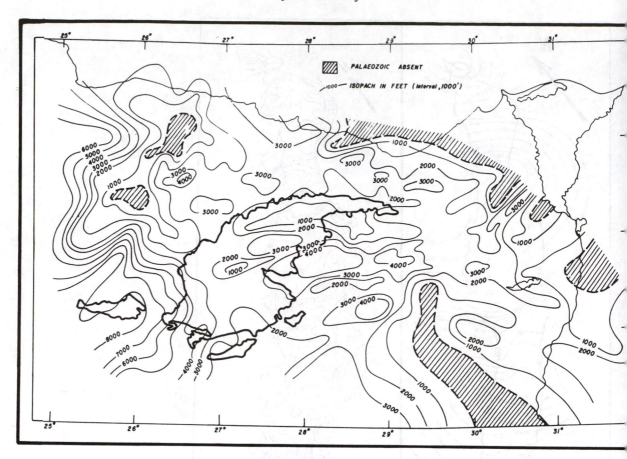

Figure 8.10 Paleozoic isopach map (after Robertson Research Intl. 1982).

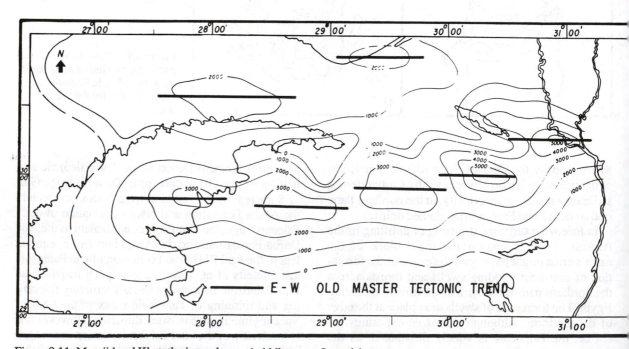

Figure 8.11 Masajid and Khatatba isopach map (middle-upper Jurassic).

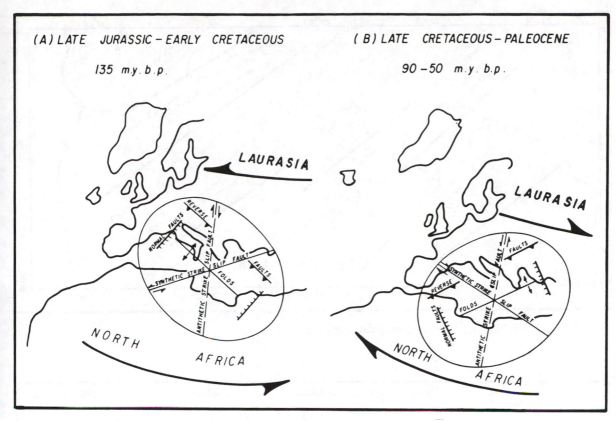

Figure 8.12 Transcurrent motion between Africa and Laurasia (modified after Smith 1971).

seems to have controlled the deposition of Paleozoic rocks (Fig. 8/9a).

Other geophysical evidence of east-west tectonic control of Paleozoic sediments comes from the regional magnetic maps of the north Western Desert (Fig. 8/8) and the Gulf of Suez (Fig. 8/9b). Both maps show strong and broad regional magnetic anomalies cutting across northern Egypt in an east-west direction.

A Paleozoic isopach map of part of the north Western Desert (Fig. 8/10) substantially covers the area termed the 'Unstable Shelf' by Said in 1962. This map shows variable thicknesses of Paleozoic sediments in various parts of the mapped area ranging from 1000 to 8000 feet, primarily along an east-west trend. The map in Figure 8/10 also shows relic north-south or north-northwest basement highs over which Paleozoic sediments are thin or absent.

Mesozoic to early Tertiary tectonic trends

The first deposit of the Mesozoic of the Western Desert was the continental Bahrein Formation of the early Jurassic. This was followed by the deposition of uniform shallow marine sediments of the middle Jurassic, an epoch of continental margin stable shelf conditions. The late Jurassic, however, was marked by the deposition of variable types of sediments with highly variable thicknesses. The isopach pattern of middle-upper Jurassic deposits (Khatatba and Masajid) shows a strong east-west tectonic trend (Fig. 8/11), indicating that the deposition of sediments from late Paleozoic until late Jurassic time was controlled primarily by east-west tectonics.

Sedimentary patterns from the late Jurassic to early Tertiary appear to have been influenced significantly by two primary tectonic forces related to Tethyan plate tectonics: (1) the sinistral shear during the late Jurassic to early Cretaceous (Fig. 8/12a), and (2) the dextral shear during the late Cretaceous to Paleocene time (Fig. 8/12b).

Moreover, it appears that the dextral shear was either partially accompanied or partially followed by north-south or north-northwest/south-southeast compressive forces (Fig. 8/23). The destruction of the Paleotethys, north and east of present day northeast Africa, was accompanied by the opening up of the Alpine Tethys, or Neotethys, as a result of the opening of the central Atlantic ocean in early Jurassic time. As a result of the Jurassic initial stage of the Alpine Tethys, Africa moved eastward relative to

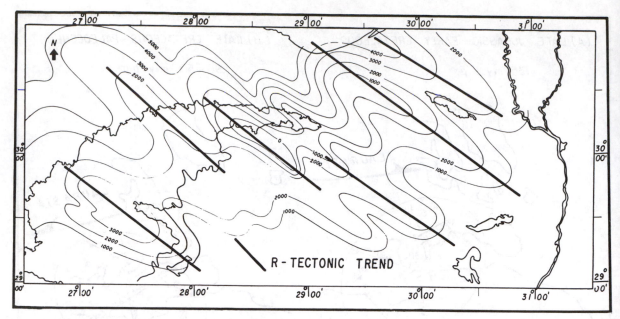

Figure 8.13 Alam El Bueib isopach map (Barremian-Necomian).

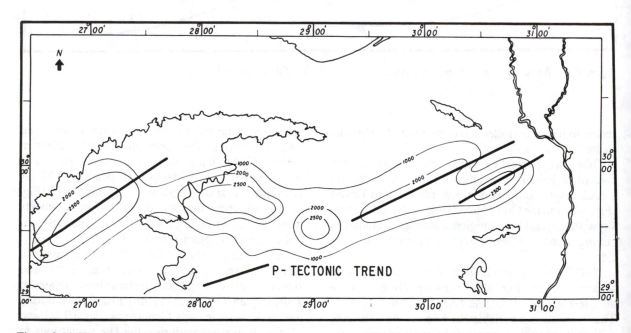

Figure 8.14 Kharita Member isopach (Albian).

Laurasia (Smith 1971). This sinistral lateral motion continued until the late Cretaceous, resulting in some 2000 km of transcurrent motion (Fig. 8/12a). Hence, during the late Jurassic to early Cretaceous, a left lateral megashear between Africa and Laurasia seems to have resulted in two main tectonic elements: west-northwest trending folds with associated thrust faults, and east-northeast trending strike-slip faults with left lateral motion (Fig. 8/12a).

The isopach map of Alam El Bueib Formation (Fig. 8/13) shows clearly that its deposition was strongly controlled by the west-northwest folding episode. This tectonic event continued until the late Cretaceous. However, during the late Cretaceous to Paleocene (90-50 my BP), the right-lateral transcurrent between Africa and Laurasia, due to the opening

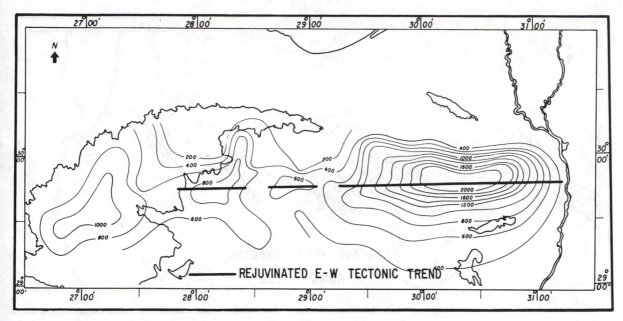

Figure 8.15 Bahariya Formation isopach (lower Cenomanian).

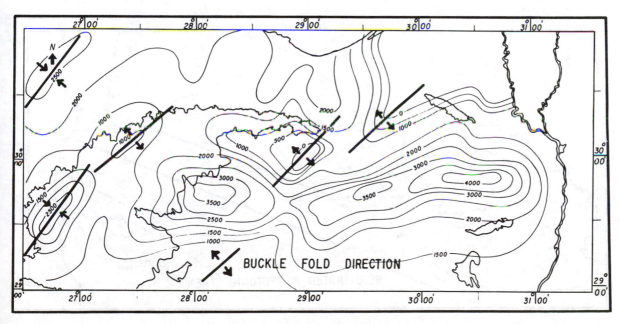

Figure 8.16 Abu Roash formation isopach map (Coniacian-upper Cenomanian).

of the north Atlantic Ocean (Fig. 8/12b), resulted in two main tectonic elements: east-northeast folding with associated thrust faults, and west-northwest strike-slip faults with right-lateral motion (Fig. 8/12b). It is believed that the east-northeast folding with associated thrusts might have resulted from wrenching along east-northeast strike slip faults formed earlier in early Cretaceous time.

The tectonic forces described above appear to have affected the deposition of the Kharita, Bahariya, Abu Roash, Khoman and Apollonia Formations, according to their respective isopach maps in Figures 8/14, 8/15, 8/16, 8/17 and 8/18 respectively.

In summary, Mesozoic tectonics resulted in two main fold systems, an older northwest trend system and a younger northeast trend system. Abdel Khalek

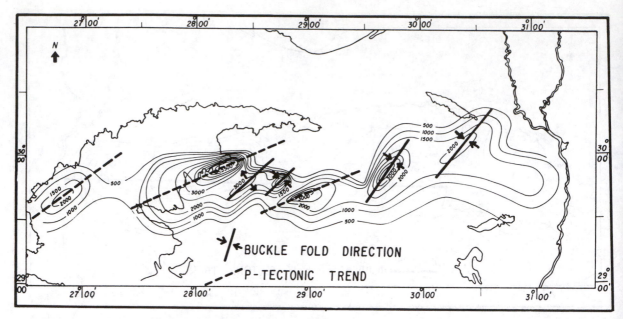

Figure 8.17 Khoman formation isopach (Santonian-Maastrichtian).

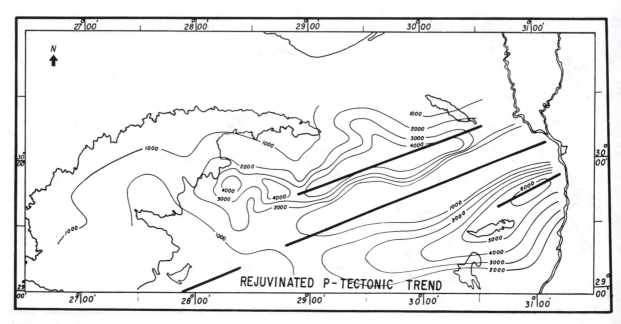

Figure 8.18 Apollonia formation isopach (Paleocene-middle Eocene).

(1979) concluded that the folding movements in the Eastern Desert, affecting the geosynclinal metamorphites, are represented by major folds trending either northwest, as demonstrated in the Hafafit and Beitan areas, or northeast, as demonstrated in the Barramiya area. He adds that in localities showing superimposed structures the northwest trending folds are older. Also the older fold system shows northwest thrust faulting.

Abdel Gawad (1970) demonstrated that the Nugrus-Hafafit shuear zone and Duwi shear zone near Quseir and Safaga are of west-northwest trend. Said (1962) describes the Duwi shear zone as a complex of ancient crystalline rocks with overlying Cretaceous and Eocene sediments.

There has been considerable controversy concerning the tectonic forces that resulted in the Mesozoic-

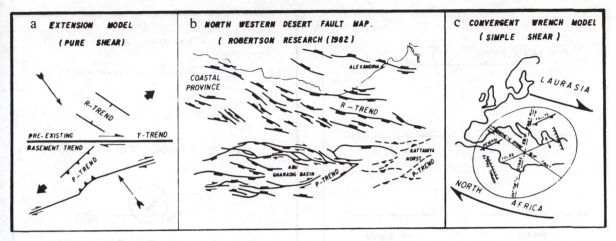

Figure 8.19 Strike disparity of faults between Abu Gharadig basin and coastal province explained by extension model a and convergent wrench model c.

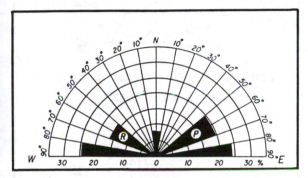

Figure 8.20 Schematic fabric diagram depicting average length for a standard (convergent) wrench system.

early Tertiary complex structural features that are mapped across Egypt.

Robertson Research International (1982) proposed a conventional wrench model with right-lateral motion (Fig. 8/19c) to explain the structural style of the north Western Desert area (Fig. 8/19b). Nelson (1986) proposed an oblique extension model (Fig. 8/19a).

From rosetta diagrams over the unstable shelf area (Fig. 8/4), showing the frequency and average length per 10° azimuth, it is evident that the tectonic fabric (or trends) is very consistent with the (R) and (P) directions shown in the schematic fabric diagram depicting average length for a standard convergent wrench system (Fig. 8/20). In wrench terminology, the (R) direction represents more frequent but shorter tectonic trends, while the (P) direction represents infrequent but longer tectonic trends.

Figure 8/20b shows the north Western Desert fault subsurface map constructed by Robertson Research

International (1982). Nelson (1986) agreed that the wrench scenario suggested by Robertson Research using right-lateral compressional force (Figs 8/19c and 8/21) explains the 2-D fault patterns as mapped (Fig. 8/19b), but not the total deformation of the area. Instead, he concludes that an extensional system or divergent wrench scenario, possibly with pre-existing east-west basement grain, best explains the data (Fig. 8/19a).

As previously mentioned, the platform regime began in Paleozoic throughout northern Africa. During the entire Paleozoic and a considerable part of the Mesozoic, the greater part of Egypt was subjected to uplift. The thrusting and/or uplifting of the north African continent against the main block of the earth's crust resulting in the formation of east-west trending swells and thrusts during Paleozoic and Jurassic times. This is revealed in both the Paleozoic (Fig. 8/10) and Jurassic (Fig. 8/11) isopach maps, which show strong east-west tectonic grain. This east-west tectonic grain set the stage for late Mesozoic-early Tertiary tectonics performed by a right-lateral compressional wrench model to act upon it. Thus a primary wrench-associated structure (Fig. 8/21) results, in which the Y shears are parallel to the master east-west wrench fault, the fabric elements 30° to the northeast of the master fault called P-shears and the fabric elements 30° to the northwest called R-shears (Fig. 8/21). Figure 8/21 also shows that buckle folds should be present.

Thus the isopach maps for formations from the Jurassic to the Apollonia (Figs 8/11, 8/13, 8/14, 8/15, 8/16, 8/17 and 8/18) and the regional magnetic map (Fig. 8.8) support the convergent wrench model of Figure 8/21. The Jurassic isopach map clearly shows the east-west tectonic grain in the early Mesozoic,

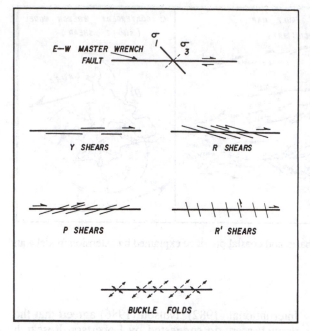

Figure 8.21 Primary wrench associated structures resulting from right-lateral compressional wrench (modified after Jamison 1983).

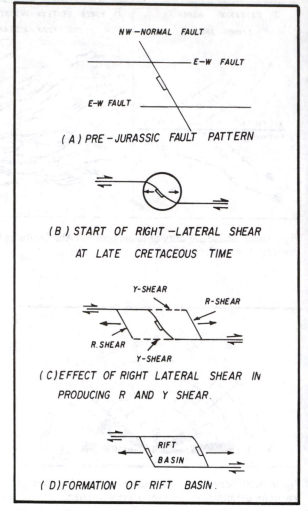

Figure 8.22 Proposed model for Abu Gharadig basin.

thus setting the stage for the primary wrench-associated structures. The Alam El Bueib isopach map (Fig. 8/13) clearly shows the development of R-trend tectonics developing about 30° to the northwest from the older master east-west tectonic trend. The Kharita isopach map (Fig. 8/14) shows the development of P-trend tectonics, about 30° to the northeast from the master east-west trend, and the Bahariya map (Fig. 8/15) shows the relative strength of the east-west trend. The appearance of the east-west trend during early Cretaceous time may be explained by the development of the Y-shear, parallel to the older east-west master fault and is interpreted as rejuvenation of the older Paleozoic-Jurassic east-west tectonic trend. Moreover, the Abu Roash (Fig. 8/16) and Khoman formations (Fig. 8/17) maps show the northeast buckle fold direction superimposed on the regional east-west master trend.

The regional magnetic map (Fig. 8/8) shows clearly a series of north-south to north-northwest/south-southeast trending positive magnetic anomalies. These are interpreted as inherited north-south Precambrian trends from the Nubian basement which may have been rejuvenated during late Mesozoic-early Tertiary times as R-tectonic trends.

Abu Gharadig basin may be explained in terms of a pulling apart of two right-lateral wrenches. The model proposed is shown in Figure 8/22. In the initial stage, in pre-Jurassic time, the main fabric grain of

the basement rocks was composed mainly of Precambrian north-northwest/south-southeast trending faults and east-west Paleozoic faults. By late Cretaceous time and the beginning of the right-lateral shears, a releasing bend was formed (Fig. 8/22b). With continuing right-lateral wrench, R-shears and Y-shears began the development of the basin (Fig. 8/22c). Thus the Abu Gharadig basin can be looked upon as a rift basin bounded on the north and south by two right-lateral shears and from east and west by northwest trending normal faults (Fig. 8/22).

Mid-Tertiary to Quaternary tectonic trends
While the greatest part of dextral, right-lateral movement occurred in the late Cretaceous and Palaeocene (90-50 ma), most of the relatively northerly component of this movement took place beginning in early Eocene time (Fig. 8/23). This northward motion of

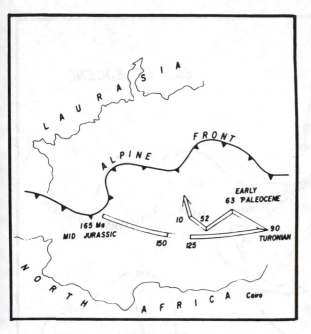

Figure 8.23 Motion of North Africa relative to Europe, Callovian to present (after Robertson Research Intl. 1982).

Africa toward laurasia during the past 50 my produced the Alpine Orogeny, causing complete closure of the Alpine Tethys (Dietz & Holden 1970). The only possible relics of the Mesozoic ocean are the Ionian and Levantine basins of the eastern Mediterranean which survived destruction along the Hellenic Trench subduction zone (Fig. 8/5).

It is evident that the northerly approach of Africa toward Laurasia has resulted in a north-south to north-northwest/south-southeast horizontal compression upon northeastern Africa, including Egypt, since early Tertiary time (Fig. 8/24). The importance of this northern pressure exerted upon Egyptian territory is noted by El Shazly (1964). Youssef (1968) suggests a shift of 10° to the northwest in the direction of the northern pressure. Meshref & El Sheikh (1973) suggest a greater shift to the northwest of 20°, with a major release in the vertical direction. They add that this north-northwest/south-southeast horizontal compressive force resulted in east-northeast uplifts or thrusts in northern Egypt during Paleocene-mid Eocene times, as shown on the isopach map of the Apollonia Formation (Fig. 8/18).

The rifting of the Red Sea and Gulf of Suez was initiated at a later time, most probably Oligocene. It was initiated as a result of the continued northwesterly horizontal pressure, but with an interchange between the intermediate and least stress axes. This is a common phenomenon in mountain building movements: as more doming and uplift takes place due to

continued horizontal pressure, the vertical direction is no longer the direction of least stress. Meanwhile the direction of least stress, or regional horizontal extension, became an east-northeast, west-southwest direction. Thus northern Africa may be regarded as having been under a regional east-northeast west-southwest extension from mid-Tertiary to present times. It is this new regional extension that resulted in the formation of the two complementary, conjugate primary shear fractures, along the north-northwest (Suez) and north-northeast (Aqaba) trends (Fig. 8/24). Usually the stress system favours one direction more than the other, and in this case it was the Suez-Red Sea direction. Thus the Gulf of Suez and Red Sea rifting was initiated around Oligocene times, while the Gulf of Aqaba was opened later in Pliocene times, though the Aqaba trending shear fractures are contemporaneous with the Suez trending fracture of mid-Tertiary age.

Basement outcrops in the Gulf of Suez region, as well as the rosetta diagrams drawn from along the Gulf of Suez and Red Sea (Fig. 8/4), suggest several tectonic trends in this region. One of these is parallel to the Gulf-Red Sea trend, while the other tectonic trends cut across the Gulf of Suez or Red Sea at varying angles. Moreover, these tectonic trends are of variable age. In view of this, a special study of tectonic trends along the Gulf of Suez-Red Sea region was carried out involving separate statistical analyses of magnetic trends (reflecting tectonic trends) for both the residual and regional magnetic maps.

It is well known that regional magnetic maps reflect deep regional basement tectonics of ages relatively older than the younger tectonics which show more strongly on the residual magnetic maps. A separate magnetic trend analysis was carried out for the northern Gulf of Suez, the southern Gulf of Suez and northern Red Sea region from both the regional and residual magnetic maps. The results of these analyses are shown in Figures 8/25 and 8/26. Figure 8/25 shows the total regional magnetic trends, both those parallel to and those cutting across the Gulf trends. These include: (1) the north-south Precambrian and east-west Paleozoic-Jurassic trends, showing relatively greater strength in the Gulf of Suez than in the Red Sea; (2) the west-northwest Darag trend of early Cretaceous age, showing greater strength in the northern Gulf of Suez; (3) the east-northeast Syrian arc, a trend of late Cretaceous to early Tertiary age, showing relatively greater strength in the northern Gulf of Suez and northern Red Sea compared to the southern Gulf of Suez, and (4) the northwest Red Sea trend, showing only in the southern Gulf of Suez and northern Red Sea.

Figure 8/26 shows the total residual magnetic trends, both parallel and cross-Gulf. These residual

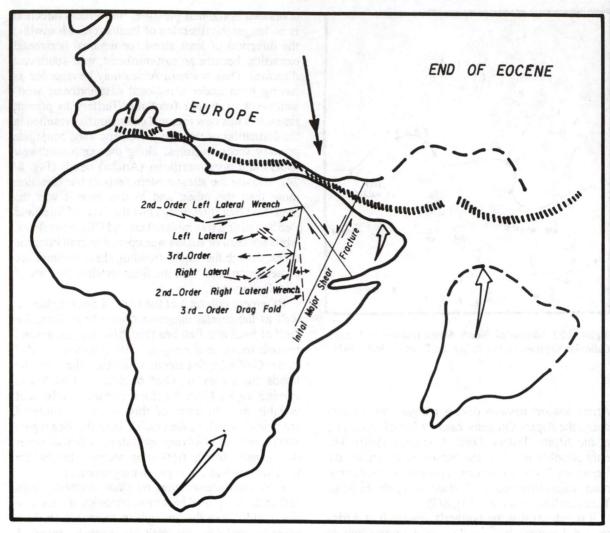

Figure 8.24 Destruction of Tethys II (adapted after Dietz & Holden 1970).

trends include: (1) the northwestern Red Sea trend in all three areas; (2) the north-northeastern Aqaba trend in all three areas; (3) the east-west trend showing relative strength in the northern Gulf, interpreted as a rejuvenation of the older east-west regional trend, and (4) the east-northeast residual, rejuvenated trend, showing relative strength in the southern Gulf of Suez and northern Red Sea.

Figure 8/27 shows the regional magnetic Gulf parallel trends. It is important to note in this figure the absence of the old Precambrian northwest trend from the northern Gulf, north of latitude 29° N. This coincides with the stable-unstable shelf boundary as identified by Said (1962). Figure 8/28 shows the residual magnetic Gulf-parallel trends and, amazingly enough, shows that the northwestern Red Sea trend is of equal strength all along the Gulf of Suez and the Red Sea. It is easy, however, to interpret this, since

this trend is of mid-Tertiary age by which time both the unstable and stable shelf areas were under the same tectonic forces of regional extension.

Figure 8/29 shows the regional cross-Gulf magnetic trends, primarily the old Precambrian north-south, Paleozoic-Jurassic east-west, early Cretaceous west-northwest, and late Cretaceous east-northeast trends. Figure 8/30 shows the residual cross-Gulf magnetic trends of younger age. Primarily among these is the north-northeast Aqaba trend with equal strength along the Gulf of Suez and the northern Red Sea. It also shows some rejuvenated older tectonic trends, such as the east-west trend in the northern Gulf region, the east-northeast trend in the southern Gulf region and the west-northwest and east-northeast trends in the northern Red Sea region.

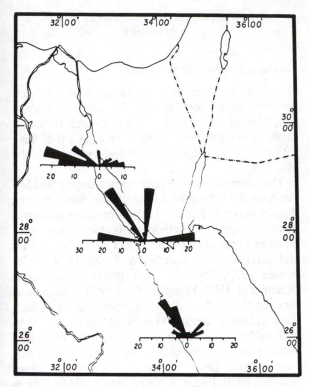

Figure 8.25 Total regional magnetic trends.

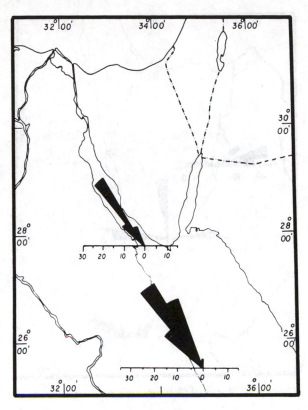

Figure 8.27 Regional Gulf-parallel magnetic trends.

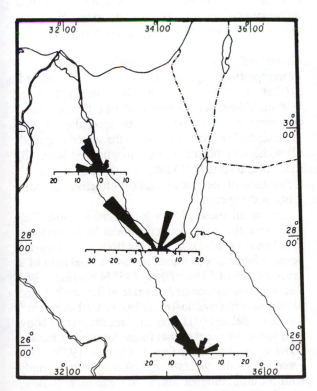

Figure 8.26 Total residual magnetic trends.

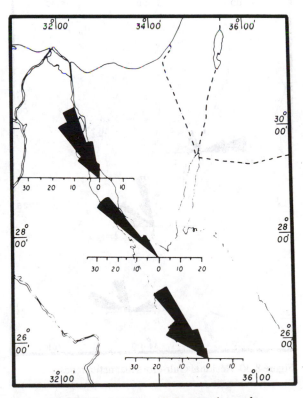

Figure 8.28 Residual Gulf-parallel magnetic trends.

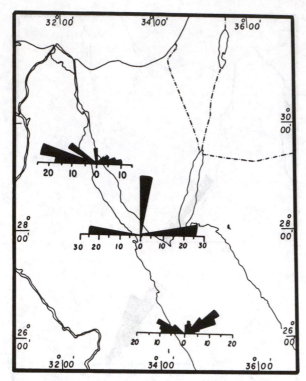

Figure 8.29 Regional Gulf-cross magnetic trends.

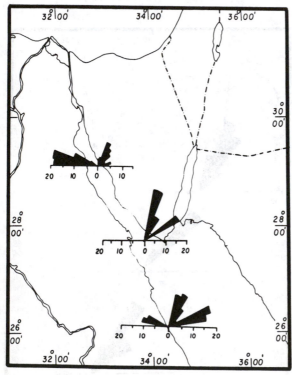

Figure 8.30 Residual Gulf-cross magnetic trends.

2 NORTHERN RED SEA–GULF OF SUEZ RIFT TECTONICS

Northern Red Sea rift tectonics

The Red Sea occupies an elongate escarpment-bounded depression between the uplifted Arabian and Nubian Shields. Stratigraphic and structural studies show that both the Red Sea and Gulf of Suez rifting began in Oligocene time and developed in the Miocene.

The extent of oceanic crust underlying the Red Sea has been the subject of a long debate. Some believe that the northern Red Sea is almost entirely underlain by oceanic crust. The evidence for this view is based on plate kinematics (Mckenzie et al. 1970), gravity and magnetic data (Girdler & Styles 1974, 1976, Rosser 1975, Styles & Hall 1980), seismic data (Knott et al. 1966, Philips & Ross 1970), and measurements of heat flow along the central Red Sea which show values of about 400 millivolts as compared to 150 millivolts on both sides. Even the measured values along the coastal areas of the Red Sea are still two to three times higher than the World mean (Girdler 1970, Evans & Tammemagi 1974). Others believe that the oceanic crust is of limited extent or non-existent in the northern Red Sea (Hutchinson & Engels 1972, Lowell & Genik 1972, Ross & Schlee 1973, Coleman et al. 1975, Lowell et al. 1975, Cochran 1983). Which viewpoint is accepted would influence one's understanding of the extent of separation or motion between Arabian and African plates.

Factors affecting sea floor spreading

In interpreting magnetic data over sea floor spread, synthetic sea floor spread models are used extensively to simulate observed magnetic data. The amplitude and wave length of the synthetic anomalies depend upon several factors: (1) depth to the sea floor spread; (2) width and thickness of the magnetized layer; (3) strike of the spread; (4) rate of spreading in cm/Y^{-1}; (5) latitude of the spread, and (6) effective susceptibility of the spread.

Taking all these factors into consideration, Hall (1979) constructed a synthetic model for the northern Red Sea (Fig. 8/31). In this sea floor spread model, oceanic layer 2 is assumed to have a thickness of 2 km at a depth of 4 km, striking 145° N, parallel to Red Sea coastlines, spreading at a rate of 0.5 cm/Y^{-1}, and having an effective susceptibility of 0.01 emu cm^{-3} for the model and 0.02 emu cm^{-3} for the central body. This value is typical of that found for oceanic basalts (Lowrie et al. 1973b, Fox & Opdyke 1973). This synthetic model yields a central magnetic anomaly with an amplitude of about 700 gammas, peak to trough (Fig. 8/31).

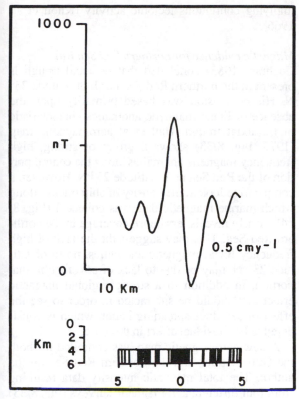

Figure 8.31 Synthetic model for recent spread over past 5 My (after Hall 1979).

Sea floor spreading history of the Red Sea

Magnetic anomalies over the Red Sea are separated into two groups on the basis of amplitude and wave length. Both groups form linear patterns and suggest two phases of sea floor spreading, an early spread characterized by relatively weaker magnetization and a recent spread with relatively stronger magnetization. Girlder & Styles (1974) show that the southern Red Sea was formed by two phases of sea floor spreading, separated by a long interval during which a huge thickness of evaporites was deposited. Figure 8/32 shows three possible synthetic models representing early spread (a) and recent spread (b and c), at different spreading rates. A comparison of observed magnetic profiles with those calculated from the synthetic models shows close agreement, thus supporting the idea of two phases of spreading, and earlier phase, 29 to 23 my (Hall et al. 1976) and a recent phase over the past 4.5 to 5 my.

Seismic reflection studies in the northern Red Sea (Knott et al. 1966, Phillips & Ross 1970) show that the upper Miocene reflector (S) is continuous across the main trough and that the upper surface of the evaporites is deformed, but with severe deformation in an area about 100 km in width in the center. Toward the margins of the main trough, the post-Miocene deposits overlying the reflector (S) tend to be much less disturbed than reflector (S), implying relative stability since the end of the Miocene. However, in the central region, the deformation continues up through the overlying sediments, giving the main trough its irregular, broken appearance, and

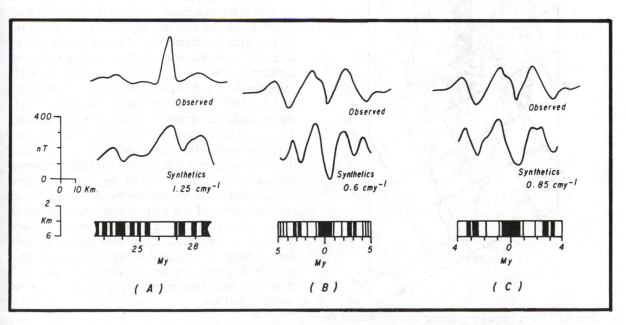

Figure 8.32 Three possible sea floor models: a. for early spreading phase, b and c. for recent spreading phase.

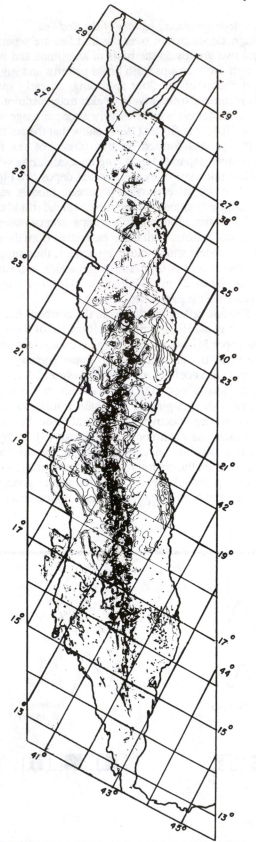

Figure 8.33 Total intensity magnetic anomaly map of the Red Sea.

implying continuous tectonic activity (Knott et al. 1966).

Magnetic evidence for northern Red Sea Rift

Cochran (1983) concluded that no axial trough is present in the northern Red Sea north of latitude 25° N. His conclusion was based primarily upon the absence of linear magnetic anomalies characteristic of the axial trough. Hall et al aeromagnetic map (1976, Fig. 8/33) shows a group of strong, high frequency magnetic anomalies along the central portion of the Red Sea up to latitude 23° N. However, a comparative look at the density of ship tracks, along which marine magnetic data was collected (Fig. 8/ 34), clearly reveals less dense coverage in the northern Red Sea. This may suggest tht the lack of high frequency linear magnetic anomalies, north of latitude 23° N, may be due to lack of coverage in that portion, in addition to a strong regional magnetic effect that should be subtracted in order to see the effect of sea floor spreading better, which is considered to be a residual effect in this case.

A new aeromagnetic map was compiled for both the Gulf of Suez and the northern Red Sea by the author. The total magnetic intensity data resulting from four different aeromagnetic surveys (Fig. 8/35) was compiled and reduced to one set of data. Due to the extreme length of the area investigated, two separate files for the gridded magnetic data were used to carry out the analyses. The first file contains the total magnetic intensity data for the Gulf of Suez, and the second file contains the total magnetic intensity for the northern Red Sea.

The total magnetic intensity was rotated to the pole in order to remove the skewness of the magnetic field vector within the area investigated, that is, to get rid of the latitude effect. The reduced-to-pole (RTP) magnetic maps for both the Gulf of Suez and the northern Red Sea are shown in Figures 8/36 and 8/37, respectively. The RTP magnetic map for the northern Red Sea (Fig. 8/37) shows various northwest trending anomalies of varying wave length and amplitude. These variations show different sources of the anomalies, as well as their depth and composition. Since the primary interest was to map the high frequency magnetic anomalies related to sea floor spreading (which seem to be overridden by the strong, broad regional effect of the surrounding crust, stretched lithosphere and upwelling mantle), it was necessary to isolate the high frequency residual anomalies from the low frequency regional anomalies by means of a suitable filtering program. Figures 8/38 and 8/39 show the residual and regional magnetic maps for the northern Red Sea.

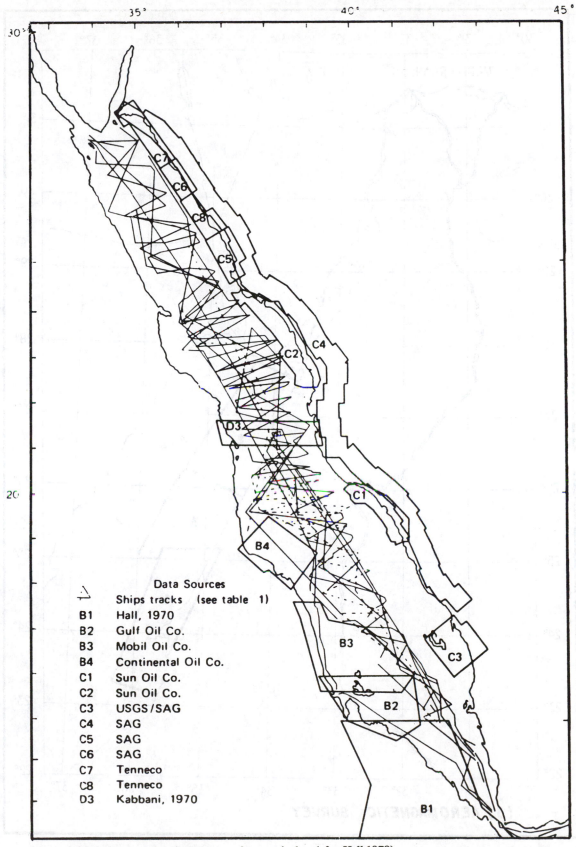

Figure 8.34 Index map showing sources of magnetic data (after Hall 1979).

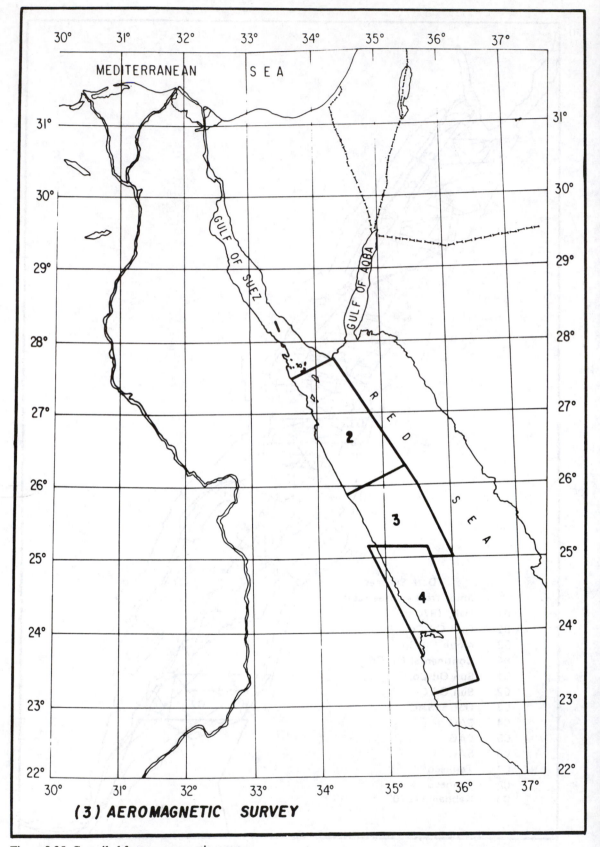

(3) AEROMAGNETIC   SURVEY

Figure 8.35 Compiled four aeromagnetic surveys.

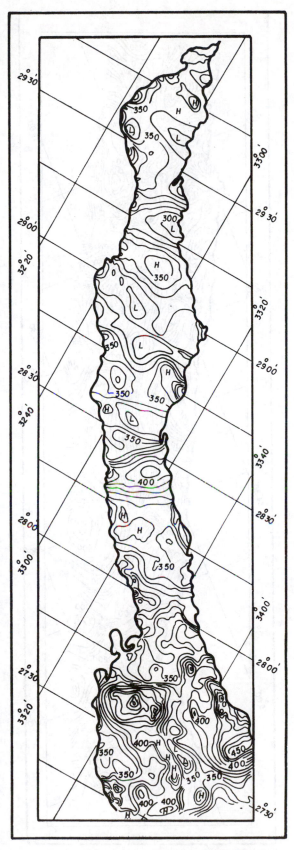

Figure 8.36 Reduced to pole magnetic Gulf of Suez.

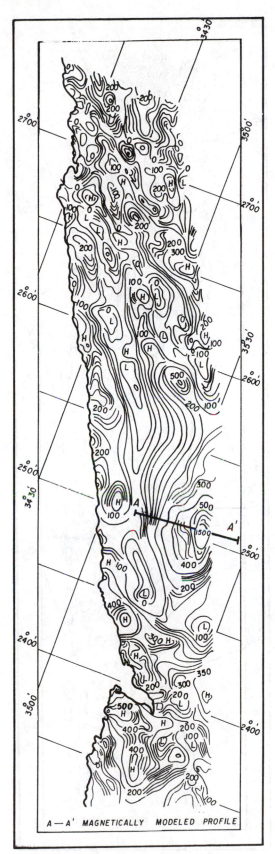

A—A' MAGNETICALLY MODELED PROFILE

Figure 8.37 Reduced to pole north Red Sea.

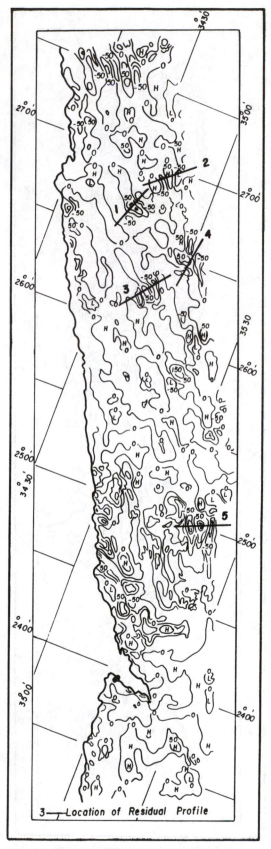

Figure 8.38 Residual magnetic north Red Sea.

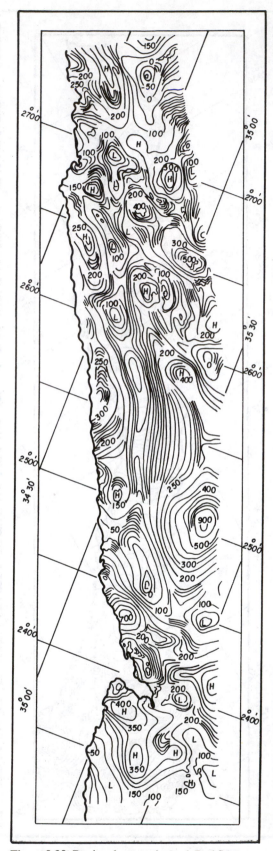

Figure 8.39 Regional magnetic north Red Sea.

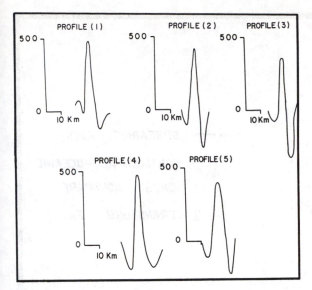

Figure 8.40 Observed residual magnetic anomalies northern Red Sea.

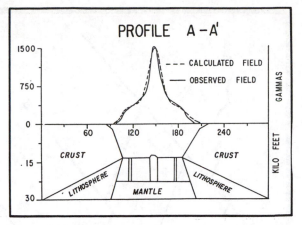

Figure 8.42 Modeled reduced to pole (RTP) magnetic profile across northern Red Sea.

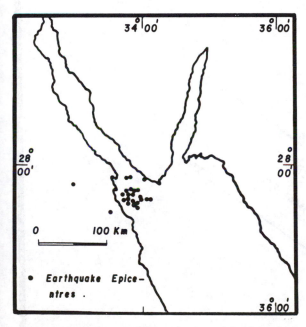

Figure 8.41 Northern Red Sea seismic activity (after Hall 1979).

Sea floor spreading

The residual magnetic map (Fig. 8/38) clearly shows two main types of northwest-trending magnetic anomalies within the offshore northern Red Sea: (a) a group of linear, parallel to coastline, magnetic anomalies of reversed polarity of very high frequency and strong amplitude (500-700 gammas), which are discontinuous in nature and very similar to those anomalies associated with the axial trough of the central and southern Red Sea, and (2) another group of relatively weak magnetic anomalies (50-150 gammas) distributed over the western portion of the northern Red Sea.

The parallelism in trend, northwest, the linear nature of both groups of anomalies, and the strong amplitude and reversed polarity of the central anomalies strongly suggest the presence of sea floor spreading in the northern Red Sea. Moreover, the strong-amplitude anomalies may be regarded as associated with the recent phase of spread, while the relatively weak anomalies are associated with the earlier phase of the spread (Girdler & Styles 1974).

Five residual magnetic profiles (Fig. 8/40), taken across the strong linear magnetic anomalies, are compared with the theoretical magnetic anomaly calculated from the synthetic sea floor spreading model for the northern Red Sea in Figure 8/31 (Hall 1979). A great similarity is seen between the observed anomalies (Fig. 8/40) and the theoretical model, both in frequency and amplitude. However, it is to be noted that the wave length of profiles 1 and 2, in the extreme northern part of the Red Sea, shows even stronger amplitude and higher frequency than that of the theoretical synthetic model, with a spreading rate of 0.5 cm/Y^{-1} (Fig. 8/31). This difference is attributed to the possibility of a higher rate of spreading, amounting to about 0.6 cm/Y^{-1} in the extreme northern end of the Red Sea, due to the proximity to the active Aqaba rift. This area is also characterized by high seismicity (Fig. 8/41). The excellent degree of fit between the observed residual and the theoretical anomalies greatly enhances the interpretation of linear magnetic anomalies as a result of sea floor spreading along the axial trough in the northern Red Sea.

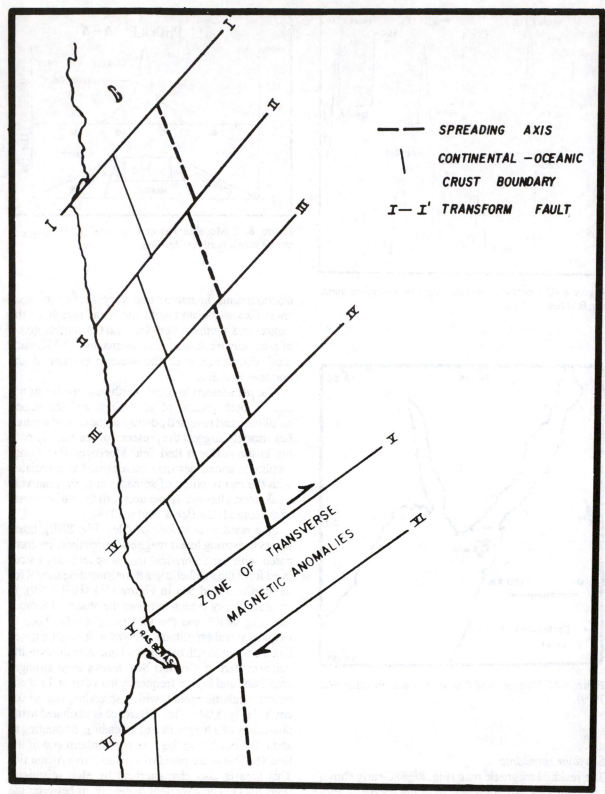

Figure 8.43 Interpretation for northern Red Sea showing spreading axis and extent of oceanic crust.

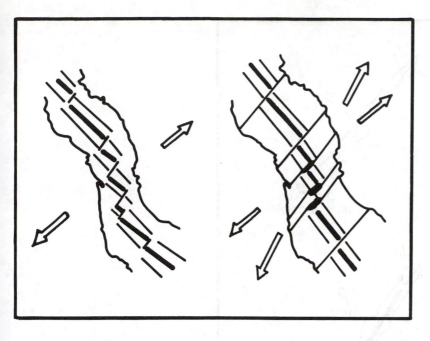

Figure 8.44 Leaky transform faults due to recent spreading along new direction (after Hall 1979).

An RTP magnetic profile (Fig. 8/38) was also modelled across the northern Red Sea, north of Ras Banas, asuming an earth crust thickness of 4-5 km over the main trough and with mantle material reaching the sea floor in the central northern Red Sea (Fig. 8/42).

Transform faults
Both residual and regional magnetic maps (Figs. 8/38 and 8/39) show several discontinuities separating the linear magnetic anomalies. These zones may represent fracture zones, transform faults or regions in which the formation of mid-oceanic ridges had not yet begun. In some places, an offset of the linear magnetic anomalies is evident (Fig. 8/38). These discontinuities or offsets are interpreted as marking the location of transform faults representing planes of differential horizontal extension across the sea floor spread. Six such faults are mapped (Fig. 8/43). The northernmost transform is considered as the southern extension of the Dead Sea-Aqaba transform. The azimuth of this transform fault is about 25°. The next three transform faults to the south show a more or less parallel trend to the northern transform, with azimuth ranging between 25 and 30°. The azimuth of these transform faults agree very well with the transform faults predicted with poles of rotation, as suggested by Freund (1970) and Girdler & Darracott (1972).

Figure 8/43 shows a large zone north of Ras Banas which is characterized by transverse magnetic anomalies (fig. 8/38) and bounded by two transform faults (V and VI) where the azimuth is greater than the other four northern transforms. This transverse zone has an azimuth of 35° between Arabia and Nubia. The zone is about 30 km wide, and magnetic patterns north and south of this zone show relative displacement, suggesting a large transform fault across the area. Hall (1979) describes such transverse zones as due to leaky transform faults (Fig. 8/44), relating the leakage of basic intrusions along the transforms to change in the motion of Arabia relative to Nubia, between the early and the recent spreading events.

A model explaining the change in direction of motion between Arabia and Nubia is presented in Figure 8/45. This model proposes that by Oligocene time the Gulf of Suez-Red Sea had been formed but that the Gulf of Aqaba had not yet been formed and was represented only by a large fracture separating the Gulf of Suez from the Red Sea (Fig. 8/45). By continued extension across the region through the Oligocene and early Miocene, the rift was initiated, resulting in the separation of Arabia from Nubia along a northeast direction. By the middle Miocene, the Gulf of Aden started to open up and the direction of extension across it was almost north-northwest-south-southeast. This new direction of regional extension across the Gulf of Aden, combined with the northeast regional extension across the Gulf of Suez and the Red Sea, resulted in a different directional vector along which the new direction of motion between Arabia and Nubia took place in late Miocene times. It is this new direction of regional extension that began the opening up of the Gulf of Aqaba and resulted, to a large degree, in left-lateral displacement along the Dead Sea-Aqaba Rift.

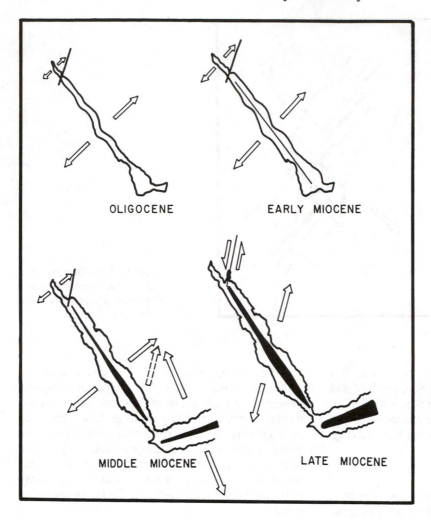

Figure 8.45 Development of Gulf of Suez, Red Sea, Gulf of Aqaba and Gulf of Aden rifts.

This model suggests that the change in direction of motion between Arabia and Nubia may have resulted in the leaky transform faults V and VI (Fig. 8/43), characterized by transverse magnetic anomalies.

Abdel Gawad (1970) correlates two shear zones on the Nubian side, the Duwi and Hafafit (Sa and Sb), with two other shear zones of the same west-northwest trend on the Arabian side, Abu Masarib (Sa) and Al Hamed (Sb) shear zones (Fig. 8/46b). He assumes that these shear zones were continuous before the development of the rift, as is clear from the restored position configuration (Fig. 8/46a). It is interesting to note that the disolocation of the second (Sb) coincides very well with the location of the zone of transverse transforms V and VI as mapped in Figure 8/43.

In summary, it is believed that there is a discontinuous spreading center along the center of the northern Red Sea, as revealed by linear magnetic anomalies and magnetic models crossing it. Spreading is believed to be at a rate of 0.5 cm/Y^{-1}, which may increase to 0.6 cm/Y^{-1} or more in the extreme northern part of the Red Sea, at its junction with the Gulf of Suez, where the Aqaba rift has been active since late middle Miocene time.

In conclusion, this area is intersected by six transform faults of NE trend. The northernmost transform coincides with the southwestern extension of the Aqaba shear. The two southernmost transforms bound a zone of transverse magnetic anomalies which are interpreted as due to leaky transform faults that seem to have separated the once continuous shear zone (Hafafit-Al Hamed).

Gulf of Suez rift tectonics

The Gulf of Suez rift comprises a northwest trending intra-cratonic basin that is separated from the Red Sea by the Aqaba transform fault. The Gulf of Suez rift is bounded on the east by the Sinai Massif and on the west by the Red Sea Hills of the Eastern Desert (Fig. 8/47).

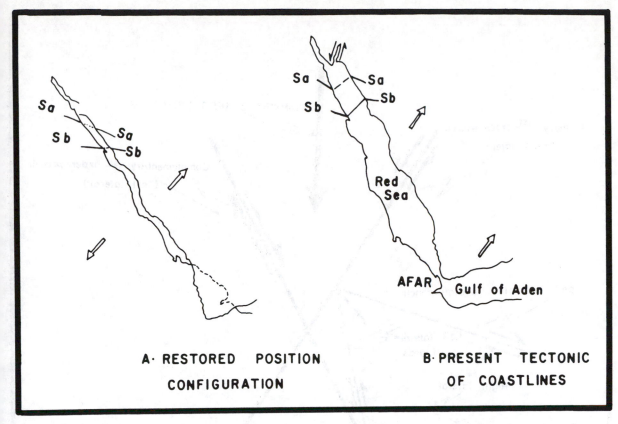

A· RESTORED POSITION
CONFIGURATION

B· PRESENT TECTONIC
OF COASTLINES

Figure 8.46 Regional tectonic setting of the Gulf of Suez (based on Abdel Gawad 1969 and Garfunkel & Bartov 1977).

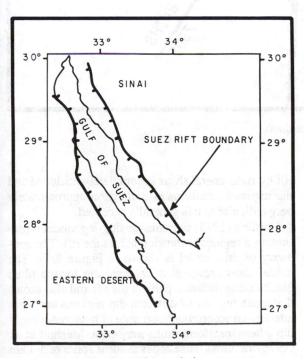

Figure 8.47 Gulf of Suez rift boundaries.

The Suez rift is considered to be the right lateral component of the two complementary shear fractures of Suez and Aqaba that resulted from a northwesterly horizontal compression (Fig. 8/48). This compressive force is believed to have started in Eocene times as a result of the northward motion of Africa towards Laurasia which destroyed Tethys II and resulted in the Mediterranean. During the Eocene, northern Africa yielded to the northwest compressive force by northeast folding. This would be expected, since the vertical direction is the direction of least stress. However, after the folding episode resulting in the Syrian arc trend, and by Oligocene time, it appears that the vertical direction was not the least stress axis due to the accumulation of sediments resulting from the folding. Hence an interchange between the northeast intermediate horizontal stress axis and vertical least-stress axis took place. This new system of stress relations resulted in the two complementary shear fractures, the Suez right-lateral shear and the Aqaba left-lateral shear (Fig. 8.49). Usually the stress field favours one shear direction over the other and, in this case, the right-lateral component was favoured, resulting in a more highly developed Gulf of Suez-Red Sea trend than the left-lateral Aquaba trend. This may

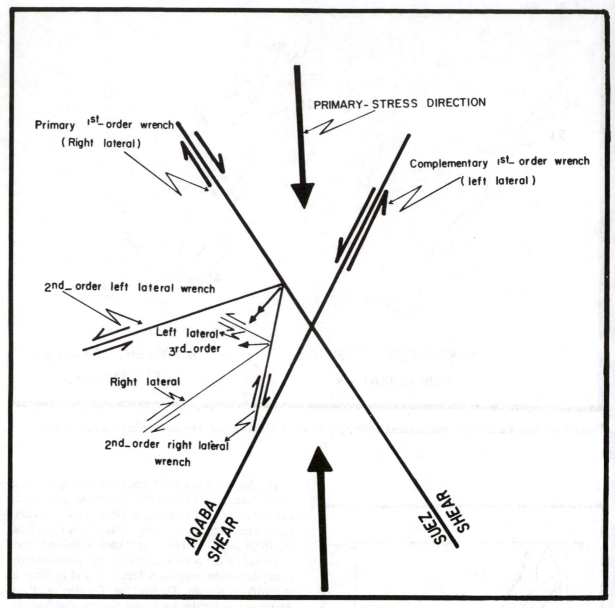

Figure 8.48 Stress model for Suez and Aqaba shears (Oligocene time).

be attributed to the fact that the northwest tectonic trend was an old Precambrian trend and that it was just rejuvenated in Oligocene time.

The two possible primary dynamic forces that could initiate the Suez rift are either the right-lateral coupled force along the shear or the regional extension across it (Fig. 8/49). The rifting due to right-lateral shear is shown in a model for Western North America (Fig. 8/50). This model resulted in basins and ranges of a northeast trend. However, the geologic evidence from the Gulf of Suez assures a northwest direction for the basins and ranges within the rift. Accordingly the possibility of initiation of the Suez

rift by right-lateral shear is completely excluded and the regional extension in a direction approximately perpendicular to it is generally accepted.

Harding (1973) presents the dog-leg model representing a regional extension across the rift. The geometry of this model is shown in Figure 8/52. The model shows reversal of regional dip for the tilted blocks along different parts of the rift and also shows that these regions of different dip regimes are separated by an accommodation zone of relatively flatter dip where transform faults may exist. Meshref et al. (1976) identifies these zones as shear zones of 5-7 km in width. The model also shows dislocation of the

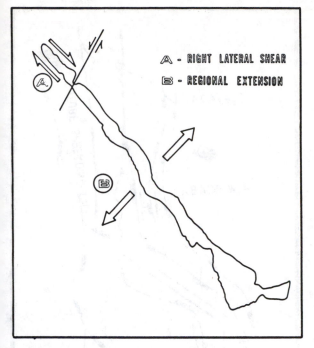

Figure 8.49 Possible dynamic forces for initiation of Gulf of Suez and Red Sea rift.

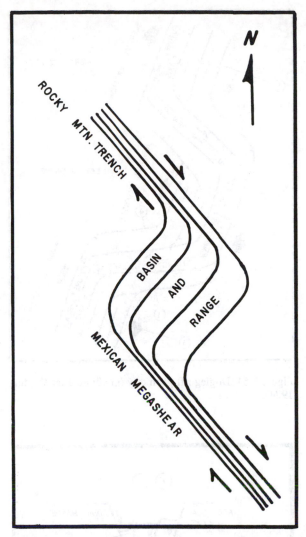

Figure 8.50 Rifting due to right-lateral shear for western North America (after Carey 1958 and Deutsch 1960).

future spreading axis of the rift from one side to another through left and right stepping respectively from south to north (Fig. 8/51). Figure 8/52 shows the attitude of the tilted fault blocks across various parts along the proposed dog-leg rift model. It is interesting to note that across the northern and southern provinces which are characterized by regional south-west dip (Fig. 8/52a), there is a possibility of local northeast dip existing in the downthrown blocks from the main eastern bounding fault of the rift. Similarly, a local southwest dip is expected to exist on the downthrown blocks from the western main bounding fault of the rift in the central province which is characterized by regional northeast dip (Fig. 8/52c).

Figure 8/53 shows the suggested dog-leg model as applied to the Gulf of Suez rift. The major feature of this model is the existence of a series of major faults that extend along the rift and bound it from both sides. It is interesting to notice that in both the northern and southern provinces there is a series of major downthrown-east faults that exist along the western boundary of the rift. Meanwhile, in the same provinces, there is only a single major downthrown-west fault that bounds the rift from the east. Thus it is expected that the combined throw of the major seires of downthrown-east faults will be more than the throw of the single major downthrown-west fault. This fact will result in regional southwest dip of the tilted fault blocks in both the northern and southern

provinces. Moreover, it is logical to assume that the future spreading axis of the rift will be located in the central part of the rift bounded by major faults of opposite throw (Fig. 8/53).

However, the central portion of the Suez rift, central province, is bounded from the east by a series of downthrown-west faults and from the west by a single downthrown-east Red Sea hills bounding fault (Fig. 8/53). Accordingly, it is expected that the central province will have a regional northeast dip. It is also interesting to note that the central province is bounded from north and south by two major accommodation zones separating it from both northern and southern provinces of regional southwest dip regime (Fig. 8/53).

Close examination of Figures 8/51 and 8/53 shows the great similarity between the dog-leg model pro-

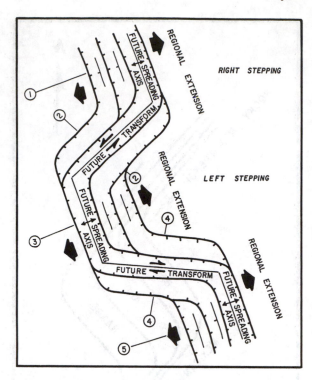

Figure 8.51 Dogleg rift geometry (modified after Harding 1984).

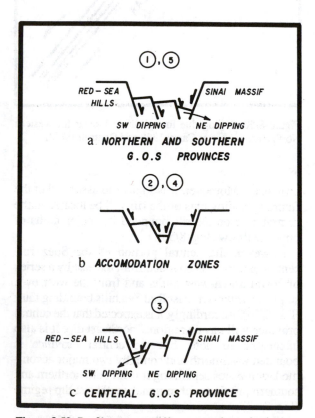

Figure 8.52 Profiles across different provinces in the proposed dogleg model for the Gulf of Suez (GOS).

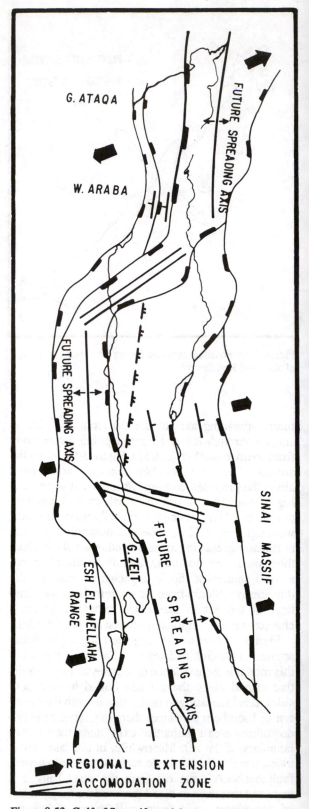

Figure 8.53 Gulf of Suez rift model.

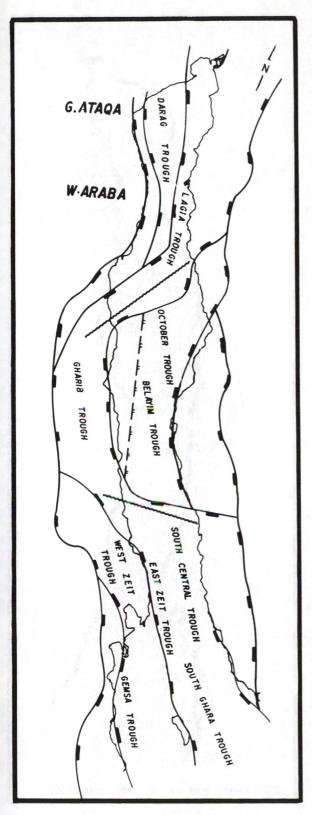

Figure 8.54 Gulf of Suez major tectonic features and troughs.

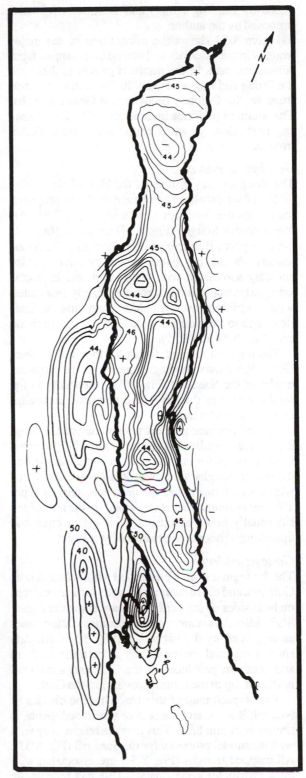

Figure 8.55 Bouguer gravity Gulf of Suez.

posed by Harding (1984) and the Suez rift model
proposed by the author.

Figure 8/54 shows the distribution of the major
troughs in the Suez rift as deduced and mapped from
subsurface data. In the northern province, these are
the Darag and Lagia troughs. In the central province
there are the October, Belayim and Gharib troughs.
The southern province includes the West Zeit, Gemsa, East Zeit, South Central and South Ghara
troughs.

Geophysical evidence

The Bouguer gravity map of the Gulf of Suez (Fig.
8/55) shows excellent agreement with the proposed
model for the Suez rift shown in Figure 8/53. All
major known troughs (Fig. 8/54) are associated with
large negative Bouguer gravity anomalies. Close attention should be given to the positive gravity
anomaly associated with Gebel Zeit and its northward extension which is most probably associated
with an uplifted basement block separating the Gharib trough to the west from the Belayim trough to the
east (Figs 8/54 and 8/55).

The regional magnetic map for the Gulf of Suez
(Fig. 8/9a) shows good agreement with the proposed
model of the Suez rift (Fig. 8/53), especially in the
southern part of the rift where magnetic anomalies
show a northwest trend.

The temperature gradient map of the Suez rift (Fig.
8/56) is in excellent agreement with the proposed
dog-leg model for the Suez rift. This map shows that
the major troughs along the rift axis are associated
with a high temperature gradient, reaching 20°F per
100 feet or more. This is explained by the axis of the
rift usually being associated with a thin crust and
upwelling of hot mantle material by convection.

Geologic evidence

The dip regime from wells drilled within the offshore
Gulf area and the measured dip from surface outcrops
on both sides of the Gulf region are shown in Figure
8/57. Also shown are the two accommodation zones
as suggested by the rift model (Fig. 8/53). All data
show a regional southwest dip in both the northern
and southern provinces of the Gulf and a regional
northeast dip in the central province of the Gulf.

The isopach map of the total Miocene clastics of
Nukhul, Rudeis, and Kareem or synrift sediments, is
shown in Figure 8/58. This map correlates very well
with the model proposed for the Suez rift (Fig. 8/53).
All mapped troughs (Fig. 8/54) are associated with
thick synrift Miocene clastics. This fact reflects the
time of development of these basins, within the lower
to middle Miocene, which is considered to be the
period of active rifting in the Gulf of Suez.

A regional Miocene stratigraphic study involving
some 300 wells was carried out by Hosny, Gaafar and

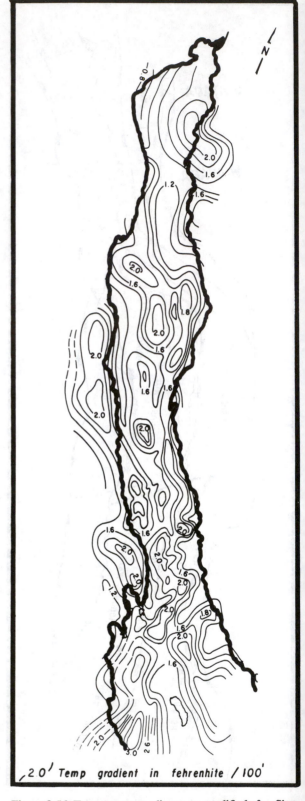

Figure 8.56 Temperature gradient map modified after Shahin & Fathi (1984).

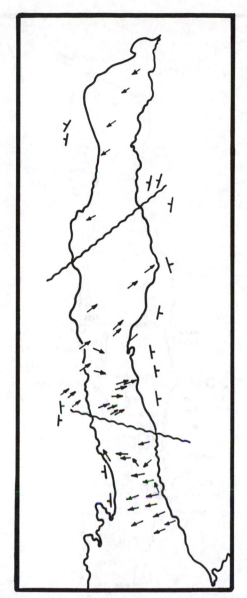

Figure 8.57 Surface and subsurface dip regimes in Gulf of Suez rift.

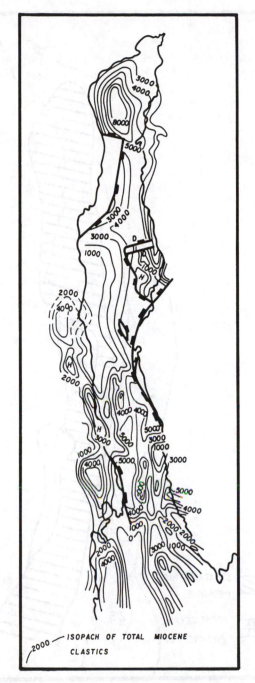

Figure 8.58 Total Miocene clastic synrift sediments iso-pach map.

Sabour (1986). This study resulted in dividing the Suez rift into two main stratigraphic provinces (Fig. 8/59) based on facies changes. Stratigraphic province 1 is characterised by Lower Miocene deposits and also by the fact that there is no sharp boundary between the Kareem and Rudeis Formations.

Stratigraphic Province 2 is divided into two subdivisions (Fig. 8/59), based on the fact that in subprovince 2A, the Markha anhydrite member is deposited separating the Kareem from the Rudeis Formation. The deposition of this anhydrite is indicative of a restricted environment. Subprovince 2B is characterized by the absence of the Markha anhydrite member,

and a continuous section of Kareem-Rudeis was deposited of sandy-shaly facies (Fig. 8/59). The two provinces seem separated from each other northward and southward by a zone characterized by the positive gravity high (Fig. 8/55) associated with the western coast of the Gulf of Suez. The very close agreement between the proposed rift model (Fig. 8/53) and

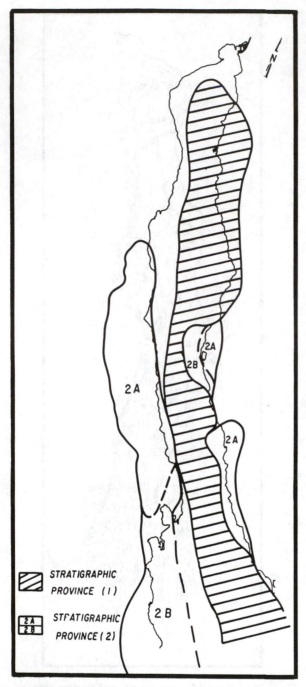

Figure 8.59 Miocene stratigraphic province in Gulf of Suez (after Hosni, Gafar & Sabour 1986).

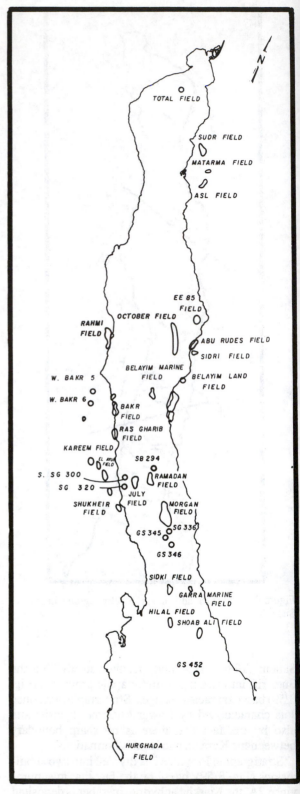

Figure 8.60 Major oil field in Gulf of Suez.

the Miocene stratigraphic provinces is indicative of the large role played by rift tectonics on Miocene stratigraphy.

The oil potential of the different basins and troughs within the Suez rift is promising; all the troughs mapped (Fig. 8/54) are petroliferous and have a good potential for oil accumulation when compared with

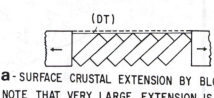

(DT)

a - SURFACE CRUSTAL EXTENSION BY BLOCK FAULTING
NOTE THAT VERY LARGE EXTENSION IS REQUIRED
BEFORE APPRECIABLE THINNING (DT) CAN TAKE PLACE
(AFTER GIBBS, 1984)

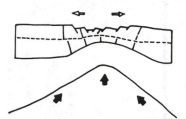

------ TRANSITION BETWEEN BRITTLE AND DUCTILE CRUST

↑ ADVECTION OF HOT ASTHENOPHERIC MATERIAL

b - CRUSTAL THINNING BY COMBINED SURFACE EXTENSION
BY NORMAL FAULTING IN THE UPPER BRITTLE CRUST
AND SUBSIDENSE OF LOWER DUCTILE CRUST.
(AFTER CHENET ET AL , 1984)

Figure 8.61 Role of subsidence in crustal thinning.

the oil show map (Fig. 8/60). It is expected that the oil type may vary from one basin to the other and in different parts of the rift depending upon the rate of rifting in the various parts of the Gulf.

Rift evolution

The Gulf of Suez is an extensional rift that comprises a northwest-trending marine basin, flanked by gravel plains that are broken by several tilted blocks such as Esh El Mellaha and Gebel Zeit to the west and Gebel Araba to the east (Fig. 8/47). The amount of extension is estimated at 25 to 50% of its original width (Garfunkel & Bartov 1977, Angelier 1985, Perry & Schamel 1985).

The evolution of the Suez rift began in eary Oligocene time by normal faulting and dyke injection, resulting in tilted fault blocks in the form of half grabens. The main bounding normal faults for the titlted blocks show an original dip angle of 60°. With active horizontal extension, more thinning of the earth's crust took place due to the rotation of blocks; this resulted in decreasing the dip angle of the main bounding normal faults, in some cases up to 40° as deduced from well data.

The direction of block rotation was not constant along the strike of the Gulf due to the regional reversal of the dip regime along the strike. Accordingly, the tilted blocks would be expected to rotate more to the southwest in the northern and southern provinces of the Gulf and more to the northeast in the central province of the Gulf.

Gibbs (1984) concluded that appreciable thinning of the crustal layer, through passive rifting by assuming block faulting alone, creates a serious spatial problem (Fig. 8/61a). Chenet et al. (1984), on the other hand, conclude that the amount of crustal thinning, as deduced from subsidence data in the southern Gulf of Suez, is much higher than the amount of horizontal extension of the upper continental crust which accounts for only some 35% of the superficial extension (Fig. 8/61b).

A new geometric model for the Suez rift was proposed by Perry & Schamel (1985, Fig. 8/62). This model is suggested for the south Gulf of Suez and assumes an initial extension in the Oligocene resulting in the formation and rotation of wide tilt blocks over an eastward-dipping low-angle listric normal fault system (Fig. 8/62a).

With continued regional extension during the early Miocene more tilting and faulting of the larger blocks

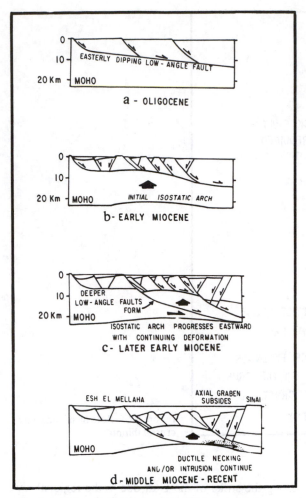

Figure 8.62 Geometric model for the Gulf of Suez rift (after Perry & Schamel 1985).

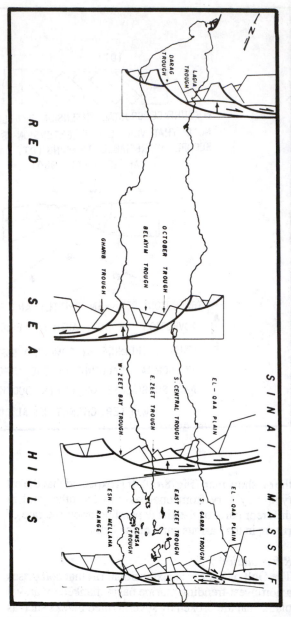

Figure 8.63 Structural cross-section showing basement blocks across different dip regimes.

took place resulting in brittle thinning of the upper crust. In response to this thinning, isostatic uplift arched the underlying low-angle fault surface (Fig. 8/62b). With continued arching, a new system of through-going faults cut the old low-angle fault system thus limiting the gravity-driven tilted blocks to the west of the newly formed faults. At the same time, it created a gulf-parallel high bend in the low-angle fault surface and the Moho under the eastern portion of the rift (Fig. 8/62c).

Later, during middle Miocene to Recent times, a system of down-dropped gravels will offset to the east of the most extended portions of the older tilt-block array (Fig. 8/62d). Thus during the early Miocene, in the southern Gulf, block faulting and subsidence, i.e. asymmetric rifting, was limited to the southwestern part of the Suez rift west of Esh El Mellaha range. But from the late Miocene on to the present, block faulting and active subsidence was concentrated in the east along the central basin trough.

Figure 8/63 shows a series of four structural cross sections along the Gulf of Suez rift. These cross sections were constructed using subsurface data, location of major faults from a proposed rift model (Fig. 8/53), and magnetic data coupled with the geometric model suggested by Perry & Schamel (1985). According to these cross sections, it would be expected that, both in the northern and southern provinces of the Suez rift, the initial rifting phase during Oligocene and early Miocene time started in the western portion of the rift as the low-angle listric faults that dip eastward. Later, by mid-Miocene, ac-

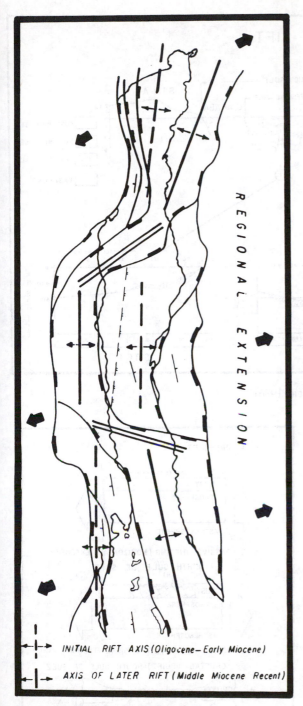

INITIAL RIFT AXIS (Oligocene—Early Miocene)

AXIS OF LATER RIFT (Middle Miocene Recent)

Figure 8.64 Gulf of Suez rift tectonics.

tive subsidence resulted in eastward movement of the rift axis toward the axial graben represented by the South Ghara and South Central troughs in the south, and Lagia trough in the north (Fig. 8/63). In the central province of the rift, in the zone of northeast regional dip regime and westward downthrown listric low-angle faults, the initial phase of rifting was limited to the eastern part of the rift, while later active subsidence took place westward along the Gharib trough (Fig. 8/63).

In summary, the Suez rift can be looked upon has having two major rift axes, a westerly axis and an easterly axis (Fig. 8/64). The western axis of the rift represents the initial rifting phase in both the northern and southern parts of the rift and the later active subsidence phases in the central portion (Fig. 8/64). The eastern axis of the rift coincides with the later active subsidence phase in the northern and southern parts of the Gulf and the initial phase of rifting in the central portion.

Stratigraphic Province 1 seems to be mainly associated with the younger phase of active rifting along the eastern part of the rift, while province 2 is associated with the western axis of initial rifting along the western part of the rift (Figs 8/59 and 8/64).

Moretti & Chenet (1986) studied the rate of rifting across the southern part of the Gulf of Suez. Figure 8/65 shows the rate of tectonic subsidence with time in three different parts of the rift. Within the offshore Gulf of Suez (Site 1, Fig. 8/65), tectonic subsidence seems to have begun about 23.5 million years ago, and the initial rate of subsidence was very slow during Nukhul time. The Rudeis phase of rifting showed the highest rate of subsidence during the period 22-28 my. Afterwards and during Kareem-Belayim times almost no significant subsidence took place within the Gulf of Suez rift. This was probably coincidental with the initiation of the Aqaba-Dead Sea rift. Since post-Belayim time and through South Gharib, Zeit, Pliocene and Quaternary times subsidence has taken place but at a very slow rate.

The subsidence rate in the area bounded by the Red Sea hills and offshore Gulf (Site 2, Fig. 8/65) seems to have initiated rifting during Nukhul and the lower part of Lower Rudeis at a rate comparatively higher than that within the offshore Gulf. During the rest of the Rudeis and Kareem time uplifting took place instead of subsidence. From Kareem onwards, another cycle of slow subsidence and uplift took place until the Pliocene-Quaternary when the rate of uplift became faster.

Across the Red Sea hills (Site 3, Fig. 8/65), no subsidence took place at all until the past 20 my where updoming started at different rates. It is to be noted from Figure 8/65 that the rate of synrift doming on both sides of the rift is proportional to the rate of subsidence within the rift. From the previous discussion, it can be suggested that the amount of doming on both sides of the rift can be taken as a measure of the degree of rift.

Figure 8/66 shows the distribution of the basement outcrops on the Nubian side of the Gulf of Suez-Red Sea rift. It should be noted that the areal extent and

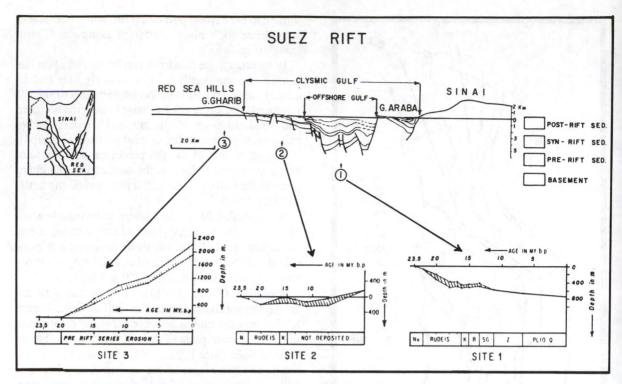

Figure 8.65 Rate of rifting across Gulf of Suez (after Moretti et al. 1986).

degree of doming of basement crustal rocks are by far greater on both sides of the Red Sea rift than that of those bounding the Suze rift. Moreover, along the Gulf of Suez rift, it can be noticed that the distribution of outcropping basement rocks is limited only to the southern half of the Gulf of Suez. This suggests that the rate of rifting in the northern Red Sea is greater than that in the southern half of the Gulf of Suez which in turn suffered more rifting than the northern half of the Gulf.

Rift model
A rifting model is proposed for the Red Sea-Gulf of Suez region in Figure 8/67. Usually rifting is initiated as a result of regional extension through a zone of extension which is bounded by already existing major fractures within the earth's crust (Fig. 8/67a). By applying horizontal extension, rifting starts by block faulting and dyke injection; this results in crustal thinning. This phase of rifting is known as the passive phase and is believed to be the present condition in the northern half of the Gulf of Suez (Fig. 8/67b). Note also that during this phase no domal uplift is recorded on either side of the rift.

With continued extension across the rift and more block faulting, continued thinning of the earth's crust takes place that results in relief of pressure on the mantle's surface, which in turn upwells as a result of

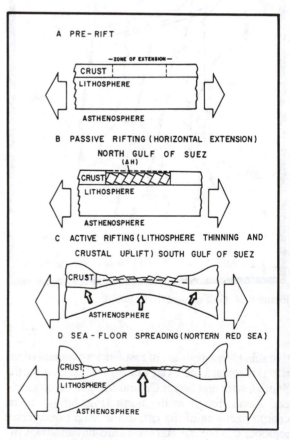

Figure 8.67 Rifting model.

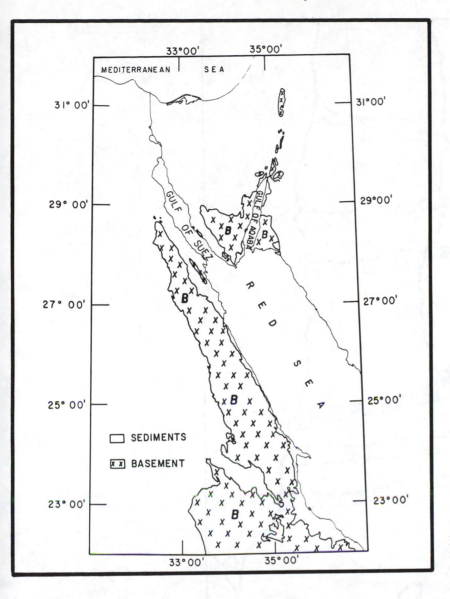

Figure 8.66 Distribution of basement rocks around Red Sea-Gulf of Suez and Gulf of Aqaba.

pressure release and convective heat. Accordingly, the lithosphere is stretched below the axis of the rift and thickness on both sides of the rift, giving rise to doming of the earth's crust on both sides of the rift (Fig. 8/67c). This stage of rifting is known as the active rifting phase and is believed to represent the degree of rifting in the southern part of the Gulf of Suez. Evidence for this stage of rifting in the southern Gulf comes from well data and magnetic modelling. Most wells drilled within the central portion of the southern Gulf and reaching the basement prove that the basement rocks are mainly of continental type. However, magnetic modelling across the southern Gulf is impossible without making two assumptions: (1) that the basement surface is regionally down-warping along the center of the Gulf (Fig. 8/68), a fact heavily supported by drilling information, and

(2) that there are more basic rocks along the central part of the southern Gulf which are close to the surface and result in assigning relatively higher magnetic susceptibility values for the basement blocks along the axis of the rift as compared to those on both sides (Fig. 8/68).

A more advanced stage of rifting takes place where the mantle material reaches the sea floor, resulting in sea floor spreading at the center of the rift and more doming and uplifting of the earth's crust on both sides of the rift (Fig. 8/67d). It is believed that this phase of rifting is the present condition in the northern Red Sea.

Figure 8/69 shows the sea floor spreading axis and the axial trough underlying the northern Red Sea. It also shows the northern extension of the rift axis along the Gulf of Suez.

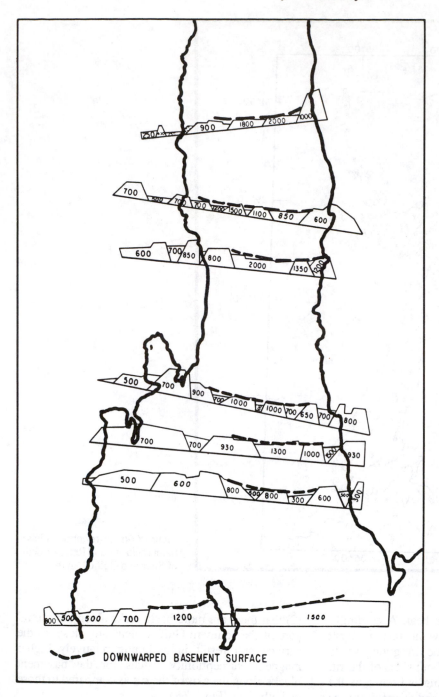

Figure 8.68 Modeled magnetic profiles across southern Gulf of Suez.

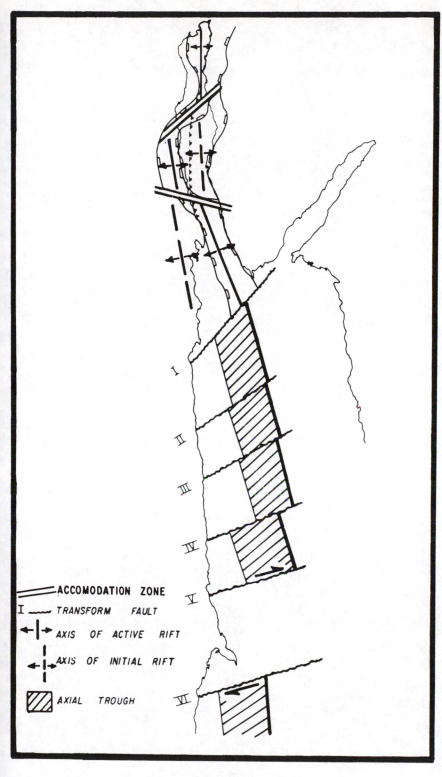

Figure 8.69 Gulf of Suez-Red Sea rift tectonics.

CHAPTER 9

Vulcanicity

MOHAMED YOUSRI MENEISY

Faculty of Science, Ain Shams University, Cairo, Egypt

During the Phanerozoic, Egypt was an exorogenic foreland receiving sedimentation affected by intermittent vulcanicity. Vulcanicity, and igneous activity at large, repeatedly occurred mainly in relation to the fracture system which originated in the Late Precambrian. Periodical reactivation of these older fracture zones throughout the Phanerozoic gave way to different types of plutonic and volcanic rock assemblages.

The Egyptian volcanic rocks of the Paleozoic and Mesozoic are diversified in their chemical characteristics, size and mode of eruption; whereas the Tertiary- and Quaternary-volcanics are largely basaltic.

Several workers studied various aspects of the Phanerozoic volcanics. Early works (e.g. Hume 1907, Barthoux 1922, Andrew 1937, Shukri 1953, Rittmann 1954, Sabet 1958) were reviewed by Said (1962). Most of these authors dealt primarily with the geologic and petrographic aspects of the different volcanics, mainly of Mesozoic and Tertiary age. Information concerning Paleozoic volcanics was then very limited.

Later works include El Hinnawi 1965, El Hinnawi & Abdel Maksoud 1968 and 1972, Sayyah & El Shatoury 1973, Meneisy & Kreuzer 1974, El Shazly et al. 1974, Meneisy et al. 1976, El Shazly 1977, Hashad et al. 1978, Abdel Monem & Heikal 1981, Kamel et al. 1981, Amer et al. 1982, Meneisy & Abdel Aal 1984 and a number of unpublished theses. These works added valuable information on the geochemistry, magma type, tectonic environment and age of these volcanics. In the light of these recent studies and the increasing availability of isotopic ages of Egyptian Phanerozoic volcanics, a meaningful temporal and spatial correlation between the different volcanic districts can be attempted.

Over the last decade, active interest in the study of the alkaline ring complexes in Egypt has provided a bulk of data concerning their origin, tectonic setting and ages. Alkaline magmatism is emphasized throughout this chapter.

Compilation of some 150 isotopic ages and a few new K/Ar ages obtained recently by the author is presented. The ages are generally considered to represent the time of formation or 'true age' of the rocks studied. The consistency of the ages obtained by different methods for the same rock is remarkable. In a few cases the ages obtained by K/Ar method are regarded as minimum ages. Also, recalculated ages using recent decay constants are indicated. The histogram approach was not attempted in the presentation of data in view of the wide spectrum of ages obtained by different methods for a wide variety of rocks. Coupled with our present knowledge of tectonic events, the isotopic age data were used to construct a sequence of the main episodes of Phanerozoic vulcanicity in Egypt. The geological timetable generally used for reference throughout this chapter is that compiled by Van Eysinga and published by Elsevier (1978). However, certain adjustments to the boundaries, especially for the Cenozoic as proposed by other authors, are taken into consideration.

Finally, a brief account of mineralization or hydrothermal activities connected with Phanerozoic vulcanicity is presented.

EARLY PALEOZOIC VULCANICITY

The earliest Phanerozoic igneous activity in Egypt was that associated with or closely related to the Pan-African tectono-thermal event. Despite current controversy as to the nature and time span of this important event (e.g. Fleck et al. 1976, Gass 1977, Rogers et al. 1978, Engel et al. 1980), it is generally accepted that the last phases of this event persisted through earliest Paleozoic times. It is with these late Pan-African phases that we are concerned here. The earlier (Late Proterozoic) phases are dealt with elsewhere in this book.

The sequence of principal Pan-African magmatic events – especially in the north Eastern Desert – has recently been outlined by Stern et al. (1984). According to his sequence, the final Pan-African phase of igneous activity took place roughly around 550-500 Ma. This is in close agreement with what was proposed by Fleck et al. (1976) for the Arabian Shield. They report ages indicating two late Pan-African thermal pulses or maxima. The first, between 610 and 580 Ma, is asociated with the major igneous activity which took place in the region at the end of the Precambrian. The second, between 540 and 510 Ma (Cambrian), was relatively less pronounced in terms of temperature and igneous activity, but was accompanied by faulting and fracturing.

Early Paleozoic (Pan-African) magmatism (550-500 Ma)

The earliest Paleozoic igneous activity reflects the emplacement of minor alkaline intrusions and granites (mainly alkali to peralkaline-G3-granites of Hussein et al. 1982) as well as 'dike swarms' common within basement rocks. Examples are Locations 1, 2, 3, 4 and 5 given in Table 1.

The alkaline ring intrusions are characteristic of the Red Sea Rift zone of Egypt and Sudan with the oldest yielding Pan-African 'Eocambrian' ages around 570-540 Ma (Razvalyayev & Shakhov 1978). The oldest known alkaline intrusion in the Eastern Desert is that of Wadi Dib with an average age of 554 Ma (Serencsits et al. 1979). This age places the Wadi Dib complex among the group of Pan-African Younger Granites with an alkaline affinity (El Ramly et al. 1982). This complex is silica over-saturated and relatively low in alkalies (as compared with the younger complexes). Structurally, it is associated with the northwest trending Najd transcurrent fault system developed at about the same time span 580-530 Ma BP (Fleck et al. 1976).

In Sinai, a ring dike complex with a central volcanic pile of alkaline rocks is reported by Shimron (1980) as associated with the 'Catherina cycle' of vulcanicity roughly dated at 550 ± 50 Ma. It is interesting to note that ring complexes in northeast Sudan yield ages in the same range. Vail (1976) reports K/Ar ages for Sabaloka granite between 476 and 540 Ma, and for Salala syenites and gabbros (411-550 Ma).

Several periods of alkaline magmatism are now known to have occurred in Egypt during the Phanerozoic and will be consecutively dealt with in this chapter.

The 'dike swarms'

The basement terrains in Egypt, especially in the Eastern Desert and Sinai, are of the most intensely dike-intruded regions known. Commonly referred to as dike swarms, these post-tectonic intrusions vary widely in composition from acidic, intermediate, basic and alkaline, and were intruded over a lengthy period of time.

The dikes are narrow, steeply dipping bodies, a few metres thick and several kilometres in length. As mapping has progressed, the distribution of these dikes has become more accurately known. Most of the dikes post-date the folding and metamorphism of the basement complex and the batholithic granites and some pre-date ring complexes.

In some areas, where paleomagnetic and isotopic age studies were carried out, the age of the dike swarms could be ascertained. K/Ar ages in the range 530-480 Ma were obtained for dike swarms located north of the Qena-Safaga Road and Um Rus area, central Eastern Desert (Nairn et al. 1980). These ages, however, should be considered as minimum ages in view of the possible argon loss due to alterations suffered by some of the dated samples. In many areas in the Nubian Shield, acidic dikes, pegmatites and aplite veins genetically and spacially related to the 'Younger Granites' are assigned to this period. A few isotopic ages of pegmatites are reported by El Ramly (1962) in the range of 530-495 Ma (Locations 3 and 4: Table 1). A brief review of relevant information concerning the age of the dike swarms in Egypt is given by Schurmann (1974).

It is important to point out here that detailed systematic studies and geochronological investigations of the dike suites in the Eastern Desert and Sinai are, as yet, insufficient.

Post Pan-African magmatic activity intermittently took place during the early Paleozoic. A number of intrusions and extrusions yielding Ordovician-Devonian ages are shown in Table 1.

In other areas, where age data is lacking, early Paleozoic ages are assigned to various subalkaline, intermediate and basic dike swarms (El Shazly et al. 1965). Rhyolite flows in Gebel Uweinat area, south Western Desert, are also assigned to this period.

Two periods of alkaline magmatism (404 and 351 Ma) are indicated by the isotopic ages obtained for Zargat Naam and Tarbtie North ring complexes corresponding to the Silurian and Devonian. The age of Zargat Naam is of particular significance since it intrudes Nubian Sandstones; a Paleozoic age is thus assigned to the latter. It should be pointed out that the formation of a ring complex took place, in some cases, through successive phases of magmatic activity. This is shown by the relative isotopic ages obtained for the different rock types of the same complex (e.g. Zargat Naam 404 and 247 Ma, see Tables 1 and 2).

LATE PALEOZOIC MAGMATISM

A number of alkaline and subalkaline volcanic rocks occur as plugs, sheets and cones as well as swarms of sills and dikes invading Precambrian formations mainly in the central and southern parts of the Eastern Desert. These volcanics are most probably related in age and genesis to the alkaline ring complexes. Based on the major chemistry statistics, these rocks include seven main types: namely, trachytes, bostonites, nepheline syenites, latites, spessartites, camptonites and keratophyres (Aly & Moustafa 1984).

Numerous age determinations (Table 1) reflect three main volcanic episodes in the Late Carboniferous, Permian, and Permo-Triassic. The rocks in the first two episodes are mainly tachytes, bostonites and nepheline syenites but also include camptonites and olivine basalts. They are widely intruded as ring complexes, sills, dikes, plugs, sheets and flows into the basement of the Eastern Desert and the sedimentary column of the Western Desert.

In the central Eastern Desert, the peralkaline volcanics of the greater Wadi Kareim area give an average Rb/Sr isochron age of 290 ± 15 Ma (Sayyah et al. 1978). This includes the age of El Atshan volcanics known for their radioactive mineralization. One isochron on Nasb El Qash trachytes gives a slightly younger age of 245 ± 15 Ma. The Wadi Kareim volcanics occur as sills, dikes and plugs of variable dimension but of limited lithologic variation. Trachytes and bostonites are the main types. The isotopic ages place these volcanics between late Carboniferous and early Permian. Their average initial Sr^{78}/Sr^{86} ratios show that they fall within the range typical for continental acidic volcanics (Hashad et al. 1981).

It is interesting to note that similar K/Ar ages are reported for volcanic dikes from Gebel Um Kibash, south Eastern Desert (El Ramly 1962). These data indicate the importance of the magmatism which occurred on the foreland of Egypt toward the late Carboniferous and early Permian time.

Permo-Triassic (230 ± 10 Ma) magmatism

Related to the initial break-up of Pangea and the closure of the Tethys, this period is characterized by rapid polar wandering (Gass et al. 1978). Records of volcanicity related to the uplift of the Aswan-Uweinat massif exist (Locations 18, 19, 20, 21 and 22, Table 1).

In the late Permian/early Triassic, the area between Gebel Uweinat and Bir Safsaf, south Western Desert, was uplifted along zones of pre-existing crustal weaknesses and these reactivated fractures gave way to the intrusion of basaltic dikes around 235 Ma as well as rhyolitic subvolcanics around 216 Ma (Schandelmeier & Darbyshire 1984). A group of K/Ar ages falling in the range of 230 ± 15 Ma is reported from northeast Sudan (Vail 1976).

During this period the massifs of Zargat Naam (partly), Bir Um Hebal and Gebel Silaia were emplaced. The granosyenite masses of Gebel Silaia and Gebel Zargat Naam fall along a major transform fault that runs approximately N 60° E, while Bir Um Hebal complex is located along a parallel transform fault somewhat to the north. The intersection of these transform faults with the northwestern faults seems to have controlled the location of the complexes. In Sinai, a major unconformity between the dominantly clastic Upper Carboniferous and Triassic deposits was detected in several wells, indicating a major movement in the Late Paleozoic (Said 1962). Recently, Meneisy (1986) reported a Permo-Triassic K/Ar age (238 Ma) for an olivine basaltic sheet from Farsh El Azraq volcanics, west central Sinai. This sheet (about 70 m thick) overlies the Upper Carboniferous rocks and is locally covered by Cretaceous Nubian Sandstone. These basaltic rocks subareally erupted along deep-seated faults and were derived from an olivine tholeiitic magma.

In a recent study of the Paleozoic rocks of Egypt, Issawi & Jux (1982) report major breaks within the Paleozoic succession especially in the south Western Desert. The breaks reflect uplift and subsidence of Arabo-Nubian Craton, echoing worldwide crustal disturbances during the Caledonian and Hercynian orogenies.

Also, in the Gulf of Suez area, Cherif (1976) concludes that most of the Carboniferous and Permian stages seem to have been subjected to the major movements of the Hercynian orogenesis, but were less strongly affected than most European and North African countries. The phases are indicated in the Gulf of Suez region by mild epirogenic movements.

To sum up, the late Carboniferous, Permain and Permo-Triassic vulcanicity is a reflection of land emergence and diastrophism occurring between the Paleozoic and the Mesozoic (Hercynian). Late Paleozoic vulcanicity in Egypt appears to have been more widespread than has previously been recognized.

MESOZOIC VULCANICITY

Mesozoic volcanic and other magmatic rocks in Egypt are abundant and diversified in size, form and composition. They include basaltic rocks, alkaline ring complexes as well as minor granitic intrusions. The largest Phanerozoic volcanic association in Egypt, namely that of Wadi Natash, southern Eastern Desert, is of late Cretaceous age. Several of the ring

Table 1. Isotopic determinations yielding Paleozoic ages.

No.	Locality	Age (Ma)	Method*	Rock	Reference
1.	W.Dib, N.ED	553 ± 11	K/Ar (b)	Qz-syenite	Serencsits et al. 1979
		558 ± 11	K/Ar (b)	umptekite	
		551 ± 11	K/Ar (b)	umptekite	
2.	N.Qena-Safaga	527	K/Ar	acidic, intermediate	Nairn & Ressetar 1980
	Road	521	K/Ar	and basic dikes	
		488	K/Ar		
	Um Rus, C.ED	480	K/Ar		
		497	K/Ar		
3.	W. Gemal, C.ED	530	Pb	allanite-pegmatite	El Ramly 1963
			Pb		
4.	W. Hafafit, S.ED.	495	K/Ar	pegmatite	El Ramly 1963
5.	Um Kroosh, S.ED	500 ± 30	Rb/Sr	andesite	El Shazly et al. 1975c
		480 ± 15	Rb/Sr	andesite	
6.	G.Abu Durba	485 ± 23	K/Ar	rhyolite	Steen 1982
7.	G.Babein, S.WD	489 ± 12	K/Ar	microgranite	Hunting 1974 (recalculated)
8.	G.Kamil, S.WD	431 ± 33	Rb/Sr (isochron)	granite	Schandelmeir & Darbyshire 1984
9.	Zargat Naam, S.ED	404 ± 8	K/Ar (h)	alkali, syenite	Serencsits et al. 1979
10.	Sharib Well no. 1, 8205 ft, N.WD	395 ± 16	K/Ar	basalt	McKenzie 1971
11.	Trabite N, S.ED	351 ± 7	K/Ar (h)	nordmarkite	Serencsits et al. 1979
12.	F.Mishbeh, S.ED	305 ± 20	Rb/Sr	Ne-syenite	Serencsits et al. 1979
13.	Um El Khors, C.ED	302 ± 15	Rb/Sr (isochron)	trachyte	El Shazly et al. 1975c
14.	W.Kareim, S.ED	300 ± 15	Rb/Sr	bostonite	Sayyah et al. 1978
15.	Hafafit Mine, S.ED	300	K/Ar	camptonite	El Ramly 1963
16.	Rabat Well no. 1, N.WD	293 ± 12	K/Ar	olivine basalt	El Shazly 1977 (reported)
17.	G.Um Kibash, S.ED	290	K/Ar	bostonite	El Ramly 1963
18.	El-Atshan, C.ED	273 ± 20	Rb/Sr	bostonite	El Ramly 1963
19.	Um Shaghir, C.ED	273 ± 15	Rb/Sr (isochron)	trachyte	El Shazly et al. 1975c
20.	Nasb El Qash, C.ED	245 ± 15	Rb/Sr (isochron)	trachyte	
21.	Farsh El-Azraq, W.C.Sinai	238 ± 3	K/Ar	olivine basalt	Meneisy 1986
		233 ± 3	K/Ar	basalt	
22.	G.Zargat Naam, S.ED	247 ± 13	Rb/Sr	syenite	Hashad & El-Reedy 1979
23.	Uweinat area, S.WD	235 ± 5	K/Ar	basaltic dike	Klerx & Rundle 1976
24.	El-Gezira, S.ED	229 ± 5	K/Ar (b)	gabbro	Serencsits et al. 1979
25.	G.Bir Um Hebal, S.ED	223 ± 9	Rb/Sr	granosyenite	Hashad & El-Reedy 1979
26.	G.Silaia, S.ED	221 ± 12	Rb/Sr	granite	Hashad & El-Reedy 1979
27.	G.El-Naga, S.ED	220 ± 20	Rb/Sr	Ne-syenite	El Shazly et al. 1975c

G. = Gebel; W. = Wadi; WD = Western Desert; ED = Eastern Desert; N. = Northern; S. = Southern; C. = Central.
h = hornblende; b = biotite.
*Whole rock, unless otherwise specified.

complexes were formed during the Mesozoic (e.g. Abu Khruq, El Kahfa, El Naga, El Mansouri, Nugrub El Fogani and Nugrub El Tahtani and Mishbeh). A few small granite bodies in the Eastern Desert are assigned to the Late Cretaceous. Basaltic dikes, sills, flows and plugs scattered in the Eastern and Western Deserts and in Sinai were intruded or extruded during the Mesozoic. The volcanic rocks of Wadi Araba and Abu Darag areas are of Early Cretaceous age.

The Mesozoic magmatic rocks can be generally related to two main phases of igneous activity in the Late Jurassic-Early Cretaceous (140 ± 15 Ma) and in the Late Cretaceous-Early Tertiary (90 ± 20 Ma).

Late Jurassic-Early Cretaceous phase (140 ± 15 Ma)

This phase of igneous activity was related to the initial rifting of the South Atlantic and the corresponding Africa-South America compression and Afro-Arabian strike slip faulting. Many blocks were affected by this event. Examples of magmatic rocks of this phase are shown in Table 2 (Locations 30-34).

Most of the masses which yield isotopic ages in the range of 140 ± 15 Ma are typically alkalic ring complexes and include those of Gebel Mishbeh, Gebel Nugrub El Tahtani, Gebel Nugrub El Fogani, Gebel El Naga and Gebel El Mansouri, in the south Eastern Desert. The isotopic age data obtained by different methods are remarkably consistent (Serenc-sits et al. 1979 and Hashad & El Reedy 1979). This 140 Ma episode of alkaline magmatism in Egypt coincided with a major episode of alkaline magmatism occurring in the areas surrounding the South Atlantic and was related to initial rifting of Africa from South America (Darbyshire & Fletcher 1979). Ring complexes in the same age range (150-130 Ma) are reported from northeastern Sudan (Vail 1976).

Examining the tectonic distribution map of the ring complexes in the south Eastern Desert after Briossov (in El Ramly et al. 1971), it is noted that these complexes are confined to a slightly curved zone of weakness that trends in a northeast direction parallel to the regional fault system of Wadi Halfa, Aswan, Mersa Alam (Aswan trend). This tectonic alignment and the close isotopic ages indicate that alkali magmas were being emplaced along this zone of weakness, which extends 200 km, during the period 130 to 150 Ma. The generation of this magma may have been triggered by some 'hot-spot' mechanism of the type postulated by Briden & Gass (1974). The zone of weakness provided the passage for the magma and the formation of these alkalic ring complexes. Further discussion of the ring complexes and their origin is presented later.

The volcanic rocks of Wadi Araba and Abu Darag, on the western side of the Gulf of Suez, are considered as related to this phase of early Cretaceous volcanicity. However, a slightly younger volcanic pulse may have followed. The rocks occur mainly as dikes and plugs cutting essentially the upper Paleozoic sedimentary series which are exposed at the core of the Wadi Araba structure and in several localities in Abu Darag area (Abdallah et al. 1973). The K/Ar ages range between 126 and 115 Ma (Meneisy & Kreuzer 1974) and are regarded as good minimum ages due to possible argon loss. The petrology and petrochemistry of these volcanics is studied by Meneisy et al. (1976). Petrographically, they are mainly nepheline-bearing and pyroclastic rocks in Wadi araba. In Abu Darag, the following volcanic association is recognized – olivine basalt-andesitic basalt and a rhyolitic variety. The Wadi Araba volcanics were extruded into the crest of Wadi Araba anticline the axis and plunge of which trend from east to west. These alkali basaltic rocks appear to belong to an active continental margin environment, being extruded within uplifted areas (Abdel Monem & Heikal 1981).

In Sinai and the Western Desert, it is difficult, as yet, to estimate the extent of volcanicity of this phase due to lack of isotopic age data.

The Late Cretaceous-Early Tertiary phase (90 ± 20 Ma)

This phase was tectonically related to the second major episode of alkaline magmatism and the large-scale strike-slip faulting in Afro-Arabia. The Late Cretaceous-Early Tertiary diastrophism referred to as 'Laramide' or Syrian arcing system has been the subject of considerable discussion (e.g. Said 1962, El Shazly 1977). The most obvious folds caused by this movement are the 'Syrian arcs' noted in northern Egypt, especially in northern Sinai and the northern Western Desert.

This is perhaps one of the most documented events of alkalic igneous activity in Egypt. The best record of this event is undoubtedly the volcanic rocks of Wadi Natash about 125 km east-northeast of Aswan, along the boundary between the Nubian Sandstones and the Precambrian basement complex. The exposed volcanics cover about 600 km^2. A number of workers have studied these volcanics (e.g. Barthoux 1922, El Ramly et al. 1971, Abul Gadayel 1974). Based on field work, El Ramly et al. (1971) report that the sequence of volcanic activity started by the formation of olivine basalt; this was followed by trachybasalts, latites and trachytes. These are considered as normal differentiation products of an olivine-basaltic magma. This magma erupted on a continental crust related to tectonic zones. The Rb/Sr isochron age of 104 ± 7 Ma obtained by Hashad & El Reedy (1979) is remarkably concordant with the 100 Ma K/Ar age reported by Abul Gadayel (1974). The isotopic ages are corroborated by Late Cretaceous leaflets found in the intercalated volcaniclastic sediments. In a recent study by Crawford et al. (1984) on Wadi Natash volcanics, the authors indicate that the age of the volcanics suggests that they were not directly associated with Red Sea rifting. Their alkaline nature may imply they were involved in a pre-rifting doming process. Alternatively, these volcanics may belong to the intraplate bimodal alkali-olivine basalt suite.

In the Uweinat area, south Western Desert, some alkaline volcanics pierce Paleozoic sandstones. The age of these rocks may be correlated with that of the Wadi Natash volcanics. Petrographically, they include alkali trachytes and alkali rhyolites (Bishady & El Ramly 1982). Potassium-argon ages are in the range 83 to 78 Ma for olivine basaltic rocks from the Darb El Arbain area (Meneisy & Kreuzer 1974a). The geology of this area is described by Issawi (1971) and the petrography of the basaltic rocks is given by Abdel Aal (1981). It should be pointed out that the volcanic rocks in this area belong to more than one phase of vulcanicity. In the neighbouring

Gilf El Kebir area, Greenwood (1969) reports ages in the same range.

The other important event during this period of igneous activity was the alkaline magmatism giving rise to Gebel El Kahfa, Gebel Abu Khruq and partly Gebel El Naga and Gebel El Mansouri ring complexes, the latter two being among ring complexes which were formed through successive phases of magmatism. Isotopic ages are fairly consistent and are roughly around 90 ± 5 Ma. It is noted that the two ring complexes of El Kahfa and Abu Khruq fall along a northwesterly trending lineament. This may imply a tectonic control for their intrusion. These complexes were related to one phase of alkali igneous activity. Wadi Natash volcanics also seem to have belonged to this phase.

Reference should be made here to the carbonatites in Sinai and the south Eastern Desert. The spatial and temporal affinities between African rift valley alkaline magmatism and carbonatites are generally accepted. In the southeastern Sinai, the Tarr albitite-carbonatite complex of Wadi Kid is studied by Shimron (1975). The complex comprises albitite masses with closely related explosion breccias, fenite aureoles, intrusive carbonate bodies, olivine dolerite and lamprophyre dikes and copper mineralization. Shimron (op. cit.) concludes that the complex and the alkaline volcanism in the northern Negev and metal-rich sediments associated with the hot brines in the Red Sea Deep represent a continuous (or staged) event localized along an accreting plate margin. The magmatic events commenced during the Cretaceous (Tarr) and have continued through the Oligocene-Miocene (basalts and dolerites) up to the present. He reports a fission track age of 103 ± 8 Ma from the fenite aureole, probably related to the emplacement of the Tarr complex.

Another aspect of the late Cretaceous igneous activity is the intrusion of small bodies of granitic composition. These Cretaceous granites have come to the attention of workers in the field during the last few years. El Shazly et al. (1973) report Rb/Sr ages around 90 Ma for yellow and red muscovite granite bodies from Abu Sawyel area, south Eastern Desert. Another example of the foreland granites assigned to the Late Cretaceous-Early Tertiary based on field evidence is reported by Samuel et al. (1983) from El Bakriya area, central Eastern Desert. They are intruded at sites of fault plane intersection and assume circular or oval outlines. They manifest contact thermal effects on the bordering Nubian Sandstone blocks.

On the origin of the alkaline complexes in Egypt

As already shown, the alkaline complexes in the Nubian Shield of Egypt range in age from 554 to around 78 Ma, i.e. Cambrian-Upper Cretaceous. The intrusive complexes of the Eastern Desert are intimately associated with major lineaments (Fig. 9/1).

They represent the northward continuation of ring complexes associated with the East African rift system. Hussein (pers. comm.) suggests that the ring complexes of Nugrub El Fogani, Nugrub El Tahtani, Mishbeh, El Naga, Gezeira and El Mansouri, together with the carbonatite bodies bordering Mansouri and extending southwestward into the Sudan, all fall on the continental trace, a zone of transform faults that extends to cut across the axis of sea-floor spreading in the Red Sea. By analogy, it may be suggested that the two complexes of Hadayib and Um Risha (El Ramly et al. 1979) together with that of Zargat Naam mark another zone of transform faults. A third zone may exist along which lie the ring complexes of Trabite North, Trabite South and Kahfa.

Phil de Gruyter & Vogel (1981) suggest that these complexes originated as a result of alkaline melts formed in the asthenosphere by shear heating caused by changes in plate motion. These melts were emplaced along reactivated Pan African fractures or pre-existing zones of weakness. The six main periods of alkaline magmatism in Egypt (554, 404, 351, 229, 145-132, 91-89 Ma) are all synchronous with changes in plate motion. It is widely accepted that alkaline magmatism has a relatively deep-seated origin and may be associated with relatively small degrees of partial melting of an enriched mantle source. The melts ascend preferentially along zones of weakness. The coincidence of alkaline magmatism in widely separated areas, as in the case of the Egyptian, Atlantic and South African alkaline provinces, indicates that these alkaline melts may have been due to worldwide, synchronous, mantle disturbances.

Although the alkaline complexes of the Eastern Desert have similar overall chemical trends, there is a distinct change in the degree of alkalinity and silica-saturation with age. The older complexes are generally silica-oversaturated and relatively low in alkalies, whereas the younger complexes are dominated by silica-undersaturated rock types and are rich in alkalies. This change in alkalinity may be due to the cooling of the lithosphere after the Pan-African event. Thus the chemical trends observed in the Egyptian alkaline complexes are consistent with decreasing thermal gradient in the Phanerozoic.

El Ramly et al. (1982) discuss the tectonic setting and petrogenesis of the alkaline ring complexes in Egypt. They suggest that under certain favourable geotectonic conditions (e.g. intracontinental hot spots) partial fusion of deeper levels in the upper mantle could produce enough heat, volatiles and

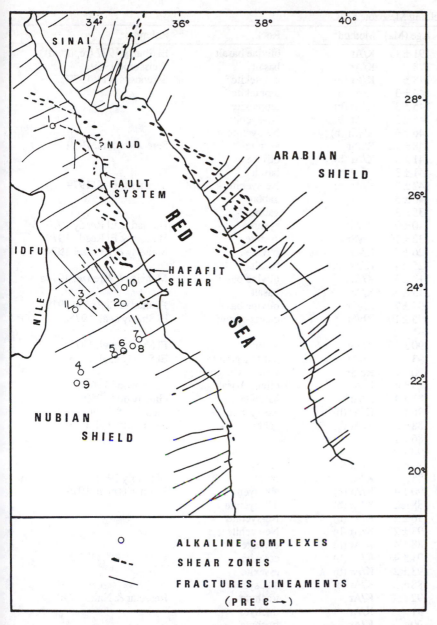

Alkaline Complexes
1. Wadi Dib
2. G. El-Kahfa
3. G. Trabite N
4. El-Gezira
5. G. El-Naga
6. G. Mishbih
7. G. Nigrub N
8. G. Nigrub S
9. G. Mansuouri
10. Abu Khruq
11. G. Trabite S

Figure 9.1 Map of tectonic lineaments (after Garson & Krs 1976).

mobile elements to melt the overlying parts of the lower crust. The two magmas evolved could produce the variety of rocks constituting the formation of these complexes.

CENOZOIC VULCANICITY

Several episodes of volcanic activity occurred in Egypt during the Cenozoic. The earliest was of Paleocene age and represented the continuation of the extensive late Cretaceous igneous activity (Table 3).

Mid-Tertiary vulcanicity was widespread and was the first to be recognized in Egypt. Numerous isotopic age determinations were carried out on Tertiary volcanics; and, in the light of age data, a number of successive volcanic pulses are indicated starting in the late Eocene with subsequent extensional phases ranging from Late Oligocene to Middle Miocene. This vulcanicity was intimately associated with the Red Sea opening. It is generally believed that the Red Sea acquired its characteristics and independence from the Tethys due to the Cenozoic tectonics, starting with late Eocene-Oligocene-Early Miocene uplift and development of a major fault system. These faults trending northwest-southeast show considerable displacement along fault planes dipping away from the Red Sea. Vulcanicity accompanied these

Table 2. Isotopic determinations yielding Mesozoic ages.

No.	Locality	Age (Ma)	Method*	Rock	Reference
28.	Kattaniya Well, N.WD	191 ± 19	K/Ar	olivine basalt	El Shazly 1977 (reported)
29.	Um Bogma, W.Sinai	178	K/Ar	basalt	Weissbrod 1969
30.	G.El. Naga, S.ED	148 ± 3	K/Ar (b)	umptekite	Serencsits et al. 1979
		146 ± 3	K/Ar (b)	umptekite	
		145 ± 3	K/Ar (b)	umptekite	
		145 ± 3	K/Ar (b)	Ne-syenite	
		146 ± 6	Rb/Sr (b)	Ne-syenite	
31.	G.Mishbeh, S.ED	148 ± 12	Rb/Sr	syenite	Serencsits et al. 1979
		141 ± 3	K/Ar (b)	olivine	
		141 ± 3		basalt	
32.	G.Nugrub El Fogani, S.ED	142 ± 2	K/Ar (b)	Ne-syenite	Serencsits et al. 1979
		140 ± 3		gabbro	
		135 ± 3		gabbro	
33.	G.Nugrub El Tahtani, S.ED	140 ± 9	Rb/Sr	syenite	Hashad & El Reedy 1979
34.	G.El Mansouri, S.ED	132 ± 10	Rb/Sr	syenite	Hashad & El Reedy 1979
35.	W.Araba, N.ED	126 ± 4	K/Ar	olivine-	Meneisy & Kreuzer 1974a
		125 ± 4	K/Ar	basalt	
36.	W.Abu Darag, N.ED	113 ± 3	K/Ar	olivine basalt	Meneisy & Kreuzer 1974a
		115 ± 3	K/Ar	basalt	
37.	W.Natash, S.ED	104 ± 7	Rb/Sr	olivine basalt	Hashad & El Reedy 1979
38.	G.El. Mansouri, S.ED (see 34)	95 ± 10	Rb/Sr	quartz-syenite	El Shazly et al. 1975c
39.	G.Zargat Naam, S.ED	90 ± 10	Rb/Sr	granite	El Shazly et al. 1975c
40.	Um Shilman, S.ED	93	Rb/Sr	red Mu-granite	El Shazly et al. 1973
		90	Rb/Sr	yellow Mu-granite	
41.	Nubia, S.ED	89 ± 4	K/Ar	olivine basalt	Greenwood 1969
42.	W.Um Hokban, S.ED	85 ± 4	K/Ar	basanite	Greenwood 1969
43.	G.Abu Khrug, S.ED	96 ± 2	K/Ar (b)	Ne-syenite	Meneisy & Kreuzer 1974b
		88 ± 5	K/Ar (b)	gabbro	Serencsits et al. 1979
		86 ± 15			
		90 ± 2			
		89 ± 2	K/Ar (b)	Ne-syenite	
		85	K/Ar	basalt	El Ramly 1963
44.	El Kahfa, S.ED	90 ± 4	K/Ar (b)	alk. syenite	Serencsits et al. 1979
		88 ± 2	K/Ar (b)	alk. syenite	
		96 ± 2	K/Ar (b)	Ne-syenite	
		93 ± 2	K/Ar (b)	Ne-syenite	
		88 ± 4	K/Ar (b)	essexite	
		91 ± 4	K/Ar (b)	essexite	
		93 ± 2	K/Ar (b)	essexite	
		88 ± 2	K/Ar (b)	alk. syenite	
45.	W.Kareim, C.ED	92	K/Ar	trachyte	Ressetar & Nairn 1980
		91	K/Ar	trachyte	
		90	K/Ar	trachyte	
	(younger generation)	74	Rb/Sr (isochron)	bostonite	Sayyah et al. 1978
46.	G.Nuhud, S.ED	78	K/Ar	trachyte	El Ramly 1963
47.	Darb El Arbain, S.WD	79 ± 2	K/Ar	olivine basalt	Meneisy & Kreuzer 1974a
		78 ± 2	K/Ar		
		78 ± 3			
		76 ± 2			
48.	Gilf El Kebir, S.WD	75 ± 3	K/Ar	basanite	Greenwood 1969
49.	G.El Naga, S.ED	86 ± 3	K/Ar	Ne-syenite	Meneisy & Kreuzer 1974b
		84 ± 3	K/Ar	Ne-syenite	

G. = Gebel; W. = Wadi; WD = Western Desert; ED = Eastern Desert; N. = Northern; S. = Southern; C. = Central.
h = hornblende; b = biotite.
*Whole rock, unless otherwise specified.

movements, which are also reflected in the sedimentary column by non-deposition or continental sedimentation. The spatial relationships between the basaltic rocks and the Suez rift are quite evident.

Quaternary vulcanicity is reported, though not well-documented in Zabargad (St John's) Island in the Red Sea and in the south Western Desert. K/Ar ages for basaltic rocks from Zabargad Island are reported by El Shazly et al. (1974) to be 1.5 Ma but actually ranging from 0.9 to 1.7 Ma indicating an early Pleistocene age. More field and isotopic age data is needed to confirm this conclusion.

In the south Western Desert, some basaltic plugs and dikes along fracture zones in the Uweinat-Kamil-peneplain and Bir Safsaf areas are assigned to the Quaternary (Almond 1979). Small-scale alkali basaltic vulcanicity occurred in this area during the Quaternary.

TERTIARY VULCANICITY

Tertiary vulcanicity is uniformly basaltic and widely distributed north of latitude 28° N (Fig. 9/2). Basaltic

extrusives cover a large area beneath the Nile delta and the adjacent parts of the Western Desert (Bayoumi & Sabri 1971, Said 1981, Williams & Small 1984). Numerous isolated outcrops also occur along the Fayum-Abu Rowash, Cairo-Suez and Tihna-El Bahnasa stretches.

In the southern parts of the Western Desert, some Tertiary basaltic occurrences are sparsely distributed. In places, they are associated with minor occurrences of acid to alkaline rocks. These volcanics vary in composition and belong to more than one phase of volcanic activity (Meneisy & Abdel Aal 1984).

Along the Red Sea coast, south of Quseir, some dolerite flows occur. A few scattered basaltic dikes and plugs intruding Nubian Sandstones in the south Eastern Desert, near Wadi Hodein, are considered of Tertiary age. Information concerning these rocks is, however, scanty. In Sinai, several minor Tertiary basaltic outcrops occur, especially in the western and central parts, and reach large dimensions in places.

The Tertiary basaltic rocks occur mainly in the form of sheets, dikes, sills which sometimes widen into small irregular stocks, plugs, cinder cones or small ridges. The geology, petrography and petroche-

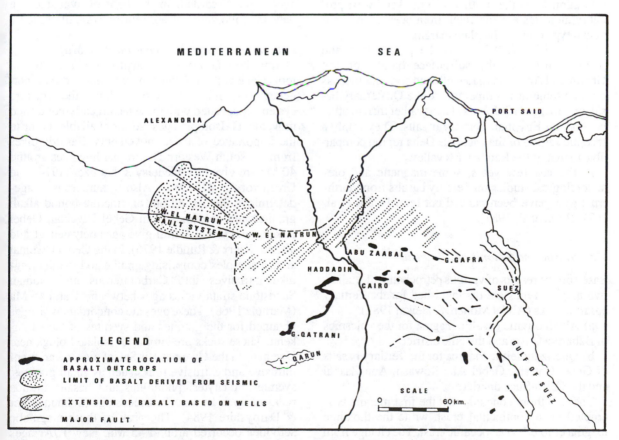

Figure 9.2 Main Tertiary basaltic occurrences in northern Egypt (after Williams & Small 1984).

mistry of these volcanics have been studied by
several workers. Petrographically, the basalts of Abu
Zaabal, Abu Rowash and Gebel Qatrani are
described as doleritic basalts; those of the Cairo-Suez
district, Tihna and El Bahnasa are mainly normal
basalts. Two varieties are distinguished at the Bahariya Oasis, namely olivine basalt and a doleritic
variety (El Hinnawi 1965, Kamel et al. 1981). Petrochemical studies suggest that most of the studied
Tertiary basalts (except those of the Bahariya Oasis)
are tholeiitic basalts (El Hinnawi & Abdel Maksoud
1972), whereas the basalts of Bahariya are alkali
basalts (El Kalioubi 1974). According to Abdel
Monem & Heikal (1981) the Egyptian Tertiary basalts exhibit a complete compositional range from
nepheline to quartz normative. They also note a relationship between the geographic latitude of the volcanic districts and the degree of silica-saturation and
increase in total alkalies. The basalts of the Cairo-
Suez district, Abu Zaabal and Abu Rowash, which
constitute the northern belt, are typically quartz-
normative, the basalts of Gebel Qatrani and the Nile
Valley are quartz-normative grading into olivine-
normative, and those of the Bahariya Oasis are
olivine-normative grading into nepheline-normative.
This systematic change in the degree of silica-
saturation from the northeast to the southwest probably indicates advancement from oceanic to continental type crust at the plate margin.

Amer et al. (1982) study the petrography and
petrochemistry of the subsurface basalts of Mit
Ghamr and Abu Hammad wells and consider them as
oceanic tholeiites comparable to the Quaternary basalt in Zabargad (St John's) Island rather than to other
continental Egyptian Tertiary basalts. They imply a
tectonic setting of the east Nile Delta region comparable to that of the Red Sea rift valley.

In the last few years, some magnetic and ore-
mineralogical studies on Tertiary basalts from northern Egypt have been carried out (e.g. Fahim et al.
1971, Basta et al. 1981).

Magma type and tectonic environment

Based on the results of various petrochemical studies,
two magma types are distinguished for the Tertiary
basaltic rocks (Abdel Monem & Heikal 1981):
a) Alkali olivine basaltic magma for the volcanics
of Bahariya Oasis and the Nile district.
b) Quartz tholeiitic magma for the Tertiary basalts
of Gebel Qatrani, Gebel Abu Rowash, Abu Zaabal
and the Cairo-Suez district.
These authors conclude that the first magma type
erupted on a continental crust, while the tholeiitic
magma erupted on an oceanic crust. The change from
the tholeiitic to alkaline character in space and time

suggests that vulcanicity migrated from an oceanic
crust environment (e.g. Cairo-Suez district in the
northeast) to a continental crust environment (e.g.
Bahariya Oasis in the southwest). Discrimination
diagrams show that all these basalts fall into the field
of 'within plate' basalts.

Recently Abdou (1983), studying volcanic rocks
in northern Egypt, indicated that in the Cairo-Suez
district the basalts of Gebel Anqabia, Gebel Um
Raqm and Gebel Abu Triefiya have some chemical
characteristics of ocean-floor basalts – a conclusion
based on clinopyroxene chemistry. However, the rest
of the mineral chemistry parameters, in addition to
the whole rock data, strongly suggest an intraplate
tectonic setting. He concludes that these rocks from
the Cairo-Suez district are closely comparable to the
Ethiopian Plateau basalts and the African basalts
attributed to zones of crustal thinning.

Tertiary volcanic episodes

The Tertiary volcanics in Egypt are well-documented
by isotopic age data (Table 9/3). Nearly all of the
age-measurements were carried out using the K/Ar
method. Results are usually reliable unless the
sample is altered. On this basis a few of the published
ages were discarded. In the light of available age
data, the following volcanic episodes are discerned.

Late Eocene-Early Oligocene (40 ± 10 Ma)

During the Late Eocene, a shallowing of the Tethys
took place and the Oligocene was marked by emergence. Volcanics developed along the fracture
systems associated with these tectonically-controlled
movements. Isotopic ages, so far available, point to
the importance of these movements. Basaltic dikes
from the south Western Desert yielding ages around
40 Ma are given by Meneisy & Kreuzer (1974) and
Greenwood (1969). Also numerous age-
determinations carried out on fracture-bound alkali
granites and volcanics from Gebel Uweinat, Gebel
Arkenu and Gebel Kamil give ages between 41 and
48 Ma (Klerx & Rundle 1975). In the Gebel Uweinat
area, a complex comprising granite and syenite ring-
dike intrusives into Carboniferous and Nubian
Sandstone strata yields ages between 41 and 45 Ma
(Marholz 1968). These ages are concordant with ages
obtained for the granites and syenites of Gebel Arkenu. These rocks are similar to isolated plugs near
Uweinat. In the Uweinat-Bir Safsaf area, a variety of
intrusive and extrusive rocks such as alkali granites,
syenites, trachytes, phonolites and basalts intrude
rocks of up to Lower Cretaceous age (Schandelmier
& Darbyshire 1984). The peak of these magmatic
activities occurred around 45 Ma. New K/Ar ages
recently obtained by the present writer for basaltic

dikes from Darb El Arbain area (45 and 46 Ma) are included in Table 9/3. They show striking similarity with the above-mentioned data. This phase corresponds to the Lutetian-Bartonian (Pyrenean movement).

A few K/Ar ages reported in Table 9/3 range between 31 and 28 Ma. These are considered as 'minimum' ages due to possible argon loss through alteration. They are thus possibly related to this phase.

Oligo-Miocene phase (24 ± 2 Ma)
Vulcanicity related to the opening of the Red Sea took place in a series of successive pulses ranging in age from the late Oligocene up to the Middle Miocene. The initial Red Sea rift, formed by continental rifting in the Late Oligocene or Early Miocene, widened through a combination of normal faulting which may have been listric in nature and of dike injection which included emplacement of some sizeable igneous bodies (Coleman et al. 1979). Isotopic dating of these bodies assists in determining the age of the stratigraphic and tectonic events.

The Oligo-Miocene phase is well-documented. In northern Egypt, several basaltic occurrences yielded ages in the range of 25 to 23 Ma (e.g. Locations 56, 57, 64, 65 and 68, Table 9/3).

In western Sinai, a number of basaltic bodies are dated at around 24 Ma (Steen 1982). In central Sinai, volcanic rocks are widely distributed. Most occurrences are of doleritic dikes, sills and plugs; flows are known near Abu Zenima and Hammam Faraoun. Some of the dikes are tens of kilometers long (e.g. Raqabat El Naam). These bodies intrude rocks of up to the Middle Eocene. The Raqabat El Naam dike trends east-west parallel to a large fault belonging to the central Sinai-Negev shear zone. It gives an isochron age around 25 Ma (Steinitz et al. 1978), and its point ages vary from around 25 to 31 Ma. This period was associated with strike-slip faulting in central Sinai.

Age data, as well as field evidence, indicate that the Oligo-Miocene vulcanicity was followed by a number of successive pulses during the early Miocene and up to the Middle Miocene. In the Cairo-Suez district, for example, a sequence of up to four successive eruptions can be distinguished in the field (M.A. El Sharkawi, person. comm.). El Sheshtawy (1979) shows that the Qatrani basalts are made up of a numbr of successive sheets, clearly distinguished in the field. This is also partly reflected by age data which range between 27 and 23 Ma. The extent and distribution of Oligo-Miocene basalts in northern Egypt can be properly envisaged when the subsurface basalts are considered. The subsurface distribution and age of basalts in the eastern part of the

Western Desert, as demonstrated by well and seismic data, are discussed by Williams & Small (1984). The Abu Zaabal (Haddadin) basalt is found in a number of wells. The type locality of this basalt is at Abu Zaabal, east of Cairo, where the thickness averages 30 m. This basalt extends eastward into the Eastern Desert, and an exceptional thickness of 328 m is noted at Mit Ghamr-1 and probably forms part of a continuous sheet through to the Maryut area (Fig. 9/2). Detailed paleontological studies of the wells which penetrated the basalt suggest an Early Miocene dating; however, seismic correlation to other nearby wells suggests that these basalts may, in part, be of Late Oligocene age.

Lower-Middle Miocene phases (20, 18 and 15 Ma)
Following the Oligo Miocene phase, successive volcanic eruptions occurred during the Miocene. Based on isotopic age data, using histograms, Meneisy & Abdel Aal (1984) propose the following sequence for the main episodes of Miocene vulcanicity, a Lower Miocene (20 and 18 Ma) and a Middle Miocene (15 Ma).

Accordingly, the youngest Tertiary vulcanicity, so far dated, is that of the middle Miocene recorded, mainly from the Bahariya Oasis. This is likely to be further substantiated when sufficient geochronological studies are carried out. It should be noted in this respect that age data for the extensive subsurface basalts in northern Egypt are as yet very scanty.

HYDROTHERMAL VOLCANIC ACTIVITIES

Of particular importance are the hydrothermal activities that accompanied magmatic eruptions and that have survived them in Egypt. These may now be represented by certain thermal springs in the country. The mineralizing effects of these hydrothermal activities are intimately related to the nature and intensity of the magmatic activity as well as to the nature of the invaded rocks. The most pronounced effects of these hydrothermal activities in Egypt are those connected with the Tertiary vulcanicity.

Mineralization connected with Early Paleozoic magmatic activity is limited. Garson & Shalaby (1976) discuss the Precambrian-Lower Paleozoic plate tectonics and metallogenesis in the Red Sea region and outline mineralization associated with some of the Late Pan-African Younger Granites, e.g. Sn-W and Nb-Ta-Sn-Be mineralizations in central and south Eastern Desert. Also Hussein & El Kassas (1980) outline radioactive anomalies distributed in different rock types including bostonites, pegmatite and aplites dikes indicating uranium or thorium mineralization. Some of these rocks may pertain to

Table 3. Isotopic determinations yielding Cenozoic ages

No.	Locality	Age (Ma)	Method*	Rock	Reference
South Western Desert					
50.	Darb El Arbain	41 ± 2	K/Ar	olivine-	Meneisy & Kreuzer (1974a)
		34 ± 1		basalt	
		46 ± 1	K/Ar	basalt	Meneisy (this work)
51.	G.Uweinat	41a	K/Ar	alkali-granite	Klerx & Rundle 1976
	G.Arkenu, G.Kamil area	to		and volcanics	Marhold 1968
		48			
52.	Uweinat approach	37 ± 2	K/Ar	hawaiite	Greenwood 1969
53.	Near pottery Hill	28 ± 2	K/Ar	olivine basalt	Greenwood 1968
North Western Desert					
54.	W.Samalut	28 ± 2	K/Ar	basalt	Meneisy & Kreuzer 1974a
		27 ± 3			
		23 ± 2	K/Ar	basalt	
		23 ± 2			
56.	G.Qatrani	27 ± 3	K/Ar	basalt	El Shazly et al. 1975c
		25 ± 1		basalt	
		24 ± 1		basalt	
		23 ± 1	K/Ar	basalt	Meneisy & Abdel Aal 1984
Sinai					
57.	G.Matulla	24 ± 1	K/Ar	basalt dike	Meneisy (this work)
58.	Raqabat El Naam	25 ± 2	K/Ar	basalt	Steinitz et al. 1978
59.	G.Iktefa 33°33'	20 ± 1	K/Ar	basalt	Steinitz et al. 1978
60.	Themed	20 ± 1	K/Ar	basalt	Steinitz et al. 1978
61.	G.Araba	31 ± 2	K/Ar	olivine basalt dike	Steen 1982
62.	W.Nukhul	22 ± 1	K/Ar	olivine basalt dike	Steen 1982
63.	W.Tayiba	21 ±	K/Ar	olivine basalt dike	Steen 1982
64.	G.Araba	18 ± 1	K/Ar	olivine basalt dike	Steen 1982
Cairo-Suez District					
65.	Abu Zaabal	23 ± 1	K/Ar	basalt	Meneisy & Kreuzer 1974a
66.	El Gafra	22 ± 2	K/Ar	basalt	Meneisy (this work)
67.	Qattamiya	22 ± 2	K/Ar	basalt	Meneisy & Abdel Aal 1984
Red Sea					
68.	South of Quseir	24 ± 1	K/Ar	basalt	Meneisy & Abdel Aal 1984
		23 ± 1	K/Ar	basalt	
		22.5 ± 1	K/Ar	basalt	K. Balovh (pers. comm., Abdel Aal)
		22.5 ± 1	K/Ar	basalt	
Bahariya Oasis					
69.	Basalt Hill	20 ± 1	K/Ar	olivine-	Meneisy & El Kalioubi 1975
	G.Maeysra	20 ± 1	K/Ar	basalt	
	G.Gel Hefhuf	18 ± 1	K/Ar	basalt	
	G.Mandisha	18 ± 1	K/Ar	basalt	
	G.Mandisha	16 ± 1	K/Ar	basalt	
	(dike)	15 ± 1	K/Ar	basalt	
70.	Zabargad (St. John's) Island	1.7?	K/Ar	basalt	reported in El Shazly 1977

G. = Gebel; W. = Wadi; WD = Western Desert; ED = Eastern Desert; N. = Northern; S. = Southern; C. = Central.
h = hornblende; b = biotite.
*Whole rock, unless otherwise specified.

Table 4. Summary of main Tertiary mineralizations

Age of rock mineralized	Replacement	Vein and other forms	Locality
Miocene (conglomerate, sandstone, limegrit, clay, gypsum)	Pb		Zug El Bohar, Abu Anz, Ras Banas
	Pb, Zn		Wizr, Um Gheig, G. Rusas, Ras Banas
	Mn, Fe	Mn, Fe	Wadi Meialik
	Mn	Mn	Halaib, Wadi El Daeit
Oligocene (sandstone)	Mn, Fe	Mn, Fe	Sharm El Sheikh
		U	Fayum
	Fe and Mn oxides impregnations with silica	Abu Roash (near Cairo); Cairo-Suez Road	
Eocene (limestone)	Mn		Near Wadi Feiran, W. El Akareb
	Fe, Mn	Fe, Mn	Bahariya Oasis
Cretaceous (limestone, dolomite, shale)	Fe, Mn	Fe	Bahariya Oasis
	Fe		Kharga Oasis
	Fe and Mn oxides impregnations with silica		Abu Roash
Jurassic (limestone, sandstone)	Fe and Mn oxides impregnations with silica		Northern and southern Galala (west Gulf of Suez)
Carboniferous (dolomite, sandstone)	Mn, Fe		Um Bogma
	Cu		Sarabit El Khadim
	Fe and Mn oxides impregnations with silica		Wadi Araba (west Gulf of Suez)
Basement complex (igneous and metamorphic rocks)		Cu	Sarabit El Khadim, Regita, Abu Nimran, Gebel Samra
		Cu, Zn	Um Samuiki
		Cu, Ni	Abu Swayel?
		Cu, Fe	El Atawi?
		Mn	Gebel Alda, Halaib, Wadi El Daeit
		Mn, Fe	Gebel Musa, Wadi Meialik

the last phases of the Pan-African event in the Early Paleozoic. Radioactive mineralization is also connected with late Paleozoic peralkaline volcanics of the Wadi Kareim area, central Eastern Desert, especially El Atshan volcanics. Copper mineralization occurs in the Carboniferous rocks in west central Sinai. These rocks are invaded by sheets and dikes of basalt and dolerite which indicate marked volcanic activity in the Post-Carboniferous history of the district. As already mentioned earlier in this chapter, some of these volcanics, e.g. at Farsh Al Azraq, are dated as Permo-Triassic.

Of particular interest is the copper mineralization and some of the hydrothermal-type alterations (fenitization in part) of the Tarr albitite carbonatite complex of the Wadi Kyd region in southeastern Sinai. The complex comprises three major albitite masses: the Tarr, Hatamiya and Samra. In his study of the complex, Shimron (1975) considers that the copper deposits resemble others that are located along crustal accretion or subduction zones.

As Tertiary vulcanicity is the most widespread, Tertiary mineralization is pronounced. The mineralizing effects include ubiquitous ferrugination, silicification and dolomitization. Ascending mineralizing solutions have also developed, by replacement and vein-filling, quite a few of the Egyptian epigenetic hypogene ore deposits. The common relation between the sites of these mineralizations and those of the Tertiary volcanic activity (extrusions and shallow intrusions) show that most of these mineralizing solutions are related to the Tertiary vulcanicity. Comprehensive accounts of Tertiary mineralization in Egypt are given by earlier workers. Table 9/4 summarizes the main Tertiary mineralizations and Figure 9/3 shows their distribution in connection with the centres of Tertiary volcanics activity. It is noted that these mineral deposits are usually situated along fault planes and in regions where volcanic activities were most pronounced. Although the basalt itself is not a good mineralizer, the hydrothermal solutions can become, during their ascent, good mineralizers through leaching and mobilizing pre-existing labile ions in the traversed rocks. More details concerning these deposits are presented elsewhere in this book.

It remains now to refer to the interesting and important subject of the hot brines and their metalliferous deposits in the Red Sea. Active interest in the study of these hot brines has culminated in a number of publications in recent years. The connection between the hot brines and vulcanicity is fairly evident. Data from geology and geophysics produce a convincing picture of the source of heat and the environment of the hot brine basins. The hot brines might have consisted of ground waters that had obtained their major salts from resolution of bedded former

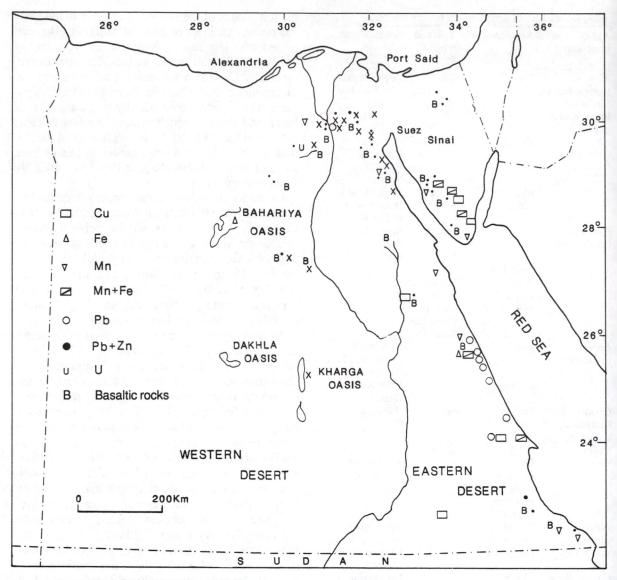

Figure 9.3 Main mineralization localities in Egypt.

evaporites and their trace elements from solutions of interbedded shales. When the ground water reached a mass of recently emplaced and still hot basalt, its dissolving power would have been enhanced; it might have received new materials dissolved in juvenile waters, and its decreased density would have caused it to rise eventually to escape as hot brine at the bottom of the Red Sea.

The metalliferous deposits of the Red Sea are characterized by the same metals that are typical of shallow, moderate and deep vein deposits with their sulphides of iron, zinc, copper, lead and accompanying trace metals. However, they are mainly disseminated in a widespread blanket of sediments. This is

exactly what one would expect when metal-bearing solutions discharge from igneous rocks through deep veins into the bottom of a deep body of ocean water. In the absence of confining walls, the solutions would spread throughout the lower levels of water body. As they cool and come in contact with sulphide ions in the brine from the veins or produced by bacterial reduction of sulfate in the original ocean water, the metals would be deposited as a widespread blanket dispersed in the enclosing carbonate and detrital sediments (Degens & Ross 1969). The process of seafloor spreading in the areas of the hot brine basins in the Red Sea and the formation of oceanic crust are continuing up to the present.

Table 5. Main phases of Phanerozoic igneous activity related to tectonic events.

Age		Magmatic phases and rock types		Tectonic events	
Cenozoic	Pleistocene				Styrian
		1.7?	Alkali basalt		
	Pliocene			Vulcanicity associated with the Red Sea opening.	
	Miocene			Red Sea rift.	
		28–15	Basalts		
	Oligocene				Pyrenean
		40–32	Basalts		
	Eocene			Red Sea doming and extension.	
	Paleocene				
		90 ± 20	Ring complexes Minor granites Various volcanics	Large scale strike-slip faulting of Afro-Arabia. Second major episode of alkaline magmatism.	
Mesozoic	Cretaceous				Late Kimmerian
		140 ± 15	Ring complexes	Initial rifting of the South Atlantic/Africa/ South America.	
	Jurassic				
	Triassic				Hyercynian
		230 + 10	Ring complexes Dikes, sheets	Overall compression and subduction of Paleo-Tethys north of Afro-Arabia.	
Paleozoic	Permian				
		290 + 10	Dikes of varied composition		
	Carboniferous				
		350 ± 10	Ring complexes	Initial break-up of Pangea.	
	Devonian				
	Silurian				
		400 + 10	Ring complex Dikes		
	Ordovician				
	Cambrian	550–500	Post-granite dikes Granites Ring complex	Pan-African final phases of tectonism and magmatism.	
Proterozoic		+ + + +		+ + + +	
		+ + + +		+ + + +	

SUMMARY AND CONCLUDING REMARKS

The main phases of Phanerozoic vulcanicity which intermittently occurred in Egypt and which were intimately related to tectonic events are summarized in Table 9/5.

The following salient features are pointed out:

1. Although based on limited data, anorogenic magmatism seems to have been more widespread in the Lower and Upper Paleozoic than had previously been recognized. It is pointed out, in that respect, that the extent or importance of a volcanic phase is not necessarily directly related to the size of age data available.

2. Six main periods of alkaline magmatism in the Eastern Desert of Egypt are defined (554, 404, 351, 229, 145-132, 91-78 Ma). Another period is recognized in the south Western Desert around 45 Ma. Accordingly, the ring complexes in Egypt seem to range in age from Cambrian to Tertiary. Similar results are known from the Sudan.

3. Age data on volcanic rocks formed in connection with the Red Sea rifting assist in fixing the main stages in its tectonic evolution. Actual rifting appears to have begun in the Late Eocene, and was well under way in the Late Oligocene to Early Miocene as shown by basaltic magmatism and by faulting. In the early Miocene the rift was already well outlined.

4. Quaternary vulcanicity of Plio-Pleistocene age is the youngest reported volcanic activity dated in Egypt. The processes of sea-floor spreading in the Red Sea are continuing up to the present (hot brines, oceanic crust).

5. The mineralizing effects of hydrothermal volcanic activities are widespread, especially those of the Tertiary.

ACKNOWLEDGEMENTS

The help of Professor R. Said during the preparation of this work is deeply appreciated. The new K/Ar ages presented here were carried out by the writer at the Federal Geological Survey, Hannover; they during a scholarship from Alexander Von-Humboldt Stiftung, which is gratefully acknowledged. I am also grateful to colleagues with whom I had profitable discussions.

PART 2

Basement complex

CHAPTER 10

The basement complex of the Eastern Desert and Sinai

SAMIR EL GABY*, FRANZ K. LIST** & RESA TEHRANI**

*Assiut University, Egypt & **Free University of Berlin*

1 INTRODUCTION

Precambrian basement covers about 100,000 km^2 in south Sinai and the Eastern Desert. Geological investigations, mostly concentrated on the central Eastern Desert, were conducted by a large number of workers (for a review see El Ramly & Akaad 1960 and Akaad & El Ramly 1960). The work of El Ramly & Akaad 1960 and Akaad & Noweir 1980 formed the basis for subsequent classifications of the Precambrian rocks of Egypt (El Shazly 1964, El Ramly 1972) which adopted the geosynclinal concept and the ensuing relationship between geotectonic and geomagmatic cycles.

The central Eastern Desert can be described as largely occupied by an intricate association of metasediments, metavolcanics, metagabbros, and serpentinites, constituting the 'ophiolitic mélange' of Shackleton et al. (1980) that is interrupted by gneisses in structural highs. This sequence is unconformably overlain by practically unmetamorphosed, intermediate to silicic volcanics (Dokhan volcanics) and molasse-facies clastic sediments (Hammamat clastics). The whole pile is intruded by a vast array of granite intrusions ranging in composition from quartz diorite to alkali-feldspar granite. The ophiolitic mélange (in a descriptive, not in a genetic sense) is highly reduced in the north Eastern Desert. It is entirely lacking in south Sinai where the basement is largely composed of granites together with lesser outcrops of high-grade gneiss, Dokhan volcanics, and Hammamat clastics or their equivalents.

The south Eastern Desert is also geologically different from the central Eastern Desert. The metamorphosed sediments in the greater Wadi Allaqi region are essentially composed of terrigenous clastic sediments, intercalated in the lower part by limestone layers that form a marked contrast to the predominantly volcanogenic sediments of the central Eastern Desert; a feature already recognized by Hunting Geology & Geophysics Ltd (1967). Moreover, the area is characterized by widespread low-pressure meta-

morphism obviously connected with the large granitic intrusions.

The basement of the Eastern Desert and Sinai constitutes part of the Arabian-Nubian shield that has been cratonized around the end of the Precambrian. A general consensus prevails among basement geologists as to the late Proterozoic volcano-sedimentary rock associations, regardless of whether they represent a eugeosynclinal association or an ophiolitic mélange, as well as to the younger rocks. A debate goes on about the participation of older rocks in the basement of the Eastern Desert and Sinai. From the structural point of view, late Proterozoic ophiolites and volcano-sedimentary rocks, metamorphosed in the greenschist facies, occur as imbricated thrust sheets (suprastructure) thrust over the early Proterozoic continental crust (infrastructure). The incorporated infracrustal rocks were subjected to mylonitization and diaphthoresis or to remobilization depending on the temperature and depth prevailing at the respective sites.

In the following, an overview of the basement geology of the Eastern Desert and Sinai is given. Since the metamorphic and tectonic evolution of this area and the metallogenic implications have been treated in some detail in a recent paper by the authors, the reader interested in these aspects is referred to that publication (El Gaby, List & Tehrani 1988). A detailed description of the major rock units as shown on the new Geological Map of Egypt at 1 : 500 000 (see chapter 3) is given in the Explanatory Notes to the map (Hermina, Klitzsch & List 1988).

2 OVERVIEW OF THE BASEMENT GEOLOGY

The basements rocks of the Eastern Desert and Sinai are grouped into:
- Pre-Pan-African rocks, comprising higher-grade metamorphic rocks, and a
- Pan-African rock assemblage, comprising ophiolites and the island arc association, a tectogenetic

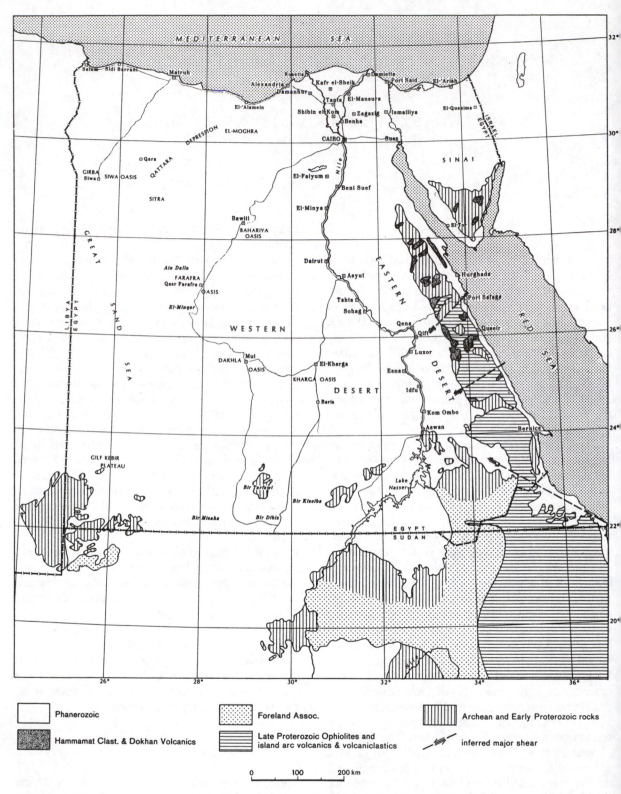

Figure 10.1 Distribution of Precambrian rocks in Egypt and northern Sudan. Rocks of Archean age are confirmed in the Western Desert only.

rock association, and
– Phanerozoic alkaline rocks.

2.1 *Pre-Pan-African rocks (infrastructure)*

The presence of pre-Pan-African rocks is a point of contention among geologists studying the Eastern Desert of Egypt. In the context of this description of the basement, it is assumed that pre-Pan-African rocks do exist in the form of the gneiss domes of the Eastern Desert. They are thought to represent a lower structural level that, although reworked by the Pan-African orogeny, is part of the older East Sahara craton exposed farther to the west in the Gebel Uweinat area.

Medium- to high-grade schist, gneiss, and granite gneiss, together with their mylonitized products, cover mappable areas:

1. along the northwestern coast of the Gulf of Aqaba (Fjord gneiss),
2. at Wadi Feiran, southwestern Sinai,
3. along the Qena-Safaga road (Barud gneiss),
4. at Gebel Meatiq,
5. at Gebel El Sibai,
6. at Gebel El Shalul,
7. at Hafafit area,
8. at Wadi Khuda and on St John's Island,
9. at Aswan, and
10. at several scattered localities to the east and southeast of Aswan (Fig. 10/1). The gneiss occurrences of Wadi Feiran, Gebel Meatiq, and Migif-Hafafit area are well researched and described (e.g. Akaad & Shazly 1972, El Ramly et al. 1983).

The rocks of the old continental crust in the Eastern Desert and Sinai were incorporated in the Pan-African orogeny. The effects of the orogeny varied according to the prevailing temperatures. At low temperature and shallow levels, the rocks of the old continental crust were subjected to plastic deformation leading to the formation of quartzo-feldspathic mylonites and blastomylonites – the 'acid gneiss' of Hume (1934). This situation can be observed at the Meatiq, Sibai and El Shalul areas where mylonites and blastomylonites enclosing relics of the original rocks constitute a carapace surrounding a less deformed kernel of gneissic granite.

At higher temperature and deeper levels, the older rocks were remobilized and mostly transformed into migmatite and granitic gneiss, commonly associated with autochthonous to parauthochthonous synorogenic granitoids of tonalitic to monzogranitic composition (G1 of Hussein et al. 1982). This situation is well displayed at Wadi Feiran in Sinai, along the Qena-Safaga road, and around Aswan in the Eastern Desert.

The spacious Hafafit culmination offers a rather complete cross-section of the infrastructural rocks. On the eastern flank of this domal structure, mica schist is diaphthorized along many small, listric thrust faults underneath a serpentinite-metagabbro thrust sheet; variably deformed gneissic granite occurs at Wadi Sikait and at the upper reaches of Wadi Nugrus within the 'psammitic gneiss' of Akaad & El Ramly (1960). Along the western bank of Wadi Nugrus, fine-grained hornblende gneiss grades westward, that is downward, into medium- to coarse-grained banded and homogenized hornblende gneiss which is intruded by parautochthonous granitoids occupying the core of this swell. On the western side of the culmination, the Wadi Shait granite is deformed and mylonitized along several major shear zones trending nearly northwest-southeast (El Gaby & El Aref 1977).

The identification and mapping of the pre-Pan-African rocks was based on structural and petrographic considerations. The infrastructural rocks comprise, as stated above.
– medium to high-grade metamorphic rocks and granites that have been deformed and diaphthorized, and
– other rocks that have been migmatized during the Pan-African orogeny.

We are confident about the identification of the pre-Pan-African rocks particularly when deformation and diaphthoresis took place below a Pan-African ophiolite/volcano-sedimentary thrust sheet. This is the general situation in the central Eastern Desert and for the Wadi Bitan gneiss in the south Eastern Desert where the contact between the rocks of the infrastructure and the overlying ophiolites is always a thrust fault.

In the case of remobilized rocks, there is always a possibility that Pan-African migmatization might have affected late Proterozoic rocks as well. However, we have found that migmatization did affect neither the overlying ophiolitic serpentinite-metagabbro thrust sheet along the eastern flank of the Hafafit culmination nor the volcano-sedimentary belt with banded iron ore layers at Gebel Um Anab to the north of the Qena-Safaga road. Nevertheless, identification of infrastructural rocks on the basis of the degree of metamorphism alone might be misleading.

Low-pressure metamorphism at the Wadi Kid area, southeast Sinai (Shimron & Zwart 1970; Reymer et al. 1984), and at Wadi Um Had in the central Eastern Desert resulted in the development of biotite, garnet, staurolite and cordierite mineralzones. These metamorphosed rocks grade through a relatively narrow anatectic zone into granite gneiss. Granite gneiss forms part of an elongated diapiric body at Wadi Kid (Reymer et al. 1984) and at Wadi Um Had. In both occurrences, thermal metamorphism affected primarily Hammamat clastics, and the anatectic rocks developed contain the critical assemblage cordierite ± sillimanite and/or andalusite ± K-feldspar ± garnet ± biotite.

In the large and composite East Aswan swell, where predominantly terrigenous sediments attain great thickness, granite intrusions pertaining to g_α are associated with wide migmatized gneiss zones and high- to medium-grade schists. Kyanite has been described from the thermal aureole at Gebel Nasb Aloba (Hussein & Rasmy 1976), and coexisting kyanite-sillimanite-andalusite from the thermal aureole at Wadi Dif. Wollastonite, indicative of the pyroxene-hornfels facies, is recorded from the marble of Abu Swayel. The high-grade gneiss and schist associated with and surrounding these granite intrusions are shown as infrastructural gneiss and schist on the new geologic map although there is a high possibility that they may include migmatized late Proterozoic rocks of the foreland association.

2.2 *Pan-African rock assemblage*

The Pan-African thermo-tectonic event (Kennedy 1964) that took place toward the end of the Protero-zoic, between 550 and 650 Ma ago (Clifford 1970), strongly affected the old cratonic margin. Ophiolites and one or several island arcs were accreted to and obducted over the continental mass and mixed with foreland-derived sedimentary, pyroclastic, and volca-nic rocks. Prolific intrusions of a granite series (Read 1955) with corresponding thermal metamorphism took place, accompanied by calc-alkali volcanism and the deposition of molasse-type sediments. This Pan-African rock assemblage dominates the base-ment of the Eastern Desert and Sinai.

2.2.1 *Ophiolites and island-arc association*

Ophiolites and rocks pertaining to an island arc asso-ciation constitute a highly deformed sequence of low-grade, regionally metamorphosed serpentinites, gabbros, volcanics and volcaniclastics. These rocks cover large areas in the central Eastern Desert and are preserved in synfolds between gneiss swells in the south Eastern Desert. They occupy narrow stretches in the north Eastern Desert but are entirely lacking in Sinai. In part they are equivalent to the 'eugeo-synclinal filling' of Akaad & Noweir (1969), the 'ophiolitic mélange' of Shackleton et al. (1980), and the 'greenschist assembly' of Vail (1983). The whole succession is dissected by many listric thrust faults causing repetitions and tectonic mixing. Neverthe-less, it can be differentiated into a lower ophiolite sequence and an upper island arc association.

Ophiolites. In general the basic to ultrabasic rocks of the Arabian-Nubian shield are currently interpreted within the framework of plate tectonics as remnants of obducted oceanic crust. The major occurrences of ophiolites in Saudi Arabia are situated at Jabal al Wask (Bakor et al. 1976), the Amar-Idsas region (Al Shanti & Mitchell 1976, Al Shanti & Gass 1983), the Bir Umq complex (Frisch & Al Shanti 1977), and at Jabal Ess. In Sudan, ophiolitic sequences of late Proterozoic age are reported from Sol Hamed (Fitches et al. 1983) and from the Wadi Onib area (Hussein et al. 1984).

In the present work, the authors support the view that those rocks that were previously classified as serpentinite, metagabbro, metavolcanics, and meta-sediments are, at least in part, components of an ophiolitic mélange in a purely descriptive sense. For-merly, the mafic and ultramafic rocks of Egypt were interpreted as geosynclinal submarine flows (Ritt-mann 1958) and as intrusive bodies emplaced into eugeosynclinal metasediments and metavolcanics (El Ramly & Akaad 1960, El Shazly 1964, Sabet 1961, 1972, El Ramly 1972, Akaad 1972, Akaad & Noweir 1980). An interpretation of the serpentinites and related rocks as ophiolitic suites is first given by Garson & Shalaby (1976).

Several ophiolite suites are reported from the East-ern Desert, e.g. from the Wadi Zeidun area (Dixon 1979), the Qift-Quseir road (Nasseef et al. 1980, Stern 1981), from the Idfu-Marsa Alam area (Shack-leton et al. 1980, Ries et al. 1983), the Gebel Mohagara-Ghadir area (Takla et al. 1982), the El Rubshi area (Khudeir 1983), and from Wadi Bitan, Wadi Rahaba and Wadi Naam (Ashmawy 1987). Ophiolites always occur as allochthonous and com-monly incoherent basic to ultrabasic bodies. The most complete ophiolite sequence is described from Wadi Ghadir (El Sharkawy & El Bayoumi 1979).

The ideal section is composed as follows, starting from the top:

5. Thinly bedded metasediments of deep facies, 100 to 200 m thick at Wadi Ghadir, are generally quite rare. They are essentially formed of pelites interca-lated by thin chert and calcareous bands.

4. Low-potash tholeiitic basalt sheets, several hundred meters thick, are sometimes pillowed, and the pillows are always right-up (Stern 1981). Alkali basalts are rare.

3. Massive diabase and sheeted-dyke complexes, 100 to 200 m thick, also of tholeiitic composition are common.

2. Metagabbro of tholeiitic composition is formed at the base of cumulate pyroxenite and gabbro, fol-lowed upward by isotropic gabbro and hornblende gabbro, frequently enclosing small dioritic bodies or being cut by appinites. Primary hornblende, when present, is of olive-brown color. Olivine,

fresh or altered, has never been recorded, so cumulate dunites cannot have developed. The contact between the metagabbro and the underlying serpentinite is always occupied by a thrust fault. Mesoscopic layering reported by El Sharkawy & El Bayoumi (1979) and Kröner (1985) is believed to be of secondary origin; it is observed in the sheared lower parts along sole-thrust planes. The thickness of the gabbro layer is about 1 km, at Fawakhir, and 2 km at Gebel El Rubshi.

1. Serpentinites, essentially after harzburgite and to a lesser extent after dunite and lherzolite, are frequently transformed into talc-carbonates, particularly along thrust faults and shear zones. They sometimes enclose boudinaged chromite lenses (El Sharkawy & El Bayoumi 1979; Khudeir 1983) and may contain enstatite, diopside, and sometimes olivine relics. They are characterized by almost constant, high Mg/Fe ratios suggestive of mantle origin.

The ophiolites are commonly dismembered and even the large serpentinite masses are frequently intercalated by thin, highly foliated pelitic layers (Shackleton et al. 1980, Khudeir 1983) marking minor thrust faults. The ophiolites are believed to have developed in back-arc basins (Stern 1981) and later obducted from the east over the old continental margin (El Gaby et al. 1984, Kröner 1985).

The proponents of an accreted island arc(s) evolutionary model for the Arabian-Nubian shield consider the ophiolites as pieces of oceanic crust obducted along destructive plate boundaries during the closure of small ocean basins. The ophiolites thus delineate, at least tentatively, tectonic sutures or arc margins (e.g. Bakor et al. 1976, Gass 1977, 1979, Shackleton 1980). Church (1976) raised doubt about this assumption, stating that the ophiolites in the Eastern Desert occupy a certain stratigraphic horizon, and that their distribution on the geologic map of Egypt is due to the presence of first-order, northwest-trending anticline structures which were eroded sufficiently deep to expose stratigraphic levels beneath the ultramafic-bearing units of the geosynclinal sequence.

The ophiolites are commonly located along major thrust faults, and particularly along the decollement surface between the obducted Pan-African ophiolites and island arc rocks (= suprastructure) and the underlying old continental margin (= infrastructure; El Gaby 1983, El Gaby et al, 1984). This would explain why ophiolite outcrops are concentrated around gneiss swells in the central and south Eastern Desert. The distribution of ophiolites might also help in unraveling the large-scale tectonics of the Eastern Desert. For example, in the Qift-Quseir area, two major thrust sheets can be recognized: a lower thrust sheet verging to the west along Wadi Atalla, and an upper thrust sheet verging to the west along Gebel El Rubshi and Gebel Um Seleimat, to the north and south of the Meatiq dome.

Island-arc association. The ophiolites are overlain and tectonically mixed with another series of weakly metamorphosed, calc-alkaline intermediate to acid volcanics essentially formed of andesite, dacite, and other volcaniclastics of comparable composition. Basalt, rhyolite and rhyodacite are subordinate. The tuff and volcanogenic graywacke is often banded, with graded bedding indicating submarine deposition. These rocks are frequently intercalated with metamorphosed iron-ore layers, particularly in the northern part of the central Eastern Desert. Farther to the south, for example at Um Samiuki, they are locally associated with stratabound base-metal sulphide deposits (Garson & Mitchell 1981, El Aref et al. 1985).

On previous maps, the volcaniclastics were generally mapped as metasediments. This volcanic association is referred to as 'younger metavolcanics' (YMV) to differentiate it from the 'older metavolcanics' (OMV) of the ophiolite association (Stern 1979, 1981). The volcanic rocks around Sheikh Shadli (Shukri & Mansour 1980), the type-locality for the 'geosynclinal metavolcanics' (El Ramly 1972), are formed of island-arc volcanics and volcaniclastics, which means that they belong to the YMV. Due to severe tectonics and tectonic mixing, it was not always possible to separate the island-arc association (YMV) from the ophiolitic metavolcanics (OMV) on the new geologic map. Nevertheless, they constitute the greater part of the volcano-sedimentary cover.

In the north Eastern Desert, the YMV are locally preserved in narrow belts bounded by faults within the major Qena-Safaga shear zone. They display the common lithologic features of the island arc association and are intercalated at Gebel Um Anab by the characteristic iron ore bands (Sabet et al. 1972).

In Sinai, the YMV appear to be entirely absent. The volcanics of the Wadi Kid and Wadi Sa'al areas, southeast Sinai, are metaphorphosed and locally deformed and sheared. Petrologically, they range from andesite to rhyolite, with the acid members prevailing; ignimbrite is quite abundant. These are the characteristic petrologic features of the calc-alkaline, Andean-type Dokhan volcanics. Moreover, the associated sediments at Wadi Sa'al are of molasse facies (Shimron 1984) akin to that of the Hammamat clastics. Bentor (1985) reported that the sediments at Wadi Sa'al are largely clastic and consist mainly of thick units of conglomerates composed of boulders and pebbles of dioritic and granitic gneiss. The presence of gneiss pebbles speaks for deposition in intracontinental basins (i.e. molasse-type Hammamat clastics) rather than in an island-arc environment.

The reported isochron age of 734 ± 17 Ma (Bielski 1982) from the metamorphosed Sa'al volcanics must be considered cautiously, particularly if one takes into account that granite-gneiss boulders from the Sa'al conglomerates yielded a Rb-Sr isochron age of 641 ± 25 Ma (Bielski 1982). Similarly, abnormally old Rb/Sr apparent ages are sometimes obtained from Dokhan volcanics (Stern & Hedge 1985).

Plutonic rocks belonging to the island-arc association are least recognized and studied in the Egyptian basement. They might include subduction-related diorite and tonalite, the plutonic equivalents of island arc andesite and dacite, as well as mantle-derived gabbro. The sheared tonalite of Gebel El-Mayit (Akaad & El Ramly 1960) which is conformably incorporated into the deformed ophiolite/island-arc mélange belongs most probably to island arc granites (s.l.). It was difficult to separate island-arc metagabbro from ophiolitic metagabbro which are petrographically almost identical. Therefore intensive field work is required for this distinction since island-arc metagabbros are intrusive and are composed to a considerable extent of metadiorite. Metagabbro to metadiorite occurs in relatively small bodies intruded into the YMV in the Sheikh Shadli region. Metagabbro/metadiorite complexes occupy wide areas between Marsa Alam and Wadi Mubarak (Hassan & Essawy 1977); two metagabbro/metadiorite intrusions in this area can be recognized on the Landsat-MSS imagery. These plutonic rocks are expected to be intimately associated with and intrusive into the YMV, and they are likely to be metamorphosed and deformed.

2.2.2 *Tectogenetic association*

The tectogenetic phase is characterized by the prevalence of granite among plutonic rocks, and rhyodacite and rhyolite among volcanic rocks, a feature that heralds the participation of sialic crust in magma generation or modification (Wyllie 1983a, b). Moreover, the volcanic activity is mainly subaerial and ignimbrites are quite abundant. The associated sediments are essentially coarse terrigenous clastics with abundant conglomerates, generally deposited in non-marine fluvial and fanglomerate environments, i.e. molasse-facies sediments. The lithology of magmatic and sedimentary rocks corroborates the end of the oceanic island-arc phase and the passage into the continental-margin Cordilleran stage. The tectogenetic associations comprise penecontemporaneous:

– subduction-related, Andean-type, calc-alkaline, intermediate to acidic igneous rocks,
– molasse-type Hammamat sediments, and
– intrusive, mantle-derived, mostly fresh spinel lherzolite, gabbro, troctolite, and meladiorite.

Calc-alkaline magmatic activity. In the period around 600 ± 50 Ma BP, the Nubian shield was the site of a pronounced calc-alkaline magmatic activity in which the silicic members prevailed. This magmatic activity is manifested in the emplacement of syn- to late-orogenic calc-alkaline granites of tonalitic to ideal-granitic composition, and in the eruption of their volcanic and subvolcanic analogues, the Dokhan volcanics and post-Hammamat felsites. Granites cover about 40% of the surface area of the basement (Hussein et al. 1982); they are more abundant in the deeply eroded north Eastern Desert and Sinai.

i. Calc-alkaline granites: Hume (1935) and Schürmann (1953) considered the Egyptian granites (apart from the deformed Shaitian granite) as of Gattarian (= late Precambrian) age, and as younger than the Dokhan volcanics and Hammamat clastics. El Ramly & Akaad (1960) divided the Egyptian granites into 'older granites' of tonalitic to granodioritic composition with unfailingly gray color, and into 'younger granites' with pink and red colors and of granitic to alaskitic composition; they are stratigraphically separated by the Dohkan volcanics and Hammamat clastics. These two granite groups are synonymous to El Shazly's (1964) 'synorogenic' and 'late orogenic' plutonites, to which he added a younger, 'post-orogenic granite' group for the Aswan porphyritic granite, probably influenced by the young K-Ar age of 470 Ma (El Ramly 1963) available at that time. El Ramly (1972) introduced the post-orogenic alkaline granites to incorporate riebeckite-bearing granites.

El Gaby (1975) postulated that the two Egyptian granite groups constitute one continuous granite series in which the granitic rocks become with time more silicic and richer in potash. He noticed that hypersolvus alkaline to peralkaline granites developed as a side-branch during the late phases of the 'younger granite' probably under subvolcanic conditions and through gas transfer. El Gaby & Ahmed (unpubl.), by making use of dike swarms to determine the relative age of granitic intrusions in the Wadi Feiran area, southwest Sinai, found that highly differentiated calc-alkaline granite intrusions were later followed by less differentiated quartz syenite, and then by the peralkaline ring intrusion of Gebel Serbal. They suggested, as adopted in later publications (e.g. El Gaby & Habib 1982) to classify the Egyptian granites into

– an older, syn- to late-orogenic calc-alkaline granite series comprising the older 'grey granites', the porphyritic granite of Aswan, and the younger, calc-alkaline two-feldspar granites, and
– a younger, post-tectonic alkaline to peralkaline granite series comprising quartz syenite, alaskite, and aegirine- or riebeckite-bearing leucocractic granites.

Greenberg (1981), recognised three possible granite clans, among the 'younger granites' which he believed to be the products of a single magmatic episode 603 to 575 Ma ago which marked the end of a 'cratonizing process' and the beginning of truly anorogenic activity.

Hussein et al. (1982) subdivided the Egyptian granites into:

– G1-granite, comprising the formerly so-called 'older', 'Shaitian', 'grey', or 'synorogenic' granites, and the Aswan porphyritic granite. G1 is a subduction-related, I-type granite formed above old Benioff zones.

– G2-granite, comprising most of the 'younger', 'pink and red' or 'post-orogenic' granites. G2 tends to be more of the S-type and is believed to have formed as a result of suturing.

– G3-granite which comprises alkaline to peralkaline granites as well as a considerable part of the 'younger granites' of El Ramly & Akaad (1960). G3 is a typical intraplate anorogenic A-type granite.

In a statistical study on Egyptian granites, El Shatoury et al. (1984) concluded that they can be subdivided into 'older or synorogenic granitoids' (Gr.A), 'younger or post-orogenic granitoids' (Gr.B$_1$), and the youngest 'alkali granites' (Gr.B$_2$); a sharp boundary exists between Gr.A on the one hand, and Gr.B$_1$ and Gr.B$_2$ on the other hand. They stated that Gr.A corresponds to the G1- and G2-granites of Hussein et al. (1982) which were formed in a compressional tectonic environment. Gr.A-granites or G1- and G2-granites correspond to the calc-alkaline granite series mentioned before.

Granites are subdivided on the new geologic map into calc-alkaline tonalite to granodiorite (g$_\alpha$), calc-alkaline two-feldspar ideal granite (g$_\beta$), and anorogenic subalkaline to peralkaline granite (g$_\tau$). Separation of the g$_\alpha$- and g$_\beta$-granites is arbitrary since transitions do occur. It is based on differences in mineralogical composition which is reflected by color and susceptibility to weathering and erosion (i.e. relief). The tonalitic and granodioritic members, representing the earlier phases of the calc-alkaline granite series, frequently grade into granite gneiss in exposed deeper crustal levels, whereas the two-feldspar ideal granites are always intrusive. The granitic members that are transitional between g$_\alpha$- and g$_\beta$-granites, which are characterized by monzogranitic composition and porphyritic texture, are grouped on the map with g$_\alpha$ when grading into granite gneiss (e.g. Aswan granite), and with g$_\beta$ if displaying clear intrusive relations.

The g$_\alpha$-granite occurs in the form of autochthonous and parautochthonous intrusive bodies elongated parallel to the regional setting of the enclosing country rocks. It constitutes the largest plutonic masses; the intrusive bodies commonly possess marginal zones with well-developed planar foliation. The early members of this group are quartz-dioritic in composition, rich in mafic minerals and strongly foliated, while the younger members are increasingly more acidic, lighter in color and commonly structureless. Enclosed mafic endogenic xenoliths decrease progressively in size and abundance in the later phases.

At deeper crustal levels, as displayed along the Qena-Safaga road (Sabet et al. 1973, Akaad et al. 1973), in the core of the Hafafit swell (El Ramly et al. 1984) and around Aswan (Gindy 1954), g$_\alpha$-granite commonly grades into homogenized granite gneiss and migmatite. It sometimes encloses large skialiths of fine-grained gneiss and amphibolite suggesting in-situ development by granitization processes. However, a low initial $^{87}Sr/^{86}Sr$ ratio reported by Stern & Hedge (1985) from the autochthonous Mons Claudianus granodiorite mass, north Eastern Desert, and Pan-African zircon ages with normal lead-isotope ratios from the Hafafit area (Gillespie & Dixon 1983) suggest only limited contamination with older crustal rocks.

The autochthonous granite (g$_\alpha$) along the Qena-Safaga road is locally enriched in K-feldspar metacrysts and thus grading into porphyritic monzogranite. Development of K-feldspar metacrysts is much more pronounced at Aswan and farther to the southeast. The porphyritic granite of Aswan is normally of monzogranitic composition; sometimes it is enriched in K-feldspar metacrysts (as well as in plagioclase, hornblende and biotite with enclosed zircon) and deprived of intergranular granitic pore liquids so that it approaches a quartz-syenite. In the Wadi Allaqi region, south Eastern Desert, orogenic stress obviously lasted longer so that g$_\alpha$-granodiorite and porphyritic monzogranite intrusions were partly deformed and locally mylonitized.

The g$_\beta$-granite bodies are always intrusive and commonly occur in the form of smaller intrusions that are also conformable with the surrounding regional setting. Marginal foliation is normally absent and the mafic endogenic xenoliths are small and few in number. Hornblende is typically absent. The Um Had pluton intruded into the Hammamat clastics, and accordingly, the most typical 'younger granite' according to El Ramly & Akaad (1960) possesses a granodioritic margin (Noweir 1968) which contains hornblende beside biotite. This means that the separation into older and younger granite groups is non-genetic and can be only applied in a descriptive manner as adopted in the present work.

ii. Dokhan volcanics: Dokhan volcanics are the surface and near-surface manifestations of the plutonic calc-alkaline magmatic activity comprising g_α and g_β. The discrimination between Dokhan volcanics and older island-arc metavolcanics on the basis of metamorphism, as adopted by El Ramly & Akaad (1960), is inadequate. It has caused serious misidentifications, for example in the cases of the Kid and Sa'al 'metavolcanics' (Shimron 1980, Bentor 1985), and the 'younger metavolcanics' at Wadi Massar and Wadi Zeidoun (Stern 1979). Bentor (1985) noticed in the Sa'al volcanics, however, that 'the high proportion of acid volcanics in such an early stage of crustal evolution is remarkable'.

The Dokhan volcanics are predominantly subaerial, distinctly silicic and intimately associated and intercalated with molasse-type Hammamat sediments. The rock types present are mainly rhyolite and rhyodacite with subordinate andesitic flows; extensive welded tuff units are particularly conspicuous. The common intercalation with molasse Hammamat sediments and the ubiquitous presence of ignimbrites and other features of subaerial volcanism indicate the dominantly terrestrial setting of this volcanic activity (Gass 1982). Geochemical investigations indicate that the Dokhan volcanics are calc-alkaline in composition and of Andean-type. They erupted along an active continental margin (Basta et al. 1980, Gass 1982, Furnes et al. 1985). However, Stern et al. (1984) believe that the common bimodality of the Dokhan volcanics expressed by andesite (56 to 62% SiO_2) in association with rhyolitic ignimbrite and related felsic rocks (72 to 76% SiO_2) indicates that the crust was actively extending and undergoing rifting.

Available radiometric ages from the Dokhan volcanics range between 639 and 581 Ma (Dixon 1979, Stern & Hedge 1985, Bielski 1982); they are highly comparable to the coeval calc-alkaline g_α- and g_τ-granites. It is worth mentioning that the metamorphosed, particularly andesitic Dokhan volcanics (recognized on the basis of their association with molasse sediments and/or with abundant silicic volcanics) commonly give older Rb-Sr ages. Abu Swayel rhyodacites gave 655 to 654 Ma (El Shazly et al. 1973) and 768 Ma (Stern & Hedge 1985), while the Nuqarah andesites gave 686 Ma and volcanics from Wadi Massar-Wadi Arak gave 632 or 690 Ma (Stern & Hedge 1985). From the metamorphosed volcanics of Wadi Sa'al, a Rb-Sr age of 734 Ma was obtained, while granite-gneiss boulders from the associated Sa'al conglomerates (= Hammamat conglomerates) gave only 641 Ma (Bielski 1982). The weakly metamorphosed volcanics from Wadi El Mehdaf, unfortunately one of the type localities of the YMV according to Stern (1981), gave 622 Ma (Stern & Hedge 1985) and therefore belong to the Dokhan volcanics; they also include proper rhyolites containing up to 2.84% K_2O (Stern 1981) which are common in the Dokhan volcanics.

Molasse-type Hammamat clastics. The Hammamat clastics are largely composed of poorly sorted clastic sediments intercalated with minor impure calcareous layers. Akaad & Noweir (1980) described the Hammamat sediments as formed at the base of an alternation of predominantly brick-red and green siltstones, the Igla formation, which grade upward into subgraywackes and volcanic arenites frequently enclosing thick conglomerate banks. The Igla formation was deposited in freshwater basins (Samuel 1977). Grothaus et al. (1979) proposed several models for the deposition of the Hammamat clastics in intermontane, alluvial and freshwater basins, whereby variation in grain-size of the clastic sediments is attributed to lateral facies differentiation and not to tectonic events.

During the Cordilleran stage, the basement was thrown into a series of open anticlinal folds and swells disposed along two geanticlines trending north/northwest-south/southeast. These geanticlines also acted as Andean-type magmatic arcs along which calc-alkaline magmas erupted. Erosional products of the Dokhan volcanics extruded along the raised magmatic arcs accumulated as molasse-type Hammamat sediments in intermontane and foreland basins. Consequently, Dokhan volcanics and Hammamat clastics are characterized by their intimate temporal and spatial association.

Hammamat clastics are best developed in the northern part of the central Eastern Desert but occupy smaller and smaller areas farther south due to the northward slope of the basement. They attain substantial development in the north Eastern Desert and Sinai, suggesting that the upheaval of the northern block along the Qena-Safaga line took place during the Cordilleran stage while eruption of the Dokhan volcanics and deposition of the Hammamat sediments were still in progress (El Gaby 1983). The model proposed by Stern et al. (1984) for the north Eastern Desert suggesting deposition of the Hammamat sediments in downfaulted grabens trending nearly northeast-southwest cannot be substantiated. We noticed that the Hammamat sediments occur on the dip-slopes of a series of faulted blocks bounded from the south by reverse (?) faults that obviously were rejuvenated during the Tertiary. In the south Eastern Desert, known occurrences of Dokhan volcanics and Hammamat clastics are very much reduced on the Geological Map of Egypt (El Ramly 1972); this point will be further treated under the foreland association (vide infra).

Younger ultrabasic to basic intrusions. A suite of

intrusive ultrabasic and basic rocks comprising spinel lherzolite, clinopyroxenite, troctolite, olivine gabbro and meladiorite occurs as small, frequently layered intrusions and sills. They were intruded coevally with the Andean-type calc-alkaline magmatic rocks. A layered gabbro is intruded by granodiorite near the mouth of Wadi Mubarak, central Eastern Desert (Kabesh et al. 1967), while the granodiorite intrusion of Wadi El Sheikh, southwest Sinai, is dissected by a meladiorite dike-like body which is in turn truncated by the more differentiated Ma'in granite. The younger ultrabasic and basic rocks are commonly unmetamorphosed, but at Abu Swayel they have clearly been subjected to the low-pressure metamorphism that affected the Wadi Allaqi region, together with the enclosing country rock.

The younger peridotite and gabbro differ from their ophiolitic and island-arc counterparts in the following aspects: fresh olivines are commonly present with marked variations in their Fe-content, indicating crystallization from a differentiating gabbroic magma. In addition, clinopyroxenes, Cr and/or Mg spinels and sometimes Cu-Ni sulphides are typically present in the peridotites. In this context we believe, until conclusive radiometric dates become available, that the peridotite of St John's (Zabargad) island belongs to this suite though Bonatti et al. (1981, 1983) are of the opinion that the peridotite represents an uplifted fragment of sub-Red Sea lithosphere (upper mantle) connected with the early stages of the development of the rift. This belief is based on the following facts:
– Bulk composition and mineral paragenesis of the Zabargad peridotite are very similar to that of spinel lherzolites described from Gebel Dahanib (Dixon 1981a) and from Gebel El Motaghayerat (Abu El Ela 1984) from the northern part of the Khuda swell, south Eastern Desert. Spinel lherzolites occur also in southeast Sinai (Shimron 1984) and at Wadi Allaqi.
– Cu-Ni sulphide mineralization associated with the Zabargad peridotite occurs also in some ultrabasic bodies in the south Eastern Desert, for example at Abu Swayel (El Goresy 1964), at Gabbro-Akarem (Bugrov & Shalaby 1973), and at El Geneina West (Kamel et al. 1980).
– The old metamorphic country rocks of Zabargad Island are composed of coarsely banded hornblende-plagioclase gneiss and amphibolite similar to the hornblende gneiss exposures widespread in the Eastern Desert (Bonatti et al. 1981, 1983) and particularly to the nearby Khuda gneiss. This signifies that the island is a detached block of the basement of the Eastern Desert, as also acknowledged by Bonatti et al. (1981, 1983), This seems more probable than to assume a vertical upheaval in the order of 30 km, particularly if we take into consideration that the

peridotite is in tectonic contact with upper Cretaceous sediments (Bonatti et al. 1981, 1983) and is unconformably overlain by basal Miocene sediments (E. Philobbos, oral comm.).

Foreland association. Hunting Geology & Geophysics Ltd (1967) noticed that the metasediments of the greater Wadi Allaqi region differ markedly from the geosynclinal metasediments described by El Ramly & Akaad (1960) from the central Eastern Desert and therefore assumed the presence of another geosynclinal basin. The sedimentary rocks are characterized by the frequent occurrence of limestone (metamorphosed into marble) in the lower part of the succession and by poorly sorted arenite and siltstone in the upper part; intercalations of fine-grained conglomerates are also not infrequent in the upper part. The facies of the clastic sediments is very similar to those of the molasse Hammamat sediments. Rhyodacite and rhyolite, though commonly metamorphosed, are quite abundant among the associated volcanics so that they should be considered pertaining to the Dokhan volcanics, as already claimed by El Shazly et al. (1973) and El Ramly (1972).

At least the lower part of the sedimentary succession of the greater Wadi Allaqi region is equivalent to the 'metasedimentary unit' of Vail (1983), deposited as a sedimentary prism on a rifted or attentuated pre-Pan-African continental margin. During the Cordilleran stage, the greater Wadi Allaqi region was transformed into a foreland basin in which a thick pile of sediments accumulated above the shelf sediments containing limestone layers. The foreland basin extends farther south across the border into north Sudan (Fig. 10/1).

Available isotopic ages from the Wadi Allaqi region are questionable or inconclusive. El Shazly et al. (1973) reported a Rb-Sr isochron age of 1195 Ma from a garnetiferous mica schist of the Abu Swayel area. However, schists in general and metamorphosed clastic sediments in particular are generally difficult to date (Jäger 1977, Hashad 1980). The model Nd age of about 1500 Ma from the metamorphosed sediments at Wadi Allaqi (Harris et al. 1984) could simply signify mixed clastic sediments from early and late Proterozoic sources. El Shazly et al. (1973) reported Rb-Sr ages of 890 to 980 Ma from intrusive quartz diorites that are based, however, on only two whole-rock data points, or even on one point, and an assumed initial ratio. Rhyodacite pertaining to Dokhan volcanics from Abu Swayel and Bir Haimur gave Rb-Sr ages of 654 and 665 Ma, respectively (El Shazly et al. 1973). Stern & Hedge (1985) reported a Rb-Sr age of 765 Ma from rhyodacite of the Abu Swayel area; they noticed, however, that weakly metamorphosed andesites frequently produced errorchrons with older apparent ages.

2.3 *Phanerozoic alkaline rocks*

About the end of the Pan-African orogeny and till the Tertiary, the basement was intermittently intruded by a number of subalkaline to peralkaline S- and A-type granite bodies, mostly as shallow cauldron complexes or as small discordant intrusions having almost circular outlines; dike-like intrusions, chimneys or volcanic necks are not uncommon. These post-kinematic silicic rocks, generally with an alkaline tendency, are lumped together on the new map as g_τ-granites that correspond to the 'younger granite complexes' of Almond (1979) and to the 'Katherina Province' of Bentor (1985). Available radiometric ages show an overlap in the time of intrusion between g_β- and g_τ-granites. S-type granites are commonly intruded in the back-arc, concurrently with the emplacement of calc-alkaline granites in the mobile belt (Pitcher 1983). The g_τ-group comprises a vast array of rocks, including two-mica granite, alkaline granite and syenite, and its members intruded over a very long time span.

This g_τ-group can be separated into several subgroups. The earlier phases of g_τ-granites are clearly potassic and unassociated with gabbroic rocks. They comprise subalkaline to peraluminous biotite granites and two-mica granites, which are difficult to distinguish petrographically from g_β-granites. These granites are obviously of the S-type with which all known Mo, and most of the Sn and W mineralizations are associated. The peraluminous character seems to be of secondary origin due to pneumatolysis and greisenization; andalusite or cordierite are never found. Quartz syenite and riebeckite-bearing alkaline granite that commonly form ring complexes occur among the early phases.

Associated volcanic rocks, namely Katherina volcanics, are rare and only described from the south Sinai as remnants of a caldera (Shimron 1980), surrounded and intruded by a large ring dike. These volcanic rocks are essentially made up of commendite and related pyroclastics (Bentor 1985). It still remains controversial, however, whether the late Precambrian/Cambrian g_τ-granites evolved from g_β-granites by differentiation and modification through gas transfer, or from an independent magmatic source. We believe in the latter alternative since we observed, particularly in Sinai, that highly differentiated g_β-granites were followed by less differentiated quartz syenite and post-kinematic biotite granite.

Mesozoic alkaline silicic igneous rocks are always sodic and frequently associated with minor gabbroic rocks, and rarely with carbonatites (El Ramly et al. 1979). They are essentially composed of syenites which sometimes carry nepheline. The associated volcanic rocks are represented by the Wadi Natash volcanics which intruded into the lower part of the Nubia sandstone. Similar volcanics also occur at Gebel Abraq in the south Eastern Desert. The Mesozoic ring complexes are apparently confined to the south Eastern Desert. Based on preliminary discrimination characteristics, namely the soda/potash ratio and an association with gabbroic rocks, it seems likely that the albite granites known as apo-granites (Sabet et al. 1973) commonly associated with Nb-Ta, Sn- and fluorite mineralizations, and the El Bakreya granite intrusion, all in the central Eastern Desert, may belong to this Mesozoic magmatic association.

CHAPTER 11

Precambrian basement inliers of Western Desert geology, petrology and structural evolution

AXEL RICHTER & HEINZ SCHANDELMEIER

Free University of Berlin and Technical University of Berlin, Germany

1 GEOGRAPHIC, GEOLOGIC AND STRUCTURAL OVERVIEW

The main part of Egypt west of the river Nile is covered by thick sequences of relatively undisturbed sedimentary strata of Paleozoic, Mesozoic and Cenozoic age. Apart from the Eastern Desert, Precambrian and younger crystalline rocks occur only in southern Egypt where they occupy an area of some 40 000 km². The majority of these rocks crop out in the Gebel Uweinat area at the junction of the borders of Egypt, Sudan and Libya. Numerous smaller basement areas are exposed between Bir Safsaf and Lake Nasser. Generally, the basement is close to the surface everywhere in that region, except between Gebel Kamil and Bir Safsaf where a downfaulted graben, the Bir Misaha trough, is filled with about 700 m of Nubian sediments (Schneider & Sonntag 1985). The basement high that separates the deep intracratonic Dakhla basin from the more shallow basins of North Sudan is referred to as the *Uweinat-Bir Safsaf-Aswan Uplift*.

This uplift is situated at the eastern margin of a large continental plate which might either be the East Sahara craton (Bertrand & Caby 1978), or the Nile craton (Rocci 1965). The generally high-grade rock assemblages of this continental plate are overlain with a marked tectonic unconformity by volcano-sedimentary-ophiolite sequences in the Aswan area (Vail 1983). Although many scientists disagree on details, it now seems a well-established fact that the Nubian shield was accreted onto this older continental margin through some kind of subduction-related processes and collisional-type tectonics during the Late Proterozoic (El Ramly et al. 1984, Kröner 1985). This late Proterozoic event, generally referred to as Pan African, had a considerable impact, particularly on the margin of the continental plate, by reworking of older crust, migmatization, resetting of isotope systems, and intrusions of voluminous granitic batholits (Vachette 1974, Kröner 1979, Vail 1983). Nevertheless, although this Pan African impact destroyed much of the older geologic record, there is still enough evidence for a pre-Pan African age of the rock units west of Aswan.

Lead isotope studies of a granite near Aswan (Gillespie & Dixon 1983) show that this rock contains distinctly more radiogenic lead than the rocks from the Eastern Desert. Abdel Monem & Hurley (1979) dated detritic zircons from arcose-type sediments derived from the Hafafit dome in the Eastern Desert. The U/Pb upper intercept age of 1770 ± 40 Ma is an indication for the age of the crustal block in the west which supplied this detritus. Nd isotope analyses of samples from migmatites from the Gebel El-Asr and Gebel Kamil areas (Schandelmeier et al. 1987) confirm a definite pre-Pan African crustal history of these rocks by their extreme ε_{Nd} values. Finally, detailed structural, petrofabric, paragenetic, and geochemical analyses of rocks from the entire uplift system (Richter 1986, Bernau et al. 1987) show that many aspects of these rocks resemble the Archean/Early Proterozoic basement of the Uweinat area in southeast Libya (Klerkx 1980, Cahen et al. 1984, Harris et al. 1984).

The uplift system consists of four large and numerous smaller basement inliers:

A. Gebel Uweinat basement inlier (35 000 km²)
  – Gebel Uweinat area
  – Peneplain area
  – Gebel Kamil complex
  – Ring-structure complex
B. Bir Safsaf basement inlier (2 500 km²)
C. Gebel El Asr basement inlier (900 km²)
D. Gebel Umm Shaghir basement inlier (600 km²).

By the regional structure of the area it can be concluded that the rock assemblages of the previously described areas seem to belong to a generally northeast-striking extensive fold belt called 'Zalingei folded zone' by Vail (1976). This fold belt is assumed by this author to be the continuation of the Kibaran belt of Central Africa that swings clockwise into North Sudan and Darfur. But available new isotopic age data argue against this view and indicate a much higher age for the Zalingei fold zone (Schandelmeier et al. 1987). Structurally, the western and eastern parts of the uplift system have to be treated separately, since the Pan African imprint was probably much more intensive in the east.

1.1 Gebel Uweinat basement inlier

In the Gebel Uweinat basement inlier the predominant trend of fold axes is northeast-southwest but locally this principal trend direction harmoniously swings to north-northwest-south-southeast or east-west. At present, fold geometry is little known. At the southeastern slopes of the Gebel Uweinat, Klerkx (1980) observed recumbent folds. The prevailing dip of the schistosity planes is moderately steep to the northwest, although it varies in respect of axial deformation of folds. A regional southeasterly vergence of axial planes is suggested by the main dip direction.

The axial deformation led to a sigmoidal bending of megafold structures. This is probably the most striking structural feature of the Gebel Uweinat basement, being conspicuous even on Landsat images. However, measurement of poles of schistosity planes show quite a large scattering. The principal trend of deformation becomes clearer by analysis of β-linears (Fig. 11/1). By means of the β-linears, a double-branched spiral geometry is obvious, suggesting a later clockwise rotation of pre-existent fold axes in combination with the crustal uplift.

According to petrographic evidence and available radiometric age data (see Chapter 3.4), the major phase of deformation in this region occurred during the Early to Middle Proterozoic, accompanied and succeeded by a regional anatectic event around 1800 Ma BP. As can be inferred from the structural data described (Richter 1986), an early orogenic phase with principal stress in a northwest-southeast direction caused fold axes trending around 55° E, and created a dual shear pattern of faults which broke up the crust into small blocks. During the succeeding mid- to late-orogenic phase, the primary fold axes were sigmoidally bent, probably due to a clockwise rotation and marginal interaction of the small blocks caused by friction. At a final stage, with increasing compression and decreasing ductility, imbrication took place causing at least part of the suggested, presumably heterogeneous crustal uplift which in turn was responsible for increasing plunge of fold axes.

This tectonic interpretation is in general accordance with the polar wandering swath inferred for West Gondwana by McWilliams (1981) for the period between 2100 and 1800 Ma BP. At that time, an initial southeasterly motion was changing to a northeasterly drift. For this reason, the clockwise rotation of small blocks within the Gebel Uweinat region must have been induced by a wide-spaced, subparallel succession of right-lateral northeast-southwest-trending mega-shears.

Landsat image interpretation showed that the fault pattern of the region is dominated by lineaments trending northeast-southwest with subordinate east-west-trending faults (Fig. 11/2). Due to assumed block rotation, the main faulting cannot be attributed to an early tectonic stage, but more likely to a late

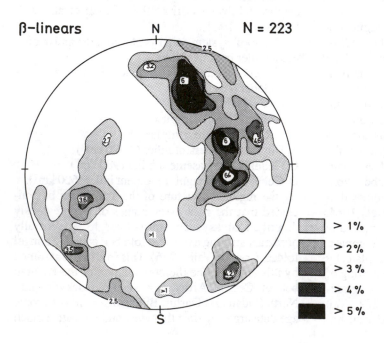

β-linears   N   N = 223

> 1%
> 2%
> 3%
> 4%
> 5%

Figure 11.1 Isoline diagram of the weighted distribution of β-linears within the eastern part of the Gebel Uweinat inlier. Numbers indicate percentage density. A two-branched spiral distribution of the maxima can be seen indicating rotation and tilting of pre-existent fold axes.

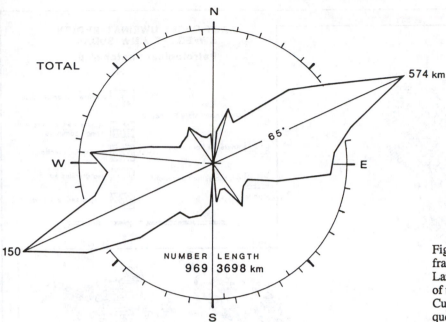

574 km

65°

150

NUMBER | LENGTH
969 | 3698 km

Figure 11.2 Rose diagram of the fracture pattern deduced by Landsat-MSS image interpretation of the eastern Gebel Uweinat area. Cumulative fault lengths and frequencies are given.

phase of the major tectono-thermal event, reactivated and probably pronounced by a Late Proterozoic intraplate stress regime.

1.2 *Bir Safsaf basement inlier*

The regional structure of the metamorphic basement in the Bir Safsaf-Aswan uplift is less clear than that of the Gebel Uweinat-Gebel Kamil area. A general east-west-trend of the foliation accompanied by varying dip directions could be recognized in all basement outcrops of that region. Deviations from this trend are very common due to the intrusion of plutons, late stage doming, and fracturing. In the Gebel El Asr and Gebel Umm Shaghir areas, mylonitic rocks with a general east-west-strike occur frequently. Their dip is low to intermediate to the south and southwest. Petrofabric investigations (Bernau et al. 1987) show that the mylonitization affecting the migmatites in a temperature interval between 300 and 500°C occurred definitely later than the migmatization.

The uplift is characterized by a dense fracture system with an east-west-trend of the major faults. The more or less parallel faults form a dextral wrench-fault zone which supposedly forms the complementary fault system to the sinistral Najd fault system of Saudi Arabia (El Gaby 1983, Bernau et al. 1987). If this interpretation is correct, and structural and geochronological arguments support this view, the Bir Safsaf-Aswan wrench-fault zone is late Pan African in age, and probably related to the late-stage collision phase during which the Nubian Shield

assemblages were thrust over the continental plate margin.

1.3 *Gebel El Asr and Gebel Umm Shagir basement inliers*

Both the Gebel El Asr and the Gebel Umm Shagir basement inliers are unconspicuous outcrops of the basement. Morphologically, these inliers present a deeply eroded landscape of low hillocks and sand-covered country rocks. They are similar in their lithology to the rocks outcropping at Bir Safsaf, so no separate description is given here.

2 GEOLOGY AND PETROGRAPHY OF THE GEBEL UWEINAT REGION

2.1 *Metamorphic rocks*

Geographically, the Gebel Uweinat basement inlier can be subdivided into four major subareas: the slopes and plains in the immediate vicinity of the Gebel Uweinat itself, the Peneplain – an elevated plateau topped by a flat denutation surface and cut by frequent wadis, the so-called 'Gebel Kamil complex', 100 km east of Gebel Uweinat – an area of rough morphology including sedimentary escarpments, basement plains and hillocks and, finally, southeast of Gebel Uweinat, an area called 'Ring-structure complex' (Richter 1986). The name of the latter refers to two dominant ring-intrusions among several megafold structures which are clearly visible on Landsat

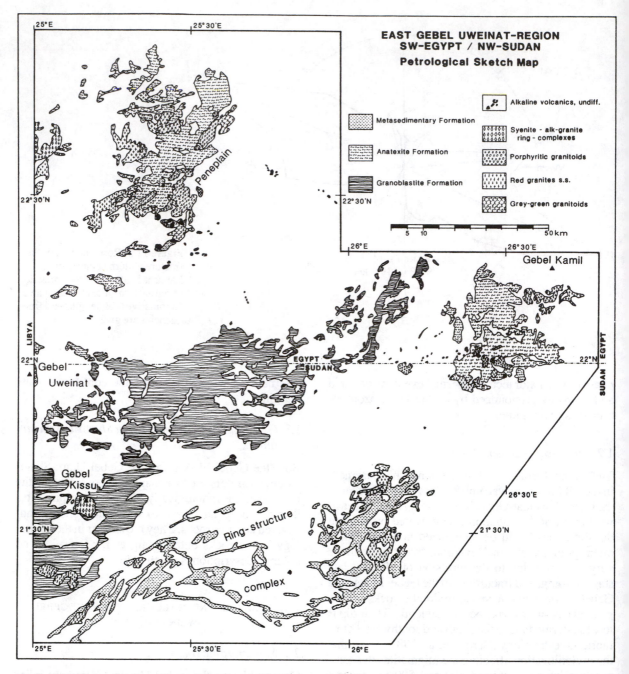

Figure 11.3 Generalized petrological sketch map of the eastern Gebel Uweinat basement inlier.

images and whose arcuate to linear crests rise to 100 m above terrain level. This part is on Sudanese territory, but has to be considered here in the interest of the geological context (see Fig. 11/3).

The earliest scientific geologic report on the Uweinat area has been published by Menchikoff (1927) who already mentioned the leptitic character of the basement along the slopes of the Gebel Uweinat massive. His early study included major-element

analyses of 12 metamorphic and magmatic rocks. For about 40 years, geological research in this area disregarded the basement until Marholz (1968) published some radiometric age data from samples collected in the Uweinat region. In the early seventies, more information was supplied by Hunting Geology & Geophysics Ltd (1974) who conducted exploration work on behalf of the Libyan government.

Klerkx and co-workers carried out field work in the

basement southeast of Gebel Uweinat during the mid and late seventies. Their results (Klerkx & Rundle 1976, Klerkx & Deutsch 1977, Klerkx 1980) improved the knowledge on this remote area, and collected fundamental information on the crustal composition of the area. Based on petrography and Rb/Sr whole rock age data, Klerkx (1980) recognized two major basement series: a) the granulitic 'Karkur Murr series' at the eastern and southeastern slopes of Gebel Uweinat, and b) the migmatic 'Ayn Dua series' which crops out along the northern, western and southwestern margin of the massive. Both series are separated by a clastic horizon, the 'Il Passo mylonite'. In addition to geochronologic work on basement rocks, Klerkx & Rundle (1976) dated a number of magmatic rocks by the K/Ar method.

The eastern extension of the area has been investigated by Schandelmeier et al. (1983), Schandelmeier & Darbyshire (1984) and Richter (1986). Mapping of the area was performed by combined Landsat-MSS image interpretation and field checks, including intensive sampling. Based on the ideas of Klerkx (1980), Richter (1986) distinguished three basement formations by their lithofacies, that is rock educt and metamorphic history. These three formations that can be partly correlated with Klerkx's (1980) series, are: a) the high grade granulitic 'Granoblastite Formation' as lower unit, overlain by b) the clearly remobilized 'Anatexite Formation', and c) the probably youngest, clearly bedded 'Metasedimentary Formation'. The latter has no equivalent on the Libyan side whereas the other two correspond to Klerkx's (1980) series in the following manner:

Ayn Dua Series   = Anatexite Formation
Karkur Murr Series = Granoblastite Formation

In this paper, the subdivision of Richter (1986) will be adopted. The generalized distribution of the distinguished rock units of the Gebel Uweinat basement inlier is depicted in Figure 11/3. Rocks of granulitic character extend in a wide northeast-southwest trending strip from the western parts of the Gebel Kamil complex to the eastern and southwestern foreland of Gebel Uweinat, including the embracement of Gebel Kissu in northwest Sudan. Rocks of anatectic character crop out on both sides of this strip, i.e. within the eastern parts of the Gebel Kamil complex (Fig. 11/3), and in the Peneplain area as well as to the south and west of Gebel Uweinat on Libyan territory. The Metasedimentary Formation builds up most of the basement in northwest Sudan. Contacts between the different Formations are mostly obscured by wide sand plains.

Petrographic mapping respectively interpretation of the basement rocks is difficult to perform on Landsat-MSS images. Sand-covered outcrops prevent development of rock specific radiometric signatures, in particular in view of the small scale lithologic variations which frequently occur in basement rocks. Even textural image features are difficult to interpret due to the morphologic character of the area which consists of flat plains of eroded basement rocks, interrupted by sand sheets. Consequently, escarpments as well as blurred filigree textures had to be used to subdivide sedimentary cover and basement. The wide or narrow-banded features of the basement represent the outcropping edges of folded gneiss and schist layers with distinctive dips. To summarize, basement mapping of the Gebel Uweinat inlier was supported by image interpretation but to a large extent it had to rely on field work, since aerial photographs of that area are not available.

2.1.1 *Granoblastite Formation (gng)*

The term 'Granoblastite Formation' is based on a paper by Winkler (1979) defining a granoblastite as a high-grade regional metamorphic rock, similar in texture to a granulite but lacking the diagnostic mineral assemblage.

The Granoblastite Formation consists in its bulk of grey-green to brownish-grey, fine to medium-grained, well-layered gneissic granulite and granoblastite. The equigranular to seriate fabric with granoblastic to flaser-texture contains as the major components quartz and plagioclase (oligoclase, andesine), whereas alk-feldspar is often missing. Typical xenoblastic quartz is aggregated in subparallel plates (lagen-quartz), including strain-oriented needles of rutile microlites. The isometric plagioclase is frequently antiperthitic and mainly saussuritic, and variably altered to muscovite, epidote/clinozoisite, zoisite, orthoclase, and albite. Xenoblastic to hypidioblastic alk-feldspar is common as microperthite, often mesoperthitic, with microcline and microclineperthite occurring as well. Sometimes, intergranular myrmekite can be observed.

In the granulite, orthopyroxene occurs, relictic at least, whereas in the granoblastite it disappeared or never was contained. However, clinopyroxene prevails throughout, paragenetic with green hornblende, pale fibrous actinolite and anthophyllite, Mg-chlorite, antigorite and talc. Pseudomorphoses of these minerals after pyroxene prove the polymetamorphic nature of the Formation. Whereas uralite is a product of low to medium grade alteration of clinopyroxene, while talc and parageneses of antigorite, Mg-chlorite, anthophyllite, and other fibrous amphiboles represent the orthopyroxene equivalent. Small hypidioblastic flakes of brown biotite are constant minor constituents, well-aligned along schistosity planes.

Frequently, they are pseudomorphosed by chlorite and sagenite.

In some layers where alk-feldspar increases and pyroxene may completely disappear, minor amounts of porphyroblastic garnet (pyralspite) appear. But like pyroxene, the grains may be replaced by a microcrystalline keliphytic belt of sericite, chlorite, green biotite, epidote and calcite. Garnet relics may be preserved in such pseudomorphoses.

There is a continuous transition from the quartzo-feldspatic rocks described to more tectonized varieties, frequently intercalated as concordant layers, but dominated by quartz and often enriched in magnetite and iron hydroxides.

The majority of the analyzed rocks are of grano-dioritic composition, with a range from tonalitic to granitic. A unique feature is the low concentrations of Rb, Th, U and K, a low value of K/Rb, relative enrichment of Ba and Sr and, relative to modern upper continental crust, a slight decrease in the ratio Ba/Sr.

To a lesser degree, ultrabasic rocks are either inter-calated in the basement as are, for instance, amphibolite, or they crop out with ill-defined relations to their hostrock-like meta-(ol)-hbl-pyroxenite, meta-px-hbl-gabbro-norite, and serpentinite east-northeast of Gebel Uweinat. The latter three are believed to be discordant. The foliated amphibolite is fine to medium-grained granoblastic, and dominated by brown and green hornblende and by unzoned saussuritized plagioclase (oligoclase, andesine). Minor amounts of ortho- and clinopyroxene are present, often altered to actinolite, talc and antigorite. Brown biotite and chlorite are further minor constituents.

The fine-grained, weakly foliated gabbro-norite shows a mineralogic composition similar to that of the amphibolite, but the zoned plagioclase is labradorite. Pyroxene within the pyroxenite is preserved in relics only and otherwise replaced by talc, whereas assumed primary olivine is completely altered to pseudomorphoses of serpentine and talc. Brown hornblende, and extremely long needles of grammatite, phlogopite and Mg-Fe-chlorite occur subordinate in this rock.

A unique feature of all these rocks is the replacement of a primary prograde mineralogy by a later retrograde facies, or even more than one. On grounds of mineralogy and rock paragenesis, a metamorphic culmination in the hornblende-granulite facies under high to medium pressure and high temperature conditions (approximately 800°C, 8.5 kbar) can be inferred. It was followed by a retrograde event within the amphibolite facies, presumably introduced by a nearly adiabatic pressure release (< 4.5 kbar), continuously or episodically followed and terminated at least in part of the area by greenschist facies condi-tions of the biotite zone (360 to 450°C, < 4.5 kbar).

2.1.2 Anatexite Formation (gna)

The characteristic feature of the Anatexite Formation is the predominance of migmatic gneiss, i.e. diatexites, metatexites and metablastites. Whereas the highly mobilised, medium-grained diatexites are of a homogeneous to nebulitic texture, rarely showing foliation, metatexites and metablastites are well-foliated and a two-phase nature – melanosome and leucosome – becomes obvious, arranged in the form of layers or schlieren and patches. Their equigranular, seriate granoblastic or porphyroblastic fabric is composed of intermediate plagioclase, alk-feldspar, quartz and biotite. Locally subordinate green hornblende and diopsidic pyroxene occur.

In general, plagioclase is idioblastic and strongly saussuritisized. Younger albite or albite-rich plagioclase is unaltered. Antiperthite has occasionally developed. Alk-feldspar is variably shaped and appears as microcline, litperthite or mesoperthitic chessboard-albite. The perthites tend to form megablasts. Quartz predominantly has grown interstitially and is rich in inclusions.

Brownish-green biotite is the dominant mafic component. A preferred orientation of the variably-sized flakes is rare in the diatexite but persistent in the metatexite. Two generations are often observable by a strong hiatus in appearance. Frequently the large, older flakes show kinking and are developed as sagenite. Replacement of biotite by intergrown, subparallel chlorite together with sphene and/or epidote is widespread. A similar alteration may be seen in subordinate green hornblende. Rarely, pale green clinopyroxene is present in relics, enclosed by biotite or hornblende. Talc, actinolite, sericite, and chlorite appear as further retrograde alteration products.

The chemical composition of the described ana-texites is in general normal-granitic, however, syenitic and (quartz-) monzonitic varieties appear as well. In comparison to most rocks of the Granoblastite Formation, a difference in significant trace elements has developed within the anatexites. Although concentrations of Th, Rb, K, and Ba are slightly to distinctly increased, if compared to the granoblastites, the K/Rb ratios are still below that of modern continental lower crust estimates (K/Rb = 249) while Ba/Sr is above the modern continental upper crust estimates (Ba/Sr = 2) after Taylor & McLennan (1981).

Within the area of the Anatexite Formation, paleosomatic gneisses not seized by anatexis, crop out everywhere. The gneisses are light to dark-grey, fine to medium-grained, and either foliated or layered. Mica-poor, blasto-cataclastic gneiss varieties show flasertexture very similar to that of granoblastites, while mica-rich varieties are more irregularly schistose. In general, the mineralogical composition is not different from that of adjacent remobilized rocks. However, in some of the samples analyzed, enrichment of SiO_2 is obvious, although paleosomes are granitic in toto. Like biotite of the granulitic rocks, biotite of the paleosomatic gneiss is brownish in most cases.

Frequently, marble, calc-silicate, quartzite, and amphibolite are intercalated with paleosomatic gneiss or occur embedded as xenoliths or domains within the anatexites, especially in the Gebel Kamil complex. As opposed to the paleosomatic gneisses, these rocks, termed 'resisters' by Mehnert (1968), escaped anatexis probably as a consequence of their composition.

The marbles are layered in centimeter to decimeter dimension. Thicker light-colored carbonatic beds are intercalated with dark, variably colored silicate beds. Frequent boudinage structures prove tectonic overprinting. Dolomite exceeds calcite in quantity. The silicate component consists of variable amounts of forsterite, diopside, pargasite, tremolite, humite, spinell, serpentinite, phlogopite, white mica and Mg-rich chlorite. An indication of previous high-grade pt-conditions that were later overprinted by low-grade metamorphism is depicted by the paragenesis of diopside-forsterite-tremolite-calcite-dolomite with tremolite rims around diopside (Winkler 1979).

A few skarn-like calc-silicate occurrences in the central Gebel Kamil complex contain either for-sterite, clinohumite and minor vesuvianite, scapolite, plagioclase, or fassaite, hornblende, clinohumite, cli-nozisite, chlorite and white mica.

Quartzite is found either as Fe-quartzite with hae-matite and goethite, or as mica-quartzite. Quartz is quite variable in shape and size. Mica orientation within the mica-quartzite is weak and quartz is poly-gonally bounded, indicating thermo-metamorphic overprinting.

Amphibolites occur either as xenoliths within dia-texites or as thin layers intercalated with paleosome. Generally, they are medium-grained and massive, but fine-grained foliated types are present as well. Their seriate granoblastic fabric is assembled by saussuritic plagioclase and green hornblende, sometimes embracing cores of uralite or clinopyroxene relics. Olive-green biotite increases to considerable amounts in some amphibolites. Epidote and sphene are present as minor constituents.

Of petrogenetic significance is a spinell-sillimanite-almandine-cordierite-gneiss from a central locality of the Gebel Kamil complex, associated with paleosomatic gneiss, marble, and or-thogneiss. The brownish, seriate fine to medium-grained gneiss is streaked with co-aligned fibrous tails of sillimanite. Subparallel to sillimanite tails, lenticular flattened garnet megablasts have grown, often poikiloblastically intergrown with cordierite, quartz, sillimanite and ore. The rockforming cordie-rite is hypidioblastic and fine-grained as it is the intermediate plagioclase. Further major components are xenoblastic alk-feldspar and quartz. While feldspar tends to form megablasts, quartz is elongated and frequently cataclastic. The minor idioblastic brown biotite flakes are aligned along the schistosity planes. Green-brown spinell is idioblastic and fine-grained. Felted pseudomorphoses might be either pyroxene replaced by talc, or pinitized cordie-rite. The paragenetic mineralogy of this gneiss indi-cates the existence of high-grade metamorphism of granulite facies, with high temperature and low pres-sure, previous to anatexis.

Summing up the metamorphic stages as indicated by the various rocks within the Anatexite Formation, a three-stage development can be inferred: a) LP-HT granulite facies (700 to 800°C, approximately 6 kbar) as culminating event, b) partial to advanced anatexis within the highest grade of amphibolite facies as a first retrograde event, c) static, i.e. tectonically inact-ive overprinting in greenschist facies (chlorite zone) as a second retrograde event. The Anatexite Forma-tion is considered as the roof of the Granoblastite Formation.

2.1.3 *Metasedimentary Formation*

Located southeast of the Gebel Uweinat, the Metase-dimentary Formation is built up of clearly ensialic clastic rocks of psammitic and subordinate pelitic character. In the northwest of its distribution area, beds of green, well-foliated, generally fine-grained and frequently cataclastic paragneiss prevail. The seriate quartz grains still show weak graded bedding, and accessories like apatite and allanite are enriched in distinct layers. The following gneiss varieties may be distinguished: hornblende-gneiss, hornblende-plagioclase-alkalifeldspar-gneiss, biotite-hornblen-de-plagioclase-gneiss, biotite-alkalifeldspar-gneiss, biotite-plagioclase-gneiss, and almandine-hornblen-de-biotite-plagioclase- gneiss. In all cases, quartz is additionally present up to considerable amounts (> 40%). Biotite is of brown or green color. Hornblende is fine-grained isometric to medium-grained prismatic, sometimes poikiloblastically in-tergrown with quartz and feldspar. Sceletal garnet is enriched in some layers.

Toward the top of the sequence in the southeast, gneiss is increasingly replaced by paraschist and quartzite which are generally weakly bounded. These very fine to fine-grained rocks are light to dark-violet, red, yellow or grey colored; their schistosity is nar-row, and a phyllitic lustrus sheen has sometimes

developed. The lepidoblastic fabric is parallel or
confuse schistose. A primary layering is obvious, and
cross-bedding is still preserved in some samples.
Beside quartz, several kinds of phyllosilicates like
white mica, sericite, biotite, chlorite, and alumoser-
pentine are present in varying amounts.

Close to a granite pluton thermal metamorphic
reactions led to crystallization of leucoxene, probably
after rutile, and knots of biotite. A third type of knots
composed of sericite, chlorite, alumosilicate and
quartz presumably represents the initial stage of cordierite blastesis.

An interesting feature of the Metasedimentary Formation is an itabiritic sequence of iron-quartzites
with intercalations of jaspilite, mica-schist, and basic
metavolcanics. This sequence crops out for several
kilometers in the central part of the Ring-structure
complex. The opaque ore grains of the quartzite
consist of haematite and goethite with relics of
(titano-)magnetite which was obviously replaced by
these two minerals.

Mica-quartzites are in general light-colored, fine-
grained and narrow schistose. The schistosity planes
are sigmoidally bent, or linear-parallel. Quite frequently, a second schistosity cuts the first one with an
angle of about 60°. Large mica flakes are kinked,
while smaller ones are mimetic crystallized. Lenticular quartz is blasto-cataclastic. At one locality, major
amounts of intergrown, well-aligned, hypidioblastic
kyanite and staurolite occur within a turmaline-
sericite quartzite.

Marble and amphibolite are subordinate embedded in the Metasedimentary Formation. Within
the contact aureole of the granite of the large ring-
structure south of the Egyptian-Sudanese border, various types of hornfels were encountered.

In contrast to the Granoblastite and Anatexite Formations, the Metasedimentary Formation has never
reached a higher metamorphic grade than almandine-
hornblende-amphibolite facies. But this metamorphic grade is only realized within the probably basal
gneiss and quartzite in the western part of the area
where they overly the retrograde granoblastites,
nearly without a break in metamorphic grade. Toward the east, presumably a continuous transition to
lower grades of metamorphism within the greenschist facies is to observe. Although facies-indicative
parageneses do not appear in the eastern parts,
according to Turner (1981) the strong schistosity of
these rocks is an indication that it came up to
pumpellyite-actinolite facies (approximately 320°C,
3.5 kbar) at least by regional metamorphism. Poly-
metamorphism of the Metasedimentary Formation is
not obvious but likely, since later low-grade events
affecting the other formations would not have resulted in new mineral phases due to the pre-existence of
stable minerals. Contact metamorphism is the youngest, probably Phanerozoic phenomenon.

2.1.4 Orthogneiss

In the Gebel Uweinat inlier, gneiss of originally
magmatic material is probably more abundant than
recognized by now. The orthogneisses occur as
members of all three formations, but are considered
together here, due to their assumed similar origin.
However, genetic relations of the orthogneiss in respect of the host rocks are still obscure, since its
extension, metamorphism, and spatial relationships
are ill-known. Its recognition is impeded by a mineralogical and textural overlap with diatexite and certain
grey-green granitoids (see 3.2.2). Thus, some rocks
of the Anatexite Formation being grouped as orthogneiss might be diatexite, and vice versa. In fact,
the term 'orthogneiss' is being used here, somewhat
arbitrarily, as a bulk term for metamorphic rocks
which deviate by their properties from the surrounding basement.

The orthogneiss bodies may occur as stock-like
xenoliths in flat, eroded basement or within sandy
plains as isolated, spheroidally weathered outcrops
suggesting a continuous complex of perhaps batholithic dimensions. At two locations, additional evidence for the presence of orthogneiss was inferred
from circular to oval features on the satellite image.
This rock group includes more or less foliated rocks
with mineral and/or bulk composition different from
the surrounding gneiss, being either more acidic or
more basic, and containing different amounts of pyroxene in respect of the host rock.

The compositional range of the orthogneiss studied varies from granitic to dioritic, and from mangeritic to noritic varieties as mentioned from Karkur
Murr by Klerkx (1980).

In some cases, textural evidence is available and
supports the classification as orthogneiss. Eutectic
and augen-textures as well as ophthalmitic, porphyritic and mortar fabrics are to observe in these rocks.
Plagioclase grains are often euhedral, tabular to prismatic with weak or pronounced zoning in contrast to
the fabric found in diatexite.

Beside plagioclase, the orthogneiss commonly
contains phenocrysts of variably structured microperthite. In the Granoblastite Formation, the perthite
may be even found in the form of mesoperthite,
microclineperthite and microcline. As mafic components, predominant biotite, frequently white mica,
amphibole, and occasionally clinopyroxene are present. The mangeritic and noritic gneisses of the Karkur
Murr also contain garnet and/or orthopyroxene in
minor amounts. As secondary components albite,
chlorite, sericite, and epidote/clinozoisite may be
present.

2.2 *Intrusive rocks*

The intrusive rocks are among the most prominent features of the Gebel Uweinat region. They penetrate the basement in all four subareas.

By their general geologic setting, macroscopic appearance and bulk composition, at least four principal suites can be recognized (Richter 1986):
1. Grey-green, calc-alkaline granitoids   (gg)
2. Red, alkaline granites s.s.   (gr)
3. Porphyritic, calc-alkaline granitoids   (gp)
4. Alkaline syenitic to alkali-granitic ring complexes   (rc)

Although radiometric age data of the intrusive rocks are scarce or not available, textural and cross-cutting relations indicate a chronostratigraphic succession as given above. While the first two rock units are believed to be of Precambrian age, a Paleozoic and Cenozoic age of the latter two has been proved by Klerkx & Rundle (1976), and Schandelmeier & Darbyshire (1984).

The exposures of grey-green granitoids in the Peneplain area are of batholitic dimensions. Here, that type of magmatites is intimately mixed with high-grade, layered paleosomatic gneiss without sharp contacts. A weak to moderate foliation is common. A similar type of rock has been found within the eroded core of a large anticline of the Metasedimentary Formation. At that locality, the granitoids appear as discordant intrusions, but the setting of grey-green granitoids in the Peneplain area indicates an origin by parautochthonous remobilization of the country rock.

In the Peneplain and elsewhere, grey-green granitoids are frequently cut by red granites s.s. of various textural appearances. Most common are unfoliated types. The intrusions took place either in the form of veins resembling agmatic textures, or as small stocks measuring some tens to hundreds of meters across.

Two large, epizonal plutons within the ring-structure complex of northwest Sudan show a diameter of approximately 20 km. They consist of coarse-grained, unfoliated red granite. These two intrusions caused contact metamorphism within an aureole several kilometers wide. Fine-grained red granites intruded as sills of considerable thickness into biotite-gneisses in the same area.

Another kind of setting can be observed in the Gebel Kamil complex. Red rapakivi granites are exposed within two batholiths. The coarse-grained rocks are surrounded by fine-grained, strongly foliated varieties. The northern intrusion grades from east to west into a quartz monzonite, shown in the map as belonging to the grey-green granitoids. At one place, the southern rapakivi-granite batholith is underlain by a hypabyssal layered intrusion of tholeiitic dolerites which are topped by a rhyolitic sill. Thus, within the Gebel Kamil complex a bimodal magmatic association is present, similar to that of the Proterozoic anorthosite-rapakivi-granite complexes quoted by Emslie (1978) as typical for incipient rifting of continental crust.

Field relations and textural differences within the red granites already give evidence for at least two genetic suites originating in different crustal levels. Nevertheless, due to the lack of age data, they are considered here as one group.

Although grey-green granitoids, and in part magmatites of the red granite group, are presumably of Precambrian age, a strong mineralogical impact of metamorphism is missing. However, more or less developed foliation indicates a pre- to syntectonic setting for at least the granitoids and some of the red granites.

The porphyritic granitoids are only present in the Gebel Kamil complex where they form medium to small-sized, irregularly shaped intrusions. They have intruded migmatites and orthogneiss.

The syenite-alkali granite ring complexes are only touched by Egyptian territory. The most prominent one is Gebel Uweinat; the intrusive part of the massive, however, is almost completely situated in Libya. Another ring complex is Gebel Barboir to the north of Gebel Uweinat, right at the border between Libya and Egypt. Ring complexes lying completely inside Libya are Gebel Arkenu and Gebel Bahari, both north-northwest of Gebel Uweinat. In the southeast, in Sudan, the ring complex of Gebel Kissu is rising as a prominent landmark above the gravel plains.

2.2.1 *Grey-green granitoids (gg)*

The rocks of the grey-green granitoid suite appear aphyric to medium-grained in hand-specimens. The microfabric resembles a distinctive granitic texture. Orientation of the components is weak, but a tectonic deformation must be inferred from interstices filled with a granitic microcrystalline phase, weak cataclasis of larger crystals, undulose quartz grains, and kinking of mica flakes. Mafic components frequently occur agglomerated.

The mineral assemblage consists of major ab-rich, mostly zoned plagioclase, minor perthitic alkfeldspar or microcline and quartz. Myrmecite has frequently developed between plagioclase and alkfeldspar. Major mafites are green biotite and green hornblende. Biotite is present in large flakes, often corroded and sometimes replaced by chlorite with additional growth of sphene. Hornblende sometimes encloses a core of uralite as evidence of primary clinopyroxene. Accessories are sphene, apatite, zircon, allanite, calcite and opaque ore.

The quartz monzonite of the northern batholith in

the Gebel Kamil complex deviates from the main rock type by an increased amount of alk-feldspar, the occurrence of clinopyroxene, and inverse zoning of part of the plagioclase.

The major element composition of three analyzed rocks from the Peneplain varies between diorite and trondhjemite. Cluster analyses based on 25 major and trace elements failed to distinguish between these three analyses and others of diatexites and meta-texites (Richter 1986). This indicates a very close genetic relation of all those rocks, particularly if considered together with field relations and the petrography of the grey-green granitoids. The Y + Nb versus Rb classification after Pearce et al. (1984) indicates VAG-type but this is doubtful in view of the assumed Precambrian age of the rock suite, an age where this classification might not be applicable.

2.2.2 *Red granite s.s. (gr)*

The mineral assemblage of the red granites (sensu strictu) is quite constant. Alk-feldspar, developed as wide-spaced vein-perthite, is prevailing, followed by quartz and minor oligoclase. Subordinately occurs green biotite as only mafic component but which may be altered to chlorite. Common accessories are sphene, rutile, apatite, zircon, opaque ore, and fluorite.

The textural appearance of the red granite is more variable than its composition. Typical of the large, epizonal ring-structure of northwest Sudan is a coarse-grained, unfoliated granite which shows an inequigranular, subhedral-granular, weakly cataclastic microfabric. A fine-grained phase of granitic composition has formed in variable amounts between larger crystals showing convex boundaries with the latter. But such a phase is enclosed in some of the megacrysts as well. Thus, a complex history of crystallization and recrystallization has to be presumed.

The granite of the two batholiths within the Gebel Kamil complex shows partly a well-developed rapakivi-texture. Large oval vein-perthites are enclosed by ab-rich euhedral plagioclase. In addition, smaller plagioclase and alk-feldspar individuals occur. Frequently the latter show weak zoning.

Adjacent to the rapakivi granite batholiths, medium to fine-grained red granite is exposed. Its well-foliated texture closely resembles metamorphic fabrics, i.e. interlobate to amoeboid bounded quartz appears in elongated to lenticular aggregates in addition to mostly elongated feldspars.

Samples of fine-grained red granite sills exhibit a typical granophyric fabric. A microcrystalline, symplectic intergrown quartz-feldspar groundmass contains medium-grained porphyrocrysts of seriticized plagioclase and alk-feldspar.

The major element composition of two samples of the rapakivi granites is granitic, while a specimen from the ring structure of northwest Sudan gave values between granitic and alk-granitic composition (Richter 1986). Among both subgroups, a clear difference is apparent for the ternary Rb-Ba-Sr relation. The granite from the ring structure shows a strong magmatic differentiation by its significantly increased Rb-content, while the rapakivi granites are only slightly enriched in Rb but contain more Sr in comparison to normal granites; therefore they have to be considered as anomalous granites according to El Bouseily & El Sokkary (1975). The calc/alkali-ratio versus SiO_2 distribution of the former is in accord with that of Nigerian 'Younger Granites', while the rapakivi granites resemble Fennoscandian rapakivis. Common to all three analyzed specimens of red granite is the WPG characteristic of the Y + Nb versus Rb-classification (Pearce et al. 1984). All samples plot within the or-field close to the thermal valley of the granitic system in the ab-an-or-qz-tetrahedron.

2.2.3 *Porphyritic granitoids (gp)*

Typical of the porphyritic granitoid suite is their common porphyric texture, although the assumed wall-rocks of the intrusion appear aphyric. The phenocrysts of the brownish-grey rock consist predominantly of feldspar up to several centimeters in size which show even macroscopic zoning. In outcrops, a magmatic fluidal orientation of the tabular feldspars can be observed as evidence for a primary magmatic origin. Under the microscope a fluidal texture of the quartz-alk-feldspar groundmass is evident by embedded, well-oriented biotite microlites surrounding seriate graded phenocrysts and xenoliths.

Within this magmatic suite plagioclase-dominated rocks with an approximately balanced content of biotite and hornblende can be distinguished from others with prevailing alk-feldspar and quartz, accompanied by biotite as the predominant mafic component. A common accessory is allanite embedded in the form of well zoned but metamict, euhedral crystals. In addition, apatite, zircon, sphene, opaque ore, and fluorite are present.

Zoned plagioclases and plagioclase aggregates, quartz xenoliths, corroded biotite and hornblende as well as symplectic aggregates of these are present as phenocrysts. With a marked hiatus, alk-feldspar phenocrysts occur. These zoned perthites frequently enclose small plagioclase grains or mafites, evidence for later overgrowth of early magmatically settled crystals. In contrast, occasional xenoliths appear to be resorbed by the groundmass. These controversial features reflect a complex history of crystallization. Within plagioclase-dominated varieties, a continuous transiton to aphyric texture can be observed by a decrease in the amount of matrix together with a coarsening of grain size, disappearance of the huge perthitic phenocrysts, and an increase in the quantity of medium-grained plagioclase.

The more acidic, quartz-rich varieties occasionally show granophyric textures. Xenocrysts of various kinds, among those plagioclase of considerable an-content (56% an), obviously acted as nuclei for the crystallization of spheroidal, eutectic symplectites of quartz and alk-feldspar.

Analyses of the chemical composition based on major oxides vary between tonalite and granite (Richter 1986). The trend of the suite in its calc/alkali- versus SiO_2-ratio coincides with that of the roughly contemporaneous Paleozoic calc-alkaline 'Younger Granites' of the Scottish Caledonides. The suite can be considered as syn-COLG, although Y + Nb versus Rb-characteristics confusingly shows VAG-rock type.

2.2.3 *Syenite-alkali granite ring complexes (rc)*

The four major ring complexes situated in Libya are described by Hunting Geology & Geophysics (1974), while the one in Sudan, Gebel Kissu, is discussed in Richter (1986). Here, only the two touched by Egyptian territory, Gebel Uweinat and Gebel Barboir, shall be considered in more detail.

The composite intrusion of the Gebel Uweinat is roughly circular with a diameter of about 23 km, and is flanked on the northern side by three smaller over-lapping intrusive rings which are aligned along a north-northeast-trending axis. The main complex is composed of deeply weathered alkaline rocks, form-ing very large well-rounded blocks. The outer rim is formed by coarse-grained quartz syenite which steeply slopes toward the surrounding basement. The rock is assembled of quartz up to 12% and major euhedral perthitic alk-feldspar beside minor aegirine.

Toward the center, in the south and west, the outer intrusive ring is followed by a highly complex zone of coarse-grained quartz syenite and coarse to fine-grained alkali granite. This zone forms the outer rim in the north. According to Klerkx & Rundle (1976), similar to the syenite the alkali granite is essentially composed of subhedral alk-feldspar but with an in-creased amount of interstitial quartz and aegirine. The relatively abundant aegirine either surrounds amphibole of the arvfedsonite group, or encloses cores of quartz and opaque ore, probably alteration products of primary amphiboles. Aegirine also ap-pears in elongated crystals along the contacts of leucocratic minerals. In the southern part of this complex zone, a riebeckite-bearing microgranite crops out.

The central depression of the ring complex is occupied by a reddish, fine-grained quartz syenite with trachytic matrix that encloses rare phenocrysts of K-feldspar up to 6 mm in length. The rock is locally cut by trachytic dikes and sills. A steep inner slope around the central quartz syenite depression is built up by alternating cone-sheets of syenite, consist-ing largely of perthite with no quartz.

The dissected oval structure of the Gebel Barboir, with its longer axis trending north-northwest, is emplaced in Precambrian gneiss and red granite. It is partly obscured by sand of the Arkenu dune. The highly complex intrusion is essentially composed of an outer quartz syenite ring-dike and a central white, medium to fine-grained, granite. The syenite shows little variation, and is a pink, coarse-grained, equigra-nular rock containing major perthitic alk-feldspar, subordinate plagioclase with minor biotite and quartz. Basic to acidic cone sheets are intercalated and, in many places, variably composed dikes have penetrated the complex.

The three other major ring complexes situated outside Egyptian territory are nearly monolithic. Gebel Bahari appears to consist of a single ring dike of medium to coarse-grained pink granite, while Gebel Arkenu is mainly made up of medium to coarse-grained nepheline syenite beside subordinate syenite with trachyte and pyroclastic phonolite in the center. Gebel Kissu is built up by a steep outer ring of coarse-grained syenite, deeply weathered to large boulders, and a center of sheeted, fine-grained quartz syenite. The complex is cut by frequent dikes of intermediate to basic composition and variable thick-ness.

2.3 *Volcanic rocks*

The Gebel Uweinat region represents a culmination in predominantly Tertiary volcanic and subvolcanic activity in northeast Africa. Various kinds of volcanic features are present, i.e. small-scale cones, necks or plugs as well as large, occasionally dissected subcir-cular to oval crater-like structures.

Circular to oval features are easy to identify in the satellite image but when they are checked in field, frequently no volcanic rocks are exposed, or exist only as relics in the center of the structures. In these cases, the circular pattern is manifested only by si-licified sediments with a higher resistance to erosion than the country rock. On the other hand, basalt cones can be easily confounded with small sandstone insel-bergs covered with desert varnish.

The widespread sills and dikes are either related to the Cenozoic volcanics or intrusive ring complexes or to the Precambrian and Paleozoic intrusive bodies. The composition of these subvolcanics ranges from basic dolerites to acidic microgranites and granophyres. Occasionally visible low-grade metamorphic impacts suggests Precambrian age for some of these rocks.

In any case, there is an obvious correlation between igneous activity and the regional fault pattern. Very often volcanic necks are situated at points of intersection of faults.

In general, the extrusives consist of alkaline, SiO_2-undersaturated to oversaturated trachytes and related rocks, as well as of olivine-basalt. To a minor extent, intermediate to rhyolitic varieties are present. However, acidic subvolcanics are more frequent than acidic volcanics.

2.3.1 *Alkali olivine basalt (vb)*

Necks and cones of fine-grained alkali olivine-basalts and associated rocks occur far beyond the boundaries of the Uweinat basement inlier. They can also be found on top of the Gilf Kebir plateau and in its eastern foreland.

In most of these rocks, dark-green olivine nodules of up to several millimeters in size are abundant. The olivine is partly or completely altered to iddingsite. The irregular groundmass shows a hypocrystalline intergranular-ophitic fabric. It consists of albite-twinned lath-shaped plagioclase microlites, in between with subordinate small euhedral pyroxene minerals (augite, titano-augite, pigeonite). Interstitially volcanic glass, analcime, zeolithes and opaque ore occur. In some rocks, titano-augite or hexagonal shaped leucite can be found in addition to olivine phenocrysts. Radial-structured aggregates of fine-grained pyroxene are occasionally embedded in the matrix.

The presence of foids and the absence of quartz indicate a general SiO_2-undersaturation. This statement is supported by some available chemical analyses.

2.3.2 *Trachyte and related rocks (vt)*

Alkali trachyte, saturated trachyte, trachy-phonolite, and phonolite are considered within this rock group which is the most common volcanic group in the Gebel Uweinat region. Typical of all is their fine-grained trachytic fabric. Some rocks are clearly fluidally developed, while others are completely unoriented. A porphyritic texture is common.

Most abundant are lath-shaped sanidine crystals beside, probably, anorthoclase. They may appear as phenocrysts or in the ground-mass. Subordinate to minor constituents of the matrix are foids like nepheline, minerals of the sodalite group and analcime, pyroxenes like aegirine and aegirine-augite, amphiboles as kaersutite and kataphorite as well as apatite, iron oxides and hydroxides. All are present in accordance with the respective bulk composition of the rock.

Nepheline appears in isometric hexagonal crystals or as short columns, and is frequently altered to sericite. In the ground-mass, it may be mistaken for similar-looking apatite. Fibrous to elongated pyroxene and amphibole has often grown in skeletal or poikilitic aggregates, caused by interstitial mineralization along the boundaries of leucocratic minerals.

At the southern margin of the Gebel Kamil complex, two sub-circular trachyte structures contain frequent xenolites of a medium-grained intrusive rock which closely resembles the grey-green granitoid suite.

When the trachytic rocks are considered together with the alkali olivine-basalt and with the related basanite, tephrite and phonotephrite outcrops from east of the Gilf Kebir plateau, a rock assemblage is defined which approximately follows the alkaline differentiation trend of Hawaii (Richter 1986). However, some caution is indicated since age datings of the extrusives (see 4) revealed Cenozoic and Mesozoic ages. This could mean that two or more such alkaline cycles have to be assumed.

2.3.3 *Rhyolitic rocks (vr)*

Extrusive rhyolitic rocks are scarce within the Gebel Uweinat region. In the Gebel Kamil complex, only one dissected ring is made up of rhyolite showing vesicular texture and phenocrysts of rounded to angular quartz and feldspar. The latter is, in general, strongly kaolinized or sericititized. Sometimes the feldspar has completely vanished by weathering leaving cavities. The phenocrysts are enclosed by a cryptocrystalline matrix without any visible texture.

Bishady & El Ramly (1980) published 13 major element analyses of volcanic rocks from the Peneplain area. The authors described most of them as alkali trachyte. However, when the analyses are recalculated by the classification method of De la Roche et al. (1980) as done by Richter (1986), quite a number of these volcanics turned out to be more rhyolitic to intermediate rocks, i.e. SiO_2-saturated, calc-alkaline rhyodacite, dacite, and quartz latite with normative quartz and hypersthene, and no nepheline. Although these rocks are found among the (alkali) trachytes possessing the same texture, they do not meet the alkaline differentiation trend of the trachytic to olivine-basaltic volcanics (Richter 1986). The genetic relation of these calc-alkaline volcanics to the alkaline suite and their general geotectonic significance is still obscure.

3 GEOLOGY AND PETROGRAPHY OF THE BIR SAFSAF, GEBEL EL ASR AND GEBEL UMM SHAGIR INLIERS

All rock associations occurring in the Bir Safsaf, Gebel El Asr and Gebel Umm Shaghir areas are petrographically more or less identical to those from the Gebel Uweinat-Gebel Kamil areas which are described in detail in the previous chapter. For this reason, only features which are significantly different will be referred to in this chapter.

3.1 *Metamorphic rocks*

Metamorphic rocks constitute the predominant rock type cropping out in the three larger complexes which form the Bir Safsaf-Aswan uplift. They are generally exposed in flat, slightly undulating plains as a patchwork of scattered outcrops separated from each other by sheets of wind-blown sand. Morphologically, Gebel Umm Shagir and Bir Safsaf complexes are lowland areas, surrounded by the topographically higher sedimentary rocks covering the Nubian group, whereas the south Gebel El Asr basement is tectonically uplifted along an east-west striking vertical fault, which now forms a linear basement escarpment. The three complexes are located within a zone marked by a significant east-west striking fracture system that can be easily recognized on Landsat images. In the field, it is obvious that the basement is exposed due to a tectonic uplift along the east-west fault planes. This holds true also for the numerous smaller basement outcrops which occur frequently in between the three major complexes.

Geologic information on these basement rocks is generally scarce, although there are a number of contributions by different Egyptian scientists which are published in the *Egyptian Journal of Geology*, and in the *Annals of the Geological Survey of Egypt*. However, most of these contributions deal with special structural, petrological or geochemical problems of selected rock units, without attempting to establish a correlation on a regional scale.

A preliminary summary, mainly of the petrology of the Bir Safsaf complex, was given by Schandelmeier et al. (1983); the first radiometric data from the Nusab El Balgum alkaline complex is reported by Schandelmeier & Darbyshire (1984). Meanwhile, the increase of geologic data from the Bir Safsaf-Aswan uplift are summarized by Bernau et al. (1987).

The majority of the metamorphic rocks consists of migmatic gneiss and migmatites of more or less granitic composition. These rocks occupy the central and southeastern part of the Bir Safsaf area (Fig. 11/4), the eastern part of the Gebel El Asr area and virtually the entire Gebel Umm Shaghir area. Supracrustal rocks like calcsilicates, marble and amphibolite generally occur as lenses or narrow bands within the migmatic gneisses; they are rarely present in the Bir Safsaf area but abundant in Gebel El Asr and Gebel Umm Shaghir. Orthogneiss of granitic composition was only found in one large outcrop of several hundred meters in diameter in the southern Bir Safsaf area.

3.1.1 *Migmatic gneiss and orthogneiss (gna)*

Migmatic gneisses very similar to those from the Gebel Uweinat-Gebel Kamil area occupy most parts of the basement complexes. Nebulitic varieties with poorly developed foliation are prevailing; metatexite is scarce.

Three major types of gneiss can be distinguished.

Gneiss of *granitic* composition occurs in all three basement inliers. Geochemically, they show no distinct element trend, and there is no chemical or mineralogical evidence for a pre-metamorphic sedimentary history of these rocks. Bernau et al. (1987) have shown that these gneiss have a magmatic educt. This is supported by its close association with partly metamorphosed S-type granites in the Gebel Umm Shaghir area.

A few larger outcrops of rather homogeneous migmatic gneiss were found in the Gebel El Asr and Gebel Umm Shaghir areas. Their composition is *trondhjemitic to granodioritic*; they resemble Archean high-alumina tonalites (Bernau et al. 1987). This magmatic gneiss contains xenoliths of calcsilicates, suggesting that it is intrusive into the granitic basement containing the calcsilicates.

The extremely homogeneous orthogneiss of the south Bir Safsaf area is of *granodioritic* composition.

3.1.2 *Calcsilicates, marble, amphibolite*

These supracrustal rocks which occur in all three complexes are practically identical to those described in Chapter (2.1.2). Calcsilicate rocks and marble from Gebel El Asr and Gebel Umm Shaghir areas were used to determine the metamorphic conditions.

Bernau et al. (1987) could show that three stages of mineral formations occurred within these rocks: the first stage of metamorphism was one of granulite facies conditions at temperatures from 800 to 1000°C, and a total pressure of 8 to 9 kbar. The second stage, which was related to the migmatization of the granitoid country rocks, took place at temperatures of 650 to 700°C, and pressures probably just below 6 kbar. The third stage is represented by a LT-LP overprint under greenschist facies conditions.

These three stages of metamorphic development could also be recognized within the metamorphic units of Gebel Uweinat (Klerkx 1980) and those of the Gebel Kamil area (Richter 1986).

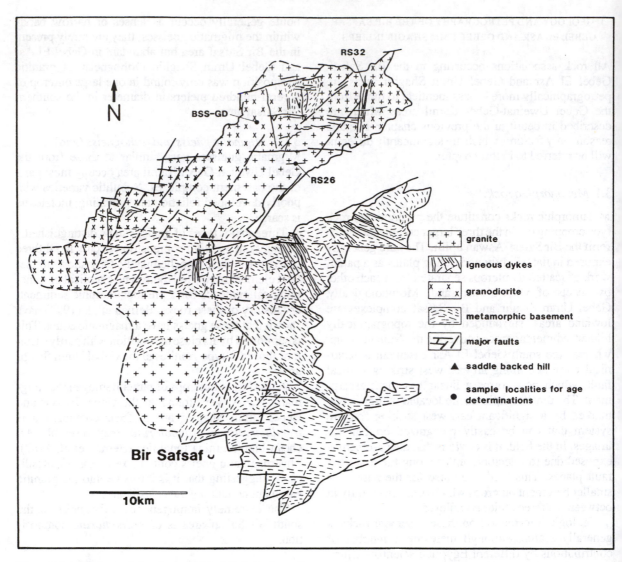

Figure 11.4 Geological sketch map of the Bir Safsaf complex.

3.2 *Intrusive rocks*

Intrusive rocks dominate the southwestern and northern parts of the Bir Safsaf complex (Fig. 11/4) where they often show complicated intrusive interrelations. Such rocks commonly crop out within extensive plains occupied by scattered hillocks of rock boulders with a more pronounced topography than the morphologic features of the metamorphic basement areas. In the Gebel El Asr and Gebel Umm Shagir areas, intrusive rocks are significantly underrepresented; they occur as smaller stocks or bodies within the migmatic gneisses.

The Bir Safsaf and Gebel El Asr areas are dissected by numerous igneous dikes which may group swarms, especially in the Bir Safsaf area. Litholo-

gically, these subvolcanics have a wide compositional range from acidic to basic. They cut both the basement gneiss and the granitic bodies, and are therefore definitely the youngest intrusive rocks in the area. The dikes are clearly related to the regional fracture system of the area.

3.2.1 *Syntectonic granite*
This type of granite occurs mainly in the Gebel Umm Shaghir area in the form of small leucogranitic plutons. They contain quartz, plagioclase, alk-feldspar, biotite, accessories, and secondary minerals like carbonate, white mica and chlorite. After Bernau et al. (1987), this rock is most likely an S-type granite in the sense of Chappell & White (1974), i.e. it is of

Table 11/1. Some radiometric age data from Precambrian rocks from the Gebel Uweinat-Bir Safsaf-Aswan uplift.

Location	Rock type	Method	Age (Ma)	Reference
Bir Safsaf, Egypt	Granodiorite	Sm/Nd WR model age	1445	Schandelmeier et al. 1987
Bir Safsaf, Egypt	Granodiorite	Rb/Sr WR isochron	581 ± 75	Bernau et al. 1987
Gebel El Asr, Egypt	Migmatite	Sm/Nd WR model ages	2399 – 1839	Schandelmeier et al. 1987
South of Gebel Kamil, Egypt	Migmatite	Rb/Sr WR regression line	673 ± 56	Schandelmeier & Darbyshire 1984
South of Gebel Uweinat, Libya	Granodioritic gneiss	K/Ar biotite	1878 ± 64	Hunting Geology & Geophysics Ltd 1974
Wadi Wahech, Uweinat, Ayn Daw series, Libya	Migmatic biotite gneiss	Rb/Sr WR isochron	1784 ± 126	Cahen et al. 1984; recalculated from Klerkx & Deutsch 1977
Ill Passo mylonite, Uweinat, Libya	Mylonite	Rb/Sr WR, botite regression line	2637 ± 392	Cahen et al. 1984; recalculated from Klerkx & Deutsch 1977
Karkur Murr series, Uweinat, Libya	Granulitic gneisses	Rb/Sr WR regression line	2556 ± 142	Cahen et al. 1984; recalculated from Klerkx & Deutsch 1977
Karkur Murr series, Uweinat, Libya	Granulitic gneisses	Rb/Sr WR model ages	2919 – 2904	Cahen et al. 1984; recalculated from Klerkx & Deutsch 1977
Karkur Murr series, Uweinat, Libya	Granulitic gneisses	Sm/Nd WR model ages	3200 – 3000	Harris et al. 1984

Table 11/2. Some radiometric age data from Phanerernozoic anorogenic rocks from the Gebel Uweinat-Bir Safsaf-Aswan uplift.

Location	Rock type	method	Age (Ma)	Reference
Gebel El Asr, Egypt	Olivine basalt	K/Ar WR	21 ± 1	Schandelmeier et al. 1987
Gebel Uweinat, Libya	Dolerite	K/Ar WR	32 ± 1	Klerkx & Rundle 1976
Gilf Kebir, Egypt	Olivine basalt	K/Ar WR	38 ± 2	Franz et al. 1987
Gebel Uweinat, Libya	Syenite	K/Ar amphibole	42 + 41 (± 2)	Marholz 1986
Gebel Uweinat, Libya	Granite	Rb/Sr feldspar	45 ± 2	Marholz 1986
Gebel Uweinat, Libya	pegmatite	Rb/Sr feldspar	45 ±) 4	Marholz 1986
Gebel Uweinat, Libya	Alkali granite	K/Ar kfsp, bio, aegirine	45 – 42 (± 1)	Klerkx & Rundle 1976
Gebel Uweinat, Libya	Phonolite	K/Ar WR	46 ± 1	Klerkx & Rundle 1976
East of Gilf Kebir, Egypt	Olivine basalt	K/Ar WR	59 ± 2	Franz et al. 1987
Gebel El Asr, Egypt	Olivine basalt	K/Ar WR and mineral	94 – 81 (± 3 – ± 2)	Schandelmeier et al. 1987
Gebel Umm Shaghir, Egypt	Olivine basalt	K/Ar WR	157 – 155 (± 4)	Bernau et al. 1986
Bir Safsaf, Egypt	Trachyte dike	K/Ar WR	193 ± 5	Bernau et al. 1986
Nusab El Balgum, Egypt	Alkali subvolcanics	Rb/Sr regression line	216 ± 5	Schandelmeier & Darbyshire 1984
Gebel Uweinat, Libya	Basaltic dike	K/Ar WR	235 ± 5	Klerkx & Rundle 1976
Gebel Kamil area, Egypt	Basalt	K/Ar WR	233 ± 9	Franz et al. 1987
Gebel Kamil area, Egypt	Trachyte	K/Ar WR	240 ± 7	Franz et al. 1987
Gebel Kamil area, Egypt	Porphyritic granite	Rb/Sr WR isochron	431 ± 33	Schandelmeier & Darbyshire 1984
Gebel Umm Shaghir, Egypt	Granodiorite dike	Rb/Sr WR isochron	529 ± 20	Schandelmeier et al. 1987
Gebel El Asr, Egypt	Granodiorite dike	K/Ar WR	541 ± 20	Schandelmeier et al. 1987
Bir Safsaf, Egypt	Various dike rocks	K/Ar WR	600 – 509 (± 25 – ± 13)	Schandelmeier et al. 1987

crustal origin. To this end, no hard evidence for a magmatic source of this rock could be found but most probably it represents a Precambrian tonalitic crust with a low LIL-element content.

3.2.2 *Late-tectonic granite and granodiorite*

3.2.2.1 *Granodiorites (gg).* In the Bir Safsaf area, a suite of granodiorites to tonalites with a few occurrences of diorites and gabbros is exposed petrographically resembling the grey-green granitoid suite of the Gebel Uweinat region. Major and trace element distributions in these magmatites suggest that they were formed through partial melting of a basic source without or with only a little subsequent fractionation (Bernau et al. 1987). They are, however, certainly influenced by crustal material as shown by their intermediate initial Sr ratio of 0.7054 which is very typical for granites that rose through old Proterozoic basement (Brown 1979). In contrast to the, in some respects, comparable magmatites of the Gebel Uweinat region (Chapter 2.2.1), the chemical composition and field evidence of the rocks from Bir Safsaf favor a post-collision granite type (Pearce et al. 1984). This is in good accord with the fact that the Bir Safsaf magmatic suite intruded during a late stage of Pan African influence.

3.2.2.2 *Biotite granite (gr).* The major part of the Bir Safsaf inlier is occupied by a coarse-grained, red biotite granite which intruded into the granodiorite (Fig. 11/4), and which itself is cut by a 521 Ma old dike (see Table 11/1). Mineralogically, it differs from granodiorite suite by preponderance of biotite instead of amphibole or pyroxene. After Bernau et al. (1987), this rock was probably formed under the same geotectonic regime as the granodorite suite but, most likely, from a more leucocratic source. Huge granite occurrences of very similar composition are described from the Nubian Desert region in north Sudan (Huth et al. 1984) and from west of the river Nile near Aswan (Engel et al. 1980) and from the Gebel Uweinat region (Chapter 2.2.2).

3.3 *Volcanic rocks*

The distribution of volcanic rocks, which occur generally as plugs, is very irregular within the three there inliers. No volcanic plugs, except in the Nusab El Balgum complex, were found in the Bir Safsaf area, whereas the Gebel El Asr area and its surroundings are extensively pierced by those. A few plugs occur in Gebel Umm Shaghir. The vast majority of the volcanic rocks is alkali olivine basalt (vb); only a few are of phonolitic-trachytic composition (vt). Rhyolitic rocks (vr) only occur within the Nusab El Balgum alkaline complex.

The mineralogical composition of these rocks which are described in detail by Franz et al. (1987) is practically identical to that of the rocks of the Gebel Uweinat-Gebel Kamil area (Chapter 3.3). It is worth mentioning that all three groups of volcanic rocks are, according to their chemical characteristics, true intraplate volcanics. Although the relationship with the regional fracture system is not always obvious, there is no doubt that the volcanic activity is closely coupled to distinct periods of initial rifting within the Uweinat-Bir Safsaf-Aswan uplift. Normal faulting predominantly acted along older, probably Pan African, zones of crustal weakness. Major reactivation of these or other fault planes occurred during the Permo-Triassic, the Upper Cretaceous and the Tertiary (see radiometric ages in Table 11/2).

4 RADIOMETRIC AGE DATA

Only a small amount of radiometric age data is available in comparison to the immense size of the investigated area. In particular, data from the basement rocks are so limited that an overall picture of the chronostratigraphy of the basement in the Uweinat-Bir Safsaf-Aswan uplift has to be necessarily tentative to some extent. Cahen et al. (1984) reported several age data from the Uweinat inlier but it has to be mentioned that some of these data, especially those from Schürmann (1974) and Marholz (1968), are either derived from single-sample analyses, or analytical details are not presented. For this reason, Table 11/1 contains only age data from Precambrian rocks which are based on reliable analytical material.

The Phanerozoic record of the presently available data is documented in Table 11/2. Since these ages are exclusively derived from anorogenic magmatic rocks, they clearly indicate the different periods of intraplate tectonic activity that reactivated the Uweinat-Bir Safsaf-Aswan uplift several times. All radiometric data quoted in Tables 11/1 and 11/2 are strictly from locations within the uplift system. Age data from the Uweinat area on Libyan territory are quoted in the interest of the geologic context.

CHAPTER 12

Precambrian of Egypt

MAMDOUH A. HASSAN* & AHMED H. HASHAD*

Faculty of Earth Sciences, King Abdul Aziz University, Jeddah, Saudi Arabia

INTRODUCTION

In Egypt, a complex of igneous and metamorphic rocks of primarily Precambrian age covers about 100,000 km² in the Eastern Desert and South Sinai, and limited areas in the south Western Desert (Fig. 12/1). The evolution of these rocks was conventionally interpreted on the basis of the classical geosynclinal orogenic cycle model. This geosyncline was considered ensialic and all its rock units were formed during successive stages of its evolution, i.e. early-, syn- and late-orogenic stages, and an epeirogenic stage (El Ramly & Akaad 1960, Sabet 1962, 1972, El Shazly 1964, Akaad & Noweir 1980).

The most important controversy among proponents of this model was concerned with whether the orogenic events represent a single sequence or whether there was more than one cycle or orogeny within the same geosyncline. Akaad & El Ramly (1960) proposed two orogenic cycles recognized by a major break between the gneisses and the overlying metasediments in Hafafit.

Recent studies in the Eastern Desert show that the evolution of this complex is better interpreted in terms of plate tectonic models. Ophiolite sequences and ophiolitic melanges are of wide occurrence in the Eastern Desert, associated with extensive metasediments of oceanic character. Calc-alkaline metavolcanics characteristic of island arcs or volcanic arcs of active continental margins were also recognized. Furthermore, episodes of granitic intrusion were correlated with events of subduction, collision and rifting. Accordingly, several plate tectonic models were proposed for the evolution of these rocks (e.g. Gass 1977, Church 1979, Hashad & Hassan 1979, Engel et al. 1980, Ries et al. 1983, Stern et al. 1984, El Bayoumi 1984). However, all these models were based on studies carried out in specific areas, and none of them is yet accepted as a unified model for the whole region.

*On leave from Nuclear Materials Authority, Cairo, Egypt.

The basement complex and Pan African

The igneous-metamorphic complex mentioned above has conventionally been referred to in the literature of Egyptian geology as the 'basement complex' or the 'basement' because it formed the base upon which younger undeformed Phanerozoic sedimentary beds were deposited. This complex was considered to be of Precambrian age (Hume 1934). Issawi & Jux (1982) reported the occurrence of Cambrian fossils in sediments overlying this complex. Geochronological studies indicate that this complex was affected by the Pan African events (Hashad 1980). On the other hand, igneous rocks of Phanerozoic age are recognized within the 'basement complex', e.g. ring complexes, dikes and sills, as well as volcanic flows and pyroclastics (Hashad & El Reedy 1979). These rocks form a very small proportion of the 'basement complex' and they will be treated separately in this volume by El Meneisy. Only basement rocks of Pan African or older age will be dealt with in this chapter.

The term Pan African Event was originally proposed by Kennedy (1964) to denote 'a specific thermo-tectonic episode nearly at the close of the Precambrian c. 550 Ma ago which caused remobilization of Archean and Proterozoic rocks, their deformation and metamorphism to higher grades as well as migmatisation, anatexis and wide scale intrusion of granites'. However, this term was later used by Kroener (1979) to include the entire depositional and thermo-tectonic evolution of the crust in northeast Africa during the time period c. 1200-c. 450 Ma. Gass (1981) used the term to describe the whole process of cratonization of ocean arc complexes and their collision and welding to the older African craton. Whether this usage is justifiable or not, the term Pan African is used in this text to mean a tectono-magmatic episode, at least in Egypt, of late Proterozoic age, with its waning stages passing into very early Phanerozoic. The term 'basement rocks' or 'basement' will be used hereafter to denote all rocks formed during the Pan African episode or older.

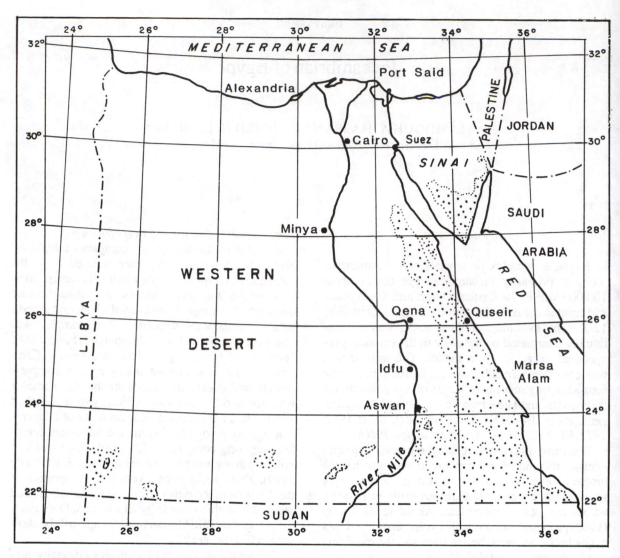

Figure 12.1 Outline of basement outcrops in Egypt.

Review of classifications

The ideas, concepts and models of the geological
evolution of the basement rocks have changed rad-
ically during the past sixty years. Hume (1934)
classified the basement rocks into ten units grouped
in four divisions and presented a model for evolution
of the basement rocks depending mainly on the concept of depth zones within the earth's crust. Schuurman (1953) added a new division to this
classification, the Shaitian granite. However, it was
later shown that the Shaitian granite is sheared older
granite and should not be considered a new division
(Akaad & El-Ramly 1963, Akaad & Moustafa 1963).
The active geological mapping of the basement rocks
carried out in the 1940s and 1950s, particularly in the
Eastern Desert, resulted in the accumulation of a
wealth of factual data on the distribution and litho-

logy of rock units, paving the way for El Ramly &
Akaad (1960) and Akaad & El Ramly (1960) to
present a lithologic succession of basement rocks
based on maps of the Eastern Desert between latitudes 24° 30' and 25° 40' N. They also presented a
model for the evolution of these rocks based on a
geosynclinal orogenic cycle.

El Ramly & Akaad (1960) arranged the basement
rock units in a chronological order, and proposed a
classification which was adopted by most workers,
and in the geologic map of Egypt published by the
Geological Survey in 1981.

Most later workers followed, in one way or
another, the original concept developed by El Ramly
& Akaad. The only significant modification is that of
Akaad & Noweir (1980), in which they arranged the
lithologic units into Formations and Groups. They

presented a map of the Egyptian basement between 25° 35' and 26° 30' N with an alternative lithostratigraphic classification, and also explained the evolution of the basement rocks according to the geosynclinal model. However, most workers have not followed this classification, due to the difficulty of correlation beyond the mapped area, and the difficulty of integrating their model with the plate tectonic models dominating the geologic literature at that time.

Lithologic succession
According to the formal geologic map of Egypt (1981 edition), the rock units of the Egyptian basement are as follows, beginning with the oldest:
1. 'Migif-Hafafit' gneisses and migmatites;
2. geosynclinal metasediments;
3. geosynclinal 'Shadli' metavolcanics;
4. serpentinite;
5. metagabbro-diorite complex;
6. older granitoids, 'syntectonic to late tectonic granite, granodiorites, undifferentiated granite and diorites';
7. Dokhan volcanics;
8. Hammamat group of 'younger sediments';
9. post-Hammamat felsite;
10. gabbro;
11. younger granitoids and post-granitic dikes.

All these units are of Precambrian age. El Ramly (1972) presented a useful summary of these units, adding three more units of Phanerozoic age:
12. Wadi Natash volcanics and trachyte plugs and ring dikes;
13. ring complexes;
14. basalts.

This chapter is restricted to the first eleven units, while the three Phanerozoic units are dealt with in chapter 9.

The primary assumptions implicit in this formal classification are:
1. The whole area of the basement was developed as one unit (the geosyncline) and all events were more or less contemporaneous throughout.
2. All sedimentary and volcanic units are autochthonous.
3. The geosyncline was ensialic, probably on a continental shelf or an epicontinental sea.

The present authors believe that new findings require a substantial revision of these formal units of classification. For example, the serpentinites, believed to be ultrabasic intrusives, are in fact largely allochthonous masses dismembered from mafic-ultramafic complexes that have been identified as ophiolite suites (in the commonly used sense of the 1972 GSA Penrose Conference on ophiolites). These serpentinites, together with substantial parts, if not all, of the Geosynclinal Metasediments, Geosynclinal Metavolcanics and the Metagabbro-Epdiorite complex of the map are believed by some workers (Shackleton et al. 1980, Ries et al. 1983, Church 1986) to be components of ophiolitic melange and do not constitute a stratigraphic sequence.

The metavolcanic components of this melange were found to have both tholeiitic and calc-alkaline affinities. The metasediments covering vast areas in the Eastern Desert were found to be a heterogeneous complex of 'distal mudstone and graphitic shale units, and more proximal turbidites, mixitites, and olistostrome units containing cobbles and blocks of ophiolitic material, dynamo-thermally metamorphosed amphibolites, chert, dirty carbonate, granitoids, felsic volcanic rocks and quartzose greywacke' (Church 1982). They also include a substantial pyroclastic component. Though these features are recorded from specific areas recently investigated, they cannot yet be generalized to the whole basement outcrops until detailed mapping reaches a more advanced stage.

One of the strongly debatable issues that emerged from the studies carried out in the last 10 to 15 years is whether the Pan African Arabo-Nubian Shield is underlain by an old sialic crust or by an oceanic substratum. Although the ensimatic island-arc evolution model calling for oceanic substratum gained more popularity for a while, it has become increasingly evident from the latest geochemical, geochronological and isotopic studies (Stacey & Hedge 1984, Harris et al. 1984, Stoeser & Camp 1985, Vail 1985, Church 1986), that a Precambrian sialic basement is present beneath a great part of the Arabian-Nubian shield. We believe that the paragneisses of the Eastern Desert do represent remnants of the pre-Pan African basement on which the Pan African rocks were overthrust, and they crop out in tectonic windows (Hassan et al. 1984).

Preview of the basement rock units
The lithologic units of the basement complex can be distinguished very generally into two main divisions: the layered sequences and the intrusive rocks. The layered sequences include all rocks of volcanic, volcano-sedimentary and sedimentary parentage that are variably metamorphosed. They are found with distinctive features in three successive major sequences: the lowermost sequence, and probably the oldest, is of mixed continental margin-oceanic character, with a significant proportion of psammitic schists and gneisses, forming a substrate for the Pan African sequences. It represents relics of the pre-Pan African basement that was remobilized during the Pan African event. The two other sequences are Pan

African, the older of which is characterized by oceanic and island arc volcanism and sedimentation, prevalence of ophiolites and ophiolitic melanges, mafic to intermediate tholeiitic to calc-alkaline volcanics, low to moderate grades of metamorphism, and weak to intense deformation. The younger sequence is dominated by felsic calc-alkaline to peralkalic volcanism, continental and molasse-type sedimentation, weak deformation and absence of, or very weak metamorphic effects.

The intrusive rocks are overwhelmingly of granitic composition and are separated into three major successive groups:

the older (1000-700 Ma) quartz-diorite-trondhjemite-granodiorite plutonites of low-K calc alkaline affinity;

the younger (700-550 Ma) more potassic, peraluminous, calc alkaline granodiorite-granite plutonites;

the post orogenic (< 550 Ma) peralkalic highly fractionated granites with 40-50% K-feldspars.

Other subordinate units include ultramafic intrusives, metagabbro-diorite, gabbro, subvolcanic felsites and dikes. Transitional rock types, such as migmatites and some subvolcanic rocks associated with volcanic sequences, occur rarely.

LAYERED SEQUENCES

Gneisses

The gneisses are a group of foliated metamorphic rocks constituting 7% of the surface outcrops of the basement rocks. Originally the group was thought to include rocks of sedimentary parentage (El Ramly & Akaad 1960, El Ramly 1972). Later, a significant portion of these were found to be of volcanosedimentary parentage (El Ramly, pers. comm.). A common source of confusion regarding the stratigraphic position of the gneisses in the Egypt basement comes from the insistence of many authors on including younger gneissified plutonic rocks with the gneisses (Hunting 1967). In this work, the term 'gneisses' is restricted to those of sedimentary and volcanosedimentary parentage.

Most early workers on the Egyptian basement considered the gneisses the oldest unit, representing the earliest phase of subsidence and infilling of the geosyncline, composed primarily of coarse clasitcs passing gradually to finer sediments with progressive subsidence and thus changing gradually upwards into metasediments (Akaad & Noweir 1980). This classic concept is now being reviewed in the light of new investigations.

Occurrence and Distribution

The geologic map shows the gneisses as occurring in more or less isolated belts in areas increasing in extent southward (Fig. 12/2). Most occur south of latitude 25° N. Some of the better studied gneisses include the Migif-Hafafit belt southwest of Marsa Alam, the Meatiq belt west of Quseir, and the Feiran-Solaf belt in Sinai. These three belts are treated in some detail. Other belts, not as well known, are mentioned briefly but should be excluded from the definition adopted.

The Migif-Hafafit gneisses

According to El Ramly & Akaad (1960) the Migif-Hafafit area consists of a succession of gneisses about 4.2 km thick, exposed in an eroded doubly pitching anticline about 55 km long and 17 km wide. These gneisses include the following divisions:

Youngest f. dark green hornblende gneisses, about 1200 m thick;

e. grey biotite gneisses, about 800 m thick;

d. pinkish and greyish psammitic gneisses, about 1000 m thick;

c. dark grey biotite-bearing and hornblende-bearing gneisses, about 500 m thick;

b. hornblende gneisses, locally migmatised, about 400 m thick;

Oldest a. migmatitic gneisses.

These gneisses are bound on the north and east by a group of metamorphosed volcanosedimentary rocks varying considerably in lithology and grade of metamorphism. The gneisses are better described as paragneisses derived from the high grade metamorphism of a succession of pelites, marls, calcareous psammopelites, pelites and marls, (El Ramly & Akaad 1960). Later, gneisses of volcanosedimentary parentage were recognized among them (El Ramly, pers. comm.). Within the Hafafit anticline, the gneisses were uplifted into several small domes with complicated structures. The northernmost of these small domes affords a very good exposure of the gneiss succession from the psammitic gneiss (unit d) downward. To the southeast a similar domal anticlinal uplift provides a fairly good exposure of the gneiss succession from the psammitic gneiss upward (Hassan 1973). Thus, the Hafafit-Nugrus area (Fig. 12/3) is a good area to study the entire succession of the Migif-Hafafit gneisses.

El Ramly et al. (1984) reinterpreted the Migif-Hafafit succession as being composed of two types of gneisses: paragneisses represented by the psammitic and some of the biotite gneiss, and ortho-gneisses

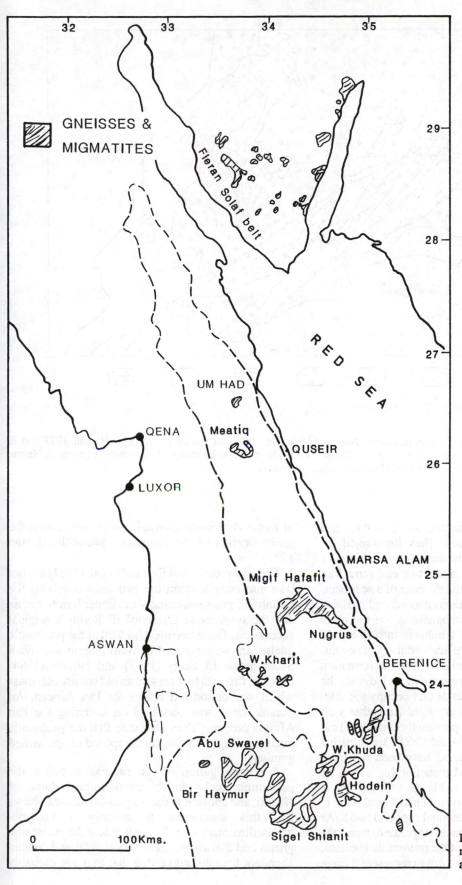

Figure 12.2 Gneisses and migmatites in the Eastern Desert
and Sinai.

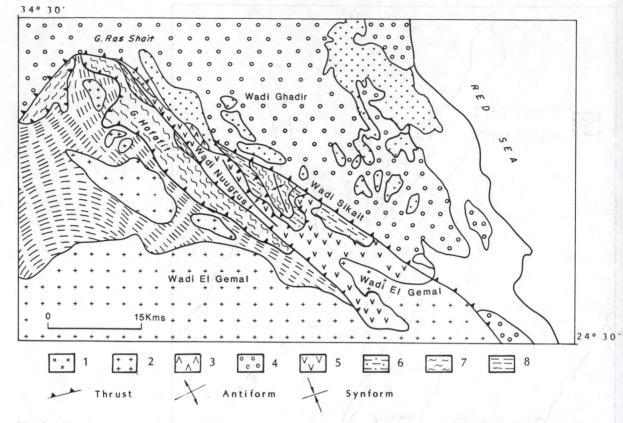

Figure 12.3 Simplified geologic map of Hafafit-Nugrus-Ghadir area (after El Ramly et al. 1984, Hassan 1973 and El Bayoumi 1980). 1. Granite; 2. Gneissose granite; 3. Gabbro; 4. Ghadir melange, horizon C; 5. Hornblende gneiss; 6. Nugrus melange, horizon B; 7. Psammitic gneiss; 8. Hafafit melange, horizon A.

represented by the migmatites, some biotite- and hornblende-bearing gneisses. They interpreted the structural evolution of this domain as includig 11 deformational events. The psammitic and some biotite gneisses once constituted the base of a sedimentary sequence that was metamorphosed and intruded by plutonic rocks. The orthogneisses, on the other hand, began as basic and ultrabasic intrusives and became progressively more acidic during the evolution. They were gneissified during the deformation events. The magmatic development ended with the emplacement of the granitoids that occupy the interiors of the domes. These granitoids, together with the associated migmatites, previously considered the oldest unit (El Ramly & Akaad 1960), are intrusive into the other paragneisses, but have been subjected to the same structural and metamorphic events as other high grade rocks. A U-Pb age of ± 680 Ma is given for the intrusive granite. Hashad et al. (1981) reported three Rb/Sr conventional ages and two K/Ar ages of ± 620 Ma on micas separated from the migmatoidal segregations of the psammitic gneisses. This age was interpreted by them to represent the age

of metamorphic-metasomatic event that caused the gneissification of the psammitic protolith (Hassan 1973).

The interpretation of El Ramly et al. (1984) did not take into consideration the two units overlying the psammitic gneisses, namely the upper biotite gneiss and the hornblende gneiss of El Ramly's original succession. These two units, as well as the psammitic gneiss, are well exposed in Wadi Nugrus and Wadi Sikait areas. El Gaby (1983) and Higazy (1984) consider this area to represent an old continental mass which was cratonized before the Pan African. An oceanic crust was obducted on it during the Pan African orogeny. They also state that the psammitic gneiss of Nugrus is largely composed of gneissified granites.

A reinvestigation of the succession below the psammitic gneiss in the northernmost dome of Hafafit, and above it in the Nugrus-Sikait area, shows that this succession is actually a volcano-metasedimentary one. The unit below the psammitic gneiss and that above it are very similar and contain abundant features suggesting that they are probably

Table 1. The rock succession in Hafafit Nugrus-Sikait area.

	PREVIOUS INTERPRETATION (EL-RAMLY 1972, EL-RAMLY AND AKAAD 1960, 1961)		PRESENT INTERPRETATION
Geosyn-clinal Units	meta sediments and metavolcanics with serpentinites ~?~?~?~		Ghadir melange }----- C major thrust or suture zone
Hafafit	Hornblende gneiss		ophiolitic gabbro ?
	Biotite gneiss		melange }----- B Thrust / Thrust
	Psammitic gneiss		metamorphosed shelf sediments ~Thrust
Gneisses	Biotite – Hb gneiss Hornblende gneiss Granitic gneiss		melange, migmatized at the bottom }----- A

melanges that were intensely foliated and metamorphosed. These two units were misidentified as biotite gneisses because they actually contain sporadic biotite schists as huge allochtonous blocks in an ophiolitic melange. Based on this and on works by Hassan (1973), Hassan et al. (1984) and El Gemmizi & Hassan (1984), an alternative interpretation of the Hafafit-Nugrus succession is shown in Table 1. From this, it follows that:

1. The granitic gneisses in the core of the domes are gneissified granitic intrusions which form zones of migmatites at their contacts with the hornblende and biotite gneisses and send apophyses, dikes and small intrusions into the overlying units. They are also similar to, and seem to be continous with, the batholithic older granites that crop out extensively south of Wadi El Gemal (Fig. 10/3). The 680 Ma age obtained by Hedge et al. (1983) on this granitic core pluton should therefore be considered as the minimum age for the paragneisses, while the U-Pb age of 1700 Ma obtained by Abdel Monem & Hurley (1979) on detrital zircon should be regarded as the maximum age for this unit.

2. The unit above the gneissified granites and below the psammitic gneiss is composed mainly of pebbly mudstones, bedded sedimentary and tuffaceous sequences, volcanic and gabbroic bodies, as well as ultramafic bodies, with bands of hornblende and biotite-rich gneisses and schists at the bottom. Although metamorphism and intense deformation have obscured the original nature of this succession to some extent, it seems to represent a metamorphosed and deformed ophiolithic melange assemblage.

3. Petrographic and mineralogical studies (Hassan 1972, 1973, El Gemmizi 1984) indicate that the psammitic gneiss is a metamorphosed sedimentary unit of a quartzo-feldspathic composition. This composition, as well as the abundance of relict detrital zircon and other heavy minerals, indicate that it was derived from a granitic provenance. The U-Pb age of 1700 Ma for this detrital zircon (Abdel Monem & Hurley 1979) indicates the presence of an exposed middle Proterozoic continental crust in the provenance of the psammitic gneiss at the time of its deposition, which took place on a continental margin. It was later tectonically emplaced at its present position, as indicated by a prominent thrust plane at the base of the psammitic gneiss.

4. The schistose unit between the psammitic gneiss and upper hornblende gneiss in the Wadi Sikait-Wadi Nugrus area contains pebbly bands, pockets of talc-carbonate rocks and serpentinized ultramafites. It shows great similarities to the biotite rich alternations occurring below the psammitic gneiss within the northern dome of Hafafit, and thus is also considered a metamorphosed and deformed melange body.

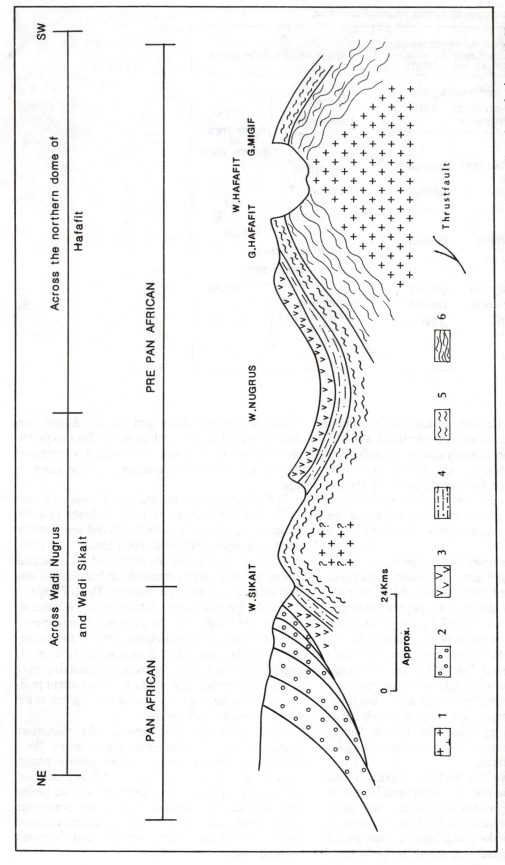

Figure 12.4 Diagrammatic cross-section (with exaggerated vertical scale) across Hafafit-Nugrus-Sikait domain. 1. Gneissose granite; 2. Ghadir melange, horizon C; 3. Hornblende gneiss; 4. Nugrus melange, horizon B; 5. Psammitic gneiss; 6. Hafafit melange, horizon A, migmatized at the bottom (see also Table 1).

5. The topmost hornblende gneiss is metamorphosed gabbroic and other basic rocks (basalts and dolerites) which was thrust over the previous succession.

6. The whole Hafafit-Nugrus domain is thus considered a tectonic window, surrounded by rocks of a distinctly lower grade metamorphism belonging to the ophiolitic melange of the Eastern Desert formed during the Pan African episode. The Hafafit-Nugrus succession was probably developed during a cycle of plate tectonic activity earlier than the Pan African event. This older cycle formed the continental mass later involved in the Pan African event. The cratonization history of this continental mass must be sought in the sedimentary cores of the gneiss domes. The intrusion of the granites at a much later date caused the uplift of the domes, probably accompanied by sliding of the layered units against each other. Figure 12/4 represents the authors' interpretation of the Hafafit Nugrus-Sikait domains.

The Meatiq gneisses
The Meatiq area lies midway between Qena and Quseir in the central Eastern Desert (Fig. 12/1). It is a more or less circular mountainous country with several prominent peaks (Fig. 12/5). The rocks in the area are thrown into a huge domal structure, composed of rocks of the Meatiq and Abu Ziran Groups, both intruded by granitic plutons (Akaad & Noweir 1969, 1980). The rocks of the Meatiq Group occur mainly on the western side of the dome in a huge semicircular structure elongated in a roughly north-south direction, with scattered roof pendants capping the igneous plutons and rims around them. This is all surrounded by metasediments, metavolcanics and serpentinites of the Abu Ziran Group with a more or less circular boundary line. Akaad and co-workers (e.g. Akaad & Shazly 1972, Akaad & Noweir 1969, 1980) defined the Meatiq Group as the oldest major lithostratigraphic unit in the Egyptian basement complex, characterized in general by marked uniformity over a considerable area within the outcrop of the group and by the virtual absence of rock types other than the predominant granulites and minor pelitic and semipelitic schists. They used the term granulite as a textural and lithologic term to denote 'a rather fine grained quartzofeldspathic metamorphic rock without conspicuous schistosity'. They identified syngenetic sedimentary structures, e.g. graded bedding, current bedding and ripple marks. They stated that the Meatiq Group represents a thick succession of sediments that are distinctly mature and siliceous, with only subordinate shaly and calcareous intercalations. This represents the slow deposition of quartzose and feldspathic sandstones, minor shale and calc-shale in an extensive and notably stable basin.

On the other hand, Akaad & Noweir (1980) described the same rocks as a large variety of high-grade siliceous gneisses and schists that appear to have originated mainly from acidic tuffs and flows.

Habib et al. (1985) indicated tht the Meatiq Group may be differentiated into five successive, almost conformable thrust sheets overlying each other. The Meatiq infrastructure represents a part of an old craton, or continent, over which the Abu Ziran suprastructure was thrust during the Pan African Orogeny. The whole succession was then regionally metamorphosed under high pressure and temperature conditions. This is inferred from the presence of almandine, staurolite, kyanite, sillimanite and cordierite in the weakly sheared rocks.

Concerning the relation between the Meatiq Group and the Abu Ziran Group, there are two contrasting concepts. Akaad and his co-workers noted that the passage from the mature psammitic granulites of Meatiq to the immature metasediments is marked by a transitional zone which was designated as a formation and named the Abu Fannani Schists, separating the two major contrasted groups, the Meatiq Group and the Abu Ziran Group.

On the other hand, several workers believe in the structural break between the Meatiq Group and the Abu Ziran Group. Habib et al. (1985) consider the Meatiq dome as an old infrastructure over which the Abu Ziran Group forms a suprastructure that was thrust from the east as a thrust sheet or nappe. El Gaby (1975) and Shazly (1971) expanded the Abu Fannani Formation to include both the upper part of the Meatiq Group, rich in amphibolite and biotite-rich intercalations, and the lower part of the Abu Ziran Group. The upper part of the Abu Fannani Formtion forms a tectonic melange zone along the thrust separating the two groups. The thrust faults dip away from the Meatiq dome and are conformable with the foliation and 'layering' of the overlying and underlying rocks. The rocks of the Meatiq Group are generally foliated in contrast to the poorly foliated Abu Ziran Group.

Dixon (1979) considers the Meatiq Group to be the best example of an older sialic sequence in the Egyptian basement. It represents a sequence of mature sediments which are exposed as a large domal structure, with a Pan African granite pluton occupying the core. The main unit of this group is a quartz-rich gneiss. The volcanic-sedimentary belt surrounding the Meatiq gneiss has been tectonically juxtaposed against the sialic dome. Indicating that the dome represents a local feature developed independently. Dixon (1979) concludes that the Meatiq gneiss could therefore represent one manifestation of an early continental margin with older crust inferred to the north, now largely obscured by remobilization of this

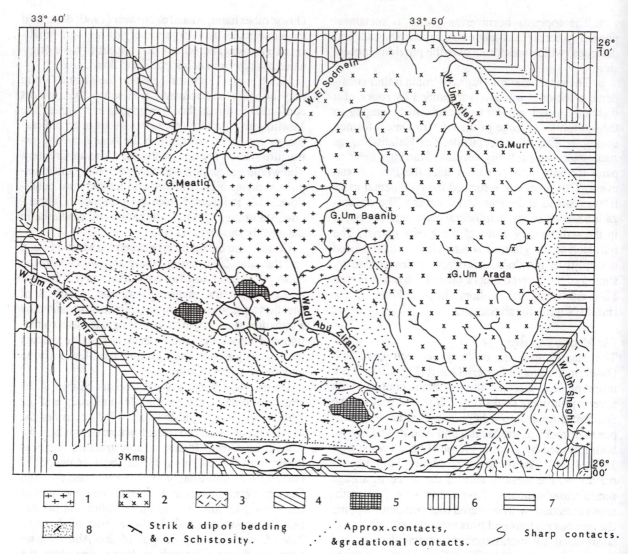

Figure 12.5 Simplified geologic map of Meatiq dome (after Shazly 1971). 1. Younger granites; 2. Um Baanib gneissose granite; 3. Grey granite; 4. Serpentinites; 5. Hornblendic rocks; 6. Abu Ziran group, with minor serpentinite lenses; 7. Abu Fannani schists; 8. Meatiq group.

sialic material during Pan African plutonism and orogeny.

Church (pers. comm.) states that in the southern part of the Meatiq Dome, the Meatiq quartzofelspathic gneisses contain abundant garnet crystals that increase in size toward the core of the dome. The rocks of the Abu Zeiran group in immediate contact with the garnetiferous gneiss are mainly serpentine melange with bands of graphitic slates that are distinctly lower in metamorphic grade. Sturchio et al. (1983) states that the rocks of the Meatiq group are quartzofelspathic mylonites, phyllonites and mafic mylonites that are derived from a granitic protolith emplaced in a mafic-ultramafic assemblage (Abu Ziran group of Akkad & Noweir 1980), 626 ± 2 Ma

ago. This granite protolith was gneissified during a metamorphic event that culminated at 614 ± 6 Ma ago. Sturchio et al. (1983) also concluded that the Meatiq group should no longer be considered the oldest exposed lithostratigraphic unit in the Eastern Desert. However, the present authors do not agree with that concept. According to our observations in the Meatiq area and with reference to Dixon (1979) and Habib et al. (1985), the lithologic, structural and metamorphic break between the Meatiq gneisses and the overlying melanges is fairly well established. This break is reminiscent of the break between the Hafafit domain and the overlying schistose metavol-canosedimentary rocks described earlier. Stuchio et al.'s (1983) 626 ± 2 Ma age is very similar to the ±

670 Ma ages obtained on micas from the psammitic gneisses by Hashad et al. (1981) and should also be interpreted as the minimum age for the Abu Ziran group.

Feiran-Solaf Migmatite-Gneiss belt
This belt lies in the southwestern part of Sinai and extends about 40 km in a northwest-southeast direction, with a width of 5 to 10 km (Fig. 12/2). The whole belt is intruded on both sides by a number of granitic plutons and dissected by a series of dike swarms. Recent studies on this belt are presented by Akaad (1959), Akaad et al. (1967), and El Gaby & Ahmed (1980). The last two authors state that field and microscopic studies suggest that the gneisses of this belt were originally sediments of arenaceous, pelitic and calcareous composition, with only minor basic magmatic intercalations that were regionally metamorphosed and variably granitized. They state that these rocks suffered anatexis, metasomatism and migmatization, causing considerable mineralogical and textural alterations in the original rock. However, they presumed that grouping several rock types into larger rock units on the basis of assumed original lithology, together with the delineation of certain major structures, has served to establish the lithostratigraphy of the gneiss belt. Hence, they classified this gneiss belt into five formations with eleven members. The migmatites, which are dominant, were formed by partial melting or anatexis (El Gaby & Ahmed 1980), or by lit-par-lit injection of granitic magma (Schurmann 1966).

According to El Gaby & Ahmed (1980), the Feiran-Solaf gneiss belt belongs to a pre-eugeosynclinal succession, its migmatization being accomplished during the same phase of regional metamorphism as that which affected the eugeosynclinal metasediments of the Egyptian basement. This phase of regional metamorphism took place c. 1100 Ma ago (El Shazly et al. 1973, Siedner et al. 1974, in Hashad 1980).

The present authors believe that the Feiran-Solaf migmatite-gneiss belt was part of a mature to semimature sedimentary succession that was metamorphosed and migmatized within a continental crust or in a core of a mature island arc, as a result of intrusion of batholithic granites of Pan African age. The ± 650 Ma Rb/Sr age reported by Bielski in Bentor (1985) on the Solaf-granite-gneiss is most probably the age of such intrusions and the accompanying migmatization of the intruded sedimentary successions. It should not be interpreted as the age of the 'gneiss' but as the age of the gneissification event or a minimum age for the gneiss.

Other gneiss belts
Several other gneiss belts appear on the geologic map of Egypt, in the Eastern Desert, the Western Desert, and Sinai. These belts are much less studied than those described above, and can be summarized as follows:

The southern belts. These occur from latitutde 24° N southwards (Fig. 12/2). Most of these belts are probably gneissified metasediments and metavolcanics, amphibolite schists and foliated plutonic igneous rocks. For example, the Wadi Khuda gneisses are spatially and genetically related to both the granodiorite of the Berenice area and to the Abu Hammamid volcanic sequence (Hunting 1967). However, Dixon (1979) states that Wadi Khuda gneisses are paragneisses and represent older sialic material with a protolith of relatively mature sediment. The metamorphic grade is considerably higher than that of the younger volcanosedimentary units.

The Wadi Kharit gneisses, occurring midway between Berenice and Aswan, were developed by infolding of geosynclinal sediments and volcanics with syntectonic granodiorites, and the imposition of gneissic structures upon them. They also show gradational change to the typical grey granite (Hunting 1967). Biotite and hornblende gneisses and schists similar to those of the Hafafit gneiss succession are suspected of being present in several less well defined domal structures (El Ramly, pers. comm.).

Um Had Gneisses. A small belt of old gneisses and migmatites occurs about 40 km northwest of the Meatiq Dome. The western part is formed of orthoamphibolites, hornblende gneisses and migmatites, whereas garnetiferous mica schists and gneisses prevail in the eastern part (El Gaby et al. 1984).

The Sinai belts. Other than the Feiran-Solaf belt, some gneiss belts occur in Sinai (Fig. 12/2), but no studies have been carried out to reveal their nature.

In addition to the above, gneisses occur in the south Western Desert. The most extensive of these outcrops is that occurring in the Uweinat region. There, they are quartz diroite-tonalite gneisses and crystalline schists intruded by granite. Ages as old as 2700 Ma were reported in these rocks (Klerkx 1980). In Gebel Kamil area, migmatic gneisses gave Rb/Sr whole rock model ages of 3123-2817 Ma (Schandelmeier & Darbyshire 1984, reported by Schandelmeier and Richter this volume, chapter 11).

In summary, the gneisses, defined as metamorphosed mature to semimature sediments, occur only in the Feiran-Solaf, Meatiq and Migif-Hafafit areas, according to present knowledge. Other gneissic belts should be investigated but there is no reason to suppose that the gneisses form an extensive base upon which other units were deposited. It is more proper to consider them remnants of deeply eroded proto-continents or mature island arcs that were involved in the Pan African Orogeny. Evidence from Hafafit-Nugrus and Meatiq areas suggest that these gneisses might have undergone an earlier phase of orogeny, thus representing a pre-Pan African continental basement. However, if they do not represent an old basement, they would represent shelf sediments which were originally deposited on a continental margin and later were involved in tectonic events during which they were metamorphosed, deformed and associated with other rock groups, thus still confirming the presence of a pre-Pan African continental basement (see discussions in Church 1979, 1981, 1983). The gneisses of the Uweinat region are believed to represent the West African Craton. It was suggested by Gillespie & Dixon (1983) that the Aswan area may be on the boundary between the Nubian Shield and an older sialic craton to the west, on the basis of isotope composition of some igneous rocks in the Egyptian Shield.

Ophiolites

Rittman (1958) was the first to identify the serpentinite masses of the Eastern Desert as ophiolites in the sense of the Steimman Trinity. Abdel Gawad (1970) drew attention to the alignment of the serpentinite masses of the Eastern Desert and North Saudi Arabia along the northwest-southeast trending belts, using this to determine the offset across the Red Sea. The first use of the term ophiolite in the context of the definition adopted by the GSA Penrose Conference (Anonymous, 1972) in the Arabian-Nubian Shield was, however, by Bakor et al. (1976) Garson & Shalaby (1976), and Neary et al. (1976). Following these works, several plate-tectonic models were suggested for the evolution of the Arabian-Nubian Shield. Most of these models (Frisch & Al Shanti 1977, Gass 1977, Abdel Khalek 1980, Schmidt et al. 1979, Stern 1979, 1981, Fleck et al. 1980, Kemp et al. 1980, 1982) were based on defining ophiolitic rocks as remnants of subduction zones and hence margins of former oceanic basins. Gass (1981) featured the Arabian-Nubian Shield as composed of a number of intra-oceanic island arcs that have been swept together, along with their sedimentary aprons and occasional slices of oceanic lithosphere, to form new continental crust. He defined five of these island arcs within the Nubian Shield of the Eastern Desert of Egypt and Sudan.

The interpretation of ultramafic rocks as 'sutures' with respect to the Eastern Desert of Egypt has been debated by Church (1979, 1982, 1983, 1986), Hashad & Hassan (1979), Engel et al. (1980) and Vail (1985). Church (1986) argued that the ophiolites formed part of a single stratigraphic unit formed during the obduction of the oceanic crust in a manner similar to that involved in the development of the western margin of the Appalachian system, and that the present distribution of the ophiolites is controlled by deformation structures. Commenting on Vail's (1985) 'palinspastic' map showing various arc 'microplates' and 'sutures', Church (1986) came to the conclusion that although many of the ophiolites of the Arabian-Nubian Shield most likely represent Pacific, Tasman or back-arc (fore-arc?) type ocean basins, their present locations do not necessarily delineate the original sites of the ocean basins and consequently do not necessarily demarcate plate boundaries.

Occurrence and distribution

Dixon (1979) estimated that ultramafic bodies account for 5.3% of all Precambrian outcrops in the Eastern Desert of Egypt. No figure is available for the combined ophiolite complexes and melanges, but it is apparently in the order of 10 to 15%. The better known occurrences in the Eastern desert lie to the south of latitude 26° N, and their distribution as indicated on the goelogical map of Egypt (El Ramly 1972) is very irregular. This could possibly be explained by their occurrence as allochthonous blocks or thrust sheets floating in a dominantly ophiolitic melange or in various types of metasediments. Shackleton et al. (1980) estimated that these masses have moved by submarine gravity sliding and later thrusting for distances of more than 200 km. This may also explain why the ophiolite complexes are found mostly as dismembered parts.

Serpentines or serpentinized ultramafics constitute the major part of these complexes, but the other units are clearly present and, in many cases, it is easy to patch them up in a complete ophiolite sequence by careful investigation. In some places (e.g. Wadi El Ghadir, El Fawakhir, El Geneina & Wadi Mubarak), complete intact sections, except for the ultramafics, are encountered.

Among the several ophiolites recognized in the Eastern Desert, two will be described in detail, El Ghadir and El Fawakhir.

El Ghadir ophiolites

Wadi El Ghadir area (Fig. 12/6) lies about 30 km southwest of Marsa Alam. In this area, the ophiolites occur as allochthonous units in a melange assemblage that has been mapped previously as part of the geosynclinal metasediments. Foliated dioritic rocks form a large fold structure with a northwest trending

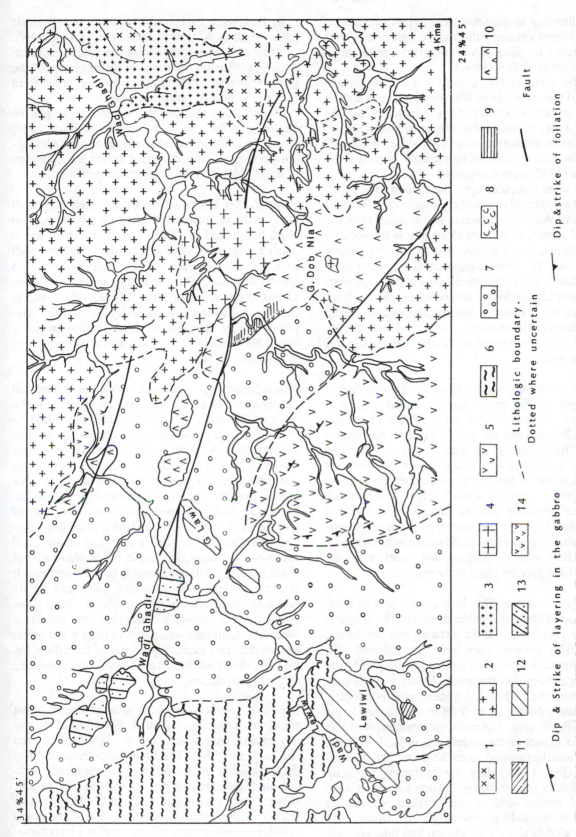

Figure 12.6 Geologic map of Wadi Ghadir area, Eastern Desert (after El Bayoumi 1980). 1. White granite; 2. Massive pink and grey granite; 3. Foliated pink granite; 4. Sheared pink granite; 5. Leucogabbro; 6. Distal melange; 7. Proximal melange; 8. Pillow basalt; 9. Sheeted diabase; 10. Gabbro complex; 11. Layered gabbro; 12. Massive serpentinized peridotites; 13. Serpentines; 14. Undifferentiated volcanic rocks.

axial direction in the southwestern part of the area. Granitic rocks occupy the eastern part of the area. Abundant later dikes of various compositions cut all other units in all directions.

El Bayoumi (1984) presented a comprehensive study of the ophiolites in Wadi El Ghadir area and the following description of El Ghadir ophiolites is based on his work. Although the areal extent of El Ghadir ophiolites is discontinuous and limited, all the formal ophiolite units as defined by the GSA Penrose Conference (1972) were recognized.

The most complete ophiolite section in Wadi El Ghadir area is well exposed at the junction of Wadi El Beda and Wadi Saudi, where all the units, from the layered gabbro at the base to the pillow basalts with associated deep-sea sediments at the top, are present (Fig. 12/6). The overall exposure is estimated to be more than 2 km. It dips moderately to the southwest. Metamorphic peridotites are, however, missing from this section, presumably due to dismembering during displacement. However, peridotites and their serpentinized equivalents occur in variably sized masses all over the area.

Metamorphici peridotites. In addition to boulders in the melange, the peridotites are found mainly as mountain-size allochthonous units in Wadi Lawi and Wadi Ghadir. They are mostly highly serpentinized and in many places are transformed into talc carbonate, cream-colored rocks. These mountainous masses of the metamorphic peridotites appear to be dismembered parts of a much larger ophiolite slab that was obducted into the orogenic zone. This is confirmed by magnetic survey measurements indicating that large peridotite and serpentinite masses of Wadi Ghdir area are rootless, and therefore are not connected by feeder pipes to the mantle.

Cumulate sequence. This includes dunite with chromitite at its base, and gabbroic rocks. The dunite is always associated with the metamorphic peridotites, presumably because they were dismembered with them due to their similar mechanical properties. Its base is defined by chromitite lenses that separate them from the underlying metamorphic peridotites. Cumulate gabbroic rocks form several masses in the Wadi Ghadir area. The main gabbroic mass of this complex occurs on the eastern side of Wadi El Beda, with a small extension in the southern side of Wadi Saudi (Fig. 12/6). Here the gabbro is distinctly layered at the bottom and grades upward into a coarse grained gabbro with a characteristic appearance known in the field as rosette gabbro. In turn, the rosette gabbro gives way upward into microgabbro. The attitude of the layers varies from horizontal to almost vertical.

In the layered gabbro zone, pockets of pegmatitic gabbro rich in feldspars (sometimes pink in color), anorthosites and trondhjemites are common. These represent injections of the later differentiates of the gabbroic magma. Pyroxenite cumulates, in the form of irregular pockets, are also present.

Several other massive and rosette gabbro masses are found in the area, at the feeders of Wadi Saudi, Wadi Lawi, and in the middle reaches of Wadi Ghadir, but these are much smaller in size and appear to be huge allochthonous blocks in the melange.

Sheeted dikes. Sheeted dikes are best seen in Wadi El Beda on the southern side of the main gabbro mass, in Wadi Saudi, and in Wadi Lawi. These dikes are in contact with each other without any 'foreign' wall rock material. For each one, the two adjacent dikes are the wall rock. The dikes range in thickness from about 50 cm to about 2 m. Although the contacts between individual dikes is hardly discernible, assymetric chilled margins were recognized in many cases. The material of the dikes is mostly fine grained to aphanetic diabase.

Pillow basalt. Pillowed basaltic rocks were observed clearly in two locations in the northwestern end of the main gabbro mass of Wadi El Beda (Fig. 12/6). In both places, the pillowed basalts are associated with sheeted dikes. In Wadi El Beda, the pillowed mass reaches a thickness of 200 m. The pillows range in size from about 20 cm to 1.5 m across with a sheath 2 to 8 cm thick depending on the size of the pillow. The pillows are best developed toward the western end of the mass, but eastward, toward the sheeted dike complex, they become smaller in size, less distinct, and grade into the sheeted dikes. In the southern side of Wadi Saudi, the pillowed mass is about 70 m thick. It is more altered, with brownish color and highly schistose sheath. Some of the pillows here show porphyritic texture, while others show vesicles. In the Wadi El Ghadir area, sediments of deep water origin are interlayering or resting on top of the pillow basalts, providing additional evidence of the submarine origin of the pillows.

Chemical characteristics. In spite of the common spilitization and the slight metamorphism of the pillow basalts, their tholeiitic nature is still recognized (El Bayoumi 1984). Their low K-content as well as their immobile trace elements patterns clearly indicate their similarity to present day oceanic basalts. However, the enrichment of Wadi El Ghadir pillow basalts in the immobile elements led El Bayoumi (1984) to consider them as being within-plate characteristics and to suggest that they have originated from a plume of oceanic basaltic magma. Comparing the chemical composition of the most primitive basaltic dikes with different models of partial melting of mantle pyrolite indicated that these rocks originated by 13% partial melting of the original pyrolite material (El Bayoumi 1984). Irregular variations in the immobile elements indicate that the fractionation processes took place in an open system in proximity to a diffuse spreading system and related fracture zones.

Tectonic setting of El Ghadir ophiolite. Based on his study of El Ghadir ophiolites and associated melange and other rocks, El Bayoumi (1984) proposed a plate tectonic model for the evolution of this segment of the Egyptian basement. According to El Bayoumi, the area represents a segment of a trench zone filled with a melange body, which is overlain by an ophiolite nappe resulting from the westward subduction of an oceanic crust under a continental margin. The geochemical studies of the ophiolites indicated that they originated from an oceanic plume basaltic magma. (Fig. 12/7). A line of hot spots (plumes) brings heat and primordial material up to the asthenosphere, causing upwelling of the lithosphere. Subsequent currents in the asthenosphere spread away from the upwelling, producing tension on the crest and compression on the bottom of the lithosphere. Crustal attenuation and fissuring then followed, combined with continuous basification of the continental crust. Further rifting caused by continuous convection currents resulted in the formation of a proto-ocean. Continued rifting with sea floor spreading caused more and more opening of the ocean and production of a new oceanic crust. Thickening of sediments on the western continental margin combined with transformation of gabbro to eclogite, led to a westward subduction of the oceanic crust under the continental margin. Subduction caused attenuation of the continental margin forming a trench zone which was filled with more sediments, was formed. During subduction, slices of the oceanic crust were detached and obducted on the continental margin. Continued subduction led to the closure of the ocean and convergence of the two continental masses with subsequent folding of the consumed continental margin and the overlying sediments. The subducted oceanic crust was then mixed with the mantle. This resulted in its partial melting followed by fractionation and the intrusion of the calc-alkaline rocks.

El Fawakhir ophiolites

The region of the Egyptian basement lying to the northwest, west, southwest and south of the Meatiq dome is occupied by extensive ophiolitic rocks and melange. The Qift-Quseir highway affords an intact and complete section of ophiolite sequence (known as El-Fawakhir ophiolites) between the Hammamat conglomerates to the west and the Meatiq dome to the east, for a distance of about 15 km. In this section, all the ophiolite units are encountered from base to top successively from west to east. The section is characterized by the abundance of basaltic and diabasic dikes traversing all its units, particularly the gabbros and the pyroxenites, and by the paucity of melange. Toward the west, this section is intruded by the Fawakhir granite, and in its middle part it is covered by a northwest-southeast belt of conglomerates defined by Akaad & Noweir (1980) as the Atud Conglomerates. These conglomerates are very similar to the Hammamat conglomerates to the west, except for its fractured pebbles.

The western contact of El Fawakhir ophiolites with the Hammamat is an easterly dipping thrust zone, while its eastern contact with the Meatiq rocks is defined by a zone of tectonic melange, mylonitized rocks and highly stretched flaser gabbro. Between these two contacts, the ophiolite belt seems to form several open folds with synclinal troughs occupied by the conglomerates. These conglomerates, as well as the Hammamat conglomerates, contain abundant pebbles and boulders of pink granite which are supposed to be derived from the older granites. However, they do not contain ophiolite fragments. A depositional unconformity surface is believed to occur between these conglomerates and the underlying serpentinites along Wadi Um Seleimat. The petrographic characteristics of this ophiolite sequence are consistent with the characteristics of oceanic crust rocks that have suffered sea floor metamorphism (Nasseef et al. 1980).

The geochemical characteristics of the massive volcanics, sheeted dikes and pillow basalts of El Fawakhir ophiolite sequence show mixed characteristics of MORB and IAT. This strongly suggests their formation in a tensional environment such as mid-oceanic ridges or back-arc basins (Hassan et al. 1983) situated between an island arc to the east and a continental mass to the west. During the succeeding stage of convergence in the back arc basin, the whole sequence was later overthrust on the eroded island arc to the east, and probably also on the continental margin to the west, as a result of continued subduction under the island arc. The final disposition of the ophiolites and the overlying sediments was shaped by the updoming of the Meatiq Group, which caused the gliding of phiolitic nappes away from the core (Habib et al. 1982). This caused the overriding of the ophiolitic nappe of El Fawakhir on the Hammamat to the West, as well as the intensification of the thrusting effects on the base of the ophiolite sequence against the Meatiq Dome.

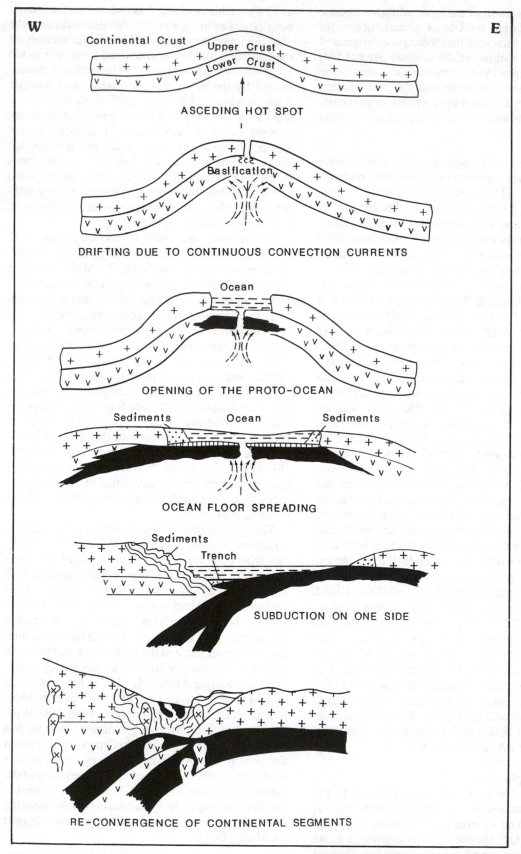

Figure 12.7 Diagrammatic model showing tectonic history of Wadi Ghadir ophiolites (after El Bayoumi 1980).

Other ophiolite occurrences

Several other ophiolite suites have been reported in the Eastern Desert, e.g. El Geneina area (Garson & Shalaby 1976), Wadi Zeidun area (Dixon 1979). At least four other suites were reported in unpublished theses: Wadi Mubarak, Wadi Bizeh (Abu El Ela 1985), Wadi El Gemal (Higazy 1984), and Gebel Mohagara (Takla et al. 1982). In addition, variably sized fragments of ophiolitic rocks are very common in the widespread Eastern Desert melange and also in the metasediments.

In Sinai, some amphibolitic gabbro bodies with cumulate ultramafic margins are exposed in the area of Wadi Feiran. Beyth et al. (1978) described two small masses of partly serpentinized harzburgite from Dhahab in the Gulf of Aqaba area. They are too small, only a few tens of meters in length, to justify any speculations as to their ophiolitic affiliation.

On the other hand, Dixon (1979) and Engel et al. (1980) showed that some of the ultramafic complexes in the Eastern Desert were originally emplaced as small sill-like bodies, intrusive into the metasediments. These ultramafic intrusives will be described in a later section.

Age of the ophiolites

In discussing the age of the ophiolites, it is necessary to distinguish clearly between the age of their formation originally as parts of oceanic crusts and the age of their emplacement in the present position. In the southern part of the Eastern Desert, undeformed volcanic rocks of 768 Ma age (Stern 1979) overlie intensely deformed allochthonous sheets of ophiolitic melange, which in turn overlie metasedimentary and metavolcanic units (Chuurch 1986). These metavolcanics gave a four-point Rb/Sr isochron age of 842 ± 22 Ma (recalculated by Hashad from the data presented by El Shazly et al. 1973). This implies that at least some of these ophiolites were emplaced between about 850 and 770 Ma.

Metasediments

This unit was first defined by El Ramly & Akaad (1960) as including a thick succession of mostly fine grained 'geosynclinal' sediments mainly in a low grade metamorphism and occasionally reaching a medium grade, surrounding the Hafafit-Migif domain. Greywackes and conglomerates are not uncommon. This unit was named the 'schist-mudstone-greywacke' series by these authors. It was later extended by various workers in other areas to incude immature, weakly metamorphosed sediments of flysch and turbidite nature and came to be known as the geosynclinal metasediments on the geological map. It was explained by most authors writing on the

Egyptian Shield as the sedimentary infilling of the geosyncline during its early stages, closely followed and overlapped by extrusion of volcanics and intrusion of serpentinites. Engel et al. (1980) believe that the metasediments are similar in most respects to the Archean greenstone belts. On the other hand, Church (1979) stated that the geosynclinal metasediments resemble the early stage 'exogeosynclinal' flysch deposits of the Appalachian and Alpine systems; such rocks have not been described from the Archean.

Except for the granites, the metasediments constitute the most extensively outcropping unit in the basement of Egypt. They occur mainly in the Eastern Desert south of latitude 26° N, with a very few outcrops north of it and in Sinai. It is a very heterogeneous unit, composed mainly of immature clastic and volcaniclastic sediments, ranging from mudstones to conglomerates. Carbonates, mature sandstones, and cherts with banded iron formations are of significant proportions. Intimate mixing with serpentinites and basic igneous rocks is very common and widespread especially around large serpentinite ranges. Metamorphic equivalents of the fine sediments and basic rocks are also common and include slates, schists, greenstones and, rarely, amphibolites.

The metasediments, as shown on the geologic map of Egypt (1981), may be divided into four assemblages as follows:

1. mature to semimature assemblage dominated by carbonates, quartzites and pelites;
2. immature assemblage dominated by flysch type volcaniclastics, mudstones and greywackes;
3. cherts and banded iron formations;
4. melange.

Each of these assemblages will be described separately and the relationships between them discussed.

Mature to semimature assemblage

This assemblage occurs mainly in the southernmost parts of the Eastern Desert (Hunting 1967, El Shazly et al. 1975). This assemblage is best developed in the Abu Swayel area. According to El Shazly et al. (1975), the metasediments in this area are composed of calc and calc-pelitic, and psammitic to psammopelitic metasediments. These are largely transformed into marbles, hornblende-biotite schists, biotite schists and quartzite and quartzofeldspathic schists. The biotite schists form the most dominant rock type, which are garnetiferous close to granitic contacts. Greywacke and conglomeratic horizons with well-shaped pebbles and cobbles are subordinate. Current and graded beddings are rare. This assemblage is considered by Hunting (1967) as suggesting an origin in shallow water approaching neritic conditions. The prevalence of intensely folded schists suggests that the southern part of the Eastern Desert represents a deeper level in the crust than the central part. Church (1982) considers the schists of Abu Swayel as representing continentally derived shelf deposits intercalated with mafic-felsic volcanics.

The exact distribution of this assemblage outside the Abu Swayel area, as well as its relation to other assemblages of metasediments, is largely unknown. It was noted by Hunting (1967) that at about latitude 24° N, there is a structural break and facies changes in the metasediments. The southern assemblage passes northward into the more immature facies of the 'schist-mudstone-greywacke' series as originally defined by El Ramly & Akaad (1960).

Immature assemblage

This covers large areas in the central Eastern Desert from latitude 24° N northward as far as 28° N. The assemblage comprises a very thick succession of clastic sedimentary and volcaniclastic rocks, mainly thinly bedded mudstones, tuffites, greywackes and less commonly, polymictic conglomerates. The assemblage represents a typical flysch-type sedimentation. The mudstones are very thinly laminated and composed essentially of quartz and feldspar flour mixed with chlorite, muscovite and subordinate clay minerals. Well-preserved sedimentary structures are very common in these rocks. They are highly compacted and indurated. Clastic fragments in greywackes include volcanic rocks, ultramafic rocks, gabbros, quartz and feldspars. Conglomerates are widely present among beds of this sequence, although in much lower proportions. The composition of these conglomerates varies from place to place; e.g. the Atud conglomerate is rich in mafic and ultramafic fragments (Akkad & Essawy 1965), while further west, the Dungash conglomerate is rich in fragments of calc-alkaline volcanics (Abu El Ela 1985).

The rocks of this sequence range from almost unmetamorphosed to highly foliated schists. The greywacke and conglomerates largely preserve their original sedimentary characters. The clastic fragments are mostly unchanged while the fine matrix is variably transformed into sericite, chlorite, epidote and actinolite. On the other hand, the mudstones and tuffites are much more affected by metamorphism. Passing from almost unmetamorphosed assemblages to completely recrystallized siliceous and argillaceous schists is common in many areas.

This assemblage possesses the features pointing to the turbidite origin of a great part of it, in moderate to deep water beyond neritic environment. Most important of these features are the graded and repetitive bedding, slump structures, rarity of cross and current bedding, and the general rarity of strongly calcareous beds.

From the association of the clasts of greywackes and conglomerates of this assemblage, some inferences may be made on its provenance. The mixing of ophiolitic fragments (ultramafics, gabbros, basalts) with sialic fragments (granites, quartzites, carbonates) in the clasts of this assemblage indicates that deposition took place in an oceanic trough with a nearby continental margin or a mature arc. This is also supported by the prevalence of calc-alkaline volcanic fragments (andesite, dacite and rhyolite) within them.

Chert-banded iron formations assemblage

This assemblage is of a very limited occurrence, being found only in the central Eastern Desert, approximately between latitude 25° 30' and 25° 50' N, in the areas of Wadi Kareim-El Dabbah and Wadi Zeidun-Arak. It can be considered as part of the immature assemblage, but has a significant proportion of siliceous and ferrugenous chemical sediments that represent a deeper water environment than the rest, probably the deepest environment within the whole metasediments. Within the areas mentioned above, three lithologic components may be distinguished (Stern 1979): (a) volcanogenic clastic metasediments; (b) marls; (c) banded iron formations.

Volcanogenic clastic metasediments are by far the most common constituents of the metasediments. They are very similar to the immature assemblage described above, but are of a dominantly volcanogenic composition. Chemical analyses of these metasediments show their affinities to andesitic and dacitic volcanic rocks (Stern 1979).

Marl layers are common in this succession, although very subordinate, occurring in bands of less than 1 m thick. They are found in the western part and are absent in the eastern part.

Banded iron formations and jaspers occur as intercalations within the wacke-rich sections. They are composed of dark iron-rich laminae alternating with jasper and other siliceous layers, and have few interbeds of volcanic, mineral and lithic fragments.

The Egyptian banded iron formations differ than those of other localities in being considerably younger in age. Banded iron formations in other localities are restricted to the Lower Proterozoic and the Archean. (Stern 1979).

Melange assemblage.

Melange will be used here as a term denoting a mappable body composed of fragments of all sizes of a variety of rock types in a pervasively deformed matrix that may be formed by sedimentary or tectonic processes or a combination of both (Penrose Melange Conference 1978). Melange occurs widely in the Eastern Desert of Egypt in association with the metasediments and metavolcanics. Ries et al. (1983) believe that the rock groups from the metasediments to the metagabbro-diorite complex, (or the Abu Ziran and Rubshi Groups of Akaad & Noweir 1980) constitute an ophiolitic melange and not a stratigraphic sequence, proposing to designate this melange as the Eastern Desert melange. Although melange is of wide occurrence in the Eastern desert, it occurs as belts within the metasediments. These belts are mostly thrust sheets or slices incorporated within the autochthonous belts of the metasediments.

A characteristic feature of the melange in the Eastern Desert is the significant proportion of serpentinites it contains, either as a matrix or as variably sized blocks. The serpentinites are commonly transformed into creamy colored talc carbonate schists that impart a very characteristic appearance to the melange outcrops.

The most common fragments and clasts in the melange are ophiolitic fragments, deep sea sediments and calc-alkaline volcanics. Less common fragments are granitic rocks, carbonate rocks, quartzites and quartz sandstones. Pebbly mudstones, with pebbles of both oceanic and continental derivation, occur as components in the melange, either as thrust slices or as olistoliths. In places, melange screens occur alternating with laminated mudstones and greywackes, probably as imbricate thrust sheets. On the other hand, the immature metasediments commonly contain isolated ophiolitic fragments of various sizes, as well as detached huge blocks of serpentine melange.

Melange of Wadi Ghadir area. The best example of melange yet described in the Eastern Desert is the Ghadir melange, first recognized and studied by El Bayoumi (1984).

The melange of this area may be divided into two facies, proximal and distal, in relation to the source of its ophiolitic components. The proximal facies is found mainly to the north and south of Wadi El Ghadir and to the east of Wadi Lawi, whereas the distal facies occurs in the northwestern part of Wadi El Ghadir (Fig. 12/6).

The proximal facies is composed of rolled and fragmented rock-debris of highly variable sizes in a matrix of scaly and schistose mudstones. Serpentinized peridotite blocks are the most abundant components in this facies. Some of these are surrounded by a sheath of schistose talc-carbonate rock due to squeezing and rolling of the blocks. Probably such blocks acted as wheels on which rock masses were moved. Second in abundance are disrupted and fragmented parts of dikes of variable sizes which are mixed and squeezed with talc carbonate rocks. Other rock types recognized among the debris of the melange are various types of volcanic rocks, greywackes, quart-

zites, chert, marble, shale, granite and other plutonic rocks, amphibolites and schistose rocks. In certain places, pebbles and cobbles of rock debris are the main components and form conglomerates and breccias. Commonly, the pebbles and cobbles are stretched due to deformation.

The distal facies is composed of low grade schists, mostly of pelitic composition. It also contains pebbles of other rock types like those in the proximal facies, but of much less abundance and smaller sizes. In several places, e.g. in the area northwest of the junction of Wadi El Ghadir and Wadi Lawi (Fig. 12/6), highly schistose talc carbonate rock of creamy color occurs as pockets and lenses within the distal facies. These pockets and lenses might have been blocks of serpentinized peridotites within a matrix of the distal facies. Metamorphism and deformation of the distal facies might have been caused by the overriding of slices of proximal facies in the form of nappes, remnants of which are common in various parts of the area.

According to El Bayoumi (1984), the Ghadir melange accumulated in a trench zone formed as a result of convergence of two plates, resulting in the subduction of an ocean crust on the east under a consumed continental margin on the west. This process resulted in the destruction and mass wastage of both ophiolites and continental margin, followed by gravity sliding and deposition of the tectonic fragments in the trench. Detached ophiolites, mostly serpentinite fragments, acted as a carpet under the moving slices. Erosion and wastage of the continental margin took place after its uplift as a result of collision. The oceanic and continental fragments are mixed in the trench zone. As a result of huge accumulations of sediments with continued westward movement, deformation and metamorphism modified the area. The whole area was then dissected by volcanic calc-alkaline dikes. Intrusion of the dikes was continuous during the formation of the melange. Finally, the area was intruded by calc-alkaline granites and leucogabbros. The calc-alkaline rocks (dikes, granites and leucogabbro) are believed to be intrusive fractions resulting from the active Benioff zone of the subducted oceanic crust. A diagrammatic representation of the melange formation is shown in Figure 12/8.

Age of the metasediments

Hashad (1980) compiled the Rb/Sr isochron dates reported from metasediments prior to 1980 from several sources. These ages range from about 1200 to 750 Ma, with initial $^{87}Sr/^{86}Sr$ ratios between 0.7023 and 0.7064. The major drawback of these ages is that sampling was carried out in the absence of an appropriate understanding of their lithostratigraphic and/or tectonic position. The wide range of these ages

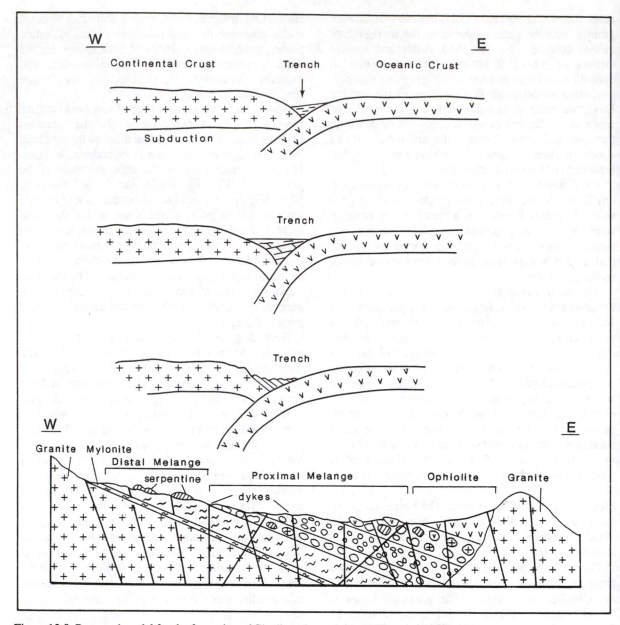

Figure 12.8 Proposed model for the formation of Ghadir melange (after El Bayoumi 1980).

most probably indicates that the samples represent at least two different events. For example, the garnet mica schist from Abu Swayel mine gave a four-point whole rock Rb/Sr isochron age of 1160 ± 144 Ma (Ri = 0.7061 ± 0.0018, MSWD = 0.02) as recalculated from the analytical data presented by El Shazly et al. (1973). Another six-point isochron on hornblende and mica schist from Wadi Haimur in Abu Swayel area produced an age of 1220 ± 62 Ma (Ri = 0.7056 ± 0.0009, MSWD = 0.48). Some other similar ages were obtained in the central Eastern Desert by El Manharawy (1977). We believe that samples giving

the older ages (c. 1200 Ma) represent pre-Pan African rocks, similar to those in the Hafafit and Nugrus areas (see section on gneisses), although they were not recognized on the present maps as such because of their lithological similarity to the Pan African meta-sediments. We believe also that a great part of the schists and amphibolites in the Abu Swayel area, which were included within the mature to semimature assemblage of the metasediments, actually belong to the pre-Pan African cycle.

The existence of a metamorphic event at about 1200 Ma recalls the Aleksod event of the Tuareg

Shield in West Africa and events of approximately the same age elsewhere in Africa (Cahen et al. 1984). Dixon (1981) reported U/Pb ages from zircons in granitic pebbles in conglomerates of the metasediments that range from 2300 to 1100 Ma, and considered these pebbles to have been derived from plutons in areas adjacent to the Egyptian Shield.

Metavolcanics

All earlier workers recognized that an important group of metavolcanics occurs among the basement rocks either partly contemporaneous with or immediately younger than the immature volcaniclastic metasediments, with which they are closely associated. These metavolcanics have been referred to as the Shadli metavolcanics, named after their type locality in the Sheikh El Shadli area about 70 km southwest of Marsa Alam (Fig. 12/9). They were explained as products of volcanic activity that took place toward the final phase of the filling of the geosyncline and continued until after the cessation of deposition. During this activity, lavas of considerable thickness were extruded and sometimes intruded at shallow depths the associated sediments (Akaad & El Ramly 1961). These volcanics are essentially of two types: basic and intermediate varieties including low grade metadolerites, metabasalts, metaandesites and metaporphyrites; and acid varieties including rhyolites, tuffstones and pyroclastics (El Ramly & Akaad 1960). They are metamorphosed into the greenschist facies and locally into the lower amphibolite facies.

To illustrate the characteristics of these metavolcanics, some of the better studied areas will be described in some detail.

Shadli metavolcanics

These cover more than 1500 km² in the area of Um Samuiki-Abu-Hammamid and around Sheikh El Shadli Tomb, after which they were named. Shukri & Mansour (1980) show that they are made up of rhyolites (32.5%), andesites (42.5%), basalts (6.5%) and volcaniclastic rocks (15%). These constitute a thick pile (10 km) of cyclic submarine volcanic flows and bedded volcaniclastics which unconformably overlie a series of metasediments, mostly of the immature assemblage. According to these authors, the Shadli metavolcanics can be differentiated into four mappable units. Beginning with the oldest, they are the Marasan Metavolcanics (1.5 km thick), Huluz Metavolcanics (3.5 km), Burrad Metavolcanics (2.8 km) and the Hammamid Metavolcanics (average thickness 2.6 km). Together with the metasediments, they form a west-northwest-east-southeast trending synclinorium, flanked by two large anticlinoria.

Searle et al. (1976) subdivided the Shadli metavolcanics in the Samuiki area into two major units (Table 2) and stressed the cyclic nature of the Hammamid Group which was formed in an island arc setting.

Kotb (1983) studied the geochemistry and petrochemistry of these volcanics in Gebel Abu Hammamid and showed that they are similar to recent volcanics occurring along immature island arcs as a result of subduction of oceanic lithosphere at convergent plate margins. The volcanics were derived by fractionation of a tholeiitic magma type and extruded successively in pulses. They are comparable to island arc volcanics in Saudi Arabia between Jabal Ibrahim and Al Aqiq. Hashad & Hassan (1979) believe that the presence of large percentage of felsic rocks with the calc-alkaline rocks of the Shadli Group does not seem to support an exclusively ensimatic island arc origin for Egyptian Shield.

Hashad (1980) reported three radiometric dates for Shadli metavolcanics, citing other authors. A K/Ar age of 825 Ma has been obtained on rhyolites of the Hammamid range. A model Pb/Pb age of 1070 Ma was obtained on galena from sulfide ore within the Abu Hammamid volcanics. Metavolcanics of the Abu Swayel area, which are considered equivalent to Shadli metavolcanics by El Shazly et al. (1973), gave a Rb/Sr isochron age of 842 ± 22 Ma, but the rocks dated are highly weathered, with very low Rb/Sr ratios.

Metavolcanics in the central Eastern Desert

Stern (1979) studied the metavolcanics in the central Eastern Desert between latitudes 25° 40' and 26° 10' N and found that they can be divided into two units: the first lying below the immature metasediments and forming a substrate to it, and the second overlying these sediments. He called the first unit the old metavolcanics and the second the young metavolcanics.

The *old metavolcanics* are composed of a thick succession of monotonous aphyric pillowed metabasalts, with rare sedimentary interbeds. They can be identified at either end of the Precambrian outcrops along the Qift-Quseir road. Their extent to the north and south is not easily determined. Northward, they seem to grade into amphibolites as the batholithic terrain of the north Eastern Desert is approached. However, unequivocal pillowed old metavolcanics occur 20 km north of El Fawakhir. To the south, greater intensity of deformation obscures their nature. Pillowed basalts of ophiolitic nature were reported as a part of ophiolitic successions in El Geneina-El Gharbia area, midway between Berenice and Aswan (Kamel et al. 1980). However, the exact extent remains to be defined. Geochemically, the old metavolcanics have strong affinities to modern mid-ocean ridges and marginal basin tholeiites (Stern 1979, Engel et al. 1980). Although the possibility that

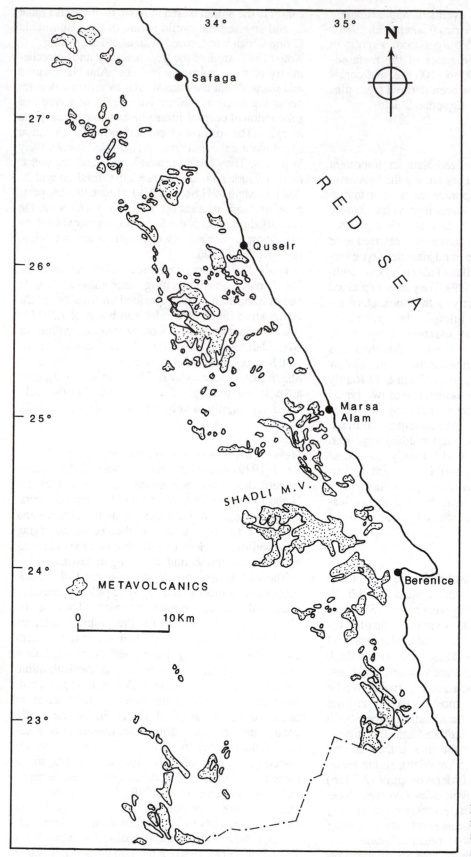

Figure 12.9 Distribution of the metavolcanics in the Eastern Desert of Egypt (after El Ramly 1972).

Table 2. Shadli metavolcanics subdivision (after Searle et al. 1978).

		DOLERITE	

SHADLI VOLCANICS	HAMAMID GROUP	CYCLE (II)		Upper banded tuffs and cherts
			acid)	Volcanic breccia
			lava)	Dacite and related intrusions
			phase)	Volcanic breccia and turbidities
			basic)	Upper pillow lavas and hyaloclastites
			lava)	
			phase)	Basic vent breccia and lavas
		CYCLE (I)		Lower banded tuffs and cherts
			acid)	
			lava)	Acid vent rocks and mineralization
			phase)	
			basic)	Microdiorite
			lava)	
			phase)	Lower pillow lavas
	Wadi Um Samuiki Volcanics			Basic and intermediate lavas with rhyolite and acid metavolcanics.

this sequence is analogous to island arc tholeiites cannot be precluded, the bulk of the data does not favor this hypothesis. They are interpreted as the products of about 20% fractional fusion of anhydrous mantle peridotite at depths of less than 60 km.

Dixon (1979) put forward the hypothesis that the old metavolcanics formed a substrate on which the young metavolcanics were erupted. These old metavolcanics, together with the associated ultramafic rocks in the Zeidun-Massar area represent late Proterozoic sea floor, thrust from the east onto a calc-alkaline volcanic edifice with intercalated volcaniclastic sediments. The sole of the thrust is marked by a prominent carbonate horizon. El Mezayen (1984) and Nasseef et al. (1980) show that the old metavolcanics in El Fawakhir Um Seleimat district are part of an ophiolite succession (see section on ophiolites). Correlatives to the old metavolcanics are found within the ophiolites of northern Saudi Arabia.

The *young metavolcanics* are composed predominantly of andesitic flows but mafic and more felsic volcanics, as well as volcaniclastic breccias and tuffs are also common. They occupy a stratigraphic position on top of the immature metasediments, but some parts may be lateral equivalents to the metasediments.

The young metavolcanics may be distinguished from the old metavolcanics in the field by the abundance of porphyritic and more felsic lithologies in the former succession. Even where the young metavolcanics are basaltic, plagioclase and pyroxene porphyries are in marked contrast to the aphyric metabasalts of the old metavolcanics. Also the young metavolca-

nics are characterized by the abundance of volcaniclastic metasediments, rarity of large pillows, and
rarity of associated serpentinites. Stern (1979) indicate that the thickness of the young metavolcanics in
the central Eastern Desert is at least 1 km. He also
suggests that the pillowed young metavolcanics to
the east are more mafic than the massive metavolcanics to the west and that their eruptions occurred in
progressively shallower water toward the west.

The young metavolcanics show the effects of low
grade metamorphism. However, their original mineralogy is generally well-preserved. The andesites are
largely plagioclase-augite porphyries. The subordinate rhyodacites are both aphyric and porphyritic.
These overlie or interfinger with the andesitic units.
Phenocrysts in the rhyodacites are typically plagioclase and quartz. The volcanic rocks have chemical characteristics similar to many contemporary
calc-alkaline sequences erupted at convergent plate
margins. Rb/Sr isochron dates on them give a range
of 610 to 700 million years, with Sr isotope ratios
between 0.7019 ± 0.0007 and 0.7030 ± 0.0001 (Stern
1979). Dixon (1979) obtained a zircon fission track
age of 639 ± 40 Ma on calc-alkaline metavolcanics in
Wadi Arak, about 75 km southwest of Quseir, which
are believed to belong to the young metavolcanics of
Stern (1979).

Metavolcanics in other areas

El Ramly et al. (1982) studied the metavolcanics of
the Kolet Um Kharit area (Fig. 12/10) and showed
that they are the first bimodal volcanics to be
described in the Egyptian shield. They are composed
of massive basalt-andesite lavas, overlain by rhyolite
and rhyolitic pyroclastics. Thin beds of banded tuffs
and cherts occur as intercalations between the rhyolites, thus marking successive periods of volcanic
activity. The metavolcanics rest on the older quartzofeldspathic gneisses and are in turn intruded by plutonic rocks. They are classified into two main rock
types: acid and basic, with minor intermediate types.
The associated pyroclastics are essentially of acidic
composition. Petrochemically and geochemically,
these metavolcanics have a mixed tholeiitic calc-
alkaline character for the basic metavolcanics and
predominantly calc-alkaline character in the acid
types. Presently, such magma type(s) are generated at
converging plate margins in island and volcanic arcs
formed over subduction zones.

At Um Khariga, metavolcanics cover about 755
km², 15 to 20 km west of Mersa Alam. They are
composed of both acidic and basic metavolcanics as
well as abundant pyroclastics, with some intercalations of tuffaceous metasediments.

Takla et al. (1982) reported a spilite association
with the ophiolite suite of Gebel Mohagara that lies to
the northwest of the Wadi Ghadir area. These rocks
occur as deformed massive bodies and as pillowed
lavas, with pillows reaching up to 50 cm across. The
spilites are composed of albite (An_{0-5}) and augite
with variable amounts of hornblende, chlorite, ilmenite, sphene and iddingsite. Varieties are albito-
phyres with very low mafic mineral content.

Dokhan volcanics

Sequences of acidic and intermediate volcanics and
equivalent pyroclastics, together with minor amounts
of volcaniclastic sediments occur widely in the Eastern Desert. Although they do not form extensive
outcrop areas as do other rock units, they have been
reported from all over the Eastern Desert from the
northernmost tip of the Precambrian rock outcrops to
the south of Wadi Allaqi, west of Abu Swayel. Earlier
records of these volcanics were reported by Andrew
(1938) and Hume (1934) in the Gebel Dokhan area,
where they were called the Dokhan Volcanics. Later
workers extended their extent on the basis of petrographic similarity and stratigraphic position.
Several workers in several areas showed that these
volcanics are almost non-metamorphosed and contain pebbles from the older syntectonic granites and,
on the other hand, supplied the Hammamat conglomerates with pebbles. Thus their time of formation is
easily bracketed. These volcanics are characterized
by the prevalence of porphyritic types and a distinct
purple color. A famous deep purple porphyritic andesite rock in Gebel Dokhan is known as Imperial
Porphyry. It was quarried as ornamental stone in
Roman times (Ghobrial & Lotfi 1967).

Dokhan volcanics in the type area

Basta et al. (1980) carried out a comprehensive study
on the Dokhan volcanics in their type locality. The
following descriptions are based on this work.

Gebel Dokhan area is covered mainly with Dokhan volcanics which cut minor exposures of metasediments and are in turn intruded by the 'younger or
post-tectonic granites' of the Eastern Desert of Egypt
(Fig. 12/11). The volcanics constitute a thick sequence of stratified lava flows of intermediate to
acidic composition, together with subordinate sheets
of ignimbrite and a few intercalations of
pyroclastics and siltstone. The contact with the granite is rather sharp and chilled, dipping 80° E. The
maximum thickness of the sequence is 1200 m, and
the flows (10 to 50 m thick), though practically
unfolded, are often tilted and crushed, particularly
near contacts with the granite. The rocks are mostly
porphyritic and amygdaloidal of different shades of
grey that frequently grades into the reddish to deep
purple color of the Imperial Porphyry. The base of

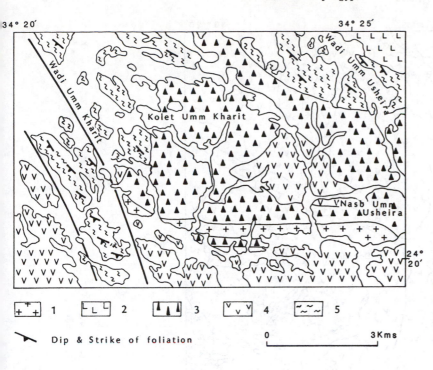

Figure 12.10 Simplified geologic map of Kolet Um Kharit area (after El Ramly et al. 1982). 1. Biotite syenogranite; 2. Metagabbro-diorite complex; 3. Acid metavolcanics; 4. Basic metavolcanics; 5. Quartzofeldspathic muscovite gneiss.

each individual flow is usually chilled, with an increase in size of the porphyritic crystals upward. Few flows, however, are non-porphyritic. Alternations of intermediate (andesite and Imperial porphyry) and acidic (rhyodacite and quartz latite) flows extend over most of the area, with a tendency for the former to prevail near the top of the formation. Intercalations of volcanic agglomerates and lapilli tuffs are locally more abundant and contain volcanic bombs up to 20 cm in diameter. A few sheets or small lenticular masses of very fine grained greyish-violet ignimbrite are also encountered.

Petrographically, seven rock types are distinguished: andesite, quartz andesite, Imperial porphyry, dacite, rhyodacite, quartz latite and quartz trachyte, as well as pyroclastics. The first three rock types are predominant. The pyroclastics are represented mainly by coarse andesitic lithic crystal tuff with a mineralogical and chemical composition approaching that of the quartz andesite of the Dokhan volcanics.

Based on the analyses of 48 samples of the Dokhan volcanics, Basta et al. (1980) showed that they compare well with the average composition of the corresponding world volcanics except for minor deviations and constitute a calc-alkalic series. Their fractionation sequence is: andesite, quartz andesite, Imperial porphyry, dacite, quartz latite, rhyodacite, quartz trachyte. The presence of latite and trachyte in such a predominantly calc-alkalic volcanic series may be attributed to a sudden potash enrichment of the magma during the late stages of crystallization. These rocks are quartz-bearing types, and they are of so low frequency compared to other rock types in the suite that the main sequence could be considered as follows: andesite, quartz andesite (including Imperial porphyry), dacite and rhyodacite. Furthermore, the Dokhan volcanics have a composition that is comparable with that of active continental margins and well-developed island arcs with a thick continental crust. Basta et al. (1980) explained the origin of Dokhan volcanics by the fractional crystallization of the basaltic magma generated by the partial melting of the upper mantle under relatively high oxygen fugacity. The change in the proportion of calc-alkalic and tholeiitic rocks with advancing development of continental-type crust may be regarded as a result of progressive depletion of the upper mantle underlying the island arc in basaltic components (Miyashiro 1973). It is also possible that casual removal of oxygen from a calc-alkalic series magma may transfer the crystallization into a tholeiitic trend, resulting in a minor proportion of tholeiitic series rocks associated with predominantly calc-alkalic ones, as is the case in the Dokhan.

Dokhan volcanics in other areas

Several workers described Dokhan volcanics in other areas, and all have agreed to their calc-alkaline nature. Ignimbrites have been identified among Dokhan volcanics west of Safaga (Gindy & Mohammed 1971). Heikal et al. (1980) described rhyolitic ig-

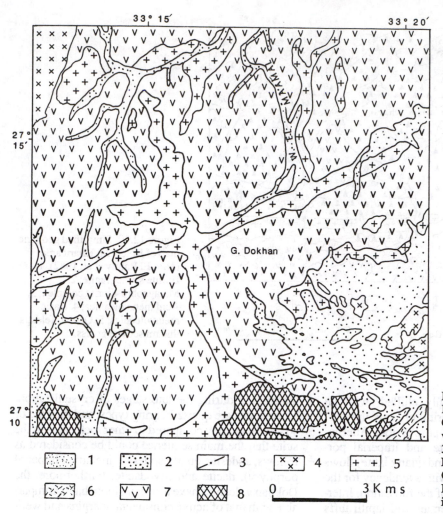

Figure 12.11 Geologic map of Dokhan area, Eastern Desert (after Ghobrial & Lotfi 1967). 1. Recent wadi deposits; 2. Gravel terraces; 3. Basic and intermediate dikes; 4. Red granite; 5. Pink granite; 6. Quartz diorite-granodiorite; 7. Dokhan volcanics; 8. Metasediments.

nimbrites in the Wassif area southwest of Safaga. The Wassif volcanics occur in the form of successive sheets of lavas and pyroclastic rocks. They are mainly andesitic and rhyolitic in composition, with minor rhyodacite. The pyroclastic rocks are lapilli tuffs, vitric crystal tuffs and lithic tuffs. The ignimbrite forms a lava-like sheet more than 25 m thick. Its main part shows a lava-like appearance of a quartz porphyry. Welding is well-developed in the central part of the flow.

In the Hammash-Sufra district, about 100 km southwest of Mersa Alam, volcanics of the Dokhan type were reported by Moustafa & Akaad (1962). They are mostly of intermediate to acidic composition, including well-bedded agglomerates. Similar occurrences were also reported in Wadi Shait and Wadi Ranga by Akaad & El Ramly (1958).

In the central Eastern Desert, Ries et al. (1983) reported the occurrence of Dokhan volcanics in several areas. In the Wadi Sodmein area north of Gebel Meatiq, they are composed of rhyolites or rhyodacites, quartz-feldspar porphyries, andesites and basalts. Characteristic rock types include plagioclase-phyric andesites, flow-banded rhyolites, agglomerates with lithophysae, quartz porphyries, together with tuffs, mudstones and some conglomerates containing pebbles of volcanic rocks and chert. A series of similar rocks, is exposed in Wadi Hammamat in a thin thrust slice between the melange and the Hammamat sediments (Ries et al, 1983). Also in Wadi Kareim, south of the Qift-Quseir road, a similar succession is encountered. Ries et al. (1983) do not differentiate between the young metavolcanics of Stern (1979) and the Dokhan volcanics, and consider both as one unit, 'calc-alkaline volcanics'.

In Sinai, two groups of volcano-sedimentary successions occur (Shimron 1984). The calc-alkaline volcanics of these two groups are believed to be equivalent to the Dokhan volcanics in the Eastern Desert.

In summary, there is enough evidence to indicate that the Dokhan volcanics were formed due to volcanic activity in mature island arcs. They were erupted in both subaerial and subaqueous environments. Although they are mostly of acidic and intermediate composition, basalts are not uncommon. According to Stern et al. (1984), the Dokhan volcanics are characterized as follows:

1. In some regions the Dokhan is a bimodal suite characterized by andesite (50 to 62% SiO_2) strongly enriched in TiO_2 (1 to 2%), K_2O (1.5 to 2.5%), P_2O_5 (0.3 to 0.6%), and light rare earth elements $[(Ce/Yb)_N = 10]$ in association with rhyolitic ignimbrite and related felsic rocks (72 to 76%) with $K_2O/Na_2O = 1$, low P_2O_5 (0.02 to 0.04%), slightly elevated TiO_2 (0.3%), and strongly negative Eu anomalies.

2. In some areas hypabyssal rhyolite porphyry is found as a magmatic facies intermediate between ignimbrite and granite, suggesting that rhyolitic volcanism, (which accompanied epizonal emplacement of granite) was an important aspect of crustal evolution.

3. Andesitic dikes clearly represent feeders for some andesitic flows.

4. Structural trends in Hammamat and Dokhan outcrops indicate that these were deposited and erupted in basins that were elongated east-west to northeast-southwest.

Stern (1979) obtained a whole rock isochron age of 602 ± 13 Ma from the Dokhan volcanics, with an initial ratio of 0.7028 ± 0.0002. The volcanic sequence in Wadi Sodmein gave an Rb/Sr whole-rock age of 616 ± 9 Ma with an intercept of 0.70271 ± 2 (2 sigma errors, MSWD = 5.0) which is interpreted as the age of extrusion of the lavas (Ries & Darbyshire in press, reported by Ries et al. 1983).

Engel et al. (1980) believe that the Dokhan volcanics represent the surface indication of the first stage of Pan African LIL-enriched plutonic activity.

Hammamat sediments

The Hammamat sediments comprise a thick succession of clastic sediments that represent the youngest unit in the layered sequences of the basement. The name 'Hammamat' is after the type locality in Wadi El Hammamat, approximately 70 km west-southwest of Quseir (Fig. 12/12). They were termed the Hammamat series (Hume 1934), younger metasediments (Sabet 1962), postgeosynclinal sediments (El Shazly 1977) and the Hammamat Group (Akaad & Noweir 1969).

These sediments crop out in isolated exposures mainly in the central and northern Eastern Desert. They are best developed in Wadi Hammamat area, their type locality, where they attain a thickness of about 4000 m (Akaad & Noweir 1980), in Wadi Kareim area with a total thickness of more than 6000

m (Abdel Aziz 1968), and in the Wadi Arak area with a thickness of more than 5000 m (Assaf 1973). These outcrops occur in down-faulted blocks or in topographic lows. The sediments are regionally folded into plunging folds and bounded by regional faults. They overlie the older units unconformably with a basal conglomerate in most cases, and in turn are intruded locally by younger granitoids and later intrusives.

Lithologic composition

According to several authors (Abdel Aziz 1968, Hafez 1970, Assaf 1973, El Ghawaby 1973, Grothaus et al. 1979, Akaad & Noweir 1980), the Hammamat sediments are composed totally of immature to semimature clastic sediments including conglomerates, greywackes, sandstones and slates. They exhibit characteristic green, purple and grey colors. In Wadi Arak area, some dolomitic bands are interbedded with the slate beds (Assaf 1973).

Conglomerates range from coarse to fine conglomerates, with the former dominating. In the fine conglomerates, the clasts are mostly subrounded to subangular and are commonly matrix supported. They are composed of a great assortment of rock fragments such as metasediments, metavolcanics, granites and, less abundantly, ophiolitic fragments. The granitic clasts are most abundant in the uppermost horizons of the succession. Some boulders of the coarse conglomerates are previously deposited conglomerates. The matrix is composed of pebbly sand size fragments of the same composition. The conglomerates are either massive without internal signs of bedding, or they are bedded, or even show graded bedding. In the case of massive conglomerates, bedding is discerned by the alternation of the conglomerates with sandstones or alternation of conglomerate beds of variable clast size. The clasts are mostly aligned parallel to the bedding, but occasionally show imbricaton.

Sandstones are composed of the same material as the matrix of the conglomerates. They form beds or lenses within the succession. Some of them are pebbly sandstones in which clast size reaches 15 cm in the longest dimension. The pebbly varieties are mostly matrix supported, but clast supported types are not uncommon. The sandstone beds and lenses are either massive or cross-bedded, especially the thick beds and lenses. Grothaus et al. (1979) distinguishes four types of cross bedding in these sandstones: low to intermediate angle tangential, scour and fill, channel fill, and angular cross stratifiction.

Siltstones occur as isolated beds and lenses through-

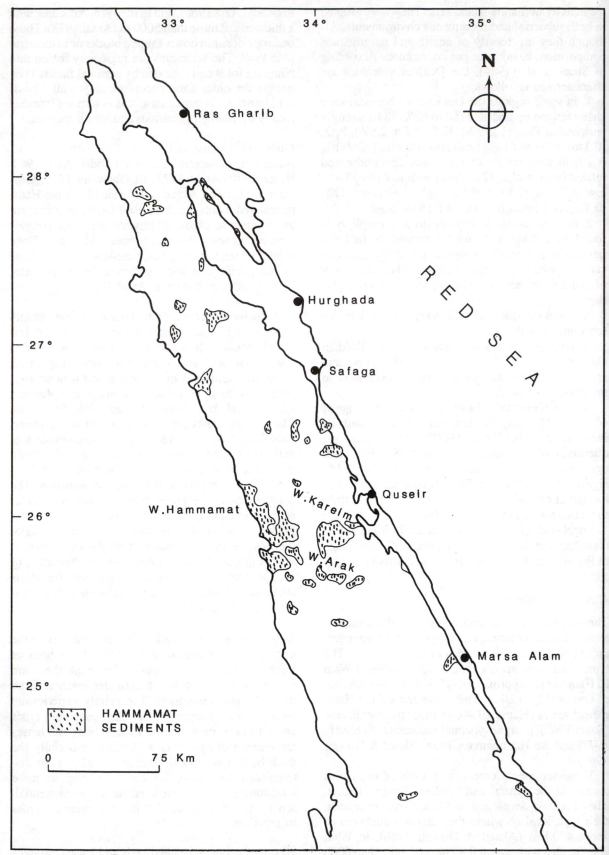

Figure 12.12 Exposures of the Hammamat sediments in the Eastern Desert.

out the Hammamat section, but mostly in the sandstone dominated areas. They are either green or red in equal proportions, although one color may dominate in a certain area. The siltstones are either massive or rippled and mudcracked. The massive siltstones occur in sections up to 100 m thick. They are usually part of a fining upward cycle grading from pebbly sandstone or coarse sandstone to siltstone. In some places, this progression is interrupted by isolated interbedded units of sandstones or conglomerates.

Slates occur as thin interbeds within the siltstone succession. They are either grey, green or red in color. Foliation is almost parallel to bedding, with pencil-like and mullion structures due to crenulation.

Assaf (1973) reported the occurrence of dolomitic bands interbedded with grey and green slate beds in the Wadi Arak area. These dolomitic bands exhibit red and brownish colors, but fresh surfaces are grey.

Classification

Various workers have used different concepts in mapping and classifying the Hammamat sediments. The most common of these is the lithologic classification in which the sediments are divided into units using rock type, texture and color as the criteria of division (e.g. Hafez 1970, Assaf 1973, El Ghawaby 1973). El Shazly (1977) termed these sediments, the 'Wadi El Hammamat Supergroup' divided into an earlier Wadi Kareem Group and a later Wadi El Mahdaf Formation. The Wadi Kareem Group was deposited before the erosional exposure of the late orogenic granites. The Wadi El Mahdaf Formation contains, among other elements, abundant granite pebbles considered to have been derived from the earliest types of the late orogenic granites.

Another stratigraphic classification of the Hammamat sediments is given by Akaad & Noweir (1980). They consider the Hammamat sediments a group composed of two formations: the Igla Formation below and the El Shihimiya Formation above. The Igla Formation is composed of sandstones, siltstones and silty mudstones with a basal conglomerate. The Shihimiya Formation is composed of conglomerates, greywackes and sandstones. Each formation was further subdivided into members.

M.L. Abdel Khalek (pers. comm.) suggests that in many cases the greywackes, sandstones, siltstones and mudstones are facies variants of one unit rather than representing different time stratigraphic units.

The Hammamat sediments are regionally unmetamorphosed (Akaad 1972). Noweir (in Akaad 1972) stated that the definite criterion of regional metamorphism, schistosity, is lacking in the largest occurrence of these sediments in Wadi El Hammamat, although the rocks show a marked degree of induration. However, the Umm Had granite pluton, which intruded the thickest succession of these sediments in Wadi El Hammamat, forms a wide contact aureole. In this aureole, the Hammamat sediments are thermally metamorphosed. Away from the aureole, the conglomerates are highly indurated and exhibit an overall appearance of a green breccia, the famous monumental 'breccia verde' quarried in ancient times. Similar metamorphism and induration effects are not uncommon in other areas. These are attributed to the effect of concealed granitic intrusions, and are not due to regional metamorphism (Akaad 1972).

Depositional environment

Most workers consider the Hammamat sediments as post orogenic molasse deposited in disconnected intermontan basins as a result of rapid uplift and erosion (Hafez 1970, Assaf 1973, El Ghawaby 1973, El Shazly 1977, Akaad & Noweir 1980). That these sediments were deposited in fluctuating basins or troughs of uneven topography is attested to by the thinning and disappearance along the strike of numerous intercalations and the presence of volcanic inliers in the midst of the conglomerates. According to Grothaus et al. (1979), the composition and textur of the Hammamat lithofacies suggest that they are the result of an alluvial fan-braided stream complex. Deposition appears most likely to have been in a series of intermontan basins, rather than one large basin. The detritus shed from the uplifted terrain was deposited adjacent to the uplifts. The finer grained facies are interpreted to have been deposited in two different environments. The thinner siltstone units in which ripple marks and mudcracks are more prevalent are interpreted as cutoff channel deposits within braided streams, whereas the much thicker units are viewed as playa or lake sediments.

Ries et al. (1983) discuss the age of the deposition of the Hammamat sediments. They state that it is bracketed between 616 ± 9 and 590 ± 11 Ma. These two ages are Rb/Sr whole rock ages given by Ries & Darbyshire (in press). The first age is of calc-alkaline volcanic rocks in Wadi Sodmein (believed to be equivalent to Dokhan volcanics) which underline unconformably the Hammamat sediments. The second age is of the Umm Had granite which intruded the Hammamat sediments in their type locality.

Questionable Hammamat conglomerates

Akaad & Noweir (1980) mapped a northnorthwest-southsoutheast trending, 30 km long belt crossing Qift-Quseir highway. This belt comprises a bedded and repeatedly alternating succession of polymictic metaconglomerates and metagreywackes, with minor metamudstones, schists and intraformational

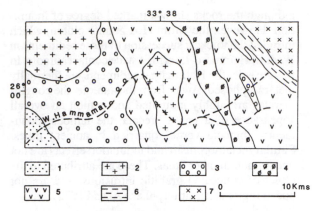

Figure 12.13 Simplified sketch map of part of Qift-Quseir road (after Akaad & Noweir 1980). 1. Nubian sandstone; 2. Younger granite; 3. Hammamat group; 4. Atud conglomerate; 5. El Fawakhir ophiolites; 6. Tectonic melange; 7. Meatiq gneisses and schists.

metabreccia. Part of this belt, north and south of the Qift-Quseir highway was investigated by the authors (Fig. 12/13). It is composed of bedded conglomerates, with a few interbeds of sandstone and greywackes. The clasts are mostly composed of acidic volcanics, pink granites, granite porphyry and chert. A characteristic feature in these conglomerates is the intense stretching, flattening and fracturing of the clasts. They are either stretched into long ellipsoidal bodies with tapering masses of the matrix at both ends, flattened into disc shaped bodies, or pervasively fractured by two conjugate shears with en echelon slippage on the shear planes. These conglomerates occupy a synclinal trough within the ophiolites and melanges. Along the eastern boundary, the lowermost few tens of meters are composed of alternating mudstones and greywackes which become coarser upward. In places, these are separated from the underlying melange by rubbly conglomerate with a talcose matrix and abundant ophiolite boulders representing the erosional surface of the melange. On the western boundary of the belt, the bedded conglomerates come in direct contact with the ophiolitic melanges and pillow basalt. It seems to be a faulted contact.

The conglomerates in this belt are very similar to the Hammamat conglomerates to the west, except for the stretching, flattening and fracturing of the pebbles. It is believed that they represent a lower part of the Hammamat sediments in this area. The same conclusion is also reached by Ries et al. (1983).

A similar situation also occurs in Wadi Zeidun area, and was reported by El Ghawaby (1973). In this area, he divided the Hammamat sediments into lower and upper units, separated by an unconformity. The lower conglomerates have the clasts flattened and stretched, which is not the case in the upper conglomerates.

INTRUSIVE ROCKS

The intrusive rocks of the Egyptian basement appear on the goelogic map of Egypt as eight units, namely:

1. Serpentinites
2. Metagabbro-diorite complex
3. Older granitoids
4. Post-Hammamat felsites
5. Gabbros
6. Younger granitoids
7. Post-granitic dikes
8. Ring complexes.

All these units are of an intrusive nature except the serpentinites, which should be divided into three categories:

1. Ultramafic members of ophiolitic sequences;
2. Serpentinite masses of ophiolitic melange; and
3. Magmatically emplaced ultramafic rocks.

The first two categories are allochthonous and have been discussed before in this chapter. The third category is described below.

The surface exposures of the metagabbro-diorite complexes are of limited extension and show irregular distribution all over the Shield. The term 'epidiorite', originally proposed by Hume (1934) for one of these complexes and then widely used in the literature on the Egyptian Shield, has been abandoned in favor of 'metagabbro-diorite complex'. As the term 'complex' implies, this unit is a heterogenous assemblage of rock types including metamorphosed gabbros, norites, dolerites and basalts. Since the response of these rocks to regional metamorphism and deformation is not uniform, the prefix 'meta' has been added to distinguish igneous rocks that are free of dominant metamorphic textures but show mineralogical reconstitution to some degree. As such, they are clearly distinguished from the relatively fresh late to post-orogenic gabbros which occur throughout the Shield.

Granitoid rocks constitute about 50% of the intrusive rock assemblage of the Shield. There is unanimity in the literature in classifying these rocks into two major groups: 1. the older granitoids, variously described in the literature as 'old', 'Shaitian', 'grey' or 'synorogenic' granites. The term 'granite' was used as a useful field term in the sense of the present usage of the term 'granitoid' as suggested by Streckeisen (1976) which includes, beside granites, granodiorites, tonalites, quartz diorites and trondhjemites. 2. The younger granitoids, variously described in the literature as 'younger', 'Gattarian', 'pink', 'red', 'late to post orogenic' granites. This group includes granodiorites, syeno- and monzogranites, alkali feldspar granites, alkalic or peralkalic granites and syenites.

Attempts to characterize each of the two groups or to distinguish them into further subdivisions were based on geochronology, mineralogy, petrochemistry, radiometry, tectonic setting and mode of emplacement (e.g. Schurmann 1966, Dardir & Abu Zeid 1972, Fullager 1981, Hashad 1980, Hussein et al. 1982, El Shatoury et al. 1984).

A detailed description of the main divisions of the intrusive rocks are given below.

Magmatically emplaced serpentinites

On the new geologic map of Egypt (1981), no distinction has been made between the magmatically emplaced 'autochthonous' serpentinites and the tectonically emplaced 'allochthonous' serpentinites which form an integral part of the ophiolitic sequences and associated melanges. These allochthonous serpentinites have been described earlier in the section dealing with ophiolite and melange.

Magmatically emplaced ultramafic rocks occur as subordinate members of largely gabbroic intrusions occurring as small masses widely within the basement rocks. Examples of these masses are Gebel Atud, Abu Ghalaga, Gebel Dahanib and Gabbro Akarem. In these masses, the ultramafic rocks are mostly fresh and rich in clinopyroxenes, in contrast to the tectonically emplaced ultramafic rocks (dunites, harzburgites and lherzolites) which are olivine-orthopyroxene rich. Three examples of such masses were given by Dixon (1979). All these bodies are characterized by abundant (85%) clinopyroxene which is dominantly a low Ca variety.

Engel et al. (1980) believe that large volumes of mafic to ultramafic magmas were intruded mainly as sills into the metasediments. The sills approximate basaltic komatiite in composition, and many parts of the serpentinites of the basement represent serpentinized products of these bodies, but specific examples of these bodies were not described or defined. Dixon (1979) cited Gebel Dahanib intrusion as an example of these komatiitic sill-like bodies. However, Church (1983) stated that it seems doubtful that komatiitic rocks are quantitatively important in the Eastern Desert. They represent only a small group of relatively fresh cumulate sequences which often carry Ni sulfides.

It is believed by the present authors that all these fresh non-ophiolitic mafic-ultramafic bodies represent late to post-tectonic intrusions which are rift related. All these bodies are grouped as one unit on the geological map of the basement (El Ramly 1972). These bodies will be treated in a later section.

Metagabbro-diorite complex

This unit of the basement complex occurs in several outcrops of limited extension all over the Eastern Desert and Sinai. The largest of these outcrops occurs west of Mersa Alam. The outcrop at the extreme southeastern part of the Eastern Desert (Geologic Map of Egypt 1981) was compiled after the Atlas of Egypt and is believed to include large masses of grey granitoids and metavolcanics (El Ramly 1972).

The metagabbro diorite complex is composed of a heterogenous assemblage of rock types. They are mainly metamorphosed basic rocks including gabbros, norites, dolerites and basalts, in which the igneous textures are partly preserved. They suffered considerable deuteric uralitization as well as injection by, and interaction with granitic material producing hybrid diorites. This interaction also resulted in the permeation of the metagabbro diorite masses by leucocratic granitic material in the form of intricate veinlets and highly irregular masses, with gradational contacts.

The metagabbro diorite masses commonly contain rafts of metasediments, and occasionally form dike-like bodies and minor intrusive masses cutting metasediments. Also, the grey granites in several localities carry xenoliths and large masses of metagabbro-diorite (El Ramly & Akaad 1960).

The spatial association of these complexes with the older granitoids, as well as the older metavolcanics and some ophiolitic sequences in the south Eastern Desert, suggests that they may represent border facies for the huge batholithic intrusions of this area. These border facies are known to comprise a wide variety of rock types with a very wide range of metamorphic responses. These facies extend for tens of kilometers, depending on the size of batholithic intrusions. In fact, the interaction of older metavolcanics, metasediments and ophiolitic sequences with the invading granitic material may represent the earliest phase of sialification of the Pan African volcanic arc systems. The extreme hybridization, uralitization and other alterations render the dating of this event rather difficult, if not impossible. A detailed description of one of these complexes in the south Eastern Desert may further clarify this statement.

Metagabbro-diorite complex west of Mersa Alam
This mass was studied by Akaad & Essawy (1964) and by Hassan & Essawy (1977). It consists of dark green, amphibole-rich rocks, characterised by marked variations in grain size and in the proportion of mafic to felsic minerals. These variations may take place within very short distances and sometimes a single outcrop can produce a large variety of hand specimens of appreciable differences discernible

only under the microscope. The hornblendic rocks of the complex are frequently crossed by quartzo-feldspathic veins, exhibiting all stages of interaction with them and ultimately producing leucocratic hybrid rocks. In localized patches, the grain size of the hybrid rocks increases considerably, leading to coarse grained rocks possessing appinitic tendencies.

Apart from xenoliths of the adjacent metasediments and metavolcanics, the rocks of the complex contain swarms of enclaves or inclusions of hornblendic rocks that represent older amphibolites. These inclusions show all stages of interaction and incorporation by the host. The ultimate stage is the formation of 'ghost' xenoliths or enclaves hardly discernible from the surrounding matrix.

The main rock types of the complex are: relic gabbros, uralitized gabbros, metamorphosed gabbros, quartz diorites, granodiorites, quartzo feldspathic veins, and amphibolites. Petrographic investigations revealed the effect of hybridization, post-magmatic alteration and regional metamorphism.

The metagabbro-diorite complex may be interpreted as a border facies assemblage formed by interaction of felsic magma of the older granitoid batholithic intrusions with older ocean floor sequences.

Older granitoids

'Older granitoids' include the assemblage of felsic plutonic rocks of essentially intermediate composition previously referred to as 'grey granite' by Hume (1935) and El Ramly & Akaad (1960), as 'synorogenic granite' by Sabet (1962) and El Shazly (1964) and as 'older granite' by Akaad & El Ramly (1960). This assemblage also includes the sheared cataclased variety called 'Shaitian granite' by Schurmann (1953) after its type locality in Wadi Shait, south Eastern Desert.

In spite of the reluctance expressed recently by some authors to the use of rock colors in the nomenclature of the basement rock units, it remains a fact that more than 95% of the exposures of older granitoids exhibit distinctly grey color. However, it was the previous usage of the word 'granite' for an assemblage composed essentially of trondhjemites, tonalites and granodiorites which is not proper. Furthermore, potential confusion may arise from the use of the world 'older' in the nomenclature of this assemblage. What will happen if an assemblage of felsic plutonic rocks older than the 'older granite' is discovered in the vaguely known areas of the southern parts of the Eastern Desert? This possibility exists in the light of the increasing evidence of a pre-Pan

African basement in the shield.

Classifications implying genetic connotation were also suggested for the plutonic rocks of the shield. The 'older granitoids' were classified as subduction related granites by Hussein et al. (1982), or plutonic rocks of the island arc stage by Bentor (1985). However, evidence of such mechanisms are still the subject of controversy (Hashad & Hassan 1979, Engel et al. 1980, Church 1982, Duyverman & Harris 1982, Stern et al. 1984, Harris et al. 1984, Vail 1985, Bentor 1985 and Church 1986).

Distribution and field occurrence

The older granitoids form large composite batholiths up to several hundred square kilometers in extent. Morphologically, they are characterized by vast sandy plains with scattered blocky outcrops with well-developed exfoliation. In such areas, for example the upper part of Wadi El Gemal and in Wadi Bizeh, the rock is mostly coarse grained, grey to whitish, highly weathered and friable. However, in certain areas, specially near the batholith borders, they form rugged terrain of high relief, as in the pluton crossed by the lower courses of Wadis Ghadir, Ereir and El Gemal. In such terrains, the granitoid rocks exhibit pinkish colors and include large rafts of the country rocks (metasediments and metavolcanics) with all grades of interaction and hybridization. Areas of older granitoids with intermediate features are also common. It is believed that these features correspond to various depths within batholiths exposed by erosion.

The batholiths are mostly concordant with their volcanosedimentary country rocks, with gradational and highly foliated contacts. The granitoids themselves are highly foliated and gneissose. They contain abundant xenoliths and rafts of the country rocks in various stages of digestion. Not uncommonly, cross cutting relations may be observed between parts of the older granitoids and their country rocks. In such cases, the cross cutting parts seem to be protruding masses from the main batholithic bodies, and the masses themselves are very coarse grained and pegmatoidal. In turn, the older granitoids are cut by pegmatites, aplites and various dikes ranging from basic to acidic in composition. In the core of the domal structure of Hafafit, a gradational zone of migamtites occurs between the batholithic granitoids and the overlying rocks.

Composition

Most of the 'older granitoids' occur as composite plutons often of batholithic dimensions. These plutons cover the compositional spectrum of trondhjemite, tonalite, granodiorite and rarely granite. In general, they have fine to coarse grained equigranular to porphyritic subsolvus textures. Plagiocalses

(An$_{20-45}$) and quartz are the essential minerals, microcline is relatively rare; mafic minerals constitute less than 20% by volume and hornblende exceeds mica which is mostly biotite. Accessory minerals are apatite, sphene, allanite and magnetite.

The chemical analyses of igneous rocks from Egypt compiled by Aly et al. (1983) show that the older granitoids have a broad compositional spectrum from highly silicic to basic (75 to 50% SiO$_2$), and typically display a unimodal distribution of SiO$_2$. With the exception of a few samples from the more silicic varieties, the K$_2$O percentage is less than 2.2. In general, they exhibit very low K$_2$O/Na$_2$O ratios and the FeO$_t$/MgO ratios are in the range of 0.5 to 3. Their contents of TiO$_2$, CaO, Na$_2$O and P$_2$O$_5$ also show wide variation. The CaO/Alk ratio varies between 0.3 and 1.7, while the FeO$_t$/FeO$_t$ + MgO ratio varies between 0.6 and 0.9.

The CIPW norms calculated for 22 of the typical older granitoid plutons show these rocks to be silica and alumina saturated. While normative quartz, orthoclase, albite and anorthite are persistently present; minor corundum shows in some analyses. The calculated anorthite content of the plagioclase is in the range of 16 to 48, but in the more silicic varieties, it is generally less than 25.

On the modified O'Connor Ab-An-Or ternary diagram (Fig. 12/14A), these samples show widely scattered spread between tonalites, trondhjemites and, to a lesser extent, granodiorites, but none in the granite field. On the K$_2$O-SiO$_2$ diagram (Fig. 12/14B), they show the same distinction when compared with the granodiorite-tonalite-trondhjemite associations of the Arabian Shield (Jackson 1986). Most of the samples of the so-called Shaitian granite fall in the trondhjemite (low-K tonalite) field. The analyses show, however, wider SiO$_2$ and K$_2$O ranges than observed in the typical island arc plutons of Solomon Islands and New Britain, and greater deviations from the magmatic arc coastal batholith of Peru. The fields of these plutons were outlined by Ramsay et al. (1986) from published data.

The calc-alkaline nature of these granitoids has been shown by Hussein et al. (1982) using the calc/alk index of Wright (1969). On the AFM ternary diagram (Fig. 10/14C), most of the analyses fall in the region drawn by Ramsay et al. (1986) from data on typical calc-alkaline intrusive provinces with similar petrographic characters. However, this common calc-alkaline affinity should not imply a comagmatic origin for the different plutons as supposed by El Gaby (1975), since the analyses represent different plutons that have been intruded throughout more than 300 Ma time span (Hashad 1980). The trend of the analyses on the AFM diagram (Fig. 10/14C) is also suggestive of compressional environment for the for-

mation of these granitoids (Petro et al. 1979). Most of these plutons show petrofabric features indicative of this compressional setting which led to their recognition as 'syntectonic'.

The trondhjemite affinities of some of the older granitoids are redemonstrated using Na$_2$O-K$_2$O-CaO diagram (Fig. 12/14D). They plot in the field drawn by Ramsay et al. (1986) from data on typical non-ophiolitic trondhjemites, such as the Sparta Complex, Oregon and the Twillingate Complex, Newfoundland. The other samples show typical calc-alkaline affinities.

Trace element data available (Aly et al. 1983, Greenberg 1981, Dixon 1979, El Gaby 1975) show these older granitoids to have low abundances of Rb, Ba, REE and other high field strength elements, a feature characteristic of primitive crust. The LIL-depleted character led Dixon (1979) to suggest a comparison of Egyptian older granitoids with the earliest known examples of continental crust, such as the Amitsoq and Univak gneisses and the tonalitic diapirs of the Barberton Mountainland. Further support for the primitive geochemical nature of these granitoids comes from the initial Sr and Nd isotopic ratios of few dated samples.

Age and tectonic setting

Available isotopic ages on the older granitoids are very limited in spite of their wide distribution all over the shield. Hashad (1980) compiled these ages; few dates have been added since then. The highest age so far available for these granitoids was by El Shazly et al. (1973) on Um Kroosh (22° 38' N to 33° 43' E) quartz diorite complex (958 ± 32 Ma, Ri = 0.7100, MSWD = 0.01) as recalculated by Cahen et al. (1984). Their age is based on mixed three point isochron, the slope of which is dominated by the ratios of a single microcline separate analysis. Such an age is not suitable for use for a sound geochronological interpretation. Several other ages in Hashad (1980) compilation are without detailed analytical data and are being revised now. These dates have been excluded from the compilation presented in Table 3.

The ages of Table 3 range between a maximum of 850 and a minimum of 614 Ma. It is difficult to interpret this 236 Ma time span as representing the duration of a 'continuous phase' or a particular tectonic setting 'stage' in the crustal evolution of the Egyptian Shield. There is no analogy in Phanerozoic plate tectonics for a similar event. While acknowledging the very limited number of ages in Table 3, it nevertheless seems possible to recognize three possible events of igneous activity during which different plutons of older granitoids were emplaced.

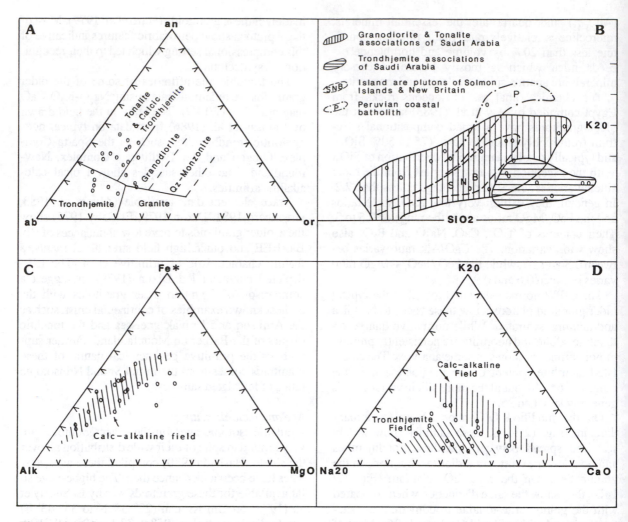

Figure 12.14 Some geochemical parameters of the older granitoids. A. Normative Ab-An-Or triangular diagram. Field boundaries are after O'Connor 1965, modified by Barker 1979; B. K₂O-SiO₂ diagram. Suit boundaries from Jackson 1986; C. AFM ternary diagram. Calc-alkaline field after Ramsay et al. 1986b; D.: NKC ternary diagram. Calc- alkaline and trondhjemite fields after Ramsay et al. 1986b. Data source on older granitoids are from compilation of chemical analyses by Aly et al. 1983.

The oldest igneous event is tentatively called the Shaitian event and is dated between 800 and 850 Ma. During this event, the Shaitian quartz diorite (801 ± 24 Ma) and Gebel Zabara foliated granite (850 Ma) were emplaced. Both plutons are among the most deformed older granitoids in the south Eastern Desert. Further support for the existence of this event is the whole rock Rb/Sr isochron age of 768 ± 31 Ma obtained for rhyodacite from Abu Swayel (Stern & Hedge 1985).

The second igneous event is tentatively called the Hafafit event and is most probably bracketed between 670 and 710 Ma. Five plutons out of nine presented in Table 3 fall within this range. Among the plutons emplaced during this event is the Hafafit tonalite that occupies the core of the Migif-Hafafit dome and

considered for a long time the fundamental basement for the overlying metamorphic succession (Hume 1934, El Ramly & Akaad 1960). However, recent studies have demonstrated the intrusive nature of the tonalite into the enveloping high-grade assemblage. Stern & Hedge (1985) believe that the 682 Ma age also dates the tectonic activity and thrusting in the surrounding area.

The third igneous event which is tentatively called the Meatiq event has been defined between 630 and 610 Ma based on two ages from the Meatiq dome and its surroundings. The dated rocks have been subjected to extreme mylonitization and the ages seem to represent the timing of major crustal movements in this part of the Eastern Desert. An event of intensive plutonism was started in this part of the world at this

Table 3. Isotopic ages of some older granitoids of the Egyptian Basement Complex.

Area	Location	Rock type	Method	Age	Remarks	Reference
Abu Ziran	26° 00' N 33° 45' E	tolanite/granodiorite	U/pb Zircon	614 ± 8	regression of chord to concordia	Stern & Hedge 1985
Meatiq	26° 05' N 33° 45' E	granite gneiss/ mylonite	Rb/Sr WR isochron	626 ± 2	Ri = 0.7030	Sturchio et al. 1982
Wadi El-mai	25° 15' N 33° 45' E	granodiorite/foliated	Rb/Sr WR isochron	671 ± 33 674 ± 13	Ri = 0.0734 Ri = 0.7027	Stern 1979 Stern & Hedge 1985
Hafafit	24° 45' N 34° 30' E	tonalite/foliated	U/Pb Zircon	682	2 point chord	Stern & Hedge 1985
W. Khreiga	22° 51' N 35° 15' E	tonalite/altered	U/Pb Zircon	709	single zircon	Stern & Hedge 1985
Dahanib	23° 45' N 35° 12' E	tonalite	U/Pb Zircon	711 ± 7	concordant ages	Dixon 1979
Shaiit	24° 45' N 34° 15' E	cataclased Qz-diorite	Rb/Sr WR isochron	801 ± 24	Ri = 0.7086	Hashad et al. 1972
J. Zabara	24° 38' N 34° 50' E	granite/foliated	Sm/Nd WR isochron	850	ENd (T) = 5.4	Harris et al. 1984

time and resulted in the formation of one of the largest granitic masses on Earth. It is perhaps more suitable to relate this event (630 to 610 Ma) to the younger granitoid events discussed later in this chapter.

The extreme isotopic homogenization expressed in the coincidence between a mixed Rb/Sr isochron age of 626 ± 2 Ma for Meatiq gneiss and U/Pb zircon age of 614 ± 8 Ma for Abu Ziran tonalite-granodiorite is worth noting. The resolution of this event into 626 ± 2 Ma age for the original crystallization of the Meatiq dome igneous protolith and 614 ± 6 Ma age for the culmination of a metamorphic event during which the minerals and some of whole rocks behaved as open systems (Sturchio et al. 1982) is difficult to accept in the light of the regional events prevailing at this perod, particularly the intensive intrusion of younger granites.

With the exception of the 0.7086 ± 0.0002 initial $^{87}Sr/^{86}Sr$ ratio obtained by Hashad et al. (1972) from the Wadi Shait quartz diorite isochron (Table 3), a very restricted range of 0.7030 ± 0.0003 is shown by the rest of the plutons. This low range has been interpreted by most workers to indicate a mantle source by analogy with ratios observed in recent mantle derived rocks. Alternatively it may imply derivation from not much older 'primitive' crust, depleted in Rb and other LIL elements and similar in this bulk composition to oceanic mafic rocks.

On the other hand, a subduction related mechanism for the generation of these granitoids has been suggested by Hussein et al. (1982), most probably in an ensimatic island arc setting (Gass 1977, Bentor 1985). Although such a speculative model is quite possible, it should be kept in mind that low initial Sr isotopic ratios in the 0.702 to 0.703 range may also characterize magmas of basic to intermediate compo-

sition that form by the melting of granulites and similar high grade metamorphic rocks at the base of the continental crust having low Rb/Sr ratios (Faure 1977).

The higher initial Sr isotopic ratio (0.7086) observed in the Shaitian quartz diorite may be the result of isotopic heterogeneity at the time of emplacement, or due to contamination of the quartz diorite magma with radiogenic ^{87}Sr derived from unexposed sialic rocks of older basement at the southwestern parts of the Eastern Desert. Such processes are quite possible through a subduction-related mechanism in an Andean-type arc setting (Hashad & Hassan 1979).

Gabbros

Gabbroic rocks in the basement complex occur in four distinct associations:

1. layered and cumulate masses of ophiolitic successions, such as the Wadi Ghadir ophiolites;

2. members of the metagabbro-diorite complex, where they are distinctly metamorphosed and considered as synorogenic intrusions;

3. small intrusive discordant masses of fresh and unmetamorphosed gabbroic rocks, collectively grouped into one unit in the basement (El Ramly 1972) and appearing as such on the goelogical map of Egypt;

4. minor units in some ring complexes, in which case they are mostly alkaline gabbros (El Ramly et al. 1971).

The first two associations were described earlier in this chapter. The fourth is beyond the scope of the discussion of basement rocks, as they are of Phanerozoic age. The third association is the subject of the present section and will be collectively called gabbros, or gabbroic rocks.

Gabbros and gabbroic rocks occur as small intrusions of a very limited extent, but of wide distribution, within the basement rocks. In a few of these masses, the gabbroic rocks are associated with ultramafic rocks, due either to differentiation in situ or due to multiple intrusions, such as Gebel Kalalat, Gebel Dahanib and gabbro Akarem masses. The rock types of these masses are for the most part extremely fresh and include various types of gabbro and norite, together with subordinate amounts of anorthosites and peridotites. These masses are usually discordant to the general trend of their country rocks. They are believed to be post-tectonic intrusions, and also post-Hammamat, since the latter does not contain any boulders of fresh gabbro that can be related to this group (El Ramly 1972). They may have been emplaced during the same episode as the younger granites, or somewhat earlier (Basta & Takla 1974). On the other hand, Sabet (1972) and Sabet et al. (1976) consider the gabbros earlier than the grey granites. Several of these gabbroic intrusions have been studied in some detail. Three examples will be described: Gebel Atud, Gabbro Akarem, and Gebel Dahanib.

Gebel Atud lies about 50 km west of Mersa Alam. According to Amin et al. (1955), it is composed of a thick sheet of gabbroic rocks ranging from olivine-rich gabbros to leucogabbros. Rhythmic layers are not displayed in this intrusion, but there is a general tendency for the olivine-rich variety to dominate in the lower parts. The sheet is intruded into volcaniclastic metasediments to the south and highly weathered uralitized metagabbros to the north, east and west. All the rock types of the sheet are quite fresh. A fresh gabbro dike occurs to the northeast of the main mass and appears to be an offshoot from it.

Gabbro Akarem is a small mafic-ultramafic complex composed of two small bodies lying midway between Aswan and Berenice (Fig. 12/1). According to Carter (1975) and Carter et al. (1978), and personal communications with A.A.A. Hussein, the complex consists of two separate bodies 1500 m apart. The total length of both bodies is 10 km in an east-northeast direction, with a width of 1 to 2 km. The two bodies are steeply dipping dike-like intrusions, with inward dipping contacts against metasediments. The main bulk of the complex is composed of noritic rocks, intruded by pipe-like bodies of peridotites in two generations, the later of which is mineralized with Cu-Ni sulfide minerals. The complex was intruded into a deep-seated fracture zone that is one of the important series of such east-northeast trending zones controlling several geologic features in the south Eastern Desert (Krs 1977). The successive intrusion of rocks of increasing basicity (norite, olivine-melanorite, peridotite and finally mineralized

peridotite) indicate the gradual deepening of a major fracture zone. The age of the complex is almost certainly later than the main episode of regional folding and metamorphism of the enclosing metasediments, but predates the surrounding grey granites.

According to Dixon (1979), Gebel Dahanib is a large (1 × 4 km) well-preserved mafic-ultramafic intrusive body in the south Eastern Desert. This body has been magmatically intruded into a metavolcanic and metasedimentary assemblage now largely preserved as amphibolite screens around the margins. The following descriptions are from Dixon.

The Dahanib body was emplaced as a large sill-like body at relatively shallow crustal levels. Over its approximately 1.5 m thickness, an overall differentiation sequence is apparent, with olivine-orthopyroxene cumulates at the base, grading upward to peridotites, pyroxenites and banded gabbros, with local horizons of anorthosite and hornblende gabbro developed near the top. Well-developed cumulate textures are common throughout. Margins of the body are unsheared and dikes from the main intrusive mass may intrude tens of meters into the country rock, ruling out any possibility of tectonic emplacement.

Surface exposures of major rock types consist of subequal amounts of gabbro, pyroxenite and peridotite, with peridotite slightly dominant. Banded gabbro is well-developed near the eastern boundary which represents the original stratigraphic top of this sill or lopolithic body, now structurally rotated toward the vertical. Anorthositic layers are locally developed near this margin but are not common. Large pegmatoidal amphiboles growing perpendicular to the main layering direction (comb-layering) were observed at one location near the central zone of the eastern margin.

The observed mineral chemistry in the Dahanib sill implies bulk chemical characteristics of the Dahanib parent magma equivalent to those deduced from the weighted compositions of the various rock types. Relative to modern ocean ridge tholeiitic basalts and most large differentiated mafic bodies, the Dahanib magma is enrichued in Ca, Mg, Ni and Cr, and depleted in Ti, Na and K. These characteristics typify basaltic or pyroxenitic komatiite magmas of the Archean, Proterozoic and Phanerozoic. However, Church (1983) stated that Gebel Dahanib cumulates are similar to the island-arc cumulates described by Stern (1979) from the Marianas.

Post-Hammamat felsites

These include effusive felsite, felsite porphyry and quartz porphyry bodies and sheets, plugs and breccias that occur in small outcrops all over the basement exposures (Akaad 1957, Shazly 1971, Dardir & Abu Zeid 1972, Essawy and Abu Zeid 1972). The largest of these bodies is the Atalla felsite intrusion forming the summit of Gebel Atalla in the central Eastern Desert. It forms an elongated body 19.5 km long and 1 to 3 km wide. It is intruded in a belt of rhyolitic flows and tuffs belonging to the Dokhan Volcanics on its western side and in basic metavolcanics and metasediments on its eastern side. The contacts are sharp with no contact effects. The felsite is fine grained to cryptocrystalline with yellow, buff and brown colors.

In many places the felsites have intrusive relations with the Hammamat sediments, and in turn were introduced by the younger granites. However, Dardir and Abu Zeid (1972) mention that there is no clear field relation between the felsites and the Hammamat sediments in the north Eastern Desert between latitudes 27 and 27° 30' N. They further state that the presence of felsite pebbles in the Hammamat conglomerates in that area may be taken as a criterion of the older age of the felsites. According to Greenberg (1981), field evidence and chemistry indicate that the felsites might be directly related to or comagmatic with the younger granites. According to the limited chemistry available, the felsites are considered to include Fe-poor alkaline rhyolite and rhyodacites (Bentor 1985). Their age, being post-Dokhan, is probably less than 580 Ma.

Younger granitoids

Occurrence

The younger granitoids are widely distributed all over the Egyptian Shield, constituting approximately 30% of its plutonic assemblage. Their relative abundance to the older granitoids increases from 1 to 4 in the south of the Eastern Desert to approximately 1 to 1 in the north (Stern 1979) and 12 to 1 in Sinai (Bentor 1985). They occur mostly as isolated, more or less equidimensional plutons of 1 to 10 km diameter, crosscutting virtually all the previously mentioned basement rock units. Some of these plutons have elongated outcrop outlines roughly parallel to the strike of the country rock's regional structures. Some other plutons particularly the very large ones have irregularly shaped boundaries. Semi-ring structures with different core-margin lithologies, concentric ring dikes and cone sheet complexes are not uncommon. A common feature of all these plutons is their high resistance to erosion as reflected in their topography.

The younger granitoids, though represented by a wide range of felsic plutonic rocks, are overwhelmingly dominated by granites which show a character-istic red to pink color caused by the prevalence of potash feldspars variably impregnated by hematitic dust. Color variations toward white, yellow and grey shades are also common. The contacts of the younger granitoids with their country rocks are mostly sharp and well-defined indicating passive epizonal emplacement. In most cases they are discordant with the regional foliation, although some of the elongated plutons have their long dimensions roughly parallel to the regional foliation. Their emplacement was most probably influenced by major tectonic lineaments and fractures. Contact effects are generally absent. However, gradational contacts and contact effects do occur in some cases, such as the effect of Um Had pluton on the Hammamat sediments producing the famous highly indurated Breccia Verde in the central Eastern Desert. A notable feature of the younger granites is the general scarcity of exogenic xenoliths.

Pegmatites are absent from most of the youngr granitoids, but quartz veins and aplite dikes and veins are quite common. In some places the quartz veins are concentrated in marginal zones where they contain appreciable amounts of ore minerals (e.g. Homr Akarim pluton).

Composition

Within individual plutons, younger granitoids shows striking uniformity in mineralogical and chemical composition but, between plutons, notable variations occur in texture, grain size and modal mineralogy. Greenberg (1981) presented modal analyses of some 20 typical 'younger granitoid' plutons of the central Eastern Desert. On Streckeisen (1976) modal digram, these analyses plot essentially in the syenogranite and monzogranite fields. A substantially lesser number of samples plot in the granodiorite and tonalite fields. Greenberg (1981) included the albite in palgioclase with modal plagioclase rather than with the modal alkali feldspar. This is perhaps the reason why alkali feldspar granite, a common rock type among the younger granitoids of the Egyptian Shield, did not plot in its proper field on a Streckeisen diagram.

A gradual change occurrs between plutons relatively rich in felsic minerals to plutons relatively rich in mafic minerals. In the more felsic plutons, quartz amounts to 40% or more, with total feldspars around 60% and very small proportions of micas. In these felsic varieties, alkali feldspars, mainly perthites, dominate over plagioclase (An_{20}), the ratio being dependent on the degree of differentiation of the pluton. On the other hand, plutons with more mafic mineral content have 30% quartz or less and plagioclase (An_{10-30}) dominates over alkali feldspars; biotite or hornblende are more abundant. In such 'mafic' plutons, exogenic xenoliths are common.

Accessory minerals common to both varieties include magnetite, apatite, zircon and sphene. Fluorite and allanite are less common.

Some of the younger granitoid plutons are peralkalic and contain appreciable amounts of pyriboles (arfvedsonite, riebeckite, barkevikite, aegirine). These plutons are more or less concentrated in the north Eastern Desert (Gebels Zeit & Gharib), and in south Sinai (Gebel Kathrina).

The textural features of the younger granitoids are somewhat complex, due to a complex history of crystallization and late deuteric and hydrothermal alterations. Greenberg (1981) noted that the more siliceous younger granites are mostly hypersolvus, indicating crystallization at high temperature and low water pressure. Transolvus textures have been developed in most plutons by the deuteric-hydrothermal addition of water. The more mafic younger granitoid plutons show mainly subsolvus textures, indicating crystallization under conditions of higher water pressure.

Variations in chemical composition reflect complex evolution. Some 40 averages and individual analyses of younger granitoids were selected from Aly et al. (1983) compilation and from the analyses presented by Greenberg (1981) and Bentor (1985) to represent: 1) the pre-Hammamat foliated 'pink granites' (Schuurman 1966) presumably representing the transitional phase between the older and the younger granitoids, 2) the normal 'pink' granites representing the main bulk of the younger granitoids, 3) the alkali granites representing the pyribole bearing granites (Schurmann 1966, Bentor 1985). These analyses show SiO_2 to range between 70 and 79%, K_2O to be generally higher than 3.8%, the K_2O/Na_2O ratios to be generally greater than 1, the FeO_t/MgO ratios to be greater than 4, and the $FeO_t/MgO + FeO_t$ ratios to be higher than 0.8. On the other hand, CaO shows very wide variation from 0.1 to 1.6%.

The calculated CIPW norms of these analyses demonstrate some overlap between the chemistry of the three groups. The pre-Hammamat granitoids and the majority of the normal granites are peraluminous with minor amounts (0.1 to 4%) of normative corundum. Some of the normal granites are metaluminous with no corundum but with normative anorthite. Only one average for Kadabora batholith (normal granite group) shows marginal peraluminous-peralkalic character demonstrated by the absence of both normative corundum and anorthite and the appearance of acmite in scarce amounts (0.06%).

Almost all the alkali granites have no corundum in their norms. They have either minor anorthite or minor acmite which reflects metaluminous to mild peralkalic character. Available analyses on these rocks seem to include some alkali-rich granites which

are not truly peralkalic (Streckeisen 1976).

On the Ab-An-Or triangular diagram (Fig. 12/15A), data points of the three groups scatter mostly within the granite field of Barker (1979). If the composition can be interpreted as equilibrium cotectic liquid, some of the pre-Hammamat granitoids may represent melts that formed at a pressure of about 8 kb. On the other hand all the alkali granites seem to have equilibrated at a much lower pressure, less than 2 kb, which is consistent with shallow depths of emplacement. The main bulk of granites show a very wide range of pressure, thus suggesting emplacement at different levels of the crust.

From the ternary variation diagrams of chemical oxides (Figs 12/15B, C and D), certain trends can be demonstrated. On the AFM diagram (Fig. 12/15B), the fields of the pre-Hammamat granitoids and the normal granites conform to the reference calkalkaline field, though there is a tendency for the two groups to be on the alkalic end of the field and for the former group to be more mafic. The alkali granites fall in a distinctive field stretched along the Alk-Fe side.

The NKC diagram (Fig. 12/15C) permits the maximum separation of granitoids having diversed affinities. In spite of the overlap of the younger granitoid groups, they are distinctly separated from the other granitoids of calc alkaline and trondhjemitic affinities. The pre-Hammamat and the main bulk of normal granites conform to the alkalic end of the refrence calc-alkalic field. The potassic nature of the younger granitoids compared to the sodic nature of the older granitoids is clearly demonstrated.

On the Cao + MgO – FeO – Al_2O_3 diagram (Fig. 12/15D), a gradual increase in alumina is obvious from older granitouds to pre-Hammamat granitoids to the normal main bulk granites. A sudden shift away from alumina is observed however among the alkali granites.

The trace elements abundance and trends of the younger granitoids have been studied by several workers (e.g. Sayyah et al. 1973, Von Knorring & Rooke 1973, El Bouseily & El Sokkary 1975, El Gaby 1975, Nagy 1978, Greenberg 1981, and Stern & Gottfried 1986). The data obtained have been used in most cases to identify magma types through correlation with averages of typical calc-alkali or alkalic provinces in different parts of the world. The data were also used to demonstrate certain differentiation trends and to classify these younger granites accordingly. Some attempts were made to use these data to elucidate the tectonic setting but were met with difficulties related to the complicated petrogenetic history of these granites involving redistribution and possible loss of elements by volatile fluxing which prevent normal equilibrium crystallization of trace

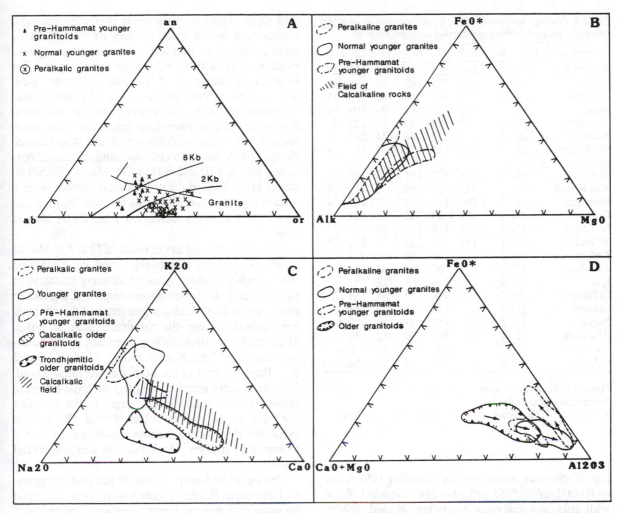

Figure 12.15 Some geochemical parameters on the younger granitoids. A. Normative Ab-An-Or triangular diagram. Field boundaries are the same as in Figure 12.14A. Water saturated peritectic compositions at 2 Kb and 8 Kb are after Whitney 1975; B. AFM ternary diagram showing fields of younger granitoids compared with calc-alkaline field given by Ramsay et al. 1986; C. NKC ternary diagram showing fields of younger granitoids compared with the fields of older granitoids given in Figure 12.14D; D. Variations in chemical composition of younger granitoids showing the abrupt change to peralkalic magmatism.

element-rich minor phases. Moreover, variable degrees of crustal contamination may have occurred particularly at the margins of intrusions and in areas relatively rich in enclaves.

In general, the calc-alkalic to alkali-rich granites forming the main bulk of the younger granitoids show wide variation in their trace element contents. The averages calculated for different intrusions (Table 4) show the more calcic plutons (CaO > 1%) to be relatively enriched in Sr, Ba and Zr compared to the more fractionated ones (CaO < 0.8%) which are enriched in Rb, Y and Nb. More enrichment in these latter elements and a significant depletion in Sr and Ba characterize the peralkalic younger granites. Zirconium is considerably low in this latter group if compared with similar rock types around the world.

The progressive transformation of an island arc terrain into true continental crust through the emplacement of such great volumes of highly evolved LIL-enriched granites in a relatively short interval of time is one of the most peculiar features of the Egyptian Shield.

Age and tectonic setting

Hashad (1980) compiled the radiometric dates of the younger granitoids. Several other ages have been obtained since then (Abdel Monem & Hurley 1980, Fullager 1980, Frisch 1982, Sturchio et al. 1983, Ries et al. 1983 and Stern & Hedge 1985). An updated compilation for the more reliable ages is presented in Table 12/5. Ages included are those based on at least 3-point isochrons. Conventional ages for rocks with

Table 4. Average content of some trace elements in younger granitoid plutons of the Egyptian Basement Complex.

Pluton name	(N)	Ref.	Ba*	Rb	Sr	Y	Zr	Nb
Raba Garrah	(101	1	106	214	21	72	115	35
Eridya	(26)	1	192	287	48	70	119	40
Um Had	(4)	1	111	291	10	189	72	35
Sibai	(3)	1	116	149	14	70	104	55
Sh. Salem	(7)	1	167	139	14	53	123	25
Kadabora	(20)	1	472	112	56	74	225	31
Hom. Gha-nem	(7)	1	330	119	52	64	163	29
G. Elgindi	(4)	1	534	137	141	53	100	16
Abu Kharif	(4)	1	345	101	88	6	115	28
Fawakhir	(23)	1	750	112	287	21	171	14
G. Ruses	(4)	1	485	62	72	54	301	7
M. Claud	(5)	1	482	110	276	20	198	27
W. Endiya	(4)	1	754	113	333	28	255	14
Kadabora	(8)	2	–	94	60	24	203	–
Abu Toyour	(3)	2	–	90	48	34	196	–
Igla	(7)	2	–	73	68	18	182	–
El Faliq	(6)	2	–	101	9	37	155	–
El Sebai	(7)	2	–	125	17	35	143	–
Nugrus	(8)	2	–	152	93	19	100	–
Abu Diab	(3)	2	–	192	30	12	30	–
Sh. Salem	(3)	2	–	201	8	37	76	–
El Ineigi	(3)	2	–	220	8	55	62	–

*concentrations are in ppm
N = number of samples
References: 1. Greenberg (1981); 2. Sayyah et al. (1973).

high Rb/Sr ratios and errors not exceeding 10% in age or 0.0002 in $^{87}Sr/^{86}Sr$ ratio are also included. Ages with unknown analytical errors are omitted. Where necessary some data were recalculated using the IUGS recommended constants (Steiger & Jager 1977) and isochron regression procedure (York 1969).

The compiled ages lie within the 620 to 530 Ma range. During this 90 Ma time span, magmatic activity in the Eastern Desert and Sinai reached its peak. This activity shows a pronounced shift from the dominantly trondhjemtic tonalitic granodioritic calc-alkalic plutonism to one characterized by predominance of more felsic, more alkalic, granodioritic and granitic rock types. The final cratonization and sialification of the Egyptian Shield was achieved in this interval. Based on the data presented, this period can be resolved into two major events spanning the time intervals 620 to 570 and 570 to 530 Ma.

During the first magmatic event (620 to 570 Ma), tentatively called the 'Dokhan event', most of the peraluminous to metaluminous calcalikalic 'pink' granitoids of the Eastern Desert and associated dike swarms were emplaced with an obvious northward increase in abundance relative to other rock units. Syenogranites and monzogranites are the essential rock types. This magmatic event is perhaps the most intense one in the entire history of the Shield as manifested by the enormous volume of magma generated in only a fifty million year time span. The extensive successions of 'Dokhan volcanics' were also erupted during this time span. The famous Aswan monumental granite was emplaced in this event. It represents the southernmost extension of the magmatic activity. The resolution of this event by Stern & Hedge (1985) into two crust-forming magmatic episodes, 625 to 610 and 600 to 575 Ma, is difficult to accept in the light of the very limited number of ages used to identify their older episode, and their restriction to the strongly deformed Abu Ziran Meatiq area.

During the second major event (570 to 530 Ma) of magmatic activity, tentatively called the Katherina event, highly evolved granites strongly enriched in alkalies and LIL elements were emplaced. Peralkalic granites and alkali feldspar granites and their extrusive equivalents are the dominant granitic types. These rocks are more or less restricted to the northernmost part of the Nubian Shield, south Sinai and the Midyan terrain of the northwest Arabian Shield. The rocks were emplaced during the post-orogenic phase (or post cratonization stage) in the complex evolution of the Arabo-Nubian Shield. During this event, the rigid Arabo-Nubian massif was subjected to tensional stresses, block faulting and differential uplift.

The initial Sr isotopic ratios of the younger granitoid intrusions (Table 5) show a wide range and argue for more complex evolution history and tectonic setting. It is true that most of the Ri values are in the low range (0.7035 ± 0.0010) which suggests derivation from upper mantle or primitive island arc protolith deficient in lithophile elements. However, there appears to have been at least some pre-existing felsic material in the source region of other intrusions in order to account for their high Ri ratios (0.7045 to 0.7085). The wide range of Ri values cannot be explained by simplistic concepts of partial fusion in the upper mantle with variable degrees of contamination by crustal material due to the lack of any supporting chemical evidences. Neither can the wide range in major and trace element chemistry be explained by a common tectonic setting for all the younger granitoids.

Taking into consideration the complicated history of the whole Shield, we may think of three possible tectonic settings and magma generating sources for the evolution of the younger granitoids:
– granitic magma generated during subduction processes in volcanic arc setting and characterized by low Ri values and trace element patterns similar to that in the more calcic granites;

Table 5. Isotopic ages of some younger granitoids of the Egyptian Basement Complex.

Pluton	Location	Lithology	Method	Age	Remarks	References
El Guwaf/Mander granite	28° 08' N 34° 22' E	Granite	Rb/Sr WR	529 ± 8	Ri = 0.7045	Bentor 1985
G. Gharib	28° 08' N 32° 06' E	A. granite	Rb/Sr MA	544	Rb/Sr = 45	Stern & Hedge 1985
G. Katharina/Iqna suite	28° 35' N 33° 58' E	A.F. granite	Rb/Sr WR	560 ± 10	Ri = 0.7062	Bentor 1985
Aswan	24° 00' N 33° 00' E	Por. granite	U/Pb	565	S.F. zircon	Abdel Monem & Hurley 1980
Raba El Garrah	25° 22' N 34° 22' E	Granite	Rb/Sr WR	568 ± 17	Ri = 0.7020	Fullager 1980
Eridya	26° 28' N 33° 30' E	Granite	Rb/Sr WR	570 ± 5	Ri = 0.7024	Fullager 1980
El Einegy	25° 12' N 34° 06' E	Granite	Rb/Sr WR	573 ± 8	Ri = 0.7044	New age
El Fawakhir	26° 12' N 33° 42' E	Qz-monzonite	Rb/Sr WR	574 ± 009	Ri = 0.7025	Fullager 1980
G. Tenassib	28° 32' N 32° 35' E	Granite	Rb/Sr WR	575 ± 11	Ri =0.7042	Stern & Hedge 1985
Kadabora Manharawy 1977	25° 30' N 34° 25' E	Granite	Rb/Sr WR	576 ± 14	Ri = 0.7038	Recal.
W. Hawashiya/ Gazalla	28° 25' N 32° 34' E	Granite/Aplite	Rb/Sr WR	577 ± 6	Ri = 0.7036	Stern & Hedge 1985
W. Kid Iqna suite	28° 21' N 34° 11' E	A.F. granite	Rb/Sr WR	580 ± 23	Ri = 0.7028	Bentor 1985
Shahamit Hill/Iqna suite	28° 57' N 33° 43' E	A.F. granite	Rb/Sr WR	580 ± 6	Ri = 0.7050	Bentor 1985
Salah El Belih	27° 11' N 32° 22' E	Granodiorite	U/Pb	583	S.F. zircon	Stern & Hedge 1985
Sheikh Salem	25° 99' N 34° 30' E	Granite	Rb/Sr WR	585 ± 16	Ri = 0.7023	Fullager 1980
Meatiq Dome	26° 06' N 33° 46' E	Granite	U/Pb	585 ± 14	S.F. zircon	Sturchio et al. 1983
G. Atawi	25° 36' M 34° 09' E	Granite	Rb/Sr WR	587 ± 11	Ri = 0.7050	El Manharawy (1977)
G. Faraid	23° 16' N 35° 22' E	Granite	Rb/Sr WR	587 ± 11	Ri = 0.7035	Stern & Hedge 1985
Aswan	24° 00' N 33° 00' E	Rap. granite	Rb/Sr WR	591 ± 10	Ri = 0.7092	Hashad et al. 1972
G. Zeit	27° 58' N 33° 29' E	Granite	Rb/Sr	592	Model age	Stern & Hedge 1985
Timna	29° 46' N 34° 58' E	Granite	Rb/Sr WR	592 ± 7	Ri = 0.7048	Bentor 1985
Homrit Ghanim	25° 48' N 34° 15' E	Granite	Rb/Sr WR	594 ± 8	Ri = 0.7061	Fullager 1980
G. Kadabora	25° 30' N 34° 25' E	Granite	Rb/Sr WR	595 ± 8	Ri = 0.7016	Fullager 1980
G. Dara	27° 55' N 33° 03' E	Granite	U/Pb	596	S.F. zircon	Stern & Hedge 1985
Um Malaq	27° 57' N 34° 11' E	Granite	Rb/Sr WR	597 ± 022	Ri = 0.7028	Bentor 1985
W. Hawashiya	27° 55' N 28° 10' E	Granodiorite	U/Pb	614	S.F. zircon	Stern & Hedge 1985
W. Dib	27° 36' N 32° 55' E	Granodiorite	U/Pb	620	S.F. zircon	Stern & Hedge 1985

– granitic magma generated as a result of arc-continent collision event and characterized by intermediate to high Ri values and trace element patterns similar to those in the more differentiated granites;

– granitic magma generated within attenuated continental crust as a result of post-cratonization rifting and characterized by peralkalic chemistry and intermediate Ri values.

Deleniation of plutons belonging to the first two groups is difficult in the absence of enough isotopic and trace element data on most of them. Extensive thrusting of ophiolitic rocks also made the delentiation of suture belts as markers of subduction zones quite misleading.

Ring complexes

Some ring complexes in the Egyptian Eastern Desert have been identified and studied (El Ramly & Hussein 1982, Hashad & El Reedy 1979). Of these, only the Wadi Dib Ring Complex is of Pan African age (554 Ma, Serencsists et al. 1979). According to El Ramly et al. (1982), Wadi Dib Ring Complex is a perfectly circular ring intrusion emplaced into older granites at the intersection of latitude 27° 34' N and longitude 32° ' E, in the north Eastern Desert. It measures about 2 km across and is made up of an outer ring composed of various feldspathoidal-bearing syenites and an inner stock that is relatively homogeneous and composed of nordmarkite. On the top of the central stock, there are some roof pendants of trachytic flows and agglomerates that might represent remnants of an older volcanic cone. El Ramly & Hussein (1982) include Wadi Dib Ring Complex in the Mishbeh type of complexes characterized by a limited range of rock types but including feldspathoidal rocks and having a poorly defined ring structure. Nevertheless, Wadi Dib Ring Complex is the only one in Egypt in which quartz bearing syenites form a stock surrounded by feldspathoidal rocks. According to Garson & Krs (1976), Wadi Dib Ring Complex lies at the intersection of a north 60° east lineament and a northwest transcurrent fault.

El Ramly et al. (1982) believe that the formation of the Wadi Dib Ring Complex is the result of intraplate magmatism, began by the extrusion of alkali trachytes forming a small stratovolcano on a gneissic granite plateau. This was followed by the introduction of the subvolcanic stock of quartz syenite by magmatic stoping, in turn followed by the stoping of the silica deficient magma along a ring fault to produce the outer feldspathoidal ring dyke. The whole structure was later intruded by dikes and cut by faults. According to Serencsists et al. (1979), the various units within individual Egyptian ring complexes, including the Wadi Dib Complex, give ages that are identical within analytical uncertainties. This limits the period of emplacement of any one complex within a few million years. The complexes were emplaced at shallow levels within the upper crust and are associated with consanguineous volcanics and dikes.

Dikes

All the basement units described above are traversed by many dike systems. From the cross-cutting relations and field setting, it seems that throughout the formation of all basement units, dikes of variable composition and dimensions were intruded into basement units and outlasted them to the Phanerozoic. With regard to age, the dikes within the basement units may be separated into three groups. The dikes of the first group are intruded before the older granites and are evidently deformed with their country rocks. The dikes of the later group span the time from the older granites to younger granites. There are cases in which the younger granites truncate some of these dikes. The third group of dikes is of Phanerozoic age, as indicated by their cross-cutting relations with the Phanerozoic ring complexes and their radiometric ages, and does not belong to the present discussion. However, in many cases it is not possible to assign a dike or dike system to the second or third group, and undeformed dikes in the basement are thus usually called post-granitic dikes.

The best documented case of the first group of dikes is in the Wadi Ghadir area (El Bayoumi 1980). Highly disrupted and deformed dikes are common in the melange. They range in thickness from a few tens of centimeters to about 2 m. Their most common composition is basaltic, with development of greenschist mineral assemblages. However, porphyritic andesitic dikes with streams of large plagioclase phenocrysts in the central zone of the dike (Borneoli effect), are not uncommon. Similar andesitic dikes were also observed in the upper parts of Wadi Nugrus and in Wadi Ramariem, south of Wadi Ghadir.

The post-granitic dikes are very common and widespread within the basement units. They have been described in nearly every work on the basement. A brief summary of their main characteristics in the central and south Eastern Desert is given by Sabet (1972). According to him, the dikes vary greatly in thickness, from less than 0.5 to more than 25 m, the average being 2 to 5 m. They are essentially rectilinear with parallel walls and vertical to very steep dips (more than 60). They vary greatly in composition, from felsites to olivine basalts and diabases. They may be distinguished into four compositional groups as follows:

1. acid dikes, comprising the granite porphyries,

felsite porphyries, porphyritic dacites, rhyolite sand plagiophyres, light in color;

2. basic dikes, commonly composed of basalts and diabases which are fine grained and dark grey in color and usually spheroidally weathered;

3. intermediate dikes, commonly composed of trachytes and andesites and rarely occurring as plugs;

4. lamprophyre and bostonite dikes, sills, sheets and plugs. This group is essentially fine grained and most, if not all, are of Phanerozoic age.

The basic, intermediate and acidic dikes show variable cross cutting relations, indicating synchronous emplacement. However, in several small areas, the basic dikes appear to be the latest.

The post-granitic dikes are especially abundant in the northern part of the Eastern Desert, forming swarms that trend east to northeast (Schurmann 1966, Kabesh & Shahin 1968, Ghanem 1973). These dike swarms are strong evidence for north-south to northeast-southwest directed crustal extension. However, the original attitude of the dikes could have been changed because of block tilting and rotation. Compositionally, the dike swarms reported by Schurmann (1966) range from basaltic (47% SiO_2) to rhyolitic (75% SiO_2) and from calc-alkalic to alkalic with a bimodal aspect (Stern et al. 1984). Stern & Hedge (1985) reported Rb/Sr ages for some acidic to intermediate dikes in the north Eastern Desert ranging between 543 and 589 Ma.

Evolution models

The foregoing description of the basement units show that the volcanic-sedimentary units are predominantly of Pan African age (1100 to 550 Ma) and have been intruded by granites, which include the synorogenic grey granites and the late orogenic and post-orogenic younger granites, including the alkali granites as well as the Wadi Dib Ring complex. Ophiolites are sporadically present within the layered sequence of the Pan African age, but mostly as dismembered fragments or allochthonous sheets. These Pan African units show evidence of evolution and cratonization from an oceanic environment, with the interaction of continental masses. Within the Pan African sequence, there are inliers of tectonic windows of gneissic terranes that seem to represent older than Pan African continental masses on which the Pan African sequence was overthrust. These continental masses represent remnants of deeply eroded proto-continents or mature island arcs which were involved in the Pan African orogeny as passive masses.

Several models have been proposed to explain the evolution of the basement complex of the Egyptian Shield. Many of these models have been proposed for the whole Pan African in northeast Africa and Arabia. A review of these models is beyond the scope of the present work and might well become obsolete by the time of its publication. New works appear constantly, indicating universal interest in the Pan African.

Broadly speaking, the evolution of the Egyptian basement complex is explained in terms of four main concepts:

1. Opening and closure of a limited ocean basin (e.g. Garson & Shalaby 1976, Dixon 1979) or development of rift ocean basins and subsequent subduction related phenomena (Church 1979, 1981).

2. Episodic and/or successive evolution of island arcs which were swept and welded together (e.g. Gass 1977, 1980).

3. Evolution in a continental margin environment similar to the present Andean-type setting (e.g. Hashad & Hassan 1979, El Bayoumi 1980, El Gaby et al. 1984).

4. Evolution of oceanic terranes lying between continental plates to the east and west (Vail 1985).

In view of the present state of knowledge, the authors believe that the Egyptian basement rocks evolved in a continental margin environment similar to the present-day Andean setting. The boundary between the oceanic and continental domains is not known due to the extensive cover of Phanerozoic sediments. However, it is believed that this boundary is not far from the western limit of the basement exposures.

Further mapping and study of the distribution and tectonic setting of the basement units are still needed to give more insight into the development of the Egyptian Shield and its relation to the Pan African belt. However, it is advisable at this time to avoid use of formal names nomenclature in the definition of rock units. Many rock units are allochthonous and formal stratigraphic nomenclature should not be applied to them. Efforts should be directed to increasing our understanding of the distribution and tectonic setting of the rocks. The authors propose a tentative classification of the rock units of the Egyptian basement. The main features of this classification are:

1. Separation of the gneisses and allied rocks as pre-Pan African, stressing the major break between them and the Pan African units.

2. Separation of ophiolites as an independent unit.

3. Division of the metasediments into three assemblages.

4. Separation of the exposures of syntectonic granitoids into high-level and deep-level bodies.

Proposed legend

Dikes. Ranging from basic to acidic, age ranges from Precambrian to Cenozoic.

Wadi Dib ring complex. The oldest of a series of ring complexes, it is the only one of Precambrian age (550 Ma).

Alkali granites. Simple small intrusions of alkalic and peralkalic granites carrying alkali mafics, discordant to the main structural trends, probably formed in extensional environment, of uncertain age.

Post-tectonic granitoids. Pink, red, white, yellow and rarely grey granitoids, mostly in the form of cross-cutting bodies of small dimensions with sharp contacts with the country rock and weak or no contact effects, potash-rich with two feldspars, abundant hydrothermal and metasomatic alterations, some associated by mineralization of Sn, Nb, Mo, U, Th ... etc., dominantly of calc-alkaline nature, poor in mafics which are mostly mica, probably formed due to collision with a marked peak in age range about 570 to 590 Ma.

Intrusive gabbro. Small discordant masses of fresh gabbros, norites and troctolites, of uncertain age, but post-Hammamat, mostly layered with ultramafic components.

Felsites. Subvolcanic intrusions of acidic composition of uncertain age.

Hammamat sediments. A thick succession of molasse-type sediments including conglomerates, sandstones, siltstones and slates, commonly folded, in part highly sheared; conglomerates contain pebbles of syntectonic granite and Dokhan-type volcanics; includes 'Igla formation' which is the basal part of the sediments composed of shale and sandstone characterized by a unique purple color.

Dokhan volcanics. Unmetamorphosed subaqueous to subaerial volcanics, mostly composed of intermediate to acidic flows and pyroclastics; basalts are rare; ignimbrites are not uncommon; porphyritic varieties including Imperial Porphyry are common; intercalations of volcaniclastic sediments are rare; include the Dokhan Volcanics in Gebel Dokhan area, and probably volcanics in Saal, Kid and Kathrine areas; represent products of volcanic activity in mature island arcs of active continental margins; probably in part representing the surface manifestation of the later phases of the post-tectonic granitoids.

Syntectonic granitoids. Undifferentiated and gradational varieties of the following two types, including highly sheared variations.

High level syntectonic granitoids. Pink to whitish and yellowish, weakly foliated to massive bodies with batholithic dimensions, gradational to country rocks through wide zones, very abundant partly digested rafts and xenoliths, form extensive dikes in the country and roof rocks, represent a high level of intrusion, gradational in part to the other type, best represented in the lower parts of Wadi El Gemal and Wadi Ghadir.

Deep level syntectonic granitoids. White to grey foliated bodies with batholithic dimensions, mostly gradational and concordant contacts, pegmatoidal in parts, with abundant migmatite zones gradational to overlying rocks, contains abundant rafts, roof pendants and xenoliths of highly metamorphosed and granitized country rocks, represent the deep level of intrusion, best represented in the upper parts of Wadi El Gemal and lower parts of Wadi Hafafit.

Metagabbro-diorite complex. Medium to small size masses with gradational contacts, composed of uralitized gabbros and granitized basic rocks and appenites, of controversial origin, probably a heterogeneous assemblage of basic and intermediate intrusions which were hybridized by the syntectonic granitoids.

Old Calc-alkalic metavolcanics. Calc-alkalic island arc or volcanic arc volcanics with associated pyroclastics, mostly of andesitic composition, bimodal in part, cyclic in part, including the Abu Hammamid Metavolcanics and equivalents, all metamorphosed in greenschist facies, locally in lower amphibolite facies.

Metasediments. Undifferentiated one or more of the following units, representing sedimentation in oceanic environment by turbidity currents, in part in a trench over a subduction zone and in part grading into or approaching continental environment.

Melange. Proximal facies composed of heterogeneous assemblage of fragmented blocks of mostly ophiolitic derivation, mainly serpentinites, together with fragments of continental derivation all in a highly sheared argillaceous matrix grading in part into pebbly schists as a distal facies.

Mature to semi-mature metasediments. Mostly psammitc, psammopelitic and calcareous metasediments, with sub-greywackes and conglomerates; garnetiferous schistose and gneissose equivalents very common in the southwest, which may belong to the pre-Pan African schistose metasediments.

Immature metasediments. Including laminated tuffites and mudstones, graywackes, conglomerates, cherts and BIF, with minor marly and calcareous bands; metamorphic equivalents in the form of micaceous and chloritic schists are common and gradational, pebbly mudstone and ophiolitic fragments are common; most extensive in the central Eastern Desert.

Ophiolites. Undifferentiated one or more of the following units, representing allochthonous fragments of oceanic crust.
   Volcanic sequence. Includes pillowed and massive basalts, sheeted dike complexes and massive diabase, with a thin veneer of pelagic sediments on basalts in some cases.
   Cumulate sequence. Includes layered dunite, layered massive and fine grained gabbros with minor pyroxenite, anorthosite and trondhjemite.
   Basal ultramafics. Serpentinized peridotites including lherzolite, harzburgite and dunite, with serpentinites and talc carbonate schists, forming allochthonous rootless masses or thrust sheets, dismembered basal parts of ophiolitic sequences.

Major break

Hornblende gneisses. A single outcrop extending in Wadi El Gemal, Wadi Nugrus and Wadi Sikait, probably a thrust sheet on top of the underlying successions within Hafafit-Nugrus window, composed of metamorphosed basic rocks (gabbro-basalt) in lower to intermediate amphibolite grade.

Upper schistose metasediments. Pelitic and calc-pelitic garnetiferous schists with abundant pebbly bands, only in Wadi Sikait and Wadi Abu Rusheid, very similar to the basal parts of the lower schistose metasediments and may be structurally equivalent.

Undifferentiated gneisses. Gneissose rocks of variable composition with questionable position, tentatively grouped with the old gneisses until verified, mostly south of latitude 24° N.

Old gneisses. Dominantly psammitic gneisses, with psammopelitic components, partly migmatized, including Migif-Hafafit psammitic gneiss, Meatiq granulites and Feiran-Solaf migmatites and their equivalents.

Lower schistose metasediments. Pelitic, calc-pelitic and psammopelitic, with serpentinites and talc-carbonate bands and lenses, migmatized becoming garnetiferous gneissose at the base, metamorphism decreases rapidly upward, occurring in Hafafit-Nugrus window forming the structurally lowermost unit, may also be represented in Abu Swayel area.

ACKNOWLEDGEMENTS

We thank W. Church, University of Western Ontario, for stimulating and fruitful discussions during our joint field trips to the Eastern Desert of Egypt. Critical comment by M.F. El Ramly, Egyptian Geological Survey, are deeply appreciated. We have benefitted from discussions with our colleagues in the Faculty of Earth Sciences, King Abdulaziz University, Jeddah, Saudi Arabia. Participants of the IGCP (Project 164) field trip to the Eastern Desert of Egypt (Winter 1983) had contributed in various ways to this work.

PART 3

Geology of selected areas

CHAPTER 13

South Western Desert

E. KLITZSCH & HEINZ SCHANDELMEIER
Technical University of Berlin, Germany

Most of southwest Egypt is a cuesta type landscape of Late Jurassic to Cretaceous clastics, of small to medium high escarpments with extensive sand and gravel sheets situated between them. From the north, this area is entered by large longitudinal dunes of the great western sandsea which forms a closed sandmass (erg) north of the Ammonite Hills.

To the east, this landscape is bounded by the late Cretaceous and early Tertiary Plateau, forming the escarpment along the Arba'in caravan road. Toward the south, the landscape becomes a monotonous sand sheet, whereas to the southeast, towards Lake Nasser, the morphology keeps the same cuesta type character which dominates the area between the oasis of the New Valley and Gilf Kebir. In some places of relatively minor extension, Precambrian Basement is exposed.

The area west of the line from Gebel Kamil to the eastern edge of Gilf Kebir, the southwestern corner of Egypt, is somewhat different. The southern and southwestern foreland of Gilf Kebir is mostly made up of Precambrian basement of irregular and broken relief overlooked by the Gebel Uweinat intrusion situated at the border between Egypt, Libya and Sudan. Gebel Kamil is an isolated hill made up of middle to late Cretaceous sandstone at the northern edge of an extensive exposure of Precambrian rocks which reaches from there to Gebel Uweinat. Gilf Kebir is a large and approximately 1,100 m high plateau made up of late Jurassic to late Cretaceous clastics (for locations, see Figure 13/1). Toward the Libyan border and northwest of Gilf Kebir, this plateau at the Akaba Passage is connected with another plateau of similar size but of different geological construction: the Abu Ras Plateau is made up of Silurian, Devonian and mainly Carboniferous clastics, overlain by a relatively thin cover of Late Jurassic to Middle Cretaceous strata. The Paleozoics are truncated toward the east and are no longer present in the Gilf Kebir Plateau (in sensu stricto), except for the Akaba area.

Strata of Paleozoic age are also present in the eastern part of Gebel Uweinat and in its eastern foreland. There are, however, no Paleozoics between there and southern Gilf Kebir as had been previously assumed (Geological map of Egypt 1981).

The areas around the New Valley, along Darb El Arba'in and between there and Lake Nasser belong to the classical areas of the so-called Nubian Sandstone. Because these strata were thought to be barren, it was only recently that the stratigraphy as well as the structural and paleogeographical development of these areas have been relatively well understood.

Early information on the geology of the southwestern desert of Egypt came mainly from four expeditions. During the Rohlfs expediton in 1872-1873, basic information about the marine strata overlying the so-called Nubian Sandstone was gathered (Zittel 1883), and a first geological map of large parts of the Western Desert was published. It was not until 1923 that important information was added by Menchikoff (1927) who visited Gebel Uweinat and discovered Carboniferous strata at the southeastern part of this isolated mountain area. Menchikoff accompanied Hassanain's 1923 expedition, which was responsible for the first scientific survey of Gebel Uweinat. Gilf Kebir was first discovered by Kamal El Din in 1926. In 1935, Sandford travelled through large areas of northeast Chad, northwest Sudan and southwest Egypt, and came back with a preliminary conception of the subdivision of strata in this very extensive empty desert. Several other expeditons, for example the ones of Bagnold, gathered much geographical and some geological information, but the geology of the area remained little known. A major breakthrough came in 1976 when the seemingly monotonous 'Nubia Sandstone' beds were subdivided and mapped (Klitzsch 1978, 1979, Barthel & Boettcher 1978, Klitzsch et al. 1979, Harms 1979, Beall & Squyres 1980, Klitzsch & List 1980).

Precambrian and younger crystalline rocks cover an area of some 40,000 km² in southwest Egypt. The majority of these rocks are exposed around the Gebel Uweinat-Gebel Kamil area at the convergence of the

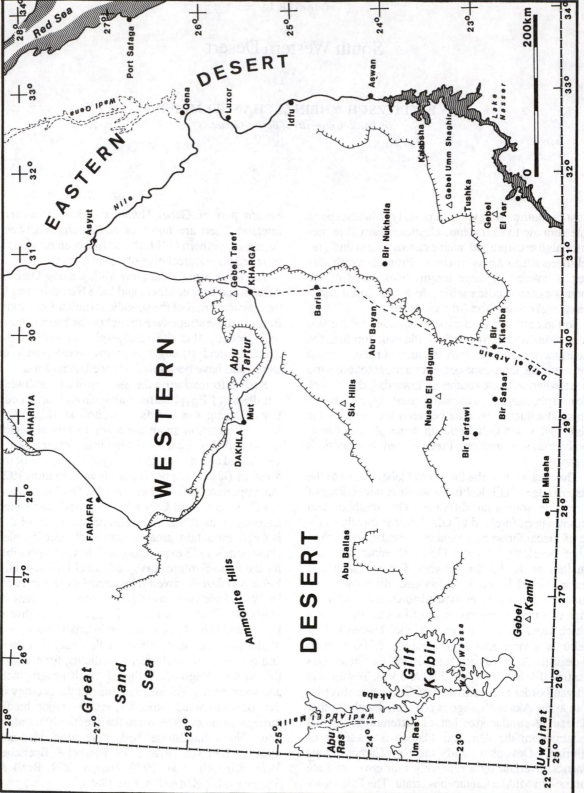

Figure 13.1 Location map of south Western Desert

frontiers of Libya, Egypt and Sudan. Smaller basement inliers like the Bir Safsaf, Gebel El Asr and Gebel Umm Shaghir complexes are exposed between Bir Safsaf and Lake Nasser. Between Gebel Uweinat and Lake Nasser, the basement is everywhere close to the surface except between Gebel Kamil and Bir Safsaf, where a probably downfaulted graben (Bir Misaha trough is filled with up to 700 m of Cretaceous sediments (Schneider & Sonntag 1985). The basement high, which separates the deep intracratonic Dakhla Basin from the more shallow basins of northern Sudan is referred to as the Gebel Uweinat-Bir Safsaf-Aswan Uplift System (Schandelmeier et al. 1983).

The basement exposures form slightly undulating plains with little relief. Outcrops are often scattered and covered by thin sheets of windblown sands. In a few cases (e.g. the peneplain area northeast of Gebel Uweinat, the south Gebel El Asr inlier) where the basement is in a topographically more elevated position than the surrounding sediments, this is either the result of tectonic uplift or, as in the case of Gebel Uweinat uplift itself or the Nusab El Balgum, the result of the intrusion of younger anorogenic bodies.

Geological information on the basement is still limited. Early information on the petrology is available from Menchikoff (1927) and Sandford (1935). More recently, the structure, metamorphism and geochronology of the basement, as well as the younger magmatic rocks, have been described by Burollet (1963), Marholz (1968), El Ramly (1972), Hunting Geology and Geophysics Ltd (1974), Klerkx & Rundle (1976), Klerkx & Deutsch (1977).

A synthesis of the Precambrian geology, however, has been developed only recently (Richter & Schandelmeier, this book).

1 PRECAMBRIAN AND YOUNGER MAGMATIC ROCKS

The Precambrian of the Western Desert of Egypt is exclusively part of the pre-Pan African East Saharan Craton of northeast Africa. The Gebel Uweinat-Bir Safsaf-Aswan Uplift is situated at the eastern fringe of this craton, and its high grade metamorphic and granitic rock associations change with marked tectonic contact into the metavolcanic-metasedimentary-ophiolite sequences of the Eastern Desert of Egypt (Nubian Shield).

Opinions still differ on details of the development of the Arabian-Nubian Shield, but there is general agreement that the shield was accreted onto the older continental plate through subduction and collisional type tectonics during the late Proterozoic.

Contemporary with the Late Proterozoic cratonization of the Arabian-Nubian Shield, the Pan African influence on the continental plate occurred in the form of reworking of older crust, resetting of isotope systems, migmatization and intrusion of voluminous batholiths (Vail 1983). These processes often obscure the older geological record, making it difficult to identify pre-Pan African rock units. The intensity of the Pan African imprint on the continental plate increases from West (Gebel Uweinat area) to East (Bir Safsaf-Aswan Uplift).

The lithology of the basement areas is typical for cratonic environment and can be generalized in the following lithostratigraphic successions:

A. Granulites and granoblastites (only Gebel Uweinat-Gebel Kamil area).

B. Migmatic gneisses and migmatites of generally granitic composition.

C. Supracrustal rocks (marbles, calcsilicates, amphibolites).

D. 'S-type' syn-collision granites and 'Caledonian I-type' post-collision granites.

E. Various kinds of dikes, ring-complexes and volcanic plugs.

The Gebel Uweinat-Gebel Kamil area is part of a metamorphic belt with a regional northeast-southwest trend, clearly visible in satellite images. Vail (1976) speculated that this belt might be the continuation of the Kibaran fold belt of central Africa which swings clockwise into Darfur and northern Sudan (see Richter & Schandelmeier, this book); but this regional structure is more likely the result of an Eburnean event. The regional stress field for this area was investigated by Richter (1986), who showed that deviations from the general northeast-southwest trend were due to clockwise rotation of crustal segments.

Structures in the Bir Safsaf-Aswan Uplift are more diffuse due to an intense Pan African overprint. Late Pan African mylonization (Bernau et al. 1987) seems to have been related to thrust fault movements which occurred when the Nubian shield assemblage collided with the cratonal margin. Migmatization and formation of S-type granites preceded this event and voluminous I-type granitoids were formed subsequently.

The regional fracture system of the Uweinat-Bir Safsaf-Aswan Uplift is probably the result of Late Proterozoic compressive stress induced onto the continental plate through accretional forces at the plate margin. The result was generally dextral strike slip movement on vertical fault planes, which were frequently reactivated during distinct periods in the Phanerozoic.

A late Archean to early Proterozoic age for the basement was directly proven for the Uweinat areas (Klerkx 1980, Harris et al. 1984), and indirectly by Nd model ages for rocks from the Bir Safsaf-Aswan

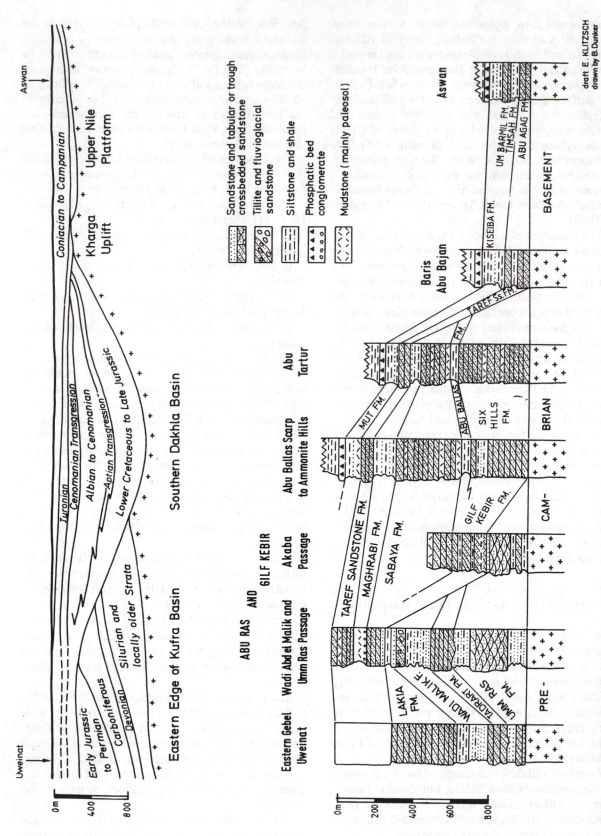

Figure 13.2 Typical sections of southwestern Egypt and basin development.

Uplift (Schandelmeier et al. in press). Middle Proterozoic rock ages were reported by Klerkx (1980) from the Uweinat area: he connected the 1800 Ma anatectic event to the regional folding along northeast-southwest trending fold axis. Analogies in the polymetamorphic development (Bernau et al. 1987) of the whole uplift system, structural investigations (Richter 1986), Nd model ages (Schandelmeier et al. in press, Harris et al. 1984) and U/Pb ages from detritic zircons in psammitic gneisses of the Eastern Desert derived from the continental block (Abdel Monem & Hurley 1979) all give a strong indication that a major crustal forming event affected the continental plate between Gebel Uweinat and Arabian-Nubian Shield in the early to middle Proterozoic.

The Pan African record, particularly in the Bir Safsaf-Aswan Uplift, is represented by migmatization and intrusion of granitoid batholiths (until 560 Ma), by uplift, erosion and block faulting of the marginal plate and by the invasion of dikes along vertical fractures ($\approx$ 520 Ma).

The Phanerozoic deformation of the marginal East Saharan Craton is exclusively one of intraplate nature. Differentiation of the continental plate into domal uplifts, as well as formation of basins and adjacent uplifts, grabens and incipient rifts are Phanerozoic phenomena, but in many cases were reactivations along older zones of weaknesses.

The oldest tectonic event in the Uweinat-Bir Safsaf-Aswan Uplift (Franz et al. 1987, Schandelmeier & Darbyshire 1984) occurred in the Permo-Triassic. Uplift block faulting and the ejection of wide-spread alkaline volcanics seem to indicate the formation of an incipient rift which failed to develop. These movements are probably related to an incipient fragmentation phase of Pangea. Jurassic volcanism of alkali-olivine basaltic nature occurs in the eastern part of the uplift near the margin of the continental plate.

In the upper Cretaceous, initial strike slip movement along the present-day Red Sea axis converted the Pan African, east-west striking shear planes into normal faults under tensional conditions. Horst and graben structures were formed especially in the Tushka depression area.

Alkali-olivine basaltic volcanism accompanied the fracturing. All the stages which subsequently contributed to the development of the Red Sea rift have their records in the continental plate. Volcanics of different ages generally occupy the same regional fractures, indicating a frequent reactivation of old fault systems (Franz et al. 1987, Bernau et al. 1987).

2 PALEOZOIC AND MESOZOIC STRATA

Paleozoic and Mesozoic rocks of Gilf Kebir-Abu Ras-Gebel Uweinat area

The best exposures of Paleozoic strata in the Gilf Kebir-Abu Ras-Gebel Uweinat area are at the southern slope of the Abu Ras Plateau, the Um Ras area (near the Libyan border), Wadi Abd El Malik (within the Abu Ras Plateau) and its side wadis and, further south, at the northeastern, eastern and southeastern edge of Gebel Uweinat including Karkur Talh and Karkur Murr. Mesozoic strata are best exposed in the Gilf Kebir area, north of the Abu Ras Plateau, at Gebel Kamil and in the middle and upper parts of the eastern flank of Gebel Uweinat.

In *Gebel Uweinat* shallow marine to fluvial sandstone of Ordovician and Silurian age forms the lower sandstone unit directly overlying the Precambrian Basement in the northeast (for example at Karkur Talh). These strata are, in parts, highly burrowed by *Scolithos* sp., and the basal part contains *Cruziana rouaulti* Lebesconte of Ordovician age (found by Monod and identified by Seilacher, both personal communications, see also Klitzsch & Léjal-Nicol 1984). The upper part of these early Paleozoic sediments is very similar to strata found north of Gebel Uweinat near the Um Ras Passage along the Libyan border and strata found southeast of Gebel Kissu. In both areas, these sediments contain *Harlania harlani* Desio and *Cruziana acacensis* Seilacher and are of Silurian age (probably Llandovery). The maximum thickness of Ordovician and Silurian strata of Gebel Uweinat is in the order of 100 m. At the southeastern edge of this mountain area, at Karkur Murr, Carboniferous strata rest directly on the Precambrian basement. At the outliers and small plateaux, directly east of Gebel Uweinat along the Sudanese border, the basal part of the strata consists of early Paleozoic strata, as around Karkur Talh.

Above the Ordovician and Silurian Karkur Talh and Um Ras Formation (Klitzsch & Léjal-Nicol 1984), 40 to 80 m of fluvial sandstone follow unconformably; these are also present in lesser thicknesses southeast of Gebel Kissu in Sudan and, in greater thickness, north of the Um Ras Passage toward Wadi Abd El Malik. This standstone is very similar in appearance to the Tadrard Sandstone of Libya (early to middle Devonian) and probably correlates with this formation which is much better known further west.

Both in the southeastern part of Gebel Uweinat at Karkur Murr and in the northeast at Karkur Talh, and also directly east of the mountain area, Carboniferous strata are present in varying thickness and facies. The lower 50 to 120 m of the Carboniferous section are mainly made up of sandstone, siltstone and some shale deposited in shallow marine, coastal flood plain or fluvial environment. Parts of the strata are intensively burrowed and brachiopods as well as plant remains of early Carboniferous age are common (Wadi Malik Formation, Klitzsch 1979, for flora and fauna, see Klitzsch & Léjal-Nicol 1984). At Karkur Murr these strata rest directly on Precambrian rocks, while at Karkur Talh, as well as east of Gebel Uweinat, they overlie older Paleozoic strata. In most locations, they more or less conformably overlie the Tadrard Sandstone Formation.

Above these partly marine strata, a unit of 40 to 120 m of very heterogeneous sediments follows. It is made up of fluvial sandstone which is interbedded with a very regularly-laminated siltstone (similar to varve sediments) in some locations, for example the section at Karkur Murr; the base of this unit is made up of several meters of a chaotic sediment similar to tillite. The direction of transport of the fluvial sandstone is to the south and southwest. At Karkur Talh and southeast of Gebel Uweinat in Sudan, the varve-type siltstone dominates the whole formation. It occupies the same stratigraphic position of the tillite and the fluvio-glacial sandstone of the northern Wadi Malik and is, therefore, called the North Wadi Malik Formation. Previously, the varve-type sediments at Gebel Uweinat were called Lake Beds (Klitzsch 1980).

Along the middle part of the eastern slope of Gebel Uweinat, the Ordovician to Carboniferous strata are overlain by a 250 to 300 m thick fluvial sandstone, partly immature and interbedded with paleosol. Direction of transport is mainly towards south to southwest. These strata must be equivalent to the Lakia Formation of northern Sudan (Klitzsch & Léjal-Nicol 1984). The Lakia Formation, like the underlying North Wadi Malik Formation, was deposited after the eastern part of North Africa was uplifted in connection with the collision between Gondwana and the northern continents (Hercynian Event). As a consequence of this, Paleozoic and older strata were eroded from most of south Egypt and transported south to southwestward. Gebel Uweinat is the only area in south Egypt where the resulting deposits are present. Further south and southwest, they cover large areas in northern Sudan (Lakia Umran Escarpment, for example) and in the Kufra Basin in southeast Libya and northeast Chad. These strata were named the Lakia Formation (Klitzsch & Léjal-Nicol 1984); they are of Permian to Early Jurassic age. The uppermost 200 to 300 m of the sedimentary section of Gebel Uweinat are made of sandstone equivalent to the Gilf Kebir Formation (Jurassic to lower Crectaceous, Klitzsch 1978).

Along the southern foreland areas of the *Abu Ras*

Plateau (including the Akaba Passage) and the Um Ras Passage, Paleozoic strata can be studied at many places. The Silurian Um Ras Formation reaches a thickness of approximately 400 m between the Um Ras area and the southwestern slope of the Abu Ras Plateau. It consists mainly of fluvial and deltaic white sandstone, intercalated with interbeds of shallow marine siltstone, some shale and burrowed sandstone. These beds contain *Harlania harlani* Desio, *Cruziana acacensis* Seilacher, *Scolithos* sp. and other trace fossils at several levels. The Um Ras Formation reaches as far east as the Akaba Passage between Gilf Kebir and the Abu Ras Plateau. Further east, it is truncated by the Gilf Kebir Formation of Late Jurassic to Early Cretaceous age. Between this area and the Eastern Desert, Mesozoic strata rest directly on basement in south Egypt.

Between the Um Ras Formation and the Wadi Malik Formation of Carboniferous age, 80 to 100 m of fluvial sandstone similar to the Tadrart Sandstone of Libya, are exposed at the southwestern rim of the Abu Ras Escarpment.

The Wadi Malik Formation more or less conformably overlies these Devonian beds in the middle part of the southwestern and western Abu Ras Escarpment. It is, however, best exposed all along the Wadi Abd El Malik and its side wadis which reach from the southern and southwestern part of this 60 × 120 km large plateau towards its relatively flat northern end, and drains this plateau in a northerly direction. The 100 to 150 m thick Wadi Malik Formation consists of marine sandstone, siltstone and shale, interbedded with fluvial, deltaic and tidal sandstone. In marine beds, brachiopods (*Camerotoechia* sp.), trace fossils (like tracks of tribolites) and starfish or burrows like *Bifungitis fezzanensis* Desio are frequent (Seilacher 1983). Several beds contain frequent plant remains including many different species of *Sigillaria* and *Lepidodendron*. The age of this flora is Visean (lower Carboniferous, see Klitzsch & Léjal-Nicol 1984).

The overlying strata, as at Gebel Uweinat are very heterogeneous. In north Wadi Abd El Malik, the Wadi Malik Formation underlies unconformably a 30 to 60 m thick bed of a chaotic sediment, consisting of clay, sand, gravel, blocks of older strata up to house size. Southward, in the middle and southern parts of Wadi Malik and its side wadis, these strata are replaced by fluvial sandstone, conglomerate and sandstone with isolated blocks of older strata (erratic boulders).

The thickness of this fluvio-glacial unit is several tens of meters. It is overlain by cross-bedded and parallel-bedded sandstone containing plants of Stephanian age in the central parts of the Abu Ras Plateau, for example *Cordaites angulostriatus*

Grand'eury (A. Léjal-Nicol pers. comm.), was found in sandstone directly overlying the fluvio-glacial North Wadi Malik Formation and underlying the basal parts of the Gilf Kebir Formation of Jurassic to early Cretaceous age.

In the upper part of the south Abu Ras Plateau and on isolated hills on the plateau in this extensive landscape, the Carboniferous strata normally underlie unconformably the Gilf Kebir Formation. At the Akaba Passage southeast of the Abu Ras Plateau, in the Gilf Kebir area and its foreland, these strata consist of fluvial sandstone, lacustrine sediments and paleosols. It contains some nearshore to possibly shallow marine, burrowed sandstone, with *Thalassionoides* sp.

At the eastern edge of Gilf Kebir toward the Abu Ballas area, the upper part of the Gilf Kebir Formation interfingers with the marine Abu Ballas Formation of Aptian (early Cretaceous) age. Within the Akaba Passage and directly north of there, as well as at the western end of Wadi Wassa and in the southern reaches of Gilf Kebir, several horizons are rich in flora: they contain ferns and other plants of Jurassic to Early Cretaceous age (among many others, *Cladeophlebis oblonga* Halle, *Phlebopteris polypodoides* Brongniart and *Weichselia reticulata* Stockes & Webb, see Klitzsch & Léjal-Nicol 1984). Shale and siltstone contain fossil insects at several locations (Schlüter & Hartung 1982).

More or less the same flora also characterizes the Gilf Kebir Formation along the northern edge of Gilf Kebir, as well as along the northern edge of the Abu Ras Plateau north and northeast of Wadi Abd El Malik. Similarly the lower part of Gebel Kamil and the Mesozoic strata between there and the northeastern foreland of Gebel Uweinat are characterized, down to some meters above basement, by this Jurassic to lower Cretaceous flora.

While in the Abu Ras Plateau, the Gilf Kebir Formation represents only the top part of the sediments and is eroded over large areas, it builds up the lower half to lower two-thirds of the whole Gilf Kebir Plateau (except for the Akaba Passage where remains of the Silurian Um Ras Formation are present at the base and above the basement).

The upper half to one-third of the Gilf Kebir Plateau is made up of younger Cretaceous strata equivalent to the Sabaya Formation (Albian to Cenomanian, here mainly fluvial sandstone) and to the Maghrabi Formation (Cenomanian, mainly paleosols and fluvial sandstone). Equivalents to these strata are also present between the northern end of the Abu Ras Plateau and Wadi Qubba. In Wadi Qubba, these sediments are overlain by fluvial and lacustrine sandstone and paleosol, probably equivalent to the Taref Sandstone of Turonian age.

Mesozoic pre-Maastrichtian strata of the area between the New Valley, Gilf Kebir and the southern border of Egypt

The large area lying in the shadow of the Limestone Plateau of the middle latitudes of Egypt and extending from the Abu Tartur and Ammonite Escarpments to the north, to Gilf Kebir to the west and the Selima Sandsheet to the south is characterized by a relatively simple structural pattern and by six to seven lithostratigraphic formations of great consistency. This area is at the southern rim of the Dakhla Basin, whose center lies to the north under the Great Sandsea. It is also the transitional area toward the shallow basins and uplifts of north Sudan and is part of a large regional high formed between the Late Carboniferous and Early Jurassic. Therefore, this area is free of older Mesozoic or Paleozoic strata (erosion and/or nondeposition between Late Carboniferous and Early Jurassic time). The principal lithology of the strata present in this area can be studied along both sides of the Kharga-Dakhla road, and the subdivision given by Klitzsch (1978), Barthel & Boettcher (1978) is valid, with some alterations, for the whole area. Previously, the whole succession was described as belonging to the Nubian Strata. The subdivisions are as follows:

The basal unit is made up of up to 600 to 700 m thick fluvial sandstone, paleosol and, toward the top, minor nearshore marine sandstone. It is called basal Clastics (Klitzsch 1978) or better Six Hills Formation (Barthel & Boettcher 1978). These clastic sediments were deposited while the area was subsiding and before the Aptian transgression advanced from the north. This development was the reverse of the previously described development of Late Carboniferous to Early Jurassic time. After Pangea began to disintegrate during the Jurassic, the area was tilted toward the north and the drainage, going south to southwest between Late Carboniferous and Early Jurassic time, reversed again and consequently fluvial sediments were prograded northward.

The first transgression which reached the area of southwest Egypt during this new structural cycle was the Aptian transgression. It is represented by the second Formation from the bottom, the Abu Ballas Formation (Barthel & Boettcher 1978, Boettcher 1982, originally *Lingula* Shale, Klitzsch 1978). It consists of up to 60 m of shale, siltstone and sandstone of a very shallow marine transgression of probably very high salinity. Parts of the strata are rich in fauna of mainly small species of lamellibranchiades, brachiopods and gastropods and they also contain plant remains including fossil fruit. An Aptian age for this transgression was also proven by palynological means (Schrank 1982, 1983).

Above a sedimentary break (formation of thick paleosol) a fluvial sandstone (the Sabaya Formation) was deposited during the Albian, or more likely, at the beginning of the Cenomanian. The Sabaya Formation (Barthel & Boettcher 1978), formerly Desert Rose Beds (Klitzsch 1978), is more than 200 m thick and represents one of the best aquifers of the Dakhla Basin, New Valley area. This is topped by sediments of a short and shallow marine transgression made up of flaser-bedded silt and sandstone, shale and, in certain areas, also paleosol and fluvial sandstone intercalations. These sediments contain small species of lamellibranchiades, gastropods, brachiopods, sea urchins, frequent fish and other vertebrate remains as well as middle Cretaceous flora. The age is most likely Cenomanian and the formation has great similarity to the Bahariya Formation of the north Western Desert of Egypt (Dominik 1985). It was originally called Plant Beds (Klitzsch 1978) and later changed to Sabaya Formation (Barthel & Hermann-Degen 1981), it is 50 to 200 m thick.

Above an unconformity, the Sabaya Formation is overlain by the next fluvial sandstone formation, the Taref Sandstone (Awad and Ghobrial 1965). This 30 to 150 m thick sandstone was deposited after some structural unrest which occurred during the Late Cenomanian or Early Turonian. In Wadi Qena a sandstone unit of similar position and facies characteristics (Um El Omeiyid) is of Turonian age (Klitzsch, Gröschke, Hermann-Degen, this book). The upper part of the Taref Sandstone Formation locally seems to be of eolian origin. The predominant continental history of the area comes to an end during the Campanian. Above the Taref Sandstone, a multi-colored unit of shale, siltstone and sandstone was deposited, originally called Variegated Shale (Said 1962) and now called Mut Formation (Barthel & Hermann-Degen 1981). These strata are more than 150 m thick and represent a coastal flood plane, estuarine and shallow marine environment. Apart from small invertebrates and plant remains, they contain remains of saurian and of other vertebrates. The upper part of these strata grades into sediments containing phosphate beds. The age of the formation is Coniacian to mainly Campanian. It is possible that sedimentary breaks (represented by paleosol) are included.

The described formations are all part of the strata which were formerly called Nubian or Nubian Sandstone (Russegger 1838, Said 1962). A complete and easily visible exposure of these strata occurs at the Abu Tartur escarpment and its southern foreland directly west of kilometer 100 on both sides of the Kharga-Dakhla road. There, the road passes the Sabaya Formation west of an extensive playa. North of the road toward the Abu Tartur escarpment, Magh-rabi Formation, Taref Sandstone and Mut Formation are exposed, while, south of the road, Abu Ballas Formation with a rich fauna lamellibranchiates and brachiopods (*Lingula* sp.) overlies the Six Hills Formation. The base of the latter is the only part of the section not exposed near the road. The Six Hills Formation reaches much further south toward Gebel Abu Bayan and the basement exposures of the Bir Tarfawi-Bir Safsaf area.

An excellent exposure of the 'Nubia Sandstone' also occurs along the road which leads to the Abu Tartur phosphate mine. At the nearby settlement of this mine, the Maghrabi Formation is exposed. Overlying the Sabaya Formation, about 1 km west of the settlement, rich flora of Middle Cretaceous age occurs within the lower part of the Maghrabi Formation (flaser beds). Between there and the main escarpment, Taref sandstone is present, followed by a well-exposed section of Mut Formation, directly overlain by the phosphate beds of the mine area. A similar section can be seen between Kharga Oasis and Gebel Taref, where at least the middle and upper parts of the Maghrabi Formation and the whole Taref Sandstone are well-exposed. Another area of easily accessibility, where parts of these strata are exposed, is along the Mut (Dakhla)-Bir Tarfawi road.

The strata described are present, with some alterations, in the extensive area between the Kharga-Dakhla road and the eastern edge of Gilf Kebir. There the Abu Ballas Formation interfingers with the upper part of the Gilf Kebir Formation. The Maghrabi Formation loses its partly marine character west of longitude 27° E. Equivalents of this formation north of Gilf Kebir consist only of fluvial sandstone, conglomerate and paleosols. Because of very long and consistent more or less east-west striking faults, transecting the southern part of the Western Desert, repetition of strata, mainly in the lower part of the section, is common. This is especially noticed along and south of latitude 24° N. Older faults were active after the Cretaceous and resulted in vertical displacements of usually not more than 10 to 100 or 200 m. But, because the dip toward the Dakhla Basin or toward local structural lows is extremely gentle, the displacement resulted in a repetition of strata. For example, the Abu Ballas Formation is exposed in several areas south of 23° N. Facies changes in this direction are to be expected. Between the Abu Tartur area and the eastern edge of Gilf Kebir, the facies is very consistent because this whole area was originally situated more or less parallel to the Aptian coast; the areas further south were closer to this coast at the rim of a fluctuating and more shallow sea. The Abu Ballas Formation of Aptian age, however, is found as far south as the upper part of the escarpment northwest of Wadi Halfa near the Sudanese border, just north of

Abu Simbel and in the Selima area in Sudan (base of strata there).

4 MESOZOIC PRE-MAASTRICHTIAN STRATA EAST OF DARB EL ARBAIN ROAD

An obvious change in age and facies of strata occurs in the areas east of the Darb El Arbain road. It seems that the eastern edge of the Dakhla Basin was along this general trend. It is possible that the so-called Kharga uplift which formed the eastern border of this basin extending from the Kharga-Baris-Bir Hussein area eastward to the area between Qena and Aswan marked the beginning of an area of erosion or non-deposition (wide uplift of platform). The older formations of the former Nubian sandstone wedge out toward this area: the Six Hills, Abu Ballas Sabaya and Maghrabi Formations are absent in this large structural high. The strata present along the southern rim of the regional high range in age from Turonian to Campanian and younger (equivalents to Mut Formation, recently called Kiseiba Formation, Klitzsch & Léjal-Nicol 1984). Between the area of Bir Kiseiba and Kalabsha, equivalents of only these younger strata are present on top of the Precambrian basement. South of the Bir Safsaf/Bir Nukheila and the Kalabsha basement exposures, the section becomes complete again. There, a more or less eastward striking trough must have been connected with the Dakhla Basin since the Late Jurassic or Early Cretaceous (Six Hills and Abu Ballas time).

It is possible, therefore, to postulate that the so-called 'Southern Nile Basin' was a structural high until the Cenomanian. After all or most of the remains of older strata had been removed from this area, it became a large northeast-southwest trending embayment by relief inversion. Fluvial, lacustrine and marine sediments of Turonian to Campanian age were then deposited. The sea remained in that area until the early Eocene and its sediments sealed this former high.

CHAPTER 14

The surroundings of Kharga, Dakhla and Farafra oases

MAURICE HERMINA

Conoco, Egypt

The oases of Kharga, Dakhla and Farafra, and their surrounding areas in the southern and central parts of the Western Desert of Egypt, afford excellent exposures of Cretaceous and lower Tertiary. This chapter deals with the geology of an area of about 90,000 km^2 within these surroundings. The areas covered comprise Kharga, Abu Tartur-Dakhla and Farafra, from the southeast to the northwest (Fig. 14/1). The geology of these areas has been the subject of many investigations since the pioneering work of Zittel (1883) and the early memoirs of Ball & Beadnell of the Geological Survey of Egypt (1900-1905). Interest in the surface and subsurface geology of these areas was revived in the 1960s and 1970s when they came to be parts of a major reclamation project, the 'New Valley Project'. The works of Hermina, Ghobrial & Issawi (1961), Awad & Ghobrial (1965), Ghobrial (1967) and Hermina (1967) deal with the surface geology and describe numerous sections in and between Kharga and Dakhla oases.

Awad and Ghobrial (1965) subdivide the Cretaceous and lower Tertiary sediments in the Kharga area into several litho- and biostratigraphic units. Issawi (1969, 1971) defines the Paleocene and lower Eocene rock units which cover the areas to the south of Kharga and which he terms the Garra-El Arba'in facies. Hermina & Issawi (1971) outline and correlate the upper Cretaceous and lower Tertiary formations in south Egypt. Issawi (1972) reviews their stratigraphy in central and south Egypt and differentiates three main types of facies. El Hinnawi et al. (1978), El Deftar et al. (1978) and Issawi et al. (1978) discuss the stratigraphy and geologic history of the Kharga and Abu Tartur areas, while Mansour et al. (1982) deal with the west Dakhla area. Other interesting publications from Egyptian universities discuss the paleontology and sedimentology of the rocks covering the Kharga and Dakhla areas. The biostratigraphy and structure of the Farafra area is dealt with in the publications of LeRoy (1953), Said & Kerdany (1961), Omara et al. (1970) and Youssef & Abdel Aziz (1971).

The first zonation and age dating on the predominantly sandstone sequences of the 'Nubia' encountered in the many wells drilled for water in the oases are given in El Shazly & Shata (1960), Barakat & Milad (1966), Helal (1966), Abbas & Habib (1971), Philip & Asaad (1972) and Barakat & Abdel Hamid (1974). These zonations are based on electric logging, heavy mineral analyses, and micropaleontology.

During the past ten years, the geology and hydrology of the oases areas have been treated extensively. Much of this activity is related to renewed interest in the regional geological mapping of these areas to serve both groundwater and oil exploration projects.

A principal achievement during this recent phase was the subdivision of the 'Nubian Sandstone' unit into several formations which have been traced in the field. More light has also been thrown on the facies changes displayed by the overlying marine sediments and their depositional history. Outstanding amongst the publications of this period are Barthel & Boettcher (1978), Klitzsch et al. (1979), Barthel & Herrmann-Degen (1981), Beottcher (1982), Bisewski (1982), Klitzsch (1984), Dominik & Schaal (1984), Hendriks et al. (1984), Hendriks (1986), Dominik (1985), Klitzsch & Wycisk (1987), Wycisk (1987), Hendriks et al. (1987), Schrank (1987) and the New Geologic Map Series (EGPC-Conoco 1987-1988).

Paleogeography and depositional framework

Figure 14/2 illustrates the relief of the basement top in parts of the south and central Western Desert as related to the Kharga-Farafra stretch. Two sedimentary basins exist on the northward-sloping African craton. These are the Dakhla basin on the west and the Assiut-Upper Nile basin on the east, with the Kharga uplift in between. The Kharga uplift, a northeastwardly plunging positive structure, delimits the southeastern rim of the Dakhla basin and was active with varied intensities during Jurassic and Cretaceous times. Uparching of the Precambrian basement

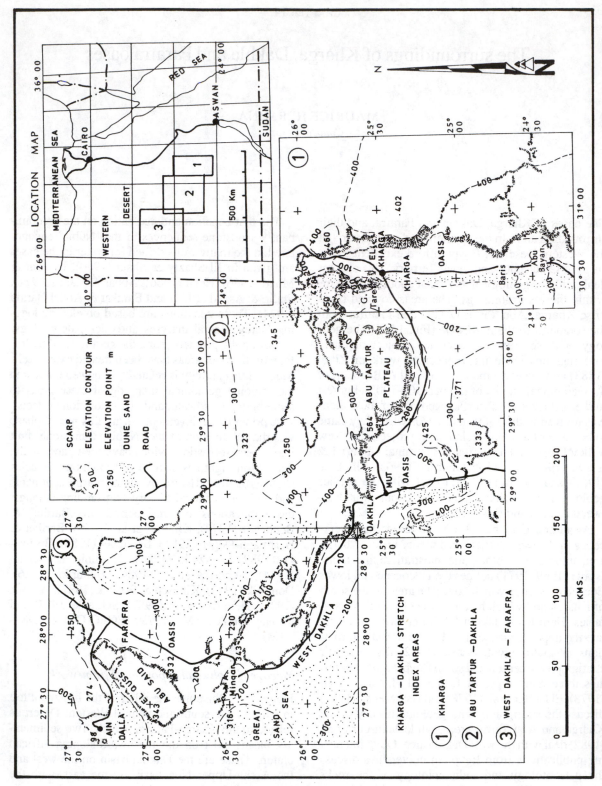

Figure 14.1 Index location map of the Kharga-Farafra stretch and its areas.

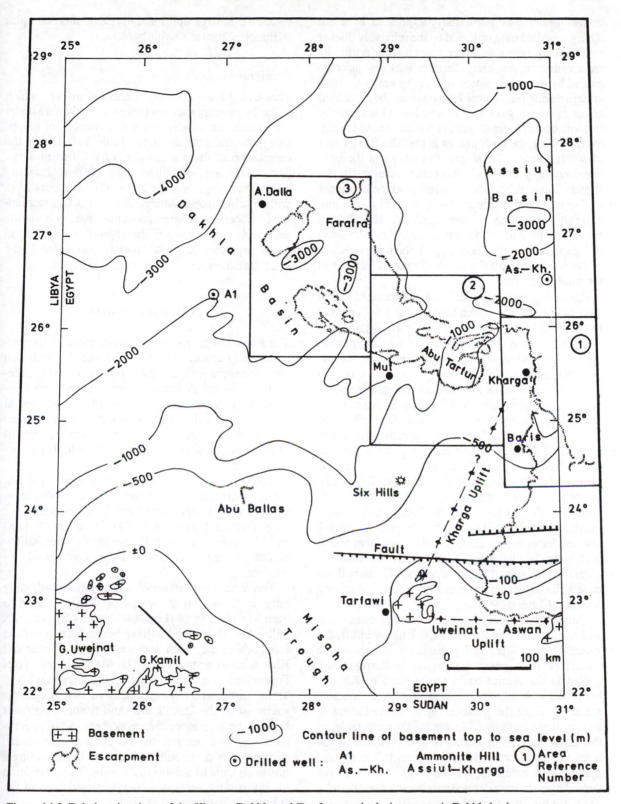

Figure 14.2 Relative situations of the Kharga, Dakhla and Farafra areas in the intracratonic Dakhla basin.

in this uplift was particularly expressed in south Kharga, while being nullified by the relatively thicker sedimentary cover in other areas to the north. Its reactivation in the early Tertiary was also accompanied by the rejuvenation of a major east-west fault system during post-lower Eocene time, the combined effect of which gave rise to small-sized and patchy tectonic occurrences of granite bodies exposed as hill masses along the fault planes in the Abu Bayan area of south Kharga. These granites represent the only basement exposures in the entire Kharga-Farafra stretch. In the subsurface, the relief on the basement top, represented in Figure 14/2, is based on drilling data and geophysical investigations. The Dakhla basin was filled up with continental and marine strata of Paleozoic to early Eocene age in the northwest and of Jurassic (or early Cretaceous) to early Eocene in the south and east.

Paleogeographically, the areas incorporated within the Kharga-Farafra stretch lie differently with respect to the Dakhla basin (Fig. 14/2). In the Kharga and Abu Tartur areas (marked 1 and 2), which represent the eastern and marginal parts of the basin, the top basement lies at 1000 to 500 m (or lower) below sea level. Paleozoic sediments are not reported from surface or subsurface data and late Jurassic continental sediments unconformably overlie the basement. Deeper in the Dakhla basin, drilling in the Mut area to a depth of 1100 m below sea level in the Jurassic reached no basement. The top of the basement is estimated to lie at about 2000 m below sea level and it is not certain whether an upper Paleozoic section underlies the Jurassic in the West Dakhla area. Further to the northwest, the Ammonite Well-1 reached basement at 2000 m below sea level and at least 860 m of a Paleozoic section (Carboniferous and older) is recognized in the stratigraphic interval between the underlying basement and the overlying Jurassic (Conoco 1978).

Outside the limits of the stretch under discussion, and to the north and east of the Kharga uplift, the Assiut-Upper Nile Valley basin is delineated. It should be mentioned that the Assiut-Kharga well, drilled in the Assiut basin to the north of Kharga, bottomed in lower Senonian at about 820 m below sea level. Here the presence of an upper Paleozoic section underlying the Mesozoic is questionable, as the basement is estimated to be at 2000 m below sea level. To the east of the Kharga uplift, the present plateau area extending to the Nile Valley was occupied by a shallow depression characterized by relatively shallow marine sediments of late Cretaceous to early Eocene age and presumably by thin development of older Mesozoic strata. Some authors propose that it was mainly a shallow platform situated be-tween the Kharga uplift and the Eastern desert uplift (Klitzsch, Chapter 13, this book).

Stratigraphy

The late Mesozoic-early Cenozoic rocks, which make the primary sedimentary cover in the area under discussion, are subdivided into a number of mappable lithostratigraphic units. Table 14/1 shows the correlation of these units across the different areas. The units are classified into: (a) a Jurassic-Campanian sequence, predominantly continental but with marine intercalations, and (b) a Campanian-lower Eocene, transgressive-regressive open marine sequence. A summary of the depositional history in the Kharga-Farafra stretch and its correlation in the south Bahariya region is illustrated in Figure 14/3.

JURASSIC-CAMPANIAN

The Jurassic-Campanian sequence includes the predominantly continental sandstone and clay beds that were formerly lumped under the term 'Nubia'. Recently, upon the detection of marginal marine clay-shale strata which occur persistently in outcrops at certain stratigraphical intervals within the sandstone sequences, its subdivision into well-defined units of cyclical continental and marine deposition has become possible.

Deposited in the continental basins are: the Six Hills Formation (late Jurassic), Sabaya Formation (Albian-early Cenomanian) and Taref Formation (? early Turonian). The Abu Ballas Formation (Aptian) and Maghrabi Formation (Cenomanian) are shallow marine deposits separating these continental sequences.

The areal distribution of these northerly-dipping units is shown on the maps of Kharga and Abu Tartur-Dakhla areas (Figs 14/4 and 14/5). The Six Hills, Abu Ballas and Sabaya Formations cover the forelands to the south and west of Abu Tartur and Kharga areas respectively. The Maghrabi and Taref Formations outcrop between north Kharga and Abu Tartur and extend over an extensive area toward the southeast of the Dakhla area and further southwest beyond the area under discussion here. A rather complete exposed section of the Jurassic-Campanian succession is described by Bisewski (1982) along a line from Qulu El Sabaya to Six Hills. The correlation of this surface section with interpreted logs in the subsurface is attempted in Figure 14/6.

Six Hills Formation (Barthel & Boettcher 1978; = Basal Clastics; Klitzsch 1978, Bisewski 1982)

			Farafra-West Dakhla	Dakhla-Abu Tartur	North Kharga	South Kharga (Baris)
early Tertiary	Eocene	middle	Naqb		Drunka / Drunka	Dungul
		early	Naqb / Farafra	Farafra / Dungul / El Rufuf	El Rufuf / Serai (Thebes)	
			Garra (A.Dalla) / Esna / Garra	Esna / Garra / Esna	E s n a	Garra
	Paleocene	late	Tarawan	Tarawan / Kurkur / Tarawan	T a r a w a n	Kurkur
		middle	D a k h l a / Khoman	Dakhla / Kurkur / Dakhla	D a k h l a	Dakhla / Kurkur
		early				
late Cretaceous		Maastrichtian	Khoman / Dakhla	D a k h l a	D a k h l a	Dakhla
		Campanian	El Hufuf / Duwi	D u w i	D u w i	Duwi
			W.Hennis / Quseir	Q u s e i r	Q u s e i r	Quseir
		Santonian				
		Coniacian				
		Turonian		T a r e f	T a r e f	T a r e f
		Cenomanian		M a g h r a b i	M a g h r a b i	Maghrabi
				S a b a y a	S a b a y a	
early Cretaceous		Albian				
		Aptian		A b u B a l l a s		Six Hills
				S i x H i l l s		
		Barremian-Neocomian				Basement
		Jurassic				

Table 14.1 Correlation of lithostratigraphic units, Kharga-Farafra stretch.

The Six Hills Formation reaches up to 500 m in thickness in its type area (Six Hills, 24° 10' N, 29° 15' E, Fig. 14/2), at about 100 km south of Mut in Dakhla, and is thus beyond the limits of the map for this dicussion. It is generally assigned to late Jurassic-early Cretaceous age (Klitzsch & Lejal-Nicol 1984). In the subsurface, based upon pollen investigations, Helal (1965) and Soliman (1977) refer the lower part of the Six Hills Formation to late Jurassic. Soliman (1977) also identifies late Jurassic foraminifera in some Kharga wells. Schrank (1987) describes pollen of middle to late Jurassic from a horizon about 300 m below the well-defined overlying Aptian shales in Ammonite Well-1. Bisewski (1982) notes the presence of marine influence at a horizon 60 m below the top of the Six Hills Formation to the south of Abu Tartur. This may be considered a prelude to the Aptian transgression and is overlain by a fluviatile series. Dominik (1987) refers at least the exposed part that crops out at the Six Hills type locality to the Aptian transgression.

Abu Ballas Formation (Barthel & Boettcher 1978; = *Lingula* Shale; Klitzsch 1978)
The type locality of this formation is at the escarpment south of Abu Ballas (24° 23' N, 27° 35\' E, Fig. 14/2). In the Qulu El Sabaya area (Fig. 14/5), the Abu Ballas sediments attain a thickness of 16 to 25 m.

The transition between the fluviatile sandstones of the Six Hills Formation and the overlying marine clays and shales of the Abu Ballas Formation is easily recognized, although an erosional surface is not found everywhere (Bisewski 1982).

Irregular intercalations of iron crusts and mud cracks indicate that periods of continental influence prevailed during the marine deposition of this formation. Fully marine conditions do not occur in these sediments except in the topmost strata which are related to the climax of the Aptian transgression. An Aptian age is proposed for this rock unit by Boettcher (1982) in spite of the absence of index fossils. Pollen and spores investigated by Schrank (1987) from the subsurface in West Mawhub prove an Aptian age for

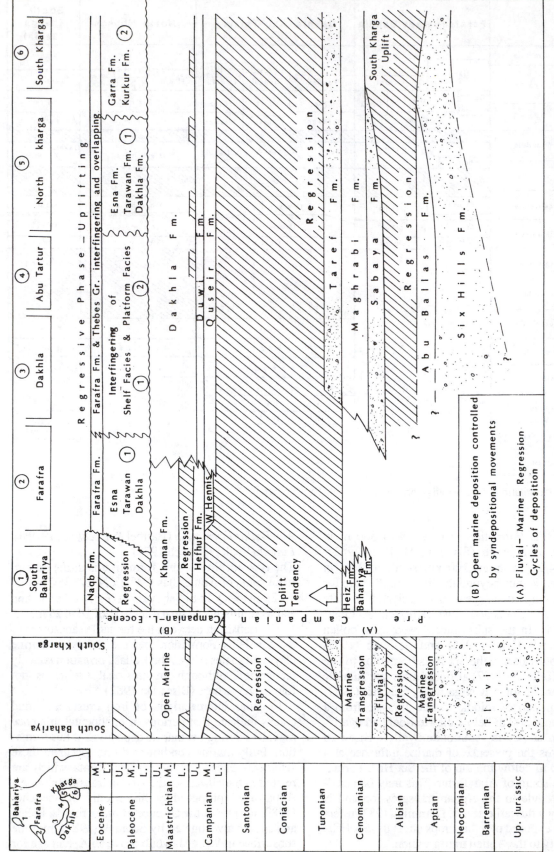

Figure 14.3 Diagram. Summary of depositional history in the Kharga-Farafra-south Bahariya stretch (modified and expanded after Dominik, oral comm.).

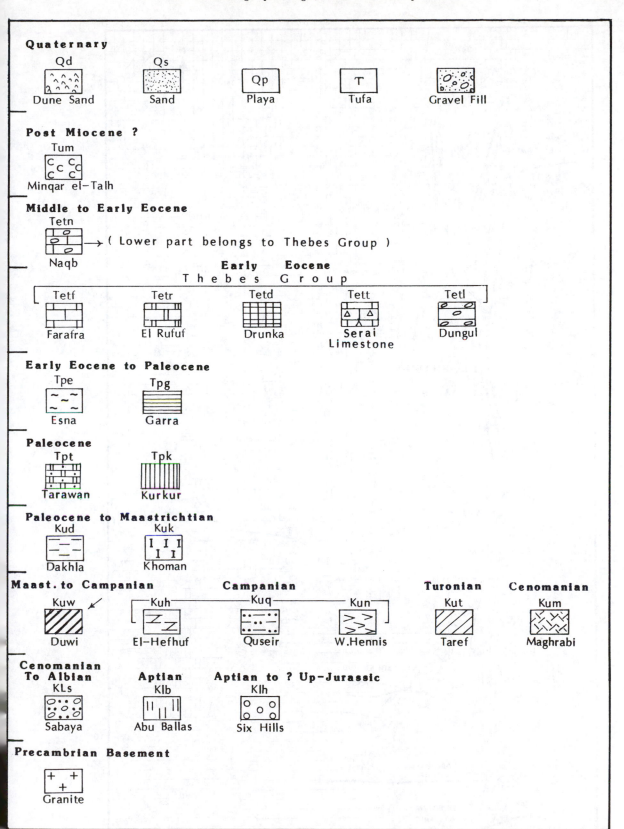

Table 14.2 Explanation of symbols appearing on maps (Figs. 14.4, 14.5 & 14.6).

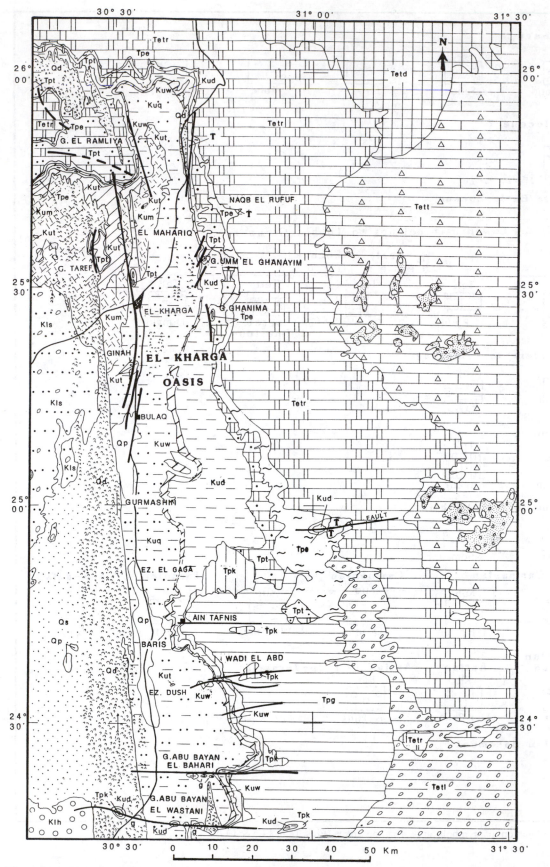

Figure 14.4 Geological map of the Kharga area (after Awad & Ghobrial 1965, El Hinnawi et al. 1978, El Deftar et al. 1978 and EGPC/CONOCO map sheet Luxor 1988). For legend, see Table 14.2.

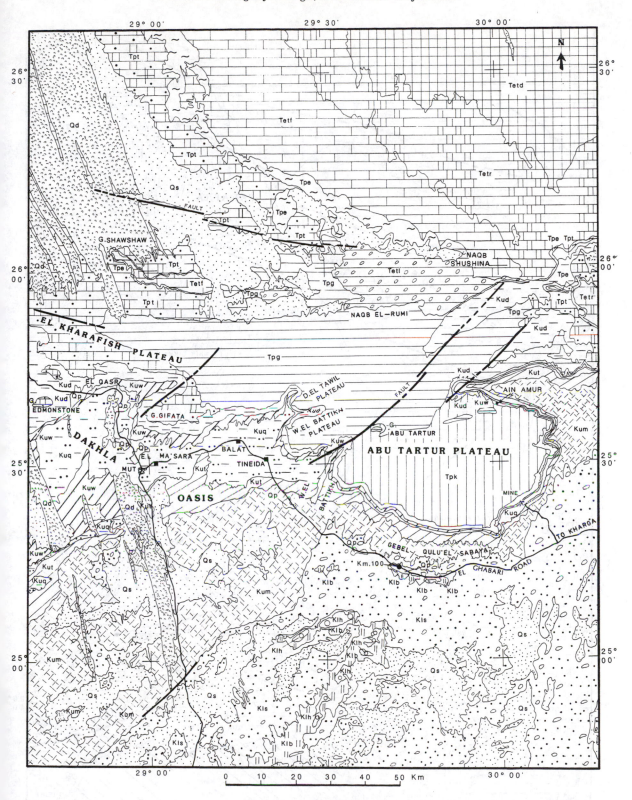

Figure 14.5 Geological map of the Abu Tartur-Dakhla area (after Hermina 1967, El Deftar et al. 1978, and EGPC/CONOCO map sheet Dakhla 1987). For legend, see Table 14.2.

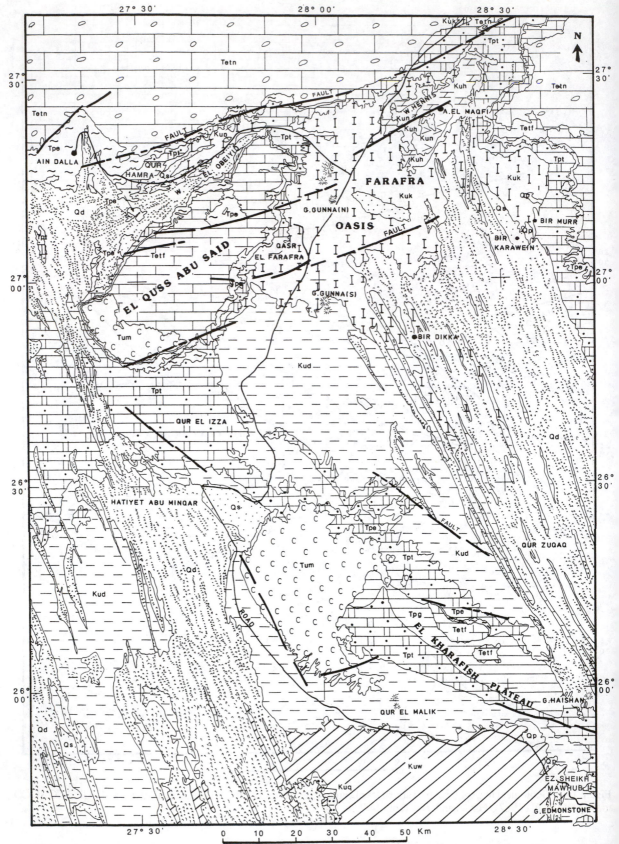

Figure 14.6 Geological map of the west Dakhla-Farafra area (after EGPC/CONOCO map sheets Dakhla and Farafra 1987) (for legend, see Table 14.2).

a horizon correlated with Abu Ballas sediments (Fig. 14/6).

Sabaya Formation (Barthel & Boettcher 1978; = Desert Rose Beds; Klitzsch 1978)

The Sabaya Formation overlies the Abu Ballas and follows the Albian regression (Fig. 14/3). The type locality of this formation is at Qulu El Sabaya hills on the Kharga-Dakhla road (25° 21' N, 29° 43' E) where it measures 170 m in thickness. In the more basinal parts of the Dakhla basin, this unit reaches a thickness of up to 200 m and may have some marine influence (Bisewski 1982). It is made up of a sequence of clearly fluviatile sediments which start at the base with an erosional surface overlain by a 30 m thick white kaolinitic paleosol sandstone rich in root remains. Due to the increased epeirogenic intensity on the south Kharga uplift, the Sabaya sandstones were probably not deposited in areas of south Kharga (according to the interpretation of logs of wells drilled in the Baris area, Figs 14/3 and 14/6).

The Sabaya Formation is most likely of Albian to early Cenomanian age. A sample investigated by Schrank (1987) from the top of a correlative unit in Ammonite Well-1 yielded a microflora of late Albian-Cenomanian age (Sample S3, Fig. 14/6). Between the continental Sabaya Formation and the following transgressive Maghrabi Formation, there is a gap in sedimentation and formation of topographic relief (Bisewski 1982).

Maghrabi Formation (Barthel & Herrmann-Degen 1981; = Plant Beds; Klitzsch 1978)

The interbedded claystones, siltstones and sandstones of the Maghrabi Formation transgressively overlap the paleorelief on top of the Sabaya Formation. The type of the unit is the exposure covering the southeastern forelands of Abu Tartur to the west of Kharga. It is about 60 m in thickness but thins out considerably over the Kharga uplift in the Baris area (Logs, Fig. 14/6). The best exposures occur in the vicinity of the phosphate mine at the foothills of Abu Tartur plateau (map, Fig. 14/5), where basal flaser-bedded sandstone at the base of the formation contains abundant remains of angiosperms (mainly leaves) and other plant remains of probable Cenomanian age (Klitzsch & Lejal-Nicol 1984). The overlying shale, siltstone and sandstone are less fossiliferous. Glauconitic sand layers contain poorly preserved remains of lamellibranchs and fish teeth from shallow marine to tidal flat sedimentation. There are horizons with root remains and carbonaceous matter.

The Maghrabi formation can be correlated with the fluviomarine sediments of the Cenomanian Bahariya Formation (Fig. 14/3). In support of this correlation, Dominik (1985) discusses the similarity of the mineralogical composition of the two formations. The Maghrabi Formation represents the southern spur of the Cenomanian transgression which reached its climax in the north.

Taref Formation (Awad & Ghobrial 1965, Barthel & Boettcher 1978, with modification of the lower stratigraphic limit)

The Taref Formation unconformably overlies the Maghrabi Formation and is comparable in composition to the Sabaya and Six Hills Formations. It attains a thickness of more than 100 m at its type locality at Gebel Taref, a conspicuous outlier in the north Kharga depression. Occasional interbeds of clay and shale occur in the sandstones of the Taref Formation in the Abu Tartur area, with leaf impressions and fragmentry wood in the section (Hermina 1967). These sediments locally bear a marine influence in the form of wood and stems pierced by marine organisms (Dominik 1985). Mansour et al. (1979) divide the Taref Formation in the Ain Amur embayment (Fig. 14/5) into two subunits. 'The sandstone of the lower subunit is very poorly cross-bedded, and its contained shale-silt intercalations yield arenaceous foraminiferal association which indicate deposition in cold water environment. The dominance of *Haplophragmoides* spp. and *Trochammina* spp. within the shale intercalations can indicate deposition in less normal salinity. A probably intertidal flat low energy environment is inferred for this subunit. On the other hand the upper subunit of the Taref Formation is characterized by the dominance of cross-bedding with foresets having an average of 30° dip, representing beach dunes dissected by most probable 'braided streams'.

The cross-stratified medium to coarse-grained sandstones of this formation mark a general regression and slight epeirogenic movements during the early Turonian which probably continued on to the early-middle Campanian. These movements are thought to have been related to the more pronounced uplifting phase in the Bahariya area. During Turonian, Coniacian and Santonian times, the Bahariya arch represented an area of erosion in this part of the Dakhla Basin. Meanwhile, the southern parts received fluvial sediments of the Taref Formation at least during the early Turonian.

The Campanian transgression

Following the tectonically-controlled early Turonian regression and the subsequent periods of intense erosion, a renewed subsidence affected the Kharga-Farafra stretch. Marine deposition commenced during the Campanian, depositing Quseir and Duwi Formations in the Kharga-Abu Minqar areas and their stratigraphic equivalents of the Wadi Hennis Formation and El Hefhuf Formation in the Farafra depres-

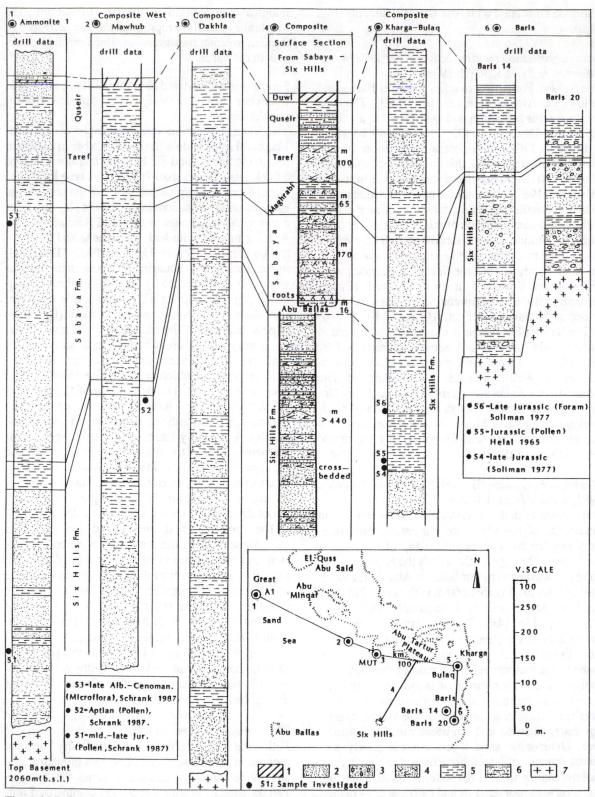

Figure 14.7 Correlation of stratigraphic subdivisions of the pre-Campanian clastics in areas of southern Dakhla basin. Logs: after 1. CONOCO well unpubl. 1978; 2. Abbas & Habib 1970; 3. Barakat & Milad 1966; 4. Bisewski 1982; 5. El Shazly & Shata 1960; 6. El Shazly & Shata 1960 and Philip & Asaad 1972. Interpretation: after Wycisk 1987, with modifications. Lithological symbols: 1. phosphate rock; 2. sandstone; 3. conglomeratic sandstone; 4. rippled sandstone; 5. clay/shale; 6. sandy clay; 7. basement.

sion area. This marine invasion continued, with regressions, throughout the period to the lower Tertiary. The successively overlying rock-units in the different areas and their correlations are illustrated in Table 14/1.

CAMPANIAN-LOWER EOCENE

Quseir Formation (Youssef *1957; = Mut Formation;* Barthel and Herrmann-Degen *1981*)

The claystones, siltstones and sandstones of the Quseir Formation overlie the sandstones of the Taref Formation and attain a maximum thickness of 70 to 90 m. They cover extensive areas in Kharga and Dakhla depressions and participate as well, with varying thicknesses, in the formation of the foothills of the bordering scarps (Figs 14/4, 14/5). In the Kharga area, the basal succession comprising clay and sandstones gives evidence of terrestrial and brackish environments, grading to a shallow shelf. The upper sediments, which are made up primarily of varicolored, mottled, silty and sandy claystones, reflect a prodeltaic shallow shelf facies (Fig. 14/7). In general, the deposits of the Quseir Formation indicate a gradual, occasionally stagnating transgression. Freshwater gastropods, remains of freshwater reptiles, as well as abundant terrestrial plant debris, and dinosaur skeletons and bones are found embedded in the basal sediments indicating limnic conditions (Awad & Ghobrial 1965, Hendriks et al. 1984). At Baris, a lense-shaped sandstone horizon in the upper part of the formation carries abundant casts of pelecypods and gastropods (Hendriks et al. 1984). The faunal assemblage in the Quseir Formation does not provide a definite age, though middle to late Campanian is assumed for it in the Kharga area as it is stratigraphically superimposed by the reliably dated Duwi Formation of Campanian or ?Campanian-Maastrichtian age. In the Abu Tartur area, outcrops of the Quseir Formation are restricted to a narrow belt along the footscarps and are formed essentially of varicolored claystones with sandstone pockets. In the Dakhla area, the Quseir Formation forms the floor of the excavated depression. It is subdivided here into a lower brick red subunit (maximum exposed thickness 30 m) and an upper subunit of alternating ferruginous-glauconitic sandstone, brown and gray sandy clay (20 to 35 m thick), with a distinctive green siltstone bed in between. The lower and upper subunits are designated as Unit I and Unit II respectively (Hermina et al. 1961), and are named the Mut member and El Hindaw member (Omara et al. 1976).

Barthel & Herrmann-Degen (1981) refrain from classifying these sediments in the Dakhla area under

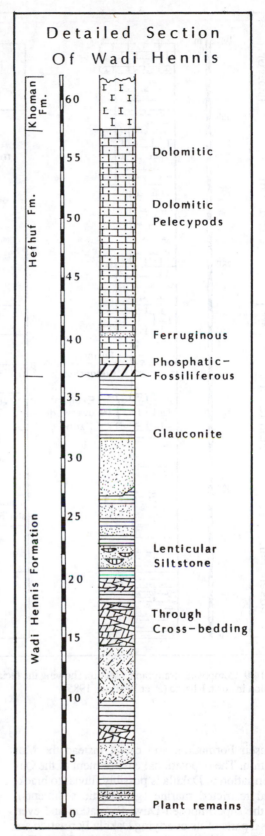

Figure 14.8 Stratigraphic section, Wadi Hennis and El Hefhuf formations (after Dominik 1985).

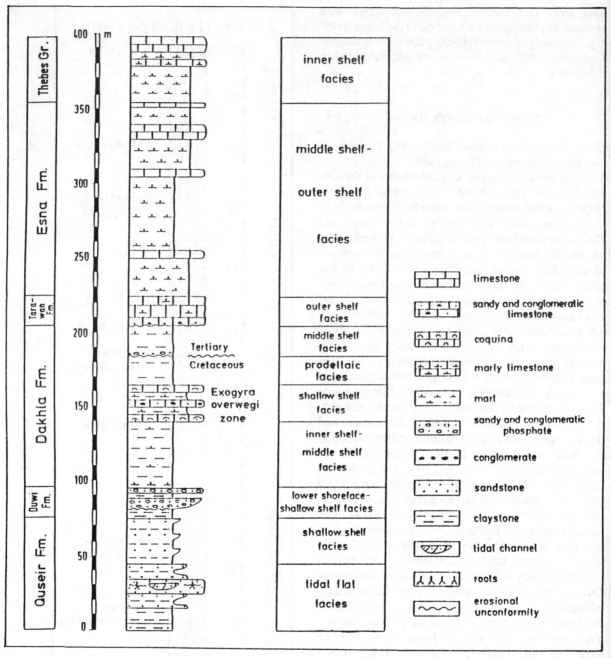

Figure 14.9 Composite stratigraphic section showing the facial interpretation of the Campanian-lower Eocene sedimentary succession in north Kharga (after Hendriks 1985).

the Quseir Formation, and call it instead the Mut Formation. The depositional environment of the Quseir Formation in Dakhla is possibly fluvial to brackish and restricted marine (glauconitic sediments). Since the superimposed Duwi Formation and even the basal parts of its overlying Dakhla Formation are of late Campanian age (see below), the Quseir Formation is assumed to have been deposited in early to middle Campanian time. In the Farafra area, a 35 m thick clastic sequence of sublittoral sediments forms the core of the anticlinal feature in the northeast of the depression between Wadi Hennis and El Maqfi. This sequence was related to the Cenomanian by previous authors. Recently, it has been reviewed and named the Wadi Hennis Formation (Dominik 1985) and is considered to be a stratigraphic equivalent of the

Quseir Formation of upper Campanian age (Table 14/1). The Wadi Hennis Formation lies unconformably below a carbonate unit dated as Campanian (El Hefhuf Formation) and consists of a section of clays and sandstones (Fig. 14/8). The clays are often bituminous; the sands are occasionally glauconitic and yield abundant fish teeth and remains of Campanian age comparable to those in the Quseir Formation of Baris. The top part of the Wadi Hennis Formation may correlate, in part, with the basal beds of the Duwi Formation which overlies the Quseir Formation in the southern oases (Table 14/1).

*Duwi Formation (*Youssef *1957; = Phosphate Formation;* Awad & Ghobrial *1965)*
The Duwi Formation is a phosphate-bearing unit that occupies a stratigraphic position at the top of the Quseir Formation and underlies the Dakhla Formation. It outcrops in a narrow belt along the extension of the bordering footscarps in Kharga and Abu Tartur. In the Dakhla area its rock succession is expressed topographically on an irregular east-west trending lower cuesta, subsidiary to the main scarp and overlooking the depresion on its north side. To the west of Dakhla, beds of this formation change their strike to the southwest. A hard limestone rock capping these beds forms the floor of a major part of the peneplain surface extending to the south of the Dakhla-Abu Minqar road (Figs 14/4, 14/5, 14/9). Lithologically the Duwi Formation consists of phosphate beds interbedded in a sequence of alternating claystone, sandstone, siltstone and conglomerate. Some sandstone beds are rich in glauconite.

The succession of the Duwi Formation is subdivided into two phosphate-bearing horizons (informally designated as the A and B Horizons) and a calcareous cap rock of restricted occurrence. The latter overlies the B Horizon in the Dakhla and west Daklha areas only (Fig. 14/10).

The A Horizon, which marks the base of the formation, is of widely variable thickness and exhibits rapid lateral changes in lithology. In the Kharga area, it is less than 1 m thick (completely missing in northernmost Kharga, Figure 14/10, section 4), and it consists of a brown siliceous or calcareous mudstone belonging to the *Neaera subcomplanata* Zone (Awad & Ghobrial 1965). The lower surface of this bed is slightly irregular with scour-and-fill structure, and may reflect the presence of para-conformity with the underlying varicolored clays of the Quseir Formation (El Hinnawi et al. 1978). In the Abu Tartur area, the A Horizon is more developed and consists of rocks exhibiting rapid lateral changes. Of economic importance is the development of this horizon into an almost solid phosphate bed (with minor clay intercalations) attaining a thickness of 4.5 to 5.0 m in some

parts of the southeastern sector of the Abu Tartur plateau. In the Dakhla and west Dakhla area, the A Horizon attains a thickness of 2 to 3 m with a fossiliferous mudstone bed (the *Neaera* bed) marking its base. In the Dakhla area, the A Horizon of the Duwi Formation lies on the upper sandstone unit of the Quseir Formation (Unit II, Hermina et al. 1961), while in the Abu Tartur and Kharga areas, this unit is missing and the Duwi Formation rests on the lower red clay unit of the Quseir Formation (Unit I). The absence of Unit II may be explained as a wedge-out of the sandstone body or, more likely, as due to its non-deposition on the uplifted areas of Abu Tartur and Kharga in the early late Campanian. In north Kharga this uplifting movement reaches its maximum. The overall thickness of the Duwi Formation here is much reduced, and in section 4 (Fig. 14/10), the entire thickness of the Duwi Formation is barely represented by less than 50 cm of brownish phosphatic mudstone bed overcrowded with borings of *Zaphellia* spp. belonging to the topmost beds of the Duwi Formation (Awad & Ghobrial 1965).

The A Horizon of the Duwi Formation is followed upward by a section of shale with intercalations of phosphate and claystone beds (B Horizon). In the Dakhla area, it attains an almost constant thickness of 15 to 20 m, while in the Abu Tartur and Kharga areas it is widely variable and is completely missing in some sections (Fig. 14/10). The top of the B Horizon is marked by the occurrence of the uppermost phosphate bed in the section. It is only in Dakhla and to its west that a 5 to 12 m thick section of fossiliferous calcareous shale, whitish marl and chalk-like sediments follows upward on top of the B Horizon and is designated the 'calcareous cap' of the Duwi Formation in Figure 14/10. Herrmann-Degen (1981) includes this calcareous interval in the basal member of the overlying Dakhla Formation and names it Qur El Malik Member. The '*Isocardia chargensis* Limestone' of Abbass & Habib (1971) is also included in this member. In the Abu Tartur and Kharga areas, this horizon is much reduced in thickness or completely missing.

From the wide variations in the overall thickness of the Duwi Formation, as well as the clear evidence of near unconformities, reworking and disturbed deposition (particularly noticeable at the contacts of the intercalating phosphate beds with the underlying shales), it can be inferred that the Duwi Formation was deposited on a highly oscillating bottom of a shallow shelf. Criteria indicating such disturbed conditions of deposition are given in Awad & Ghobrial (1965) and Hermina (1967). In the Bahariya and Farafra areas, the El Hefhuf Formation (Akkad & Issawi 1963) is considered the stratigraphic equivalent of the Duwi Formation in the southern oases

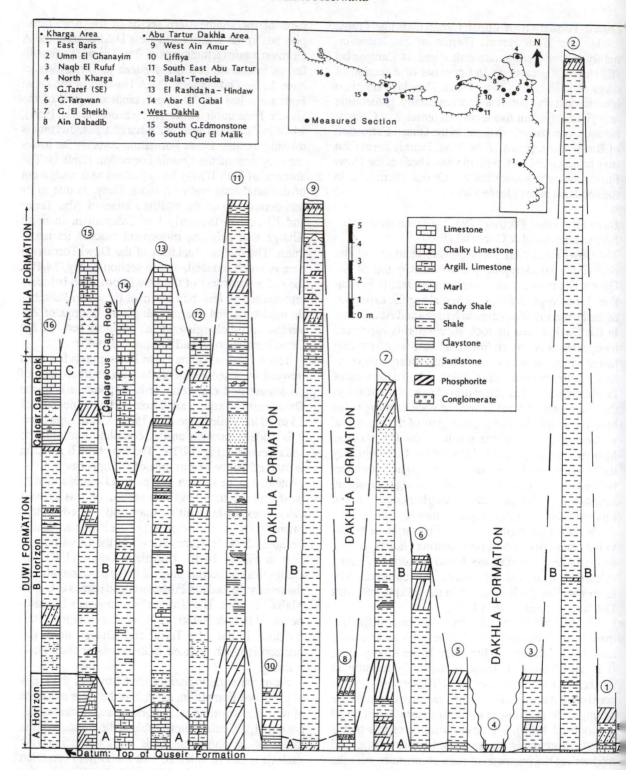

Figure 14.10 Correlation of the Duwi Formation, after 1. El Hinnawi et al. 1978; 2–7. Awad & Ghobrial 1965; 8–11. Hermina 1967; 12–14. Hermina 1959–1960, Geol. Surv. Egypt internal reports; 15 and 16. El Deftar et al. 1970, Geol. Surv. Egypt internal report 14/69.

(Table 14/1, Fig. 14/3). Ammonite fauna prove the upper Campanian age of El Hefhuf formation (Dominik & Schall 1984). The El Hefhuf Formation and its underlying Hennis Formation constitute the oldest rocks exposed in the Farafra area. They compose the core of the northeast-southwest trending anticline within the area between Wadi Hennis and Ain El Maqfi (Fig. 14/9), in alignment with the Bahariya anticlinal feature to its northeast. Dolomite, sandy dolomite and sandy to dolomitic chalk, devoid of clastics, are the constituent rocks of the El Hefhuf Formation in Farafra. The base of the formation is phosphatic, sandy, and contains limonitic concretions. Youssef & Abdel Aziz (1971) record the presence of *Isocardia chargensis*, *Cardium* spp., and other fossils from a calcerenite bed in this formation at Ain El Wadi. Omara et al. (1970) introduce the name 'Ain El Wadi Limestone' for this unit. Barakat & Abdel Hamid (1973) recognize this unit in the subsurface as 'the compact dolomitized limestone' overlain by the Maastrichtian chalk of the Farafra depression and underlain by 615 m of the 'Nubia group' (Fig. 14/13). They correlate the 'compact dolomitized limestone' with the El Hefhuf Fomration, and the top 100 m of the underlying clastics with the bottom-most clastic outcrops of Ain El Wadi (Wadi Hennis Formation).

In the area of Farafra it is possible to establish a tectonic uplift related to the Bahariya arch at the Campanian-Maastrichtian boundary. Foraminiferal investigations (Said & Kerdany 1961) and Abdel Aziz (1971) give evidence of this movement. In the Dakhla and west Dakhla areas, Campanian-Maastrichtian sedimentation was continuous. In these areas, Barthel & Herrmann-Degen (1981) note the presence of uppermost Campanian foraminifera in the 'Qur El Malik Member' which forms the top horizon of the Duwi Formation. They emphasize the upper Campanian fauna of the overlying shales and marls of the Dakhla Formation. The 'Qur El Malik Member' is missing in the areas around Abu Tartur and Kharga where the B Horizon of the Duwi Formation is directly overlain by shales of the Dakhla Formation of Maastrichtian age (*Globotruncana gansseri* Zone, Luger in Hendriks et al. 1984).

From the above considerations, it is possible to conclude that the phosphate-bearing Duwi Formation in the Dakhla and west Daklha areas was deposited in continuity with both the underlying and overlying Campanian Formations of Quseir and Dakhla. In Farafra, it is assigned to the Campanian and unconformably underlies the Maastrichtian. In the Abu Tartur and Kharga areas, it is apparently unconformable with both the underlying middle to late Campanian Quseir Formation and the overlying middle Maastrichtian Dakhla Formation.

Dakhla Formation (Said *1961*)

The Dakhla Formation consists of shale, marl and clay with intercalations of calcareous, sandy and silty beds. It forms the major thickness of the succession that overlies the Duwi Formation and underlies the Paleocene limestone beds along the scarp face from south Kharga to Abu Minqar. It also covers parts of the plain to the west of Dakhla and is exposed in the shallow depressions in the Abu Tartur-El Kharafish plateau (Figs. 14/4, 14/5 and 14/9). The tripartite subdivision of theDakhla Formation in the Kharga area (Awad & Ghobrial 1965) keeps to gross lithological aspects throughout the various areas in the Kharga-Abu Minqar stretch, and reflects repeated sea level fluctuations during its deposition. A general increase in the sandstone-siltstone component in its sections is noticeable in the areas of south Kharga and to the west of Dakhla toward Qur El Malik and Abu Minqar reaches.

Kharga area. In north Kharga, the basal subdivision of the Dakhla Formation, the 'Mawhoob shale member', consists of marl and papery shale, of inner to middle shelf deposition. In south Kharga, it becomes more silty and sandy and contains phosphatic conglomerate beds, reworked gastropods and abundant plant debris, indicating shallowing conditions of deposition. From the wide variations in the thickness of this member, from 5.5 m in the northwest scarp to 64.5 m in central Kharga and from 6 to 49 m in the south, it appears that deposition took place in semi-detached basins separated by ridges. A conspicuous basin occupied the central part of the present depression, flanked on the north and south by high areas. Another local but deep sub-basin developed in the high shelf to the east of Baris. Foraminifera of the *Globotruncana gansseri* Zone are present in the north, indicating an early middle Maastrichtian age for this member (Hendriks et al. 1984), and inner to middle shelf facies deposition (Fig. 14/7). Arenaceous forms, among them *Haplophragmoides* spp. and *Ammobaculites*, are present in south Kharga. The middle subdivision (Baris oyster mudstone Member) attains its maximum thickness (54 m) in the sections to the east of the Baris area (Fig. 14/11) and consists of a number of fossiliferous calcarenites and mudstone beds interbedded in a clay-shale section. Among the abundant fossils in the calcarenite layers are *Exogyra overwegi* and *Libycoceras ismaeli*. These calcarenite beds are commonly replaced by siltstone and contain intraformational conglomerate of phosphatic pebbles as well as small vertebrate remains. In the north parts of Kharga, the thickness of this member ranges from 2 to 15 m, with maximum development along the northernmost scarps (Fig. 14/11). This middle member is attributed to the upper

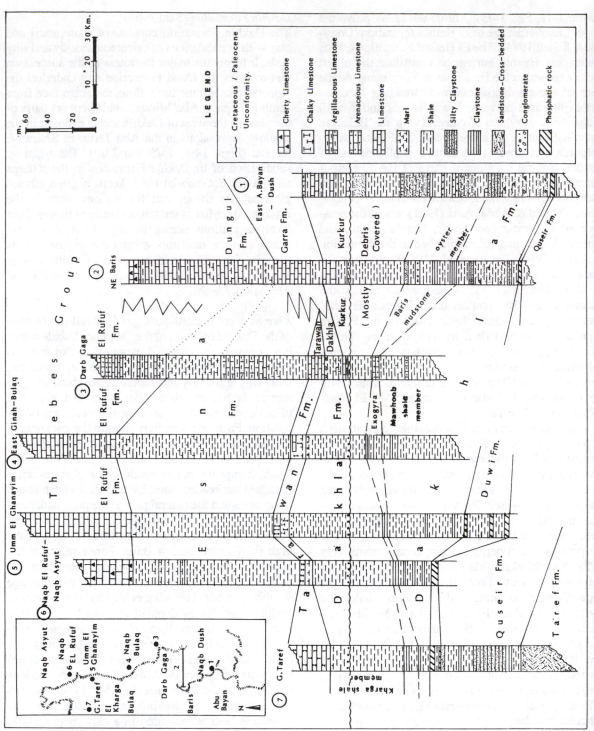

Figure 14.11 Correlation of stratigraphic sections in El Kharga area, after 1–3 El Himawyi et al 1978; 4–7 Awad & Ghobrial 1965, with modifications

part of the *Globotruncana gansseri* Zone (Middle Maastrichtian) and represents a shallow shelf facies of deposition (Fig. 14/7).

The upper subdivision (Kharga shale Member) has an overall thickness of 50 to 70 m, with no marked variations as exhibited in the lower two members. The lower part (10 to 40 m) of this division consists of claystone and shale of nearshore prodeltaic facies of deposition. It contains poor faunal remains of arenaceous forms. A phosphatic conglomerate (0.5 to 2.0 m) marks the top of this lower interval and an erosional hiatus separates it from the upper part of the Kharga shale member (of Paleocene age). This hiatus contains reworked Maastrichtian fossils (*Libyco-

ceras ismaeli) and also includes the lower Paleocene. In south Kharga the Paleocene part of this subdivision is partly or completely replaced by reefal limestone facies of the Paleocene 'Kurkur Formation'. The Paleocene part of the Kharga shale member, in areas of north Kharga, consists of fossiliferous marl and shale which contain *Globorotalia compressa* and *G. pseudobulloides*, while the uppermost marls belong to the *Morozovella angulata* Zone (early to middle Paleocene).

Abu Tartur-Dakhla area. The members of the tripartite division of the Dakhla Formation identified in the Kharga area (Awad & Ghobrial 1965) can be fol-

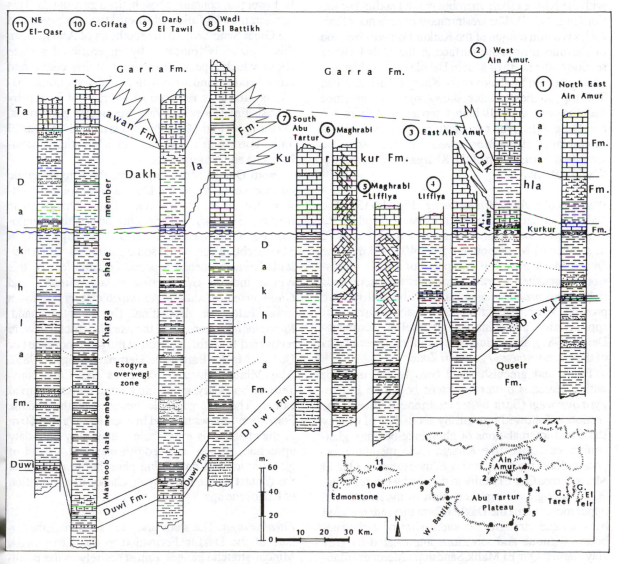

Figure 14.12 Correlation of stratigraphic sections along Abu Tartur-Dakhla scarp, after 1–6. Hermina 1967; 7–11. Hermina et al. 1962, with adoption of rock unit nomenclature.

lowed in the Abu Tartur-Dakhla area. Foraminiferal assemblages of the *Gansserina gansseri* Zone are identified from the Mawhoob shale member. The overall thickness of the Dakhla Formation generally increases to a maximum of more than 250 m in Dakhla. Figure 14/12 shows the variations in thickness and the correlations of the different members. Differential movements of the sea bottom during the Maastrichtian controlled the deposition of the Dakhla Formation in the Abu Tartur area (Hermina 1967). These movements show a similar history to that prevailing in the Kharga area at the same time. Of interest in the Abu Tartur area also is the thick development of the carbonate complex of the 'Garra El Arba'in facies' above the upper Cretaceous-Paleocene unconformity. The Paleocene lower members completely replace or variously interfinger with the Kharga shale member of the Dakhla Formation (Fig. 14/12). The westernmost occurrence of this facies is a thin tongue of the Kurkur Formation on top of the unconformable surface in the Gebel Gifata section to the north of Mut in Dakhla oasis, where it is again overlain by shales of the Kharga shale member. The top of these shales shows signs of disturbed bedding and a small hiatus separates it from the overlying Paleocene carbonate rocks. Foraminiferal assemblages of the *Morozovella uncinata* Zone are identified from the top of the Kharga shale member at Wadi El Battikh.

West Dakhla-Abu Mingar area (Fig. 14/13). The Dakhla Formation starts with its basal member, the Mawhoob shale member. It attains a thickness of 72 m at Abar El Mawhoob and from 40 to 60 m at the Qur El Malik area. It is composed of a shale succession intercalated by rare calcareous mudstone beds rich in *Inoceramus regularis*. Marl beds at its base provide *Nostoceras* cf. *helicinum* and *Libycoceras* spp. of late Campanian age (Barthel & Herrmann-Degen 1981), and include foraminiferal assemblages of the *Globotruncana gansseri* Zone.

Black and greenish shale beds follow upwards with calcareous oyster mudstone beds rich in *Exogyra overwegi* (Baris oyster mudstone Member). The section is followed upward by about 100 m of grey shale and intercalations of phosphate-bearing, glauconitic sandstone (the Kharga shale member), exposed along the scarp from Edmonstone to Qur El Malik. From Qur El Malik to the northwest, for about 90 km along the scarp face, most of the middle and upper members of the Dakhla Formation are replaced by a unique stratigraphic succession of carbonate-bound siltstone and sandstone intercalated by grey clay, named Qur El Malik Sandstone Member (Mansour et al. 1982). An equivalent sandstone unit, termed the Ammonite Hill Member, is known from

the Ammonite Hill scarp (26° 10' N, 27° 05' E) to the west of Abu Minqar and beyond the area presently under discussion (Barthel & Herrmann-Degen 1981). The top of this sandstone member in the type area is exposed above the last occurrence of *Exogyra overwegi* beds. The sandstones are heavily bioturbated and rich in fossils. This quantitative increase of sand and sandstone toward the west from Dakhla is also well remarked in other members of the Dakhla Formation. Enclosed in the sand/shale section of the upper member of the Dakhla Formation is a horizon of conglomeratic phosphatic and glauconitic bands (Bir Abu Minqar Horizon) which occurs in the areas to the west of Dakhla (Abbass & Habib 1969, Mansour et al. 1982, Barthel & Herrmann-Degen 1981). Bir Abu Minqar Horizon is about 1.5 m thick and consists of sandy, glauconitic and ferruginous marl. Its lower part contains phosphatized gastropods. This horizon may be followed toward the southwest into the Great Sand Sea. To the southeast of Abu Minqar, this horizon is truncated by an erosional surface above which a peculiar section of fairly coarse and cross-bedded sandstone with partially kaolinitic matrix occurs; it is replaced or overlain at its top by whitish limestones carrying freshwater gastropods. On the map (Fig. 14/9), this unit is named 'Minqar El Talh Formation', and an undifferentiated post-Miocene age is assigned to it.

From Qur El Malik eastward, the conglomeratic horizon continues to the Dakhla area, followed upward by the upper division of the Kharga shale member (Fig. 14/13). The occurrences of this conglomeratic horizon contain ammonites, suggesting a Maastrichtian age, while some gastropods and nautiloids have Paleocene aspects. Foraminifera of this region include *Globigerina pseudobulloides* and *Globigerina triloculinoides* which clearly point to a lower Paleocene deposition. On the other hand, *Rugoglobigerina scotti* of the late Maastrichtian were recovered from this horizon at the scarp northeast of Qur El Malik (Barthel & Herrmann-Degen 1981). Abu Minqar Horizon expresses the Cretaceous-Tertiary boundary, involving a small hiatus of varied extent. The Kharga shale member of the Dakhla Formation follows upward in a section of about 60 m in thickness; it is grey shale, rather silty and slaty upsection, ending in a reddish clay and a band of glauconitic sand containing phosphorite nodules at the contact with the overlying chalky-like formation of Paleocene age.

Farafra area. The shale and marl section characteristic of the Dakhla Formation in the Kharga-Abu Minqar stretch changes almost entirely in the northern part of the Farafra depression into a chalky limestone unit named the Khoman Formation. This is a white to light tan chalky calcilutite. A type locality is proposed by Norton (1967, unpublished report) at Ain Khoman scarp, southwest Bahariya Oasis, where it attains a thickness of 50+ m. Where exposed at Qasr El Farafra in the Farafra depression (Fig. 14/9), it attains a thickness of about 4 m with the base unexposed. At the northern scarp and in the cliffs of Wadi Hennis it reaches thicknesses of 50 and 80 m respectively. Subsurface evidence suggests it may be as much as 160 m thick (Barakat & Abdel Hamid 1974, Fig. 14/13). A shale section (17 m) overlies the Khoman chalk and separates it from the overlying Tarawan Formation. This shale separation is believed to be equivalent to the uppermost part of the Kharga shale member in the south. From both subsurface and surface data, the Khoman chalk in Farafra is considered to range in age from uppermost Campanian to lower Maastrichtian (*Globotruncana lapparenti tricarinata* Zone to the *G. gansseri* Zone and probably younger (Kerdany, unpubl. thesis 1969). The overlying shale section is a tongue of the Dakhla Formation of lower to middle Paleocene (*Globorotalia trinidadensis* to *G. angulata* Zones), which is overlain by the chalky limestone, 'Tarawan Formation', of upper Paleocene age (*Globorotalia pseudomenardii* Zone). It is evident that there is a facies change in the *Globotruncana gansseri* Zone from shale in the south to chalky limestone in Farafra.

The intertonguing of the Dakhla and Khoman is probably due to paleotectonic events. An unconformity of different magnitudes occurs in Farafra at the upper Cretaceous-lower Tertiary boundary (Le Roy 1953, Said & Kerdany 1961, Said & Sabry 1964). It is more accentuated in the Maqfi area where the *pseudomenardii* Zone lies directly on the upper Cretaceous *Globotruncana gansseri* Zone.

Kurkur, Garra, Tarawan and Esna Formations
Following the major regression of the upper Cretaceous-Tertiary, the areas of south Kharga and the Abu Tartur-El Kharafish plateau stood as high platforms in the shallow shelf environment of the late early Paleocene transgression. On these elevated platforms, a succession of reef-like limestone with shale intercalations of near-shore environment was deposited, including the Kurkur Formation at the base and the Garra Formation at the top. In other areas, outward from these two platforms, sediments of more shale facies were deposited in middle to outer shelf environments. Their succession includes the Paleocene part of the Dakhla Formation overlain by the Tarawan and Esna Formations. Sediments of the platform and the deeper facies display gradual lateral and vertical changes in lithology and exist in interfingering or overlapping relationships.

Kurkur Formation. This unconformably overlies the Dakhla Formation and forms the top of the scarp to the south of Baris, as well as the lower bench of the plateau to its northeast as far as Gaga locality. It also forms the core of two small domal structures in association with the east-west faulting that cuts across the plateau to the east of Baris (Fig. 14/4). In the Abu Tartur area, the Kurkur Formation makes the topmost part of the scarp and the plateau surface which is semi-encircled between Ain Amur embayment and Wadi El Battikh (Fig. 14/5).

Lithologically, the Kurkur Formation is characterized by its reef-like, earthy brown, hard, thick-bedded limestone which is sandy in parts, with intercalations of shale. The limestone is rich in fossils: *Cardita wegneri*; *Cardita tenedensis*; *Ostrea orientalis*; *Turritell* spp. The base of its bottommost beds is invariably conglomeratic, including ferruginous pebbles in clay or calcareous matrix. In south Kharga, it overlies the Dakhla Formation in thicknesses decreasing gradually to the north from 39 m in the east Abu Bayan area to 6 m east of Gaga (Fig. 14/11), where it forms the northernmost tongue in the Dakhla Formation. Around the southeastern and southern scarp of Abu Tartur plateau, it is of residual thickness due to erosion, but a maximum of 110 m is recorded in a borehole drilled some 15 km to the south of Ain Amur on the plateau (Issawi et al. 1978). At Wadi El Battikh scarp, it measures about 40 m in thickness and replaces part of the Dakhla Formation shale above the unconformity. To the west, it gradually loses thickness to become completely missing in the sections to the northwest of Mut (Fig. 14/12, sections) where its base is taken over by the thin conglomerate erosional surface (Bir Abu Minqar Horizon).

The foraminiferal assemblage of the Kurkur Formation in south Kharga and in Abu Tartur includes *Globorotalia uncinata*, *Globorotalia pseudobulloides* and *G. trinidadensis*, which assign it to an early Paleocene age and suggest an inner shelf to tidal flat deposition.

Garra Formation. The shallow flooding of the sea in the early Paleocene gave way to a deeper sea in late Paleocene, and the Garra Formation was deposited during the late Paleocene-early Eocene.

In south Kharga, the Garra Formation covers the middle bench of the plateau conformably overlying the Kurkur Formation. It consists of well-bedded, massive white limestone separated from the underlying brownish Kurkur limestone by a clay bed marker. The Garra thickness decreases gradually to the north, from about 50 m in east Baris to 18 m at Gaga. Further northward, it passes laterally into the Esna /Tarawan Formation and partly into the top of the Dakhla Formation (Fig. 14/11). Its foraminiferal

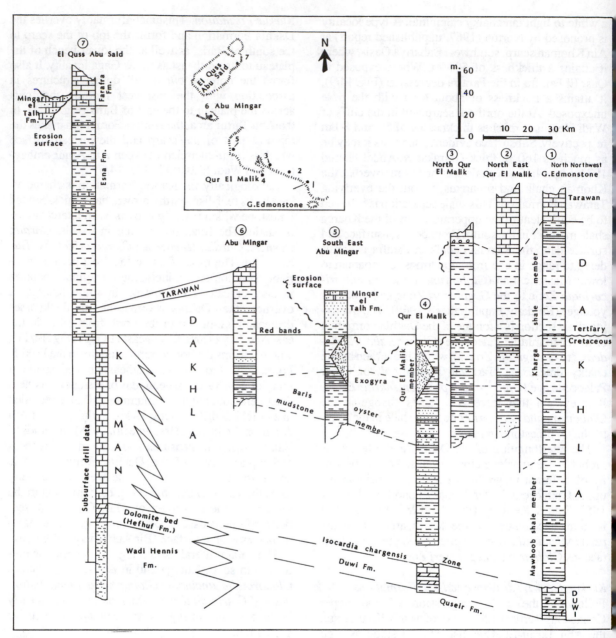

Figure 14.13 Correlation of the stratigraphic sections along west Dakhla scarp with the surface and subsurface sections at Farafra, after 1–6. Barthel & Herrmann-Degen 1981 and El Deftar et al., Geol. Surv. Egypt internal report 14/69; 7. surface section, El Kerdany, unpubl. PhD thesis 1969, Ain Shams Univ., Cairo; subsurface section, Barakat & Abdel Hamid 1974.

assemblage includes, from bottom to top: *Morozovella uncinata*; *M. angulata*; *M. pseudomenardii* and *M. velascoensis*, assigning it to late Paleocene at the top.

In the areas of the Abu Tartur-El Kharafish plateau, the Garra Formation covers major parts of the surface (Figs 14/5, 14/9). It overlies different Paleocene rock units: Kurkur, Dakhla or Tarawan, as illustrated by the sections (Figs 14/12, 14/14). In these areas, it maintains the same lithologic types as in south Khar-

ga, with common variations in the ratios of the carbonate and shale components. At the scarp north of Wadi El Battikh, it is represented by 42 m-thick, highly fossiliferous limestone beds of which the bottom 9 m are marly and contain *Ananchytes fakreyi*, characteristic of the Tarawan Formation. The underlying shales belong to the topmost Dakhla Formation and contain *Globorotalia pseudomenardii* foraminifera. The macro and microfaunal assemblage associated with the Garra Formation in these areas is

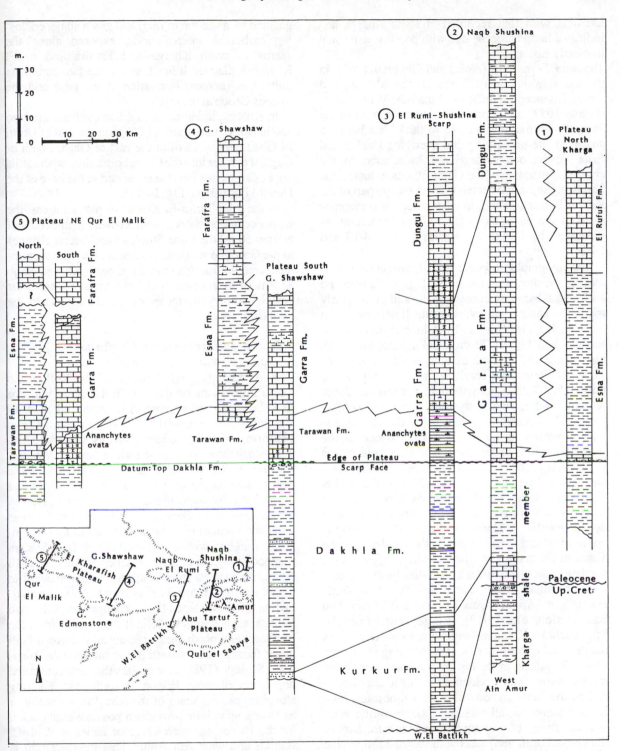

Figure 14.14 Stratigraphic sections and lithofacies changes of the lower Tertiary rock units on Abu Tartur-El Kharafish plateau, after 1. El Deftar et al. 1978; 2–4. Hermina 1967 and El Deftar et al., Geol. Surv. Egypt internal report 14/69.

identical with that recorded in south Kharga and indicates late Paleocene age with possible extension into early Eocene.

Tarawan Formation (Awad and Ghobrial 1965). In the areas from north Kharga to Abu Minqar, and where Paleocene sediments of the Nile Valley facies (Issawi 1972) are deposited, a carbonate unit of the Tarawan Formation, 4 to 45 m thick, lies between shales of the underlying and overlying Dakhla and Esna Formations respectively. In relation to the coeval equivalents of the Garra El Arbain facies, the Tarawan Formation correlates with the top part of the Kurkur Formation and, in other places, it is incorporated in the basal part of the Garra Formation, as illustrated by the sections (Figs 14/11, 14/12 and 14/14).

The geographic occurrences of the Tarawan Formation are shown on the maps (Figs 14/4, 14/5 and 14/9). In Kharga, it consists of fossiliferous, partly marly or chalky, yellowish white limestone of an outer shelf facies (Fig. 14/8). A bioturbated conglomeratic bed (10 to 20 cm thick), associated with phosphatic and calcareous units at the base of the Tarawan Formation, marks a *hiatus* on top of the underlying Dakhla Formation. It contains vertebrate remains and reworked dwarfed fauna and solitary corals. The occurrence of this bed is due to a regression of the sea during middle to late Paleocene. The hiatus horizon is traced throughout the whole area of north Kharga and at several localities in Dakhla, west Dakhla and Farafra. Around Abu Minqar, the Dakhla Formation is succeeded upward by a very peculiar facies which consists of sandy marl and vermetid debris (Barthel & Herrmann-Degen 1981) which is possibly related to this hiatus. In the Farafra area, the Tarawan Formation has the same lithology as in the southern oases, and forms a chalk-like limestone bed, 4 to 5 m thick, cutting across the Dakhla and Esna Formations. Around Qasr El Farafra (El Quss Abu Said section, Fig. 14/13), it is separated from the Maastrichtian Khoman Formation, to which it bears much lithologic similarity, by a tonguing shale bed, about 17 m thick, of the top Dakhla Formation. From microforaminiferal evidence in different occurrences, the Tarawan Formation is not time transgressive but seems to fall within the *Globorotalia velascoensis* zone. It (or its equivalent beds at the base of the Garra Formation) rests with varied stratigraphic gaps on the underlying Paleocene part of the Dakhla Formation. Strougo (1986) relates this stratigraphic relationship in the south and central Western Desert and in other areas to a large-scale syndepositional tectonic disturbance, including faulting, which resulted in marked changes of depositional patterns.

Esna Formation. The Esna Formation is distinguished by a section of marl and green shale, enclosing carbonate intercalations, exposed along the scarps in north Kharga and Farafra and in El Kharafish plateau. It lies between the two carbonate units, the Tarawan Formation at the base and the Thebes Group at the top.

In Kharga, it decreases in thickness from north to south, from a maximum of 160 m to the east of Umm El Ghanayim to 45 m to the east of Gaga. South of Gaga, it merges into the Garra Formation, whereas its upper part passes into the shale part at the base of the Dungul Formation (Fig. 14/11).

In the Abu Tartur-El Kharafish plateau areas, the Esna Formation shows (with an average thickness of around 50 m) the same lithology and lateral changes to the Garra Formation, as illustrated by the sections measured across this tract from north Kharga-Naqb El Rhumi in the east to Gebel Shawshaw and west Gebel Haishan (northeast of Qur El Malik) to the west (Fig. 14/14).

Paleocene facies changes in Abu Tartur-El Kharafish plateaux and in Farafra

The interchanging relationships of the Kurkur and Garra Formations on the Abu Tartur scarp with the Dakhla Formation on west Ain Amur and Wadi El Battikh scarps (Fig. 14/12) are due to a change from platform to shelf deposition probably influenced by a syndepositional northeast-trending fault system during the middle to late Paleocene. Traces of these faults show on the surface (map, Fig. 14/5). Similar fault-controlled depositional effects are expressed by the changes of the Garra Formation into the Tarawan and Esna Formations. The interchange of facies coincides roughly with a sub-latitudinal fault system mapped along the northern reaches of Abu Tartur-El Kharafish plateau from Naqb Sushina westward to Gebel Haishan (maps, Figs 14/5 and 14/9). The association of these depositional changes with the two fault systems seems to delimit the plateau from the southeast and north, and suggests an upthrown fault block forming this plateau in the middle-late Paleocene. Strougo (1986) emphasizes the fault-controlled deposition of these Paleocene sediments. Faulting along the eastern scarp of the Abu Tartur plateau or its strong uparching provides a possible explanation for the Paleocene interchange of facies at Wadi El Battikh and west Ain Amur. This interpretation is partly in accord with Hermina (1967) who visualizes the Abu Tartur plateau as a submarine swell 'flanked on its eastern and western sides by subsiding lands sloping towards Kharga and Dakhla respectively'.

In Farafra, the Esna Formation has the same stratigraphic position as in the southern occurrences. It forms the shale-marl sections exposed on the slopes of the El Quss Abu Said plateau (70 to 176 m thick),

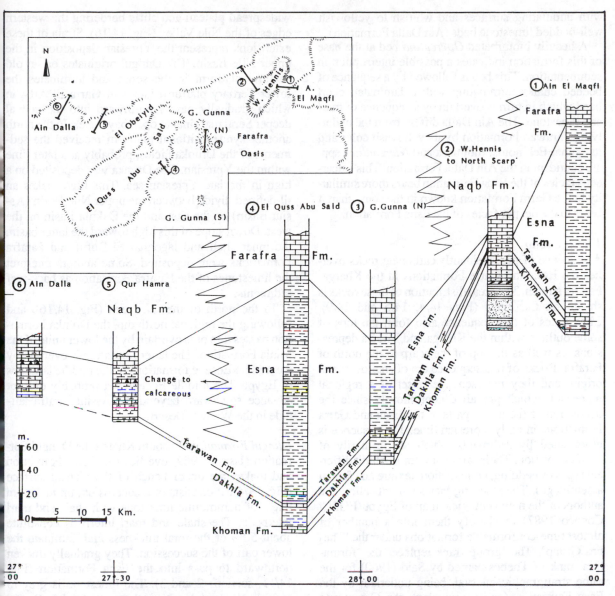

Figure 14.15 Correlation of Farafra stratigraphic sections, after 1 and 3 Youssef & Abdel Aziz 1971; 2. Dominik 1985; 4. El Kerdany, unpubl. PhD thesis 1969; 5 and 6. Barthel & Hermann-Degen 1981.

the eastern scarp (El Maqfi, 150 m), and the outlier of Gebel Gunna North (71 m). Lithologically it consists of greenish shale and marl with intercalations of nummulitic biocalcarenite. At its base and in transition to the Tarawan Formation, beds of hard limestone sometimes occur. In the south corner of El Quss Abu Said the Esna-Tarawan contact is calcareous. Upward in section, the Esna Formation is more calcareous and partly coralline. It grades into the overlying Farafra Formation. It thins out to the northeast of Gebel Gunna (North), and is completely missing on the northern scarp due to its later erosion on the most uplifted part of the Farafra anticline (Section

2, Fig. 14/15). Along the northern scarp, the Tarawan Formation is directly overlain by the Naqb Formation. The Esna Formation in Farafra, as in the southern occurrences, is assigned an upper Paleocene age for its lower two-thirds (*velascoensis* zones) and an early Eocene for its top part. To the west of the Farafra depression, the Tarawan Formation is overlain by a calcareous facies of the Esna Formation. At Qur Hamra (outlier of the north scarp) and along the Ain Dalla scarp (Fig. 14/15), its section, around 47 m thick, consists of chalky limestone and marl. This succession is described by Barthel & Hermann-Degen (1981) as a succession of alternating grey marl with undulating surfaces and whitish to yellowish well-bedded limestone beds (Ain Dalla Formation).

A heavily bioturbated *Operculina* bed at the base of this formation indicates a possible interruption in sedimentation. This bed is followed by a sequence of bedded chalks terminating with a laminated chert band which in turn is overlain by a sequence of marl and limestone. The Ain Dalla differs from the underlying Tarawan Formation by its yellowish color and its fauna. Echinoids, mollusca and *Nummulites* spp. are abundant in the Ain Dalla Formation. This calcareous facies of the Esna Formation bears more similarity to the Garra Formation known in the south than it bears to the deeper facies of the Esna Formation.

Thebes Group

A sequence of predominantly carbonate rocks overlies the Esna and Garra Formations in the Kharga-Farafra stretch. The areal distribution of these rocks is shown on the maps (Figs 14/4, 14/5 and 14/9). Limestones of this sequence also form the tops of some outliers within the Kharga and Farafra depressions, as well as the top of the scarp to the north of Farafra. Rocks of this sequence are of shallow water origin, and they represent the onset of a regional regression which prevailed continuously since the deposition of the upper parts of the Esna and Garra Formations in early Ypresian time. The sequence is represented by sediments which are primarily of shelf deposition. Their carbonate strata exhibit different types of bedding, composition, texture and faunal assemblages. The resulting facies variants allow the authors of the new geological map of Egypt (EGPC/ Conoco 1987) to classify them into a number of almost time-conformable formations under the 'Thebes Group'. The 'group' rank replaces the 'formation' rank of Thebes named by Said (1960) for the same stratigraphic interval, being underlain by the Esna Formation (or its equivalent, the Garra) and overlain by the Minia Formation.

The formations of the Thebes Group along the Kharga-Farafra plateau comprise: the Serai (=Thebes s.str., see Said, Chapter 24, this book), Farafra, El Rufuf, Dungul and Drunka. A regional map showing the areal distribution of these formations in the Kharga-Farafra plateau and its wider extension to the east, is compiled from the new geological map (1987) in Figure 14/16. A schematic interpretation of the different facies patterns accompanies the map whereby the Serai is interpreted as marginal bay, the Farafra as lagoonal, the El Rufuf as inner shelf and the Dungul and Drunka as platform deposits.

Outward from the Kharga-Farafra stretch, the extension of the Dungul Formation to the south and southeast and that of the Serai Limestone and Drunka Formation to the east forms the erosive surface of the widespread plateau and cliffs bordering the western edges of the Nile Valley (Fig. 14/16). Strata of these extensions represent the Ypresian deposition in the upper Nile Basin. The Dungul originates on an old shallow platform in the south and terminates the lower Tertiary platform facies of Garra El Arba'in which gradually passes northward into the coeval deeper Serai marginal to open bay facies. In the north another syndepositional platform received the sediments of the Drunka facies, probably at a later time within the Ypresian. The Drunka was deposited on a high in the late Ypresian sea. This high makes an ill-defined divide between the upper Nile Basin (Assiut Basin) on the east and the Dakhla Basin on the west. Down slope of this high toward the latter basin, the inner shelf and lagoonal El Rufuf and Farafra Formations were deposited. Some authors interpret the limestones of the Farafra Formation as back-reef sediments.

To the north of the map area (Fig. 14/16), and following the regional north dip, the Drunka Formation passes into or is overlain by the lower units of the Minia Formation. The latter formation is overlain by the middle Eocene formations of the middle latitudes of Egypt. The middle Eocene sea probably did not advance southward beyond the Assiut-Farafra latitude in the Western Desert.

Dungul Formation. In south Kharga the Dungul Formation (Issawi 1969) overlies the Garra Formation and makes the upper bench of the plateau surface (Fig. 14/4). It consists of a succession, up to 130 m thick, of nummulitic limestone with shale and marl interbeds. The shale and marl interbeds constitute about 50% of the total thickness and dominate the lower part of the succession. They gradually thicken northward to pass into the Esna Formation (Fig. 14/11, sections 2 and 3). The limestone is grey to pinkish, cherty, dolomitic at the top and rich in *Ostrea multicostata*, gastropods, *Nummulites* and *Operculina* spp. In places the top 15 m of the formation are reef-like carbonates. A shallow platform environment is indicated for this formation.

El Rufuf Formation. Northward the Dungul passes into the El Rufuf Formation which crops out along a belt 10 to 25 km wide, building the upper part of the scarp on top of the Esna Formation (Fig. 14/4). The type locality of the formation is at Naqb El Rufuf on the Kharga scarp to the east of El Mahariq, and consists of marls which are progressively replaced upward by marly limestone and thick-bedded limestone (3 to 15 m bed thickness) with an increasing content of brown chert bands. The ratio of clastic constituents is much more reduced compared to that in the Dungul Formation. A thickness of 145 m is

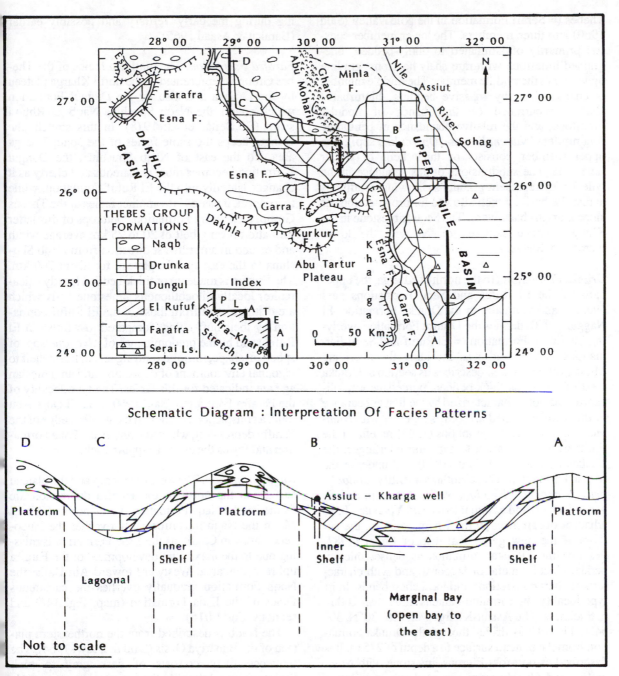

Figure 14.16 Geological map of the Thebes Group in Kharga-Farafra plateau and its eastern extension (compiled from EGPC/CONOCO geological map of Egypt 1987).

attained at Gebel Um El Ghanayim. An inner shelf environment is proposed for these strata. Abundant gastropod casts, *Lucina thebaica*, *Conoclypeus dela-nouei* and *Nummulites* spp. are present in the limestone. The shale beds yield *Globorotalia esnaen-sis* and *G. gracilis* of Ypresian age.

Serai Formation (= Thebes s.str). Outward, on the eastern extension of the Kharga scarp, the plateau surface is largely covered by thinly-bedded shelf chalk, chalky limestone, and cherty limestone of the Serai Formation. Oyster and Nummulite layers, possibly allochtonous, are common. Facies of the Serai limestone indicate their deposition on a shelf of deeper and lower energy environment than that of El Rufuf and Dungul. Snavely (1979) subdivides the

Thebes (= Serai) Formation in the Nile Valley (Said 1960) into three members. The lower member consists primarily of laminated to thinly-bedded, fine-grained limestone with rare shaly horizons, grading upward into the middle member. The middle member is characterized by massive bedding, bioturbated chalks, occurrence of thin bands of nodular limestone, and the relative abundance of benthonic foraminifera (*Nummulites* and *Operculina* spp.). The upper member consists of thick beds of oyster limestone. The subdivisions of this sequence in the Nile Valley represent gradual shallowing conditions across the basin, which allowed for progradations of slope margins basinward. The Serai Limestone in the Kharga plateau can be equated with the lower member of Snavely.

Drunka Formation. To the north of Kharga, the upper parts of the El Rufuf and Serai Formations pass laterally and vertically in the Drunka Formation. El Naggar (1970) defines the Drunka Formation overlying his Luxor Formation (= El Rufuf and Serai Formations in this study) as including the overlying Minia Formation, and describes its section at Drunka west of Assiut as thick-bedded, porcellaneous, siliceous limestone characterized by the first appearance of the spindle-shaped alveolines together with *Nummulites*. Mansour & Philobbos (1983) redefined the limits of this formation to cap (and interfinger) the Thebes Formation of Said (1960) and underlie the Minia Formation. These authors identify *Conoclypeus delanouei*, *Alveolina*, *Nummulites* spp. and algae in its beds and assign it to the late Ypresian. In its adjacent occurrences in the Kharga plateau, the authors of the new geological map of Egypt (1987) describe the Drunka Formation as dense, thickly-bedded, locally reefal or lagoonal, and with characteristic chert concretions and local chert bands. In its type locality, the formation attains more than 200 m in thickness. The Assiut-Kharga well (26° 30' N, 30° 54' E, Fig. 14/16), drilled through the Drunka Formation from the plateau surface to a depth of 275 m. It is described as non-fossiliferous limestone with *Nummulites* and *Alveolina* spp. recorded only in its lower horizons (Barakat & Asaad 1965). A succession of fossiliferous marly limestone with shale intercalations follows for another 155 m below the Drunka. It includes *Nummulites* and *Operculina* spp., and seems to correlate with the El Rufuf Formation. The cumulative thickness of the Thebes Group in this well is 430 m. It is underlain by a 218 m thick Paleocene section belonging to the Esna Formation. These increased thicknesses of the Paleocene and lower Eocene, compared to the equivalent thicknesses in the Kharga area, point to their accumulation in a down-faulted basin which was continuously subsiding during the early Tertiary and possibly earlier (Barakat & Asaad 1965).

Farafra Formation. The platform facies of the Thebes Group is represented in the north Kharga plateau by the Drunka and also by the Dungul Formation which covers the Naqb Shushina-Naqb El Rhumi scarp (El Deftar et al. 1978). In this stretch, the Dungul keeps the same facies of the south Kharga area. To the east of Naqb Shushina, the Dungul Formation becomes more calcareous and cherty as it changes laterally into the El Rufuf Formation, while to its north, it passes into another facies of the Thebes Group, the Farafra Formation. Outcrops of this latter formation form a belt of 20 to 25 km average width and extend in a northwest direction from Naqb Shushina to the east Farafra plateau for about 200 km. The Farafra is made up of thick-bedded (partly calcarenite) locally allochthonous limestone beds which are distinguished from those of the El Rufuf Formation by their buff color (compared to whitish in El Rufuf) and by the predominance of *Alveolina* spp. of spherical shape, miliolids and algae. An inner shelf to lagoonal environment of deposition and an Ypresian age are indicated for this facies. The type locality of the Farafra Formation (Said 1960) is at El Quss Abu Said, a conspicuous outlier on the western edge of the Farafra depression, where it overlies the Esna Formation and forms the eroded capping rock.

Naqb Formation. The top of the scarp and the plateau surface to the north of the Farafra depression are formed of the Naqb Formation. To the north of Ain El Maqfi the Naqb unconformably overlies the Paleocene Tarawan Chalk, as the Esna Formation is missing due to the maximum development of the Farafra uplift in this area. Westward toward Ain Dalla, the Naqb Formation gradually overlies the calcareous facies of the Esna Formation (map, Fig. 14/9 and sections, Fig. 14/15).

The Naqb is described from the northeastern plateau of the Bahariya Oasis (Said & Issawi 1964). The sequence there consists of dark grey to pink limestone and dolomitic limestone with minor shale intercalations. The top part is sandy and carries pelecypod and gastropod shells, *Operculina*, elongate *Alveolina*, *Nummulites* spp. and coralline algae. In the northern plateau of Farafra, the Naqb is made up of pinkish, dense and nummulitic platform limestones with local flint bands. To the east of Farafra, it extends over a large part of the plateau as far as Ghard Abu Moharik where it passes laterally into the Minia Formation in the area to the east of Ghard Abu Moharik, but the passage is not clear in the dune-covered areas.

POST-LOWER EOCENE

Following the retreat of the sea and the uplifting phase in post-early Eocene time, terrestrial conditions prevailed over the Kharga-Farafra stretch since that time. Alluvial, lacustrine and eolian sediments are described from several localities.

Gravel Fills. These are alluvial deposits which occur as well-developed mounds which cover patch areas, ranging from 2 to 20 km², on the plateau surface east of Kharga (Fig. 14/4). They stand out as heights, sometimes exceeding 10 m, above the lower Eocene limestone of the Thebes Group. They consist of poorly bedded gravel of dark brown limestone and chert embedded in pale brown silty matrix. The gravels are derived from the underlying Thebes limestone. El Hinnawi et al. (1978) and El Deftar et al. (1978) suggest that these sediments were transported short distances by flood waters during pluvial periods and deposited in low topographic depressions. Said (Chapter 25, this book) suggests that these gravel fills represent inverted wadi deposits of a post-Eocene drainage system and correlates them with the Oligocene Nakheil Formation on top of the faulted Eocene blocks in the Quseir-Safaga area, Red Sea region.

Minqar El Talh Formation. A sequence of cross-bedded sandstones, in a partially kaolinized matrix, topped by fresh-water gastropod-bearing limestone, forms the summits of the west Dakhla and Abu Minqar scarps (Fig. 14/9 and sections Fig. 14/13). It rests on detritic carbonate and reddish shale or sandstone of the late Maastrichtian and Paleocene substrata. The same sequence is recorded truncating the Paleocene-early Eocene marls and limestones of the Esna and Farafra Formations at the southern end of El Quss Abu Said plateau (map 14/9). It is not possible to date these sediments by their terrestrial and fresh water gastropods (Barthel & Herrmann-Degen 1981).

A similar sandstone and sandy limestone sequence resting on the middle Eocene limestone of the Mokattam group is described by Lebling (1919) from the plateau north of Bahariya. In the Siwa area, Zittel (1883) records several localities with ?Neogene deposits carrying fresh water fauna and flora. The authors of the new geological map (1987) classify this sequence, named Minqar El Talh Formation, with the post-Miocene deposits on stratigraphic grounds. At the type locality in the southwest Qattara depression, the sequence is described as 'light-colored, continental to lacustrine sandstone with root marks; yellow siltstone, capped by lacustrine limestone with borings and gastropods'. Said (Chapter 25, this book)

attributes this sequence to the late Pliocene drainage system or systems which developed over 'the exhumed surface of the late Miocene erosional episode'.

Gravel terraces. The gravel terraces cover large parts of the foot scarps of the Kharga depression, Abu Tartur and other areas to the west. They consist of pebbles and boulders of limestone and chert embedded in a clayey matrix. They were formed during the Pleistocene pluvial episodes after the excavation of the depressions. Strong torrents cut deep gullies in the face of the scarps and carried the detrital wash depositing the coarse gravels at the foothills (Awad & Ghobrial 1965). The finer clay material was further transported by the torrents to the numerous lakes which occupied the lower areas of the excavated depressions where they were deposited (see playa deposits, *vide infra*).

Playa deposits. Among the large playas in the area under discussion are those to the west of Um El Ghanayem (200 km²), around Baris (125 km²), at Qulu El Sabaya, west Mawhoob and Farafra. There are several others of lesser dimensions. On the plateau surface near Gebel Abu Tartur, other playas are found covering areas ranging from 10 to 100 km² (El Deftar et al. 1978). The playa deposits are made up of horizontal, alternating bands of soft, friable sand, clay and silt with frequent plant remains. Some artifacts are found in them, particularly near their edges. Gastropod shells are of common occurrence in the playa deposits. Morphologically the playas form isolated well-eroded hummocks or narrow elongte ridges in heights of a few meters to 20 m above the depression floor.

On the top surface of some playas, a salt crust of capillary origin forms a thin veneer 10 to 20 cm thick. Salt crusts also intercalate, in places, the shale and silt bed rocks of some parts of the depressions. Sodium chloride is the major salt constituent in the crust, about 82%. For a discussion of these deposits, their age and genesis the reader is referred to Said (Chapter 25, this book).

Traverine and Tufa. These are massive or porous fresh water limestones with abundant plant remains. The thickness of these deposits in the southern oases areas is variable, is within the range of 10 m. They spread over the scarp face to the east and north of Kharga, south of Ain Amur embayment, and in other localities to the west. They also occur over the plateau surface to the east of Garmushin in Kharga (El Hinnawi et al. (1978) and to the west of El Ramliya promontory (El Deftar et al. 1978).

These occurrences and others in the Kharga-

Dakhla stretch are comparable to similar deposits of spring activity in the south Western Desert, described by Said (1969), Said & Issawi (1964) and Issawi (1969, 1971). The age of these deposits is difficult to ascertain, although most authors classify them with the Pleistocene. Said (Chapter 25, this book), however, emphasizes that the massive tufas on top of the Kharga plateau are of pre-late Miocene age, though younger generations of tufas are known.

Sand dunes. Sand dunes cover about 20% of the total area under discussion. The southeastern extension of the conspicuous Abu Moharik dune belt, which originates on the plateau east of Bahariya Oasis, enters the plateau north of Kharga and continues with minor breaks in nearly the same southeastern direction for about 150 km, delimiting the western side of the Kharga depression. Patches of sand dunes and drifts exist at the foot scarps of Kharga at the Baris and Mahariq areas, and along the slopes of Abu Tartur-Abu Minqar scarps. A deluge of parallel, northwest-southeast longitudinal dunes overwhelms large parts of the area from east of the Farafra depression to the El Kharafish plateau. Occasional inter-dune corridors expose Tarawan or Dakhla Formations. The extension of this sand mass invades in discontinuous narrow belts the western side of the Dakhla depression (Figs 14/5 and 14/9). The Great Sand Sea, formed of similar northwest-southeast running dunes, mainly of the seif type, borders the Abu Minqar-Farafra stretch on its west and blankets various rock units ranging from the Quseir and Duwi Formations in the south to the Esna Formation in the north (Fig. 14/9).

SUMMARY OF DEPOSITIONAL HISTORY

Precambrian basement rocks in the Kharga stretch are restricted to scattered tectonic occurrences which form the small-sized granite hills at Abu Bayan. They owe their exposure to uplift and faulting movements, mainly of post-Eocene age. The sedimentary cover in the Kharga and Dakhla areas represents the infill of the Dakhla Basin from late Jurassic to early Eocene, with a total thickness ranging from less than 500 m on the south Kharga uplift to about 2000 m in Dakhla and north of Kharga. In the west Dakhla-Farafra area, the Dakhla Basin is partially Paleozoic and the Paleozoic-early Eocene sediments there increase in thickness to about 3000 m. The depocenter of the Dakhla Basin falls under the Great Sand Sea further to the northwest, beyond the limits of the area under discussion. Late Jurassic sediments are the oldest recorded in the stretch, as revealed by surface investigations and the available subsurface drilling data. Two main facies are distinguished: a Jurassic to Cam-

panian facies, predominantly continental but with shallow marine invasions in the Aptian and Cenomanian, and a Campanian to early Eocene facies of transgressive-regressive open marine facies. Late Jurassic to early Turonian continental and marine sequences wedge out, or are lacking on the Kharga uplift, and the entire stretch was emergent from middle Turonian to early Middle Campanian time. The Campanian transgression which followed terminated the predominantly continental deposition in the Dakhla Basin. At the Campanian-Maastrichtian boundary, phosphate-bearing sediments were deposited in association with sea bottom oscillations of varied intensities, developing in basin-ridge configuration. While the Campanian to Maastrichtian deposition was continuous in deep parts of the basin (Dakhla and west Dakhla areas), it was interrupted by diastems in the Kharga and Farafra areas. Shale and mudstone are the main lithotypes of Maastrichtian sediments in the Kharga-Abu Minqar stretch. Sand content in the shale increases generally in areas of south Kharga and between west Dakhla and Abu Minqar. In the Farafra depression, the Maastrichtian shale changes to chalk, free of clastics. Stratigraphic intervals of shallowing depositions are repeatedly reported within the Maastrichtian sediments, particularly in its middle members, and eventually a well-documented erosion surface terminates the Maastrichtian deposition. The marine Paleocene sediments which followed on the Maastrichtian with varying stratigraphic gaps display two distinct types of facies: marginal marine limestone facies (Garra El Arba'in facies) which developed in the south Kharga and Abu Tartur-El Kharafish plateau, and deeper, middle to outer shelf facies in the other basinal areas (Nile Valley facies). Interfingering relationships between different formations of the two facies are well-displayed. Shallowing conditions started again toward the close of the Paleocene and the sea completely withdrew to the north after the deposition of the Thebes Group limestones during the early Eocene. The latter limestones are classified into a number of interchanging formations of platform to variable shelf environments of deposition. Since the post-early Eocene regression and the following uplifting movement and erosional phase that affected Egypt, terrestrial conditions have prevailed over the Kharga-Farafra stretch as well as in other areas of south Egypt.

DESCRIPTION OF GEOLOGIC STRUCTURES

The structural elements of the Kharga-Farafra stretch are the result of typical stable shelf tectonics. Faults and, to a lesser extent, large scale gentle folds are reflected on the surface and these indicate differential block movements in the basement. Most probably, the type and intensity of the resulting tectonics and deformations in the overlying strata are governed by the thickness and lithology of the rocks which differently constitute the sedimentary cover in the various areas. In Kharga, where the sedimentary cover is relatively thin, and particularly in its southern parts, faults are the dominant tectonic feature and are of greater density and persistence. On the other hand, in other areas to the west and northwest, including the Abu Tartur and Dakhla areas, the role of broad warpings and undulations is more prominent. The Farafra area is distinguished from Dakhla by its relatively more intense tectonics, as it represents the southern extension of the Syrian Arc system, more fully developed in Bahariya to its northeast.

The structural features of each of the structural sections of the stretch of territory under investigation are described. Location names and fault lines are shown on the maps in Figures 14/4, 14/5 and 14/9.

South Kharga sector

This is the only sector in the Kharga-Dakhla stretch where outcrops of crystalline basement are found with a thin sedimentary cover and where the uparching of the basement rocks is pronounced.

Faulting

The structural setting of the south Kharga sector is determined primarily by a group of parallel faults running in a mainly east-west direction for distances of up to 50 km. These major faults divide the sector into parallel blocks of variable width (Fig. 14/4). The vertical displacement of rocks on the fault lines ranges from 10 to 50 m and, in some faults, the vertical displacement is accompanied by horizontal shifts, as in Abu Bayan El Bahari and Abu Bayan El Wastani faults. Along the course of the major east-west faults, small diagonal faults are observed. El Hinnawi et al. (1978) describes in detail a number of the major faults dissecting the rocks in the depression and plateau areas of south Kharga (Fig. 14/4). The Abu Bayan El Wastani fault affects the Campanian Quseir Formation in the depression. Three centroclinal basins, formed of beds belonging to the Dakhla, Kurkur, Garra and Dungul Formations, as well as a granite body, are arranged along the fault line. On the plateau surface, the limestone beds of the Kurkur and Garra Formations are highly inclined and much brecciated, and the fault has a vertical displacement of 20 m to the south. To its north, a nearly parallel fault at Abu Bayan El Bahari is marked by three granite hills along its course in the depression,

and it displaces the Paleocene rocks on the scarp and plateau with 25 m of downthrow to the north. At Wadi El Abd, on the scarp and plateau to the east of Dush, two faults of nearly the same east-west trend affect the Paleocene limestone beds of Kurkur and Garra Formations and enclose a graben. A domal structure is associated with the northern fault, with Kurkur beds forming its core. To the east of Ain Tafnis, a similarly trending fault of 20 m throw to the south dissects the Paleocene beds on the scarp and plateau surface. Ain Tafnis spring represents a structural trap formed due to the displacement of Kurkur limestones against the clays of the Dakhla Formation. The northernmost extension of this east-west trending fault system is represented by a fault on the plateau approximately along latitude 25° N cutting across a domal structure with beds of Dakhla Formation at the core and of Tarwan, Esna and el Rufuf on the flanks. El Hinnawi et al. (1978) describe a dark grey travertine which covers the eastern extremity of the fault line as a deposit of an old spring located on the fault plane.

This dominant east-west fault trend is traced for several hundred kilometers to the east and west, beyond the Kharga sector. To the west it extends as far as the eastern forelands of El Gilf El Kebir in the south Western Desert, where its age seems to be mainly early Mesozoic. At a locality about 120 km to the west of Abu Bayan area, this faulting has dropped a narrow block of lower Eocene carbonate into the Jurassic Six Hills Formation with an estimated throw of about 1000 m. From there on to the east, this fault system should be at least partially of post-Eocene age.

Folding

Gentle anticlinal undulations coincide with the small bays along the scarp face with dips ranging from 2 to 5° on the flanks. As mentioned above, some centroclinal domes are associated with the east-west faulting on the plateau surface. The area of these folds ranges from 1 to 25 km² and their beds are highly tilted near the fault lines. Along Abu Bayan El Wastani fault, two centroclinal basins occur. Lower Eocene limestones of the Dungul Formation crop out at the center of the western basin, being flanked on the north and south sides by older rock units which dip at 15 to 30°.

North Kharga sector

The north Kharga structural sector, situated to the north of latitude 25° is structurally characterized by predominance of north-south trending faults of the normal type. They extend over the depression and along the eastern and northwestern escarpment as well. Slight azimuth deviations from the north-south trend to the north-northwest or to the north-northeast are common along the courses of some of these faults. The east-west trending faults characteristic of the south Kharga sector are very subordinate in the northern sector. Ghobrial (1967) describes in detail the different fault patterns in north Kharga; a summary of which is given here (Fig. 14/4).

Taref-Teir fault. This fault extends for at least 30 km in the western limits of the Kharga depression and truncates along its course the western sides of Gebel El Teir and Gebel Tarawan. The sandstones of the Taref Formation on the west side of the fault are brought to a level higher than that of the Duwi Formation in Gebel El Teir and even higher than that of the upper shale beds of the Dakhla Formation as in Gebel Tarawan. The maximum displacement is about 225 m to the east in the vicinity of the latter hill. A group of minor faults run for short extensions parallel to this major fault of Gebel El Teir, with 5 to 20 m vertical displacement to either east or west. On the western side of the major Taref-Teir fault another group of lesser faults of nearly the same trend are found on the western side of the conspicuous outlier of Gebel Taref. On both sides of the major Teir-Taref fault, two north-south, doubly plunging basinal structures are developed in the area to the north of El Kharga town. The one to the east encloses the outliers of Gebel El Teir and Gebel Tarawan, and the one to the west encloses the outlier of Gebel Taref and the small hill to the west of it, Gebel Sheikh. The rocks which form the basinal parts of these outliers belong to the Duwi, Dakhla and Tarawan Formations. It is only in Gebel El Teir that eroded remnants of the Esna Formation and Thebes-El Rufuf Formation are preserved as capping rocks.

Quarn Ginah-Boulaq faults. A group of isolated hillocks in the depression area at Ginah to the north of Boulaq are intensely folded and are separated from each other by a group of inferred faults which form two horsts separated by a graben. The extensions of these faults range from 4 to 20 km following a trend of average azimuth 13°, and with amounts of downthrow from 20 to 160 m. Several domal and basinal structures with steeply inclined flanks are developed in association with this group of faults. Most conspicuous of these folded structures is Quarn Ginah, a doubly plunging domal structure developed around a north-south trending axis, and fault-truncated on its western side. The folding of Quarn Ginah could have been initiated in pre-Campanian times as indicated by the occurrence of a joint system within its formative sandstone layers belonging to the Taref Formation, which is noted to be older than a younger joint

system in the lower Eocene limestones on the plateau (Ghobrial 1967).

Down-to-east step faults along the northern escarpments. Two faults are mapped along the western and eastern scarp faces and foothills which bound El Mahariq embayment in the northernmost Kharga area (Fig. 14/4). The western fault extends for a distance of about 22 km and trends at an azimuth of 34°. Its maximum downthrow is estimated to be in the range of 110 m to the east as inferred from elevations on top of the Duwi Formation on both sides of its course (Ghobrial 1967). The eastern fault runs for a distance of about 22 km along the scarp on the east side of the embayment in an almost northerly direction (azimuth 6°). The downthrow is toward the east with a maximum vertical displacement in the range of 120 m, as indicated by elevations in the Taref sandstones on the western side and those on top of the Tarawan Formation on the eastern sides. The above two faults are two step faults, both having their downthrow to the east.

Umm El Ghanayim-Ghanima faults. A group of faults affect Gebel Umm El Ghanayim and Gebel Ghanima, outliers of the eastern scarp in north Kharga, as well as the areas in their vicinity (Fig. 14/4). These faults extend for distances of 5 to 10 km. Two of these faults truncate the western sides of the two outliers and they both have the downthrow sides to the east with vertical displacements of 73 and 86 m at Umm El Ghanayim and Ghanima respectively. The fault cutting the area in between has a downthrow also to the east, of 30 m. Exceptionally, the fault to the northwest of Umm El Ghanayim has a downthrow to the west of about 130 m. Similar to the Teir-Tarawan-Taref basins, the Umm El Ghanayim and Ghanima outliers represent basinal structures truncated by faults on their western sides.

It is quite certain that block faulting is the basic characteristic feature in the Kharga area in general. Faulting took place in several phases during its geologic evolution, and it is also evident that the tectonic faults were activated later, when new ones as well occurred. The activity of the majority of these faults can be established as taking place in post-lower Eocene times. The present day morphological configuration of the Kharga depression and its eastern bounding escarpment in an almost north-south direction, and its parallelism to the dominant fault trends in north Kharga, is noteworthy. Over the entire stretch from Abu Minqar to Abu Tartur at the western approaches of Kharga, the retreat of the scarp face has been progressively affected to the north, along the gentle to almost horizontal regional dip, and the present scarp face on the northern sides of the depressions acquire an almost east-west trend. The remarkable and abrupt change of the scarp alignment to the south in north Kharga developed along the dominant step-faulting system in an area of a flat regional dip. Another feature that draws attention in north Kharga is the direct relationship between the folded structures, domal and basinal, and the north-south system of faults. 'Nearly all observed folded structures are truncated on their west sides by faults. Also, the intensity of folding is directly proportional to the amount of downthrow of the respective faults' (Ghobrial 1967).

Abu Tartur-Dakhla-Abu Minqar sector

This structural unit is characterized by simple geologic structure and very gentle northwardly regional dips. The central part of this sector, situated in the Dakhla area, is occupied by a major northeast-plunging syncline which forms part of the regional Dakhla Basin. The Dakhla syncline is thrown into a number of small anticlinal and synclinal undulations of different intensities. The axes of the anticlinal undulations coincide with the embayments of the scarp at the northeast of El Qasr, along Darb El Tawil and at Wadi El Battikh (Fig. 14/5). On the other hand, the axes of the synclinal undulations coincide with the promontories of the scarp which lie at Mut and Teneida. In spite of the subdued surface expression of the Dakhla syncline and the superimposed fold undulations, the structures continue at depth with little or no crestal shifting as the subsurface data indicate (Barakat & Milad 1966). The folding movements date back to Jurassic-early Cretaceous times.

The southeast flank of the Dakhla syncline occupies the present Abu Tartur plateau, with minor gentle undulations and a deeper closed subsyncline located in the center of the plateau. A structural uplift area occupies the desert peneplain between west Kharga and the Abu Tartur plateau. The structurally higher setup of this area, relative to the lower Dakhla area to its west, dates back to pre-Campanian times, as this high area probably formed a northern extension of the Kharga paleo-uplift. As the prevailing dips are very gentle in degree, the differential structural relief between the areas of east Abu Tartur and Dakhla are revealed only by systematic structure contouring. Datum elevations carried out (Hermina 1967) on the base of the Duwi Formation in the Abu Tartur area and by Hermina et al. (1961) in the Dakhla area show a gentle drop from 450 m in the east to less than 100 m around Edmonstone in west Dakhla. These datum elevations should have attained higher values to the east of Abu Tartur in the Kharga area, if it were not for the later downthrow effects of the north-south fault system to the east. Elevations on

the same datum in Kharga do not surpass the 400 m on the highest parts of the Kharga central north-south elongated anticlines, and they reach down to below 100 m along the dips toward the bounding escarpment (Ghobrial 1967).

Faults around Abu Tartur are of diverse orientation and the intersection of many minor faults appears to divide the area into small echeloned sections. These are best revealed by the use of aerial photographs or satellite imagery. Some other minor, normal faults extending irregularly in different directions in the depression and along the scarp faces can be directly traced on the ground. Two sets of northeast faults with sublatitudinal trends are partially and discontinuously traced on the scarp face, but more completely on the Abu Tartur-El Kharafish plateau. They affect the Paleocene limestone beds. As discussed above, their origin is attributed to syndepositional tectonism at that time and they are responsible for the development of the near-shore 'Garra El Arabain' facies on the upthrown blocks.

The relatively most important structure affecting the northerly monoclinal dip in western Dakhla is a north-south trending syncline enclosing the Edmonstone outlier. Similar to the basinal structure in the Kharga depression, the Edmonstone syncline is located near the downthrown side of a north-south trending fault indicated by some authors to its west side. The amount of downthrow of this fault is estimated to be 80 m.

The possibility of the existence of a group of major faults around this syncline cannot be eliminated. This is strongly indicated by the abundance of minor faults throughout the surrounding area (Hermina et al. 1961). A fault or a group of faults is inferred to exist along the southern footscarp of Gebel Edmonstone, whose cumulative downthrow is in the range of 100 m to the southwest. The *Iscordia chargensis* limestone bed, which caps the Duwi Formation, is noted to drag at 10° near the faults of south Edmonstone. Other sets of faults of a general west-northwest trend can be traced or inferred for distances of 10 to 15 km along the face of the northern main scarp and in the area between it and Edmonstone, with vertical displacements of 20 to 50 m, primarily to the southwest. Another fault runs parallel to the scarp face north of El Qasr. It is indicated by the abundance of minor faults in this area, whose downthrow sides are mostly towards the north. The existence of this fault is also suggested by the occurrence of a group of hot sulphurous water springs along its inferred course, at the foothills of the scarp.

A characteristic feature of the area to the northwest of Edmonstone is the lack of major faults (Fig. 14/9). A series of minor normal faults are observed along the scarp face west of Qur El Malik where they dip to the northeast. The morphological shape of the escarpment southeast of Abu Minqar suggests the presence of a fault running parallel to it. This fault is much concealed below continental deposits of younger age (Minqar El Talh Formation). The west-northwest trending faults recorded on the El Kharafish plateau and to its west, fall almost on the same trend of the similar faults previously described on the Abu Tartur plateau. They also appear to be a result of the Paleocene syndepositional tectonism which controlled the distribution of the facies characteristic of its rocks.

Farafra sector

The Farafra structural sector is distinguished from the Dakhla-Abu Minqar sector by its relatively more intense tectonics. Its structural elements may represent the transitional stage between the typical stable and mobile shelf conditions. Faults, joints and folds indicate movements in the basement, but these are faintly distinguished on the surface due to the relatively thick sedimentary cover (Barakat & Abdel Hamid 1973). The sedimentary record in this sector reflects the occurrence of a number of tectonic movements at different times. These movements produce four gently folded structures. Two anticlines (Farafra & Ain Dalla anticlines) alternate with two synclines (El Quss Abu Said and El Ghard synclines). Rather dense faulting and very prominent jointing resulted from these tectonic disturbances. A prominent set of faults of northeast strike is detected parallel to the northern escarpment of the depression. Although individual faults have been traced for a considerable distance, no one fault is found to extend completely across the whole scarp.

The Farafra main (central) anticline is a doubly-plunging anticline, trending northeast-southwest with the steeper plunge on the northeast. The flanks show gentle dips, 2 to 3°, increasing to 7° near the fault. Clastic rocks of the Wadi Hennis Formation (Campanian) are the oldest rocks exposed on the

central part of the anticline in the area of Ain El Maqfi and Wadi Hennis (Fig. 14/9). Maximum weathering effects are observed along the southwestern extension of the axis in a direction toward Qur El Izza and Abu Minqar. Numerous faults and a dense system of joints are developed in the areas around North Gunna, an area characterized by numerous springs. Survey studies around Qasr El Farafra in the depression show a very marked consistency of joint patterns in chalk beds of the Khoman Formation which swing gradually from northwest north of Qasr El Farafra to the northeast south of there. El Ramly (1964) remarks that 'these joints which are perpendicular to the bedding planes are joined by (en echelon) gash joints filled by calcite and which stand out above the ground surface as low ridges'. He adds that this jointing is certainly partly contemporaneous with the folding.

Faults are rather difficult to detect due to sand cover. Some faults of a northeast trend cut obliquely across the Farafra anticline in the area between Gunna North and Ain El Maqfi (Fig. 14/9). The eastern flank of the Farafra anticline constitutes a huge monoclinal structure covered by limestones of the Thebes Group, which persistently dip gently to the southeast. El Ramly (1964) shows a huge syncline to the southeast of Farafra anticline whose axis also trends northeast-southwest in the dune area between Bir Karawein and Qur Zugag (El Ghard syncline).

The northwest flank of the Farafra anticline is defined by a parallel-trending and double-plunging syncline of El Quss Abu Said, which in turn is defined by the Ain Dalla minor anticline on its northwest flank. The latter, however, is an ill-defined structure, due to the vast cover of sand in the area.

Omara et al. (1970) define an upthrown-faulted block to the south of El Quss Abu Said. The bounding faults extend in a northwest-southeast direction on the plateau areas from Abu Minqar toward El Kharafish. It is possible that this block faulting is related to the previously discussed system that affected the Paleocene rocks to the east.

CHAPTER 15

North Western Desert

GAMAL HANTAR
Consultant, Cairo, Egypt

The northern part of the Western Desert forms an almost featureless plain which, with the exception of the small folded and faulted Abu Roash complex to the north of Giza pyramids, offers few prominent topographical or geological features that would reflect its intricate geological history. Most of the surface is covered with gentle-dipping Neogene starta of reasonable lithological uniformity. There are a few lines of major faults, and the few folds noted are minor rolls with gentle dips and large amplitude. Topogrpahically the monotony of the plains is cut by occasional low questas, the great Qattara depression, Siwa oasis and the Wadi Natrun hollows.

Deep drilling in this desert, however, has shown that this apparently geologically simple structure made by the thin cover of later sediments conceals beneath it an intricate geological structure made up of a large number of swells and basins. The sedimentary column is thick. In the Abu Gharadig basin it reaches between 8 and 9 km, while to the north it may reach 3 to 6 km. The complicated structure and the great thickness of the sedimentary column, when compared with the areas to the south, justifies the use of the terms Stable and Unstable Shelves which were introduced by Said (1962) for the areas north and south of the Abu Gharadig basin.

Recent literature covering this area include: Abdine (1974), Abdine & Deibis (1972), Abu El Naga (1984), Awad (1984, 1985), Bayoumi & El Gamili (1970), Ezzat & Dia El Din (1974), Deibis (1976), Dia El Din (1974), El Gezeery et al. (1972), El Sweifi (1975), Elzarka (1984, 1986), Gindy & El Askary (1969a, b), Khaled (1975), Marzouk (1969), Khalil & El Mofty (1972), Metwalli & Abdel-Hady (1975), Norton (1967), Omara & Ouda (1969, 1972), Riad (1977), Salem (1976) and Sestini (1984). Basing its work on the wealth of material resulting from oil exploration work in the north Western Desert, Robertson Research International (RRI 1982) carried out a comprehensive study of the geology and the oil potential of this region.

Tectonics and structure

The eastern Mediterranean basin evolved as a result of plate motions responding to the opening of the Atlantic Ocean starting from the Jurassic and resulting in the destruction of the Paleotethys and the opening of the Neotethys. The north Western Desert forms part of the African plate. At present it is characterized by having a narrow continental shelf (15 to 50 km wide) which is bound by a steep continental slope representing a major fault or hingeline separating the continental crust from the continental margin. In the past, the north Western Desert was intermittently submerged by epicontinental seas. The entire sedimentary section, except in limited areas and for short durations, is of subaerial or shallow marine origin excluding a continental margin model and putting the north Western Desert well south of the Neotethys suture.

Several tectonic events affected the north Western Desert. The early Paleozoic (Caledonian) and the late Paleozoic (Hyrcenian) events were mild and are represented by regional uplifts of moderate magnitude producing disconformities within the Paleozoic and between the Paleozoic and the Jurassic. The presence of widely-spread continental Jurassic indicates that the late Paleozoic event could not have produced major structural or topographic irregularities. During the Jurassic, which was accompanied by major plate movements including the separation of the Apulian microplate, many of the emergent lands of north Egypt became submerged by the newly formed Neotethys. The end of the Jurassic witnessed a major orogenic movement which resulted in the emergence of the land.

The most important tectonic event occurred during the late Cretaceous and early Tertiary and was probably related to the movement of the north African plate toward Europe. It resulted in the elevation and folding of major portions of the north Western Desert along an east-northeast west-southwest trend (Syrian arcing system) and in the development of faults of considerable displacements.

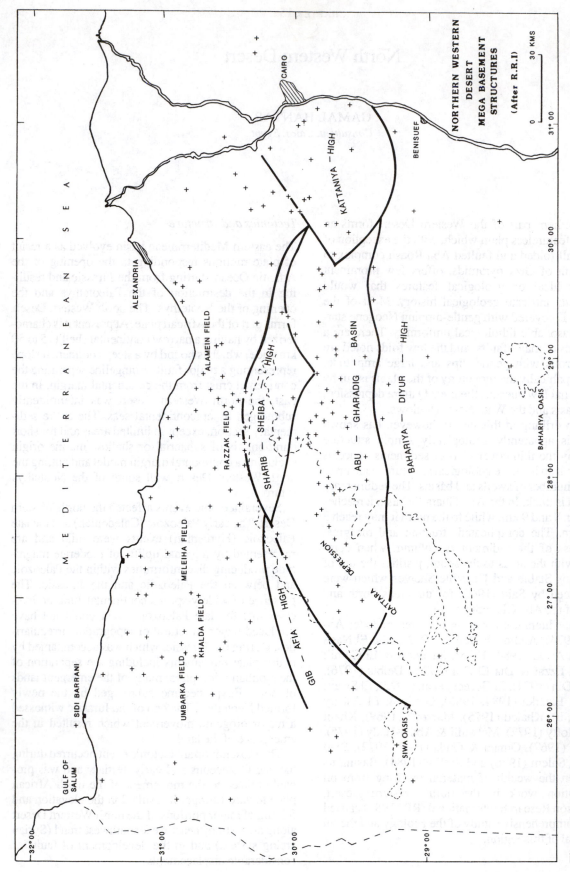

Figure 15.1 Mega-basement structures (after Robertson Research International (RRI) 1982).

Basement trends

The tectonic evolution of north Africa was affected among other factors by the reactivation of the basement fractures. The trends of these fractures must have had an important influence on the sites and orientation of basins and other mega and small structures. The basement trends can be seen in the basement outcrops. They include:

1. A north-south trend charachteristic of the Nubian Shield. This trend extends northward and underlies the area between the Nile Valley and the Qattara depression.

2. A northeasterly trend characteristic of the Tibesti Massif and the Pelusium line which were probably genetically related. This trend extends below the western part of the Western Desert.

3. An east northeast-west southwest trend. This trend results from the interaction of the above-mentioned trends marking a major structural discontinuity. The expression of this trend is along a line that runs from Siwa oasis to Alexandria.

4. An east-west trend. This trend does not show on the surface but is strongly indicated from the aeromagnetic measurements of the Western Desert. This trend was most probably responsible for the shaping of a number of the megastructures that extend along latitude 30° N.

Basement mega-structures

Information obtained from drilling, aeromagnetic and gravity measurements and seismic interpretations indicate the presence of basement mega-structures which are aligned in a more or less east-west direction. These structures lie along the Stable-Unstable Shelf contact and seem to have affected the sedimentary history and geologic evolution of the region. The following are the most important elements of these structures (Fig. 15/1).

Bahariya-Diyur high. The southern part of the area is dominated by a basement high where the basement is always shallower than 3 km. To the east, however, lies the Gindi (Fayum basin) in which the basement lies below 5 km.

Gib Afia high. This high runs in a northeasterly direction from the western side of Siwa oasis to the northwestern edge of the Qattara depression. The basement lies at a depth of 3 to 4 km.

Sharib-Sheiba high. This high has an east-west trend. It extends from the western corner of the Gib Afia high to longitude 30° E.

Kattaniya high. This high has a northeast-southwest trend and is centered about 50 km from Cairo. It is 25 to 30 km wide and forms a horst block bounded by faults with a displacement of about 3 km.

Abu Gharadig basin. This basin has a thick sedimentary section and great hydrocarbon potential (Awad 1984, 1985).

Structures

The north Western Desert structure is dominated by faults many of which can be identified from seismic and well data (Fig. 15/2). The majority are steep normal faults and most have a long history of growth. Some of the normal faults suffered strike slip movements during part of their history. Strike slip movements seem to have affected the orientation of many of the fold axes. The strike slip movements were probably related to the lateral movements which the African plate underwent during the Jurassic (sinstral) and late Cretaceous (dextral).

Faults with displacements in the magnitude of 1500 to 3000 m are limited to the Kattaniya horst (northeast-southwest) and the Abu Gharadig graben (east-west). The greatest subsidence along these faults took place in post-Turonian time and greatly affected the isopachous distribution of the late Cretaceous and early Tertiary sediments.

Faults with displacements in the magnitude of 750 to 1500 m are present in the northern parts of the region but are widely spaced. These faults have an east-west trend with their end acquiring a northeasterly or northwesterly direction.

Faults with less than 750 m throw are more frequent. Their trend is east-west in the Abu Gharadig basin, northeast-southwest in the Kattaniya high and northwest-southeast over the rest of the north Western Desert. Faults of north-south trend are known only in the area to the southwest of Matruh. There are also a large number of hanging faults affecting the shallower parts of the section and usually of limited throw. These faults are common in the northern part of the region.

Most folds owe their origin to compressional movements which affected the area during the late Cretaceous-early Tertiary tectonic event. These folds have a northeast-southwest trend and a periclinal geometry. Abu Roash is a classic example of this type of fold. In addition, there are other folds which owe their origin to normal or horizontally displaced faults. These folds are usually confined to fault blocks. Their axes are parallel, oblique or perpendicular to the fault block trend depending on the magnitude of the strike slip component of the movement.

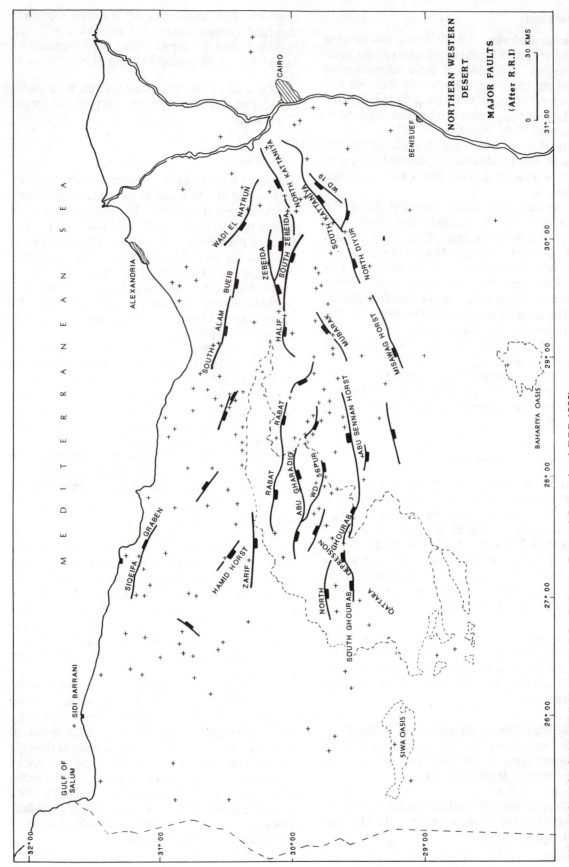

Figure 15.2 Major faults and displacements (after Robertson Research International (RRI) 1982).

Stratigraphy

For most of its geological history, the greater part of the north Western Desert formed a platform characterized by relatively mild subsidence; it was situated near actively subsiding basins or depocenters.

During the Paleozoic most of the area lay to the east of the more active Paleozoic basin occupying the Siwa-Kufra, Libya area (the Kufra Basin). During the Jurassic substantial tilting shifted the center of the basin to northeastern Egypt leaving part of the Western Desert in the form of a platform. With the onset of the early Cretaceous and up to the Recent, the active part of the basin shifted to the north occupying the present Mediterranean offshore area parallel to the present shore line. During these times, the north Western Desert formed a platform which lay to the south of the offshore basin to the north.

During different periods, however, local depocenters of limited dimensions developed in different places over this platform. Among these mention is made of the narrow pullapart basins that straddle latitude 30° N; they were given the names Betty, Abu Gharadig and Gindi (Fayum) basins. These basins were particularly active during late Cretaceous-early Tertiary times. The Matruh Basin deserves special mention. It has a north-south trend and was particularly active during the early Cretaceous (and probably earlier). Its origin is not yet fully understood. However, it is thought by different authors to represent either a failed arm of an early Mesozoic crustal rift, a normal graben or simply a submarine erosional canyon that was filled later with lower Cretaceous clastics.

As has been already stated, the north Western Desert, with the exception of the small outlier of Abu Roash, is a plateau covered with Neogene sediments. Recent active oil exploration work including drilling, seismic, gravity and aeromagnetic measurements has revealed the presence of a subsurface stratigraphic column which ranges in age from the Paleozoic to the Recent. The sediments occur in a number of basins with varying degrees of subsidence. There is no formal and universally accepted nomenclature for the subsurface rock units encountered in the north Western Desert. The formational names used in this chapter are mainly those suggested by Norton (1967) and later followed, with minor modifications, by the authors of the RRI report (1982). In a few instances it was necessary to deviate from this nomenclature to avoid correlation problems with other areas of Egypt.

PALEOZOIC

Paleozoic sediments are reported from 31 wells, five of which are uncertain. The identification of the Paleozoic is based on the stratigraphic position of the strata, the presence of a few megafossils separated from the raised cores and cuttings and wire line log correlation. Although these data do not permit definitive results regarding the exact boundaries of the different stages, they were, in many cases, authenticated by later comprehensive paleontologic studies.

The Paleozoic sediments of the north Western Desert are of monotonous composition and are made up of interbedded sandstone and shale with a few carbonate beds. This monotony makes the identification of workable rock units difficult. Norton (1967) attempted to extend the use of rock unit names in Libya to Egypt: Gargaf (Cambro-Ordovician), Acacus (Silurian) and Gara Dalma Formations (Carboniferous). These formations were difficult to distinguish in the Egyptian section and the nomenclature is not followed in this chapter.

Figure 15/3 gives the isopachs of the Paleozoic. The maximum thickness of Paleozoic strata was penetrated in Faghur-I (2500 m), Zeitun-I (2416 m) and Siwa-I (2406 m). Gravity and magnetic modelling suggest that in the western part of the north Western Desert, basins containing more than 4000 m of Paleozoic strata may well exist (Abu El Naga 1984).

Cambrian

Cambrian strata were reached in 14 wells and were fully penetrated in five of these wells. Only in four wells (Bahariya-I, Ghazalat-I, Gib Afia-I and Kahraman-I) was the Cambrian definitely proven by paleontological means. The Yakout I section, which was originally assigned on stratigraphical grounds to the Cambrian by the operator (Shell), is given a Silurian to Devonian age by RRI (1982) on the ground that it is devoid of palynomorphs. From a paleogeographic point of view, it seems reasonable to assign this section and other undifferentiated Paleozoic sections to the Cambrian rather than to younger ages. The Cambrian strata of the Bahariya, Gib Afia and Ghazalat wells lie on old basement highs and seem to represent remnants of a very widely spread sea.

The Cambrian strata are made up of sandstones of various light colors, sometimes glauconitic, and shales of reddish, brick red and grey colors. In Rabat-I well the section contains thin beds of coarse gravel. Definite fossil-bearing Cambrian rocks reach a maximum thickness of 860 m (Bahariya-I well). Undiffe-

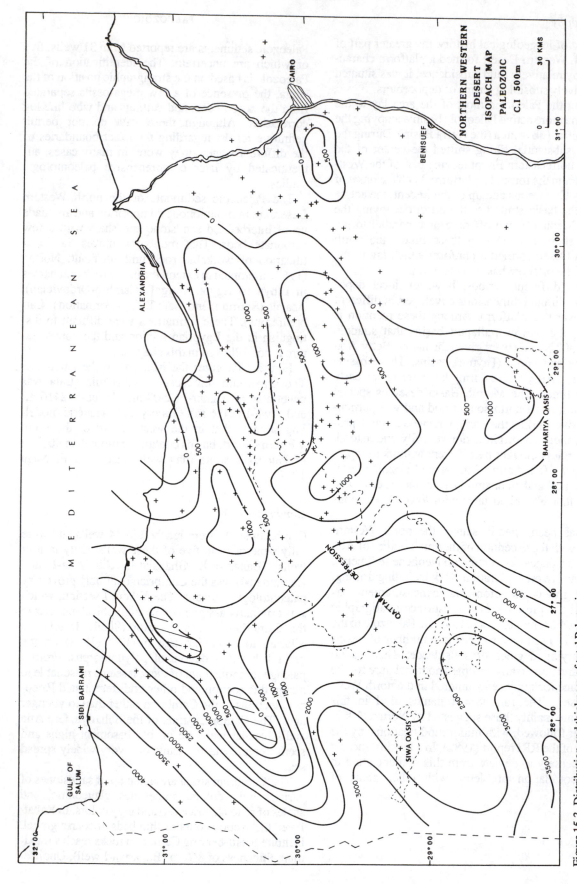

Figure 15.3 Distribution and thickness of total Paleozoic.

rentiated pre-Devonian strata ascribed questionably to the Cambrian may reach a thickness of 1240 m (Siwa-I well).

The presence of trilobites, brachiopods and acritarchs in the Cambrian of Bahariya, Ghazalat, Gib Afia, Kahraman and Yakout wells point to a marine environment of deposition. The abundance of clean sands suggests a near-shore high energy environment. On the other hand, the presence of brick red shales indicates either a non-marine environment for the intervals when these shales were deposited or a proximity to a source of hematitic mud. The undifferentiated clastic sequences assigned to the Cambrian point to a fluvial to continental environment.

Cambrian strata rest unconformably over the basement rocks which provide a clear boundary. The upper boundary, however, is less certain and is usually marked by an arbitrary stratum of Silurian, Devonian, Carboniferous or younger age.

Ordovician

No fossil-bearing strata of Ordovician age were identified in the region. However, a number of operators assign a Cambro-Ordovician age to the barren clastic section penetrated in a number of wells below the identifiable Devonian.

Silurian

Paleontologically identified beds of Silurian age were penetrated only in Sheiba-I well on the Sharib-Sheiba basement high. Here the section is made up of shale, siltstone and thin limestone beds intruded by a gabbroic sill.

Devonian

Fossiliferous Devonian strata were identified in 10 wells and questionably in two. With the exception of Yakout-I, all the wells lie to the east of longitude 27° E. The Devonian section is made up of a lower sandstone unit with minor shale interbeds and an upper shale unit with minor siltstone and sandstone interbeds. The sandstone is fine to coarse-grained and its color ranges from white to brown or pink. The shale is mainly grey or greenish grey. The thickness of the Devonian strata is consistent throughout the region and is in the range of 900 to 1000 m. The lower and upper boundaries are poorly defined and are usually arbitrarily marked. In the few wells where the Devonian rests below the Carboniferous, a disconformity marks the upper boundary.

The presence mainly in the upper shale unit, of marine foraminifers, ostracods, conodonts, acritarchs, brachiopods, bryozoans and echinoderms suggests a marine environment of deposition for at least this part of the section. The sands of the lower unit may have been deposited under fluvial conditions.

Carboniferous

Strata of Carboniferous age were identified in 11 wells. With the exception of the Abu Roash-I well, all other wells lie to the west of longitude 27° E. Although the dominant lithology is clastic, the Carboniferous sediments exhibit lateral variation of facies from north to south. In the north (e.g. Sidi Barrani-I) 20 to 30 m thick, light-colored, fine-grained and dolomitic interbeds are abundant. Southward (e.g. NWD 302-I) the carbonate interbeds disappear and are replaced by grey to brick-red shales. In the south (e.g. Bahrein-I) the section is made up entirely of light colored to reddish, fine to coarse-grained sandstone. The maximum thickness reported for the Carboniferous is 910 m in the Siwa-I well. Other thicknesses range from 680 to 780 m.

Because the Carboniferous was partially penetrated in only five of the 11 wells in which it was reported, the lower boundary is difficult to define. With the exception of the Abu Roash-I well where the Carboniferous rests directly over the basement, the lower boundary is marked by the disconformable contact with the underlying Devonian strata. The upper boundary is marked by the unconformable contact with the overlying marine or continental late lower or early middle Jurassic strata. In only two wells the Carboniferous underlies the Permian.

The presence of rich micro- and macro-fossil assemblages points clearly to the marine nature of the sediments (Said & Andrawis 1961, Abd El Sattar 1983). The Carboniferous shows the first indication during the Paleozoic of the increased influence of marine over terrestrial conditions.

Permian

Permain strata were identified with certainty in two wells: Misawag-I (south of Abu Gharadig basin) and Faghur W-I, and questionably in Agnes-I well. The lithology of the Misawag occurrence is similar to the other Paleozoic sediments; the strata are made up essentially of sandstone, thin shale interbeds and a few coal seams. The lithology of the Faghur W-I occurrence, however, is different; the strata are made up of dolomites and dolomitic limestones with a few thin shale and sandstone interbeds. The thickness of the Misawag occurrence, which was not fully penetrated, is more than 1320 m while that of the Faghur W-I, which rests disconformably on the lower Carboniferous, is only 55 m. In the Misawag well the

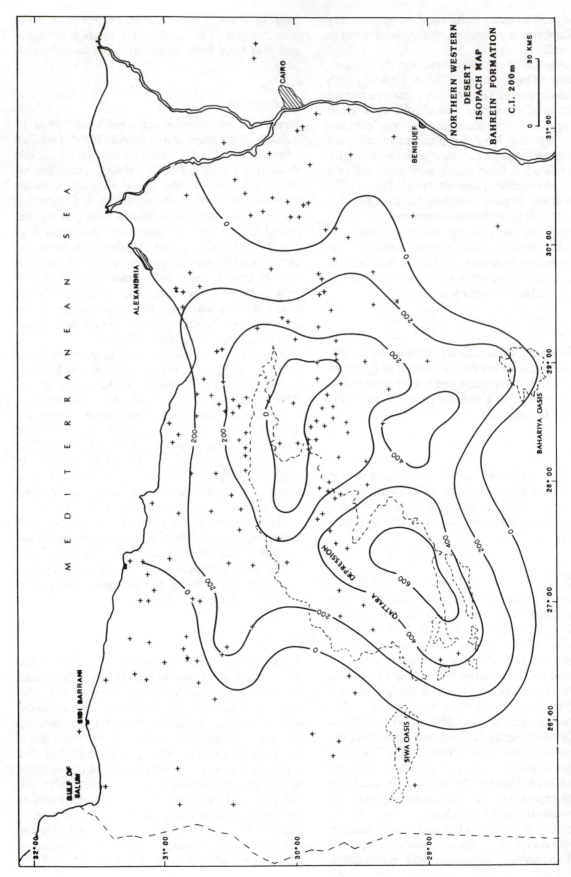

Figure 15.4 Distribution and thickness of Bahrein Formation.

Permian rests unconformably below the Bahrein Formation (continental Jurassic). In the Faghur W-I it rests unconformably below the upper Cretaceous Bahariya Formation.

The Permian occurrences seem to have been deposited in littoral to sublittoral environments. In the case of Misawag, marginal marine supralittoral conditions seem to have prevailed.

MESOZOIC

With the close of the Paleozoic, the north Western Desert, like most other parts of Egypt, formed a positive area until the first Mesozoic transgression of the middle Jurassic. No Triassic or early Jurassic marine sediments are known in the region in spite of the fact that early Jurassic continental sediments are recorded in most parts of the region. The wide-spread nature of these latter sediments indicates most probably that the north Western Desert, along with many other parts of Egypt, formed an area of low relief. As has already been mentioned, the first marine Mesozoic transgression in north Egypt occurred simultaneously with the opening of the Neotethys and the separation of the Apulian microplate. This must have been accompanied by movements that tilted the land of Egypt, elevated the actively subsiding Paleozoic basin of the western part of the north Western Desert and formed the subsiding basin of the Delta region and Sinai. Most of the north Western Desert formed a platform where a shallow open marine environment prevailed for a long time. Minor lateral variations are noted.

During Mesozoic time the clastic supply seems to have come from the east and south. In Sidi Barrani well, the whole Mesozoic section is made up of carbonates in spite of its proximity to the Paleozoic high which stood to the west. This conclusion is based on the assumption that carbonate sedimentation becomes dominant when the clastic supply is interrupted or when the site is far from the sources of this supply.

Jurassic

The subsurface Jurassic deposits of the north Western Desert are classified into the following units (from bottom to top).

Bahrein formation

This new formational name was introduced by WEPCO for the continental Jurassic (and possibly Triassic) sediments which underlie the marine Jurassic strata in many parts of the desert. The name replaces the Eghi Group which was proposed by Norton (1967) for the continental section above the Carboniferous. The name is gaining acceptance and was used in the RRI report (1982).

The type section is the interval between 3888 and 4437 m in the Betty-I well (29° 40' N and 27° 45' E). The name Bahrein was proposed because the name Betty was preoccupied. The formation is made up of red color clastics. In the type well the formation consists of fine to coarse quartzose sandstone with thin pebble interbeds, siltstones and shale, occasionally carbonaceous or pyritic. The formation is remarkably uniform over the whole area. In Yakout-I, a few anhydrite beds are present.

The formation unconformably overlies different units of the Paleozoic or even the basement. The contact with the Paleozoic is easily defined when the latter contains identifiable fossils. In cases where the Bahrein lies on barren Paleozoic sediments, its lower boundary is arbitrarily defined on the basis of the grain size of the clastics which is usually coarser, on the color which is darker and on the shale content which is usually smaller than in the Paleozoic.

The Bahrein Formation lies unconformably below the marine Khatatba Formation. The upper boundary of the Bahrein is easily identified because the Khatatba includes, in its lower part, many argillaceous and carbonate members. In places along the southern and western stretches of the area, the Bahrein rests unconformably below the dominantly clastic lower Cretaceous Betty Formation or the Alam El Bueb member. In these places the upper boundary of the Bahrein is recognised on the basis of the darker color of its clastics.

Toward the eastern and northern parts of the area, the Bahrein Formation grades laterally into the marine formations of Wadi Natrun and Khatatba. To the west, the Bahrein pinches out against the Paleozoic high and grades locally into the Alam El Bueb member.

The isopachs of the Bahrein (Fig. 15/4) show that it thins toward the north, east and west and on predepositional highs. The maximum thickness is reported in Betty-I (550 m).

The Bahrein Formation is of early to middle Jurassic age and possibly older. In Bahrein well the formation is of Callovian age. Along the southern and eastern reaches of the area, the formation may range into the late Jurassic.

The lithology of this dominantly coarse red sandstone deposit points to a subaerial environment of deposition. Sedimentological studies on this formation from the Betty well indicate, however, a possible littoral high energy environment. The presence of anhydrite in Yakout-I well may point to a supratidal to lagoonal environment.

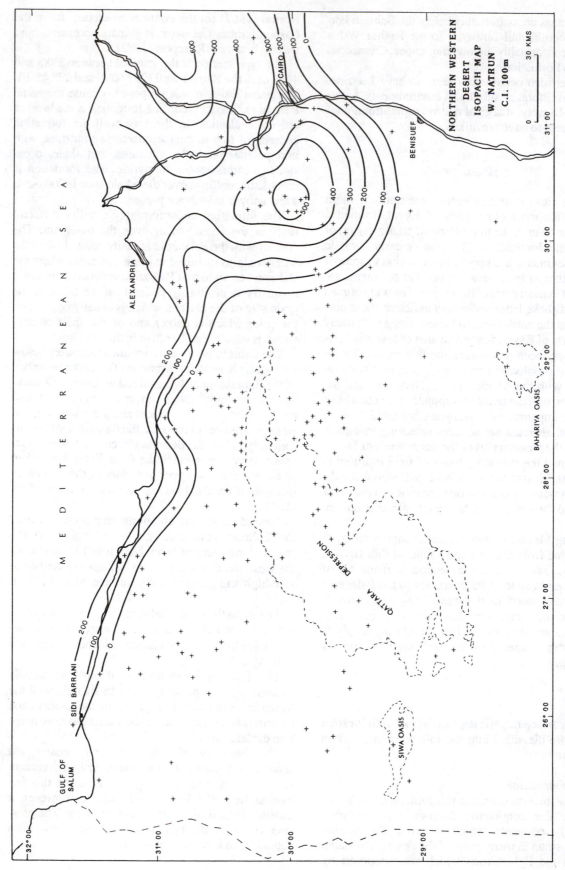

Figure 15.5 Distribution and thickness of Wadi Natrun Formation.

Wadi Natrun Formation

The name Wadi Natrun Formation was proposed by Norton (1967) to include the marine carbonate-shale sequence of middle Jurassic age. The type locality is the interval 3594 to 4056 m in the Wadi Natrun-I well (30° 23' 27" N and 30° 18' 31" E) according to Norton and the interval 3594 to 3620 m according to the RRI report (1982). The carbonates of the section are mostly dolomitic and are more frequent in the upper part of the section. Anhydrite is present in the Gebel Rissu and Kattaniya wells.

The formation rests unconformably over the basement (e.g. Wadi Natrun-I) or the Paleozoic (e.g. Abu Roash-I) or conformably over the Bahrein Formation (e.g. Kattaniya-I). The contact is usually sharp and clear. The Wadi Natrun is always overlain by the Khatatba Formation. The contact between these two formations is drawn on paleontological evidence and on the fact that the Wadi Natrun Formation includes more limestone beds in its top part than the lower part of the overlying Khatatba Formation. As has already been stated the formation grades laterally to the west and south into the Bahrein Formation.

The age of this formation is middle Jurassic although the lowermost part could be of early Jurassic age. The age is based on the presence of a palynomorph assemblage including among others: *Deltoidospora* spp., *Corrollina meyeriana*, *Spheripollenites* spp., *Dichadogonyaulax stauromatos*. Available paleontological and sedimentological evidence indicates that the formation was deposited in a shallow marine low energy environment.

The Wadi Natrun Formation has a limited distribution and is known only in the eastern part of the area and along its northern borders (Fig. 15/5). The maximum thickness reported is 833 m (Natrun T-57-I well).

Khatatba Formation

The name Khatatba Formation was proposed by Norton (1967) for the marine middle Jurassic dominantly clastic section of the north Western Desert. The type locality is the interval 335 to 1536 m in Khatatba-I well (30° 13' 44" N and 30° 50' 07" E) according to Norton and the interval 442 to 1536 m according to the RRI report.

The clastic section of the Khatatba Formation has a few limestone interbeds and is made up of sandstone and shale. The sandstone is fine to medium-grained and is brown in color. The shale is grey to brownish grey in color. The limestone interbeds become thicker and more frequent near the upper part of the section especially in the eastern and northeastern parts of the area. They are typically microcrystalline, argillaceous and in places carbonaceous. Thin coal seams are present at different levels of the section.

The formation rests conformably over the Wadi Natrun Formation in the northeastern and eastern parts of the area and over the Bahrein Formation in all other parts of the area. It underlies conformably the Masajid Formation in most areas except in the south where it underlies unconformably the lower Cretaceous Burg El Arab Formation. The contact with the Masajid is sharp and is marked by the change of facies from the dominantly clastic section of the Khatatba to the more calcareous section of the Masajid. The upper contact, in areas where the Masajid is absent, is difficult to draw. As has already been stated, the formation grades laterally into the continental Bahrein Formation toward the south and west. The age of this formation is middle Jurassic. The environment of deposition is shallow marine except at the southern and western margins where the formation interfingers the Bahrein clastics and becomes more continental in character.

The formation is thickest in the eastern part of the area where a maximum thickness of 1375 m was penetrated in Natrun T-57-I well (Fig. 14/6). The formation thins appreciably along the Alexandria-Ghazalat line and disappears to the west of longitude 27° E.

Masajid Formation

This is typically a massive limestone sequence of middle to late Jurassic age. The name was proposed by Al Far (1966) who designated the surface exposure in Wadi Masajid, north of Bir Maghara, Sinai, as a type locality. The carbonate sequence is cherty in the northeast and central parts of the north Western Desert. Pyrite and carbonaceous matter occur locally. In the northern and eastern parts of the region shale interbeds become more common especially near the base and top of the formation. The formation is clearly marked in most of the region. It overlies conformably the clastic Khatatba Formation and underlies unconformably the clastic lower Cretaceous sequence. In the northeastern part of the region, where the whole Mesozoic section is made up of carbonates, the formation is less marked and its identification is possible only by paleontologic evidence. The Masajid grades laterally toward the south and west into the Bahrein and possibly, at least in part, into the Khatatba.

The formation is thickest in the Wadi Natrun area (Fig. 15/7). It assumes a thickness of more than 450 m in Wadi Natrun-I well. The extent of the Masajid sea was smaller than that of the Khatatba; the 0-isopach contour lies north of latitude 29° N. The Masajid is known to thicken greatly to the east of the Nile delta region. A thickness of 840 m was reported from Q-72-I well. With the exception of the Rabat high, where it is missing, the Masajid has a thickness

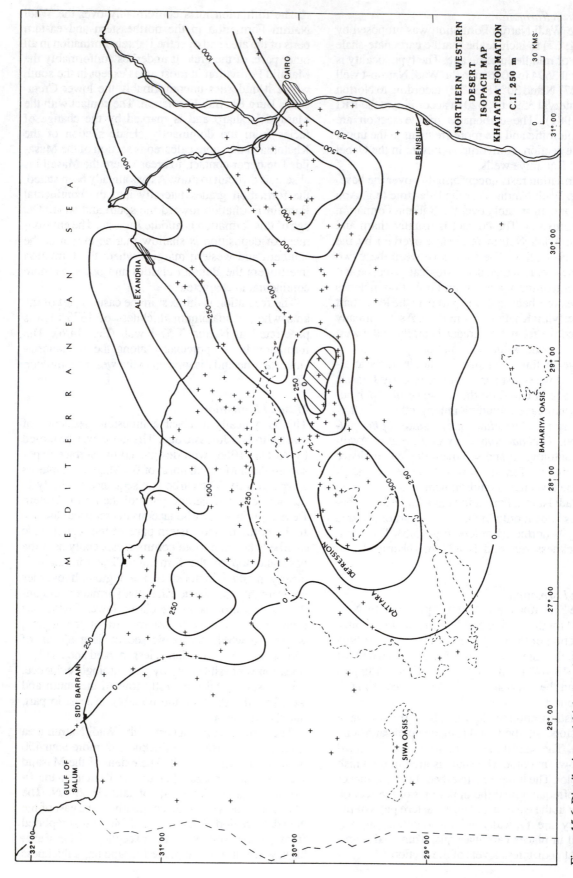

Figure 15.6 Distribution and thickness of Khatatba Formation.

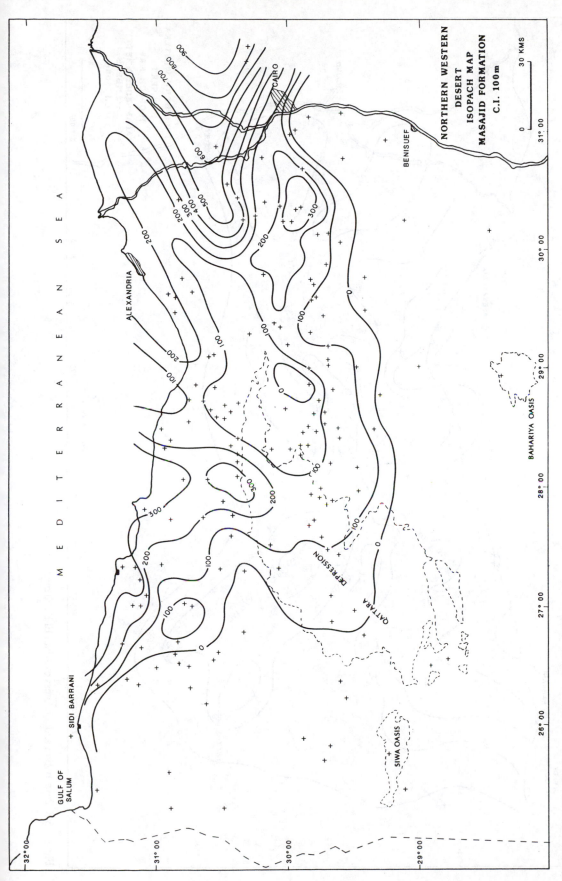

Figure 15.7 Distribution and thickness of Masajid Formation.

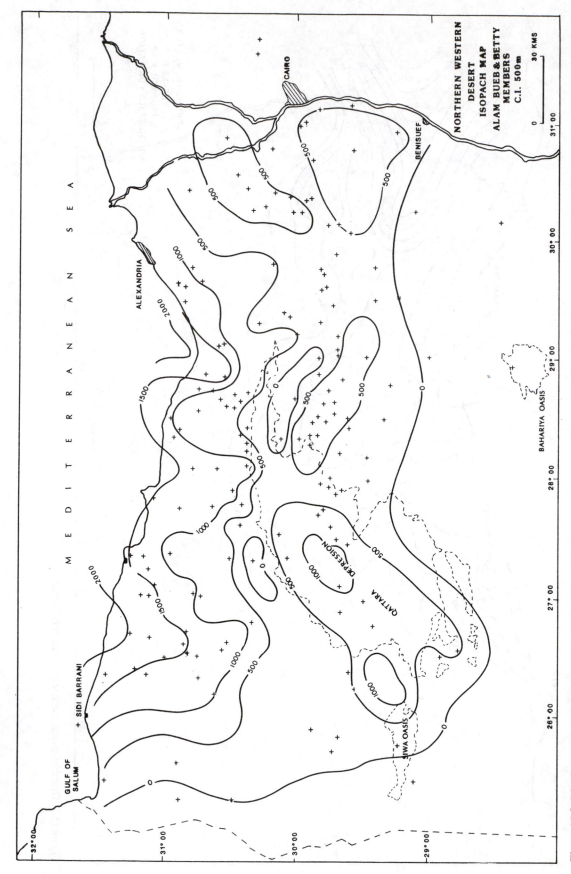

Figure 15.8 Distribution and thickness of Alam El Bueb member.

which ranges from 200 to 400 m.

The formation is of shallow marine origin deposited in a basin which must have undergone mild subsidence under relatively low energy conditions. The age of the Masajid is Callovian in most areas. In the north and east its age ranges from Bathonian to Oxfordian.

Sidi Barrani Formation

The Sidi Barrani is a thick carbonate section of middle Jurassic to early Cretaceous age. Its Jurassic part is coeval with the Masajid Formation, while its upper part is the equivalent to the lower Cretaceous clastics (Betty Formation or part of Alam El Bueb Formation). The type locality is the interval 1899 to 4301 m of the Sidi Barrani well (30° 41' 39" N, 25° 28' 00" E). The carbonates are mainly dolomitic. A few interbeds of sandstone, shale and anhydrite occur at the base of the formation. The formation is easily defined. It overlies conformably the clastics of the Khatatba Formation and underlies the clastics of the Alam El Bueb member of the Burg El Arab Formation.

The Sidi Barrani Formation is limited in its distribution to the northwestern part of the region. It was penetrated in less than 10 wells nearly all of which are located between Matruh and Sidi Barrani. The maximum thickness reported is 2404 m in the Sidi Barrani-I well.

The formation is of open marine origin. It was deposited in a low to moderate energy environment. The rarity of clastics and the abundance of organic material indicate an environment free from turbidity. According to the interpretations of some of the seismic lines, RRI (1982) reports the possible presence of reefal developments in this formation in spite of the fact that none has so far been recognized in the wells drilled in the area.

CRETACEOUS

The Cretaceous is divided into a lower unit made up primarily of clastics and belonging to the lower Cretaceous and an upper unit made up mainly of carbonates and belonging to the upper Cretaceous. The lower unit includes an important carbonate bed of great areal extent, the Alamein dolomite, which provide the reservoir rock for three important oilfields in the region. Except for the local depocenters of the Abu Gharadig, Betty, Shoushan and Gindi areas, the north Western Desert formed a platform of uniform sedimentation with little facies or isopachous variations.

Lower Cretaceous

Burg El Arab Formation

The lower Cretaceous is represented by the Burg El Arab Formation made up of a dominantly thick sequence of fine to coarse-grained clastics. The formation is divided into two units, the Alam El Bueb and Kharita, separated by a widespread carbonate unit, the Alamein, and a less distributed shale unit, the Dahab. The type locality of this formation is the interval 2305 to 4054 m in the Burg El Arab well (30° 55' 20" N, 29° 31' 28" E). The formation has a sharp contact when it overlies conformably the carbonate Masajid Formation. When it overlies unconformably the Wadi Natrun, the Bahrein, the Paleozoic or the basement the contact is less sharp. The formation conformably underlies the Bahariya Formation which is characterized by having a carbonate bed at its base providing an excellent boundary and a valid seismic marker.

Except in the local depocenters of Matruh, Abu Gharadig, Gindy and Betty, the thickness of the formation ranges from 500 to 2000 m. There is a regional increase of thickness toward the north. The maximum thickness is reported from Matruh-I well (3057 m).

The formation is divisible into four members, from bottom to top: Alam El Bueb (or its lateral equivalent the Matruh), Alamein, Dahab and Kharita.

Alam El Bueb member. The Alam El Bueb member underlies the well-marked Alamein dolomite member. It is a sandstone unit with frequent shale interbeds in its lower part and occasional limestone beds in its upper part. The limestone beds become thicker and especially abundant in the northwest. The type locality is the interval 3927 to 4297 m of the Alam El Bueb-I well (30° 38' 39" N, 29° 08' 37" E). RRI (1982), however, proposes Almaz-I well (interval 2902 to 3712 m) as the type locality for this member.

The Alam El Bueb member includes units that were given different names by different operators such as Matruh Group, Aptian clastics, Alamein shale, Dawabis, Shaltut, Umbaraka, Mamura and operational units A, B, C, D1, D2, E, F1 and F2. Also some of the reservoirs in the Umbaraka oilfield were given depth assignments (e.g. 10,700' sand). The lower part of this unit was also distinguished as a separate formation, the Betty.

The member ranges in age from Barremain to Aptian with at least one time gap identified within this unit. The environment of deposition was shallow marine with more continental influence toward the south.

The isopachous distribution of this member is

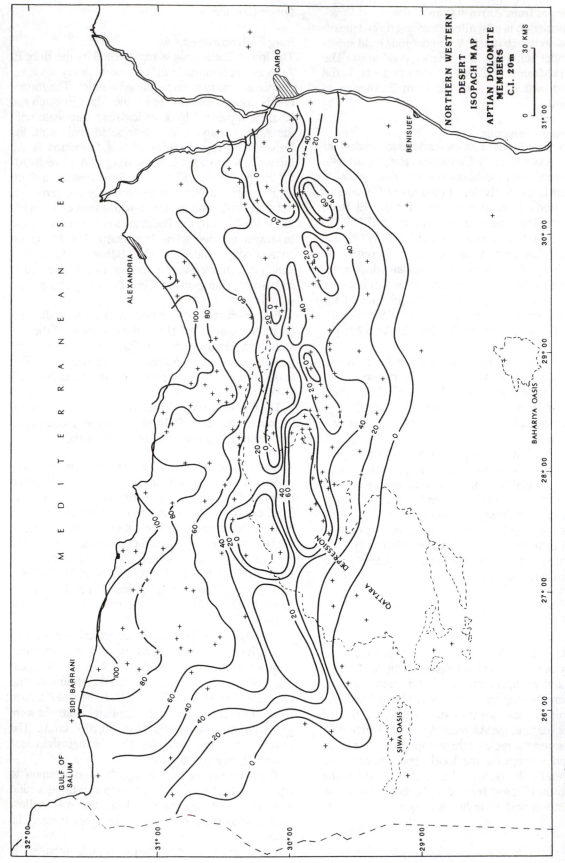

Figure 15.9 Distribution and thickness of Alamein Dolomite member.

shown in Figure 15/8. The unit is missing south of latitude 27° N and in the Rabat area, Sharib-Sheiba high and in Zarif well. The maximum thickness reported is 1820 m (Alamein-I well). In the rest of the area, the thickness ranges from 200 to 1200 m.

Matruh shale member. This is a dark brown to dark grey pyritic calcareous shale unit which includes carbonaceous and lignitic layers in its upper part. Very few carbonate and sandstone beds are reported. The unit is coeval with the Alam El Bueb Member and is limited in distribution to the Matruh area. The type section is the interval 2585 to 4572 m in the Mersa Matruh-I well. The base of this unit was reached only in the Siqeifa-I well where it rests conformably over the Sidi Barrani Formation. The age of this formation is Neocomian to Aptian. The environment of deposition was shallow marine with more continental influence near the top.

Alamein dolomite member. This is a widely spread unit known all over north Africa and Arabia. It is made up of light brown hard microcrystalline dolomite with vuggy porosity. A few thin shale interbeds are present. The unit contrasts with the clastics of the Alam El Bueb and Dahab members which underlie and overlie it respectively. The type section of this member is the interval 2489 to 2573 m of the Alamein-I well (30° 36' 39" N, 28° 43' 52" E). The Alamein dolomite grades laterally into shales along a narrow belt that runs to the north and parallel to latitude 30° N between longitudes 31 and 27° E and in the Matruh area.

The isopachous distribution (Fig. 15/9) shows a uniform deposit with thicknesses in the range of 20 to 80 m over most of the area except in the north where maximum thicknesses are reported, Kanayis-I (97 m) and Alamein E-I (92 m). The unit seems to have been deposited in a shallow marine, low to moderate energy environment. The age of the Alamein dolomite is Aptian to Albian in the type area. In the Umbaraka area, RRI (1982) assigns a Neocomian to Barremain age to this member and suggests, therefore, to give this unit a new name, the Umbaraka dolomite, a suggestion which has not been followed by any of the operators.

Dahab shale member. This is a grey to greenish grey shale unit with thin interbeds of siltstone and sandstone. It rests conformably over the Alamein dolomite and under the sandy Kharita member. The type section is the interval 3180 to 3354 m of the Dahab-I well (30° 48' 24" N, 28° 45' 25" E).

The Dahab Member is relatively thick in the Alamein area. The maximum thickness is at the type well (174 m). In other areas, it is usually thin. The thickness of the Dahab shale is of relevance to oil exploration work. Faults with a throw exceeding the thickness of the Dahab shale will adversely affect the underlying Alamein reservoir. The age of the Dahab shale is Aptian to early Albian.

Some operators lump the Dahab with the Alamein dolomite in one unit, the Alamein Formation. Earlier classifications (Metwalli & Abdel-Hady 1975) lumped the Dahab and the overlying Kharita under the name Abu Subeiha.

Kharita member. This is a unit of fine to coarse-grained sandstone with subordinate shale and carbonate interbeds. The carbonate interbeds increase in frequency and thickness toward the northwest. A certain amount of amorphous silica is always associated with the section and helps characterize this unit. The type section of this member is the interval 2501 to 2890 m of the Kharita-I well (30° 33' 48" N, 28° 35' 32" E). The Kharita formed the upper part of the now defunct Abu Subeiha Formation.

This member rests conformably over the Dahab shale or unconformably over older units from lower Cretaceous to basement. It underlies everywhere the widely-spread Bahariya Formation at the easily recognisable contact provided by the limestone bed which is persistently present at the base of the Bahariya Formation.

The maximum thickness of this unit was reported from the Mersa Matruh-I well (1100 m); it must have accumulated in a local graben. Most other thicknesses are considerably less. Figure 15/10 is an isopach map of the combined thicknesses of both the Dahab and Kharita members. It clearly shows, for the first time, the presence of a number of depocenters extending parallel to the south of latitude 30° N including the Gindi, Abu Gharadig and Betty areas. It also shows a better definition of the Shoushan and Bahariya basins.

The Kharita is assigned an Albian to Cenomanian age. In the Umbaraka area, where the underlying dolomite is given a Neocomian to Barremian age, the unit may be of Aptian age at least in its lower part. The unit was deposited in a high energy shallow marine shelf. In the extreme north, the unit seems to have been deposited in deeper water, while in the south it was under the influence of continental conditions.

Upper Cretaceous

In Egypt the upper Cretaceous marks the beginning of a major marine transgression which resulted in the deposition of a dominantly carbonate section (Middle Calcareous Division of Said 1962). In the north Western Desert, the mainly calcareous sediments of

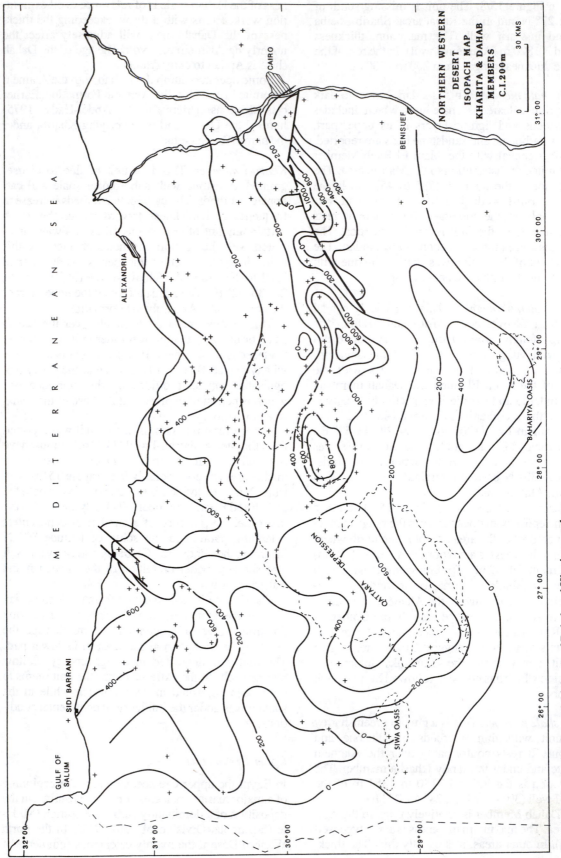

Figure 15.10 Distribution and thickness of Dahab and Kharita members.

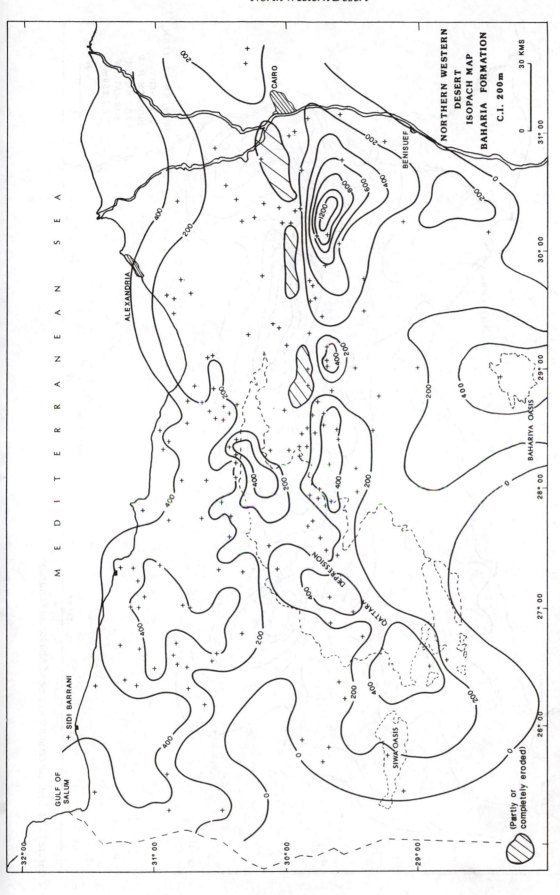

Figure 15.11 Distribution and thickness of Bahariya Formation.

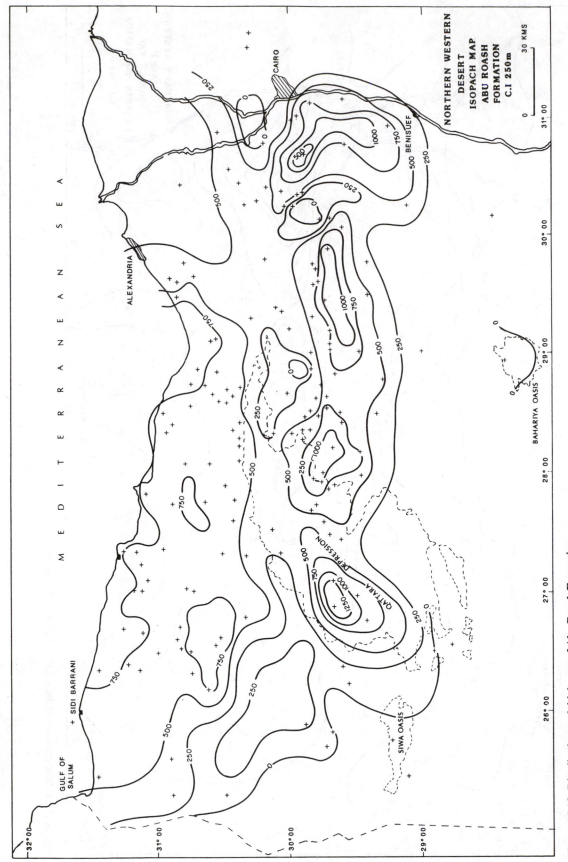

Figure 15.12 Distribution and thickness of Abu Roash Formation.

the upper Cretaceous are particularly developed in the Abu Gharadig basin where they form a number of oil reservoirs. These sediments are divided into three rock units which are from bottom to top: Bahariya, Abu Roash and Khoman.

Bahariya Formation

The Bahariya is exposed along the floor and both sides of the Bahariya depression. It has been recently subjected to a comprehensive study by Dominik (1985). The exposed section measures at least 170 m and is divisible into three members (from bottom to top): Gebel Ghorabi, Gebel Dist and El Heiz. The more calcareous El Heiz was first described and given formational status by Akkad & Issawi 1963. In many wells of the north Western Desert, member 'G' of the overlying Abu Roash Formation could well be equivalent to the El Heiz Formation. The Bahariya Formation is of late Cenomanian age and according to the extensive studies of Dominik (1985), was deposited first under fluviatile conditions (Gebel Ghorabi Member), then under estuarine conditions (Gebel Dist Member) and finally under lagoonal conditions (El Heiz Member). The Gebel Ghorabi Member is made up of cross-bedded, coarse-grained, seemingly non-fossiliferous sandstones, while the Gebel Dist Member is made up of fine-grained, well-bedded, ferruginous clastics carrying a large number of fossils including vertebrates in the lower levels and an assortment of oysters and other fossils in the middle and upper levels. The El Heiz is made up of dolomites, sandy dolomites and calcareous grits rich in fossils. The base of the formation is unexposed and its downward limit in the Bahariya-I well is questionable. RRI (1982), assigns the upper 216 m to this formation, while Norton (1967) assigns the top 170 m to it.

The thicknesses given in Figure 15/11 are for the Bahariya Formation without the El Heiz Member which is considered in this chapter as forming part of the overlying Abu Roash Formation. The maximum thickness occurs in the Kattaniya-I well (1143 m). Over the rest of the area, the thickness ranges from 50 to 500 m.

Operators had previously given several names to this formation: Razzak sand, Meleiha sand or Medeiwar Member of the Abu Subeiha Formation.

Abu Roash Formation

This is mainly a limestone sequence with interbeds of shale and sandstone. The type locality is the classic Abu Roash structure to the north of the Pyramids of Gizeh, which has made the subject of a large number of studies since the classic work of Beadnell (1902). The unit is divided into seven members designated from bottom to top: G, F, E, D, C, B and A. As has been already stated, the lowermost member is probably coeval with the El Heiz Member of the Bahariya Formation. Members F, E, D, C and B are probably the lateral equivalents of the exposed units in the Abu Roash structure described under the names: Rudistae series, limestone series, *Acteonella* series, flint series and *Plicatula* series respectively (Said 1962). Member A is not recognised in the surface in the Abu Roash structure and was probably eroded away. In the subsurface the members are well defined and are easily traced. Members B, D and F are relatively clean carbonates while members A, C, E and G are largely fine clastics. The limestone is cryptocrystalline, grey, brown or white, locally argillaceous and dolomitic. Dolomite is more frequent in the north and east. The shales are grey to green, sometimes richly glauconitic and pyritic. They grade into carbonates to the north. Minor sandstone interbeds are reported and these usually grade into shales rather abruptly. The sandstone is fine-grained, argillaceous and calcareous. The clastic intervals as well as members D and F form the reservoir rock of the Abu Gharadig basin. Members D and F develop secondary porosities due to dolomitization.

The Abu Roash Formation overlies conformably the Bahariya Formation. In the northern part of the area, where most of the formation is made up of limestone, the contact is sharp and easily recgonised. In the south, where the lower unit G (El Heiz) is made up of interbedded carbonates and clastics, the boundary is taken at the top of the thick sand interval of the Bahariya Formation. The Abu Roash underlies the Khoman Formation where the contact is determined, in the absence of fossils, by the change of lithology from the crystalline limestones of the Abu Roash to the chalky limestones of the Khoman.

The age of member G is late Cenomanian, while the age of the overlying members ranges from the Turonian to Santonian. Many sedimentary gaps exist. With the exception of unit G, which is of lagoonal origin in the south, the formation was deposited in an open shallow marine shelf during several sedimentary cycles which developed in response to oscillations of sea level.

Like the Bahariya Formation, the Abu Roash Formation is best developed in a number of basins that run parallel and to the south of latitude 30° N, namely the Gindi, Abu Gharadig and Betty. In the WD-19-I well in the Gindi basin the thickness is 1916 m, while in the Ghourab-I well in the Betty basin the thickness is 1814 m. In the Abu Gharadig basin the thickness is more than 1000 m. Over the rest of the area the thickness ranges from 250 to 750 m (Fig. 15/12).

Khoman Formation

The Khoman Formation is a distinctive unit of snow-

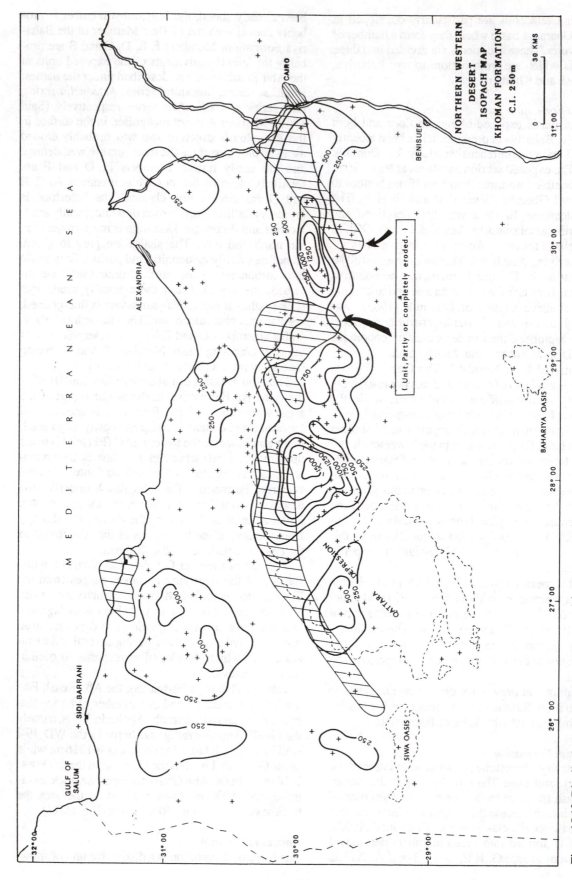

Figure 15.13 Distribution and thickness of Khoman Formation.

white chalk and chalky limestone with abundant chert bands. The type section is the scarp west of Ain Khoman to the southwest of Bahariya Oasis (27° 55' N, 28° 30' E). The formation has in many places a thin shale bed at the base. This bed interfingers the limestone and forms a thick unit of interbedded shales and limestones in the Abu Gharadig basin, justifying the division of this unit into two members: A and B.

The formation overlies unconformably different units of the Abu Roash or the Bahariya and is easily distinguished by virtue of its lithology and stratigraphic position. The formation underlies unconformably the Apollonia or Dabaa Formations. The contact between the Khoman and the overlying formations is again easily recognised due to the contrast between the chalky nature of the Khoman and the dense crystalline nature of the Apollonia.

The distribution and thickness of this formation (Fig. 15/13) are structurally controlled and seem to have been affected by: (1) the type of block movements which took place at the close of the time of deposition of the underlying Abu Roash Formation producing a number of high and low fault blocks that straddled latitude 30° N; and (2) the active subsidence of many syndepositional faults which developed under the weight of the thick Khoman sediments in the Natrun, Gindi, Betty and Abu Gharadig basins. The maximum thickness of the Khoman is reported from the WD-7-I well (+ 1644 m). To the north of these basins, the Khoman is either missing or in the range of 20 to 100 m.

The Khoman Formation is of Maastrichtian age. It was deposited in open marine outer shelf conditions. The deposition of the Khoman was associated with a rise of the sea level which extended the sea to the south of the north Western Desert and brought about a substantial change of facies.

CENOZOIC

During the Paleogene, open marine conditions continued to extend further south into upper Egypt. Paleocene deposits are mainly mudstones which, in north Egypt, were reduced or mainly removed during the early to middle Eocene. The distribution of the Eocene rocks bears evidence of this syndepositional tectonism. Thinner shallow-water cherty bioclastic limestones or dolomitic limestones (Apollonia Formation) were deposited on or around the highs of the north Western Desert, with chalky limestones in the basins.

During the late Eocene-Oligocene, thick open marine calcareous shales (Dabaa Formation) were

deposited in the northern reaches of the Western Desert. To the south, at the latitude of the Fayum, littoral marine to continental clastics (Qasr El Sagha and/or Abu Muharik Formations) were deposited during the late Eocene and were followed by the deposition of continental to deltaic deposits (Qatrani Formation) during the Oligocene.

The Neogene opened with a recession and the deposition of neritic shallow water calcareous shales (Mamura Formation) in the extreme northern reaches of the north Western Desert. These grade to the south into the delta-front turbiditic clastics of the Moghra Formation and the fluvial sediments of the Gebel Khashab redbeds. The middle Miocene shallow water carbonates of the Marmarica Formation cover a large part of the north Western Desert. No late Miocene sediments are recorded.

Marine Pliocene deposits fringe the coast of the Mediterranean and are in the form of shallow marine pink limestones or their lateral equivalent, the littoral to lagoonal sandy limestones and evaporites of the Hagif Formation. A continental to lacustrine deposit (Minqar El Talh Formation) is reported along the west Qattara rim and is probably related to the old drainage of the Pliocene. During the Pleistocene, calcareous ridges of kurkar were formed along the coast in response to fluctuating sea levels (Alexandria Formation).

In the following paragraphs a description is given of the Paleocene to middle Miocene rock units from bottom to top.

Apollonia Formation

This is a Paleocene to middle Eocene limestone unit with subordinate shale members. The type section is at the hills south of the village of Apollonia (Libya). The type section is 250 m thick and is made up of massive siliceous limestones with numerous chert bands. In the north Western Desert, the section consists predominantly of white, light grey or brownish grey nummulitic limestone and minor shale beds. Glauconite, pyrite and chert nodules are common particularly in the lower part of the section. The limestone is dolomitized in the northeastern parts of the region.

Over the largest part of the north Western Desert the formation overlies unconformably the Khoman chalk. A significant hiatus covering the entire Paleocene (and sometimes the early Eocene) exists in most sections. In the Natrun, Gindi and Abu Gharadig basins, however, the relationship of this formation to the underlying Khoman is conformable. In these areas the age of the Apollonia ranges from Paleocene to middle Eocene. The Paleocene section is usually made up of limestone alternating with thin layers of shale especially toward the deeper parts of the basins

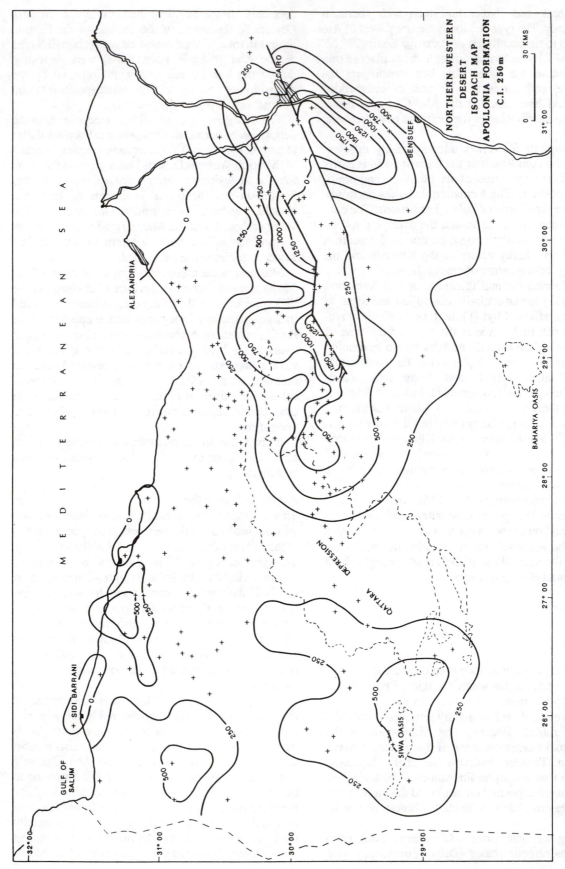

Figure 15.14 Distribution and thickness of Apollonia Formation.

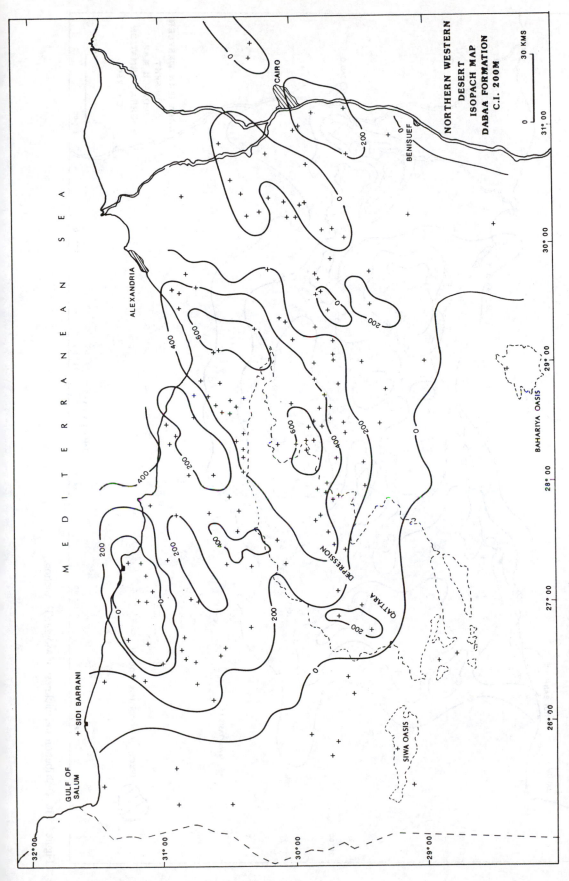

Figure 15.15 Distribution and thickness of Dabaa Formation.

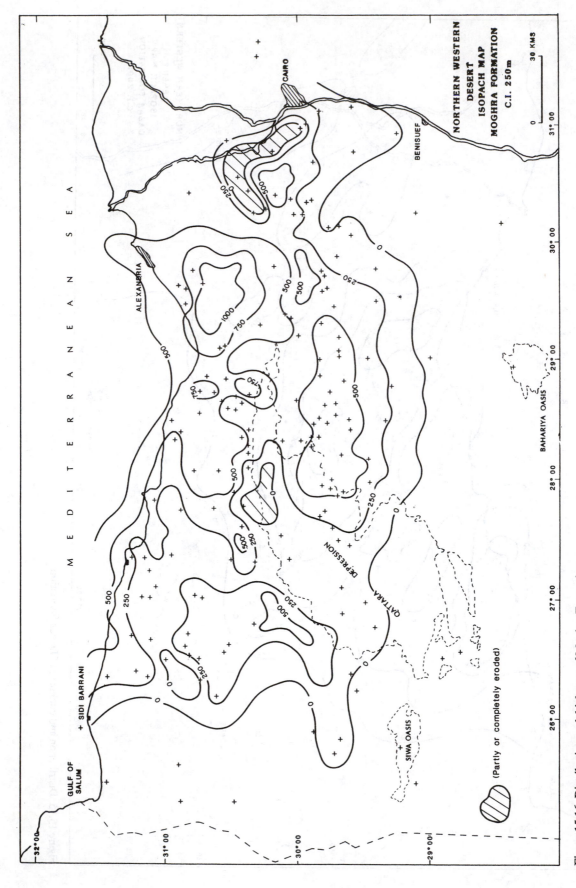

Figure 15.16 Distribution and thickness of Moghra Formation.

(Elzarka 1986). The formation is conformably overlain by the Daba'a Formation.

Figure 15/14 gives the thickness distribution of the Apollonia. The formation is particularly thick in the Gindi Basin (1788 m), in a number of depocenters straddling latitude 30° N (up to 1300 m) and along the westernmost part of the area (up to 550 m). Over the rest of the area, the formation is usually less than 100 m in thickness. It is missing over the Kattaniya high, west of the northern delta, Matruh and Sidi Barrani areas.

The formation has been previously described as the Gindi Formation or as the Esna, Thebes and Mokattam Formations. The name Apollonia is probably suitable for the shelf limestones of the north Western Desert. The thicker deposits of the basinal areas probably deserve a more refined classification.

Daba'a Formation

The Daba'a Formation is a subsurface unit of upper Eocene-Oligocene marine shales. The name was proposed by Norton (1967) and has been widely used since then. The name El Daba'a given by Abdallah (1967) to the exposed Pliocene pink limestones of the Mediterranean littoral will be considered a homonym. This formation has been previously given various names: Qasr El Sagha, Maadi, Birqet Qarun and Gehannan.

The type section is the interval 579 to 1021 m of the Dabaa-I well (31° 01' 19" N, 28° 29' 42" E). RRI proposes, however, the interval 729 to 999 m of the Almaz-I well. The formation is of uniform lithology throughout the north Western Desert. It is made up of light grey to greenish grey shales with subordinate thin beds of limestone.

The formation rests on the Apollonia Formation with a minor disconformity and with a sharp contact. It is conformably overlain by the early Miocene, the Moghra or the Mamura Formations. The environment of deposition was inner shelf to littoral which became estuarine toward the top in many areas. Toward the south the Dabaa grades laterally into the well-known littoral to deltaic deposits of the Fayum region.

Figure 15/15 gives the isopachous contours of this formation. It shows that the northern part of the region was shaped, for the first time, into a number of lows and highs rather than forming a platform. A northeast-southwest depocenter is formed between Mariut and Abu Gharadig and a less conspicuous center is found to the south of Mileiha. Highs are found in the Ganayin, Mileiha and Matruh-Zayid areas. The maximum thickness of this formation is reported from WD-7-I well (828 m) and from Fayad-I (723 m). Over the rest of the area the thickness ranges from 200 to 400 m.

Moghra Formation

The Moghra is a clastic fluvio-marine delta-front sequence of early Miocene age which grades laterally to the north and west into the marine Mamura Formation and to the south into the genuine fluviatile redbeds of Gebel Khesheb. The type section is the classic surface section of Moghra at the extreme eastern tip of the Qattara depression where Said (1962) describes a 230 m section (base unexposed) of sandstone, siltstone and calcareous shale with vertebrate remains and silicified tree trunks. In the subsurface, this widely distributed unit includes, in places, more shale and anhydrite beds. The Moghra overlies the Dabaa and underlies the Marmarica. The thickness distribution of this formation is given in Figure 15/16.

Mamura Formation

The Mamura is a limestone and calcareous shale sequence which is the marine equivalent of the Moghra. The name was proposed by Marzouk (1970). The type section is the interval 114 to 401 m of the Mamura-I well (31° 30' 04" N, 26° 15' 21" E). The formation is of uniform lithology. It rests above the Dabaa Formation and is conformably overlain by the middle Miocene Marmarica Formation.

The Mamura sea seems to have submerged the Matruh-Bir Khamsa area which formed a positive area during the late Eocene-Oligocene time. The Alam El Bueb depocenter, which started forming during the late Eocene-Oligocene time, became conspicuous during Mamura time. The maximum thickness reported for this formation is in Dahab-I well (964 m).

Marmarica Formation

The Marmarica is a limestone, dolomite and shale sequence of middle Miocene age. The type locality is the scarp to the north of Siwa where the surface section is 78 m thick (base unexposed) and is made up of limestone and marl with several oyster banks and rich neritic and reefal assemblages (Said 1962). Drilling near the type locality and in the Diffa plateau shows that the formation overlies the Mamura. In all other areas, it overlies the Moghra Formation.

The formation covers the Marmarica plateau of the north Western Desert and is overlain, along the Mediterranean littoral, by a thin shallow marine unit of pink limestones which was named by Abdallah (1967) the El Daba'a Formation. To the southeast of the Gulf of Arabs enclave, the pink limestone unit grades into a more marginal unit with anhydrite, the Hagif Formation.

CHAPTER 16

Wadi Qena: Paleozoic and pre-Campanian Cretaceous strata

E. KLITZSCH, M. GROESCHKE & W. HERRMANN-DEGEN
Technical University Berlin

The geology of Wadi Qena is characterized by an evident northward increase of marine influence within the middle to late Cretaceous strata. Paleozoic rocks are present in northern Wadi Qena and between there and Wadi Dakhal. They are the remains of a Paleozoic cover whose distribution was more or less controlled by older pre-Cenomanian structural events (Fig. 16/1).

The situation along the west side of the north Red Sea and the Gulf of Suez is similar, but exposures there are less complete, due to the structural development which formed this graben system. In both areas, marine Cenomanian strata interfinger increasingly southward with deltaic and fluvio-continental sandstone. In the same direction, strata also become younger, for example, the thick Cenomanian (to possibly Albian) sandstone of Wadi Kibrit at Gebel Zeit and of northern and central Wadi Qena thins out toward the Qena-Hurghada Hinge Line and is replaced by a fluvial sandstone unit of seemingly Turonian to Campanian age. The Aptian transgression and its prograding underlying sandstone units, which are present over large areas in the Western Desert, is absent in the whole Eastern Desert and the Red Sea-Gulf of Suez area. In early Cretaceous time, these areas were part of a large regional high.

A PALEOZOIC STRATA OF NORTH WADI QENA

1 *Araba Formation (Cambrian)*
At Somr El Qaa (around 28° 13' and 32° 22'), the Precambrian basement is unconformably overlain by a very heterogeneous sequence of clastic sediments filling a relatively uneven morphological relief. The basal Araba Formation is up to 25 m thick and consists of immature arkosic conglomerate (mainly near the base), fine-grained to very coarse beach sand, silty and/or sandy shale (partly bioturbated) and fine- to medium-grained well-bedded marine sandstone. The base contains rare to very abundant trace fossils, mainly different species of *Cruziana*

(trilobite tracks). This fauna is very similar to that in the basal strata at Umm Bogma where there is a very rich assemblage of *Cruziana* sp. According to A. Seilacher (pers. comm.), this fauna is of early Cambrian age.

2 *Somr El Qaa Formation (Carboniferous)*
In the area of Somr El Qaa, up to 150 m of sandstone unconformably overlie the Araba Formation. These strata consist of cyclical sequences of fine- to coarse-grained fluvial to deltaic sandstone, shallow marine sandstone with *Scolithos* bioturbations and mudstones. They were evidently deposited in a shallow and fluctuating sea. South of the Somr El Qaa area, this sequence is only present (together with thin remains of the Araba Formation) in local patches. It is likely that its exceptionally large thickness at Somr El Qaa (and north of there) is due to a large and distinct Precambrian fault striking west-northwest to northwest.

Toward Wadi Dakhal, both the Araba and Somr El Qaa Formations disappear locally, and in some locations, Wadi Qena sandstone of Cretaceous age rests directly on the basement. In the southwestern part of Wadi Dakhal, however, the basal part of the Somr El Qaa Formation contains plants of early Carboniferous age: *Bothodendron* aff. *depereti* Vaffier, *Knorria* sp., *Lepidodendron spetsbergense* Nathorst, *Lepidodendron veltheimii* Sternberg, *Lepidophylloides* sp., *Leptophlocum rhombicum* Dawson, *Precyclostigma blakaense* Lejal, *Precyclostigma* sp., *Pseudolepidodendropsis scobiniformis* (Meek) Lejal, *Pseudolepidodendropsis klitzschii* Lejal.

These plant-bearing strata are overlain by approximately 50 m of white and very mature sand (mined as glass sand at Wadi Dakhal). Some of these beds are full of *Scolithus* structures, but the majority are cross-bedded fluvial sandstone. They are overlain by 10 to 15 m of well-stratified fine-grained sandstone, siltstone and silty shale, containing impressions of Vrachiopodes (*Productus* sp., *Spirifer* sp.), trace fossils (*Scolithos*, *Chondrites*, tracks of sea urchin) as

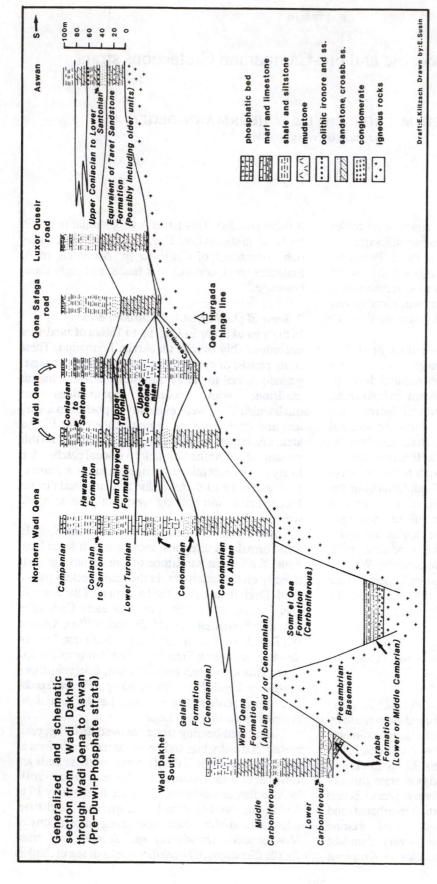

Figure 16.1   Generalized and schematic sections from Aswan to Wadi Qena (pre-Duwi phosphate strata).

well as fossil plants like *Nothorhacopteris argentinica* Archangelsky. The latter indicates a middle Carboniferous age (Late Viséan to Early Namurian).

These strata are overlain by white sandstone of early to middle Cretaceous age (Wadi Qena Formation).

B: PRE-CAMPANIAN CRETACEOUS STRATA BETWEEN WADI DAKHAL AND ASWAN

In the southern part of central Wadi Qena (between 27° N and 27° 20' N), marine strata of late Cenomanian age (Galala Formation) overlie a very thin sheet of fluvial sandstone (in general 5 to 15 m) and, in some locations, the marine sediments rest directly on Precambrian basement. Toward the north, the thickness of the lower sandstone gradually increases (Wadi Qena Formation), and in Wadi Dakhal the upper part of this fluvial to deltaic sequence interfingers with marine sediments.

South of approximately 27° N, this fluvial to deltaic sandstone pinches out between Precambrian basement and marine late Cenomanian strata. South of approximately 26° 53' N, the latter strata disappear and are replaced by a sandstone unit of Turonian age which directly overlies the Precambrian basement (Umm Omeiyid Formation which is an equivalent of the Taref Sandstone). As during Cenomanian time, continental influence increased southward during Turonian to Campanian time. It seems that the most extensive transgression in the Eastern Desert-Red Sea area was the transgression of the Coniacian. In late Coniacian, a short marine interval reached as far south as to the Aswan area. Oolitic iron ore and oolitic sandstone were found at approximately the same stratigraphic level and within near-shore environment in many places between north Wadi Qena and the Aswan area. At Wadi Qena as well as at Aswan, paleontological evidence suggests a Coniacian age for this episode.

Some general characteristics of the area between north Wadi Qena and Aswan are very similar to those along the Gulf of Suez-Red Sea coast (see Figs 16/2 and 16/3).

The thick basal Cretaceous Sandstone Formation of Wadi Kibrit (Gebel Zeit) pinches out southward. Overlying strata of late Cenomanian to Campanian age show increasing fluvial influence toward the south. It seems that there was a distinct margin (for example, a hinge line) in Cenomanian time, striking northeast from the southern part of Wadi Qena toward the area north of Hurghada. The areas south of this line were areas of erosion until late Cenomanian time. After the Cenomanian, they became sites of sedimentation. North of the hinge line, a thick sequence of sandstone was deposited during Albian and Cenomanian time (coastal flood plain and deltaic environment).

1 *Wadi Qena Formation* (? Albian to Cenomanian)
This formation forms a distinct escarpment on the western side of Wadi Qena from approximately 27° 27' N northward to the area of Somr El Qaa. Between there and the Wadi Dakhal area, this sandstone is underlain by Paleozoic strata and can be confused with parts of these older sediments, especially because much of the landscape is an irregular type cuesta. The formation is probably coeval with the Malha Formation known from Sinai and the Gulf of Suez.

In Wadi Umm Omeiyid, Wadi Qena (approximately 27° 34' N and from there east and westward), the Wadi Qena Sandstone Formation rests unconformably on Precambrian basement and is conformably overlain – in most places – by marine strata of the Galala Formation. The thickness of the formation at this location is in the order of 80 to 100 m, increasing northward and decreasing southward. The lower 50 to 70 m are made up of fine- to coarse-grained white sandstone with some conglomeratic beds. The sandstone is very mature: most of it is tabular to partly trough cross-bedded with some convolute bedding, and it is mainly of fluvial to deltaic origin. Some thin intervals are burrowed (*Scolithos*-type bioturbations). The direction of transport is mainly between north and west. Silicified wood and other plant remains are locally frequent in the lower part of this unit.

The upper 20 to 30 m consist of parallel bedded fine-grained sandstone, partly silty, with intercalations of tabular cross-bedded sandstone. Some of this sandstone is highly burrowed and contains fragments of bones and silicified wood. This upper part represents transitional fluvial to nearshore marine environment at the beginning of the Cenomanian transgression.

Further north, for example, southwest of Somr El Qaa, this upper part of the Qena Sandstone is totally marine. There, the underlying fluvial strata in its upper part are intensively and deeply burrowed, which indicates more or less continuous sedimentation. The Wadi Qena Formation, therefore, is most likely of Cenomanian (possibly Albian to Cenomanian) age. It is possibly an equivalent of the Sabaya Formation (Desert Rose Beds) of the Western Desert.

2 *Galala Formation (late Cenomanian to early Turonian)*
The Galala Formation represents the fully marine environment of the Cenomanian transgression. It

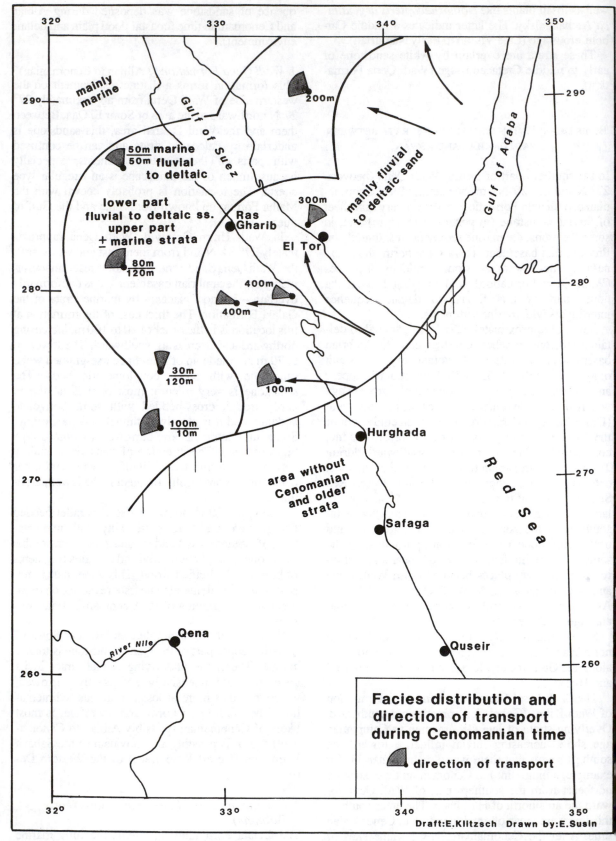

Figure 16.2 Facies distribution and direction of transport during Cenomanian time.

thickens northward along Wadi Qena and toward Wadi Dakhal and pinches out not far south of 27° N. Along 27° 34' N, it is only 20 to 30 m thick, while southwest and west of Somr El Qaa in north Wadi Qena, it attains a thickness of 70 to 90 m.

The Galala Formation is made up of silty shale, sandy marl, fossiliferous limestone and locally up to 18 m thick oyster beds (mainly near the top west of Somr El Qaa).

The basal part contains intercalations of highly burrowed fine-grained shallow marine sandstone (*Teichichnia* sp.). The presence of ammonites at different levels in south-central and central Wadi Qena indicates a late Cenomanian to early Turonian age. Fauna from the Galala Formation directly above the Wadi Qena Sandstone in south-central Wadi Qena between 27 and 27° 20' N includes: *Angulithes mermeti* (Coquand), *Metengonoceras acutum* (Hyatt), *Neolobites brancai* Eck, *Exogyra africana* Lamarck, *E. conica* Sowerby, *E. olisponensis* Sharpe, *E. suborbiculata* Lamarck, *Heterodonta* sp., *Ostrea (Crassostrea) isidis* (Fourteau), *Ostrea biauriculata* Lamarck, *O. ouremensis* Choffat, *Plicatula auressensis* Coquand, *Pterodonta* sp., *Strombus incertus* d'Orbigny, *Tylostoma* sp. (age: late Cenomanian).

A similar fauna is found along 27° 34' N: *Angulithes mermeti* (Coquand), *Metengonoceras acutum* (Hyatt), *Neolobites* sp., *Exogyra africana* Lamarck, *Modiolus* sp., *Neithea aequicostata* (Lamarck), *Pinna* sp., *Strombus incertus* d'Orbigny (see: late Cenomanian). According to Hendriks et al., the upper part of the Galala Formation in this area contains an ammonite fauna of early Turonian age.

In north Wadi Qena, southwest of Somr El Qaa, the lower part of the Galala Formation contains among others: *Exogyra africana* Lamarck, *E. conica* Sowerby, *E. flabellata* Goldfuss, *E. olisiponensis* Sharpe, *E. suborbiculata* Lamarck, *Ostrea (Crassostrea) delletrei* Coquand. At this locality, a limestone bed overlying the oyster reefs in the upper part of the Galala Formation yielded *Mammites* sp. and *Pseudoaspidoceras* sp., which indicate an early Turonian age.

A large part of the fauna collected has not yet been identified. The material investigated, however, proves the given age. The Galala Formation is an equivalent of the Bahariya and Maghrabi Formations of the Dakhla Basin (Western Desert).

3 Umm Omeiyid Formation (late Turonian)

In Wadi Omeiyid, the central part of Wadi Qena, along approximately 27° 34', the Galala Formation is overlain by the Umm Omeiyid Formation which is made up of a lower and an upper trough and tabular cross-bedded sandstone unit (both 15 to 20 m thick)

and a middle unit of shale, silt, marl and marly sandstone (25 to 30 m thick). The sandstone units represent fluvial fans with a northwest and north (locally also east and southeast) direction of transport. The middle part of the formation is of marine origin. It increases in thickness northward until it replaces the fluvial deposits. At the type area at Wadi Umm Omeiyid, most of the sandstone and marly sandstone is highly burrowed (*Thalassinoides* sp., *Teichichnia* sp.), some of the sandstone beds contain plant remains and marly sandstone to sandy limestone beds with frequent ammonites: *Coilopoceras* cf. *multicostatum* Lewy, *C. requienianum* (D'Orbigny). The ammonites indicate a late Turonian age.

It is very likely that the Umm Omeiyid Formation is replaced, in the south Wadi Qena area and between there and Aswan, by the basal fluviatile sandstone formation overlying basement which is known locally as the Abu Aggag Formation. In Aswan, the Abu Aggag is overlain by Coniacian (to possibly early Santonian) marine sediments with beds of oolitic iron ore which are also present above the Umm Omeiyid Formation at its type area. The Umm Omeiyid is probably an equivalent of the Taref Sandstone Formation of the Western Desert.

At present, both the Umm Omeiyid and the Hawashiya Formation (see below) are under detailed investigations by teams of Assiut University and Technical University of Berlin. Results will be published soon.

4 Hawashiya Formation (Coniacian to Santonian)

Along 27° 34' N (central Wadi Qena) the upper sandstone of the Umm Omeiyid Formation is overlain by an approximately 55 m thick sequence of shale, silty shale, marl, sandy marl and some sandstone intercalations. Thin oyster beds are typical and, together with small sea urchins and gastropods, ammonites appear in the upper part of the section. Near the top, *Ceratodus* sp. is also present, indicating shallowing of the sea and regression. The fauna contains among others: *Metatissotia fourneli* (Bayle), *Ostrea (Crassostrea) boucheroni* (Coquand), *O. (C.) costei* (Coquand). The ammonite indicates a middle to late Coniacian age (see also Hendriks et al. 1987).

In north Wadi Qena, southwest of Somr El Qaa, the Hawashiya Formation contains among others: *Lopha dichotoma* (Bayle), *Ostrea (Crassostrea) heinzi* Peron & Thomas.

In Wadi Dakhal, equivalent strata above the sandstone of the Umm Omeiyid Formation contain the following fossils: *Forbesiaster gaensis* Jux, *Lopha dichotoma* (Bayle), *Plicatula ferryi* Coquand, and *Ceratodus* sp.

Further south in Wadi Abu Aggag northeast of

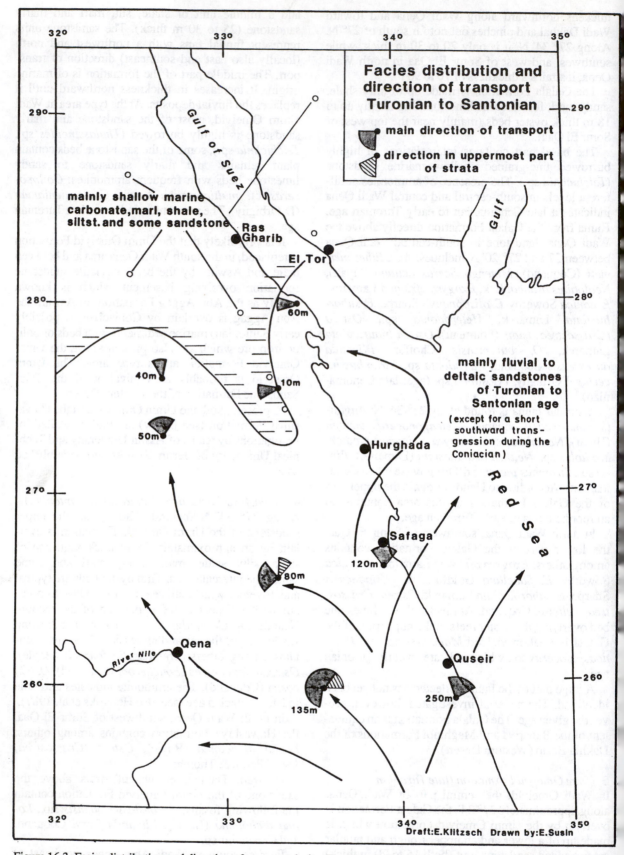

Figure 16.3 Facies distribution and direction of transport during Turonian to Santonian time.

Aswan, an equivalent unit (Timsah Formation) carries the following fossils: *Inoceramus (Platyceramus) cycloides* Wegner and *I. (Volviceramus) balli* Newton. Frequent fossil plants occur also in intercalations of fluvial sandstone and in silty shale: *Brachyphyllum crassicaule* Fontaine, ?Magnolia obtusata Heer, *Magnoliaephyllum* sp. In several places in Wadi Abu Aggag and also southeast of Aswan, tracks of Tetrapoda are present in this fomration.

Along the western side of Wadi Qena, as well as in the Wadi Dakhal area and near Aswan, the Hawashiya Formation and its equivalents are characterized by the occurrence of oolitic sandstone and/or beds of oolitic iron ore.

Before the fauna of Wadi Abu Aggag was given a Coniacian to early Santonian age (E. Seibertz, pers. comm.), the Timsah was given a Cenomanian age and was correlated with the Bahariya and Maghrabi Formations of the Western Desert. Paleontological evidence and correlation between the Aswan and the Wadi Qena area, however, suggest that the Timsah Formation is more or less an equivalent of the Hawashiya Formation. Consequently, the underlying Abu Aggag Formation of the Aswan area (= Aswan. of Klitzsch & Léjal-Nicol 1984) might be an equivalent of the Taref Sandstone.

5 *Santonian, Campanian, Maastrichtian*

All along the Eastern desert between the Aswan area and Wadi Dakhal this more or less marine interval of Coniacian to possibly early Santonian age seems to be followed by a sedimentary break or by the deposition of fluvial sandstone (Aswan area) or by lacustrine deposits (*Ceratodus* at different places in Wadi Qena and Wadi Dakhal). This sedimentary break or continental interval of Santonian age, however, is followed in the north Wadi Qena by marine to fluvial or coastal flood-plain sediments of Campanian age. These strata contain phosphatic horizons at different levels (Duwi Phosphate). This, in turn, is overlain by the well-known Sudr Chalk or Dakhla shale mainly of Maastrichtian age. Within all these formations, an increasing continental influence toward the south and southeast is evident.

CHAPTER 17

Nile Delta

J.C. HARMS & J.L. WRAY
Consultants, Littleton, Colorado, USA

The Nile Delta played an important and early role in the development of geologic concepts. The term 'delta' was first applied to the subaerial portion of its deltaic plain by Herodotus nearly 2500 years ago because of its resemblance to the shape of the Greek letter. Herodotus also recognized that the thin increments of sediment laid down by floods implied a considerably long time span for the deposition of this large feature and considerable antiquity for the earth. In spite of this historical significance and its general recognition as the type delta, the Nile Delta has received relatively less attention in the published literature than several other deltas. The geologic history and internal morphology of the Nile Delta have remained less well-known than deserved because subsurface and geophysical data have been somewhat sparse and proprietary. Certainly, the geology of north Egypt including the Delta was synthesized by Said (1962) and somewhat revised by Salem (1976). Rizzini et al. (1978) defined three sedimentary cycles within the delta province ranging from mid-Miocene to Holocene. Ryan (1978) described the late Miocene Messinian salinity crisis along the southeastern margins of the Mediterranean Sea and its general effects on erosion and deposition. Barber (1981) elaborated on this topic, in particular, the subaerial modification of the Nile Delta region related to the Messinian erosional surface. Sestini (1976), Zaghloul et al. (1979), and Montasir (1937) discussed the geomorphology and later history of the Nile Delta. Ross & Uchupi (1977) defined the bathymetry, structure, and sedimentary history of the area offshore of the Delta in the southeast Mediterranean. Said (1981) has contributed the most complete treatment of the evolution of the Nile River and Delta, elaborating in particular the Pliocene and Pleistocene history in detail. For the Quaternary history of the Delta and a complete reference to existing literature, the reader is referred to that extensive work and to Chapter 25 of this book.

This chapter provides an interpretation of the stratigraphy and structure of the pre-Quaternary Delta area. Eighteen deep wells and a regional grid of seismic lines (Fig. 17/1) have been used as the primary data base to develop this interpretation. In each well, lithologies were identified based on cuttings, descriptions and log responses, ages were assigned from calcareous nannofossils and depositional environments interpreted on the basis of faunal constituents, mainly benthic foraminifera, and lithologies. Standard Cenozoic low-latitude nannofossil zones (Martini 1971) can be distinguished in marine and marginal marine Nile Delta sediments (Table 17/1) and these provide the high-resolution biostratigraphic framework, particularly in the Miocene, within which to interpret depositional history of the region. Formation names have not been applied to these time units because formations were defined earlier on the basis of other information or do not exist for some units; a review of Miocene and younger formation names is given by Said (1981, Appendix B). The wells were tied to seismic sections using velocity information and synthetic seismograms, and reflection events were used to interpret structure and stratigraphy. The data base shown in Figure 17/1 is still sparse considering the size of the Delta which is 270 km wide at the shoreline and 160 km from the head of the distributary system to the sea. Some additional wells were used in constructing the maps shown in Figures 17/8 to 17/12. Data from some wells along the shoreline and offshore were not available for this study.

REGIONAL TECTONIC SETTING

In a tectonic sense, the coastal area has behaved mainly as a passive continental margin. The oldest clearly defined seismic event which can be traced regionally is a prominent reflection that marks the top of the Cretaceous (or thin overlying Eocene) carbonate units. This reflector, which can be tied to wells in the south, dips northward and can be traced into the mid-Delta area where the rate of dip, in terms of time,

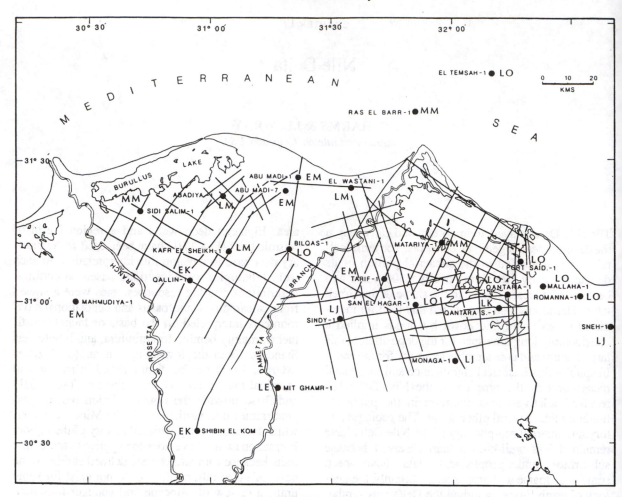

Figure 17.1 Index map of the Nile Delta showing wells and seismic lines used in this study. For each well, the stratigraphically deepest penetration is indicated: J – Jurassic; K – Cretaceous; E – Eocene; O – Oligocene; M – Miocene; P – Pliocene, and E, M or L for Early, Middle or Late.

increases (Fig. 17/2). Associated with this increase in dip is a line north of which the top of the Cretaceous reflection complex appears to lose amplitude and continuity. Still farther northward on Figure 17/2, beyond the zone of steeper dips, no clear continuous reflections can be traced. These features are interpreted as a major facies change where well-bedded platform carbonates to the south change to finer-grained slope and basinal facies to the north. This facies change probably overlies the transition from the continental crust of Africa to the oceanic crust of the Mediterranean, as suggested by deep refraction data in Sinai. The carbonate shelf area of Figure 17/2 corresponds approximately to Said's (1981) South Delta Block and the slope and basin is his North Delta Embayment. Although faulting may accentuate the flexural form of the zone separating these two areas, it is mainly the result of original facies and bathymetric influences. It appears that from at least Jurassic

time onward, the coastal zone subsided and a thick pile of relatively little deformed sediment accumulated, suggesting thermal subsidence and isostatic loading over a continental-oceanic crustal transition.

This zone, which lies in line with the projected continental margin in Sinai to the east and the Western Desert to the west, had a continued influence on later stratigraphic and structural characteristics. Thick Tertiary strata lie only to the north of this zone. Deep water deposits represented by bathyal facies of Oligocene to Pliocene age also lie north of this zone. Relatively large rotated fault blocks of Miocene strata occur only to the north of this earlier Cretaceous platform.

Two additional tectonic influences overprint this simple, subsiding margin scenario. First, a belt of apparently compressional folds extends regionally from Sinai to the Western Desert suggesting a late Cretaceous-Eocene event that may reflect collision of

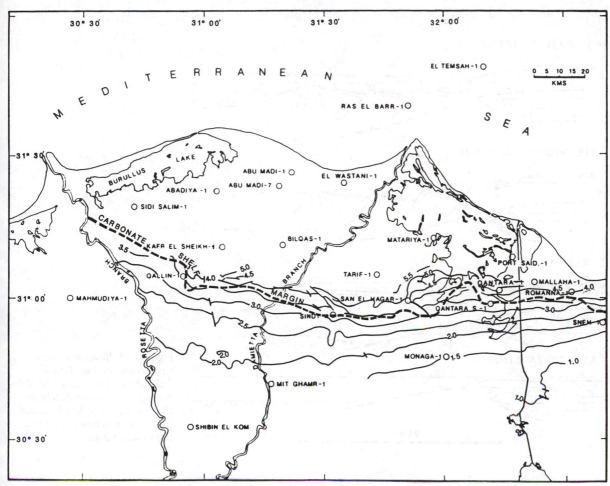

Figure 17.2 Contour map on the top of the Cretaceous (locally thin Eocene) in two-way travel time. Contours are spaced at 0.5 s intervals. The dashed line indicates the probable edge of the Cretaceous massive shelf carbonates, based on the amplitude and continuity of the seismic event. Thick shallow-water carbonate strata lie south of this line and slope and deeper water carbonate or marl lie to the north.

the northward-moving Africa with another plate. Second, uplift, rifting, and transform faulting occurred in late Oligocene to Miocene time in the Gulf of Suez-Red Sea area and appears to affect structural trends in the eastern Delta near the Suez Canal and offshore. A north-south uplift through the central Delta of about middle Miocene age is aligned with this trend and may be related to Gulf of Suez events because of its age and orientation, which is at odds with the general structural grain of a continental margin.

SUMMARY OF STRUCTURAL DEVELOPMENT

The evolution of major structural features for a typical north-south cross section of the delta is shown in Figure 17/3. From Jurassic through Cretaceous

time, a carbonate platform developed south of an east-west line through the mid-Delta. Based on interpretation of seismic lines and the model of north Sinai, the carbonate platform is replaced by basinal facies toward the Mediterranean. This major facies belt is thought to overlie the transition from African continental crust to Mediterranean oceanic crust and, if this is true, represents the most fundamental structural boundary in the delta area. The crustal blocks can of course be depressed by deposition, but isostacy dictates that the continental crust remain relatively high, receiving shoal-water sediment, whereas the oceanic crust can be a deeper water depositional basin.

Several other structural features closely parallel this zone. Following an episode of shelf progradation in earliest Miocene times, caused possibly by lowered sea level and regional upwarping, a thick

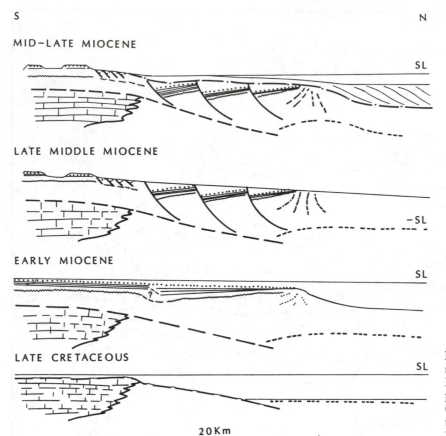

S   N

MID–LATE MIOCENE

SL

LATE MIDDLE MIOCENE

–SL

EARLY MIOCENE

SL

LATE CRETACEOUS

SL

20 Km

Figure 17.3 Simplified north-south cross-section of the Nile Delta showing the sequence of deposition and deformation from Cretaceous to mid-late Miocene time. Area was emergent during late middle Miocene.

wedge of strata deposited from Oligocene to middle Miocene time failed as a series of blocks bounded by northward-dipping normal faults. The northern edge of the early Miocene shelf-slope complex may have provided the north-inclined free surface which promoted this failure.

Southward-rotating fault blocks 4 to 8 km wide are typical of the area north of the underlying Cretaceous platform margin (Fig. 17/4). Similar but much smaller fault blocks occur to the south, in many places at the edge of a buried mesa topography. Although the faults are obvious at the level of the middle Miocene to late Oligocene, rarely are underlying Cretaceous beds broken, suggesting that the faults are listric in form and flattened into bedding at depth. At the northern margin of those fault blocks, reflectors of early Miocene age disappear; there may be a zone of diapiric upward flow of overpressured Oligocene shales, suggested diagrammatically by faint high angle lines on Figure 17/3.

Following fault movement, erosion formed a broad angular unconformity in late middle Miocene to early late Miocene time. Paleontologic data indicate a hiatus which encompasses progressively greater periods of time southward in the delta. This angular unconformity which truncates rotated fault blocks is clearly evident on seismic records throughout the mid-Delta. The rocks below the unconformity have been definitely dated as early to middle Miocene in the eastern delta. In the western delta, similar appearing fault blocks, as at the Kafr El Sheikh well (Fig. 17/4), have not been paleontologically dated where penetrated. However, we infer these blocks to be some part of the Miocene section.

A large north-south trending uplift through the central delta (Fig. 17/4) developed at the same time as the large normal faults. This uplift was also truncated so that rocks perhaps as young as late Miocene rest on truncated early and middle Miocene strata along the flanks of the uplift and upon Oligocene along the axis. At the Bilqas well, paleontologically undated strata about 600 m thick occur at this unconformity, so that precise age relationships of sediment related to the uplift are not everywhere well-known. The trend of this structure is nearly at right angles with all other structural trends in the delta area, which suggests that the uplift may be related to the Gulf of Suez and Red Sea structural elements.

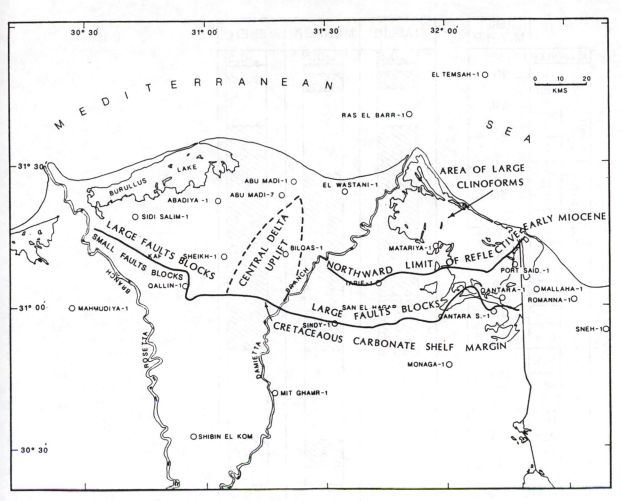

Figure 17.4 Areal distribution of major tectonic features, based on seismic interpretation.

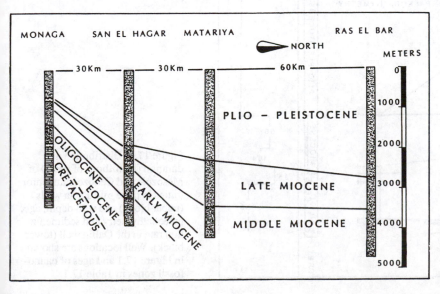

Figure 17.5 North-south cross-section of the eastern Delta showing the thickness of major time units and average well penetration in the area.

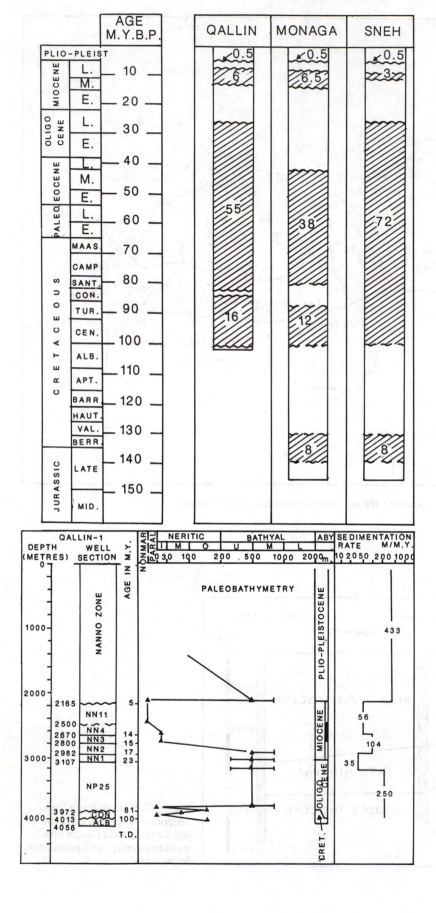

Figure 17.6 Distribution and duration (in millions of years) of Mesozoic and Cenozoic unconformities in three mid-Delta wells (upper block) and well depth, age, paleobathymetry and sedimentation rate in the Qallin well (lower block). Well locations are shown in Figure 17.1 and ages of nannofossil zones in Table 17.1.

Near the beginning of late Miocene time, sea level rose and marine sediments were deposited far to the south over the previously eroded unconformity. The major structural movement during Tortonian-Messinian time was a considerable subsidence of the northeastern quadrant of the delta (Fig. 17/4). An impressive clinoform sequence 1 to 3 km thick was deposited in the Wastani-Matariya area, as shown diagrammatically in the top section of Figure 17/3.

The delta area was exposed during the Messinian salinity crisis, a valley was entrenched, and a thick Pliocene to Pleistocene deltaic wedge was deposited when sea level was restored to near present position. General subsidence of the delta area allowed deposition of up to 3500 m of sediment in the last four or five million years, perhaps mainly because of isostatic loading. Local structural movement, reflected by small faults or low amplitude folds, was rare and minor. A few very small normal faults cut the base of the Pliocene in the mid-Delta and low domes like Abu Madi formed, presumably as minor adjustments to compaction or subsidence.

MESOZOIC STRATIGRAPHY

Mesozoic beds are penetrated in only a few wells across the middle and southern part of the Nile Delta because of the northward thickening Tertiary section (Fig. 17/5). Figure 17/6 shows the distribution in time of rock units and unconformities from three wells, including Sneh, Monaga, and Qallin. Two of these wells encountered late Jurassic rocks at total depth and had fairly complete early Cretaceous sections. Unconformities of considerable duration separate the Jurassic and early Cretaceous and eliminate nearly all of the late Cretaceous. Because of the sparse Mesozoic well control, no thickness and facies maps within the Delta area are presented in this chapter. However, the Mesozoic rocks are mainly carbonate or marl and were deposited in inner to outer neritic environments (Fig. 17/6). In this respect the rocks resemble depositional environments encountered in North Sinai and the adjacent Western Desert. For most of the time represented, environments across the mid-Delta ranged from platform shoals to open shelf lagoons or open platform margins. Within the late Cretaceous Senonian interval, lithologies are dominated by pelagic lime mudstone that infers a deeper outer shelf margin. Although no penetrations exist across the north Delta, Mesozoic strata were probably deposited in slope and deep oceanic environments north of the line shown in Figure 17/2.

TERTIARY STRATIGRAPHY

The depositional history of the Nile Delta area can be fairly completely reconstructed for the Miocene and Pliocene, but is less well-known for the early Tertiary. Latest Cretaceous, Paleocene and Eocene rocks have either not been reached or are absent in most wells in the Nile Delta. Eocene strata have been penetrated in only two wells in the mid-Delta area, the Mit Ghamr and Monaga, and in those wells are relatively thin. It appears that a late Cretaceous to early Eocene orogeny, well-known from folds formed across North Sinai and in the Western Desert, influenced the Delta area significantly and caused either non-deposition or erosion during this time span where well control exists (Fig. 17/6). Oligocene sediments have been penetrated by eight wells, mainly in the middle and eastern delta. The Oligocene beds are difficult to date and subdivide; most appear to have been deposited in depth ranges typical of the upper or middle slope. Toward the south, remnants of basalt of Oligocene age cap Oligocene or Eocene strata and represent subaerial flows that occurred during an Oligocene low sea-level stand.

The Neogene history of the Delta area is much better known than that of older units. Two major

Table 17/1. Standard Cenozoic low-latitude nannofossil zones (after Martini 1971) and age assignments used in this article.

Age	my	Calcareous nannoplankton zones
Pliocene		NN12 *A. tricorniculatus*
	5.5	
		NN11 *D. quinqueramus*
M L	8.0	
		NN10 *D. neohamatus*
	9.0	
I		NN9 *D. hamatus*
	10.0	
		NN8 *C. coalitus*
O	11.0	
		NN7 *D. kugleri*
M	13.0	
C		NN6 *C. miopelagicus*
	14.5	
		NN5 *S. heteromorphus*
E	16.0	
		NN4 *H. ampliaperta*
	17.5	
N		NN3 *S. belemnos*
E	19.0	
		NN2 *D. druggii*
E	23.0	
		NN1 *D. deflandrei*
	23.5	
Oligocene		NP25 *C. abisectus*

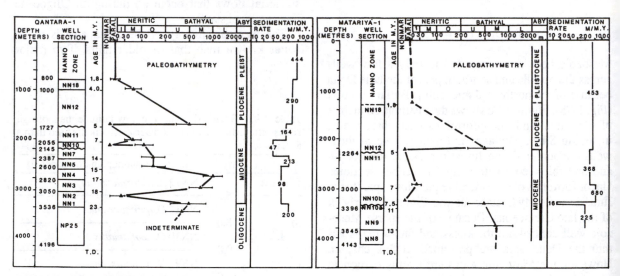

Figure 17.7 Distribution and duration (in millions of years) of Neogene unconformities in 4 wells in the middle or northern Delta (upper block). Well depths, age, paleobathymetry and sedimentation rate in the Matariya and Qantara wells in the lower blocks.

unconformities of regional extent subdivide the Miocene and Pliocene intervals. A major unconformity with basin-diminishing duration separates middle and late Miocene strata (Fig. 17/7). This unconformity has the longest duration in wells in the southern part of the Delta and a significantly briefer duration northward. A second unconformity separates late Miocene and Pliocene strata and is widespread over the entire Mediterranean area, where it represents the Messinian desiccation event. These unconformities

mark periods of major change in paleobathymetry and sedimentation rates. The ages, paleobathymetry and sedimentation rates are shown for three representative delta wells in Figures 17/6 and 17/7. The age and nannofossil zone numbers and names used for the late Oligocene to early Pliocene are shown in Table 17/1.

A series of maps (Figs 17/8 to 17/12 and 17/14) illustrate the shifts in depositional environment and the thickness of sediment deposited and preserved in the middle and northern delta area through Miocene and Pliocene time.

Early Miocene

Early Miocene depositional facies range from non-marine in the south to shelf and slope to the north (Fig. 17/8). During this time period, sea level first rose, then fell slightly, then rose continuously world-wide (Haq et al. 1987). Probably as a result of this eustatic rise in sea level, marine waters transgressed far to the south in the Gulf of Suez region during the early Miocene. Shelf sands and carbonates were deposited at Monaga while slope conditions prevailed at Qallin and San El Hagar, although a shift to inner shelf carbonate deposition occurred during the early part of the interval, possibly related to the eustatic drop in sea level. A similar history was recorded at Qantara, although a final shift to shallow water conditions here and at El Temsah may be due to tectonic

uplift. A middle or early Miocene sandy section at Kafr el Sheikh and Bilqas may have an associated bathyal fauna and could represent proximal turbidites deposited on a slope or shallow sediments deposited during the low stand in sea level, if the indications of bathyal fauna are related to caved samples.

The thickness of early Miocene beds is highly influenced by rotational block faulting in the east and west-central parts of the Delta. Early Miocene strata probably extended far to the south, although the beds were thin, and the thickness is more influenced by erosion of the high parts of the fault blocks during the late middle Miocene unconformity than by structural growth during deposition. Early Miocene beds were also erosionally removed from the central part of the delta around Bilqas by the mid-Miocene unconformity. The thickness contours show general structural trends; true structure is more complicated. Paleontological confirmation of the early Miocene age of beds in the west Delta has been established only at the

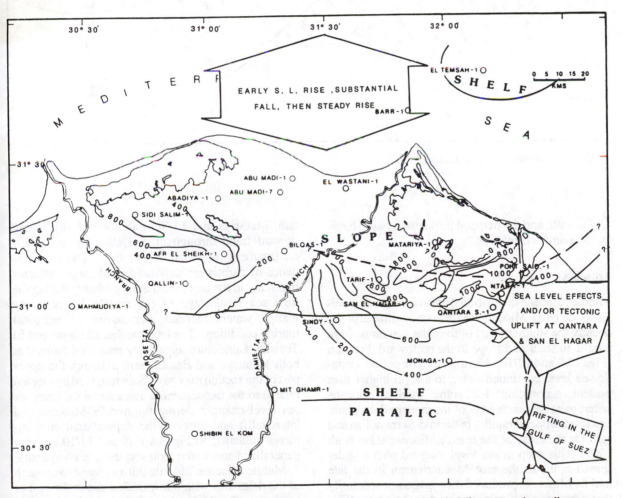

Figure 17.8 Early Miocene facies and thicknesses. Thickness contours are in meters. Early Miocene marine sediments are known from the Cairo-Suez district (paralic = marginal marine and associated environments).

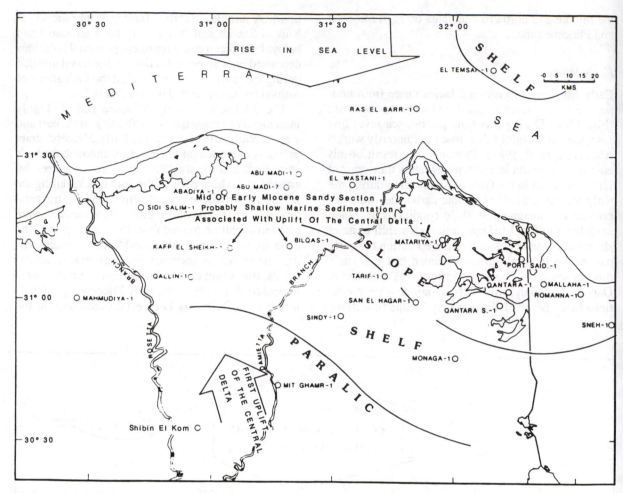

Figure 17.9 Middle Miocene (Langian) facies.

Middle Miocene

Qallin well, and the mapped thickness is based large-ly on seismic correlation.

Middle Miocene

Middle Miocene depositional environments are simi-lar to the early Miocene in that non-marine deposits occupy the southern part of the delta and range from paralic to shelf and slope in the northward direction (Figs 17/9 and 17/10). During this time period, eusta-tic sea level continued rising to a level higher than present accounting for some marine Miocene outcrops at the south edge of the modern delta plain. Sea level dropped rapidly in the mid-Serravallian and again at the close of the middle Miocene (Haq et al. 1987). This drop in sea level, coupled with a wide-spread uplift in the east Mediterranean in the late middle Miocene, produced the marked unconformity that separates the middle and late Miocene (Fig. 17/7). Paleobathymetric data at San El Hagar, Qan-

tara, Matariya and El Temsah show a shallowing upward trend through time. Qallin lay in a shelf environment for much of this time. The predomi-nance of carbonates deposited at Monaga indicates that it lay in the inner shelf zone probably during the high sea level stage of the middle Miocene. As marine waters receded, non-marine or marginal marine conditions developed at San El Hagar and El Temsah. Conditions apparently remained bathyal at both Matariya and Ras El Barr, although the upper part of the section may have been removed by erosion related to the unconformity. Because of the complex sea level changes during the middle Miocene and later uplift and erosion, the depositional environ-ments indicated on Figures 17/9 and 17/10 are very generalized and do not represent the true complexity.

Middle Miocene beds are thin or absent over much of the delta area with a few thick areas in the east or northeast. The thickness patterns are related mainly to structural movement. In the central part of the delta

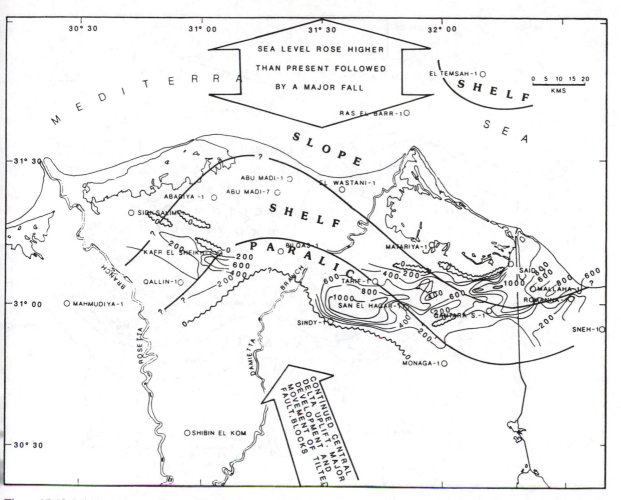

Figure 17.10 Middle Miocene (Serravalian) facies and total middle Miocene thicknesses. Thickness contours are in meters.

around Bilqas, broad uplift apparently caused the erosion of previously deposited beds during the late middle Miocene. In other areas, rotated fault blocks greatly affect the thickness of the middle Miocene sediments, probably mainly because of erosion of the higher parts of the fault block during the development of the broad late middle Miocene unconformity. Facies patterns do not suggest that fault movement occurred during deposition.

Late Miocene

Late Miocene facies are shown in two maps, one for the Tortonian (Fig. 17/11) and one for the Messinian (Fig. 17/12). During this time span, there was an overall progradation that caused paralic and shelf facies to advance from the mid-delta to the northeast delta area. The interval is most noteworthy for the deposition of a sequence of large clinoforms which can be recognized on seismic data in the northeast

part of the delta. These steeply inclined sedimentary units were deposited in deep water north and east of the early Tortonian shelf-slope break. The sequence of inclined beds prograded toward the east, probably under the influence of easterly longshore drift (Fig. 17/11). The source of these sediments is still problematic, but they may indicate the first appearance of a major integrated river system whose mouth lay just east of the uplift in the Bilqas area. During the Tortonian, this river may have built a small delta which prograded eastward bringing non-marine or marginal marine conditions to San El Hagar and eventually to Matariya at the end of this interval. Farther west in the Bilqas area, sedimentation began under sandy shelf conditions and became increasingly shaly through time, becoming bathyal at both the Abadiya and Bilqas wells. This transgression was probably the result of a eustatic rise in sea level during the Tortonian (Haq et al. 1987). Shoreline or near-shore conditions must have persisted throughout

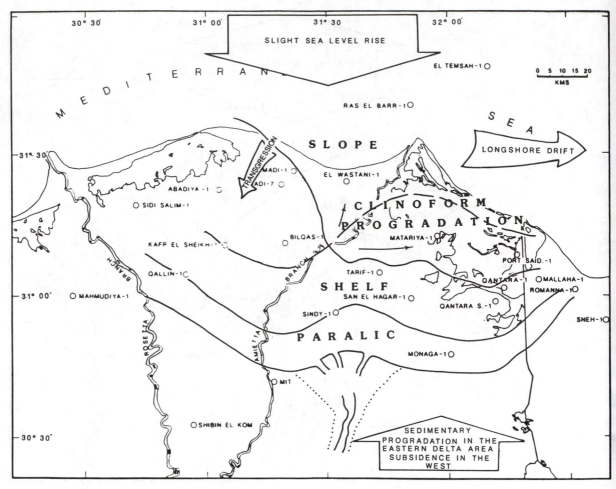

Figure 17.11 Late Miocene (Tortonian) facies.

the interval at Qallin, based on paleobathymetric indicators and the presence of an anhydrite bed. Abrupt bathymetry changes from upper bathyal to inner shelf or non-marine and back to upper bathyal in the El Temsah and Ras El Barr area, indicated by foraminiferal assemblages, suggest local tectonic uplift and renewed subsidence northeast of the delta.

During the Messinian, the deep-water clinoform sequence recognized on seismic records continued to prograde toward the east. Non-marine conditions and sand deposition persisted at San El Hagar and Monaga. Toward the north, in the Matariya and Qantara area, depositional environments fluctuated between marginal marine and inner shelf. The northwest quadrant of the delta shows a progressive shoaling through the Messinian which probably reflects both a lowering of the Mediterranean sea level and progradation of the shelf. At the end of the Messinian interval, global sea level fell significantly, and the isolated Mediterranean was drastically lowered by

the 'Messinian salinity crisis' (Hsu et al 1978). There is widespread evidence of non-marine, shallow marine or evaporite deposition during the latest Messinian in most wells in this area. Anhydrite beds common in many areas at this stratigraphic level show as prominent seismic reflectors.

Rifting and pronounced subsidence occurred in the Red Sea and the Gulf of Suez areas toward the end of the Miocene and may have caused the focus of deposition of the ancestral Nile to lie in the east delta. The local tectonic events at El Temsah and Ras El Barr offshore may perhaps also be related to the projection of the Gulf of Suez trend.

The thickness of late Miocene beds is shown on Figure 17/12 at the end of the Miocene progradation. The area of greatest deposition was in the northeast part of the delta region, related to the prograding massive clinoform sequence. The position of the integrated river system shown on Figure 17/11 and 17/12 is problematic but is placed toward the east

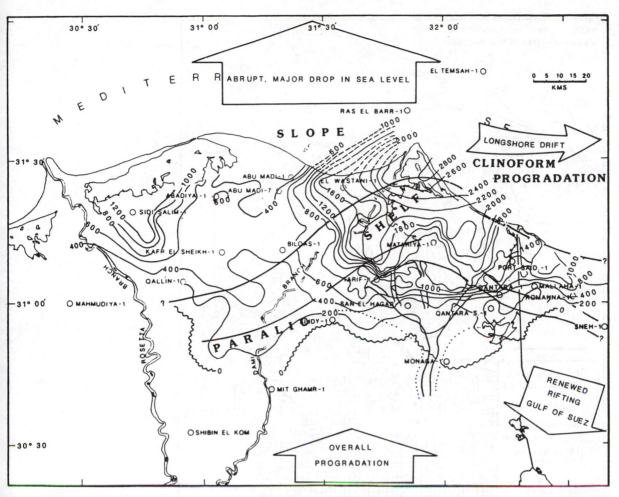

Figure 17.12 Late Miocene (Messinian) facies and total late Miocene thicknesses. Thickness contours are in meters.

because deposits are thickest in the eastern part of the delta. The late Miocene is thin in the north-central delta area partly because of low rates of subsidence over the underlying structural high and probably also because of the erosional removal of the late Miocene during the Messinian lowering of sea level. Rotational block faulting affected thickness variations in the east- and west-central parts of the delta area. Underlying topography was locally irregular following the mid-Miocene unconformity, and continued rotational movement of some fault blocks during the late Miocene apparently also influenced sedimentation. Erosion of the topographically higher parts of many fault blocks during the Messinian lowered sea level also strongly modified the pattern of late Miocene isopachs.

The dramatic drop in Messinian sea level occurred at about five million years before present, based on nannofossil zonation, and caused a hiatus throughout most of the Nile Delta area with a duration of about

half a million years. The subcrop pattern beneath the unconformity (Fig. 17/13) was interpreted by Barber (1981). The deepest erosion shown here trends from Cairo north to between the Shibin El Kom and Mit Ghamr wells, a position 50 km or more west of the implied river positions in Figures 17/11 and 17/12. These positions are necessarily speculative, but represent depositional distributaries rather than the entrenched river system which dominated the post-Messinian history. Barber interpreted this surface as developed in the early Messinian, but nannofossil data suggest the unconformity is slightly younger. Therefore, the interpretation made here suggests that Pliocene sediments rest on this unconformity throughout most of the delta and that latest Messinian sediments younger than the unconformity occur only along the axis of the deep river canyon and were deposited during the recovery of sea level in very latest Messinian time. Thus the Abu Madi sand facies of Rizzini et al. (1978) is interpreted as fluvio-deltaic

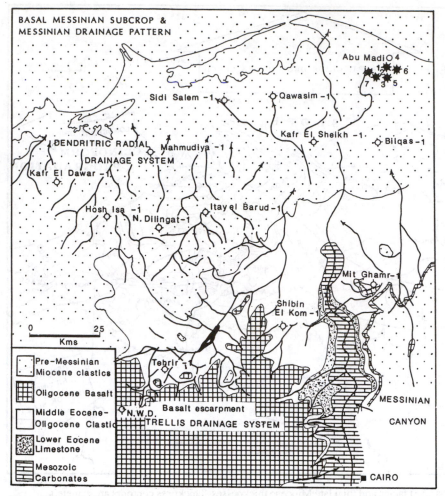

Figure 17.13 Subcrop at the Messinian unconformity in the western Delta (after Barber 1981).

and related to transgression at the close of the Messinian. This facies is as much as 350 m thick along the axis of the canyon.

Pliocene

The coastal deposits of the Nile area are truly deltaic only in the Pliocene and Pleistocene, in the sense of being caused by a large focused supply of sand and mud from an integrated river. Before the Pliocene, sediment is more typical of a trailing-edge (Atlantic-type) continental margin, and streams were relatively small and not integrated into a major regional drainage like the present Nile. Indeed, there have been suggestions that the major pre-Pliocene ancient river system across north Africa trended westward toward the Benue trough (McCauley et al. 1986). However, because of the very low stand of the Mediterranean in latest Messinian time, a north-trending major integrated stream system developed and eroded a ca-

nyon 1.5 km or more deep at the latitude of Cairo. This is the Eonile of Said (1981). The topography of the latest Miocene erosion surface is indicated by an isopach map of the combined Pliocene and Quaternary thicknesses (Fig. 17/14) which is based on the regional grid of seismic lines and well control. A somewhat more detailed unconformity map was presented for the western part of the delta area by Barber (1981, Fig. 6) and for the canyon near Cairo by Bentz & Hughes (in Said 1981, Fig. 68). The thick wedge of sediment indicated in Figure 17/14 was deposited after sea level rose to its present position and as the Pliocene delta prograded over the subcrop surface shown in Figure 17/13. The initial transgression flooded the ancient Nile canyon as far south as Aswan where Pliocene sediments within the incised area are marine. Pliocene facies within the delta area indicate an overall depositional regression. Deeper-water muddy sediments occur near the base of the Pliocene section and are overlain by inclined beds of

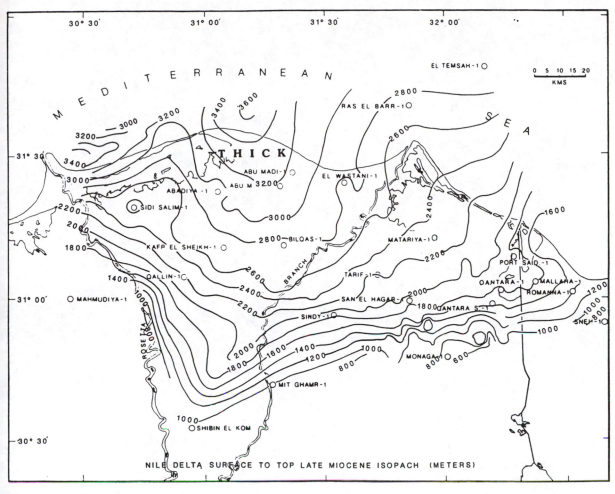

Figure 17.14 Pliocene-Quaternary thicknesses. Contours are in meters.

prodelta muds and finally fluvial and shoreline sandy sediments. Deltaic facies within the Pliocene section have not been mapped in detail because age dating is difficult, lithologic data from logs are rather general, and the section is too shallow to be prospective for hydrocarbons. Offshore the deltaic wedge ties into the shelf slope and Nile cone sediments discussed by Ross & Uchupi (1977).

ACKNOWLEDGEMENTS

The authors gratefully acknowledge AGIP, Conoco and the Marathon Oil Company for support and permission to publish this article. This work was conducted at the Denver Research Center, Marathon Oil Company in 1980-81. S.J. Derksen, H.N. Cappel, K. Byerley, C.H. Ellis, W.H. Lohman and M.E. Heiman provided significant technical contributions to the study.

CHAPTER 18

Red Sea coastal plain

RUSHDI SAID

Consultant, Annandale, Virginia, USA and Cairo, Egypt

The Red Sea range abuts against the Red Sea coast in the south of Egypt leaving a narrow maritime plain where pre-Miocene sediments were eroded. In the Quseir-Safaga district, however, strike faults gave rise to a remarkable topographical complexity in which the pre-Miocene (Cretaceous and Eocene) strata are preserved forming limestone plateaux of which Gebel Duwi is conspicuous. The Duwi range consists of a long sharp ridge which drops precipitously to the southwest and slopes gently to the northeast. Other crests similar in character are seen both to the southeast and north. In these the limestone and sandstone beds reappear in the same succession and are sharply cut off by the dark hills which surround them. The Eocene-capped Gebel Hamadat is conspicuous by reason of the dark crystalline ranges which, from a distance, appear to enclose it on all sides.

Thus while on the western side of the Red Sea hills the Cretaceous and Eocene beds are more or less connected, on the eastern side they consist of a series of widely separated outliers that owe their preservation to the action of faults.

The sedimentary rocks of this region are separable into two great divisions: the pre-rifting Cretaceous-Eocene group and the post-rifting Oligocene and later sediments group. The latter division exhibits a continuous succession from the middle Miocene onward.

In the Quseir-Safaga district the Cretaceous-Eocene deposits occupy the troughs of synclinal-like folds within the crystalline hill ranges. In this district, as on the western side of the Red Sea hills, marine upper Eocene and Oligocene deposits are absent. Hence, it seems that the region must have undergone elevation during these two epochs. The fundamental complex of ancient crystalline rocks with its overlying mantle of Cretaceous and Eocene sediments was then subjected to intense deformation which resulted in a series of highly tilted fault blocks running along axes trending north-northwest or northwest. Most of the synclinal folds were due to the drag of the western

parts of the sediments over the underlying basement complex (Akkad & Dardir 1966). During this continental episode the area underwent considerable denudation before it was finally submerged in the Miocene by the sea which was connected first to the Mediterranean and later to both the Mediterranean and the Indian Ocean. This sea was the prototype of the present-day Red Sea.

CRETACEOUS AND EOCENE ROCKS OF
QUSEIR-SAFAGA DISTRICT

The succession of this older divison of rocks is not different from that in many other parts of south Egypt. In the isolated outcrops of Gebel Duwi and their continuation at Gebels Atshan and Hamadat (Fig. 18/1), as well as in the Safaga district (Fig. 18/2), the succession is as follows from top to bottom: Nakheil, Thebes, Esna, Tarawan, Dakhla, Duwi (Phosphate) Formations and Quseir and other Formations of the Nubia Sandstone Group (Fig. 18/3 & table 1).

The sections at Quseir and Safaga are described by many authors including Barron & Hume (1902), Ball (1913), Beadnell (1924), Youssef (1957), Faris & Hassan (1959), Akkad & Dardir (1966, 1969), Abdel Razik (1967, 1972), Issawi et al. (1969, 1971), Tarabili (1966, 1969) and Gindy et al. (1973, 1976).

Nubia group
The oldest sedimentary beds, belonging to the well-known Nubia group, rest unconformably over the basement complex. The beds occupy many of the topographic lows in the Quseir-Safaga area and are divided into two units. The lower unit is 200 m thick and is made up of seemingly non-fossiliferous sandstone beds with intercalations of mudstones. The upper unit is 70 m thick and is made up of variegated shales of the Quseir Formation. In the following paragraphs a description is given of these units from the oldest to the youngest.

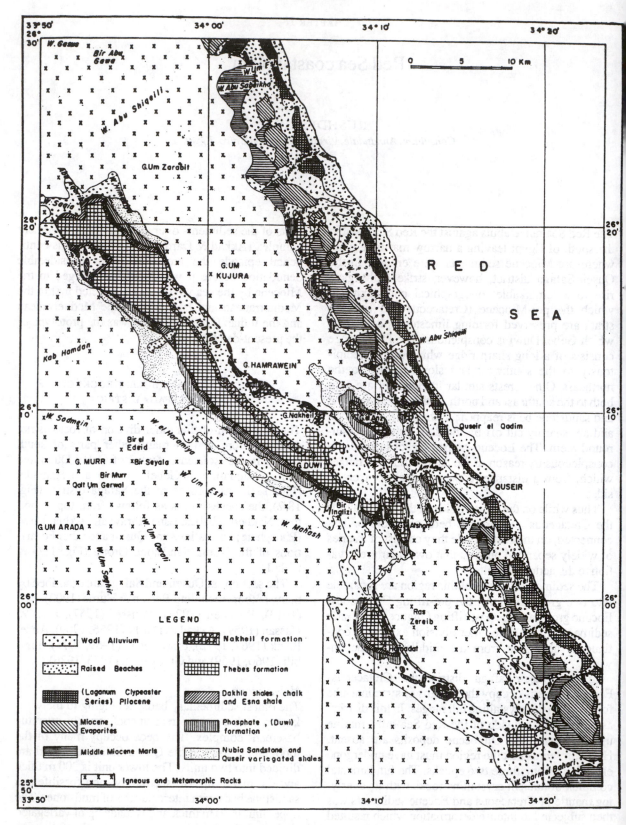

Figure 18.1 Geological map of Quseir district (after Youssef 1949 with modifications).

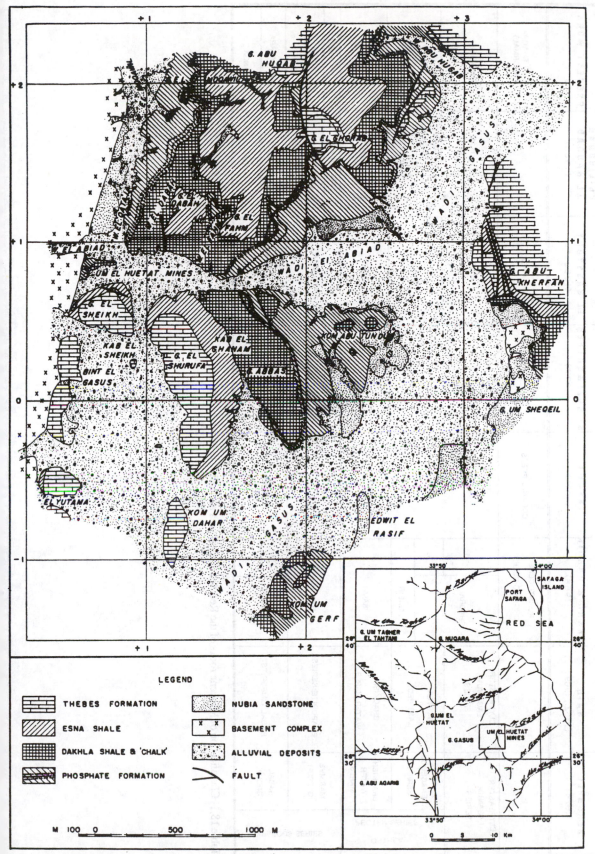

Figure 18.2 Geological map of Safaga district (after Hume 1927 with modifications).

BEADNELL (1924)	SAID (1962, 1969)	AKKAD & DARDIR (1966)	ISSAWI (1971)	SAMUEL & SALEEB (1977)	EL BASSYONY (1982)	PHILIBBOS & EL HADDAD (1983)	PRESENT WORK
CLYPEASTER LAGANUM SERIES	CLYPEASTER LAGANUM SERIES	PLEISTOCENE ORGANIC REEFS	ORGANIC REEFS		UM GHEIG		WADI SHAGRA (INFORMAL)
		SHAGRA U			WIZR	SHARM EL ARAB	SHAGRA
ARENACEOUS SERIES	OYSTER CAST BEDS	SHAGRA L	GASUS			DISHET EL DABAA	
		GABIR		SAMH	GABIR	ABU SHIGEILI	GABIR
BRACKISH WATER MARLS	BRACKISH WATER MARLS	SAMH			SAMH	WIZR	SAMH
MASSIVE GYPSUM	EVAPORITE SERIES	GYPSUM	GYPSUM	UM GHEIG	ABU DABBAB	UM GHEIG	UM GHEIG
				ABU DABBAB		ABU DABBAB	ABU DABBAB
						SYATIN	
BASAL GROUP	BASAL LIMEGRITS	G.RUSAS U	G.RUSAS	UM MAHARA	ABU HAMRA/GHADIR	UM MAHARA	UM MAHARA
		G.RUSAS L		RANGA	G.RUSAS — RANGA	RANGA	RANGA

(Rotated group labels: BEADNELL — OSTREA-PECTEN SERIES, GYPSEOUS SERIES; PHILIBBOS & EL HADDAD — SHAGRA, MERSA ALAM)

Table 18.1 Classification of Miocene rocks of the Red Sea coastal area according to different authors.

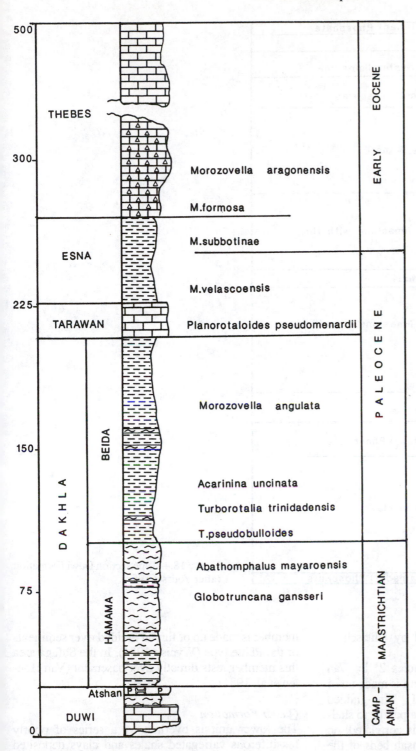

Figure 18.3 Columnar section Gebel Duwi (foraminiferal zones modified after Krasheninnikov & Abdel Razik 1969).

The lower unit of the Nubia group forms the subject of the recent studies of Ward et al. (1979) and Van Houten et al. (1984). These authors divide this unit into three members (described as 'facies'). According to these authors these 'facies' reflect Cretaceous regressions and transgressions of the Tethys across northeast Africa.

The lowest member termed 'facies 1' by Van Houten et al. (1984) is 21 m thick and is made up of non-marine, lenticular, trough cross-bedded sandstones and minor conglomerates. This member seems to represent a fluvial regressive phase. Although non-fossiliferous, the formation is correlated, on stratigraphic grounds, with similar beds in Aswan

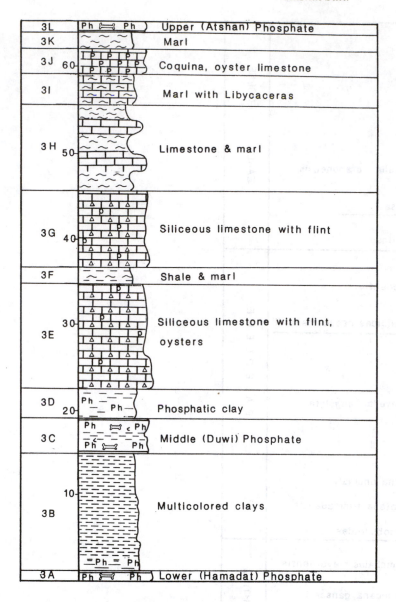

Figure 18.4 Type section Duwi Formation (after Youssef 1957).

(the Abu Aggag) which are dated by Klitzsch & Léjal-Nicol (1984) as (?)Turonian.

The middle member termed 'facies 2' by Van Houten et al. (1984) is 23 m thick and is made up of parallel-bedded mudstone and fine-grained sandstone. This member represents a paralic to shallow marine transgressive phase. It is correlated, on stratigraphic grounds, with the Timsah beds of the Aswan area which are dated Coniacian by Klitzsch & Léjal-Nicol (1984).

The upper member forms the lower part of 'facies3' of Van Houten et al. (1984). It is correlated with the Um Barmil Formation of the Aswan area. It is 86 m thick and is made up of non-marine, tabular cross-bedded sandstone. This member represents a regressive phase during Santonian (?) time. This

member is made up of fluvial braided river sediments of the Platte type (Wycisk 1984). In the Safaga area this member rests directly over basement (Van Houten et al. 1984).

Quseir Formation
The lower unit is overlain by a series of poorly fossiliferous variegated shales and clays deposited under non-marine to marginal marine conditions. They reach about 70 m in thickness in Gebel Atshan. A comparable section is noted in many parts of south Egypt above the sandstones of the Nubia group. These shales, which were previously classified with the 'Nubia Sandstone', are separated by Youssef (1957) and given the name Quseir Variegated Shales. TheQuseir is correlated with the Mut Formation of

the Western Desert and is assigned a Campanian age.

Duwi (Phosphate) Formation

The Quseir variegated shales are overlain by the Duwi (Phosphate) Formation (Fig. 18/4) made up, in the Quseir-Safaga reach, of three phosphate horizons separated by beds of marl, shale and oyster limestone with flint. Youssef (1957) and Ghanem et al. (1970) include all the phosphate horizons within the Duwi irrespective of their thickness or position within the section. Many field geologists, however, map the Duwi as including only the thick and mappable phosphate horizons of this formation.

The lowermost of the phosphate horizons, the Hamadat (= Abu Shigeila or the 'C' horizon of Said 1971), is separated from the middle Duwi or 'B' horizon by a section of variegated shales of non-marine to marginal marine origin similar to those of the underlying Quseir Formation. In areas where the Hamadat Horizon is represented by thin stringers, it is common practice among many field geologists to lump the Hamadat with the Quseir Formation. The Hamadat assumes some thickness and is mined in the Abu Shigeila, Gihaniya and Hamadat areas. In the Safaga district, Faris & Hassan (1959) record the ammonite *Mannambolites aff. julieni* from a bone bed within the variegated shales of this horizon. The age of the Hamadat is Campanian.

The middle horizon (the Duwi or 'B' beds) is the best developed and most consistent of the phosphate horizons. It is exploited in the Duwi and Hamadat mines as well as in the Safaga mines. It is made up of several phosphate beds separated by thin shale, marl and silicified limestone beds. The exploited phosphate bed in Gebel Duwi is 150 cm thick and has a tricalcic content of about 70%. The Duwi Horizon has a thickness of 4 to 10 m. In Hamadat, the exploited phosphate bed is 3 m thick and has a tricalcic phosphate content of 40 to 60%. Here the phosphatic rock is dark in color and has silicified phosphatic nodules. An important marker in this horizon is the thick oyster limestone bed which caps and intercalates this horizon.

The Duwi carries a large number of vertebrate remains, coprolites, fish teeth and heteromorph ammonites. These form the subject of a recent study by Dominik & Schall (1984). According to these authors, the faunal assemblage belongs to the *Bostrychoceras polypolocum* zone whch assigns the age of this formation to the late Campanian.

The upper phosphate horizon (the Atshan or 'A' beds) is exploited in the Atshan, El Daba'a, Younis and El Nakheil mines in the Quseir area. When well-developed, the thickness of this horizon ranges from 4 to 10 m and is made up of several phosphate beds (between 3 and 7) which are separated by shales and marls. In Atshan the exploitable phosphate bed is between 160 and 170 cm in thickness and has a tricalcic content of between 65 and 70%. In areas where the Atshan is not well-developed, its place is usually taken by one or two thin phosphatic conglomerates which intercalate the marine lower marly member (the Hamama) of the Dakhla Formation. The age of the Atshan phosphate is Maastrichtian.

Dakhla Formation

The Duwi Formation is overlain by a series of marls and shales which vary in thickness from one place to another. The formation is stratigraphically comparable to the Dakhla shales which are developed in many parts of south Egypt. In Gebel Duwi the formation reaches its maximum thickness and development. Here it is 175 m thick and is divisible into a lower marl member (the Hamama) and an upper shale member (the Beida).

The lower member is 75 m thick and is made up of marl beds in the Quseir and Safaga areas. These lower marls (named the Hamama Member by Abdel Razik 1972) include, as in the Nile Valley and the oases, *Isocardia chargensis* and *Pecten farafrensis* as well as other invertebrate fossils comparable to those found in equivalent beds in other parts of south Egypt. The Hamama overlies the Atshan phosphate beds when these become mappable and well-developed. Otherwise the lower boundary of the Hamama is extended to the top of the Duwi Member. In this latter case, the Hamama includes one or more phosphatic conglomerate stringers marking the Atshan Horizon. The age of the Hamama member above this phosphate horizon is Maastrichtian.

The upper shale member (named Beida by Abdel Razik 1972) is 100 m thick. Its age is early Paleocene. A biostratigraphic analysis of this section is given in Krasheninnikov & Abdel Razik (1969), and although these authors report a minor disconformity between the Hamama and Beida Members (corresponding to the Cretaceous-Tertiary boundary), the section seems to represent a complete Maastrictian-early Paleocene sequence, possibly indicating an unbroken sedimentary record (Masters 1984). According to the latter, no other section in Egypt offers such a complete sequence.

In other areas of Quseir and in Safaga the sequence is less complete. In Gebel Anz (Quseir area) the Dakhla is 90 m thick and is made up only of the lower Maastrichtian marly member. No early Paleocene is recorded in this area. In Safaga, the Dakhla Shales are very thin and are represented by the lower Hamama Member made up of thickly-bedded marls, marly limestones and hard limestones weathering to a characteristic pinkish color (Akkad & Dardir 1966). The thickness of the Dakhla in the Safaga area ranges from 23 to 30 m and its age is Maastrichtian.

A notable feature of the Quseir, lower Dakhla and the intervening Duwi Formations is the presence of oil (black) shales within and above the phosphatic beds. Some of these shales are reported to have caught fire in the past. The average oil content of some weathered samples of these shales is 19 gal/t (Troeger 1984). Geochemically, they display differences with regard to their $CaCO_3$, organic matter and trace elements contents although they are not significantly enriched in trace elements (Gindy & Tamish 1985). These differences suggest deposition under marine, lacustrine, brackish and freshwater euxenic to well-oxygenated environments. Some of the shale beds are rich in casts of plant remains, and these may have been deposited in mangrove swamps.

The varied environments of deposition of the part of the section covering the Quseir, the Duwi and the lower Hamama Member of the Dakhla Formation are related to repeated synsedimentary tectonism. This tectonism seems to confirm the hypothesis of a tectonically controlled phosphate concentration in basins which formed below the wave base (Hendriks & Luger 1987).

Shafik (1970), Sadek & Abdel Razik (1970) and Abdelamlik (1982) study the Maastrichtian and early Paleocene nannoplankton from the Dakhla Shales of the Quseir area. Schranck (1984) studies the microfloras of the cores of the Atshan Phosphate Horizon and the overlying beds of the Dakhla Formation. These cores were raised from the north Younis mine (Quseir area). The cores yielded microfloras which reflect the course of a progressive marine transgression. The near shore phosphate beds carry land-derived pollen and spores some of which reach large grain size. The younger Dakhla beds are dominated by a succession of dynoflagellate cyst assemblages which characterize a progressive transgression from near-shore to genuine open marine environments.

Tarawan Formation

The Dakhla Shales are overlain by a carbonate bed which is recognised with ease in the Quseir area due to its sharp contact with the overlying and underlying dark green shales. In the Safaga district, where the Dakhla is more calcareous, this unit is difficult to distinguish. This carbonate bed is equivalent to the Tarawan Chalk of the Western Desert oases. The formation is made up of marl and marly limestone and ranges in thickness from 6 to 18 m. Megafossils are rare and, when found, are poorly preserved. The microfauna point to a late Paleocene age.

Esna Formation

The Tarawan is overlain by a well-developed unit of grey laminated shales that are similar in lithological characteristics to the Esna Shale of the Nile Valley. In Gebel Duwi the Esna is 47 m thick whereas in the Quseir-Safaga area it varies from 16 to 54 m. In Safaga the beds are about 50 m in thickness. This unit conformably underlies the Thebes Formation and has, like its equivalents in many parts of Egypt, a middle limestone member that runs consistently through it. The Esna Shale is rich in microfauna of late Paleocene to early Eocene age. El Dawoody & Barakat (1972) describe calcareous nannofossils from the late Paleocene-early Eocene rocks of Gebel Duwi.

Thebes Formation

The Esna Shale is overlain by a unit of limestone with flint which forms the important topographical features of Gebels Duwi, Atshan, Anz and Hamadat. This unit forms bold white cliffs facing southward. To the north these cliffs run sharply to ground level with a dip slope of 15 to 20°. This unit is equivalent to the Thebes Formation described from the Nile Valley. It attains about 285 m in thickness in the Duwi range, about 250 m in Hamadat and 166 m in Gebel Anz. It consists of thinly-bedded limestones with a noticeable development of flint bands or flint nodules and chert concretions interbedded, toward the top by a marly bed which contains *Lucina thebaica*. The limestone with flint is overlain by a thinly-bedded chalky limestone unit rich in *Operculina libyca*, *Nummulites* spp. and numerous *Turritellae*. The age of the Thebes is early Eocene according to many authors. However, Strougo & Abul Nasr (1981) separate from a 25 m-thick nodular limestone bed 90 m below the top of the section middle Eocene fauna, confirming an earleir contention (Youssef 1957) that the Thebes in Quseir could be of early to middle Eocene age.

Snavely et al. (1979) divide the Thebes Formation of Gebel Duwi into three 'informal members' representing different stages of deposition. The lower member rests conformably on the Esna Formation and is consistently about 40 to 60 m thick throughout the Red Sea area. It is composed of laminated to thinly-bedded limestone and chalk as well as massive bioturbated chalk beds. Chert is abundant in the upper portion. This member indicates a deep-water phase of predominantly pelagic deposition. The middle member is made up of thinly-bedded limestone and chalk interbedded with nodular limestones, intraformational conglomerates, reworked and *in situ* skeletal limestones, benthonic foraminiferal lime sands, cross-bedded planktonic foraminiferal lime sands and rare chert. This unit is about 100 to 120 m thick according to Snavely et al. (1979) and 200 m thick according to Strougo & Abul Nasr (1981). This heterogeneous accumulation of carbonate lithologies is indicative of progressive shallowing and outward progradation of shallow-water depositional environments. The upper member is thin (10 to 25 m thick) to non-existent. When present, it is made up of fine-grained limestones locally interbedded with cross-bedded oyster and shell hash. The distribution of this member probably indicates a period of local uplift coupled with a period of rapid sea-level drop.

Nakheil Formation

The Thebes Formation is overlain by a thick section of breccia beds interbedded with fine-grained lacustrine sediments made up of varicolored limestones, clays and sandstones (mainly yellow, red and green). The breccia beds are made up of large blocks derived from the Eocene and Cretaceous rocks. The only rounded components are the flint boulders which retain their original shape before transportation. This formation is named the Nakheil by Akkad & Dardir (1966). The formation is widely spread in the Quseir-Safaga district and may reach a thickness of 60 m or more. The formation is made up of two types of deposits: very coarse breccia beds and fine-grained lacustrine deposits. The coarse breccia beds do not include Precambrian fragments. The Nakheil occupies the slopes of the Thebes formation on the down-thrown side of the faults. It is closely associated with the early faulting which must have taken place after the deposition of the Thebes. The formation is seemingly non-fossiliferous. Most authorities give the formation an Oligocene age (Akkad & Dardir 1966).

MIOCENE AND LATER SEDIMENTS

The Miocene and later sediments form a strip along the coast. They are essentially littoral in character exhibiting marked lithological changes laterally and vertically. They rest unconformably and with a depositional dip on older rocks. The sediments form the subject of the classical work of Beadnell (1924) and Cox (1929). More recent workers include Souaya (1963), Akkad & Dardir (1966), Said (1969), Issawi et al. (1971), El Bassyony (1970, 1982), Tewfik & Burrough (1976), Tewfik & Ayyad (1982), Philobbos & El Haddad (1983), Samuel & Saleeb (1977), Samuel & Cherif (1978), Abou Khadrah & Wahab (1984) and Khedr (1984). Beds of Miocene and later sediments fall into the following units of which the evaporites, both topographically and geologically, are the most important and conspicuous (Fig. 18/5).

8. Reefs and raised beaches   Pleistocene

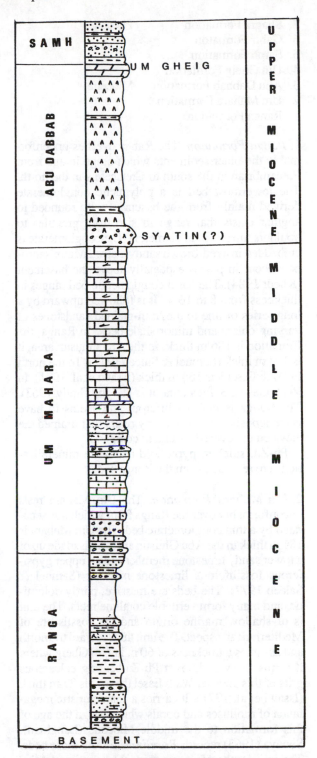

Figure 18.5 Miocene section of Red Sea coast (after Samuel & Saleeb-Roufaiel 1977 with modifications).

7. Shagra Formation   Pliocene
6. Gabir Formaton
5. Samh Formation
4. Um Gheig Formation
3. Abu Dabbab Formation   Miocene
2. Um Mahara Formation
1. Ranga Formation

1. *Ranga Formation.* The Ranga overlies unconformably the older sediments which range in age from Precambrian in the south to Cretaceous in the north. The lowermost bed is a polymictic conglomerate derived mainly from the basement with rounded to angular clasts that range in size from granules to boulders (up to 1 m in diameter). The conglomerate is embedded in a red-brown sandy matrix. Minor sapropels occur in places especially above the basement (Khedr 1984). The basal conglomerate bed ranges in thickness from 8 to 16 m. It is followed upward by a long series of fine to medium-grained sandstones of varying colors and minor shale beds. In Ranga, the formation is 186 m thick. In the Abu Ghusun area, it is 103 m thick (Samuel & Saleeb 1977). To the north in Wadi Essel it is 166 m thick (Issawi et al. 1971). In Wadi Gasus, Safaga area, it is 123 m (Gindy 1963). The Ranga is non-fossiliferous and seems to have been deposited subaerially by rivers that drained the elevated uncovered basement complex.

Pb-Zn, sulphur, pyrite and marcasite mineralizations are recorded from the Ranga beds.

2. *Um Mahara Formation.* The Um Mahara rests unconformably over the Ranga from which it is separated by a thin conglomerate bed. The Um Mahara is 181 m thick in the Abu Ghusun area and is made up of a lower sandy limestone member and an upper gypsiferous fossiliferous limestone member (Samuel & Saleeb 1977). The beds are massive, partly dolomitic, and many form veritable coralline reefs. The unit is of shallow marine origin and the fossils are of Mediterranean aspect. The unit thins toward the north and assumes a thickness of 60 m in Um Gheig where it forms the well-known Pb-Zn-bearing calcareous grits of this area. In Wadi Essel the unit is 27 m thick (Issawi et al. 1971). It carries a characteristic megafauna of molluscs and corals which assign the age of this formation to the middle Miocene (Langhian). Souaya (1963) records *Borelis melo* from these beds. The unit is, therefore, younger than the basal Miocene beds of the Gulf of Suez region. Mineralizations of Mn, S, Pb-Zn and pyrite are recorded from this unit in many localities.

The Ranga and Um Mahara were first described as one unit, the Gebel Rusas Formation, by Akkad & Dardir (1966a). Tewfik & Burrough (1976) subdivide the formation into two units, a lower clastic unit

and an upper carbonate unit. These were named by Elbassyony (1982) as the Ranga and the Abu Hamra Formations. Elbassyony further distinguishes the coralline limestone beds, the remnants of which overlie the basement in many areas in the south, into a separate unit, the Ghadir. In the opinion of most authors the Ghadir is part of the Um Mahara complex.

3. *Abu Dabbab Formation.* (=Evaporite series of older authors). Evaporite deposits extend for hundreds of kilometers along the coastal plain of the Red Sea. Their distribution is patchy and their thickness varies from one place to another being in the range of 90 to 400 m in the onshore areas. An excess of 3000 m of rock salt, equivalent to the onshore gypsum and anhydrites, was penetrated in the offshore wells (Tewfik & Ayyad 1982). In a few localities the formation rests upon the flanks of the basement complex, but more often the formation overlies conformably the Um Mahara. In the Abu Ghusun area it is separated from the underlying limestone beds of the Um Mahara by a thick conglomerate bed, about 120 m thick (Tewfik & Ayyad 1982). This unit may be equivalent to the calcareous sandstone unit recorded between the Um Mahara and the Abu Dabbab by Philobbos & El Haddad (1983) and named the Syatin Formation. The Abu Dabbab consists of solid white gypsum, weathering to a hard coralloid-like hackly surface of characteristic yellowish brown color. Intercalated shales are rare and generally confined to the base of the formation, while sands and gravels are practically absent. The only intercalations commonly associated with the gypsum are irregular masses and lenticles of hard, compact and sometimes dolomitic or semicrystalline limestone, often full of small isolated cavities. The gypsum usually exhibits distinct traces of bedding planes, the dip being at low angle averaging about 5° toward the coast. The formation is non-fossiliferous but its age is assumed to be middle to late Miocene on the basis of its similarity with the evaporites of the Gulf of Suez.

4. *Um Gheig Formation.* Overlying the Abu Dabbab is a characteristic 8 to 10 m thick, hard, ledge-forming dolomite bed which Samuel & Saleeb (1977) classify as a unit, the Um Gheig, because of its areal extent and usefulness as a map marker. The Um Gheig forms the basal bed of the Samh as described by Abou-Khadrah & Wahab (1984). The bed is a grainstone (mud-free carbonate rock) rich in crinoids, oncoids, algae and bioclasts and seems to have been deposited in agitated shallow water above the wave base. This unit is frequently referred to in the literature as the oil-tainted limestone; the rock when fractured emits an odor of petroleum.

5. *Samh Formation*. The Samh overlies unconformably the Abu Dabbab Formation or the Um Gheig bed. It is made up of a lower 5 m thick green to grey shale and fine-grained variegated sandstones, a middle 2 m-thick escarpment maker bed of hard sandstone and an upper 15 to 20 m-thick limestone bed with occasional conglomerates. The limestones are recrystallized and form the bulk of the formation. They carry, in addition to many poorly preserved fossils, casts of species of the freshwater *Melania* and others. The presence of these fossils indicates deposition in brackish, if not fresh, water environment. Abou-Khadrah & Wahab (1984) advocate deposition in intertidal to supratidal flats. These authors base their argument on the petrographic characteristics of the sediments. The age of the Samh is probably late Miocene.

The exact type locality of this formation is not known as Wadi Samh does not appear on the published topographic maps of the area. For this reason Elbassyony (1984) designates Wadi Wizr as the type locality; he describes from this area a 53 m-thick section of sandstones, marls and shales. Philobbos & El Haddad describe a non-marine unit of fine-grained clastics and limestones which they name the Wizr Formation and which may be correlatable with the Samh.

6. *Gabir Formation*. (= oyster and cast beds of Beadnell 1924). The Gabir overlies the Samh with seeming conformity. It was originally described by Akkad & Dardir (1966) as a unit made up mainly of sandstone which is overlain conformably by the lower member of the overlying Shagra Formation. El Bassyony (1982) believes that the lower Shagra member (as described by Akkad & Dardir) belongs to the Gabir; both are of Pliocene age and are conformable. The Gabir, thus emended, is 124 m thick and is separated from the overlying Shagra by a surface of unconformity (Fig. 18/6). It is made up of a lower 44 m-thick succession of sand stones (80%) and an upper 80 m-thick unit of sandstone (40%), marls (29%), reefal limestones (10%), calcareous grits and gravel beds. Many beds are rich in casts of molluscs. Several oyster interbeds carrying the large *Ostrea gingensis* intercalate this succession. The most common fossils in the cast beds are *Metis papyracea* and *Pecten erythreanus*. Souaya separates from these beds *Archaias aduncus*. This formation marks the first proper marine invasion after the Abu Dabbab Formation. The fauna encountered in this unit is of Pliocene age and Indo-Pacific origin (Cox 1929). The transgression must have come from the south across the Bab El Mandab strait. Philobbos & El Haddad (1983) are unable to identify the Gabir. Instead they

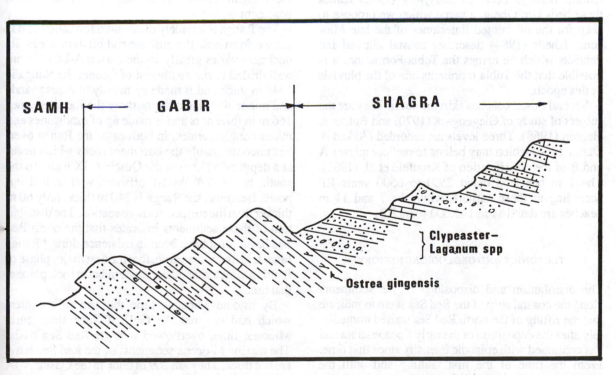

Figure 18.6 Diagrammatic cross-section in Wadi Wasaat showing relationship of Gabir and Shagra Formations (after Elbassyony 1982).

describe two units: the Abu Shiqeili and Dishet El Daba'a which may be correlatable with the two members of the Gabir. It is interesting to note here that, in the Quseir district, Issawi et al. (1971) are not able to differentiate the post-evaporite formations (the Samh, Gabir and Shagra) from one another and propose to lump them into one formation, the Gasus.

7. *Shagra Formation.* (= *Clpeaster-Laganum* series of Beadnell 1924, upper member of Shagra Formation of Akkad & Dardir 1966, Wizr Formation of Elbassyony 1982). The Gabir is followed unconformably by a succession of littoral deposits containing mainly arkosic sandstones and minor marls rich in *Clypeaster scutiformis* and *Laganum depressum*. The fauna is Indo-Pacific (Cox 1929) and is assigned to the Pliocene. The thickness of this formation is 22 m. This formation is probably coeval with the sandstones of the Sharm El Arab described by Philobbos & El Haddad (1983) from the Quseir area.

8. *Pleistocene reefs and raised beaches of Wadi Shagra.* Akkad & Dardir (1966) describe from Wadi Shagra a 34 m section made up of a succession of four organic reefs separated by conglomerate and gravel beds. These latter beds are interpreted as representing pluvial episodes which interrupted an otherwise dry climate. It is certain that the section described by Abou-Khadrah & Wahab (1984) under Shagra Formation belongs here. El Bassyony (1982) names these beds Um Gheig, a name which we propose to keep for the oil-tainted limestones of the late Miocene. Khedr (1984) describes coastal alluvial fan deposits which he names the Tubia Formation. It is possible that the Tubia represents one of the pluvials of this epoch.

Several raised beaches skirt the coast. They are the subject of study of Giegengack (1970) and Butzer & Hansen (1968). Three levels are recorded (Akkad & Dardir 1966) which may belong to reef complexes A and B of the classification of Kronfeld et al. (1982). The 1 m beach is dated 2500 to 6500 years BP according to these authors, while the 7 and 11 m beaches are dated 81 to 141,000 years BP.

TECTONICS AND GEOLOGICAL HISTORY

The distribution and disposition of the sediments along the coastal strip of the Red Sea seem to indicate that the rifting of the north Red Sea started immediately after the deposition of the early Eocene strata and has continued with episodic intensity since that time. From the time of the first faulting and until the beginning of the middle Miocene, only non-marine sediments were deposited indicating that the earliest

trough must have been perched and must have had no access to the sea.

The oldest of these sediments is the Nakheil, a lacustrine deposit with many interbedded gravity breccias, the clasts of which are of local derivation and include no Precambrian pebbles. This deposit seems to have filled the valleys and topographic lows of this emerging landscape. Subsequent erosion destroyed the greater part of the sedimentary cover which is now known only in a few of the infolded and faulted hills of the Quseir-Safaga reach. Similar sediments to those which make up these hills must have covered the entire Red Sea range. Since none of these early sediments is recorded in any of the wells hitherto drilled in the offshore areas of the north Red Sea, it is feasible to believe that immediately after the accumulation of the Nakheil Formation intensive erosion must have taken place transporting these sediments away from the Red Sea basin. This phase of erosion continued until the early Miocene when the basement was uncovered and pebbles from it were incorporated in the conglomerates of the Ranga Formation.

The presence of Precambrian clasts in the Ranga conglomerates indicates that the Red Sea hills must have been stripped of their sedimentary cover by Ranga (?early Miocene) time. Before their uncovering the Red Sea hills must have formed a formidable mountain range. During Ranga time the north Red Sea basin became for the first time a receptacle receiving the detritus transported from the uncovered basement complex.

The Ranga is a widely distributed formation and is known from both the onshore and offshore areas. Its thickness varies greatly. In the Quseir A-IX offshore well drilled to the southwest of Quseir, the Ranga is 1546 m thick and is made up mainly of coarse sands and minor shales. On the onshore the Ranga ssumes 166 m in thickness and is made up of sandstones and minor conglomerates. In both cases the Ranga overlies unconformably the basement rocks which occur at a depth of 5039 m in the Quseir A-IX well. To the south, in the RA West-1 offshore well drilled opposite Berenice, the Ranga is 240 m thick, only 60 m thicker than the surface onshore section. The distribution of these sediments indicates that the north Red Sea basin must have been in existence during Ranga time. It is likely, therefore, that the first major phase of faulting which shaped the modern basin took place at that time.

By middle Miocene time the arm of the area, which had covered the Gulf of Suez since early Miocene time, overflowed into the Red Sea basin. The marine Miocene sediments of the Red Sea basin are not thick. They are 209 m thick in the Quseir A-IX well, as against 31 m on the coastal plain. In the south the section becomes more calcareous. A 72 m-thick

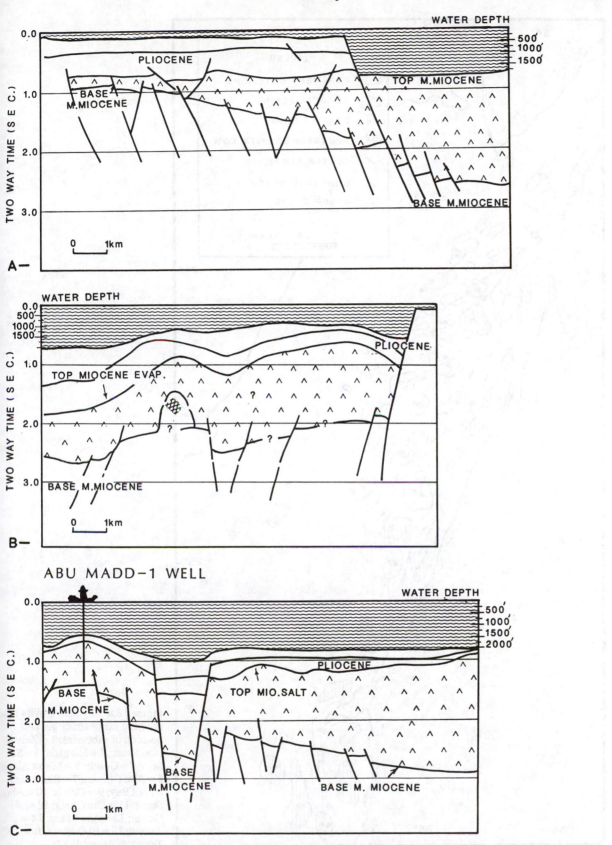

Figure 18.7 Red Sea offshore seismic profiles showing A. near-shore homoclinal structures; B. shallow structure with Miocene sediments and C. basinward fault block structure underneath massive salt (adapted from Tewfik & Ayyad 1982).

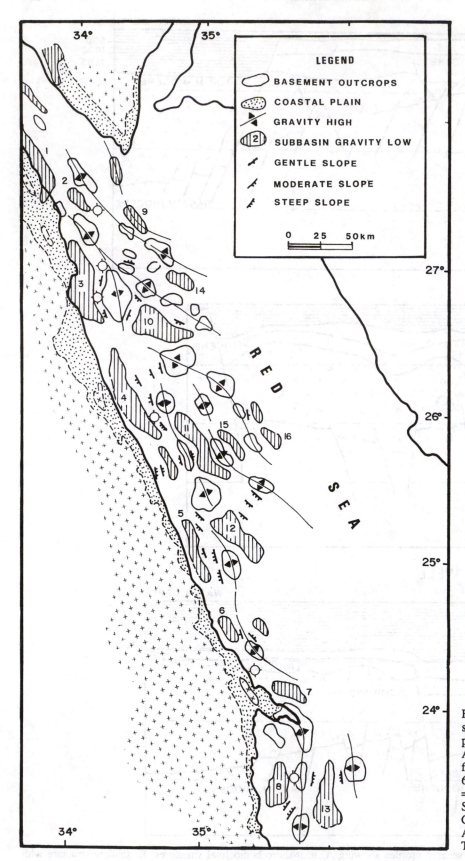

Figure 18.8 Regional gravity map showing major structures and position of subbasins: 1 = Zeit-Abu Shaar; 2 = Hurgada; 3 = Safaga; 4 = Quseir; 5 = Mersa Alam; 6 = Abu Ghusun; 7 = Ras Banas; = Foul Bay; 9 = Giftun; 10 = Abu Soma; 11 = Um Gheig; 12 = Gemal; 13 = Abu Madd; 14 = Atshan; 15 = El Qadim (after Tewfik & Ayyad 1982).

solid carbonate unit is recorded from RA West-1 well and appears on the log of this well as 'Nullipore rock'. In the onshore area in the same general region, the section is 180 m thick and is fossiliferous and reefal.

During the late Miocene the lowering of the sea level and final desiccation of the Mediterranean, with which the Gulf of Suez was connected, severed the Red Sea from the world's oceanic system and converted it into a lake or a series of lakes where evaporites were deposited. The evaporites are widely distributed and are known from the onshore areas (mainly in the form of anhydrites) and from the offshore areas (mainly in the form of rock salt). Thicknesses vary greatly. In some offshore areas, thicknesses of up to 4000 m are known. The thickness was probably governed by the bottom topography. The gravity map (Fig. 18/8) shows the presence of numerous subbasins which probably have deeper columns of brine and thicker evaporites.

A short pluvial followed in which the clastics and carbonates of the Samh Formation were deposited. These were followed by the Pliocene marine sediments of the Gabir Formation rich in Indo-Pacific fossils. The transgression came from the south. It is not certain whether the transgression was the result of the global rise of sea level or the result of a tectonic movement that opened up the Bab El Mandab Strait. The northerly extent of the Pliocene transgression is difficult to determine, but it certainly did not cover the greater part of the Gulf of Suez.

The Pleistocene section of the Red Sea coastal strip seems to have been deposited in a continuously subsiding basin. The sequence, 60 m thick, consists of alternating coral reefs and sands and conglomerates deposited in littoral to beach environment during alternating arid and pluvial episodes. Several polymictic conglomerates interbedded with coral reefs are described from Mersa Alam in the early and middle Pleistocene of Wadi Shagra area. The late Pleistocene is represented by three raised beaches: 11, 6 to 8 and 1 m in elevation above the modern sea level. The 1 m beach has a Uranium series date of 6500 to 8500 years BP. It is remarkable that the climatic fluctuations suggested by the distribution of these sediments coincide to a large extent with the curve suggested by Cita et al. (1973) for the Mediterranean region.

The outcropping sediments along the Red Sea coast have a regional eastward dip to the sea. Dip angles range from 35° at the inland basement contact to nearly horizontal near the shore. Faults with northwest-southeast trends are the main structural elements in the area. Cross faults perpendicular to this trend occur. Recent geophysical data (Tewfik &Ayyad 1982) show that these surface faults extend eastward into the main Red Sea depression. They also show that the Red Sea marginal homocline dips gently eastward from the shore and is made up of many horsts and fault blocks (Fig. 18/7). Away from the shore and along the shelf many of the anhydrite and salt beds are folded, probably because of the mobility of the basal evaporite section (Phillips & Ross 1970). Salt domes and diapirs are common in the Red Sea offshore areas (Fig. 18/7). Further offshore the seismic reflection data indicate a relatively narrow northwest trending zone of pre-middle Miocene horsts and tilted fault blocks. In many cases the structural axes are shifted laterally, probably as a result of east-west cross faults (Fig. 18/7).

The gravity maps show northwest trending negative gravity anomalies which are believed to represent subbasins within the Red Sea. These subbasins trend parallel to the coast line and are separated by structural ridges (Fig. 18/8). They seem to be of Miocene (pre-evaporite) age and probably controlled the isopachous variations of the evaporites. Khattab (1985) studies the geophysical data of the Benas basin (basin 9 in Figure 18/8) and concludes that the configuration of the basin was controlled by pre-middle Miocene and Quaternary faulting which, according to him, supports the hypothesis of Red Sea arching and subsequent faulting at marginal shelves.

CHAPTER 19

North and Central Sinai

DAVID A. JENKINS

Conoco Oil Company, Houston, Texas, USA

1 INTRODUCTION

The Sinai Peninsula covers an area of approximately 61,000 km² and is separated geographically from Egypt by the Suez Canal and the Gulf of Suez. Highly dissected igneous and metamorphic mountains, whch rise to a height of 2675 m (Gebel Musa), form the southern tip of the peninsula.

The central part of the peninsula consists of subhorizontal Mesozoic and Tertiary sediments, creating the plateaux of Gebel El Tih and Gebel Egma, which are drained by the northerly flowing affluents of Wadi El Arish.

North of latitude 30° N, the topography comprises low alluvial plains which are broken by large uplifted Mesozoic domes and anticlines, such as Gebel Yelleq (1090 m), Gebel Halal (890 m) and Gebel Maghara (735 m). Northward these 'Syrian Arc' structures sink seaward, due to Tertiary down-to-the basin faulting, and are hidden under the Quaternary coastal plain and continental shelf deposits.

North of Gebel Maghara and extending nearly to the Mediterranean coast is a broad tract of sand dunes, some of which attain heights of 91 m above sea level.

2 TECTONIC SETTING

The Sinai Peninsula (Fig. 19/1) is wedged between the African and Arabian plates the boundaries of which are defined by the Gulf of Suez and Gulf of Aqaba-Dead Sea rift systems.

In the south, exposed pre-Cambrian igneous and metamorphic rocks form the Arabo-Nubian shield. Field and petrographic evidence indicate that the shield consists of a series of island arcs which were cratonized during the late Proterozoic-early Paleozoic (1200 to 500 my BP) Pan-African orogeny (Gass 1981). The peneplaned paleosurface of the shield dips gently northward with the overlying sediments,

ranging from Cambrian to Recent, thickening northward.

In central Sinai the Gebel El Tih-Egma plateaux, 914 m above sea level, represent a thin sedimentary cover, which is affected only by faulting. This region has been described in detail by Shata (1956) and Said (1962).

An east-west trending shear zone of dextral strike slip faults with up to 2.5 km of displacement has been recognized in central Sinai by Steinitz et al. (1978) (Fig. 19/1).

Northward from the Raqabet El Naam dextral wrench fault (Fig. 19/1), the style of deformation becomes increasingly complex and consists predominantly of 65° N to 85° E oriented anticlinal folds and monoclinal flexures expressed mainly in the Cretaceous strata. This belt of folds extends offshore into the southeast Mediterranean Sea.

The individual anticlines, which increase both in size and in amplitude northward, culminate in the extremely large, overturned, thrusted structures such as Gebel Maghara and Gebel Halal (Fig. 19/1). Shata (1956) and Said (1962) have described the individual structures in great detail.

The Syrian Arc structures attain a more northerly trend aligning themselves with the sinistral Dead Sea fault system and the Pelusium line (Fig. 19/1), to the east and northeast of Sinai. In these regions, the folds appear to be reminiscent of fault plane drag.

There are two proposed models for the tectonic evolution of north and central Sinai. The central issue is the timing of the initiation and development of the fold belt.

Some authors consider that the folding was initiated in the late Paleozoic when the embryonic Gulf of Suez rift was created. The folding was then reactivated throughout the Mesozoic and the deformation climaxed in the Oligocene (Agah 1981).

The alternative theory is that the fold system is closely related to the compressional stresses created when the Tethys Sea, which was located between the Afro-Arabian and Eurasian land masses, began to

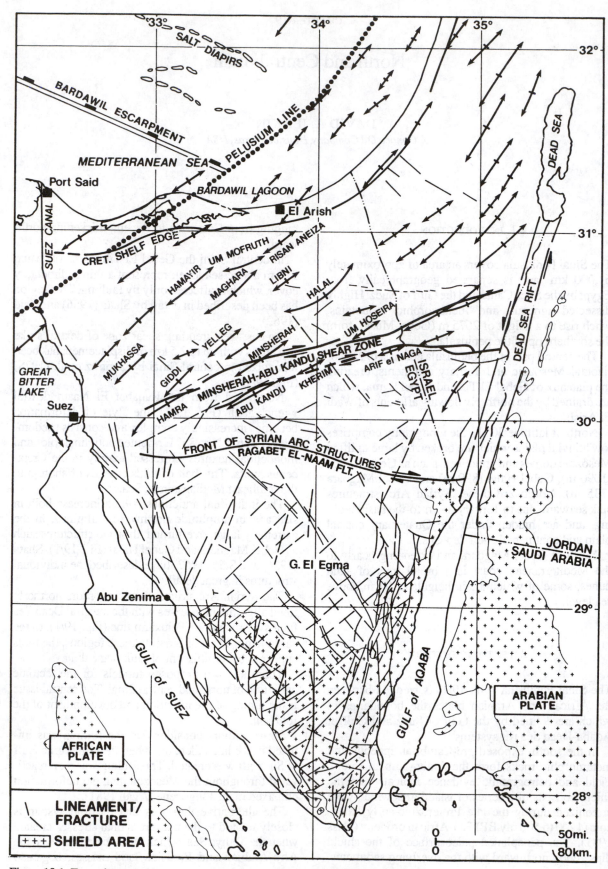

Figure 19.1 Tectonic map of north and central Sinai (after Neev 1975 and Agah 1981).

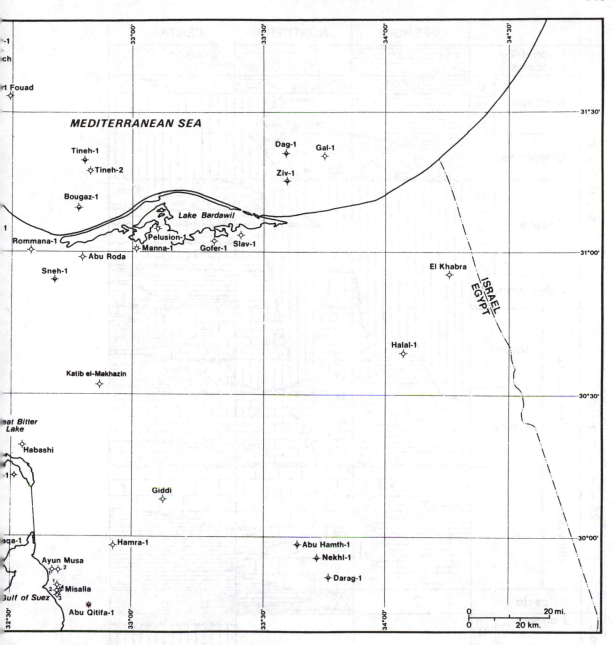

Figure 19.2 Location of boreholes, north and central Sinai.

close, as a result of northerly subduction, during the late Cretaceous (Senonian) period. The closure of this ocean had far-reaching consequences throughout the Middle East, including the widespread emplacement of southwesterly directed nappes along the entire length of the Arabian plate from Oman (Coleman 1981) to south Turkey, the formation of wrench basins and the reactivation of old structural grains (Murris 1980). Bartov et al. (1980), in a detailed study of the stratigraphic and structural history of Gebel Arif El Naga (Fig. 19/1) postulated that folding

continued throughout the Coniacian-Maastrichtian with regional differential uplift during the post-Eocene period (40 to 14 my BP).

In northernmost Sinai and the offshore area, the sedimentary cover increases in thickness from less than 1829 m to in excess of 7620 m. Ginzburg & Gvirtzman (1979) postulate that this large increase in sediment thickness and the development of the Mesozoic platform margin facies (Fig. 19/8) is related to the transition from continental to oceanic crust. These platform margin facies are interpreted to

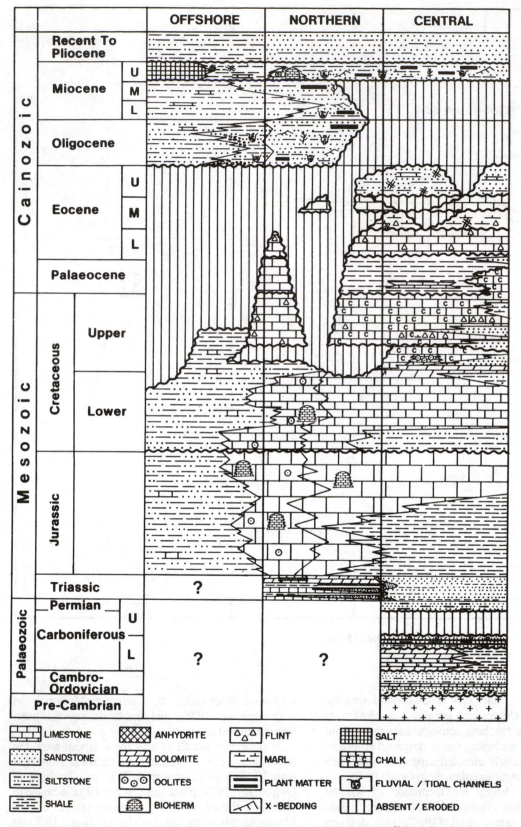

Figure 19.3 Generalized chronostratigraphic column for central, northern and offshore Sinai.

define the fossil Mesozoic continental margin which bordered the Tethys Ocean.

Neev (1975) proposed that the oceanic-continental crustal boundary is represented by the northeast-southwest trending Pelusium line, which lies 22 to 25 km offshore (Fig. 19/1). The Pelusium line is interpreted to represent a transcontinental megashear suture extending southwest through the Kufra basin in central Sahara and into the Benue trough. Rhys-Davies (1984) is of the opinion that the Pelusium line had a profound effect on the distribution and type of faulting which occurred in the offshore region during the late Tertiary.

The exact relationship between the Pelusium line, the oceanic-continental crustal boundary and the Mesozoic platform margin facies requires additional research.

In terms of reconstructing plate movements, the shifting of the Arabian plate by as much as 105 km in a right-lateral sense along the Dead Sea fault, coupled with a clockwise motion, necessitates, and dictates, a westerly and clockwise translation of the Sinai plate in order to avoid crowding and overlapping of the continental crust in the Dead Sea-Gulf of Aqaba shear zone. In turn, the bulk of the Sinai's gross westerly motion must then be taken up along and within the Gulf of Suez rift, which is a well-defined structural boundary of the Sinai cratonic element, separating it from the African plate. This motion could be further resolved into two net components: one parallel (left-lateral) along, and the other normal (compression) to the rift trend. This sequence of plate motions can simulate compressional and left-lateral shear stresses capable of partially closing the Gulf of Suez. Thus the reverse sense of deformation, i.e. ductile extension of the crust (McKenzie et al. 1970, Freund 1970, Cochran 1981) and right-lateral strike-slip, since at least the late Oligocene time, could have caused the formation of the Gulf.

This model of plate interaction and reconstruction shows that the closing of the Gulf of Suez by a minimum of 25 km normal to its present northwest trend and left-laterally shifting the Sinai side of it by a distance of 30 to 50 km relative to its Nubian side satisfies the required 105 km of motion along the Dead Sea shear. It would also match the structures of north and central Sinai with those of northeast Egypt, across the rift. This reconstruction also matches the pre-rift Proterozoic Najd fault system of the Arabian and Nubian shields across the Red Sea, as well as the structurally controlled coastlines of the Red Sea and the Gulf of Suez, with a remarkable accuracy (Agah 1981).

3 STRATIGRAPHY

The Sinai is covered by sediments which were deposited on a predominantly shallow platform and range from Cambrian to Recent in age. A generalized lithostratigraphic column for the Sinai is illustrated in Figure 19/3.

3.1 *Paleozoic*

The Paleozoic sediments are only exposed in the southern central parts of the Sinai, primarily in the Um Bogma area east of Abu Zenima and at Abu Durba.

These sediments have been studied in detail by Abdallah & Adindani (1963), Hassan (1967), Weissbrod (1969), Said (1980) and Issawi & Jux (1982). However, the lack of diagnostic fossils has made accurate age-dating and correlation extremely difficult and additional confusion has been created by the use of various formation nomenclature.

A composite lithostratigraphic section of the Paleozoic in Sinai is illustrated in Figure 19/4.

The basal Cambro-Ordovician sequence comprises the Araba, Naqus and Wadi Malik Formations of Issawi & Jux (1982), which are equivalent to the Yam Suf group and the Netafim Formation of Weissbrod (1969).

These clastic sediments, deposited on the peneplained surface of the pre-Cambrian basement, are dominated by grits, siltstones, subarkoses and conglomerates, with a few dolomite beds.

Issawi & Jux (1982) have interpreted some of the conglomerates to be of fluvio-glacial origin, suggesting that some of these sediments may be of Silurian age whilst the upper parts are interpreted as Devonian. However, these age datings are not based on direct faunal evidence and therefore may not be totally correct.

Said (1980) reports that a carbonate bed at Abu Durba has yielded algal stromatolites which have been identified as belonging to the species *Vetella ushbasica,* known from the lower Cambrian of Kazakhstan, USSR.

These basal Paleozoic sediments were deposited under fluvial-paralic conditions and probably represent relics of a very large sediment body which originally extended over a large area, but was deeply eroded before the deposition of the upper Paleozoic strata.

The lower Carboniferous Um Bogma Formation (Fig. 19/4) provides the first reliable date for the Paleozoic sequence in Sinai. These marine carbonates are richly fossiliferous, especially in foraminifera (Tetrataxidae, Archadiscidae, Endothyridae) and have been dated as Tournaisian-Visean. Ferro-

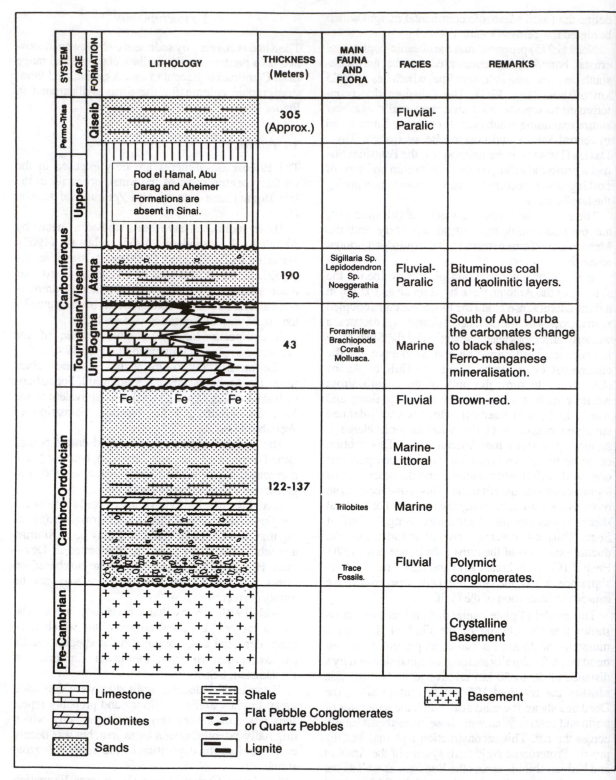

SYSTEM	AGE	FORMATION	LITHOLOGY	THICKNESS (Meters)	MAIN FAUNA AND FLORA	FACIES	REMARKS
Permo-Trias		Qiseib		305 (Approx.)		Fluvial-Paralic	
Carboniferous	Upper		Rod el Hamal, Abu Darag and Aheimer Formations are absent in Sinai.				
	Tournaisian-Visean	Ataqa		190	Sigillaria Sp. Lepidodendron Sp. Noeggerathia Sp.	Fluvial-Paralic	Bituminous coal and kaolinitic layers.
		Um Bogma		43	Foraminifera Brachiopods Corals Mollusca.	Marine	South of Abu Durba the carbonates change to black shales; Ferro-manganese mineralisation.
Cambro-Ordovician			Fe Fe Fe			Fluvial	Brown-red.
				122-137		Marine-Littoral	
					Trilobites	Marine	
					Trace Fossils.	Fluvial	Polymict conglomerates.
Pre-Cambrian			+ + + + + + + +				Crystalline Basement

Legend:

- Limestone
- Dolomites
- Sands
- Shale
- Flat Pebble Conglomerates or Quartz Pebbles
- Lignite
- Basement

Figure 19.4 Lithostratigraphic section of the Paleozoic in central Sinai.

manganese minerals within the basal part of the carbonates at Um Bogma have been extensively mined for many years. South of Abu Durba the Um Bogma Formation is represented by black marine shales (122 m thick) which have yielded similar fauna.

The basal 132 m in the Ataqa-1 well has been assigned to the Um Bogma Formation.

Unconformably overlying the Um Bogma Formation are thick cross-bedded, very fine to fine grained sandstones of the Ataqa Formation. Visean flora and fauna have been identified within these sediments by Horowitz (1973). Kaolinitic layers are present within the Ataqa Formation and these are exploited commercially in the Wadi Abu Natash area. At Um Bogma the Ataqa Formation contains bituminous coal seams and is capped by a basaltic sill. Shales and siltstones which were penetrated in the Ataqa-1 (226 m) and the Ayun Musa-2 (240 m) wells are possible facies equivalent of the Ataqa Formation; however the lack of diagnostic fauna precludes a definitive correlation.

Upper Carboniferous sediments were previously thought to be absent in Sinai. However, Issawi & Jux (1982) have reported the presence of the Aheimer Formation at Um Bogma.

In central and south Sinai, the Permian is represented by the continental sediments of the Budra or Qiseib Formation, which will be discussed later. In north Sinai there is neither surface nor subsurface data available on the Permian, although it is highly probable that marine conditions existed in this area during this period.

3.2 *Mesozoic*

3.2.1 *Triassic*

In the Sinai, the only outcrop of Triassic sediments is in the core of Gebel Arif El Naga (Awad 1946, Karcz & Zak 1968) (Fig. 19/5), whilst in the subsurface they have only been penetrated in the Halal-1 borehole.

As a result of the lack of data on the Triassic system and in order to understand the Triassic depositional environments, it is necessary to extrapolate and integrate the subsurface data and interpretations from the Negev contained in an excellent report by Druckman (1974).

During most of the Triassic period, shallow marine conditions prevailed throughout north Sinai, with up to 914 m of sediments being deposited, and these conditions were governed by the Tethys Sea which lay to the northwest. Transitional environments, such as tidal flats and deltas moved northward and southward over north Sinai throughout the Triassic period. In central and south Sinai, continental, mainly fluvial,

conditions predominated with up to 762 m of sediment being deposited.

Druckman (1974) subdivided the Triassic system into five formations, and these are the marine equivalents of the continental Qiseib Formation of central and south sinai. The lower two formations are recorded from the subsurface (Halal-1 well).

The lowermost formation consists mainly of brown to dark grey, fissile shales, grey biomicrites, biosparites, marine sandstones with dolomites and fine-medium grained, deltaic sandstones becoming increasingly abundant southward. The lithology and fauna of this formation, which is 196 m-plus thick at the Halal-1 well, indicates a broad northerly lime-mud carbonate shelf with southerly deltaic complexes and localized supra-tidal conditions. It is postulated that during this period the shoreline lay in an east-west direction just north of the Halal-1 well.

The overlying Formation consists entirely of biomicrites and micrites (50 m thick at Halal-1) indicating a low energy carbonate shelf resulting from a southerly marine transgression with a postulated east-west shoreline located slightly north of the Hamra-Abu Hamth wells.

During middle Anisian time the shoreline prograded to the north and northwest with a corresponding increase in clastic sedimentation. In the subsurface, tidal flat facies grade into clean winnowed sands in the nearshore region, with fine clastics deposited offshore and low energy carbonates further north on the open shelf. The exposed section at Arif El Naga is named the Arif El Naga Formation (Fig. 19/5). The lower 50 m-thick unit of this section consists of multi-colored, coarse-grained well cemented sandstones, variegated siltstones and shales carrying plant remains. Paleocurrent analysis of the festoon-bedded sandstones indicate a southeasterly source for the clastics. This unit has the informal name Arif El Naga 'A' beds (Said 1971). Paleocurrent analysis of the festoon-bedded sandstones indicate a southeasterly source for the clastics. Overlying this unit is a 19 m thick unit of argillaceous micrites, biomicrites and biosparites which is given the informal name Arif El Naga 'B' beds or the *Beneckia*-bearing beds (Said 1971). Both the 'A' and 'B' beds are named Gevanim by the Israeli geologists.

The overlying formation, Arif El Naga 'C', represents the main marine transgression during the Triassic period and it extended southward well into central Sinai. Marine carbonates of this formation have been identified in the Ayun Musa-2 (69 m), Hamra-1 (100 m), Abu Hamth (38 m) and Nekhl (35 m) wells. The 'C' beds are named Saharonim by the Israeli geologists.

This formation consists of biomicrites, biosparites, and shales in the lower part grading upward into

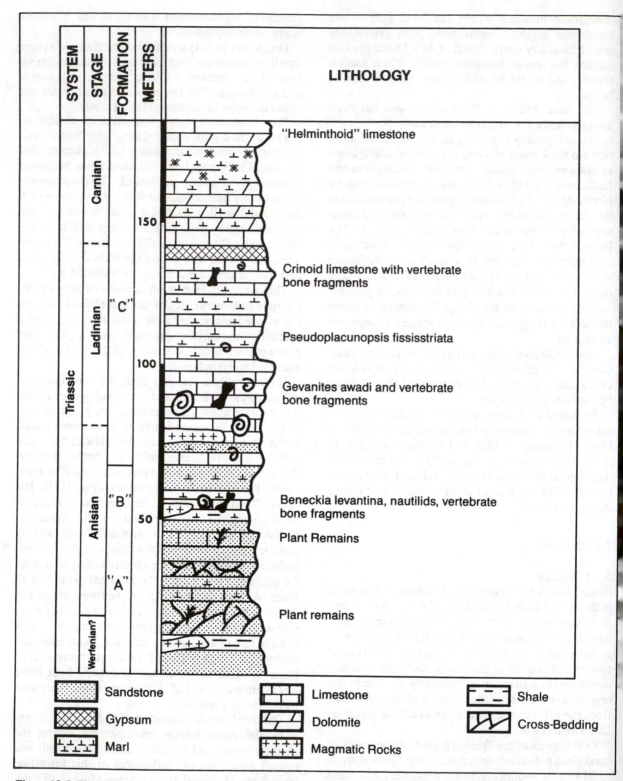

Figure 19.5 The Triassic sequence of Gebel Arif-el-Naga (after Karcz & Zak 1968 with modifications).

oobiomicrites, micrites, pelmicrites, dolomicrites, algal stromatolites, calcareous and dolomitic shales and flat pebble conglomerates. At Gebel Arif El Naga the Saharonim Formation is 116 m thick, whilst in the Halal-1 well it is 275 m thick.

The lithofacies suggest that a subtidal shallow marine environment existed on the open shelf to the north, changing southward into a wide expanse of tidal flats which incurred occasional prolonged marine incursions. The presence of thick evaporites in the subsurface of the Negev indicate the presence of localized salina and sabkha conditions, which were subject to occasional flooding and it is highly probable that similar localized conditions existed in Sinai.

The uppermost Triassic sequence is represented by the Mohilla Formation which consists of dolomicrite, dolomitic shales, dolomitic limestones, algal stromatolites and anhydrite. In the Halal-1 well this forma-tion is 50 m thick, but is absent at Gebel Arif El Naga, suggesting that this structure may have been emer-gent at this time. The overall depositional environ-ment is interpreted to be supratidal to sabkha condi-tions as indicated by the lack of fossils, algal mats, and birdseye structures, with localized hypersaline lagoons. At Halal and to the north shallow marine conditions existed.

In central and south Sinai, the entire Triassic se-quence is represented by the fluvial sediments of the Budra or Qiseib Formation. These sediments consist of brown, purple and grey, fine-coarse grained sandstones, variegated shales and siltstones with channel and overbank deposits, abundant ripple marks, cracks, cross-bedding and silicified tree-trunks, several metres in length.

3.2.2 Jurassic
During the Jurassic period, north and central Sinai

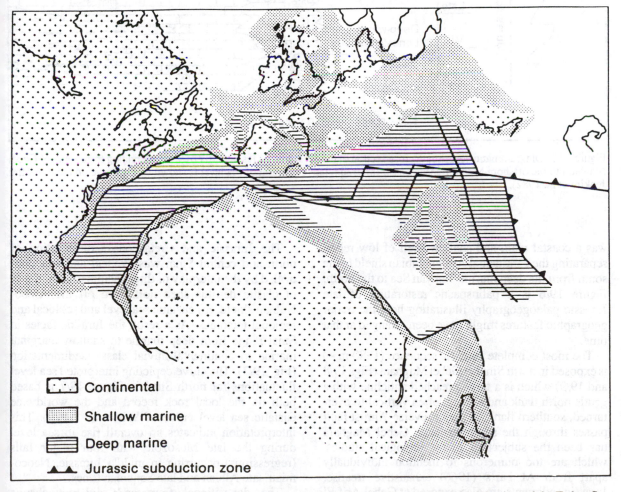

Figure 19.6 Palinspastic restoration of late Jurassic paleogeography showing the young central Atlantic, and the Tethys Sea between Africa and Europe. Uncertain areas are shown without pattern and include a probable deep marine basin north of Egypt (after Eyal 1980).

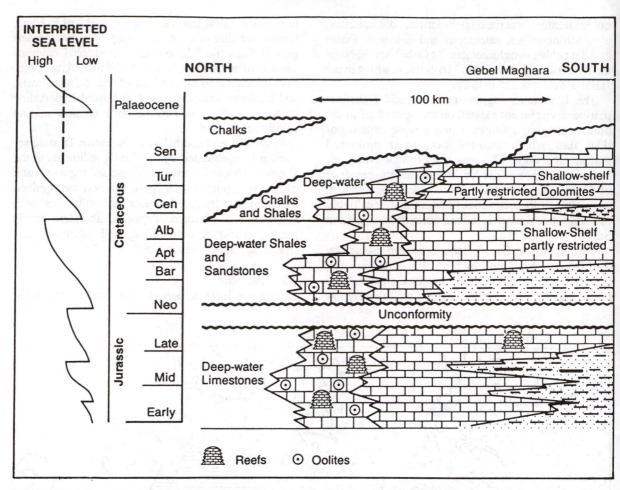

Figure 19.7 Diagrammatic north-south cross-section of the Mesozoic facies of northern Sinai, extending from south of Gebel Maghara to the offshore Mediterranean coast. Inferred relative position of sea level is shown by the curve on the left (after Jenkins et al. 1982).

was a coastal plain and shallow shelf of low relief, separating the continental Arabo-Nubian shield to the south from the deep marine Tethyan Sea to the north. Figure 19/6 is a palinspastic restoration of late Jurassic paleogeography illustrating how the major geographic features might have been arranged at that time.

The most complete Jurassic sequence (1980 m +) is exposed in north Sinai at Gebel Maghara (Figs 19/8 and 19/9) which is a large breached anticline with a gentle north flank and a steep, often vertical to over-turned, southern flank. A major reverse-?thrust fault passes through the southern flank. Gebel Maghara has been the subject of many geological studies, which are too numerous to mention individually apart from Al Far's (1966) exhaustive treatise. Jurassic sediments are also exposed at Gebel Arif El Naga (141 m), Gebel Minshera (80 m) and have been penetrated in 13 boreholes.

The Jurassic and Cretaceous sedimentation of north and central Sinai was cyclical and was con-trolled by the interplay of three factors (a) the supply of clastic detritus shed northward off the Arabo-Nubian shield, (b) eustatic sea level and (c) local and regional tectonics. As a result the Jurassic facies in Sinai range from deep marine to shallow-marginal marine to continental-fluvial clastic sedimentation (Fig. 19/7). The curve depicting interpreted sea level fluctuations for north Sinai in Figure 19/7 is based both on the local rock record and the worldwide eustatic sea level curve of Vail et al. (1979). This interpretation indicates an overall rise in sea level during the late Mesozoic with four major falls (regressions) occurring in middle Jurassic, Neoco-mian, late Albian and late Senonian times.

The depositional framework did not change greatly through the Jurassic and Cretaceous periods and the dominant sediments are limestones and do-

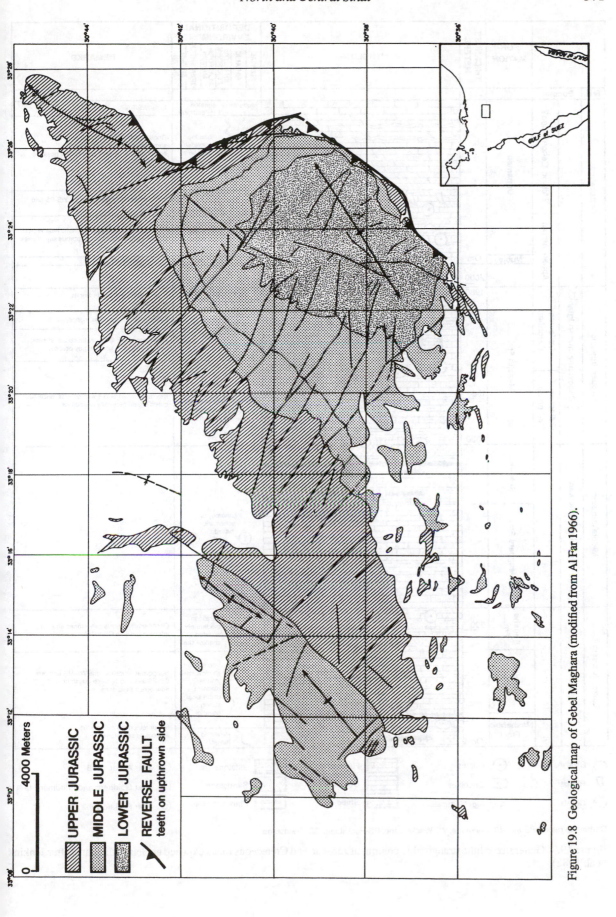

Figure 19.8 Geological map of Gebel Maghara (modified from Al Far 1966).

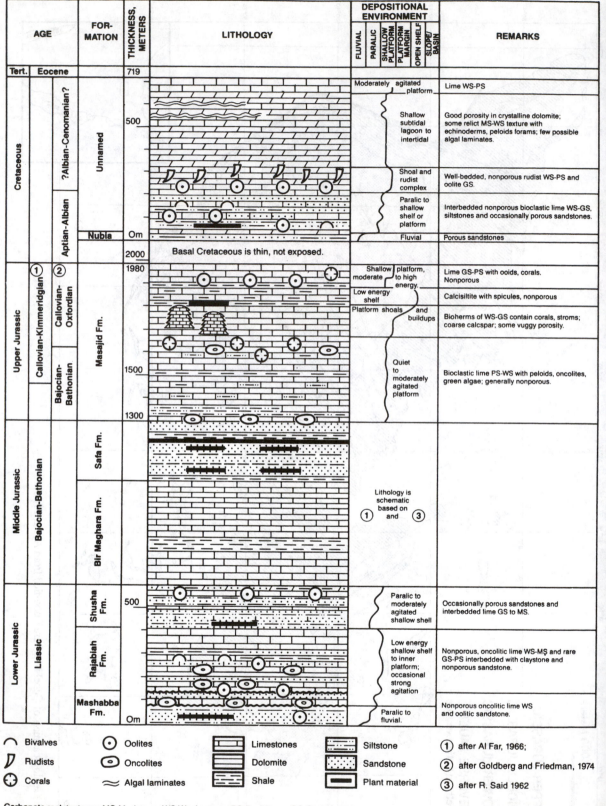

Figure 19.9 Generalized lithostratigraphic column of Jurassic and Cretaceous rocks exposed at Gebel Maghara (after Jenkins et al. 1982).

lomites of shallow, relatively low energy origin. These interfinger intimately with clastic sediments to the south and change to deep marine clastics and carbonates to the north. The belt of high energy platform-margin sediments depicted in the north is based both on subsurface control and inference from the geology of coastal Israel.

3.2.2.1 *Lower Jurassic*.
The lowermost Jurassic sediments exposed at Gebel Maghara belong to the Mashabba Formation (Fig. 19/9). These clastic sediments are of Liassic age and were deposited by northerly-flowing braided streams carrying detritus shed off the Arabo-Nubian massif. The dominance of continental clastics was probably the result of low eustatic sea level coupled with tectonic uplift to the south.

The thin basal fluvial sandstones containing large wood fragments are succeeded by interbedded shallow marine carbonates and nearshore marine clastics of the Rajabiah and Shusha Formations. The carbonates are rich in lime mud having been poorly winnowed in low energy environments. Rare interbeds of well-sorted oolitic limestone are present, suggesting higher energy conditions, commonly with quartz sand grains indicating proximity to the shoreline.

The interbedding of carbonates and clastics might have been caused simply by the shifting of different depositional facies across a low energy shelf.

The Halal-1 well penetrated a very similar sequence of lower Jurassic sediments which attained a thickness of 2240 m.

At Gebel Arif El Naga, lower Jurassic sediments have a thickness of 141 m and consist of variegated sandstones, ferruginous silty shales, limestones, sandy dolomites and marls. These sediments are interpreted as being deposited in saline-brackish lagoons bordering on a continental-fluvial environment.

Lower Jurassic sequences of sandstones and shales have been penetrated in the Nekhl (162 m) and Abu Hamth (162 m) wells. A thick clastic sequence in the Ayun Musa-2 (622 m) and Hamra-1 (534 m) wells has been assigned a middle-lower Jurassic age, whilst the El Khabra well has an undifferentiated Jurassic section (1430 m) of sandstones, shales and limestones.

3.2.2.2 *Middle Jurassic*.
The middle Jurassic sediments at Gebel Maghara can be subdivided into a lower carbonate unit (Bir Maghara Formation) and an upper clastic unit (Safa Formation) which attain a thickness of approximately 701 m (fig. 19/9). These clastics contain sub-bituminous coals which are interpreted as having been deposited in lakes or lagoons adjacent to the coastline.

This clastic-carbonate sequence is also exposed at Gebel Minshera and has been penetrated in the Halal-1 (780 m), Katib El Makhazin (502 m +) and Giddi-1 (805 m +) wells.

The undifferentiated middle-lower Jurassic clastic sediments penetrated in the Ayun Musa-2 and Hamra-1 wells indicate that continental conditions existed throughout central Sinai during the lower-middle Jurassic period.

Detailed petrographic studies of the middle Jurassic sediments in boreholes drilled along the present-day coastline (Oesleby 1981) indicate that a high-energy carbonate platform margin existed in this area. These sediments are predominantly ooid and bioclastic lime grainstones and packstones.

North of this platform margin a deep-water environment existed (Fig. 19/7).

3.2.2.3 *Upper Jurassic*.
The upper Jurassic sediments (Masajid Formation) are dominated by carbonates (Fig. 19/10) representing a southerly marine transgression at the end of Bathonian-Callovian times (Fig. 19/7).

At Gebel Maghara these carbonates are 680 m thick and consist of bioclastic, oncolitic, peloidal packstones and wackestones with isolated biohermal development (Fig. 19/10). The bioherms, which range in size from 8 to 24 m in height and 15 to 200 m in length, contain a wide variety of coarse-grained organic components including corals, stromatoporoids, sponges, crinoids, and skeletal blue-green algae set in a poorly sorted, peloidal, bioclastic matrix with associated debris-flow conglomerates (Fig. 19/10). They are interpreted as small patch reefs that developed in well-circulated, relatively shallow water.

During the upper Jurassic period, the Sinai was a shallow carbonate shelf with continental-fluvial facies existing to the south, a high energy platform margin trending ?east-west along or adjacent to the present-day coastline and a deep marine facies further north.

This interpretation has been confirmed by petrographic studies of the various boreholes (Oesleby 1981). In the north the Manna-1 wells encountered 102 m plus of skeletal, peloidal wackestones, packstones and grainstones with some corals and ooids present. The Gofer-1 well penetrated 150 m of ooidal lime packstones and grainstones, which are overlain by a thrusted sequence of Cretaceous and Jurassic deep marine shales, suggesting that a deep-water environment was proximal at the time of deposition.

To the south shelfal carbonates have been rec-

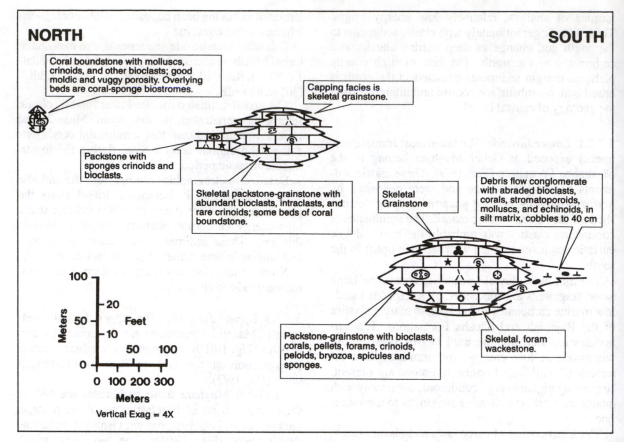

NORTH   **SOUTH**

Coral boundstone with molluscs, crinoids, and other bioclasts; good moldic and vuggy porosity. Overlying beds are coral-sponge biostromes.

Capping facies is skeletal grainstone.

Packstone with sponges crinoids and bioclasts.

Skeletal packstone-grainstone with abundant bioclasts, intraclasts, and rare crinoids; some beds of coral boundstone.

Skeletal Grainstone

Debris flow conglomerate with abraded bioclasts, corals, stromatoporoids, molluscs, and echinoids, in silt matrix, cobbles to 40 cm

Packstone-grainstone with bioclasts, corals, pellets, forams, crinoids, peloids, bryozoa, spicules and sponges.

Skeletal, foram wackestone.

Meters — 100 / 50 / 0
Feet — 20 / 10
0 / 50 / 100
0   100   200   300
Meters
Vertical Exag = 4X

Figure 19.10 Bioherms in the Masajid Formation. Spatial distribution is approximate. Laterally equivalent beds are poorly exposed but include mainly bioclastic lime wackestones and packstones with some oolitic grainstones and minor marls.

ognized in the Halal-1 (214 m), Ayun Musa-2 (121 m), Hamra-1 (65 m) and Katib El Makhazin (388 m) boreholes.

3.2.3 Cretaceous

3.2.3.1 Lower Cretaceous
The basal fluvial-continental lower Cretaceous sediments unconformably overlie the shallow marine upper Jurassic carbonates as a result of a major eustatic fall in sea level.

In central Sinai the pre-Cenomanian section is composed entirely of 'Nubian-type' sandstones which attain a thickness of 780 m at Um Bogma. The basal part (282 m) of these sands have been assigned a Carboniferous age, whilst the upper 498 m is thought to range from Triassic to lower Cretaceous in age. These sandstones are exposed along the scarp of Gebel Tih.

'Nubian-type' sandstones of early Cretaceous age have also been penetrated in the following boreholes, Abu Hamth (370 m), Nekhl (247 m), Ayum Musa-2

(149 m), Hamra (366 m) and Kabrit (103 m).

Further north similar sandstones outcrop at Gebel Maghara, Gebel Arif El Naga, Gebel Halal, Gebel Yelleq, Gebel Giddi, Gebel Kherim and Gebel Minsherah.

At Gebel Maghara the contact between the Jurassic and Cretaceous sediments is obscured by wadi fill. The basal exposure of the Cretaceous consists of thin, cross-bedded, very fine to coarse-grained sandstones and conglomerates (Fig. 19/9). These represent the major pulse of late Mesozoic land-derived sediments in north Sinai which were deposited by northerly-flowing streams. Overlying these sands is an interbedded series of argillaceous clastics and carbonates of Aptian-Albian age, which represent fluvial-paralic to shallow shelf facies deposited in a low energy system, with occasional high energy episodes, indicated by the presence of rudists and oolites (Fig. 19/9).

A similar sequence outcrops at Gebel Yelleq (Fig. 19/11) where a basal conglomerate (5 to 15 m) is overlain by cross-bedded sandstones (160 to 168 m)

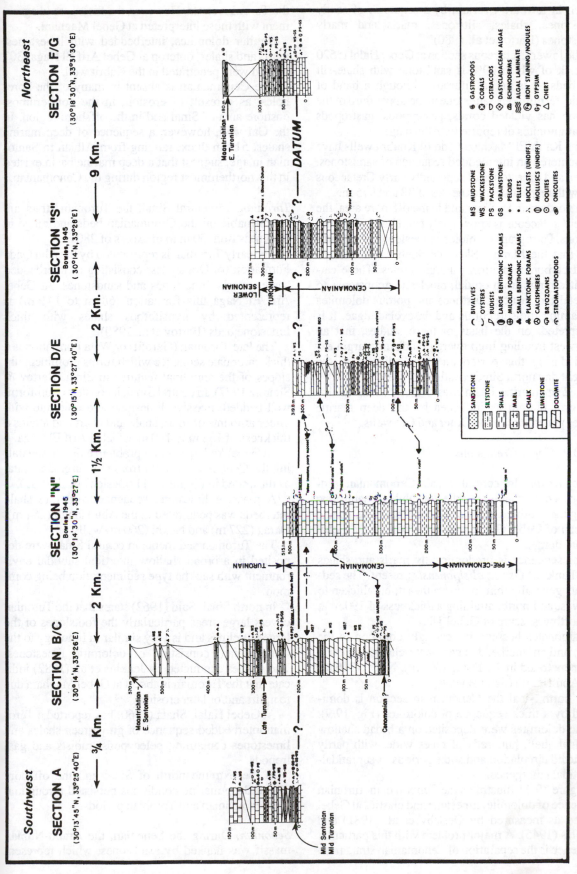

Figure 19.11 Cretaceous sections along the southern flank of Gebel Yelleq, north-central Sinai (after Bowles 1945 and Oesleby 1981).

which in turn are overlain by a sequence of sandy limestones, shales, siltstones, marls and marly limestones (Bartov et al. 1980).

The lower Cretaceous section at Gebel Halal is 520 m thick of which 90% is sandstone with shale-silt interbeds. It is unfossiliferous, although a band of oolitic, ferruginous sandstone in the upper third of the section has yielded corals, pelecypods, gastropods and ammonites of reportedly Albian age.

The Katib El Makhazin and El Khabra wells have encountered an interbedded sequence of sandstones, shales and limestones of reportedly early Cretaceous age with respective thicknesses of 32 and 975 m.

In northernmost Sinai and in the offshore area, the lower Cretaceous is known only from borehole penetrations. Petrographic studies by Oesleby (1981) indicate that the Manna, Slav and Sneh wells encountered carbonate platform margin facies. These carbonates range from ooidal, bioclastic grainstones to skeletal bioclastic packstones and porous dolomites containing rudistid corals and dascyclad algae. It is interpreted, on the basis of these studies, that an east-west trending high energy platform margin existed during this period along or adjacent to the present day north Sinai coastline (Fig. 19/7).

A deep marine basin was present north of this platform margin as indicated by the deep marine shales penetrated in the Gofer and Gal wells.

3.2.3.2 Upper Cretaceous

Cenomanian. In central Sinai, Cenomanian sediments conformably overlie the Nubian sandstone along the southern scarp of Gebel Tih, on the upland plateau of Gebel Gunna and in the core of the Gebel Somar dome.

The sequence is predominantly marls and shales with banks of *Ostrea olisiponensis* present. The sediments generally thin to the southeast, but thicken to the west and north, attaining a thickness of 190 m at the northwest scarp of Gebel Tih.

Carbonates become increasingly common to the north and an interbedded carbonate-clastic sequence was penetrated in the Darag (307 m), Nekhl (316 m) and Abu Hamth (326 m) wells.

In north Sinai the Cenomanian section is dominated by a thick sequence of dolomites (Fig. 19/9). These dolomites were deposited on a broad shallow subtidal shelf, hundreds of miles wide, with partly restricted circulation and some periods of supratidal-intertidal emergence.

Figure 19/11 illustrates the Cenomanian-Turonian sequence of dolomite, limestone and clastics at Gebel Yelleq as measured by Oesleby et al. (1981) and Bowles (1945). A major problem with this particular sequence is the separation of Cenomanian strata from

the Turonian strata. However, the facies are in agreement with those interpreted at Gebel Maghara.

Similar dolomites, interbedded with limestones, marls and shales, outcrop at Gebel Arif El Naga (307 m) and were penetrated in the Kabrit well.

The Cenomanian is absent in many of the boreholes, as a result of erosion, in the northernmost onshore area of Sinai and in the offshore region. In the Gal well, however, a sequence of deep marine shales, 512 m thick, ranging from Albian to Santonian in age, suggest that a deep marine basin existed in this northernmost region during the Cenomanian.

Turonian. In central Sinai, the Turonian strata are conformable on the Cenomanian beds and range in thickness from 50 m to in excess of 280 m.

The early Turonian is represented by the Abu Qada Formation (or Ora shales) consisting of interbedded shales, marls, limestones and sandstones. At Gebel Arif El Naga this formation (68 m to 141 m) is represented by gypsiferous shales with thick limestone beds (Bartov et al. 1980).

The late Turonian (Gerofit) or Wata Formation is a thick carbonate sequence which forms prominent dip slopes of the anticlinal features in Sinai (Bartov & Steinitz 1977). In central Sinai, it consists of uniform, well-bedded, massive limestone and dolomite with minor amounts of marl, shale and chert and attains a thickness of 132 m to 140 m at Gebel Arif El Naga.

At Gebel Yelleq, a major problem exists in separating the Cenomanian strata from the Turonian strata, as illustrated in Figure 19/11 (Oesleby et al. 1981).

A massive limestone sequence with rare shale interbeds was penetrated in the Abu Hamth (254 m), Darag (227 m) and Nekhl (260 m) wells.

The Turonian sediments in central Sinai were deposited in a broad shallow intertidal-subtidal environment with sabkha type sedimentation being common.

In north Sinai, Said (1962) states that the Turonian covers large areas particularly the footslopes of the large anticlines and is very similar in lithology to the Cenomanian, consisting of dolomitic limestones. The studies conducted by Jenkins et al. (1982) indicate that the Turonian is absent at Gebel Maghara due to uplift and/or later erosion.

At Gebel Halal, Shata (1960) has reported a Turonian interbedded sequence of grey-green shales and limestones containing pelecypods, rudists and gastropods.

In the extreme north of Sinai and the offshore region deep marine conditions persisted throughout the Cenomanian and Turonian periods.

Senonian. During the Senonian, the Arabo-Nubian massif was flanked by sandstones, which represent

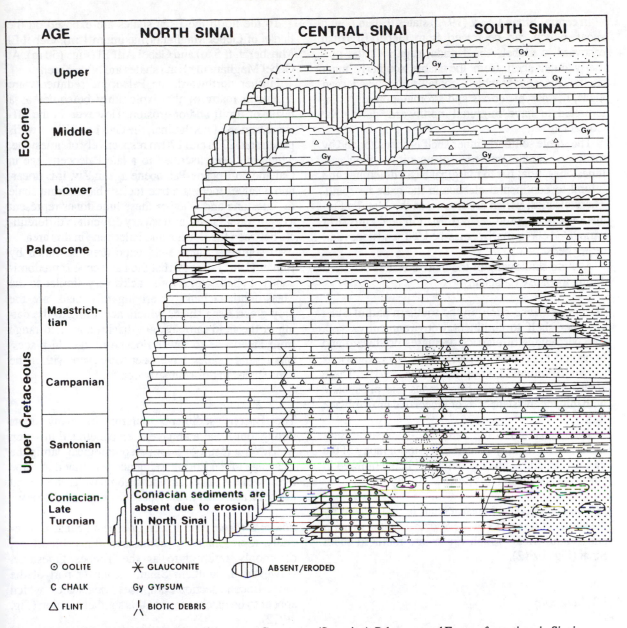

AGE		NORTH SINAI	CENTRAL SINAI	SOUTH SINAI

Figure 19.12 Lithofacies relationships of the upper Cretaceous (Senonian), Paleocene and Eocene formations in Sinai.

⊙ OOLITE   ✳ GLAUCONITE   ⬭ ABSENT/ERODED

C CHALK   Gy GYPSUM

△ FLINT   ⊥ BIOTIC DEBRIS

terrigenous clastic pulses derived from the south, indicating either a temporary regression or a relative increase in the supply of clastics. Northward from the massif these clastics prograde into carbonates which were deposited in increasingly deeper water northward (Fig. 19/12).

In north and central Sinai, it is often very difficult to subdivide the Senonian stage due to the monotonous sequence of chalks present. These chalks are well-exposed along the wall of the Egma tableland and outcrop in the lowlands between the major anticlinal structures. It is absent from many of the

structures due to uplift and/or later erosion.

In many of the boreholes the Senonian section is undifferentiated, but it has been penetrated in the Darag (235 m), Nekhl (33 m), Abu Hamth (326 m) and El Khabra (192 m) wells.

Coniacian. The Sinai comprised three sedimentological regions during the Coniacian period (Fig. 19/12).

In north Sinai a rapidly subsiding outer shelf was developing with marly and chalky limestones being deposited.

In central Sinai, Lewy (1975) states that an oolitic shoal, which can be traced from Gebel El Hitan eastward to Gebel Abu Kandu, separated the northerly outer shelf from the stable southerly inner shelf. This shoal, which contains dascycladacean algae, hermatypic corals and stromatoporoids, is interpreted to have migrated northward during the Coniacian.

The stable southerly inner shelf is characterized by medium coarse-grained, glauconitic, bioclastic limestones, tidal flat to marginal basin dolomites and littoral to continental clastics in the extreme south. These clastics, which are known as the Matulla Formation, are exposed at Wadi Matulla and consist of variegated clays, marls, cross-bedded sandstones and oyster beds.

A slight shallowing of the marine waters occurred during the late Coniacian.

Santonian-Campanian. In the southern part of Sinai an inner shelf to continental environment existed. The percentage of clastics decreases northward and in central Sinai thick white, shoft chalks, known as the Sudr Formation, are present (Fig. 19/12). These chalks are occasionally bituminous and phosphatic with thin inconsistent beds of brecciated brown-black flint.

The deposition of the Santonian-Campanian sediments was probably related to the topography of the various structural units which were being uplifted in parts of Sinai during this period.

Maastrichtian. The Maastrichtian is represented by the snow-white chalks of the Sudr Formation in north and central Sinai, which grade into marls in south Sinai (Fig. 19/12).

3.3 Cenozoic

3.3.1 Paleocene

Regional data indicates that the Paleocene was deposited in the lows between the major structural highs in Sinai and its distribution closely follows that of the underlying upper Cretaceous chalk unit.

The Paleocene section, which is known as the Esna shales, has a uniform lithology of greenish-grey shale, although in the south a basal chalky facies is present (Fig. 19/12).

In central Sinai it is a thin but clearly mappable dark unit in the Gebel Egma cliffs (35 m thick) and forms extensive plains on the northern flank of Gebel El Mineidra El Kebira and outcrops in the core of the Nekhl and Darag domes. In the Nekhl and Darag boreholes the Esna shales have a thickness of 59 m and 38 m respectively.

To the north the Esna shales are exposed on the flanks of Gebel Maghara (maximum 1 m), Gebel El Minsherah (65 m) and Gebel Arif El Naga (50 m). At Gebel Maghara the Esna shales are often absent.

Further northward, the Paleocene sediments are absent in many of the exploratory boreholes as a result of uplift and/or erosion. However, in the offshore region of north Sinai, the Gal-1 and Ziv-1 wells penetrated 27 m and 170 m respectively of limestone, which has been ascribed to a late Paleocene age in Gal and an Eocene-Paleocene age in Ziv. It is uncertain whether there is a true facies change from shale to limestone and whether these limestones represent a deep-water facies, as it is very difficult to differentiate between the Eocene and Paleocene in this area.

In southern Sinai, field exposures examined by Bunter (1981) suggest that the Paleocene is predominantly a chalky limestone facies very similar to the Maastrichtian chalk. Overlying this chalk are the grey-green Esna shales which, according to the dating of the field samples, are diachronous and range from Paleocene to lower Eocene in age. In places, small sand bodies have been observed within the Esna shales in this southern area.

3.3.2 Eocene

The Eocene sediments outcrop in many areas throughout Sinai and have been penetrated in seven wells. As a result of poor paleontological dating the exact age of the Eocene sequence penetrated in some of the wells is uncertain and in some wells an interval has been assigned to the Eocene solely on lithology.

Lower Eocene. The lower Eocene in Sinai is generally represented by a massive flinty limestone, commonly referred to as the Thebes limestone. However, in southern Sinai, the lower part of the lower Eocene section comprises Esna shales, which appear to be diachronous in this southern region (Fig. 19/12).

At Gebel Egma in central Sinai this flinty limestone, known locally as the 'Egma Limestone', covers the extensive tableland of the Egma plateau. It varies in thickness from 125 m to 240 m and northward the basal part has a development of yellowish marly limestones (Safra beds).

In north Sinai this lower Eocene limestone occupies the broad synclinal lowlands to the north and east of Gebel Halal, as well as between Gebel Halal and Gebel Yelleq, Gebel Maghara and Gebel Yelleq and between Um Hoseira and Kherim. It exhibits uniform lithology but has large thickness variations being very thin to absent in places.

In the Habashi, Gal, El Khabra and Manna wells, a chalky limestone interval has been dated as Eocene. It is uncertain whether this is the deeper water equivalent of the lower Eocene Thebes limestone or the middle Eocene Mokattam limestone.

Middle-upper Eocene. At Wadi Nukhul in central Sinai, the lowermost part of the middle Eocene section is represented by the Darat and Khaboba Formations. These comprise green-brown shales, marls, limestone stringers and gypsiferous shales with thin flint bands. These beds have also been observed in south Sinai but are absent in north Sinai.

Overlying these predominantly clastic formations is the Mokattam or Plateau limestone which is crowded with *Nummulites gizenhensis* and has a widespread distribution in Sinai.

The uppermost part of the Eocene epoch is a regressive period. In the south it is represented by the thinly bedded gypsiferous marls of the Tanka beds. Northward these beds grade into the sandy limestones and grey, yellow, sandy shales of the Maa-di Formation, which has not been recognized in north Sinai.

The middle to upper Eocene sequence in north and central Sinai is characterized by large unconformities with parts of the sequence missing in certain localities (Fig. 19/12).

3.3.3 *Oligocene*

The distribution of the Oligocene sediments, which are known only from boreholes drilled in the extreme north of Sinai and in the offshore area, is related to the relationship of three geological components.

Firstly, the late Cretaceous compressional event which probably commenced at the end of the Turonian and resulted in the thrusted, assymmetrical, domal anticlines which are recognized in both offshore and onshore north Sinai. Differential uplift of these structural highs from the Eocene through to the

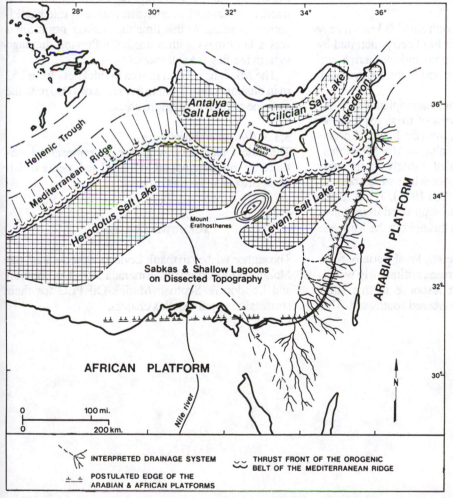

Figure 19.13 Paleogeographic map of the eastern Mediterranean during the upper Miocene Messinian (after Gvirtzman & Buchbinder 1977).

lower Miocene period controlled the distribution and nature of the sediments.

Secondly, uplift of the granitoid basement of Sinai from late Eocene times resulted in the northward retreat of the Eocene seas to a position similar to that of the present-day Mediterranean.

Thirdly, the differential movement of the Sinai and the Levantine basin created a major faulted hinge line parallel and sub-parallel to the Pelusium line (Fig. 19/1).

Thus the Oligocene drainage system flowed north-west from the emergent, southerly granitoid massif and debouched sediment within northerly prograding deltaic and submarine fan systems along the major faulted hinge line. These depositional systems have been recognized in the subsurface from borehole data in the extreme north of Sinai and in the present-day offshore area.

The thickness of the Oligocene sequence ranges from less than 152 m in the northernmost Sinai to in excess of 1800 m in the offshore region.

3.3.4 *Miocene*

The Miocene sequence in north Sinai is known only from the subsurface where it has been penetrated by exploratory boreholes. In central and southern Sinai, Miocene sediments are restricted to the eastern bank of the Gulf of Suez.

Detailed paleobathymetric, nannoplankton biostratigraphy and facies analyses of these exploration boreholes have revealed that from the lower Miocene period through to the beginning of the Messinian, four main facies belts prevailed, namely continental-fluvial, paralic, shelfal and slope with their associated clastic sedimentation (Marathon 1981). These facies belts migrated northward and southward in response to eustatic sea level changes throughout the Miocene epoch.

During the early Miocene sea level initially rose, then fell slightly and finally rose continuously world-wide (Vail et al. 1979). In response to this latter change, the facies belts transgressed southward. This eustatic sea level rise continued during the middle Miocene until Serravallian times when a sudden drop occurred. This lowering of sea level resulted in a northerly migration of the facies belts and it is also responsible for the marked unconformity between middle and late Miocene sediments. Gvirtzman & Buchbinder (1977) have interpreted this sea level drop as the cause of the middle Miocene desiccation phase.

Eustatic sea level rose during Tortonian time with a corresponding marine transgression. An east-west trending reefal carbonate belt developed in northern-most Sinai at this time and these high energy carbonates have been identified in the Gofer, Manna, Pelusion, Slav, Sneh and Abu Roda wells.

During uppermost Miocene 'Messinian' time, the Mediterranean basins were cut off from the global ocean system, resulting in the production of endorheic salt lakes and the 'Messinian Salinity Crisis'. These salt lakes developed north of the northwest-southeast Bardawil fault scarp and the Pelusium Fault Zone (Fig. 19/13). The evaporites were later remobilized, as a result of overburden pressure, into halokinetic structures. At this time the onshore north Sinai was a land mass with a northerly-flowing drainage system feeding the salt lakes.

The Miocene section ranges in thickness from 152 m in onshore north Sinai to in excess of 1000 m in the extreme north and offshore areas.

3.3.5 *Pliocene-Quaternary*

The Plio-Quaternary sequence is represented on-shore by thin continental to littoral sediments which are approximately 3000 m thick.

ACKNOWLEDGEMENTS

The author wishes to thank Conoco Incorporated, the Marathon Oil Company, Amerada Hess Corporation and London & Scottish Marine Oil PLC for their permission to publish this chapter.

CHAPTER 20

Structural characteristics and tectonic evolution of north Sinai fold belts

ADEL R. MOUSTAFA & MOSBAH H. KHALIL

Faculty of Science, Ain Shams University and Gulf of Suez Petroleum Company, Cairo, Egypt

This chapter discusses the structural characteristics and tectonic evolution of north Sinai fold belts. The study is mainly based on the detailed stereoscopic study of aerial photographs of scales 1:20,000, 1:30,000 and 1:40,000, and the analysis of Landsat images of scale 1:250,000. Field checks were carried out in several key localities throughout the study area. A detailed structural map has been prepared covering north Sinai structures (Fig. 20/1).

Several east-northeast and northeast oriented doubly plunging anticlines form a distinctive tectonic province in north Sinai. These folds are the main topographic highs of the region. They are of different sizes including large folds (Gebels El Maghara, Yelleq and El Halal), medium folds (Gebels El Minsherah, Kherim, Araif El Naga, etc.), and many other small folds which are less than 2 km in length. Folds of the first two categories are generally oriented east-northeast to northeast and are asymmetric. The northwest flanks dip 5 to 20° while the southeast flanks are steeper and sometimes vertical or overturned and deformed by northwest dipping thrusts such as in Gebels El Maghara, Um Mafruth, Araif El Naqa, etc. These folds were the subject of the study of Sadek (1928), Shata (1959), Said (1962), and Youssef (1968).

In the following paragraphs, a description is given of the structural characteristics of some north Sinai folds.

1 Gebels El Maghara, Um Mafruth and El Amrar (20/2, 20/3).

The middle part of Gebel El Maghara represents the highest part of the structure which has the form of an asymmetric, east-northeast oriented doubly plunging anticline. The core of this fold is dome-like. The northwest flank dips about 25° while the southeast flank is very steep, vertical or overturned. The latter flank is bounded by a major thrust where the Jurassic rocks ride over the lower Cretaceous section (Fig. 20/2).

Gebel El Maghara anticline was formed due to the drag of the upthrown side of the thrust. The lower Cretaceous rocks in the downthrown side were also dragged forming a foot wall overturned syncline. The clearlest exposure of the two dragged sides of the thrust is at Wadi El Sagaan and south of Gebel El Rukba where the thrust is crossed by Wadi El Maghara. Toward the southwest the downthrown side of the thrust is further deformed by several second order folds which have a right-stepped en-echelon arrangement. Some of these folds are overturned and all the axes plunge southwest to south-southwest and form an east-northeast elongated belt which extends from Gebel El Mahash through Gebel Um Latiya to Gebel El Urf. This elongated belt of en-echelon folds was probably thrust over the topographically low area lying south of it.

Gebel El Rukba lies to the north of Gebel El Maghara and is a northward plunging, east-facing monocline. To the east of this monocline is another anticline that plunges northeastward and represents a hanging wall anticline on the north segment of the main thrust of Gebel El Maghara. The Cretaceous rocks at Gebel Manzour are folded into a negative box fold composed of two monoclines facing each other and lying on its eastern and western sides. The western monocline is in the downthrown side of the north part of Gebel El Maghara thrust, while the eastern monocline is probably in the downthrown side of another thrust. The large northeast plunging anticline lying east of Gebel Manzour is probably in the upthrown side of the latter thrust.

The north part of Gebel El Maghara thrust is offset twice by two east-west oriented right-lateral strike-slip faults. On the upthrown side of the displaced part of the thrust in Gebels El Torkmaniya, Um Mitmam, and El Mistan is another northeast elongated belt of right-stepped en-echelon folds that plunge south to southwest. This belt is similar to that extending from Gebel El Mahash to Gebel El Urf in the southern part of Gebel El Maghara.

Gebel Um Mafruth is an east-northeast oriented

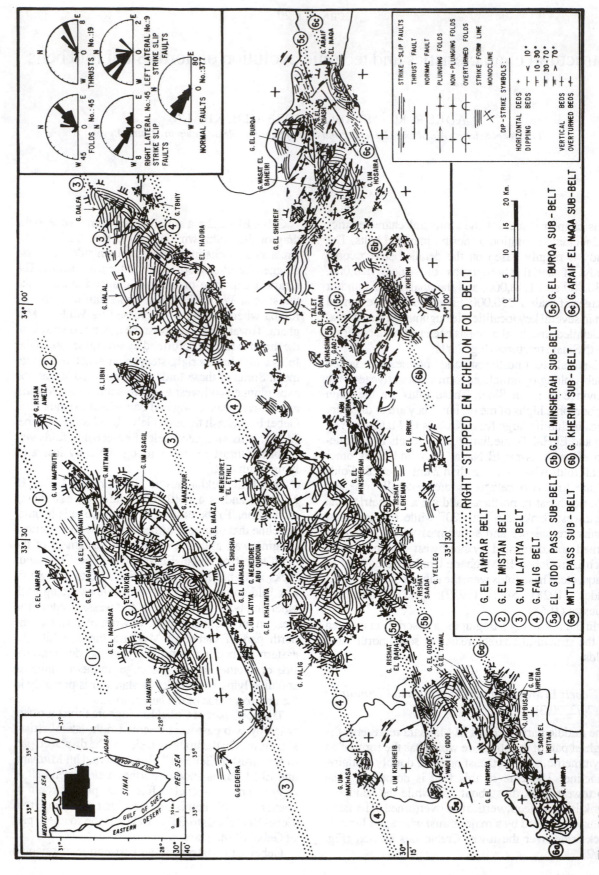

Figure 20.1 Structural map of north Sinai showing the en-echelon fold belts.

doubly plunging anticline that is dissected by a west-northwest oriented right-lateral strike-slip fault. This fold probably extends southwestward to Gebel El Lagama.

Gebel Hamayir lies to the northwest of Gebel El Maghara and is on the upthrown side of another northeast striking thrust. The fault plane dips 54° northwest and has diagonal-slip slickensides which plunge 36° in the S 78° W direction indicating diagonal-slip displacement with right-lateral strike-slip component. The Jurassic rocks of Gebel Hamayir are folded by three south-southwest plunging folds.

Gebel El Amrar lies to the north of Gebel El Maghara structure and is itself an east-northeast oriented doubly plunging anticline with Jurassic rocks exposed in its core. This fold is on the upthrown side of a northeast oriented thrust.

Several northwest oriented normal faults affect the area shown in Figure 20/2. They are transverse to the fold axes and probably originated as extension fractures with dip-slip displacement during the uplift of the structure. Most important among the faults of the area are the east-west oriented right-lateral strike-slip faults in the north part of Gebel El Maghara.

2 Gebels Yelleq and El Minsherah

Gebel Yelleq is the largest among the north Sinai folds (Fig. 20/1). It is oriented northeast and has moderate dips on its northwest side and very steep dips on its southeast side. The southwest one-quarter of Gebel Yelleq is different from the rest of the mountain. It is composed of two southwest plunging anticlines and an intervening syncline. Northeastward, the two anticlines combine into one structurally high block that has the form of an asymmetric positive box fold with a steep southeast limb. In the central part of this box fold the Cretaceous rocks have a horizontal attitude. The whole structure is dissected by several northwest oriented normal faults similar to those of Gebel El Maghara. These transverse faults probably originated as extension fractures with dip-slip displacement during the uplift of the structure.

Juxtaposing the north side of Gebel Yelleq is an east-northeast elongated belt of right-stepped, en-echelon, doubly plunging anticlines. These anticlines are oriented northeast (Gebel Meneidret El Etheili, Gebel Meneidret Abu Quroun, Gebel Falig, and other unnamed ones especially south of Gebel Meneidret El Etheili). Another east-northeast elongated belt including two right-stepped en-echelon folds bounds the south part of Gebel Yelleq. The two folds are Gebel El Minsherah doubly plunging anticline and another unnamed anticline that lies about 9 km southwest. Both structures are dissected by east-northeast

oriented diagonal-slip (right-lateral thrust) faults. In Gebel El Minsherah the diagonal-slip fault strikes N 68° E and dips 65 to 82° SE. It has diagonal-slip slickensides that plunge 46° in the S 84° E direction indicating diagonal-slip movement with right-lateral strike-slip component and thrust component. The western part of the fault juxtaposes the upper Cretaceous chalk against the Jurassic rocks. This fault continues east-northeastward to Gebel Abu Suweira and west-southwestward to Rishat Leheman. In Gebel El Minsherah core are two small east-northeast oriented, doubly plunging en-echelon anticlines. This belt of en-echelon folds continues toward the west-southwest and includes a third minor anticline affecting the Jurassic rocks of this locality, a plunging syncline between Gebel El Minsherah and Rishat Leheman, and an anticline in the latter area. This belt of en-echelon folds is probably related to Gebel El Minsherah fault and its continuation on both sides.

About 9 km southwest of Gebel El Minsherah is another large anticline that lies at the southwest part of Gebel Yelleq. It is bounded by an east-northeast oriented diagonal-slip (right-lateral thurust) fault that probably changes its direction of dip at its southwest part. Like Gebel El Minsherah fault, this fault is accompanied by some second order folds that make an acute angle with it, e.g. Rishat Saada.

3 Gebels El Halal and Libni

Gebel El Halal is the third largest structure in north Sinai (Fig. 20/1). This asymmetric doubly plunging anticline is structurally simpler than the Gebel El Maghara and Gebel Yelleq structures. The axial trace of the fold is curved and its northeast part (Gebel Dalfa) plunges northeast while its southwest part plunges west-southwest to due west. The asymmetry of the fold is not as pronounced at Gebel Dalfa as it is in the central and southwestern parts which are highly asymmetric. Vertical to overturned beds are found on the south and east sides of El Hadira. The areas to the northeast and southwest of this steeply-dipping area are refolded by second order folds that plunge due south to south-southwest and form two east-northeast elongated belts where the second order folds are en-echelon and right-stepped. The steeply dipping and overturned area may be bounded by a concealed northwest dipping thrust on its east side.

Gebel El Halal is dissected by several transverse (northwest oriented) normal faults. They are similar to the transverse faults of Gebel el Maghara and Gebel Yelleq and probably originated as extension fractures with later dip-slip displacement. One of these faults offsets the axis of Gebel El Halal anticline and has a right-lateral strike-slip component.

Gebel Libni lies to the northwest of Gebel Halal

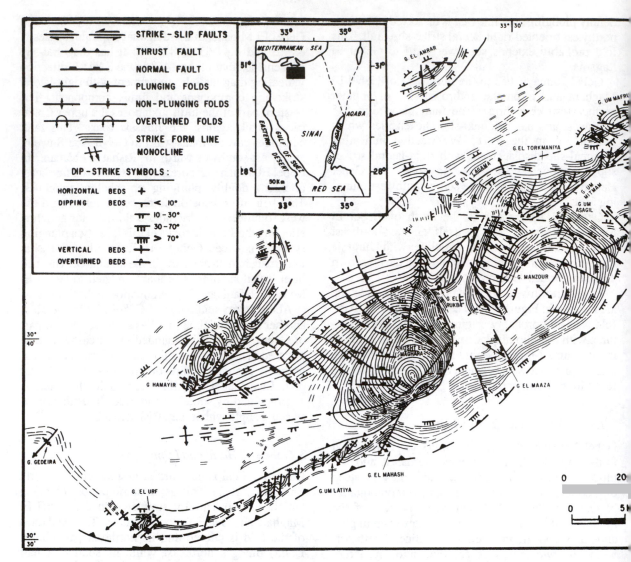

Figure 20.2 Structural map of Gebels El Maghara, Um Mafruth and El Amrar.

and is an asymmetric northeast oriented doubly plunging anticline. It is also dissected by some transverse normal faults.

4 Folds of the Giddi and Mitla passes area

Despite the small size of the folds in the Giddi and Mitla passes area (Fig. 20/1), they are very important in unravelling the nature of the structural deformation and its age. The Mitla pass area contains an east-northeast elongated belt of right-stepped, en-echelon, doubly plunging anticlines. These folds are oriented northeast and have an average length of 5 km and an average width of 3 km (Gebel Homra, Gebel El

Hamraa, Gebel Um Busal, Gebel Um Hreiba, and another unnamed one to the east). These folds affect most of the upper Cretaceous rocks. The upper Cretaceous chalk seems to be unfolded or slightly folded. The Eocene rocks capping the plateau south of the en-echelon fold belt (Gebel Sadr El Heitan plateau) are not folded. This indicates that the folding event reached its climax in late Cretaceous time and most probably continued through the Maastrichtian and early Tertiary time but with a smaller magnitude. Most probably this deformational event ended by middle Eocene time, at least in the Mitla and Giddi passes area.

The Giddi pass area also shows folded Cretaceous

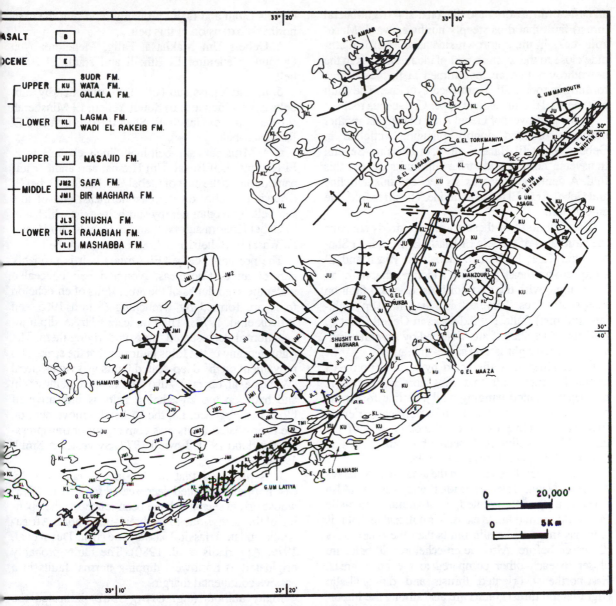

BASALT ☐ **B**

OCENE ☐ **E**

┌─ UPPER	**KU**	SUDR FM. WATA FM. GALALA FM.
└─ LOWER	**KL**	LAGMA FM. WADI EL RAKEIB FM.

┌─ UPPER	**JU**	MASAJID FM.
├─ MIDDLE	**JM2**	SAFA FM.
	JM1	BIR MAGHARA FM.
└─ LOWER	**JL3**	SHUSHA FM.
	JL2	RAJABIAH FM.
	JL1	MASHABBA FM.

Figure 20.3 Structural and geologic map of Gebels El Maghara, Um Mafruth and El Amrar.

rocks unconformably covered by the horizontal Eocene rocks of Gebel Um Khisheib plateau. In the pass itself, narrow folds affect the Cretaceous rocks forming two narrow east-northeast oriented folds on the south side of an east-northeast oriented right-lateral strike-slip fault. To the west of this fault, several small en-echelon folds affect the Cretaceous rocks. The area between the Giddi and Mitla passes is folded by larger northeast oriented folds. Two east-northeast oriented small folds lie on the west side of Gebel El Tawal.

In Gebel Um Makhasa, the Cretaceous rocks are folded by a northeast oriented anticline. Like the Giddi and Mitla passes area, the nearby Eocene rocks are not folded.

5 Gebels Araif El Naqa, El Burqa, and Kherim area

Several folds exist in the area around Gebels Araif El Naqa and Kherim (Fig. 20/1). In this area the exposed Triassic, Jurassic and Cretaceous rocks are intensively folded while the upper Cretaceous (Maastrichtian) chalk is slightly folded.

Gebel Araif El Naqa lies in the eastern part of this area. Mesozoic rocks in this mountain are steeply folded by an east-northeast oriented doubly plunging anticline. The southeast (steeper) flank is bounded by an east-northeast oriented diagonal-slip (right-lateral thrust) fault that dips steeply northwestward. Mesozoic rocks in the upthrown side have a vertical attitude close to the western part of the fault. Beds in the downthrown side are overturned in the same area forming a foot wall overturned syncline. The fault juxtaposes Triassic against upper Cretaceous rocks.

To the northwest of Gebel Araif El Naqa is another northeast oriented doubly plunging anticline in Gebels Wasat El Baheiri and Um Hosaira. Several transverse normal and strike-slip faults dissect this fold. A gentle syncline exists between this anticline and Gebel Araif El Naqa anticline. The three folds are en-echelon.

The area north of these en-echelon folds contains the structures of Gebels El Riash, El Burqa, El Shereif, and Talet El Badan. Gebels El Riash and El Burqa are northeast plunging anticlines that make an acute angle with the east-northeast oriented fault on their south sides. This fault is a diagonal-slip (right-lateral thrust) fault. Folded rocks in Gebels Talet El Badan and El Shereif are dissected by several northwest oriented right-lateral strike-slip faults.

Gebel Kherim is dissected by a main northeast oriented thrust fault that dips northwestward. A northeast oriented hanging wall anticline forms the main part of Gebel Kherim and a northeast oriented foot wall syncline exists on the other side of the thrust. The syncline is dissected by some east-west oriented right-lateral strike-slip faults.

Several small folds exist in the area around Gebels Araif El Naqa, Talet El Badan, and Kherim. Also several northwest oriented, right-lateral, strike-slip faults and normal faults exist. It is noticeable that this area has more strike-slip faults than the other areas discussed before. Also the en-echelon fold belts are closer to each other compared to the other areas. East-northeast oriented thrusts and diagonal-slip (right-lateral thrust) faults are also among the important structures of this area.

STRUCTURAL CONFIGURATION AND TECTONIC SYNTHESIS

The north Sinai folded area contains several east-northeast elongated belts of right-stepped en-echelon folds of intermediate size (2 to 10 km long). The three major fold ranges (Gebels El Maghara, Yelleq, and El Halal) lie between some of these belts of en-echelon folds (Fig. 20/1). These belts are from north to south:

1. Gebel El Amrar belt.
2. Gebels El Torkmaniya, Um Mitmam and El Mistan belt.
3. Gebels El Urf, Um Latiya and El Mahash belt.

Gebels Libni and Dalfa probably represent the east-northeast extension of this belt.

4. Gebels Um Makhasa, Falig, Meneidret Abu Quroun, Meneidret El Etheili and south El Halal belt.

5. a) Giddi pass sub-belt. Two other nearby sub-belts exist which are: b) South Yelleq, El Minsherah sub-belt; and c) Gebels El Shereif, El Burqa and El Riash sub-belt.

6. a) Mitla pass sub-belt including Gebels Hamra, El Hamraa, Um Busal, Um Hreiba, and another unnamed fold to the east of Gebel Um Hreiba. Gebel El Bruk is probably on the east-northeast end of this sub-belt. Two other nearby sub-belts exist which are: b) Gebel Kherim sub-belt; and c) Gebels Um Hosaira, Wasat El Baheiri and Araif El Naqa sub-belt.

The northern (Gebel El Amrar) belt is probably part of an east-northeast elongted belt concealed under the sand dunes of the area. Belts of en-echelon folds are formed by wrenching (Smith 1965 and Wilcox et al. 1973). They indicate a strike-slip rejuvenation of deep-seated faults underlying them. The right stepping of the folds indicates that the strike-slip movement on the deep-seated faults was right-lateral (Wilcox et al. op. cit.). Also the existence of thrusts in and between the en-echelon belts is indicative of block convergence, i.e. the strike-slip movement on the deep-seated faults was convergent or transpressive (Harland 1971, Lowell 1972, Sylvester & Smith 1976).

It is proposed that north Sinai is underlain by east-northeast oriented deep-seated faults (Fig. 20/3) whose origin seems to have been related to the opening of the Tethys due to the break-up of north Africa-Arabia in late Triassic-Liassic time (Biju-Duval et al. 1979, Argyriadis et al. 1980). The faults probably originated as northwest dipping normal faults in a passive continental margin.

Many authors (e.g. Orwig 1982) indicated the existence of east to east-northeast oriented faults in the basement rocks of north Egypt. These separate the craton (stable platform) area to the south from the structurally low area to the north where sedimentary units abruptly increase in thickness (miogeosyncline).

The distance between each of the northern five deep-seated faults is about 15 km. On the other hand, the distance between the two southernmost faults is about one-half this value (Fig. 20/4). The blocks lying between the deep-seated faults are also deformed by the right-lateral transpression. Therefore, Gebel El Maghara belongs to the deformed block lying between the second and third deep-seated faults; Gebel El Halal belongs to the deformed block between the third and fourth faults, and Gebel Yelleq belongs to the deformed block between the fourth

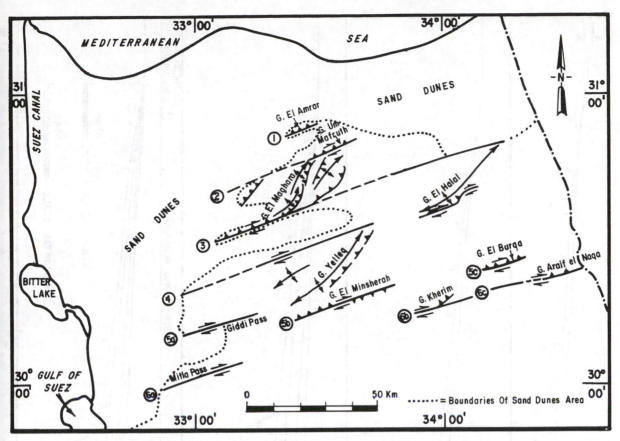

Figure 20.4 Proposed deep-seated faults in north Sinai. Numbers correspond to those mentioned in Figure 20.1.

and fifth faults. The axes of these three major structures and the accompanying thrusts make acute angles with the deep-seated faults (Fig. 20/4). Generally, all the thrusts affecting these major mountain ranges dip northwestward. The steep southeastern side of each of these structures probably represents the dragged leading edge of a thrust sheet or is probably draped over an underlying thrust fault similar to the Rocky Mountains structures (Fanshawe 1939, Berg 1962, Prucha et al. 1965, Sales 1968, Stearns 1971, among others). As these positive structures popped up they were affected by local extension parallel to their axes. Therefore, the folded rocks yielded by developing transverse extension fractures with later dip-slip displacement. The east-west to west-northwest oriented right-lateral strike-slip faults dissecting the north Sinai structures represent one of the two conjugate sets of strike-slip faults that developed by the right-lateral wrenching. They are equivalent to the Riedel shears or the synthetic strike-slip faults of Wilcox et al. (1973). The other conjugate set of strike-slip faults (conjugate Riedel shears) always has little chance of developing (Tchalenko 1970).

The area between the two southernmost deep-seated faults in Figure 20/4 (fifth and sixth faults) is relatively narrow. Therefore, large structures similar to Gebels El Maghara, Yelleq, and El Halal did not develop but smaller folds developed instead. This area also includes several diagonal-slip (right-lateral thrust) faults.

During late Cretaceous-early Tertiary (Laramide) deformation these pre-existing deep-seated faults were rejuvenated by right-lateral transpression (Fig. 20/5). Smith (1971) concludes from Atlantic spreading data that Africa moved west-northwest relative to Eurasia in late Cretaceous to late Eocene time (Smith's op. cit., Fig. 8). This motion would produce a right-lateral shear couple between north Africa and Eurasia. This shear couple probably caused the right-lateral rejuvenation of the deep-seated faults in north Egypt. It is important to notice that the west-northwest oriented shear couple makes an angle with the deep-seated faults which probably accounts for the convergent nature of the right-lateral wrenching.

Study of the mesostructures in some of the north Sinai folds and similar ones in Palestine is in agreement with this conclusion. The maximum principle stress axis ($\sigma 1$) that formed the mesostructures in the

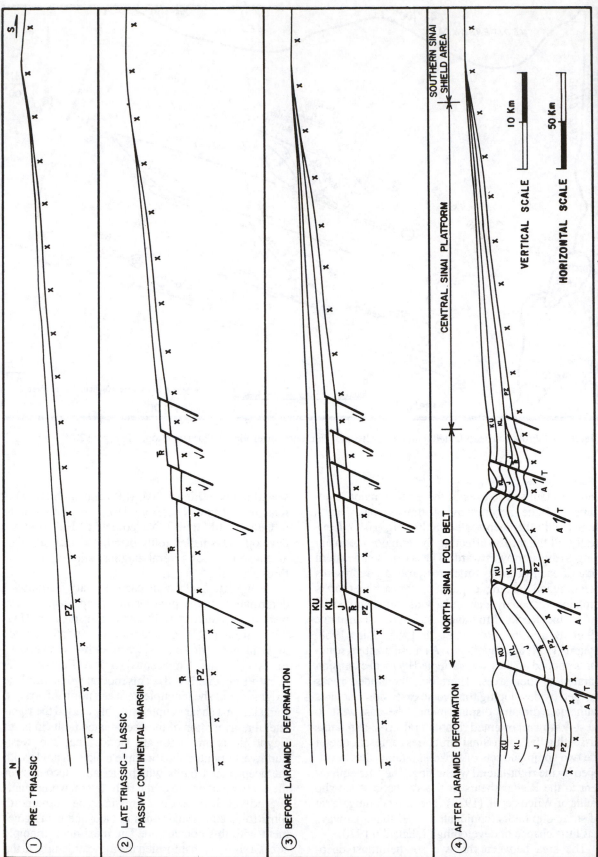

N

S

① PRE - TRIASSIC

PZ

② LATE TRIASSIC- LIASSIC
PASSIVE CONTINENTAL MARGIN

R

R

PZ

③ BEFORE LARAMIDE DEFORMATION

KU
KL
J
R
PZ

KU
KL
J
R
PZ

④ AFTER LARAMIDE DEFORMATION

NORTH SINAI FOLD BELT

CENTRAL SINAI PLATFORM

SOUTHERN SINAI
SHIELD AREA

KU
KL
PZ
J
R
A
T

KU
KL
J
R
PZ
A
T

KU
KL
J
R
PZ
A
T

KU
KL
J
R
PZ
A
T

10 Km

VERTICAL SCALE

50 Km

HORIZONTAL SCALE

Cretaceous and Eocene rocks was oriented (N 287° ± 3°) (Eyal & Reches 1983). This stress direction probably accounts for the right-lateral convergent rejuvenation of the pre-existing east-northeast oriented deep-seated faults of north Sinai.

The tectonic evolution of north Sinai indicates that late Triassic-Liassic rifting formed east to east-northeast oriented normal faults which were reactivated during the closure of the Tethys by right-lateral transpression. Such structural inversion of old structures is probably similar to the deformation of the Atlas Mountains. The high and middle Atlas Mountains were formed by the inversion of an earlier Triassic rift structure in late Cretaceous-early Tertiary time (Stets & Wurster 1982). This deformation was related to the closure of the Tethys. Laubscher & Bernoulli (1979: 17) mentioned that the compressive movements in the Alpine systems were accompanied by dextral movements. This adds further evidence to the right-lateral transpression in north Sinai which was certainly related to the Alpine deformation.

SUMMARY AND CONCLUSIONS

Structural study of north Sinai shows the abundance of northeast to east-northeast oriented doubly plunging anticlines including large (tens of kilometers long, e.g. Gebels El Maghara, Yelleq and El Halal), intermediate (several kilometers long, e.g. Gebels El Minsherah, Kherim, Araif El Naqa, etc.), and small folds (less than 2 km in length). The large and intermediate folds are asymmetric. The northwest flank dips 5 to 20°, while the southeast flank is steeper and may be vertical or overturned. The latter flank is in quite a number of cases affected by thrust faults.

Six east-northeast elongated belts of right-stepped en-echelon folds are recognized. These east-northeast oriented belts probably overlie deep-seated faults rejuvenated by right-lateral transpression in late Cretaceous-early Tertiary time (Laramide deformation). The blocks between each of the northern five deep-seated faults were affected by the right-lateral transpression and were occupied by the large structures of north Sinai: Gebels el Maghara, Yelleq, and El Halal. The steep (southeast) flank of each of these folded structures represents the dragged leading edge of a thrust sheet or draping over an underlying thrust fault similar to the Rocky Mountains structures. These folded ranges and the accompanying northwest dipping thrusts make acute angles with the deep-seated faults. They are dissected by east-west to west-northwest oriented second order, right-lateral strike-slip faults (Riedel shears) and northwest oriented (transverse) normal faults that probably originated as extension fractures.

The east-northeast deep-seated faults were probably formed due to late Triassic-Liassic rifting in north Africa-Arabia that formed the passive continental margin of the Tethys in the east Mediterranean region. They were rejuvenated by Laramide right-lateral transpression. The rejuvenation probably took place due to the west-northwest movement of Africa-Arabia relative to Eurasia creating a right-lateral shear couple between the two plates. This west-northwest oriented shear couple was not parallel to the deep-seated faults, therefore the right-lateral wrenching was convergent.

Similar structural inversion of old structures is reported in the Atlas Mountains, and the closure of the Tethys is also documented to have been achieved by right-lateral convergence in other areas in the Alpine system.

ACKNOWLEDGEMENTS

We thank Professor M.I. Youssef, Department of Geology, Ain Shams University, for reviewing an advance copy of the manuscript. Aerial photographs were provided by Ain Shams University. Field work was sponsored by the Gulf of Suez Petroleum Company. M. Reda kindly drafted the illustrations.

PART 4

Discussion

CHAPTER 21

Paleozoic

E. KLITZSCH

Technical University, Berlin, Germany

Until recently, surface exposures of Paleozoic strata occupied a small area on the geological map of Egypt. These included the well-known Carboniferous exposures along the Gulf of Suez and in the Gebel Uweinat area in the southwest corner of Egypt. Subsurface Paleozoic occurrences have also been reported from the northern part of the Western Desert. Pre-Carboniferous Paleozoic strata, however, were not clearly identified until relatively recently. In 1942, Picard already suggested that the lower part of the Paleozoic strata of the Um Bogma area of Sinai is of Cambrian age. Later Omara & Conil (1965), Omara (1972) and Weissbrod (1969) reached more substantial conclusions on the Cambrian age of some of the Paleozoic strata of Sinai. In 1981, Said suggested the presence of a large exposure of 'undifferentiated Lower Paleozoic' in the southwest corner of Egypt.

The work on the extent and subdivision of these and other Paleozoic strata in southwest Egypt between Gebel Uweinat and the Abu Ras Plateau west of Gilf Kebir was carried out by Klitzsch (1978, Klitzsch & Léjal-Nicol 1984). There, strata of Ordovician, Silurian, Devonian, Carboniferous and Permian to Triassic age of a combined thickness of up to 1000 m thickness are present at the eastern edge of the Kufra Basin and the southwest edge of the Dakhla Basin. These remote areas comprise the largest and most complete exposure of Paleozoic sediments in Egypt. In the north Western Desert Paleozoic (Cambrian to Permian) strata of up to several thousand meters in thickness are known from the subsurface.

The regional distribution of the Paleozoic strata in Egypt shows that there is a thick sequence of these strata in northwest Egypt and a thinner sequence in the Gulf of Suez/Sinai area and in southwest Egypt. Paleozoic strata are missing over most of the remaining parts of Egypt: there are no proven surface exposures of Paleozoic strata south of latitude 27° N except in the areas west of longitude 26° E.

It is likely that some areas which are free of Paleozoic strata today were originally covered by some of the Paleozoic formations. Due to intensive erosion following a regional updoming of much of southern, central and eastern Egypt towards the end of Carboniferous time (late Paleozoic-early Mesozoic event) pre-Carboniferous strata were stripped from large parts of Egypt (Klitzsch 1986). In many of these parts of Egypt, continental strata of late Jurassic to Cretaceous age rest directly on Precambrian basement, indicating this structural high position over long periods after this event.

In the case of the southwestern occurrences, an attempt is made to summarize the results of the most recent work carried out in Libya and Egypt by the author. The fieldwork in southwest Egypt was mainly carried out during the years 1976-1980 within the Mapping Project of the Continental Oil Company and within research projects of the German Research Foundation. At present the reconstruction and interpretation of Paleozoic strata is still under discussion and is not completely finalized.

Until recently, the Paleozoic strata of southwest Egypt were undifferentiated due to insufficient stratigraphical evidence. One of the major problems of stratigraphy in cratonal areas is the lithological similarity of strata over long periods of time. The uniformity of sedimentary environments and organism behavior makes identification of the different trace fossils difficult and subtle. Some of the recent discussions on Paleozoic formations in the Eastern Desert (Aswan area, south Wadi Qena) are based on ill-identified ichnofossils. Much of the distribution of Paleozoic sediments was controlled by a northnorthwest striking structural relief formed as the result of extensional movements in Early Paleozoic time (Klitzsch 1986, Schandelmeier et al. 1987 and Fig. 21/1).

CAMBRIAN

Sediments of Cambrian age are exposed in the Um Bogma area of Sinai and at several places between

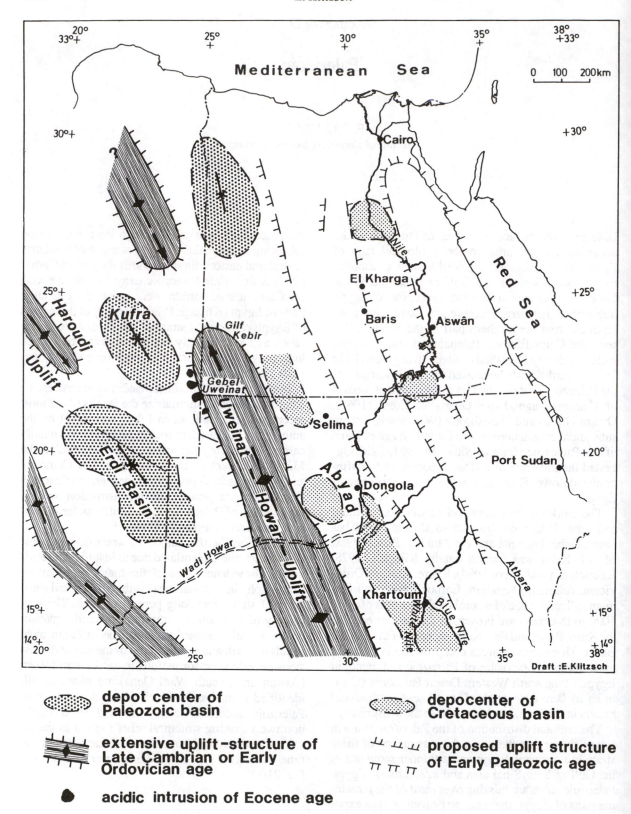

Figure 21.1 Structural relief of late Cambrian to early Carboniferous time, which controlled sedimentation processes during most of the Paleozoic and which was partly reactivated in late Jurassic to Cretaceous time (after Schandelmeier et al. 1987).

there, Wadi Feiran and Abu Durba. These seemingly non-fossiliferous clastic strata which rest above basement and below the well-dated Carboniferous carbonates of the Um Bogma Formation were considered to be part of the Carboniferous section by most authors until Omara & Conil (1965), Weissbrod (1969) and Omara (1972) proved a Cambrian age for the lower part of this clastic section. These strata are examined by Seilacher (this book, Chapter 32), who confirmed an early Cambrian age for them.

Similar sediments were found by Seilacher and the author at the northern end of Wadi Qena (Eastern Desert) in the surroundings of Somr El Qaa and between there and some kilometers south of the western end of Wadi Dakhal (Klitzsch 1986 and this book). These sediments were proven to be of an early Cambrian age by Seilacher.

Cambrian strata are also present in the subsurface of northwest Egypt. There, clastics of Cambrian age are proven in at least four wells in the Bahariya area and between there and the Qattara Depression. These sediments belong to a larger Paleozoic basin, which is in connection with well known Paleozoic occurrences in the subsurface of northeast Libya. Earlier records of the existence of possible Cambrian strata in southwest Egypt (Said 1971 after Burollet 1963 and Conant & Goudarzi 1964) are not substantiated.

Cambrian of Sinai

In the area around Umm Bogma, the Precambrian basement is overlain by a sequence of clastic sediments, which represent several cycles of fluvial to near-shore marine environment. The basal 5 to 20 m are made up of mainly coarse grained to conglomeratic sandstone, filling the pre-existing relief. All cycles above are relatively homogeneous and of consistent thickness over several kilometers and are made of sandstone, siltstone and silty shale. The marine parts of the sequence contain numerous trilobite tracks and other trace fossils: *Cruziana aegyptica*, C.cf. *nabataeica*, *C. salomonis*, *Bergauneria sucta*, *Dimorphichnus* cf. *obliquus*, *D.* cf. *quadarfidus*, and others (Seilacher, this book).

The *Cruziana* species clearly indicate a lower Cambrian age of this sequence, at least for the lower three quarters of the section. The uppermost 20 to 26 m seem to be barren, consist of crossbedded sandstone, partly similar to beach sand and possibly belonging to a later sedimentary cycle.

Weissbrod (1969) subdivided this sequence into six formations which he correlated with the type sections of the south Negev Desert. He gives them an early Cambrian age. For the uppermost formation (our uppermost 20 to 26 m), however, he leaves the age open (Netafim). It is overlain by dolomitic sediments of Lower Carboniferous age.

In order not to introduce new names and not to correlate over too long distances, we prefer to call the lower three quarters of the section Araba Formation (Hassan 1967, Said 1971), and the upper 20 to 30 m of the Um Bogma area we call, after the same authors, Naqus Formation.

It must be mentioned that Soliman & Fetouh (1970) and later Kora (1984) gave the Cambrian strata of the Um Bogma area the following names from base to top: Sarabit El Khadim, Abu Hamata, Nabib and Adedia Formation and gives it a Cambro-Ordovician age. For the time being, the author will follow the nomenclature adopted by Said (1971) for the regional use: Araba Formation.

In the area south of Wadi Feiran, strata of Cambrian age containing trace fossils similar to those of the Umm Bogma are exposed directly west of the Precambrian basement and south of the Abu Rudeis-St. Catharine road. Here, the Araba Formation is only 20 to 40 m thick and consists of coarse multi-colored sandstone and conglomerate at the base, overlain by well bedded fine- to coarse-grained sandstone and clayey to silty fine-grained sandstone. It is overlain by 120-150 m of fluvial cross-bedded sandstone interbedded with shallow marine sandstone full of *Scolithos* bioturbations and rare remains of *Cruziana* sp. and *Bergauneria sucta*. These strata possibly also belong to the Araba Formation. They are followed by approx. 150 m of fluvial sandstone equivalent to the Naqus Formation probably also of Cambrian age. The top is marked by a very ferruginous sand- and siltstone of up to 6 m in thickness.

In the area of Abu Durba the Precambrian basement is overlain according to Omara (1972) by some meters of limestone containing stromatolites and archeocyathids of most likely Early Cambrian age. This important discovery is in accordance with Seilachers interpretation of trilobite tracks from the Um Bogma, Wadi Feiran and north Wadi Qena area (Seilacher, this book, Chapter 32). During our own fieldwork in the Abu Durba area, however, we could not find the basal limestone of Omara, but we found approx. 80 m of typical Araba Formation. It is overlain by 150-200 m of mainly fluvial sandstone equivalent to Naqus Formation. The basal 30 m contain *Scolithos* layers similar to those characteristic for large parts of the Araba Formation. The top is marked by very ferruginous sandstone beds like at Wadi Feiran. Above (and below the Abu Durba shale) follow approx. 50 m of fluvial sandstone and marine sand- and siltstone, partly full with trace fossils similar to those known from the Abu Thora or Ataqa Formations of early Carboniferous age. The thickness of the various formations reported in earlier literature in general seems to be overestimated.

Cambrian of the Eastern Desert

In the area of north Wadi Qena and between there and south of the western end of Wadi Dakhal, strata of Cambrian age are exposed, which are very similar to the Cambrian sediments at Wadi Feiran and Um Bogma. The basal 5 to 25 m normally consist of coarse sand and conglomerate which fill the pre-existing relief and which partly are forests of deltaic environment. This basal unit is followed by 15 to 25 m of brown and red silty or clayey sandstone and siltstone and by planar sandstone beds, which partly contain abundant trilobite tracks. Seilacher identified *Cruziana aegyptica, C. salomonis, Dimorphichnus* cf. *quadarfidus*.

This fauna was found 5 to 10 m above basement at the eastern side of Gebel Somr El Qaa and at a point 22 km to the north in the same beds which strike northward toward Wadi Dakhal. South of Somr El Qaa and in some isolated areas between Somr El Qaa and Wadi Dakhal as well as from Wadi Dakhal northward, Cambrian strata are eroded or were not deposited. Between Somr El Qaa and Wadi Dakhal, however, in most areas the Cambrian strata are overlain by plant and *Scolithos*-bearing sandstone of Carboniferous age. In some areas, however, sandstone of Albian to Cenomanian age rests directly on Precambrian basement.

Cambrian of north Western Desert

Cambrian subsurface strata of Cambrian age are reported in wells drilled in the western part of the Western Desert (Bahariya and northward). They include sandstone, frequently glauconitic, siltstone, grey and red shale, basal conglomerate and rare carbonate. The thickness ranges between 600 to 1000 m. Age determination is based on the identification of acritarchs, brachiopods, trilobites and (indirectly) on the absence of palynomorphs.

It is likely that most of north and possibly also some of central Egypt has been transgressed by the sea in early Cambrian time. The southern edge of this Cambrian sea, however, cannot be identified with any degree of precision because Paleozoic strata have been eroded from large areas in Egypt after the Hercynian event. Southwest Egypt, however, was not reached by the sea in Cambrian time (for a tentative interpretation, see Fig. 21/2).

ORDOVICIAN

Strata of Ordovician age have not been identified in north Egypt, neither at the surface nor in wells. Strata described as Cambro-Ordovician in some publications from Sinai and in oil company reports are either of Cambrian age or of younger age than Ordovician. The only location in south Egypt, where strata of Ordovician age are identified was in Karkur Talh in the northeastern part of Gebel Uweinat at the Egyptian-Sudanese border. There, Monod (pers. comm.) found trilobite tracks in shallow marine sandstone directly above Precambrian basement. Seilacher (pers. comm.) identified *Cruziana rouaulti* Lebesconte from these beds and gave the sediments an Ordovician age. This formation – which we call Karkur Talh Formation (Klitzsch & Léjal-Nicol 1984) – consists of several ten meters of fluvial and shallow marine sandstone. It is unconformably overlain by sandstone of Silurian age, which in other areas of southwest Egypt rests directly on Precambrian basement.

The Ordovician of Gebel Uweinat is certainly part of the extensive blanket of Ordovician strata, which cover large areas in south Libya and north Chad. The Gebel Uweinat occurrence must have represented the eastern edge of an Ordovician transgression, which came from the northwest or west. Most of Egypt seems to have been a positive area during the Ordovician.

SILURIAN

The Silurian is reported from the subsurface of the north Western Desert (Schrank 1984, Hantar this book, Chapter 15). Surface exposures cover large areas in southwest Egypt (Klitzsch & Léjal-Nicol 1984). As in Ordovician time, Egypt was during the Silurian near the eastern edge of the sea, which covered large areas of North Africa and reached its maximum extension during Llandovery time. This sea seems to have transgressed into the southwest, west and northwest Egypt (Fig. 21/2).

From two wells (Foram 1 and Sheiba 1) in northwest Egypt Silurian sandstone, siltstone and shale is proven, containing palynomorphs of Silurian age. Trace fossils typical for Silurian of Libya have also been found (e.g. *Harlania harlania* Desio). The real thickness of Silurian sediments in northwestern Egypt is not known, because stratigraphical control is insufficient. The minimum thickness seems to be in the order of 200 to 300 m.

In southwest Egypt in the area between Gebel Uweinat and the Abu Ras Plateau west of Gilf Kebir, Silurian sediments reach a thickness of approximately 400 m, around the Umm Ras Passage south of the Abu Ras Plateau near the Libyan border. There, fine to medium and partly coarse-grained white sandstone of fluvial and deltaic origin is interbedded with nearshore marine sandstone, beach sand and

SILURIAN

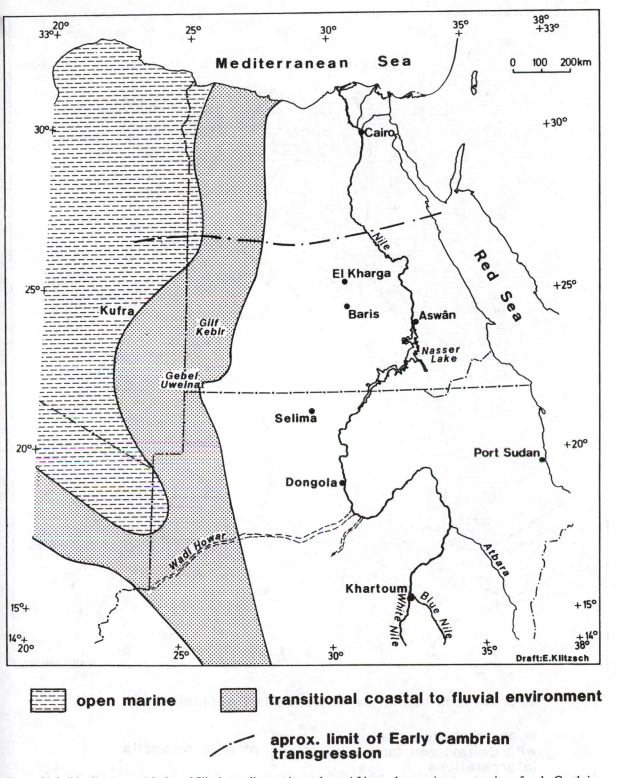

Figure 21.2 Distribution and facies of Silurian sediments in northeast Africa and approximate extension of early Cambrian transgression (after Klitzsch & Wycisk 1987).

LOWER CARBONIFEROUS (AND DEVONIAN)

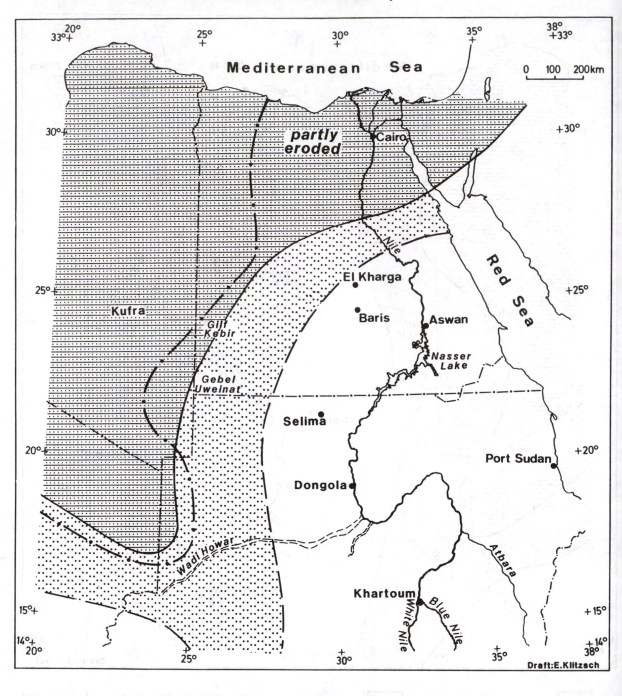

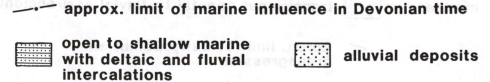

Figure 21.3 Distribution and facies of lower Carboniferous sediments in northeast Africa and approximate extension of marine influence in Devonian time (after Klitzsch & Wycisk 1987).

silty shale to clayey silt-stone. Several beds are intensively burrowed by *Scolithos* sp., others contain *Harlania harlania* Desio and *Cruziana acacensis* Seilacher. The trilobite tracks are typical for the Silurian, the other ichnofossils and the sedimentological characteristics indicate that southwest Egypt was near the edge of the Silurian (probably Llandovery) transgression (see also Klitzsch 1978, 1979, 1981, 1983). Silurian *Cruziana* are abundant in some beds near the base of the formation only some meters above Precambrian basement as well as near the top of the formation. This formation is named Um Ras Formation (Klitzsch & Léjal-Nicol 1984). It unconformably overlies Precambrian basement and is unconformably overlain by fluvial sandstone of Devonian age.

Approximately 180 km south of the type area at the northeastern part of Gebel Uweinat, the Umm Ras Formation rests unconformably above the Karkur Talh Formation of Ordovician age and is unconformably overlain by fluvial sandstone as in the type area (probably Devonian). Further to the south, these early Paleozoic strata are truncated by Carboniferous sediments: at Karkur Murr in the northwesternmost corner of Sudan, Carboniferous strata rest directly on basement. The Silurian Um Ras Formation, however, is present again southeast of Gebel Kissu, in Ennedi and in Gebel Tageru of northern Darfur.

DEVONIAN

No surface Devonian exposures are known in north Egypt, the Sinai, the Gulf of Suez or the Eastern Desert. Devonian sediments are referred to from at least 12 wells in northwest Egypt in unpublished oil company reports (Hantar, this book, Chapter 15). These occurrences belong – like the underlying Silurian strata – to a Paleozoic basin which extends over large areas in northeast Libya and northwest Egypt; southward it is in connection with the Kufra Basin of Libya and Chad. This situation was reconstructed by the author (Klitzsch 1970). The only published proof of Devonian strata in northwest Egypt is by Schrank (1984), who identified numerous palynomorphs of Late Emsian to Early Givetian age (late Lower to early Middle Devonian) from the Foram 1 well.

The other wells contain remains of brachiopods, bryozoans, echinoderms as well as foraminifera, conodonts, ostracodes and acritarchs. The environment is at least partly marine with a southward increase of continental influence. Most of the sections consist of light colored porous sandstone and of siltstone with some intercalations of partly black shale. These shale intercalations can reach a thickness of 150 to 200 m. Total thickness of Devonian strata is in the order of 900 to 1200 m in the northwest, thinning southward toward the Kufra Basin.

Devonian strata are exposed on surface along the western and southwestern edge of the Abu Ras Plateau near the Libyan border in southwest Egypt and at the northeastern part of Gebel Uweinat. These occurrences are part of the Devonian blanket described above and represent the surface exposure of Devonian strata reaching further south toward the Ennedi Mountains in northeast Chad. Unfortunately, no paleontological proof has been found until now within southwest Egypt. The stratigraphical interpretation is based on the position of these sediments within the section (in both areas unconformably underlain by strata of proven Silurian age and more or less conformably overlain by strata of proven early Carboniferous age). Moreover, the presence of very similar strata of Devonian age in neighbouring areas of Libya and Chad back this interpretation (Klitzsch & Léjal-Nicol 1984).

The Devonian sediments of the western Abu Ras Plateau and of northeast Gebel Uweinat are compared with the Tadrart Sandstone Formation of Libya (early to middle Devonian). In these areas, this formation consists of 50 to 70 m of mainly tabular cross-bedded fine- to coarse-grained sandstone, partly slightly conglomeratic, frequently with convolute bedding. It is certainly a fluviatile sediment deposited in the southern and eastern to southeastern foreland of one (or several) Devonian transgressions.

CARBONIFEROUS

Of all the Paleozoic sediments of Egypt, strata of Carboniferous age were the first to be recognised. They have been discussed the most and are exposed in large areas, both in north Egypt (Sinai, Gulf of Suez) and in the southwest at the Abu Ras Plateau west of Gilf Kebir and in Gebel Uweinat. The Carboniferous strata of Egypt differ more in facies than other Paleozoic sediments. The Carboniferous sediments range from fully marine carbonate, shallow and deep marine clastics, deltaic and continental fluviatile sandstone to lacustrine and fluvio-glacial deposits as well as tillite. The main reason for the very differentiated appearance of Carboniferous strata is the structural development of that time: until late Visean or Namurian time, Egypt was at the southern edge of a more or less shallow sea which transgressed parts of the country (Fig. 21/3). At the same time, northward draining rivers from surrounding areas in the south and southeast filled depressions with fluviatile sediments. During the late Carboniferous (probably in Namurian or Westphalian time) and

UPPERMOST CARBONIFEROUS TO LOWER JURASSIC

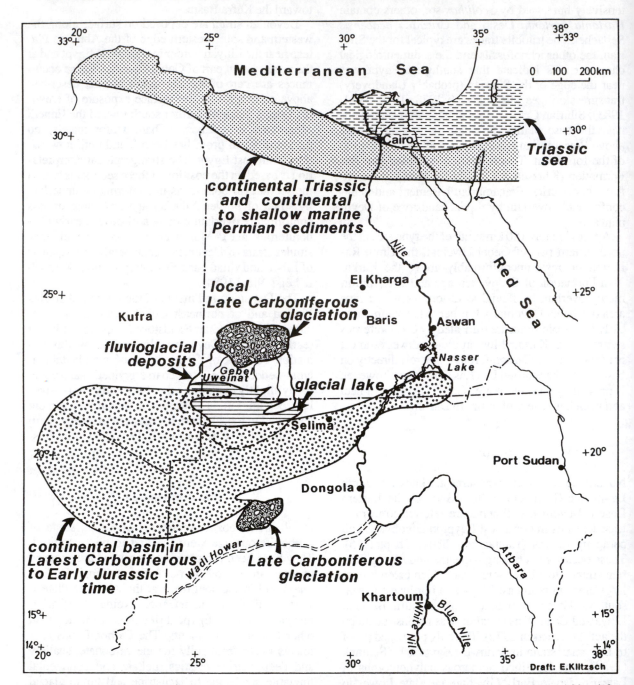

Figure 21.4 Distribution and facies of upper Carboniferous and Permian to Triassic sediments in northeast Africa (after Klitzsch & Wycisk 1987).

in connection with the Hercynian structural event, large parts of central and south Egypt were uplifted, the sea retreated and in higher parts of southwest Egypt, glaciers developed. Moreover, due to the structural development, the drainage system of central and south Egypt was reversed. From latest Carboniferous until early Jurassic time, the drainage was toward the south and southwest. Large areas in central and south Egypt were elevated and the Paleozoic strata – where present – were now subjected to erosion. The oldest sediments, for which this new situation can be proved, are the glacial deposits of the

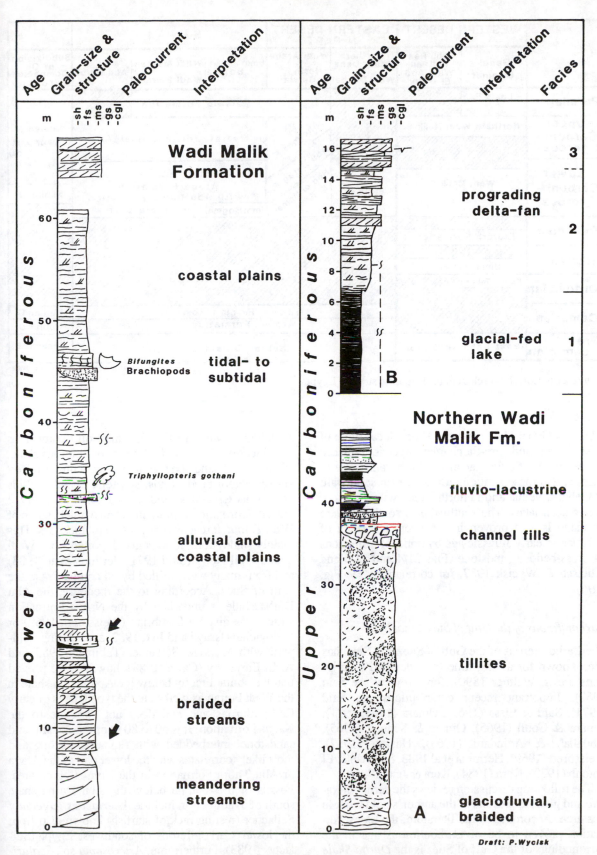

Figure 21.5 Carboniferous strata of the Wadi Abd El Malik (lower Carboniferous and lower part of upper Carboniferous section) and of Gebel Uweinat (upper part of upper Carboniferous) (after Klitzsch and Wyask 1987).

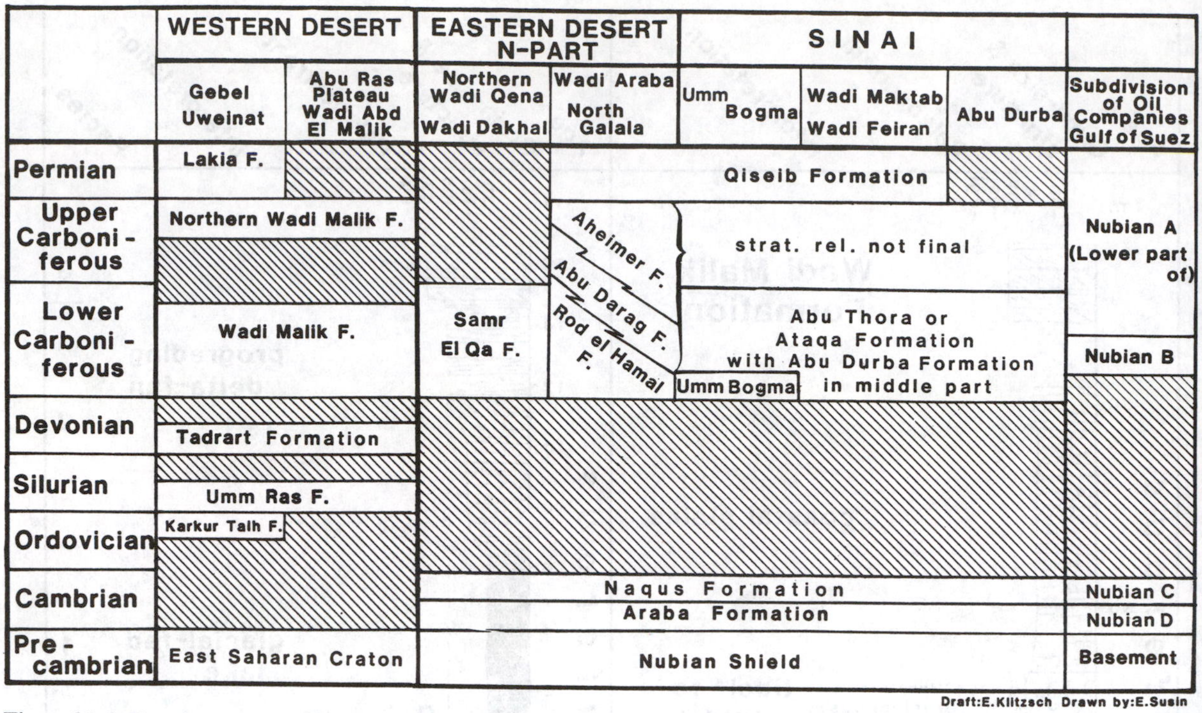

	WESTERN DESERT		EASTERN DESERT N-PART		SINAI				
	Gebel Uweinat	Abu Ras Plateau Wadi Abd El Malik	Northern Wadi Qena Wadi Dakhal	Wadi Araba North Galala	Umm Bogma	Wadi Maktab Wadi Feiran	Abu Durba	Subdivision of Oil Companies Gulf of Suez
Permian	Lakia F.				Qiseib Formation			Nubian A (Lower part of)
Upper Carboni-ferous	Northern Wadi Malik F.			Aheimer F. / Abu Darag F.	strat. rel. not final			
Lower Carboni-ferous	Wadi Malik F.		Samr El Qa F.	Rod el Hamal F.	Abu Thora or Ataqa Formation with Abu Durba Formation in middle part			Nubian B
					Umm Bogma			
Devonian	Tadrart Formation							
Silurian	Umm Ras F.							
Ordovician	Karkur Talh F.							
Cambrian					Naqus Formation			Nubian C
					Araba Formation			Nubian D
Pre-cambrian	East Saharan Craton				Nubian Shield			Basement

Draft:E.Klitzsch Drawn by:E.Susin

Figure 21.6 Correlation chart of the Paleozoic strata of Egypt.

Abu Ras-Gebel Uweinat area. There, local tillite of post-Visean and pre-Stephanian age is replaced southward by fluvio-glacial sandstone and conglomerate, and these strata interfinger again southward at Gebel Uweinat and in north Sudan with varves of peri-glacial lakes. The southward to southwestward drainage is also proven in the overlying strata of Permian to early Jurassic age by transport directions of cross-bedded sandstone (Figs 21/4 and 21/5 and Klitzsch & Wycisk 1987, for correlations see Fig. 21/6).

Carboniferous of the Gulf of Suez-Sinai area

The Carboniferous of the Gulf of Suez and Sinai has been known for a long time (first detailed descriptions are in Walther 1890), for a review see Said 1962). Important recent contributions are Said (1971), Said & Eissa (1969), Omara (1965, 1971), Omara & Conil (1965), Omara & Schultz (1965), Abdallah & Adindani (1963), Hassan (1967), Weissbrod (1969), Hermina et al 1983, Soliman & El Fotouh (1970), Kora (1984), Kora & Jux (1986).

The following discussion reviews the earlier literature and gives the results of the author's field work in this area. According to the literature, the oldest formation proven to be of Carboniferous age in the surroundings of the Gulf of Suez is the *Durba Shale Formation* (Hassan 1967), which is identical with the

so-called 'Nubian B' of the oil companies classification (Beets 1948). It is a fossiliferous shale, siltstone and sandstone sequence, containing different species of productoids, spirifers, rhynchonellids as well as *Conularia* and other fossils. In the type area, the formation is more than 160 m thick, south of Wadi Feiran, it reaches about 40 m in thickness. This mainly shaley formation is also reported from Wadi Araba Well No. 1 (Said 1971, Hermina et al. 1983) and from many wells drilled by oil companies in the Gulf of Suez. According to the literature, the Abu Durba shale is underlain by the Naqus Formation which was given a Cambrian (Weissbrod, 1969) or Ordovician (Issawi and Jux, 1982) age. During fieldwork with A. and M. Seilacher (Tübingen, 1983) and A. El Barkooky (Cairo, 1988), however, we proved that the strata directly below the Abu Durba shale in the Wadi Feiran as well as in the type area is of early Carboniferous age and does not correlate to the Naqus Formation. It is 50 to 80 m of marine silt- and sandstone, interbedded with fluvial sandstone and most likely correlates with the lower part of the Ataqa or Abu Thora Formation in the Umm Bogma area. Fossils of these strata below the Abu Durba shale south of Wadi Feiran include *Asteriacites gugelhupf* Seilacher (resting rack of starfish) known also from the lower Carboniferous of southwest Egypt (Seilacher 1983). Furthermore, *Asterosoma* sp., *Scalarituba* sp., *Scolicia* sp., *Conostichus* sp., *Curvolithus* sp., *Trichophycus* and other ichnofossils indicating shallow marine environment and partly proving Carboniferous age (Seilacher pers. comm.).

Schrank (pers. comm.) has also identified palynomorphs of lower Carboniferous age from mudstone exposed approximately 15 m below the described trace fossil horizon. It is therefore certain that at least some of the strata deposited above the Cambrian Araba Formation and below the Durba Formation are of early Carboniferous age. The Durba shale must be an equivalent of shale and siltstone units more or less of the middle part of the Abu Thora or Ataqa Formation in the Umm Bogma area and it should be definitely younger than the Umm Bogma Formation. At the type area, this formation consists of dolomite, which is mainly sandy and conglomeratic near the base, as well as sandy marl and intercalations of silt and sandstone. Some beds are very fossiliferous and contain a rich fauna of Visean age. Kora (1984) and Kora & Jux (1986) recently investigated the fauna of the type area in detail, published a long list of megafossils and added a number of newly discovered brachiopods, bryozoans and corals to the fauna already published by Klebelsberg (1911) and Ball (1916). The name Umm Bogma is now reserved to this carbonate unit only. Previously the name was used for this unit and the underlying clastic section (Kostandi 1959, Said 1962).

The thickness of the Umm Bogma Formation in the type area ranges from some meters to more than 55 m. The formation overlies the manganese ore lenses of the Um Bogma area, which indicates at least a discontinuation in sedimentation between the underlying Naqus Formation and the dolomitic Um Bogma Formation. It is conformably overlain by the *Ataqa Formation* (Kostandi 1959, Weissbrod 1969), recently called *Abu Thora Formation* (Kora 1984, Weissbrod 1976).

In the Umm Bogma area as far south as Wadi Mukattab (some kilometers north of Wadi Feiran), the Ataqa Formation consists of almost 100 to over 200 m of interbedded fluvial sandstone with nearshore marine sandstone, siltstone, shale and – locally – thin coal seams which also are present in the Wadi Dakhal area west of the Gulf of Suez. The upper part of the formation contains basalt sills at several localities. At some sandstone beds impressions of brachiopods are frequent (*Rhynchonella* sp.), others contain ichnofossils like *Spirophyton, Asterosoma, Cruziana costata, Planolites, Margaritichnus reptilis* and *Scolithos* (Seilacher pers. comm.).

Together with this fauna we found in the same part of the Ataqa sequence fossil plants: *Cyclostigma pacifica, Lepidodendron rhodeanum, Lepidodendropsis africanum* and *Nothorhacopteris* sp., which are of Visean age (Léjal-Nicol this book). At Wadi

Mukattab, a very rich layer of fossil plants was found between beds full of marine ichnofossils and impressions of brachiopods. According to Léjal-Nicol (pers. comm.), the flora contains – in addition to the same flora mentioned above – *Lepidodendron veltheimi, Lepidodendropsis venestrata, Tomiodendron kemerovense* and several new species. The flora is of Visean age.

On the other side of the Gulf of Suez – between northern Wadi Qena and Wadi Dakhal at the southern Galala Plateau – the Cambrian strata (where present) are overlain by up to 150 m of fine- to coarse-grained sandstone, partly massive and clean glass sand, partly well-bedded with abundant *Scolithos* structures and with intercalations of silty shale and shale. At Somr El Qaa (north Wadi Qena) cross-bedded sandstone dominates. Shale is almost totally absent and well-bedded sandstone is full of *Scolithos* structures. Northward toward Wadi Dakhal, shale and siltstone increase, only some layers are burrowed by *Scolithos* (Wadi Dakhal) and in the middle of the Carboniferous section occurs a massive 40 to 60 m thick white, slightly kaolinitic quartz sand consisting of beach sand and large deltaic forests near the top. Below this glass sand member, cross- to parallel-bedded sandstone contains fossil plants like *Lepidodendron veltheimii, Bothrodendron* aff. *deperati, Pseudolepidodendropsis scobiniformis, Precyclostigma blakaense* and several others (Léjal-Nicol, chapter 29 this book). Fine-grained marine sandstone and siltstone above the glass sand contains different brachiopods, ichnofossils and also plant remains, all of early Carboniferous age. At this level, also very thin coal seams are present. This lower Carboniferous formation is most likely equivalent to the Umm Bogma and the Ataqa Formation of Sinai, but marine influence is less than at Sinai. We call these Carboniferous strata *Somr El Qaa Formation* after Gebel Somr El Qaa in the northern part of Wadi Qena. The formation rests above a fossil soil and above fossiliferous clastics of lower Cambrian age (Araba Formation) at Somr El Qaa and on Precambrian basement at Wadi Dakhal. Between the two locations, Cambrian strata are present in most areas below the Somr El Qaa Formation. At the Somr El Qaa area and at Wadi Dakhal, the Somr El Qaa Formation is overlain with a very slightly unconformity by massive white sand (partly also glass sand) of Albian to Cenomanian age (Wadi Qena Formation), containing plant remains like *Paradoxopteris stromeri*. The stratigraphic and paleogeographic interpretation of the whole strata below marine Cenomanian has been very different in the past (Jux & Issawi 1982).

Further north in the Wadi Araba area and at the southern and eastern foreland of the north Galala Plateau, the Carboniferous increases rapidly in thickness. Above approximately 100 m of beds coeval with the Abu Durba Shale Formation (see above), follow 360 m of thick sandstone, siltstone and limestone with abundant crinoids, brachiopods, gastropodes (*Bellerophon* sp. sp.) and other marine fossils. These strata are called the *Rod El Hamal Formation* (Abdallah & Adindani 1963). They are overlain by the *Aheimer Formation*, which consists of 280 m of successions of shale, siltstone, sandstone and silty to sandy marl and limestone, partly full of mainly shallow marine fossils (brachiopods, gastropods, pelecypods, crinoids and others). For a full description of the Wadi Araba occurrence, reference is made to Abdallah & Adindani (1963).

The age of the Rod El Hamal Formation is late Carboniferous and that of the Aheimer Formation is late Carboniferous to early Permian (Said & Eissa 1969). Because the Aheimer Formation is overlain by the Qiseib redbeds now proven to be of lower Permian age according to its rich flora (*vide infra*), it is possible that the Aheimer Formation is of late Carboniferous age only. From own fieldwork together with A. El Barkooky, I have the impression that subdivision, correlation and exact dating of Carboniferous strata in the Wadi Araba – northern Galala area, is still not fully understood. Recent research of Cairo University (El Barkooky, Barakat) is supposed to improve this gap in Egyptian stratigraphy.

Carboniferous of the subsurface in the north Western Desert was already reported by Said (1962) and is treated by Hantar (this book, Chapter 15). In the Faghur well situated near the Libyan border, a rich fauna is described (Said & Andrawis 1961). Carboniferous strata are also known from at least 12 wells *west* of longitude 27° E and from several wells to the east of this longitude. Toward the east, the occurrences have less marine influence and are recognised solely on the basis of stratigraphic evidence.

In the western areas, the strata are more or less fully marine, consist of sandstone, siltstone and shale with thin limestone intercalations, mainly in the upper part and with several at least partly biohermal limestone or dolomitic units of up to 30 m in thickness. Southward, the limestone gradually disappears and fluvial influence increases. This seems to be even more evident eastward, but control in this direction is very weak and in large areas Carboniferous strata seem to be missing to the east of 27° E. West of there, thickness is generally in the order of 670 to slightly over 950 m.

Schrank (1984) identified a rich microflora of Visean age from the middle part of the Carboniferous sediments of the Foram 1 well (near the Libyan border, approximately 160 km south of Siwa). He compares it with the microflora from Visean strata of

the Gulf of Suez identified by Said (1962, 1964) and Sultan (1976).

The Carboniferous of the Abu Ras-Gebel Uweinat area in southwestern Egypt was first reported from the Karkur Murr in northwest Sudan by Menchikoff (1926). Later, Carboniferous strata were described by several authors from bordering areas in Libya, for example Burollet (1963) and Hecht et al. (1963), but the geology of the Carboniferous sediments of the southwest corner of Egypt was only recently partly worked out (Klitzsch 1979, 1983. Klitzsch & Léjal-Nicol 1984). In this paragraph, the results of this work, which has not yet been finalized, are given.

The area where Carboniferous strata can best be studied in southwest Egypt is Wadi Abd El Malik, a south to north wadi system with many tributaries. The whole system is deeply cut into the Carboniferous and the overlying younger sediments of the Abu Ras Plateau. This plateau is situated along the Libyan border to the northwest of the equally large, but better known Gilf Kebir Plateau (made of strata of late Jurassic to mainly early to middle Cretaceous age). The two plateaux are in contact at the Akaba passage, a very narrow remnant of an originally much larger plateau.

It is very difficult to reach Wadi Abd El Malik from the east. That is why its geology remained unknown until recently. The easiest access is from the Kufra area (Libya) or from the north through Wadi Qubba. The approach of Wadi Qubba from Egypt is through the sand sea, which can be traversed either from the Siwa Oasis southward through the parallel southward-striking channel system of longitudinal dunes or from Abu Minqar (northwest of Dakhla) westward across the dunes. It is very difficult to reach Wadi Abd El Malik from the south. The south Abu Ras Plateau can be reached by car through the Akaba pass, but from there it is almost impossible to enter Wadi Abd El Malik, unless one goes more than 100 kilometers north to north-northeast into the sand sea and then goes westward across the westernmost part of the sand sea.

The strata of Wadi Abd El Malik and its side valleys including the western and southern scarps at the Abu Ras Plateau, are of Carboniferous age and form the type section of the Wadi Malik Formation. They rest more or less conformably upon the Tadrart Sandstone of Devonian age. They are made up of 100 to 150 m of alternating silty shale, siltstone, fine- to coarse-grained and partly conglomeratic sandstone, all of fluviatile to shallow marine to beach and intertidal environment. Cross-bedded fluvial sandstone increases upward and southward as well as eastward. The lower part of the section contains rare impressions of brachiopods, some trace fossils and plant

remains like *Eremopteris whitei* of early Carboniferous age. The middle part is rich with trace fossils in
some layers, some of which also contain frequent
Camerotoechia sp. Seilacher (1983) identified
among many others *Conostichus broadhedi* (a feeding burrow of actinians frequent in the Lower Carboniferous), *Asteriacites gugelhupf* (resting track of
stalleroids), *Bifungites fezzanensis* (dwelling burrows well known from Late Devonian and Early
Carboniferous strata in Libya). In the central parts of
Wadi Abd El Malik and some meters below the above
fossil beds, a rich flora of early Carboniferous age
was found, including *Archaeosigillaria minuta*,
Heleniella costulata, *Lepidodendron veltheimi*, *Lepidodendropsis africanum*, *L. fenestrata*, *L. hirmeri*, *L.
vandergrachti*, *Lepidosigillaria intermedia*, *Prelepidodendron rhomboidale*, *P. lepidodendroides*, *Rhacopteris orata*, *Triphyllopteris gothani* and several
others (see also Léjal-Nicol, this book, Chapter 29).
Similar flora was also found at the eastern edge of the
Abu Ras Plateau, there it is truncated by late Jurassic
to early Cretaceous strata containing abundant ferns.

At central Wadi Abd El Malik and along the eastern edge of the Abu Ras Plateau, the partly marine
Wadi Malik Formation is overlain disconformably by
conglomerate, conglomeratic sandstone and cross-
bedded sandstone, locally containing large erratic
blocks. About 3 km west of Wadi Abd El Malik and
at the major east-west striking side wadi along its
western scarp, the upper part of the Wadi Malik
Formation consists of 30 m of cross-bedded
sandstone, topped by 2 m of fine-grained, white sand-
and siltstone, which contains *Triphyllopteris gothani*
(lower Carboniferous) among others.

Above an erosional contact follow 4 to 5 m of
conglomerate with angular boulders (partly from the
underlying siltstone and sandstone). These are followed by 25 m of heterogeneous conglomeratic
sandstone, which are topped by 4 to 5 m of mudstone
with large cracks filled with conglomerate (probably
paleosol with ice cracks). This bed is overlain by
20 m of medium- to coarse-grained, cross-bedded
fluvial sandstone. The uppermost beds contain plant
remains of *Rhodea*. Above a thin conglomerate bed
Cordaites angulostrictus occurs. The latter should be
an indication for Stephanian (uppermost Carboniferous) age (Léjal-Nicol pers. comm.). This 50 to 55 m
of mainly coarse sediment above the erosional contact is the *Northern Wadi Malik Formation*. It is
interpreted as a fluvio-continental sequence of
Namurian or Westphalian to Stephanian age. Northward, toward the northern end of Wadi Malik, this
unit interfingers with a sixty or more meters-thick
sediment, which consists of white clayey sand and
gravel, including boulders and blocks of sediments
up to house size. This sediment is interpreted as of

glacial origin. The underlying sandstone and siltstone
is folded and compressed, probably as a result of the
movement of glaciers.

At Gebel Uweinat and its southeastern and eastern
foreland as well as at the northeastern escarpment,
the lower Carboniferous Wadi Malik Formation is
also present. There, it is up to 120 m thick, consists of
shallow marine sandstone and siltstone with only thin
shale intercalations, but with thick units of cross-
bedded, fluvial sandstone. Apart from trace fossils in
some marine beds and poorly preserved impressions
of brachiopods, plant remains are frequent, including
Cyclostigma ungeri, *Lepidodendropsis vandergrachti*, *L.* aff. *rhomboformis*, *Precyclostigma*
and others. We found similar strata also further south
in the Sudan, where marine influence disappears. At
the northeastern Uweinat region, the Wadi Malik
Formation rests on the Devonian Tadrart Sandstone
Formation. At Karkur Murr on the Sudanese side of
Gebel Uweinat, from where Carboniferous strata
were first described by Menchikoff (1926), the Carboniferous overlies unconformably the Precambrian
basement.

In Gebel Uweinat, the North Wadi Malik Formation assumes a distinct facies which differs from that
of the type area: it is mainly a fluvial sandstone
showing south and southwestward transport directions; a large part of the section is made up of varve-
type siltstone with very regular and very thin layering
of fine-grained sand. At Karkur Murr, the basal part
contains a tillite bed of several meters of thickness.
Southward and southeast of Gebel Kissu in Sudan,
this formation consists almost exclusively of varve-
type siltstone, which is interpreted as indicating the
presence of a very large glacial lake at late Carboniferous time. Thickness of the north Wadi Malik Formation is between some 10 and 120 m (Karkur Murr).
The formation is overlain by Permo-Triassic sediments (Lakia Formation) in the Gebel Uweinat area
and by late Jurassic to early Cretaceous sediments
further north.

PERMIAN

Permian of the Gulf of Suez-Sinai area can be referred to under the name *Qiseib Formation* (Abdallah
& Adindani 1963). The original authors describe a 50
m thick sequence of red shale and brown to red
friable sandstone from Wadi Malha in the north Galala Escarpment, overlying marine fossiliferous
strata of Aheimer Formation. The same strata in
similar thickness overlies marine fossiliferous strata
of Carboniferous age in the Umm Bogma area west
of the Gebel Tih Escarpment. The Qiseib is also
present in Wadi Araba between the Zafarana-Cairo

road and St. Anthony as well as from 9 km west of the St. Anthony junction westward (or approximately 41 km west of Zafarana). At this location, we collected from red silty sandstone directly next to and north of the road the following flora (see also Léjal-Nicol 1986) of lower Permian age: *Asterotheca* aff. *leeukui-lensis*, *Callipteris conferta*, *Thinnfeldia* aff. *decurrens*, *Spenophyllum* sp., *Lobatannularia* sp., *Dory-cordaites* sp. and several others just added to this collection.

In the foreland of Gebel Tih on Sinai, these strata are overlain by a 60 to 70 m thick sequence of mainly well-bedded brown lacustrine and fluvial sandstone, mostly likely of Triassic age. At the type locality, the overlying strata (Malha Formation) are of Cretaceous age, as most likely also in Wadi Araba.

The flora found in the Qiseib Formation of Wadi Araba finalizes the discussion about the age of this formation formerly interpreted to be Permo-Triassic. It also backs our interpretation of a Carboniferous (and not Permian) age of the Aheimer Formation.

Not many details are known about the Permian in the subsurface of northwest Egypt. Thickness and facies of the Permian on the few locations from which Permian strata are reported, varies from 50 to 60 m of dolomite, sandstone and shale of a shallow marine environment north of Siwa and more than 400 m of light to pink sandstone with shale and thin coal intercalations east of the Qattara Depression.

Permian strata of *southwest Egypt* – if present at all – seem to be restricted to the Gebel Uweinat area. South of the Sudanese border, large areas are covered by immature fluvio-continental and lacustrine deposits of Permian to early Jurassic age. These strata are the result of erosion of older strata in large parts of south and central Egypt, following uplift of these areas in late Carboniferous time and southward transport by rivers.

The only area where this *Lakia Formation* (Klitzsch & Léjal-Nicol 1984) is present in Egypt is above the Northern Wadi Malik Formation in northeast Gebel Uweinat. There, 350 to 400 m of cross-bedded and partly clayey sandstone and mudstone form a major part of the eastern Gebel Uweinat Escarpment above late Carboniferous and below late Jurassic to Cretaceous strata. This unit was probably meant by Burollet (1963) with Cima Sandstone. The stratigraphic proof for the Lakia Formation is weak. At Lakia Arbian fossil wood of Triassic age is present in middle and higher parts of the section. Thirty kilometers south of Selima, the top of the formation is of lower Jurassic age. The Permian age for the lower part of the formation is suggested because no discontinuity has been observed between the Northern Wadi Malik Formation of late Carboniferous age and the Lakia Formation.

CHAPTER 22

Mesozoic

M. T. KERDANY & O. H. CHERIF

Conoco Egypt, Inc. and Faculty of Science, Ain Shams University, Cairo, Egypt

Mesozoic rocks are known in Egypt since the earliest phases of geologic research in this country. The first recognised Mesozoic sequences are the Cretaceous rocks which crop out extensively. Jurassic rocks were later identified in northern Sinai (Barthoux & Douvillé 1913) and in the Eastern Desert in the district between Gebel Ataqa and Gebel Galala El Bahariya (Sadek 1926), while Triassic exposures were found to be restricted to Gebel Arif El Naga in eastern Sinai (Awad 1946).

The distribution of Mesozoic rocks in Egypt indicates that each of the three systems recognized reflects a distinct paleogeographic setting. The Triassic seas encroached only over a limited region of the extreme northeast part of the country. The Jurassic sea seems to have covered northern Egypt, giving the impression that an embayment existed over the site of the present Gulf of Suez during that time. The late Cretaceous transgression, witnessed one of the widest encroachments of the sea over the country in Phanerozoic times. This extensive distribution of late Cretaceous seas seems to correspond to a worldwide change in sea level (see Morgan, Chapter 7, this book).

Intensive deep drilling operations in the course of exploring for oil, particularly after World War II, increased our knowledge of the distribution of the Mesozoic rocks in the subsurface, particularly in the northern parts of the country, where they are generally covered by extensive Tertiary blanket deposits masking the pre-Cenozoic geologic history.

Mesozoic strata crop out in southern Egypt and in northern Sinai where an almost complete sequence from Triassic to Cretaceous is known. In the northern Western Desert, however, Mesozoic strata are buried beneath younger sediments and are known only from the subsurface.

Figure 22/1 shows the major structures of north Egypt. It is based on subsurface, seismic and well data. In the northern Western Desert the structure is dominated by west-northwest faults which abut in the southeast against a set of east-northeast faults constituting the Kattaniya horst. In the northern Eastern Desert and in Sinai, similar elements to those of the northern Western Desert are recognized on the surface. In Sinai, however, the major 'anticlinal' structures (Yelleg, Halal, Risan Aneiza, Maghara, etc.) follow the same trend as the Kattaniya horst. These structures are affected by the younger faults of the Gulf of Aqaba trend (northeast-southwest) and of the Gulf of Suez (Erythrean) trend (northwest-southeast) related to the rifting of the Red Sea and to the development of the Levant Shear (Gulf of Aqaba and Dead Sea system of transcurrent faults).

Figure 22/1 also depicts the structural elements which played an important role in the development and evolution of the various Mesozoic sedimentary basins in Egypt. These lie in the continental shelf area of north Africa, which must have been affected by the various movements of the microplates of the Tethyan realm. A full understanding of the tectonic evolution of Egypt during the Mesozoic will necessarily, therefore, touch upon the paleogeographic evolution of the Mediterranean region during the Mesozoic and the tectonic implications of various plate movements in this realm (Fig. 22/2).

Various models for plate movements in the Mediterranean have been proposed in the last decade (for a review of these and other aspects of the geological evolution of the east Mediterranean, the reader is referred to Dixon & Robertson 1984). In spite of many differences most models agree on many points. The following are particularly significant for the comprehension of the Mesozoic geologic history of Egypt.

1. The rifting-off or separation of a 'Turkish' (Apulian) microplate from the Egyptian continental mass in mid-Jurassic time (Figs 22/2, 22/3).

2. The movement of Africa with respect to Eurasia as a result of the stresses acting on east-west fractures generated on the northern part of the African continental mass. This movement is related to the successive stages in the opening of the Atlantic ocean. The initial mid-Jurassic opening of the mid-Atlantic

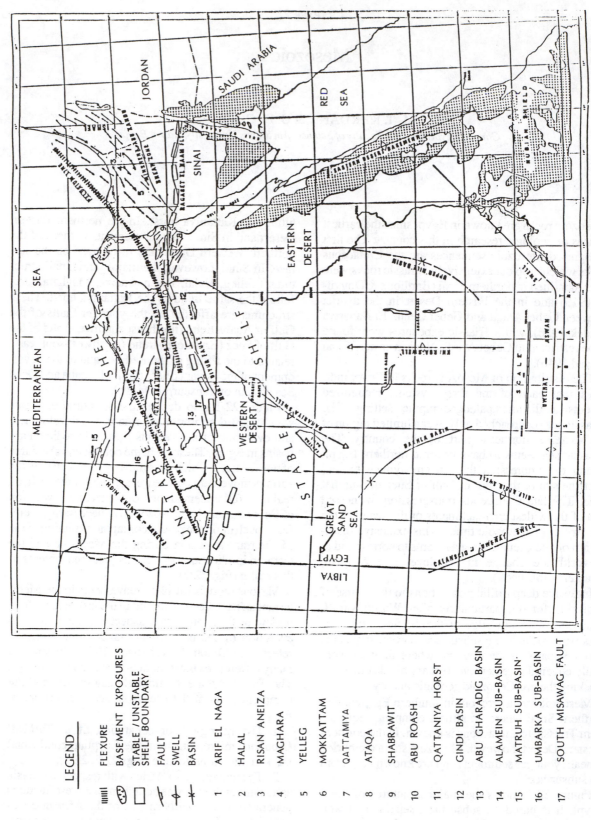

LEGEND

||||||| FLEXURE

▨▨ BASEMENT EXPOSURES

□ STABLE / UNSTABLE
  SHELF BOUNDARY

⊢ FAULT

⊣ SWELL

✕ BASIN

1   ARIF EL NAGA
2   HALAL
3   RISAN ANEIZA
4   MAGHARA
5   YELLEG
6   MOKKATTAM
7   QATTAMIYA
8   ATAQA
9   SHABRAWIT
10 ABU ROASH
11 QATTANIYA HORST
12 GINDI BASIN
13 ABU GHARADIG BASIN
14 ALAMEIN SUB-BASIN
15 MATRUH SUB-BASIN
16 UMBARKA SUB-BASIN
17 SOUTH MISAWAG FAULT

Figure 22.1 Map showing the distribution of the various tectonic elements and sedimentary basins of the Mesozoic in Egypt.

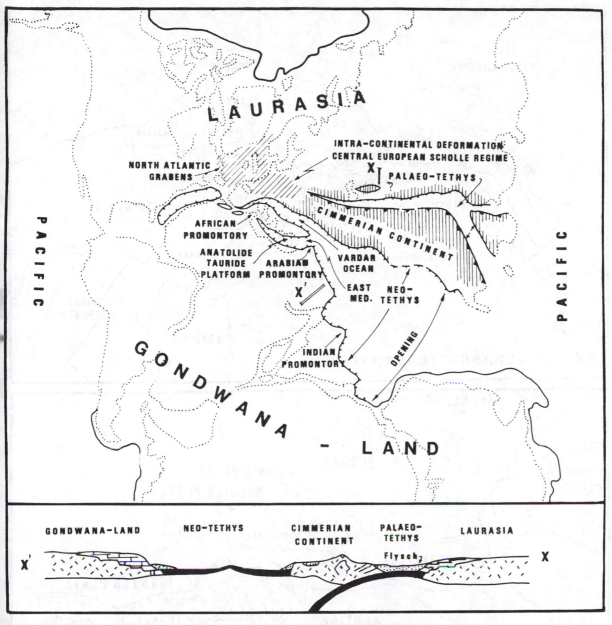

Figure 22.2 Latest Triassic-early Jurassic reconstruction of the Tethyan realm (after Sengor 1979).

(Dewey et al. 1973) caused Africa to move eastward with respect to Eurasia, generating a sinistral shear on the various fractures of the north African continental mass (Fig. 22/3a). The late Cretaceous (probably Turonian) opening stage caused Eurasia to move eastward with respect to Africa, generating a dextral shear (Fig. 22/3b). Many authors think that this dextral shear is still active at the present time (Pitman & Talawani 1972).

3. The separation of the 'Turkish' microplate from Africa is related to the formation of the Neotethys (Sengor 1979) which succeeded the older Paleotethys which was closed by the gradual drifting of the Turkish microplate in mid-Jurassic time (Sengor & Yilmaz 1981). Since this time, the movement of Africa toward Eurasia generated compressive stresses acting more or less in a north-south direction in the realm of the Neotethys. A significant acme of the action of these stresses occurred in late Cretaceous times (Laramide movements in the Alpine realm, mentioned by early authors, in Said 1962). Such compressions continued to be active until the early Eocene.

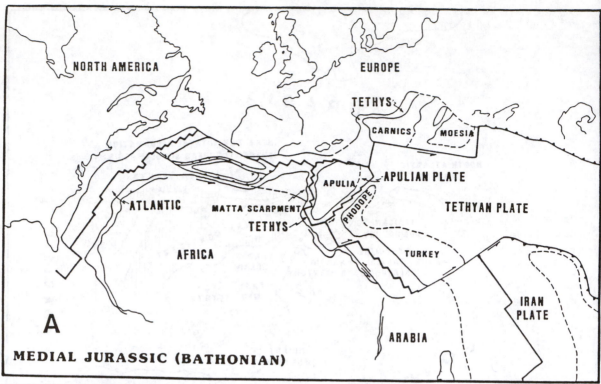

MEDIAL JURASSIC (BATHONIAN)

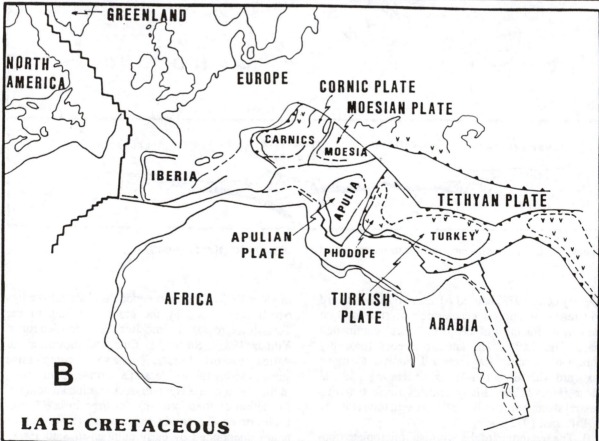

LATE CRETACEOUS

Figure 22.3 Some important phases of the evolution of the Mesozoic Tethys (after Dewey et al. 1973): a. Medial Jurassic (Bathonian); b. Late Cretaceous.

STRUCTURAL SETTING OF EGYPT DURING THE MESOZOIC

Egypt forms part of the large north African continental platform which extends to the west through Libya, Saharan Algeria and Morocco to the Atlantic Ocean. The Mediterranean coasts of Egypt are bordered by a narrow continental shelf (15 to 50 km wide) ending at a steeply faulted continental slope (Sestini 1984). To the east of the Nile Delta, in north Sinai, the continental slope probably lies very near to the shore, but is masked by the sediments of the Nile cone. During the Mesozoic, the African, Sinai and Arabian plates were connected into a single continental mass. The steeply faulted continental slope along the west Mediterranean coasts of Egypt referred to as the 'hinge line' (Orwig 1982), may be a passive margin of the African plate. The area between this 'hinge line' and the line of underthrusting of the Aegean arc and its eastward continuation through Cyprus and the Tauride, seems to grade from abnormal (thin) continental crust in the west to normal continental crust in the east, where continental collision occurs between the Arabian plate and Anatolia (Fig. 22/4).

Said (1962) subdivides the continental platform area of Egypt into a northern 'Unstable Shelf' and a southern 'Stable Shelf' (Fig. 22/1). The 'Stable Shelf' borders the 'Nubian Shield' (Sestini 1984). The 'Unstable Shelf', comprising the northern Western Desert, northern Eastern Desert and northern Sinai, shows an outstanding tectonic complexity. It is characterized by a deeper and more mobile Precambrian basement and by thicker Phanerozoic sediments. The 'Unstable Shelf' of Egypt was particularly affected by the various tectonic stresses generated by the plate movements in the Tethyan realm. These stresses were strongest during pre-Tertiary times, as can be attested by the complicated subsurface geology of northern Egypt when compared with the simple surface geology of the Tertiary times. The Tertiary geology becomes, however, more compli-

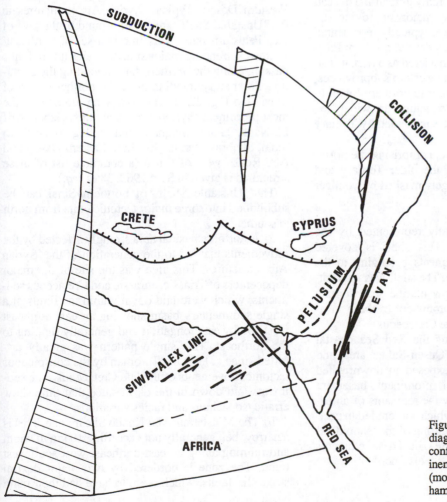

Figure 22.4 Schematic block diagram showing the major plate configurations in relation to prominent structures in the region (modified after Nur & Ben Avraham 1978).

cated in the eastern part of the 'Unstable Shelf' which is strongly affected by the late Paleogene and Neogene tectonic episodes, of the Gulf of Suez and Red Sea rift and the Levant shear.

The eastern part of the 'Unstable Shelf' stands in a structurally higher position than the western part during the Tertiary. Mesozoic rocks are often exposed in the east of the Eastern Desert (Shabrawet, Ataqa and the Galalas) and many areas in northern Sinai. This suggests that the subcrust in front of the 'Unstable Shelf' of the Western Desert yielded to more uniform subsidence of this region during the Neogene while that of the northern shores of Sinai was more rigid, allowing many Mesozoic structures to remain exposed to the present time (Fig. 22/4).

The apparent tilt of the continental mass of Egypt to the west is also reflected in the 'Stable Shelf' region by extensive basement exposures in the east.

Thus, Egypt may be subdivided into the following broad Mesozoic structural provinces (Fig. 22/1):

1. The 'Unstable Shelf' with complex structures, strongly affected by the shear system generated by the plate movements in the Tethyan realm. This can further be subdivided into two major subprovinces:

a) An Eastern subprovince, typically represented by northern Sinai. It also includes the northern Eastern Desert which may be considered as a transitional zone between the eastern and western subprovinces, but which is tectonically closer to northern Sinai. In this subprovince, Mesozoic structures are widely exposed on the surface and are surrounded by Tertiary sediments.

b) A Western subprovince, comprising the northern Western desert, where the older Tertiary and Mesozoic structures are largely masked by younger Tertiary blanket deposits.

2. A 'Stable Shelf', typically represented by the southern Western Desert. This shelf is covered mainly by continental sediments belonging to the Paleozoic and the Mesozoic. The shelf was intermittently overlapped by shallow marine seas coming from the north. The most prominent of these invasions occurred during the late Cretaceous.

Mesozoic sediments along the Red Sea coastal areas of the Eastern Desert (Quseir-Safaga area) and in west central Sinai, are exposed in downfaulted blocks in the vicinity of prominent basement outcrops. They are probably the remnants of an extensive sedimentary cover, which was eroded in post-Oligocene times after the uplifting of the 'shoulders' of the Gulf of Suez and Red Sea rift. These sediments are considered as part of the 'Stable Shelf'.

UNSTABLE SHELF

A. EASTERN SUBPROVINCE

1 *Structural setting*

This region lies to the east of the 'Pelusium line' (Neev 1975) (Fig. 22/1), and includes the northern parts of the Eastern Desert and Sinai. In northern Sinai the southern limit of the Unstable Shelf is drawn along the Ragabet El Naam east-west trending dextral wrench fault. This fault separates a northern region with complex structural pattern related to the 'Syrian Arc' trend from a southern region with relatively simple structures. In the northern Eastern Desert, the boundary between the 'Stable' and 'Unstable' shelves is more difficult to trace, but may be tentatively drawn along a line of east-west disposed escarpments extending from Cairo to Suez and marking the northern boundary of a number of topographic highs: Gebels Mokattam, Qattamiya and Ataqa. The westward extension of this line is marked by a sequence of major faults constituting the southern boundary of the Kattaniya horst in the northern Western Desert. Typical 'Syrian Arc' structures in the 'Unstable Shelf' are mostly found to the east of the 'Pelusium line'. This line is a set of northeast-southwest trending major sinistral transcurrent faults passing along the northern fault bounding the Kattaniya horst and extending up to the northern coast of Sinai (see Fig. 22/1). The following are among the most prominent 'Syrian Arc' structures: Gebels Arif El Naga, Halal, Maghara and Yelleg in northern Sinai, Shabrawet in the northern Eastern Desert and Abu Roash west of Cairo (a complete list of these structures is given in Said 1962: 38 *et seq.*).

The 'Unstable Shelf' of northern Sinai can be subdivided into three major tectonic units from north to south:

a) A northern basinal area strongly affected by the movements that led to the generation of the 'Syrian Arc' structures. This area was the site of the major depocenters of Triassic, Jurassic and Cretaceous sediments, which were laid down within the limits of a single sedimentary basin showing an east-northeast west-southwest elongation and generally dipping to the northeast. This single pattern was greatly distorted since at least the Turonian by the conspicuous tectonic movements of the late Cretaceous. The sediments laid down in the basin since that time show erratic isopachous and facies variations.

b) The Minsherah-Abu Kandu Shear zone. This is a narrow belt generally not exceeding 10 km in width and running along an east-northeast west-southwest trend. The zone is bordered by two major dextral strike slip faults, which seem to have had their main activity in post-Mesozoic times as they do not affect the facies distribution of the Mesozoic sediments.

c) To the south of the shear zone lies an area slightly affected by the 'Syrian Arc' folding, but which became a pronounced depocenter in late Turonian times as indicated by the isopachous and lithofacies distribution of the Wata Formation.

The Mesozoic geology of the northern Eastern Desert seems to be a continuation of the northern basinal area of Sinai (Table 22/1). The facies distribution of the Mesozoic sediments in this part of the 'Unstable Shelf', however, shows a gradual change to the west approaching the facies displayed in the northern Western Desert.

2 Triassic

2.1 Sedimentation and facies

The oldest Mesozoic sediments known in Egypt belong to the Triassic. These are exposed in northeastern Sinai at the core of Gebel Arif El Naga. This Gebel is one of the highest structures in Sinai which lies within the Minsherah-Abu Kandu shear zone and along one of the more prominent Syrian arcs. The stratigraphic column of the Gebel ranges from middle Triassic at the core of the structure to rocks of Neogene age at its outermost periphery. Earlier marine Triassic rocks are known only from the subsurface at Halal-1 well which lies to the northwest. Gebel Arif El Naga itself is an assymetrical anticline trending east-northeast west-southwest. A fault with the same trend traverses the southern flank of the structure. The Triassic rocks of Gebel Arif El Naga were first described by Awad (1946). The earlier geological mapping was conducted by the Geological Survey of Egypt; later mapping was conducted by the Geological Survey of Israel. Several paleontological studies were carried out including among others: Spath (in Awad 1946), Eicher (1946, 1947), Kummel (1960), Lerman (1960), Eicher & Mosher (1974), Druckman (1974), Parnes (1964, 1975) and Hirsch (1976) who established the ammonite biostratigraphy of this important Triassic exposure. Karcz & Zak (1968), Heller-Kallai et al. (1973), Druckman (1974), Druckman et al. (1975), Bartov et al. (1980) worked on different aspects of the sediments.

Said (1971) names the Triassic beds exposed in this structure the Arif El Naga Formation and subdivides them into three members: the 'A' or the lower 'Nubia'-type sandstone, the 'B', or *Beneckia* beds and the 'C' or genuine marine *'Ceratites'* beds. The 'A' and 'B' members are given the name Gevanim and the 'C' member the name Saharonim by Zak (1968). Two other formations, Zafir and Ra'af (Scythian-Anisian), are described from the subsurface below the 'A' beds from Halal well no. 1 where they attain a thickness of 196 and 50 m respectively (Zak 1968).

The 'A' and 'B' members (Gevanim) are exposed over a small area (0.3 km^2) in the core of the Arif El Naga anticline where they attain a thickness of 68 m. The 'A' member is 50 m thick and is made up of fluvial to fluviomarine deposits consisting of a lower unit, 23 m thick, of non-fossiliferous quartzitic sandstone with minor clay and an upper unit 27 m thick, of sandstone, limestone and clay beds carrying vertebrate remains, plant imprints and fossil wood. Among the fossils are the bivalves: *Trigonodus tenuidentatus* Lerman and *Unionites fassaensis* (Wissman). The 'B' member is 18 m thick and is made up of limestone and clay beds carrying, among others, the ammonite *Beneckia levantina* Parnes and the bivalve *Neoschizodus orbicularis* (Bronn). The age of the 'A' and 'B' beds is middle to early late Anisian (middle Triassic). The faunas of these beds, like their predecessors in the late Paleozoic, are endemic and are difficult to relate to the Germanic or north Tethyan associations. The 'C' member (Saharonim Formation) consists of 117 m of fossiliferous limestones, marls, shales and a few beds of dolomite and gypsum. The formation is subdivided into three mappable lithostratigraphic units by Zak (1964). The lowest is a fossiliferous limestone unit, 35 m thick, marked at its top by a 6.5 m thick bed of hard gray limestone (the so-called mottled scar bed of Awad 1946). An olivine basalt dike intrudes this sequence. Parnes (1975) recognizes three informal ammonite zones within this lower unit which are from bottom to top: *'Paraceratites binodosus'* (Eicher), *Gevanites inflatus* Parnes and *G. awadi* Parnes.

The middle limestone and marl unit is 31 m thick and carries the ammonite *Gevanites epigonus* Parnes, the bivalve *Pseudoplacunopsis fossistriata* (Winkler) and other molluscs and vertebrates. The upper limestone and gypsum unit is 51 m thick with many dolomite and marl intercalations. The macrofossils include many poorly preserved nautilii, gastropods, crinoid stems and echinoid spines. Most diagnostic is the conodont *Pseudofurnishius murcianus* Boogard which points to a Ladinian age for this succession.

The age of the 'C' beds is late Anisian to late Ladinian or early Carnian. It was deposited in a shallow marine environment with local hypersaline conditions developing in late Ladinian time. The faunas form a distinct biofacies different from that of the Alpine Tethys and resemble that of many south Mediterranean countries and Spain. Hirsch (1976) names this association the sephardic (Spanish) biofacies.

In the Gulf of Suez region and in central Sinai the Triassic is represented by the paracontinental Qiseib

CHRONOSTRATIGRAPHY					SINAI OFFSHORE	EXTREME N. SINAI	NORTHERN SINAI
MESOZOIC	CRETACEOUS	LATE	SENONIAN	MAASTRICHTIAN	SUDR	SUDR	SUDR
				CAMPANIAN			
				SANTONIAN			
				CONIACIAN			
			TURONIAN		DEEP MARINE CHALKS & SHALES		WATA
			CENOMANIAN				HALAL
		EARLY	ALBIAN			REEFAL L.S.	CORAL BE
			APTIAN		DEEP WATER SHALE & S.S.		RISAN ANEIZA
			BARREMIAN				NUBIA S.S.
			NEOCOMIAN				
	JURASSIC	LATE	PORTLANDIAN				
			KIMMERIDJIAN				
			OXFORDIAN				AROUSIA
		MIDDLE	CALLOVIAN				KEHAILIA
			BATHONIAN				SAFA
			BAJOCIAN				BIR MAGHARA
			AALENIAN				SHUSHA
		EARLY	LIAS		DEEP WATER L.S	REEFAL L.S.	RAJABIAH
							MASHABBA
	TRIASSIC	MIDDLE LATE	NORIAN				QISEIB
			CARNIAN		?	?	
		MIDDLE	LADINIAN				ARIF "C
			ANISIAN				EL NAGA "E
		EARLY	SCYTHIAN				"A
PALEOZOIC	PERMIAN						CONTINENTA
	CARBONIFEROUS						MARINE
	DEVONIAN						CONTINENTA S.S
	ORDOVICIAN/SILURIAN						
	CAMBRIAN						MARINE
	PRECAMBRIAN						BASEMENT

Table 22.1 Main Mesozoic stratigraphical subdivisions of northeastern Egypt.

GEBEL ABRAWET	GEBEL ATAQA	NORTH GALALA	CENTRAL SINAI	WADI QENA	QUSEIR/ SAFAGA
RA EL BAHARI ~?~	MOGHRA EL BAHARI		SUDR	SUDR	DAKHLA
	ADABIYA		THELMET ? RAJIM	HAWASHIYA	DUWI & QUSEIR
			MATULLA MAGMAR		TAREF
RA EL HADIDA	MOGHRA EL HADIDA		WATA	UM	? ? ? NUBIA
			ABU QADA	OMIEYED	
ALALA	GALALA ? ? ?	GALALA	RAHA		
AN ANEIZA			"MALHA"		
? ?		MALHA	? ? ?	MALHA	
	NOT EXPOSED	KHASHM EL GALALA	? ? ?	? ?	
T EXPOSED		? ?		? ?	
		QISEIB	QISEIB OR BUDRA		
		MARINE	MARINE		
			NAQUS		
			? ?		
		BASEMENT	BASEMENT	BASEMENT	BASEMENT

Formation. In its type locality along the western coastal plain of the Gulf of Suez the Qiseib is 43 m thick. It increases in thickness southward toward Wadi Araba to 140 m. The formation is made up of a lower thick unit of redbeds mainly variegated brown, purple and gray shales and siltstones and minor sandstones and an upper thin carbonate clastic unit. The Qiseib is a characteristic unit by virtue of its color and lithology. The formation was first described by Abdallah & Adindani (1963) and given a Permo-Triassic age because of its stratigraphic position above the Aheimer Formation which was then believed to be of late Carboniferous age. The lower clastic part of the formation carries fossil tree trunks and other plant remains (identified as Permian, see Klitzsch, this book, chapter 21) while the upper carbonate part includes in the Abu Darag area marine fossils that belong to the middle Triassic gastropod *Naticopsis* and to the pelecypods *Hoernesia* and *Pleuromya* spp. (Weissbrod 1969). Beds similar to

the Qiseib are described from the opposite side of the Gulf in the Um Bogma area above the early Carboniferous Ataqa Formation. They are named the Budra Formation by Druckman (1974).

In the subsurface, the Qiseib is recorded in many wells where it is referred to as the 'red shale series' or the 'upper Ataqa Formation' in many oil company reports. In the well Abu Hamth-1, the Qiseib is 376 m thick; the upper 36 m are made up of limestones rich in middle Triassic marine fossils (Druckman 1974).

The lower clastic redbeds include many thin coal seams from which were separated the palynomorphs: *Sulcatisporites kraeuseli, Verrucosisporites applanatus, Corollina meyeriana, Cyathidites minor* and *Concavisporites jurienesis* (Horowitz 1970) suggesting an early to middle Triassic age. In both the wells Ayun Musa-2 and Abu Hamth-1 remains of *Clamites* spp. are recorded. Similar sections are known from the wells Nekhl-1 and Hamra-1.

During the Triassic central Sinai and the Gulf of

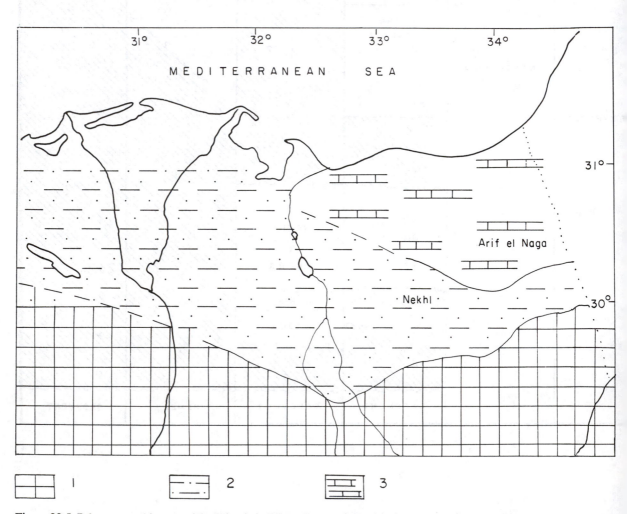

Figure 22.5 Paleogeographic map of the Triassic in Egypt (1 = continental facies, 2 = tidal flats, 3 = marine).

Suez region seem to have been the site of the continental fluvial deposits of the Qiseib Formation with its characteristic channel and bank deposits, abundant ripple marks, mud cracks, cross-bedding and silicified tree trunks of several meters length. A marine middle Triassic tongue of the sea terminated this fluvial episode.

2.2 *Triassic paleogeography*

The marine Triassic transgression followed the late Paleozoic (Hercynian) tectonic event and affected only the structurally lower areas of northeast Egypt. During the middle Triassic, an arm of the sea advanced to the south covering central Sinai and parts of the Gulf of Suez region to the locations of wells Ayun Musa-2, Hamra-1, Abu Hamth-1 and Nekhl-1. During most of the Triassic, however, Sinai and the Gulf of Suez regions were under the influence of tidal flat and deltaic sedimentation when the deposits of the Qiseib Formation were laid down. This formation lies unconformably over the late Paleozoic Aheimer Formation and conformably below the marine middle Triassic rocks.

The isopachs of the Triassic are regular reaching a maximum of +900 m toward the north (Jenkins, Chapter 19, this book). The Triassic sediments seem to have been deposited in a shallow shelf which was affected by several regressive cycles which caused the tidal flat-deltaic sediments to the south to interfinger the marine sediments of the north (Fig. 22/5). Toward the end of the middle Triassic a major regression caused hypersaline conditions to develop.

The faunas of the Egyptian Triassic are different from the Alpine Tethyan forms and resemble those of many south Mediterranean countries such as Turkey and Greece. This may be taken as evidence that the Turkish and Greek microplates still formed part of the shores of the southern Tethys and that migration along these shores could not have extended to the Eurasian shores.

3 *The Jurassic*

3.1 *Sedimentation and facies*

Jurassic rocks are exposed in northern Sinai at Gebel Maghara, Giddi, Minsherah and Arif El Naga. The thickest and most complete exposure in Egypt is that of Gebel Maghara which formed the subject of numerous studies. Al Far (1966) gives a review of the earlier literature and describes in detail the stratigraphy and structure of this important occurrence (see also Jenkins, Chapter 19, this book). The following table summarizes the thickness and lithological composition of the different units of the Maghara Jurassic.

Thickness and lithologic composition of Jurassic formations, Gebel Maghara.

Formation		Thickness (m)	sd/sh/ls ratio
Masajid	Arousiah	443	0–0–100
	Kehailia	132	2–14–84
Safa		215	34–37–29
Bir Maghara	Bir Maghara	216	1–68–31
	Moweirib	133	2–64–34
	Mahl	93	2–19–79
Shusha		271	52–36–12
Rajabiah		292	2–20–78
Mashabba		100	50–23–27

The section is made up of alternating genuine marine shelf deposits (Rajabiah, Bir Maghara and Masajid) and tidal flat-deltaic-fluvio-marine sediments (Mashabba, Shusha and Safa). The latter are made up of cross-bedded sands, fine clastics and plant beds which become veritable coal seams in the Safa Formation. Adindani & Shakhov (1970) give a description of the viably economic coal beds of Gebel Maghara.

The fauna of the Gebel Maghara Jurassic is described by Douvillé (1916), Arkell (1952, 1956), Said & Barakat (1958), Al Far, Hagemann & Omara (1965), and Hirsch (1976). The top of the Rajabiah Formation yields the earliest ammonite in the succession which is assigned to the Toarcian genus *Grammoceras* with some uncertainty by Arkell.

The middle and lower members of Bir Maghara Formation (the Mowerib and Mahl members respectively) yield the typical middle Bajocian species: *Bucardyomia orientalis*, *Dorsetensia* sp. and *Otoites* sp. The upper marls of Bir Maghara Formation yield a rich late Bajocian fauna including: *Ermoceras*, *Telermoceras*, *Normannites*, *Kosmoceras*, *Leptosphinctes*, *Trimargina*, *Thamboceras* as well as brachiopods, pelecypods and gastropods.

The lowermost beds of the Safa Formation carry *Thambites* spp. which assign this formation to the early Bathonian.

The middle part of the Masajid Formation (the lower part of the Arousiah) carries an *Erymnoceras* fauna (Callovian) while the upper part of this member yields the Oxfordian *Euaspidoceras* fauna. The Kimmeridgian seems to be missing in Gebel Maghara.

The ammonite fauna is essentially Ethiopian in realm; Tethyan elements appear only during the Bathonian and make an important element of the fauna from then on.

The 80 m Jurassic exposure at Gebel Minsherah (Farag & Shata 1954) carries a typical *Ermoceras* fauna (late Bajocian).

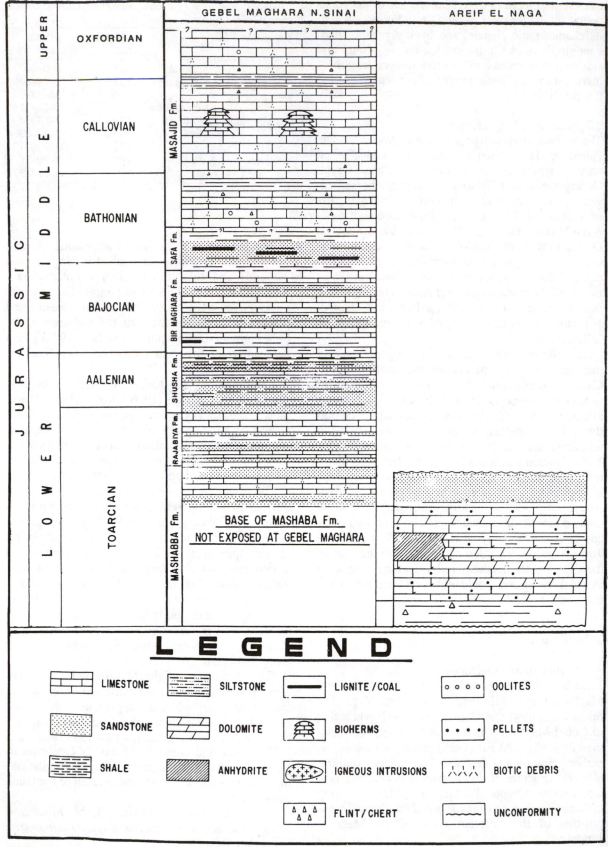

Figure 22.6 Composite section of the Jurassic system in northern Sinai.

The Arif El Naga exposes the oldest Jurassic rocks (? early Lias) in Sinai. These are represented by a 21 m-thick ferruginous silty shales, alternating with limestones and dolomites, marls and variegated sandstones, with a basal silty pisolitic, ferruginous shale (the so-called 'Flint Clay'). The section is described by Zak (1964) and Bartov et al. (1980). The sequence seems to have been deposited over a shallow shelf in saline or brackish lagoonal environment. It probably represents the initial phase of the Jurassic transgression which peaked in Toarcian time. In Arif El Naga this sequence, and the overlying 120 m-thick section of reddish variegated sandstone with a few layers of brown and dark gray shales, are probably equivalent to part of the Mashabba Formation exposed at the base of Gebel Maghara (Fig. 22/6).

Subsurface Jurassic occurrences in Sinai are recorded in Khabra-1 (+1430 m), Halal-1 (3234 m), Katib El Makhazin-1 (890 m), Hamra-1 (706 m), Abu Hamth-1 (475 m), Ayun Musa-2 (743 m), Nekl-1 (162 m), Sneh-1 (+120 m), Slav-1 (617 m), Manna-1 (+313 m), Gofer-1 (+100 m) and Tineh-1 wells (+77 m).

There are few surface Jurassic outcrops in the Gulf of Suez, the best known of which is that of Khashm El Galala first described by Sadek (1926). Here the Jurassic strata make the lowermost 170 m of this fault block mountain situated at the northeastern corner of the north Galala plateau. At the base of the section there are about 50 m of fluviatile sandstones and calcareous shales rich in fossil plants of Rhaetic-Infraliassic age that include *Equisites, Phlebopteria, Zamites* and *?Chaldophlebis*. This flora is identical to remains found in the Kohlan sandstone of Yemen (Carpentier & Farag 1948). Immediately above the plant beds lie the marine Jurassic beds: a long series of marls and thin limestones, sandy limestones and sandstones with, at least, twenty species of pelecypods, brachiopods, and (more rarely) gastropods. Shells such as *Trigonia pullus, Grevillia waltoni, Astarte* sp., *Nucula* sp. and others occur in profusion in certain thin ferruginous bands; the tests are preserved intact. Other slabby limestones are crowded with crushed rhynchonellids. Unfortunately no ammonites have been observed, but the assemblage is essentially upper Bathonian (Arkell 1956). Above the Bathonian, lie about 150 m of seemingly non-fossiliferous sandstones of presumed early Cretaceous age on which, in turn, rest Cenomanian marls and late Cretaceous limestones.

Several thinner Jurassic outcrops are known to occur along the western coast of the Gulf. To the south of Khashm El Galala at Ras El Abd on the Red Sea coastal road a section (over 100 m thick) of marine marls, limestones, and sandstones with Bathonian pelecypods and brachiopods is known. Farag (1948) attributes some of the pelecypods to the Bajocian; but the bulk of the fauna, according to Arkell (1956), is Bathonian.

In the Abu Darag area, Nakkady (1955) and Abdallah & Adindani (1963) record several Jurassic outliers in the midst of the Permo-Carboniferous Abu Darag exposure. These downfaulted blocks are thin, about 28 m in thickness, made up of argillaceous sandstones and yellow limestones with *Nucula variabilis, Trigonia pullus, Thamastraea crateriformis, Rhynchonella asymmetrica* and others.

In the subsurface, at Ayun Musa wells, the Jurassic rests directly below the Miocene. Here the Jurassic section has a thickness of 955 m and a clastic content of 85%. There are abundant coal seams (Adindani & Shakhov 1970). In Ataqa well a Jurassic section (580 m thick) is recorded; it has coal seams and a clastic content of 72%. The beds rest below early Cretaceous sandstones and are reached at a depth of 156 m. Borings in Hamra reached Jurassic at depth, where a 100 m section with a clastic content of 85% is reported.

Other subsurface Jurassic occurrences in the north Eastern Desert and the Gulf of Suez area are present in Abu Sultan-2 (+244 m), Ataqa-1 (585 m), GS9-1 (questionable Jurassic or Triassic 689 m), GS24-1 (+651 m), Q71-IX (+816 m), Q72-IX (+884 m), Sukhna-1 (+94.5 m), R69-1X (+1585 m), Abu Hammad-1 (+2347 m), Monaga-1 (+103 m) wells. Jurassic sediments are recorded in the subsurface as far south as Wadi Matulla (118 m) and Baba (370 m) (Kostandi 1959).

3.2 *Jurassic paleogeography*
The thickest exposed Jurassic sequence known in Egypt is that of Gebel Maghara (+2000 m). During this period the northern basinal area of Sinai was the site of a major sedimentary basin extending over the present Nile Delta area up to a region to the north of the Kattaniya horst where the thickest Jurassic sequence in the Western Desert is recorded (± 1000 m) (Fig. 22/7).

The first marine transgression of the Jurassic on the Egyptian shelf was restricted to north Sinai and did not extend beyond the region of the northern basinal area. This occurred during the Toarcian as attested by the marine algal limestones and marls of the Rajabiah Formation overlying the fluviomarine Mashabba Formation.

The limited Toarcian transgression was followed by a regressive phase represented by the fluviomarine Shusha Formation. This, in turn, is followed by a more important transgressive phase during the Bajocian which was responsible for the deposition of the marine carbonates and clastics of the Bir Maghara Formation. This Bajocian transgression reached the

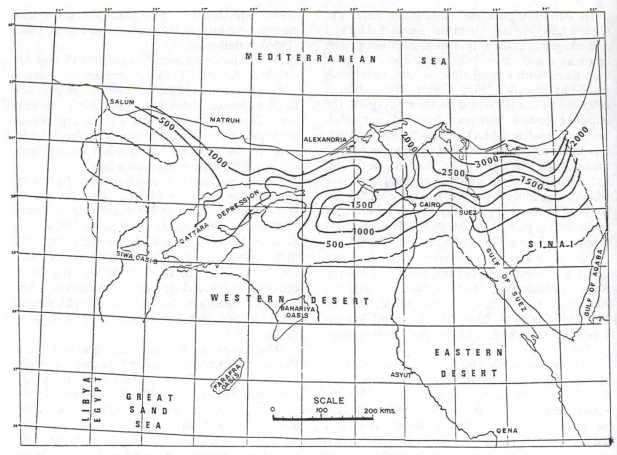

Figure 22.7 Isopach of the Jurassic (contour intervals in meters).

northeastern part of the northern Western Desert where the clastic to calcareous marine Wadi Natrun Formation was synchronously deposited (Table 22/2). Close to the boundary between the Bajocian and the Bathonian a marked regression is indicated in Gebel Maghara by the fluviomarine Safa Formation. The two transgressions of the Toarcian and Bajocian seem to have followed the same path as the Triassic transgression. A significant transgressive episode is also indicated in the Maghara area in late Bathonian times. This transgression is characterized by widespread shelf carbonate deposition. It reached the northeastern part of the Western Desert in early Callovian time and spread over most of the eastern part of the western 'Unstable Shelf' area in the mid-Callovian. This episode corresponds to the deposition of the Masajid Formation. The carbonates of this formation continued to be laid down in the Maghara area until the Oxfordian. In post-Oxfordian times, the sea regressed from northern Sinai. The late Jurassic regression seems to have started earlier in the Western Desert.

It is possible that the east-northeast west-southwest trending Jurassic sedimentary basin extending over northeast Egypt was generated by the sinistral Mediterranean shear acting on the northern African platform in mid-Jurassic times. These times also witnessed pronounced magmatic activity. Weissbrod (1969) describes a basalt sill in the Um Bogma region dated 178 m.a. (middle Jurassic). The widespread regression of the sea recorded in Egypt at the end of the Jurassic corresponds to a global eustatic change in sea level (see also Morgan, chapter 7 this book). It may also be attributed to the tectonic events that produced the numerous basalts dated as late Jurassic-early Cretaceous (140 ± 15 m.a., Meneisy, Chapter 9, this book).

4 Early Cretaceous

4.1 Sedimentation and facies of the early Cretaceous and Cenomanian

During Triassic and Jurassic times, eastern Egypt was in a structurally lower position than the western parts of the country and was thus affected by the consecutive transgressions of these periods. This structural setting seems to have been reversed since early Cretaceous times. During the Cretaceous, western Egypt became structurally lower, thus receiving the transgressions of this period earlier than the eastern parts. This situation continued during Tertiary times, thick Tertiary sediments cover pre-Tertiary structures in the northern Western Desert.

In northern Sinai, at Gebel Maghara the contact between Jurassic and Cretaceous strata is obscured by wadi fill. Here, the lowermost exposed Cretaceous strata consist of thin cross-bedded very fine to coarse-grained sandstones and conglomerates. These represent the last pulse of land-derived fluviatile sediments before the oncoming of the major marine transgression of the Aptian. They could be of Barremian age;

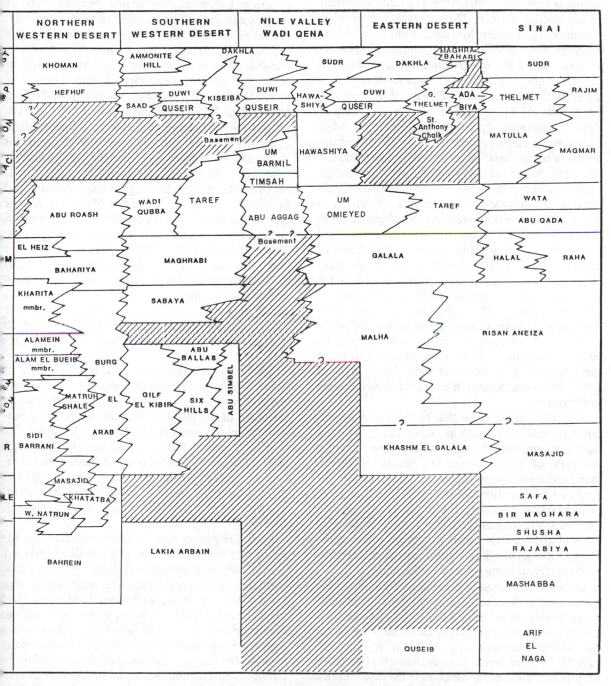

Table 22.2 Correlation of Mesozoic lithostratigraphic units in significant localities in Egypt.

late Jurassic to early Cretaceous strata seem to be missing.

The first major marine transgression in Cretaceous times occurred during the Aptian in response to a world-wide rise in sea-level (Vail et al. 1977). It covered both the north and south Western Desert as well as the margins of the Arabo-Nubian craton along northern Sinai and the Isthmus of Suez.

In the marginal areas of the Arabo-Nubian craton in northern Sinai and in the Isthmus of Suez, the Aptian is represented by marine to paralic sediments which overlie a lower 'Nubia' type sequence of sandstones at Gebels Maghara, Yelleg and Halal (northern Sinai) and Shabrawet (Isthmus of Suez). In all these areas the lower sequence of continental sediments grade upward into similar strata deposited under marine to paralic conditions and belonging to the Albian. At Risan Aneiza, a hill to the north of Gebel Maghara, the Aptian-Albian deposits consist of argillaceous clastics interbedded with carbonates which seem to have been deposited under fluvio-paralic to shallow marine conditions in a low energy system, with occasional high energy episodes (indicated by the presence of several oolitic bands). The Aptian-Albian section overlying the Maghara structure is named the Risan Aneiza Formation (Said 1971). At the type locality on the northern flanks of the structure at Bir Lagama, the formation is 110 m thick. The lowest calcareous sandstone bed of the section carries the Aptian Orbitolina lenticularis (Said & Barakat 1957). The lowest of the oolitic ferruginous limestone beds of the upper part of the section carries Knemiceras sp. as well as Douvilleceras mammilatum. These are considered to be of early Albian age (Mahmoud 1955, Lewy & Raab 1976). In these latter beds the pectinid Nithea syriaca is diagnostic.

A similar sequence to that of the Risan Aneiza outcrops is that exposed at Gebel Yelleg where a basal conglomerate (5 to 15 m thick) is overlain by a sequence of sandy limestones, shales, siltstones, marls and marly limestones (Bartov et al. 1980). The Aptian-Albian section of Gebel Halal is 520 m thick of which 90% is sandstone and minor shale and silt interbeds. It is generally non-fossiliferous, although a band of oolitic ferruginous sandstone in the upper third of the section has yielded corals, pelecypods, gastropods and ammonites of Albian age.

In central Sinai and in the Gulf of Suez, the oldest Cretaceous rocks are represented by alluvial near-shore 'Nubian type' sandstones which rest unconformably over older strata. These sandstones are named the Malha Formation by Abdallah & Adindani (1963). The type area is from the western coast of the Gulf of Suez. In central Sinai, as in the type area, the Malha overlies disconformably the Qiseib red beds of Permo-Triassic age. In the Um Bogma area the Malha assumes a thickness of 70 to 130 m and includes red to grey, fine-grained, partly kaolinitic sandstone and siltstone beds. The formation contains conglomerate beds in the lower part, as well as stringers of pebbles high in the section. Thin lenses of kaolin occur sporadically. The thickness of the Malha varies greatly from one place to another ranging from 70 to 130 m in the Gulf of Suez area. In Khashm El Galala (western coast of the Gulf of Suez), as well as in many boreholes, the Malha overlies beds of Jurassic age. In the Abu Hamth and Nekhl boreholes, central Sinai, the Malha is penetrated at depth and is 398 and 245 m thick respectively. The Malha is non-fossiliferous although in Wadi Malha (Gulf of Suez) and along the entrance of Wadi Qena, it is intercalated in its upper part by thin fossil-bearing marl beds of Albian age (Van der Ploeg 1953, Attia & Murray 1952). The age of the Malha is usually given as early Cretaceous although there is evidence that it is mostly of Albian age. Further north, similar sandstones outcrop below the marine upper Cretaceous strata at Arif El Naga, Giddi, Kherim and Minsherah. In Arif El Naga the Malha consists of grey, white and variegated sandstones, cross-bedded and quartzitic in part, with some interbeds of silty, limonitic crusts and shale beds, mainly in the lower part of the section (Bartov & Steinitz 1977). Close to the base of the section, basaltic flows occur, mainly olivine basalt, with a composite thickness of about 15 m. The flows are separated by reddish paleosols. The Malha rests over a 5 m thick conglomerate bed with quartzite clasts, some of which reach 30 cm in diameter.

The Malha is believed to be mainly the deposit of rivers which carried clastic materials during a low stand of sea level from the positive areas formed in the wake of the late Jurassic tectonic event. It probably filled low areas in an irregular topography. The rivers seem to have been braided and of high-energy and low-sinuosity. The presence of thin marine fossil-bearing strata of Albian age within the upper part of the formation may point to minor marine incursions. The Malha forms part of facies 1 of Van Houten et al. (1984).

Early Cretaceous rocks are also encountered in the subsurface in northern Sinai in Katib El Makhazin-1 (32 m), Khabra-1 (975 m) and Halal-1 (406 m) wells. In offshore areas, early Cretaceous sediments are also known from borehole penetrations. In Manna-1 (164 m) and Sneh-1 (815 m) wells, these sediments are carbonates which seem to have been deposited in a high energy platform margin which stretches along or adjacent to the present-day north Sinai coastline. In the northern Eastern Desert, early Cretaceous sediments are also known from the subsurface in R-69-1

(409 m) and Q-72-1X (415 m) wells.

Cenomanian strata cover large stretches of north and central Sinai. They form the bulk of Gebels Halal and Yelleg, and a large part of Gebels Giddi and Libni and other anticlinal structures in northern Sinai. In central Sinai they skirt the entire Tih scarp.

Wherever seen, Cenomanian strata lie above the Malha or Risan Aneiza with no break in sedimentation. They form a distinctive unit, which in the north is made up of carbonate rocks such as fossiliferous limestones, dolomites and marls with minor shales. This unit is named the Halal Formation by Said (1971). In the south, the unit has a greater clastic component and is named the Raha Formation by Ghorab (1961). On the west coast of the Gulf of Suez along the Galala scarps, Abdallah & Adindani (1963) and more recently Mazhar et al. (1979) describe a unit similar to the Raha which they name the Galala Formation.

Good descriptions of the Halal are given by Shata (1956, 1960). Recently these strata were subjected to detailed studies by Bartov & Steinitz (1977), who correlate them with the Hazera Formation (Albian-Cenomanian) of southern Israel. These authors divide the Hazera into five members which they claim can be traced into Sinai.

At its type locality in Gebel Halal, the formation is 450 m thick and is made up of carbonate rocks, white to light grey, medium hard to hard crystalline limestones and dolomites with a few marl bands making less than 8% of the total thickness. Many fossiliferous beds are present and include a variety of megafossils such as echinoids, ammonites, rudists, oysters and others. The ammonites are particularly common in the upper levels; the top bed carries the ammonite *Neolobites vibrayeanus*.

At Arif El Naga, Bartov et al. (1980) describe a late Albian to Cenomanian section which belongs to the Hazera Formation (= Halal). It overlies the Malha Formation and is 310 m thick. The dominantly carbonate section is interrupted by two thin marl and marly limestone beds, about 16 and 63 m thick, which divide the section into five members. The lower member, 141 m thick, carries *Eoradiolites liratus*, *Strombus incertus*, *Exogyra flabellata*, *Hemiaster* cf. *julieni* and is given a late Albian age. The overlying two members carry a rich assemblage of fossils: *Heterodiadema libyca*, *Holectypus larteti*, *Pterodonticeras diffisi*, 'Dosinia' *delettrei*, *Exogyra flabellata*, *E. columba* and are assigned an early Cenomanian age. The upper two members carry the fossils *Exogyra flabellata*, *Toucasia carinata*, *Chondrodonta joannae*, *Praeradiolites biskarensis*, *Nerinea gemmifera*, *Polytremacis chalmasi* and are assigned a late Cenomanian age.

In central Sinai and the Gulf of Suez, the Cenomanian overlies the Malha with seeming conformity. The sediments are thinner and include more marl and shale beds; sandstone beds occur only in east central Sinai and in the south. The latter are named the Raha. At Sheikh Attiya, the Raha is 100 m thick with 50% of the total thickness made up of clastics. At the Somar dome, the Raha is 140 m thick with more than 40% of the total thickness made up of clastics. Along the Tih escarpment and in Gebel Gunna, the section is 190 m thick. In central Sinai wells Darag, Nekhl and Abu Hamth, the Raha was penetrated at depth and is made up of an interbedded carbonate-clastic sequence with thicknesses of 310, 316 and 326 m respectively.

In the Eastern Desert, Mazhar et al. (1979) describe an 80 m thick section of Cenomanian strata from the Galala plateaux which they correlate with the Galala Formation. This formation is made up of a lower interbedded marl-shale-sandstone unit and an upper carbonate unit.

The Cenomanian seems to form a continuous transgressive event of relatively long duration. The sediments were deposited in a wide shelf without major regressions or influx of detrital material.

4.2 *Sedimentation and facies of the Turonian to Maastrichtian*

The Turonian is represented in Sinai and the Gulf of Suez by two distinct formations: the Abu Qada and Wata which are of early and late Turonian age respectively. In the Galalas the two units are classified under the name Maghra El Hadida Formation (Akkad & Abdallah 1971), although the lower 55 m of this formation in the Ataqa type section could be ascribed to the Abu Qada, while the upper 87 m of the same section could be ascribed to the Wata.

The Abu Qada is coeval with the Ora Shales of Israel (Bartov & Steinitz 1977). It is made up of a lower unit of alternating green-gray soft shale and hard yellowish marly limestone rich in early Turonian ammonites (the so-called ammonite bed of Said 1962); a middle unit of hard dolomites, nodular limestones, chert, with minor marl and shale; and an upper unit of marl-shaly facies in east Sinai and sandy facies in most of Sinai and the Gulf of Suez. The lower unit becomes more calcareous in northwestern Sinai, the Galalas and in the Gulf of Suez. The sandstones of the sandy facies are cross-bedded and form a blanket over most of Sinai. In contrast to the Halal, the Abu Qada exhibits great isopacheous variations, the maximum thickness is in central north Sinai where basins developed in response to the first pulse of the Syrian arching movements.

In Gebel Arif El Naga, the Abu Qada is 68 to 141 m thick. The lower beds carry ammonites: *Mammites nodosoides*, *Neoptyches cephalotus*, *Fagesia* and

Thomasites spp. Also present in these beds is the characteristic fossil of the sixth ammonite biozone *Choffaticeras luciae trisellatum* which represents the peak of the early Turonian transgression (Freund & Raab 1969). The absence of the earlier Turonian biozones (T1 to T5) indicates that the area was a relative structural high during the time. The middle layers of the Abu Qada Formation in Arif El Naga are made up of thinly-bedded limestone beds, 'the paper limestones' containing oysters, *Cerithium* sp. and many *Caryocorbula* spp., indicating shallow marine to brackish conditions. These are overlain by marly limestone beds containing *Coilopoceras* aff. *jenksi*.

In central Sinai (e.g. Sheikh Attiya), in the Isthmus of Suez (e.g. Gebel Ataqa) and in the Galalas, the lower ammonite bed carries the ammonites of the earlier biozones: *Pseudoaspidoceras* (T2) and *Choffaticeras meslei* (T3).

The Wata Formation is late Turonian in age and is coeval with the Gerofit Formation of Israel. It is a hard cliff-forming carbonate unit which overlies conformably the Abu Qada. The topmost beds of this formation mark a most useful mappable horizon. The Wata is 102 m thick in the type area in Wadi Wata, it is made up of massive limestone and dolomite sequence with very minor amounts of marl and shale beds. The upper layers carry the gastropod *Nerinea requieniana*. Compared with the underlying and overlying units, the Wata is poor in clastics. It is only in the southeastern coastal strip of the Gulf of Suez at Gebels Qabeliyat and Abu Durba that minor amounts of sandstone occur.

The Senonian deposits in Sinai, Gulf of Suez and the Galalas are generally marine in nature. During the Coniacian, however, fluvial sediments alternate with the marine deposits of the Gulf of Suez, south Galala Plateau and north Wadi Qena. In these areas the Senonian is represented by three units, the Matulla (Coniacian-Santonian), Gebel Thelmet (Campanian) and Sudr (Maastrichtian). In Wadi Qena the Coniacian-Santonian beds are named Hawashiya. The Campanian beds of the Gulf of Suez form a distinct unit of dark-colored limestone beds which overlie the Matulla and which are named the 'Brown Limestone' beds by oil geologists.

The Matulla Formation (Ghorab 1961) overlies the hard crystalline carbonates of the Wata Formation. It is coeval with the Magmar Formation proposed by Al Far (1964) for central Sinai and the Zihor Formation of the Israeli geologists. The Matulla Formation attains 170 m in thickness in its type locality. It is made up of three distinct units: an 80 m-thick lower unit of fluvial (?) cross-bedded sandstones with thin argillaceous limestone and shale interbeds, a 58 m-thick middle unit of varicolored glauconitic shale with thin limestone interbeds and a 32 m-thick upper unit of

chalky, occasionally nodular, limestone with a few glauconitic shale interbeds. On paleontological grounds, the middle unit is of Coniacian age. It carries the ammonites of the CA3 Zone in the middle layers and the CA4 and CA5 Zones in the upper layers. The upper unit is assigned a Santonian age. Abdallah & Eissa (1971) name the upper unit the St Anthony Chalk.

At Gebel Magmar, the Matulla is 52 m thick. Among the few ammonites recorded in this section, Lewy & Raab (1977) mention the presence of *Barrosiiceras* sp. of the third Coniacian ammonite biozone (CA3) in the middle levels and *Heterotissotia* sp. of the fourth Coniancian biozone (CA4) in slightly higher levels. The section at Gebel Magmar is poor in ammonites. According to Lewy (1975) and Lewy & Raab (1977) the section could be of late Turonian to Santonian age. A complete record of the five Coniacian ammonite biozones is reported from sections cropping out in western Sinai. In Gebel Minsherah the Matulla is 148 m thick and is made up of a lower marly sequence followed by a chalky limestone sequence with sandy glauconitic interbeds.

The Gebel Thelmet Formation (Abdallah & Eissa 1971) is 225 m thick at the type locality and is made up of yellow argillaceous richly fossiliferous limestones partially dolomitic and phosphatic with thin shale and sandstone interbeds. This formation is coeval with the Rajim Formation suggested by Al Far (1964) for northern Sinai sections and the Menuha and Mishash Formations (or their equivalent the Sayyarim) of the Israeli geologists. It also corresponds to the so-called 'Brown Limestone' unit, used by oil company geologists to designate the lower phosphatic part of the overlying Sudr Formation. This unit is correlated with the Duwi Formation of the Stable Shelf areas of Egypt.

In Gebel Ataqa, the Campanian rests directly over the Cenomanian Galala Formation. There, the Campanian strata assume 235 m in thickness and are named the Adabiya Formation (Akkad & Abdallah 1971). The formation is made up entirely of dolomites and limestones and is poorly fossiliferous.

The Sudr Formation (Ghorab 1961) is a widely distributed chalk rock unit which rests over the Thelmet Formation with seeming conformity. This unit has been previously lumped with the underlying Santonian St Anthony Chalk into one formation and given the name 'Chalk' by Said (1962). In the present work the widely distributed upper chalk sequence which covers large areas of Sinai and the Gulf of Suez is treated as a distinct unit, the Sudr Formation.

The formation is made up of snow-white and chalky limestones. The greatest recorded thickness is at El Themed (250 m) and at El Markha plain along the eastern coast of the Gulf of Suez (220 m).

Thicknesses up to 150 m are reported from Sudr, Homra and Gebel Somar. The Sudr is usually poor in megafossils but rich in microfossils. The age of this formation is Maastrichtian.

4.3 *Cretaceous paleogeography and tectonism*

At the end of the Jurassic and in the early Cretaceous the eastern part of the 'Unstable Shelf' of Egypt stood probably in a higher structural position than the western part, as it was reached by the early Cretaceous marine transgression in relatively later times. No conspicuous tectonic movements have been recorded in northeastern Egypt until the end of the Cenomanian. This is reflected in the uniform isopacheous distribution of rock units belonging to this geologic time span. Except for the Aptian transgression, the early Cretaceous marine transgressions did not extend far beyond the limits of the Unstable Shelf. They resemble in this respect the transgressions of the Triassic and Jurassic. Extensive transgressions covering the 'Stable Shelf' areas are known only since the Cenomanian. The complex distribution of the rock units of the Turonian over Sinai, may suggest a pronounced tectonic activity leading to tectonic differentiation of the Shelf area of northeastern Egypt.

The most well-defined late Cretaceous phase of tectonic activity seems to have taken place in late Coniacian times. Late Coniacian sediments are absent from the 'Unstable Shelf' area of northern Sinai (Lewy 1975). This suggests that north Sinai was uplifted in late Coniacian times, while the central part of the peninsula remained under sea level. In some of the uplifted parts erosion was so strong as to remove the Turonian sediments (e.g. Gebel Maghara). This situation seems to have continued until late Cretaceous times. The isopachous variations of the Sudr Formation, show a conspicuous sedimentary basin in central Sinai. The northern rim of the 'Unstable Shelf' of Sinai stood as a relatively high area, where the Sudr Formation was erratically deposited in relatively low places. Some highs were possibly never reached by the seas of this period. In the well Boughaz-1, the Turonian is directly overlain by the Oligocene. In the wells Manna-1 and Sneh-1, Tertiary sediments rest on early Cretaceous sediments, while in the well Slav-1 they rest on Jurassic sediments. In the northern Eastern Desert at Gebel Shabrawet, the Maghra El Bahari continental red beds overlie the Turonian Moghra El Hadida Formation while in Gebel Ataqa they overlie the Campanian Adabiya Formation.

All these considerations suggest strongly that the uplifting of the 'Unstable Shelf' area of north Sinai and the north Eastern Desert took place during late Coniacian time although it could have been initiated earlier and is related to the generation of the 'Syrian arc' system.

B. WESTERN SUBPROVINCE

1 *Structural setting*

This subprovince covers the 'Unstable Shelf' of the northern Western Desert and extends from the Mediterranean coast southward to an arbitrary line which separates the 'Stable' from the 'Unstable' shelf. This line runs in a north-northeast direction parallel to the faults delimiting the prominent Kattaniya structural high (= Khatatba-Wadi Khadish block in Said 1962: 215). The 'Unstable Shelf' areas to the north of this line have a deep basement, a complex structure and a thick column of sediments. The areas to the south have a high basement and less than 1000 m of sediments. As an exception, the Fayum forms a basin (the Gindy basin) within the 'Stable Shelf' as delimited here. It has, in places, more than 2000 m of sediments.

The major structural elements to the south and southeast of the Kattaniya block follow a predominantly north-northeast direction, while those to the north and northwest of this line follow a north-northwest direction. The interaction of these two trends was probably responsible for the generation of the frequent east-west fractures of this subprovince. Two of these east-west fractures delineate three main structural features which are from south to north: the Abu Gharadig basin, the Qattara ridge and the northern basinal area.

The western part of the northern basinal area is a structural high, here termed the Faghur-Maamura high. It was elevated in late Paleozoic time after having formed a major Paleozoic basin. The axis of this high runs along a line passing through the Faghur-1 well and a point midway between Sidi Barrani and Matruh on the Mediterranean coast. To the east of this high lies the Umbarka subbasin, and to the northwest lies the subbasin of Sidi Barrani. Both these subbasins are separated from the conspicuous subbasin of Matruh which lies to the east by an arch which seems to be a prolongation of the Qattara ridge. The Matruh subbasin is bound on its eastern edge by a high which separates another subbasin, the Alamein.

In Abu Gharadig basin major marine transgression and regression cycles together with numerous tectonic cycles resulted in a highly deformed, thick sedimentary cover (Awad 1984, 1985) with considerable hydrocarbon potential.

The various lithostratigraphic subdivisions of the Mesozoic of the north Western Desert are given in

426   *M.T. Kerdany & O.H. Cherif*

			SIDI BARRANI	FAGHUR MAAMURA HIGH	UMBARKA BASIN	MATRUH BASIN
CRETACEOUS	U.	MAASTRICHTIAN				
		CAMPANIAN	KHOMAN A	KHOMAN A	KHOMAN A	KHOMAN
		SANTONIAN CONIACIAN			KHOMAN B	KHOMAN
		TURONIAN	ABU ROASH	ABU ROASH	ABU ROASH	ABU ROASH
		CENOMANIAN	BAHARIYA	BAHARIYA	BAHARIYA	BAHARIYA
	L.	ALBIAN	KHARITA	KHARITA	KHARITA	KHARITA
		APTIAN		KHARITA		ALAMEIN Do.
		BARREMIAN	ALAM EL BUEIB	ALAM EL BUEIB	ALAMEIN Dol. / UMBARKA Dol.	MATRUH SH
		NEOCOMIAN			ALAM EL BUEIB	
JURASSIC	U.	MALM	SIDI BARRANI		ALAM EL BUEIB	
	M.	CALLOVIAN	SIDI BARRANI			
		BATHONIAN	KHATATBA			
		BAJOCIAN	WADI NATRUN			
	L.	LIAS				
TRIASSIC						
PERMIAN						
CARBONIFEROUS			LITTORAL TO INNER SHELF			
DEVONIAN			? ?			
ORDOVICIAN – SILURIAN			?	LITTORAL TO INNER SHELF		
CAMBRIAN			? ?			
PRECAMBRIAN						

Table 22.3 Main Mesozoic stratigraphical subdivisions of northwestern Egypt.

MEIN SIN	QATTARA RIDGE HIGH	ABU GHARADIG BASIN	KATTANIYA HORST HIGH	GINDI BASIN	N. MARGIN STABLE SHELF BAHARIYA
MAN A	KHOMAN A	KHOMAN A		KHOMAN A	KHOMAN A
		KHOMAN B			
		?			
ROASH					
	ABU ROASH	ABU ROASH		ABU ROASH	EL HEIZ
ARIYA	BAHARIYA	BAHARIYA		BAHARIYA	BAHARIYA
ARITA AB SH. EIN Dol. EL BUEIB	KHARITA	KHARITA DAHAB SH. ALAMEIN Dol. ALAM EL BUEIB		KHARITA	
EL BUEIB		ALAM EL BUEIB	ALAM EL BUEIB	ALAM EL BUEIB	
EL BUEIB		ALAM EL BUEIB	ALAM EL BUEIB	ALAM EL BUEIB	
?		MASAJID	MASAJID	MASAJID	
	MASAJID				
	KHATATBA	KHATATBA	KHATATBA	KHATATBA	
	BAHREIN	BAHREIN	WADI NATRUN	W. NATRUN	
			BAHREIN	BAHREIN	
?		?			
		NON MARINE			
?	NON MARINE				
?					

	SIDI BARRANI BASIN	FAGHUR MAAMURA HIGH	UMBARKA BASIN	MATRUH BASIN
LATE CRETACEOUS (SENONIAN)		DISAPPEARANCE OF SIDI BARRANI BASIN	GENERATION OF UMBARKA BASIN	DIFFERE
EARLY CRETACEOUS TO TURONIAN				
LATE LIAS TO LATE JURASSIC	CONSPICUOUS MARINE BASIN	NORTHERN	BASINAL AREA IN	RELATIVELY HIC
TRIASSIC–EARLY LIAS			POORLY	UNC

Table 22.4 Table summarizing the tectonic evolution of the various sedimentary basins and 'highs' in the northern Western Desert during the Mesozoic (arrows show direction of movement; subsidence or uplifting).

Table 22/3. This table also shows the chronostratigraphy of each unit together with the stratigraphical relations of the various facies.

2 The Triassic-early Jurassic

2.1 Sedimentation and facies
The whole of the Western Desert of Egypt seems to have been dry land suffering erosion during most of the Triassic and earliest early Jurassic times, as no sediments belonging to these periods have been found in the region. The youngest widespread definitely marine sediments underlying the first datable Mesozoic rocks in the Western Desert belong to the Carboniferous and are restricted to the western part of the area. Non-marine Permian? strata have been drilled below continental to shallow marine Liassic sequences in the southeastern part of the northern Western Desert in Misawag-1 well, and

marine Permian in Faghur West-1 well. This suggests that the end of the Paleozoic witnessed a widespread regression followed by a period of tectonic stability and non-sedimentation. The Paleozoic regression must have ended within Permian times.

2.2 Paleogeography
There is no evidence for any tectonic activity during the Triassic. Some authors, however, suggest that the west-northwest east-southeast faults noted in the northern basinal area of the Western Desert may have been generated during the Jurassic or earlier. These faults constitute the framework of the present faulted ridges and basins, which resemble, according to Sestini (1984), the block faulting of the Atlantic and African margins which took place during the Triassic and early Jurassic. These may be related to the crustal rifting which caused the separation of the 'Turkish' microplate from Africa. The olivine basalt drilled in the Kattaniya-1 well and dated late Triassic to early

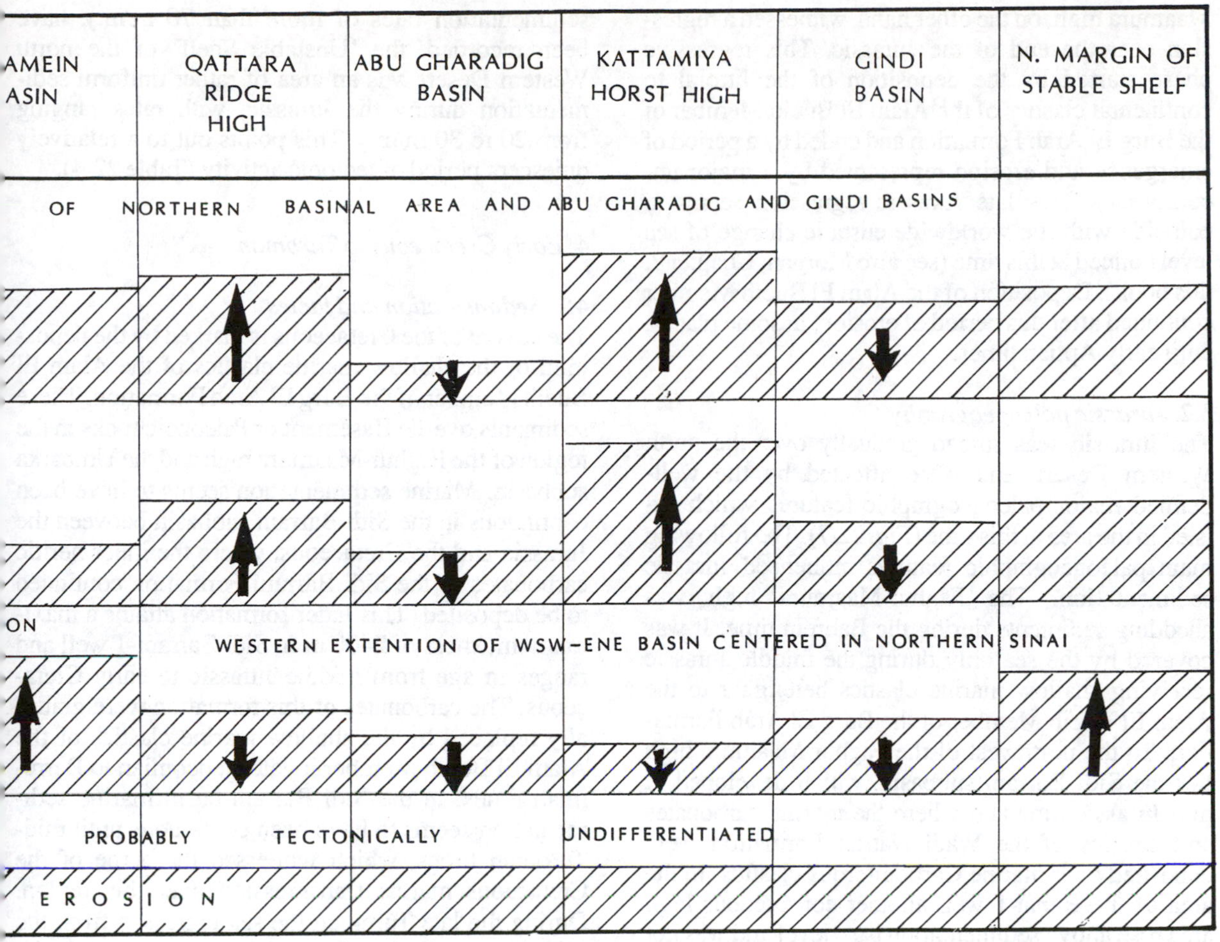

AMEIN ASIN	QATTARA RIDGE HIGH	ABU GHARADIG BASIN	KATTAMIYA HORST HIGH	GINDI BASIN	N. MARGIN OF STABLE SHELF

OF   NORTHERN   BASINAL   AREA   AND   ABU   GHARADIG   AND   GINDI   BASINS

WESTERN EXTENTION OF WSW–ENE BASIN CENTERED IN NORTHERN SINAI

PROBABLY   TECTONICALLY   UNDIFFERENTIATED

E R O S I O N

Jurassic by K-Ar methods (191 ± 19 m.a.) may represent the basaltic extrusions corresponding to this stage of rifting along the Mediterranean coast of Egypt.

3 Middle and late Jurassic

3.1 Sedimentation and facies

The basal beds of the middle and late Jurassic in the north Western Desert of Egypt are non-marine clastics with few fossil bearing horizons: the Bahrein Formation (EGPC-RRI, special report 1982). In places late Lias fossils are recorded in the base of this formation (see chapter 15, this book). The non-marine clastics of the Bahrein Formation are overlain by the marine Jurassic clastics of the Khatatba or Wadi Natrun Formations. The base of the marine Jurassic differs in age according to the time of the transgression over a particular area (Tables 22/3 and 22/4).

Sedimentation during most of the Jurassic was mainly clastics, derived probably from the elevated northeastern corner of the African craton.

These clastics are represented by the mainly continental Bahrein red beds and the overlying marine clastics and carbonates of the Wadi Natrun Formation and the predominantly marine clastics of the Khatatba Formation. These clastic units were laid down during late Lias to Bathonian times. During the late Callovian, and sometimes earlier, the shallow marine carbonates of the Masajid Formation were deposited in the northern basinal area, Abu Gharadig basin, Kattaniya horst and Fayoum (Gindi) basin (Fig. 22/1). This represents the acme of the Jurassic transgression over the eastern parts of the area. In the extreme northwestern part, at the Sidi Barrani subbasin, deposition of marine carbonates (Sidi Barrani Formation) was dominant. This formation may be a relatively deeper marine equivalent of the Masajid Formation. The Umbarka subbasin, east of the Faghur-

Maamura high, on the other hand, witnessed a regression near the end of the Jurassic. This regressive phase started by the deposition of the littoral to continental clastics of the Alam El Bueib Member of the Burg El Arab Formation and ended by a period of emergence and erosion represented by a major unconformity. This late Jurassic regression seems to coincide with the worldwide eustatic change of sea level noticed at this time (see also Morgan, Chapter 7, this book). Deposition of the Alam El Bueib Member continued after this period of erosion, in some places, until early Aptian times.

3.2 Jurassic paleogeography

The Jurassic seas spread gradually over the north Western Desert, and were affected by the well-defined major paleogeographic features which existed in the area at this time (Fig. 22/7). The following main paleogeographic features influenced Jurassic sedimentation. The Faghur-Maamura high was shedding sediments during the Bahrein time. It was covered by the sea only during the middle Jurassic receiving shallow marine clastics belonging to the Alam El Bueib Member of the Burg El Arab Formation. To the northwest of the Faghur-Maamura high lies the Sidi Barrani subbasin, which developed in mid-Jurassic time and where the marine carbonates and clastics of the Wadi Natrun Formation were deposited over marine Carboniferous clastics. In the area of the Rabat-1 well another conspicuous high area rose above sedimentation base level and was not inundated until early Paleogene times. The lowest tectonic area developed in the Jurassic is in the region of the Kattaniya horst. This low was submerged in early Bajocian time (where the Wadi Natrun Formation was deposited over the Bahrein Formation). It seems that this low area is a continuation of the major Jurassic sedimentary basin extending over the northern part of the Sinai and the north Eastern Desert (Fig. 22/7).

The Faghur-Maamura and Rabat highs follow generally a north-northeast south-southwest trend which may have been inherited from pre-Mesozoic times. This is the same trend as that of the north Kattaniya (Pelusium) line. The major east-northeast west-southwest trend is not apparent in Jurassic times. It seems that during the Jurassic very little movement occurred along most of the major faults in the northern Western Desert. Seismic and well data indicate, however, that the south Misawag and north Diyur faults were active in pre-Jurassic and Jurassic times respectively. These two faults lie approximately near the boundary between the 'Stable' and 'Unstable' shelves (Fig. 22/1).

With the exception of the areas of the Kattaniya horst and the Sidi Barrani subbasins, where Jurassic sedimentation rates of more than 70 m/m.y. have been recorded, the 'Unstable Shelf' of the north Western Desert was an area of rather uniform sedimentation during the Jurassic, with rates ranging from 20 to 30 m/m.y. This points out to a relatively quiescent period of tectonic activity (Table 22/4).

4 Early Cretaceous to Turonian

4.1 Sedimentation and facies

The advent of the Cretaceous is marked by the deposition of the shallow marine clastics of the Alam El Bueib Member of the Burg El Arab Formation. These sediments overlie Basement or Paleozoic rocks in the region of the Faghur-Maamura high and the Umbarka subbasin. Marine sedimentation seems to have been continuous in the Sidi Barrani subbasin between the Jurassic and the Cretaceous, where the inner neritic carbonates of the Sidi Barrani Formation continued to be deposited. This latter formation attains a maximum thickness of 2400 m in Sidi Barrani-1 well and ranges in age from middle Jurassic to early Cretaceous. The carbonates of this formation were gradually replaced by the shallow marine clastics of the Alam El Bueib Member in late Neocomian to Barremian times. In the Sidi Barrani basin marine sedimentation seems to have been continuous until mid-Turonian times, which witnessed the acme of the Cretaceous marine transgression over this region. During the late Turonian the sea regressed from the Sidi Barrani subbasin which together with the Faghur-Maamura was elevated above the sea. In contrast the Umbarka subbasin started to develop and was overlapped by the sea.

Another site of continuous marine sedimentation from Jurassic to Cretaceous is the Matruh subbasin, where a fault bounded trough started to develop during the Neocomian. This trough has been interpreted by Prior (1976) as a possible 'failed arm' of a crustal rift, and by Taher (1976) as a fossil submarine channel cut by turbidity currents. In this trough, the Neocomian to Aptian marine Matruh shales (the Mersa Matruh Formation of Norton 1967) were deposited above the Sidi Barrani Formation. El-zarka (1983) mentions that the thickness of the pre-Aptian strata in the northern Western Desert ranges from 53 to 913 m. The Matruh shales are overlain by an Aptian carbonate unit named the Alamein Dolomite Member of the Burg El Arab Formation. According to Elzarka (1983) the thickness of this unit varies from 62 to 584 m. The Alamein dolomite is overlain by marine clastics with some carbonates of Albian to Cenomanian age and belonging to the Kharita Member of the Burg El Arab Formation and to the Bahariya Formation. A marked increase in the depth of the sea is reflected in the deposition of the inner neritic calcareous Abu Roash 'G' unit of the Abu Roash Formation in late Cenomanian time.

Generally in the basins of the north Western Desert the Cenomanian is of marine character. The thickness varies from 138 to 1153 m with a sand/shale ratio exceeding 1. The depocenter is around the Kattaniya well. The Cenomanian seems to have been deposited in a deep water basin with syndepositional contemporaneous subsidence (Elzarka 1983). The Cenomanian is represented by a lower clastic unit (the Kheima Member) and an upper clastic-carbonate unit (the Mellaha Member) of the Kharita Formation. Metwalli & Abd El-Hady (1975) include the Cenomanian and Albian of the northern Western Desert strata in the Abu Subeiha Formation.

4.2 *Early Cretaceous to Turonian paleogeography*

The paleogeographic evolution of the region can be related to the tectonic activity which was responsible for the development of the different paleogeographic features during this time.

The late Jurassic to early Neocomian tectonic event (Neocimmeian) is expressed by a widespread unconformity over most of the area, with the exception of the Matruh and Sidi Barrani subbasins, and by fault movements leading to the appearance of the Qattara ridge, the Abu Gharadig basin and the Matruh subbasin. In late Barremian to early Aptian times a tectonic phase is expressed by another widespread unconformity, except in the Matruh and Sidi Barrani subbasins, and by fault activity. During this phase the differentiation of the Abu Gharadig basin from the Qattara Ridge was further accentuated. The late Turonian tectonic phase is also expressed by a widespread unconformity, except in the Umbarka subbasin and Abu Gharadig basin. This phase is also characterized by a general period of fault activity which may have started in Cenomanian times. The late Turonian movements led to: (a) the disappearance of the Sidi Barrani subbasin, which became a single structural entity with the Faghur-Mamura high, (b) the differentiation of the Umbarka subbasin and (c) the further development of a large basinal area south and southeast of the Qattara Ridge comprising the Abu Gharadig and Fayum (Gindi) basins and the Kattaniya horst.

5 *Senonian*

5.1 *Sedimentation and facies*

The Senonian is characterized by a widespread transgression and deepening of the seas over Egypt. In the northern Western Desert, chalk (the Khoman) was deposited. Complete sections of the Khoman are found in many of the basinal areas (Abu Gharadig, Umbarka) where sedimentation was continuous, from the Turonian to the Senonian.

In these areas the Khoman is divided into a lower or 'B' member and an upper or 'A' member. Outside the basins, it is frequent to see the upper Khoman overlying the eroded Turonian with an angular unconformity. In many instances, the latest Maastrichtian is missing from this sequence. This is indicated by a pronounced unconformity. This is probably a reflection of the late Senonian to Paleocene tectonic movement (Laramide tectonic phase mentioned by various authors, in Said 1962) which affected northern Egypt. This unconformity is also observed in northern Sinai, where it is caused by late Coniacian movements representing an acme in the generation of the 'Syrian Arc'.

5.2 *Paleogeography*

The paleogeographic pattern during the Senonian was established through fault activity which enhanced the development of the early Mesozoic sedimentary basins and ridges. In latest Cretaceous times, however, some basins disappeared as a result of 'basin inversion'. The most conspicuous basins that became highs since late Cretaceous time are the Sidi Barrani, Umbarka, Matruh and Alamein subbasins. The Kattaniya horst, which was also a basinal area during most of the Mesozoic became a high after the late Cretaceous.

THE STABLE SHELF

1 *Structural setting*

The Stable Shelf extends over southern Egypt and is generally covered by less than 1000 m of sediments. It surrounds the basement exposures of the Arabo-Nubian craton in eastern Egypt. To the west it constitutes a widespread tectonically 'low' area (Fig. 22/1). This area may be subdivided into two major, generally north-south trending intracratonic basins: The Dakhla Basin to the west and the Upper Nile Basin to the east (Fig. 22/1) (Beall & Squyres 1980, Bisewski 1982). These basins are delimited to the southwest by the Calanscio-Uweinat-Gilf El Kebir high and to the south by a sequence of uplifts, the Uweinat-Aswan High (Klitzsch 1978), extending eastward to the Eastern Desert basement exposures. This high, which is capped by a thin Mesozoic sedimentary cover, is thought to have been formed in late Paleozoic to early Mesozoic time by uplifting along major east-west faults. This phase of uplifting was accompanied by alkali magmatic intrusions, as those described from Gebel Nusab el Balgum by Schandelmeiyer & Darbyshire (1984), and dated 216 m.y. (late Triassic). The Uweinat-Aswan High suffered considerable erosion during these times. The eroded material was mainly transported southward, where thick Permo-Triassic to early Jurassic fluvio-marine sediments were laid down in west Sudan. Klitzsch (1984) advocates that the drainage of the material provided from the Uweinat-Aswan High toward the south ceased in mid-Jurassic time, when the Arabo-African platform began to tilt markedly toward the north. It seems that this time corresponds to the beginning of sedimentation of the continental Bahrein Formation in the northern Western Desert. In north-west Sudan, it is possible to recognize a sequence of continental deposits of Permian to Middle Jurassic age (the Lakia Arba'in Formation), belonging to a southward drainage system. This sequence is overlain by another continental succession of Jurassic to early Cretaceous age (the Gilf El Kebir Formation and equivalents), extending northward into the Egyptian Dakhla and possibly the upper Nile basins, suggesting a northward drainage system.

The upper Nile basin is delimited to the east by the Eastern Desert exposures. Northward it is open to the Unstable Shelf. Along Wadi Qena, it is possible to follow in outcrop the gradual north-south changes in the sedimentary facies of the various Mesozoic marine transgressions coming from the north. The upper Nile basin is separated from the Dakhla basin by the Kharga High. This High is an anticlinal basement structure plunging to the north and ending approximately to the south of Abu Tartur between Kharga and Dakhla. The Kharga High is covered by a relatively thin sedimentary suite. This High has influenced the sedimentation of Mesozoic strata until Aptian time (Abu Ballas Formation). In post-Aptian times it is not possible to differentiate between the Dakhla and the upper Nile basins in south Egypt (Bisewski 1982). Thorweihe (1982) considers the upper Nile basin as a continuation of the Dakhla basin, the former being characterized by relatively thinner sediments.

The northern parts of the Stable Shelf of Egypt were markedly influenced by the stresses acting in the north African realm during the end of the Cretaceous. These stresses caused the generation of the major structures; the Bahariya Swell which bounds the Dakhla basin from the northeast and the Wadi Araba Horst in the northern part of the upper Nile basin. The first structure is characterized by the generation of northeast-southwest trending faults, probably induced by a tensional tectonic phase acting during the Campanian and interrupting a period of compressive phase which may have started earlier than the Campanian, but which continued in late Cretaceous and early Tertiary times (El Bassyony 1978). These movements induced the generation of macroscopic gentle folds in this area, which differ from the more steeply folded structures belonging to the Syrian arc of northern Sinai.

The Wadi Araba Horst also seems to have been generated in late Cretaceous times, due to the action of the same shear system that produced the Bahariya and Syrian arc structures. This horst seems to recall structurally the Kattaniya Horst of the northern Western Desert.

2 Sedimentation and facies of Mesozoic sediments

The earliest Mesozoic sediments recognized in the Stable Shelf area are continental, fluviatile sandstones and siltstones of the Six Hills Formation, ranging in age from late Jurassic to early Cretaceous (Bisewski 1982). This formation overlies unconformably the basement rocks in the south of the Dakhla-Kharga area (at Nusab El Balgum). The Six Hills Formation attains its maximum development in the axial region of the Dakhla Basin.

The first recognizable Mesozoic marine transgression that extended over the Dakhla Basin occurred in Aptian time and is represented by the marine clastics of the Abu Ballas Formation, overlying the Six Hills Formation (Boettcher 1982, 1985) (Table 22/2). The clastics are mainly gray, white, red and brown shales intercalated with siltstones and sandstones. They contain a fauna of pelecypods, gastropods, brachiopods and other faunal elements (Barthel & Boettcher 1978, Boettcher 1982) as well as plant fossils in certain layers. These assign the age of the formation to the Aptian. Further south at Gebel Kamil and in the stretch to the south and southwest of this location, coastal sediments of the Aptian sea are encountered, they include abundant Thalassinoides and other bioturbations as well as silicified tree trunks, ferns, root structures and paleosols, partly associating the coastline with a swamp environment (Klitzsch 1984).

In southern Egypt, in the Gilf El Kebir region, the Aptian is represented by clastics which are not differentiated from the underlying late Jurassic-early Cretaceous rocks, both of which are collectively named the Gilf El Kebir Formation. The formation is 250 m thick at the type locality at the Aqaba Pass (latitude 23° 25' N, longitude 25° 45' E). The sediments are fluviatile, deltaic and partly near-shore sandstones resting unconformably over early Paleozoic rocks and are overlain by sandstones of Albian age. Fossil flora identified from this formation point to a late Jurassic-Aptian age. The formation is the lateral equivalent of both the Six Hills and the Abu Ballas Formations. Northward the Abu Ballas Formation changes into the marine Aptian Alamein Dolomite Member of the Burg El Arab Formation, widely distributed in the northern Western Desert.

The marine Abu Ballas Formation and its equivalent is followed in the Stable Shelf area by a regressive phase represented by the fluviatile and continental Sabaya Formation. These deposits consist of tabular cross-bedded fluvial sandstones with many paleosols. In the Dakhla Basin they overlie the Abu Ballas Formation with seeming conformity and assume a thickness of about 160 m. The Sabaya Formation is assigned an Albian age and may be a lateral equivalent of the marine Kharita member of the northern Western Desert.

The Maghrabi Formation (Barthel & Herrmann-Degen 1981) overlies unconformably the Sabaya Formation. It is made up of fine-grained bioturbated sandstones and claystones. The type section is to the south of the Abu Tartur plateau (Kharga-Dakhla road) where the formation assumes a thickness of 60 m. The basal part of the formation consists of flaser-bedded sandstones with abundant plant remains (mainly leaves of angiosperms) which is overlain by shale and sandy glauconitic beds carrying abundant pelecypod shells as well as fish teeth. The remains of a dinosaur were found in these beds in a location south of Amonite Hill. The flora and fauna found are of Albian-Cenomanian age. The Maghrabi Formation loses its marine character toward the south and west of the Dakhla depression where it is replaced by non-fossiliferous sandstones and claystones.

To the north the Maghrabi Formation changes laterally into the Bahariya Formation of the northern Western Desert. The type section of the Bahariya Formation forms the floor and scarps of the oasis depression and has an exposed thickness of 209 m (Soliman & El Badry 1970). The formation includes many fossil-bearing beds; the so-called dinosaur bed is followed by a bed rich in *Ceratodus* sp. and a sandy thinning upward sequence with numerous paleosols rich in *Exogyra flabellata*, *E. columba* and a host of other characteristic fossils. The Bahariya is an estuarine deposit (Dominik 1985) recording intermittent aggradations and repeated development of paleosols. The base of the formation is encountered in the Bahariya well no. 1 at a depth of 725 m. The total thickness of the Bahariya Formation is 934 m of which the lower part seems to belong to the Aptian and Albian. The top 20 m beds include more carbonate members and were separated by Akkad & Issawi (1963) into a separate unit, the El Heiz. This unit is made up of marls and shale of late Cenomanian age.

The Maghrabi Formation is overlain in the stable shelf areas of southern Egypt by a continental sandstone sequence, almost devoid of fossils: The Taref Formation. The Taref is a thick tabular-planar cross-bedded sandstone unit, which was deposited during Turonian-Santonian time. This is the typical 'Nubian' type sandstone described in the literature. It belongs to the Platte type of fluvial deposits as described by Wycisk (1984). It was probably formed in a prograding alluvial plain in closely-spaced non-sinuos distributary streams (Harms et al. 1982, Van Houten et al. 1984). Klitzsch (1984) mentions the existence of marine intercalations yielding Turonian fossils within this formation toward the north between Qena and Wadi Araba. This formation seems to indicate a gradual regression toward the north, a regression which was culminated in Santonian time. The Taref is probably coeval with the shallow marine Abu Roash Formation of the northern Western Desert.

In the south at Aswan, the section overlying the eroded and leached basement rocks (Fig. 22/8) is correlated with the Taref Formation of the Western Desert. It is divided into three units (Attia 1955) reflecting a transgressive phase between two regressive phases of the Tethys during late Cretaceous time (Van Houten et al. 1984). The three units are named (from bottom to top) the Abu Aggag, Timsah and Um Barmil by El Naggar (1970). In this chapter these units will be treated as members of the Taref Formation.

The lower unit, the *Abu Aggag*, is given different names by different authors: El Kanaiess by Abdel Razik (1972), Aswan by Klitzsch (1984) and facies 1 by Van Houten et al. (1984). The unit has a wide distribution in southern Egypt. It overlies the basement rocks with an angular unconformity. In places the basal bed of this formation is made up of gruss embedded in a violet red sandy clay matrix, 0.5 to 10 m thick. This bed, which represents the weathering products of the underlying basement rocks, is described as a soil and distinguished as a separate unit named Ibyan by Philobbos & Hassan (1975). In Aswan the Abu Aggag is 40 m thick and is made up of a lenticular fining upward sequence of coarse-grained kaolinitic trough cross-bedded quartzose sandstones and conglomerates with many paleosols.

The Abu Aggag attains greater thicknesses to the north and northeast. At the foot of the temple of Deir El Kanaiess, along the Idfu-Mersa Alam road, it is 175 to 250 m thick (Abdel Razik 1972). In Wadi Natash (Fig. 22/9), it attains a thickness of 120 to 155 m (Said et al. 1976) and its upper part interfingers with volcanics which are dated 84 to 100 Ma (Ressetar & Nairn 1980, Hashad et al. 1982). The age of the Abu Aggag, therefore, is late Cretaceous. Issawi & Jux (1982), however, give a Paleozoic age to this formation in Aswan. These authors base their dating on Zaghloul et al.'s (1983) record of the Paleozoic ichnofossil *Bifungites* in the overlying Timsah Member above the lower iron ore horizon. Since

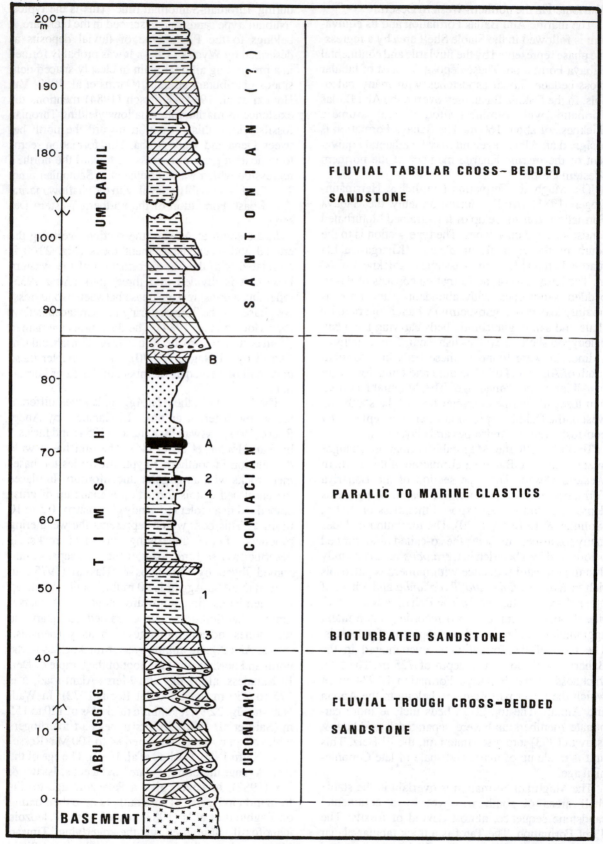

Figure 22.8 Generalized section at Aswan. A and B. lower and upper iron ore beds; 1 and 2. lower and upper plant beds; 3. *Skolithos* horizon, 4. *Isocardia-Inoceramus* horizon.

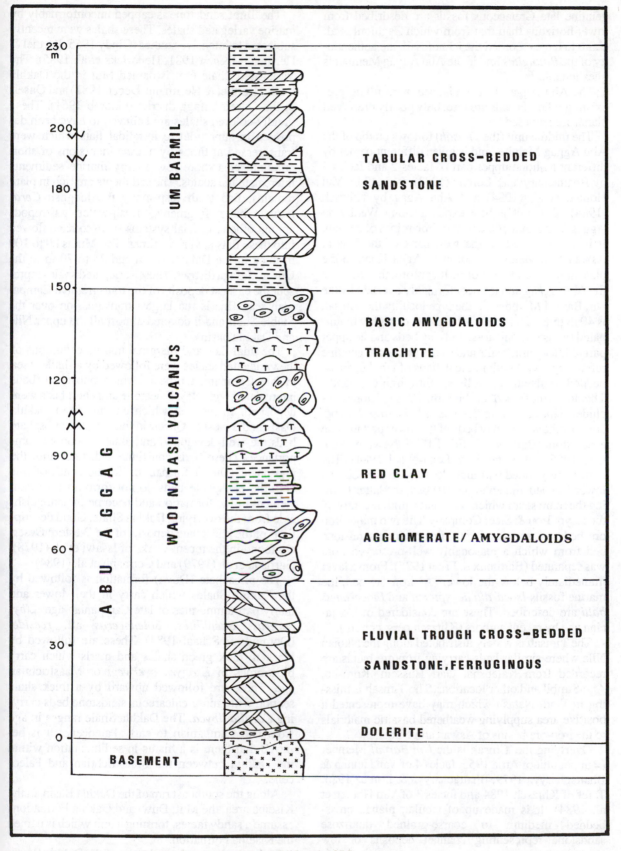

Figure 22.9 Generalized section at Wadi Natash (after Said et al. 1976).

genuine late Cretaceous fossils are desdribed from lower horizons than that from which Zaghloul et al. claim to have recorded the ichnofossil, the authenticity of the *Bifungites* find in the Abu Aggag Member is questionable.

The Abu Aggag is a fluvial sediment filling pre-existing relief. Fossils are rare; only poorly preserved plants are recorded.

The middle unit (the *Timsah*) follows on top of the Abu Aggag Member and is given different names by different authors: upper part of facies 2 and facies 3 by Bhattacharyya & Lorenz (1983), facies 2 by Van Houten et al. (1984) and Abu Agag by Klitzsch (1984). The unit is best exposed along Wadi Abu Aggag in the area of the former iron mines of Aswan. It is also exposed on the west bank of the Nile at Aswan from below the tomb of Agha Khan to the plateau north and west of it. It conformably overlies the Abu Aggag and is unconformably overlain by the Um Barmil Member. At the type locality the Timsah is 40 m thick. Its lower part consists of highly bioturbated sandstone, siltstone and clay beds and its upper part of fine-grained silts and shales (some are genuine refractory clays) with intercalations of tabular, cross-bedded sandstone as well as oolitic iron ore bands. The lowermost bed of bioturbated sandstones includes *Skolithos, Teichichnius, Thalassinoides* spp. among others. The clay beds of the lower part include a rich flora (Barthoux & Fritel 1925, Newton 1909, Cox 1956 and Klitzsch & Lejal-Nicol 1984). The upper fine-grained part includes two iron ore beds: a lower (A) bed and an upper (B) bed, the latter forming the main seam which was mined until recently by the Egypt Iron & Steel Company. The two major iron ore beds are separated by a fine-grained sandstone bed from which a reasonably well-preserved flora was separated (Barthoux & Fritel 1925). From a layer immediately below the lower 'A' iron ore bed, the marine fossils *Isocardia aegyptiaca* and *Inoceramus balli* are described. These are considered of Coniacian age by recent authors (Klitzsch pers. comm.).

The Timsah is widely distributed along the Nubian Nile where similar beds carrying iron ore bands are recorded from Kalabsha, Garf Hussein, Korosko, Abu Simbil and other locations. The Timsah is missing in Wadi Natash which may have represented a positive area supplying weathered basaltic materials to the iron-ore basins of Aswan and Nubia.

Overlying the Timsah is the *Um Barmil* Member (= upper unit of Attia 1955, facies 4 of Van Houten & Bhattacharyya 1979, Bhattacharyya & Lorenz 1983, Taref of Klitzsch 1984 and facies 3 of Van Houten et al. 1984). It is made up of tabular, planar, cross-bedded medium to coarse-grained quartzose sandstone representing channel deposits of low sinuosity streams.

The Taref sandstone is capped unconformably by marine variegated shales. These shales were recently subjected to intensive studies (Gindy 1965, Barthel & Herrmann-Degen 1981, Hendricks et al. 1985). The Variegated Shale unit is named Mut in the Dakhla Basin (Barthel & Herrmann-Degen 1981) and Quseir in the Quseir-Safaga district (Ghorab 1961). These vividly-colored shales are believed to have been deposited in supratidal to intratidal flats which were transgressed at times by marine incursions of short duration. Nearshore low energy marine sediments intercalate the shales. The sediments are rich in plant remains, fish teeth (especially the lungfish *Ceratodus*), bone fragments, fresh-water gastropods among others. Fluvial systems seem to have flowed into these flats at various times. The Mut is 80 to 100 m thick in the Dakhla Basin and 25 to 70 m in the Quseir-Safaga district. These variegated shales represent the earliest deposits of the transgressive Campanian sea. This is the largest transgression over the Stable Shelf and it covered almost all the upper Nile and Dakhla basins.

The tidal flats and marginal marine conditions of the variegated shales were followed by a shallow sea in late Campanian time in which phosphate beds intercalated with clay, limestone and chert beds were deposited. In the belt which extends from Dakhla Oasis in the west to Quseir in the east, the phosphate beds are well-developed and assume economic importance. These beds constitute a distinct unit, the phosphate (Duwi) formation. To the south of this belt, the phosphate beds do not form an important element in the formation and become indistinguishable from the overlying Dakhla Shale. Good descriptions of the phosphate deposits of the Western Desert are found in the recent works of Issawi et al. (1978), Garrison et al. (1979) and Germann et al. (1984).

The phosphate (Duwi) formation is followed by the Dakhla Shales which carry in their lower and marly part ammonites of late Campanian age: *Libycoceras, Baculilites, Solenoceras* aff. *reesidei* (Dominik & Schaal 1984). These are followed by black to dark green shales and marls which carry biostromes of *Exogyra overwegi* of Maastrichtian age. These are followed upward by a thick shale section with minor calcareous sandstone beds carrying *Cardita libyca*. The Dakhla Shale ranges in age from late Campanian to early Paleocene. It is believed that there is a hiatus in sedimentation within the Dakhla between the Maastrichtian and Paleocene.

Along the southeast rim of the Dakhla Basin, in the Kiseiba area, the Mut, Duwi and Dakhla Formations assume a sandy facies, forming a unit which is named the Kiseiba Formation.

In the northwest along the rim of the Dakhla Basin,

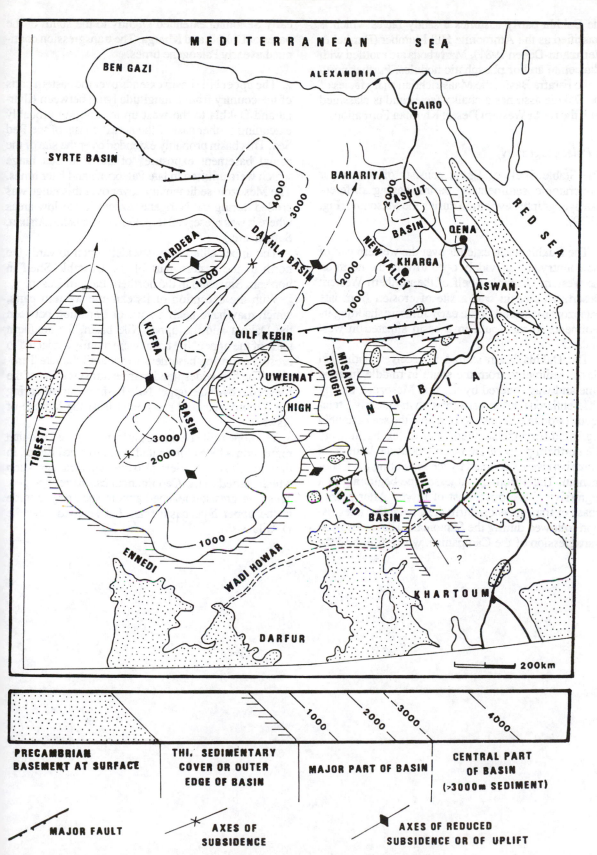

Figure 22.10 Structural interpretation of part of the eastern Sahara (after Klitzsch 1984).

the Dakhla partly assumes a sandy facies which is classified as the Ammonite Hill Member (Barthel & Herrmann-Degen 1981). Many beds are mottled with glauconite and/or phosphatic nodules. Further north in the Farafra Basin, the Maastrichtian shaly facies of the Dakhla assumes a chalky facies and is classified with the north Western Desert Khoman Formation.

3 *Paleogeography*

The Stable Shelf of Egypt consists of two major intracratonic sedimentary basins having different paleogeographic evolution during the Mesozoic (Fig. 22/10):

1. The Dakhla basin extends over the western part of the country and is widely open toward the north on the western Unstable Shelf of the northern Western Desert. This basin was a site of erosion from late Paleozoic to the end of the early Jurassic shedding its sediments to the south in a basin situated in west Sudan.

The Dakhla basin started to receive continental clastic sediments from the south in middle Jurassic time and was covered by the first Mesozoic, clearly differentiated marine transgression in the early Cretaceous (Aptian). In post-Aptian time the sea definitely regressed from the southern parts of this basin. Only the northern parts of the basin (about the latitude of Bahariya) were occupied by the sea during the Cenomanian. The Turonian witnessed the beginning of a regressive phase over most of Egypt. After a brief phase of transgression durig the Coniacian, a regression followed during the Santonian. The widespread transgression of the Campanian, which followed the early Senonian extended slightly to the south of the oases of Dakhla and Kharga. The transgression continued in early Paleogene times.

2. The upper Nile basin extends over the eastern parts of the country from a longitude lying between Kharga and Dakhla to the west up to a region probably extending farther east of the actual coasts of the Red Sea. This basin probably extended over the site of the actual basement exposures of the Red Sea range which were uplifted in late Paleocene and later times. The Mesozoic sedimentary cover over this range was eroded during the Neogene, except in the low areas where it was preserved (e.g. Duwi, Hamadat, Atshan, Safaga, etc.).

The upper Nile basin is widely open towards the north to the eastern part of the Unstable Shelf in northern Sinai and in the northern Eastern Desert.

With the exception of its extreme western parts, which may have been a zone of transition between the Dakhla and the upper Nile basin, the southern parts of the upper Nile basin were mainly subjected to erosion exposing basement rocks until the late Turonian. The first widespread marine transgression of the Mesozoic over the upper Nile basin occurred during the Cenomanian and reached at least the latitude of Aswan.

The upper Nile basin was affected by the Turonian regression which culminated, as in western Egypt, in the early Senonian. Here also the major transgression which started in the Cenomanian can be recognized. This transgression reached a more southerly latitude in the upper Nile basin than that reached over the Dakhla basin.

CHAPTER 23

Cretaceous paleogeographic maps

RUSHDI SAID

Consultant, Annandale, Virginia, USA and Cairo, Egypt

An attempt is made to depict the paleogeographic evolution of Egypt during the Cretaceous period by a series of maps. The maps are based on the results of recent surveys of the exposed rocks of this period and the subsurface records reported from the numerous boreholes drilled mainly in north Egypt and the Gulf of Suez. The maps compliment Chapter 22 on the Mesozoic contributed by Kerdany & Cherif.

The maps are intended to give a generalized view of the evolution of the land of Egypt during the Cretaceous period. The limits of the accuracy and use of the maps are numerous. The data upon which the maps are based is not of equal density in the different areas of Egypt; subsurface data in the middle latitudes, for example, are scarce. Furthermore, most of the data refer to formations whose ages differ from one place to another and which, in many instances, have a longer range of age than that depicted on the map. Thus the Sabaya formation is considered of Albian age for the purpose of the map construction although the upper part of the formation could well be of Cenomanian age. The same is true of the Maghrabi formation considered here to be of Cenomanian age although the top parts could be of Turonian age. In the case of the Abu Roash and Khoman formations, arbitrary divisions are made of these formations which cover the interval from late Cenomanian to Maastrichtian. This was necessary in order to use the data in constructing the numerous maps which cover this long interval. The divisions may not fit the interpretations of some paleontologists. Under each map are listed the formations used in the construction of the maps.

A glance at the maps shows that the Cretaceous period witnessed four transgressive cycles in Egypt. The Aptian, Cenomanian and Coniacian cycles brought very shallow seas to the passageway between the elevated Nubian and Kufra massifs. The passageway, which changed position as the massifs eroded, was filled by marginal marine sediments of intratidal, supratidal, estuarine and swamp environments frequently alternating with alluvial sediments.

The Campanian-Maastrichtian transgression, on the other hand, brought shallow open marine conditions to large parts of Egypt. Figure 23/9 depicts the eustatic sea levels in Egypt as compared with those of the global sea levels.

Early Cretaceous (pre-Aptian) (Fig. 23/1)

The late Jurassic-early Cretaceous tectonic event affected Egypt in a most marked way. It affected the east-west oriented northern basin which covered the entirety of north Egypt during most of the Jurassic period. This basin which grew from a small nucleus in the northeast during the Triassic to cover the Unstable Shelf area was destroyed by the uplift of the Arabo-Nubian massif which elevated the Eastern Desert, the Red Sea, the Gulf of Suez and south and central Sinai (Fig. 23/1). These areas became the sites of intense erosion during the late Jurassic and most of early Cretaceous time. The eroded sediments of this high were deposited in the western basin in the form of fluvial deposits, the Six Hills formation in the south and the Alam El Bueb member of the Burg El Arab formation in the north. Both units are made up of clastics which accumulated under predominantly fluvial conditions although it is certain that the northern Alam El Bueb unit was influenced by shallow marine incursions. Genuine marine conditions persisted only in the extreme northwestern part of Egypt at Sidi Barrani and Mersa Matruh areas where the Sidi Barrani and Matruh formations were deposited.

The Six Hills formation underlies the well-dated marine Aptian clastics in the Dakhla basin. To the west, outside this basin, the entire early Cretaceous sedimentary column, including the Aptian, does not include any marine intercalations and is named the Gilf Kebir formation (see Klitzsch, this book, Chapter 13). Both formations include silicified tree trunks, ferns and other plant remains which indicate a ?late Jurassic to early Cretaceous age. The sands are coarse-grained and show tabular cross-bedding. The Aswan-Uweinat High was active at the time.

Palynological assemblages from the late Jurassic-

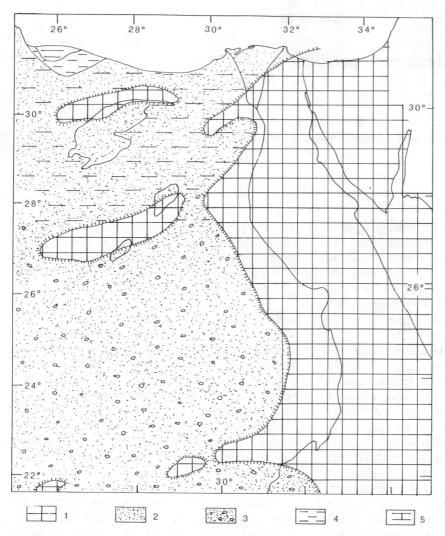

Figure 23.1 Egypt during late Jurassic-early Cretaceous (pre-Aptian) time. 1. positive areas; 2. shallow marine deposits alternating with alluvial sediments; 3. alluvial deposits; 4. shales of the open marine Matruh basin; 5. limestones of the open marine Sidi Barrani basin.

Neocomian formations of the Foram and ammonite wells are purely terrigenous and contain high percentages of pteridophyte spores associated with common gymnosperm pollen (Schrank 1984, 1987). Similar palynological assemblages are described from the subsurface strata of Kharga Oasis (Helal 1966, Soliman 1975, 1977 and Saad & Ghazaly 1976).

In the north Western Desert, the Alam El Bueb member is made up of clastics with minor shale interbeds. The member was deposited under shallow marine conditions affected by outpours of fluvial sediments which were supplied from the highs surrounding the basin from the east or standing in its midst, the Qattara ridge and the Kattaniya arch.

Aptian (Fig. 23/2)

The Aptian witnessed a transgression that brought parts of north Sinai and a large part of the south Western Desert under the influence of a shallow sea (Fig. 23/2). The Alamein and Abu Ballas formations represent thin marine intercalations in the continental section of lower Cretaceous clastics in northern and southern Egypt respectively. The thin fossiliferous red and green shales of the Abu Ballas formation are interpreted as prodelta deposits of a vast, very shallow epicontinental sea (Boettcher 1985). The prodelta sediments are underlain and overlain by fluvial and delta-front sand- and siltstones and deposits of coastal swamps. The delta front sediments lay in front of the river mouths and carry a small number of species which could tolerate this extreme environment (e.g. *Crassostrea* association). The prodelta sediments lie to the north and cover a large area. Because of the very gentle slope of the sea floor, the high energy zone was situated far off the coast, dividing the prodelta into three parts (Fig. 23/2). The upper part was the closest to the shoreline and is

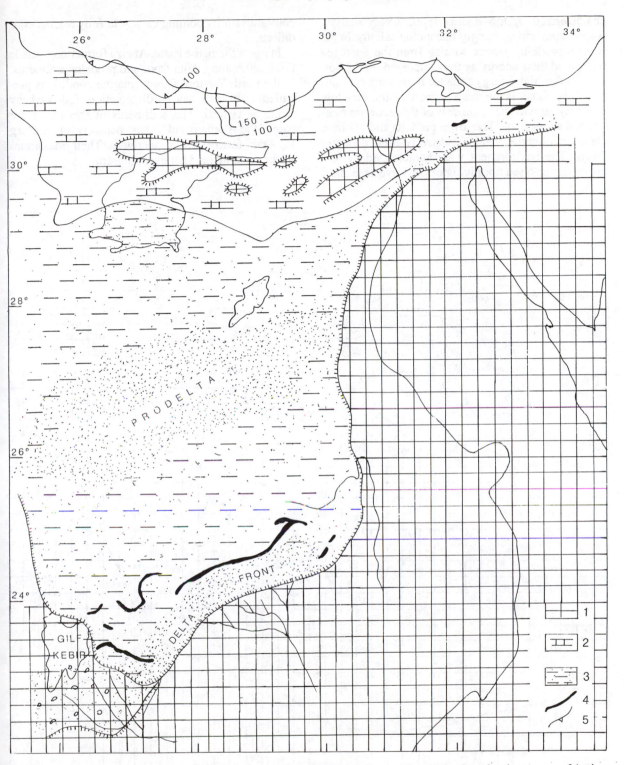

Figure 23.2 Egypt during Aptian time. 1. positive areas; 2. open marine sediments; 3. prodelta deposits; 4. outcrops of Aptian formations; 5. isopach contours.

characterized by low energy level, a very shallow water depth and a strongly fluctuating salinity. In the middle prodelta, waves coming from the open sea dissipated their energy as they impinged on the bottom. In this high energy zone the burrowing brachiopod *Lingula* sp. is the main element of the very low diversity fauna. The sediments of this zone are fine-grained sandstones. The lower prodelta and offshore clayey sediments yield a relatively high diversity fauna. The environment was characterized by low energy level, a sea floor lying below wave base and a

constant salinity coming closest to fully marine conditions.

In north Sinai, the Risan Aneiza formation is made up of alternating delta front and prodelta sediments. In the north Western Desert, marine conditions prevailed and carbonate sedimentation followed by shales is typical. The sediments of this realm are named the Alamein and Dahab members (of the Burg El Arab formation) respectively. Their maximum thickness is in the Alamein subbasin.

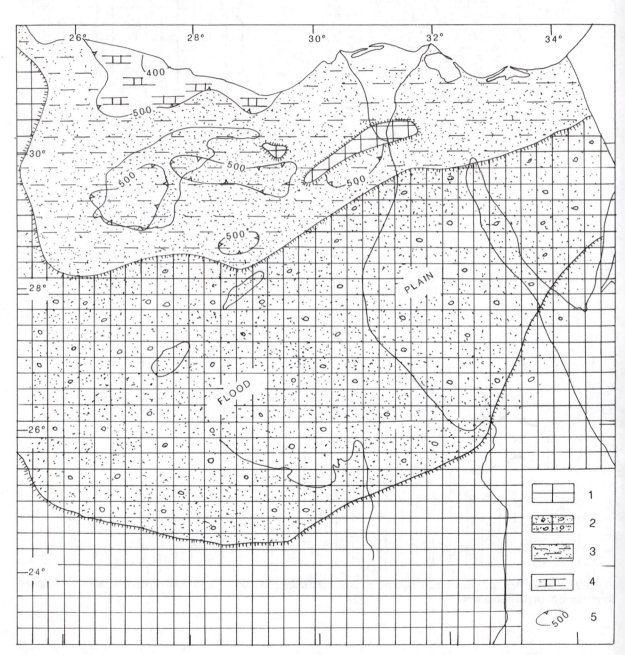

Figure 23.3 Egypt during Albian time. 1. positive areas; 2. alluvial deposits; 3. marine deposits alternating with alluvial sediments; 4. open marine deposits; 5. isopach contours.

Albian (Fig. 23/3)

The Albian represents a regressive phase in which the sea retreated northward. The northern part of the elevated Eastern Desert, as well as a large part of the Western Desert, formed a receptable receiving the fluvial detritus of the rivers emanating from the erod-ing elevated massif to the south. In the Gulf of Suez, central Sinai and Wadi Qena thick beds of fluvial sandstones of the Malha formation accumulated unconformably above older beds ranging in age from Precambrian basement to late Jurassic. In the Western Desert the sandstones of the Sabaya Formation

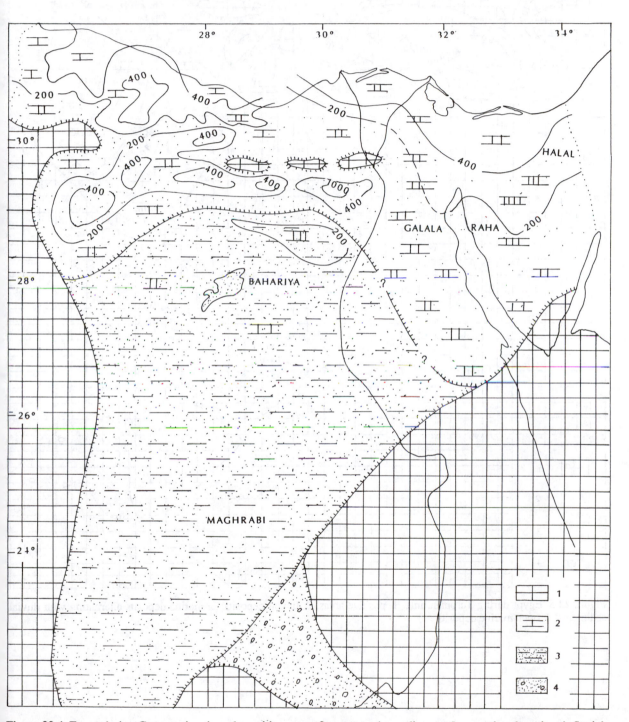

Figure 23.4 Egypt during Cenomanian time. 1. positive areas; 2. open marine sediments; 3. estuarine deposits; 4. fluvial deposits.

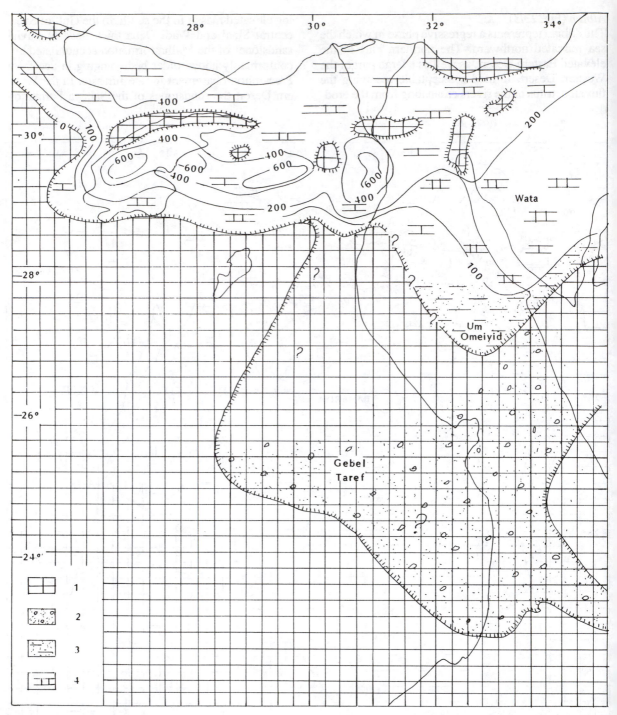

Figure 23.5 Egypt during Turonian time. 1. positive areas; 2. fluvial deposits; 3. marine deposits alternating with fluvial sediments; 4. open marine deposits.

accumulated above the Aptian.

To the north shallow marine fine- to coarse-grained sandstones accumulated in the north Western Desert above the Dahab Shales. These constitute the Kharita member (of the Burg El Arab formation). To the northwest, around the Matruh basin, carbonates form a dominant part of the section. In this area thicknesses of more than 500 m are reported.

Cenomanian (Fig. 23/4)

During the Cenomanian a marine transgression covered most of Sinai, the Gulf of Suez and northwest Egypt. In late Cenomanian time, the transgression pushed southward to form a narrow passageway which lay between the Arabo Nubian massif and the elevated Kufra basin. This passageway seems to have formed a veritable estuary in which marginal marine conditions prevailed.

The Gulf of Suez was differentiated for the first time, its eastern and western embankments were the sites of the shallow marine deposits of the Raha and Galala formations respectively. In north Sinai deeper marine conditions prevailed; thicker and more calcareous sediments of the Halal formation were deposited. Genuine marine conditions are also reported from the northwestern corner of Egypt.

The passageway lying to the south is filled in the north by the Bahariya formation made up of tidal flat, estuarine to fluviatile deltaic deposits with frequent marine intercalations (for a recent description of the Bahariya deposits, see Dominik, 1985). The Bahariya formation is divided into a lower fluviatile member, the Gebel Ghorabi, a middle estuarine member, the Gebel Dist and an upper marine member, the Heiz. The Heiz is equivalent to the 'G' member of the Abu Roash formation of the subsurface north Western Desert.

Further south the Bahariya grades into another formation, the Maghrabi, attributed, like the Bahariya, to an abrupt marine transgression flooding the Dakhla basin (Hendriks 1986). The Maghrabi differs from the Bahariya in showing lesser marine influence. It is made up of a clastic sequence which contains brachiopods (*Lingula* sp.), rare vertebrate remains (fish teeth, dinosaur bones and turtle plates) and abundant plant remains (angiosperms) which, near Kharga Oasis, form thin coal beds in the upper part of the formation. The lateral and vertical arrangement of the distinctive sedimentary facies of the Maghrabi characterize a tidal flat environment which was repeatedly cut by meandering estuarine channels. An angular unconformity and paleosol formation separate the Maghrabi from the underlying Sabaya formation.

Turonian (Fig. 23/5)

During the Turonian genuine marine conditions prevailed over a larger part of north Egypt. The marine Turonian beds cover north Egypt and the embayment of the Gulf of Suez. In the Gulf the beds are made up of a lower shale, the Abu Qada formation and an upper solid limestone unit, the Wata formation. In the north Western Desert, the subsurface Turonian is represented by the Abu Roash D, E and F (equivalent to the outcropping Rudistae and Acteonella series of the Abu Roash structure to the northwest of Cairo, Said 1962). Thick sections are recorded in the Betty, Abu Gharadig and Gindi (Fayum) basins.

The Turonian witnessed an important pulse of the Laramide movement which elevated the coastal areas of Sinai, the Sidi Barrani coastal area, the Qattara ridge and numerous structures across Sinai. It also elevated the Bahariya arch. Most authors date the carbonate section overlying the Bahariya formation (the El Hefhuf) as Campanian.

The great estuary that had covered the Dakhla basin during the Cenomanian disappeared. A thin strip of fluvio-marine deposits skirts the southern shores of the sea in the Wadi Qena area. These deposits belong to the Um Omeiyid formation and are made up of cross-bedded fluviatile sandstone beds with intercalations of marine pelitic and marly sandstone beds.

The plains of south Egypt including Aswan seem to have received fluviatile deposits during the Turonian. These are represented by the Taref formation and its lower member in Aswan, the Abu Aggag.

Coniacian (Fig. 23/6)

The Coniacian represents a transgressive phase which brought the sea inland as far as Nubia and beyond covering the entire Nile basin. Genuine marine conditions prevailed in the northern parts of Egypt and in the Gulf of Suez area where the coniacian is represented by the 'B' and 'C' members of the Abu Roash formation (= flint and *Plicatula* series of the outcropping Abu Roash structure, Said 1962) and by the Matulla formation respectively. In the Nile embayment and in the Wadi Qena area shallow marine epicontinental deposits alternate with fluvial sediments of the Timsah and Hawashiya formations respectively. These formations are described in other chapters of this book (see Kerdany & Cherif, Chapter 22 and Klitzsch et al., Chapter 16).

A pulse of the Laramide tectonic event activated a number of structures including the Suez-Cairo-Kattaniya high, the Qattara ridge, the Bahariya arch and the Sinai coastal area. Numerous basins developed in the Gulf of Suez and the north Western Desert.

Santonian (Fig. 23/7)

The Santonian represents a regressive phase during which the sea occupied only the tectonic basins of north Egypt that became clearly distinguished. In the Gulf of Suez, the Santonian is represented by the St Anthony Chalk and by the upper beds of the Matulla formation. In the north Western desert the Santonian is represented by the 'A' member of the Abu Roash formation which is recorded only in the basins intervening the major highs which became clearly distinct. The 'A' member is not present in the outcropping Abu Roash structure which became a high during Santonian time.

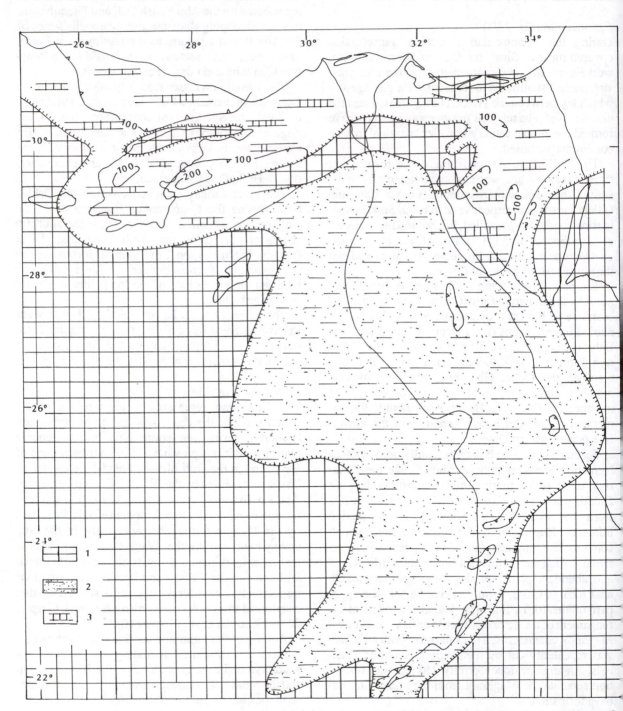

Figure 23.6 Egypt during Coniacian time. 1. positive areas; 2. shallow marine deposits alternating with fluvial sediments; 3. open marine sediments.

Campanian (Fig. 23/8)

A major transgression took place during Campanian time. During the earliest part of this transgression (?middle Campanian) the area was covered by a very shallow sea which was affected by tidal currents. Deposits of supratidal to intratidal mud, sand and mixed flats alternate with marsh and estuarine deposits. The deposits are pelitic to psammitic in texture and are varicolored. They belong to the Quseir Variegated Shale and the Mut formation. Bone beds carrying abundant fresh-water and marine fish remains, fresh-water turtle and an ornithiscian dinosaur point to deposition in near shore mixed environment. The Mut and Quseir formations belong to the middle Campanian (*Canadoceras cottreaui* and *Manambolites pivetaui* ammonite zones) and are known only along a belt, in south Egypt, underlying the Duwi (phosphate) formation which is of late Campanian age (heteromorph ammonite (*Nostoceras*) zone).

Figure 23/8 is a paleogeographic map of Egypt during the late Campanian when a shallow sea covered the largest part of the country. It was a sea which apparently formed an optimum environment for the formation of phosphate deposits. It received lesser amounts of detritus. In the belt extending from the Quseir-Safaga reach along the Red Sea to the entrance of Wadi Qena, the Nile valley and the Kharga-Dakhla stretch, thick and rich phosphate beds are common. These beds form the Duwi formation. In Wadi Qena, where shales are more common, the name Rakhiyat formation was proposed to designate the phosphate beds at the entrance of the Wadi (Hendriks & Luger 1987). In the Gulf of Suez, the Campanian marl section includes phosphatic beds and was named Gebel Thelmet formation (also named the 'Brown Limestone' by oil company geologists). In Bahirya Oasis the sequence overlying the Turonian carries phosphate beds and was named El Hefhuf. It is partly dolomitic. Allam (1986) proposes separating the phosphate-bearing part of this section to form a new formation, the Ain Giffara and to retain the name Hefhuf to the dolomite section which he considers, on stratigraphic grounds, to be of Turonian age.

In north Egypt the Campanian is represented by chalky limestones, the Rajim in north Sinai and the

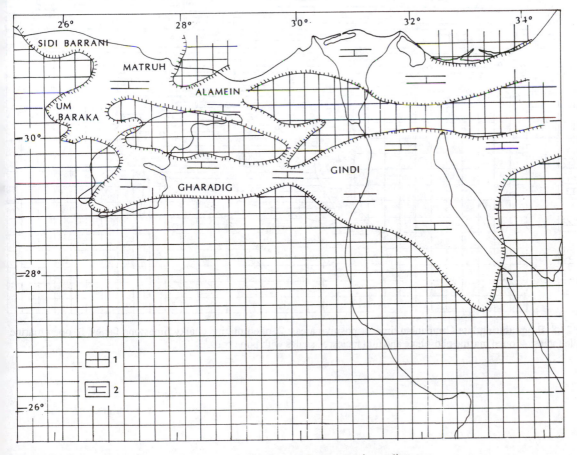

Figure 23.7 Egypt during Santonian time. 1. positive areas; 2. open marine sediments.

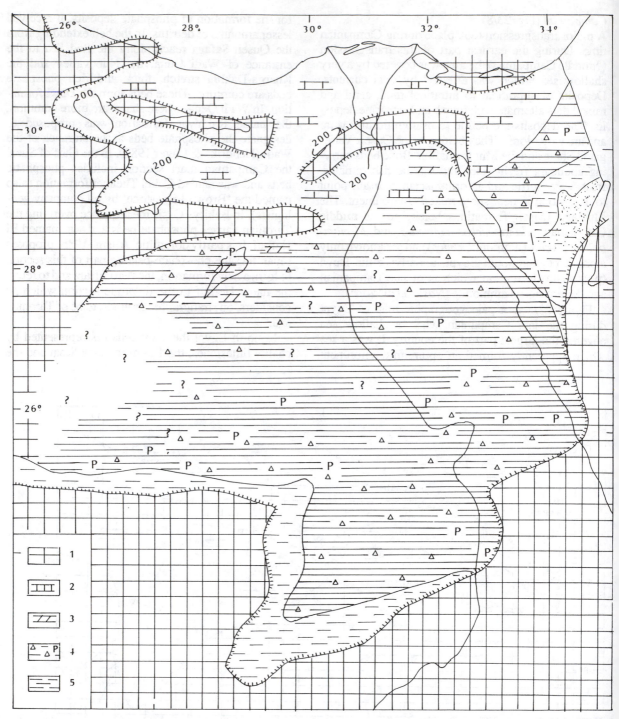

Figure 23.8 Egypt during late Campanian time. 1. positive areas; 2. open marine chalky limestone facies; 3. open marine dolomite facies; 4. open marine phosphate and flint facies; 5. shales with bone beds.

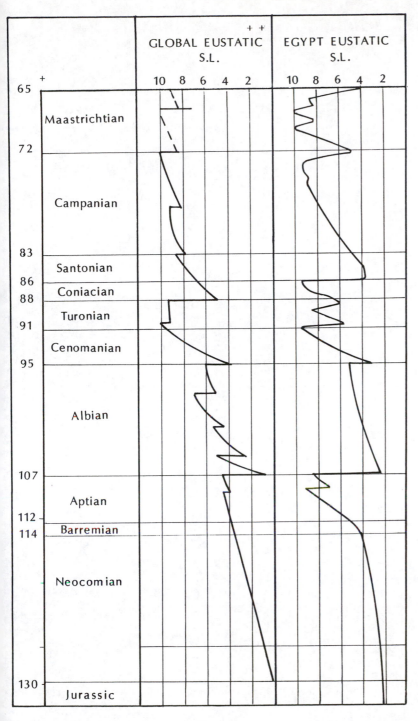

Figure 23.9 Eustatic sea levels
during Cretaceous period.

lower part of the Khoman formation, the so-called
'Khoman B'. In the latter case the section includes
numerous shale intercalations and is known only in
the basinal areas.

Maastrichtian
After a short regressive interval during the earliest

Maastrichtian, the sea advanced and covered larger
areas of Egypt than at any other time. The sea became
considerably deeper. In the north pure chalks of the
Sudr (Sinai and Gulf of Suez) and Khoman (north
Western Desert) accumulated. In the south, shales
(lower part of the Dakhla formation) were deposited.

CHAPTER 24

Cenozoic

RUSHDI SAID

Consultant, Annandale, Virginia, USA and Cairo, Egypt

The two major divisions of the Cenozoic, the Paleo-gene and the Neogene, are separated from one another in Egypt by dramatic events which changed the landscape of Egypt, initiated the process of formation of the Red Sea rift, raised mountains and activated volcanoes. The sediments of these two periods cover large areas of Egypt. They differ from one another with regard to their extent, stratigraphic setting and the type of basins in which they accumulated. In the following paragraphs a description is given of the sediments of each of these periods and an attempt is made to reconstruct their paleogeography and paleoecology.

PALEOGENE

Paleogene rocks lie unconformably over upper Cretaceous or older rocks in most areas of Egypt. The nature of this contact differs in the two major tectonic provinces of Egypt, the Stable and the Unstable Shelves. The transition from the Cretaceous to the Tertiary in the south Stable Shelf areas was not accompanied by intense tectonic disturbances; the sediments of the Tertiary lie disconformably over the Cretaceous and are separated from it by a thin intra-formational conglomerate carrying reworked late Cretaceous fossils. In these areas the Cretaceous-Tertiary boundary lies within the Dakhla Formation, a mappable unit of great areal extent in south Egypt. The extent of the Cretaceous-Tertiary hiatus can best be measured by the study of the planktonic foraminiferal assemblages along the Cretaceous-Tertiary contact. Most Cretaceous sections lack the topmost *Abathomphalus mayaroensis* Zone. Figure 24/1 shows the foraminiferal zones recorded in some Paleocene sections of the Stable Shelf. In the Dakhla area, the uppermost Cretaceous is overlain by the Paleocene P1b Zone (Barthel & Herrmann Degen 1981) whereas in the Kharga-Baris stretch it is overlain by the P2 Zone (Luger 1985). In Naqb Assiut the oldest beds of the Paleocene belong to the P4 Zone (Strougo

1986). In the Gebel Aweina section (Nile Valley) the uppermost Cretaceous is followed by the P1b Zone. In the northern parts of the Stable Shelf (e.g. Farafra) the Cretaceous-Tertiary boundary lies within the Khoman Chalk (Gunnet El Bahariya) or at its top (Ain Maqfi). In Gunnet El Bahariya the Cretaceous is followed by the P1b Zone whereas in Ain Maqfi it is followed by the P3 Zone. In the Quseir area Krasheninnikov & Abdel Razik (1969) record the most complete and probably the only section in Egypt in which there seems to be no paleontological break; the uppermost Cretaceous is followed by the P1a Zone. Luger (pers. comm.) claims that similar relationships to that of the Quseir area exist between the Cretaceous and Paleocene in Gebel Qreiya, entrance of Wadi Qena.

The relationship of the Cretaceous and Paleocene in the Gulf of Suez area is shown in Figure 24/2.

In the north Unstable Shelf areas of Egypt the late Cretaceous and early Tertiary rocks exhibit varied relationships. These relationships seem to have been governed by the paleorelief inherited from the late Cretaceous tectonism which affected the Unstable Shelf and upon which the early Tertiary sediments were deposited. In the synclinal areas of this paleorelief more complete sections of the late Cretaceous and early Tertiary are found. Here the hiatus is relatively small and the boundary lies either on top of the Sudr Chalk or else cuts across the overlying Esna Shale. Said & Kenawy (1956) discuss the impact of this paleorelief on the distribution of foraminifera in two adjacent sections in Sinai. In the more positive areas of this paleorelief great breaks are known as a result of the non-deposition of total systems or stages.

PALEOCENE

The Paleocene is represented by open marine sediments of varying lithologies reflecting the frequent epeirogenic movements and/or changes of sea level

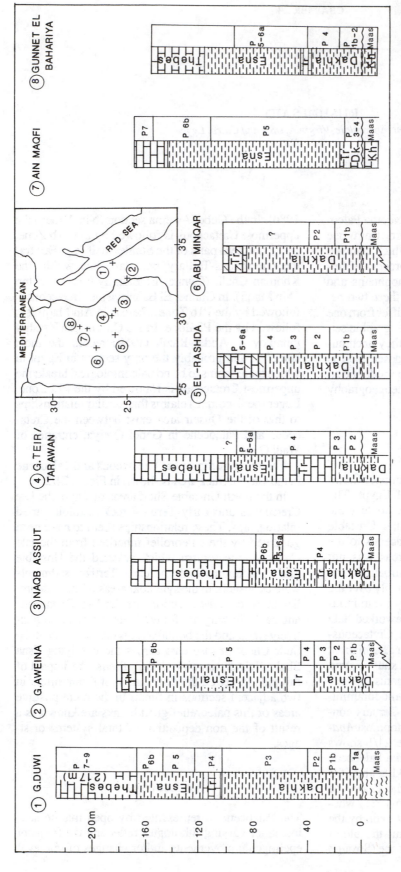

Figure 24.1 Correlation of some Paleocene sections from the Stable Shelf. Dk = Dakhla; Kh = Khoman; Th = Thebes; Tr = Tarawan. Data of sections from 1. Krasheninnikov & Abdel Razik 1969; 2. Said & Sabry 1964; 3. and 4. Hewaidy 1983 as quoted in Strougo 1986; 5. and 6. Eissa 1974; 7. Youssef & Abdel Aziz 1971 and 8. Strougo 1986.

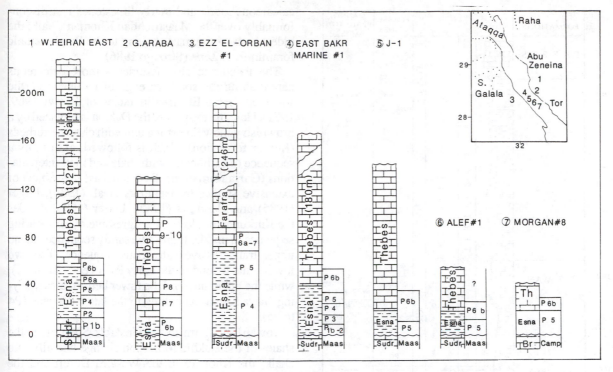

Figure 24.2 Correlation of some Paleocene sections from the Gulf of Suez. Br = Brown Limestone; Th = Thebes. Assembled by Masters, 1984. Data of well (3) from Barakat & Fahmy 1969; data of all other wells from Kerdany & Abdel Salam 1970.

which affected Egypt during this epoch. Strougo (1986) singles out the late Paleocene (the so-called *Velascoensis* time) as a significant episode of tectonism which affected many areas in south Egypt. It was marked by the deposition of burrowed bioclastic limestones in many places. A typical Paleocene section of the Stable Shelf is that of Gebel Aweina, a 450 m high outlier lying about 8.5 km to the northeast of the Sebaiya railway station. This hill is the type locality of the Esna Shale rock unit.

The Gebel Aweina section (Fig. 24/3) is made up primarily of shales interrupted by a middle carbonate bed (the Tarawan). The shales are of different colors and are of decided marine origin showing the effect of euxinic conditions at certain levels. Stringers of benthonic foraminiferal packstone associated with a large number of shells are frequent throughout the section. Toward the top bioclastic carbonates characterize the transitional layers to the massive limestone beds of the Eocene. The whole succession rests with seeming conformity on the oyster limestone and associated phosphatic beds of the Duwi formation which cover the lowlands to the south. Figure 24/3 gives the lithology, thickness and nomenclature of the different units of the section as well as the ages of these units based on their planktonic foraminiferal and calcareous nannoplankton content (Youssef

1954, Said & Sabry 1964, El Naggar 1966, Perch-Nielsen et al. 1978, Masters 1984).

The Paleocene sediments of the Stable Shelf areas of Egypt are of great areal extent and are found in areas as far apart as Dakhla and Quseir. In this stretch they are represented by the upper part of the Dakhla Shales, the Tarawan and the lower part of the Esna Shales. In Quseir the upper part of the Dakhla is made up of a solid shale unit which contrasts with the lower and more marly part. The upper part is named the Beida Member.

In Farafra the Paleocene section is lithologically similar to other sections of the Stable Shelf. Masters (1984) and Strougo (1986) reinterpret the units of the Ain Maqfi section previously described by Said & Kerdany (1961) and Youssef & Abdel Aziz (1971). They equate the Ain Maqfi Member, a thin limestone bed intercalating the lower part of the shale section, with the Tarawan chalk and divide the section (previously relegated to the Esna Shale unit) into a lower Dakhla (6 m thick), a middle Tarawan (6.5 m thick) and an upper Esna (137 m thick). According to this interpretation, the Ain Maqfi Paleocene is made up of similar units to those of the Dakhla Paleocene.

West of Farafra (Ain Dalla) the Paleocene is of a more calcareous nature and is named the Ain Dalla formation (Barthel & Herrmann-Degen 1981). These

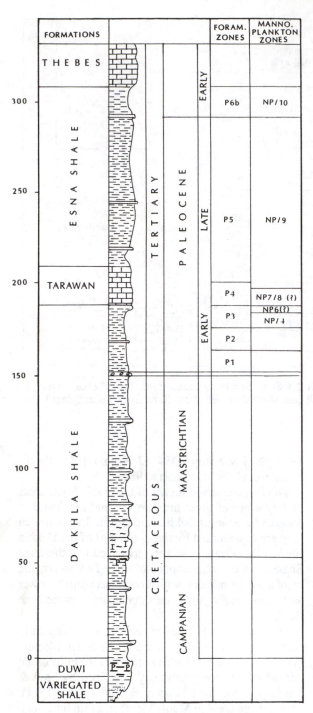

Figure 24.3 Gebel Aweina section; see table 24.1 for foraminiferal zones; mannoplankton zones after Martini (1971).

calcareous marls and marly limestones rest disconformably over the Maastrichtian Khoman Chalk; the oldest beds of the marls belong to the P4 planktonic foraminiferal Zone (Strougo 1986).

The Paleocene also assumes a more calcareous nature along the southern edge of its outcrops (the so-called Garra El Arba'in facies of Issawi 1969, 1971). Here the top part of the Dakhla is replaced by a brownish yellow limestone unit with clastic interbeds (Kurkur formation) which is followed, in turn, by a sequence of carbonates with shale and marl intercalations (Garra formation). Both units are the subject of extensive studies by Hendriks et al. (1984), Luger (1985) and Hendriks (1985). Luger (1985) divides the Kurkur into a lower transgressive oyster-bearing sequence, a middle regressive sandy sequence and an upper transgressive calcareous sequence. The two lower units belong to the early Paleocene (P1 to P3), while the upper unit and the lower beds of the overlying Garra formation belong to the late Paleocene (P4 to P5).

Toward the north in the Unstable Shelf areas the shales of the Dakhla formation change laterally into chalk; the Khoman in the Western Desert and the Sudr in the Eastern Desert and Sinai. These chalk units are of different ages, but most are of late Cretaceous age replacing only the lower late Cretaceous part of the Dakhla. The upper part of the Dakhla, the Tarawan and the Esna, which are of Paleocene to early Eocene age in the Stable Shelf areas, are represented by the Esna Shale in the Unstable Shelf areas. This unit is of wide extent and is distinct by virtue of its position between two mappable carbonate units: the Sudr below and the Thebes above. In many areas in the north the Esna is cut across by a characteristic limestone unit reminiscent of the Tarawan formation of the Stable Shelf areas. Hendriks et al. (1987) propose to name this unit the Sharib. The Sharib is 37 m thick in its type in the north Wadi Qena area, and its age extends from the Maastrichtian to the early Eocene.

Table 24/1 correlates the Paleocene rock units, while Figures 24/1 and 24/2 show the space and time relationships of these units.

EOCENE

Eocene outcrops cover about 21% of the surface area of Egypt. They bound the Nile valley, form the plateaux of the middle latitudes and build mountain scarps along both shores of the Gulf of Suez and central and north Sinai. The Eocene rocks assume several thousand meters in thickness and are made up almost exclusively of carbonates occasionally mixed

	Zones	KHARGA–UPPER NILE QUSEIR	SW ARBA'IN	UNSTABLE SHELF
EOCENE	P7	THEBES	DUNGUL	THEBES
	P6a,b	ESNA	GARRA	
PALEOCENE	P5	ESNA		ESNA
	P4	TARAWAN	K U R K U R	
	P3	D A K H L A		S U D R
	P2			
	P1a-d			
MAASTR			SHAB Member of KISEIBA FM	

Table 24.1 Paleocene rock units. Planktic Foram. Zones: Pla, b= *Globigerina eugibina, Turborotalia pseudobulloides, T. trinidadensis*; P2= *Acarinina uncinata*; P3= *Morozovella angulata*; P4= *Planorotaloides pseudomenardii*; P5= *Morozovella velascoensis*; P6a, b= *M. acuta, M. subbotinae*; P7= *M. formosa.*

with varying amounts of clastics. This lithologic monotony and the non-fossiliferous nature of large segments of the section make the recognition of regional lithostratigraphic units difficult. In addition, the persistent tectonic disturbances of this epoch resulted in many breaks and produced many sedimentary environments each with a distinctive biota making correlation and biostratigraphic zonation difficult. Nowhere in Egypt, whether in outcrop or in drillhole, can the complete section of the Eocene be examined. The section is at best composed from numerous localities.

No overall classification of the Eocene rocks of Egypt is universally accepted. Most classifications start from Zittel's major divisions: the Libysche and Mokattamstufe to which Said (1960) gave the formational names Libya and Mokattam groups respectively. The Libya group was subdivided into the Esna, Thebes (= Zittel's lower Libyan) and Minia (= Zittel's upper Libyan) formations. The name Mokattam formation was retained for Zittel's lower Mokattam and the name Maadi formation was coined to cover Zittel's upper Mokattam. Subsequent workers accepted this framework, elaborating and refining on it. Many new formational names were suggested to differentiate certain facies or to designate rock types. An attempt to sort out these names and revise the stratigraphy within this general framework is attempted by Boukhary & Abdelmalik (1983) and Strougo (1985, 1986). The authors of the new Geological Map of Egypt (1987) also distinguish within this framework several rock units which succeed and/or interfinger one another, upgrading the Thebes and Mokattam to group status. In the present work the Eocene rocks are divided into the following major rock units (from top to bottom):
— Maadi group comprising from top to bottom the Wadi Hof, Wadi Garawi and Qurn formations and their equivalents
— Mokattam group s.l. divided into:
    a lower Mokattam formation s.str. comprising the *Nummulites gizehensis*-bearing formations of the Cairo area and their equivalents in other parts of Egypt
    an upper Observatory formation for the carbonate sections of the Cairo area and their equivalents in other parts of Egypt
— Minia formation
— Thebes group comprising the Thebes formation s.str. of the Stable Shelf areas and its equivalents in other parts of Egypt
— Esna formation (top part) and its equivalents
Despite the numerous paleontological studies which have been carried out in recent years (Hamam 1975, Strougo 1977, 1979, 1985, 1986, Boukhary, Blondeau & Ambroise 1982, Blondeau, Boukhary & Shamah 1984, Omara & Kenawy 1984, Shamah & Blondau 1979, Strougo & Boukhary 1987, Ziko 1985, etc.) the exact age of many of the Eocene formations has not yet been determined. The only taxon which is of value in inter-regional correlation is the planktonic foraminifera, but these foraminifers are restricted to the pelagic facies of the Eocene which is not of wide distribution either in space or in time. The ubiquitous macro-invertebrate fossils and the *Nummulites, Operculina, Alveolina* and other larger foraminiferal tests are still of limited value in regional correlation despite the numerous studies that have recently been conducted. Ample evidence had accumulated from previous paleontological work to allow dating the Eocene rocks as early, middle and late corresponding respectively to the European stages, the Ypresian, Lutetian and Priabonian and Bartonian. Strougo (1985) advocates the use of local names as stage units

and suggests reverting to Zittel's nomenclature Libyan and Mokattamian.

In the following paragraphs a description is given of the different rock units of the Eocene. In organizing these notes I have benefitted greatly from discussions with Professors Strougo and Boukhary who have been conducting extensive research on the Eocene rocks of the Nile valley. To them I owe a note of gratitude.

ESNA FORMATION AND THEBES GROUP

The earliest Eocene sections of the Stable Shelf areas overlie the late Paleocene disconformably and with no break in sedimentation; the contact lies within the Esna Shale rock unit. The upper part of the Esna carries the early Eocene planktonic foraminiferal zones P6 and P7 (*Morozovella subbotinae* and *M. formosa* zones respectively). From these shales Kerdany (1970), Sadek (1971) and Perch-Nielsen et al. (1978) describe several calcareous nannofossils which belong to the early Eocene NP10 and NP11 zones. On top of the Esna Shale lies a limestone unit with flint, the Thebes formation, so named after its type locality near Luxor.

The name Thebes gained acceptance and wide usage despite the fact that in many areas the lower Eocene rocks assume a different lithology from that of the type section. In recent years several local names have been coined to designate these different lithologies. In the following discussion the Thebes will be retained as a formational name to designate the lower Eocene facies of the Nile valley, Quseir, Gulf of Suez and Sinai despite the fact that this name has been elevated to Group level by the authors of the new geological map of Egypt to include all the facies variants of the lower Eocene rocks.

Thebes formation

The scarp-building limestone unit of south Egypt and central Sinai was named Thebes formation by Said (1960). The type section of this formation is at Gebel Gurnah, Luxor, behind the famous temple of Deir El Bahari where it lies disconformably above the Esna formation. The section is 290 m thick and is divided into three members (Snavely et al 1979). The lower member is 135 m thick and is made up of thin (5 to 10 cm thick) alternating beds of indurated limestones and friable chalk with scattered or banded nodules of chert; the upper part of this member is marly. Fossils are relatively rare. Occasional intact nautiloids weather out of the chalk interbeds, and the pelecypod *Lucina (Anodontia) thebaica* is found in the upper marly beds of this unit.

The middle member is 125 m thick and is made up mainly of thinly-bedded fossiliferous chalk beds with nodular limestone interbeds, *Nummulites* and *Operculina* banks, and massive bioturbated chalks. Numerous echinoids and pelecypods are recorded from this unit.

The upper member is 30 m thick and is made up of reworked shell hash; oysters, echinoids and alveolines are common. The upper member assumes 95 m in thickness in Gebel Shaghab.

The age of the Thebes formation is early Eocene belonging to the planktonic foraminiferal Zones P8 and P9 (*Morozovella aragonensis* and *Acarinina pentacamerata* respectively). Other fossils that are of value in regional correlations are: *Lucina (Anodontia) thebaica* which occurs in the marly top layers of the thinly-bedded limestone-chalk lower member, *Nummulites* and *Operculina* spp. which occur as embankments in the nodular limestones of the middle member, and oyster biostromes which become more common higher up in the section. Reworked *Alveolina* tests, oysters and echinoids (e.g. *Conoclypeus delanouei*) are characteristic for the upper member. Hamam (1975) and Blondeau, Abdelmalik & Boukhary (1982) describe several characteristic species of the larger foraminifera collected from the embankments of the middle part of the section. Among the *Nummulites* described are: *N. praecursor*, *N. atacicus* and *N.* aff. *subplanulatus*. The first two species relegate the middle layers of the Thebes to the earliest Eocene.

The Thebes is widely distributed. It makes the table lands of the middle latitudes of Egypt, the bulk of the Egma plateau, Sinai and the scarps of the mountains of the Gulf of Suez. In places, the Thebes is replaced laterally or overlapped by rock units which differ in appearance due to the preponderance of one or more lithologies other than the predominant thin-bedded limestone-chalk with flint lithology of the Thebes. In recent years several local names have been coined to designate these different lithologies. Table 24/2 lists and correlates some of these names.

El Rufuf. The name El Rufuf was proposed by the authors of the new Geological Map to designate the well-bedded platform limestones of the Kharga scarp. Many fossil-bearing limestone beds intercalate the succession. In the lower levels *Nummulites* and *Operculina*-carrying beds are common. These are followed by oyster, alveolinid and echinoid embankments. The top of the El Rufuf is marked by a silcrete bed, about 30 m thick, which weathers in longitudinal furrows forming the well-known Kharafish landscape. The thickness of the El Rufuf is in the range of 110 to 140 m. In most areas it overlies the Esna Shale with a marked disconformity.

Plank. Foram Zones	Other Fossils		Main Facies Luxor	Wadi Qena	S W Facies Arba'in	W Facies Kharga	N Facies Assiut-Minia	Farafra
P9	Corals, Orbitolites	T H E B E S	/////	/////	/////	/////	///// DRUNKA	/////
	Gryphaea & Ostrea Shell banks		Shaghab	/////	DUNGUL	/////		/////
	Nummulites–Operculina Assemblage	THEBES	M Dabbabia	Serai		EL RUFUF		
P8	Lucina Thebaica	THEBES	L Hamidat	Abu Had				FARAFRA
P7	Alveolina pasticillata	ESNA	E S N A	ESNA	GARRA	/////		
P6	A.primaeva	ESNA				ESNA		

Table 24.2 Lower Eocene rock units. Planktic Foram. Zones: P8= *M. aragonensis*; P9= *Acarinina pentacamerata*.

The formation can well be correlated with the middle and upper members of the Thebes formation.

Farafra limestone. This name was proposed by Said (1960) to designate the alveolinid limestone facies of Gebel Gus Abu Said, west Farafra. The enclosed *Alveolina* spp. differ from those recorded from a higher level, the Minia, in that they are, on the whole, spherical in shape rather than elongate. These species are believed to be of earliest Eocene age (Ilerdian according to Hottinger 1960). In its type section, at Gebel Gus Abu Said, the Farafra is a 34 m thick massive, tan to buff limestone unit. The lower part is argillaceous. Most authors consider the Farafra a back reef development in the neritic zone of the early Eocene sea. The Farafra is known in the stretch between Farafra and Dakhla. In addition, the Farafra is recorded in the Ezz El Orban well (Gulf of Suez) where it assumes 230 m in thickness (Barakat & Fahmy 1969). In the nearby Gebel Thelmet, an isolated bute of Gebel Southern Galala, it assumes a thickness of 33 m (Abdel Kireem & Abdou 1979).

The Farafra limestone belongs to the P7 planktonic foraminiferal zone in Farafra, to the P6a to P7 Zone in Ezz El Orban well and to the P6b Zone in the Gebel Thelmet exposure. In all three areas the Farafra overlies the P5 Zone of the Esna formation. The Farafra is believed, therefore, to replace the upper part of the Esna Shale which is dated early Eocene in other parts of Egypt. The upper age limit of the Farafra is difficult to determine; the upper part of the thick Ezz

El Orban section does not carry any planktonic foraminifera (Fig. 24/2) and is overlain directly by the Miocene. In Gebel Thelmet, however, the Farafra is overlain by uppermost middle Eocene beds of the P13 Zone. In the north and south Galala, the Farafra includes a rich alveolinid fauna belonging to the lower *A. primaeva* zone and an upper *A. pasticillata* zone. These zones are of early Eocene age (see, however, Bandel & Kuss 1987, where these zones are considered of Paleocene age).

Dungul formation. The southern edge of the lower Eocene outcrops (Fig. 24/7) is made up of the upper part of the Garra Formation and a scarp-forming unit, the Dungul. Issawi (1969) describes the Dungul as made up of shale with limestone beds at base and a massive limestone unit with flint nodules at top. The lower beds are laden with *Nummulites* and *Operculina* species and intercalated by oyster banks. The Dungul assumes a thickness of 127 m at the type locality (Wadi Dungul) and considerably lesser thicknesses toward the west where it extends up to longitude 31° 10' E. In these latter areas, the average thickness is about 75 m of which only 16 m belong to the upper limestone unit. At the isolated hill of Barqat El Shab, a 5 m limestone bed belonging to the Dungul formation overlies a thick 'Nubia sandstone' section. This isolated occurrence carries *Lockhartia* sp. (Luger 1985) and represents the most southerly record of the Eocene in Egypt.

The upper beds of the Garra formation carry the

characteristic fossil *Lucina (Anodontia) thebaica*, whereas the lower shale beds of the Dungul formation carry *Nummulites* and *Operculina* spp. and the upper limestone beds carry oyster banks. The succession of these fossil associations leads one to believe that the upper Garra is coeval with the lower member of the Thebes, while the Dungul formation is coeval with the upper members.

Drunka formation. The name Drunka is used by the authors of the new Geological Map of Egypt to designate the massive limestone beds of the middle latitudes. The type section is at Gebel Drunka to the west of Assiut. It is made up of a lower unit, about 360 m thick, of massive and poorly fossiliferous limestones and an upper unit, about 175 m thick, of nodular and more fossiliferous limestones rich in Nummulites, alveolines and others. The lower and upper units may be synonymous respectively with the Assiuti and Manfalout formations proposed by Bishay (1961). The lower unit is composed of three beds of which the lower bed is made up of chalk, the middle (270 m thick) of massive silicified limestone and the upper of grey to yellowish limestone.

The base of the Drunka is unexposed in the type locality, but in the Qena-Safaga road the Drunka rests over the upper Thebes. The fauna described (Bishay 1961, Kenawy & Baradi 1977) points to an early Eocene age although the upper unit's fauna includes common elements with the overlying Minia formation.

MINIA FORMATION

The Minia formation follows on top of the Thebes in the Nile valley with seeming conformity. Like the Thebes, the Minia is mainly made up of carbonates of different composition, texture and structure, attesting to the varied environments in which these primarily shallow water sediments were deposited.

The name Minia formation was proposed by Said (1960) to designate the alveolinid snow white limestones which underlie the first *Gizehensis*-bearing beds of the overlying Samalut formation. The type section is at Zawiet Saada, opposite Minia. The unit was originally given an early middle Eocene age, but recent paleontological work suggests that it could be assigned a late early Eocene age (Boukhary, Blondeau and Ambroise 1982). The following is a section of the type locality as worked out by these authors:
Top:
– marly yellow limestone with *Nummulites pomeli* (A and B) and ostracods, 0.5 m
– white limestone with *N. rollandi*, 2.5 m

– white limestone with *N. partschi* (A and B) and *N. irregularis* (A and B) at the base and *N. rollandi* at the top, 22 m
– friable white limestone with algae, 13 m
– dolomitic limestone with elongate *Alveolina* sp. and algae, 41.5 m
– limestone with flint, non-fossiliferous, 10.5 m
– grey, hard to friable limestone with *N. praecursor ornatus*, *Orbitolites* spp., rich in corals and other shell debirs; forms ledge; 20 m

The lower bed is exceptionally rich in *Nummulites* spp. and other reefal forms (corals, *Orbitolites*, *Alveolina* spp. etc). The bed is massive and forms a characteristic mappable ledge to which the name Maabda member was given.

Philobbos & Keheila (1979) study the sedimentary structures and attributes of the Minia formation and conclude that it was deposited in bar and ramp coastal sub-environments. The deposits of the bar subenvironment (Beni Hassan member) are in the form of lencitular cross-bedded bodies characterized by the abundance of oolites and skeletal debris resulting in the formation of grain-supported fabric (grainstones, packstones and rudistones). The deposits of the ramp sub-environment are in the form of huge carbonate bodies which were formed under moderately agitated waters down gentle paleoslopes. They are characterized by packstone and grainstone textures and abundant bryozoa and pelecypoda.

Naqb formation. The Minia formation has a restricted areal distribution and is known only in the Nile Valley. Toward the west, in the area of the Bahariya Oasis, the Minia becomes considerably thinner and is replaced by a dark grey to pink sequence of dolomitic limestone beds which Said & Issawi (1963) name the Naqb formation. The Naqb is 29 m thick at the type locality. It rests with an angular unconformity on the Cenomanian Bahariya formation. The lowermost bed is thin (1 m thick). It carries abundant *Operculina discoidea* and is followed by a 4 m thick bed of shelly limestone rich in pelecypods and gastropods. On top of this there is a 5 m thick bed of sandy limestone rich in calcareous algae (*Dasycladacae*), elongate *Alveolina* and *Discocyclina* spp. etc. This is followed by a 3 m thick bed rich in *Nummulites ramondi*, *N. irregularis* and others which in turn is followed by a non-fossiliferous limestone bed about 8 m thick and topped by a bed rich in coralline algae.

The iron ore deposits of the Bahariya oasis are layered lenticular bodies which interfinger the Naqb (Said & Issawi 1963). In a few places they rest over the *Operculina discoidea* bed, but in all others they rest above the Cenomanian Bahariya formation. The ore beds carry calcareous algae and *Nummulites* spp.

	GROUP	SUB GROUP	Plank. Foram. Zones	NUMMULITE ZONES	GEBEL MOKATTAM	HELWAN	GIZEH PYRAMIDS PLATEAU	GEBEL ATAQA	NILE	FAYUM	NORTH BAHARIYA	GULF OF SUEZ
LATE	MAADI		P15-17	N(A)	MAADI	Wadi Hof 65m 25m		/////	Maadi	Qasr El Sagha 180m 30m		Bir Heleifiya
				N(B)		Wadi Garawi 25m 10m		/////	110m	Gehannam		Tanka
			P13-14	N(C)		Qurn 52m 97m 13m		/////	Beni Suef 86m 50m			68m
MIDDLE	MOKATTAM	OBSERVATORY			Giushi 33m	Observatory 136m	Giushi (?) Concealed	El Ramiya 78m	El Fashn 70m	Gharaq 40m Sath El Hadid 10m		Khaboba
			P12	N(D)	Upper Building Stone 70m	Gebel Hof 121m	Observatory 16m	Suez 224m	Qarara	Midawara 80m	Hamra 43m	93m
				N(E)	Gizehensis 6m		Mokattam 90m	? ?	170m			
		MOKATTAM		N(F)	Lower Building Stone 27m				Maghagha 60m	Muweilih 30m	Qazzun	Darat
		S.str.	P10-11		/////				Samalut 160m			98m
	MINIA			N(G)					Minia 115m	Naqb 29m		

Table 24.3 Middle and Upper Eocene rock units.
Planktic Foram. zones: P10= *Turborotalia bullbrooki/Hantkenina aragonensis*; P11-12= *Globigerinatheka subconglobata/ Morozovella lehneri*; P13-14= *Truncorotaloides rohri*; P15-17= *Turborotalia cerroazulensis*.
Nummulites Zones (after Strougo and Boukhary, 1987): N(A)= *N. fabianii, N. incrassatus, N. striatus*; N(B)= *N. cyrenaicus, N. cf vicaryi, N. striatus, N. aff pulchellus*; N(C)= *N. lyelli, N. discorbinus, N. cyrenaicus, N. beaumonti, N. striatus, N. aff pulchellus, N. bullatus decrouezae*; N(D)= *N. gizehensis, N. discorbinus, N. beaumonti, N. somaliensis, N. bullatus, Op. schwageri*; N(E)= *N. gizehensis, N. discorbinus*; N(F)= *N. delaharpei, N. praediscorbinus, N. aff obesus*; N(G)= *N. rollandi, N. pomeli, N. partschi, N. praecursor ornatus*.

in a succession similar to that of the surrounding outcrops.

The Naqb is a sublittoral deposit made up of intensely bioturbated dolomitized lime wackestones. Its areal extension is limited to a belt stretching from north of Bahariya Oasis westward along the presumed ancient shoreline of the late early Eocene. Although there is no concensus as to the origin of the iron ore, it is possible that it could have been deposited in lagoons developing along this ancient shoreline.

MOKATTAM GROUP, s.l.

The Mokattam group, s.l. comprises the section ori-

ginally described by Zittel (1883) under the term Unter Mokattamstufe. Although Zittel did not designate a type section for this rock unit, most authors accept Gebel Mokattam to the east of Cairo as the type locality. Here, the upper limit of the Mokattam group is clear and is marked by change of facies from solid white to yellowish carbonates to more friable yellow to brown marls and sandy limestones of the Maadi group. The lower limit of the Mokattam is not exposed, but to the south the contact between this unit and the underlying Minia becomes clear. Strougo (1986) suggests the section facing the village of Sawada, Minia, as the lower boundary reference section of the Mokattam group.

The Mokattam group comprises a number of units which could be broadly classified into a lower bedded

carbonate unit carrying in profusion large Nummulites of the gizehensis type and to which the name
Mokattam, s.str. subgroup is given and an upper solid
and massive carbonate unit with minor occurrences
of the large nummulites. To this latter unit the name
Observatory was suggested by Strougo & Boukhary
(1987). Table 24/3 is an attempt to correlate the
formations of the Minia, Mokattam and Maadi
groups.

In the following paragraphs a description is given
of the different rock units of these groups.

Mokattam s.str. subgroup
The oldest unit of the Mokattam subgroup is exposed
at the village of Sawada where a bedded limestone
unit crops out above the Minia formation. This unit,
named *Samalut*, is easily distinguished from the underlying Minia formation by its creamy color and by
virtue of the fact that it carries the first large nummulites of the gizehensis type. A composite thickness
of 160 m is recorded from the type locality by its
author (Bishay 1961). The following is a section
measured at Minia:
Top:
6. hard, dolomitic laminated limestone with abundant burrows and reworked shells; 6 m.
5. Gizehensis bank full of *Nummulites gizehensis*; 7 m.
4. rose-colored limestone with numerous banks carrying *Nummulites gizehensis*, 45 m.
3. pink limestone bed rich in *N. champollioni*, 4 m.
2. limestone packed full with specimens of *N. zitteli*
   and *N. delaharpei*; 30 m (bed quarried on large
   scale in the Samalut area).
1. cavernous, laminated to thinly-bedded, creamy
   fine-grained limestone with scattered flint nodules; occasional bands rich in *N. cailliaudi*, *N.
   pachoi* and *N. rollandi* occur; 25 m.
Boukhary, Blondeau & Ambroise (1982) and Boukhary & Abdallah (1982) confirm this zonation and add
a higher zone of *N. discorbinus*.

The environment of deposition of the Samalut
formation is worked out by Philobbos & Keheila
(1979). According to these authors, the lowermost
thinly-bedded limestones of the Samalut formation in
the Minia region overlap the bar and ramp sediments
of the Minia formation and seem to have been formed
in an open bay environment. These limestones were
designated the Meshagig member. This member is
followed by beds packed with skeletal remains which
seem to have been formed in a reefal environment.
The most characteristic of these remains are the
Nummulites and the *Discocyclina* reefs to which the
names Beni Khaled and Dhasa members were proposed respectively. Many other organic remains such
as coralline algae, bryozoa, pelecypoda, gastropoda
and others are associated with these reefs.

The vast receding plain of the Nile Valley which
lies to the north of Samalut is made up of a sequence
of well-bedded yellow and light grey limestone and
marl with minor reddish clay interbeds and a few
nummulite embankments at the base. This sequence
was given different names by different authors: the
Maghagha by Bishay (1966), Sheikh Fadl by Khalifa
(1974) and Tihna by Omara et al. (1973). The Maghagha, Sheikh Fadl and Tihna are considered coeval
with the Samalut, although the top beds of the Maghagha could be slightly younger (Strougo 1986).

Passing northward along the Nile valley, the
younger Mokattam beds exhibit varied facies. This
prompted the establishment of a large number of
formational names. The relationships of these formations is complex; the establishment of the proper
sequence of the rock units depends to a large extent
on proper paleontological correlations. The work of
Strougo (1985, 1986) and Strougo & Boukhary
(1987) was of help in sorting out the different names.

Following on top of the Samalut and its facies
variants, is a 170 m thick limestone section which
was named *Qarara* after the distinctive butte by that
name from the plain opposite Maghagha. The Qarara
is made up of a basal 20 m thick shale bed grading
upward into siltstones, with occasional carbonaceous
bands. The upper part is made up of recurring massive embankments of *Nummulites gizehensis*. From
the top of the basal bed Strougo & Azab (1982)
describe a brackish water molluscan assemblage.
This fauna as well as the structure and texture of the
sediment point to the deposition of this bed in a tidal
flat environment. This seems to confirm the enormous environmental changes that affected the
Mokattam group as a result of its tectonism.

The Samalut-Qarara succession can easily be
traced in the Fayum desert. Here the Samalut is
represented by a 30 m thick (base unexposed) section
of limestones packed full with *Nummulites zitteli* and
N. delaharpei. The formation, named *Muweilih* by
Beadnell (1905), is correlatable with the type Samalut section; the lower beds carry the same *Nummulites* spp. of the upper beds of the Samalut formation.

The *Midawara* (Beadnell 1905) overlies the
Muweilih and may well be correlated with the Qarara. It is made up of a lower 15 m thick shale unit that
is occasionally lignitic, a middle carbonate bed rich
in *Nummulites gizehensis* and an upper bed rich in
macroinvertebrate fossils.

To the north along the Nile valley in the larger
Cairo area, correlations become difficult. Figure 24/4
correlates three sections of the larger Cairo area. It is
based on the work of Strougo (1985) and the biozones he established.

At Gebel Mokattam, east Cairo, the lowermost

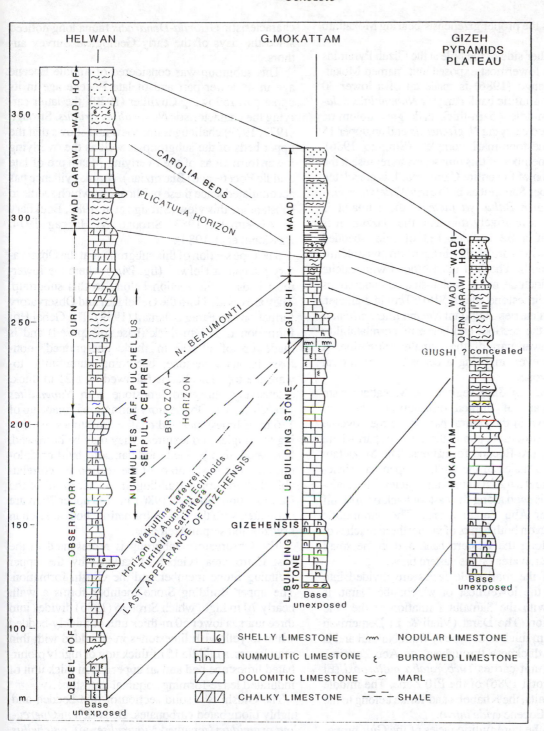

Figure 24.4 Greater Cairo area sections; data from Strougo 1986.

exposed two units of the section, the lower Building Stone and the *Gizehensis* 'Horizons' of Said & Martin's classification (1964), do not seem to have an equivalent in Helwan. The lower Building Stone Horizon starts by a 2.5 m-thick conglomerate which is followed by a 25 m-thick section of detrital nummulitic limestone. The *Gizehensis* bed is 6 to 10 m thick. These are followed by two massive limestone units: the upper Building Stone and the Giushi. The lower Building Stone and the *Gizehensis* bed can be classified as the proper gizehensis-bearing Mokattam beds.

On the other side of the Nile at the Gizeh Pyramids plateau, the lowermost exposed unit, named Mokattam by Strougo (1986) is made up of a lower 30 m-thick nummulitic bank carrying *Nummulites delaharpei*, a middle 45 m-thick dark grey dolomitic limestone bed carrying *N. gizehensis* and an upper 15 m-thick limestone-marl complex (Strougo 1986). The lower member of this unit seems to represent the oldest section of the entire Cairo area. It is correlated with the upper Samalut as both carry *N. delaharpei*.

In the *north Bahariya plateau*, the Samalut is replaced by a relatively thin unit, the Qazzun. It is made up of a 32 m-thick bed of thinly-bedded limestones which are chalky in the north and dolomitic in the south. The bed is cavernous with calcite filling and include many melon-shaped concretions of siliceous limestone, about 50 to 60 cm in diameter. The Qazzun carries abundant *Nummulites cailliaudi* throughout the section, and hence its correlatability with the lower part of the Samalut formation. It differs, however, in being a deposit of the littoral (intertidal) zone.

In the *Gulf of Suez* region, the Mokattam s.str. formations are of limited distribution. They are mostly known in the central part where they overlie the Thebes formation, and in the northern part where they overlie pre-Eocene formations. The Mokattam assumes two facies in the Gulf, an open bay shale-marl facies and a reefal nummulitic facies. The shale-marl facies is seen along the eastern bank of the Gulf in the larger Abu Zeneima area. The nummulitic facies is seen on both banks of the northern reaches of the Gulf, along the western bank and in the small Belayim-Feiran area on the eastern bank.

Rocks of the marl-shale facies are divided into three units the lowermost of which, the Darat, is correlated with the Samalut formation of the Nile Valley section. The Darat (Viotti & El Demerdash 1969) is 98 m thick at its type in Wadi Nukhul and is of uniform thickness throughout the Abu Zeneima basin. This unit carries *Turborotalia bullbrooki* (El Heiny & Morsi 1986) of the P10 Zone. The middle and upper units, the Khaboba and Tanka belong to the late middle Eocene (*vide infra*).

Rocks of the nummulitic facies of the Gulf proper belong mostly to the Samalut formation. They are followed unconformably by the Miocene sediments.

Observatory subgroup

The Observatory subgroup comprises massive limestone beds in the Cairo area. Many of the formations of this subgroup carry distinctive fossil assemblages and can, therefore, be traced for long distances. The top beds of the subgroup carry the characteristic *Gisortia-Dendracis* fauna long noticed since the days of the early Geological Survey authors.

This subgroup was considered of middle Eocene age in its lower part and of late Eocene age in its upper part following Cuvillier (1930), the latter carrying the characteristic *Nummulites striatus*. Strougo (1977, 1979) challenges this view and shows that the upper beds of the subgroup as well as the overlying lower formations of the overlying Maadi are of late middle Eocene age (Biarritzian). Ample evidence has accumulated since these publications which seems to substantiate this view (Strougo et al. 1982, Boukhary & Abdelmalik 1983, Strougo & Haggag 1974, Bassiouni et al. 1984).

The type section of this subgroup is at the Observatory plateau at Helwan (fig. 24/4) where the lowermost beds of the section belong to this subgroup. They are divided into the Gebel Hof and Observatory formations by Farag & Ismail (1959). The Gebel Hof formation is 121 m thick (base unexposed) and is made up of a 100 m thick fine-grained, non-fossiliferous limestone, becoming nummulitic toward the top. This bed is followed by a 21 m thick, extensively burrowed limestone with *Nummulites gizehensis*. The Observatory formation is made up of 136 m of limestones and chalky limestones of varying lithologies and textures. They may be burrowed, laminated, thin-bedded, nodular, soft, hard or dolomitized. The lower 66 m of the section are correlatable with the upper Building Stone 'Horizon' of the Cairo section (Strougo 1986) and the upper 70 m are correlated with the Giushi formation; both are rich in bryozoa and serpulid remains.

The Observatory subgroup is represented in the east Cairo area (Gebel Mokattam) by the upper Building Stone member and the Giushi formation. The upper Building Stone member forms a wall, nearly 70 m high, which Strougo (1985) divides into three units: a lower 50 m-thick unit of thickly-bedded white to yellowish limestones interbedded with thin marl layers, a middle 15 m-thick unit of highly bioturbated limestone bed and an upper 1.5 m thick unit of indurated ledge-forming coquinal limestone.

The Giushi is a solid section of thin-bedded and highly bioturbated carbonates. Beds rich in *Operculina pyamidum* (renamed *Nummulites* aff. *pulchellus* by Strougo 1986), serpulid and bryozoan remains are characteristic of this formation.

In the Gizeh Pyramids plateau which lies on the southern edge of the Kattaniya-Abu Roash High (*vide infra*), beds equivalent to the Observatory subgroup are very thin. It is questionable whether beds belonging to the Giushi are present; they may be represented by the covered beds overlying the Mokattam formation.

The Observatory beds seem to build the major part of the scarps of the Cairo-Suez district. In Gebel Ataqa, overlooking the city of Suez, Akkad & Abdallah (1971) divide the Eocene section, which rests unconformably over late Cretaceous-early Tertiary continental sediments, into a lower 224 m-thick section of dolomitic limestones and marls which they named the Suez formation and an upper 78 m-thick unit of clastics and marls conformably overlying the Suez formation which they named El Ramiya formation. The top layer carries the characteristic large gastropod *Gisortia*. The Suez and the El Ramiya are correlatable with the Observatory subgroup of the Cairo area.

To the south of Cairo, along the Nile Valley, the Observatory subgroup becomes difficult to follow as the valley opens up in the Maghagha-Beni Suef stretch where, as a result of the relatively easy weathering of the rocks, the valley becomes exceptionally wide. Here the rocks show greater variation in lithological composition both in space and in time. The sediments have, on the whole, a large ingredient of clastics. They usually disconformably overlie the older sediments and show a number of diastems. A typical section of this subgroup is exposed at El Fashn. Here the Qarara is overlain disconformably by a very distinctive unit made up of hard, nummulite-bryozoan limestone with chert bands and nodules which caps many scarps in the region. This unit, clearly correlatable with the Giushi, forms the lower part of El Fashn formation as conceived by Bishay (1966). Strougo (1986) re-examines the type locality of this formation at Wadi El Sheikh opposite El Fashn town and finds that the formation 'consists of two strikingly different lithologic ensembles'. The lower ensemble is correlated with the Giushi. It is proposed here to restrict the use of the name El Fash formation to the lower member of this formation as conceived by Bishay (1966).

In the Fayum desert, the Sath El Hadid and the Gharaq formations may be correlatable with the Giushi. The Sath El Hadid caps the Midawara formation and is made up of a 10 to 25 m thick limestone bed rich in large gastropods, corals and bryozoa. The Gharaq is a carbonate section of about 40 to 150 m in thickness and is poorly fossiliferous. However, the upper beds of the formation are rich in bryozoa (Strougo 1986) and may thus be correlatable with the upper beds of the Giushi formation.

To the southwest of Fayum, in the *north Bahariya plateau*, the Observatory subgroup becomes exceptionally thin and is represented by the middle Eocene beds of Gebel Hamra. These beds, which formed the lower part of the Hamra formation as conceived by Said & Issawi (1963), now form the type of the Hamra formation as emended by the authors of the new Geological Map of Egypt. These authors relegate the late Eocene upper part of Gebel Hamra to the Qasr El Sagha formation. The Hamra, as emended, is 43 m thick and is made up of yellowish brown limestone with abundant detrital grains and numerous skeletal remains. It is rich in *Nummulites* spp. The section shows many diastems represented by soil development and intraformational conglomerates. The top beds carry a large number of the characteristic large gastropod *Gisortia* and is thus believed to be coeval with the Giushi.

In the eastern side of the *Gulf of Suez*, Eocene sediments correlatable with the Observatory subgroup are of limited distribution. They are represented by the bathyal deposits of the Khaboba formation (Viotti & El Demerdash 1969). The Khaboba is made up of interbeds of gypsiferous shales, marls, limestones and chalky limestones with flint bands, the carbonates increasing upward. The formation is 93 m thick at its type locality. It carries the planktonic foraminifera of Zones P11 and P12 (El Heiny & Morsi 1986).

MAADI GROUP

The Maadi group sediments are of more clastic nature than the underlying Mokattam, s.l. sediments. They are made up mostly of shales with intercalated limestones. The shale is greyish green, highly calcareous, fossiliferous and partly sandy. The limestones are light to dark brown, highly argillaceous, medium hard and locally limonitic. The sediments were deposited in a retreating shallow sea.

The Maadi group forms a distinct unit which contrasts with the underlying massive white limestone of the Mokattam group by virtue of its yellowish color and receding line. The unit is exceptionally developed in the larger Cairo area (Fig. 24/4). It thickens to the south of Cairo and is divided in the Helwan area into three formations: the Qurn, Wadi Garawi and Wadi Hof (Farag & Ismail 1959). In the Gizeh Pyramids Plateau, where the Qurn is poorly developed, Hassan, Cherif & Zahran (1981) recognise the upper two formations, Wadi Garawi and Wadi Hof; they gave them the names Heit El Ghorab and Giran El Ful respectively. Previously, the Maadi group was believed to be of late Eocene age. However, recent studies have shown that the lower two units, the Qurn and Wadi Garawi, can best be classified as middle Eocene (Strougo & Boukhary 1987). These authors show that the Nummulite and macro-invertebrate faunas of the lower two units have close affinities to the late middle Eocene (Biarritzian) faunas of Europe. The characteristic *Nummulites* spp. of the different units of the Maadi group are listed in Table

24/3. According to this classification, the late Eocene is restricted to the upper beds of the Maadi group which carry the index fossil: *Carolia placunoides*. The characteristic *Nummulites* spp. of the late Eocene are: *N. fabianii* and *N. incrassatus*.

Middle Eocene formations of the Maadi group
The middle Eocene formations of the Maadi group in the Cairo area are exceptionally thin in the Gebel Mokattam and the Gizeh Pyramids plateau sections (Fig. 24/4). To the south, however, the Qurn and Wadi Garawi formations become well-developed. In the Helwan area, the Qurn follows on top of the Observatory formation. It is made up of a 97 m-thick sequence of marly and chalky limestones alternating with shales, sandy marls and shell banks rich in *Osrea reili*. *Nummulites beaumonti*, *N. striatus* and *N.* aff. *pulchellus*, which are recorded in the underlying Giushi formation, are also common. The Wadi Garawi formation is a 25 m-thick poorly fossiliferous sandy shale section with a hard highly fossiliferous middle bed carrying *Plicatula polymorpha*, *Nicaisolopha clotbeyi* and others. In many places this bed becomes phosphatic.

To the south, along the Nile Valley, the Qurn becomes well-developed and is easily recognised as it usually follows on top of the solid limestones of the underlying Observatory subgroup. The Wadi Garawi formation, on the other hand, becomes difficult to separate from the underlying Qurn. In Beni Suef the Qurn forms the upper member of Bishay's El Fashn formation. As originally described this member is made up of a set of strata which differs from the lower set by its softer nature – whence forming a retreated second scarp – and by the lack or scarcity of chert horizons which are common in the lower set. This unit is 86 m thick. It consists primarily of sparingly fossiliferous sandy marl and marly limestone with clay and sandstone beds at the top. The authors of the new Geological Map elevated this member to a formational status, the Beni Suef. In the Beni Mazar area Omara et al. (1977) mapped comparable beds as El Merier formation. The formation is mapped as Beni Suef in the new Geological Map and as Qurn in the Geological Survey Cairo map (1983).

In the Fayum desert, the unit is 50 m thick and named Gehannam formation. This formation is made up of a sequence of marls and sands with occasional bands of limestones. These beds were named by Beadnell (1905) the Ravine beds and are 50 m thick. They yielded the famed *Zeuglodon* remains from the lower part. Strougo & Haggag (1984) desdribe a rich macro-invertebrate and planktonic foraminiferal fauna from the Gehannan formation and date the formation middle Eocene rather than upper Eocene as was previously held (Abdou & Abdel Kireem 1975).

On the eastern side of the *Gulf of Suez*, the Tanka formation (Hume et al. 1920) may be correlatable with the Qurn and Wadi Garawi formations. The Tanka is 68 m thick at the type locality and is made up of alternating thin beds of chalky limestone and shale. It is dated middle Eocene by El Heiny & Morsi (1986) and is reported to carry planktic foraminifera of the P13 to P14 foraminiferal zones (*Truncorotaloides rohri*).

Late Eocene formations of the Maadi group
It has already been pointed out that the sediments of the Maadi group were deposited in a retreating shallow sea which was receiving detritus from the exposed hinterland. The influx of detrital material became exceptionally high during the late Eocene when large rivers seem to have discharged in the sea. One of the surviving deltas of these rivers is the well-known Fayum delta exposed to the north of the depression at the foot of the Qasr El Sagha temple.

The deltaic and interdeltaic sediments of the upper Eocene are exposed to the north of Birket Qarun, Fayum, where they form the Qasr El Sagha formation. The formation is 180 m thick at the type locality and is made up of four succeeding and/or interfingering facies (Vondra 1974). The lowermost arenaceous bioclastic facies is made up of bioturbated (mainly by the crustacean *Callianassa*) glauconitic and fossiliferous calcareous sandstone. This facies carries abundant complete and disarticulated remains of shelly marine invertebrates, abraded bones and teeth of marine and transitional marine vertebrates and carbonized wood fragments up to 10 cm in diameter and 2 m in length. From this lower facies came the well-preserved *Prozeuglodon isis* (Moustafa 1974). This facies formed as offshore bars and barrier beaches along the upper Eocene shoreline. Interbedded with this facies is a gypsiferous and carbonaceous laminated claystone and siltstone facies which seem to have been deposited in back bar open and restricted lagoons. It is made up of pale yellowish-brown to dark grey laminated claystones and argillaceous siltstones with abundant thin sheets of gypsum. This facies is very carbonaceous and contains small frgments of carbonized wood and leaves. Engelhardt (1907) identifies 22 species from the leaf prints of which eight are *Ficus* spp. Other authors recognise only four tree leaf species, one water lily leaf species, three fruit species (including a palm), one species of wood and three algae. The presence of *Ficus* spp. and the resemblance of the four tree leaf species to the present-day rain forest forms indicate an exceedingly wet climate.

Interfingering with the above facies is the interbedded claystone, siltstone and quartz sandstone facies which form the bulk of the cliffs behind the

Qasr El Sagha temple. The facies consists of 10 to 35 cm-thick interbedded units which represent a fining upward cycle of very fine to fine-grained white quartz sandstone, pale yellowish brown and dark grey siltstone and claystone. Each unit constitutes a foreset of a large-scale planar cross-strasatification set. This facies was deposited in a rapidly prograding delta front environment in quiet shallow brackish marine waters. Cut into and interfingering this facies is the quartz sandstone facies which seems to represent distributary channel deposits. From this facies came a rich fauna of fossil mammals, reptiles and fish of marine, transitional marine and terrestrial habitat. It forms the *Moeritherium-Pterosphenus* zone.

Toward the southwest in the direction of Bahariya, the Qasr El Sagha formation is about 30 m thick and is made up of richly fossiliferous arenaceous bioclastic carbonates which must have been deposited along the late Eocene shoreline. In the north Bahariya area, Said & Issawi (1963) record along this shoreline a series of saucer-like reefal structures which have inward dips of more than 40°. These reefs were built by pelecypods and gastropods and seemed to have formed part of the offshore bar and barrier beach landscape of the late Eocene.

In outcrop, the late Eocene sediments are easily distinguished by virtue of the fact that their dominant lithology distinguishes them from the underlying massive limestones of the middle Eocene and the overlying sandstones and quartzites of the continental Oligocene. Toward the north, where both the Eocene and Oligocene become buried under more recent cover, the distinction becomes less obvious as the Oligocene becomes marine in character and assumes a similar facies to that of the upper Eocene. For this reason, the upper Eocene and Oligocene sediments are usually lumped under the name Dabaa (*vide infra* under Oligocene).

OLIGOCENE

Oligocene deposits overlie late Eocene sediments disconformably. They assume two distinct facies: a fluviatile facies of sands and gravels and an open marine facies of shales and minor limestone interbeds. The distribution of these sediments was governed, to a large extent, by the volcanicity, geyser activity and tectonism which affected the Red Sea regions as well as the belt of highs between the Stable and Unstable Shelves during the Oligocene. With the exception of a questionable small outcrop recorded at the core of the Salum cliff close to the Libyan-Egyptian border on the Mediterranean Sea, the marine facies is known only from the subsurface. Here the facies is similar to that of the underlying upper Eocene sediments; both become indistinguishable from one another except through paleontological work.

Fluviatile sediments

Oligocene fluviatile sediments crop out along a narrow belt extending from Suez to Fayum via Cairo and onward into the Western Desert. Small and isolated outcrops of this facies are also known from east of Beni Suef and from Bahariya Oasis.

To this facies probably belong the small and scattered gravel mounds which lie over the top of the Ma'aza limestone plateau of the middle latitudes of Egypt and also the lacustrine to paracontinental deposits of the Red Sea coast and the Gulf of Suez. These deposits are difficult to date and they are classified with the Oligocene on stratigraphic evidence only.

Gebel Ahmar sands and gravels. A typical example of the sands and gravels of the Cairo-Suez district is in Gebel Ahmar, to the east of Cairo, made famous by the descriptions of Barron (1907) and Shukri (1954). The Gebel Ahmar sands are vividly colored, crossbedded and coarse-grained. The gravels are mostly of pebble size and are made up of banded flint usually blackened by exposure to weathering. Apart from a large number of tree trunks, the formation is nonfossiliferous. One specimen of the fresh-water snail, ?*Planorbis*, is reported to have been found within the gravels of Gebel Yahmoum El Asmar, Cairo-Suez road. The lithology and texture of these sediments indicate their deposition in laterally aggrading meandering streams. As the sediments do not include mudstones, it must be assumed that these rivers did not overflow their banks.

The Gebel Ahmar has disappeared under the foundations of Nasr City, northeast Cairo since it was last described; it has been renamed Gebel Akhdar. The Gebel Ahmar exhibits one of the rare manifestations of the action of geysers; vividly-colored silicified tubes traverse the sands and gravels. These are described by Shukri (1954). Close to Gebel Ahmar lies the conical hill known as the Rennebaum 'volcano', considered as a gas maar and resulting from a phreatic explosion creating a vent through which fumarolic activity took place (Tosson 1954).

The thickness of the Gebel Ahmar sands and gravels in the Cairo-Suez district is in the range of 40 to 100 m. In Gebel Yahmum El Asmar recent quarrying for gravel exposed more than 50 m of sands and gravels. Here the section consists of alternating thin gravel and sand laminae. Lateritic soils cap the deposit in places inducing red coloration to the underlying sediments. El Sharkawi & Khalil (1977) believe

that the fluids which caused the coloration and silicification of the Oligocene sands and gravels of the
Cairo-Suez district were exogenic. Shukri (1954)
believes that these fluids ascended along faults.

The silicified tree trunks which the Oligocene
clastics carry become concentrated in certain areas
where they are commonly referred to as petrified
forests (Gebel El Khashab); the most famous of these
is the one which occurs to the east of Maadi, Cairo.
This has attracted the attention of travellers, naturalists and scientists for a long period of time. Most
workers believe that the tree trunks (some of which
reach 30 m in length) were transported for long
distances. The alignment of these tree trunks and the
absence of twigs, fruits or any other soft parts are
cited as evidence of the long journey they made.
Kortland (1980) subscribes to this view and believes
that the presence of these trunks cannot be taken as an
indication of the climate of the area in which these
trunks are presently found. Simons (1972) and Bown
et al. (1982), however, believe that the fossil trees
must have accumulated near the place of their growth
and infer that the Oligocene climate must have been
tropical and the area resplendent with tall trees. The
silicification of the tree trunks has also been a subject
of controversy. Most authors believe that the silicification took place after the transportation of the
trees to their present place. El Sharkawi (1983),
however, believe that the silicification took place at
the place of the growth of the trees in a lake environment rich in sodium silicate minerals.

Gebel Qatrani formation. A considerably thicker deposit of the Oligocene sands occurs along the northern and western scarps of the Fayum depression.
These fluvial deposits, known as the Gebel Qatrani
formation, have become world famous because of the
unique vertebrate fauna they carry (for a discussion
of these, see Chapter 30, this book).

The Gebel Qatrani formation is the subject of the
classical work of Beadnell (1905) and, more recently,
of Bowen & Vondra (1974), Bown (1982) and Bown
et al. (1982). The formation consists of 110 to 340 m
of fluvial sandstone (80%), siltstones and claystones
(18%), minor carbonate lenses (1%) and carbonaceous shale (less than 1%). The individual beds are
lenticular and grade laterally and vertically into one
another.

The Gebel Qatrani formation disconformably
overlies the upper Eocene Qasr El Sagha formation.
The basal 100 m are characterized by a complex of
large scale trough cross-stratified channel lag and
point bar sands indicating deposition in a loosely
sinuous, low gradient, medium velocity stream. The
upper part of the formation is characterized by point
bar, flood plain splay and channel fill deposits point-

ing to an overloaded and more tightly meandering
stream. The point bar deposits form most of the
sandstones of the section. They are very fossiliferous
in the lower as well as in the upper parts of the
formation, containing abundant silicified logs and the
fossil vertebrates for which the Oligocene of Egypt is
noted. Nearly all the sandstones show diagenetic
alteration reflecting ancient pedogenesis. The few
limestones and dolomitic marl interbeds were probably deposited in shallow, perhaps ephemeral, flood
plain ponds and contain fresh-water ostracodes and
charophytes which indicate low salinity. The carbonaceous shales are flood-plain swale deposits.

A nearshore setting for Gebel Qatrani is indicated
by the presence of shark teeth, ray mouth parts,
brackish-water molluscs and abundant mangrove rhizoliths. Among the 95 species of fossil vertebrates
recorded from Gebel Qatrani, the following are abundant: crocodilians, turtles, browsing artiodactyls and
hyracoids, arboreal quadrupedal anthropoid primates, carnivorous mammals and phyomid rodents
(Simons 1968, Simons & Gingerich 1974, El
Khashab 1977 and also chapter 30, this book). The
ichnofossils and rhizoliths are abundant, well-
preserved and diverse in form (Bown 1982). The
ichnofauna contains traces of probable annelid, insect, crustacean and vertebrate origin. These include
fossil nest structures and gallery systems of subterranean termites. Rhizoliths associated with the ichnofauna document a variety of small wetland plants,
coastal mangroves and large trees. Kortland (1980)
believes that the Gebel Qatrani flora indicates a drier
(Sahelian) climate. It differs from that of Qasr El
Sagha in containing no identified leaves and in having only two fruit species (a waterlily and a relative of
the palms). The 23 species of wood identified by
Kräusel (1939) are believed to be of trees that were
transported to Egypt from far south.

The Gebel Qatrani formation is unconformably
overlain by the Widan El Faras basalt (Bowen &
Vondra 1974), the lower part of which is dated at $31 \pm$
1 my (Fleagle et al. 1986). The basalt occurs in three
sheets. The lower and upper sheets are amygdaloidal,
vesicular and intensely altered, whereas the middle
sheet is massive, compact and fresh. Layering is
noted in the upper sheet. The three sheets are similar
petrologically and are of tholeiitic nature (Heikal et
al. 1983).

Radwan formation (Bahariya). A succession of ferruginous grit, quartzite and sandstone beds cap
several of the conical and flat-topped hills which litter
the Bahariya depression and its northern plateau. The
beds are non-fossiliferous and coarse-grained. In
most areas they overlie unconformably the Cenomanian Bahariya formation, but in Gebel Hamra in the

north plateau they overlie the upper Eocene sediments. These beds were named Radwan (Akkad & Issawi 1963). The spotty distribution of the Radwan is probably due to the fact that its resistant beds were affected by the early Miocene volcanicity of the oasis. The Radwan outcrops, therefore, are probably the remnants of a more continuous sheet of the deposits of the meandering river systems which drained the northern land of Egypt during the Oligocene. The thickness of the formation at the type locality is 40 m, but in most other areas it is in the range of 1 to 10 m.

Similar outcrops are also known on top of the upper Eocene exposures in the east Beni Suef area.

Ma'aza plateau gravel mounds. The gravel mounds which litter the limestone plateaux of the middle latitudes of Egypt (the Ma'aza plateau) are discussed in Chapter 25 (Figure 25/4). They are small elongate mounds which are interpreted as inverted wadis of a defunct drainage system which must have occurred prior to the excavation of the oases depressions. They are considered part of the deposits of the Oligocene drainage system, although they could be somewhat earlier or later than the Oligocene.

Nakheil formation. The Nakheil formation is recorded from the synclinal troughs of the coastal areas of the Red Sea (see Chapter 18). It consists of gravity breccias interbedded with fine-grained lacustrine sediments. It is dated Oligocene on stratigrpahical grounds and is interpreted as accumulations along the slopes and in the lows of the incipient relief of the elevated Red Sea region.

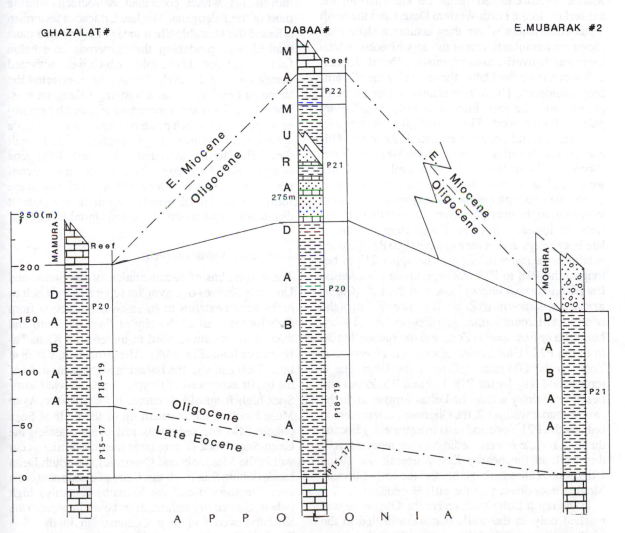

Figure 24.5 Correlation of some Oligocene sections from the north Western Desert. P15-17= *Globigerapsis semiinvolutus / Turborotalia cerroazulensis*; P18-19= *Cassigerinella chipolensis / Pseudohastigerna micra*; P20= *Globigerina ampliapertura*; P21= *Turborotalia opima*; P22= Globigerina ciperoensis. Data of sections reinterpreted from Sheikh & Faris 1985.

Tayiba redbeds. The Tayiba (also referred to as the Abu Zeneima formation in some works) overlies unconformably the late middle Eocene Tanka formation in the Abu Zeneima-Feiran area, Gulf of Suez. It consists of a 5 m-thick basal conglomerate, a 15 m-thick sequence of coarsening upward siltstone beds and an upper 15 m-thick porcellaneous limestone bed. The lower beds include reworked *Nummulites* derived from the uplifted Eocene outcrops. The formation is topped by basalt dikes which are of presumed Oligocene age. The formation could well be coeval with the Nakheil and of similar origin, although the dominant red color and the absence of large boulders may indicate deposition in more permanent lakes in areas with low relief.

Marine sediments

Marine Oligocene sediments are known from the subsurface in the north Western Desert and the north Delta embayment where they assume a shale-marl facies very similar to that of the upper Eocene which they usually overlie disconformably. This shale-marl unit was named the Dabaa formation by the oil company geologists. The differentiation between the Oligocene and the late Eocene is only possible by paleontological work. Figure 24/5 gives the planktonic foraminiferal zones of the late Eocene and Oligocene sections of three wells in the Western Desert (Sheikh & Faris 1985). Of these wells, the Dabaa well no. 1 has the most complete Oligocene section; it covers the upper part of the Dabaa formation and the lower part of the overlying Mamura formation. In this well, the lower 33 m of the Dabaa formation are of late Eocene age and belong to the P15 to P17 (*Globorotalia cerroazulensis*) Zone. The upper 213 m belong to the P18 to P19 (*Cassigerinella chipolensis/ Pseudohastgerina micra*) Zone and the P20 (*Globigerina ampliaptertura*) Zone. The lower 275 m of the overlying Mamura formation belong to the P21 (*Globorotalia opima opima*) Zone and the succeeding 30 m to the P22 (*Globigerina ciperoensis ciperoensis*) Zone. In the Ghazalat well no. 1 the Oligocene is represented only by the P18, P19 and P20 Zones and lies in its entirety within the Dabaa formation. In the east Mubarak well no. 2, the Oligocene is represented only by the P21 Zone and rests over the early Eocene directly; no late Eocene sediments are reported from this well. In the nearby East Mubarak no. 1, the Oligocene is absent altogether and the early Miocene Moghra rests directly on the early Eocene.

In the north Delta embayment the Oligocene was reached only in the wells that were drilled in the hinge zone where the Oligocene rests on the Cretaceous rocks directly (Sneh, Monaga & Qallin wells). The most complete Oligocene section is found in the Monaga well no. 1 where all the zones are represented. In both the Qallin and Sneh wells the Oligocene is represented only by the P21 and P22 Zones.

DISTRIBUTION, PALEOGEOGRAPHY AND PALEOECOLOGY OF THE PALEOGENE

The Paleogene was ushered in by a transgression which pushed its way across the southern borders of Egypt into north Sudan. The maximum transgression occurred during the late Paleocene. After the Paleocene and all through the Cenozoic the sea kept retreating toward the north almost continuously except for short intervals. The sediments that this sea left behind were affected to a large extent by the relief which was inherited from the late Cretaceous tectonism and the movements which continued episodically during most of the Paleogene. The late Crtaceous tectonism affected the Unstable Shelf areas of Egypt in a most marked way producing the numerous en-echelon folds of that belt. These folds, which were activated during most of the early Tertiary time, affected the shape and outline of the advancing Paleogene seas. The Stable Shelf areas were also affected by epeirogenic processes which produced structures with more symmetrical outlines, large amplitudes and small dips. The result was that while the Paleogene outcrops of the Unstable Shelf are thin and disconnected, those of the Stable Shelf are thick and in the form of extensive tablelands. This marks the clearest distinction between the Stable and Unstable Shelves.

Suez-Cairo-Kattaniya high

The two realms of sedimentation of the Stable and Unstable Shelves of Egypt are separated by a belt of highs which extends in an east-west direction from Gebel Raha, east of the city of Suez, to Cairo and from there southwestward to include the Khatatba-Kattaniya high (Fig. 24/8). Along this belt, which at times included also the Bahariya high, occur most of the basalt extrusions of Egypt. The Kattaniya-Cairo-Suez high is an old structure. It includes the Ayun Musa block at the northern tip of the Gulf of Suez (where Miocene sediments rest over Jurassic), the Cairo-Suez block (where large unconformities occur within the Mesozoic and Cenozoic), the south Delta block (where thin to absent Cenozoic sediments rest over Cretaceous) and the Khatatba-Kattaniya high (where Quaternary sediments rest over Jurassic in the Khatatba well and over Cenomanian in the Abu Roash wells, where very thin Paleogene sediments rest over lower Cretaceous in Kattaniya, Gebel Rissu, Wadi Khadish and Mubarak East no. 1, and where lower Miocene beds rest over lower Cretaceous in Natrun T57 well).

During the Paleogene, the limits of this belt changed by erosion and/or by tectonic activation. It was at its highest during the Paleocene; no sediments of this epoch are recorded anywhere along this belt. During the early Eocene its peripheries were eroded but it still included the Gebel Ataqa and Gebel Raha blocks; both have middle Eocene rocks resting directly over the Cretaceous. The southern boundary of the Cairo-Suez block during this age seems to have been the faults extending eastward from Tura to Treifiya and then southeastward along the Gebel Ataqa western scarp. At the junction of these faults at Treifiya occurs the longest basalt in the Cairo-Suez district. Figures 24/6 to 24/9 show the position of this belt which remained high throughout the Paleogene. It was partially covered during the middle Eocene. In Kattaniya and the Gebel Rissu wells, the middle Eocene rocks are 28 and 21 m thick respectively.

Stable Shelf

A large part of the Stable Shelf was overlapped by the

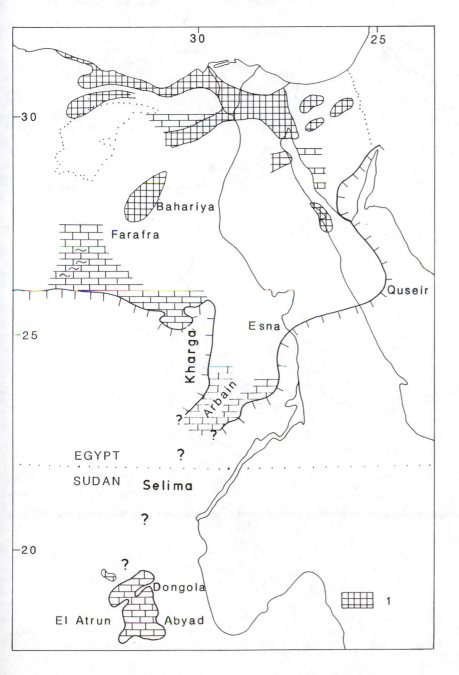

Figure 24.6 Paleocene paleo-geographic map. 1= positive areas.

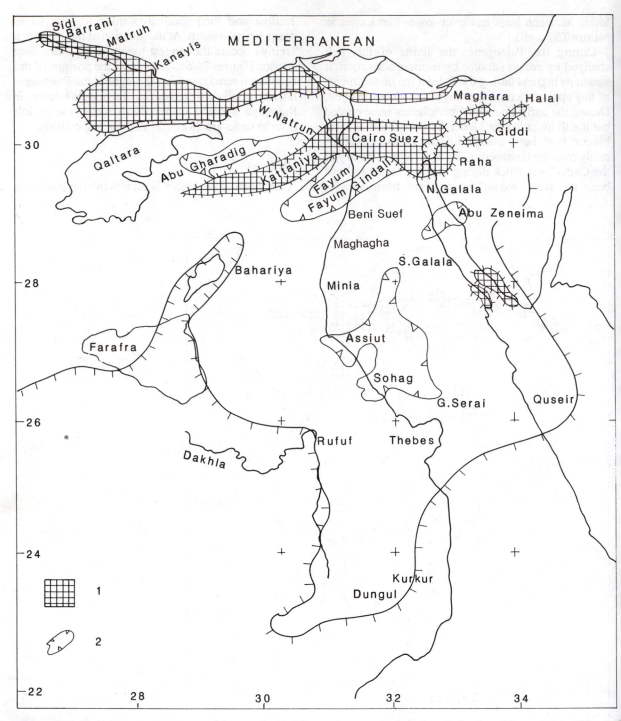

Figure 24.7 Early Eocene paleogeographic map. 1= positive areas or areas of erosion; 2= basins with more than 500 m of sediments.

sea during the Paleocene when the maximum transgression took place. The discovery of a marine Paleocene outcrop in Gebel Abyad, north Sudan (Barazi 1985), pushes the shoreline of that epoch about 400 km to the south (Fig. 24/6). The Gebel Abyad facies is similar to the Arba'in facies of the Paleocene of south Egypt and it is possible that both were formed in one basin. This basin seems to have formed a shallow and well-aerated open marine embayment which received minor amounts of clastics. It is of interest to note that the sediments of the peripheral parts of the Paleocene sea are more calcareous than those of the center of the basin. This may indicate deposition in an arid environment where the numerous intervening highs and islands of that sea were the source of the fine clastics of the Esna Shales.

During the lower Eocene, the shoreline receded to south Egypt where relatively thin (less than 500 m-thick) and uniform sediments covered this great embayment. Thicker than 500 m sediments occur only in a few basins:

1. The Fayum-Bir Gindali basin: The thickest Eocene section recorded in Egypt occurs in this basin which lies on the southern edge of the Suez-Cairo-Kattaniya high (Fig. 24/8). In the Gindi well no. 1 the Eocene section is 1,788 m thick of which the lowermost 924 m belong to the Paleocene and lower Eocene (Elzarka & Zein El Din 1985). In the Bre 3, Bre 6 and Bre 27 wells drilled in this basin, the Eocene is 1640, 955 and 940 m thick respectively. The Eocene section of all these wells forms one rock unit, the Apollonia. Paleontological work shows that the section belongs to the Paleocene, early and middle Eocene. In the Helwan area the exposed Eocene

section is more than 400 m thick; the unexposed base is of middle Eocene age. The nearby section at the Pyramids of Gizeh plateau is considerably thinner; it lies on the edge of the active Cairo-Kattaniya high.

The Fayum basin extends eastward across the Nile to cover the Bir Gindali area. Along this stretch more complete and thicker sections of the Eocene are encountered. The Fayum-Gindali basin forms the last area of the Stable Shelf which continued to be covered by the retreating middle and late Eocene seas. All the other areas of the Stable Shelf were elevated by the end of the early Eocene.

2. Assiut-Sohag basin: This basin covers the Assiut and Sohag areas where thick early Eocene sections with unexposed bases are known. The boundaries of this basin (Fig. 24/7) are inferred and are taken to include the gravity low of Wadi Qena (see Kamel, this book, Chapter 4). This basin consists of early Eocene sediments only and seems to have been elevated by middle Eocene time.

3. Abu Zeneima basin: This limited basin includes almost all the thicker than 400 m Paleogene sections of the Gulf of Suez. At north Markha, El Heiny and Morsi (1986) describe a lower-middle Eocene section 555 m thick. In Lagia no. 3 well the Eocene is 462 m thick and is made up of a thin lower Eocene (54.5 m thick) and a thick middle Eocene (367.8 m thick) section (Elzarka & Zein El Din 1985). With the exception of the small Oligocene outcrop of lacustrine to paracontinental deposits of Wadi Tayiba, no late Eocene or Oligocene sediments are known in the Gulf of Suez area which must have represented a positive area throughout that time.

The earliest Eocene is represented by the upper

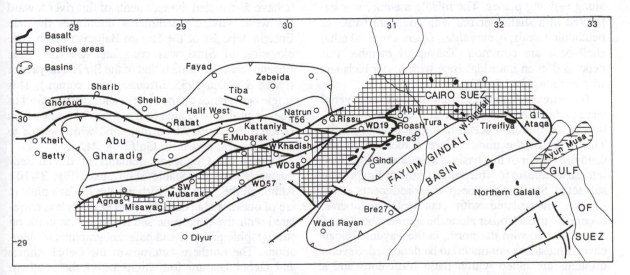

Figure 24.8 The Suez-Cairo-Kattaniya High and the bordering Abu Gharadig and Fayum-Gindali basins. Subsurface structures of Western Desert adapted from Awad 1984.

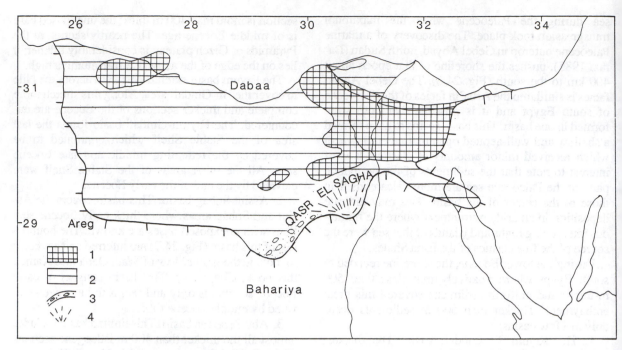

Figure 24.9 Late Eocene paleogeographic map. 1= positive areas; 2= shelf sediments; 3= shelf sediments showing effect of fluviatile sedimentation; 4= deltaic sediments (Qasr El Sagha Formation).

beds of the Esna shale and the Farafra limestone. These give way to the carbonate rocks of the lower Eocene. The main facies of these rocks, the Thebes, indicates deposition in a continuously shallowing sea. The lower member of the Thebes was deposited in a low energy deep hemipelagic environment. The enclosed chert nodules and bands are probably of secondary origin; the dissolved silica of this deep environment was reprecipitated as nodules or bands along bedding planes. The middle member was deposited in a shallower sea with abundant banks of benthonic fossils; *Nummulites, Operculina* and other shell beds are common. The upper member was deposited in an intertidal zone with oyster beds and frequent reworked shell hash.

The retreat of the sea toward the north uncovered large parts of the Stable Shelf and brought the shoreline of the middle Eocene to the latitude of Minia. The facies of the middle Eocene rocks along the southern border of this basin is unique attesting to the intense tectonism to which many parts of Egypt were subjected. Shallow and deep-water sediments, occasionally intercalated with continental sediments, occur next to each other along the southern borders of this reach. Toward the north, in the Fayum-Gindali basin, limestones continued to be deposited. Several diastems are noted within these formations and a major event took place after the deposition of these sediments which brought about the elevation of

Gebel Ataqa and large parts of the Cairo-Suez block beyond the reach of the middle and late Eocene seas.

With the advent of the late Eocene (Fig. 24/9), the Stable Shelf was almost completely uncovered with the exception of the Fayum-Gindali basin where shallow marine sediments accumulated. Toward the southwest of this basin a great river must have debouched into the sea forming the Qasr El Sagha delta. The longshore currents of the late Eocene seem to have distributed the sediments of this river toward the west; fluviatile sediments interfinger the late Eocene deposits of the Fayum Baharirya reach. The elevation of Sinai was complete. The only late Eocene record in Sinai is that of the Bir Heleifiya area (Farag & Shata 1955, Strougo, pers. comm.). This single occurrence on the eastern shores of the Gulf of Suez is probably related to the arm of the sea which covered the south Ataqa plain and which seems to have extended across the Gulf (Fig. 24/9).

During the Oligocene, remains of fluvial sediments littered the Stable Shelf areas (Fig. 24/10). Although the remains overlying the Ma'aza plateau are of questionable date, they are nevertheless correlated with the Oligocene sediments of the north on stratigraphic grounds and paleogeographic considerations. The northern outcrops of the Gebel Ahmar and Gebel Qatrani formations cover the southern borders of the Cairo-Suez-Kattaniya high in an almost continuous outcrop. They are mainly sands and gravels deposited by laterally aggrading streams. Numerous similar and isolated outcrops are recorded in the Bahariya (Radawn) and east Beni Suef areas. This indicates that the Gebel Ahmar and Qatrani fluvial deposits must have extended farther to the south. The present outcrops owe their preservation to the fact that they were consolidated under the effect of the late Oligocene-early Miocene episode of volcanic activity.

Ideas differ as to the conditions under which the Qatrani formation was deposited. Bown et al. (1982) advocate deposition in aggrading streams along a coastal lowland with mangrove swamps that gave way to at least a partly forested interior. Large tropi-

cal to subtropical trees existed in much of the Fayum area. Kortland (1980) believes, however, that the Qatrani formation was formed in a sahelian climate with no tall trees: a mosaic vegetation of medium-height, quite open dry forest and various kinds of small tree savannas, woodlands, thickets, bushlands and grasslands, with narrow strips of medium-height forest or woodland along rivers. Kortland accepts the view that the fossil trees found in the Qatrani formation were driftwood from a tropical forest lying to the south. The flora recorded offers no proof of the presence of a rain forest in Egypt during the Oligocene.

The lacustrine to paracontinental deposits found

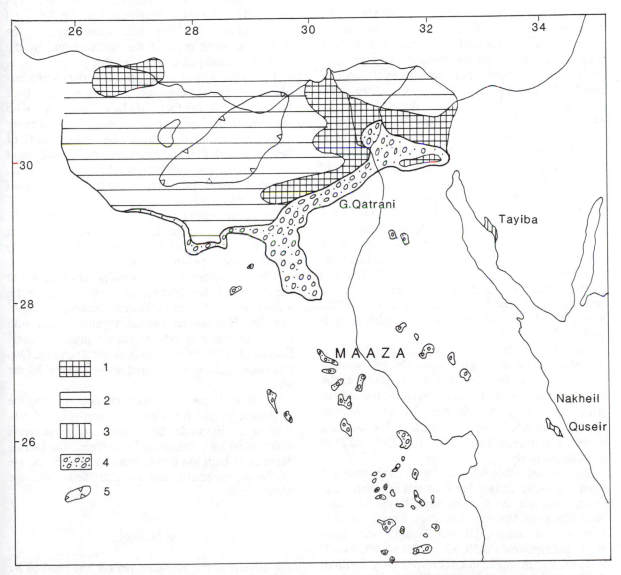

Figure 24.10 Oligocene paleogeographic map. 1= positive areas; 2= shelf sediments (Dabaa Formation); 3= lacustrine to paracontinental sediments (Nakheil and Tayiba Formations); 4= fluviatile sediments (Gebel Ahmar, Gebel Qatrani and Radwan Formations); 5= depocenters.

on both sides of the Gulf of Suez (the Nakheil and Tayiba formations) indicate that the Red Sea graben must have started forming and that the Red Sea hills were elevated during the Oligocene. The highs of the Bayuda and Nubian deserts, north Sudan, were also elevated; and it must be assumed that the drainage which produced the deltas along the shores of the Oligocene emanated from these highs. It is interesting to note that the Ma'aza plateau gravel mounds lie midway between the Nubia-Bayuda complex and the Gebel Qatrani fluvial deposits. The Ma'aza mounds could be the last remains of this defunct drainage system (see also discussion in Chapter 25).

Unstable Shelf

The Paleogene sediments of the Unstable Shelf differ from those of the Stable Shelf in being thinner and more patchily distributed. They are also of more uniform lithology and are divided into two major units: a lower marl-limestone unit of Paleocene-middle Eocene age, the Apollonia, and an upper marl-shale unit of late Eocene-Oligocene age, the Dabaa. The thickest sections of the Apollonia formation lie in the elongate east-west Abu Gharadig basin which lies immediately to the north of the Kattaniya high (Fig. 24/8). Five wells drilled in the shadow of the major northern fault which bounds this high have thicknesses of more than 1000 m. In the wells WD32 no. 1, WD33 no. 1, Natrun T56 no. 1, WD12 no. 1 and east Mubarak no. 2, the Apollonia is 1180, 1312, 1286, 1447 and 1086 m thick respectively. To the north in the Natrun South, NWD343, Zebeida, Tiba, Halif West, Fayad, Rabat, WD8, WD7, WD5, Abu Sennan, Bre23 no. 2 and the Abu Gharadig wells, the Apollonia is 886, 665, 715, 750, 854, 787, 577, 700, 839, 529, 528, 521 and 636 to 676 m thick respectively.

The Apollonia is absent or exceptionally thin along the Mediterranean coastal areas. In the western stretch of this coastal belt it is totally absent in the Kanayis, Mamura and Matruh wells. In the eastern stretch it is very thin. In Burg El Arab, Mariut, Shaltout, Dahab, Dabaa, Almaz and Abu Subeiha wells the Apollonia is 50, 33, 69, 42, 47, 47 and 21 m thick respectively.

Exceptionally thin Apollonia sections are recorded from the wells drilled between the Mediterranean coastal belt and the Qattara ridge. In the Mansour, Nasr, Marzouk, Meleiha, Shoushan, Razzak, Mingar, Kasaba, Um Baraka, Kharita, Washka and Fadda wells, the Apollonia is 59, 80, 96, 71, 66, 85, 94, 71, 72 to 88, 91, 28 and 72 m thick respectively. Medium thicknesses are noted in the western reaches, Siwa (463 m), Faghur West (411 m), Khalda (420 m) and Abu Tunis (572 m) wells.

In north Sinai, the Eocene plateau limestones are divided into a lower unit of limestone with flint (the Thebes) and an upper unit of chalky limestones (the Mokattam). The lower unit occupies some of the broad synclinal lowlands to the north and to the east of Gebel Halal as well as the areas between Halal and Yelleg, between Maghara and Yelleg, and between Um Hoseira and Kherim. The unit is similar in lithological characters to the Thebes formation of the Stable Shelf. However, the unit is much thinner when compared with its counterparts in the Stable Shelf. In Qussaima it does not exceed 30 m in thickness, and to the north of that village it assumes 50 m in thickness. In places, the Thebes overlies conformably the Esna shale and, in others, it becomes either very thin or missing as in the northwestern flanks of Gebel Arif El Naga, Ras El Gindi, Maghara, Um Khsheib and Raha. In Giddi, the Esna Shale is overlapped by the Thebes in some parts of the structure and by the Mokattam chalky limestones in other parts.

The upper unit of the plateau limestones of Sinai (the Mokattam) consists, in places, of hard limestones and, in others, of chalky limestones with many shale and marl intercalations. It has a more extensive distribution than the lower unit of limestone with flint and forms many of the tableland regions of the lowlands between the major structures. It covers almost the whole of the lowland occupied by the Ismailiya-Abu Augeila road, and it forms the tableland of Gebel um Khsheib. The Mokattam formation of Sinai is thin. It does not exceed 100 m in thickness except in a few places (e.g. Ain Gedeirat) where it attains 340 m in thickness.

The late Eocene saw the withdrawal of the sea to the north and the elevation of Sinai as well as the western part of the north Western Desert (Fig. 24/9). The late Eocene sea formed an embayment with fewer islands and positive areas than the early Eocene. The Mediterranean coastal belt and the Qattara ridge highs were lowered and inundated by the sea.

The late Eocene embayment continued during the Oligocene except for a minor retreat (Fig. 24/10). Both seas were under the influence of great rivers which must have debouched into them. The marine deposits of both the late Eocene and Oligocene are similar lithologically and are classified under the name Dabaa.

NEOGENE

The advent of the Neogene period was marked by intense tectonic movements which had a great effect on the present-day structural framework of Egypt. The eastern part, including Sinai and the Red Sea

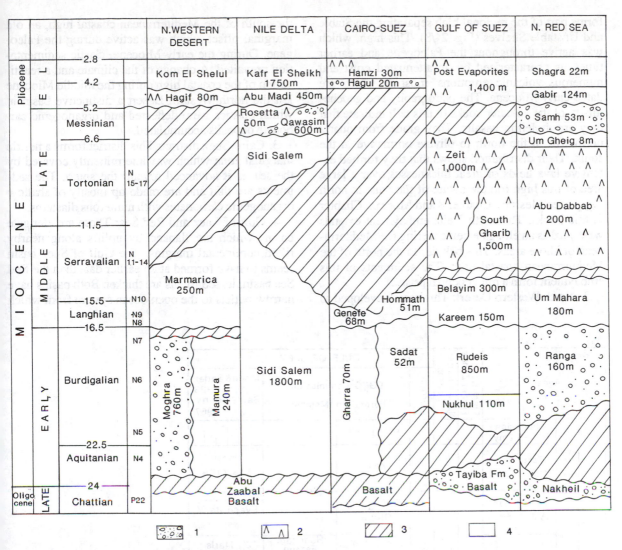

		N.WESTERN DESERT	NILE DELTA	CAIRO-SUEZ	GULF OF SUEZ	N. RED SEA

Figure 24.11 Correlation of Miocene and Pliocene rock units.

region, which had been elevated since at least the late Eocene, was affected in a most marked way. Early Miocene rifting produced the Gulf of Suez and the Red Sea grabens in more or less their present shape. Contemporaneous movements affecting the belt separating the Stable and Unstable Shelves were accompanied by intense volcanic activity.

The Miocene was terminated by a spectacular event which caused the desiccation of the Mediterranean basin, the lowering of the baselevel and the beginning of an immense erosional period which shaped the modern face of Egypt.

MIOCENE

The Miocene is the subject of a large number of publications. For a review of earlier work the reader is referred to Sadek (1959) and Said (1962). More recent work is reviewed by El Gezeery & Marzouk (1974) and El Heiny (1981, 1982). Detailed descriptions of the sediments of the Gulf of Suez, the Red Sea coastal plain and the Delta are given in other chapters of this book.

The Miocene sediments exhibit great facies variations and have a large number of unconformities reflecting the nature of the tectonically formed basins in which they were deposited (Fig. 24/11). Four distinct tectonic provinces can be distinguished:

1. The north Delta embayment: The northern part of the Nile delta forms a basin with a thick Neogene section. This basin, named the north Delta embayment by Said (1981), lies between the east Mediterranean oceanic basin and the south Delta block which

forms part of the regional high separating the Stable and Unstable Shelves (Fig. 24/8). This high, which was active throughout the Paleogene and earlier times, is characterized by an attenuated crust and numerous volcanic eruptions, most of which are dated early Miocene. The petrography and geochemistry of the volcanic rocks of this belt indicate that they are tholeiitic basalts. They grade from quartz normative in the north to olivine normative in the south. Their distribution shows that they formed extensive lava fields of great thicknesses over a large area of this high (on the nature of the crust in north Egypt, see Meshref, Chapter 8 and Morgan, Chapter 7, this book). The north Delta embayment extends westward as an elongate belt covering the Mediterranean offshore areas. This belt lies to the north of the Mediterranean coastal high. The sediments of this embayment form a miogeoclinal prism.

2. North Western Desert: This basin developed to the south of the Mediterranean coastal high, an old marginal offset which was active during the Paleogene. During the early Miocene, clastic sedimentation prevailed. A change of the climate and a reactivation of the coastal high during the middle Miocene left the north Western Desert a distinctive basin in which clastics were deflected and organogenic carbonate deposits accumulated.

3. Cairo-Suez district: This district forms a neritic marginal zone which was intermittently covered by the sea as it advanced toward the south. The sediments are thin and are made up mostly of shallow organogenic carbonates with numerous diastems.

4. Gulf of Suez and Red Sea: These are elongate basins which are flanked by uplifts along nearby rifted continental margins. The Gulf of Suez basin seems to have formed at an earlier date than the Red Sea basin; its sediments are thicker. Both basins have narrow outlets to the open ocean system from which

Ages			Planktic Foram. Zones	Nannoplankt. Zones	Group	GULF OF SUEZ		Moon & Sadek 1923 Said & Heiny 1967
						EGPC Commission Formation \ Member		
MIOCENE	Late	Messinian (6.6–5.2my)						
		Tortonian (11.5–6.6my)				Zeit		Upper Evaporite
	Middle	Serravalian (15.5–11.5my)			Ras Malaab	South Gharib		Evaporite V
		Langhian (16.5–15.5my)	N11	NN5		Belayim	Hammam Faraun	Δ Marls
							Feiran	Evaporite IV
							Sidri	γ Marls
			N10				Baba	Evaporite III
			N9		Ayun Musa	Kareem	Shagar	β Marls
							Markha	Evaporite II
								α Marls
								Evaporite I
	Early	Burdigalian (22.5–16.5my)	N8	NN4	Gharandal	Rudeis	U	Mreir
								Asl
			N7				L	Hawara
				NN3				Mheiherat
			N6	?		Nukhul		
			N5					
		Aquitanian (24–22.5my)	N4					
OLIGO-CENE		Chattian	P22			Tayiba redbeds		

Table 24.4 Miocene rock units of the Gulf of Suez.
Planktic Foram. Zones: P22= *Globigerina ciperoensis*; N4= *Globorotalia kugleri*; N5= *Globigerinoides primordius*; N6= *G. altiaperturus*; N7; *G. trilobus trilobus*; N8= *G. sicanus/Praeorbulina glomerosa*; N9= *Globorotalia peripheroronda*; N10= *G. peripheroacuta*; N11= *G. sikaensis*.
Nannoplankton Zones: NN3= *Sphenolithus belemnos*; NN4= *Helicosphaera ampliaperta*; NN5= *Sphenolithus heteromorphus*.

they were separated by sills, the Suez (Ayun Musa) high in the north and the Bab El Mandab in the south.

Geochronology

The numerous radiometric dates of the Neogene volcanics are of little use in establishing a geochronological scale for the Neogene. Most of the dates are for non-layered volcanic rocks which fall mostly within the Oligo-Miocene interval. The geochronology of the Neogene is based, to a large extent, on paleontological as well as stratigraphical evidence. The large collection of macroinvertebrates recorded from the Miocene of Egypt (Blanckenhorn 1900, 1901, Fourtau 1920, Sadek 1959, Said & Yallouze 1955, etc.) has not been successfully used to zone the Miocene rocks. Mention has frequently been made of the cephalopod *Aturia aturi* as an index of the Langhian. Successful taxa used in dating the Neogene rocks of Egypt are the planktic foraminifera and calcareous nannoplankton. Following on the early attempts to zone the Miocene section of the Gulf of Suez by planktic foraminifera by Said & Heiny (1967), Wasfi (1969) and Beckmann et al. (1969) make the first comprehensive attempt to establish standard zones applicable to Egypt. El Heiny & Martini (1981) and Andrawis (this book, Chapter 31) refine the zonation by emending some zones and inserting a few additional ones. Table 24/4 gives the ranges of these zones as proposed by these authors. Wasfi and Gupco staff (as quoted in Scott & Govean 1984), however, give different ranges for these zones. Thus the Kareem and Belayim formations belong to Zone N9 according to Wasfi and Gupco staff and Zones N9 to N11 according to El Heiny & Martini. The Nukhul belongs to Zone N4 to N5 according to Wasfi and to Zone N6 according to El Heiny & Martini.

For the calcareous nannoplankton, the works of El Heiny & Martini (1981) and Harms & Wray (this book, Chapter 17) establish the standard zones pertinent to Egypt (Table 24/4).

For the neritic sediments which do not carry planktic foraminifera, attempts were made to establish the contemporaneity of some of the characteristic benthic foraminiferal species which they carry with the standard planktic zones (Souaya 1961, 1963, Sadek Ali 1968, Cherif 1974). Souaya established the *Miogypsina globulina*, *intermedia*, *cushmani* succession to which a lower zone of *M. tani* was added by Sadek. These zones were equated with N7, N8, N9 and N4 planktonic foraminiferal Zones respectively. Cherif also suggests that *Heterostegina costata costata* could be a good index fossil for the Burdigalian and that *H. costata politatesta* could be an index fossil for the Langhian.

EARLY MIOCENE

The earliest Miocene sediments of Aquitanian age are of limited areal distribution. They are recorded with certainty in the north Delta embayment wells. Wherever their base was reached they were found to rest conformably over the marine Oligocene sediments (Bilqas, San El Hagar, Qantara and El Temsah wells). Aquitanian strata are also recorded in one locality (Gebel Homeira) in the Cairo-Suez district (Sadek Ali 1968). Here they assume a reefal carbonate facies and carry the characteristic fossil *Miogypsina tani*.

The maximum marine transgression of the Miocene epoch occurred during the Burdigalian when the sea covered large areas of north Egypt and overflowed into the newly-formed Gulf of Suez (Fig. 24/12). A large part of the transgressing sea was under the influence of fluvial sedimentation forming a wave-dominated delta plain covering the eastern part of the north Western Desert. In the following paragraphs a description is given of the sediments of the different provinces.

North Delta embayment (Sidi Salem formation)

The most complete section of the early Miocene is found in the north Delta embayment and its extension in northwest Sinai to the north of the Pelusium line. Here the sediments assume thicknesses of 600 to 1000 m and are made up of a solid clay unit. They rest conformably over the marine Oligocene and are rich in calcareous nannoplankton (NN1 and NN2 Zones) and planktic foraminifera (Zones N4 and N5). They show breaks in sedimentation in some wells toward the end of the early Miocene (see Harms & Wray, Chapter 17, this book).

North Western Desert (Gebel Khashab and Moghra formations)

A thick clastic sequence carrying a mixed fluvio-marine fauna overlies the long basalt of the Abu Roash-Fayum stretch and extends northward into the north Western Desert. It forms the lower part of the wall of the Qattara depression where at Moghra, at the eastern end of the wall, Fourtau (1918) separated and described the well-known vertebrate fauna of this locality. The section at Moghra forms the type locality of this formation. Said (1962) assigns an early Miocene age for this sequence and shows that it grades laterally into other facies. The name Moghra was originally proposed to cover all the facies exhibited by the early Miocene sediments of the north Western Desert. Said (1962) distinguishes an 'estuarine', marine, reefal and open bay facies. In addition, a genuine fluviatile facies, given the name Gebel Khashab Redbeds, is also recognized. Marzouk

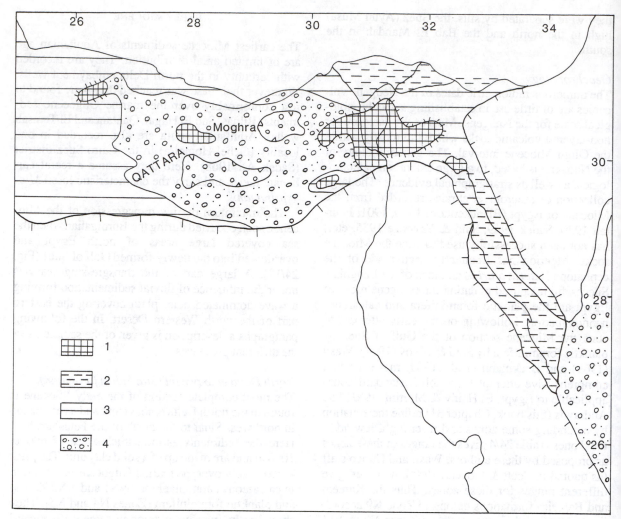

Figure 24.12 Early Miocene paleogeographic map. 1= positive areas; 2= bathyal sediments; 3= shelf sediments (Mamura and Gharra Formations); 4= shelf sediments influenced by fluviatile action (Moghra and Ranga Formations).

(1969) restricted the use of the name Moghra to the outcropping 'estuarine' clastics and proposed the name Mamura for the marine, reefal and open bay sediments. These sediments occur in the subsurface in wells drilled to the west of longitude 27° E where the influence of the fluvial sediments of the Moghra delta is not felt. The type section of the Mamura formation is from depth 114 to 399 m, Mamura well no. 1. The basal 40 m made up of sandy fossiliferous limestones are given the name Shushan by Omara & Ouda (1972), while the upper shale-limestone section is given the name Khalda by these authors. The sediments of the Mamura formation are of shallow marine origin. The Mamura rests over the marine Oligocene (Fig. 24/5) and is marked by a number of sedimentational breaks. The lower part of the section carries the characteristic fossil *Miogypsina tani* and

belongs to the Aquitanian (see Hantar, Chapter 15, this book).

The Gebel Khashab Redbeds and the Moghra seem to have been deposited under one regime belonging to a river system which must have flowed into the sea from the north of Fayum. The sediments of this system cover about 70,000 km². The Gebel Khashab Redbeds are made up of coarse non-fossiliferous clastics which were probably deposited in aggrading streams. These sediments grade northward (at the latitude of Wadi El Faregh, southwest of Wadi Natrun) into delta plain deposits of the Moghra formation. The Moghra is made up of finer-grained clastics which carry, in places, shark teeth and other terrestrial vertebrates.

The Moghra seems to have been deposited in a high-energy wave-dominated delta. Strandline sands make the bulk of the delta plain, with local occurrences of distributary channel sands. The sands form elongate bodies whose orientation parallels depositional strike. Since nearshore marine fossils are associated with the deposits of the delta, it is assumed that the progradation must have occurred seaward and along the entire delta front. The fossils separated from these sands are ostracodes, euryhaline foraminifera, echinoids, molluscs, shark teeth, turtles, mastodons and many others. Occasional silicified tree trunks are found. These seem to have been rafted.

The thickness of the Moghra formation is in the range of 200 m but to the north, where it becomes buried under younger sediments, it is thicker. At Dabaa, Almaz wells the Moghra is 738 and 651 m thick respectively. The size of this delta and the thickness of the sediments clearly point to a river of great competency terminating in a shelf area. The earlier Oligocene Gebel Qatrani river did not build a delta of comparable dimensions and its sediments were distributed on the submerged Suez-Kattaniya high which seems to have acted as a barrier to the open sea. The sediments of the Moghra and Gebel Qatrani rivers are similar mineralogically; the two rivers seem to have had the same source and were

probably formed under similar climatic conditions. It is certain that the climate was tropical to subtropical but it is not certain how much rain the area itself received. The pedologic evidence points to wetlands which could have developed in delta expanses under different climatic conditions. Most of the *in situ* biota recorded points to a tropical climate and African affinities. The tree trunks could well be drift wood (see Kortland 1980 and Bown et al. 1982 for a discussion on this controversial subject). The paleogeographic evidence points to a river system which derived the largest part of its waters from a source farther to the south, probably the elevated Nubian Massif (see discussion in Chapter 25, this book). This conclusion is based on the fact that no fluvial sediments accumulated in any of the basins of Egypt during these two epochs other than the basins which received the deposits of these two river systems.

Cairo-Suez district (Gharra formation)

Shallow marine shelf deposits are recorded along the Cairo-Suez road (Fig. 24/13). They are made up of detrital limestones carrying *Operculina, Heterostegina, Miogypsina* spp., and other reef-building forms. The early Miocene part of the section belongs

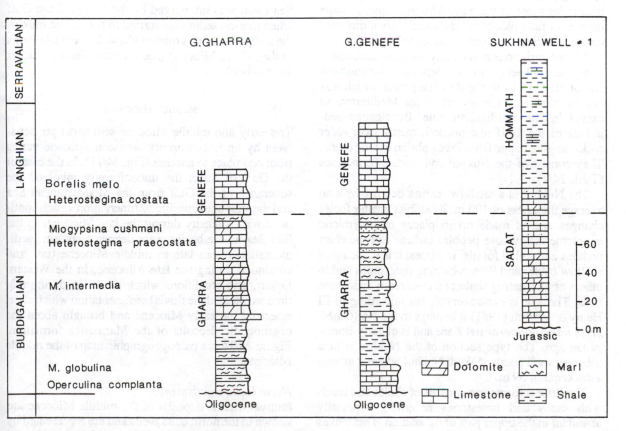

Figure 24.13 Correlation Miocene sections along Cairo-Suez Road.

to the Gharra and Sadat formations. The 120 m thick Gebel Gharra section is the subject of a study by Souaya (1961, 1963) who is able to distinguish three *Miogypsina* zones: *M. globulina*, *M. intermedia* and *M. cushmani*. These fossils allocate the Gharra to the early Burdigalian, late Burdigalian and Langhian respectively (Cherif 1974). The Gharra Burdigalian strata are also rich in *Operculina complanata* and the Langhian strata carry *Hetrostegina praecostata*. Burdigalian strata are also reported from Dar El Beida, Homeira, Agrud and Sadat areas of the Cairo-Suez district.

No Aquitanian is recognised from the Cairo-Suez district (El Heiny 1982) although Ali Sadek (1968) dates the lowermost beds of Gebel Homeira (where specimens of *Miogypsina tani* were separated) as Aquitanian.

To the east toward the entrance of the Gulf of Suez, the early Miocene sediments assume a carbonate facies, the Sadat formation (Abdallah & Abdel Hady 1968).

Gulf of Suez (Nukhul and Rudeis formations)
With the advent of Burdigalian time, the Gulf of Suez was submerged by the waters of the Mediterranean Sea. The free flow of the waters across the Suez sill resulted in an aerated bottom and clean bathyal sediments for most of the early Miocene time. A large number of fault blocks build the Gulf. Their different rates of subsidence contributed to the large thickness and facies variations of the early Miocene sediments. A regional sill seems to have separated the southern end of the Gulf from the Red Sea proper which was not overflown by the waters of the Mediterranean except in late Burdigalian time. Burdigalian sediments of the Gulf rest unconformably over older rocks ranging in age from Precambrian to Oligocene. They consist of the Nukhul and Rudeis formations (Table 24/4).

The Nukhul is a shallow marine deposit with an average thickness of 100 m. It exhibits greater facies changes and is made up in places of polymictic conglomerates whose pebbles include Eocene chert nodules and rolled fossils. In places, it is made up of shallow oyster and Pecten-bearing limestones and in others of alternating shale and calcareous sandstone beds. The age is controversial, but according to El Heiny & Martini (1981) it belongs to the N6 (*Globigerinoides altiaperturus*) Zone and is of early Burdigalian age. The type section of the Nukhul is in a tributary wadi south of Wadi Nukhul where it attains a thickness of 59 m.

The Rudeis formation consists of shales and marls with sands and limestones becoming especially abundant in the upper part of the section. This makes possible the subdivision of the Rudeis into a lower and an upper unit; the upper usually rests unconformably ovr the lower. The Rudeis is thick; thicknesses of 2000 m are common. It is richly fossiliferous and belongs to the N7 and N8 *Globigerinoides trilobus trilobus* and *G. sicanus/Praeorbulina glomerosa*) Zones (El Heiny & Martini 1981).

In Gebel Zeit-Esh El Mellaha area, a unique sequence of the early Miocene is recorded and named Abu Gerfan by Ghorab & Marzouk (1967). It is made up of coarse to very coarse calcareous conglomerate boulder beds alternating with gritty coralline limestone beds. The thickness is 20 m at the type locality at Gebel Zeit. This formation may be correllatable with Scott & Govean's (1984) 'basal carbonate' bed (classified with the Nukhul) which they record along the south Suez coast.

Red Sea (Ranga formation)
During the early Miocene, the Red Sea seems to have been separated by a sill from the Gulf of Suez and seems to have been filling up with a thick clastic deposit, the Ranga. This formation is made up of a basal conglomerate, about 8 m thick, followed by a long series of vari-colored sands and shales. Thicknesses of 180 m are common. The formation is clearly made up of subaerially deposited alluvial fans. Toward the end of the early Miocene the Red Sea basin was submerged by the waters of the Gulf when marine sediments started to form. The bulk of the sediments of this marine phase, however, belongs to the middle Miocene (see for more detail, Chapter 18, this book).

MIDDLE MIOCENE

The early and middle Miocene sediments are separated by an unconformity whose magnitude varies from one place to another (Fig. 24/11). In the case of the Gulf of Suez, the unconformity involved the severance of the Gulf from the Mediterranean Sea and the start of evaporitic sedimentation which continued with intensity during the late Miocene. In the Red Sea whose basins were deeper, evaporitic sedimentation began late in middle Miocene time and continued during the late Miocene. In the Western Desert, arid conditions which prevailed during that time terminated the fluvial sedimentation which characterized the early Miocene and brought about the organogenic deposits of the Marmarica formation. Figure 24/14 is a paleogeographic map of the middle Miocene.

North Delta embayment
Bathyal marine deposits of the middle Miocene are known in the north delta wells and are represented by shales of more than 1000 m in thickness (San El

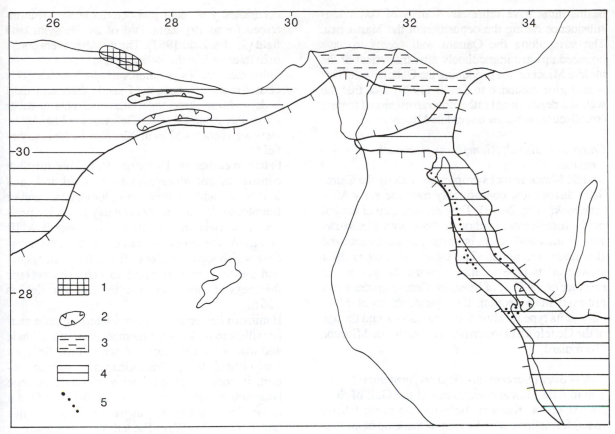

Figure 24.14 Middle Miocene paleogeographic map. 1= Positive areas; 2= depocenters; 3= bathyal sediments; 4= shelf sediments; 5= Nullipore reefs (Hammam Faraun Member and Hommath Formation).

Hagar well). This well seems to lie at the foot of the great south wall which borders the north Delta embayment (Said 1981). In the Monaga well, barely 20 km to the south, the middle Miocene becomes thin and assumes a carbonate facies similar to that known from the Cairo-Suez district.

North Western Desert (Marmarica formation)
Middle Miocene sediments of the north Western Desert follow unconformably on top of the Moghra formation. They are represented by the Marmarica limestones which cover the entire north Western Desert (Fig. 24/14) extending from the Libyan borders to Siwa eastward along the Qattara wall. The formation is made up of an upper white limestone fossiliferous member, a middle snow-white chalk member and a lower member made up of alternating cross-bedded carbonates and fissile shales and marls. These members are named El Diffa plateau, Siwa and Oasis respectively by Gindy & El Askary (1969). At the type section to the north of Siwa, Said (1962) describes a 78 m thick section the lower 32 m of which include some marl intercalations. The section is

richly fossiliferous including among others the characteristic Langhian fossil *Borelis melo*.

The thickness of the Marmarica is more or less uniform throughout except at a few localized areas where it becomes exceptionally thick. The average thickness outside these areas is less than 200 m. Wells drilled in the narrow strip overlooking the Qattara wall from Sanamein to Marzouk have exceptionally thick sections. In the Sanamein, Tarfa, Ghanem, Garf, Obeidalla and Marzouk wells, the thickness of the Marmarica is 802, 753, 714, 1070, 530 and 642 m respectively. To the north along another narrow and elongate strip exceptional thicknesses are recorded along an east-west line passing through the Kheima, Ganayen, Yidma and Washka wells. Thicknesses recorded in these wells are 639, 900, 1113 and 975 m respectively. It is remarkable that in the narrow strip separating these two basins thicknesses of less than 100 m are recorded. Among other localized areas with exceptionally thick Marmarica sediments is the Um Baraka basin where it attains a thickness of more than 500 m. The only explanation that I can offer for these exceptional thicknesses is that these elongate basins must have represented areas of continuous subsidence during the deposition of the Marmarica. The strip along the Qattara wall seems to have bounced upward immediately after the retreat of the middle Miocene sea producing the Qattara wall. This would give credence to Said's (1979) idea that the wall is a depositional rather than an erosional feature (see discussion below under late Miocene).

Cairo-Suez district (Genefe and Hommath formations)

Middle Miocene rocks cropping out along the Cairo-Suez district rest conformably over the early Miocene rocks (Fig. 24/13). They are made up of detrital richly fossiliferous carbonate beds with *Heterostegina costata* and others belonging to the Genefe and Hommath formations. The Genefe is a 69 m-thick section at its type locality. Toward the east at the entrance of the Gulf of Suez, the Genefe grades into a grit-marl-shale section, the Hommath about 51 m thick at its type locality. Some authors extend the age of the Genefe and Hommath to the early late Miocene (Tortonian).

Gulf of Suez (Kareem and Belayim formations)

The middle Miocene sediments of the Gulf of Suez consist of the Kareem, Belayim and south Gharib (pars) formations. The Kareem is made up of clastics with one or more interbeds of anhydrite and occasional limestone. It is 260 m thick at the type locality (Gharib north well no. 2). In most areas it averages 110 m in thickness. It is subdivided into two members, a lower member, the Markha and an upper member, the Shagar. The Markha is made up of a basal anhydrite bed, a middle shale-marl sequence and an upper anhydrite-sand-shale sequence. These sequences correlate with Evaporite I, α interevaporite marls and Evaporite II of Said & Heiny's classification (1967). The Markha is developed along the western banks of the Gulf where evaporite sedimentation seems to have started. In many of the eastern parts of the Gulf where Evaporite I is not formed, the separation of the Rudeis formation from the overlying Kareem formation is difficult; the upper Rudeis and the Kareem marls and sands form one unit which is named the Ayun Musa formation. The Shagar member is a shale-marl unit with sand and conglomerate interbeds which become common toward the shoulders of the graben. The Kareem carries Pay Zones IV-A and V of the Belayim land oilfield.

The Belayim formation is 302 m thick at the type section (Belayim well 112-12). It is made up of four members which are from bottom to top:
- Baba member (= Evaporite III): This member is made up of anhydrite and salt with a clastic interbed made up usually of shale and marl and occasionally of sand and conglomerate. This interbed forms Pay Zone IV-I of the Belayim land field (Said & Zaki 1967). The member averages 51 m in thickness in the Belayim field.
- Sidri member (= γ interevaporite marls): This clastic member consists of sand, shale and marl beds with rare limestone interbeds. This member forms Pay Zone IV of the Belayim land field. The member averages 57 m in thickness in the Belayim field.
- Feiran member (= Evaporite IV): This member consists of anhydrite with shale, marl and sand interbeds and rarely conglomerates and/or limestones. In this unit several pay zones are present in the Belayim land field: Pay Zones II-A, III and III-A. In the eastern part of the field the anhydrite which separates Zones II and II-A pinches out and the two zones merge into one. The average thickness of this member in the Belayim field is 156 m.
- Hammam Faraun member (= δ interevaporite marl or nullipore rock): This member consists of shale and marl or shale, marl and sand in the Belayim land oilfield. In the Wadi Gharandal surface section, it consists of massive gypseous limestone beds rich in *Lithothamnion*. In the Belayim field it carries Pay Zone II. The average thickness of this unit in the field is 50 m. The nullipore rock assumes a carbonate facies along the western coast of the Gulf of Suez (Fig. 24/14).

The Kareem and Belayim formations are of Langhian age. They belong to the planktic foraminiferal zones N9 to N10 and N11 respectively.

In the Gebel Zeit-Esh El Mellaha area a unique facies of the middle Miocene is recorded and named Gharamul by Ghorab & Marzouk (1967). The Gharamul formation is made up of algal unbedded limestone, occasionally dolomitic. The thickness at the type section at Gebel Gharamul is 126 m. The Gharamul may be the reefal equivalent of the Hammam Faraun member of the Belayim formation.

Red Sea (Um Mahara formation)

The Red Sea basin was inundated during the late early Miocene and middle Miocene times when it received the sediments of the Um Mahara formation. This formation is made up of a lower sandy member and an upper fossiliferous gypseous limestone member, the latter grading laterally into coralline limestones. The formation is of middle Miocene age at least in its upper part. This part carries the fossil *Borelis melo* which characterizes many of the Langhian sections of Egypt. The Miocene sediments of the Red Sea basin are dealt with in more detail in Chapter 18.

LATE MIOCENE

A continuous withdrawal of the sea from Egypt took place during the late Miocene. By Messinian time not only was the land of Egypt completely uncovered, but so was the whole Mediterranean Sea as it severed its connection with the world oceanic system. The impact of this event was enormous in shaping the modern landscape of Egypt. The Nile excavated its modern course and the oases and other depressions were formed in adjustment to the new lowered base-level of the Mediterranean. The late Miocene was an episode of erosion with few sediments preserved. These are mostly evaporites which accumulated in the Gulf of Suez, Red Sea and north Delta embayment. Coarse grained clastics accumulating in front of the forming Nile are also recorded from the subsurface.

North Delta embayment

Late Miocene sediments are recorded from most of the wells drilled in the embayment. They rest over the sahles of the middle Miocene Sidi Salem formation forming a wedge of alternating beds of poorly sorted coarse-grained quartzose sandstones and fine-grained shales. The coarse-grained layers include chert pebbles and rolled fossils which seem to have been derived from the elevated Cretaceous and Eocene table lands of Egypt. The formation was named Qawasim and is interpreted as forming the deltaic fan of the Eonile (Said 1981). The type section of the Qawasim formation is the interval between 2800 and 3765 m of the Qawasim well. The formation is thin over the hinge zone of the delta which separates the embayment from the south Delta high (Qantara well (420 m), San El Hagar well (513 m), Mahmudiya well (250 m), etc.) and thicknest at the footslopes of this zone (Kafr El Sheikh well (1313 m), Bilqas well (1336 m), Sidi Salem well (1275 m), etc.).

The age of the Qawasim formation is difficult to ascertain because of the lack of diagnostic fossils, but it is most likely of late Tortonian to early Messinian age. This age assignment is inferred from the stratigraphic position of the formation which overlies unconformably the well-dated early Tortonian sediments of the Sidi Salem formation and underlies the evaporite beds which are correlatable with similar deposits recorded in the Mediterranean basin and dated as Messinian.

The evaporites which overlie the Qawasim were named the Rosetta formation (Rizzini et al. 1978). This formation is encountered in most of the wells drilled in the embayment at the same level recorded beneath the Mediterranean Sea by the Deep Sea Drilling Project Cruise leg 13 at depths ranging from 3963 m (Kafr El Sheikh well) to 2059 m (El Tabia well) below sea level. The type section of this formation is the interval 2678-2720 m of the Rosetta well no. 2. Here the formation is made up of layers of anhydrite interbedded with clay. In the Qawasim well, the Rosetta covers the interval 2651-2800 m. It is made up of a basal 7 m-thick gypsum bed followed by a sequence of quartzose sands, shales and evaporites. The age of the Rosetta evaporites is Messinian. They are interpreted as having been formed as sabkhas on a dried sea floor.

North Western Desert

The sheet of halite which fills the deeper southwestern part of the Qattara depression (Fig. 2/4) has been recently interpreted as a remnant of a sheet of evaporites which probably formed during the Messinian in a perched basin (Said 1979). The basin seems to have formed in the shadow of the rising Qattara ridge which received the thickest middle Miocene sediments.

Gulf of Suez

The evaporite section which overlies the Belayim formation is partly of late Miocene age, although most of the section is of late middle Miocene age. The section consists of the south Gharib and Zeit formations. The surface outcrops of these formations are considerably thinner than their counterparts in the subsurface. The thickness of the evaporite formations increases from zero in the north and on local structural highs to more than 3000 m in the southern part of the Gulf (Fawzy & Abdel Aal 1984).

The south Gharib formation overlies the Belayim formation. It is made up of anhydrite with shale intercalations at the periphery of the Gulf and mainly of salt in the depocenters. The type section of this formation is the south Gharib well no. 2, where it is represented by a 701 m-thick anhydrite bed with salt and shale intercalations. The formation is mainly non-fossiliferous although some of the lower shale intercalations carry foraminifera (?Serravalian affinities). The thickness of the formation varies from zero in the northern part of the Gulf to more than 1800 m in the south. Isopachous variations are due, to a large extent, to the paleorelief inherited after the post Belayim movements which activated many of the blocks that made up the Gulf. Fawzy & Abdel Aal (1984) divide the formation into a lower unit of salt with minor intercalations of anhydrite, shale and sands. This member attains its maximum thickness toward the south partially due to the plastic flow of the salt. The upper unit is made up of thinly-bedded salt and anhydrite interbeds.

The Zeit formation follows on top of the south Gharib formation. It is made up of shale, sand and sandstone beds with anhydrite and salt interbeds. The type section of this formation is the Gebel Zeit well no. 1 where the formation attains a thickness of 941 m. The formation is non-fossiliferous and the age is determined on stratigraphic grounds.

Red Sea

The late Miocene sediments of the Red Sea are reperesented by an evaporite sequence which was named Abu Dabbab by Akkad & Dardir (1966). The type section of this formation is at Wadi Abu Dabbab, north of Mersa Alam, where the formation rests over the Um Mahara formation. The thickness of the formation varies from 30 to 400 m. It is made up of massive gypsum beds with minor carbonate and shale intercalations. In the offshore areas of the Red Sea, the formation may exceed 4000 m. The formation is usually capped by a dolomitic limestone bed, the Um Gheig formation, referred to in the older literature as the oil-tainted limestone.

The Abu Dabbab and the Um Gheig formations are overlain by a sequence of fine-grained clastics of the Samh formation. The sequence is about 53 m thick in Wadi Wizr. It carries a few poorly preserved fossils of fresh water habitat which are not age diagnostic. The formation is attributed to the late Miocene on stratigraphic and paleogeographic evidence. The sediments could represent the fresh water lacustrine deposits of the retreating late Miocene sea. The detailed description of these formations is given in Chapter 18, this book.

PLIOCENE

The advent of the Pliocene epoch was marked by the flooding of the Mediterranean basin across the Gibralter Strait and the gradual inundation of the north Delta embayment, the northern coastal areas and the Eonile canyon by this sea. The Gulf of Suez and the Red Sea, which had been isolated from the Mediterranean, were connected with the Indian Ocean across the Bab El Mandab Strait. The late Pliocene saw a withdrawal of the seas and a dramatic climatic change that brought local rains. In the following paragraphs a brief description is given of the sediments of each of these realms.

Nile valley

The Pliocene sediments of the Nile valley consist of a lower marine sequence of early Pliocene age and an upper fluviatile sequence of late Pliocene age.

Early Pliocene marine sequence

The rising sea level of the early Pliocene brought the Mediterranean into the excavated Nile canyon transforming it into a narrow gulf reaching as far south as Aswan. In the north, with the rise of sea level, the waters overflowed the banks of the canyon and covered large tracts of the areas surrounding the delta and the eroded south Delta block. The early Pliocene sediments of the north Delta embayment are thick sand-shale deposits which carry a rich open marine fauna. The planktic foraminifera belong to the *Sphaeroidinellopsis* spp. zone. The sediments follow on top of the late Miocene or older deposits with a marked unconformity which can be traced on all the seismic records with ease. The sediments include more sand members in their lower part (the Abu Madi formation) and are made up of shales in the upper part. These early Pliocene shales form the lower part of a thick and uniform sequence of shales named the Kafr El Sheikh formation. The type locality of this formation is the depth interval 975 to 2735 m of the Kafr El Sheikh well. The lower 855 m belong to the early Pliocene planktic foraminiferal zones N19 & N20. The upper 905 m include a brackish water foraminifera and belong to the late Pliocene (*vide infra*).

In the Nile gorge itself the deposits of the deeper parts are in the form of sands (18 to 59%) and montmorillonitic clays with thin lenses of fine-grained polymictic sands and sandy loams rich in authigenic minerals: glauconite, pyrite and siderite. Toward the peripheries of the canyon the deposits are in the form of sandy limestones, marls and coquinal beds carrying shallow marine fossils; the beds abut against the bounding cliffs of the canyon. The fossiliferous beds of this episode (best exposed in Kom El Shelul, south of the Gizeh Pyramids) rest on the slipped masses of the Eocene or, more frequently, on the Eocene bedrock itself. An intervening band of conglomerate or breccia up to 3 m in thickness separates the marine Pliocene from the underlying bedrock. Little (1936) and Said (1981) map the Pliocene beds of this gulf. According to Little, the average width of the gulf was 12 km but the arms of the sea extended inland to some distance on either side (see map given by Said 1981 for the Nile valley).

The type section of the marine deposits of the Nile gulf is at Kom El Shelul where it is made up of a basal 10 m oyster bed made almost exclusively of *Ostrea cucullata* shells. This is followed by a 2 m sandstone bed crowded with *Pecten benedictus* and *Chlamys scabrella*. Upon this lies a sandstone bed of 50 cm thick full of remains of *Clypeaster aegyptiacus*, casts of large gastropods (*Strombus coronatus, Xenophora infundibulum,* etc.), the same pectens in the underlying bed, and many other fossils. Then follows a 10 m bed of non-fossiliferous yellow quartzose sand and brownish sandstone with large flint pebbles.

The most southern occurrence of the marine Pliocene of the Eonile canyon is reported from Aswan. Chumakov (1967) describes from the cores raised from the Nile bed at the site of the Aswan High Dam a marine Pliocene sequence between depths 170 and 260 m. The sequence is made up of grey montmorillonitic clay with thin lenses of fine-grained maicaceous sand and sandy loam with abundant plant detritus and a few specimens of ostracodes.

Late Pliocene fluviatile sequence

The onset of more humid conditions in late Pliocene time converted the marine gulf of the Nile into a veritable river channel to which Said (1981) gave the name Paleonile. The sediments belonging to this river system consist of a long series of interbedded red-brown clays and thin fine-grained sand and silt laminae which crop out along the banks of the Nile. In the north Delta embayment sediments belonging to this river system are reported from the subsurface. They form the upper part of the Kafr El Sheikh formation. This part carries a fluvio-marine fauna which seems to have been deposited under the effect of fresh water.

To the same river system belong the fluvio-marine deposits of Wadi El Natrun which seem to have formed the earliest delta of the Paleonile before this river built its complex delta system into the Mediterranean offshore (Said 1981). The Wadi Natrun fluvio-marine deposits attracted the attention of many authors (Blanckenhorn 1901, Lyons 1906, Sromer von Reichenbach 1905, James & Slaughter 1974, etc.). The Wadi Natrun fluvio-marine section is best exposed at Gar El Muluk, a prominent topographic feature in the flat surrounding terrain of Wadi Natrun where it attains a thickness of 33 m. The formation is made up of a sequence of shales, calcareous sandstones and thin limestone interbeds. It includes a rich vertebrate fauna (*Hipparion, Hippopotamus* and *Hippotragus* spp. and others). The lowermost bed of the succession carries the middle Pliocene fossil *Ostrea cucullatta*. The Gar El Muluk beds carry a rich ostracode fauna of late Pliocene age.

Coastal areas of north Western Desert

The distribution of the early Pliocene sediments shows that, in addition to the delta and the Nile canyon, the Pliocene sea covered large tracts of the lands around the modern delta especially along its western edge, but did not overlap except very small fringes of the present day coast of Egypt. Along this coast, a 25 m-thick pink dolomitic limestone sequence overlies the Marmarica and underlies the oolitic limestone Kurkar ridges of the Mediterranean coast (see also Chapter 25). Fossils are rare but include a few benthonic foraminifera and abundant

stromatolitic calcareous algae. The formation was named Dabaa by Abdallah (1966). This name is now reserved for the late Eocene-Oligocene sequence encountered in the subsurface in many wells of the north Western Desert (*vide supra*). The authors of the new goelogical map of Egypt named this formation Hagif, after the type to the northwest of Wadi Natrun. In this locality the formation becomes thicker (80 m) and includes workable gypsum beds.

Cairo-Suez district

In the Cairo-Suez district and in the depression between Gebel Ataqa and Gebel north Galala, a 20 m-thick sequence of calcareous sandstone beds overlies unconformably the marine middle Miocene. The sequence, previously named non-marine Miocene (Shukri & Akmal 1953) is now named Hagul. It is non-fossiliferous although it carries a large number of rolled fragments of silicified tree trunks. The type locality of this formation is at triangulation point 92, Wadi Hagul (Abdallah & Abdel Hady 1968). Early authors dated the formation as (?) late Miocene but in the present work it is given, together with the overlying Hamzi formation, a late Pliocene age. Since the Hagul is a fluviatile deposit it is difficult to conceive it as forming during the arid Tortonian age (when evaporites were forming) or during the wet Messinian age (when downcutting and erosion took place).

The Hagul is capped in many places by a sequence of white porcellaneous limestone beds with layers of flint. These beds also occur across the Nile in the area to the west of Cairo where at Gebel Hamzi, they reached their maximum thickness of 35 m. The formation, named Hamzi (Said 1971), is usually thin. Poorly preserved fossils, belonging mostly to the fresh water snail *Pirenella*, occur in places. The fossils are not age diagnostic but indicate deposition in fresh water lagoons. The top 2 m-thick sandy limestone bed with flint bands, which caps the Hagul at its type locality, may be equivalent to the Hamzi formation. This may show that the Hagul and Hamzi form one sequence that could have developed during the wet phase of the late Pliocene.

Gulf of Suez

A section of clastics and thin evaporites follow on top of the Zeit formation. It is dated Pliocene to Recent. Isopachous and lithological variations are large. In Wadi Gharandel and Gebel Zeit surface sections, the lower part of the post evaporite section carries the large *Ostrea gryphoides*, several mollusc and echinoid spp. of decided Indo-Pacific affinities. The upper part of the surface sections is made up of clastics of seeming terrestrial origin. The lower part belongs to the early Pliocene when the rising waters of the Gulf came from the south; the upper part was deposited in the regressive phase of the late Pliocene and later times.

Considerably thicker sections of more than 1000 m are encountered in the wells drilled in the Gulf. They are of marine to lagoonal origin. Fawzy & Abdel Aal (1984) distinguish two facies in the Gulf which they named Gihan and Morgan. The Gihan facies is made up mainly of sands with thin and minor anhydrite beds. The Morgan facies consists primarily of a lower unit of thin anhydrite and shale beds, a middle salt bed and an upper unit of pisolitic limestones and sands. The Gihan sand facies seems to have developed as submarine delta fans in front of the mouths of the wadis that seem to have debouched into the Gulf during Pliocene and Pleistocene times.

Red Sea

Pliocene sediments of the Red Sea coastal areas consist of the marine calcareous sandstone sections of the Gabir and Shagra formations. These formations are characterized by the appearance of the first elements of Indo-Pacific faunas. After severing its connection with the Mediterranean, the Red Sea basin was inundated by the waters of the Indian Ocean which flowed across the Bab El Mandab Strait. The Gabir is 124 m thick in its type locality and is made up of a lower 44 m-thick section of sandstones followed by an 80 m-thick section of sandstone, marls and reefal limestones. It is rich in corals, oyster and Pecten shells and other fossils. The Shagra is 22 m thick at its type locality and is made up of arkosic sandstones and minor marls rich in the echinoids *Clypeaster scutiformis* and *Laganum depressum*. The detailed description of these formations and their mode of origin is given in Chapter 18, this book.

CHAPTER 25

Quaternary

RUSHDI SAID

Consultant, Annandale, Virginia, USA and Cairo, Egypt

The Quaternary sediments of Egypt have recently been subjected to intensive studies (see Said 1975, 1980, 1981, 1983, Haynes 1980, 1982, Wendorf & Schild 1976, 1980, 1986, Pachur & Braun 1986 and the abstracts of papers of symposia on Quaternary and development, Mansoura 1982-1987). They lie unconformably over the Pliocene or older sediments in the Nile Valley and the surrounding deserts. Of these two environments, the Nile trough possesses the more complete record of the Quaternary in Egypt where the sediments assume great thicknesses and are divisible into units which are unconformable with one another. In the deserts, however, which are the sites of intense erosion, the Quaternary sediments are thin and incomplete. The correlation of the sediments of the different environments is difficult because of the presence of great gaps in the sedimentary record and because the precise age of most of the sediments is unknown. With the exception of the youngest sediments whose age falls within the range of radiocarbon dating, the sediments do not include materials amenable to radiometric measurements. Fossils are sparse and long-ranging. In the following paragraphs a description is given of the Quaternary sediments of the Nile valley and the deserts of Egypt.

QUATERNARY SEDIMENTS OF NILE VALLEY

Basing his work on field studies, a large amount of well and borehole data, seismic reflection profiles and the results of archeological expeditions which worked out in detail the young sediments of the Nile (de Heinzelin 1968, Butzer & Hansen 1968, Wendorf & Schild 1976, Said 1981) gives an outline of the geological evolution of the Nile River within the boundaries of Egypt. Figure 25/1 is a composite section of the sediments which fill the through of the Nile, while figures 25/2 and 25/3 show cross sections across the Nile and the delta.

The Nile trough is filled with alluvial sediments which are divisible into five units each of which is unique with regard to its texture, structure and mineral composition (Fig. 25/1). Each of these units seems to have been formed by a river system which was unique with regard to its sources, competency and regimen. Indeed the Nile can be conceived as having passed through five main stages since its valley was cut down in late Miocene time. Each of these stages was characterized by a master river system. Toward the end of each of the first four stages (the last is still extant) the river seems to have declined or ceased entirely to flow into Egypt.

The first of the rivers, the Eonile, was a late Miocene feature which cut its course to a great depth in response to the lowered base level of the desiccating Mediterranean. It formed a canyon longer and deeper than the Grand Canyon, Arizona. The Eonile canyon was traced by the use of seismic data (Bentz & Hughes 1981, Barber 1981) in north Egypt and by the use of gravity data in upper Egypt (El Gamili 1982). In upper Egypt the width of the Eonile canyon ranges from 2 to 20 km, and the thickness of the riverine sediments ranges from 170 to 900 m. In the delta region, however, the thickness exceeds 3 km in the extreme northern parts. Several water falls seem to have blocked the Eonile; a significant one was at Maghagha which formed a constriction along the path of the Eonile (El Gamili 1982).

The deposits of the Eonile are known only in the subsurface in the north Delta embayment. They are made up of a lower unit of coarse-grained sands and gravels derived from the eroded Cretaceous and Eocene rocks of Egypt (the Qawasim) and an upper unit of evaporites (the Rosetta) which is correlated with the evaporite suite recorded beneath the bottom of the modern Mediterranean (Hsu et al. 1973).

The canyon was transgressed by the advancing Mediterranean as it started filling up during the early Pliocene. There is evidence that this ingression extended inland in the form of an elongate gulf up to the latitude of Aswan (Chumakov 1967). In the north, with the rise in sea level, the waters overflowed the banks of the canyon and covered the peripheries of

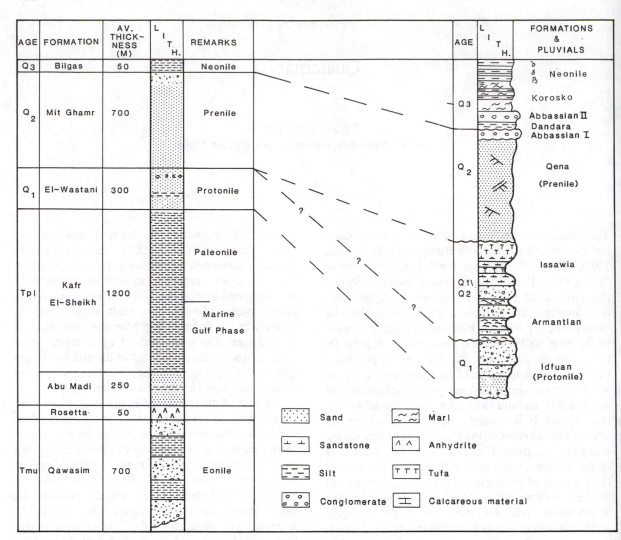

AGE	FORMATION	AV. THICK-NESS (M)	LITH.	REMARKS
Q3	Bilqas	50		Neonile
Q2	Mit Ghamr	700		Prenile
Q1	El-Wastani	300		Protonile
Tpl	Kafr El-Sheikh	1200		Paleonile / Marine Gulf Phase
	Abu Madi	250		
	Rosetta	50		
Tmu	Qawasim	700		Eonile

AGE	LITH.	FORMATIONS & PLUVIALS
Q3		Neonile / Korosko / Abbassian II / Dandara / Abbassian I
Q2		Qena (Prenile)
Q1\ Q2		Issawia / Armantian
Q1		Idfuan (Protonile)

Sand   Marl
Sandstone   Anhydrite
Silt   Tufa
Conglomerate   Calcareous material

Figure 25.1 Composite columnar section of Nile sediments; left column of subsurface sediments based on delta well logs; right column of surface sediments based on measured sections mainly along the banks of the Nile Valley in upper Egypt; scale of surface section enlarged $2\frac{1}{2}$ times in comparison to that of the subsurface section.

the Nile and delta (Kom El Shelul formation).

The effect of fresh water on the marine sediments of the gulf increased drastically in late Pliocene time converting it first into an estuary and then into a veritable river channel. The sediments of this river, the Paleonile, consist of a long series of interbedded red-brown clays and thin fine-grained sand and silt laminae which crop out along the banks of the valley and many of the wadis which drain into it (Madamud formation or Blanckenhorn's *Melanopsis* Stufe). The sediments are also known from the subsurface in practically all the boreholes drilled in the valley and delta (Kafr El Sheikh formation). The Paleonile sediments make about 20% of the section of riverine deposits of the valley and delta. By the end of Paleo-

nile sedimentation, the Eonile canyon was completely filled up and an immense wedge of delta front sediments filled part of the embayment which lay in front of the river.

The northern part of the modern delta, therefore, extends beyond the continental margin. It started forming during the late Pliocene in the embayment (named North Delta Embayment by Said 1981) which lay in the shadow of the elevated southern part of the delta (named South Delta Block by Said 1981).

The lithology and mineral composition of the Paleonile sediments indicate that they must have been derived from areas receiving sufficient precipitation and having an effective vegetation cover. The

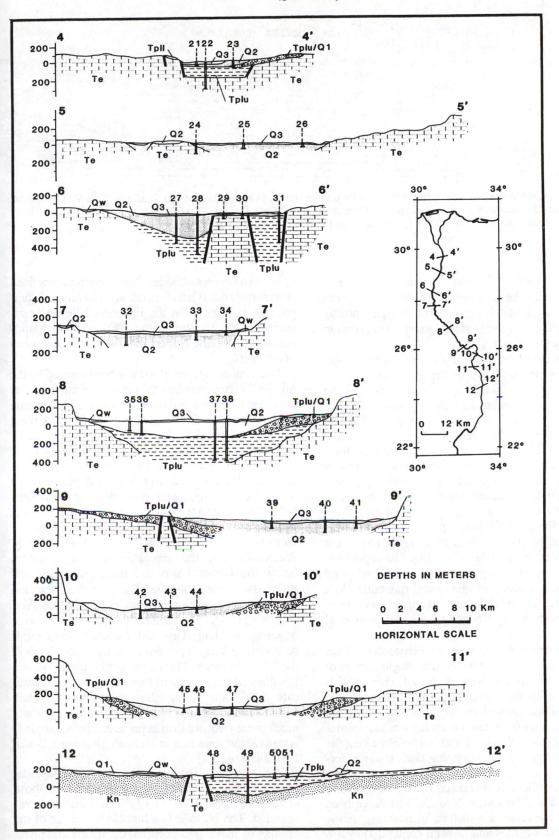

Figure 25.2 Cross-sections in the Nile Valley based on borehole data (boreholes 21–51) and surface measurements. Te = Eocene; Tpll, Tplu = lower and upper Pliocene; Q1, Q2, Q3 = lower, middle and upper Quaternary (after Said 1981).

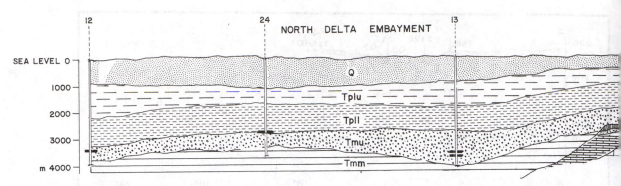

Figure 25.3 Longitudinal section in Nile delta. 12 = Baltim well no. 1; 24 = Abadiya well no. 1; 13 = Kafr El Sheikh well no. 1; 14 = Shebin El Kom well no. 1; J = Jurassic; K = Cretaceous; Te = middle Eocene; To = Oligocene; B = Basalt; Tmm, Tmu = middle and upper Miocene; Tpll, Tplu = lower and upper Pliocene; Q = Quaternary (after Said 1981).

absence of central African freshwater faunal elements in these beds suggests that these sediments must have been largely derived from Egypt, probably from highlands in the Eastern Desert. This points to extremely wet climates.

The advent of the Pleistocene epoch in Egypt was marked by dramatic events the most important of which, relative to their impact on the history of the Nile, were those related to climate and tectonism. The advent of the Pleistocene brought to Egypt a pattern of aridity that set the tone of the climate prevailing in Egypt, with minor fluctuations, throughout the Pleistocene. The earliest Pleistocene was an episode of great aridity in which Egypt was converted into a veritable desert. Not only did aridity set over the country, but the Paleonile stopped flowing into Egypt. This long episode of aridity was interrupted by the intrusion of a highly competent river, the Protonile, in the Nile valley. The deposits of this river are made up of cobble and gravel-sized sediments composed of quartz and quartzites (Idfu formation). These sediments seem to have been derived from deeply leached terrain and from local sources.

At a later time during the early Pleistocene a short pluvial occurred during which the conglomerates of the Armant formation were deposited. This was followed by an episode of spring activity and the deposition of bedded and/or massive tufas. These tufas are associated with thick talus breccias which accumulated during an episode of high seismicity along the piedmont slopes of the bounding cliffs (Issawia formation).

The Protonile was succeeded by two other rivers, the Prenile and the extant Neonile. The deposits of each of these rivers are distinct in lithology, stratigraphic relationships and mineral content. They are separated from each other by an unconformity and a long recession. The deposits of the Prenile are made

up of massive cross-bedded fluvial sands interbedded with dune sands (Qena formation). The mineral composition indicates that the Egyptian Nile was connected for the first time with the Ethiopian highlands across the elevated Nubian massif by way of a series of cataracts.

This connection seems to have been severed by the middle Pleistocene when the channel of the river was occupied by ephemeral rivers depositing polygenetic conglomerates derived from the uncovered basement complex of the Red Sea hills. During this crisis, which lasted probably more than 200,000 years, a hyperarid phase interrupted this period during which time the Nile flowed from the south depositing silts which are indistinguishable from those of the modern Nile. These silts are named Dandara formation (= α Neonile silts). The conglomerates below and above these silts are called Abbassia I and II respectively. Previously only the generation of conglomerates below the Dandara silts had been recognised and given the name Abbassia. We owe this new and revealing discovery to the work of Paulissen & Vermeersch (1987). The Abbassia conglomerates were deposited in a humid interval, the Abbassian I and II respectively. They represent conspicuous horizons in the Nile sequence. The composition, thickness and lithology of the gravels of these conglomerates indicate deposition during short wet intervals when winter-season cyclonic cloud bursts were intense and much more frequent than at present. The Abbassia II conglomerates are rich in archeological materials of Acheulian age.

The deposits of the last of the rivers, the Neonile, which is still extant, are indistinguishable from those of the present-day river. The African connection was resumed. The Neonile is a humble successor of the Prenile. Its harbingers occurred during the arid phase which interrupted the Abbassian Pluvials when the Dandara (α Neonile) silts were deposited. Since that

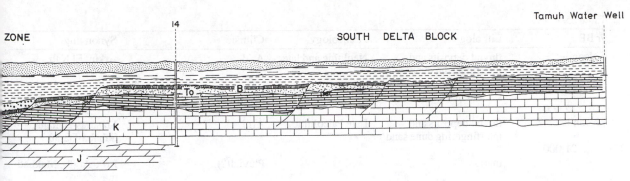

time and for a long period (?200,000 to ?60,000 BP) the floods seem to have been erratic; only during a few intervals were the floods high enough to leave behind thin silt terraces. Most of the time the floods were smaller than usual and the Nile was lowered down to levels which cannot be determined. The recessional deposits which followed the Abbassia II gravels, named the Korosko formation, include the deposits of two major pluvials during which great geomorphological changes took place over the valley. In the earlier Pluvial (Saharan I) Mousterian remains become abundant. In the second pluvial, the Saharan II Aterian sites are recorded. Two or more silts occur in the midst of the recessional deposits.

This long recessional period was followed by three aggradational phases which were interrupted by minor recessional episodes. The aggradational deposits are named β, γ amd δ Neonile.

Results of recent work in Egypt (Wendorf & Schild 1986, Hassan 1986, Paulissen & Vermeersch 1987) necessitate the redefinition of these units. The β, γ and δ deposits (younger Neonile deposits) are represented by massive structured silts with interfingering dune sand. The lowermost β silt carries late middle Paleolithic artifacts of the Khormusan tradition. It is exposed along the Nubian Nile which has been downcutting its channel since its inception and is buried beneath the modern silts of the aggrading downstream Nile. In Nubia, it is given the name Dibeira-Jer by de Heinzelin (1968). As the Nubian lands drowned under the rising waters of the High Dam, the only location where it is still preserved is in Wadi Kubbaniya to the northwest of Aswan. It is described as middle Paleolithic silt by Wendorf & Schild (1986). The age of this unit is not known with certainty, but it probably ended about 40,000 years BP (i.e. during oxygen isotope stage 3).

The succeeding γ Neonile deposits are recorded in many areas in upper Egypt. They form a complex which includes all the units which were previously described under the names Masmas-Ballana, Deir El Fakhuri and Sahaba-Darau (Wendorf & Schild 1976, Said 1981). These units are now believed to form one major aggradational cycle with minor recessional intervals. The deposits carry late Paleolithic artifacts and are dated between 21,000 and 12,000 BP. The cycle is known to have been terminated by a series of high floods (Vermeersch 1970).

The δ Neonile silts are the deposits of the youngest of the aggradational cycles of the Nile. They started to form about 10,000 BP and are separated from the underlying silts by recessional deposits which belong to the Nabtian pluvial. The silts are named Arkin (de Heinzelin 1968) or El Kab (Vermeersch 1970, Hamroush 1986).

The α Neonile silts (the Dandara or the older Neonile deposits) lie unconformably over the Abbassia I gravels. Their age is unknown; the archeology points to a pre-Acheulian age (probably 400,000 BP during oxygen isotope stage 7).

The following table (p. 492) summarizes our present understanding of the events of the Neonile.

In conclusion it can be stated that the connection of the Nile with its African sources is new. The earliest Niles derived their waters from local sources, the Red Sea range and the Nubian and southwest Egyptian highs. It was only during the middle Pleistocene that the first African connection occurred. It is almost certain that the Ethiopian connection preceded the equatorial African connection; the White Nile, according to the work of Salama (1987), formed a series of interiorly-drained lakes until relatively recent time.

SEDIMENTS OF DESERTS

The Quaternary sediments of the deserts of Egypt are

Table 25/1.

yr BP	Lithology	Archeology	Climate	Synonymy
δ Neonile	fluvial deposits and interfingering dune sand	Neolithic and later	Arid	Arkin, El Kab
10,000				
	silt and marl		Erratic floods	Dishna-Ineiba, Birbet
12,000				
γ Neonile	fluvial deposits and interfingering dune sand	Late Paleolithic	Arid	Masmas-Ballana, Sahaba-Darau
21,000				
	marls		Pluvial(?)	Korosko(?), Makhadma
40,000				
β Neonile	fluvial deposits and interfingering dune sand	Khormusan	Arid	Dibeira-Jer, M. Paleolithic silt
?				
		Aterian	Saharan II pluvial hyper-arid	Gerza
Korosko	silt, marl and dune sand	Mousterian	Saharan I pluvial arid (?)	Ikhtiariya
?200,000				
Abbassia II	polygenetic conglomerate	Final Acheulian	Abbassian II pluvial	
?				
α Neonile Dandara	silt	?Final Acheulian	Arid	
?				
Abbassia I	polygenetic conglomerate		Abbassian I pluvial	
?400,000				

varied and complex. In the following paragraphs a description is given of some of these deposits which were selected because of their significance to the discussions which will follow on the past climates of Egypt and the evolution of its landscape.

1 Eolian deposits

Sand dunes

Sand dunes cover a large part of the surface area of Egypt (see Fig. 2/1). They have attracted the attention of many authors and make the subject of the classic basic research of Bagnold (1933, 1941) and Bagnold et al. (1939). The sand sea, the largest of the dune-covered areas of Egypt, lies to the south of Siwa Oasis extending to the southern borders of Egypt and into north Sudan. The sea is covered with belts of sand dunes of immense length and comparatively small breadth running from a north-northwesterly direction in the north and to a south-southeasterly direction in the south. In the north and under the influence of varying wind regimes, the sand mass is sinuous forming barchanoid ridges and coalescing dunes (McKee 1979). In the south linear (seif) dunes are common. They form ridges separated by avenues (Clayton 1933). The Abu Moharik dune belt is another conspicuous belt in the Western Desert. It extends for a distance of 300 km in length and only a few kilometers in width from west of Bahariya Oasis to the north end of Kharga and continues with minor breaks and nearly in the same direction for another 150 km within Kharga depression. Here the chain is made up of barchan dunes (Embabi 1982). Dunes move at rates that have been variously estimated from 10 to 100 m per year (Bagnold 1941, Embabi 1982). The rate is proportional to the length of duration of the effective wind, as well as to the size, angle of slope and length of the windward side of the dune.

Two major wind regimes are known in Egypt: predominant westerlies along a narrow corridor of the Mediterranean sea coast and generally north-northwesterlies throughout the Western Desert (El-Baz & Wolfe 1979). The accumulation of modern dunes is governed by the frequency of the effective (sand moving) rather than the dominant winds and by the alignment of the scarps and other topographic features which lie in the path of the wind.

Dunes arise wherever sand-laden wind deposits sand on the windward slopes of a random patch. Under regimes of constant wind, barchan dunes develop. The mounds grow to a height of 10 to 15 m when a slip-face is established by avalanching on the sheltered leeward side. As the dune migrates, the extremities, offering less resistance to the wind than the summit region, advance more rapidly until they extend into wings of such length that their obstructive power becomes equal to that of the middle of the dune. The resulting crescentic form then persists with only minor modifications of shape and size so long as the wind blows from the same quarter. The width of the barchan is commonly about ten times its height.

Gabriel & Kroepelin (1985) describe a type of barchan, the parabolic dune, where the middle moves faster than the extremities. The resultant dune's extremities, therefore, do not point to the direction of the wind. The parabolic dune seems to form under wet climatic conditions. They are reported from Wadi Howar, northwest Sudan, where they are taken as important indictors of past Holocene climates.

The modern moving dunes are of Holocene age. They overlie unconformably surfaces covered by Neolithic soils and occupations. Relicts of earlier dunes of pre-Holocene arid phases are of limited areal extent. They have been preserved as small patches of indurated dunes since the Acheulian. They formed favored sites for ancient settlements (Wendorf et al. 1976, 1977).

2 Fluvial deposits

Many of the wadis of the deserts having external drainage are covered with Holocene alluvial fill which seems to have accumulated in response to the rising sea level during this episode. Records of earlier deposits and terraces are known along many of the wadis (see, for example, Shata 1959 and Sneh 1982 for a description of Pleistocene alluvial sediments of Wadi El Arish). In the Western Desert, fluvial sediments are difficult to recognise; they were either eroded away or modified by wind beyond recognition. In many instances the wadis themselves were entirely covered by sand. It is only recently, through intensive field studies, that alluvial deposits in the form of inverted wadis and gravel mounds have been discovered and the nature of the sand sheets deciphered.

Sand sheets
Sand sheets form large and flat stretches of the Arba-'in desert. They probably owe their flatness to wind scour which was limited by the depth of the water table (Haynes 1982). The sand sheets had been traditionally classified among the eolian deposits by most authorities (Sandford 1935, Bagnold 1933, Said 1975, Haynes 1980, Maxwell 1982). In the light of new data obtained from drilling and seismic profiling, Said (1980) advances the view that these sheets represent the relics of a complex drainage system. They probably represent deposits of braided streams which spread from the highlands to the south and west.

The upper 20 to 50 cm are made up of unconsolidated laminations of medium to coarse-grained sand. These rest over layers of slightly more consolidated sand conspicuously browner in color and representing a paleosol which seems to be of Neolithic age as suggested by the associated archeology. Drilling and recent seismic work indicate that these layers have a thickness varying from a few centimeters to 20 m. Beneath these layers of poorly consolidated sand, there is a 50 to 200 m-thick layer of indurated and weathered coarse-grained sand grading into gravels. The lithology thus revealed shows clearly the alluvial nature of these deposits.

The age of the sand sheet deposits is not known with certainty. The sheets took their modern shape by Acheulian time and are topped by a soil complex the young parts of which are of Neolithic age. The sheets are overlain unconformably by the modern dunes. The sheets rest over an erosion surface which was carved in late Miocene (*vide infra*). They are probably of Pliocene age.

The Selima sand sheet covers an area of 4000 km^2 in southwest Egypt and north Sudan. Its center is in Bir Tarfawi where it is strewn with kankar, root drip and carbonate. The carbonate layer displays polygonal cracks and is almost continuous around the Bir Tarfawi low. It forms a crust over the bevelled surface of the underlying sand sheet and is closely related to it. The layer seems to have been formed by calcified root drip and other vegetative materials which must have covered the Bir Tarfawi area during higher water table levels accompanying the Pleistocene pluvials.

Inverted wadis
Inverted wadis are elongate, sinuous and sometimes branching gravel ridges which stand out above the rock-cut surfaces. They are first noted by Giegengack (1968) in the Nubian desert (Fig. 25/4). They represent ancient thalwegs of the old drainage system which, in the case of the Nubian desert, seems to have had some relationship with the ancient Nile system. The inverted channels are indurated wadi deposits of poorly sorted gravel and coarse sand. They include rolled Acheulian artifacts and are related to the early Pleistocene Armant pluvial (*vide infra*).

Older (?Oligocene) inverted wadis (Fig. 25/4) occur above the middle limestone plateau in the form of mounds of gravel ranging in size from 2 to 20 km^2. They lie unconformably over the underlying early Eocene limestones. The mounds are made up of alternating poorly-bedded gravel beds, sometimes reaching thicknesses exceeding 10 m, and thin coarse sand-pebble beds (El Hinnawi et al. 1978). The gravels are mostly dark brown limestone and chert. They range in size from pebbles to cobbles, are poorly

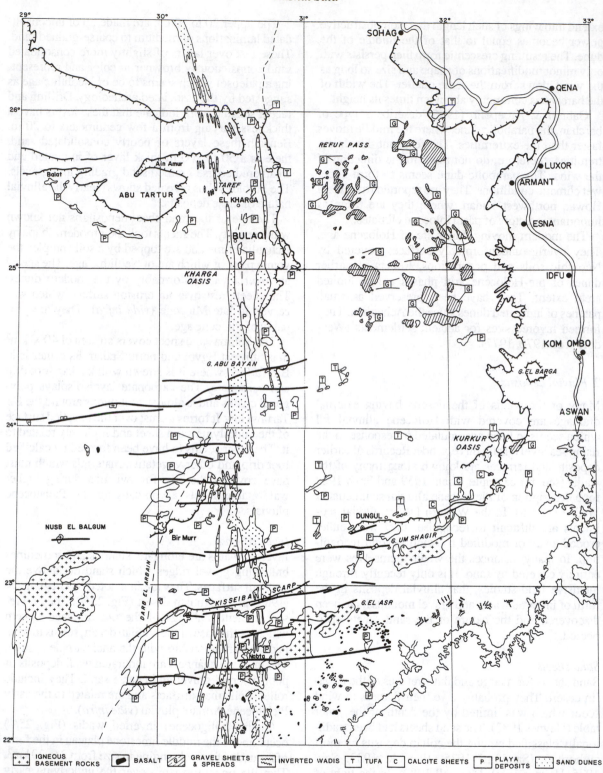

Figure 25.4 Map of Kharga oasis region showing distribution of surficial deposits (after Said 1980).

sorted, rounded to subrounded and sometimes disc-shaped. The gravels are embedded in a pale-brown sandy matrix. The mounds lie unconformably over the limestone plateaux on both sides of the Nile in its middle latitudes.

3 Spring and solution features and deposits

Spring deposits

Spring deposits cover a small area. They feature clean marlstones, black soils, peat deposits, tufas and salt encrustations. They are described from Kharga (Caton-Thompson 1952), Kurkur (Said & Issawi 1964, Butzer 1964, Issawi 1969, 1971), Dungul (Said 1969) and the Nile cliffs (Sandford 1934, Said 1981). There seems to have been several generations of spring activity. The oldest are represented by the massive tufas that lie on top of the limestone plateau in Kharga, Kurkur and Dungul regions. They are 10 to 20 m thick and are made up of solid to vesicular crystalline stone without bedding. They include

pockets, up to 1 m in diameter, containing boulders of local derivation. The tufa mass is composed of hard, crystalline carbonate rock precipitated around various species of plants of which the stems and internal structures are still preserved. The age of these massive tufas is difficult to ascertain, but they were certainly formed prior to the excavation of the oasis depressions and are, therefore, of pre-late Miocene age.

Several other generations of tufas are known over the plateau, along the scarps or within the depressions. Many overlie or intercalate boulder beds, marlstones or peat beds. The tufas range in thickness from 1 to 20 m. At least four generations are recognised in the Kurkur and Dungul regions (Figs 25/4 and 25/5). Tufas associated with the depressions are all connected with boulder beds and fine-grained spring deposits. They seem to have been formed under wetter and most probably colder climatic conditions, in closed basins with a rich mat of vegetation. The youngest tufas are of middle Paleolithic to Neoli-

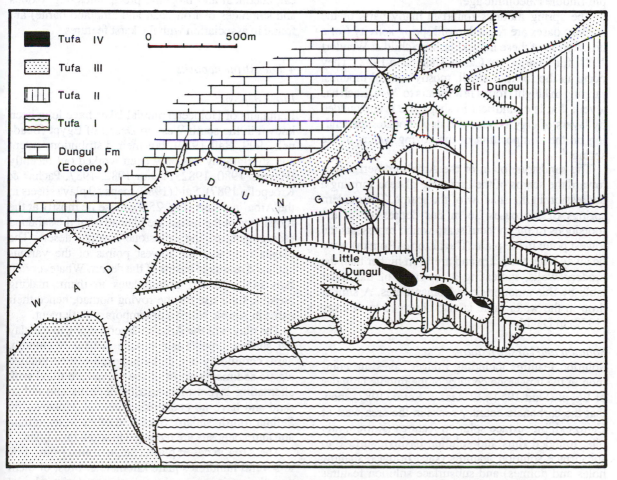

Figure 25.5 Tufa deposits of Dungul oasis classified from older (Tufa I) to younger (Tufa IV) (after Said 1969).

thic age judging from the implements included within the boulder beds (Hester & Hobler 1969). It is
difficult to establish the age of the old tufas. A
discussion of the depositional environment and the
limnological conditions prevailing in the basins is
given in Said (1969).

Old spring activity is inferred from geomorphological evidence in many areas that have no source of
water today. Among the features that may belong to
this old activity are the shallow circular to oval
depressions which are seen all along the sandstone
plains of the Arba'in desert. The diameter of these
depressions varies from 15 to 100 km, and the difference in height between the lowest and the highest
points in them never exceeds 40 m. Although the
origin of these depressions is problematical, it is
highly likely that they resulted from the settlement of
the water-bearing sandstone beds after the drying of
the springs that must have been active all over the
sandstone plains of the desert. In one of these depressions, the dike area, deflated playa deposits with
middle Paleolithic artifacts are found (Wendorf &
Schild 1980), indicating that some of them are of
pre-middle Paleolithic age.

The spring mounds found in many parts of the
modern oases are related to old spring activity (for a
description of these in Dakhla, see Schild & Wendorf
1981; for Kharga, see Caton-Thompson 1952). They
occur in the form of small, mostly conical mounds,
ranging in height from a few meters to 20 m. Each has
a vent that is cylindrical in shape and is surrounded
by layers of clay, ocher, sand or carbonates, depending on the composition of the oozing waters of the
spring.

The sandstone plain of the Arba'in desert slopes
toward the scarps of the limestone plateau forming a
depression in the shadow of the scarps. The modern
springs of the oases lie in this depression at the
footslopes of the modern scarp which represents the
lowest point in the plain. The historic (Roman) wells
lie away from the scarp and are 10 m higher than the
modern wells, indicating a lowering of the water
table by this amount since Roman times. A similar
drop in the water table is also observed by Murray
(1952) in Dakhla.

Solution and karstic features
Solution features including midget caves are common in the limestone country of the middle latitudes
of Egypt (Said 1954). El Aref & Refai (1987) show
that the carbonate country rocks in the area west of
Cairo exhibit surface solution features (Karren, rain
pits, rounded rims and solution basins), surface to
subsurface solution features (karst bridges, shallow
holes and dolines) and subsurface solution features
(opening and cavities).

Haynes (1982) reports patches of *terra rosa* soils
in the limestone plateau overlooking the Kharga
oasis; these extend downward as pipe-like fillings of
cavities in the bedrock. These cavities are interpreted
as karstic features. Sinkholes are reported in the
western part of the Qattara depression and in the Gib
Afia water well between depths 228 and 246 m (El
Ramly 1967). Exploration wells drilled on the Abu
Tartur plateau lost circulation in the limestone of the
Kurkur formation, and some are reported to have
gone through virtual sinkholes. Stringfield, Lamoreaux & LeGrand (1974) propose that the solution of
carbonate rocks is responsible for the development of
depressions in which oases occur. The oases of
Kurkur and Dungul lie on top of the limestone plateau
and seem to owe their origin to karstic processes.
Kurkur lies at the confluence of several wadis which
have no outlet. Wadi Dungul seems to have cut its
outlet in relatively recent time.

El Aref et al. (1986) report conekarst and karst
ridge landforms with minor surface and subsurface
solution features in the Miocene limestones and evaporites of the coastal plain of the Red Sea. Mechanical, chemical and biogenic precipitates (e.g. oxides
and sulphides of iron, lead and zinc and barite) are
found in association with the karst features.

4 Lacustrine deposits

Playa deposits
Remnants of Holocene pluvial lakes have long been
known in the south Western Desert of Egypt (Beadnell 1909, Ball 1927), but their distribution, extent
and stratigraphy has only been worked out recently
(Haynes 1980, 1982, Pachur 1982, 1987, Pachur &
Kroepelin 1987). Said (1980) maps 50 playa sheets in
the Arba'in desert (Fig. 25/4); these lie mostly at the
footslopes of the limestone plateau or of the subdued
escarpments of the sandstone-covered desert. The
playas represent the lowest points of the various
enclosed drainage basins of the desert. Whatever rain
falls in the desert accumulates in them, making
possible their use for the roving nomad, hence their
name in Arabic *Hattiya* or temporary settlement.

The playas of the Arba'in desert belong to the clay
flat surface type. Their surface is usually smooth,
hard, commonly dry, and composed of fine-grained
clastic sediments. most of the playas are filled with
alternating layers of lacustrine deposits and eolian
sand and have no economic potential. However, in
Selima (north Sudan), Sheb and other wells, nitrate
and alum deposits are recorded in small quantities.

The Arba'in desert playas range in area from a few
to several hundred square kilometers. Most of them
owe their origin to surface discharge during pluvial

episodes and are not related to springs. Many of the Arba'in desert playas formed during the terminal Paleolithic-Neolithic pluvial period. In most of them the water table is usually deep. During their formation, they were active sites of human habitation. In the Nabta playa, one of the few playas with a shallow water table, several cultural layers succeeded one another during the entire late Paleolithic-Neolithic pluvial (Wendorf & Schild 1980).

Some of the playas of the Kharga region owe their origin to spring activity. Many survived up to Roman times; the younger section of the playa deposits is made up of drifting sand which was carried into the lakes by the wind. The earlier part of the playa contains, like most other surface-discharge playas, fine-grained clastics which seem to have slid into the lakes by rain.

Apart from these playa deposits of the Arba'in desert, lacustrine sediments are also reported from Wadi Feiran, Sinai (*vide supra*).

Armored playas

Armored playas are extensive sheets of playa deposits which are veneered by a layer of white nodular chalcedony cobbles up to 15 cm in diameter embedded in a reddish brown matrix. Issawi (1971) was the first to record these in the Darb El Arba'in area. He maps them as chalcedony sheets and relates them to doming movements and structural lines. Recent work, however, has shown that the chalcedony cobbles seem to have been formed in standing bodies of water rich in sodium carbonate and silica having a pH of 9.5 or higher (Haynes 1980). The age of these playas is not known, but they are certainly older than the Neolithic. They overlie bedrock of Cretaceous age and could well be older than the Quaternary.

Jux (1983) advocates the view that the *Liban Desert Silica Glass* (LDSG) is of sedimentary origin and that it was formed in a playa. The origin of the LDSG has remained a mystery since its discovery by Clayton and Spencer in 1932 (for a review of the work carried out on the LDSG, see Giegengack and Issawi, 1974). Until the publication of the work of Jux, the glass was believed to have been formed under great temperature generated by an extra-terrestrial event (Cohen 1959, 1961; Kleinmann 1969; Barnes & Underwood 1976; Underwood & Fisk 1980; Frischat et al 1982 etc.). Research on the origin of the LDSG before the publication of the work of Jux centered on the finding of meteoritic impact features and astroblemes which could be related to the event that produced the glass. Several concentric outcrops were described from southwest Egypt and the Kufra area and were claimed to be impact features linked to the glass. In fact, work along these lines continued after publication of Jux by advocates of the fusion process

that presumably formed the glass. As this book went to press Giegengack, Underwood and Weeks sent a manuscript for inclusion in this book in which they attempt to explain the process which produced and sustained for a long time the high temperature under which the glass was formed and concluded that of the "variety of imaginative hypotheses thus far advanced to explain the origin of the glass, none has yet met the constraints imposed by existing field and laboratory investigations". These authors are, therefore, convinced that the glass "is a unique material that owes its origin to a unique process, an event thus far undescribed from any part of the geologic record".

Figure 25/6 is a generalized geological map of the silica glass area while figure 25/7 is a cross-section along one of the main avenues separating the longitudinal dunes covering southwest Egypt where the glass is found. It shows the avenue to be covered by a thin veneer of gravel embedded in a red matrix. This gravel seems to extend beneath the dune which is made up of a lower indurated part and an upper active and moving part. The lower part of the dune is topped by a red soil. Abutting against the lower part of the dune are lacustrine sands and clays. Haynes (1982) found clasts of the glass in the gravel bed and Jux (1983) found them in the lacustrine deposits and in the crevices underneath. This led Jux to conclude that the glass must have been formed in the lake as a result of the coagulation of the silica gel when the lake water came in contact with the underlying ground water which seemed to have had a high pH value. The sedimentary origin of the glass was further substantiated by the discovery and separation of palynomorphs and organic compounds within the glass.

The idea that the LDSG is of sedimentary origin and was formed in playas suggests to the present author that the glass could be of similar origin to the "chalcedony" fragments which are strewn over the armored playas and which are assumed to have been formed in waters of similar composition. Field work shows that the two have a similar stratigraphic context and it remains to be seen whether the chemical and mineralogical composition of the fragments of the armored playas are comparable to those of the LDSG. The age of the glass is difficult to ascertain, but it must be of pre-Acheulian age, for the glass itself was used in the making of artifacts of this age (Roe et al. 1982). If one accepts the 28.5 million year fission-track age for the LDSG (Storzer & Wagner 1971), then the glass could not have been deposited in the lake with which they are associated today as suggested by Jux. It is possible that the glass was deposited in a considerably older lake and was later transported to its present place during pluvial episodes related to the formation of the gravel bed and the lacustrine sediments described by Haynes and

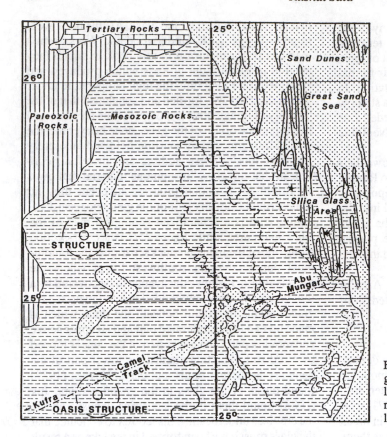

Figure 25.6 Generalized geological map of silica glass area, southwestern Egypt. Asterisks indicate locations of discovered in situ glass, the southernmost location is in B-C avenue. Circles show locations of reported astroblems (after Jux 1983).

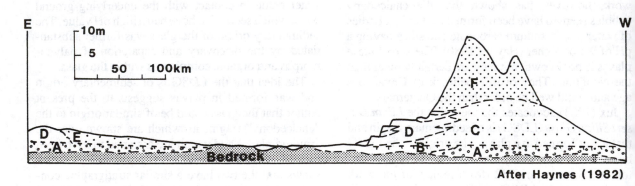

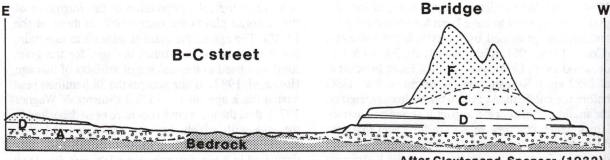

Figure 25.7 Cross-section along B-C avenue, silica glass area, southwestern Egypt after the interpretation of Haynes (1982) (top) and Clayton & Spencer (1932) (bottom). A. Red gravel, fossil aquifer; B. Pebbly silty sand; C. Stabilized dune; D. Lacustrine deposits; E. Eolian sand; F. Active dune (after Jux 1983).

Jux. The gravel bed forms part of the surface of the ?Neogene sand sheet described above.

5 Coastal deposits of the Mediterranean coast west of Alexandria

Kurkar ridges (Alexandrian formation)

The Mediterranean coastal plain to the west of Alexandria is characterized by the presence of a number of elongated ridges which run parallel to the coast, separated by longitudinal depressions (Fig. 25/8). The ridges are composed mainly of oolitic limestone (Hume & Hughes 1921, Shukri, Philip & Said 1956, Butzer 1959). They seem to represent successive fossil off-shore bars that were formed in the receding Mediterranean during the Pleistocene. The nearest three ridges to the shore, the 10, 25 and 35 m high ridges (named the Coastal, Abu Sir and Maryut bars) can be followed for long distances along the coast. The succeeding five inland ridges, the 60, 80, 85, 90 and 110 m high ridges (named the Khashm El Eish, Alam El Khadem, Mikheirta, Raqabet El Halif and Alam Shaltut), are less conspicuous and do not form continuous ridges. The depressions which separate the bars contain lagoonal deposits (evaporites and marls). In Gharbaniyat and El Omayid economic gypsum deposits are known in the lagoon separating the Gebel Maryut and Khashm El Eish bars (Said, Philip & Shukri 1956, Abdel Aziz et al. 1971, Adindani et al. 1975).

QUATERNARY CHRONOLOGY, PAST CLIMATES AND CLASSIFICATION OF SEDIMENTS

The age of the Pleistocene stage boundaries, and particularly the transition from the Tertiary to the Quaternary, is uncertain. The Plio-Pleistocene boundary is placed at about 1.85 my BP at the Olduvai magnetic interval by some authors (Berggren & van Couvering 1974, Haq et al. 1977) and at about 2.8 my BP at the Kaena magnetic interval by others (Beard et al. 1976, 1982). Those who advocate a younger age emphasize the stratigraphic evidence obtained from the classic stratotypes of the Neogene of the Mediterranean region, while those who espouse an older age give a climatic implication to this boundary. The older age seems to have more support. The International Union on Study of the Quaternary (INQUA) and the Committee of Mediterranean Neogene Stratigraphy (CMNS) now put the boundary at 'the first appearance of cold water north Atlantic immigrants (species) in the Plio-Pleistocene sequence of Calabria, Italy'. The numerical age of the Pleistocene stages adopted in this book follows, with minor modifications, that of Beard et al. (1976, 1982). Figure 25/9 gives the adopted geomagnetic time scale of the Pliocene and Pleistocene and the accompanying climatic events as revealed from the history of the Nile.

The classification proposed for the Egyptian Quaternary (Said 1983) is based on the subdivision of the sediments into climato-stratigraphic units which were formed during several pluvial episodes separated by interpluvial episodes (Fig. 25/10). The pluvials are represented by spring, torrential or lacustrine deposits as well as paleosols, while the interpluvials are represented by dune accumulation and/or salt and sabkha formation. At least seven major pluvials are recognised. They correspond, to a large degree, to the world-wide episodes of higher sea level and warmer climates. The two oldest pluvials, the Idfuan and the Armantian, are of early Quaternary age and are not succeeded or preceded by interpluvial deposits, hence their questionable position within this sequence. The five more recent pluvials, the Abbassian I, Abbassian II, Saharan I, Saharan II and Nabtian, are recorded in a stratigraphic sequence and their relative position within the middle and late Quaternary is better established.

Evidence obtained from the geomorphology and distribution of archeological material in the deserts of Egypt leads one to believe that the effect of the pluvials was felt only in the southern reaches of the country and that during the Quaternary Egypt was subjected to two different climatic models, the south being affected by the northward migration of the Sudano-Sahelian savanna belt and the north by the Mediterranean pre-Sahara steppe. Indeed it can be assumed that a northward shift the present-day climatic belts of about 15° of the latitude they now occupy would bring an increase in precipitation in south Egypt and would set a climate similar to that which must have prevailed during the Quaternary pluvials.

The Quaternary chronology of the northern reaches rests solely on evidence obtained from ancient shorelines. The altimetry of the Mediterranean Kurkar ridges skirting the coast provide evidence of at least eight transgressions during the Quaternary. The five oldest bars achieving altitudes in the approximate range of 50 to 120 m can be reasonably considered early Quaternary on the basis of marine mollusca and foraminifera which they carry (Shukri et al. 1956). The three youngest bars were probably formed during the last three major interglacials in the middle and late Quaternary. No archeology is associated with these deposits, but the second bar (Abu Sir) carries a fauna of Tyrhennian age or early late Quaternary. The first and younger (coastal) bar is probably of late Quaternary age, while the third and

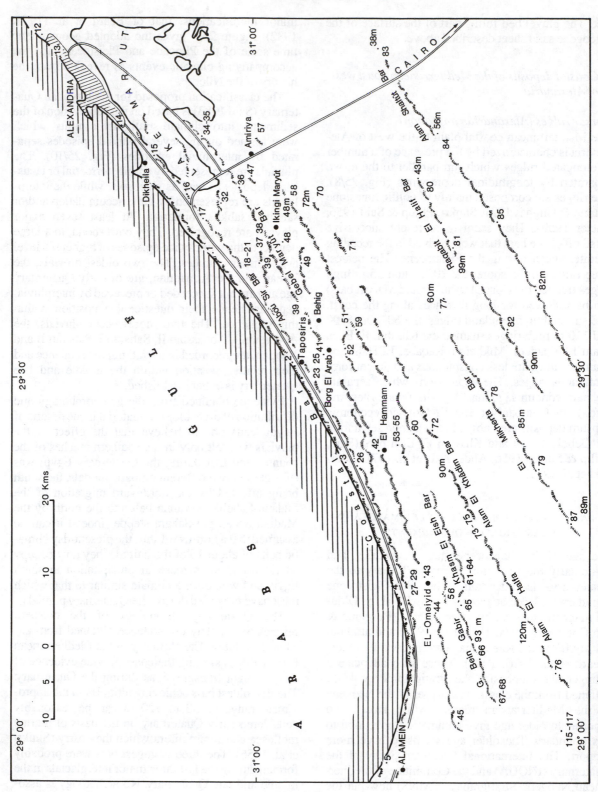

Figure 25.8 Distribution of kurkar ridges west of Alexandria (after Shukri, Philip & Said 1956).

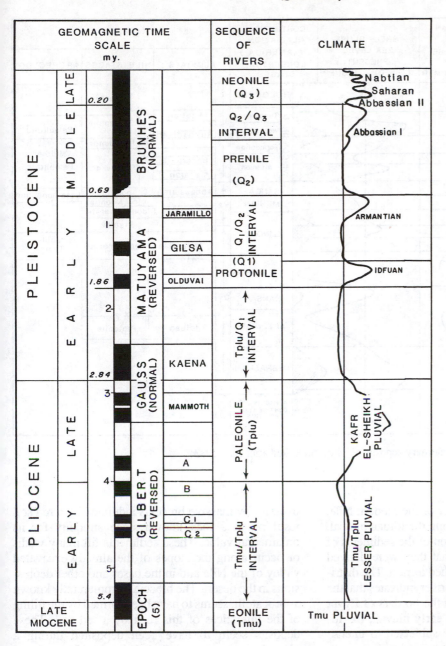

Figure 25.9 Geomagnetic scale, dates of different Niles and accompanying climatic events; vertical scale of upper part starting from 0.69 my BP doubled (modified after Said 1981).

older Maryut bar is probably of middle Quaternary age.

The poorly developed drainage lines, the absence of playa deposits, the poor to non-existent archeology (Hassan 1978) in the northern reaches of Egypt and the preservation of old salt deposits within the Qattara depression – all point to the long periods of aridity in these reaches. The coastal Kurkar ridges are made up of wind-blown sands which were deposited under water. The lagoonal deposits which were formed during episodes of lower sea levels behind and between the bars are evaporites and marls of definite arid origin. Other sediments in the northern reaches of Egypt are in the form of dunes, sabkha and salt crusts of seemingly young age. Bar-Yosef & Phillips (1977) and Sneh (1982) report no significant pluvial deposits in the reconstructed Quaternary column of Gebel Maghara and Wadi El Arish in north Sinai.

Pluvials of early Quaternary. Two major units belonging to this stage are recognised in the Nile Valley. The earlier is made up of cobble and gravel sediment which was deposited by a highly competent

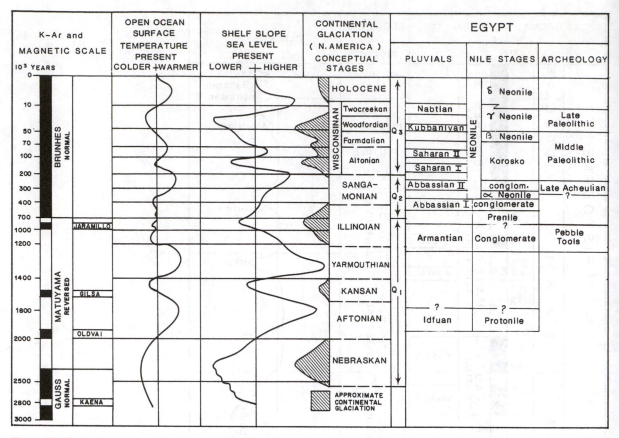

Figure 25.10 Classification of Quaternary deposits of Egypt (modified after Said 1983).

river, the Protonile, a precursor of the modern Nile, which occupied the valley during the Idfuan pluvial. The simple mineral composition of the sediments of this river attests to the fact that they were derived from sources with deeply leached terrain. The distribution of the sediments of this river indicates that the source of the river was within the borders of Egypt and north Sudan. During this early pluvial, the Idfuan, these source lands were subjected to intense chemical disintegration leaving behind surfaces with siliceous cobbles and pebbles only. This seems to indicate a vegetative cover and a wetter climate through the year to account for the weathering of the non-stable minerals. There is no archeology associated with the Idfuan deposits of the Nile. In the deserts, deposits belonging to the Idfuan pluvial are difficult to identify. Some of the old inverted wadis littering many parts of the south Western Desert may belong to this pluvial. The age of many of these wadis is difficult to ascertain.

The later early Quaternary pluvial deposits are in the form of bedded and massive or vesicular travertines which overlie and are interbedded by sheets of gravel and other locally-derived detritus. These are overlain by massive breccias indurated by a red soil resulting in a rock known since antiquity for its ornamental value. These sediments fill many wadis or occur along the slopes of the already excavated valley of the Nile and in the oases and other depressions in the desert. The breccia, commercially known as brocatelli, seems to have been formed by the filling of the interstices of this rock by a red soil. These deposits seem to have been deposited during a pluvial, the Armantian. Early human pebble tools are recorded from the gravels of the Armant formation near Luxor by Biberson et al. (1977).

Pluvials of middle Quaternary. The advent of the middle Quaternary was marked by important events which led to the forming of a new river, the Prenile, with new connections to the Ethiopian highlands. It represents the first river with an African connection since the Oligocene.

The middle Quaternary was terminated by a pluvial, the Abbassian I, which was characterized by intense rains over Egypt resulting in the accumulation of thick, locally derived gravel deposits which rest unconformably over the eroded surface of the

Prenile and earlier sediments of the Nile valley. The gravels are polygenetic and contain abundant crystalline rocks and feldspathic sands representing the first generation of gravels derived from the uncovered basement rocks of the Eastern Desert of Egypt.

Pluvials of late Quaternary. The late Quaternary is represented in the Nile valley by the deposits of the Neonile which seem to have been laid down in four aggradational episodes (α, β, γ and δ) separated by recessional episodes. The oldest of the aggradational deposits, the Dandara, is separated from the succeeding aggradation by a long recession and a marked unconformity. The earlier part of this recession is marked by a lower bed of gravel which overlies the Dandara and is similar to the Abbassia I gravel bed underlying the Dandara. This bed represents the deposits of a pluvial, the Abbassian II. This pluvial occurred after the first Neonile sediments reached Egypt. The deposits of this pluvial are rich in archeological material of Acheulian age.

In the deserts of Egypt, rich late Acheulian sites are associated with wadi deposits accumulating at the piedmonts of the major gebels such as Gilf Kebir (Myers 1939, McHugh 1975, 1982); spring deposits such as Kharga (Caton-Thompson 1952), Dakhla & Bir Sahara-Terfawi areas (Wendorf & Schild 1980). In Kharga, Kurkur and Dungul oases, the slope travertines and gravel sheets belong to this stage. Detailed work in the Bir Sahara-Terfawi area reveals the presence of two submaxima for this pluvial separated by an arid interval, the latter being associated with the final Acheulian. While correlation is extremely difficult, it is possible that these two submaxima correspond to the Abbassian I and II respectively.

The later part of the Nile recession is marked by the sediments of two pluvials, the Saharan I and Saharan II, which are separated from each other by a number of thin silt beds and dune sand. The sediments of the Saharan I and II pluvials carry Mousterian and Aterian artifacts respectively.

In the desert, where detailed studies were carried out on the artifact-carrying deposits of these pluvials in the Bir Sahara-Terfawi area, Wendorf & Schild (1980) distinguish these two pluvials clearly with at least five submaxima. In Kharga Oasis, the Saharan I pluvial deposits carry Mousterian tools in gravel beds alternating with bedded spring travertines (Caton-Thompson 1952, Wendorf & Schild 1980). In Dungul and Kurkur Oases similar Saharan I deposits carrying Mousterian tools are recorded (Hester & Hobler 1969). In the Bir Tarfawi area, the Saharan I deposits are in the form of lacustrine deposits (Wendorf & Schild 1980) and deflated playa sediments with lag artifacts in many areas of the south Western Desert. Saharan II pluvial sediments carrying Aterian artifacts are recorded in lacustrine deposits in Bir Terfawi, and in spring deposits in Kharga and Dungul oases.

Absolute dates for the long recession which separated the α Neonile from the succeeding β Neonile are not available. It is safe to state that its duration was long covering the Acheulian, Mousterian and Aterian periods in Egypt. It probably lasted about 300,000 years or close to three-quarters of the time of the duration of the Neonile itself. During this recession the floods were erratic, mostly on the low side; and only during a few short intervals were the floods high enough to deposit silt embankments.

The last of the pluvials, the Nabtian, is well-documented in the playas of the Western Desert. Sediments belonging to this pluvial (10,000 to 5000 y BP) carry terminal Paleolithic and Neolithic artifacts. In the Nabta playa, where abundant archeological work was carried out (Wendorf & Schild 1980, 1982), three short dry episodes interrupted this pluvial (between 8300 and 8150 y BP; 7700 and 7600 y BP and 6600 and 6400 y BP). The Nabtian pluvial is represented in the Nile valley by the deposits of the recessional episode between the γ and δ Neoniles.

The Nabtian playas which occur in profusion in southwest Egypt and northwest Sudan are the subject of the extensive and elaborate studies of Hassan (1986), Pachur (1974, 1982, 1987), Pachur & Braun (1980, 1986), Pachur & Kroepelin (1987), Pachur & Roepper (1980) Kroepelin (1987), Haynes (1982), Haynes & Haas (1980), Brookes (1983) and others.

ORIGIN OF LANDSCAPE

The geomorphological evolution of the deserts of Egypt has attracted the attention of a large number of workers. The extensive work of the Bardai Research Station in the Tibesti mountains (Hagedorn 1971, Jaekel 1971, Pachur 1974 and others) has contributed considerably to our understanding of the desert processes and the evolution of its landscape. The Egyptian Sahara has also long been the subject of research by many scientists, the majority of whom believe that the present landscape was due to wind deflation. Early members of the Geological Survey of Egypt advocate the idea that wind was capable of shaping slopes, excavating the enormous depressions of the Sahara and forming the pediments which cover enormous stretches of the desert. However, with the exception of sand dune deposits which are obviously the result of wind action, there are few features or deposits which could be related with definitive evidence to the action of wind, such as yardangs, eolian corrosion features, sorted sand sheets, etc. Many recent authors allude to the action of water as a factor

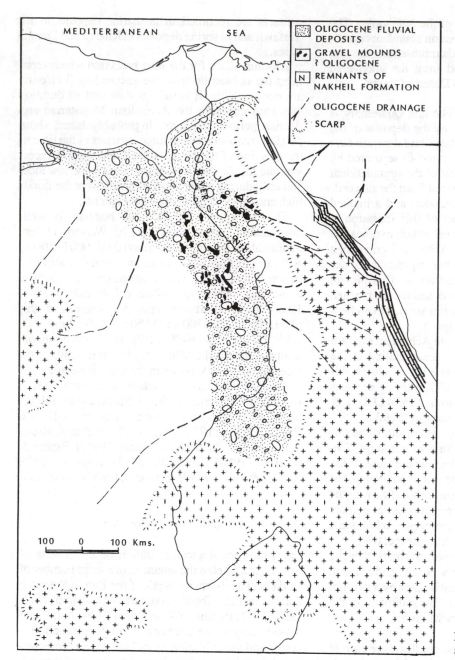

MEDITERRANEAN   SEA

⬚ OLIGOCENE FLUVIAL
    DEPOSITS
🝙 GRAVEL MOUNDS
    ? OLIGOCENE
Ⓝ REMNANTS OF
    NAKHEIL FORMATION
--- OLIGOCENE DRAINAGE
⌒ SCARP

100   0   100 Kms.

Figure 25.11 Hypothetical drainage system during the Oligocene.

in shaping the landscape. Said (1960) and Haynes (1982) believe that the alternations of past climates is the cause of the present landscape although both authors speak of the great depressions as being 'deflated'. McCauley et al. (1982) emphasize the importance of running water in shaping the Gilf Kebir highlands. Said (1980, 1983) attempts to relate the present landscape to defunct drainage systems which seem to have been related to major events that overtook the desert after its elevation in late Eocene time. Credence to this theory comes from recent

work on the paleoclimates of Egypt showing that Egypt enjoyed a wetter climate throughout most of the Tertiary and that aridity was almost exclusively a Pleistocene feature, and even then the arid episodes were of short duration in comparison to the wetter episodes (*vide supra*). The lithology, faunal and floral remains of the late Eocene to early Miocene deposits indicate either an exceedingly wet climate (Bown et al. 1982) or at least savanna-type scrubland (Kortland 1980).

Using the shuttle imaging radar (SIR) signals

which are capable of penetrating the obscuring top layers of loose sand, McCauly et al. (1982, 1986) are able to confirm the presence of a drainage network in the east Sahara. Some of its trunk channels are about 10 to 30 km wide and hundreds of kilometers long. Although the source and direction of these channels are not clear from the radar images, these authors assume that the rivers originated from the Red Sea hills and passed to the Bodele Chad basin in Africa.

Albritton et al. (1985) accept the presence of an old drainage system and propose that 'the Qattara depression is a segment of an old stream valley dismembered by karstic processes during the late Miocene, and afterward modified by mass wasting as well as by eolian and fluviatile processes'.

Sequence of drainage systems

Said (1983) proposes a sequence of drainage systems which influence the present landscape of Egypt based on the study of the stratigraphy of the fluvial sediments which seem to have formed after the retreat of the sea from the land of Egypt in late Eocene and subsequent times.

1 *Late Eocene-Oligocene drainage system*
This system developed in the wake of the great tectonic movements which elevated the Arabo-Nubian massif. The late Eocene sea was marked by its regression toward the north and by the start of the influx of detrital materials into the dominantly carbonate sediments of the Eocene. The shore-line of the late Eocene sea from Fayum to Siwa is littered with scattered vertebrate remains which become exceptionally abundant and layered in certain localized areas such as Fayum or the extreme western tip of the Qattara depression. The process which started in the late Eocene and which led to the introduction of detrital materials continued during the Oligocene with greater vigor as the tectonic movements became more pronounced.

The Oligocene was an epoch of erosion in Egypt. The most pronounced unconformity in the sedimentary section in the Gulf of Suez and north Egypt belongs to that epoch. The exposed sediments of the Oligocene are in the form of alluvial sands which crop out along a strip bordering the footslopes of the elevated lands of Egypt, extending from Suez to the western tip of the Qattara depression (Fig. 25/11). These fluvial outcropping sediments are second-cycle, well-sorted and well-rounded orthoquartzitic sands which seem to have been derived from the sandstones to the south. To the north of that strip, the Oligocene becomes buried underneath the Miocene and younger sediments and becomes finer-grained and of marine character. In the Fayum stratified del-

taic deposits occur carrying an African fauna and flora (see also Chapters 24 & 30, this book).

The late Eocene-Oligocene drainage system is difficult to reconstruct. It seems to have been almost totally exhumed during the late Miocene erosional episode. Only a few relics of this system are preserved in the form of inverted wadis which have recently been described from the top of the middle limestone plateau (*vide supra* under inverted wadis). Figure 25/11 is an attempt to depict the general pattern of this drainage in a generalized hypothetical manner. It is based on the assumption that the drainage must have emanated from the elevated Arabo-Nubian massif and the Bayuda desert, as well as from the elevated plateaux of the Libyan desert to debouch into the Tethys where late Eocene-Oligocene deltaic deposits are known. Although it is difficult, at present, to delineate the channels of this old drainage system, it is certain that its trunk channel was directed to the north toward the deltas of the time which carry rich African faunas.

The age of the inverted wadis is post-Eocene. The wadis are certainly related to a landscape that preceded the excavation of the oases depressions and the bevelling of the sandstone plains of south Egypt. The gravels of these wadis resemble, in many ways, the colluvial deposits of the Nakheil formation known on the eastern side of the Red Sea range on top of the faulted Eocene blocks of the Quseir-Safaga reach (Chapter 18, this book).

The correlation of these gravel mounds with the Nakheil is of interest. The Nakheil is made up of coarse breccia, alternating with fine-grained lacustrine sediments made up of limestone, clay and variegated, brightly coloured siltstones. Thus during the Oligocene the eastern side of the newly elevated Red Sea range was occupied by lakes which apparently developed on the peripheries of the incipient Red Sea graben. On the western side of the range, however, fluvial conditions prevailed, the channels carrying boulders and pebbles of similar lithological nature to those of the Nakheil.

The Nakheil and the plateau gravel mounds seem to have been related to the faults which were responsible for the elevation of the Red Sea range and the formation of the Red Sea graben. The coarse breccia in the east and the conglomerates in the west seem to have been derived from the Eocene and Cretaceous rocks which covered the elevated Red Sea range. In the east they were not transported for long distances and thus do not have a great extent, while in the west they were transported over some distance due to the expanse of the land. The age of these formations is definitely post-early Eocene and pre-middle Miocene. Middle Miocene conglomerates overlie unconformably the Nakheil along the Red Sea coast. Since these formations were closely associated with the faulting movements which led to the formation of the Red Sea range and graben, it is justifiable to refer the age of these deposits to the first faulting episode during the early Oligocene.

2 *Late Miocene drainage system*
The most dramatic event which helped shape the modern landscape of Egypt occurred during the late Miocene when the Mediterranean Sea was isolated from the world oceanic system and then desiccated (Ryan et al. 1973). The land of Egypt was subjected to intense erosion. The modern Nile valley was excavated and developed as a subsequent stream to the numerous consequents emanating from the elevated lands to the east. These consequents were deeply incised in the plateaux which surrounded the river on both sides and which can still be seen to this day (Said 1981). While the drainage lines of the elevated plateaux of the Eastern Desert are still preserved, those of the western part of Egypt have been almost completely obliterated by later erosion and/or deposition of younger sediments. During the late Miocene, the Western Desert, like the Eastern Desert, must have had a drainage system which caused a major lowering of the land in response to the lowering of the Mediterranean sea level. Active sheet wash erosion resulted in the levelling of the sandstone plains of the Nubian desert and the excavation of the southern oases depression.

3 *Late Pliocene drainage system*
The late Miocene drainage system of the Western Desert is buried under the cover of later sediments or the rising waters of the sea. Buried channels of this system were discovered by the use of geophysical methods along the Marmarica plateau in Libya (Barr & Walker 1973) and along stretches of south Egypt and north Sudan (McCauley et al. 1982). The sediments which cover these buried channels seem to have been laid by a drainage system or systems which developed over the exhumed surface of the late Miocene erosional episode. The deposits were later bevelled to form the great sand sheets of the south Western Desert (*vide supra*).

No fossils have been recorded from these sheets so far, but to the north in the Siwa area, Zittel (1883) records several localities with (?)Neogene deposits carrying fresh water fauna and flora. These deposits have recently been mapped by the authors of the New Geological Map of Egypt and given the name Minqar El Talh formation. The formation is described as 'light-colored continental to lacustrine sandstone with root marks; yellow siltstone, capped by platy lacustrine limestone with borings and gastropods'. The age is given as undifferentiated post-Miocene.

Although it has not been demonstrated in the field that these deltaic deposits are connected with the extensive sand sheets of the south Western Desert, it is likely that the two are related. They could have been formed during the late Pliocene pluvial.

The above discussion makes it more fitting to classify the east-west channels described by McCauley et al. (1986) from the east Sahara with the late Pliocene drainage system rather than with the earlier system as claimed by these authors. According to McCauley et al. the channels represent a trans-Saharan drainage network which drained the Red Sea hills into sub-Saharan African basins and which, according to these authors, ceased when it was beheaded by the late Miocene Eonile. If the east-west direction of this system is accepted, it is possible that the entire drainage formed part of the tributary system of the late Pliocene Paleonile (*vide supra*). As has already been pointed out, a westward direction for the drainage seems most improbable in Oligocene-Miocene times for many of the north African mountains were elevated impeding any drainage from the Red Sea hills to the African basins. In addition, the presence of deltaic Oligocene deposits with rich African fauna and flora along the shores of the Tethys indicates a northward drainage system.

The sand sheets and the underlying gravel beds cover large stretches of the south Western Desert. Their absence from the oases depressions may indicate a later lowering of these deposits during a post-Pliocene pluvial. The sheets form coalescing alluvial cones which seem to have been fed by channels emanating from the highlands to the west and south. Some of these may have formed a feeder of the old Paleonile debouching into the Nubian Nile. It is also possible that others coalesced to form the great trunk channel extending from Gilf Kebir to Siwa Oasis and beyond. It is along this trunk channel that the modern eolian sand dunes of the great sand sea occur, and it is possible that the sand of the modern dunes was derived from the underlying sand sheets as they were bevelled by wind action.

4 *Pleistocene drainage systems*
Whereas the late Pliocene (?) sand sheets seem to have filled the lowlands produced by the late Miocene erosion, there is indication that the oases depressions were the sites of further lowering by a drainage system which seems to have been connected to the Mediterranean via the Nile. Recent drilling in the Tushka area, Nubia, has shown the presence of deep channels filled with early Pleistocene gravels which seem to have drained into the Protonile which carried similar gravels (Haynes 1980). The idea that the southern oases depressions were connected to the Nile was suggested by Ball (1927), but it was rejected

on the ground that no alluvial sediments or beach features had been described from these depressions. In addition, the southern oases depressions form interior basins with no external drainage, hence the theory that they were wind-excavated. With the discovery of a number of buried channels filled with Protonile sediments in Tushka, situated at the Nubian arch which separates the Nile from the oases depressions, Ball's idea gains credence. The oases depressions seem to have passed through a stage in which they did not form interior basins.

PART 5

Economic mineral deposits

CHAPTER 26

Mineral deposits

ABDEL AZIZ A. HUSSEIN
Geological survey of Egypt

(SECTIONS 2.5 AND 2.6 BY M. A. EL SHARKAWI)
Faculty of Science, Cairo University

1 INTRODUCTION AND HISTORICAL REVIEW

Mineral deposits, particularly those of gold, copper and gemstones, were known and exploited by the ancient Egyptians since pre-Dynastic time. The Egyptians were certainly able to smelt gold and copper and to produce bronze about 2500 BC. The amazing colors in the tombs of Thebes were produced by artists using the green of malachite, the blue of the turquoise and the purple of the amethyst. Craftsmen produced ovoid and even faceted beads of agate, beryl, chalcedony and garnet.

With increasing demand for gold, copper, pigments and gemstones, economic geology had its beginning in the recording of the mode of occurrences of these deposits, the formulation of crude theories of origin, and the organization of expeditions for the discovery and exploitation of ores. The first economic geologist known to us is the Egyptian Captain Haroeris who in about 2000 BC led a successful expedition to Sinai, prospecting for turquoise. Three months after setting out, he discovered and extracted large quantities of the mineral (Jensen & Bateman 1981).

Mineral deposits known to occur in Egypt include gold, copper, tin, tungsten, lead, zinc, nickel, chrome, iron, titanium, beryllium, talc, barite, asbestos, graphite, phosphate, marble and alabaster. Almost all of these have been exploited at one time or another but, at present, only iron, phosphate, talc and some ornamental and building stones are exploited commercially.

The systematic study of the mineral deposits of Egypt began in this century with the pioneering work of Hume, who made a comprehensive list of mineral occurrences in association with Precambrian rocks, with notes on stratigraphy, mode of occurrence and genesis of some deposits (Hume 1937). He grouped the mineral deposits into: occurrences of gold, occurrences of silver, copper, zinc, molybdenum, tungsten, iron, chromium, nickel, lead, tin, platinum and graphite, and occurrences of precious and semiprecious

minerals and ornamental stones. Hume noted that mineral deposits are not associated with Protarchean rocks, that gold deposits are of Metarchean age, and that the Samuiki type copper deposits are associated with the Eparchean period. He related the tin, tungsten and molybdenum ores to the quartz veins forming late injections in the Gattarian.

Amin (1955) classified the mineral deposits of the Eastern Desert in seven groups, six of which are of Precambrian age, while the seventh includes those of Miocene and younger ages. The seven groups, in descending order of age, are:

7 Lead-zinc mineralization, at the base of the Miocene deposits in the Red Sea coastal plain;

6 Tin-tungsten mineralization associated with post-Gattarian quartz veins;

5 Gold mineralization, hypogene, epigenetic auriferous quartz veins associated with post-Gattarian dikes;

4 Ilmenite mineralization, associated with some gabbroic intrusions;

3 Steatite mineralization, associated with epidiorites and other intrusions of basic to intermediate composition;

2 Chromite-magnesite-asbestos mineralization, in association with serpentine and talc-carbonate rocks;

1 Marble-graphite-magnetite mineralization, associated with schists, amphibolites and mudstones.

El Shazly (1957) classified the mineral deposits of Egypt on the basis of time relations and their supposed mode of formation. In his classification the mineral occurrences are grouped under Precambrian, Cretaceous, Miocene and Pleistocene to Recent deposits, with a further classification of the Precambrian group into early and late subgroups. The main features of El Shazly's classification are as follows:

Pleistocene-Recent

– Beach and eluvial placers (including black sands)
– Evaporites

Miocene

– Red Sea coast Pb-Zn and related ochre deposits

– Sulfur deposits
– Evaporites
– Manganese-iron deposits in Sinai and the Eastern Desert

Cretaceous

– Aswan iron ores
– Phosphate deposits

Late Precambrian

– Hydrothermal replacement deposits (steatite and talc, zinc and copper, copper)
– True hydrothermal fissure vein deposits (tin-tungsten, molybdenum, gold, barite)

Early Precambrian

– Deposits formed by magmatic segregation (ilmenite)
– Deposits related to old pegmatites (asbestos, vermiculite, beryl)
– Deposits formed from ultrabasic intrusions (chromite, peridot, nickel, magnesite, talc)
– Metamorphosed sedimentary deposits (bedded iron ores and graphite).

This classification by El Shazly was adopted by Said (1962), who introduced 'minor modifications to make the discussion in harmony' with the concepts introduced in his work.

In the years following publication of *The Geology of Egypt* by Said (1962), many refinements took place regarding mineral deposits in Egypt. The available data on mineral deposits were compiled and presented by Moharram et al. (eds) (1970), and new types of deposits were discovered as a result of the intensive exploration programs undertaken by the Egyptian Geological Survey. Hussein (1973) reported the identification of some of these new types, such as the tin-tungsten and beryllium disseminations in the albitized and greisenized parts of some Younger Granite masses (e.g. Muelha), anomalous concentrations of U, Mo, Pb and Nb in some of the alkaline ring complexes (El Naga) or in some granitic rocks (Abu Rushaid), and the probable existence of porphyry copper deposits at Hamash. Bugrov et al. (1973) introduced, in Egypt, the term apogranite to describe the autometasomatically altered granites in Egypt containing the paragenetic association Sn, W, Nb, Ta, Be, Mo and F.

Invanov & Hussein (1972), and later Ivanov et al. (1973), adopted Bilibin's concept (Bilibin 1955, 1968) for a metallogenetic classification of mineral deposits in Egypt. According to these works, three metallogenetic epochs are recognizable: Pregeosynclinal, Geocynclinal and Platform epochs. They found that data on the Pregeosynclinal epoch is lacking, but that nevertheless, some pegmatite veins in gneisses (e.g. Hafafit) may belong to this epoch. The Geosynclinal metallogenetic epoch was further divided into initial, synorogenic and terminal phases.

With the initial phase, Ivanov & Hussein (1972) included the metamorphosed iron ore deposits (Wadi Karim), ilmenite, chromite, and Zn-Pb-Cu-talc deposits (Samuiki). The synorogenic phase was considered to be either not very much mineralized, or else very deeply eroded. Still, it includes some quartz veins bearing copper or lead-zinc sulfides. The terminal phase includes mainly copper and gold mineralization in association with propylitized subvolcanic parts of the Dokhan volcanics (Hamash). The platform metallogenic stage includes Nb, Ta, Be, W, Sn, Mo, U and Li mineralization associated with the Younger Granites, the phosphate deposits and Aswan iron ores within the Nubia formation, the Nb, Th, U, V, Zr and Cu in ring complexes, as well as the Pb, Zn, Mn, Cu, S and Fe related to the Red Sea rifting and believed to be of Miocene age.

Garson & Shalaby (1976) presented a plate tectonic model for the evolution of the Arabian-Nubian shield, and related the mineral deposits of Egypt to this model of crustal evolution. According to them, the mineral deposits of the country can be categorized in three groups of varying geotectonic environments:

1. Mineralization associated with ophiolitic belts;
2. Mineralization associated with calc-alkaline porphyries;
3. Mineralization associated with granitic intrusions.

The mineralization associated with ophiolitic belts includes podiform chromites, asbestos, talc and Cu-Ni-Co mineralization associated directly with ophiolite suites, and Cu-Ni-Co mineralization and massive ilmenite in layered ultramafic complexes along probably transverse tectonic structures. The metamorphosed iron ore deposits of the Eastern Desert are closely related to ophiolitic rocks, since they were formed from a volcanic exhalative source in association with the development of frontal arcs comprising slices of ophiolitic rocks.

With the calc-alkaline and siliceous volcanic and subvolcanic intrusions, Garson & Shalaby (1976) associated the porphyry type Cu-Mo mineralization and alteration as well as the Kuroko type massive sulfides. As to the mineralization related to granitic rocks, these authors differentiated between the barren synorogenic granites and granodiorites on the one hand and the Younger Granites of the Pan African orogeny that are often mineralized on the other. The mineralization includes Sn-W and Sn-Mo in quartz sheets and Nb-Ta-Sn-Be mineralization in apogranitic phases.

Hilmy & Hussein (1978) felt the need for a new classification of the mineral deposits in Egypt to serve as a frame for the detailed study of these deposits and to facilitate their correlation with world-wide deposits. Such a classification would take advantage of the accumulation of geological information during the Geological Survey of Egypt's active decade, 1968-1978, the discovery of new types of deposits in the country, and the development of the plate tectonic concept in crustal evolution with its reflections on ore genesis and distribution of mineral deposits. They thus proposed a classification following the now widely accepted notion (Stanton 1972) that mineral deposits are integral parts of the petrological associations with which they occur, and that they may have formed in all the ways that ordinary rocks have formed. According to Hilmy & Hussein, mineral deposits and occurrences in Egypt may be classified in the following groups:

1. *Ores of mafic-ultramafic association*
   1. chromium-nickel-platinoid association
   2. iron-nickel-copper sulfide association
2. *Ores of felsic association*
   1. carbonatite-alkaline complexes association
   2. anorthosite iron-titanium oxide association
   3. quartz monzonite-granodiorite copper-molybdenum sulfide association (porphyry type deposits)
3. *Stratiform sulfides of marine-volcanic association*
4. *Stratabound ores of sedimentary association*
5. *Ores of vein and disseminated association*
6. *Ores of sedimentary affiliation*
   1. iron deposits
   2. manganese deposits
   3. true sedimentary deposits
      a) biogenic-chemical deposits
      b) clastic and placer deposits
      c) evaporites
7. *Ore deposits of metamorphic affiliation*

Pohl (1984) presented an extensive study of the metallogeny in the Pan African in east Africa, Nubia and Arabia. According to him, types of mineral deposits with probably economic potential in the Arabian-Nubian shield include:

1. Ores of mafic-ultramafic association
   1. Alpine type chromites, the largest of these being in the Ingessana hills, Sudan
   2. ilmenite-magnetite (e.g. Abu Ghalaga, Egypt) and Cu-Ni sulfides (e.g. Gabbro Akarem, Egypt) in layered gabbros
   3. massive magnetite lenses in gabbroic intrusions (Ankor, Sudan)
2. Syngenetic stratiform deposits
   1. banded iron oxide ores (Umm Nar, Egypt; Sawawin, Saudi Arabia); these are usually small and may be of distal volcanogenic-hydrothermal origin
   2. magnesite of evaporitic origin, forming a large lens at Jabal Rokham, Saudi Arabia

3. Volcanogenic deposits associated with island arc volcanism, including
   1. proximal massive sulfide deposits enclosed within acid volcanics and their pyroclastic equivalents, comparable to Canadian and Kuroko deposits and containing mainly copper with some zinc and lead and traces of gold and silver (Jabal Sayid, Saudi Arabia)
   2. distal lenticular massive Zn-Cu-Pb sulfide deposits with many sedimentary features (Nuqrah, Saudi Arabia)
   3. quartz veins with Au-Ag mineralization associated with acid subvolcanic intrusions (Mahd ad Dahab, Saudi Arabia)
4. Mineralization related to plutonic rocks
   1. albite-muscovite granites of calc-alkaline to peralkaline nature containing disseminated Ta-Nb-Sn (Abu Dabbab, Egypt) and W (Jabal Eyob, Sudan)
   2. pegmatites genetically related to the late granites, which may be mineralized with Sn, Be, Mo, U, Th, Nb, REE, with very limited economic significance
   3. quartz (-carbonate) veins with Au, Sn, W, Mo, Be and F, related to granite cupolas (Igla, Egypt); some veins with dominant Cu, Pb-Zn, Sb, F, Ba and Sr in Saudi Arabia may be associated with Najd faulting

2 CLASSIFICATION OF MINERAL DEPOSITS IN EGYPT: GENERAL STATEMENT

In the present synthesis, a framework classification is presented for the categorization and description of mineral deposits of Egypt. While this classification follows the general outlines of that presented by Hilmy & Hussein (1978), substantial modifications have resulted from the intensive and careful review of mineral deposits as a whole, and those of the country in particular, by the author. Within this proposed classification, mineral deposits are grouped and a list of deposits pertaining to each group is given along with a review of the geology and economic potentials of the more important ores. Also, the mode of formation and geotectonic environment of each group has been developed, in harmony with the crustal evolution models suggested for the shield in other chapters of this work.

As will be clear in the following paragraphs, most groups of mineral deposits are associated with the hard rocks of the Precambrian. However, the fifth and sixth groups of the classification given below, namely the stratabound ores of sedimentary association and the ores of sedimentary nature, are all found enclosed within the cover rocks of Paleozoic and younger ages, formed under epicontinental environments following cratonization of the shield. Thus, the classification followed here goes along the following lines:

1. Mineral deposits associated with mafic-ultramafic assemblages:
   1. in ophiolite sequences
      a) chromite deposits
      b) Cu-Ni-Co sulfide deposits
      c) asbestos, vermiculite, corundum, talc and magnesite deposits
   2. in layered mafic-ultramafic intrusions
      a) Cu-Ni deposits
      b) Ti-Fe oxide deposits
      c) Ni-bearing veins and peridot
2. Mineral deposits in felsic association
   1. in Ring Complexes
      a) mineralization in carbonatites
      b) aluminium raw material
      c) mineralization in ferruginated nepheline syenites
      d) radioactive veinlets
   2. Cu-Au porphyry type mineralization
   3. mineralization related to granites
      a) disseminated and vein molybdenum mineralization
      b) disseminated and vein tin mineralization
      c) vein tungsten mineralization
      d) disseminated and vein Nb-Ta mineralization
      e) beryllium mineralization
      f) fluorite mineralization
      g) uranium mineralization
3. Stratiform volcanogenic massive sulfide deposits and related talc
4. Precious and base metal vein type deposits
   1. dominantly gold veins
   2. dominantly base metals
   3. barite veins
5. Stratabound deposits in sedimentary sequences
   1. zinc-lead deposits
   2. stratiform copper
   3. sulfur deposits
   4. barite in sedimentary rocks
6. Ores of sedimentary nature
   1. iron ore deposits
   2. manganese ore deposits
   3. true sedimentary ores
      a) phosphorites
      b) coal deposits
      c) carbonates
      d) clastic and placer deposits
      e) evaporites
      f) weathering products
      g) sedimentary uranium deposits
7. Mineral deposits in metamorphic association
   1. metamorphogenic deposits

2. metamorphosed mineral deposits
   a) banded iron ore deposits
   b) marble deposits
8. Miscellaneous

2.1 MINERAL DEPOSITS ASSOCIATED WITH MAFIC-ULTRAMAFIC ASSEMBLAGES

2.1.1 *In ophiolite sequences*

a) *Chromite deposits*

Small and irregular masses of podiform chromites are frequent within serpentinized members of ophiolitic sequences in the Eastern Desert, mostly south of latitude 26° N. Almost all of the known occurrences have been worked out, and are described hereafter as a matter of academic interest. The best-known occurrences of chromites are tabulated below (table 26.1).

Other, less important sites include: Ras Shait (24° 51', 34° 34'); Wadi El Nakari (24° 51', 34° 50'); Wadi Khashab (24° 22', 34° 22'); Um Kabu (24° 34', 34° 56'); Wadi Gerf (24° 55', 34° 48'); Gabal Korabkansi (very high grade massive ore found as float by author in 1970), and many others of negligible importance.

The country rocks enclosing chromite lenses are serpentinites or serpentine-talc-carbonate rocks (products of alteration of the serpentinites) formed after harzburgite and dunite, at the very base of the cumulate ultramafic rocks of the ophiolitic sequences in the areas noted.

Anwar et al. (1969) studied some of the chromite occurrences in the Eastern Desert, showing that the chemical composition of these chromites do not show wide variations. Ivanov & Hussein (1972) concluded that these chromites are of the Alpine-podiform type, on the basis of the petrological assemblage with which they occur, the shape and size of the individual pods, and the chemistry of samples analyzed (high MgO/FeO, low Fe_2O_3, low Al_2O_3/Cr_2O_3, and high Cr/Fe ratios). No analyses are available for Pt in Egyptian chromites as yet, but Ni is always present in both the chromites and associated serpentinites.

It is believed that these chromites were formed through early crystallization followed by crystal settling from basic magmas at spreading centers, during the formation of new oceanic crust. This crust, with its enclosed chromites, was tectonically emplaced during accretion, prior to cratonization.

Production of chromite is very insignificant in Egypt, only a few hundred tons annually. It reached a climax of 998 tons in 1977, then dropped to its present negligible level due to exhaustion of the known lenses. Efforts must be concentrated on the exploration for new chromite deposits interstratified

Table 26.1

Occurrence	Lat.	Long.	Remarks
Gabal Moqassem	22°08'– 22°12'	33°55'– 34°10'	Six small lenses, 180 tons, medium grade ore at G. Moqassem; 9 small (1 × 10 m) lenses of small tonnage and medium grade at Um Domi
Um El Tiyur	22°15'	34°35'	14 lenses in talc-carbonate rocks of Gabal El Adrak, 1500 tons of medium grade ore
Sol Hamid	22°19'	36°11'	13 lenses, 630 tons; 48% Cr_2O_3
Wadi Allaqi area	22°40'	33°48'	
Um Krush			16 lenses, 1100 tons, 49% Cr_2O_3
Dyniyat El Gueleib			6 small lenses of high grade ore
Wadi Haimour			1 lens, 550 tons high grade ore
Wadi Arayes	23°35'	34°51'	33 small lenses enclosed in talc-carbonate rocks of G. Arayes
Abu Dahr	23°39'	35°08'	1 lens 2 × 10 m high grade (53.9% Cr_2O_3); some small lenses at G. Mastura and Wadi Betan
Wadi Ghadir area	24°44'– 24°57'	34°42'– 34°49'	
Wadi Ghadir			8 lenses, 4800 tons medium grade ore
Wadi Um Hegari			1 lens low grade ore (25% Cr_2O_3)
Wadi Lawi			10 lenses in serpentine-talc-carbonate rocks, medium grade ore (35% Cr_2O_3)
Dungash area	24°58'	33°24'	8 sites, several lenses in each, about 1000 tons medium grade ore
Abu Mireiwa	25°01'	33°52'	4 lenses, 195 tons, 36% Cr_2O_3
Wadi Um Khariga	25°02'	34°42'	4 lenses, 350 tons, 35% Cr_2O_3
Barramyia-Um Salatit area	25°06'– 25°07'	33°46'– 33°54'	9 sites, some 84 lenses at periphyry of serpentine-talc-carbonate rocks
Wadi Sifein	25°06'	34°47'	2 lenses, 250 tons, 35% Cr_2O_3
Kolet Um Homr	25°45'	34°15'	17 lenses, small reserves, around 42% Cr_2O_3
Gabal El Rabshi	26°09'– 26°15'	33°36'– 33°55'	18 sites, more than 100 lenses of massive chromite, more than 2700 tons, averaging 44% Cr_2O_3

with the lowermost parts of layered mafic-ultramafic intrusions, as well as the serpentinized segments of the ophiolite belts.

b) *Cu-Ni-Co sulfide deposits*

Copper-nickel-cobalt sulfide deposits are known to occur in two petrological assemblages of the mafic-ultramafic association, namely in the gabbroic members of ophiolite sequences and in layered intrusions. Both types are represented in Egypt by Abu Swayel and El Geneina for the former, and Gabbro Akarem for the latter. In this section, we only deal with those believed to be directly associated with ophiolitic sequences, the Abu Swayel and El Geneina deposits.

Abu Swayel deposit. This deposit is located at about 185 km south of Aswan, near the head of Wadi Haimour. A number of smaller but similar occurrences are known in the vicinity along Wadi Haimour. The area was worked by the ancients for copper and malachite. The main studies in Abu Swayel are those by Hume (1937), El Shazly (1957), Bassyouni (1960), DEMAG (1960), IPCO (1963), El Goresy (1964) and El Shazly et al. (1965). A shaft was sunk to a depth of 69 m and a total length of 1205 m diamond core drilling was carried out. The Aswan Mineral Survey Project restudied the area, emphasizing geophysical and geochemical exploration (Ivanov & Hussein 1972).

The orebody includes both massive and disseminated mineralization hosted in a lens-like body of amphibolite, 500 m long, 30 m wide, striking northwest-southeast with dips at 60 to 80° northeast. The amphibolite lens is surrounded by biotite-garnet schist of basic derivation. The amphibolite and biotite-garnet schist may represent the metamorphosed equivalents of the gabbro and the basalt of a dismembered ophiolite suite respectively.

The ore minerals include pyrite, pyrrhotite, chalcopyrite, pentlandite, bravoite, violarite, cubanite and ilmenite, with brochiantite, chalcanthite and malachite in the oxidation zone.

Ore reserves were estimated at 85,000 tons of ore containing 2.8% Cu and 1.53% Ni as well as minor amounts of Co. The copper-nickel sulfides were formed as a result of liquid immiscibility from the silicate melt during solidification of the basic magma into an oceanic crust.

El Geneina occurrence (23° 57' N, 34° 37' E). A gossan with copper and nickel secondary minerals was discovered in 1973 during a geochemical exploration program undertaken by the Aswan Mineral Survey Project. Detailed study of the area, including some shallow diamond core drilling, showed that it is an occurrence with no economic potential (Garson & Fredrickson 1975). Malachite and garnierite-stained gossans are associated with thrust slices of mafic-

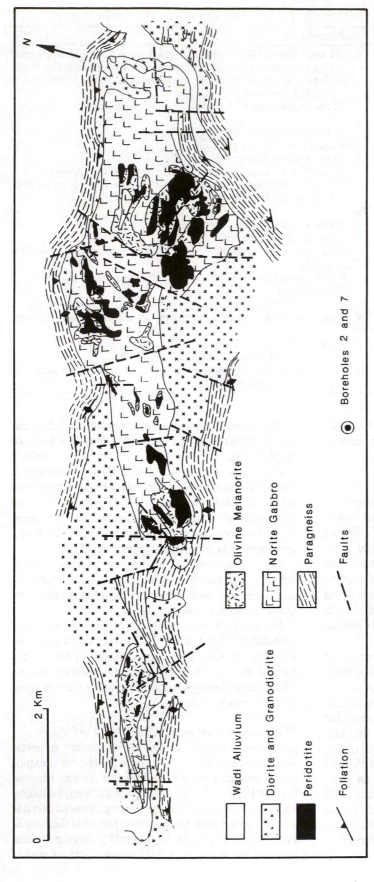

Figure 26.1 Geological map of Gabbro Akarem mafic-ultramafic complex (after Carter 1975).

Wadi Alluvium

Diorite and Granodiorite

Peridotite

Olivine Melanorite

Norite Gabbro

Paragneiss

Faults

Foliation

Boreholes 2 and 7

ultramafic rocks that include peridotite, pyroxenite and gabbros. Fresh ore minerals are represented mainly by pyrite, phyrrhotite, chalcopyrite and pentlandite. Core assay indicated 0.17% Cu and 0.38% Ni, but ore reserves are insignificant.

c) *Asbestos, vermiculite, corundum, magnesite and talc deposits*

Asbestos and vermiculite. Chrysotile asbestos is known to occur as very small veinlets, 1 to 2 mm in width, criss-crossing the serpentinized parts of the ultramafic masses almost wherever they crop out. None of these occurrences attains the grade or tonnage to warrant economic consideration. Nevertheless, anthophyllite asbestos occurs in association with vermiculite in a number of occurrences spread over an area of about 15 km^2 between 24° 28' to 24° 29' N and 34° 27' to 34° 47' E at Hafafit. The main occurrences are those of Wadi Shidani, El Duwaig, Um Groof, Um Kuhl, Um Fahm, Um Kisbash, Wadi El Hisa, and north of Bir Hafafit (Amin & Afia 1954). In the last-mentioned occurrences, anthophyllite has been mined since 1944 with an average production of 500 tons per year, together with smaller amounts of vermiculite.

Anthophyllite and vermiculite are restricted to serpentinized ultramafic masses (350 × 150 m at the north of Bir Hafafit occurrence) embedded in the gneisses and believed to represent blocks in a melange zone (Hassan, pers. comm. 1985). The ore is developed only where these serpentinite masses are intruded by pegmatite veins and veinlets. Here, and at the periphery of the pegmatite veins, vermiculite followed by actinolite, then anthophyllite and lastly talc, are developed within the serpentinite mass with the transformation of the pegmatite into a quartz-free, feldspar-mica rock which may occasionally bear corundum.

The formation of anthophyllite-vermiculite at Hafafit was attributed by Rasmy (1974) to bimetasomatic reactions between pegmatitic material and the ultramafic rocks, where alkalis and silica supplied by the pegmatites infiltrated the ultrabasic mass, producing successive, almost monomineralic zones of vermiculite, actionlite, anthophyllite and talc away from the pegmatite intrusion. The relative abundance of alkalis to silica determines the degree to which each of the anthophyllite or vermiculite zones are developed.

Corundum. Corundum occurrences are known in Egypt only at Hafafit, where it was first recorded by Amin & Afia (1954). The main occurrences are at Abu Nimr, Abu Merikhat, Um Karaba, Wadi El Hema and Abu Fahm. In all of these, corundum is restricted to plagioclasite pegmatite cutting through the serpentinite masses enclosed within the Hafafit gneisses.

At Abu Nimr a serpentinite mass is in contact with hornblende gneiss. The contact is followed by pegmatite veins, one of which, 80 m long and 80 cm wide, bears corundum, with the complete absence of quartz. Away from the serpentinite mass, similar pegmatite veins are quite rich in quartz. The ultramafic rock is altered into actinolite and vermiculite. Corundum constitutes between 5 and 60% of the vein, and forms either colorless crystals or crystals with pale pink or blue tints.

According to Rasmy (1974) the plagioclasite-pegmatite and its associated corundum were formed through a process involving the progressive bimetasomatism between the uprising pegmatitic material (probably the product of sweating from the gneisses) and the ultrabasic mass. The removal of silica and alkalis from the pegmatitic material during its ascent is the reason for the development of corundum.

Magnesite. Ophiolite related magnesite is known to occur at Semna, Khor Um El Abas, Sagia, Bir Mineih, Um Salatit, Gabal El Mayiet, Ambaout, Zargat Naam and Wadi Eikwan. It forms thin veinlets, stockworks and pockets in the serpentinized ultramafic masses, seldom exceeding a few meters in length and a few centimeters in width. The most important occurrences, exploited at one time or another, are khor Um El Abas, Sagia, Um Salatit and Ambaout. Total production reached a maximum of 800 tons in 1974, then dropped to almost nil at present due to the exhaustion of reserves. Magnesite was formed during the process of serpentinization to accommodate the excess Mg released during that process.

Talc. Talc forms as a product of metamorphism or hydrothermal alterations of Mg-rich rocks, especially ultramafics, dolomites and some tuffs. In Egypt, the mineral is always found in association with serpentine minerals, carbonates, silica and tremolite in the carbonatized serpentinites formed after the ultramafic members of the ophiolitic suites, locally known as Barramiya rock. This type of occurrence is nowhere susceptible to exploitation in the country. Most of the talc production in Egypt (Atshan, Darhib, etc.) comes from the intensively altered tuffs and volcanics in the Keel zones to volcanogenic massive sulfide deposits, even where the ore bodies themselves are not so well developed. Consequently, talc deposits will be dealt with in Section 2.3, in relation to the stratiform volcanogenic massive sulfide deposits.

1.2 *In layered mafic-ultramafic intrusions*

a) *Cu-Ni sulfide deposits*

At present, Cu-Ni sulfide deposits in layered intrusions are only known in the area of Gabbro Akarem in the south Eastern Desert of Egypt.

Gabbro Akarem prospect. In 1972, Gabbro Akarem (latitude 24° 00', longitude 34° 17') was discovered in a regional geochemical survey undertaken by the Aswan Mineral Survey Project (Bugrov & Shalaby 1973), and later studied in greater detail by the members of that project (Carter 1975). Here, Cu-Ni sulfide mineralization, both massive and disseminated, occurs within a mass of norite, melanorite and peridotite. The intrusion consists of two separate masses, together extending some 11.5 km in an east-northeast direction (Fig. 26/1). The main eastern body that carries the mineralization is 7.5 km in length and ranges in width from 1 to 2 km. It was previously thought of as a layered lopolith consisting of alternating layers of peridotite and gabbro (Bugrov & Shalaby 1973) but further work revealed that it is built up of successive intrusions of mafic-ultramafic rocks (Carter 1975). The rock types encountered are norite, olivine melanorite, peridotite, pyroxenite and fine grained basic dike rocks. Emplacement began by the intrusion of the olivine-poor rocks (norite and olivine melanorite) followed by two phases of olivine-rich rocks (peridotites), an earlier non-mineralized peridotite and later mineralized peridotite. Pyroxenite and basic dikes represent the last phase of intrusion. Gabbro Akarem is believed to be younger than the main episode of regional folding and metamorphism, and is definitely older than G2 granite, dikes of which cut through the mafic-ultramafic complex in many places.

On the surface, the mineralization is expressed in the form of three zones of gossans within the peridotites and is possibly related to massive sulfide bands. The sulfide mineralization occurs as disseminations or as massive bands and the sulfide grains are molded around the silicate crystals with no signs of replacement. This indicates that the sulfides are magmatic and constituted an integral part of the original magma. Primary, sulfide assemblage includes pyrrhotite, pentlandite, chalcopyrite, and cubanite. Pyrrhotite and pentlandite are replaced partially by a secondary sulfide assemblage that includes pyrite, marcasite, violarite and mackinawite (Rasmy 1982). Secondary minerals developed in the gossans include malachite, chyrsocolla and garneirite.

Investigations of Gabbro Akarem have included geophysical, geochemical and geological studies, supplemented by an intensive program of diamond core drilling. On the basis of these investigations, reserves were estimated at 700,000 tons of mineralized peridotite at a grade of 0.95% combined Ni and Cu, of which 270,000 tons of grade 1.18% Ni+Cu are in the drill proven category. It is evident that neither the grade nor the tonnage permit consideration of exploitation under present circumstances (Carter 1975). It was also concluded that the ore was formed as a result of pre-intrusion segregation, followed by the emplacement of successive phases, starting with norite and ending with the mineralized peridotite which represents the residual sulfide bearing fraction of the primary magma.

Gabbro Akarem was emplaced in association with a deep-seated transverse tectonic structure trending east-northeast. It is therefore suggested that, like most of the layered mafic-ultramafic intrusions in the world (Eckstrand 1984), Gabbro Akarem was formed from a mafic magma, mantle-derived in most cases, which was emplaced quiescently in multiple phases at higher crustal levels in a tensional rift environment.

b) *Ti-Fe oxide deposits*

A number of Ti-Fe oxide deposits are known in Egypt, in association with mafic-ultramafic masses that include rocks ranging in composition from melagabbro-melanorite to anorthosite. These occurrences are: Hamra Dome, Abu Dahr, Wadi Rahaba, Um Ginud, Wadi El Miyah, Um Effein and Abu Ghalga areas. In all of these occurrences, the ore is present as massive lenses or disseminations of magnetite, hematite, ilmenite, rutile and apatite. In addition to Fe and Ti, Cr and V are the principal minor constituents, sometimes with traces of Cu. Many studies have been made of these deposits (Hume 1937, Amin 1954, 1955, Attia 1950, Holman 1954, Nakhla 1954, Amer & Abdel Tawab 1958, Abdel Tawab 1961, Khairy et al. 1964, Basta 1970, Basta & Girgis 1968, 1969, etc.). In all these deposits, with the exception of Abu Ghalga, the tonnage is very limited (a few hundred thousand tons), and the TiO_2 content is relatively low (16 to 22%). We shall present Abu Ghalga as an example.

Abu Ghalaga. The Abu Ghalaga deposit occurs in a hill overlooking Wadi Abu Ghalga, 20 km west of the port of Abu Ghosun. The host rocks include meta-gabbro, noritic gabbro and anorthosite that show primary banding or layering. The gabbroic mass is emplaced within older volcanic and pyroclastic rocks of a dominantly andesitic composition and is intruded by G2 granite. Thus, the age of Abu Ghalga is most probably similar to that of Gabbro Akarem.

The mineralization occurs as bands or lenses of massive ore intercalated with the gabbro layers, or it forms disseminations gradational between the massive ore bands and enclosing gabbro. The main band extends 350 m in a northwest-southeast direction and is 50 m wide. It dips at 45° to the northeast. The massive ore includes ilmenite, magnetite, hematite, rutile, goethite and anatase representing about 70% of the ore, silica minerals 28% and sulfides 3% (Basta & Takla 1974). The chemical composition of the ore was given by Moharram et al. (1970) as:

TiO_2	36.36–49.9%	V_2O_5	0.52%
FeO	24.5–28.59%	CaO	0.1%
Fe_2O_3	17.82–28.30%	MgO	2.18–2.93%
Al_2O_3	0.61–3.40%	SiO_2	2.20–7.70%
P_2O_5	trace	S	0.03–0.99%

The deposit was reinvestigated by the Geological Survey in 1974-75, resulting in detailed geological maps and a total of 3000 m of diamond core drilling (Naim 1978). This study showed that the ore averages 150 m in thickness and that it extends beyond the limits of the exposed bands. Ore reserves are given as more than 45 million tons, with an average grade of about 35% TiO_2. The area is under consideration for exploitation.

Similar to Gabbro Akarem, the Abu Ghalga deposit is located along another of the transverse tectonic

structures running east-northeast (Garson & Shalaby 1976), and is related to rifting. The iron and titanium oxide phases were separated by crystal settling or filterpressing during crystallization of the gabbroic magma to form syngenetic bands and segregations of massive ore. The discordant dike-like orebodies, especially at Abu Dahr, were formed through the separation of late stage, Fe-Ti-P rich immiscible liquid and its intrusion into the lithified parts of the gabbroic mass.

At Wadi El Miyah (25° 17', 34° 00'), and in a gabbroic mass ranging in composition from normal to olivine gabbro, two small lenses of ore are known to occur. The lenses are of olivine, clinopyroxene and amphibole, with 33 to 35% opaques. The opaques form the matrix to the silicates and contain up to 38% ilmenite, 14% magnetite, and a substantial amount of apatite (Basta 1970).

c) Ni-bearing veins and peridot

At St John's Island, off Ras Banas in the Red Sea, Ni-bearing and peridot veins are known to occur (Moon 1923, Nassim 1949). The island presents a problem, since it is not clear whether it constitutes a part of an ancient ultramafic mass caught up in a transverse tectonic structure or if the peridotites are

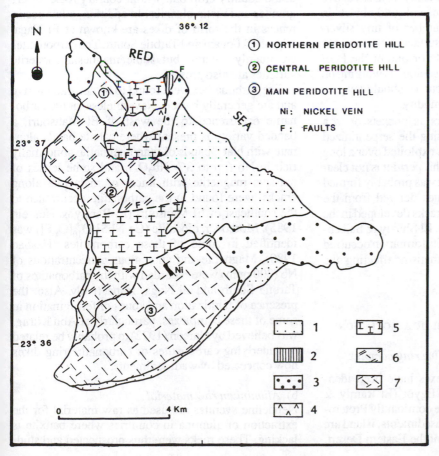

① NORTHERN PERIDOTITE HILL
② CENTRAL PERIDOTITE HILL
③ MAIN PERIDOTITE HILL
Ni: NICKEL VEIN
F : FAULTS

Figure 26.2 Geological map of St John's Island. 1. Younger coral reef and alluvium; 2. Basalt; 3. Older coral reefs; 4. Evaporites; 5. Detrital calcareous sediments; 6. Basic metavolcanics; 7. Ultramafic metavolcanics.

younger intrusions related to sea floor spreading. Garson & Shalaby (1976) placed it in their mineralization groups with layered ultramafic complexes within transverse tectonic structures, and this point of view is adopted here. The island (Fig. 26/2) is made up essentially of serpentinized peridotite, surrounded by coral reefs and covered in places by Recent basalt (El Shazly & Saleeb-Roufaiel 1972).

Two garnierite-bearing veins striking east-northeast extend for about 50 m with a width varying from 0.6 to 2 m. The ore consists of garnierite and a mixture of iron oxides and hydroxides, with a Ni content of 5 to 9%. These veins were exploited in 1937-1938. Analysis of the vein material gave the following composition:

Ni	4.86%	Cu	0.25%
Fe	12.25%	Au	0.19 ppm
Pt	0.93 ppm	Ag	6.2 ppm

Reserves were estimated at 5000-6000 tons of ore.

The secondary ore minerals were presumably leached from the peridotite mass, where they may have existed as disseminated primary sulfides, and redeposited in open fractures.

It is interesting to note that the only recording of platinum in Egypt is that in the analysis listed above for the Ni ore. Nevertheless, several gold objects in the Cairo Museum show a number of tiny silver-white metallic specks on the surface. They are not silver, but most probably are Pt or one of the PGE (Hume 1937). Prospectors and geologists working on the ultramafic masses in the country should keep an eye open for this valuable commodity.

Peridot (gem olivine) occurs as pockets or in a network of veinlets criss-crossing the serpentinized peridotite mass. It was formerly exploited over a long period of time. The genesis of this peridot is not clear but, according to Said (1962), it was probably formed by the crystallization of silicates derived from the partly consolidated melts into cracks developed in the lithified parts of the same mass. El Shazly & Saleeb-Roufaiel (1979) suggested that its formation occurred through a process of metasomatism affecting the ultrabasic rocks.

2.2 MINERAL DEPOSITS IN FELSIC ASSOCIATION

2.2.1 *In carbonatites and alkaline ring complex*

Some 15 alkaline ring complexes have been identified, mapped and studied in Egypt (El Ramly & Hussein 1985). They intrude the dominantly Proterozoic basement of gneisses, metasediments, island arc volcanics and older granitoids of the Eastern Desert.

They are circular or elliptical in plan, a few kilometers in diameter, and include a wide variety of rocks with syenites and equivalent volcanics dominant. They range in age from Cambrian (554 Ma for Wadi Dib) to late Cretaceous (89 Ma for Abu Khrug). All the complexes show alkaline affinity and carbonatites are well-developed only at El Manosuri, the southernmost of the ring complexes in Egypt.

During prospecting activities of the Aswan Mineral Survey Project, U, Nb, Mo and Cu geochemical anomalies were encountered in the ring complexes of El Naga, Nigrub El Fogani and Abu Khrug, but no deposits of any significance were discovered. Nevertheless, the presence of very pronounced zones of alteration, as well as radiometric and geochemical anomalies suggest the possibility of the future location of ore deposits now concealed.

Ring complexes, carbonatites and associated mineralizations are rift related features. They were formed in association with the rifting episodes that followed cratonization at the end of the Pan African event (El Ramly & Hussein 1985).

a) *Mineralization in carbonatites*

To the southeast, north and west of the El Mansouri ring complex, the largest occurrences of carbonatites in the country crop out, within country rocks of G1 granites and metavolcanics. In addition, other occurrences in the form of dikes are known at El Naga, Nigrub El Fogani and Tarbtie South. The carbonatites are mainly sövites, but some magnesite-forsterite varieties are also present.

The carbonatites were found to be non-radioactive and are generally barren. Nevertheless, at the carbonatite occurrence to the west of El Mansouri, a banded variety is present where calcite bands alternate with black bands of titanomagnetite abnormally rich in Nb (up to 3% Nb_2O_5), and some lenses of massive magnetite with traces of malachite along cracks, were found in the carbonatite occurrence to the southeast of El Mansouri (El Ramly & Hussein 1985). Moreover, niocalite ($Ca_4NbSi_2O_{10}[O, F]$) was identified in some of these carbonatites (Hashad 1980). Magnetite and anomalous concentrations of Nb (300 ppm) were recorded with the carbonatites of Tarbtie South and El Naga, respectively. Also, the presence of fenitization and zones of ferrugination in some of these complexes, particularly at Abu Khrug, was believed by Garson (1972) to probably be related to underlying carbonatites or carbonatite ring dikes now concealed by wadi sediments.

b) *Aluminium raw material*

Nepheline syenites are used as raw material for the extraction of alumina in countries where bauxite is lacking. These rocks were thus prospected and stud-

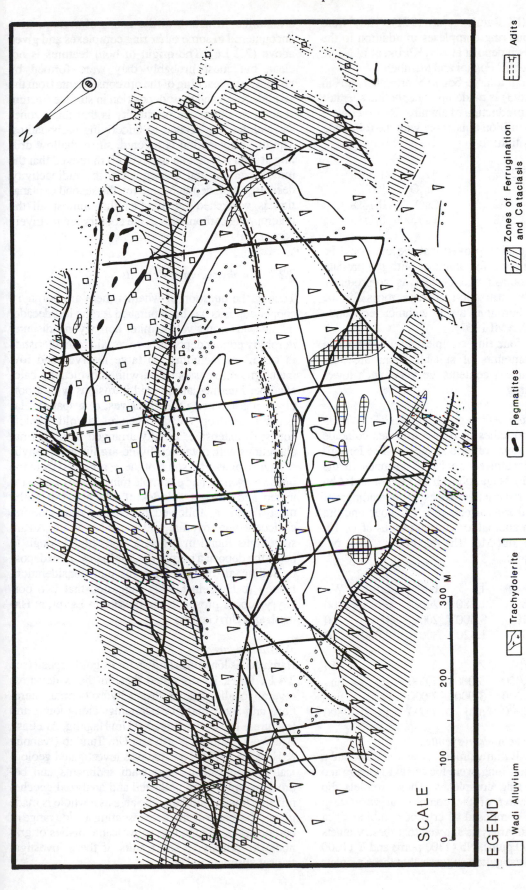

SCALE

| | | | | | |
0   100   200   300
M

LEGEND

Wadi Alluvium

Aegirine-augite Foyaite

Biotite Aegirine-augite
Foyaite

Dykes
Trachydolerite

Tinguaite

Bostonite

Pegmatites

Zones of Zeolitisation

Zones of Cataclasis

Zones of Ferrugination
and Cataclasis

Faults

Xenoliths

Adits

Figure 26.3 Geological map of prospected part of the nepheline syenite at Abu Khrug ring complex (after El Ramly et al. 1970).

ied in Egypt (El Ramly et al. 1971). They were identified in four ring complexes in addition to the previously known deposit at Abu Khrug, at El Naga, El Khafa, Nigrub El Fogani and Mishbeh.

At Abu Khrug, a large body covering 0.7 km² in outcrop (Fig. 26/3) is made up of nepheline syenites suitable for the production of alumina. Reserves were estimated at 26 million tons of ore with the following overall composition:

SiO_2	55.24%	Al_2O_3	21.63%
Fe_2O_3	4.80%	Na_2O	9.6%
K_2O	4.94%	CaO	0.74%
MgO	0.15%	TiO_2	0.11%

Technological testing proved that the ore could be successfully sintered for alumina, with the production of soda, potash and cement as co-products. Accordingly, the utilization of these rocks for the production of aluminium awaits technical-economic consideration (VAMI 1967).

In the other four ring complexes with nepheline syenites, the amounts of suitable rocks are very limited and alumina contents are very much lower than at Abu Khrug.

c) *Mineralization in ferruginated nepheline syenites*
Radiometric anomalies in ring complexes coincide with extensive zones of reddish, gossan-like ferruginated and feldspathized nepheline syenites in the inner parts of El Naga, Nigrub El Fogani and Abu Khrug. These radioactive zones gave readings of several hundred micro-roentgens per hour and are associated with anomalous concentrations of U, Th, Pb, Nb, Zr, Be and Mo. The maximum values obtained are, in ppm:

	U	Th	Pb	Zn	W
A. Khrug	560	1,180	200	na	100
El Naga	510	8,400	200	450	200
N. Fogani	130	1,420	300	2,000	200

	Nb	Zr	Y	Be	Mo
A. Khrug	2,000	3,000	1,000	20	na
El Naga	3,000	3,000	2,000	na	na
N. Fogani	3,000	maj	1,000	1,000	800

d) *Radioactive veinlets in syenites*
Systems of small thin veinlets, a few cm thick and a few tens of meters long, were located at El Gezira and Tarbtie North ring complexes. These veinlets dip towards the center of the complex at very steep angles and are composed of cryptocrystalline silica and iron oxides. Analysis showed that these vienlets contain U (1000 ppm), Nb (1300 ppm) and Y (1600 ppm). Field investigation suggests that these vienlets

are similar in nature to the zones of ferrugination encountered in some other ring complexes and given above (2.2.1.c). The origin of both features is not clear, but most probably they were formed by groundwater leaching of the ore constituents from the alkaline rocks and their deposition in surface fracture zones. An alternative possibility is that these zones and veinlets are related to underlying carbonatites. The first explanation is favored, since shallow drilling at Abu Khrug to a depth of 40 m proved that the mineralization, and consequently the radioactivity, decreases with depth. Moreover, the second explanation is weakened by the fact that almost all the outcropping carbonatites in the country are relatively barren.

2.2.2 *Porphyry copper deposits*

Though the question of whether there are porphyry copper deposits has been pursued for the last decade, the answer is not yet definite, nor beyond dispute. Porphyry copper deposits have certain characteristics in common, including: a) large reserves and low metal grade; b) association with subvolcanic, calc-alkaline, intermediate to acid intrusions with a porphyry phase among the intrusives; c) a spatial relationship to deep crustal fractures or old Benioff zones; d) extensive, zonally arranged hydrothermal alterations with potassic-phyllic-argillic and propylitic zones, or at least some of these, arranged from the inside outwards, and e) simple mineralogy of disseminated and stringer sulfides of which pyrite is the most abundant, followed by chalcopyrite, bornite, chalcocite and covellite, and where Mo and/or Au are sometimes found in concentrations high enough to name the deposit a porphyry Cu-Mo or Cu-Au deposit. Taking all of these features into consideration, Ivanov & Hussein (1972) suggested that two porphyry copper prospects are present in Egypt, at Hamash and Um Garayat.

a) *Hamash area*
Hamash has long been known for its gold deposit (see 2.4.1). Later, it was noted that, in the wider area, several localities have strong hydrothermal alterations and show malachite stainings along joints and fractures. These were noted at Um Hagalig, Ara East, Ara West, Hamash North and Um Tundub (Ivanonv & Hussein 1972). The area was investigated geologically, geophysically and wadi sediments and bedrock samples were collected and analyzed geochemically. It was found that the area as a whole is made up of different volcanics representing a wide range of composition, intruded by subvolcanic bodies of granodiorite porphyry. The results of these investigations is summarized here.

At Um Hagalig, a zone of intensive propylitization, sericitization, kaolinization and silicification occurs within the granodiorite porphyry. A number of quartz veins with malachite cut the granodiorite porphyry and were excavated in the past. In the geochemical samples, anomalous values for copper (up to 0.5%) were found in close vicinity to the veins, but did not extend far beyond them into the granodiorite porphyry. Mo values (up to 50 ppm) were obtained in a few scattered samples. No pronounced electrical conductors were found in the geophysical surveys.

At Ara East and Ara West, quartz-sericite-pyrite zones are characteristic. Analyses showed isolated patches of anomalous copper (200 ppm) and Mo (80 ppm) concentrations (Bugrov 1972). Most of the two areas show values near to background. The drill holes penetrated vertically through rocks with up to 30% pyrite, but with no other sulfides, to a depth of 200 m.

Hamash North is more interesting geologically, with abundant quartz veins with pyrite-chalcopyrite, as well as the presence of gossans and zones of hydrothermal alterations up to 500 m long and 100 m wide, within the andesite-granodiorite porphyry. Analysis of samples from the zones of alteration yielded anomalous concentrations of Cu (200 to 5000 ppm) and Mo (10 to 50 ppm) (Bugrov 1972).

Um Tundub was chosen by Ivanov (1975) as the most promising locality at Hamash, and it was hence studied in more detail. Here, alterations are most pronounced and two vertical drill holes, 250 m each, passed through a zone of intensive development of pyrite. The zone of hydrothermal alteration is almost circular and measures about 2000 × 1700 m, excluding the outermost propylitized rocks. Within this zone, concentric, more or less complete, subzones are recognized with propylitization to the outside, pyrophyllitization-sericitization in the middle, and an inner subzone of silicification, alunitization and development of a small amount of hydrobiotite (Ivanov 1975). Drilling showed that pyrite is disseminated and fills cracks and fractures in the whole length of the core down to a depth of 250 m. This body of pyrite is estimated to contain 500 million tons of pyrite, but other sulfides are almost completely absent. It is established that in porphyry–type deposits, copper is always located towards the inner margins of the pyrite zone. At Um Tundub, the drilling of vertical holes in the pyritic zone, with sites selected before the pattern of hydrothermal alteration was fully understood, did nothing to help reach conclusive results as to the presence or absence of copper mineralization. Thus, for the various sites of Hamash, taking the positive indications of porphyry copper deposits noted above into consideration as well as the failure of this drilling to encounter copper minerali-

zation, four possibilities are suggested:

a) economic porphyry-type mineralization never developed, due to the virtual absence of copper from the mineralizing solution;

b) the present level of erosion is above the zone of disseminated and stringer copper mineralization, implying that mineralization is present at depth but that the drill holes (of the 1970s) were unfortunately located in the wrong places and drilled vertically in the wrong direction;

c) the intrusion and its associated mineralization, of Proterozoic age, has been eroded down below the level of mineralization.

Field evidence, analytical results and the presence of Cu-Au bearing quartz veins in the area, among many other factors, make the present author favors (b) above.

b) *Um Garayat*

Um Garayat (22° 34' N, 33° 24' E) was located when a very pronounced airborne electromagnetic anomaly (Lockwood 1968) coincided with a quartz porphyry stock some 3 km to the southeast along the strike of the Au-bearing quartz veins exploited at Um Garayat gold mine (see 2.4.1). Here (Fig. 26/4), the area is made up of different assemblages of granodiorite and quartz-andesite porphyries. The granodiorite porphyry intrusion shows concentric, almost complete zones of hydrothermal alteration, and very rich sulfide mineralization, now in the oxidized state and forming gossan-like bodies. The central core of quartz porphyry (granodiorite) suffered intensive silicification, sericitization, pyrophyllitization, and development of minor hydrobiotite. These are followed outwards by kaolinization and propylitization. Pyrite mineralization is superimposed on these alterations, which pass gradually into less altered rocks, then fresh rocks. Within the silicified core, minor quartz veins and lenses with apatite, tourmaline and occasional specks of native gold are encountered.

Geophysical investigations revealed a central conical electrically conductive body surrounded by concentric zones of varying electrical conductivity, corresponding to the zones of different hydrothermal alterations (Krs 1972). Moreover, samples selected from shallow pits over the electrically conductive bodies showed anomalous concentrations of As, Mo, Cu, Au and Ag (Krs 1972). The sequence of events leading to the mineralization of Um Garayat, as visualized by Ivanov (1975) must have begun with the emplacement of andesite-granodiorite porphyry in a subvolcanic environment, as indicated by the textures and lack of tuffs. The parent magma was intermediate to acid in composition and rich in volatiles. The emplacement was immediately followed by propylitization of a wide area caused by circulat-

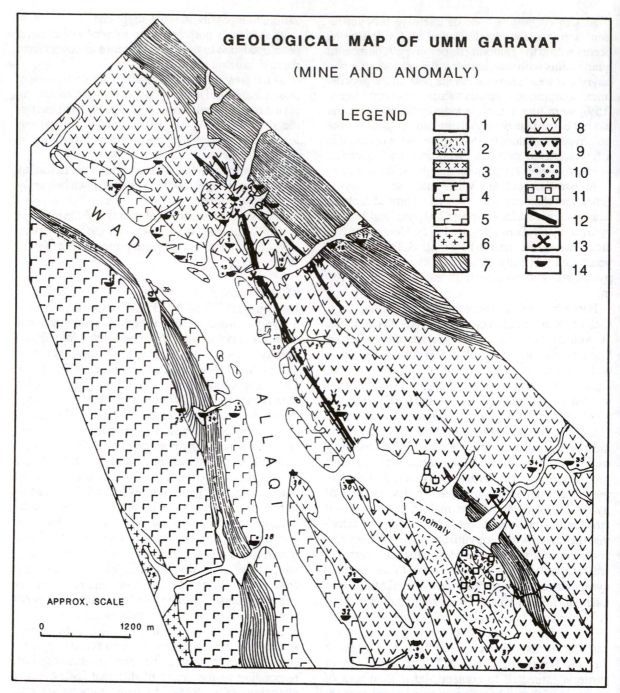

Figure 26.4 Geological map of Um Garayat (modified after Ivanov & Hussein 1972). 1. Wadi alluvium; 2. Quartz porphyry; 3. Granodiorite porphyry; 4. Gabbro; 5. Fine-grained gabbro; 6. Granite; 7. Metasediments; 8. Layered rocks; 9. Andesites; 10. Zones of silicification; 11. Zones of argillic alteration; 12. Gold-bearing veins; 13. Old workings; 14. Sampling sites.

ing groundwater heated by the magmatic bodies. The magmatic phase of hydrothermal alterations started by the formation of a little hydrobiotite or orthoclase (potassium metasomatism), high temperature si- licification and pyrophyllitization. This stage was accompanied by the apatite and tourmaline-bearing

quartz lenses, and small amounts of pyrite, pyrrhotite and chalcopyrite. Another phase of silicification fol- lowed, and with it most of the pyrite and some of the native gold were introduced. The third, low tempera- ture phase of silica introduction was accompanied by most of the gold and silver, mainly in the form of veins in the mined area (Ivanov 1975). Again, as is the case at Hamash, the absence of copper in surface exposures is explainable in two ways: (a) copper was not introduced, being very poor in the mineralizing solutions or (b) it was formed but subsequently leached out almost completely from the zone of oxidation but is present deeper, probably with a supergene enrichment zone.

The nature of the parent magma, the pattern and types of hydrothermal alterations as described by Lowell & Gilbert (1970), as well as the geophysical and geochemical results, favor alternative (b), though the answer awaits properly located deep drilling.

It is interesting to record here that when the report by Ivanov & Hussein (1972) was given for review by the UNDP to Dr R. Sillitoe, expert and consultant on porphyry copper deposits, he agreed with the author's conclusions that both Hamash and Um Garayat represent porphyry copper deposits worthy of proper exploration (Sillitoe, pers. comm. 1978).

2.2.3 *Mineralization related to granites*

In Egypt, granitoids constitute about 40% of the Proterozoic shield rocks, cropping out mainly in the Eastern Desert, south Sinai and the southernmost part of the Western Desert. Hussein et al. (1982) classified these granites into three groups: G1, G2 and G3. According to these authors, G1 granites are subduction related, formed in old Benioff zones, and characterized by being calc-alkaline in nature, closely associated with island arc volcanics, formed under compressional environments, and are I-type magnetite series granites. Moreover, these G1 granites usually form large intrusions that include rocks ranging in composition from diorites to granites, with SiO_2 contents not exceeding 65%. Their Nb and REE contents are always less than 20 and 50 ppm respectively and their radioactivity in the field does not exceed 10 μ R/h. This group of G1 granites has been proven to include most of the granite masses previously referred to as 'Old', 'Shaitian', 'Grey', 'Synorogenic' or 'Aswan' granites. On the other hand G2 granites were formed as a result of suturing and are characterized by being S-type, ilmenite series, calc-alkaline granites, formed under environments of compression. They tend to form relatively small plutons with a narrow range of composition, always granitic, with SiO_2 contents higher than 65%. They may contain accessory monazite, zircon and cassiterite. Their Nb is less than 50 ppm, REE content low, and radioactivity between 10 and 20 μ R/h. They cover most of the 'Younger', 'Gattarian', 'Pink', 'Red' or 'Postorogenic' granites referred to by earlier workers. The third group, G3, includes the small, simple intrusions of alkali to peralkaline intraplate anorogenic granites.

These are characterized by being S-type ilmenite series granites formed under extensional environments and have a very limited range of compositions, with SiO_2 between 70 and 75%. The accessory minerals may be monazite, zircon and cassiterite. These granites are rich in Nb and REE (higher than 50 ppm) and in outcrop, their radioactivity is higher than 40 μ R/h. They are frequently partly greisenized and/or albitized.

Many mineral deposits in Egypt are associated with these granitic rocks, covering a wide variety of ore mineral types and spread over the central and northern parts of the Eastern Desert, where granitic rocks predominate. With G1 type granites, no appreciable mineralization is known to occur, apart from some scattered minor base metal bearing quartz veins. Most of the Sn, W, Mo, Nb-Ta, REE, Be, and F deposits are associated with the G3 granites, but some of them may also be associated with G2 granite masses. In the following paragraphs, the mineral deposits are dealt with according to the dominant ore present, though one or more of the other elements of this paragenesis is always present.

a) *Disseminated and vein type molybdenum mineralization*

Hume (1937) reported the occurrence of quartz veins with molybdenite traversing 'Pink granite' at Gabal Gattar, Um Harba, Wadi Dib, Abo Marwa, Gabal Um Disi and Gabal El Dob, in the northern part of the Eastern Desert. In all six localities, the quartz veins trend mainly in a submeridional direction and dip steeply eastwards. Some of these veins extend for more than 700 m but their thicknesses are always very limited (10 to 20 cm), and Mo content in them is always low. These same occurrences were restudied by Dardir and Gad Allah (1969) and the following descriptions are based on their work.

At Gabal Gattar, mineralization occurs at the intersection of 27° 05' 29" N and 33° 16' 10" E, and is in the form of a zone of veins, 25 m wide and 500 m long, cutting through the pink granites of Gabal Gattar. Individual veins vary from 1 mm to 10 cm, and they pinch and swell, bifurcate and join other veins. They generally strike in the N 10° E and dip at about 80° to the east. The mineralized zone persists with the same attitude and thickness in underground workings at depths of 46 and 136 m below the surface. Molybdenite occurs in the quartz veins and as disseminations in the enclosing granite. It forms coarse crystals, 1 to 3 cm in size, concentrated in the veins or arranged along their contacts with the country rocks. Some thin veinlets are made wholly of ore. Disseminations in the granite extend for 10 to 15 cm from the veins in both sides. Molybdenum was assayed in an 8 m wide zone and its values ranged from 0.27 to

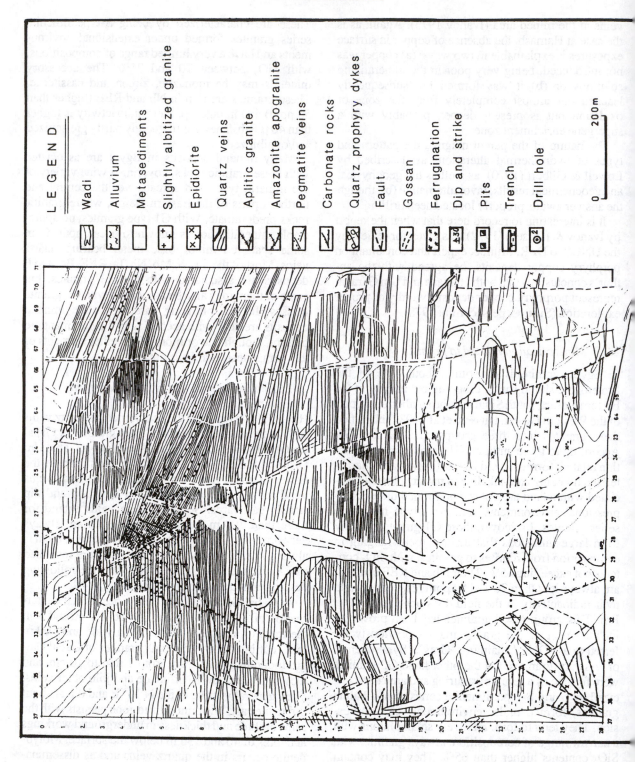

LEGEND

	Wadi
	Alluvium
	Metasediments
	Slightly albitized granite
	Epidiorite
	Quartz veins
	Aplitic granite
	Amazonite apogranite
	Pegmatite veins
	Carbonate rocks
	Quartz prophyry dykes
	Fault
	Gossan
	Ferrugination
	Dip and strike
	Pits
	Trench
	Drill hole

0   200m

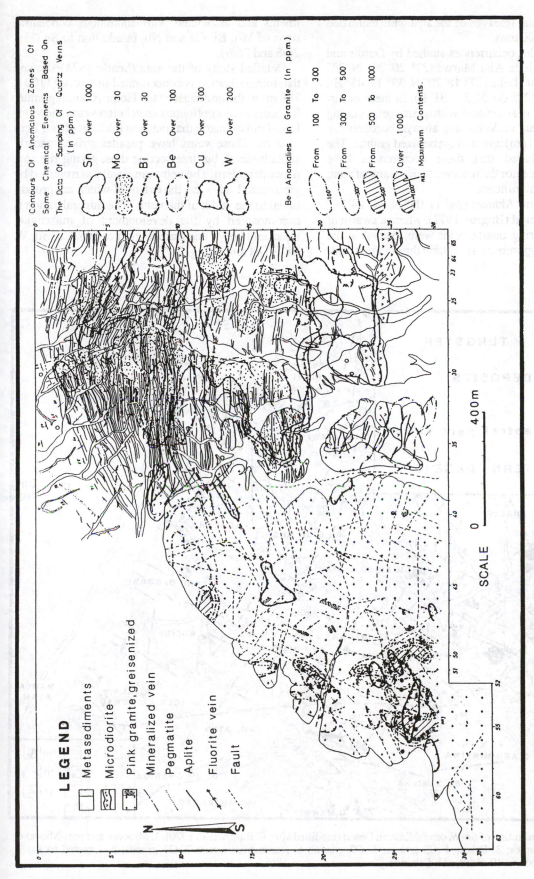

Figure 26.6 Geology and geochemistry, Homr Akarem (after Bugrov 1972).

2.25% Mo. Total reserves of the area were estimated to be about 4500 tons.

Other, smaller occurrences studied by Dardir and Gad Allah include Abu Marwa (27° 20' 20" N, 33° 05' 45" E), Abu Harba (27° 18' 20" N, 33° 12' 45" E), and Um Disi (27° 00' N, 33° 31' E). In these occurrences, the ore is associated with quartz veins cutting across red granite. Veins are always bordered by narrow (5 to 10 cm) zones of greisenized granite. The authors concluded that these occurrences have neither the grade nor the reserves to make any of them economically significant.

In 1969, Homr Akarem (24° 11' 08" N, 34° 04' 39" E) was discovered (Bugrov 1972), where a swarm of cassiterite-bearing quartz veins were located in the vicinity of the granite mass, cutting through metasediments and associated with anomalous concentrations of Mo, Bi, Cu and Nb, in addition to Sn (Fig. 26/5 and 26/6).

Detailed study of the area (Searle 1974) showed that mineralization is concentrated in a zone (1100 × 800 m) to the northeast of the Homr Akarem granite. It occurs in discontinuous en echelon veins and veinlets of milky quartz that range in thickness from 1 cm to 2 m. These veins have parallel strikes, mostly sublatitudinal, but intersecting dips, giving rise to a network pattern. The veins are always surrounded by greisenized selvages, the extent of which are proportional to the width of the veins. The mineralization is accompanied by the development of muscovite, topaz and beryl, and introduction of Mo, Cu, Sn, W, Bi, F, OH and S (Fig. 26/7).

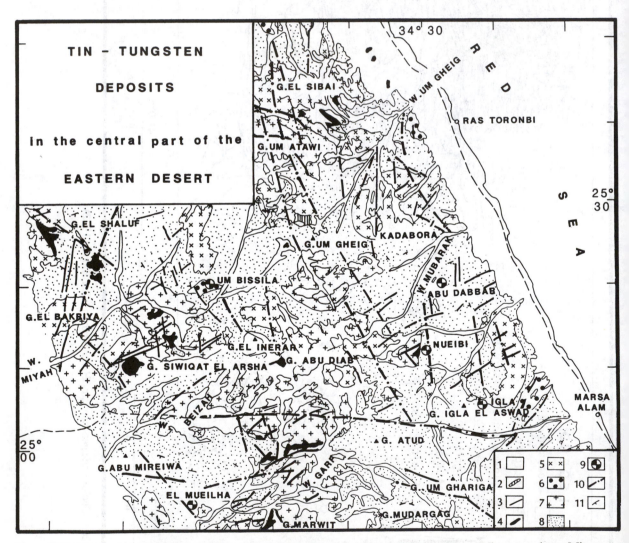

Figure 26.7 Distribution of Sn-W, central Eastern Desert (modified after El Ramly et al. 1970). 1. Miocene and post-Miocene; 2. Tertiary volcanics; 3. Dikes; 4. G3 granite; 5. G2 granite; 6. Hammamat series; 7. G1 granite; 8. Layered rocks; 9. Tin-tungsten deposits; 10. Faults; 11. Foliation.

Mineral zoning is evident in the field, and is confirmed by mineralogical and chemical analysis of the core samples. Here, predominantly Mo mineralized core is surrounded by a cassiterite zone, followed by a border zone with Cu mineralization. Mo mineralization is not extensive on the surface and occurs mainly as flakes enclosed within the quartz, or as paint on slicken-sided surfaces. The greatest concentration of molybdenite is in thin veinlets filling fractures in the veins. On the surface, powellite takes the place of molybdenite. Cassiterite, chalcopyrite and trace amounts of W and Bi are present, fluorite is more common in the Sn zones, and Be is present as a geochemical anomaly in all the three zones. On the surface, oxidation products are powellite, limonite and various secondary copper minerals.

Homr Akarem was investigated geologically, geochemically, and geophysically. Five drill holes totalling 1357 m explored the most interesting and promising anomalies. On the basis of these studies, Searle (1974) concuded that the effective area of mineralization measures about 240×400 m, in which molybdenite is the only mineral of economic importance, but it is not evenly distributed in the area or within individual veins. The average ore grade was calculated by him to be 0.031 Mo, and the reserves at about 8,000,000 tons. Thus, Homr Akarem can be classified as a subeconomic deposit.

In addition, vein and disseminated Mo mineralization was discovered in the granite mass of Wadi Hafia (24° 56' 40" N, 34° 07' 46" E). Here, a small mass of 'Younger Granite' (500×300 m) is emplaced in a country rock of schist, acid volcanics and gabbro-amphibolite. The granite is greisenized in part, and with it is associated a swarm of quartz veins that extend for more than 1 km across the mass and its enclosing rocks. In the veins, Mo is present as powellite. Samples collected from the veins and greisenized granite showed anomalous concentrations that reached 300 ppm for Mo, 80 ppm for Sn and 200 ppm for each of Cu, Be and Bi (Bugrov 1974).

b) *Disseminated and vein-type tin mineralization*

Tin is one of the metals searched for and exploited by the ancient Egyptians, who used it for the production of bronze. It was most probably obtained primarily from the Muelha mine, as indicated by the hieroglyphic inscriptions still preserved on the rocks along the wadi leading to the mine. Until recently, tin as cassiterite was known only in quartz veins, sometimes with wolframite, mainly at Igla, Nuweibi, Abu Dabbab, and Muelha. Most of these locations were described by Amin (1947), who stated that the mineralized veins are made up of massive milky quartz with cassiterite, wolframite, topaz, beryl, yellow mica, tourmaline, fluorite, sericite, and chlorite, traversing various rock types that include metavolcanics, metasediments, diorites and granites (Amin 1947). The veins are always `confined to the margins of granitic masses and their contacts with the country rocks. The veins are believed to be genetically related to muscovite granite masses, always with signs of post-magmatic alterations, mainly greisenization. Data available on the cassiterite and/or wolframite veins were summarized by El Ramly et al. (1970) in the table reproduced below (table 26/2). The location of these localities was given by El Ramly et al. (1970) (Fig. 26/7). It is interesting to note that in all these deposits the veins are well-defined fracture fillings, with greisen rims along their contacts. Most of them strike northeasterly and dip steeply to the southwest. Zones of mineralization, usually with tens of veins, are about 500 m wide and extend for more than a kilometer. The veins vary in thickness from a few centimeters to 1.5 m. The ore minerals identified are cassiterite, wolframite, beryl, chalcopyrite, chalcocite, scheelite, huebnerite, columbite, tantalite, malachite, iron and Mn oxides, covellite, azurite, chrysocolla and Bi-bearing ochre (El Ramly et al. 1970). Cassiterite and/or wolframite are always present along the margins of the veins, either directly embedded in the quartz or associated with the micas of the greisen zones. The gangue minerals present, in addition to quartz, are zinnwaldite, muscovite, sericite, fluorite, tourmaline, topaz, orthoclase, albite, calcite, chlorite and kaolin. Wall-rock alterations associated with these veins are mainly greisenization, whereby the country rocks, granitic or otherwise, are changed into flakes and aggregates of mica, quartz, some fluorite and topaz. In some of the occurrences (Igla, for example), monomineralic fluorite or beryl veins may be present. The Sn content in most of these veins is relatively low, ranging at Igla between 0.04 and 0.94%, with limited reserves.

It was only recently (Bugrov et al. 1973) that cassiterite, in association with W, Nb-Ta, Be and other lithophile elements, were discovered as disseminations in autometasomatically altered granites (apogranites). During the verification of an airborne radiometric anomaly, this type of mineralization was found at Gabal Muelha, some 3 km to the southeast of the old mine. This was followed by the discovery of similar situations at Igla, Abu Dabbab, Nuweibi and many other sites. This suggests that at almost all the occurrences cited above, both vein and disseminated types of deposits are present.

Gabal Muelha (24° 52' N, 34° 00' E) is a dome-like mass of biotite granite intruded into metasediments. The granite is albitized especially at its southwestern contact, where it is altered into an albite-microcline-quartz-Li mica rock. Greisens form lenses and veins of quartz and mica with some impregnations of

Table 26.2. Mineralogical composition and mode of occurrence of the tin-tungsten deposits

Deposits	Abu Hammad	Fatira El Beida	Abu Kharif	El Dob	Maghrabiya	Umm Bissilla	Abu Dabbab	Nuweibi
Ore minerals	Wolframite Huebnerite	Wolframite	Wolframite	Wolframite Fe and Mn oxides	Wolframite	Wolframite	Cassiterite Wolframite	Cassiterite Wolframi(te) Columbit(e) Tantalite Malachite
Gangue minerals	Quartz Mica Orthoclase	Quartz	Quartz Fluorite Mica Orthoclase	Quartz Mica Orthoclase	Quartz Mica	Quartz	Quartz Yellow mica Fluorite Topaz	Quartz Yellow m(ica) Fluorite
Country	Red granite between two dolerite dykes	Granite	Grey granite near felsite and andesite dykes	Grey, white, pink granites, near basic dykes	Granite	Hammamat greywackés	Muscovite granite and green stone	Greenston(e) near smal(l) nite intrus
Strike, angle and direction of dip	E-W, < 90°	NE, < 30° NW	Older lodes: E-W, < 30° N Younger: NW or NE, < 70° NE or NW	25°-30° NE < 30°-70° SE	NE, < 80° NW N-S, < 90°	NW, < 26°-35° NE	N-S and NE, and E-W, 40°-75° SE	NE, N-S (and) E-W, < 40 SE E and
Number of ore bearing veins	14 veins	A great number of veins		Series of veins	Several veins (3 main ones)	Series of veins (3 main ones)	Series of veins	Numerou(s) veins

fluorite, cassiterite, powellite and iron-copper sulfides. These greisen zones are closely associated with the main fractures, the best developed zone being that found at the intersection of the main northeast and northwest trending faults that gave rise to a central depression in the mass, where total greisenization of the rock took place. Cassiterite, columbite and beryl were identified in panned samples and as disseminations in albitites and greisens. Geological, geochemical and radiometric surveys showed that the greisenized and albitized parts are radioactive and associated with anomalous concentrations of Sn (500 to 1000 ppm), Nb (500 to 800 ppm), Mo (up to 2000 ppm), Bi (up to 300 ppm) and Be. The latter occurs as euhedral crystals, a few centimeters long, in thin pegmatite veins, especially at the western contact of the mass. The mineralization at Gabal Muelha is not economic as such. Nevertheless, wadis draining the mass were investigated (Fig. 26/9) and trenched. Panned samples were found to contain cassiterite, magnetite, zircon, columbite, garnet and ilmenite. Cassiterite content is about 500 gm/m³, and the total volume of alluvium is around 100,000 m³, so a small deposit of alluvial cassiterite is there, about 500 tons.

The Muelha mine (24° 54' N, 33° 55' E) was exploited in the past, where cassiterite was obtained from the quartz veins and panned from alluvium. Here, a mass of intensively greisenized granite covers an area of 0.15 km² and intrudes a country rock of metamorphosed sediments and volcanics (Fig. 26/8). Seventy quartz veins were mapped (El Ramly et al. 1959), of which 23 carry both cassiterite

...gla	El Muellha	Zargat Nåm	Gash Amer
Cassiterite Wolframite	Cassiterite Wolframite Beryl	Wolframite	Wolframite
Chalcopyrite Chalcosite Scheelite	Chalcopyrite Scheelite		
Columbite ? Tantalite ? Malachite Fe and Mn oxides Covellite Azurite Chrysocola	Malachite Azurite Bismuth-bearing ochres		
Quartz Yellow mica Fluorite Muscovite Tourmaline Enstatite	Quartz Yellow mica Fluorite Quartz-II Muscovite	Quartz	Quartz
Contact zone of pink granite and metavolcanics	Muscovite granite and metasediments	Granite	Granite
3° NE up to < 85° SE, large cellular stockworks	NE, < 25°-65° SE	i	Almost horizontal
Over 20 veins	About 70 veins		Series of veins

was concluded that the area is not economic because of the limited reserves and low grade of ore. The presence of more extensive zones of greisens below the metasediments may alter the picture, but further drilling is required (Bugrov 1972).

The Igla mine (25° 06' N, 34° 39' E) was also exploited in the past, the ruins of a tin smelter still being there. Here, a terrain of acid to intermediate volcanics and tuffs is intruded by a small mass of apogranite with disseminated cassiterite. Quartz veins related to this apogranite, and occurring in association with its contact, carry cassiterite, wolframite and beryl. There are more than 60 such veins, all striking northeast and dipping to the southeast. They are always bordered by greisen selvages of zinnwaldite, tourmaline and fluorite. Cassiterite crystals are restricted to the sides of the veins, while wolframite, if present, is usually embedded in the inner parts of the veins. Some chalcopyrite and secondary malachite may be present. Panned alluvial samples from the wadis draining the apogranite contain cassiterite, tantallite-columbite, thorianite and monazite (Kochin & Bassiuni 1968). Samples representing the mineralized veins gave the following range of metal contents:

Sn	$0.001 - 0.5$ %
W	$0.0005 - 0.06$%
Be	$0.001 - 0.6$ %
Li	$0.001 - 0.1$ %
Fluorite	$1.0 - 5.0$ %
Topaz	$1.0 - 2.0$ %

Proven reserves were estimated at about 1700 tons of ore containing some 7 tons of metallic tin. Probable reserves may provide more than 60 tons Sn.

Abu Dabbab (25° 20' 27" N, 34° 32' 20" E) is another deposit where vein type cassiterite has been known for some time. Recent studies (Sabet et al. 1976) showed that the veins are related to a mass of apogranite that contains essentially Ta-Nb mineralization. This is dealt with in section 2.3.d below. In this area, a small apogranite stock is intruded into metasedimentary country rocks. Some 58 cassiterite-bearing quartz veins and veinlets are known, ranging from 50 to 600 m in length and from 10 to 50 cm in thickness. The veins are made up of massive quartz with cassiterite, some wolframite, zinnwaldite, fluorite and topaz. Ore reserves in the veins are very limited (16,000 tons of ore), but there are some 500,000 tons of alluvium at 0.1% Sn in the wadis draining the mass.

At Nuweibi (25° 12' N, 34° 30' E), Nb-Ta mineralization impregnates zones in the Nuweibi apogranite mass, while Sn is present in quartz cassiterite veins. The Nb-Ta mineralization is dealt with in section

and wolframite, 30 carry only cassiterite, and 17 are barren. The quartz veins strike northeast and dip to the southeast at angles between 25 and 65°. They contain cassiterite, wolframite, scheelite, chalcopyrite, beryl, fluorite, Li-mica and supergene malachite and limonite. The greisenized granite is sometimes banded and contains cassiterite and wolframite as disseminated minute grains or as clusters and pockets. Analyses showed that Sn values range from 0.1 to 0.3%. Other elements present are Nb (up to 300 ppm) and Be (100 to 1000 ppm). Anomalous values of W were sporadically spread, and high values of Bi (100 to 1000 ppm) were found only in quartz veins with sulfide mineralization. The zone of mineralization was checked with three shallow holes (45 m each) and Sn was found to average 0.1%. It

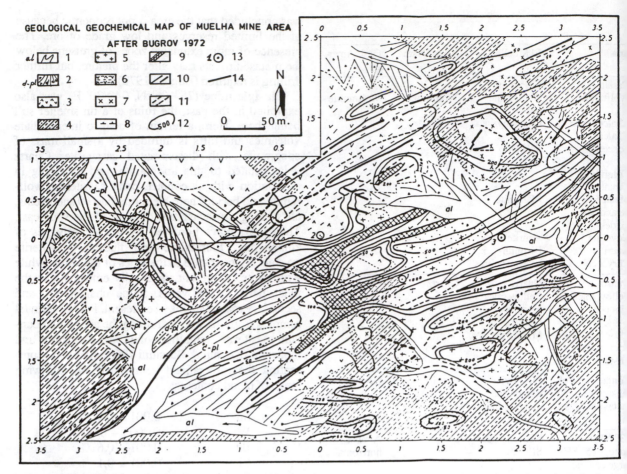

Figure 26.8 Geological-geochemical map of Muelha mine area (after Bugrov 1972). 1. Wadi alluvium; 2. Deluvium-proluvium thickness; 3. Metavolcanics; 4. Metasediments; 5. Greisenized granite; 6. Metavolcanics and metasediments; 7. Diorite; 8. Serpentinite; 9. Greisen and greisenized zones; 10. Acidic (granitic) dike; 11. Quartz vein; 12. Tin anomalies, based on spectrometry of bedrock; values in parts per million; 13. Numbered drill hole; 14. Quartz vein with Sn over 1000 ppm.

2.3.d. The zone of cassiterite veins occurs at the east of the apogranite mass, and includes 72 veins ranging in length from 100 to 800 m and in thickness from 25 to 30 cm. They strike northeast and dip to the southeast. The veins have thick mica selvages and are made up of massive milky quartz with cassiterite, wolframite, some molybdenite, tantalite, beryl, malachite, azurite and limonite. Distribution of cassiterite in the veins is highly irregular, and Sn grades vary from 0.003 to 0.3%. Vein and alluvial tin reserves were given at 94,000 tons of ore containing 0.05% Sn (Kochin & Bassiuni 1968).

In addition to the localities reviewed above, potential deposits of Sn may be expected in many of the sites listed in table 26/3, with which pronounced Sn anomalies are associated.

c) *Vein-type tungsten mineralization*

Tungsten occurs together with Sn in many of the deposits cited in 2.3.b above, where it is always of secondary importance. Nevertheless, there are a number of deposits where W is the main, if not the only, commodity. These are reviewed below.

At Abu Hamad (27° 31' N, 33° 20' E), 14 veins are associated with a red granite mass cut by dolerite dikes. The veins are almost vertical and of a latitudinal trend, and range in thickness from 2 to 30 cm. They are composed of quartz, feldspar and mica with wolframite irregularly distributed. Scheelite was identified microscopically in some sample of these veins. Chemical analysis of the wolframite itself gave a WO_3 content ranging from 46.4 to 69.05%, but no estimates of reserves or grade are available (Dardir & Abu Zeid 1972).

At Fatira El Beida (26° 48' N, 33° 20' E), a number

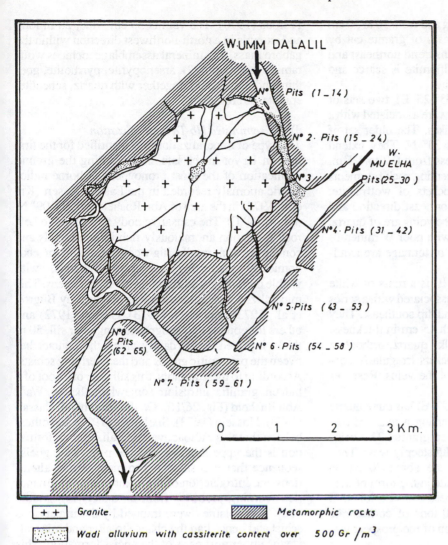

Figure 26.9 Distribution of tin-bearing alluvium and location of pits dug around Gebal Muelha (after Bugrov 1972).

Table 26.3 List of sites with Sn

Locality	Lat.	Long.	Remarks
Abu Zarb	25°13'N	34°12'E	Pegmatoidal greisenized granite (0.0003–0.0006% Sn)
El Ineigi	25°13'N	34°09'E	Endocontact of El Ineigi granite mass (0.001–0.0006% Sn)
Um Backra	25°16'N	34°08'E	Muscovite-albitized granite with fluorite (0.0003–0.001% Sn)
Thettabi	25°21'N	33°46'E	Mass of muscovite granite (0.005–0.002% Sn)
El Abbadiya	25°23'N	33°41'E	Medium grained biotite granite (0.0003–0.006% Sn)
Hamdallah	25°33'N	33°54'E	A mass of biotite granite (0.005–0.006% Sn)
Sheikha Oteyfa	25°19'N	33°51'E	Muscovitized granite (0.0003–0.0006% Sn)
Atshan	25°09'N	33°34'E	Mass of muscovite granite (0.0003% Sn)
El Arsha	25°11'N	33°51'E	Muscovite granite (0.001% Sn)
Rod Aid	25°13'N	33°55'E	Muscovite granite (0.0003–0.002% Sn)
Um Naggat	25°30'N	34°15'E	Greisenized and albitized marginal zone of Um Naggat granite (0.015–0.2% Sn)
El Shalul	25°27'N	33°40'E	Albitized muscovite granite (0.008% Sn)
El Backriya	25°18'N	33°41'E	Radioactive and altered zones in biotite granite (0.0004% Sn)
Rod Ishab	25°08'N	34°06'E	Pegmatite body (0.001% Sn)
Um Safi	25°20'N	34°08'E	Chalcopyrite and pyrite (0.003–0.02% Sn, 0.006% W)

of quartz veins up to 500 m long and from 10 to 30 cm thick, are associated with a mass of granite cut by numerous basic dikes. The veins trend northeast and dip 30° to the northwest. Wolframite is scarce and scattered inrregularly in the veins.

At Abu Kharif (26.48' N, 33° 25' E), two sets of wolframite-bearing quartz veins are associated with a granite mass cut by felsitic dikes. The older set of veins strikes east and dips at 30° N. They extend about 150 m, range in thickness from 0.2 to 1.5 m, and are made up of quartz cementing rock fragments and irregularly distributed pockets of wolframite. The younger set strikes in a northeast direction and dips very steeply northwest. The veins are of quartz, orthoclase, fluorite and mica, with poor W mineralization. No estimates of grade or tonnage are available.

El Dob (26° 27' N, 33° 28' E) is a mass of white granite, greisenized in places, associated with a series of veins that strike northeast and dip southeast. They extend for 1 km and may reach 45 cm in thickness. The veins are made up of milky quartz, orthoclase and mica, with wolframite pockets irregularly concentrated in the central parts of the veins. Reserves seem to be very limited.

Maghrabiya (26° 22' N, 33° 27' E) has three quartz veins bearing wolframite traversing a mass of partially greisenized and kaolinized granite. The veins strike north or northeast and dip steeply west. They extend for some 150 m and are about 30 cm in thickness. They carry wolframite in the form of irregularly distributed pockets. These were exploited in 1938-39, producing some 180 tons of concentrate (52% WO_3). No later evaluation of the prospect was made.

Um Bisilla (25° 21' N, 34° 01' E) is located in the Hammamat sediments and associated volcanics that are intruded by a small mass of muscovite-albite apogranite. Some 50 small mineralized veins strike mainly in a northwest direction and dip northeast. The veins are made up of quartz with some calcite, chlorite, muscovite and barite, and include wolframite, magnetite, pyrite, chalcopyrite and malachite. These veins were assayed at 0.006 to 0.01% W, but the reserves are very limited (El Ramly et al. 1970).

Zargat Naam (23° 46' N, 34° 41' E) is a very limited occurrence, where three or four quartz veins carrying W mineralization occur in the marginal zone of a Younger Granite mass.

Gash Amer (22° 18' N, 36° 12' E) is another area of very limited W mineralization where a granite mass is traversed by almost horizontal fissure filling quartz with some wolframite.

At the base of Gabal Atud a gold-bearing quarz-wolframite vein was reported by Saleeb-Roufaiel & Fasfous (1970). The vein is 30 cm wide, 20 m long and extends in a north-northwest direction within the gabbro mass. The mineral assemblage includes wolframite, tungstinite (?), arsenopyrite, pyrrhotite, goethite and native gold, together with quartz, scheelite, epidote and rutile.

d) *Disseminated Nb-Ta mineralization*

This type of mineralization was identified for the first time in Egypt in the late sixties during the ground verification of the most pronounced airborne radiometric anomaly recorded in the Eastern Desert (Krs et al. 1973) in the area of Abu Rusheid (24° 37' 09" N, 34° 46' 04' E). The causative body was found to be a rock formation anomalously rich in rare metals and containing abnormally high concentrations of economically interesting accessory minerals, with visible columbite and zircon in hand specimens. This mineralized rock formation was believed by Bugrov et al. (1973), Krs et al. (1973), Garson (1972) and others to represent an apogranite tongue or sill, 50 m thick, intruded concordantly along the contact between the psammitic gneiss and the overlying schists. According to these authors, this sill is an offshoot of a hidden granitic intrusion somewhere below Wadi Abu Rusheid (Fig. 26/10). On the other hand, Hassan (1973), Hussein (1973), Snelgrove (1972) and others believe that this radioactive mineralized rock formation is the uppermost part of the psammitic gneiss sequence that was subjected to metasomatic alterations and introduction of ore components by emanations coming probably from an underlying granite body. The volatiles were trapped by the impervious schist and hence had the chance to alter and mineralize the uppermost parts of the gneiss formation. This dispute has not been settled. The mineralized rock is a quartz-albite-microcline rock with the following mineralogical composition (Bugrov 1972):

Quartz	15–40%
Albite	25–35%
Microcline	10–15%
Li-mica	5–15%
Ore minerals	1–05%

Detailed studies included radiometric survey, geochemical bedrock sampling, and geological mapping (Fig. 26/11). This was supplemented by diamond core drilling of six holes totalling 407 m, two of which were to trace the mineralization down the dip, and the other four to check pronounced anomalies. The study showed that the ore minerals present are columbite (Nb_2O_5:Ta_2O_5, ratio 5:1 to 8:1), cassiterite, monazite, xenotime, fluorite, zircon, thorite-thorogummite, and sulfides of iron, zinc, copper, lead and molybdenum (Knorring 1971). Dril-

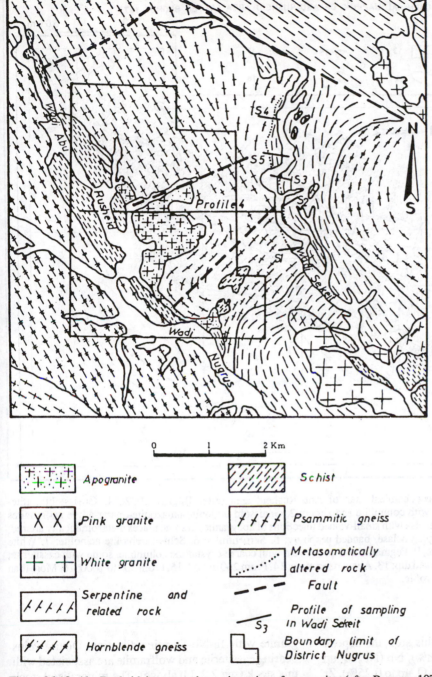

Figure 26.10 Abu Rusheid, interpreted as an intrusion of apogranites (after Bugarov 1972).

Legend:

Apogranite

Pink granite

White granite

Serpentine and related rock

Hornblende gneiss

Schist

Psammitic gneiss

Metasomatically altered rock

Fault

Profile of sampling In Wadi Sekeit — S_3

Boundary limit of District Nugrus

ling indicated that the mineralization decreases gradually downwards, and the thickness of the mineralized section ranges from 45 to 50 m in the central part of the body, to 25 to 30 m in the southernmost part, and then into thin zones of 1 to 5 m. This is clearly shown below:

Average contents of some minerals in core samples (%)

Depth (m)	Nb_2O_5	Ta_2O_5	W	Sn	Zr	Li
4.6– 21.0	0.250	0.023	0.013	0.054	0.200	0.090
21.0– 38.6	0.022	0.013	–	0.030	0.200	0.065
38.6– 48.7	0.064	0.008	–	0.017	0.090	0.020
48.7– 60.2	0.066	0.007	–	0.006	0.200	0.040
60.2–173.0	0.010	0.004	–	0.010	0.067	0.014

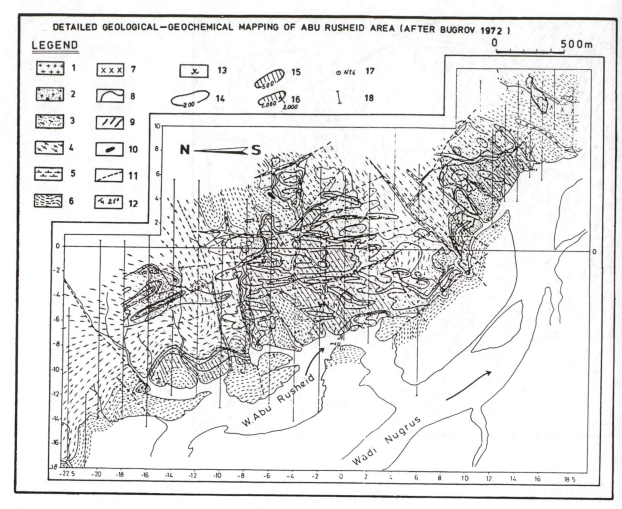

Figure 26.11 Detailed geological-geochemical map of Abu Rusheid area (after Bugrov 1972). 1. Gneiss, lithionite-microcline-quartz-albite apogranite with columbite and zircon; 2. Gneissoid lithionite-amazonite-quartz-albite apogranites with columbite and zircon; 3. Orthogneiss with small-veined injections of apogranite, in some places metasomatically altered; 4. Orthogneiss: mica-quartz, mica-plagioclase, banded porphyry; 5. Serpentinite; 6. Schist; tremolite-acinolite; 7. White granite; 8. Small layers of apogranite; 9. Pegmatite and quartz vein with coarse-crystalline columbite, some with cassiterite; 10. Quartz plug; 11. Fault; 12. Strike and dip; 13. Abandoned mine; 14. From 200 to 500; 15. From 500 to 1000; 16. More than 1000; 17. Numbered drill hole; 18. Profile.

Thus, it was concluded that, in this area, the deposit contains Ta (Ta_2O_5, up to 0.0328%), Nb (Nb_2O_5, up to 0.3%), Sn (up to 0.3%), Li (Li_2O, up to 0.25%), Zr (1%), U (up to 0.86%), Th (up to 1.43%), and traces of Cu, Zn, Pb and Mo. Of these, the Nb and Ta are the most interesting. Abu Rushaid is therefore an ore body containing 90,000 tons of Nb_2O_5 and 13,000 tons of Ta_2O_5, at a cutoff grade of 0.02% Ta_2O_5. If and when worked, some Sn, Zr, Li and other elements will be obtained as byproducts.

Abu Dabbab (25° 20' N, 34° 32' E) is in a country rock of paraschists that include quartz-biotite and quartz-chlorite schists with intercalated beds of tuffs and agglomerate, with an intruded stock of apo-

granite with Ta-Nb-Sn mineralization. Quartz veins bearing cassiterite and wolframite are associated with this stock (see 2.2.3.b above). The mineralized stock covers an area of 3.2 km² and is made up of quartz-albite or quartz-albite-amazonite lithionite. Mineralization is most intensive in the northeastern endocontact zone of the mass, which is 30 m wide. This mass has been studied in detail, drilled, and assayed. Ta_2O_5 ranged from 0.0095 to 0.075%, with an average of 0.028% (cutoff grade, 0.01% Ta_2O_5). Nb_2O_5 values varied from 0.002 to 0.029%, average 0.008%. Abu Dabbab is thus primarily a tantalum deposit, as the Ta_2O_5:Nb_2O_5 ratio averages 3:1. Proven ore reserves were calculated at ten million tons, containing 2800 tons Ta_2O_5 and 800 tons Na_2O_5 (Sabet et al. 1976).

At Nuweibi (25° 12' N, 34° 30' E), the country rocks are mainly quartz-biotite schists. Nb-Ta mineralization occurs as disseminations in the apogranite part of the Nuweibi mass. Quartz-cassiterite veins with greisen selvages characterize the exocontact zone (see 2.2.3.b). Mineralogical analysis of apogranites from the most intensively albitized and mineralized parts of the mass revealed the presence of columbite, cassiterite, magnetite, ilmenite, zircon, topaz, barite, and trace amounts of apatite, rutile and monazite. Analyses show that Ta_2O_5 contents vary from 0.018 to 0.024%, and Nb_2O_5 from 0.005 to 0.009%, with the Ta_2O_5:Nb_2O_5 ratio ranging from 5:1 to 2:1. Drilling and assay of cores indicate that the lower levels of the mass show a sharp decrease in Ta and Nb contents, to 0.004% Ta_2O_5 and 0.002% Nb_2O_5. Detailed mapping and assessment (Sabet et al. 1973) reveal that the eastern part of the apogranite mass is potentially important. There is a zone 1000 m long, 450 m maximum width, with mineralization down to 50 m. Average Ta_2O_5 and Nb_2O_5 are 0.018% and 0.009% respectively, and the Ta_2O_5:Nb_2O_5 ratio is 2:1. These figures allow calculation of proven reserves of 20 million tons of ore containing 3600 tons Ta_2O_5 and 1800 tons Nb_2O_5.

In addition to the three deposits described above, Nb-Ta mineralization has been identified in a number of localities, some of which also contain Sn or W. The main occurrences are listed in table 26/4.

e) *Beryllium mineralization*

Beryl, as emerald, was known and exploited in many locations in ancient times, as is evidenced by the excavations, dump sites and ruins of Medinet Nugrus and Sikait. The same old mines were reopened repeatedly, but there is no exploitation at present. Beryl, including the gem variety emerald, occurs at many localities in the Eastern Desert in pegmatite lenses and quartz veins that cut across mica and talc schists, as well as in the schists themselves. It was later discovered as pockets, lenses and disseminations in greisenized and albitized granite masses and also, together with fluorite, as a major constituent in the greisen selvages associated with Sn-W bearing quartz veins in almost all localities with this type of mineralization. Beryl thus occurs in two geological environments (Hassan & El Shatoury 1976), in schistose rocks associated with the contacts of psammitic gneiss, and as pegmatite veins and lenses in certain granite masses which also carry disseminated beryl.

The first mode of occurrence is restricted to a 50 km long northwest-southeast striking belt that extends from Zabara in the northwest to Um Kabu in the southeast. Beryl mineralization in this belt is restricted to micaceous rocks following very closely the contact of the quartzofeldspathic gneiss. The belt includes Zabara, Nugrus-Sikait and Um Kabu-Um Debaa areas (Basta & Zaki 1961).

At Zabara (24° 45' N, 34° 41' E), beryl mineralization occurs in a zone 1 to 20 m thick of mica rock with frequent quartz stringers, following the sharp contact between the quartzofeldspathic gneiss and the overlying mica, talc and amphibolite schists. Beryl forms well-developed six-sided crystals a few centimeters in width that range from emerald (transparent deep green color) to ordinary beryl. The gem quality tends to occur more frequently in the mica schist than in the quartz stringers. Greisenized and albitized parts of the psammitic gneiss of this locality show 80 ppm Be. Moreover, Mo, Sn and W have been found in thin quartz veinlets in the area (Bugrov 1972).

The Nugrus-Sikait area was extensively mined by the ancient Egyptians. Here, Be mineralization is also restricted to the mica rock along its contact with the quartzofeldspathic gneiss of Wadi Abu Rushaid (2.2.3.d) and Wadi Nugrus. Beryl, and associated

Table 26.4 Localities with Nb-Ta mineralization.

Locality	Lat.	Long.	Remarks
Um Naggat	25°30'N	34°15'E	Albite-microcline- quartz apogranite in marginal zone of the Um Naggat mass (0.02% Ta and 0.03–0.15% Nb)
El Shalul	25°27'N	33°40'E	Altered muscovite granite (traces of Ta and 0.01% Nb)
El Backriya	25°18'N	33°41'E	Altered zone in the N endocontact of El Backriya granite (0.01% Nb, traces of Ta)
Um Bisilla	25°21'N	34°00'E	Muscovite-albite- quartz apogranite (0.005% Nb)
Um Salim	–	–	Muscovite-microcline- albite-quartz apogranite, 5 m wide endocontact zone (0.005% Nb, traces of Ta)
Rod Ishab	25°08'N	34°06'E	Sheet-like pegmatite body (0.001–0.008% Nb)
Muelha mine	24°54'N	33°55'E	Greisen with disseminated cassiterite (0.01% Ta and 0.01% Nb)
Um Dubr	22°40'N	35°50'E	Zones of amazonite granite (panning yielded columbite and xenotime; radioactive zones associated with quartz amazonite pegmatites and altered zones in granites; sampled pegmatites and altered zones (. . .–0.1% Ta_2O_5 and 0.5–1.1% Nb_2O_5)

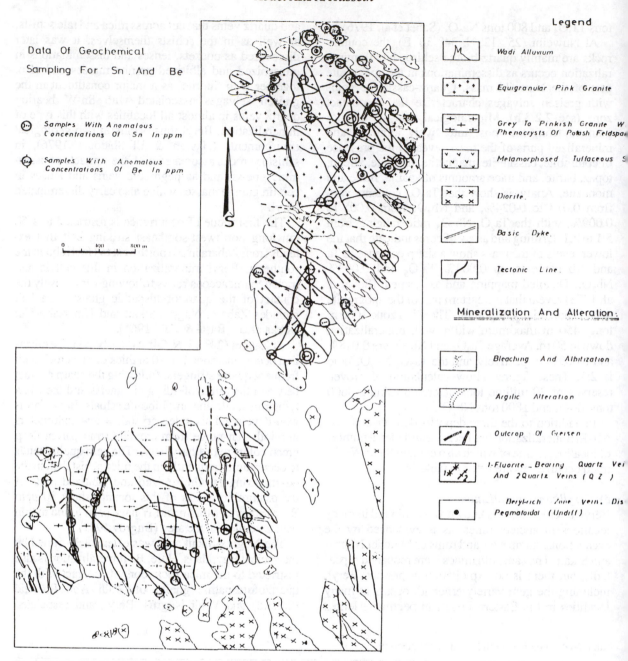

Data Of Geochemical
Sampling For Sn And Be

⊙⁰⁻ Samples With Anomalous
     Concentrations Of Sn In ppm

⊕ Samples With Anomalous
     Concentrations Of Be In ppm

Legend

M	Wadi Alluvium
	Equigranular Pink Granite
	Grey To Pinkish Granite W... Phenocrysts Of Potash Feldspa...
	Metamorphosed Tuffaceous S...
	Diorite
	Basic Dyke
	Tectonic Line

Mineralization And Alteration:

	Bleaching And Albitization
	Argilic Alteration
	Outcrop Of Greisen
	1-Fluorite - Bearing Quartz Vei... And 2 Quartz Veins (Q Z)
	Beryl-rich Zone vein, Dis... Pegmatoidal (Undiff)

Figure 26.12 Geologic map of Gebel Homret Mikbid (after Bugrov 1972).

emerald, occurs in the mica rock and the quartz
stringers that traverse it.

The situation at Um Kabu-Um Debaa area is rather
different, as quartzofeldspathic gneiss does not crop
out. Beryl occurs in the mica schist and the lenticular
quartz bands enclosed within it.

The second mode of occurrence is where beryl
occurs in association with granite masses that show
signs of greisenization and albitization, either as pegmatite pockets, lenses and veins, disseminations, or

in the greisen selvages of mineralized quartz veins.
The best examples of this mode are at Homret Mikpid
and Homr Akarem.

Homret Mikbid (24° 10' N, 34° 23' E) is a mass of
granite intruded into a country rock of basic metavolcanics, diorites and granodiorites (Fig. 26/12). The
mass is split into two parts by a latitudinal fault. It is
altered in parts with zones of intensive greisenization
and amazonitization. Frequent quartz veins and some
fluorite veins are present. Beryl occurs as dissemina-

tions in the altered parts of the granite in association with amozonite and fluorite. Also, in the northern part of the mass, beryl occurs in small pegmatite veins of an east-west strike, 5 to 30 cm in thickness, and about 90 m along the strike. The mineral forms well-developed crystals of greenish or bluish shades. Along the northern contact, beryl occurs together with amazonite and fluorite in quartz veins that follow the contact between the granite and the metavolcanic country rocks.

At Homr Akarem (2.2.3.a and b), mineralization occurs in small pegmatoidal lenses and pockets (20 × 80 cm) in the northern parts of the granite mass. It is also present, together with orthoclase, in small pegmatite veins and in the greisen selvages of the quartz-cassiterite veins.

f) *Fluorite mineralization*

Fluorite is always present, as an essential guangue mineral, in association with practically all of the autometasomatically altered granites (apogranite), especially with zones of greisens within the granites where it is an essential constituent, and also in the selvages of the Sn, W or Mo bearing quartz veins. Good examples of this mode of occurrence are the central depression in Gabal Muelha and the vein system at Muelha mine (2.2.3.b).

Fluorite may also form veins, monomineralic or together with quartz. Some of these veins are known in the area of Gabal Ineigi (25° 13' N, 34° 09' E), where they have been worked out. The quartz-fluorite vein is 2 to 3 m in thickness, and occupies a northwest trending fracture zone across the granite mass. It passes northwards into a number of quartz veins with minor fluorite and galena, then into veinlets of pure quartz. Similar veins are known from El Bakriya area (Saleeb-Roufaiel et al. 1982), and also just to the north of Hamret Mikpid, extending northwards from the granite contact. At Homr Akarem also, some fluorite veins are found to the north of the granite mass. They are of very limited reserves and are currently being exploited.

g) *Uranium mineralization*

Uranium mineralization is known in varied environments in Egypt. It is known in association with some Carboniferous and Cretaceous black shales, and in phosphorite deposits. It was also discovered in the Oligocene sandstones and associated rocks at Gabal Qatrani, where uranium of up to 0.3% U_3O_8 is concentrated in the intersitital spaces between sand grains (Said 1962). In this section, we will concentrate on the uranium mineralization related to granite masses, where it occurs either as disseminations in

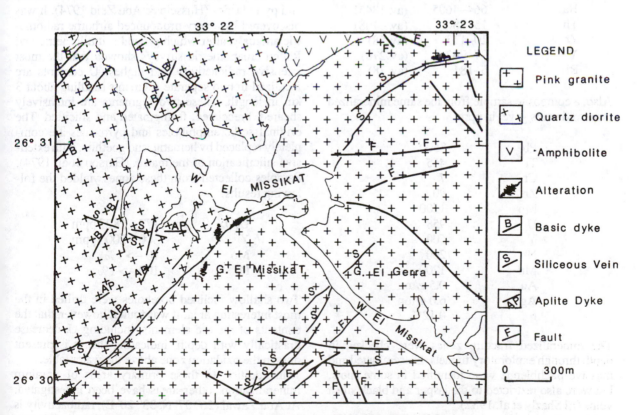

Figure 26.13 Geologic map of northern part of Gebal El Missikat (after Hussein et al. 1986).

the autometasomatically altered parts (greisens and albitites), or where it forms veinlets and stringers across granite masses (Hussein et al. 1986). The more important occurrences are reviewed below.

El Missikat (26° 22-26' 39" N, 33° 14-33' 40" E. Fig. 26/13) was located by an airborne radiometric survey (El Shazly et al. 1982). Upon verification, it was found that the anomalous radioactivity is related to uranium mineralization associated with highly brecciated black and jasperized silica veinlets, confined to minor faults and fractures striking east-northeast. Mineralization is represented on the surface by urnaophane and soddyite, which were observed visually as thin films or aggregates of acicular crystals or as a paint on fractures and cavities. It is always associated with sulfides (pyrite, chalcopyrite, galena, sphalerite and molybdenite), and the gangue minerals are mainly fluorite and Fe-Mn oxides (Attawiya 1983). The granites show signs of silicification, sericitization and kaolinization in the vicinity of the U-bearing veinlets.

Attawiya (1983) reported an analysis of black and jasperoid silica veinlets as follows (in ppm):

U	497–8856	(av. 1845)
Pb	968–17615	(av. 6509)
Nb	32–485	(av. 296)
Sr	5–70	(av. 28)
Ba	664–4605	(av. 1845)
Th	15–280	(av. 108)
Zr	10–113	(av. 58)
Y	97–962	(av. 270)
Rb	5–181	(av. 78)

Also, a composite sample from the same mineralized jasperoid veins gave, in ppm:

Pb	3500
Ti	465
Cr	7
Zr	90
Mo	1200
U	958
Y	500
Nb	200
Mn	1200
Au	0.5 g/ton
Ag	6.0 g/ton
F	0.12%

The mineralized fracture zones were checked in depth through exploratory tunnels. Disseminated and massive pitchblende were identified in some spots, but were also restricted to the jasper and black silica veins (El Shazly et al. 1982).

El Erediya uranium deposit (26° 20' N, 33° 28' E)

is in a very similar situation to that of El Missikat: the mineralization is directly associated with jasperoid veins occupying faults and fractures that strike at varying degrees between north and east (Fig. 26/14). Uranium minerals identified on the surface are uranophane, beta-uranophane, dossyite and renardite, but uraninite is the primary mineral in the subsurface. The mineralization is closely related to intensive brecciation, silicification and kaolinization of the granite host rocks (El Tahir 1985).

El Atshan is one of the earliest discovered and studied uranium deposits in the Quseir area. Here, pitchblende and atshanite occur filling fractures in the contact zone between a bostonite (microgranite?) dike and its enclosing rocks of slates and low grade schists (Hussein et al. 1970, El Kassas 1974). It is believed that the hydrothermal U-bearing solutions represent a late stage of magmatic differentiation, probably the same magma that gave rise to the bostonite sills.

Um Doweila (22° 17' N, 33° 26' E) is the most striking of these U-bearing microgranite dikes. It extends in a northeast direction for 10.6 km and ranges in thickness from 2 to 20 m. The northern non-altered and non-mineralized part shows that it is made up of grey alkaline microgranite with aegirine, riebeckite and arfvedsonite. It cuts through country rocks of practically non-metamorphosed volcanics and pyroclastics (Hussein & Abu Zeid 1974). It was discovered by a very pronounced airborne radiometric anomaly, after which ground scintillometric and Radon emanometric surveys showed that the most intensive radioactivity and highest U contents are localized in two separate segments totalling about 3 km in length. These two segments are intensively sheared, brecciated, ferruginated and silicified. The original sodic amphiboles and pyroxenes are completely replaced by hematite and goethite, with extensive silicification of the matrix (Garson et al. 1974). Samples collected from these zones yielded the following results:

U_3O_8	177 – 1827 ppm
Y	300 – 2000 ppm
Mo	10 – > 1%
Pb	100 – > 1%

Two shallow inclined boreholes were drilled in the most interesting area. It was found, however, that the contents of the ore elements, including U, decrease drastically with depth, indicating that it is present only in the oxidation zone above the water table.

In addition to these primary areas of uranium mineralization, other areas have been investigated. At Abu Zawal (26° 37' N, 33° 20' E), radioactivity is associated with pegmatite veins in altered pink gra-

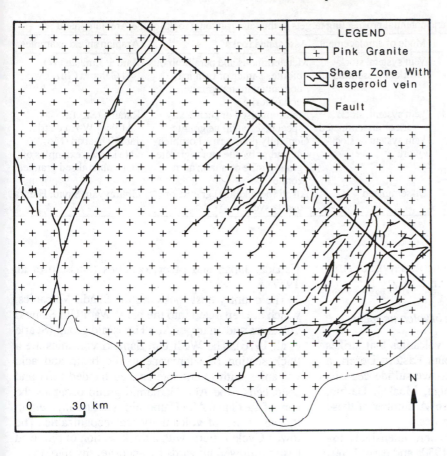

Figure 26.14 Silicified shear zones of El Erediya uranium occurrence (after Hussein et al. 1986).

nites, and is related to the presence of allanite in these veins (El Tahir 1978).

In Wadi Hawashia-Gabal Dara area, U, Th, and Nb mineralization is restricted to some veins of pegmatites cutting through the granite mass (El Shazly et al. 1982).

In the wider area of Wadi El Gemal, several radioactive anomalies proved upon verification to be related to concentrations of allanite in granite pegmatites associated with the white granites of the area (Bugrov & Krs 1972).

At Um Shilman (22° 37' N, 33° 49' E), a radioactive apogranite zone occurs at the margin of the Um Shilman pluton. It contains disseminated secondary uranium minerals in addition to anomalous concentrations of Pb, Zr, Y, Nb and Ta (Bugrov 1972). In fact, many of the autometasomatically altered granites (apogranites), where silicification, albitization and greisenization are intensive, carry accessory thorite, uranothorite, zircon and kasolite, together with their abnormal concentrations of Sn, Bn, Ta, Zr, Li and REE (cf. 2.2.3.b and d). The same sort of alterations and mineralizations are associated with the psammitic gneiss formation of the Abu Rushaid area (2.2.3.d). Uranium may be recovered as a byproduct if these deposits are exploited.

h) *Granites and mineralization, a discussion*

Granites are broadly classified into magnetite series and ilmenite series granites (Ishihara 1981). The former is usually associated with sulfide mineralization while the ilmenite series granites are related to the lithophile elements such as Sn and W. In Egypt, the magnetite series are represented by the G1 granites, known to be generally barren or associated with minor grains of Cu, Zn or Pb sulfides. G3, and to some extent G2, granites are ilmente series and hence are likely to be associated with Sn-W mineralization (Hussein et al. 1982). A similar study was later carried out in Saudi Arabia (Jackson et al. 1983) and the results show that most of the central Higaz granites are of the magnetite series and are likely to host Cu-Mo sulfide mineralization. In addition, a small number of intrusions are of the ilmenite series and, in a number of them, W and/or Sn are known to occur. Moreover, the same authors recognized a potential for four types of granitoid related mineralization in the Arabian Shield. Similar work is yet to be done on the Nubian Shield in Egypt. The four types are given by Jackson et al. in the form of a table:

Summary of mineralization types and potential prospecting targets in the felsic plutonic rocks of the Arabian Shield

Type	Mineralization	Mode of emplacement	Associated rocks	Arabian shield examples
W-Sn	W and/or Sn	Vein system, stock-work or disseminated	Metaluminous and peraluminous biotite, 2-mica or Li-mica syenogranite, monzogranite, or granodiorite	Baid al Jimalah (W) Al Gaharra (Sn)
Plumasitic	Ta, Nb, Sn, Li, Be, F, W, Au	Vein system, stock-work or disseminated	Metasomatised or chemically specialised alkaline granite	Ratama (Sn, Ta) Al Wajj W-Au district
Mo-Cu	Mo and/or Cu (Au, Ag, Pb, Zn, Bi)	Vein system or stock-work	Generally porphyritic hornblende and/or biotite morzogranite, granodiroite or quartz monzonite	Jabal Thaaban (Mo) No known examples of Cu-mineralised granites
Agpaitic	Nb, Zr, Y, Ta, Sn (REE, U, Th)	Disseminated	Alkali (peralkaline) granite or quartz syenite	Ghurayyah, Sayid, Hamra

2.3 STRATIFORM VOLCANOGENIC MASSIVE SULFIDE DEPOSITS AND RELATED TALC

2.3.1 *Massive sulfide deposits in volcanic rocks*

Within a 60 km long belt of volcanics that strikes northwest-southeast in the south Eastern Desert, a number of generally small massive sulfide deposits are known: Um Samuki, Helgit, Maakal, Darhib, Abu Gurdi, Egat and El Atshan. A number of these have been identified as talc quarries.

Some of these occurrences were intensively explored and studied in the late 1950s and early 1960s and the concensus of opinion was that they represent epigenetic mesothermal deposits formed by replacement of talc and other rocks in shear zones (El Shazly 1957, El Shazly & Afia 1958, Shukri & Basta 1959). Um Samuki is the largest in reserves, the best in ore grade and the best-studied area amongst these occurrences. It is thus described here in detail as representative of this group of ore deposits in Egypt.

Um Samuki (24° 14' N, 34° 30' E) is a Zn-Cu-Pb deposit in an area of very rugged topography amidst the belt of calc-alkaline island arc volcanics (andesites and their pyroclastic equivalents with lesser amounts of basalts and rhyolites). Copper was discovered and exploited in the area in ancient times, as indicated by old workings and slag heaps. Two underground workings are known and referred to as the Eastern and Western Mines.

The area was studied by Hume (1937), El Shazly & Afia (1958), Kovacik (1961), Mansour et al. (1962) and others. All of these authors concluded that the mineralization is epigenetic, introduced by hydrothermal solutions along a shear zone, and developed by replacement of pre-existing rocks. In 1974, the area was restudied by the Aswan Mineral Survey Project and it was concluded that the deposit is a volcanogenic stratiform massive sulfide deposit, similar in many respects to those described from

Canada, Japan and other localities (Hussein et al. 1977).

The country host rocks are the Shadli Volcanics, which are separated into the basal Wadi Um Samuki volcanics and an upper Abu Hamamid group (Searle et al. 1976). The Wadi Um Samuki volcanics are a thick section of submarine cyclic basic and acid volcanics with minor intercalated banded tuffs and chert beds. The Abu Hamamid group occupies the core of the Gabal Abu Hamamid syncline and, within it, two cycles of volcanicity are recognizable. The lower, Cycle I, starts with a thick section of pillowed basalt followed upwards by andesite, rhyolite, rhyolite breccia, and vent rocks. The rhyolite breccia is the footwall to the massive sulfide ore bodies, while their hanging wall is of banded and graded bedded tuffs. Cycle II is marked by basic vent breccia and pillowed basalt, followed by banded tuffs and chert.

The massive sulfide ore bodies at both the Eastern and Western Mines occur along a specific stratigraphic horizon, namely that separating the brecciated rhyolite and vent facies on one hand, and the banded graded bedded tuffs on the other. The upper contacts of the massive sulfide ore bodies are sharp and well defined, while on the footwall side, an extensive pipe or funnel of alteration is present. Here, intensive magnesium metasomatism, silicification and dissemination of sulfide minerals (mainly pyrite) are encountered. Magnesium metasomatism resulted in the development of talc, antigorite and tremolite on the expense of the footwall rocks and even the bottom parts of the ore bodies. The ore lenses are banded with local development of sedimentary textures. Moreover, they show some zonation whereby Zn increases towards the hanging wall, while Cu increases towards the footwall. The mineralogy was given by Rasmy (1982).

The ore body at the Western Mine attains some 90 m in strike length. Its width is in the range of a few meters, and its ore assays Cu 2.2%, Zn 21.6%, Pb

0.5%, and Ag 109 g/t. Ore reserves are estimated at slightly less than 200,000 tons (Searle 1975). At the Eastern Mine, three lenses are known, the largest of which is 35 m in strike length and up to 2 m in width. The ore quality is poorer than that in the Western Mine, assaying Cu 1.8%, Zn 13.8%, and Pb 3.4%. Its ore reserves are also less than those of the Western Mine.

In addition to the Eastern and Western Mines, the wider area of Um Samuki includes two other occurrences called Helgete and El Maakal. At Helgete, a thin lens, less than 1 m in width, extends about 70 m along strike. Its assay showed Cu 2.9%, Zn 13.6%, and Pb 11.4%. Reserves were estimated at 15,000 tons. The occurrence at El Maakal includes a small number of lenses enclosed within the talcose footwall rocks. Samples of the massive ore assay Cu 3.5%, Zn 17% and Pb 2%. The ore reserves are very small.

Contrary to the epigenetic-hydrothermal deposition and replacement mode of origin advocated by the earlier authors above, it is here believed that the massive sulfide ore bodies were deposited during the Abu Hamamid volcanic episode, on top of a submarine volcanic vent system, to be sedimented conformally with the enclosing rocks at the interface between the volcanic pile and sea water. The ore bodies overly a stockwork of altered rocks resulting from intensive metasomatic effects induced by the ascending volcanic exhalations on the channelways through which they ascended (Hussein et al. 1977). The hot concentrated metal-bearing solutions continued their metasomatic role after the precipitation of most of their metal contents. Metasomatism (introduction of magnesium, silica and disseminated pyrite) was most intensive below the ore bodies as a result of the ceiling effect of the latter. Products of these metasomatic alterations are talc, tremolite, chlorite, and carbonates. The talc thus formed is now being exploited economically at Darhib and El Atshan talc mines.

2.3.2 *Talc deposits hosted in volcanic rocks*

Talc (including steatite, soapstone and phyrophyllite) is known in some 35 sites in the Eastern Desert and Sinai. The most important occurrences are those associated with the belt of island arcs hosting the Zn-Cu-Pb volcanogenic massive sulfide deposits of Um Samuki and the like. The main talc producers in Egypt are the Darhib talc mine and a number of surface and underground workings around El Atshan, near Hamata.

At El Atshan (latitude 24° 15', longitude 35° 13'), the host rocks are volcanic and volcaniclastic rocks of basaltic to rhyolitic composition (Fig. 26/15). They enclose a number of small lenses of the massive

sulfides, exposed as gossans on the surface and inducing a greenish tint to the talc in their vicinity. Talc occurs, in the footwall rocks to the sulfide bodies, in association with serpentine minerals tremolite, epidote, quartz, dolomite and magnesite, in an alteration zone 30 m wide that extends for some 500 m in an east-west direction.

The situation is similar at Darhib, and many small insignificant lenses of the sulfides were encountered in the lower levels of the underground workings.

The origin of talc was related by previous workers to a process of hydrothermal alteration acting upon lenses and bodies of ultrabasic rocks enclosed within the volcanic pile (El Shazly 1957). The very low Ni, Cr and Co contents of these rocks preclude such an ultrabasic origin (Gad et al. 1978). Here, it is believed that talc and the associated minerals (serpentine, tremolite, chlorite, magnesite) were formed through a process of intensive Mg-metasomatism associated with the volcanic exhalative episode responsible for the formation of the massive Zn-Cu-Pb deposits, and hence their close spatial association. Similar deposits, where intensive Mg-metasomatism resulted in the development of talc and related minerals are known from the Canadian Shield, in association with massive sulfides (Sangster 1972).

Production of talc in Egypt amounts to 12,200 tons per year (1983-84 figures), most of it is used as a filler in local industries, though 1500 tons were exported in the same year (Geological Survey Egypt 1984).

2.4 PRECIOUS AND BASE METAL VEIN-TYPE DEPOSITS

2.4.1 *Dominantly gold (silver) vein deposits*

a) *General statement*
Gold was very highly prized by the ancient Egyptians. It was no doubt the driving force behind most of their mining activities, and they were quite successful in prospecting for the noble metal and its extraction from open pits and underground mines in spite of their primitive technology. Some 95 gold occurrences are known in the country, spread over the Eastern Desert, and most of them were discovered and exploited by the ancient Egyptians.

Gold occurrences are in most cases restricted to quartz veins occupying pre-existing fractures, and are made up of massive quartz with disseminated gold and sulfide minerals. Nevertheless, there are some occurrences where the ore is associated with dikes (mainly felsites), either as stockworks of minor quartz veinlets or contained in pyrite grains disseminated throughout the whole mass of the dike (Geolog-

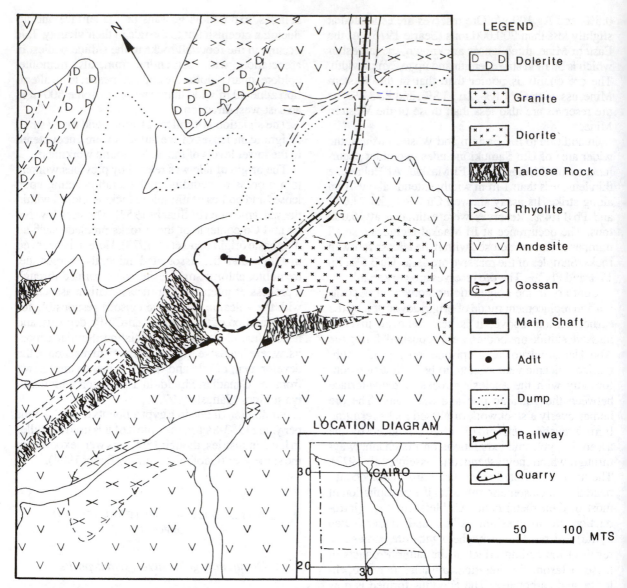

Figure 26.15 Geological map of El Atshan talc mine (after Gad et al. 1978).

ical Survey Egypt 1986). This type of dike ore is known in Fatiri, Um Mongul, Abu Mereiwat, Marahib, Sagi, Kurdeman and Sabahia.

The country rocks hosting mineralized veins are quite varied and include serpentinized ultramafics, metamorphosed volcanic and sedimentary rocks, intrusive gabbros and different granitoids, including the latest phases, yet most of the veins are associated with diorite-granodiorite plutons and their contacts with schists. The mineralized veins are structurally controlled, being fissure fillings, confined to fault planes, or zones of intensive fracturing. They may be arranged as a series of en echelon veins with pinches and swells. The main veins are always accompanied

by parallel veinlets and stringers to form ore zones of considerable thickness as compared to the veins.

At Barramiya, Atud, Hangaliya, Um Rus and Sukari, this zone ranges between 15 and 20 m in thickness and reaches up to 100 m at El Sid. Along the strike, the zones extend for some hundreds of meters and may continue for considerable distances along the dip, for example, 455 m at El Sid. The main veins vary in width, generally between 0.6 and 1.5 m, but may reach up to 5 m as at Semna, Sukari and Um Egat. The vein zones are associated with hydrothermal alterations that extend for 1 to 2 m on both sides of the zone, and are represented by sericitization, chloritization and introduction of pyrite. The veins are made up mainly of massive, milky or grayish-white quartz. In many cases the quartz represents at least two generations, an older brecciated milky quartz which is usually barren, and a younger gray quartz that cements the fragments of the older phase and is usually gold-bearing. Calcite may also be present, and some veins (e.g. at Korbiai) are made up of a mixture of quartz and calcite. Gold is found mostly in the native form and, otherwise, is carried in auriferous pyrite. Gold contents are very variable, even within one vein. An appreciable increase in tenor has been noted where the veins or their enclosing rocks carry finely disseminated graphite. Gold generally averages 11 to 30 g/t. Silver is always present in association. The veins carry appreciable amounts of sulfide ore minerals, mainly pyrite, arsenopyrite, sphalerite, chalcopyrite, galena and pyrrhotite (Kochin & Bassiuni 1968).

The features reviewed above are common to almost all of the gold occurrences in the country, and hence the occurrences defy any grouping on the basis of any set of commonly used parameters. Hunting (1967) and El Ramly et al. (1970) classified the gold occurrences in geographical groups: northwestern, northern, central, southeastern and southwestern, each including a number of deposits, as shown in Figure 26/16, reproduced here after El Ramly et al. (1970).

b) *Main occurrences*

In the following paragraphs, some of the more important occurrences are briefly described as examples, followed by a synthesis of the mode of formation offered by the present author.

Um Rus (latitude 27° 28', longitude 34° 34'). At Um Rus, a number of mineralized quartz veins are spread over an area of about 7 km², where they cut through a granodiorite mass intruded into layered gabbros. The veins are enclosed mainly within the granodiorite, but may extend into the gabbro where they pinch out. The veins generally strike northeast and dip northwest at gentle angles. They vary in thickness but are generally about 40 cm thick, and are formed of massive milky quartz with occasional feldspars and carbonates. They carry minor amounts of pyrite and arsenopyrite. Gold is present in the native form, but about 4% of the total gold content is locked in pyrite (Amin 1955). The wall rocks to the veins are strongly altered, where the feldspars are sericitized and the ferromagnesian minerals are changed into chlorite. Pyrite is disseminated in the vicinity of the veins and is changed into red zones with hematite on alteration. The total assured reserves are 16,000 tons assaying 11 g/t gold (Amin 1955).

El Sid (26° 00' 17" N, 33° 35' 42" E). This is the largest and richest gold deposit in the country. It was exploited in ancient times, and recently. From 1944 to 1958, 120,000 tons of ore containing 27.9 g/t gold were mined to produce 2653 kg of fine gold over a period of 15 years. Mining operations totalled 4000 running meters of shafts, drifts and winzes (Geological Survey Egypt 1986).

The country rocks hosting gold-bearing quartz veins are an assemblage of schist and amphibolite of mafic-ultramafic derivation intruded by the Fawakhir granodiorite pluton. All the veins are confined to the western part of the pluton, but a few extend for some distance into the schist. The majority of the veins strike almost east-west and dip at 25 to 65° to the south.

The main vein in the area extends 450 m along strike, 455 m along the dip, and varies in thickness from a few centimeters to 1.5 m. It occupies a shear zone that cuts across the granodiorite in the west to the schistose serpentinite in the east. Within this zone, the vein is not a simple continuous one, but is made up of a number of lenticular veins arranged en echelon and is associated with a series of parallel veinlets and offshoots. It is of milky quartz with irregular disseminations of free gold and rare sulfides (galena, pyrite, chalcopyrite, pyrrhotite and sphalerite). Some calcite may be associated with the quartz. The average gold grade is 27.9 g/t.

Another relatively large vein occurs about 18 m to the north of the main vein. This northern vein was also exploited in ancient and recent times, and it seems to be about 0.4 m thick and of a high gold content. A number of other veins and veinlets occur in the vicinity of these veins, making a vein zone 80 to 100 m wide.

In addition to the El Sid deposit, the wider El Sid area (6 × 3 km) contains more than 170 quartz and quartz-calcite veins and veinlets and zones of silicification. All are fracture fillings, controlled by two systems of fractures of approximately east-west and northwest trends. An interesting zone of silicification with frequent veins and veinlets occurs at about 0.5 km to the southeast of the El Sid deposit. The zone is 1300 m long, 130 m thick, and includes a number of individual veins 20 cm or less each. The average gold content in these veins is 8.8 g/t (El Ramly et al. 1970). A similar zone occurs 2 km to the northeast of the El Sid deposit.

Currently, the El Sid is included in an exploration license granted to a joint venture company reassessing the area in preparation for exploitation.

Barramiya (25° 04' 24" N, 33° 47' 16" E). The Barramiya gold deposit is relatively accessible, situated midway on the Mersa Alam-Idfu road. The

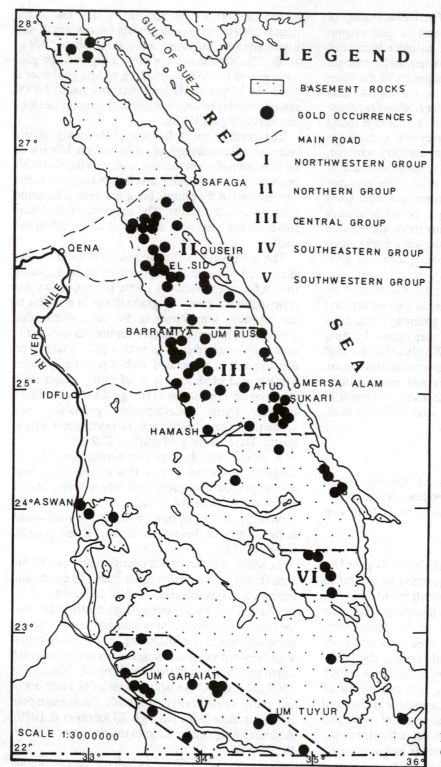

Figure 26.16 Gold deposits, Eastern Desert (after El Ramly et al. 1970).

country rocks enclosing the auriferous veins are graphitic or calcareous schists associated with serpentinites and related rocks of probable ophiolitic nature. These are cut by several systems of fractures giving rise to fracture zones mainly conformable with the regional schistosity. Quartz diorite dikes (20 m thick) were reported from underground workings.

The auriferous veins are restricted to these zones of fractures and occur mainly in the graphitic schist. They are gray or dark blue gray quartz with fine disseminations of gold and small amounts of sulfides (pyrite and chalcopyrite). Veins and veinlets trend east-west, north-south or northeast, with the east-west trend dominant. Main veins are usually associated with thin veinlets that occupy fractures and planes of schistocity. The quartz veins are sometimes cut by younger carbonate veins. The schists in close vicinity to the veins are sometimes impregnated with gold and pyrite.

The gold deposit at Barramiya was divided by miners who exploited it into four lodes, referred to as the Main Lode, Taylor's Reef, Caunter Lode and New Caunter Lode. The Main Lode is the eastern part of the area of mineralization and is associated with a zone of intensive fracturing in the graphitic schist. The lode extends for 1000 m in an east-northeast direction and dips steeply (80°) northwards. The average thickness is 1 m, but the vein pinches, swells and splits into smaller veins and is accompanied by a number of parallel veinlets and stringers within the fracture zone. The ore still present in this lode is estimated at 20,000 to 27,000 tons with a gold content that varies according to different estimates from 2.2 to 15.5 g/t. Taylor's Reef does not crop out but was encountered underground to the north of the Main Lode. It extends 75 m along an east-west strike and is almost vertical. Its average thickness is 0.6 m and the average assay is 16.5 g/t, with occasional richer spots. One of these 'bonanzas', containing 105 kg/t, was picked out (Hume 1937). Reserves are estimated at 13,800 tons.

On the other hand, the Caunter and New Caunter Lodes represent two branches of the mineralization zone. Their general strike is north-south and they dip steeply to the southeast. The former was mined for 130 m along the strike and 75 m along the dip. It consists of a zone 60 cm thick of quartz veinlets enclosing pieces of country rocks. Gold is present in both veinlets and fragments, grade ranging from 18.69 to 43.09 g/t. Reserves are 2740 tons (Kochin & Bassiuni 1968). The New Caunter Lode was worked for 90 m along the strike and 115 m along the dip. Its average thickness is 1 m and the average grade is 8.4 g/t, but both thickness and grade decrease with depth.

In addition to the reserve estimations given from these four 'Lodes', some 54,000 tons of auriferous sand and slime representing the tailings of previous operations are present in the area. Their average gold content has been estimated at 5.67 g/t. These should be taken into consideration in future evaluations.

Atud (25° 01' 10" N, 34° 24' 10" E). At Atud, a number of gold-bearing quartz veins are concentrated mainly in three localities, referred to as the Main Atud, Atud East I, and Atud East II. The first is the most important, occurring on the eastern footslopes of Gabal Atud. Most investigations, including an intensive underground exploration program, have centered on this locality.

The Main Atud is associated with a shear zone extending north-northwest for almost 1500 m across the Atud gabbros. The quartz vein is exposed in a number of separate outcrops and is not persistent. It shows frequent pinches and swells and in places takes the form of sheared en echelon lenticular bodies, sometimes accompanied by a series of parallel veinlets and offshoots. The average thickness of the vein is 70 cm and the average grade is 16.65 g/t in the northern part of the deposit, known as the main lode, decreasing in the southern part.

Atud East I occurs about 2 km to the east of Main Atud. Here, a single vein occurs in metadiorites. It strikes northeast and dips westward at 20°. It extends more than 200 m and averages 2.71 g/t, a value which increases with depth.

Atud East II is located 3 km south of Atud East I. In this area, two parallel veins, 200 m apart, crop out. They can be traced for 500 m along the strike and vary in thickness from 15 to 45 cm. Gold values in samples from dumps and ancient workings range from 2.3 to 31.9 g/t (Geological Survey Egypt 1986).

The principal lode is thus the Main Atud, with 19,000 tons ore grading 16.276 g/t. In addition, 1600 tons of dump with 12.4 g/t gold are piled in the area.

Sukari (latitude 24° 56', longitude 34° 42' 27"). This is an ancient working that was reopened and exploited this century. A gold-bearinjg quartz vein occupies the contact zone between a granophyre and the enclosing schists. The vein varies from 1 to 4 m in width, dips east (± 45°) and extends in a north-south direction for about 500 m. The average grade is reported to be 52 g/t gold and 18 g/t silver (Hume 1937).

Samut (latitude 24° 41' longitude 33° 52'). A mineralized quartz vein strikes N-10-E and dips at 80° E. It extends almost a kilometer and is 0.5 m wide, with frequent horsetailings, pinches, and swells. The vein cuts across a granodiorite mass intruded into schists and basic volcanics. Propylitization and listwanitization are observed in the near vicinity of the vein and anomalous concentrations of pyrite, arsenopyrite and chalcopyrite are characteristic of the vein and its enclosing rocks.

The vein was worked to a depth of 60 m but no data as to the grade or reserves are available. The information given by Bugrov (1972) indicates that gold occurs not only in the quartz vein but also in the hydrothermally altered rocks in close vicinity.

Hamash. In the wider area of Hamash, gold is known in the localities called Hamash, Um Hagalig and Um Hamr. The gold-bearing quartz veins strike northeast or east and dip northwest or north. They are a few hundred meters in strike length and 50 to 70 cm thick. The country rocks are granodiorites intruded into metavolcanics. On the surface, the veins carry malachite, azurite and chrysocolla, which grade into chalcopyrite, chalcocite and covellite with increasing depth. Workings went down to a depth of 60 m or more, but no estimates of reserves or grades are available (Moustafa & Hilmy 1959).

c) *Genesis of the gold deposits*
Amin (1955) and El Shazly (1957) related gold mineralization in Egypt to Gattarian times, in a probable genetic association with the Gattarian granites. Selim (1966), in his age dating of galena from various occurrences, reported an age of 500 Ma for galena associated with a gold-bearing quartz vein at Fawakhir. Ivanov & Hussein (1972) stated that at least one generation of gold is chronologically and genetically related to the Dokhan volcanics.

In the present study, and on the basis of the general characteristics of the deposits given above, it is believed that these belong to the precious metal vein type deposits (epithermal or bonanza type) developed in association with Andean type calc-alkaline felsic volcanics. The general characteristics of this type were given by Sillitoe (1978). They correspond surprisingly well to those of the vein deposits of Egypt, particularly in the following respects.

1. The veins occupy pre-existing tension fractures and mineralization is sporadic in the veins.

2. The deposits contain principally precious metals, but base metal sulfides in various proportions are also present. Gold occurs in native form, but is usually argentiferous.

3. The country rocks show propylitic or advanced argillic alterations, with sericitization, silicification and dissemination of pyrite restricted to the immediate vicinity of the veins.

Thus, these deposits may be related to subduction in a continental margin environment and are hence associated with the Dokhan volcanics or their co-magmatic batholiths. The model of formation suggested here is that of a convective cell, wherein water, sea or meteoric, circulates downwards to be heated by granodoritic intrusions and then rises through the volcanic-volcanicalsic pile of rocks, leaching precious and base metals, to deposit them in the already opened fracture system at the cold end of the cell near the surface. The same model has been suggested for some of the precious and base metal vein type deposits in the Arabian Shield (Babattat & Hussein 1983). The mechanism might have been repeated successively with later, younger intrusions emplaced within volcanic-volcaniclastic piles.

d) *Placer gold deposits*
Recent alluvial placer deposits of commercial value should not be expected in the country under prevailing arid conditions. Nevertheless, fossil placers, as auriferous conglomerates, might be present at paleosurfaces of erosion and unconformities. Examples of probable sites of such occurrences are the base of the Hamamat, within Atud conglomerates, and on the Shield-Nubia formation unconformity. In fact, some placer gold has been reported in a buried stream in the area of Abu Dahr (Gad 1980, pers. comm.). No data are available on this occurrence, but it (and similar environments) deserves further investigation. In addition, beach placers of black sands were reported to contain minor amounts of gold particles accumulated as a result of wave action on Nile silts (El Gemmizi 1985).

2.4.2 *Dominantly base metal veins*
Some of the quartz veins occurring in environments similar to the gold-bearing veins reviewed above are not auriferous. Instead, they carry base metal sulfides, either copper or lead and zinc. They are of no economic significance. Veins with mainly copper mineralization are described from Regeita, Abu El Nimran and Samra in Sinai (El Shazly et al. 1955, Moustafa & Hilmy 1959). The veins are fissure fillings in various rocks that include schists, granodiorites and granites. Copper mineralization is represented by chalcocite, covellite and cuprite. Similar occurrences are known from El Atawi, in the Eastern Desert, where the mineralization occupies a shear zone cutting through amphibolites. The copper minerals present are secondary minerals, mainly malachite, associated with some hematite with no primary sulfides (El Shazly & Sabet 1955).

Veins with lead-zinc sulfide mineralization occur in many parts of the Eastern Desert, good examples being Um Gheig (west of the sedimentary hosted mineralization), Wadi Sitra, Wadi Siwiqat El Soda and Wadi Hamad. In these areas, quartz or chalcedony veins carry sphalerite and galena and cut across a country rock of mainly granodiorite and schist. The veins vary from 0.1 to 1 m in thickness and extend up to 200 m. The oxidation zone contains limonite, cerussite, calamine and some malachite. The age of these veins is determined to be around 650 Ma for galenas from Wadi Siwiqat El Soda and Wadi Sitra (Selim 1966).

2.4.3 *Barite veins*

Many barite veins are encountered here and there in Egypt. They include the barite-galena veins cutting across the pink granite east of Aswan, those at the mouth of Wadi Hamash, El Gara El Soda, and Wadi El Gerera in the Elba region. In the latter area, some 16 barite veins are known. They cut across basement rocks and some of them also cut the Miocene sediments. The veins are composed of coarsely crystallized barite, with some calcite and occasional sulfides. The reserves of this area were estimated to be 250,000 tons. At El Hudi, near Aswan, several veins of barite are known, classified by Saleeb-Roufaiel et al. (1976) into veins in granitic rocks and veins in metamorphic rocks. The veins in granitic rocks are composed of reddish barite with quartz, microcrystalline silica, and a few orthoclase grains. A composite sample showed Ba 52.9%, Mg 0.17%, Sr 0.70%, Fe^3 0.95%, and Si 2.95%. The barite veins in metamorphic rocks are economically more important. In these, the barite is colorless or white and is accompanied by appreciable amounts of sulfide minerals, mainly galena with a little sphalerite and chalcopyrite. No estimation of reserves has been made for this area.

The base metal and barite veins are believed by the present author to have been formed in the same manner suggested for the auriferous veins. The differences between gold-bearing and non-auriferous veins may be attributed to variations in the metal contents of the volcanic pile permeated by leaching solutions, as well as to differences in physico-chemical conditions (T, P, Eh, pH, etc.) prevailing during the processes of leaching and deposition.

2.5 STRATABOUND DEPOSITS IN SEDIMENTARY SEQUENCES

This group includes a number of deposits, each restricted to a certain stratigraphic horizon, occurring with the Phanerozoic sedimentary cover. The deposits described here are the zinc-lead, stratiform copper, sulfur, and barite deposits.

2.5.1 *Zinc-lead deposits*

Seven zinc-lead occurrences are known in the Eastern Desert between the latitudes of the Red Sea coastal towns of Quseir and Ras Banas: Zug El Bohar, Essel, Wizr, Um Gheig, Abu Anz, Gabal El Rusas and Ranga (El Shazly 1957). Zinc and lead ore materials occur in the lower Gabal El Rusas formation, which rests unconformably on the peneplaned Precambrian rocks at Zug El Bohar and Essel. In the other occurrences, mineralization is associated with the upper Abu Dabbab formation and may extend into younger sediments. Primary sulfide minerals include galena, sphalerite, pyrite and marcasite. On surface exposures, these are extensively altered into cerussite, anglesite, smithsonite, hemimorphite, hydrozincite, jarosite and limonite. Wulfenite was also recorded from Um Gheig mine. The mineralogy and geochemistry of the ore deposits have been extensively studied – Barakat & El Shazly (1956), Soliman & Hassan (1969), Hilmy et al. (1972), Rasmy & Montasser (1982), El Aref & Amustutz (1983), El Aref (1984). The two major occurrences, at Um Gheig and Gabal El Rusas, have been the subject of intensive geophysical and geochemical investigations (Youssef et al. 1966, Bayoumi & El Dashlouty 1967, Soliman & Hassan 1971). Drilling revealed extension of the sulfide mineralization in depth, predominantly rich in the iron sulfides (Akkad & Dardir 1966, El Shazly & Hassan 1962).

Many opinions have been expressed regarding the origin of these zinc-lead deposits, but a genesis through replacement of 'limegrit' and 'conglomeratic limegrit' by hydrothermal 'telethermal to leptothermal' solutions has been agreed upon by many authors (Amin 1955, El Shazly 1957, 1968, El Shazly et al. 1956, Moharram 1970, Sabet et al. 1976). A syn-sedimentary origin was suggested by El Ramly et al. (1970), and an exhalative sedimentary origin is favored by Hilmy et al. (1972).

The most recent work (El Aref & Amstutz 1983) on Um Gheig, Wizr, Essel and Zug El Bohar, and by El Aref (1984) on the Ranga deposit, shows that these deposits can be classified in two groups:

a) lead-zinc deposits of the filling type, occurring in the Abu Dabbab 'Formation' and represented by the Um Gheig, Ranga and Wizr deposits (the latter is a typical karst ore); and

b) stratiform to stratabound galena in sandstone, confined to the lower beds of the Gabal El Rusas formation and represented by the Zug El Bohar and Essel occurrences.

Surface and subsurface mapping at Um Gheig workings revealed that the ore is confined to a fill mass developed along a major northwest-southeast rift fault apparently affiliated to an intercontinental

rift zone (Mitchell & Garson 1981), and not necessarily related to magmatism during rifting.

A prevailing *sebkha* environment during the deposition of the Abu Dabbab 'Formation' is apparent,
which contributed to the deposition of zinc and lead
sulfide minerals. Youssef (1986) recognized a relatively high lead-zinc-copper concentration in the
stromatolitic and nodular dolomites of the Abu Dabbab 'Formation' south of Mersa Alam on the Red Sea
coast.

2.5.2 *Stratiform copper deposits*

In the Phanerozoic sediments, minor copper occurrences are noted in the Paleozoic sediments of west
central Sinai, the site of extensive copper production
in Pharaonic times (Lucas 1927). Copper occurs in
the form of secondary minerals, predominantly
malachite impregnating clastic sediments (Hilmy &
Mohsen 1965). In some occurrences, copper
minerals occur admixed with manganese ore deposits.

The Sinai copper occurrences are quite similar to
the copper deposits mined at Timna in the Negev and
Wadi Dana in Jordan, where the deposits occur in
lower Cambrian sediments exposed on both sides of
the Aqaba-Dead Sea transcurrent fault.

From the Mesozoic, minor copper mineralizations
are reputed to occur in Cretaceous sediments in Wadi
Araba. So far, nothing has been published regarding
the nature of these deposits.

2.5.3 *Sulfur deposits*

Sulfur has been produced from two mines, one at
Gemsa, just north of Hurghada on the Gulf of Suez
(Shukri & Nakhla 1955) and at Ranga, near Abu
Ghusun on the Red Sea coast (El Shazly & Abdallah
1964). In both occurrences, native sulfur is found as
lenses and bands replacing gypsum or dolomite of
middle Miocene age. Gemsa sulfur deposit is believed by Shukri & Nakhla to have been formed by
ascending hydrocarbons carrying sulfur in solution,
judging from lithologic controls and the structural
features. Schnellmann (1959), however, regarded the
formation of sulfur as a result of the reduction of
calcium sulfate by inorganic processes at ordinary
temperatures in the presence of an unidentified cata-
lyst.

The origin of sulfur at Ranga is related by El Aref
(1984) to biogenic action, whereby sulfur is deposited and accumulated by rhythmic crystallization differentiation during the diagenetic stage in a 'partly
closed system'. Iron sulfides and galena are common
associates of sulfur and are confined to the Abu

Dabbab formation. They show the following geometric types (El Aref 1984):

a) stratabound rhythmic type of pyrite/marcasite
associated with calcite of algal origin, barite, and
quartz;

b) stratiform to stratabound rhythmic type of sulfur associated with anhydrite, calcite and bitumen;

c) stratabound anthigenic galena occurring as a
cement of karst fillings and associated mainly with
cryptalgal calcite and anhydrite.

Apparently, the original deposition of the Abu
Dabbab formation and associated ore minerals was
under a *sebkha* environment modified by diagenetic
and epigenetic 'karstification processes'.

A minor sulfur occurrence at Um Reiga area to the
south of the Um Gheig lead-zinc deposit, in Miocene
sediments, has been described (El Shazly & Mansour
1962).

2.5.4 *Barite in sedimentary sequences*

In addition to the barite occurring in veins, described
in section 2.4.3, it is a common association of stratabound ore deposits. Barite is recorded in the Mn-Fe
ore deposits of west central Sinai, the Mn ores of
Halaib district, the Zn-Pb deposits of the Red Sea
coast, and the iron ore deposits of Ghorabi and El
Gedida in the Western Desert.

Mining excavations at the El Gedida iron mine
revealed the occurrence of barite in association with
iron ores. It occures only in the western sector of the
open pit, associated with high grade iron ore, in the
form of pockets or lenses. Individual crystals may
reach up to 20 cm in length, with well-developed
crystal faces. Fragmental barite and barite in siliceous concretions are occasionally observed in the
clastic sediments overlying iron ore beds. Unfortunately, barite has been discarded during iron mining
operations. At Ghorabi, exploratory drilling revealed
the occurrence of barite at the base of the iron ore, in
addition to the surface exposure of barite in the
western sector of the area (Akkad & Issawi 1963).

2.6 ORES OF SEDIMENTARY NATURE

2.6.1 *Iron ore deposits*

Economioc iron ore deposits occur in Cretaceous and
Eocene sediments northeast of Aswan and north of
the Bahariya Oases, respectively, and are being
mined there. Recently, mining operations have concentrated upon the Bahariya occurrence after operations were discontinued on the Aswan deposit.

a) *Aswan iron ore deposit*

Iron has been produced from an area northeast of Aswan since Pharaonic times (1580 to 1350 BC). About seven million tons were produced between 1956 and 1973 to supply the Helwan Iron and Steel Complex south of Cairo. The estimated reserves are 121 to 135 million tons (Attia 1955). The ore is a bedded oolitic type of Senonian age in the form of two bands interbedded with ferruginous sandstone and clay capping Precambrian rocks. The thickness of the bands varies from 20 up to 350 cm. The ore minerals are mainly hematite, with minor goethite. Gangue minerals include quartz, gypsum, halite, glauconite and clay minerals.

The iron content of this ore ranges from 31.2 to 62.3%, averaging 46.8% Fe, SiO_2 ranges from 5 to 31%, averageing 14.1%, P from 0.04 to 3.5%, Mn up to 1.3%, and S up to 0.3%. The total reserves of a 54 km^2 area are estimated at 121 to 135 million tons of ore, with 20 million tons proved reserves (Attia 1955). The ore is considered by most writers to have been formed under sedimentary lacustrine conditions during the deposition of Senonian sediments.

b) *The Bahariya iron ores*

Workable and potential iron ores are confined to the lower part of the middle Eocene limestone (El Naqb formation) in four major occurrences north of Bahgariya Oases. These occurrences are called El Gedida, Ghorabi, Nasser and El Harra, extending over 11.7 km^2, and the ore thickness varies from 2 to 25 m, averaging 9 m. The ore is thought to be localized in the crests of two major anticlines trending in a northeasterly direction. El Gedida and El Harra ore deposits are localized on the eastern anticline, while Ghorabi & Nasser are on the western anticline (Akkad & Issawi 1963). The structural nature of the Bahariya Oases is not yet well understood but it is apparent that most of the reported folds were generated by faulting affiliated with the Pelsuium megashear, along which the Bahariya Oases are located (Neev et al. 1982).

The mineralogical and geochemical characteristics of the ore deposits, particularly those of Ghorabi and El Gedida, have been reported by many authors (Gheith 1955, Nakhla 1961, El Hinnawi 1965, Basta & Amer 1969, Tosson & Saad 1972, Kamel 1971, Nakhla & Shehata 1967, El Sharkawi & Khalil 1977). The ore consists mainly of hematite and goethite with occasional pockets of sooty ochre and pyrolusite. Gangue minerals include barite, kaolinite, glauconite, alunite and silica minerals represented by chert and tripoli.

Generally, four types of ore are distinguished based on constituents, texture and chemical composition. These are called the hard, friable, banded-cavernous, and pisolitic types. It is necessary to blend the various types to obtain Fe 53%, SiO_2 7.5%, Cl 0.7% and MnO 1.98%, for use in the metallurgical plants at Helwan. The ore reserves and average chemical composition of the four main ore occurrences are given below.

Area	Reserves M tons	Fe %	SiO_2 %	Mn %	S %	P %	Cl %
El Gedida	126.7	53.6	8.9	2.3	0.9	0.2	0.6
Ghorabi	57.0	48.0	9.0	3.0	0.7	0.9	0.8
Nasser	29.0	44.7	6.7	3.9	0.6	0.1	1.3
El Harra	56.6	44.0	12.5	2.9	1.0	0.1	0.8

Ambiguity arises regarding the genesis of the iron ores in the Bahariya Oases area. Attia (1950) favored a shallow water lacustrine origin during Oligocene time. Deposition of leached iron under lagoonal environment and subsequent replacement of the underlying middle Eocene and Cenomanian beds is envisaged by El Shazly (1962), and a similar origin is posited by Akkad & Issawi (1963). Evidence of replacement is apparent where most of the calcareous fossils, especially the diagnostic nummulites of the middle Eocene, are almost completely replaced by iron oxides (Gheith 1955, Nakhla 1961).

Contrary to these opinions, Tosson & Saad (1974) suggested that the ores were formed by metasomatic replacement associated with impregnations and cavity filling from ascending solutions affiliated with volcanic activity. They considered the oolitic and pisolitic iron ore outcropping in the Ghorabi area to be sygenetic, the iron being supplied by weathering processes. Based on combined geological, mineralogical and geochemical data, Basta & Amer (1969) considered the ore to have been formed by metasomatic replacement extending to the lower Bahariya formation and the upper Radwan formation. Most authors agree with the observation that the high grade ores exist in the crests, and that low grade ores are localized in the limbs of the anticlinal structures.

Recent studies based on sequential observations of the blasted faces at El Gedida mine, led by El Sharkawi et al. (1984) resulted in distinguishing three genetic types:

1. Genetic type I: Represented by the high central area in El Gedida mine. The ore is high grade, with high iron and sodium chloride contents, low silica, and variable Mn and Ba contents. High traces of Zn and Cu are present. The mineralized middle Eocene Naqb formation is brecciated, 'possibly karsitifed', and metasomatically replaced by hydrothermal solutions ascending along northeast-southwest trending fractures. The deposited iron ore is a hydrothermal-metasomatic type and of a massive nature.

2. Genetic type II: Following the emergence and

faulting of the mineralized middle Eocene block, the generated depressions received reworked rocks including high grade ore from the high central area. Fresh water lakes occupied the depressions where remobilization of iron and manganese and their redeposition were effected, possibly through biogenic interference. Tripoli earth and kaolinite were authigenetically deposited with the debris. Detrital barite is a common associate. Abrupt change in grade characterizes the iron ore of this genetic type. The iron ore is usually cavernous, ochreous or massive.

3. Genetic type III: This follows type II in age and is tied to the post-middle Eocene glauconitic succession which caps the reworked iron ore of type II. Enrichment of the marine depositional basin in iron and potassium promoted the formation of glauconite. Cyclic deposition of glauconitic clays and sands was interrupted by intermittent emergence followed by lateritic weathering of the glauconitic sediments. Profound changes in the mineralogy of these sediments took place, resulting in the deposition of a low grade iron ore characteristically poor in Mn and Ba. The iron ore is usually oolitic or pisolitic in texture.

2.6.2 *Manganese ore deposits*

Workable manganese deposits are known in the Um Bogma district in Sinai and in the south Eastern Desert in the Halaib district close to the Sudan border. At present there is minor production in the latter area. In addition, minor occurrences are known in Wadi Mialik, near Abu Ghosun in the Eastern Desert, and Wadi Sharm El Gibli to the north of Hurghada.

a) *Um Bogma district*

The geology of this district attracted the attention of many geologists even before the work of Ball (1916) on the geography and geology of west central Sinai was published. Extensive workable manganese deposits contributed significantly to the Egyptian economy up to 1967, when the mines were abandoned. Reopening the best mines is being considered and evaluation of newly discovered occurrences is in progress. It is also planned to produce ferromanganese products at the plant installed at nearby Abu Zeneima, a port on the Gulf of Suez.

Manganese ore deposits occur in Paleozoic sediments of Carboniferous age (see for a discussion of the age of this section chapter 21, this book).

The manganese ore is a stratiform type occupying more or less the same stratigraphic horizon in the dolomite member of the Um Bogma formation which caps the Adedia formation. The ore bodies usually show abrupt contacts with the dolomite and are frequently found to fill depressions in the underlying

Adedia formation clastics. According to El Shazly (1957), the ore occurs in the form of lenses or lenticular beds varying in dimensions and extent. The thickness varies from tens of centimeters to 8 m and the extent of the beds may reach hundreds of meters. Attia (1956) observed that the small lenses are richer in manganese than the lenticular beds, where manganese occurs admixed with iron.

The normal mineralogy of the ore includes pyrolusite, manganite, psilomelane, cryptomelane, and less commonly ramsdellite, with common associates of goethite and hematite. Gangue minerals include dolomite, barite, calcite, gypsum, clay minerals and alunite. Multistage formation of the manganese minerals is noticed especially in the regeneration and recrystallization of pyrolusite (El Shazly et al. 1963). Mart & Sass (1972) differentiated three zones in the large ore bodies, characterized by certain mineral assemblages:

1. The inner zone: essentially composed of psilomelane and pyrolusite with rare manganite, hausmanite, polianite and pyrochroite. Hematite and clay usually do not exceed 25%. The structure is massive, but concretions of pyrolusite may be present.

2. The intermediate zone: consists of psilomelane, pyrolusite and hematite with up to 15% goethite, quartz, barite and clays. The ore is massive and constitutes the main ore reserves of Um Bogma.

3. The outer zone: composed mainly of hematite and goethite with minor psilomelane. Detrital quartz is common and spherulitic concretions are frequent.

This zoned pattern indicates a low pH/high Eh condition at the rims, and high pH/low Eh in the cores. The mineralogy indicates a low temperature of formation in a sedimentary environment.

It is evident from the isopach and facies contour maps of Um Bogma formation prepared by Mart & Sass (1972) that the manganese mineralization isopach contours run in a general northeast-southwest direction, and that two facies prevail in the area, a silty facies to the southeast and a dolomite facies to the northeast.

A shallow marine origin of the manganese ore deposits is supported by El Shazly et al. (1963) and Mart & Sass (1972), and this view is favored by exploration geologists working in the area. Within this framework, the behavior of certain trace elements during the formation and diagenesis of the manganese ore body was studied by Margaritz & Brenner (1979), following and refining the sedimentary origin as first proposed by Barthoux (1924).

On the other hand, advocates of a hydrothermal-metasomatic origin as proposed by Ball (1916) include Fennine (1931), Attia (1956) and Gill & Ford (1956). They tie the mineralization to volcanic activity, faults, and replacement of the dolomitic host rocks.

Current studies favor sedimentary origin, as there is no link between the ores and faults and where manganese deposits are present in faults, they are introduced as fillings from above. The late Professor G. S. Saleeb-Roufaiel (pers. comm.) was convinced that deposition of insoluble manganese and iron oxides followed chemical weathering of the dolomite. Moreover, he noted that manganese oxide minerals occur, pseudomorphing coral reefs at some mines. El Sharkawi, El Aref and Abdel Moteleb are currently involved in studying the genesis of manganese mineralization within the framework of the controlling factors of karst topography. The induced effect of weathering of the clastic rocks capping the manganese-dolomite layer is thought to have epigenetically added turquoise, malachite and alunite to the mineralogy of the manganese ore deposits.

b) *Halaib 'Elba' region*

Manganese occurs in sedimentary rocks of probable Miocene age in 24 areas within a featureless plain in the Halaib region, situated in the southern extremity of the Egyptian Eastern Desert near the Red Sea coast. In a few cases, manganese deposits occur as fracture fillings in basement rocks, especially granites.

Manganese minerals occur either in veins trending within a range of 118 to 130° in a belt about 70 km long and less than 7 km wide, or occasionally replacing the Miocene conglomerates and limegrits. The mineralogy of the ore (El Shazly & Saleeb-Roufaiel 1959, Basta & Saleeb 1971) includes pyrolusite, psilomelane, cryptomelane, ramsdellite, todorokite, and nsutite, in addition to occasional goethite and hematite. The gangue minerals include quartz, black calcite, barite, opal and chalcedony.

El Shazly (1957) envisaged an origin by weathering of the Precambrian rocks and supergene deposition in fissures accompanied by replacement along the walls of the fissures. An epigenetic low temperature origin is proposed by Basta & Saleeb (1971), based on the predominance of stable higher oxides of manganese and the absence of manganese silicates, carbonates and sulfides, which reflect near-surface deposition of the ore.

Manganese mining is restricted to elementary operations due to the remoteness and unfavorable conditions of the area. Extraction of manganese began in 1955 and since then only about 60,000 tons of high grade ore have been produced.

c) *Other manganese occurrences*

Minor manganese occurrences are recorded from Wadi Mialik, near Ras Banas on the Red Sea coast of the Eastern Desert. The ore occurs as fillings in fault zones and fissures in Precambrian amphibolites (El Shazly 1957).

In the Miocene sediments of Abu Shaar El Qibli, in the southern part of the Esh El Mellaha range north of Hurghada, Ghobrial (1963) investigated the workability of a thin manganese deposit (about 50 cm thick). The mineralogy and geochemical aspects of this deposit are reported by Hassan & Sabet (1979). Nothing has been produced from this occurrence.

2.6.3 *True sedimentary ores*

a) *Phosphorites*

Superphosphates are produced for agricultural use in Egypt and for export. Phosphorites are currently produced in two districts. The Red Sea coast phosphorites are exported primarily to southeast Asia, while the Nile Valley phosphorites are produced for domestic use. Production from a major occurrence of phosphorites at Abu Tartur plateau in the Western Desert awaits the results of a lengthly feasibility study. The distribution of the phosphate-bearing rocks is shown in Figure 26/17.

1. *Red Sea coast phosphorites*. The phosphatic deposits in the Quseir-Safaga district are present in three members within the Duwi formation (Youssef 1957). The Cretaceous and Eocene sediments are tilted by faulting in different directions, forcing mining through inclined shafts. The economic phosphorites occur in the form of lenses or beds and are mined at many localities in the Quseir group of mines (Hamadat, Atshan, Duwi, Anz, Abu Tundub, Hamrawein), and at the Safaga group (Um El Howeitat, Gasus, Wasif, Mohamed Rabah). Facies changes in the phosphate bearing sediments are recorded at the Safaga mines group. The P_2O_5 content ranges from 22 to 30%, reaching 34% in rare cases.

The intimate association of the phosphorites with black shales and the presence of pyrite and organic matter indicate that phosphate deposition was in a basin with negative Eh. Youssef (1957) linked the formation of phosphorites to upwelling in synclines, while Tarabili (1969) favored a model of deposition on the western slopes of a northwest trending gulf parallel to the present Red Sea, with the Safaga area representing deeper marine conditions than the Quseir area. Reworking of the phosphatic matter was noted by Ramzy (1964), in which collophane grains were transported from deeper to shallower levels.

Mining of the Red Sea coast phosphorites began in 1910, for export to the Far East. The phosphorites are subjected to calcination to upgrade the product. The largest mine operating at present is that at El Hamra-

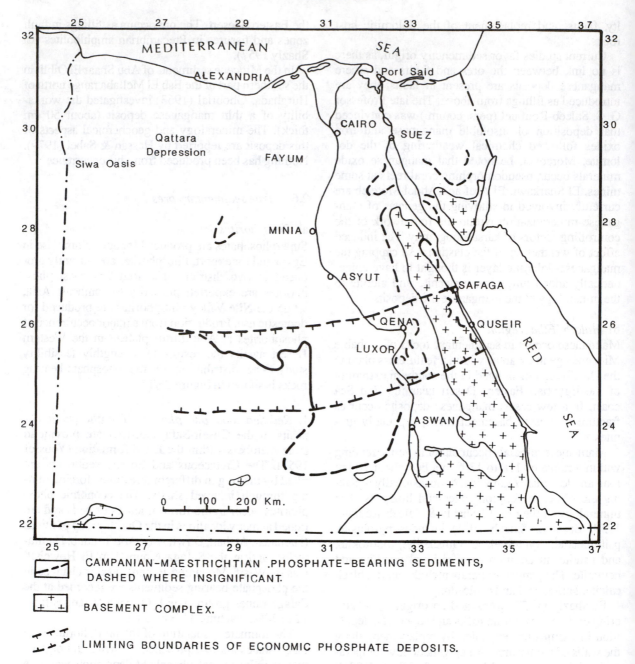

Figure 26.17 Phosphate-bearing rocks (after Spanderashvilli & Mansour 1970).

wein, but this operation is encountering serious problems due to the low price of phosphates and the costs of mining.

2. *Nile valley phosphorites.* Domestic needs for phosphorites for the production of superphosphates are supplied from the Nile Valley deposits between the latitudes of Esna and Edfu. They were the first phosphorites discovered in Egypt and have been exploited for some 75 years. The phosphate bearing

Duwi formation includes three phosphate units intercalated with marl and oyster limestone beds (El Bassyouni et al. 1970). The middle and lower units are the most economic and are mined at several localities. Mineralogically, the phosphorites consist of collophane and francolite, with small amounts of quartz, calcite, goethite, chlorite and, rarely, zircon (Kotb et al. 1978). The environment is shallow marine with strong agitation. The mineralogical and chemical compositions are modified by diagenesis.

Silicified marl and chert bands appear to have formed during diagenesis.

3. *Abu Tartur phosphorites*. Phosphorites are known to occur in the scarp face bounding the Dakhla-Kharga depression in the Western Desert (Ball 1900, Beadnell 1901). The thick and relatively high grade phosphorites are recorded in the Maghrabi-Liffiya sector of the Abu Tartur plateau, about 60 km west of El Kharga town (Hermina 1967) were explored by surface mapping, drilling, and exploratory mines by the Geological Survey of Egypt, (Said, 1971, Wassef 1977). The phosphate bearing formation of Campanian-lower Maastrichtian age (Said 1962) overlies unconformably the variegated shale unit of the Nubia formation and is in turn capped conformably by the Dakhla shales. The phosphate formation is divided into three members, the lower member being 4 m thick and the most economic, averaging 26.5% P_2O_5. Fresh phosphorites are dark grey to black due to the presence of carbonaceous matter and pyrite, and acquire a creamy to yellow color on weathered surfaces.

The phosphatic matter occurs in the form of pellets, oolites, coprolites, organic remains, nodules and concretions. The phosphatic minerals include carbonate apatite 'collophane' and crystalline francolite-dahllite with subordinate wavellite and manganapatite (Kamel et al. 1977). The non-phosphatic minerals are represented by dolomite, ankerite, montmorillonite, calcite, gypsum-anhydrite, iron oxides, pyrite-melnicovite, marcasite, glauconite, quartz and carbonaceous matter. El Kammar (1977) drew attention to the relatively high rare earth elements content, which averages 2034 ppm with a high yttrium contribution to this content.

The current reserve estimates in the exhaustively investigated area runs in the order of a billion tons. Unfortunately, the occasional presence of iron sulfides seriously affects the workability of these tremendous reserves. It is estimated that though about L.E. 235 million have been spent on infrastructures (roads, drilling, feasibility studies) since 1968, the present low price of phosphates discourages mining operations.

b) *Coal deposits*

Exposed coal deposits are known in two areas of Sinai, the Maghara district and Um Bogma district. Subsurface coal seams and coaly sediments have been recorded in oil exploration wells in the Gulf of Suez region and in the Western Desert (Fig. 26/18). A subsurface coal deposit at Ayun Musa (Sinai, Gulf of Suez region) has been thoroughly explored by core drilling. At present, the only deposit considered economic is that of Maghara in north Sinai.

1. *Maghara coal deposit*. Al Far (1966) reported the occurrence of coal seams in the Bathonian sediments on the northwestern side of the Maghara anticline. Some 84 wells were drilled up to 1970 and more wells have since been drilled to examine its workability. Ten coal seams are known, of which two, the main and the upper, are of commercial value. The estimated reserve is about 51.8 m.t. in a 30 km^2 area. The coal is predominantly vitrinite and clarain, long-flamed coal, consisting of:

Carbon	70.0 – 80.0%
Hydrogen	5.6 – 6.6%
Nitrogen	1.04%
Oxygen	8.2 – 9.2%
Sulfur	2.7 – 3.5% (main pyrite)
Ash	4.0 – 8.0%
Volatiles	51.0 – 59.0%

The technological properties of Maghara coal are described by Adindani & Shakhov (1970). Production in 1970 was projected as 150,000 tons per year, increasing to 300,000 tons per year. An exploratory adit was opened at El Safa area.

2. *Um Bogma coal deposit*. Carboniferous coal deposits have recently been recorded from oil wells drilled in the Gulf of Suez region at east Gharib, east Bakr and Belayim Marine sites. Similar deposits were discovered in outcrop in the area around Um Bogma in west-central Sinai at Thora, Beda, Allouga and Abu Zarab. The age assigned to these coaly deposits, based on palinologic analysis, is upper Visean (Petrascheck & Nakhla 1961). Extensive exploratory work was conducted in 1958-59 in the Thora and Beda areas and the results were reported by Moustafa & Yarovoi (1961). Thora, more promising than Beda, has coal seams 10 to 80 cm thick and extending over a distance of 15 km, dipping 4° N. The coal reserve at Thora is about 1.5 m tons and the average chemical composition is as follows:

Ash content	39.0 – 49.0%
Volatile recovery	17.0 – 27.0%
Moisture	2.0 – 3.8%
Fixed carbon	30.0 – 44.0%
Sulfur	0.4 – 1.0%

3. *Ayun Musa coal deposit*. Subsurface coal deposits similar in age to the Maghara coal (Jurassic) occur at a depth of 400 to 600 m. The deposit has been explored during oil exploration in the area 14 km southeast of Suez. Since its discovery in 1946, the Egyptian Geological Survey has drilled 26 wells in an area 30 km^2 to calculate reserves and the nature of the coal seams (Moustafa & Auslender 1961).

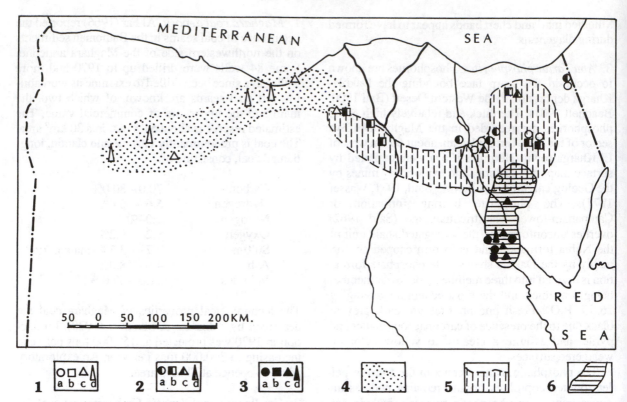

Figure 26.18 Distribution of coal formations in Egypt (after Adindani & Shakhov 1970). 1. Proved presence of coal in lower Cretaceous rocks, a. carbonaceous rocks, b. lenses and coal intercalations up to 0.7 m thick, c. coal seams 0.71 to 2.0 m thick, d. coal seams more than 2.0 m thick; 2. Same, for middle Jurassic rocks; 3. Same, for lower Carboniferous rocks; 4. Area of proven maximum coal content present in lower Cretaceous rocks; 5. Area of proven maximum coal content present in middle Jurassic rocks; 6. Area of proven maximum coal content present in lower Carboniferous rocks.

Two coal horizons some 40 to 60 m apart are present. The upper horizon is comprised of up to 10 coal seams of variable thickness, reaching a maximum of 120 cm. The lower horizon includes one coal seam varying in thickness between 20 and 120 cm. The reserve is about 36.8 m tons.

The deposit is regarded as non-commercial due to the non-persistence of the coal seams, complicated structures, and the great depths which would require special mining skills.

c) *Carbonates*

1. *Limestones.* For building and road construction purposes, the extensive Eocene limestone exposures satisfy the needs of the country. The Cretaceous limestone exposed at Abu Roash is extensively quarried, especially for road construction materials, while dimensional limestone blocks are obtained from the Eocene of Gabal Mukkatam for building needs in Cairo. Limestone blocks are produced for building purposes from the Pleistocene oolitic limestone ridges west of Alexandria.

Lime production needs are met from the middle Miocene limestone exposed in the Cairo-Suez district and magnesium-poor limestones of Cretaceous and Eocene age. The limestones of the Pleistocene ridges west of Alexandria are also used for production of lime and soda ash.

The Eocene limestones exposed near Helwan and south of Suez are extensively quarried for the cement industry. The sintering plant at the Helwan Steel Complex used limestone for sintering ore fines, acting as a fluxing agent. Certain limestones having high compressive strength (about 600 kg/cm²) are suitable for making concrete blocks.

High grade limestone is quarried from the Eocene succession at Samalut, in upper Egypt. The $CaCO_3$ content ranges from 96.4 to 98.8%, $MgCO_3$ is 1%, and NaCl is 0.2 to 0.7%. This product is highly valued and finds uses in many industries.

2. *Alabaster.* This is a special type of calcium carbonate displaying colloform banding, similar to world alabaster (hydrous calcium sulfate) in appearance. It is expected to be present in limestone plateaux, especially those of Eocene age. The most famous and workable deposit, producing alabaster since Pharaonic times, occurs in the vicinity of Beni Suef in the Nile Valley. Since alabaster only develops in fractures in limestone plateaux, its genesis is thought to be through calcium-secreting algae that flourished extensively in karstified and jointed limestones. The carbonate minerals acquired with aging the characteristic honey yellow color. Another alabaster occurrence is known in Wadi El Assiuty, near Assiut in upper Egypt (Akaad & Naggar 1963).

3. *Dolomite.* Economic dolomite is currently quarried from Eocene strata at Ataqa, about 20 km south of Suez. A product with more than 18% MgO is needed for the glass, ceramic, and iron and steel industries. Hard dolomitic rocks are used in furnace linings, harbor construction and shore protection.

Eocene dolomite has been quarried in the Abu Roash district west of Cairo, at Giran El Foul. Dolomitization of Cretaceous limestones is a common feature along fault planes and fractures, requiring selective quarrying at El Gigia and near Abu Roash village.

d) *Clastic and placer deposits*

1. *Gravel.* The composition of gravel varies with provenance. Gravels quarried from the Red Sea coastal plain are predominantly reworked basement rocks of various composition. Gravels in northern Egypt are thought to form in depressions created by post-Eocene tectonics. The main contribution to such basins comes from the Cretaceous (Turonian and Maastrichtian) cherts which are rich in silicified Cretaceous microfauna and phosphatic matter in addition to lower Eocene silicified limestones rich in nummulites and alveolins.

More than 95% of the sieved gravels are composed of chalcedony, chert and silicified limestone. The rest includes quartz pebbles, felsites, metavolcanics and granitic pebbles derived from older formations. The gravels in Wadi El Natrun district may enclose large boulders of basement rocks, mostly granite and rhyolite, pointing to the possible presence of basement exposures shedding debris in the basin, or a possible direct connection with the basement rocks exposed in the Eastern Desert prior to the carving of the present day Nile course in northern Egypt.

Since the formation of the Oligocene gravels, these were either preserved by capping basaltic sheets or were reworked to newly formed basins. Reworking took place during the lower Miocene, Plio-Pleistocene and in Recent times.

Most of the workable large gravel deposits are those of Oligocene age, which escaped diagenetic deterioration of gravels by biogenic means or subjection to arid climate which produces splintery gravels. Caliche-cemented gravels pose problems in gravel quarrying. Such products seriously affect the quality of the clinker.

Faults do occur in gravels and can be detected, if with difficulty, with the aid of aerial photographs wherein fault planes assume a zigzag configuration. Abrupt transitions from gravel dominated areas to sandy areas is a common finding in gravel quarrying, and is commonly related to faults.

Gravels of good quality are currently produced in the Cairo-Suez district, while inferior quality gravels occur in the Western Desert, south of Cairo, and near Wadi El Natrun.

2. *Sand.* Most of the quarried sands come from Oligocene and Recent sediments. The main use of sand in Egypt is in the manufacture of concrete, precast walls, and sand bricks. Size-graded sands are used for water purification and filtration.

3. *Glass sands.* Many active quarries operated by public and private sector firms are operating in Abu Darag and Wadi Dakhal near Zaafarana on the Red Sea coast. Sands satisfying the production of high quality glass are found at Wadi El Dakhal. The deposit belongs to the Malha formation of Cretaceous age.

Extensive glass sand 'belt' outcrops are found in central Sinai about 50 km to the northeast of St Catherine. The high kaolinite percentage in this sand, about 80%, is regarded as a hindrance to the workability of the deposit. Preliminary processing of the raw sand, however, produced a high quality glass product (A. A. Omar, pers. comm.). This may encourage a reconsideration of the specifications currently required by the glass industry. Production of good quality glass in Jordan using similar raw materials has proved successful. Glass sand for inferior quality glass products is produced from a quarry east of Maadi, south of Cairo, and from another quarry 80 km east of Edfu.

In addition to its chemistry, grain size is a sensitive parameter in ranking sand deposits for use in glass production. The fraction between 0.2 and 0.5 mm should be about 98%. Current production from Wadi El Dakhal assays 99.2 to 99.5% SiO_2 and 0.02% Fe_2O_3, free of alumina and titania, and from Abu Darag 98.5% SiO_2 and 0.3 to 0.5% Fe_2O_3, with traces of titania. These are both used in the production of good quality glass. Inferior quality glass sand, such as that of the Maadi cocurrence, which assays 97%

SiO_2 and 1% Fe_2O_3, can be used in the manufacture of window glass. The grain size of the high quality glass sand produced from Wadi El Dakhal is as follows:

0.850 mm	0.5%
0.600 mm	0.5%
0.425 mm	2.0%
0.250 mm	75.0%
0.150 mm	17.0%
0.125 mm	1.0%
< 0.125 mm	4.0%

4. *Quartzite*. This name is currently applied to diagenetically silica cemented sands and sandstones. Such sandstone is a tough, hard rock usually in colored shades of brown, red, and even black, as a result of the incorporation of iron oxides in the silica cement. The best-known locality of quartzite is Gabal Ahmar, on the northeastern border of Cairo, developed in exposed Oligocene sands and gravels. The source of the cementing silica is thought to be through the action of geysers percolating in the clastics of Gabal Ahmar. Shukri (1954) related the hydrothermal solutions to Oligocene volcanic activity. Silica cementation, however, could also be induced under surface conditions and at normal temperatures, and a silica source from above is certainly a possibility. Quartzite bricks and blocks have been quarried for decorative purposes, and are traditionally used as millstones.

5. *Brick-making clay*. The governmental ban on the use of topsoil for brick-making has activated an intensive search for suitable clays in Phanerozoic sediments. Most clays used at present create problems for automated machinery, being substituted for the traditional hand methods. Crushed basalts and granites are added to montmorillonitic clays to avoid bricks cracking during drying and firing. Gypsum and halite, common associates of Cretaceous, Eocene and Miocene clays, generate acidic gasses during firing that seriously affect the bricks' quality for construction purposes. Kaolinitic clay and bituminous clay are the most suitable for brick-making. Grain size, not less than 2 microns, and plasticity are critical factors. Calcium carbonate can be tolerated only up to 7%, as lime causes blowing and misshapenness in the product. It is customary to use an electromagnet to eliminate iron objects that contaminate the clay during quarrying, in order to prevent damage to the machinery. Almost every governorate in Egypt has established its own brick-making factory to meet urgent needs for this product.

6. *Kaolin*. Most of the kaolin supplied to the porcelain and ceramic industries in Egypt comes from Aswan and from west central Sinai. Production of low grade kaolinite from the Abu Darag area has dropped sharply since the discovery of the Kalabsha kaolin southwest of Aswan (Issawi 1969, Said & Mansour 1971).

At Kalabsha, 105 km southwest of Aswan, kaolin occurs between two members of the Nubia facies tentatively assigned to the Senonian. The upper sandstone member encloses thin beds and lenses of kaolin. Three types of kaolin are present, the pisolitic, non-pisolitic 'plastic', and concretionary kaolin. The thickness of the kaolin members revealed by drilling ranges from a few centimeters up to 9.15 m. In the upper sandstone member, kaolin may reach 30 cm in thickness. The proven reserves are about 16.5 million tons.

In Sinai, kaolin occurs interbedded with sandstones of Cenomanian age and is mined in many localities. Wadi Abu Natash and Naqb Burda produce low plastic kaolin, while non-plastic kaolin is being mined at Musaba Salama and also occurs in Wadi Kabrit, Ras Dereib Agrag-El Tih, Wadi Abu Rigim, and Farsh El Ghazlan. The low grade kaolin of Abu Darag is used in the production of white cement.

7. *Ball clay*. All clays in the Abu El Rish Qibli district are suitable for the production of earthenwares, due to their high plasticity and good sintering properties. Alumina content reaches about 33% and it is planned to use this material for the production of alum. Ball clay is added to the low and non-plastic kaolins to increase their plasticity.

8. *Black sands*. The Mediterranean coast of Egypt between Alexandria and El Arish is spotted with black sand concentrations in many places, the most economic of which is that near the mouth of the Rosetta branch of the Nile. The geomorphology of the deltaic coastal plain between Rosetta (Rashid) and Port Said is presented by Said (1958) and the factors controlling sedimentation of black sands were elucidated by Rittmann & Nakhla (1958) and Said (1958).

Abu Khashaba black sands. The Rosetta black sands have been the subject of many articles and dissertations, as well as technical reports by private firms. The Abu Khashaba area, 5 km north of Rashid on the eastern side of the Rosetta mouth, was proven to be the most economic site. The 8 km² area was explored through close-grid boreholes and the upper 12 m proved to contain 7.5 million tons of black sands containing 707,000 tons of economic minerals, of which 287,000 tons are ilmenite (Kamel et al. 1973). Other minerals of economic interest are present, including monazite, zircon, rutile and xenotime. Gold was recently reputed to occur in the black sands. Zircon and monazite do not exceed 1 to 2% and monazite is characterized by a low thoria content of 5% (Abdel Monem & Hurley 1980).

e) *Evaporites*

1. *Gypsum.* Gypsum deposits of Miocene and younger ages are known in many areas all around the Mediterranean and Red Sea coast. Low grade gypsum occurs in Fayoum (Abdallah & El Kadi 1974). High quality gypsum is produced at El Ghorbaniat and Deir El Birqat-El Hagf to the west of Alexandria. At El Ghorbaniat, the gypsiferous beds reach 5 m, capped by 1 m of fossiliferous limestone (Tosson 1963, Abdallah 1967). In the newly discovered gypsum deposit in the Deir El Birqat-El Hagf area, gypsum occurs as two lenticular bodies, an upper and a lower, interbedded with Miocene sediments including sandstone, marl and limestone. The upper lens averages 5.1 m, while the lower lens averages 6.3 m. The proven reserves, estimated by El Shazly et al. (1976), amounts to 50 million tons. The product assays 95 to 98% gypsum and is used in agriculture, plaster of paris, and insulation materials.

The Miocene sediments in the Gulf of Suez district yield gypsum-anhydrite products suitable for the cement industry and the manufacture of plaster of paris. It is produced at Ras Malaab on the east coast of the Gulf of Suez, and in an area south of Zaafarana on the west coast.

Granular surface gypsum is deposited in shallow lakes through evaporation at El Ballah, north of Ismailia, as well as at Mariut and Manzallah lakes near the Mediterranean coast. At El Ballah, gypsum is produced from a 1 m-thick bed admixed with sodium chloride and calcium carbonate. It is used as a retarder and in the production of plaster of paris.

2. *Rock salt.* Rock salt is obtained by scaling crusts artificially deposited through evaporation of sea and lake waters in salines. The main production comes from Al Max, near Alexandria, Rashid, Mersa Matruh, Idku and Baltim.

3. *Alum and magnesium sulfate.* These are deposited in Dakhla and Kharga Oases as thin banks in black shales of Nubia sandstone. The salts are believed to have been formed through supergene oxidation of original syngenetic sulfides and their subsequent leaching and deposition through the evaporation of ground waters (Said 1962, 1975).

f) *Weathering products*

1. *Alunite.* Alunite, the hydrous sulfate of aluminium and potassium, was recently identified in the Phanerozoic sediments of Egypt (El Sharkawi & Khalil 1977). This mineral was previously thought of as a product of endogenetic processes, especially during hydrothermal alteration of rhyolites. Alunite is difficult to identify in the field and is therefore often overlooked in sedimentary successions, and may be mistaken for kaolinite.

Extensive alunite deposits are known in the El Gedida area, Western Desert, in post-middle Eocene sediments. Alunite genesis at El Gedida mine was during weathering of glauconitic sediments capping the Eocene iron ores (El Sharkawi & Khalil 1977). The workable alunite sands developed capping and intercalated with the iron ore.

Minor occurrences of alunite are known in west central Sinai, developed in association with the manganese ore deposits (El Shazly et al. 1963) and in Oligocene sediments of Gabal Ahmar near Cairo (Shukri 1954). The mineral is also observed in association with ochreous deposits at Um Greifat, near Quseir, and is locally known as 'the potatoes'. At Ras Gemsa, north of Hurghada, it is exposed intercalated with green clays, sulfur and gypsum in Miocene sediments. Recently, Hilmy et al. (1983) described natro-alunite in sedimentary rocks of Tertiary age between Cairo and Fayum.

Alunite is a prospective raw material for aluminium production. It also finds applications in the ceramic industry, and in alum production. It is used in the specialized production of greenhouse glass.

2. *Ochre.* Yellow and red ochre have been mined intermittently on a small scale from Um Greifat, about 50 km south of Quseir on the Red Sea coast. The deposit occurs in middle Miocene sediments including conglomerates, marl, sandstone and green clay, resting on peneplaned basement rocks. It appears in the form of lenticular bands about 2 m thick, capped with green clay. Alunite is occasionally found admixed with the ochre. The yellow ochre contains 46.5% Fe and the red ochre, 58.7% Fe.

The ochreous deposit was first described by Attia (1950) and its genesis is considered by Kabesh et al. (1970) to be by oxidation of a primary siderite replacement deposit, while Tosson & Saad (1972) favor a shallow marine sedimentary origin with later alteration. The green clay suffered repeated weathering, resulting in the formation of alunite and deposited ochreous iron.

g) *Sedimentary uranium deposits*

In Phanerozoic sediments, radioactivity is related to

the presence of thorium and/or uranium bearing refractory minerals or to secondary uranium minerals. Radiogenic potassium may produce similar effects. The basement-sediment contact is the target for prospection for radioactive minerals.

The Carboniferous sediments 30 km east of Abu Zeneima, west central Sinai, were investigated by trenching and adits, especially the sandstone bed belonging to the Lower Sandstone Series with radioactivity of 60 mR/hr. The principal mineral causing this radioactivity is thought to be the radiogenic mineral xenotime (Hussein et al. 1971).

The Cretaceous sediments, especially the black shales and phosphorites, display high radioactivity. In the black shales, radioactivity is related to organic matter, while in phosphorites it is related to uranium substitution in the apatite structure. Uranium concentration in Egyptian phosphorites has been treated in many publications (e.g. Higazy & Hussein 1955, Zaghlul & Mabrouk 1964, Hassan & El Kammar 1975). The Western Desert phosphorites are poorer in uranium than the Red Sea and Nile Valley phosphorites, which average 100 ppm.

In the Oligocene sediments of Gabal Qatrani, north of Fayoum, uranium mineralization occurs in various rock types including phosphatic sandstone, carbonaceous clay, limestone, and fossil wood (El Shazly 1974). Uranium is abnormally high in francolite separated from the phosphatic sandstone, about 0.6% U. A hot brine activated by Oligocene vulcanicity is the model adopted to explain the origin of abnormal uranium concentrations.

The beach placers at Rashid (Rosetta) on the Mediterranean Sea coast might yield the radiogenic minerals zircon and monazite.

2.7 MINERAL DEPOSITS IN METAMORPHIC ASSOCIATION

2.7.1 Metamorphogenic deposits

Ores that form as a result of metamorphic processes include graphite, sillimanite, kyanite, andalusite, garnet and others. None, with the exception of graphite, is known as deposits in Egypt, probably due to the absence of high grade metamorphic assemblages in the country. Thus, this group is represented here by only graphite.

a) Graphite

Graphite was first discovered in Egypt in 1938 at Gabal Um Selim in the Barramyia area and later, in 1943, at Bent Abu Geraiya and Wadi Sitra (El Alfy 1946). Other occurrences have been recorded from Wadi Dendikan (Hamata area), Wadi Ghadir and Wadi Haimour. In all these occurrences, graphite is present in thin seams of graphite schist intercalated with other paraschists, in close association with serpentinites. The paraschists are mainly chlorite-tremolite-biotite schists. The graphite schist is strongly foliated and has a soapy feel and is composed of chlorite and tremolite with finely dispersed graphite dust. Chemical analysis of graphite schists from Wadi Bent Abu Geraiya showed graphitic carbon in the range of 1.6 to 1.84%, and ash between 92 and 94% (Said 1962). At Wadi Sitra, some 20 to 25 seams of graphitic schist are present in close association with paraschists and serpentinites.

The origin of graphite is believed to be that suggested by Said (1962) who attributed the development of graphite to the low to medium grade metamorphism of the carbonaceous matter contained in the original sediments.

Ore reserves of graphite in Egypt were estimated at 200,000 tons by El Alfy (1946) with a contemplated annual production of 1000 tons concentrate containing 65% graphite.

2.7.2 Metamorphosed ore deposits

a) Banded iron formation (BIF)

In the Eastern Desert of Egypt, and in a restricted area between latitude 25° 15', 26° 31' N and longitude 33° 22', 34° 20' E, banded iron formation (BIF) occurs in a number of localities: Abu Marawat, Wadi Karim, Wadi El Dabbah, Um Khamis El Zarga and Um Nar. In all of these areas, the ore is present in the form of bands and lenses of magnetite, martite and hematite with a gangue dominantly of quartz. The bands may extend for some kilometers along the strike, and vary in thickness from a few centimeters to 10 m or more. The overall chemical composition of these deposits is as follows:

Fe	31.9 – 52.3%	SiO_2	19.3 – 32.4%
P	0.17 – 0.61%	Mn	0.23%
Ti	0.27%	S	0.10%

These deposits resemble the Algoma type BIF known from the Archean (and younger) greenstone belts (Sims & James 1984), since they occur in successions of volcaniclastic rocks and intercalated lava flows.

1. *General characteristics.* The general characteristics of these deposits as described by Sims & James (1984) and El Ramly et al. (1971) are as follows:
– The BIF represents sharply defined stratigraphic units within layered volcanic-volcaniclastic se-

quences of calc-alkaline nature and andesitic composition.
- Individual bands range from a few centimeters to more than 10 m in length and are frequently faulted and folded with steeply dipping limbs.
- Microbanding occurs on a scale of a centimeter or less, where iron-rich bands alternate with bands of jasper or, sometimes, of carbonates or silicates.
- Most of the iron is present as magnetite (altered in places to martite) concentrated in steel-black bands alternating with red jasper or with iron-poor grey or greenish bands; hematite is less frequent. The gangue minerals present are mainly quartz, chlorite, biotite and clay minerals.
- Frequent contemporaneous folding, faulting, brecciation and slump structures are found.
- Greenschist facies metamorphism, with the development of chlorite, sericite and the iron silicate stilpnomelane and possibly minnesotaite occur. On the contact with intrusives, local metamorphism may reach amphibolite facies with the recrystallization of the iron minerals and silica and development of epidote and garnet.

Detailed ore evaluation, including drilling in most of these areas, gave the following results (El Ramly et al. 1971).

	AM	WK	WD	UKh	GH	UN	Total
Reserves (M.t.)	6.5	17.7	6.0	5.6	3.6	13.7	53.1
Fe% surface	44.4	44.6	38.2	44.6	45.7	45.8	43.7
Fe% subsurface	–	43.0	34.9	42.1	45.0	41.8	–
Fe% in concentrate	–	56.4	53.5	59.7	69.0	61.0	55.3
Expected conc. - res.	–	10.0	3.2	3.6	2.6	7.0	26.4

The following paragraphs give some notes on each of the main occurrences of BIF in the Eastern Desert.

2. Main occurrences

Abu Marawat (26° 31' N, 33° 22' E). Ore bands occur in a country of regionally metamorphosed volcaniclastics (tuffs) with intercalations of dolerite and andesite flows. These bands range in thickness from 3 to 17 m and are folded in a series of anticlines and synclines with folding axes in a northwest-southeast direction, superimposed on a broad anticlinal structure with axial plane trending northeast-southwest. The bands are arranged in groups and are concentrated on the upper parts of Gabal Abu Marawat. They dip steeply either east or west. Hematite is the main constituent in outcrop, probably due to martitization. Magnetite, siderite, quartz, calcite, tremolite and chlorite are present. Total reserves, to the wadi level, are estimated at 6.5 m.t. of ore containing 44.4% Fe.

Wadi Karim (25° 54' N, 34° 09' E). Four major ore seams called A, B, C and D, alternate with metasediments of a volcanic derivation in a 130 m-thick section. Seam A is the most important, extending for more than 3 km along the strike (northwest) and 220 m down dip (northeast). Its thickness varies from 3.5 to 12 m. The seams are folded into a recumbent anticline and show extensive microfolding. They are cut by numerous faults. The ore is mainy of magnetite and, to a lesser extent, hematite, both finely intergrown with silica. Ankerite and goethite are also present and the gangue is mainly of jasper, chalcedony and quartz as well as some rock material. Mineable ore (bands greater than 1.5 m thick and containing 40% or more Fe) is estimated at 17.8 m.t. and is assayed as:

Fe	44.0 – 45.0%
SiO_2	23.3 – 23.5%
CaO + MgO	8.7%
P	0.2 – 0.3%

Wadi El Dabbah (25° 49' N, 34° 08' E). A number of conformable ore bands up to 10 m thick occur within a succession of fine tuffaceous metasediments over an area of about 6 km^2. The ore bands gather into three groups, the central of which is most important. They extend 5 km along the strike and form part of a broad anticline with axis striking north-south. Minor folds strike east-west and plunge east. Faulting is pronounced, with a major north-south fault coinciding with the anticlinal axis and some minor faults in the east-west or northeast-southwest direction (Fig. 26/19). Two varieties of ore are present, a magnetite-rich or black ore and a hematite-rich or red-violet ore, the latter being higher in grade. The ore minerals are hetatite and/or magnetite, with quartz, chlorite, actinolite, epidote and garnet gangue. Calcite and pyrite are present as secondary minerals. Martitization is pronounced near the surface. In the vicinity of the granite contact, veinlets of epidote and porphyroblasts of garnet are developed in the ore. Mineable ore reserves (more than 1.5 m thick bands with greater than 35% Fe, to 100 m depth) are given as 6.1 m.t., and the average analysis given by Akaad & Dardir (1983) is as follows:

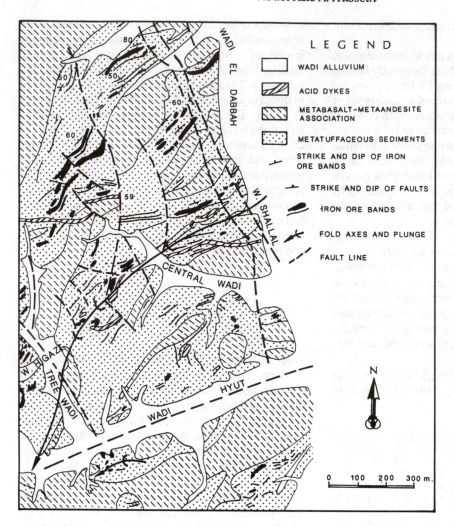

LEGEND

☐ WADI ALLUVIUM

▨ ACID DYKES

▨ METABASALT-METAANDESITE ASSOCIATION

⋯ METATUFFACEOUS SEDIMENTS

⊢ STRIKE AND DIP OF IRON ORE BANDS

⊢ STRIKE AND DIP OF FAULTS

▬ IRON ORE BANDS

⊢ FOLD AXES AND PLUNGE

- - - FAULT LINE

0   100   200   300 m.

N

Figure 26.19 Geological map of Wadi El Dabbah iron ore deposit (after Akaad & Dardir 1983).

	Hematite ore	Magnetite ore
Total Fe	43.8	39.0
FeO	2.1	6.2
Fe_2O_3	59.6	48.48
SiO_2	25.1	30.57
Al_2O_2	1.7	5.9
CaO	5.0	4.5
MgO	0.94	0.95
P	1.0	1.54
MnO	0.04	–
S	0.7	–
L.O.I.	0.4	–

Um Khamis El Zarga (25° 32' N, 34° 17' E). A number of bands of small thickness are found in greenschists folded into an east-west anticline with cross folding and frequent faults. Most of the bands are unmineable (less than 1.5 m in thickness) and are less persistent along strike than in the other areas. The ore is mostly magnetite with extensive martitization

that continues to considerable depths. Chalcedony is the main gangue and along the granite contact, garnet and epidote are developed in the ore. Ore reserves are given to the wadi level as 3 m.t., with the following analysis:

Fe	40.00 – 41.00%
SiO_2	26.41 – 28.05%
P	0.40%
Al_2O_3	2.50 – 7.20%
CaO	3.60%
MgO	1.00%

Gabal El Hadid (25° 20' N, 34° 10' E). Here, ore bands alternate with bands of chert within a predominantly volcanic sequence of submarine lavas and tuffs metamorphosed into chlorite actinolite schist. The area is folded and frequent faults truncate the bands. The main ore is a black magnetite ore, but carbonate bearing or jaspilitic ores are also present.

The gangue minerals are quartz, hornblende, feldspar, chlorite, sericite, clay minerals and secondary carbonates. Mineable reserves (bands greater than 1.5 m thick with higher than 40% Fe) are 2.15 m.t. to the wadi level and 1.4 m.t. below it. Assay of the ore showed:

Fe	43.00 – 47.00%
SiO_2	23.50 – 25.00%
P	0.26 – 0.29%
CaO	2.98 – 5.40%
MnO	0.60%
P	0.26 – 0.29%
S	0.15 – 0.40%

Um Nar (25° 18' N, 34° 15' E). Within a volcanosedimentary sequence of 2200 m thickness, nine groups of ore bands are found, separated by quartzitic or biotite schists. The sequence is folded and extends for a 7 km strike length. The ore is schistose and has the coarsest grain size of all the occurrences. It is made up of magnetite, lesser amounts of hematite and quartz, chalcedony, hornblende and biotite gangue. Epidote and garnet develop close to the granite contact. The mineable ore reserves are given as 13.7 m.t. of the following average composition:

Fe	40.51 – 45.50%
SiO_2	27.50 – 32.27%
P	0.21 – 0.22%
Al_2O_3	0.40 – 0.20%
CaO	3.74 – 5.70%
MgO	0.17 – 0.35%
S	0.10%

3. *Genesis of banded iron formation (BIF)*. The genesis and mode of formation of these deposits summarized by Sims & James (1984) is accepted here. They state that these deposits originated as chemical sediments of exhalative fumarolic source in a number of separate small basins developed between intraoceanic island arcs. They were accumulated, with volcaniclastic tuffs and intercalated laval flows in shallow submarine environments during periods of quiescence in submarine volcanic activity. After diagenesis and lithification, they were folded more than once, faulted, and regionally metamorphosed into greenschist facies.

b) *Marble deposits*

Marble, in the correct sense of the word, is known to occur in some localities within the shield rocks in Egypt. Many quarries produce 'Phanerozoic marbles', hardened limestones subjected to advanced stages of diagenesis with no metamorphism. These are described in Section 2.6.3 above by El Sharkawi.

True marbles, metamorphosed calcareous rocks, occur as bands or lenticles enclosed within schists of clastic or volcaniclastic derivation at a number of localities in the shield area, particularly within the southern assemblage (Hunting 1967). Main occurrences north of latitude 25° N are those at Wadi Dib, and Wadi Dagbag and Gabal Rokhan off Wadi Mia. More extensive deposits are known at Wadi Maryia, the wider area of Abu Swayel and Wadi Allaqi, south of Aswan. In these southern localities the marble is white, bluish grey or black in color, sometimes with very attractive banding. Sculptural grades may be found (Hunting 1967).

Though the reserves in the southern areas are substantial, the remoteness of the quarries limits exploitation. Construction of the planned Wadi Halfa railway, or Aswan to Khartoum highway, to the east of the Nile might encourage more extensive working of these deposits.

2.8 MISCELLANEOUS

In any classification, there are usually some items that defy assignment to specific pigeonholes. The present classification is no exception, and some minerals and rocks must be placed here, under none of the above headings. These include some gemstones, as well as some of the building and ornamental stones that have been and still are exploited in Egypt. Some of these are of special historic importance, such as the peridot of St John's Island, emerald of Zabara, monumental granite of Aswan, Chephren diorite, Hammamat breccia verdiantico, and the Imperial Porphyry of Gabal Dokhan.

Peridot is the gem variety of olivine. It has been used and praised since the eighteenth dynasty of ancient Egypt. Its occurrence is restricted to St John's Island (see 1.2.c) and is not exploited at present.

Emerald is green transparent beryl. The ancient Egyptians mined it in Nugrus, Sikait, Zabara and other localities (see 2.2.3.e).

Garnet is abundant in metamorphic rocks. It was used for beads and necklaces in ancient Egypt and was probably obtained from the mica schists of Gabal Mitiq or Wadi Gemal (Hume 1937). No record of gem quality garnet is known in Egypt at present.

Amethyst is the violet transparent variety of quartz. It was used extensively in jewellery in ancient Egypt. Considerable workings are known near Gabal Abu Diyeiba, Wassif area, where amethyst occurs lining cavities in drusy veins cutting across red granites (Murray 1914).

Amazonite, a beautiful bluish-green microcline feldspar, occurs in coarse pegmatite veins intruded into the gneisses of Gabal Migif and Wadi El Gemal, and also as a constituent of the metasomatically altered gneisses of Abu Rushaid (see 2.2.3.d).

Malachite, in addition to its use as an ore of copper and as a pigment, was used extensively by the ancients for necklaces and carved scarabs. It was obtained from the gossans covering copper mineralizations, as in Abu Swayel.

Turquoise occurs as fissure fillings in Carboniferous formations in Sinai. It was exploited on a very small scale until recently for the local market.

Lapis lazuli was used in ancient Egypt but no occurrence of the stone is known in the country. Ibrahim (1949) believed it was obtained from the vicinity of Oweinat Oasis in the extreme southwestern corner of Egypt and otherwise, was imported from Persia or Afghanistan.

Ornamental stones, i.e. any stone that takes a fine polish and has an attractive appearance, are of many types in Egypt (Ibrahim 1949). The famous Monumental Granite of Aswan and the hornblende granite known as the granite of Mons Claudianus have been cut and polished into sarcophagi, obelisks and statues since ancient times. At present, a limited amount of granite is obtained from quarries at Aswan.

Diorite and gabbro occur in many localities. The famous Chephren Diorite was quarried in the Western Desert 65 km northwest of Abu Simbel. The gabbros of Wadi Semna, Eastern Desert, were also worked by the Romans (Hume 1937). Imperial Porphyry, characterized by its beautiful purple color with rose-white phenocrysts, occurs only at Gabal Dokhan. It was cut into columns, blocks and vases and transported to Rome. Serpentinites in various shades of green, Breccia Verdi antico and Breccia Brocatelli, are outstanding ornamental stones used in ancient times but almost completely neglected at present.

Fresh hard basalts are regarded as a national strategic raw material. Active quarries are found around Cairo (Abu Zaabal, El Yahmoum, El Haddadin, Qatrani), and it is planned to use weathered basalts and dolerites in the manufacture of bricks. Most basalt extents may be concealed under a thin Quaternary cover. Recent seismic data from the western Delta area proved the subsurface extension of the El Haddadin basaltic sheet (Williams & Small 1984) of Oligomiocene age exposed at Tel El Zallat and Tel El Haddadin to the north of the Abu Ruwash structure. Similar extensive sheets are recorded covering the area between Abu Treifiya and Nasuri in the Cairo-Suez district. Other workable basalts occur near El Bahnasa and Darb El Arbaien in upper Egypt. Extensive sheets and dikes are known in the Bahariya Oasis area in the Western Desert and in south Sinai.

3 CONCLUDING REMARKS

In conclusion to this brief review of the mineral deposits and occurrences of Egypt, there are two relevant questions that require discussion. The first concerns the metallogeny of these deposits and their relation to the models of crustal evolution suggested for the country, and the second concerns the prospects of finding new metallic deposits in the shield rocks.

3.1 METALLOGENETIC CONSIDERATIONS

There are three approaches to dealing with a country's mineral deposits: a) as commodities, such as copper, zinc, gold deposits, etc.; b) as products of, and in association with, crustal evolution in a plate tectonics model; and c) assigning them to the petrological associations or rock assemblages of which they are characteristic and form integral parts.

The first approach is obviously suitable for miners and those dealing with extractable metals as such, regardless of the geology, genesis or mode of occurrence of the deposits. This approach of course places geologically widely diverse deposits under one heading, such as cupriferous pyrites in oceanic crust to stratiform deposits in sedimentary association to porphyry copper deposits, three types with nothing in common except their having copper as their most valuable metal.

The author follows the basic principles of metallogeny that state that mineral deposits faithfully follow crustal evolution (Bilibin 1968), both being characteristic products of the geotectonic environment prevailing. Also, plate tectonics are considered by him to be the only explanatory model for the evolution of both (Mitchell & Garson 1981, Sawkins 1984). Nonetheless, to deal with the mineral deposits on the basis of these principles is to rely heavily upon interpretation, probably neglecting the simple geologic facts observable in the field.

The choice thus adopted here is the third, assigning the mineral deposits to the petrological assemblages with which they are associated and of which they are characteristic.

Regarding the evolution of mineral deposits in relation to crustal evolution and in the framework of one or another of the plate tectonic models suggested for the country, a few concluding remarks are in order. All these models agree in their essential elements; an old continental mass; belts of melange and obducted slabs of oceanic crust; island arcs; magmatic arcs on continental margins; intraplate magmatism in a cratonized crust, and sedimentary sequences formed in epicontinental seas. No important mineral deposits are known to occur associated with the old continental mass, exposed only in small tectonic windows (such as at Hafafit), except corundum, beryl and the like, associated with a few small bodies of pegmatites. The obducted oceanic crust hosts chromite occurrences as well as the Cu-Ni sulfide deposits of the type found at Abu Swayel. On the ocean floor, and as a result of exhalative processes associated with the early stages of island arc formation, the deposits of banded iron formation of the central Eastern Desert were formed. The mature island arc stage was accompanied by the formation of volcanogenic massive sulfide deposits such as those of Um Samuiki. With continuing subduction, under a continental margin environment, the porphyry copper prospects (Hamash) and the precious metal (Au, Au-Ag) vein deposits were formed. The intraplate hot spot and early rifting activities were associated with a number of deposits, including the lithophile elements (Sn, W, Mo, Nb, Ta, Be, Li, U) deposits in G3 granites, the porphyry molybdenum deposits (Homr Akarem), ring complexes and carbonatites, the Cu-Ni sulfides (Gabbro Akarem), and the Fe-Ti oxides (Abu Ghalaga) found in layered mafic-ultramafic intrusions. Intraplate sedimentary environments are responsible for the stratabound Zn-Pb deposits in limestone (Um Gheig), the true sedimentary ore deposits such as phosphorites, the iron ores, and the various limestones and other construction materials.

3.2 PROSPECTS OF FINDING NEW METALLIC DEPOSITS

It is clear from the review presented here that despite the large number of occurrences and prospects of metallic mineral deposits found in the shield rocks, practically none of them is large or rich enough to warrant large or even medium scale mining operations. This is not peculiar to the shield in Egypt, but is also the case in all the Pan-African Arabian-Nubian Shield exposed in Saudi Arabia, Egypt, Sudan, Ethiopia and Somalia. Total metals produced from this shield amounted to only a few million dollars worth in the period 1978-1982 (Pohl 1984). The reasons for this situation, which has long been tantalizing yet economically unfortunate, have been the subject of extensive investigation: a) is this due to the lack of adequate mineral exploration programs in the shield terrains? b) was the time of the formation and evolution of the shield not a particularly rich metallogenetic epoch? c) could the situation be attributed to erosion, whereby many mineral deposits were eroded together with their enclosing rocks during the long time span since their formation in the upper Proterozoic?

Many mineral exploration programs have been carried out on some parts of the shield. The Aswan Mineral Survey project, undertaken jointly by the UNDP and the Geological Survey of Egypt during 1966-1976, was very successful both technically and scientifically. Nonetheless, it concluded with the identification and delineation of an additional number of mineral occurrences, most of them are at best sub-economic (Hussein 1973). Similarly, in Saudi Arabia, particularly during the period 1975-1985, very extensive, though by no means exhaustive, exploration and evaluation programs were conducted. The results are very similar to those in Egypt, a great number of occurrences with virtually none of them economic under prevailing conditions. The situation is not much different in the Sudan. A joint Sudanese-Soviet mineral survey project covered the area of the Red Sea Hills (northeast Sudan) during the period 1971-1978. No deposit was discovered during, or as a result of, this project (Hassan Ali 1987, pers. comm.). Thus, lack of exploration programs is not alone enough to account for the absence of valuable deposits.

To assume that the period of formation and evolution of the shield was a particularly poor epoch in ore generation is somewhat unrealistic, since the period was some 600 million years and included all geotectonic processes – rifting, spreading, subduction, accretion and intraplate activity – with which ores are generated.

Before considering the role that might have been played by erosion, it must be noted that most of the ore deposits expected in the shield areas are those hosted by obducted oceanic crust, arc related volcanic-volcaniclastic-sedimentary formations, the subvolcanic protrusions of the I-type G1 granite batholiths, and skarn zones-copulas-veins fields associated with the top parts of high level S-type intraplate (G3) granites. All these deposits form within the upper 2 km or so of the crust. In addition, orthmagmatic ore deposits, particularly those formed by magmatic segregation, accumulate in the lowermost parts of batholiths or layered intrusions, i.e. at a depth of some kilometers within the crust. Thus, in the exposed shield, where erosion of the upper few kilometers must have taken place since cratonization and uplift in the late Precambrian, it is not surprising to find that most of the arc volcanic-volcaniclastic and sedimentary rocks as well as those formed in subvolcanic environments and in the roof zones of high level intrusions have been eroded away together with the deposits they host. It is interesting to note that the only volcanogenic massive sulfide deposits known in Egypt at present are those that escaped erosion, being hosted in a down-faulted synclinal block (at Abu Hamamid).

The same factor, erosion, is used by Sawkins

(1984) to explain the rarity of porphyry copper deposits in Paleozoic and older rocks, stating, 'Erosion levels are of major significance, for compressive arcs tend to form thick crustal roots and stand high, and their uppermost few kilometers, where porphyry deposits form, are susceptible to removal. Few of the porphyry deposits of the Andes or those of the Philippines, for example, would survive 20 million years into the future, given continuation of current uplift and erosion rates in these areas' (Sawkins 1984: 22).

On the other hand, this level of erosion in the shield is not deep enough to expose deposits related to early stages of rifting, buried below very thick sequences of sedimentation, or those associated with the basal parts of batholiths and layered intrusions.

In conclusion, it is believed that a number of metallic mineral deposits must have formed during the evolution of the shield, but most of them have been unfortunately lost to erosion. Some hope still remains, however. A number of metallic deposits may be discovered in areas where shield rocks were protected structurally or below early Phanerozoic cover, as for example in the Western Desert. It appears that the mineral occurrences of Egypt can be exploited only through continuing small scale mining operations, though deposits enclosed within Phanerozoic sedimentary cover deserve further attention.

CHAPTER 27

Petroleum geology

MOSTAFA K. EL AYOUTY

Consultant, Cairo, Egypt

Oil seeps have been known in Egypt throughout recorded history but modern petroleum exploration is scarcely 100 years old (see Royds, Mason & Eicher 1975, Egyptian General Petroleum Corporation 1986). Commercial accumulations of oil and gas are so far known in four provinces: Gulf of Suez, north Western Desert, the Nile delta and north Sinai. A description is given below of the petroleum geology of each of these provinces.

GULF OF SUEZ PROVINCE

The Gulf of Suez graben was formed as a result of tectonic movements initiated in the Oligocene which continued with intensity until post-Miocene times. Tensional faults of considerable displacements determine the configuration of the graben and its boundaries. Within the confines of these boundary faults the graben area is dissected into numerous pre-Miocene fault blocks of various sizes. The tilt of these blocks is predominantly to the southwest in the northern and southern parts of the Gulf, and to the northeast in the central part. The different blocks were subject to varying degrees of erosion. The Miocene transgression covered an eroded and topographically uneven surface of pre-Miocene rocks. Accordingly, Miocene sediments rest unconformably over rocks ranging in age from Precambrian to late Eocene. The rugged topography of the pre-Miocene surface is evident from the presence of both shallow and deep water marine environments side by side in the early stages of the Miocene transgression. These conditions were favorable for the development of reefs on the uplifted blocks and/or organic-rich clastics on the downthrown blocks.

The movements continued during the Miocene and later times, rejuvenating old lines and producing an environment where great lateral facies changes occurred. These Miocene and younger events affected the oil habitat of the Gulf, the 'cooking' conditions of source rocks, the oil migration process, and the type and size of the traps.

The Gulf of Suez is the main oil province in Egypt. Oil is present in and is produced from sediments belonging to the Basement, Paleozoic, Mesozoic and Cenozoic. The fields are mainly structural traps, though some are stratigraphic.

Reservoir rocks

The bulk of the oil in the Gulf of Suez is housed in sandstones of Paleozoic (Carboniferous and probably older), Cretaceous (Cenomanian, Turonian and early Senonian) and Tertiary (Miocene) age. Some oil is present, however, in fissured and cavernous limestones of late Cretaceous, Eocene and Miocene ages. In the last few years oil accumulations have been discovered in porous basement rocks in several structures.

Basement reservoirs

Basement rocks with some porosity are found in a few localities in Egypt. The porosity is of the secondary type, either due to fracturing or to weathering of the basement.

So far oil productive basement rocks are present in four localities in the Gulf of Suez basin: Shoab Ali, south Geisum, G.S. 304 and Zeit Bay fields. Of these, the last field has the largest basement reservoir. It consists of fractured basement topped by basement wash, both of which are productive. Production at the other three localities comes from fractured basement reservoirs.

Paleozoic reservoirs

The Paleozoic pay in the Gulf of Suez consists of the lower part of the 'Nubian Sandstone' section. It was first recognized in the Hurghada oilfield. Here the lower part of the Nubian type sandstones overlying the basement yields plant remains of Devonian age (Van der Ploeg 1953). This is followed by a section of Carboniferous age. The Paleozoic sandstones house the bulk of the oil of this field. The same sandstones are also present in the Ras Gharib field where they form the main oil reservoir. Because of the lack of diagnostic fossils in the Nubian Sandstone, the section is subdivided according to the heavy mineral assemblages into an upper Nubia 'A' which is considered Cretaceous, a middle Nubia 'B' which is mainly made up of black shales of Carboniferous age, and a lower thick Nubia 'C' which is oil-bearing in Ras Budran, July, Ramadan and Sidky fields.

Cretaceous reservoirs

Early Cretaceous Nubia 'A' sands are oil-bearing in the Hurghada, Ras Gharib, Bakr, July and October oilfields. Cenomanian sands are productive in the Belayim Marine and October oilfields. Turonian sands and limestones form reservoir rocks in the Belayim Marine, Bakr, Amer, Ras Gharib, Kareem, July and Ramadan fields. The early Senonian sands are oil-bearing in July, Ramadan, Belayim Marine, October and Sidky fields.

Eocene reservoirs

The Eocene reservoir in the Gulf of Suez province consists of fractured and/or cavernous limestones of middle, and in some cases, early Eocene. These Eocene limestone reservoirs are present in Sudr, Asl, Ras Matarma, Bakr, west Bakr, Kareem and Shoab Ali fields. The completion of the wells as producers from these limestone reservoirs includes acidization.

Miocene reservoirs

The Miocene reservoirs are the most prolific producers in the Gulf of Suez. They house the bulk of the reserves and are present in many fields both on land and offshore. A number of reservoirs are present within the Zeit-South Gharib sequence, but the main oil-bearing levels are present in the Belayim, Kareem, Rudeis and Nukhul formations. With few exceptions, the Miocene reservoir rocks are sandstones.

The Zeit and South Gharib formations comprise a number of pay sands in Belayim Land field. These have erratic lateral extensions, due to the rapid changes of depositional environments.

The main Miocene sand reservoirs lie within the Kareem-Rudeis formations as in Morgan, Belayim Land, Belayim Marine, July, Shoab Ali and Zeit Bay fields, and in Belayim formation as in Belayim Land, Belayim Marine, Morgan and Shoab Ali fields. Less conspicuous production is obtained from these formations in many other fields in the province.

The lowermost Miocene rock unit, the Nukhul formation, is oil-bearing in many fields. It was first recognized as a pay in Rudeis and Sidri fields, where it is the only oil-bearing horizon. The Nukhul reservoir consists of sandstones in Shoab Ali, Rudeis,

Sidri and GS-173 fields but in a few other fields, some production is obtained from carbonates such as in El Ayun and Kareem (Saoudi & Khalil 1984).

Miocene limestones are oil-bearing in a few fields within the Belayim and Rudeis formations (e.g. Sudr, Asl and Matarma). In Ras Gharib field the limestone pay section is in the form of a reef, the Nullipore Rock.

The Miocene reefal limestones were the producing reservoir in the depleted Gemsa field. It is difficult to determine the stratigraphic position of the pay section in the field. The only published information concerning the geology of the field is that of Bowman (1931). The well logs shown in Plates 24 and 25 of that report indicate that the oil was present in limestone and dolomitic limestone interbeds within a mainly evaporite section which probably belongs to the Ras Malaab group (?South Gharib formation).

Pliocene-Pleistocene reservoirs

So far, sediments belonging to the Pliocene-Pleistocene have been found to be oil-bearing only in the Abu Durba field. In the only published report on this field the reservoir in many of the wells is attributed to the Nubian sandstone section (Bowman 1926, Van der Ploeg 1953). However, a careful examination of the data indicates that the field is actually a surface seep along faults which impregnated the surface material along the foot of Gebel Araba. This conclusion is based on the fact that the oil is housed in reservoir rocks consisting of conglomerates with igneous pebbles, calcareous sands, shales and limestones. This type of lithology does not tally with the known lithology of the Nubian Sandstone section but rather with sediments of post-Miocene age. However, the two most easterly situated wells, Abu Durba 1 and 3, seem to have penetrated at the bottom a thin oil-bearing section of what can be considered Nubian Sandstone. These two wells are the nearest to the Cretaceous outcrops in the area.

Cap rocks

Sealing beds are abundant throughout the geological column, especially in the Cretaceous and Miocene. In these stratigraphic levels, beds of shale, compact limestone and evaporites could seal off any lower hydrocarbon pool.

The most effective of these sealing beds, however, belong to the Miocene. The evaporite beds within the Nukhul, Kareem, Belayim, South Gharib and Zeit formations are associated with the oil accumulations of the Miocene sequence. They, together with pre-Miocene sealing beds, form an effective sealing cover for accumulations in the pre-Miocene sequence (Ras Gharib and Hurghada). Miocene evaporites, shales and marls are also effective in sealing (or trapping) the oil laterally, by virtue of being very thick in the grabens and depressions where most of the oil accumulated.

Traps

The rifting process which caused the formation of the Gulf of Suez graben left a marked imprint on the tectonic framework and geological history of the province. The sequence of events involved in the process greatly affected the distribution of the younger Tertiary sediments, their thickness and facies, as well as their structural setting.

The rift faults which caused block faulting of the sediments in the form of horsts and grabens were rejuvenated, causing the blocks to be tilted, the grabens to be deepened, and the horsts to be elevated. This was followed by erosion of the highs and deposition of markedly thicker late Tertiary, particularly Miocene, sediments in the lows. The continuous movements produced marked lithological variations reflecting the rapidly changing environments of deposition. Moving away from the depocenter, to the margins of the basin, sediments change from fine to coarse clastics and from salt to anhydrite. In addition, and as a result of these movements, unconformities and/or depositional breaks occur at different levels of the section. The geological record indicates that differential erosion took place after the rifting started, so that when the Miocene sea transgressed from the north marine sediments were deposited on a variety of older formations ranging from pre-Cambrian basement to late Eocene (Abdine 1981).

In contrast, the geological history of the Gulf of Suez area in pre-Miocene times was characterized by movements which were not as intensive and continuous (Said 1962, Gilboa & Cohen 1979). Pre-Miocene sedimentation was not interrupted except during the late Paleozoic, early Mesozoic, late Mesozoic and early Tertiary tectonic events. The Paleozoic and Mesozoic sediments in the Gulf of Suez area are, therefore, lithologically uniform laterally and do not display the marked lithological changes of the Miocene sequence. This rather regular deposition makes the formation of stratigraphical oil traps unlikely.

On the other hand, the late Cretaceous-early Tertiary tectonic event which affected the Paleozoic and Mesozoic sequences is related to most of the oil occurrences in Egypt. This event formed the northeast-southwest trending lines of highs (folds) of the Syrian arcs. In the Gulf of Suez they occur only in the northern part of the Gulf. Although oil accumulations have been found in some of these structures in the Western Desert, no oil has been found so far in features related to it within the northern part of the

Gulf of Suez province. Another tectonic event was the Tertiary rifting which is displayed by the more conspicuous Clysmic faults and the less prominent Aqaba faults. The result was the formation of fault blocks, with a main northwest-southeast trend, which consist of Paleozoic and/or Mesozoic sequences on which Tertiary formations overlapped. The oil accumulations in the Paleozoic and Mesozoic sediments are trapped in fault blocks which are sealed off laterally by juxtaposition against impervious pre-Miocene and Miocene sediments in the surrounding lows. Paleozoic and/or Mesozoic oil accumulations are trapped, therefore, by fault (structural) conditions rather than by any other type of trapping mechanism.

Aeromagnetic and Bouguer gravity studies suggest that most of the oil discoveries in the Gulf of Suez province are associated with basement uplifts. Three major north-northwest trending uplifted basement belts separated by two structural lows are recognized in the Gulf (Meshref, Refai & Abdel Baki 1976).

The Tertiary rifting movements constitute the prime factor in determining the habitat of oil in the Gulf of Suez province. The northwest-southeast trending lows lying between the highs and fault blocks received thick and deep water organic-rich clastics (potential source rocks). The nearby higher features (of Miocene and pre-Miocene sequences) acted as reservoirs for the oil after migration. Moreover, as a result of the irregular topography in late Eocene and Miocene times, the conditions of deposition changed laterally from one location to the other, thus allowing for the marked lateral facies changes which led to the formation of stratigraphic traps in the Miocene reservoirs.

Faulting is the main factor which determined the geological setting, and hence the trapping processes. The anticlines or half anticlines described from the Gulf are associated with faults. According to Said (1962) they were 'either produced by the bending of the strata before breaking or by movements that caused the less rigid sediments, especially the Miocene, to bend in anticlinal or synclinal folds'. Another mechanism suggested for the formation of these folds is 'that the topographic configuration of the pre-Miocene surface was inherited by younger sediments due to both differential compaction and to more subsidence of pre-depositional troughs relative to the highs' (Hantar 1967). This mechanism, according to Hantar, 'gave rise to a number of inherited anticlines above pre-depositional highs and synclines over erosional lows and, consequently, the pre-Miocene topography was to a great extent responsible for the present structural configuration of the Miocene and younger sediments . . . it follows from the foregoing argument that such structures have been formed at the same time in which the sediments were being deposited' (*ibid.*).

These folds, together with the faults which are associated with traps in which stratigraphic factors are involved in the accumulation of oil, establish the combination of both structural and stratigraphical elements in determining the category of traps referred to as the structural-stratigraphical (or combination) traps.

It is difficult in many cases to differentiate between the three types of traps. In many of the multi-reservoir traps the oil in one pool or more might be trapped by the effect of the structural element on the reservoir and sealing formations, whereas the oil in another pool might be trapped due to stratigraphic conditions only (lensing or non-permeability trapping). Even in the case of some single pool traps, oil may be present in a lens which is later on affected by faulting or folding or both.

The 'pure' structural traps are rather limited in number in the Gulf of Suez province and are almost totally confined to the fields in which oil is housed in pre-Miocene reservoirs. This is due, as mentioned above, to the fact that the pre-Miocene in the Gulf of Suez province was characterized by continuous and laterally uniform sedimentation and thus the possibility for lensing, depositional breaks and lateral facies changes was minimal.

In the Gulf of Suez province oil is structurally trapped in 'Nubian-type sandstone' reservoirs belonging to the Paleozoic-Mesozoic (including Cenomanian, Turonian and Senonian) sandstones and in fissured and fractured limestones of Eocene age. The most prominent structural traps in the Gulf of Suez district include the Paleozoic and/or Mesozoic pools in Hurghada, Ras Gharib, Bakr, Kareem, Belayim Marine, Ramadan, Sidky, G.S-391, October, Shoab Ali and Ras Budran oilfields. This category of traps also comprises the Eocene pools in Sudr, Asl and Matarma, as well as a few Miocene pools mainly within the lower part of the sequence. In a few recently discovered reservoirs, oil is stored in pre-Cambrian fractured basement rocks (Shoab Ali, Zeit Bay and south Geisum structures). The fracturing responsible for the secondary porosity in which the oil is stored in these structures was due to movements and rejuvenation of movements related to the creation of the Gulf of Suez structural faults.

The purely stratigraphic traps are not very numerous in the Gulf of Suez province. Due to the intense and continuous tectonic movements which the region witnessed after late Eocene times, it is hard to conceive of an oil trap in the region which would owe its presence to stratigraphical conditions only, with no structural element involved. One field may be excepted, the Gemsa oilfield along the west coast of the Gulf of Suez near its junction with the Red Sea. The well information indicates that there is a basement horst which is capped by Miocene and younger sediments. Within the lower part of the Miocene section, reefal limestones developed where oil accumulated in four horizons. The Miocene reefs in question are surrounded by fossiliferous shales which in turn surround the basement horst (Bowman 1931). The oil should have migrated to the reef porosity from the surrounding organic-rich shales which are assumed to be the source rocks. There is no evidence to indicate that the trap owes its presence to post-depositional tectonic movements. Another trap considered to be of the stratigraphic type is a Miocene reefal limestone which is present on the eastern flank of the Ras Gharib field.

In contrast, the majority of the oil discoveries in the Gulf of Suez province can be assigned to the stratigraphic-structural combination type of traps. This applies to most of the Miocene pools in the Gulf which display marked lateral changes in facies, permeability and depositional breaks, reefal development and irregular evaporite deposition, in addition to the influence of the tectonic movements.

Source rocks

The earliest suggestions regarding the source rocks of oil in the Gulf of Suez province were based on mere geological associations. The subject was treated on grounds of regional geological observations, stratigraphic and structural relationships and petrographic examination of sedimentary units, especially shales. According to Weeks (1952), the Miocene sediments are the source, while Van der Ploeg (1953) suggests the *Globigerina* marl of the Miocene section as the source. He also stressed that 'the chalk series of the upper Cretaceous and Eocene have been found to be slightly bituminous all over Syria, west Jordan, Israel and Sinai, and therefore they have been regarded by several geologists as the proper source rock of the Egyptian oil'. Ghorab (1961) suggests the early to middle Miocene *Globigerina* marls as source rocks as well as possibly the late Cretaceous-Eocene carbonates.

In the mid-1960s some oil companies operating in the Gulf of Suez region began to make quantitative studies of source rocks. Generally, the approach was based on the relationship between hydrocarbon occurrences, the organic richness of sediments, the type of kerogen present, the temperatures to which the sediments were subjected, and their age. The results of work undertaken by some oil companies on these subjects are reported in Barakat (1982) and Shahin & Shehab (1984).

Organic richness is a parameter reflecting the

potential of a rock to generate hydrocarbons; it is measured by the percent by weight of the organic carbon the rock contains (Total Organic Carbon, TOC). The organic richness of sediments is considered 'poor' when TOC is less than 0.5%, 'fair' when it is between 0.5 and 1.0%, and 'good' when it is more than 1.0%. However, there are differences in the conclusions reached. Rohrback (1980b) rates the organic richness of the early Miocene (Kareem and Rudeis formations) as 'good', the rest of the Miocene formations as well as the Paleocene and Cretaceous as 'fair' and the Carboniferous as 'low'. Barakat (1982) rates all the formations studied (Miocene, Eocene/Paleocene, upper Cretaceous, lower Cretaceous and Paleozoic) as 'good' source rocks. Shahin & Shehab (1984) rate the lower Miocene clastics as 'poor' to 'good' source rock and the pre-Miocene as 'good' to 'excellent'.

Though the conclusions of the authors differ, because they emphasize various parts of the sequence, they all agree that the early Miocene (Kareem and Rudeis), the Eocene/Paleocene and the late Cretaceous sediments contain enough organic carbon to generate oil.

The type of organic matter, or kerogen, in a rock is closely related to the type and amount of hydrocarbons generated from the rock. Kerogen consists of organic matter in sedimentary rocks not soluble in organic solvents of either marine or terrestrial origin. Kerogen of marine origin consists of the remains of marine algae and amorphous organic matter. Being rich in hydrogen, this kerogen type is known to be oil-prone. Kerogen of terrestrial origin consists of pollen, spores, woody and coaly materials and is known to be gas-prone.

The kerogen types are recognized visually by petrographic means through microscopic examination in both transmitted and reflected light, by which the degree of maturation of the source rock is also determined.

According to Rohrback, the lower Miocene (Kareem and Rudeis formations) and Eocene limestones and shales are oil-prone. Their kerogen content is more than 80% amorphous (marine) material. He considers the South Gharib formation and Eocene limestones and shales as oil and gas-prone, and the Paleocene, Cretaceous and Carboniferous as gas-prone. Barakat rates the Belayim, Kareem and Rudeis formations and the upper Cretaceous shales as oil-prone, the Eocene and Paleocene as oil and gas-prone and the lower Cretaceous and Paleozoic as gas-prone. Shahin & Shehab consider the Miocene clastics as oil and gas-prone and the pre-Miocene as oil-prone.

These findings indicate that oil-prone rocks are present at many levels within the geological succession, rather than in a single stratigraphic unit. Among the potential oil-prone source rocks are the Kareem and Rudeis formations of the Miocene, the Eocene shales and limestones and the late Cretaceous carbonates and shales. Possible oil and gas-prone source rocks are the Belayim formation and the Paleocene. The generation of oil from the oil-prone source rocks depends upon their stages of maturation.

The degree of maturation of sediments is determined by visual analysis of kerogen using one of two methods. The first is the study of the coloration of spores and pollen grains, as color reflects the degree of thermal maturation of the rock. The second is the vitrinite reflectance method in which the fraction of incident light reflected from polished surfaces of wood fragments (vitrinite) in sediments is measured.

The conclusion reached by Rohrback (1980b) is that the degree of maturation of kerogen based on spore coloration and vitrinite reflectance shows that all sediments studied, both Miocene and pre-Miocene, are of sufficient maturity to generate oil. Barakat (1982), on the other hand, finds that only the lower Miocene and upper Cretaceous sediments are mature in parts of the Gulf of Suez. Shahin & Shehab (1984) conclude that all Miocene and pre-Miocene samples are immature or in the early generation stage.

Some geochemical studies have been undertaken to correlate possible source rocks with the crude oils produced in the Gulf of Suez province. Barakat (1982) mentions this subject briefly, while Rohrback gives more details (1980, 1982, 1983). The geochemical techniques used include gas chromatography, stable carbon isotope measurements and gas chromatography-mass spectroscopy. Both authors conclude that the Gulf of Suez oils seem to be of the same genetic family, a fact indicating a common (or similar) source rock. According to Rohrback, the Eocene and lower Miocene sediments correlate well with the Gulf of Suez crudes, with more crudes correlating with the Eocene sediments as suggested by combined gas chromatography-mass spectroscopy. Although Barakat suggests that the lower Miocene and upper Cretaceous sediments are sufficiently mature to generate oil in parts of the Gulf of Suez, he also suggests that the lower Miocene, rather than the upper Cretaceous, sediments are the source rocks of the oil. His argument is that since all of the major fields are undersaturated (that is, they do not have gas caps) and since mature rather than less mature sediments give gas, and since the lower Miocene sediments are less mature than the upper Cretaceous sediments, then the lower Miocene sediments must be the source of the 'undersaturated' oils in question.

The preceding discussion indicates that (a) there are not yet enough comprehensive studies of Gulf of Suez oils to provide conclusive answers to the many questions posed; (b) more than one source for the crudes is suggested; (c) more than one source suggestion fits the complex geological setting of the region in which a formation or formations were subjected to diverse geological conditions (depth of burial, lithology and facies variations, organic richness and depositional environments); and (d) that the lowr Miocene (Kareem and Rudeis formations) and the upper part of the pre-Miocene (Eocene and upper Cretaceous shales and carbonates) may be the source of the Gulf of Suez crudes.

The geological setting in the Gulf of Suez province furnished the conditions for the primary migration of oil (from source rocks to reservoir rocks). Block faulting caused the juxtaposition of source and reservoir rocks. The intercalation of porous reservoir beds within source rock sections allowed for the squeezing of the oil generated into reservoir rocks. Likewise, sand lenses, when confined within source rock sections, trapped oil expelled from surrounding organic-rich materials.

On the other hand, the lateral lithological changes displayed by the Miocene formations must have had an adverse effect on long distance secondary migration owing to the permeability barriers caused by such changes in facies. Moreover, vertical secondary migration must have been minimized by the flowage of Miocene rock salts, caused by the effect of overburden, to areas of less overburden and hence pressure. By this process, the potential vertical avenues for migration such as fractures and faults are sealed, and hence migration of oil through them is obstructed. Wherever salt was not deposited, vertical secondary migration might have taken place. This situation appears in the eastern part of Belayim East (Land) field where oil accumulations are present in a number of rather thin sandstone intervals within both the Zeit and South Gharib formations. Study of the stratigraphy and structural setting of the field indicates that these pools are in no way connected to any of the potential source rocks which could have supplied the oil through primary migration. The only way to account for such accumulations is to assume that the oil migrated to the pools in question from lower accumulations along fractures and faults in the eastern part of the field where no salts had been encountered in the wells (Hantar 1967).

The conditions adverse to horizontal and vertical secondary migration in Miocene sections do not seem to be pronounced in pre-Miocene formations. Secondary migration might have taken place in that older part of the sedimentary sequence in the absence of salt in the section and the less pronounced lateral lithological variations. The numerous surface oil seeps found on both sides of the Gulf of Suez simply represent vertical secondary migration along fault planes.

Geology of the oilfields

Oil is found both onshore and offshore in the Gulf of Suez province. Many of these occurrences are in the form of shows or small quantities which are categorized as non-commercial. Others remain to be evaluated. Of the discoveries developed and put into production some are now depleted. Following is a summary of the geology of some of these oilfields that show the different types of traps.

Gemsa oilfield (1909)

The Gemsa field, 1 km^2, produced from reefal limestones and dolomites interbedded with Miocene evaporites (anhydrite and salt). The section sits on a southwest dipping basement horst. The average depth of wells is 1000 feet. The Miocene section encountered in the wells seems to belong mainly to the South Gharib formation. Porosity is 16% and the API gravity ranges from 32 to 41°.

Abu Durba oilfield (1918)

Interest in this area began very early this century due to reports of oil seeps along the western foot of Gebel Araba on the east coast of the Gulf of Suez by geologists of the Petroleum Research Board of the Egyptian Geological Survey.

Subsurface information collected from 23 wells drilled indicate that there is practically no trap, structural or stratigraphical, in which oil is stored. The well logs show that there is no sealing (cap) rock on the oil-bearing sands (Bowman 1926) indicating that the Abu Durba 'field' is actually an open field, making it perhaps unique among fields in the country.

According to Bowman (1925) and Van der Ploeg (1953) the reservoir consists of Nubian sandstones which are outcropping on the surface. However, the reservoir rocks as reported by Bowman consist of sands with igneous pebbles and conglomerates with calcareous sands, shales, limestones and igneous rock fragments – a lithology which does not conform with the known lithology of Nubian sandstones but conforms more with that of Pliocene and Pleistocene sediments. The only possible Nubian sands in the 'field' is a short sand section at the bottom of two wells (Abu Durba 1 and 3) along the extreme eastern side of the 'field' next to the nearest Cretaceous outcrops in the area. The accumulation seems to be the result of seepage of oil upward along the major fault zone bordering the 'field' in the east into young surface sediments. The 'field' was depleted in the early 1940s.

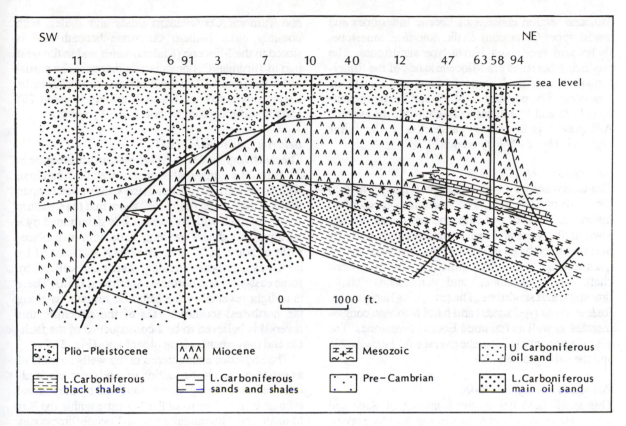

SW   NE

11   6 91   3   7   10   40   12   47   63 58 94

sea level

0   1000 ft.

Plio–Pleistocene   Miocene   Mesozoic   U. Carboniferous oil sand

L. Carboniferous black shales   L. Carboniferous sands and shales   Pre–Cambrian   L. Carboniferous main oil sand

Figure 27.1 Cross-section through Ras Gharib field (after Van der Ploeg 1953).

Ras Gharib oilfield (1938)

The first two exploration wells in the area were drilled between 1921 and 1923 on a surface structure and they were both completed as dry holes. During 1936-37 a Torsion Balance Pendulum survey was conducted in the area; it showed a gravity high some 3.5 km to the southeast of these dry holes. This was tested by the Ras Gharib discovery well (Morgan & El Barkouky 1956).

According to Van der Ploeg (1953) the structure consists of a fault bounded, partly denuded pre-Miocene upthrown block unconformably covered by the Miocene anhydrite-shale sequence which by renewed movement of the block in Pliocene time was warped into a dome-like structure that plunges steeply to the northwest, southwest and southeast, and gently to the northeast (Fig. 27/1).

The section consists of Plio-Pleistocene grits and conglomerates which are followed by a thick shale-evaporite sequence of Miocene age which grades eastward to reefal limestone beds (the Nullipore Rock) on the flank. The organic-rich part of the Miocene section (Kareem and Rudeis formations) is not encountered in the majority of the wells. The Miocene section is underlain by a sequence consist-

ing of upper Cretaceous limestones, lower Cretaceous 'Nubian' sandstones, Carboniferous black shales and Carboniferous 'Nubian' sandstones.

The Ras Gharib reservoirs are three Carboniferous and two Cretaceous horizons, all consisting of 'Nubia-type' sandstones with a porosity of 16 to 18% and a permeability of 100 to 200 MD, and one Miocene horizon on the eastern flank of the structure consisting of reefal limestones with a porosity of 16% and a permeability of 32 MD. The reefal reservoir belongs to the Belayim formation and is known as the Ras Gharib Nullipore Rock. The average API gravity is 24°.

Sudr oilfield (1945)

The discovery well was drilled to test a gravity high, and the wells drilled indicate that the structure consists of two blocks, a narrow productive pre-Miocene upthrown block dipping toward the west and bounded to the south and west by faults and a small productive Miocene block in the north bounded to the south by a post-Miocene eastwest transverse fault and to the east by the major boundary fault. Both blocks are overlain unconformably by an anhydrite-shale sequence (Van der Ploeg 1953). The pre-

Miocene section consists of Eocene limestones and marls, upper Cretaceous chalk, dolomite, limestone, shales and sands and Nubia-type sandstones. The producing horizons are Miocene sands of the Rudeis formation and an Eocene fissured and cavernous limestone. The porosity is 24.5% and the permeability is 1620 and 38 MD respectively, while the average API gravity is 23 and 20° respectively. The sulfur content is about 1.8% by weight.

Asl oilfield (1947)

The discovery well was drilled on a gravity high to the south of Sudr field. The structure consists of an upthrown block of pre-Miocene sediments dipping towards the west and broken into separate parts by northwest-southeast trending faults. The Miocene sequence is like that of Sudr oilfield, resting on Eocene chalk and limestone, and Cretaceous chalk, limestone and sandstone. The producing horizons are Rudeis sands (Asl sands) and basal Miocene conglomerates as well as fractured Eocene limestones. The average porosity is 23%, the average API gravity 22° and the sulfur content 2%.

Ras Matarma oilfield (1948)

This small field lies to the southeast of Sudr and northwest of Asl. It was also recognized by gravity. The structure consists of an upthrown block of pre-Miocene sediments dipping towards the west and bounded to the east and south by northeast to southeast and east-west trending faults respectively. The pre-Miocene block is covered by a Miocene section which is similar to that known in Sudr and Asl. The producing horizons belong to the Miocene (Rudeis sands) and Eocene fractured limestone. The reservoir characteristics and crude quality are more or less the same as those of Asl field.

Feiran oilfield (1949)

This was the first oilfield in Egypt to have been discovered by the reflection seismic surveying method. Well data indicate that the structure is a down-faulted block with its eastern side determined by a major northwest-southeast bounding fault which brought Miocene (Rudeis-Nukhul) sediments against the basement, downthrowing to the west. A number of northwest-southeast trending step faults cut the structure parallel to the boundary fault; these are also throwing to the west.

The section begins with post-Miocene clastics which unconformably overlie the Miocene section of both the Ras Malaab and Gharandal groups. This section is down-faulted along the eastern side against the basement. Older rocks are encountered in the wells drilled in the western part of this structure. The pre-Miocene section consists of Eocene limestones and Turonian/Cenomanian sands and shales, with possibly older Nubian sandstone beneath. Oil is stored in the Miocene (Rudeis) sands and in the west part in Turonian/Cenomanian sandstones. The porosity ranges from 13% in the Rudeis sands to 15% in the Cretaceous reservoir. The API gravity is about 23° and the sulfur content is about 2.2% by weight.

Belayim land oilfield (1955)

The discovery well was drilled on a seismic feature in the form of a northwest-southeast trending anticline, with its western part beneath the Gulf of Suez water. Wells drilled in the field (166) show that the structure is a fault block with its eastern side determined by a northwest-southeast trending fault with a displacement of some 8000 feet. The block is dissected by northwest-southeast trending faults as well as by some east-west and northeast-southwest faults. There is a slight reversal of the Miocene formations along the northwest-southeast axis of the structure; this reversal is believed to be a consequence of the faulting and movements along older faults (Fig. 27/2).

The sequence encountered in the wells consists of a post-Miocene clastic section of sands, gravels, multicolored clays and gypsum, followed by the normal sequence of rock units of the Miocene within the Ras Malaab and Gharandal groups. Eocene limestones were encountered in a few wells belonging to the middle Eocene (Ayouty 1961).

In this field oil is housed in 11 pays within the Miocene. Ten of these pays consist of sandstones whereas the remaining pool consists of flint and limestone conglomerate. The shallowest pay was encountered at a depth of 1700 m and the deepest at 2786 m. The top three pays are sandstone interbeds within the Zeit/South Gharib formations. The succeeding six pays are also sandstone intervals within the Belayim formation. The following pay is the flint and conglomerate reservoir described above and is present in the Kareem formation. The lowermost reservoir is a conspicuous sandstone interval within the Rudeis formation. Many of these pools show lateral lithological variations from sandstones to shales which pinch out completely. This is why the 11 pools are not found in every well in the field and why the field displays quite well the structural-stratigraphical or combination type of trap.

The oil-water contact of most of the pools was recognized in many wells. Occasionally, two levels for the oil-water contact were recognized in one and the same pay. This difference in level is attributed to effective separation by faulting after the trapping of the oil.

There are two main producing horizons in the field, one in the lower part of the Belayim formation (Pay IV) and the other in the Rudeis formation (Pay

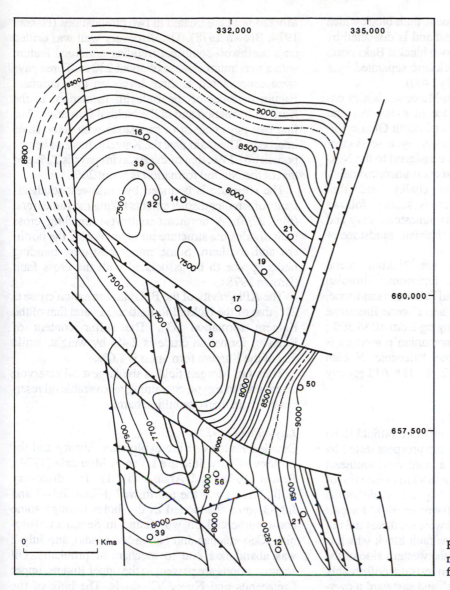

Figure 27.2 Structure contour map, top Zone IV, Belayim land field (after El Mahdy 1985).

V). Porosities range from 21 to 25% and permeability can be more than 1000 MD. The API gravities vary from 17° in the lowermost pay to 24° upward. All the crudes obtained from the different pays are undersaturated.

Rudeis-Sidri oilfield (1957)

The seismic prospect which proved oil-bearing was found to represent a structure consisting of two fault blocks with a Nukhul formation reservoir in both of them. The normal succession of post-Miocene and Miocene sediments are encountered in the wells, with a pronounced thickening of the Rudeis formation. The Nukhul formation is developed and is the oil-producing horizon in the area. It consists of basal conglomerates and sands which rest on a section of

Red Beds, probably of Eocene age. The porosity is about 13% and the permeability is some 270 MD. The API gravity ranges between 20 and 23°. The sulfur content is about 2% by weight.

Bakr oilfield (1958)

The general area of Bakr witnessed relatively active gravity survey exploration efforts in the 1940s leading to the drilling of six exploratory wells between 1940 and 1945, all of which were plugged and abandoned as dry holes though there were some oil shows in both the Miocene and Cretaceous. After a seismic survey was concluded in 1957, a location for a new wildcat was chosen in the light of combined gravity and seismic data; it was completed as an oil discovery in 1958.

The Bakr field is a pre-Micoene fault block which has a northwest-southeast trend and is dissected by transverse cross faulting into two blocks, Bakr north and Bakr south. The two blocks are separated by a non-productive graben (Hadidy 1960).

The section consists of post-Miocene clastics underlain by a Miocene sequence of evaporites and clastics belonging to the Zeit and south Gharib formations, followed unconformably by a section of reefal limestones which may be assigned to the Nukhul formation. The Miocene section unconformably overlies Eocene marls and chalky and flint limestones. An upper Cretaceous section follows which consists of chalk, marls, limestone, clay and sandstone, in turn overlying a 'Nubian' sandstone of Paleozoic age.

The producing horizons are Nukhul reefal limestone, Eocene fractured limestone, Turonian limestone and Cenomanian and Paleozoic sandstone. The porosity of the Miocene and Eocene limestone reservoirs is rather erratic, ranging from 10 to 30%; that of the Turonian and Cenomanian reservoirs is about 15%, and that of the Paleozoic Nubian sandstone ranges from 18 to 21%. The API gravity ranges between 20 and 23°.

Belayim Marine oilfield (1961)

This was the first completely offshore oilfield to be discovered in Egypt. The seismic prospect tested by the discovery well showed a northwest-southeast trending anticlinal structure some 9 km to the west of Belayim Land field. Later drilling indicated that the Miocene sediments which show reversals to the east and west away from the northwest-southeast axis of the field overlap a pre-Miocene fault block which is tilted to the northeast. Along the western side of the field, a normal Miocene section rests unconformably on the Carboniferous 'Nubian', and eastward it overlaps younger pre-Miocene sediments. In the eastern side of the field the Miocene lies unconformably on Eocene limestones. Below the Eocene limestone a Paleocene, upper Senonian, lower Senonian, Turonian, Cenomanian and Carboniferous 'Nubian' sandstone sequence follows.

The oil-bearing horizons are within the Miocene and Cretaceous. Two sandstone pays are in Belayim, one in Rudeis formation and another in the lower Senonian-Turonian-Cenomanian sands and limestones. The porosity ranges from 17 to 22% and the API gravity of the oil in Belayim formation pays is 26.5°, while the lower pays have an API gravity of 30°. The sulfur content ranges from 1.6 to 2.4% by weight.

Morgan oilfield (1965)

The history of exploration and the geology of the Morgan field are treated in two publications (Hassan 1974, Brown 1978). The discovery well was drilled on a northwest-southeast trending anticlinal feature with a maximum closure of about 230 m. Three pays were encountered at three levels, in the Hammam Faraun member of the Belayim formation, in the Kareem formation and in the Nukhul formation. The first two pays consist of sandstone with some shale interbeds. The main pay, the Kareem, has an average porosity of 23%, that of the Belayim pays is 22% and that of the Nukhul formation is about 20%.

The structure is bounded by northwest-southeast faults of smaller displacements taking the same trend (Fig. 27/3). A northeast-southwest trending cross fault divides the structure into two blocks, the northern and southern. Some minor east-west trending faults appear to be associated with the cross fault (Brown 1978).

The API gravity of the Belayim formation crude is 21°, that of the Nukhul formation 31° and that of the Kareem formation 33°. The sulfur content of Belayim formation crude is 2.4% by weight, while that of the Kareem formation is 1.6%.

Thus far, Morgan field is the largest oil reservoir discovered in Egypt, containing recoverable oil reserves of m ore than one billion barrels.

July oilfield (1973)

Detailed reviews of the exploration history and the geology of July field are given in Moustafa (1974), Brown (1978) and Abdine (1981). The discovery well (July-4) was the fourth well drilled. July-1 and July-3 were abandoned as dry holes (though some non-commercial oil was found in Senonian, Turonian, Cenomanian and Nubia 'A' sands), and July-2 was abandoned due to mechanical problems. Oil accumulations are present in the upper Rudeis, upper Cretaceous and Nubia 'C' sands. The bulk of the reserves is contained in the lower Rudeis and Nubia 'C' sands which have an average net pay of 158 and 207 m respectively.

The structure consists of two main fault blocks which are tilted to the northeast and are separated by a major cross fault. It is bounded on its western up-dip edge by a large down to the west fault (Figs 27/4 and 27/5). This fault represents a large displacement not recognized by seismic studies until relatively recently. The failure to recognize the fault caused unexpected results during the exploration and early development operations in the field. July-4 (the discovery well) made a lower Rudeis discovery instead of the 'Nubian' sand target which was found wet, whereas the delineation well which followed made a 'Nubian' sand discovery though the well was drilled as an upper Rudeis pay delineation well. Moreover, the well data indicate that in the southern block of the

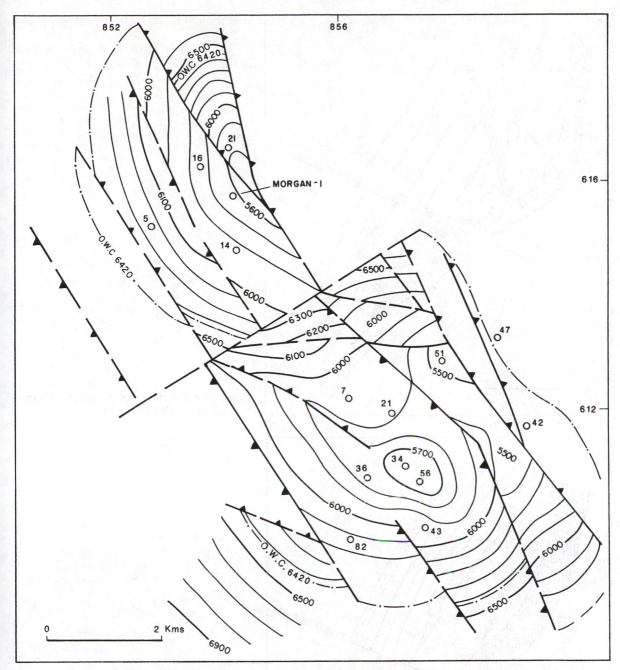

Figure 27.3 Structure contour map, top Kareem, Morgan field (after Brown 1978).

field, to the south of the major cross fault, the Nubia 'C' sands are absent along the crestal part of the block, and that the Rudeis sands are absent along the crestal part of the northern block. To explain this situation, two major periods of activity are suggested to have taken place along this fault. The first movement took place in Oligocene time and caused the downthrowing of the northern block relative to the southern. The second movement took place along the

same cross fault at the time of deposition of the late Rudeis sediments, with the relative direction of movement opposite to the first one, that is, in a down to the south direction. Erosion cut deep in both the uplifted blocks, resulting in the above-mentioned missing section on both sides of the fault (Brown 1978).

The porosity of the Rudeis pay sands is 24%, that of the Cretaceous sands ranges from 13 to 17% and

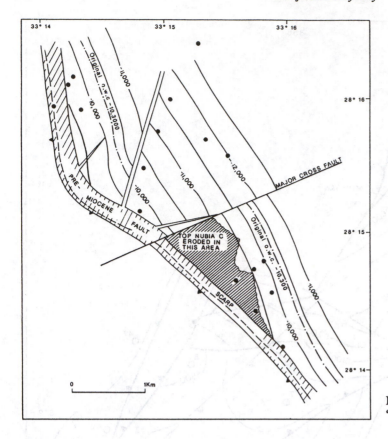

Figure 27.4 Structure contour map, top Nubia 'C', July field (after Brown 1978).

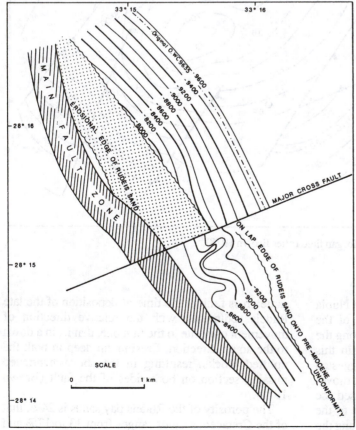

Figure 27.5 Structure contour map, top Rudeis, July fields (after Brown 1978).

that of the Nubia 'C' sand from 18 to 22%. The average API gravity is 34°.

Ramadan oilfield (1974)

The exploration history and the geology of this field are treated in detail by Brown (1978) and Abdine (1981).

The first well drilled in the area, A-1 or Alef-1, was a dry hole. In light of the dipmeter data and information from more advanced seismic acquisition, it was found tht Alef-1 well was drilled on the downdip eastern flank of a pre-Miocene structure and that the structure below the bottom of the evaporites was offset to the west relative to the top of the evaporite structure. Thus, well GS 303-1 (later called Ramadan-1) was spudded 1.5 km to the west of Alef-1 to test the pre-Miocene high, bearing in mind that only a few insignificant stringers of sands were found in the Kareem and Rudeis sections in Alef-1.

Ramadan-1 encountered a total of 446 m of net oil sand between −3246 and −4040 m (T.D.). Of these, 49 m are within the Turonian-Cenomanian, 23 m in the uppermost Nubian sandstone and 363 m in the massive Carboniferous Nubia 'C' sandstone. The average

net oil sand for the field is 230 m.

The Ramadan field structure consists of a series of northeast dipping pre-Miocene blocks separated by northwest-southeast trending down to the west normal faults (Fig. 27/6). Except for the most westerly block, which dips to the west, the northeasterly dip prevails. The drillinjg results confirmed the seismic interpretation that the structure at the top evaporites is offset to the structures at the base evaporites and the underlying pre-Miocene horizons. The structure at the base evaporites is almost coincident with the deeper pre-Miocene structure (Abdine 1981).

The porosity of the Cretaceous (Turonian/ Cenomanian) reservoirs, which consist mainly of sandstone with some limestones, ranges from 13 to 17% and that of the Paleozoic Nubia 'C' sandstone from 16 to 18%, with a permeability of 100 to 200 MD. The API gravity of the crude produced from this field is 39.9°.

October oilfield (1977-78)

This field consists of a pre-Miocene tilted fault block which trends northwest-southeast to the west of Abu

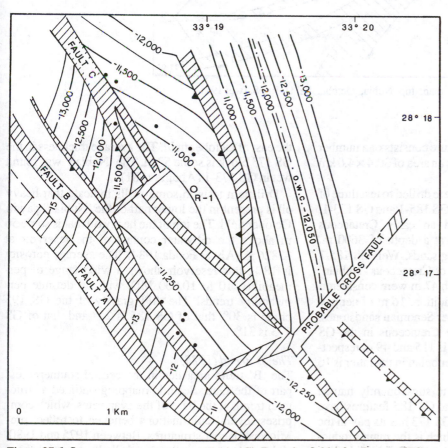

Figure 27.6 Structure contour map, top Nubia 'C', Ramadan field (after Brown 1978).

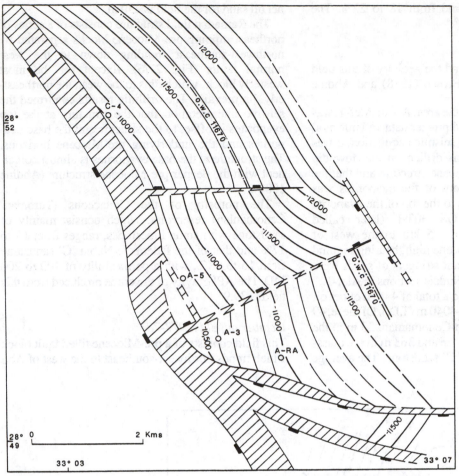

Figure 27.7 Structure contour map, top Nubia, October field (after Safwat 1985).

Rudeis land oilfield. The feature consists of a number of culminations and occupies an area of 30.4 × 4.0 km (Fig. 27/7).

Three exploration wells were drilled to test three of the culminations, GS 195-1, GS 185-1 and GS 173-1. Well GS 195-1 penetrated an early Cretaceous 'Nubian' sandstone section at a depth of 3600 m, with 115 m of net oil bearing sands. Well GS 185-1 encountered a total of 78 m of Cretaceous 'Nubian' sandstone gross pay, of which 47 m were considered to contain movable oil. In addition, 16 m of net pay were encountered in the lower Senonian sandstones. The average net pay of the Cretaceous in the GS 195-1 and 185-1 discoveries is 115 and 49 m respectively, and that of the lower Senonian in the latter is 16 m.

These oil discoveries were subsequently named the October field. GS 195 and GS 185 features have the same oil-water contact. GS 173 has its pay in the Nukhul formation (35 m) which is the main producing horizon in the nearby Rudeis-Sidri field on land to the east of October field. The average thickness in the GS 173 block is some 52 m. The 'Nubian' was found wet in GS 173-1 (Abdine 1981).

Drillstem tests in some wells indicate that heavy oil is present in the lower part of the 'Nubian' pay in October field. The top of the heavy oil mat is some 56 m above the oil/water contact of the reservoir at –3826 m (Aly & Fouda 1984). The average porosity is 18%; this reservoir shows a wide range of permeability, 10 to 10,000 MD, with no definite permeability trends. The API gravity of the GS 195 crude is 30°, that of GS 185 is 25° and that of GS 173 is 31°.

The 'B' trend (1975)

The 'B' trend area lies in the central southernmost part of the Gulf. Seismic mapping outlined a structural trend, at the base of the evaporites, which comprises several culminations believed to reflect pre-Miocene deeper structures. Between 1975 and 1980, eight exploratory wells were drilled to examine these structural features. Oil was discovered in four of the culminations in the four blocks, GS 365, GS 373, GS 382 and GS 391. Of these, only GS 382 has been in production since 1977, while the development projects of the remaining three discoveries are still in progress.

The GS 382 oilfield (Sidky oilfield) is a southwest dipping pre-Miocene fault block with faults and structure trending northwest-southeast (Fig. 27/8). The deep pre-Miocene structure in the B-trend, in general, is associated with an overlying salt section varying in thickness from 2300 m in the central area to 655 m on the flanks. This salt section caused severe problems in seismic mapping, and the deep events are considered unreliable.

The main producing horizon which was put on production consists of 272 m of 'Nubia C' sandstones; another pay is present in the lower Senonian where 26 m of sandstones are also oil-bearing. The average porosity of the 'Nubia C' sands is 15% and that of the lower Senonian sands is 20%. The API gravity of the oil produced from the 'Nubia C' sands is 34° and the gas/oil ratio is 1954.

Shoab Ali oilfield (1977)
This field was discovered in 1977 and was then called 'Alma Field' during the occupation of Sinai. The field was on production when it was taken over by the

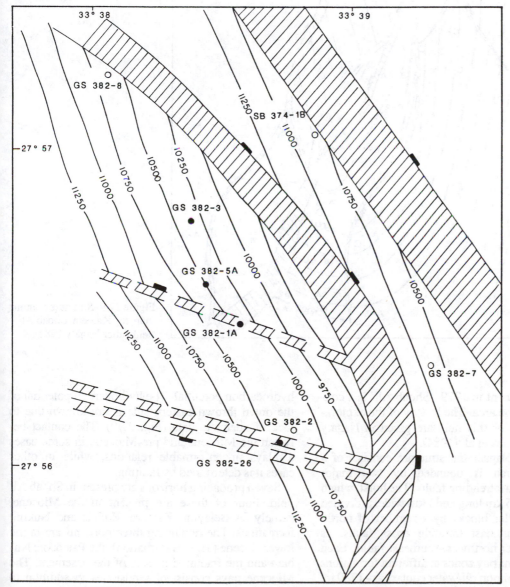

Figure 27.8 Structure contour map, top Nubia 'C', Sidky (GS 382) field (after Maghoub & Mohamed 1985).

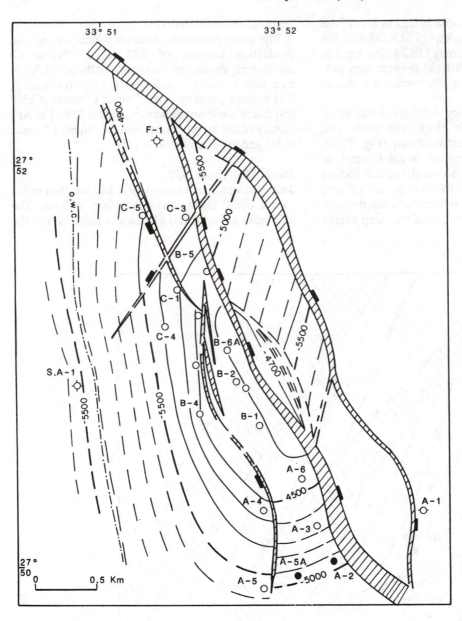

Figure 27.9 Structure contour map, top Kareem, Shoab Ali field (after Nagaty 1982).

Egyptian government in 1979. 'Shoab Ali' is a conspicuous reef in the area. The history of the exploration and geology of this field are given in Nagaty (1982) and Hassouba et al. (1984).

According to Nagaty the structure consists of a horst block which is bounded by two major northwest-southeast trending faults. The horst, which is approximately 5 km long and 1 km wide, is divided into several smaller blocks by a number of north/northwest-east/southeast trending cross faults, as well as some other northwest-southeast faults. Each fault block contains pay zones in different formations but all share a common oil/water contact at −1780 m. The downthrown side to the west of the horst has

hydrocarbon potential. The hydrocarbon potential of the down thrown eastern side is not known due to lack of adequate data (Fig. 27/9). The contact between the Miocene and pre-Miocene in some cases displays unconformable relations, while in other cases it is determined by faulting.

Seven producing horizons are present in Shoab Ali field. Four of these are present in the Miocene, namely, in Belayim, Kareem, Rudeis and Nukhul formations. The remaining three horizons are in the lower Eocene (Thebes formation), the Paleozoic Nubian and the fractured granite of the basement. The Miocene pays consist of sandstones in addition to some limestone reservoir beds in the Belayim and upper Rudeis formation. The Thebes formation reservoir consists of limestone.

The bulk of the reserves is in the Miocene reservoirs (85%) and the Paleozoic sands (12%). A small accumulation is found in lower Senonian sandstones (0.7% of the field reserves), but is not on production.

Oil is not obtained from all pay zones in every well. The Rudeis reservoirs are the main producing horizons. Porosity ranges from 22 to 26%, the average oil gravity is 32° API, and the average gas/oil ratio is 450 standard cubic feet per stock tank barrel.

Ras Budran oilfield (1978)

This field lies some 9 km to the northeast of GS 195 culmination of the October field. The structure is pre-Miocene fault block which is less prominent than the GS 195 feature (Fig. 27/10). The reservoir occupies all the Cretaceous 'Nubian' section as well as the upper part of the Paleozoic Nubian sands. The dis-

covery well contains a gross oil column of 426 m of which 406 m of pay is Paleozoic 'Nubian'. The oil/water contact is 230 m lower than that of the nearby October field. The average porosity is 14%, the average gas/oil ratio 160 and the average API gravity 26.4°.

Ras Fanar oilfield (1978)

A seismic high was recognized in Block KK 84 three km off the west coast of the Gulf of Suez near Ras Gharib. The high is an elongated horst block trending north/northwest-south/southeast; it is 7 km long and 1 km wide (Fig. 27/11) and is tilted to the northeast. The block is bounded by faults on both the east and west sides, with some faults cutting into the horst itself. The wells show that the Miocene Nullipore carbonates were deposited on top of the truncated and eastward-dipping Sudr chalk (upper Senonian), Esna Shale and Thebes limestone (Eocene) of the pre-

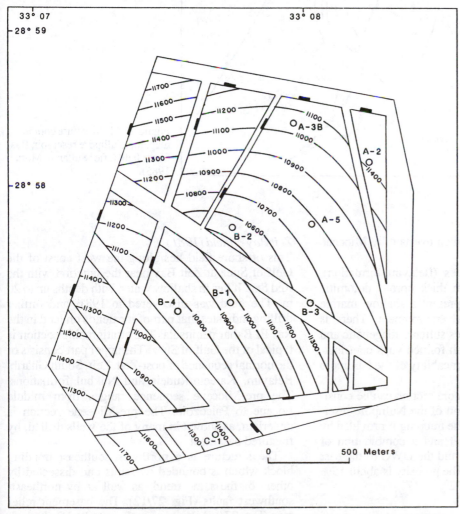

Figure 27.10 Structure contour map, top Nubian, Ras Budran field (after Rose et al. 1983).

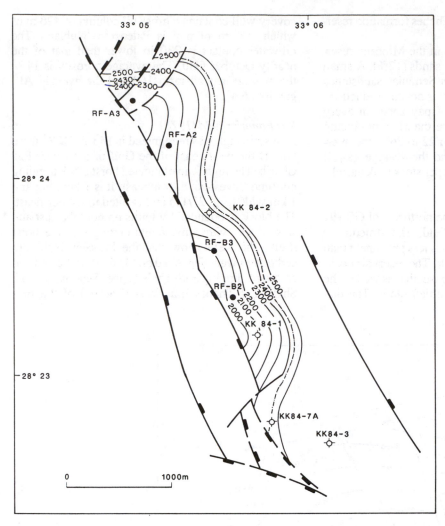

Figure 27.11 Structure contour map, top Nullipore reservoir, Ras Fanar field (after Sultan & Moftah 1985).

Miocene high; they were then overlain by Miocene evaporites.

The Nullipore carbonates (Belayim formation), consisting of some 213 m thick porous dolomitic limestone to dolomite, represent a shallow marine normal saline depositional environment. There is evidence that the carbonates suffered diagenetic and dissolution processes which formed vugs contributing to the porosity and permeability of the Nullipore section (Kulke 1982).

Oil trapped in the Nullipore section can be considered an eastward extension of the Nullipore reservoir of Ras Gharib field. The trapping is provided by the western boundary fault and a combination of permeability deterioration and the OWC to the east (Sultan & Moftah 1985). The porosity is about 15% and the API gravity about 22°.

Zeit Bay oilfield (1981)
This offshore field lies along the west coast of the Gulf of Suez at Zeit Bay near the junction with the Red Sea. It lies in shallow water, with depths up to 20 m. The field was discovered in 1981, and further drilling indicates that part of it extends on land in the Ras El Bahar peninsula. The stratigraphic section is typical of the Gulf of Suez. The upper part consists of the normal sequence of post-Zeit, Zeit, South Gharib, Belayim, Kareem/Rudeis and Nukhul formations. The pre-Miocene sequence ranges from middle Eocene to Paleozoic. The pre-Miocene section is underlain, as shown in many of the wells drilled, by fractured basement.

The structure is a northwest-southeast trending block which is bounded by faults and dissected by others of the same trend, as well as by northeast-southwest faults (Fig. 27/12). The basement relief map (Fig. 27/13) shows a southwest tilted fault block with Nubian sandstones offlapping the basement. The basal Miocene sediments (Nukhul) transgress over the Nubian/basement surface. The reservoirs in Zeit Bay field belong to the basement, Nubian sandstone, basal Miocene Nukhul and Rudeis/Kareem formations.

The crystalline basement was penetrated in almost all the wells of the field. Oil is obtained from granitic and non-granitic intervals which are intensely fractured. Log porosities in the granites vary from 1 to 5%. On top of these fractured basement rocks a section of altered and weathered basement rocks is encountered in the form of oil-bearing basement wash with a thickness ranging from 6 to 16 m. The average porosity is 8% and the permeability ranges from 0.1 to 10 MD.

The 'Nubian' sandstone reservoir varies in thickness from 0 to 230 m. The section consists of medium to coarse-grained sandstones which are occasionally conglomeratic. The porosity ranges from 12 to 22% (average 18%) and permeabilities of up to 5 darcies were measured, though the typical range is 500 to 300 MD.

The basal Miocene reservoir consists of dolomitic sandstones, quartzitic dolomite and sandstones. In the upper part of the section the dolomites are anhydritic. The thickness ranges from 10 to 32 m, the porosity from 10 to 26 with an average of 20%, and permeabilities range from 0.1 to over 1 darcy.

The Rudeis/Kareem reservoir ranges in thickness from 100 to 230 m. The reservoir is mainly carbonate and is restricted to the western flank of the field. The porosity reservoir is mainly of the intercrystalline-microvuggular type; this accounts for nearly half the storage capacity of the field. Vuggy and solution-enhanced inter-granular porosity contributes, to a

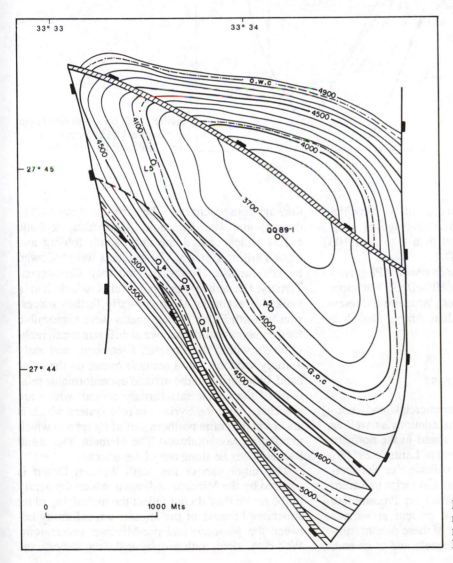

Figure 27.12 Structure contour map, top Basal Miocene Zeit Bay field (after Sultan, Moftah & Hafez 1985).

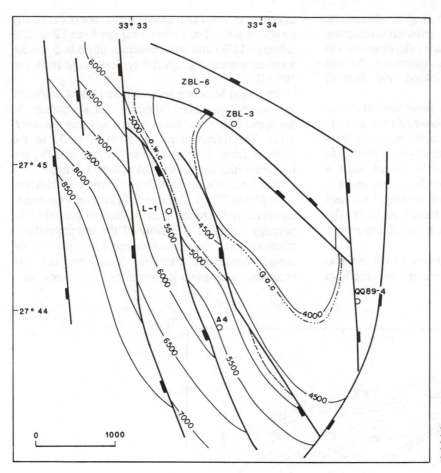

Figure 27.13 Basement relief map Zeit bay field (after Sultan, Anthon, Moftah & Hafez 1985).

lesser degree, toward effective porosity. Porosities range from 17 to 22% with an average of 19%. Permeabilities range from less than 1 to over 100 MD, with an average of 50 MD.

The Zeit Bay crude is sulfur-free with API gravity of 34°. The initial GOR is 680 scf/bbl. For more details on the exploration history, structure and reservoirs, refer to the paper of Sultan, Anton, Moftah & Hafez (1985).

WESTERN DESERT

Up to the present, all commercial and non-commercial hydrocarbon accumulations as well as oil and gas shows have been found in the northern part of the Western Desert, north of Latitude 29° 00'. Surface and subsurface data indicate that the sedimentary succession ranges from Cambrian to Recent with the oldest sediments resting on Precambrian basement. Unconformities are present at various stratigraphic levels. A number of these unconformities are known within the Paleozoic and at its end;

they are consequences of the early Paleozoic (Caledonian) and late Paleozoic (Hercynian) tectonic events which caused the north-south folding and block faulting systems. These events were followed by late Jurassic (Nevadian) and then Cretaceous-Tertiary (Laramide) tectonic events which had a marked effect on the Mesozoic-early Tertiary succession in north Egypt. These events were responsible for the unconformities known at different stratigraphic levels within the Jurassic, Cretaceous and early Tertiary sections. The Laramide event, on the other hand, brought about the marked unconformable relations of the Cretaceous-Tertiary contact which are associated with the Syrian arc fold system which is known in the entire northern part of Egypt and which trends northeast-southwest. The Alamein, Yidma and Razzak fields lie along one of these arcs.

The major part of the north Western Desert is covered by the Miocene sediments which dip gently to the north; they do not reflect the underlying older structures because of the angular unconformity between the Miocene and pre-Miocene successions. Well data, along with geophysical data, indicate the presence of a number of sedimentary basins of varying dimensions within the Paleozoic, Jurassic, lower Cretaceous, upper Cretaceous and early Tertiary. The geology of these basins is treated by many authors (Shata 1951, Kostandi 1959, 1963, Barakat 1970, Abdine & Debies 1972, Adland & Hassan 1972, El Gezeery, Mohsen & Dia El Din 1974, Ezzat & Dia El Din 1974, Khaled 1974, Debies 1976 and Abu El Naga 1984).

Exploration work began in the Western Desert in 1940 and has continued since then with only a few interruptions. In addition to the surface geological, land magnetic, aeromagnetic and gravity surveys, some 150,000 km of seismic work was carried out and about 223 exploratory wells were drilled up to the end of 1983.

The first oil discovery in the Western Desert was the Alamein field in 1955. Since then a number of discoveries have been declared commercial. Oil shows and non-commercial accumulations were found in many wells at different stratigraphic levels in the Paleozoic, Mesozoic (Jurassic and Cretaceous) and Tertiary (Eocene), with shows confined to areas where relatively dense exploration drilling was carried out. Most of the producing pools in the Western Desert are confined to the Cretaceous. Deeper pools were encountered in both Um Baraka and Razzak fields, assigned to the Paleozoic and Jurassic respectively.

A commercial gas accumulation was discovered in the lower Cenomanian Bahariya formation in Abu Gharadig field. The gas pool is separated and sealed off from the overlying oil pays of the Abu Roash units by a thin limestone and shale sequence. It is presently in production.

Other gas discoveries recently made in the Badr El Din and Sitra concessions found gas in the Albian-Cenomanian sands, similar to the 1967 gas discovery in the Bahariya sands in Abu Senan structure to the southeast of Abu Gharadig field. Other small gas finds have been made recently in the general area of Abu Gharadig basin.

Reservoir rocks

In the Western Desert oil (and gas) are stored in dolomites, dolomitic limestones and sands of Cretaceous age.

Carbonate reservoirs

The carbonate reservoirs are present in both the Aptian and Turonian. The most important of these is the Aptian dolomite first discovered in Alamein field. Other less conspicuous carbonate reservoirs are present in the Abu Roash 'D' and 'F' units of Turonian

age and the Abu Roash 'G' unit of upper Cenomanian age.

The Aptian carbonates consist of an upper unit, the Alamein formation, and a lower unit. The two are separated by a shale-dolomitic sandstone member (Hamed 1972). The carbonates consist of dolomite, limestone and dolomitized limestone of different degrees of dolomitization.

According to Abdine & Debies (1972) the lower carbonate unit consists of a limestone-dolomite sequence with clastic interbeds. The thickness ranges from less than 30 m in the south to more than 170 m in Umbaraka-Shushan wells in the east Faghur area. The clastic unit between the lower and upper carbonate units, the intercarbonate unit, consists of sands and shales with interbeds of carbonates of 265 m maximum thickness. The upper carbonate unit (Alamein formation), which is the main Aptian producer, consists of dolomite and dolomitic limestone with fair to excellent intergranular, vuggular and fractured porosity (Hamed 1972). The Alamein formation is an easily recognizable and well-defined rock unit extending widely in the northern part of the Western Desert where it is the most prominent seismic reflector. It is one of the most important targets in exploration drilling. This formation is productive in Alamein, Yidma and Razzak.

The Cenomanian/Turonian carbonate reservoirs owe their porosity to fracturing and are oil-bearing in Abu Gharadig, W.D. 33, W.D. 19, Alamein and Razzak fields.

Sandstone reservoirs

The Cretaceous sand and sandstone reservoirs belong to different stratigraphic levels starting from the Barremian in Um Baraka field (the 3500 m sand), the upper Aptian Dahab sand in Alamein field, the Albian Kharita sands in W.D. 19 and Badr El Din fields, the lower Cenomanian Bahariya sands in Abu Gharadiq, Meleiha, Alamein, Razzak and W.D. 19 fields and the Turonian Abu Roash 'C' and 'E' sands in Abu Gharadig and W.D. 33 fields.

The younger of these sandstone pays consist of rather thin sands and sandstone beds which display discontinuous lateral extension. This applies to many of the Turonian-Cenomanian sand pays which tend to pinch out laterally or to be of the channel type. Older Cretaceous sands (Albian and older) have marked wider lateral extension. Examples are the Kharita sand of Albian age and the Bahariya sands.

Cap rocks

More than one potential sealing cap rock can be identified. The shales and compact limestones and dolomite beds of the Cretaceous and Eocene can be

efficient cap rocks. For example, the upper 5 m of the
Alamein formation which consist of tight dolomitic
limestone with a thin shale bed below, together with
the overlying shale beds of the lower Albian, can be
the cap rock for the Alamein formation reservoir in
Alamein field. Likewise, shales and limestones of
Turonian and younger Cretaceous units are believed
to be sealing off any oil trapped within the
Cenomanian-Turonian porous sequence.

The lowermost producing pool in the Western
Desert, the 3500 m sand in Um Baraka field, is
protected from any upward dissipation by the sealing
effect of the overlying shales and carbonates of the
lower Aptian.

Traps

So far, all the hydrocarbon discoveries in the Western
Desert have been drilled as structural prospects,
either in the form of three or four-way closure structures or as fault block structures. The development of
the finds indicates that the structural element was the
main factor determining the trapping of oil in almost
all of the discoveries. However, in some fields the
stratigraphic element in hydrocarbon trapping is evident in the pinching out of some sand pays in the

Cenomanian-Turonian sequence, as well as in the
facies changes from clastics to carbonates.

Most of the Western Desert structures in which oil
and/or gas accumulations were discovered indicate
that late Cretaceous-early Tertiary movements were
instrumental in their formation. The accumulation of
hydrocarbons in these structures took place after
early Tertiary times. This conclusion is confirmed by
the study undertaken on the burial history of a
number of horizons in the Mersa Matruh area which
show that oil generation (and consequently accumulation) must have taken place some time after the end
of the Cretaceous (Taylor 1984).

Source rocks

So far, two studies have been published on the source
rock potentialities of the different subsurface formations in the Western Desert (Urban, Moore & Allen
1975, Parker 1982). The subject is also treated in the
study undertaken by Robertson Research International for the Egyptian General Petroleum Corporation, finalized in 1982.

The first work is a study on the thermal alteration
and source rock potential of samples from three wells
in the Western Desert. The report of Parker includes a

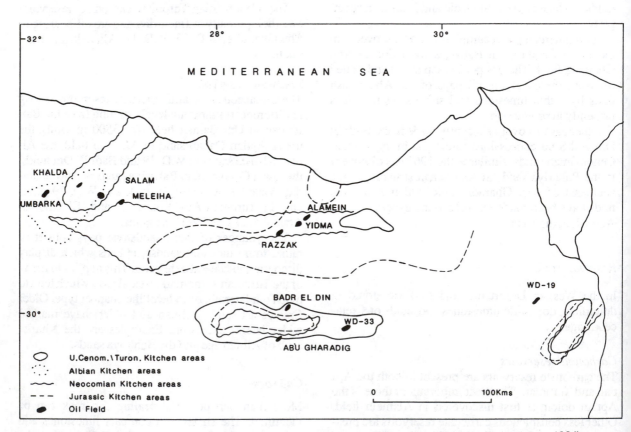

Figure 27.14 Distribution of hydrocarbons 'kitchen' areas in the northern Western Desert (after Schlumberger 1984).

synthesis of the geochemical reports previously made by oil company laboratories on 35 wells drilled in the Western Desert, as well as analyses of ten oil samples and six source rock extract samples. The Robertson Research study includes a synthesis of available source rock studies on Western Desert wells, geochemical analyses and source rock studies on nine additional wells. The Parker and Robertson Research studies include the construction of regional maps showing facies and source rock type distribution for the stratigraphic units of the Mesozoic, temperature gradient maps and organic maturity maps at or close to the structural levels of the top Cenomanian, top Alamein and top Jurassic. Based upon the source rock type distribution maps, maturity maps and reservoir distribution maps, the hydrocarbon generating areas (kitchens) are delineated and indicated on 'kitchen maps'. Figure 27/14 shows in one map the delineated kitchen areas in the Western Desert.

The kitchens indicated are:

a) A pronounced hydrocarbon kitchen covers the area extending from Sitra wells to Gindi well and includes the Abu Gharadiq and Gindi basins. In this kitchen area, Type II upper Cenomanian-Turonian source rocks have oil generating potential and Type III pre-Alamein lowr Cretaceous and Jurassic sequences have gas-generating potential.

b) In the region extending between longitude 25° and 33° 10' E and latitude 31° 10' and 30° 50' N, a hydrocarbon kitchen is present in which Type II and Type III lower Cretaceous Neocomian (pre-Alamein) and Khatatba and lower Masajid (Jurassic) sequences have oil generating potential. Type II upper Cenomanian/Turonian sequences may have generated oil in the area between east Alamein and Zebeida.

c) The general area between Mersa Matruh and Barakat-Louly in the western part of the Western Desert is another kitchen area with Type II source rocks in the thick Mersa Matruh shale and Type II/III source rocks in the Neocomian.

d) A fourth kitchen area lies to the west of Um Baraka area around NWD 302-1 well, where Devonian rocks are believed (Robertson Research) to have generated oil.

Oil and gas (commercial, non-commercial and shows) were encountered in association with these four hydrocarbon kitchens, but no oil has so far been encountered in the extreme western area. The general areas of Western Desert oil and gas fields seem to have had more than one source rock contributing to hydrocarbon accumulation.

Geology of the oilfields

Following is a brief review of the geology of some of the oilfields in the Western Desert. This review shows that almost all of the fields have more than one reservoir within the Cretaceous and that the majority of the fields are faulted anticlines or faulted blocks which are related to late Cretaceous-early Tertiary movements.

Alamein oilfield (1966)

The location of the discovery well was on top of a seismic high showing a four-way dip closure. At the Aptian level the area of the closure is some 8×6 km; this represents the largest areal closure so far found productive in the Western Desert (Fig. 27/15). The Alamein structure is one of a number of high features which constitute a ridge extending between Alamein and Qattara, the Qattara ridge. The ridge is part of a regional northeast-southwest trending line of highs, one of the Syrian lines of arcs, caused by the late Cretaceous-early Tertiary (Laramide) event. The field is affected by some of the cross faults which are more or less perpendicular to the Qattara ridge. The displacements are minor and do not affect the closure (Abdine 1974).

Oil is present in five horizons, the most important of which is the Alamein dolomite of Aptian age. The other horizons are the Razzak sands (Cenomanian) and the Dahab sands above the Alamein dolomite (Albian or Aptian), as well as two less conspicuous horizons, the Abu Roash 'G' at the base of the Cenomanian carbonate unit and a thin Aptian sand just below the Alamein dolomite. The average porosity of the Abu Roash 'G' unit, which is a dolomite in this field, is 30%, that of both Razzak sands and Dahab sands is 25%, that of the Alamein dolomite is 11%, and that of the Aptian sands is 10%. The porosity of the Alamein dolomite is mainly secondary in the form of vugs and fractures. The average API gravity of Alamein crude is 30°.

Um Baraka oilfield (1969)

The discovery well was located on top of a northeast-southwest trending seismic structure closed by dips from all directions. The succession is similar to that encountered in many of the wells in the northern part of the Western Desert, especially in the Matruh area. Marked thickning in the upper Cretaceous section is noted, and the pre-Aptian section is mostly clastics.

Only one pay is on production, namely the 3500 m sand of Aptian age, some 525 m below the Alamein dolomite. In Um Baraka-3 oil was also found in a sandstone section at around 3870 m, which could be Paleozoic. The Um Baraka crude is highly waxy compared to crudes from other fields. The average API gravity is 43°.

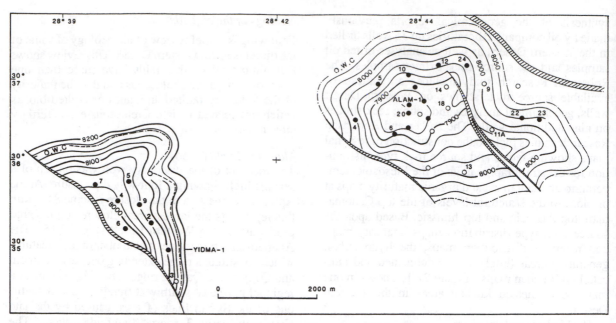

Figure 27.15 Structure contour map, top Alamein Dolomite, Alamein and Yidma fields.

Yidma oilfield (1971)

The Yidma field lies some 6 km to the southwest of Alamein field on the Qattara ridge. The closure is markedly smaller than that of Alamein field (Fig. 27/15), covering an area of 3 × 2 km and determined by dips on all sides except to the southwest where the feature is closed against a fault with a throw of 80 m to the southwest gradually decreasing to the southeast. The dips are gentler than those determining the Alamein field closure. West Yidma-1 dry hole in the down-thrown side of the fault indicates that the faulting took place in the late Cretaceous-early Tertiary (Eocene) time. Since faulting is believed to be responsible for closure to the southwest, this would mean that the closure was not complete until that time, hence the final entrapment of the oil. This raises the possibility of secondary migration of oil in both Alamein and Yidma fields.

The section encountered in Yidma is the same as that in Alamein field except that the middle and lower Eocene section is thicker in Yidma than in Alamein. The producing horizon is the Alamein dolomite which has a secondary porosity of 8% compared to 11% of the same pay in Alamein field. The original oil/water contact was at −2680 m which is lower than that of Alamein field, indicating that the two features are separate. The API gravity of the crude is 38.5°.

Abu Gharadig oil and gas field (1969)

This field represents the first commercial oil discovery in the late Cretaceous-early Tertiary Abu Gharadig basin (Adland & Hassan 1972). In addition to the

oil production obtained from the Turonian-upper Cenomanian horizons, the field produces natural gas from a separate pool below the oil pools, which belongs to the lower Cenomanian. Highly developed Tertiary and upper Cretaceous sections are present. The thick Santonian-Senonian section rests unconformably on the eroded tilted fault blocks of the Turonian and the uppermost Cenomanian (Abu Roash formation) which lie unconformably on the eroded lower Cenomanian (Bahariya formation).

The structure is a highly faulted anticlinal feature. It is highly deformed and dissected by faults into a number of fault blocks (Fig. 27/16). The faults belong to three generations of fracturing. The Cenomanian movements were responsible for the Bahariya formation tilted fault blocks which were later subjected to erosion; thus, the Abu Roash formation lies unconformably on them. The Turonian episode of deformation further complicated the structure by more faults, though these were of small displacement. The effect of these deformations is shown by the different oil/water contacts of the Abu Roash pays in different fault blocks. The late Cretaceous-early Tertiary movements caused relatively intense rejuvenation of movement along old faults so that the older Bahariya formation fault blocks were subjected to further movements relative to each other, thus adding to the differences in hydrocarbon/water contact in the field (Dia El Din 1974).

The gas in the Bahariya formation in this field is believed to be an exogenous hydrocarbon which migrated from the deep depositional center of the

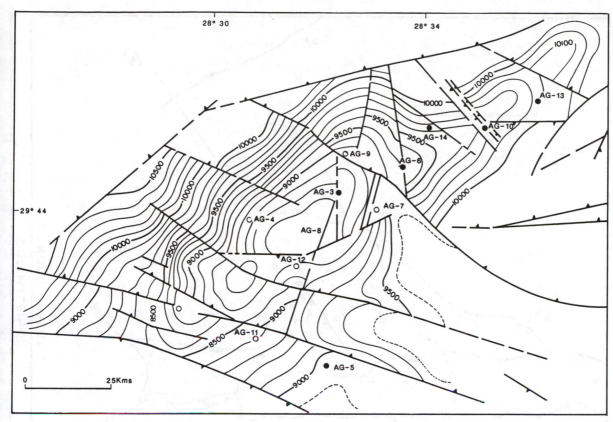

Figure 27.16 Structure contour map, Abu Roash 'C', Abu Gharadiq field (after Ezzat & Dia El Din 1974).

Abu Gharadiq basin into the Bahariya tilted fault blocks sealed by the Abu Roash shale and limestone section. Migration continued into the Abu Gharadig deformed structure until late Turonian-Santonian time. The detailed correlation of the Abu Roash 'C' and 'E' oil sands shows a stratigraphic pinchout of the sandstones toward the southwest flank of the structure. This fact, together with source rock analysis data, indicates that the Abu Roash 'C' and 'E' pays are mostly stratigraphic traps with an indigenous type of oil (Dia El Din 1974).

Razzak oilfield (1972)
Oil was discovered in seven separate reservoirs in the structure which lies on the northeast plunging anticlinal nose of one of a series of structural highs forming the Qattara Ridge (Ezzat & Dia El Din 1974). The sediments range from upper Jurassic to Recent. The structure is dissected by a number of faults, both parallel to the structure and perpendicular to it, mostly without large displacements.

In 1978 one of the culminations along the Qattara ridge a few kilometers to the northeast of the Razzak field was recognized by seismic. It turned out to be a faulted anticlinal closure structurally separated from the main Razzak field. Oil was found in the Bahariya formation sands; a less conspicuous pay is the Abu Roash 'G' dolomite unit. The Alamein dolomite, which was the main objective in drilling the exploratory well (Razzak-15) was found wet.

There are seven oil-bearing levels in the Razzak field: Abu Roash 'G' dolomite, Bahariya sands (both of Cenomanian age), upper Aptian sands (only 7 m), Alamein dolomite, lower Aptian sands (8 m); Neocomian sands (2875 m sand) and Betty sands (lowermost Cretaceous or Jurassic). The porosities of these seven reservoirs are, from top to bottom, 33, 23, 18, 7.5, 19, 22.5 and 11%. The average API gravity of the crudes of the seven reservoirs, beginning with the youngest reservoir, is 40, 31, 35, 37.5, 41, 54 (condensate) and 43°.

Meleiha oilfield (1980)
Oil was encountered in two Cenomanian sands with markedly low resistivity (1.5 to 0 ohms) due to the silty character of the sand. The structure trends east/northeast-west/northwest and is a broad relief fold which is crossed by faults of small vertical displacements. The succession encountered resembles that known in Alamein, Yidma and Razzak fields, with a

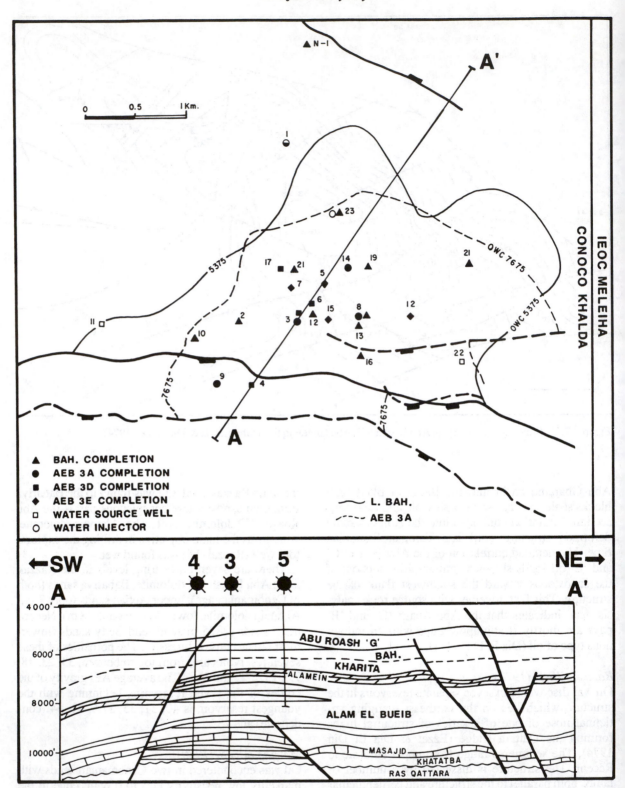

Figure 27.17 Salam field, lower Bahariya and Alam El Bueb closure and cross-section.

thicker section of upper Cretaceous, a thinner section of Alamein dolomite and a greater affinity of the pre-Alamein dolomite section toward the Matruh shales facies.

W.D.-33 oilfield (1972)
This field is small and lies to the east of Abu Gharadig field within the Abu Gharadig basin. The structure is a northeast-southwest trending faulted anticline believed to have been formed by the late Cretaceous-early Tertiary Syrian arcs movement. The structure is dissected by a number of cross faults, producing a number of smaller blocks within the main feature. The stratigraphic sequence is the same as that of Abu Gharadig field. The productive levels belong to the 'C' and 'E' units of the Abu Roash formation and both consist of sandstones. The porosity ranges from 12 to 24% and the API gravity averages about 30°.

W.D.-19 oilfield (1972)
This small field lies at the edge of the Abu Gharadig field in the easternmost part of Abu Gharadig basin. The structure is a northeast-southwest faulted anticline within a horst block with the same trend. The stratigraphic sequence is different from that of Abu Gharadig and W.D.-33 fields in that in the W.D.-19 structure, the Moghra formation (Miocene) rests unconformably on upper Senonian without the Oligocene-Eocene sequence which is some 1150 m thick in the Abu Gharadig field. Evidently this is due to the effect of the Laramide movement in late Cretaceous-early Tertiary times.

Oil is obtained from the 'F' limestone unit of Abu Roash formation (Turonian), the Bahariya sands (lower Cenomanian) and the Albian sands. The API gravity of the oil produced from Abu Roash 'F' limestone is about 16°.

Hayat oilfield (1986) (After Khalda Pet. Co.)
Hayat field consists of a four-way dipping northeast-southwest trending anticlinal feature. It lies due south of the Salam field in the Alam El Bueb formation, a double anticilinal closure exists independent of any faulting; however, at top Bahariya formation level, several northwest-southeast faults cut a complex anticlinal structure. The discovery wells tested oil from six zones in the Alam El Bueb and Bahariya formations.

Safir oilfield (1986) (After Khalda Pet. Co)
Safir field structure is formed by a four-way dipping north/northeast-south/southwest trending anticlinal closure. In the Alam El Bueb formation a double anticlinal closure exists separated by an east-west fault; however at top Bahariya level, an east/northeast-west/southwest horst block bounds the

structure to the north. The discovery well tested oil from the Bahariya, and Alam El Bueb formations; small amounts of oil also appear present in the Kharita formation.

Tut oilfield (1986) (After Khalda Pet. Co)
The Tut field structure is an anticlinal feature, bounded in the north and south (Bahariya formation only) by major east-west to east/northeast-south/southwest trending faults. It lies due north of the Salam field. The discovery wells encountered reservoir in the Bahariya formation and three units in the Alam El Bueb formation; minor gas was also tested from the Khatatba formation.

Khalda oilfield (1984) (After Khalda Pet. Co.)
The Khalda field is an oval domal structure trending north-south and cut by northwest-southeast step faults in the north and horst block in the south.

The western flank of this domal structure shows more gentle dip than the steeply dipping and partly fault-bounded eastern flank. The discovery well tested oil from the Bahariya formation.

Salam oilfield (1985) (After Khalda Pet. Co)
The Salam field (Fig. 27/17) lies at the center of major north/northeast-south/southwest trending ridge, apparently formed by deep-seated normal faults: later east-west to east/northeast-west/southwest trending stitch-slip faults have created associated closures of which Salam field is the biggest (also Tut and Hayat fields). At the top Bahariya formation level, Salam field is an anticlinal structure, but by minor faulting and bounded to the south by a major east-west fault; however, in the Alam El Bueb formation this fault has limited influence, although the anticlinal structure is bisected by a severe east-west fault.

The discovery well, Salam-3, encountered several prospective reservoirs, namely, the Bahariya formation, four units in the Alam El Bueb and the Khatatba formations (the last containing both oil and gas sands).

BED-1 Oilfield (1980)
(Information provided by Badr El Din Oil Co.)
The BED-1 field is located in the Western Desert some 9 km nortwest of the Abu Gharadig field in the central part of the Abu Gharadig basin. The field is an elonagted northwest-southeast tilted fault block approximately 8 km long and 1-2 km wide with structural dip toward the northeast and bounded to the southwest by a major normal fault (800 m throw). Based on the Abu Gharadig model the primary objectives of well BED 1-1 were the upper Creatceous sands of the Abu Roash and Bahariya formations.

However, the well missed the primary objectives which were found cut out by the major normal fault but the Kharita sands (lower Cretaceous) were found hydrocarbon-bearing. Testing of the well proved 130 m oil column in the Kharita formation (tested 9500 b/d of 39 API oil) with a 30 m gas cap (7 MMscf/d). The Kharita reservoirs consist of a stacked sequence of channel fill deposits alternating with lagoonal shales. The reservoir quality is generally moderate to good with average porosities of 15%. Permeability varies between 50 & 400 md. Two appraisal wells (BED 1-2 & 1-3) were drilled (both wells planned to penetrate the Kharita reservoirs in a similar structural setting to that of the discovery well) and both wells found the Kharita formation hydrocarbon-bearing. The western extension was proven by 3 appraisal wells (BED 1-4, 1-5 & 1-8) and the pressure data indicate good horizontal communication. The average API is 37039 and the GOR is 1500 scf/stb.

Beside the proven commercial oil accumulation in the lower Creatceous Kharita Formation, hydracarbons have been evaluated and/or tested in the upper Cretaceous Abu Roash and Bahariya formations (which constitute the main producing formations in the adjacent Abu Gharadig field) by two wells BED 1-6 & 1-7. These wells tested oil from the Abu Roash 'C', 'D' & 'F', 23-32 API and GOR 40-200 scf/stb.

BED-3 gas/condensate and oilfield (1983)
(Information provided by Badr El Din Oil Co.)
The BED-3 gas and oilfield is the main hydrocarbon field discovered to date in the western part of the Abu Gharadig basin and is located some 60 km to the west of the Abu Gharadig field. The BED-3 structure is essentially a north northwest-south southeast trending horst block, the development of which probably began in the Albian and was largely completed by early Campanian time. The filed was discovered in 1983 by well BED 3-1 (Shell Winning N.V) which penetrated two separate gas accumulations in the Albian Kharita formation. Production testing gave 30 MMscf/d & 950 b/d condensate from the upper accumulation and 25 MMscf/d & 540 b/d condensate from the lower accumulation. Detailed pressure measurements over the Kharita formation proved a gas-water contact in the lower accumulation but left open the possibility of an oil leg below the upper gas accumulation.

Appraisal well BED 3-2 was drilled along the eastern extension of the structure with the prime purpose of investigating the possible oil leg below the upper gas accumulation. This well proved field continuity but failed to prove an oil leg because of lateral shaling of part of the reservoir sequence. The extensive production testing in BED 3-2 supported and matched the high potential flow rates of BED 3-1. A second appraisal well, BED 3-3, was drilles to appraise the downdip area of the field to finally determine the presence/absence of an oil rim. The RFT results and testing proved the presence of a 33 m oil column below the upper accumulation which flowed 6,000 b/d 29 API oil on test.

THE RED SEA

With the exception of the Hurghada field, no commercial hdyrocarbon accumulations have been found in the Red Sea graben starting from its junction with the Gulf of Suez and Gulf of Aqaba southward.

Only about 45 exploration wells have been drilled in the whole depression. A few oil and gas shows were recorded in some of these wells. Most famous of the non-commercial finds is the gas/condensate Barqan discovery, offshore of Saudi Arabia opposite to Gemsa on the western coast of the Gulf of Suez. Drilling in the Barqan feature encountered four gas-bearing sandstones within the '*Globigerina*' group (Ahmed 1972). The wells also indicate that the Miocene lies directly on the basement.

In the Egyptian part of the Red Sea graben only nine offshore wells have been drilled, all plugged and abandoned as dry holes. A number of shows were reported in some of these wells at some Miocene levels which are oil-bearing and producing in the Gulf of Suez province.

The geology and oil potentialities of the Egyptian Red Sea was discussed by Tewfik & Burrough (1976), Tewfik & Ayyad (1982) and Barakat & Miller (1984). Prior to the latest round of activities of oil companies in the region, beginning in 1974, a number of oil company reports covered the geology of the coastal strip in this district. Other studies were conducted by the Geological Survey of Egypt (for a review of this work, see Said, chapter 18, this book)

Beginning in 1974, oil companies ran aeromagnetic surveys covering both coastal land strips and offshore areas. These were followed by marine geophysical surveys which included marine magnetic, marine gravity and marine reflection seismic. The aeromagnetic survey covered 14,300 km, the marine magnetic survey 19,333 km, the marine gravity survey 19,173 km and the marine seismic work 22,579 km.

The seismic survey data coupled with data obtained from other geophysical surveys indicate the presence of more than one high trend parallel to the direction of the depression. The interpretation of seismic data was a difficult task due to the lack of well and surface data to be correlated with the seismic events.

The offshore drilling data indicate that the sections drilled consist of Miocene clastic and evaporite intervals with a post-evaporite Pliocene to Recent section. The pre-Miocene was not reached in any of the nine offshore wells. It should be recalled, however, that pre-Miocene sections (Eocene and Cretaceous) are known to exist on the surface. Reservoir rocks in the form of sands and sandstones were penetrated in most of the offshore wells within the Miocene evaporites and the pre-evaporites clastic sections.

Oil and gas shows were found in some of the sandstone intervals within the evaporite and pre-evaporite clastic sections in seven of the nine offshore wells including the Abu Maad-1 well in the Ras Banas area where the shows consisted of methane and heavier gas fractions (Tewfik & Ayyad 1982). Onshore, another oil find, though not commercial, was reported in Ras Abu Soma-1, not far from Hurghada. In this well, a total of 11 barrels of black oil (29 to 32° API) were swabbed from a silty limestone (probably basal Belayim equivalent) at a depth of 855 to 863 m (Barakat & Miller 1984). The evaporites, which were encountered in all the wells drilled, represent a good and effective cap for hydrocarbon accumulations, as is the case in the Gulf of Suez depression.

The geophysical surveys conducted in the district show that there are many structural anomalies following the Gulf of Suez-Red Sea trend, which might represent structural traps in the form of horsts and tilted fault blocks similar to many features in the prolific Gulf of Suez province. Moreover, sedimentation conditions during the Miocene would have allowed for reefal development, lateral lithological changes and pinch-outs, thus furnishing the conditions for stratigraphic trapping of hydrocarbons.

These oil and gas shows, as well as those reported from other parts of the Red Sea depression (Ahmed 1972) indicate that oil was generated not only in the Egyptian part of the Red Sea but in other parts of the depression including offshore northwestern Saudi Arabia, offshore Sudan, offshore Ethiopia and offshore Yemen. The source rocks of these shows could be in the Miocene and possibly in pre-Miocene organic-rich sediments which have not so far been reached by drilling. Barakat (1982) and Barakat & Miller (1984) suggest that the lower Miocene Rudeis shales and to a smaller extent the Paleocene-upper Cretaceous shales and carbonates may be potential source rocks in the northern part of the Red Sea depression. This conclusion is based on measurements of samples in investigations dealing with organic richness, organic matter type and maturation. For the central and southern parts of the Red Sea, no such studies are so far available. However, Ahmed (1972) is of the opinion that in the Red Sea the only source rock of great interest is the fossiliferous Miocene section of the *Globigerina* group which includes the Rudeis shales of the Gulf of Suez and northern part of the Red Sea.

Hurghada oilfield (1913)

The discovery well was located on the axis of a surface anticline which encountered oil at a depth of 200 m in 1913. Later on, deeper pays were encountered as development drilling continued. The depth of the production in the field ranged between 200 and 820 m. Development drilling continued until 1949 and the number of wells drilled reached 139, of which 107 were completed as producers. The well data indicate that the subsurface structure is much more complicated than that displayed by surface geology. The data show that the basement is overlain by a Paleozoic-Cretaceous sequence which is strongly faulted and which is unconformably overlain by the Miocene sequence (Fig. 27/18). The Paleozoic section consists of shales and 'Nubia-type' sandstone, principally of Carboniferous age. The shales predominate in the upper part of the sequence and the sandstones are more pronounced in its lower part. The shales and sandstones contain an appreciable amount of plant remains. At the base of the succession a boulder bed is reported which contains rounded and angular Precambrian boulders. The upper Cretaceous overlaps the eroded surface of the Carboniferous and consists of 'Nubia type' sandstones and shales, with upper Cretaceous shales and marls on top.

The Miocene succession unconformably overlies the pre-Miocene erosion surface. The erosion effect is more pronounced along the axis of the structure. Moreover, the basal section consists of assorted lithology (as products of erosion), overlain by a *Globigerina*-rich marl and shale section (possibly Rudeis formation). According to Kostandi (1961), 'Rejuvenation of the movements along the pre-Miocene clysmic faults must have taken place during the time of deposition of the *Gloigerina* marls since the thickness of these marls increases noticeably on the downfaulted flanks'. The marl and shale section is followed by a dolomitic limestone section with a few anhydrite intercalations. This section is overlain by a sequence of evaporites (anhydrite) alternating with shales and sands.

The Miocene section is followed by a clastic section which consists of marls, variegated sands and marls with grits, calcareous sandstones, oolitic limestone beds and *Pecten*-rich calcareous sandstone beds. This section is attributed to the Pliocene.

Pleistocene-Recent beds exposed on the surface consist of Pleistocene gravels and reefs with younger alluvial deposits of gravels, sand and clay.

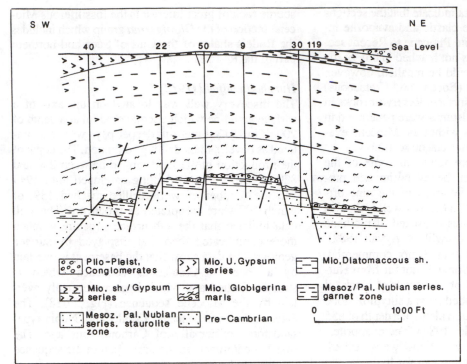

Plio-Pleist,
Conglomerates

Mio. sh./Gypsum
series

Mesoz. Pal. Nubian
series. staurolite
zone

Mio. U. Gypsum
series

Mio. Globigerina
marls

Pre-Cambrian

Mio. Diatomaceous sh.

Mesoz/Pal. Nubian series.
garnet zone

0   1000 Ft

Figure 27.18 Cross-section
through Hurghada field (after
Kostandi 1961).

The Hurghada structure consists of two main horsts, Hurghada West and Hurghada East, with a non-producing horst, the Abu Minqar horst, in the middle. The three structural elements are determined by northwest-southeast trending faults. Other faults trending north-south to northeast-southwest cut the field (Kostandi 1961).

The oil reservoirs belong mainly to the Paleozoic-Mesozoic Nubian sandstones with porosity ranging from 16 to 18% and permeability from 100 to 200 MD. They are also found in Basal Miocene clastics and the younger Miocene dolomitic beds, with a porosity of 13% and a permeability of 260 MD. The APE gravity is 22.0°.

SINAI

The following discussion of the petroleum geology of Sinai excludes the southwestern area which was previously treated with the Gulf of Suez province. The discussion, in fact, is limited to north Sinai as the south of Sinai, apart from the Gulf of Suez coastal strip, is occupied primarily by basement rocks with relatively thin sediments along their northern edge.

Exploration surveys and drilling began in the 1940s by a number of oil companies. The work included surface geological surveys as well as magnetic and gravity surveys. In light of these sur-

veys, four exploration wells were drilled, three of them on surface structures in central Sinai (Nakhl-1, Darag-1 and Abu Hamth-1) and the fourth in northeast Sinai (El Khabra-1), drilled to test a prospect delineated by gravity. Some oil shows were found in the Cretaceous (Cenomanian) in Nakhl-1 and Abu Hamth-1.

In the early 1960s, some experimental seismic work was conducted in a few locations in north Sinai, and a regional offshore seismic survey was conducted to cover the north Sinai offshore area in the Mediterranean.

In the late 1960s, exploration efforts were resumed in north Sinai and its offshore area during and after the Israeli occupation of the peninsula. Fifteen exploration wells were drilled during the Israeli occupation, four of which were offshore (Gal-1 and Ziv-1). A gas accumulation was found in one of the land wells, the Sadot field to the southwest of Rafah, and was exploited. It is now almost depleted. The gas is in a Cenomanian dolomite and limestone reservoir. Oil shows were found in the Aptian in Ziv-1.

A few discoveries have been made lately offshore in the Mediterranean, such as the oil find in Tineh structure in the Oligocene and the gas finds in both Wakar and Port Fuad structures to the north of Port Said. The structures, in fact, lie within the Nile delta basin which extends eastward to include the western part of the offshore area of north Sinai.

The most prominent tectonic event that could have had a bearing on the oil potential of north Sinai is the movement connected with the late Cretaceous-early Tertiary northeast-southwest trending Syrian arc folding (Jenkins, Harms & Oesleby 1982). This is the same tectonic event which is related to some of the oil-bearing structures in the northern part of the Western Desert. Many of the highs along these lines of arcs could have been suitable traps for oil and/or gas accumulations.

Reservoir rocks are present within the Mesozoic and Paleozoic sections in the form of sandstones and carbonates, if fractured or cavernous. Eocene limestones may also be suitable reservoirs if enough secondary porosity is developed. The storing capability of the Eocene limestones is very well demonstrated in the Ein Gudeirat-Kuntilla area along the eastern border of Sinai where the fractured Eocene limestones store huge amounts of fresh water.

The cap rocks could be present within the Paleozoic, Mesozoic and early Tertiary in the form of shales and compact carbonates. The Triassic section at Aref El Naqa in east central Sinai includes some anhydrite intervals toward its upper part. Such anhydrites, wherever found, can seal off any hydrocarbon accumulations at levels below these capping intervals.

So far, the potential source rocks in north Sinai are rather speculative. The conclusions regarding this subject are based only upon analogy with comparable formations in the Western Desert. That hydrocarbons have been generated in the area is established by the presence of the Sadot gas accumulations and by the numerous oil and gas shows in the Cretaceous and Eocene in some of the wells drilled both onshore and offshore. The Mesozoic organic-rich shales and carbonates may be potential source rocks. The source rocks of the oil and gas found in wells in the northeastern offshore part of the Nile delta basin could also be present in parts of north Sinai. The latter source rocks are believed to be organic-rich shales in the upper and lower Oligocene.

NILE DELTA

Exploration began in the Nile delta region in 1964. The first discovery was the Abu Madi field in 1967. Before 1964 practically nothing was known about the deep subsurface geological conditions of the Nile delta. Since then some 25,000 km of land seismic and some 27,000 km of offshore seismic lines have been run and about 60 exploration wells have been drilled. So far, the outcome of these activities has been the discovery of two commercial gas/condensate accumulations presently on production, the Abu Madi and offshore Abu Qir gas fields. Other onshore and offshore accumulations have been found but have not been proven commercial. Almost all these latter finds in the delta region are gas/condensate, though two of the finds contain oil rather than gas condensate. The first oil find in the delta is recorded in the Tineh structure to the northwest of Port Said, while the other find is in a clastic horizon below the gas/condensate reservoir in the offshore Abu Qir field.

The literature concerning the geology of the Nile delta basin, both onshore and offshore, is limited. Deep subsurface data were not available before the mid-1960s and most data are still scattered. Efforts to collect and synthesize these data include the works of Barber (1981), Poumot & Bouroullec (1984), Deibis (1982), Elf Aquitaine (1982), Hantar (1975), Rizzini et al. (1976), Lacaze & Dufresne (1982), and Said (1981). For a review, see Harms & Wray (chapter 17, this book).

The literature indicates (1) that the downcutting of the Nile river began in late Miocene time, (2) that regional structural movements affected the sedimentation process in the Tertiary in general and the Neogene in particular, and (3) that the section in the northern part of the Nile delta became deltaic only in post-Miocene time. Sedimentation prior to the Miocene, and particularly in Cretaceous and Eocene times, took place under marine conditions in shelf, slope and deep basin environments. Most studies stress that no growth faults are noted within the Nile delta basin in contradistinction to other hydrocarbon producing deltas in other parts of the world. On the other hand, growth faulting is reported by Deibis (1982) in the Abu Qir area.

Hydrocarbon habitat

All the hydrocarbon-bearing reservoirs so far found in the Nile delta basin consist of sandstones belonging mainly to the Pliocene and Miocene. The main gas/condensate producing section is the Abu Madi formation. Other gas/condensate bearing sections belong to both Sidi Salem and Kafr El Sheikh formations. Due to their deltaic environment of deposition some of the sand levels do not have wide lateral extension even within the boundaries of the structure.

The geological setting in the Nile delta region displays the elements required for the formation of both structural and stratigraphic traps. Late Cretaceous-early Tertiary events, as well as uplifting and faulting in Miocene and later times would form folding and doming together with block faulting of Tertiary sediments, providing a possibility of traps. The deltaic nature of the late Miocene and younger

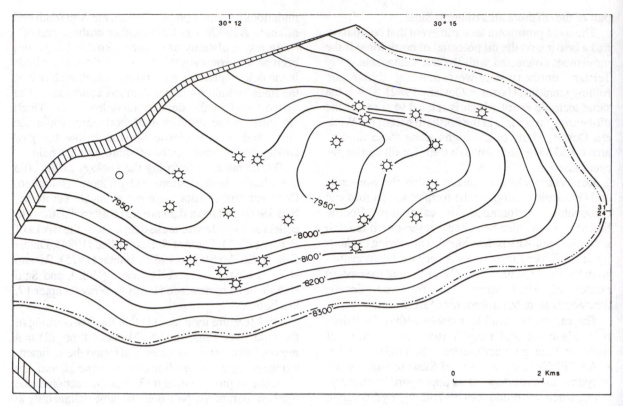

Figure 27.19 Structure contour map, top Abu Madi formation, Abu Qir gas field (after Deibis 1982).

sediments allows for marked changes in sedimentary facies which would form different types of stratigraphical traps. In pre-Miocene sections, reefal development along the edge of the carbonate platform occupying the southern part of the Delta may host oil or gas by forming a stratigraphic trap.

So far, the traps delineated by enough wells have been found to be combination traps in which the stratigraphic element plays a major role. This is established by the lenticular nature of the sands (and shales) in many of the discoveries so far made, whether commercial or not yet proven commercial. At the same time, the finds are confined within structures which owe their presence to earth movements. For example, the Abu Madi field is a low relief dome, with faulting to form blocks. Within one and the same structure, the stratigraphical trap-forming element is rather prominent. The Abu Qir field, on the other hand, is an east-west elongated anticline in which the lateral extension of the individual producing levels differs from one level to another due to lateral facies changes.

Studies devoted to source rock potential in the Delta are likewise limited. The subject is touched upon by Deibis (1982), who suggests that the Kafr El Sheikh formation (shale) is the source of the gas in both Abu Madi and Abu Qir gas fields. The International Egyptian Oil Company (AGIP) tends to consider the source of the gas/condensate accumulations to be lower Miocene and Oligocene and possibly deeper horizons (pers. comm.). These conclusions must be regarded as tentative as work on the subject is still in the early stages.

Abu Qir gasfield (1969)

This is the first gas/condsensate field to be discovered offshore in the Mediterranean. It lies in Abu Qir Bay 10 km north of the coast and 40 km northeast of Alexandria. The field lies within the Nile delta basin and hence the stratigraphic sequence encountered in the Abu Qir wells is the same as that penetrated by wells drilled in the embayment. The section consists of massive clastics with rare limestone interbeds of Miocene, Pliocene, Plio-Pleistocene and younger ages (Deibis 1982). The structure is an east-west trending anticlinal closure which is bounded along its northern and southern sides by east-west faults, claimed by Deibis to be of the growth fault type (Fig. 27/19).

The gas/condensate accumulations are present in four pay sands. The upper three are located within the Abu Madi formation and are referred to as the upper middle and lower Abu Madi sands. The fourth and lowermost is within the Sidi Salem/Qawasim formations. Recently an oil leg was discovered in the Sidi Salem formation within the western extension of the field.

CHAPTER 28

Nubian Aquifer system

ULF THORWEIHE
Technical University, Berlin

1 INTRODUCTION

Approximately 95% of the Egyptian landscape is desert area. Only the coastal regions, the delta, the Nile valley and the oases can be used for agricultural utilization. Here more than 40 million people live with a population density of more than 1000 km². For that reason, new areas of utilization had to be developed within the desert area where the groundwater reservoir could be used. Thus, at the end of the fifties, President Abdel Nasser started the New Valley Project for groundwater development in the oases Kharga, Dakhla, Farafra and Bahariya, which are situated in a long stretched-out depression in the Western Desert more or less parallel to the River Nile. Since that time, there has been an extensive artificial effect on the groundwater reservoir and it became necessary to thoroughly clarify the formation and origin of groundwater.

Since Ball (1927), the assumption in the hydrogeological models has been that the groundwater of the Nubian aquifer in the Western Desert is recharged by a regional groundwater movement from southern or southwestern areas and that the groundwater reservoir is in steady state (Sandford 1935, Hellström 1939, Ezzat 1959, 1974).

This chapter will show that a regional movement exists but that the groundwater of the Nubian aquifer was formed mainly by local infiltration. It is a fossil groundwater reservoir which gets recharged from other fossil reservoirs (Kufra basin, northern Sudan) but, compared to the discharge, the recharge is small, so that the Nubian aquifer is an unloading system. For projects of groundwater use, exact calculations are necessary to avoid wasteful groundwater mining.

2 HYDROGEOLOGICAL FRAME

The Nubian aquifer system in Egypt is formed by pre-Tertiary sediments and is part of the Nubian aquifer system of the east Sahara, which also includes the Kufra basin in southeast Libya and northeast Chad and parts of north Sudan. The Nubian aquifer system in Egypt is bounded to the north by the saline-freshwater interface which stretches roughly along the 29th latitude (BGR 1976) and to the east by the basement complex of the Eastern Desert (Fig. 28/4). In the south, there is the Gebel Uweinat-Aswan uplift system which is crossed by a southward-trending graben between Gebel Kamil and the Bir Tarfawi area, where sediments reach up to 700 m in thickness and ensure a hydraulic groundwater connection to north Sudan. To the west, the system is not bounded, so that the Libyan border is to be assumed for any calculation of the groundwater in Egypt.

The aquifer area in Egypt falls into three structural units which are not completely independent of one another.

The northwestern basin, which is north of the Cairo-Bahariya uplift and is formed by the unstable shelf. Only the southernmost strip of this unit stretches south of the saline-freshwater interface, so that this basin is of minor importance for the Nubian aquifer system.

The Dakhla basin is the largest and most important unit in Egypt. The southwesternmost corner is related to the Kufra basin. In the area of the Abu Ras plateau, fine- to coarse-grained Paleozoic sandstone crops out; it has a good permeability and reaches more than 1500 m in thickness. Only part of the Silurian sediments are of marine origin and have a low permeability.

Excepting this southwesternmost corner, the southern part of the Dakhla basin is filled with mainly fine- to coarse-grained continental sandstone of lower Cretaceous and Cenomanian age; it has a good permeability and comes up to just under 1000 m in thickness. This sediment pile, formerly called Nubian sandstone, is intercalated by two regional formations of low permeability: the marine Abu Ballas formation (Barthel & Böttcher 1978), which is formed by a few 10 m of poorly permeable shale, increasing in thickness towards the north, and the

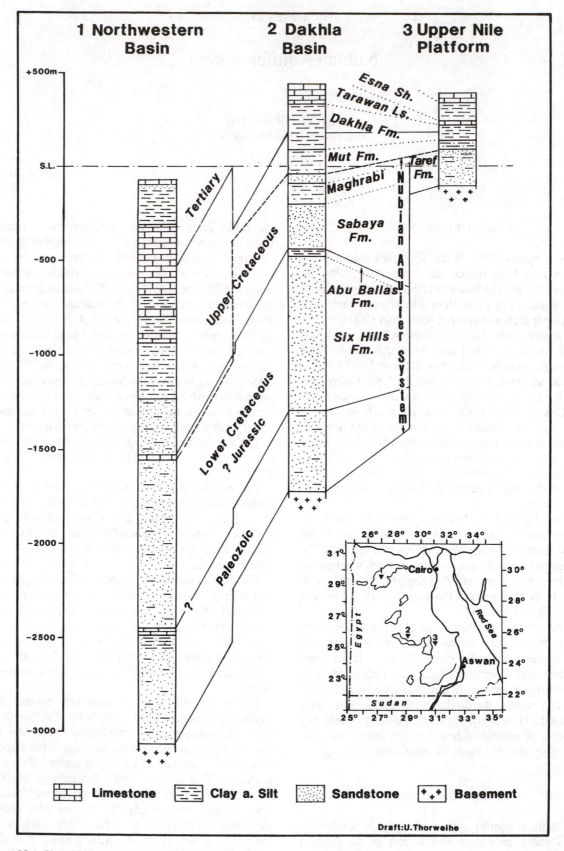

Figure 28.1 Simplified sections of the Nubian aquifer system.

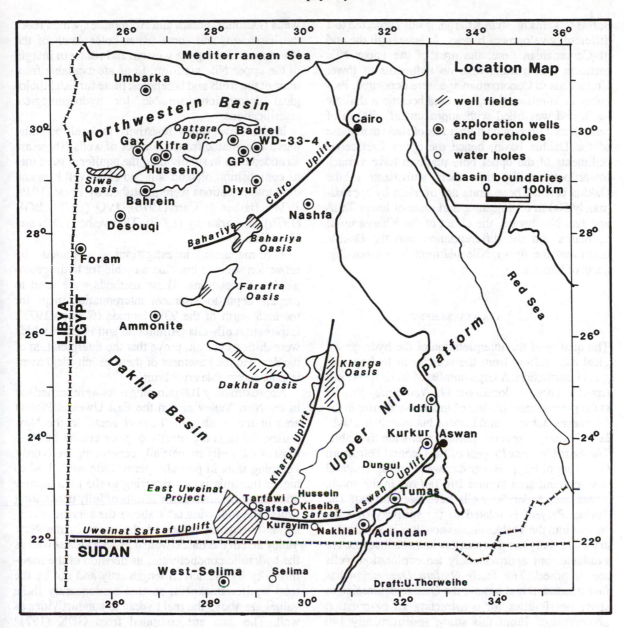

Figure 28.2 Location map.

deltaic to shallow marine Maghrabi formation (Barthel & Herrmann-Degen 1981), which reaches up to 200 m in thickness and is formed by interbedded shale and sandstone with fair permeability.

In the northern part of the Dakhla basin, just beyond the 25th latitude, the Cretaceous sediments overlie Paleozoic sediments and are covered by the variegated shale (Mut formation, Barthel & Herrmann-Degen 1981) and the Dakhla formation (Said 1962), which reach some 100 m in thickness and form the confining bed of the Nubian aquifer system. The Paleozoic sediments are not studied in

detail because of missing outcrops. But the few exploration wells which fully penetrate the Paleozoic sediments indicate that they should reach up to 2400 m along the Libyan border and up to 800 m along the longitude of the Dakhla and Bahariya oases. They are mainly continental sandstone with minor intercalations of shale, dolomite and limestone. They represent the approximately fair permeable lower part of the Nubian aquifer system.

The upper Nile platform is situated east of the Dakhla basin and is separated from it by the Kharga uplift which strikes from the basement outcrop of Bir

Safsaf towards north to Kharga. It was subjected to a different development history. At least until the end of Cenomanian time, the area of the upper Nile platform was a high, thus no sediments of lower Cretaceous to Cenomanian age were deposited. Probably in Turonian time, this area became a shallow basin and was filled with approximately 500 m of sediments with a much higher rate of shale than those of the Dakhla basin, hence the upper Cretaceous sediments of the upper Nile platform have a much lower permeability than their equivalents in the Dakhla basin. These strata are overlain by approximately 500 m of limestone and shale of lower Tertiary age. Northward, the effect of the Kharga uplift decreases and the differentiation into the Dakhla basin and the upper Nile platform is not possible north of Kharga.

3 DATA ASSESSMENT

The quality of the interpretation of the hydrogeological data suffers from the scarcity of hydrogeological information. A large number of wells is concentrated in very few locations: *The New Valley Project* is the predominant project of groundwater use. In the four oases Bahariya and Farafra, but mainly in Dakhla and Kharga, several hundred wells are available. *The East Uweinat Project* of the General Petroleum Company of Egypt is a groundwater and agricultural development area around Bir Tarfawi in the south, where just under 40 wells have been drilled. *The Toshka Project* is related to the spillway of Lake Nasser into the Toshka depression. It is in a preliminary stage: only six piezometers in a small square are available, but approximately ten exploration wells are proposed. *The Study Qattara Depression* has investigated in and around the Qattara depression to study possibilities for constructing an evaporation powerstation. From this study, approximately 140 boreholes are available, they are all located north of the 29th parallel, i.e. north of the Nubian aquifer system. Furthermore, there are three exploration

wells (Desouqi, Foram and Ammonite), which reach the basement, and some water holes south of the Kharga oasis along the western and southern margin of the upper Nile platform. Logs are available from most of the wells and boreholes: these furnish lithological descriptions suitable for hydrogeological classifications.

In the New Valley resistivity and self-potential logs are available from a number of wells. These are an added help in subdividing the aquifer system into its composition of sand, silt and clay. Well logs and their interpretations were published by Ezzat (1959, 1974), Barber & Carr (1976), JVQ (1978), BGR (1976), El Barkouky (1979), Thorweihe (1982) and Nour (1984).

Aero-magnetic investigations, supported by refraction seismic lines are suitable for hydrogeological interpretations. These methods were used to prepare depth-to-basement interpretation maps for the area north of the 23rd latitude (Conoco 1977). Exploration oilwells (Ammonite and Foram), which were drilled later on, prove that the interpretation of the depth to the basement of these geophysical maps is in an acceptable error limit.

Approximately 100 pumping tests were carried out in the New Valley and in the East Uweinat Project area in the south (Bir Tarfawi area). In the New Valley, the tests are mostly of poor quality because most of the wells are partially penetrating wells only. Pumping tests in partially penetrating wells lead to higher transmissivities referring to the total aquifer thickness compared to the results of fully penetrating wells. The following table shows the average pumping test transmissivities of the oases in the New Valley and of the East Uweinat Project area as well as the hydraulic conductivities as division of the transmissivity by the screen length only and not by the total aquifer thickness, so that consequently these values are above the real hydraulic conductivities as well. The data are compiled from GDK (1971) (Bahariya), BGR (1976) (Farafra), Barber & Carr (1976) (Dakhla and Kharga) and Nour (1984) (East Uweinat).

Table 28/1. Average transmissivities and hydraulic conductivities of the New Valley oases and the east Uweinat project area.

Region	Number of pumping tests	Aquifer thickness (m)	Average screen-lengths (m)	Transmissivity (m^2/sec)	Hydraulic conductivity (m/sec)
Bahariya	10	1880	196	8.7×10^{-3}	4.4×10^{-5}
Farafra	8	2600 (?)	221	1.0×10^{-2}	4.7×10^{-5}
Dakhla	21	1850	181	1.1×10^{-2}	6.1×10^{-5}
Kharga	59	1280	205	5.9×10^{-2}	2.9×10^{-5}
East Uweinat	6	410	185	2.4×10^{-2}	1.3×10^{-5}

4 GROUNDWATER GENESIS

It is important to understand the origin and movement of the groundwater in the aquifer system before responsible groundwater utilization can be made. Different sources for the groundwater of the Nubian aquifer system in Egypt are possible.

1. Groundwater recharge by the Nile river in that section where the alluvial deposits are in contact with the groundwater-bearing strata which is the case upstream of Qena. In the area between Aswan and Qena, on which we have little information, the piezometric head of the Nubian groundwater is above the elevation of the Nile river, so that groundwater recharge is not possible there. On the contrary, groundwater discharge is to be assumed. However, the artificial Lake Nasser raised the water table by approximately 100 m, reversing the piezometric gradient and causing groundwater recharge to take place along the lake.

2. Groundwater influx from south and southwest, from north Sudan and from the Kufra basin. It is a fact that the piezometric head is increasing in the southern direction, so that a groundwater influx from north Sudan is to be assumed. From the southwestern direction, no piezometric measurements are available but, from the hydrogeological point of view, groundwater influx from the Kufra basin is very likely.

3. Groundwater formation by local infiltration during wet periods in the past. From the western and central part of the north Sahara, it is known that the groundwater was formed mostly in former times (Gonfiantini et al. 1974, Said 1975, Klitzsch et al. 1976, Sonntag et al. 1978). The same can be expected in Egypt.

To clarify the paleoclimatic situation in Egypt, no less than 150 groundwater samples of the Nubian aquifer system were analyzed for stable- and radio-isotope contents (Sonntag et al. 1978, 1979, 1982, Haynes & Haas 1980, Thorweihe 1982, Thorweihe et al. 1984, Schneider & Sonntag 1985).

Stable isotopes deuterium and oxygen-18

The pattern of deuterium and oxygen-18 of groundwaters of the north Sahara shows a significant decrease from west to east (Sonntag et al. 1978). The interpretation of this fact is that the groundwater of the north Sahara was formed by wet air masses which were transported by a western drift from the Atlantic ocean across the continent. This is in contrast to the south Sahara, south of approximately 20° latitude, where the groundwater is of tropical convective origin (Sonntag et al. 1978). Across the continent, the content of the heavy stable isotopes deuterium and oxygen-18 in the water vapour decreases, because these isotopes condense preferably by precipitation and remain preferably in the fluid phase by evaporation, as compared to the light isotopes. This continental effect can be observed in recent European groundwater, formed by western drift, where the ratio of deuterium (D) and oxygen-18 (^{18}O) follows the Meteoric Waterline:

$\delta D = 8 \times \delta^{18}O + 10$.

The content of deuterium and oxygen-18 in the groundwater of the north Sahara fits approximately with this Meteoric Waterline: the ratio has the same slope, but the deuterium-excess is smaller due to a lower water vapour pressure in the paleo-air masses of the north Saharian latitudes, which led to a lower evaporation rate.

The content of deuterium and oxygen-18 of the groundwater of the oases in the New Valley is greatly decreased on account of the remote Atlantic ocean. The fact of the isolines pattern in Egypt as the eastern part of the total pattern of the north Sahara prove the continental effect, which leads to the assumption that the groundwater of the Nubian aquifer system was mainly formed by wet air masses transported by a western drift (Sonntag et al. 1978).

Other paleoclimatic evidences show that Egypt is in the reach of the south Sahara as well, where the wet air masses in humid phases are of tropic convective origin. After Said (1983) in southern parts of Egypt late Paleolithic and Neolithic flora and fauna of African savanna were found which prove that at least during this time this region was affected only by summer rains from the south. That led to the assumption that the south has been affected by the northward migration of the Sudano-Sahalian savanna whereas the north by the Mediterranian pre-Saharian steppe. Northward shift of the present day climatic belts to about 15°, the latitude they now occupy would bring an increase in precipitation in south Egypt and set a climate similar to that which must have prevailed during the Quaternary (Said pers. comm.). It is not proven whether these climatic conditions took place even during the late Pleistocene, but it is likely.

Table 28/2. Average content of the stable isotopes deuterium and oxygen-18 of the four oases in the New Valley, as per mill-deviation from the Standard Mean Ocean Water (SMOW).

	Kharga	Dakhla	Farafra	Bahariya
δD (SMOW ‰)	−83.4	−83.8	−79.1	−80.4
$\delta^{18}O$ (SMOW ‰)	−10.9	−11.5	−10.5	−10.5

Radiocarbon groundwater ages

For groundwater age-dating, it is not possible to sample a piece of water as it would be with archeological or pre-historic material, but a groundwater sample is the result of a complex flow situation, which varies in space and time. Furthermore, normally it is not possible to get water samples from defined positions of the aquifer, but the samples are mixtures along the lengths of the screens. Thus, it is simple to measure the radiocarbon content of the water, but the calculation of the water age depends on a model of groundwater flow. The calculated water ages are apparent ages.

In depressions, which are always discharge areas, even in wet periods, an upward flow condition of the groundwater takes place, so that all ages of groundwater formation are represented there. In such areas, the distribution of the apparent groundwater ages depends on the mean life of the radioisotope used for age-dating. For example, ^{14}C will determine the age of younger fraction, ^{36}Cl that of the older one.

In high areas, which were recharge areas in wet periods, a downward flow condition takes place, so that there the latest groundwater formations are represented predominantly. In such areas, the calculation of water ages does not depend so much on the mean life of the used radioisotope. Thus, if the detectable limit of the radioisotope is large enough to catch all groundwater formations of the high area, the age calculation should lead to a real age distribution of groundwater formation in the past.

In Egypt, groundwater samples are available from the New Valley depression, as well as from the high area of the Gebel Uweinat-Bir Safsaf uplift. In the New Valley, the groundwater samples only have apparent radiocarbon ages of more than 20,000 years. In the Gebel Uweinat-Bir Safsaf uplift area, however, the groundwater showed radiocarbon ages younger than 14,000 years (Schneider & Sonntag 1985), which proves a period of groundwater formation at that time. But apparent ages older than 18,000 years also occur, calculated from samples of either continuously flowing wells, or from wells located in a graben system crossing the uplift area and filled by approximately 700 m of sediment. It is remarkable that the distribution of the radiocarbon ages of groundwater in Egypt fits more or less with the results of recent Quaternary geological investigations in the same region (Wendorf et al. 1977, Pachur & Braun 1980). These studies found a wet period younger than 10,000 years and evidence of several wet climates older than 26,000 years.

The distribution of the radiocarbon-dated groundwater in Egypt does not show any gradient of age-increase in any direction. This may be due to the fact that the admixture of groundwater of different formations does not allow any age-gradient to show. The

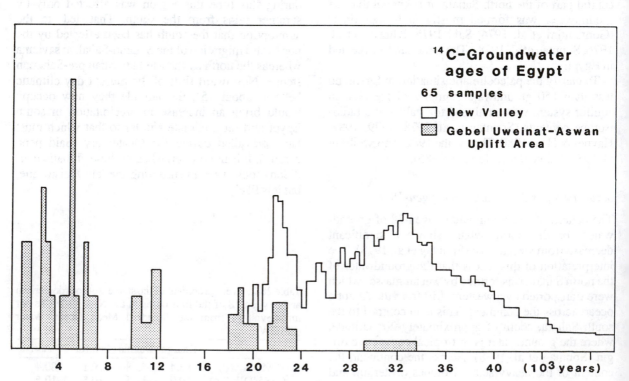

Figure 28.3 Distribution of radiocarbon-dated groundwater ages.

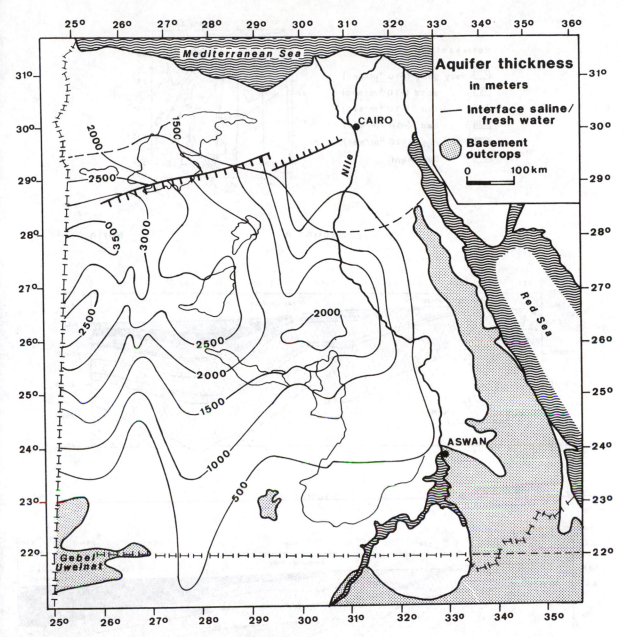

Figure 28.4 Map of thickness of the Nubian aquifer system.

time scale of the detectable limit of radiocarbon (in water approximately 40,000 years) is too short to recognize significant groundwater flow, where the flow velocity can be 3 m each year only, and consequently the flow distance not more than 100 km within the detectable limit. Further radioisotope methods, where the mean life has a suitable value, like chlorine-36, will probably show significant trends.

5 GROUNDWATER MASS CALCULATION

Using radiocarbon age-dating and the stable isotopes deuterium and oxygen-18, it is proven that the groundwater in the Nubian aquifer system in Egypt is fossil, no recent formation is detectable. This does not exclude that groundwater influx from adjacent reservoirs of fossil groundwater takes place in Egypt and recharges the groundwater. Hence, it is important to calculate the groundwater mass to get a basic value for groundwater development projects.

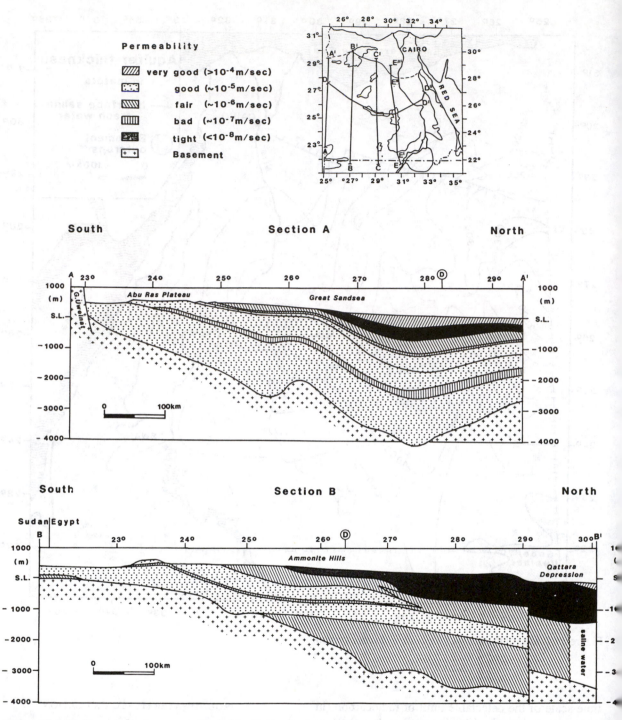

Figure 28.5 Hydrogeological cross-sections of the Dakhla basin.

The area of the Nubian aquifer system within Egypt has an extension of approximately 630,000 km². The position of the basement top is the best known parameter in this calculation, due to aero-magnetic interpretation maps (Conoco 1977) supported by fully penetrating exploration wells. The position of the top of the groundwater-bearing strata, i.e. the base of the Mut formation, or of the Dakhla formation, is less known. A limited amount of data is available from exploration wells, resistivity logs and geological field investigations, so that interpolation over large areas has to be carried out. The proven

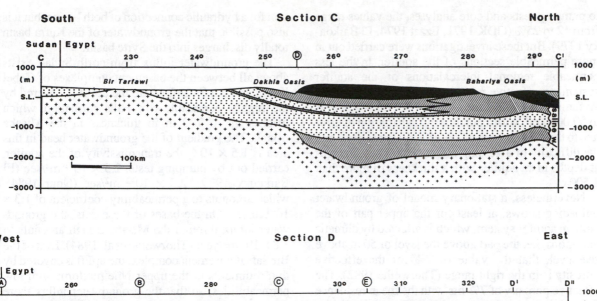

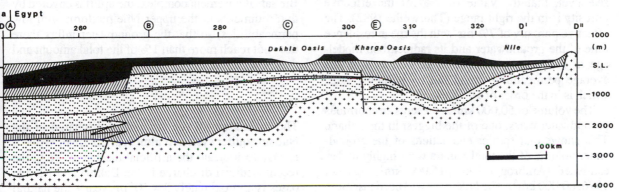

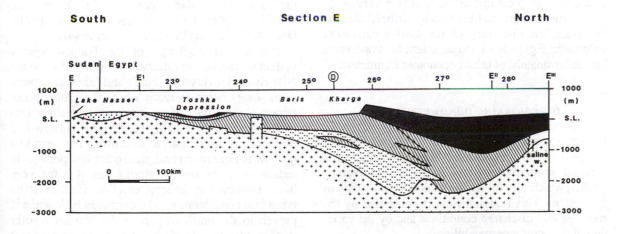

piezometric head of the groundwater in the Nubian aquifer system ranges from 250 m above sea level in the east Uweinat project area to 150 m in Farafra and Dakhla, 120 m in Bahariya and Toshka and down to 50 m in particular areas of the Kharga oasis. For the mass calculation, an average piezometric head of

150 m is assumed. The influence of this parameter does not significantly affect the error of the mass calculations. Based on these parameters, the volume of the Nubian aquifer system comes up to 780,000 km³.

The most sensitive parameter in the groundwater mass calculation is the effective porosity. According to pumping tests and core analyses, the values range from 12 to 25% (GDK 1971, Ezzat 1974, El Barkouky 1979). But these investigations were carried out in more permeable sections of the aquifer. In the less permeable regional intercalations of the aquifer system, an effective porosity is estimated to be 7%. Assuming this value, the groundwater mass amounts to 50,000 km^3, with a potentially high margin of error due to the arbitrary use of 7% effective porosity – and the difficulty of assessing effective porosity changes at depths in an aquifer with a thickness of more than 3,500 m.

Nevertheless, a stationary model of groundwater unloading shows, at least for the upper part of the Nubian aquifer system, which is affected by climatic variations, i.e. the part above the level of 50 m above sea level, that the value of 7% for the effective porosity is in the right range (Thorweihe 1982). The effective porosity of 7% fits with the mean residence time of the groundwater and its radiocarbon model-age of the major pluvial time as well as with the average permeability found out by pumping tests, which is in the order of 10^{-5} m/sec.

The volume of 50,000 km^3 represents an enormous groundwater mass, one of the biggest in the Sahara. This shows that former evaluations of the ground-water volume of the total Sahara were highly under-estimated (Ambroggi 1966: 15,000 km^3, Gischler 1976: 60,000 km^3). The Nile river will need just more than 500 years to discharge the water mass of the Nubian aquifer system in Egypt. But, of course, from economic and technical aspects, only a very small part of this amount can be mined by artificial dischar-ge. Thus, the discovery of the total groundwater volume in Egypt is of major scientific importance but, unfortunately, of minor economic significance.

6 REGIONAL GROUNDWATER FLOW

Piezometric measurements certainly prove a regional gradient of the groundwater head and therefore a regional groundwater flow exists in the Nubian aquifer, which superposes the stationary unloading of the system. This regional flow is controlled by the recharge and discharge conditions and by the varia-tion of the aquifer permeabilities.

Recharge

The position of the groundwater head is only known from the wells mentioned, so that groundwater influx from north Sudan is proven. But any groundwater influx from the Kufra basin is not provable because no piezometric data from the southwestern part of the aquifer area exist. There is the geological precondi-tion for a hydraulic connection of both basins, but it is also possible that the groundwater of the Kufra basin totally discharges into the Syrte basin.

The groundwater influx from north Sudan exists above all between the basement complexes of Gebel Kamil and Bir Safsaf, where the uplift is covered by the highly permeable Six Hills formation which reaches up to 700 m in thickness in the Misaha trough. The gradient of the groundwater head in this area is 1.5×10^{-4}, the transmissivity of the aquifer, carried out by pumping tests, 2.3×10^{-2} m^2/sec (El Barkouky 1979), 2.5×10^{-2} m^2/sec (Nour 1984), which amounts to a permeability coefficient of 1.3×10^{-4} m/sec. On the bases of these data, the ground-water influx through the Misaha trough amounts to 3.8×10^7 m^3/year (Thorweihe et al. 1984). East of the Bir Safsaf basement complex, the uplift is covered by thin sediments of the upper Nile platform with low permeabilities, so that the groundwater influx there does not reach more than 1% of the total amount and thus is negligible.

Discharge

The calculation of the amount of discharge of the Nubian aquifer system is almost impossible. While it is easy to indicate the artificial discharge – e.g. the recent artificial discharge in the Dakhla and Kharga oases is approximately 3×10^8 m^3/year – the natural discharge can be specified by groundwater flow models only. Such models were established during the last decade by Barber & Carr (1976), Ezzat et al. (1977), Amer et al. (1981) Chow & Wilson (1981), Heinl & Holländer (1984) and Nour (1984).

The natural discharge of the Nubian aquifer system takes place in different ways: overflow to the Nile river, delivery of springs and lakes in depres-sions, evaporation in areas where the groundwater table is close to the surface, discharge by leakage into roof deposits and probably the regional outflow in a northern or northeastern direction. The regional outflow is not proven and, due to the low permeable sediments of the northwestern basin, it is not very likely. However, in the long term this effect has to be taken into consideration. The discharge by leakage is proven in the northern part of the Nubian aquifer system only (JVQ 1978), but it has to be assumed that leakage takes place in all areas where the aquifer is confined. In desert areas of unconfined groundwater, the evaporation affects at least the uppermost 30 m of the groundwater-bearing strata. This effect has been shown in stable isotope analyses, where the ground-water is enriched in deuterium and oxygen-18 by evaporation. This effect is closely related to the depth of the groundwater table. Schneider & Sonntag (1985) calculated an evaporation rate of 0.2 mm/year at a depth to the groundwater table of 18 m and for an area of 6,900 km² around Bir Tarfawi an evaporation amount of 2 to 4×10^6 m³/year. In areas of artesian pressure the groundwater forms springs and lakes. Such areas exist only in the New Valley and are decreasing rapidly due to artificial groundwater use. The amount of natural discharge in these areas will not be calculable because artificial and natural discharges are not separable. Groundwater overflow to the Nile river was described by Ball (1927) in the Dakka area, which is presently covered by Lake Nasser. As mentioned above, the gradient of the groundwater table in the Toshka area certainly proves an eastern overflow, but a calculation in quantity has not been made so far.

The above survey of the recharge and discharge situation in the Nubian aquifer system shows that a regional groundwater flow is proven: but, for exact calculations, available data are lacking. Only groundwater flow models, which comprise the entire area of the Nubian aquifer system of the east Sahara – a first step was taken by Heinl & Holländer (1984) – will furnish sufficient data for calculations in quantity.

7 CONCLUSION

In the Nubian aquifer system, Egypt has an enormous groundwater mass which can be used for agricultural development and mining industry. But it is a fossil groundwater reservoir which has to be carefully developed. The groundwater influx from southern regions is one order of magnitude lower than the artificial groundwater use in the New Valley and a major discharge of the influx amount by evaporation in the Gebel Uweinat-Aswan uplift area is likely.

It will take several hundred thousand years for the groundwater influx to reach the New Valley. During that time, frequent wet periods would have formed groundwater by local infiltrations, thus the regional flow seems to be of minor importance.

The groundwater yield differs regionally and vertically: the average transmissivity in the Dakhla basin is distinctly higher than that of the upper Nile platform, which means that from the hydrogeological point of view, Kharga is not a suitable area. Vertically, the Six Hills formation and the Sabaya formation are the sections in the aquiver system with the highest permeabilities. Future projects of groundwater use should be developed in areas where these formations occur with a high thickness and where the piezometric head is close to the surface. This is to some extent the case in Dakhla: but, from the hydrogeological point of view, the most advisable area is the one around Farafra. Even in this area it is unavoidable that, in the long term, the groundwater head will be lowered so much that any project will be economically endangered.

PART 6

Paleontological notes

CHAPTER 29

Fossil flora

ANNIE LEJAL-NICOL
Université Pierre et Marie Curie, Paris, France

The importance of paleobotany in dating continental sediments is unanimously admitted. Plants are generally fossilized in place (except woods which may be transported). The study of fossil flora contributes to our understanding of biostratigraphy, paleoecology and paleoclimatology. The recent works published on Egypt prove that importance. Since 1978 systematic studies on Egyptian fossil flora have led to important results on the biostratigraphy of many of the seemingly non-fossiliferous beds and to our understanding of the evolution of the flora from the Devonian to the Quaternary (table 29, 1 & 2).

In Egypt the first fossil land plants were described in 1870 by Carruther's from the petrified forest near Cairo. Numerous works were later published: Woenig (1897), Fourtau (1898), Bonnet (1904), Engelhardt (1907), Renner (1907), Seward (1907, 1935), Couyat (1910), Hirmer (1925), Fritel (1926), Cuvillier (1928), Loubiere (1935), Kraüsel (1939), Carpentier & Farag (1948), Chandler (1954), Jongmans & van der Heide (1953, 1955), Jongmans (1955), Webber (1961), Schürmann, Burger & Dijkstra (1963), Lejal-Nicol (1981, 1987), Gregor & Hagn (1982) and Klitzsch & Lejal-Nicol (1984).

1 PALEOZOIC STRATA

The Paleozoic flora of Egypt has been studied by Jongmans & Koopmans (1940) and Jongmans & Van der Heide (1953, 1955). In Ras Gharib and Ayun Musa, on both sides of the Gulf of Suez, these authors describe Lycophyta and Pteridophylla with new species of *Lepidodendropsis*, *Cyclostigma*, *Sublepidodendron* and *Sphenopteris*. This flora presents similarities to the Mississipian flora of Spain, Donetz basin, United States, Libya, China and Peru and is of early Carboniferous age. Schurmann, Burger & Dijkstra (1963) report Permian sediments from Wadi Araba carrying one typical *Cordaites* sp. and palynoflora.

Recently (1978-1985), new Paleozoic flora has been reported from the southwestern part of Egypt

(from Gilf Kebir to Gebel Uweinat area) and from Wadi Qena and the Gulf of Suez.

Late Devonian-Tournaisian flora

In southwest Egypt (Wadi Malik area), the following species were found:
Archaeosigillaria minuta Lejal
Sublepidodendron fasciatum Jongmans
Cf. *Heleniella costulata* Lejal
Pseudolepidodendropsis klitzschii Lejal-Nicol
Lepidodendropsis sinaica Jongman, Gothan and Darrah
Lepidodendron veltheimii Sternberg

Visean flora

In Wadi Malik area, an association of Lycophyta and Pteridophylla of Tournaisian to Visean age is found:
Prelepidodendron lepidodendroides Lejal
P. rhomboidale Corsin
Rhacopteris ovata Walkom
Triphyllopteris gothani Daber
In the Wadi Malik formation, in the central part of Wadi Abdel Malik, the following Lycophyta and one Pteridophylla of Visean age are found (Lejal-Nicol 1981, 1987, Klitzsch & Lejal-Nicol 1984):
Lepidodendropsis lissoni Jongmans
L. africanum Lejal
L. fenestrata Jongmans
L. hirmeri Lutz
Lepidodendron veltheimii Sternberg
Eskdalia malikense Lejal-Nicol
Caenodendron primaevum Zalessky
Lepidosigillaria intermedia Lejal
Nothorhacopteris sp.
This association presents similarities to the flora found in the early Carboniferous from other localities in north Africa, in particular in the Djado basin in Niger.

Directly west of the Abu Ras Plateau, within the lowermost shale of the Wadi Malik formation, Pteri-

Table 29/1. Stratigraphical ranges of fossil plant taxa from Paleozoic of Egypt and north Sudan (A. Lejal-Nicol 1987, loc. cit., p. 196)

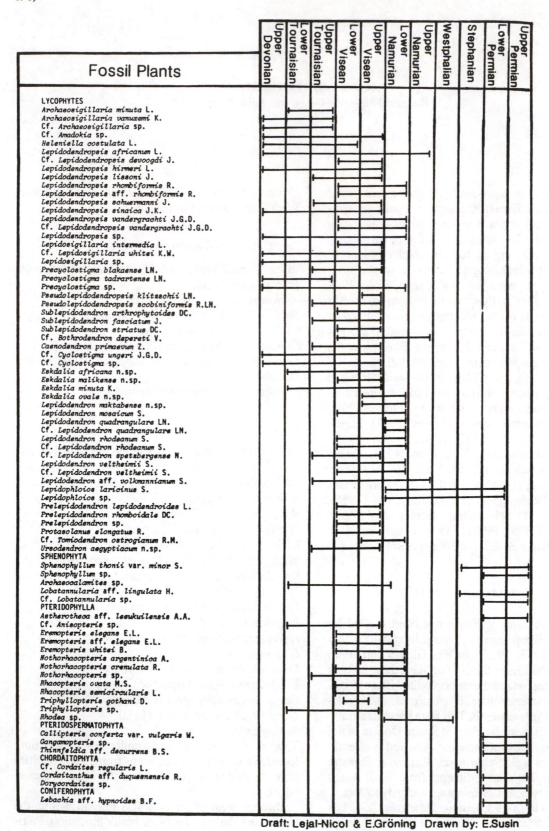

Draft: Lejal-Nicol & E.Gröning   Drawn by: E.Susin

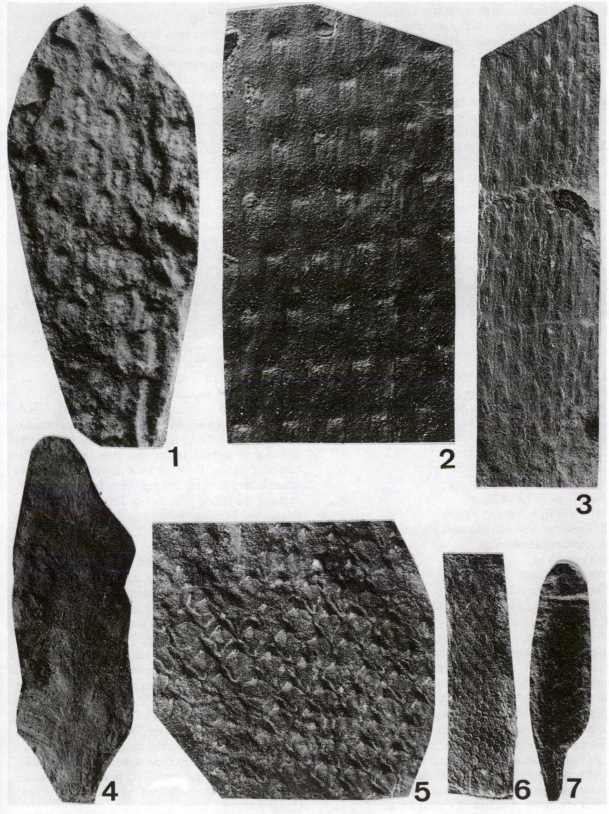

Plate 29.1 1. *Lepidodendron* aff. *volkmannianum* Sternberg. Lower Carboniferous – Gilf Kebir (×2,2); 2. *Prelepidodendron rhomboidale* Corsin. Visean – Wadi Malik (×2,7); 3.*Lepidodendron veltheimii* Sternberg.Visean – Wadi Mukattab (×1,4); 4. *Thinnfeldia* sp. Permian – Wadi Araba (×2,7); 5. *Lepidodendron mosaicum* Salter. Visean – Wadi Mukattab (×1,5); 6. *Archaeosigillaria minuta* Lejal-Nicol. Visean – Wadi Malik (×2,7); 7. *Graminae inflorescens*. Upper Cretaceous – Bahariya (×5,5).

dophylla is found and is represented mainly by *Eremopteris whitei* Berry which corresponds to a species from the lower Carboniferous rocks of the Gulf of Suez which Jongmans (1955) named *Sphenopteris whitei*.

In Gebel Uweinat, the Wadi Malik formation contains the following flora of early Carboniferous age:
Cyclostigma ungeri Jongmans
Lepidodendropsis aff. *rhombiformis* de Rouvre
Cf. *L. vandergrachti* Jongmans, Gothan and Darrah
Precyclostigma Lejal-Nicol
On the western side of the Gulf of Suez in Wadi Dakhal area, Lycophyta and Pteridophylla of early to mid-Carboniferous age, mostly of Visean age, are found:
Lepidodendropsis africanum Lejal
Sublepidodendron arthrophytioides Danzé-Corsin
Lepidodendron rhodeanum Sternberg
Protasolanus elongatus Radczenko
Eskdalia minuta Kidston
Rhacopteris ovata Walkom

Visean to Namurian flora

In Wadi Mukattab area in Sinai, Lycophyta, Sphenophyta and Pteridophylla are found:
Lepidodendron mosaicum Salter
L. maktabense Lejal-Nicol
Lepidodendropsis lissoni Jongmans
L. rhombiformis de Rouvre
Caenodendron primaevum Zalessky
Lepidophloios laricinus Sternberg
Eskdalia ovale Lejal-Nicol
Bothrodendron depereti Vaffier
Cf. *Tomiodendron ostrogianum* Meyen
Cf. *Archaeocalamites* sp.
Rhacopteris sp.
R. ovata Walkom
R. semi-circularis Lutz
Nothorhacopteris sp.
N. crenulata de Rouvre
Cf. *Triphyllopteris* sp.
This assemblage seems to indicate that the climate was warm with varying degrees of humidity. It is possible to distinguish: lowlands (water swamps) with only Lycophyta, midlands (flood plains) with Lycophyta and fragile Pteridophylla and uplands (sunny zones) with Lycophyta and Pteridophylla with resistant lamina leaves.

At Gebel Uweinat Cf. *Artisia* sp.? and Pteridophylla (*Rhodea lontzenensis* Stockmans and Williere, *Rhodea* sp.) are found; they are of Namurian age. At higher levels the presence of *Cordaites angulostriatus* Grand'Eury seems to indicate a Stephanian age.

Stephanian flora

In the Gulf of Suez near Bir Quseib, north Galala, a flora of Stephanian age is identified:
Lepidodendron posthumi Jongmans and Gothan
Sigillaria icthyolepis Presl in Sternberg sensu Weiss
Syringodendron sp.
Cf. *Tunguskadendron* sp.
Equisetites sp.
Sphenopteris aff. *souichii* Zeiller
Walchia sp.
Lebachia hypnoides Florin

Permian flora

In the Suez area, a typical early Permian flora with an association of Sphenophyta, Filicophyta, Peltaspermaceae and Cordaitophyta is found:
Callipteris conferta (Sternberg) Brongniart var. *vulgaris* Weiss
Astherotheca leeukuilensis Anderson and Anderson
Cf. *Gangamopteris* sp.
Thinnfeldia sp.
Sphenophyllum thonii var. *minor* Sterzel
Cf. *Lobatannularia* aff. *lingulata* Halle
Cordaitanthus aff. *duquesnensis* Rothwell
Lebachia aff. *hypnoides* Florin
An association of Euramerican (*Callipteris*), Gondwanan (*Thinnfeldia*) and Cathaysian (*Lobatannularia*) flora was hitherto unknown in north Africa. The presence of such an association would seem to indicate that during Permian time Egypt occupied an intermediary position among the paleofloristic provinces of the world (like Libya during the Triassic). During the Permian the climate was not as warm as during the Carboniferous and light was not as intense. Sphenophyta and Cordaitophyta are found in the lowlands, Pteridophylla in the flood plains and a few Conifers with small leaves in the highlands.

In conclusion, it can be stated that the study of the fossil flora has helped to date the continental and seemingly non-fossiliferous beds of the Paleozoic of southwest and north Egypt. Upper Devonian to Namurian paleoflora are widespread. No Westphalian paleoflora is found whereas Stephanian to Permian flora are reported from a few areas. In Egypt two main paleofloristic periods can be distinguished during the late Devonian to Carboniferous time. The first, corresponding to late Devonian/Tournaisian, Visean and early Namurian times, is dominated by Lycophyta with locally a few Sphenophyta and Pteridophylla. The second, extending from late Namurian to Permian, is characterized by Pteridophyta and Gymnosperms.

The flora of the earlier Carboniferous

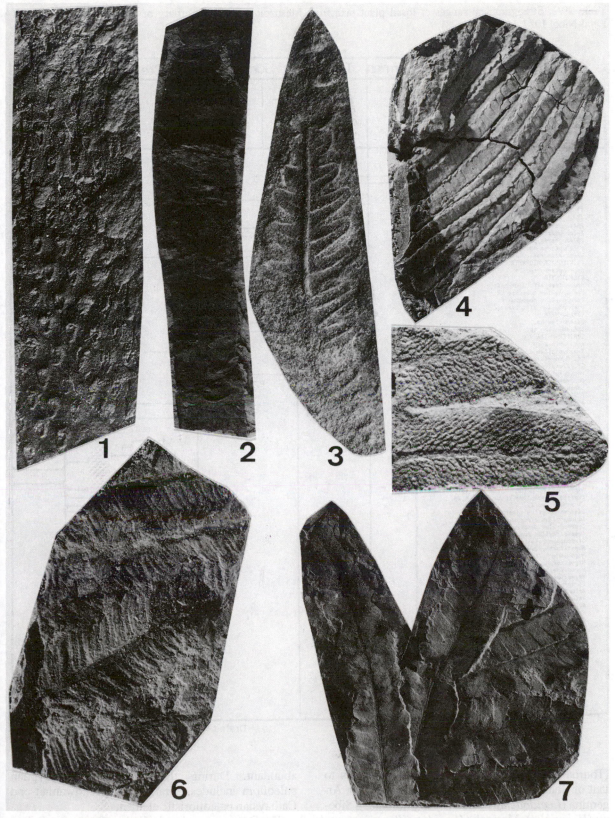

Plate 29.2 1. *Lepidosigillaria intermedia* Lejal. Tournaisian – Wadi Malik (× 2,7); 2. *Cordaites* sp. Permian – Wadi Araba (× 1,1); 3. *Cladophlebis oblonga* Halle. Jurassic – Wadi Malik (× 2,5); 4. *Weichselia reticulata* Stockes & Webb. Cenomanian – Bahariya (× 0,7); 5. *Weichselia reticulata* Stockes & Webb. Cenomanian – Bahariya (× 10); 6. *Astherotheca* aff. *leeukuilensis* Anderson & Anderson. Permian – Wadi Araba (× 2,6); 7. *Callipteris conferta* Brongniart. Permian – Wadi Araba (× 1,5).

Table 29/2. Stratigraphical ranges of fossil plant taxa from Mesozoic and early Tertiary of Egypt and north Sudan (A. Lejal-Nicol 1987, loc. cit., p. 198)

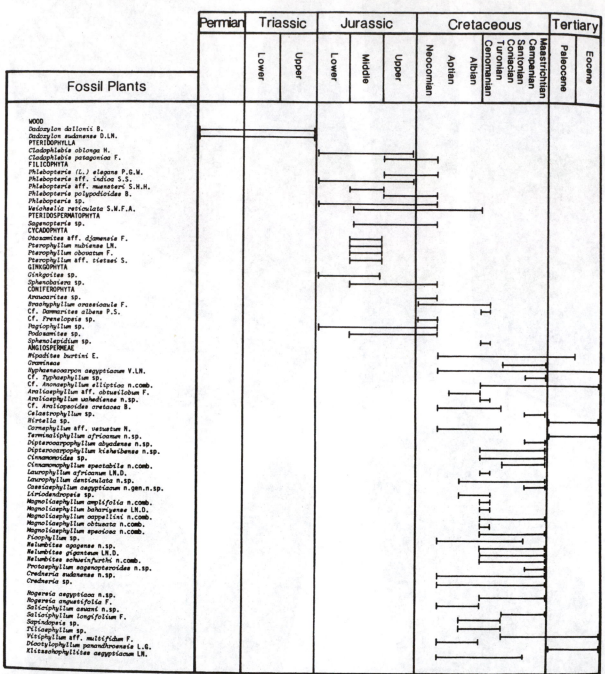

Draft: Lejal-Nicol & E.Gröning   Drawn by: E.Susin

(Tournaisian-Visean) display strong similarities to that of South America, Chile (*Lepidosigillaria*), Argentina (*Lepidodendropsis cyclostigmatoides*), Siberia (*Heleniella*), Mongolia (*Lepidosigillaria intermedia*) and Himalaya (*Archaeosigillaria minuta*). From the Namurian to the Stephanian the paleoflora is not

abundant. During the Permian the Egyptian paleoflora includes Euramerican, Gondwanian and Cathaysian paleofloristic elements.

The Devonian and early Carboniferous fossil flora indicate a generally warm climate. Pteridophylla were sphenopsid forms with lamina leaves more or less incised, having thin cuticula and numerous nervs, indicating rather wet climatic conditions. Among Lycophyta, those with long leaves and parichnos were living on the river sides or around lakes; the others with spiny leaves without parichnos were growing on drier sunny slopes. The few Pteridophylla were growing under the shade of the high lycophitic plants.

2 MESOZOIC STRATA

Permo-Triassic-early Jurassic flora

In Egypt the only area from which Triassic continental flora is reported is El Galala El Bahariya on the western border of the Gulf of Suez (Carpentier & Farag 1948). It includes:

Equisetites laevis Halle
E. scanicus Sternberg
Phlebopteris aff. *muensteri* Schenk
Todites williamsoni Brongniart
Cladophlebis sp.
Zamites gigas Lindley and Hutton
Z. schmiedellii Sternberg

Late Jurassic-early Cretaceous flora

Fossil flora separated from the type locality of the Gilf Kebir formation at the Akaba pass (Gilf Kebir) indicate the presence of late Jurassic to early Cretaceous strata. Pteridophylla and Gymnosperms are found with:

Cladophlebis oblonga Halle
Phlebopteris elegans Presl
P. aff. *indica* Sahni and Sitholey
P. polypodioides Brongniart
Podozamites Braun
Pterophyllum obovatum Fakhr
Weichselia reticulata Stockes and Webb
Ginkgoites Seward
Dadoxylon (Araucarioxylon) dallonii Boureau
Cycadophyta fructifications
In the south of Gilf Kebir and the northeastern foothills of Gebel Uweinat, a similar flora is present with new elements such as:
Cladophlebis aff. *patagonica* Frenguelli
Pagiophyllum Heer
Phlebopteris aff. *muensteri* Schenk
Xylopteris Frenguelli
Podozamites sp.
This association of Filicophyta, Pteridophylla, Cycadophyta and Gymnosperms lacking Angiosperms is of Jurassic age. The presence of *Weichselia* does not exclude this interpretation since Boureau in 1958 found *Weichselia* with typical Jurassic flora in north

equatorial Africa (Algeria and Libya).

The climate was rather warm with short rainy seasons. Pteridophylla are found in floodplains and Conifers in highlands.

Middle to late Cretaceous flora

From the Aswan area, Fritel (1926) describes a Cretaceous flora with ferns (*Weichselia*) and Angiosperms dicotyledonae (*Laurus, Cinnamomum, Magnolia, Nelumbium*) and monocotyledonae (*Zosterites, Sabalites*).

In Wadi Zeraib, south of Quseir, Seward (1935) reports dicotyledonae leaves (*Dipterocarpophyllum, Dicotylophyllum* and *Nelumbium*). This assemblage, which he dates Cretaceous to Tertiary, contains neither *Magnoliaeophyllum* nor *Laurophyllum*.

Studying mainly fossil woods, Kraüsel (1939) identifies numerous dicotyledonous and monocotyledonous angiosperms, ferns, and gymnosperms of middle to late Cretaceous age. He reports *Weichselia* and Nympheaceae from Wadi Araba and Bahariya; *Dadoxylon* from Gebel Hefhuf; Dipterocarpaceae and Nympheaceae from between Esna and Wadi Halfa; Gymnosperms, Proteaceae, Sterculiaceae and Ebenaceae from between Kharga and Dakhla.

New results

Aptian-Albian flora
The following flora is found in the type locality of the Abu Ballas formation (Lejal-Nicol, 1981) considered of Aptian age by Boettcher (1982) and Schrank (1982):
Klitzschophyllites aegyptiacum Lejal-Nicol
Leguminocarpon abuballense Lejal-Nicol
Sphenolepis Texeira
Weichselia reticulata Stockes and Webb
In Wadi Shait, north of Aswan, *Ficophyllum* sp., *Araliaephyllum* sp. and *Araliaephyllum* aff. *obtusilobum* Fontaine are found.
Fruit of Palmaceae (*Hyphaeneocarpon aegyptiacum* Vaudois-Miéja and Lejal-Nicol) is reported from the Aswan area (Vaudois & Lejal 1987).
The type area of the Maghrabi formation is the southeastern foreland of the Abu Tartur plateau, west of Kharga. In the basal part of the formation there are angiosperms:
Cf. *Araliopsoides cretacea* Berry
Cornophyllum aff. *vetustum* Newberry
Magnoliaephyllum obtusata Heer
Cf. *Rogersia angustifolia* Fontaine

Cenomanian-Santonian flora
From the Bahariya oasis Dominik (1985) reports

Plate 29.3 1. *Credneria* sp. Cretaceous – Wadi Halfa (× 1,4); 2. *Laurophyllum africanum* Lejal-Nicol & Dominik. Cenomanian – Bahariya (× 2,8); 3. *Saliciphyllum aswani* Lejal-Nicol. Cenomanian – Bahariya (× 1,2); 4. *Hirtella* sp. Eocene – Gulf of Suez (× 2,5); 5. *Pagiophyllum peregrinum* Lindley & Hutton. Jurassic – Wadi Malik (× 3,4); 6. *Hyphaeneocarpon aegyptiacum* Vaudois-Mieja & Lejal-Nicol. Upper Cretaceous – West Nasser Lake (× 2,4).

numerous *Weichselia reticulata* Stokes and Webb associated with angiosperms of Cenomanian age (Lejal-Nicol & Dominik 1987):

Nelumbites giganteum Lejal-Nicol and Dominik
Magnoliaephyllum bahariyense Lejal-Nicol and Dominik
Cornophyllum aff. *vetustum* Newberry
C. distense Lejal-Nicol and Dominik
Liriophyllum farafraense Lejal-Nicol and Dominik
Laurophyllum africanum Lejal-Nicol and Dominik
Rogersia longifolia Fontaine
Cf. *R. angustifolia* Fontaine
Typhaephyllum sp.
Vitiphyllum aff. *multifidum* Fontaine

On the west bank of the Nile, the beds overlying the Abu Aggag formation include plant remains such as angiosperm leaves with Cf. *Rogersia angustifolia* Fontaine and numerous fruits similar to those of Aceraceae. This association of fruits and leaves (work in progress) confirms the middle Cretaceous age given to these beds by Bartoux & Fritel (1925).

The basal part of the Kiseiba formation, northeast of Wadi Halfa, includes a rich flora of Cenomanian to Maastrichtian age:

Dipterocarpophyllum kisheibiense Lejal-Nicol
Rogersia aegyptiaca Lejal-Nicol
Magnoliaephyllum obtusata Heer
Nelumbites schweinfurthi Fritel
Typhaephyllum sp.

Campanian to Maastrichtian flora

In the Kiseiba formation a flora of Campanian to Maastrichtian age is found:

Cassiaephyllum aegyptiacum Lejal-Nicol
Credneria sp.
Ficophyllum sp.
Nelumbites giganteum Lejal-Nicol and Dominik
Tiliaephyllum sp.
Typhaephyllum sp.

In conclusion one can state that from the beginning of the Mesozoic to the end of the Jurassic, Pteridophylla and Cycadophyta were numerous. During the Cretaceous the first angiosperms appeared and were represented in Egypt by forms similar to the Virginia flora (Aptian-Albian). From the late Jurassic to the Cenomanian numerous Weichseliaceae flourished. In the late Cretaceous monocotyledonous and dicotyledonous angiosperms were prevalent (leaves, flowers, fruits and woods). The floristic association during the Jurassic and Cretaceous in Egypt indicates that the climate was generally warm. In the lowlands (deltaic zone, swamp areas and lakes) there were Palmaceae (*Hyphaeneocarpon*), Nympheaceae (*Nelumbites*) and Typhaceae (*Typhaephyllum*); in the flood plains or river sides there were Magnoliaceae (*Magnoliaephyllum*), Cornaceae (*Cornophyllum*), Proteaceae (*Rogersia*), Platanaceae (*Credneria*), Moraceae (*Ficophyllum*) and Dipterocarpaceae (*Dipterocarpophyllum*); and in the highlands there were Lauraceae (*Laurophyllum*) and Araliaceae (*Araliaephyllum*). Up to the Cenomanian there were Weichseliaceae in the arid areas. The climate was warm with dry and rainy seasons. In late Cretaceous time few areas had a more temperate climate.

3 TERTIARY STRATA

Chandler (1954) describes the first fossil fruits from the early Tertiary (Paleocene) sediments: *Paleowetherellia schweinfurthi* Heer from Farafra and *Nipa*, *Anonaspermum* and *Icacinicarya* from Quseir.

Gregor & Hagn (1982) identify the following fruits from the Cretaceous-Paleocene boundary (Danian of Abu Munqar, Farafra-Dakhla road):

Cf. *Coryphoicarpus globoides* Koch
Nipa burtini (Brongniart) Ettinghausen
Cupulopsis klitzschii Gregor
Erythropalum sp.
Stizocaryopsis bartheli Gregor
Munqaria kraüeseli Gregor

Most of the Tertiary fossil flora of Egypt is in the form of woods. Kraüsel (1939) describes a diversified angiosperm woods from the Eocene, Oligocene, Miocene and Pliocene strata. The Eocene flora (Sterculiaceae, Rubiaceae, Moraceae, etc.) was not very different from that of the Cretaceous. The early Oligocene was characterized by gymnosperms, Leguminosae, Sterculiaceae, Palmaceae, Sapindaceae and Ebeneceae. New species of angiosperms (*Palmoxylon, Ficoxylon, Leguminoxylon, Guttiferoxylon*, etc.) appeared during the late Oligocene to Miocene. In the middle Pliocene several species disappeared and new forms marked the advent of the Quaternary.

Lejal-Nicol records the following Eocene flora from the Gulf of Suez area:

Ficophyllum sp.
Dicotylophyllum panandhroensis Lackanpal and Guleria
Cf. *Sparganiophyllum* sp.
Cf. *Anonaephyllum* sp.
Terminaliphyllum africanum Lejal-Nicol
Cf. *Hirtella* sp.
Cf. *Platanophyllum* sp.
Cf. *Tiliaephyllum* sp.
Cassiaephyllum aegyptiacum Lejal-Nicol

This flora is an association of Cretaceous forms with younger ones.

In conclusion, the Egyptian Tertiary paleoflora appears to have been diversified. Louvet (1973) studies the Tertiary fossil woods from Libya and compa-

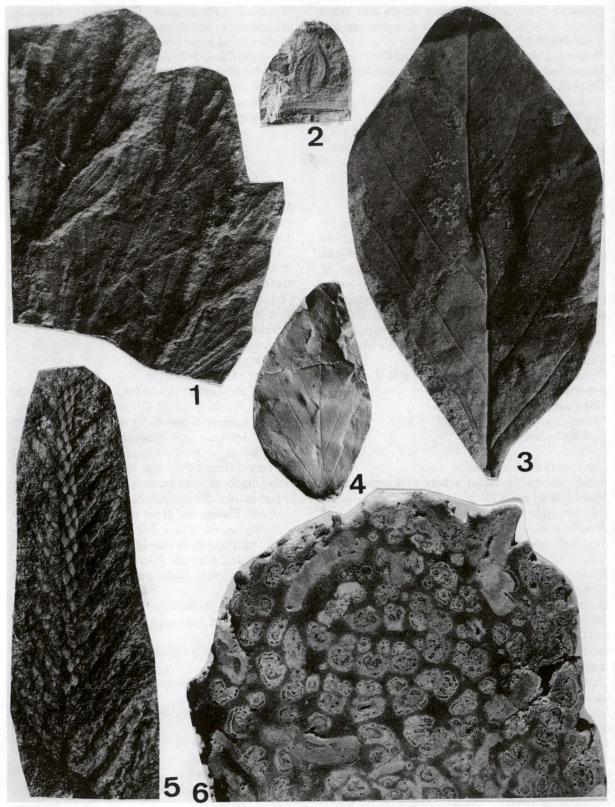

Plate 29.4 1. *Triphyllopteris gothani* Daber. Visean – Wadi Malik (× 3,5); 2. *Paulliniacarpon* sp. Upper Cretaceous – Wadi Abu Agag (× 1); 3. *Magnoliaephyllum cappelini* (Heer) (Höllick) Lejal-Nicol. Cenomanian – Kharga (× 2,5); 4. *Magnoliae-phyllum* aff. *obtusata* Heer. Upper Cretaceous – Wadi Abu Agag (× 1); 5. *Brachyphyllum crassicaulis* Fontaine. Upper Cretaceous – Wadi Dakhal (× 1,5); 6. *Weichselia reticulata* Stockes & Webb. (*Paradoxopteris stromeri*) Cenomanian – Bahariya (× 3).

res them to those of surrounding areas including Egypt. He shows that during Eo-Oligocene time the zone of tropical rain forest moved from south to north forming at times marshy lands which bounded the stretch of Mediterranean littoral between Tunisia and Egypt. At the end of the Miocene and during the Pliocene a change of climate led to the disappearance of the tropical flora and the beginning of desert conditions in the Sahara. During the Quaternary, after the big glaciations of the northern hemisphere, a Mediterranean flora appeared in the mountainous areas of the Sahara relics of which remained today. The desert as we know it today was established.

ACKNOWLEDGEMENT

Grateful acknowledgement is made to Professor E. Klitzsch and his colleagues for supplying the material upon which this study is based and to Mr Kleeberg for the photos which accompany this work.

CHAPTER 30

Vertebrate paleontology of Fayum:
History of research, faunal review and future prospects

ELWYN L. SIMONS & D. TAB RASMUSSEN

Duke University, Durham, North Carolina, USA, and University of California, Los Angeles, California, USA

HISTORY OF RESEARCH

The celebrated fossil vertebrate fauna of the Fayum depression in Fayum Province, Egypt, stands as the earliest well-known land mammal fauna of Africa. Scattered finds of contemporaneous or earlier African mammals have been made in Libya (Savage 1971, Wight 1980), Algeria (Sudre 1979, Coiffait et al. 1984, Mahboubi et al. 1986) and Angola (Pickford 1986) but fossils at these sites are scanty or remain largely undescribed. Many of Africa's Eocene fossil-bearing deposits are marine in origin, such as the Qasr El Sagha formation of the Fayum. Vertebrate fossils in these formations are few in number and only a very limited variety of terrestrial mammals are preserved. In contrast, the extensive badlands developed in the Oligocene continental sediments that overlie the Qasr El Sagha formation are rich with an abundance and diversity of fossil vertebrate remains. These provide us with an unrivalled view of the evolution of Africa's early Tertiary plants and animals.

The Fayum depression is believed to have been excavated by wind erosion in the Pliocene and early Pleistocene. At some point, perhaps only a few thousand or tens of thousands of years ago, overbank flooding from the Nile filled in a lake, Birket Qarun, in the center of this depression that gives the region its name (the Greek word *phiom* means 'lake'). From the north shore of the lake a series of flat-lying benches with intervening escarpments rise from below sea level to an elevation of over 300 m. Today, as for millenia in the past, these benches and escarpments are subjected to the erosive action of frequent windstorms and occasional flash flooding. Most of the present day exposures are either cliffs or relatively level surfaces covered with scattered gravel called desert pavement, or *serir*. It is in Oligocene river bar deposits on these serir-covered benches that fossil vertebrates occur.

The lowest of the prominent escarpments on the northern side of the Fayum depression is a steep cliff formed by the upper part of the Eocene Qasr El Sagha formation. Earlier Eocene sediments below the cliffs are partly obscured by Plio-Pleistocene and Recent deposits from Birket Qarun. Above this lowest escarpment lies the Oligocene Jebel Qatrani formation, which begins as a series of relatively flat lands broken by rounded hills and littered with petrified logs and the fossilized remains of vertebrates. Many of these sediments are point bar deposits of one or several major rivers which in early Oligocene times were flowing from highlands to the east, usually in a westerly direction toward a large southerly embayment of the ancestral Mediterranean, or Tethys Sea (Bown & Kraus 1988). These bar deposits consist of largely unconsolidated gravels and sands of varying degrees of coarseness. The Jebel Qatrani formation also includes variegated green and red overbank clay deposits and occasional cliff-forming channel deposits. At frequent intervals there are limestones bearing fossils of freshwater ostracods that confirm the presence of ancient freshwater lakes (Bown et al. 1982). At about mid-section of the Jebel Qatrani formation is a widespread sandstone termed the barite sandstone (Bown & Kraus 1988) that forms a second low escarpment separating the 'Lower Fossil Wood Zone' of earlier authors from the 'Upper Fossil Wood Zone'. Recently, the subdivisions of the Jebel Qatrani formation have been revised by Bown & Kraus (1988) and their terms 'upper sequence' and 'lower sequence' will be used in place of the old terms 'Upper and Lower Fossil Wood Zones'. Above the barite sandstone lies a second areally smaller bench bearing the principal quarries yielding small Fayum vertebrates. Above this bench rises a third and final escarpment, the Jebel Qatrani, meaning the 'tar hills'. This escarpment is capped by a series of four basalt flows, the oldest of which has recently been dated at 31 ± 1.0 million years (Fleagle et al. 1986). The entire thickness of the Jebel Qatrani formation from its lower contact with the Qasr El Sagha formation to the base of the basalts is 340 m, an interval that probably represents several million years of deposition.

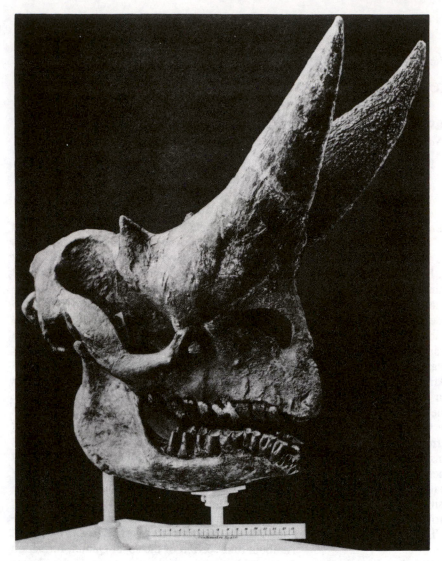

Figure 30.1 Skull of an adult male *Arsinoitherium* found in Egyptian Oligocene deposits. This skull is more than a meter long from the occiput to the tip of the horn cores. In life, the horns that covered these cores were probably much longer than the cores themselves. The largest individuals of *Arsinoitherium* were larger than a present-day white rhinoceros, which is the third largest living land mammal.

Fossil vertebrates were first located in the Fayum by the German geologist and explorer Georg Schweinfurth who began a detailed geologic study of the Eastern Desert of Egypt in 1877. In 1879, while conducting geological reconnaissance he crossed to an island near the center of Birket Qarun called Gezirit El Qorn. From the Eocene sediments there he recovered shark teeth and cetacean bones that were given to the paleontologist W. Dames. Later, with additional specimens found by Schweinfurth in the Qasr El Sagha escarpment, Dames (1894) diagnosed the whale as a new species, *Zeuglodon osiris* (now placed in the genus *Dorudon*).

In 1898, the Egyptian Geological Survey began new geological exploration of the Fayum depression under the direction of a British geologist, H.J.L. Beadnell. Mapping and section measurement that year took the Beadnell party in a traverse from east to west across the north side of Birket Qarun and at that time some fossil vertebrate bones were found in the Eocene marine deposits there. In April 1901, Beadnell's team was joined by C.W. Andrews of the British Museum of Natural History who had come to collect modern Egyptian mammals of the region. The Andrews-Beadnell group, while ascending the Qasr El Sagha escarpment discovered a late Eocene fauna of marine vertebrates in shoreline or estuarine deposits. The group began a prolonged collecting campaign that yielded remains of the large proboscideans *Barytherium* and *Moeritherium*, the serenian *Eosiren*, as well as giant sea snakes.

About this time, a camelman brought fragments of primitive elephant teeth into the camp from a site he had found to the north in the Jebel Qatrani formation's continental deposits. His finds were the first discovery of the land mammal fauna of the Egyptian Oligocene. Andrews (1901) reported that more of the elephant tooth row was recovered at the camelman's site. These teeth became the type specimen of the oldest known elephant, *Palaeomastodon*. Discovery of the continental Oligocene vertebrates brought about a burst of scientific activity by Andrews and Beadnell who then collected in the continental beds of the 'fossil wood zone' during the winters of 1901-1904, and published a series of descriptive articles on these new mammals. In short preliminary notes the two authors described the proboscideans already mentioned, new genera of giant hyracoids, and the huge, exotic four-horned ungulate, *Arsinoitherium*, for which Andrews (1906) established a new mammalian order (Fig. 30/1). Each of these two authors summarized their work in monographs (Beadnell 1905, Andrews 1906).

After Andrews and Beadnell completed their work in the region, the Fayum Oligocene localities became available to other collectors. A German private collector named Richard Markgraf established himself in the Fayum and began collecting the larger vertebrates for sale to various museums, principally in Germany. When it became clear with the publication of Andrew's work that the order Proboscidea had originated in Africa, the prominent American student of elephant evolution, Henry Fairfield Osborn, organized an American Museum expedition to the Fayum with the sponsorship of the Egyptian Geological Survey. When the American Museum party reached quarry A in early February, Osborn was accompanied by a large group of relatives and guests including his nephew Fairfield Osborn, much later noted as a conservationist. Fairfield Osborn lived to be the last survivor of these early expeditions into the Fayum region. After a short time, H.F. Osborn left the field camp putting Walter Granger in charge of the collecting program with George Olsen as assistant. Markgraf, who had been collecting independently in the region for E. Fraas of Stuttgart, was hired by Osborn to join the two Americans as a collector and to turn over the collections he had already made to the American Museum.

The American Museum party, with Markgraf, remained in the field until June 1906. Andrews' and Beadnell's discoveries had included large numbers of giant mammal skulls and jaws belong to *Arsinoitherium* and *Palaeomastodon*. The remaining mammals that they had found were mostly large or moderately large creodonts, hyracoids and anthracotheres. These early scientists apparently were not trained to find the remains of small mammals, or perhaps only looked for large fossils. Finding small fossils was to be the accomplishment of the Ameri-

can Museum expedition but only because of the superior collecting abilities of Richard Markgraf. This scientist is one of the mystery men of paleontology. Early in 1906 he had discovered a 'pocket' in what has come to be known as the upper sequence, north of where the other members of the expedition were collecting, that contained some species different from those found in the lower half of the formation. In the collections he made in this upper sequence 'pocket', small mammal remains appeared for the first time. For instance, in the 1906 season he found the jaw of a primate that was later described by Osborn (1908) as the type of *Apidium phiomense*. This was the first Oligocene anthropoidean primate ever discovered.

The American Museum party returned to New York with over 500 specimens that included a few small mammals such as the primate mentioned already, the type of a new rodent, *Phiomys andrewsi*, and also a strange, 'giant' insectivoran, *Ptolomaia* (Osborn 1908). The larger hyracoids and proboscideans in the American Museum collection were studied much later by Matsumoto (1922, 1923, 1924, 1926). Meanwhile Markgraf continued to collect on his own from 1906 to 1911 and materials he sold to the Stuttgart Natural Cabinet were included in a monograph on the Fayum mammals published by Schlosser (1911), and one on the anthracotheres by Schmidt (1913). Schlosser (1910, 1911) described three additional small primates that had been found by Markgraf. These were the types of *Parapithecus fraasi*, a monkey-like form resembling Osborn's *Apidium*, and two primitive hominoids whose jaws resembled those of gibbons, *Propliopithecus haeckeli* and *Moeripithecus markgrafi* (now also placed in the genus *Propliopithecus*). Other interesting small fossils in Markgraf's collection included the humerus of a bat, and the oldest known specimen of an elephant shrew, *Metoldobotes*. Markgraf's collecting activities in the Fayum came to an end shortly after the beginning of World War I. He died in 1916 at his home in the village of Sinnuris near the southeastern end of Birket Qarun. (In 1961, Simons was shown 'the house of the foreigner' in this village.)

Disinterest, two wars, and changing political conditions left the Fayum uninvestigated for over 30 years, and it was only in 1947 that paleontologists again collected in the Fayum bone beds, this time by members of the 'Pan African Expedition' directed by Wendell Phillips from the University of California at Berkeley. This expedition made a small collection of the various mammals but without finding any primates or other small vertebrates.

Another period of inactivity ensued until 1961 when Elwyn Simons organized and led the first of a new series of expeditions. By this time everyone who had been connected with collecting in the Fayum before the First World War was deceased except for the nephew of Osborn, and so there was very little transfer of knowledge as to localities and collecting techniques. Another major factor affecting fossil hunting had changed considerably. The bones of large fossil vertebrates in the Fayum are typically whitish and are often associated with accumulations of chalky white coprolites. With the use of field glasses, such fossil clusters can sometimes be seen from more than a kilometer away. This high detectability means that all the large vertebrate fossils exposed by weathering were cleaned out by the early workers. For this reason, few specimens of the giant Fayum mammals have been found since 1961. It is estimated that wind erosion takes off less than 3 cm per century from the serir surfaces in the Fayum, and therefore no new large fossils have yet been exposed. (The basis for estimating this erosion rate is that a 4000-year-old 12th Dynasty road made of fossil wood that traverses the Jebel Qatrani formation now stands nearly 1 m above the surrounding surface.)

The principal purpose of the 17 expeditions in the Fayum conducted between 1961 and 1986 has been to collect small mammals such as primates, rodents, smaller creodonts and hyracoids. In the course of this work, several bats, tiny insectivores, a marsupial, three prosimian primates, and a variety of small birds have also been found. The success at finding these small vertebrates has partly resulted from the development of a new method for harvesting fossils that utilizes the power of the frequent Sahara windstorms. Large areas of serir-covered surface are swept with brooms to remove the over-burden of resistant materials that have built up over the centuries. In the succeeding year, winds may remove 15 cm or more of the loosely consolidated sands from the swept areas, leaving fossils behind to be collected from the surface. This method is especially important in the Fayum because the fossils there are very poorly mineralized and cannot withstand exposure on the surface for periods of many years, and so small specimens are rarely found outside the freshly swept quarry areas. Other methods used for fossil collecting in the Fayum include screening of fine sands, and active quarrying with hammers, chisels and brushes of more resistant sandstones and mudstones.

The collections made on these expeditions number tens of thousands of specimens. They have been divided between the Cairo Geological Museum, which also houses all type specimens, and the two collaborating American institutions – Yale Peabody Museum from 1961 to 1967, and the Duke University Primate Center from 1977 to the present. The new discoveries made by these expeditions of the last 25 years have been described in a large number of

publications covering many aspects of the Fayum fauna and flora. Many of the publications are concerned with the small mammals because it is among these that the greatest number and diversity of new species have been discovered, including ten new primates. Important advances have also been made in knowledge of the non-mammalian fauna, such as ichnofossils (trace fossils) that result from the burrowing and nesting behavior of various organisms (Bown 1982) and the diverse bird fauna (Rasmussen et al. 1987). Plant fossils and rhizoliths have also been investigated (Wing & Tiffney 1982, Bown 1982). Knowledge of these important components of the plant and animal life, along with detailed geological studies, have proven conclusively that the Fayum environment during the Oligocene was wet and tropical, with swamps and marshland bordered by forests and cut by many meandering rivers (Bown et al. 1982, Olson & Rasmussen 1986, Bown & Kraus 1988). This increase in knowledge of the Fayum fauna and environment shows no signs of slowing down. Figure 30/2 shows the cumulative annual rate of discovery of new mammal species, and indicates that new species are still being discovered steadily even after the many years of expeditions. A complete annotated bibliography of Egyptian vertebrate fossils covering all work up through 1980 was published by El Kashab et al. (1983). The most recent faunal lists that have been published are ones by Simons (1968) and Bown et al. (1982).

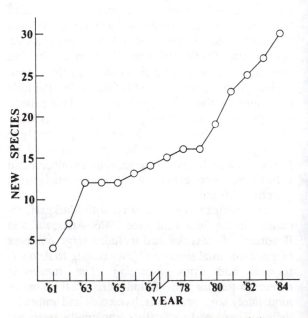

Figure 30.2 Chart showing the cumulative rate of discovery of new species of mammals from the continental Oligocene deposits of the Fayum. This indicates that the rate of discovery of new forms has climbed rapidly in recent years.

FAUNAL REVIEW

The result of this extensive work on the diverse Fayum fossils is that the Fayum stands as the best known Paleogene site in Africa. It provides by far the most complete view of the endemic African fauna before the significant faunal interchanges of the Miocene resulted in a major influx of other groups such as the many families of rodents, artiodactyls, perissodactyls, and fissipeds that are typical of modern Africa. Like the mammals, the Fayum reptiles also include many taxa unknown in Africa today. In contrast, the fish and birds of the Fayum do resemble, in general, those of modern tropical Africa.

Species of reptiles present in the Fayum assemblage (Andrews 1906, El Kashab 1977) but absent from modern Africa include the gigantic extinct aquatic snakes, *Pterosphenus* and *Gigantophis*, most common in the Eocene Qasr El Sagha formation. The large false gavial, *Tomistoma*, is today restricted to Borneo in southeast Asia but during the Oligocene occurred sympatrically in the Fayum with both long- and short-snouted species of *Crocodilus*. One group of reptiles present in both the Fayum Oligocene and modern Africa are boid snakes, a primitive group of constrictor. The most common and diverse of the reptiles in the Fayum are chelonians (turtles) represented by at least four genera (*Testudo*, *Podocnemis*, *Stereogenys* and *Pelomedusa*), including both river turtles and tropical land-tortoises. The fossilized remains of turtle shell and crocodilian scutes are scattered abundantly in the Fayum's riverine sand bar deposits. Fish are represented by a great abundance of siluroid catfish and lungfish similar to forms that inhabit rivers and swamps of modern sub-Saharan Africa (Andrews 1906, Bown et al. 1982). At certain quarries in the lower sequence there are numerous remains of sharks and rays (*Carcharinus*, *Myliobatis*, *Aetobatis* and others). The reptile and fish faunas remain the most poorly studied components of the Fayum fossil assemblage, and future work is sure to yield valuable information about these two important vertebrate groups.

The bird fauna of the Fayum is very diverse and represents the best known Paleogene record of the class in Africa. There is only a single described bird specimen from Africa that predates the Fayum assemblage – the sternum of a large pelecaniform bird from Eocene marine sediments in Nigeria (Andrews 1916). In contrast, the Fayum has yielded fossil remains of 13 bird families. Two of the Fayum bird families are now extinct, while the other 11 are still represented by living species. One of the extinct birds is a large ratite known only by fragmentary fossils recovered during the early part of the century (Andrews 1906, Lambrecht 1933) that were believed to be similar to the extinct elephant birds of Madagascar, although this conclusion must be considered tentative given the known fossil material. Two generic names have been given to the ratite specimens (*Stromeria* and *Eremopezus*) but it is possible that the fossils actually represent only a single species, *Eremopezus eocaenus* (Rasmussen et al. 1987). The other Fayum bird belonging to an extinct family is a newly described species that has been allocated to a family all its own (Rasmussen et al. 1987). This bird is a large, robust form with a distinctly curved bill (Fig. 30/3) that is probably most closely related to herons (Ardeidae).

The remaining 11 bird families are all represented by modern species that allow, by analogy, precise inferences regarding the paleoenvironment of the Fayum. The most abundant and diverse of these Fayum families are the herons (Ardeidae) and jacanas (Jacanidae). Jacanas, or 'lily-trotters', are long-toed water birds that walk on lily pads and other floating vegetation in fresh water swamps. They are represented in the Fayum by three species in two genera, including two giant species that are more than 30% larger in linear dimensions than the largest living jacanas. The Fayum herons include night herons that are indistinguishable in known elements from the modern genus *Nycticorax*, and also typical herons or egrets similar to modern *Ardea* and *Egretta*. Herons forage along water edges and in swamps and marshes for fish, amphibians and other small vertebrate prey.

Other water birds identified in the Fayum avifauna are rails (Rallidae), cranes (Gruidae), flamingos (Phoenicopteridae), storks (Ciconiidae), cormorants

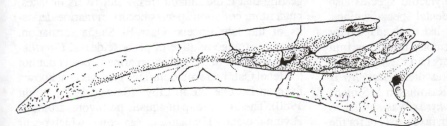

Figure 30.3 Enlarged drawing of the rostrum of a large heron-like bird known only from the Fayum. The recovery of delicate bird bones from the Fayum's fluvial sediments has increased dramatically in recent years. Length of this specimen from tip to base is 8.5 cm.

(Phalacrocoracidae) and the earliest fossil record of the gigantic aberrant family, Balaenicipitidae, the shoebilled storks Shoebills are shy, rare birds of dense African swamp that locomote on floating or submerged vegetation while hunting for catfish and lungfish like those that are so abundant in the Fayum quarries, which are then snapped up in the enormous bill.

Two birds of prey, or raptors, have also been recovered from the Fayum, including the world's earliest record of ospreys (Pandionidae). Today, one species of osprey still survives, and it is strictly a fish-eater, diving into the water for its prey from great heights. The second raptor is a large eagle-like bird (Accipitridae) possibly related to the fish eagles of the genus *Haliaeetus* (including the familiar bald eagle of North America). The only true arboreal birds known from the Fayum are two specimens representing turacos (Musophagidae), a group today restricted to tropical Africa. These bright, frugivorus canopy birds shared the trees with the Fayum primates, and may have competed with them for fruit. All 11 families of Fayum birds that are still alive today can be found together only in a limited area of Uganda bordering Lake Victoria and the upper Nile river (Olson & Rasmussen 1986). As indicated by this assemblage, the climate of the Fayum during the Oligocene was warm, wet and tropical. The environment consisted of dense vegetation-choked freshwater swamps and areas of open water bordered by forest.

In contrast to the birds, the Fayum mammal assemblage was unlike any other known mammalian fauna, living or extinct. The mammal community consisted of a mixture of endemic African groups that have now become extinct or greatly reduced in diversity, plus some important immigrant groups from Eurasia. The dominant terrestrial herbivores were hyracoids of the family Pliohyracidae, represented by eight genera (Meyer 1978, Rasmussen & Simons 1988). The Fayum hyracoids include small species similar in size to modern hyraxes (*Saghatherium*, *Thyrohyrax*) and also very large species bigger than modern tapirs (*Titanohyrax*, *Megalohyrax*, *Pachyhyrax*). A rare species, *Titanohyrax ultimus*, had molars that are much larger than those of the contemporary Fayum anthracotheres and the proboscidean *Moeritherium*. In addition to being highly variable in size, the Fayum hyracoid species also differed from one another in dental specializations. Some forms had bunodont, pig-like teeth adapted for crushing food items such as roots, tubers or fruit (*Geniohyus*, *Bunohyrax*). Others had highly selenodont teeth indicating a folivorus diet (*Titanohyrax* and a new genus described by Rasmussen & Simons 1988). The dentition of *Thyrohyrax* was similar to modern grazing and browsing hyraxes, except for the

retention of a full eutherian complement of teeth (3.1.4.3). The most common of the large hyracoids was *Megalohyrax eocaenus*, which may have been a tapir-like resident of the Fayum swamps or riverbanks (Fig. 30/4). These diverse hyracoids filled most of the herbivorous niches that were later usurped by suids, bovids and other ungulates that arrived from Eurasia by the early Miocene. The Fayum is the only fossil locality where the magnitude of this ancient hyrax radiation has been documented.

Three other kinds of large terrestrial herbivores occurred sympatrically with the hyracoids – embrithopods, proboscideans, and anthracotheres. Fossils of the order Embrithopoda are found only from the Fayum and a poorly-known site in Angola (Pickford 1986). Two species are present in the Fayum, *Arsinoitherium zitteli* and *A. andrewsi* (Fig. 30/1). These massive animals stood about 1.75 m high at the shoulder and carried two pairs of bony horns on their heads, a small pair on the frontal and a huge pair on the nasals, that were probably covered in life with a horny sheath (Andrews 1906, Tanner 1978). The extremely hypsodont, full eutherian dentition lacking any diastemata is unlike that of any other known group of ungulates. The relationships of *Arsinoitherium* cannot be surely determined without older fossils, but it probably shares a common ancestor with the other endemic African paenungulate orders – the hyracoids, proboscideans and sirenians.

Smaller than *Arsinoitherium* were the early elephants, *Palaeomastodon* and *Phiomia*, that gave rise to the later diverse radiation of proboscideans in Africa, Eurasia and North America (Andrews 1906, Matsumoto 1924). The Fayum elephants had long, low skulls and longer necks with respect to modern elephants, and were much smaller in size. They bore two pairs of long tusks, one from the upper jaw and another from the lower. Of all the archaic African ungulate groups, elephants were most influential in later faunas of other biogeographic regions. Another proboscidean that was present in the Fayum fauna is the partially aquatic *Moeritherium*, an animal with inflated, lophodont molar cusps not unlike those of *Palaeomastodon* and a long, stout, short-limbed body (Andrews 1906, Matsumoto 1923). One species of *Moeritherium* is occasionally found in the Jebel Qatrani formation's lower and upper sequences, suggesting that it did inhabit freshwater rivers or lakes, but a more common species occurs in marine deposits of the late Eocene Qasr El Sagha formation. Another strange relative of proboscideans, *Barytherium*, is found in the Eocene marine deposits but not the Jebel Qatrani formation. This genus is best known from an Eocene site in Libya (Savage 1971, Wight 1980). The final group of aquatic herbivores from the Fayum are early sirenians, or sea-cows, which occur in warm coastal regions of both hemispheres today.

One family of large ungulates from the Fayum, Anthracotheriidae, is not related to the African paenungulate radiation, but rather represents the first kind of artiodactyl known to have reached Africa from Eurasia (Schmidt 1913, Black 1978). This marks the beginning of the ungulate invasion that eventually replaced most of the endemic taxa. The Fayum anthracotheres, which are in need of taxo-nomic revision, were stout-bodied forms that may have been aquatic like their distant relative, the hippopotamus.

The lumbering Fayum herbivores described above were preyed upon by large hyaenodontine creodonts (*Apterodon*, *Pterodon*, *Hyaenodon*; see Andrews 1906, Savage 1978). These animals had large, blade-like carnassial teeth highly specialized for meat-eating and bone crushing. *Pterodon africanus* was a

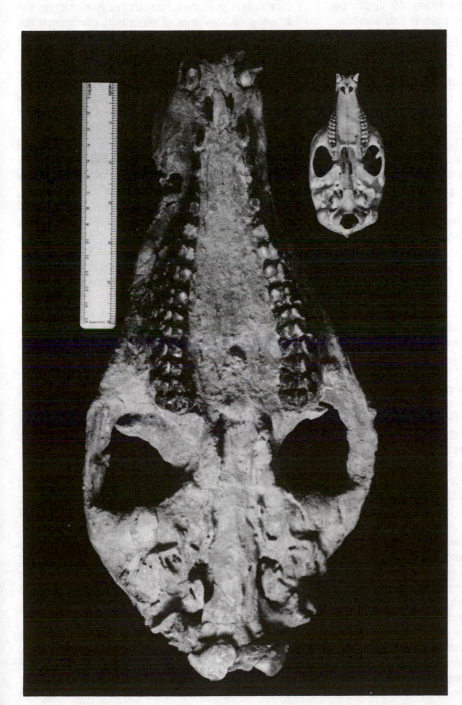

Figure 30.4 Ventral view of the skull of a large Oligocene hyrax from the Fayum, *Megalohyrax eocaenus*. The small skull next to it is from a modern African hyrax (*Heterohyrax brucei*). Numerous species of hyracoids evolved in Africa during the Oligocene that differed dramatically from each other in size and dietary speciali-zations. Photo by Asenath Bern-hardt.

large species possibly capable of preying upon the most robust Fayum ungulates. In addition to the large hyaenodonts, two genera of smaller proviverrine creodonts were also present, *Metasinopa* and *Masrasector* (Simons & Gingerich 1974). Unlike the hyaenodontines, these genera retained relatively broad talonids on their lower teeth. Puncture marks on the skeletal elements of various Fayum taxa, including the primates, have been shown to be bite marks of creodonts, rather than puntures due to scavenging by crocodiles (Gebo & Simons 1984). Fissipeds are completely absent from the Fayum and therefore presumably did not enter Africa until the late Oligocene or early Miocene.

Another group of carnivore-like mammals from the Fayum is the Ptolomaiidae, an enigmatic family of oversized pantolestan insectivores including the genera *Ptolomaia*, *Qarunavus* and a new, undescribed form (Osborn 1908, Simons & Gingerich 1974, Simons & Bown 1987). The dog-sized jaws of *Qarunavus meyeri* lack the sharp, shearing crests of strict meat-eaters, and it may well have been a racoon-like omnivore that preferred crustaceans, frogs, or other river-side prey. Several genera of smaller, more typical insectivorans that seem to be related to European forms have recently been recovered from the Fayum quarries and are currently being described by Bown & Simons. These include fossil tenrecs, a group now restricted to the island of Madagascar. A small insectivorous didelphid marsupial, *Peratherium africanus*, that closely resembles species of *Peratherium* in the Eocene and Oligocene of Europe, has also been recovered from the Fayum (Bown & Simons 1984). Other small insect-eating species include a variety of bats being studied by Thomas Bown, and the early elephant shrew, *Metoldobotes* (Schlosser 1911, Patterson 1965). Fossils of a pangolin, a large ant- and termite-eater, have also been found (Gebo & Rasmussen 1985), as well as fossilized ant or termite nests that show evidence of being burrowed into (Bown 1982) probably by the pangolin in search of food. Ecologically, three recently described prosimian primates should also be mentioned with this insectivorous group (see below). It is clear that the Fayum fauna contained many small, insectivorous mammals, some of which have been found only in recent years, and it is likely that more remain to be discovered.

Only eight species of rodents have been described from the Fayum, all belonging to a single family (Phiomyidae) of lophodont mouse- and rat-sized forms that may be near the ancestral stem of the modern African genera *Petromus* and *Thyronomys* (Wood 1968). The genus *Phiomys* comprises three species distributed in the upper and lower sequences. A larger rodent, *Paraphiomys simonsi* of the upper

sequence, may be descended from one of the lower sequence species of *Phiomys*. An allied lineage is represented by *Metaphiomys schaubi* of the lower sequence and *M. beadnelli* of the upper. *Phiocricetomys minutus* of the upper sequence is a very small species with a reduced cheek dentition, and *Gaudeamus aegyptius* from the lower sequence has distinctive three-crested teeth very similar to those of the modern cane rat, *Thryonomys*. Common rodent families known from the Miocene of south and east Africa are completely absent from the Fayum despite the discovery of hundreds of phiomyid specimens; the missing families include Bathyergidae, Pedetidae, Sciuridae, Cricetidae, Gliridae, Ctenodactylidae, and Anomaluridae. The Fayum Oligocene evidently had an impoverished rodent fauna with all species probably descended from a single common ancestor arriving in Africa not too long before deposition of the Jebel Qatrani formation (Wood 1968).

The most thoroughly studied group of fossil mammals from the Fayum are the primates. Those from the lower sequence are rare and remain very poorly known. These are: 1) *Oligopithecus savagei*, known from one jaw containing five lower teeth and about 12 isolated teeth (Simons 1962, 1971, 1972, Rasmussen & Simons in press.); 2) *Qatrania wingi*, known from three tiny jaws (Simons & Kay 1983); and 3) an unnamed tarsioid prosimian known only by two teeth that belongs in Omomyidae, a diversified family otherwise found in the Eocene of Eurasia and North America (Simons et al. 1987). *Oligopithecus* appears to be allied with the hominoids of the upper sequence as it shares with them the same dental formula (premolars reduced to two) and a similar structure of the premolars and, to a lesser extent, the molars (Rasmussen & Simons in press.). However, the molars have sharper crests, less bunodont cusps, higher trigonids and a slightly different pattern of wear facets than do the Fayum hominids, all features that are somewhat prosimian-like. Thus, *Oligopithecus* presents, in a general sense, an intermediate dental morphology between Eocene prosimians and Oligocene anthropoideans. The other lower sequence anthropoidean, *Qatrania wingi*, is about the size of the tiny South American marmosets and tamarins. The mound-shaped bunodont cusps without sharp cresting are unusual for such a small primate and indicate that the diet may have lacked a strong insect component. The morphology is similar to that of the larger parapithecids of the upper sequence, and *Qatrania* has been placed in that family.

In contrast to the lower sequence anthropoideans, those from the upper sequence are known from many hundreds of specimens representing 11 species. The family Parapithecidae includes monkey-like primates of about the size and proportions of modern squirrel monkeys (*Saimiri*). One of these, *Apidium phiomense*, is the most common of the primates and along with the hyracoid *Thyrohyrax domorictus*, is also the most abundant mammal of the upper sequence. Like New World monkeys, the dental formula is 2.1.3.3. Tooth cusps are bulbous and both the upper and lower molars have unusual central cusps. The face of *Apidium* was rather short, the frontal bones were fused with no trace of a metopic suture, the mandibular symphysis also fused early in life, and there is evidence of complete postorbital closure; all of these are morphological characteristics of modern Anthropoidea (Simons 1959, 1972). The canines were sexually dimorphic in size indicating by analogy with modern primates that *Apidium* probably lived in polygynous social groups (Fleagle et al. 1980). The distal tibiofibular articulation shows extensive apposition, which together with other evidence from the post-cranial anatomy indicates that *Apidium* was a specialized arboreal leaper with an ankle stabilized against lateral movements (Fleagle 1980, 1983, Fleagle & Simons 1979, 1983). Another species, *A. moustafai*, is significantly smaller and occurs at a lower stratigraphic level than *A. phiomense* (Simons 1962).

The closely related genus *Parapithecus* also contains two species, *P. grangeri* of the highest upper sequence quarries (Simons 1974) and the poorly known *P. fraasi* thought to have been found at a different stratigraphic level. (The generic name *Simonsius* has been applied to *P. grangeri* (Gingerich 1978, Fleagle & Kay 1985) but the diagnosis given for the proposed new genus is not sufficient to distinguish it from *Parapithecus*.) *Parapithecus* differs from *Apidium* primarily in the weaker central cusps on the molars, larger canines, smaller third molars, and the complete absence of lower central incisors in

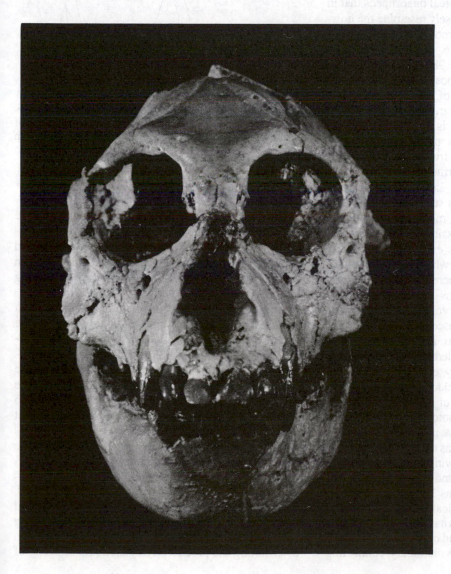

Figure 30.5 Facial view of the skull of *Aegyptopithecus zeuxis* from the Fayum. The cranium is almost complete but the upper incisors and mandible are restored from other specimens. Distance across the orbits is about 5 cm.

adults, a feature unique among primates (Kay & Simons 1983, Simons 1986). In size and morphology, the teeth and jaws of *Parapithecus* are comparable to those of the smallest living Old World monkey, *Cercopithecus talapoin*, and further resemble it in showing a primitive bilophodont molar organization. These dental resemblances and also some details of tarsal morphology (Gebo & Simons 1987) suggest a possible close relationship between *Parapithecus* and the Old World monkeys that can only be fully tested by the discovery of fossil monkeys between 20 and 30 million years ago.

The dawn apes of the Fayum include *Aegyptopithecus zeuxis* (Fig. 30/5) and four species in the genus *Propliopithecus* – *P. chirobates*, *P. ankeli*, *P. haeckeli* and *P. markgrafi*, the latter two known from only one specimen each collected by Richard Markgraf at the beginning of the century (Schlosser 1910, 1911, Simons 1965, Simons et al. 1987). Both of these genera were frugivorous arboreal quadrupeds that in size and proportions most closely resemble the howling monkeys (*Alouatta*) among living primates (Simons 1967, Conroy 1976, Fleagle & Simons 1978, 1982, Kay & Simons 1980). The two genera differ from each other in the position of the cusps on the molars, the relative sizes of the cheek teeth, and other aspects of dental anatomy (Kay et al. 1981, Simons et al. 1987). *P. ankeli* is a large species with relatively broad, expanded premolars, while *P. chirobates* is smaller and differs from *P. haeckeli* and *P. markgrafi* in details of tooth structure.

These two genera are the earliest known hominoids, although some authors suggest that they predate the divergence between modern apes and true Old World monkeys (a possibility that does not change their hominoid status except in the eyes of those cladistic systematists who are intolerant of paraphyletic taxa, and who therefore hold the untenable position that organisms must be classified according to the eventual evolutionary branching patterns of their far distant descendants). *Aegyptopithecus* resembles *Afropithecus*, *Proconsul* and other Miocene dryopithecines in dental and cranial anatomy (Simons 1987, Leakey et al. in press). No morphological features exclude *Aegyptopithecus* from being a direct ancestor of great apes and humans. The Fayum hominoids differ from later pongids in the retention of a number of primitive prosimian-like features, such as the longer snout with large, vertical premaxillary wings, the absence of a tubular ectotympanic bone, and the relatively small brain (Radinsky 1973, Simons 1987). The shape of the proximal humerus indicates more restricted movement at the shoulder than in later apes (Fleagle & Simons 1978). For these and other reasons, *Aegyptopithecus* forms a perfect evolutionary connecting

Figure 30.6 Enlarged drawing of the mandible of *Afrotarsius*, a small tarsioid prosimian, superimposed on a sketch of its probable appearance. The actual specimen is only 1.5 cm long.

link between the prosimians of the Eocene and the apes of the Miocene. Like *Apidium*, the Fayum hominoids were sexually dimorphic in body size and canine size indicating that they probably lived in social groups characterized by a polygynous mating system (Fleagle et al. 1980).

Among the most recent additions to the Fayum primate fauna are three species of prosimians. The first of these to be identified was *Afrotarsius chatrathi* (Simons & Bown 1985), a small tarsier-like primate known from one lower jaw collected in the upper sequence (Fig. 30/6). Modern tarsiers are small, nocturnal, insectivorous leapers restricted to the islands of southeast Asia. *Afrotarsius* is the first known fossil of this family, and in known anatomy is hardly distinct enough to warrant separation from the modern genus *Tarsius*. Tarsiers are evidently phyletically allied with the diverse and common omomyids of the Eocene (Simons & Russell 1960, Simons 1961, Rosenberger 1985), and further fossil finds of *Afrotarsius*, especially specimens preserving the diagnostic anterior dentition, may prove crucial in determining the precise phylogenetic relationship of these two groups. An omomyid has been identified in the Fayum's lower sequence, but it is presently known from only two isolated teeth (Simons et al. 1987). Its affinities lie with the morphologically primitive subfamily Anaptomorphinae, rather than the more tarsier-like Microchoerinae. A third species of prosimian recently identified by a single isolated molar apparently belongs in Lorisidae, a modern family including the bushbabies, lorises and pottos of Africa and Asia (Simons et al. 1987). The Fayum specimen is the earliest record of the family and, in fact, is the earliest known member of the tooth-combed prosimian group that includes not only Lorisidae but all of the diverse Malagasy prosimians as well. These new prosimian finds make the Fayum fauna the most diverse in terms of primates of any place in the world, living or fossil. This diversity, along with its crucial temporal and geographic position, makes the Fayum the most important known site for studying the evolutionary transitions from the archaic prosimians of the Eocene to the modern anthropoideans, tarsiers and toothcombed prosimians.

The ichnofossils (trace fossils) of the Fayum are among the best preserved and most diverse that have yet been recognized in fluvial sediments from anywhere in the world. These are of particular interest to geologists as several ichnofossils have potential value in Egypt for pinpointing the ages of sediments where vertebrate bones are not preserved. The 15 types of ichnofossils described by Bown (1982) fall into four main categories – large communal nests and passageways of social insects (termites and ants), dwelling burrows of various invertebrates, pellet-filled tunnels of oligochaete worms, and vertebrate burrows and excavations. The most common of the social insect ichnospecies are spherical or oblate masses (10 to 75 cm in diameter) of anastomosed galleries (3 to 7 mm in diameter) that have been designated *Termitichnus qatranii*. These resemble in size and architecture the structures formed by modern subterranean fungus-growing ants and termites. Other invertebrates are represented by a variety of burrows including vertical cylindrical ones probably excavated by freshwater crabs or crayfish, meniscate burrows similar to ones formed by crustaceans, mollusks, and insects, and teardrop-shaped burrows attributable to solitary wasps. The presence of giant tropical oligochaete worms (earthworms) up to almost 2 cm in diameter is indicated by horizontal, unbranched, pellet-filled burrows preferentially cemented with calcite, as in modern earthworm burrows (Bown & Kraus 1983).

Most notable of the vertebrate ichnofossils are clusters of large burrows 15 to 20 cm in diameter and up to 2 m or more in length that indicate the presence of 'villages' of a social burrowing mammal. None of the phiomyid rodents were large enough to construct these burrows. Among the Fayum mammals, the small creodont *Masrasector* and the ptolomaiids are of approximately the right size to have constructed the burrows, but neither are known to have foot or forelimb anatomy that could prove fossorial adaptations, and neither has any close living relative that could offer a behavioral analogy. It may be significant, however, that at least in dental anatomy and size *Masrasector* resembles the social, burrowing viverrids *Mungos*, *Helogale*, *Cynictis* and *Suricata*. These African mongooses dig communal burrows and live in groups of about 12 individuals, although some groups may grow as large as 40 individuals (Ewer 1973, Rood 1975, 1978, Rasa 1977). Another type of vertebrate excavation in the Fayum is the preserved remains of large pits (20 to 40 cm deep and 16 to 50 cm wide) that always occur with *Termitichnus* and are probably attributable to digging by the Fayum pangolins in search of their insect prey (Bown 1982, Gebo & Rasmussen 1985).

Fossil plants from the Fayum include fruits, seeds, leaves, logs (Wing & Tiffney 1982, Bown et al. 1982) and the best examples of rhizoliths in the world (Bown 1982). The most commonly encountered fossil fruits in the Jebel Qatrani formation are the large compound fruits of *Epipremnum*, a liane, that is today restricted to tropical forests of southeast Asia and the western Pacific islands. Fossil leaves include the floating aquatic plants *Salvinia* and *Nelumbo*, a cattail-like plant resembling modern *Typha*, fan palms of the Ochnaceae and Sapotaceae and, in the lower part of the formation, *Cynometra*, a genus that includes several mangrove species. Also present is the fern *Acrostichum* that today is restricted to the landward side of mangrove forests. Mangrove rhizoliths occur near the base of the Jebel Qatrani formation. Rhizoliths of other small and medium-sized plants are abundant throughout the formation but the plant taxa responsible cannot be identified. The largest fossil rhizoliths ever recognized in the world are stump casts in the lower sequence that attain a diameter of 1.2 to 2.2 m. Forests of large trees are indicated by the most obvious of the plant fossils, the giant fossil logs that litter many areas of the Fayum badlands. The fossil plant evidence thus strongly supports other lines of evidence that prove deposition of the Jebel Qatrani formation by a swampy river system bordered by tall liane-draped forest.

FUTURE PROSPECTS

The areal extent and stratigraphic thickness of the continental Oligocene sediments in the Fayum are so great that future exploration for new sites can continue almost indefinitely. For instance, from 1981 to 1984 more than 40 new fossil mammal localities were discovered. The western end of the Fayum depression is still largely unexplored. Oligocene beds north and east of Baharia Oasis show great promise for yielding new sites of interest. Searching should continue for new vertebrate localities because each separate site generally has a different faunal assemblage. Mammals rare or absent in a particular site or area may be common elsewhere. For example, all of the presently known primate specimens from the lower sequence come from but a single quarry. Scientists might expect to add very significantly to the number of primates and other small mammals if other productive sites can be found in the lower part of the section. Recently, Thomas Bown and Mary Kraus have discovered some continental and beach deposits in the upper part of the late Eocene Qasr El Sagha formation that offer the potential of yielding land mammals.

Further afield, the Oligocene of the Eastern Desert and the late Eocene and the Miocene of the Western Desert may be expected to yield more vertebrate fossils as paleontologists survey these regions. The Miocene exposures stretching from Siwa Oasis along the northern rim of the Qattara depression and east ward to known fossil fields lying near Hatiyet Moghra may also provide new information about the Tertiary of Egypt. This will be particularly true if localities rich in small mammals and birds like those of the Jebel Qatrani formation can be found somewhere in this vast area of exposures. Such discoveries could prove to be valuable in documenting the transition from the Fayum fauna composed of many endemic African groups to the African Miocene faunas that are dominated by immigrants from Eurasia.

The fossil primates of the Egyptian Oligocene include the first worldwide appearance of undoubted anthropoidean primates, and also the first records of tarsiers and lorisoids. Because of their relationships to humans, these ancient relatives receive the greatest attention from scholars. However, none of them is particularly well-known skeletally, and only one, *Aegyptopithecus zeuxis*, is known from skulls as well as skeletal parts. Knowledge of several of the Fayum primate species is restricted to only one or two fragmentary lower jaws. The unnamed lorisoid and omoyid are known only by isolated teeth. In order to gain richer insight into the anatomy and adaptations of all these primates, it will be necessary to collect again and again from the essentially inexhaustible quarry sites from which the known parts have already been obtained.

One of the most valuable conclusions to be drawn from the century of paleontology in the Fayum is that repeated, persistent fossil collecting in the same productive areas will continue to unveil new fossil species and new information about paleoenvironments, biogeography and the evolution of all variety of plant and animal life. Often the rarest or hardest to find specimens, such as delicate bird bones, invertebrate burrows, or tiny insectivore jaws, are of the greatest interest and value to geologists and evolutionary biologists. The Fayum presents the earliest broad view of the African continent in the Tertiary, and thus serves as an important foundation for studies of the later Miocene, Pliocene and Pleistocene sedimentary deposits of the continent. Many future advances in knowledge can be expected with further exploration, discovery and study of the Egyptian continental sediments and fossils.

ACKNOWLEDGEMENTS

Field expeditions to the Fayum organized by Simons and the Egyptian Geological Survey have been going on now for more than a quarter of a century. So many people have contributed to the success of these expeditions that it is impossible to acknowledge all of them. We wish to take this opportunity to thank especially those who have made valuable contributions in the early 1980s when many of the findings reviewed here have been made. Work in the Fayum would not have been possible without the support and assistance of the Egyptian Geological Survey, the Geological Museum, Cairo, and the General Petroleum Company of Egypt. We especially thank the Egyptian geologists and paleontologists who have regularly participated in field operations: Baher El Kashab, Yousry Attia, Magdy Zakaria, Abd El Ghani Ibrahim, and Ahmed El Awadi Kandil. Prithijit Chatrath has organized and managed field operations, and is curator and preparator of the collections at the Duke Primate Center. Among the many skilled field crew members from the USA, Egypt, France and the Malagasy Republic, we would like to especially acknowledge these participants of the 1980s: Friderun Ankel-Simons, Asenath Bernhardt, Thomas Bown, Jeff Brown, Inderdeep Chatrath, Tom Churchill, Bert Covert, John Fleagle, Dan Gebo, Phillip Gingerich, Marc Godinot, Greg Gunnell, Wahid Ibrahim Hassan, John Kappelman, Richard Kay, Mary Kraus, Andrew and Bruce McKenna, Casey McKinney, Rick Madden, Charles and Karen Messenger, Alex van Nieveldt, John Oakley, Ernestine Rahorimavo, Holly Smith, Michael Stuart, Lloyd Tanner, Laura Vick, J.P. Waters, Scott Wing and Mark Wolf. We thank Thomas Bown for comments on this manuscript. The Fayum project has been financially supported in recent years by NSF grants in anthropology.

CHAPTER 31

Tables of foraminiferal biozones

SAMIR F. ANDRAWIS

Consultant, Cairo, Egypt

The following tables show the main stratigraphical divisions of time and biostratigraphic units based on foraminifera. These are compiled in a correlative scheme for the practical use of the oil and other related industries. The tables are followed by lists of the assemblages of species associated with these zones.

The scheme is based on many years of applied work in oil exploration and on the results of a large number of published and unpublished works. It is useful in establishing ages and correlating strata over wide areas of Egypt. Eighty-seven zones are introduced based on both the planktonic and benthonic foraminifera. Some of these zones, especially those based on the planktonic foraminifera, are of world-wide distribution but many others are of local appli-cation to Egypt. Some of the standard world-wide zones are not recognized in Egypt on account of latitudinal differences or facies conditions. Numer-ous attempts were made in the past to zone different parts of the stratigraphic column by the use of forami-nifera. Of these, special mention is made of the pioneering work of Beckman et al. (1969). Other valuable publications can be found in the bibliogra-phy under Abdel Sattar, Andrawis, Ansary, El Heiny, Kenawy, Nakkady, Omara, Said, Souaya, Wasfy and others.

Acknowledgement. The author wishes to acknowledge the help of Professor R. Said for his suggestions and critical review.

SYSTEM	SERIES	STAGE	ZONE N°	BLOW 69 (LETTERS)	ZONES	UNSTABLE SHELF			STABLE SHELF		EASTERN DESERT & GULF OF SUEZ
						N.WESTERN DESERT	NILE DELTA	N.SINAI	S.WESTERN DESERT	NILE VALEY	
QUATERNARY	HOLOCENE (RECENT)		87	N 23	Globigerina calida calida (PARKER)		✓				
QUATERNARY	PLIESTOCENE		85	N22	Globorotalia truncatulinoides (D'ORBIGNY)		✓				
QUATERNARY	PLIESTOCENE		86	N22	Borelis schlumbergeri (REICHEL)	✓					
TERTIARY	PLIOCENE	LATE PIACENZIAN	84	N21	Globorotalia crassaformis (GALLOWAY & WISSLER)		✓	✓			
TERTIARY	PLIOCENE	MIDDLE ZANCLEAN	83	N21	Globigerinoides obliquus extremus BOLLI	✓	✓	✓			
TERTIARY	PLIOCENE	EARLY ZANCLEAN	82	N20	Globorotalia margaritae BOLLI & BERMUDEZ	✓	✓	✓			
TERTIARY	PLIOCENE	EARLY ZANCLEAN	81	N19	Sphaeroidinellopsis Spp.		✓	✓			
TERTIARY	MIOCENE	LATE TORTONIAN/MESSINIAN	80	N 17/18	Globorotalia dutertrei (D'ORBIGNY)		✓				
TERTIARY	MIOCENE	LATE TORTONIAN/MESSINIAN	79	N15/16	Globorotalia acostaensis BLOW		✓				
TERTIARY	MIOCENE	MIDDLE SERRVALIAN	77	N13/14	Globorotalia mayeri CUSHMAN & ELLISOR		✓				
TERTIARY	MIOCENE	MIDDLE SERRVALIAN	78	N13/14	Borelis melo (FICHTEL & MOLL)	✓		✓			✓
TERTIARY	MIOCENE	MIDDLE LANGHIAN	75	N11/12	Globigerinoides ruber (D'ORBIGNY)		✓	✓			✓
TERTIARY	MIOCENE	MIDDLE LANGHIAN	76	N11/12	Heterostegina costata costata D'ORBIGNY			✓			✓
TERTIARY	MIOCENE	MIDDLE LANGHIAN	76	N10	H.costata politatesta PAPP and KUPPER			✓			
TERTIARY	MIOCENE	MIDDLE LANGHIAN	74	N10	Globorotalia fohsi peripheroacuta BLOW & BANNER	✓					✓
TERTIARY	MIOCENE	MIDDLE LANGHIAN	73	N9	Globorotalia fohsi peripheroronda BLOW & BANNER	✓	✓				✓
TERTIARY	MIOCENE	EARLY BURDIGALIAN	71	N8	Praeorbulina glomerosa (BLOW)		✓				✓
TERTIARY	MIOCENE	EARLY BURDIGALIAN	72	N8	Miogypsina cushmani VAUGHAN / Heterostegina praecostata PAPP & KUPPER	✓					✓
TERTIARY	MIOCENE	EARLY BURDIGALIAN	70	N8	Globigerinoides sicanus DE STEFANI- Praeorbulina transitoria (BLOW)	✓					✓

SYSTEM	SERIES	STAGE	ZONES N°	BLOW 69 (LETTERS)	ZONES	N. WESTERN DESERT	NILE DELTA	N. SINAI	S. WESTERN DESERT	NILE VALLEY	EASTERN DESERT & GULF OF SUEZ
						UNSTABLE SHELF			STABLE SHELF		
TERTIARY (cont.)	MIOCENE (cont.) — EARLY MIOCENE	BURDIGALIAN (cont.)	68	N 5 / 6 / 7	Globigerinoides subquadrata BRONNIMANN– Globigerinoides diminuta BOLLI						✓
			69		Miogypsina intermedia DROOGER	✓	✓	✓			✓
		BURDIGALIAN	66	N4	Globigerinoides quadrilobatus primordius BLOW & BANNER						✓
			67		Miogypsina globulina DROOGER	✓	✓				✓
		AQUITANIAN	65		Miogypsina tani DROOGER	✓	✓				✓
			64		Nonion granosum D'ORBIGNY						✓
	OLIGOCENE — LATE	CHATIAN	62		Turborotalia kugleri BOLLI	✓	✓				
			63	P22	Miogypsinoides complanata SHLUMBERGER — Lepidocyclina spp.	✓					✓
			61		Globigerina ciperoensis ciperoensis BOLLI		✓				
	MIDDLE	RUPELIAN	60	P21	Turborotalia opima opima BOLLI	✓	✓				
			59	P20	Globigerina ampliapertura BOLLI	✓	✓				
	EARLY	SANNOISIAN	58	P18/19	Cassigerinella chipolensis (CUSHMAN & PONTON)		✓				
			57		Globigerina selli (BORSETTI)	✓					
EOCENE — LATE		BARTONIAN	55	P16/17	Turborotalia cerroazulensis (COLE)	✓			✓		
			56		Nummulites striatus (BRUGUIEREY) N. chavannassi DE LA HARPE — N. fabianii (REVER)	✓					✓
			54	P15	Globigerapsis simiinvolutus (KEIJZER)	✓	✓				
MIDDLE EOCENE		LUTETIAN	53	P13/14	Truncorotaloides rohri BRONNIMANN & BERMUDEZ						
			52		Nummulites striatus (BRAGIERE) = Sphaerogypsina globula (REUSS)	✓				✓	
			51		Morozovella lehneri CUSHMAN & JARVIS						✓
			50	P12	Nummulites striatus (BRUGUIERE) = Dictyoconus aegyptiensis (CHAPMAN)	✓				✓	✓

SYSTEM	SERIES	STAGE	ZONE N°	BLOW 69 (LETTERS)	ZONES	UNSTABLE SHELF			STABLE SHELF		EASTERN DESERT & GULF OF SUEZ
						N.WESTERN DESERT	NILE DELTA	N.SINAI	S.WESTERN DESERT	NILE VALLEY	
TERTIARY (cont.)	EOCENE (cont.)	MIDDLE (cont.) / LUTETIAN (cont.)	48 / 49 / 47	P11	Globigeropsis kugleri BOLLI / Nummulites gizehensis gizehensis (FORSKEL) / Turborotalia bullbrooki BOLLI	✓	✓	✓	✓	✓	✓
			45 / 46	P 10	Hantkenina aragonensis NUTTAL / Nummulites gizehensis zitteli DE LA HARPE	✓	✓	✓			
	EARLY / YPRESIAN		42 & 43 / 44 / 41	P9	Acarinina esnaensis (LE ROY) or Morozovella aragonensis NUTTAL / Nummulites irregularis DESHAYES — N.rollandi FICHEUR / Nummulites planulatus (LAMARCK)	✓			✓	✓	✓
	PALEOCENE / LATE / LANDENIAN		39 / 40 / 38	P5	Morozovella formoza BOLLI — M.REX MARTIN / Nummulites burdigalensis (DE LA HARPE) / Morozovella velascoensis (CUSHMAN)	✓	✓	✓	✓		
			37	P4	Planorotalites pseudomenardii BOLLI	✓			✓	✓	✓
		MIDDLE / MON-TIAN	36	P3	Morozovella angulata (WHITE)	✓			✓	✓	✓
			35	P2	Morozovella uncinata (BOLLI)	✓			✓	✓	✓
		EARLY / DANIAN	34	P1	Globostica daubjergensis BRONNIMANN — Morozovella trinidadensis BOLLI	✓			✓	✓	
			33		Globigerina eugubina LUTERBACHER & PREMOLISILVA	✓					
CRETACEOUS / LATE / SENONIAN		MAASTRICHTIAN	32		Abathomphalus mayaroensis (BOLLI)	✓	✓				
			31		Gansserina gansseri (BOLLI)	✓	✓	✓			✓
		CAMPANIAN	30		Globotruncana tricarinata (QUEREAU)	✓	✓	✓			✓
			29		Globotruncanita elevata elevata (BROTZEN)	✓	✓				
		SANTO-NIAN	28		Lacosteina maquawilensis ANSARY & FAKHR			✓			✓

SYSTEM	SEREIS	STAGE	ZONES N°	ZONES	UNSTABLE SHELF			STABLE SHELF		EASTERN DESERT & GULF OF SUEZ
					N.WESTERN DESERT	NILE DELTA	N.SINAI	S.WESTERN DESERT	NILE VALLEY	
(cont.)	(cont.)	SANTONIAN (cont.)	27	Globotruncana angusticarinata GANDOLFI	✔	✔				
			26	Dicarinella concavata concavata (BROTZEN)	✔	✔				
		CONIACIAN	25	Dicarinella concavata cyrenaica (BARR)	✔	✔				
			24	Marginotruncana sigali REICHEL	✔					
C R E T A C E O U S	L A T E		23	Discorbis turonicus SAID & KENAWY	✔					✔
		TURONIAN	22	Ceratobulimina aegyptiaca SAID & KENAWY	✔					
			21	Ammomarginulina ovoidea SAID & KENAWY	✔		✔			✔
			20	Heterohelix globulosa (EHRENBERG)	✔					✔
		CENOMANIAN	19	Thomasinella punica SCHULMBERGER	✔		✔			✔
			18	Thomasinella fragmentaria OMARA	✔		✔			✔
		ALBIAN		BARREN						
	E A R L Y	APTIAN	16	Hedbergella infracretacea (GLAESSNER)	✔		✔			
			17	Orbitolina discoidea GRAS — Choffatella decipiens SCHULMBERGER	✔		✔			
		BARRAMIAN	15	Globigerina graysonensis	✔					
		NEOCOMIAN HAUTERVIAN								
		VALANGINIAN		BARREN						
		BERRIASIAN								
J U R A S S I C	LATE MALM	PORTLANDIAN	14	Kurnubia morrisi REDMOND	✔	✔				
		KIMMERIDGIAN								
		OXFORDIAN	13	Steinekella steinekei REDMOND	✔	✔				✔
	MIDDLE DOGGER	CALLOVIAN	12	Kurnubia palastiniensis HENSON — Trocholina palastiniensis HENSON	✔	✔				✔

SYSTEM	SERIES	STAGE	ZONES N°	ZONES	UNSTABLE SHELF			STABLE SHELF		EASTERN DESERT & GULF OF SUEZ
					N.WESTERN DESERT	NILE DELTA	N.SINAI	S.WESTERN DESERT	NILE VALLEY	
JURASSIC (cont.)	MIDDLE (cont.) DOGGER (cont.)	CALLO-VIAN (cont.) / BATHO-NIAN	11	Pfenderina spp. assemblage	✓	✓				✓
		BAJOCIAN / TOARCIAN		BARREN						
	EARLY LAIS	SINEMU RIAN	10	Problematina liassica (JONES)	✓					
		CHARMOU TIAN								
		HETTAN – GIAN		BARREN						
TRIASSIC				BARREN						
CARBONIFEROUS	MIDDLE	WESTPHA LIAN	9	Hyperammina earlandia assemblage			✓			✓
		NAMURIAN	8	Hemigordius simplex REITLINGER	✓					✓
			7	Ozawainella umbonata BRAZHNIKOVA	✓					
			6	Eostafella postmosquensis KIREEVA	✓					✓
	EARLY	VISEAN	5	Archaediscus krestovnikovi RAUSER	✓					
			4	Endothyranopsis crassa (BRADY)	✓		✓			✓
			3	Tetrataxis conica EHRENBERG	✓		✓			✓
		TOURNAIS IAN		BARREN						
DEVONIAN	LATE		2	Plavskina piriformis REITLINGER	✓					
	EARLY		1	Thurammina deformens IRELAND	✓					

Assemblages associated with foraminiferal zones

Zone 87 *Globigerina calida calida* (PARKER)
Nonion asterizans (Fichtel & Moll); Cibicides rhodiensis (Terquem); Spiroloculina sp.; Quinqueloculina spp. and Triloculina sp.

Zone 86 *Borelis schlumbergeri* (REICHEL)
Amphistegina triloba (D'Orbidny); Borelis costulatus (Eichwald); Borelis reicheli Souaya; Cymbaloporella squamosa (D'Orbigny); Peneroplis pertusus (Forskal); Schlumbergerina alveolinoformis (Brady); Elphidium macellum (Fitchtel & Moll) and Quinqueloculina spp.

Zone 85 *Globorotalia truncatulinoides* (D'ORBIGNY)
Globigerina tosaensis Takayanagi & Saito; Globigerina falconensis Blow; Asterigerina planorbis D'Orbigny; Elphidium crispum (Linne); Cibicides refulgens Montfort; Gypsina vesicularis (Parker & Jones) and Nonion scapha (Fichtel & Moll).

Zone 84 *Globorotalia crassaformis* (GALLOWAY & WISSLER)
Orbulina bilobata (D'Orbigny); Globigerina druyri Akers; Globigerina woodi Jenkins; Globigerina falconensis Blow; Elphidium crispum (Linne) Streblus beccarii (Linne); Streblus punctatogranosus (Seguenza); Marginulina filicostata Fornasini and Amphistegina radiata (Fichtel & Moll).

Zone 83 *Globigerinoides obliquus extremus* BOLLI
Globigerinoides obliquus obliquus Bolli; Globigerina foliata Bolli; Globigerina nepenthes Todd; Globigerinoides trilobus immaturus Le Roy and Eponides haidingeri (Brady).

Zone 82 *Globorotalia margaritae* BOLLI & BERMUDEZ
Globigerinoides obliquus obliquus Bolli; Globigerinoides trilobus immaturus Le Roy; Globigerina foliata Bolli; Globigerina nepenthes Todd; Eponides haidingeri (Brady) and Pullenia sphaeroides (D'Orbigny).

Zone 81 *Sphaeroidinellopsis* spp.
Sphaeroidinellopsis subdehiscens (Blow); Sphaeroidinellopsis kochi (Caudri); Sphaerodinellopsis grimsdalei (Keijzer); Sphaeroidinella subdehiscens Blow; Sphaeroidinella dehiscens (Parker & Jones); Globigerinoides conglobatus (Brady); Globigerina riveroae Bolli and Gloigerina nepenthes Todd.

Zone 80 *Globorotalia dutertrei* (D'ORBIGNY)
Globigerinoides kennetti Keller & Poore; Globigerinoides conglobatus (Brady); Globigerina foliata Bolli; Pulleniatina obliquiloculata (Parker & Jones); Orbulina universa D'Orbigny and Cibicides dutempli (D'Orbigny).

Zone 79 *Globorotalia acostaensis* BLOW
Globihgerina venezuelana Hedberg; Globigerina continuosa Blow; Globigerina concinna Reuss; Globigerina pachyderma (Ehrenberg); Siphonina reticulata (Czizek) and Textularia neurugosa Thalmann.

Zone 78 *Borelis melo* (FICHTEL & MOLL)
Quinqueloculina seminula (Linne); Quinqueloculina pulchella D'Orbigny; Amphistegina lessonii D'Orbigny; Elphidium macellum (Fichtel & Moll); Pyrgo bulloides (D'Orbigny) and Nonion boueanum (D'Orbigny).

Zone 77 *Globorotalia mayeri* CUSHMAN & ELLISOR
Globigerina nilotica Viotti & Mansour; Globigerina foliata Bolli; Globigerinoides trilobus immaturus Le Roy; Orbulina universa D'Orbigny; Orbulina bilobata (D'Orbigny) and Nonion scapha (Fichtel & Moll).

Zone 76 *Heterostegina costata costata* D'ORBIGNY/ *Heterostegina costata politatesta* PAPP & KUPPER
Miliolids spp.; Amphistegina lessonii D'Orbigny; sponge spicules, algal and bryozoan remains.

Zone 75 *Globigerinoides ruber* (D'ORBIGNY)
Globigerinoides ruber pyramidalis (Van den Broek); Globigerinoides trilobus altiaperturus Bolli; Globigerinoides trilobus immaturus Le Roy; Hastigerina aequilateralis (Brady); Orbulina universa D'Orbigny; Orbulina suturalis Bronnimann and Bulimina elongata elongata D'Orbigny.

Zone 74 *Globorotalia fohsi peripheroacuta* BLOW & BANNER
Globigerinoides trilobus immaturus Le Roy; Streblus beccarii (Linne); Orbulina universa D'Orbigny; Orbulina suturalis Bronnimann and Orbulina bilobata (D'Orbigny).

Zone 73 *Globorotalia fohsi peripheroronda* BLOW & BANNER
Globigerinoides trilobus trilobus (Reuss); Globigerina venezuelana Hedberg; Bulimina elongata elongata D'Orbigny; Bulimina elongata subulata Cushman; Cancris auriculus auriculus (Fichtel & Moll); Cassidulina cryusi Marks and Cibicides variolatus (D'Orbigny).

Zone 72 *Miogypsina cushmani* VAUGHAN/*Heterostegina praecostata* PAPP & KUPPER
Operculina carpenteri (Silvestri); Amphistegina lessonii D'Orbigny; Sphaerogypsina globula (Reuss); sponge spicules and algal and bryozoan remains.

Zone 71 *Praeorbulina glomerosa* (BLOW)
Globigerinoides trilobus trilobus (Reuss); Globorotalia siakensis (Le Roy); Orbulina suturalis Bronnimann and Rectuvigerina tenuistriata (Reuss).

Zone 70 *Globigerinoides sicanus* DE STEFANI/ *Praeorbulina transitoria* (BLOW)
Globigerinoides trilobus trilobus (Reuss); Streblus beccarii (Linne); Sphaeroidina bulloides D'Orbigny; Cyclammina acutidorsata (Hantken) and Eggerella compressa (Andreae).

Zone 69 *Miogypsina intermedia* DROOGER
Operculina complanata Defrance; Operculina carpenteri (Silvestri); Miogypsina complanata Schlumberger; Amphistegina lessonii D'Orbigny; Eponides repandus (Fichter & Moll); sponge spicules; algal and bryozoan remains.

Zone 68 *Globigerinoides subquadrata* BRONNIMAN/ *Globigerinoides diminuta* BOLLI
Globigerina concinna Reuss; Globigerina bulloides Blow; Cyclammina acutidorsata (Hantken); Nonion pompilioides (Fichtel & Moll); Uvigerina

costata Bieda; Cancris auriculus primtivus Cushman & Todd; Textularia carinata D'Orbigny and Bulimina pupoides D'Orbigny.

Zone 67   *Miogypsina globulina* DROOGER

Operculina complanata Defrance; Operculina carpenteri (Silvestri); Amphistegina lessonii D'Orbigny; Elphidium advena (Cushman); Streblus spp.; sponge spicules; algal and bryozoan remains.

Zone 66   *Globigerinoides   quadrilobatus   primordius* BLOW & BRANNER

Globigerinoides trilobus trilobus (Reuss); Turborotalia kugleri Bolli; Globigerina ciperoensis ciperoensis Bolli; Uvigerina semiornata D'Orbigny; Uvigerina venusta Franzenau and Valvulineria complanata (D'Orbigny).

Zone 65   *Miogypsina tani* DROOGER

Operculina complanata Defrance; Amphistegina lessonii D'Orbigny; Elphidium flexuosum (D'Orbigny); sponge spicules; algal and bryozoan remains.

Zone 64   *Nonion granosum* D'ORBIGNY

Cibicides ellisi ellisi Souaya; Elphidium advena (Cushman); Nonion spp.; Streblus spp. and Bolivina shukrii hintei Souaya.

Zone 63   *Cyclammina cancellata deformis* GUPPY

Robulus submamilligera (Cushman); Cyclammina acutidorsata (Hantken); Bulimina striata D'Orbigny and Globigerina turritilina turritilina Blow & Banner.

Zone 62   *Turborotalia kugleri* (BOLLI)

Globigerina continuosa Blow; Globigerinita martini martini Blow and Coryphostoma sinusum (Cushman).

Zone 61   *Globigerina ciperoensis ciperoensis* BOLLI

Globigerina woodi Jenkins; Globigerina tripartita tripartita Koch and Globigerina praebulloides praebulloides Blow.

Zone 60   *Turborotalia opima opima* BOLLI

Globigerina sinilis Bandy; Globigerina venezuelana Hedberg; Globigerina tripartita tripartita Koch; Globigerina praebulloides leroyi Blow & Banner; Karreriella spihonella (Reuss) and Turborotalia opima nana Bolli.

Zone 59   *Globigerina ampliapertura* BOLLI

Viguninella pertusa Reuss; Eggerella spp.; Uvigerina minuta Cushman & Stone and eponides praecinctus Karrer.

Zone 58   *Globigerina selli* (BORSETTI)

Globigerina rohri Bolli; Globigerina yeguaensis pseudovenezuelana Blow & Banner; Robulus macrodiscus (Reuss) and Haplophragmoides carinatum Cushman & Renz.

Zone 57   *Cassigerinella chipolensis* (CUSHMAN & PONTON)

Globigerina venezuelana Bolli and Hastigerina micra (Cole).

Zone 56   *Nummulites striatus* (BRUGUIERE)/*Nummulites chavennesi* DE LA HARPE/*Nummulites fabianii* (REVER)

Nummulites pulchellus De la Harpe accompanied by the three zonal species.

Zone 55   *Turborotalia cerroazulensis* (COLE)

Globigerina pseudoampliapertura Blow; Globi-

gerina yeguanensis Weinzierl & Applin and Globigerinita unicava (Bolli, Loeblich & Tappan).

Zone 54   *Globigerapsis semiinvolutus* (KEIJZER)

Porticulosphaera mexicana (Cushman) and Turborotalia centralis (Cushman & Bermudez).

Zone 53   *Nummulites   striatus   (BRUGIERE)/ Sphaerogypsina globula* (REUSS)

Nummulites beaumonti D'Archiac & Haime and Nummulites discorbinus (Schlotheim).

Zone 52   *Truncorotaloides rohri* BRONNIMANN & BERMUDEZ

Acaranina spinuloinflata (Bandy); Morozovella spinulosa (Bandy) and Globigerina linaperta Finlay.

Zone 51   *Nummulites striatus* (BRUGIERE)/*Dictyoconus aegyptiensis* (CHAPMAN)

Nummulites beaumonti D'Archiac & Haime; Gypsina carteri Silvesteri and Somalina stefaninii Silvesteri.

Zone 50   *Morozovella lehneri* (CUSHMAN & JARVIS)

Hantkenina mexicana Cushman; Morozovella renzi (Bolli); Globigerina senni (Bechmann); Bulimina jarvisi misrensis Ansary and Chiloguembelina cubensis (Palmer).

Zone 49   *Nummulites gizehensis gizehensis* (FORSKAL)

Nummulites discorbinus (Schlotheim); Nummulites beaumonti D'Archiac & Haime and Gypsina carteri Silvesteri.

Zone 48   *Globigerapsis kugleri* BOLLI

Truborotalia broedermanni (Cushman & Bermudez); Acaranina spinuloinflata (Bandy) and Globigerapsis index (Finlay).

Zone 47   *Turborotalia bullbrooki* (BOLLI)

Turborotalia broedermanni (Cushman & Bermudez); Turborotalia triplex Subbotina; Bulimina jacksonensis Cushman and Robulus trompi Ansary.

Zone 46   *Nummulites gizehensis zitteli* DE LA HARPE

Nummulites discorbinus (Schlotheim) and Nummulites beaumonti D'Archiac & Haime.

Zone 45   *Hantkenina aragonensis* NUTTAL

Acaranina spinuloinflata (Bandy); Morozovella spinulosa (Cushman); Planulina sinaensis Ansary and Bolivina moodysensis Cushman & Todd.

Zone 44   *Nummulites irregularis* DE SHAYES/*Nummulites rollandi* FICHEUR

Nummulites ataticus Leymerie; Nummulites ornatus Schaub; Nummulites globulus Le Roy; Nummulites rotularis De la Harpe and Assilina laxispira De la Harpe.

Zone 43   *Acaranina esnaensis* (LE ROY) or *Morozovella* and 42   *aragonensis* NUTTAL

Acaranina gravelli (Bronnimann); Morozovella quetra (Bolli); Morozovella planoconica (Subbotina); Acaranina esnaensis (Le Roy); Turborotalia soldadoensis (Bronnimann); Globigerina turgida Finlay and Turborotalia triplex Subbotina.

Zone 41   *Nummulites planulatus* (LAMARCK)

Nummulites subplanulatus Hantken & Madarasz; Nummulites praecursor De la Harpe and Nummulites aticicus Leymerie.

Zone 40   *Nummulites burdigalensis* (DE LA HARPE)

Nummulites praecursor De la Harpe and Nummulites deserti De la Harpe.

Zone 39 *Morozovella formosa* (BOLLI)/*Morozovella rex* (MARTIN)
Morozovella formosa gracilis (Bolli); Morozovella quetra (Bolli); Acaranina esnaensis (Le Roy); Morozovella aequa (Cushman) and Acaranina wilcoxensis (Cushman & Ponton).

Zone 38 *Morozovella velascoensis* (CUSHMAN)
Morozovella velascoensis parva Rey; Morozovella acuta (Toulman) and Morozovella occulosa (Loeblich & Tappan).

Zone 37 *Planorotaloides pseudomenardii* (BOLLI)
Acaranina convexa (Subbotina); Morozovella aequa (Cushman & Renz); Bolivinopsis kneblei (Le Roy) and Zeauvigerina aegyptiaca Said & Kenawy.

Zone 36 *Morozovella angulata* (WHITE)
Morozovella angulata abundocamerata (Bolli); Morozovella colligera (Schwager); Pseudoclavulina farafraensis Le Roy; Turborotalia ehrenbergi Bolli and Guadryina laevigata Francke.

Zone 35 *Acaranina uncinata* (BOLLI)
Turborotalia ehrenbergi (Bolli); Acaranina pseudobulloides (Plummer); Zeauvigerina aegyptiaca Said & Kenawy and Bolivinopsis dentata (Alth).

Zone 34 *Globostica daubjergensis* Bronnimann/Morozovella trinidadensis (Bolli); Turborotalia compressa (Plummer); Osangularia convexa Le Roy and Chiloguembelina subtriangularis Beckmann.

Zone 33 *Globigerina eugubina* LUTERBACHER & PREMOLI SILVA
Globigerina anconitana Luterbacher & Premoli Silva; Globigerina fringa (Subbotina); Turborotalia californica (Smith) and Turborotalia globosa (Hantken).

Zone 32 *Abathomphalus mayaroensis* BOLLI
Globotuncana contusa (Cushman); Globotruncanita stuarti (De Lapparent); Pseudotextularia elegans (Rezhak) and Rugoglobigerina spp.

Zone 31 *Gansserina gansseri* BOLLI
Globotruncanita stuarti (De Lapparent); Globotuncana conica White; Globotruncana arca (Cushman); Heterohelix reussi Cushman; Bolivina incrassata lata Egger and Bolivina incrassata limonensis Cushman.

Zone 30 *Globotruncana tricarinata* (QUEREAU)
Rosita fornicata fornicata (Plummer); Globotruncana ventricosa White; Globotuncanita stuarti (De Lapparent) and Heterohelix ultimatumida (White).

Zone 29 *Globotruncanita elevata elevata* (BROTZEN)
Rosita fornicata fornicata (Plummer); Rosita fornicata manaroensis Gandolfi; Globotruncana stephensoni Passagno; Stensioina exculpata Reuss and Bolivinoides decoratus decoratus (Jones).

Zone 28 *Lacosteina maquawilensis* ANSARY & FAKHR
Bulimina prolixa Cushman & Parker; Neobulimina canadensis Cushman & Wickenden and Anomalina sp.

Zone 27 *Globotruncana angusticarinata* GANDOLFI
Globotruncana marginata (Reuss); Rosita fornicata manaroensis Gandolfi; Stensioina praexculpata keller and Heterostomella mexicana Cushman.

Zone 26 *Dicarinella concavata concavata* (BROTZEN)
Globotruncana marginata (Reuss); Globigerinelloides ehrenbergi (Barr) and Praeglobotruncana hilalensis Barr.

Zone 25 *Dicarinella concavata cyrenaica* (BARR)
Globotruncana coronata Bolli; Haplophoragmoides calculus Cushman & Waters; Discorbis minuts Said & Kenawy and Kyphooxya undulata Lootterle.

Zone 24 *Discorbis turonicus* SAID & KENAWY
Discorbis minutus Said & Kenawy; Haplophragmoides eggeri Cushman and Cuneolina conica D'Orbigny.

Zone 23 *Marginotruncana sigali* REICHEL
Globotruncana coronata Bolli and Haplophragmoides gracilis Said & Kenawy.

Zone 22 *Ceratobulimina aegyptiaca* SAID & KENAWY
Ammomarginulina curvatura Ansary & Tewfik; Ammobaculites turonicus Said & Kenawy; Nonion beadnelli Said & Kenawy and Discorbis turonicus Said & Kenawy.

Zone 21 *Ammomarginulina ovoidea* SAID & KENAWY
Ammomarginulina spp.; Ammobaculites turonicus Said & Kenawy and Ostracod species Cytherella tuberculifera Alexander.

Zone 20 *Heterohelix globulosa* (EHRENBERG)
Praeglobotruncana helvetica (Bolli); Bulimina prolixa Cushman & Parker, with maximum development of the zonal species.

Zone 19 *Thomasinella punica* SCHLUMBERGER
Cribrostomoides sinaica Omara; Cribrostomoides parallens Ansary & Tewfik; Nezzazata simplex simplex Omara; Flabellammina compressa (Bissel) and Rotalipora cushmani Morrow.

Zone 18 *Thomasinella fragmentaria* OMARA
The zonal species is occasionally found with rare planktons such as Hedbergella washitensis (Carsey).

Zone 17 *Orbitolina discoidea* GRAS/*Choffatella decipiens* SCHLUMBERGER
Orbitolina spp. and Pseudocyclammina sp.

Zone 16 *Hedbergella infracretacea* (GLAESSNER)
Hedbergella spp.; Biglobigerinella sp. and Orbitolina spp.

Zone 15 *Globigerina graysonensis* TAPPAN
Nannoconus steinmanni Kamptner and Hedbergella spp.

Zone 14 *Kurnubia morrisi* REDMOND
Pseudocyclammina ammobaculitiformis Maync; Ammobaculites natrunensis Ebeid and Conicospirillina trochoides (Berthelin).

Zone 13 *Steinekella steinekei* REDMOND
Trocholina intermedia Henson; Pseudocyclammina sulaiyana; Steinekella creusi Redmond; Verneuilinoides minuta Said & Barakat and Kurnubia trunkata Ebeid.

Zone 12 *Kurnubia palastiniensis* HENSON/*Trocholina palastiniensis* HENSON
Nautiloculina circularis Said & Barakat; Nautiloculina oolithica Mohler; Trochammina callamina Loeblich & Tappan; Lenticulina muensteri (Roemer) and Ammobaculites jurassica Henson.

Zone 11 *Pfenderina* spp. assemblage
Pfenderina gracilis Redmond; Pfenderina neoco-

miensis (Pfender); Sandrella laynei Redmond; Pfenderina inflata Redmond; Nautiloculina oolithica Mohler and Meyendorffina ghorabi Ebeid.

Zone 10 *Problematina liassica* (JONES)
Triplasia bartensteini Loeblich & Tappan; Ammomarginulina sp. and Trochammina sablei Tappan.

Zone 9 *Hyperammina/Earlandia* assemblage
Earlandia pulchra Cummings; Earlandia spp.; Hyperammina spp.; Glomospira simplex Harton and Thuramminoides sphaeroidalis Plummer.

Zone 8 *Hemigordius simplex* REITLINGER
Hemigordius discoideus Brazhnikova; Endothyra bradyi Mikhailov; Bradyina pauciseptata Reitlinger; Ammovertella sp. and Fusulina fallsensis Thomson, Verville & Looke.

Zone 7 *Ozawainella umbonata* BRAZHNIKOVA
Ozawainella angulata (Colani); Ozawainella parationi Manikalova; Millerella pressa Thompson and Millerella concinna Potievskaya.

Zone 6 *Eostaffella postmosquensis* KIREEVA
Eostaffella ikensis Vissarionova; Eostaffella mutabilis Rausser; Eostaffella tortula Zeller; Eostaffella pinquis Thompson and Glomospira vulgaris Lipina.

Zone 5 *Archaediscus krestovnikovi* RAUSER
Archaediscus convexa Grozdilova & Lepideva; Tetrataxis digna Grozdilov & Lepideva; Tetrataxis paraminima Vissarionova; Endothyra bradyi Mikhilova and Climmacamina deckerelloides Lipina.

Zone 4 *Endothyranopsis crassa* (BRADY)
Earlandinella cylindrica (Brady); Glomospira simplex Harlton; Palaeotextularia davisella Cummings; Climmacamina ferra Cummings; Tetrataxis conica Ehrenberg and Valvulinella youngi (Brady).

Zone 3 *Tetrataxis conica* EHRENBERG
Tetrataxis minuta Brazhnikova; Tetrataxis eominima Rauser; Earlandia elegans (Rauser); Bigenerina sp. and Palaeotextularis davisella Cummings.

Zone 2 *Plavskina piriformis* REITLINGER
Caligella spp.; Saccammina spp.; Moravammina segmentata Pokorny; Tolypammina jacobschapolensis Conkin; Tolypammina cyclops (Gutschick & Trechman) and Tolypammina bulbosa (Gutshick & Trechman).

Zone 1 *Thurammina deformens* IRELAND
Thurammina tubulata Moreman; Thurammina elliptica Moreman and Tolosina sedenta Ireland.

CHAPTER 32

Paleozoic trace fossils

A. SEILACHER
Tübingen University, Germany and Yale University, USA

1 INTRODUCTION

Trace fossils, sedimentary structures resulting from animal activity, cannot compete with body fossils when it comes to taxonomic resolution. Commonly it is impossible to tell apart quite unrelated species or genera of worms, or bivalves, from the tracks and burrows they have left behind. On the other hand such traces may provide detailed information about life activities, behavioral programs and nutritional strategies, that could never be derived even from the most perfectly preserved body fossils. Traces, or 'Lebensspuren', also have the advantage that they cannot be reworked to become redeposited outside the original habitat or as ghost fossils in deposits of much younger age.

It is for these properties that as biogeologists we use trace fossils primarily to register changes in environmental parameters such as water depth, oxygenation, turbulence or productivity through stratigraphic time and space. This can be done on a broad scale to distinguish shallow marine shelf and epicontinental deposits from flysch type sediments that have been deposited on oceanic crust. Or one can work on a finer scale to map local environmental patterns, or biotic responses to episodic events (storms, floods), in the framework of basin analysis.

In Egypt, such finely tuned environmental analysis has been successfully applied to the Tertiary of the Nile basin (Aigner 1982). In older rocks it is less needed for two reasons: firstly, because the area has remained cratonic throughout Phanerozoic time (oceanic sediments from the opening Red Sea have not yet been incorporated into the rock record). Secondly, detailed facies analyses are, at the present stage of reconnaissance, less needed than gross stratigraphic correlations. This is particularly true for the Paleozoic sediments which, in many cases, have been lumped with the 'Nubian Sandstone' – a unit that is much younger, if defined in a stratigraphic sense. This error is pardonable, however, because a 'Nubian Facies', dominated by fluvial and shallow marine silicoclastic sediments, had existed in this setting long before the upper Cretaceous transgression. This clastic facies is notoriously poor in classical index fossils, because it does not favor the preservation of calcareous hard parts (molluscs, brachiopods, trilobites, ostracods, foraminifera, etc.). Nor does the lithology give us a clue to the antiquity of a rock, because in the absence of a huge overload not even the oldest sandstones did turn into quartzites – a fact that also left the sand available for repeated recycling.

Progress in the unravelling and dating of Nubian facies rock sequences during the last decades has come from lithostratigraphic correlations and from fossils that are less prone to early diagenetic dissolution: acritarchs, plant remains, vertebrate bones and trace fossils. Age determinations derived from such substitute index fossils usually do not provide resolution down to the level of biozones and subzones. Nevertheless they serve as provisional measuresticks that can be further improved by standardization in other areas of the world, where an orthostratigraphic scale is available.

In the Paleozoic of Egypt, the biogeological value of trace fossils is twofold. Even the longest-ranging trace fossils (except for 'non-descript' forms) allow the distinction between terrestrial, fresh water and marine environments. Thereby they allow the recognition of marine ingressions within seemingly uniform clastic sequences. The traces that trilobites have produced abundantly as part of their sediment-feeding activities serve as markers of marine conditions. By their high degree of fingerprinting they are also better stratigraphic indices than any other group of trace fossils (with the possible exception of tetrapod tracks), although their resolution still does not reach beyond the stage level. Unfortunately, this valuable tool becomes rather blunt – due to the trilobites' decline – in later parts of the Paleozoic.

Accordingly, we shall subdivide our presentation into two parts: the first one deals with the interpretation of trilobite traces as a prerequisite for their biostratigraphic use, while actual ichnofaunas will be discussed in stratigraphic order in the second part.

Throughout this discussion we should be aware that cratonic rock sequences, at all scales, tend to be full of major stratigraphic gaps, which are difficult to recognize because of lithologic and diagenetic monotony and the lack of angular unconformities. In stable cratonic settings, stratigraphic history generally presents itself as a series of sedimentary episodes, during which short marine ingressions were able to extend over tremendous areas of the low-relief basement surface. We should also remember that in the absence of strongly subsiding basins the sediment veneer could at any place reach only a limited thickness, so that deposition had to be compensated by erosion of earlier deposits. In such a cannibalistic scenario, additionally favored by low grade diagenesis, continuity of the stratigraphic record cannot be expected – neither in a vertical, nor in a horizontal sense. What we do find is a puzzling stratigraphic patchwork. It is only thanks to the tremendous size of the Egyptian territory and for the recent efforts of devoted field geologists to explore inaccessible desert areas, that in spite of this patchiness the Paleozoic time scale can now be fairly well filled with records from disparate regions of this country.

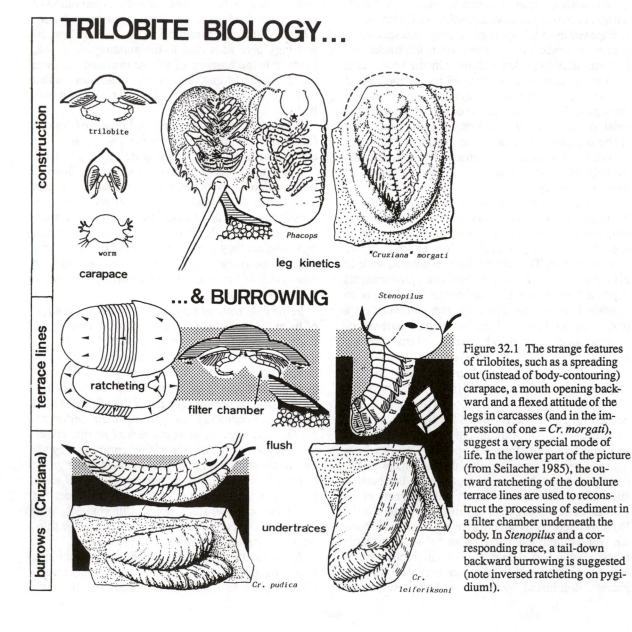

Figure 32.1 The strange features of trilobites, such as a spreading out (instead of body-contouring) carapace, a mouth opening backward and a flexed attitude of the legs in carcasses (and in the impression of one = Cr. morgati), suggest a very special mode of life. In the lower part of the picture (from Seilacher 1985), the outward ratcheting of the doublure terrace lines are used to reconstruct the processing of sediment in a filter chamber underneath the body. In Stenopilus and a corresponding trace, a tail-down backward burrowing is suggested (note inversed ratcheting on pygidium!).

2 ANALYSIS OF TRILOBITE TRACES

Trilobites, the classical guide fossils of the early Paleozoic, are hardly ever present in the carbonate-poor Nubian facies. But fortunately carapaces are not the only record that these strange organisms have left behind. As active deposit feeders that either browsed the surface or processed the upper sediment layer in a filter chamber underneath their carapace hood (Fig. 32/1), trilobites were also very productive as trace-makers. Their shallow diggings are usually preserved on the sole faces of sandstone beds as bilobed casts (hence the old name 'bilobites'). They are here united under the ichnogeneric name *Cruziana* D'Orbigny, whether made in a stationary (coffee bean-shaped 'rusophyciform' expression) or in a bulldozing manner (band-shaped 'cruzianaeform' expression). In addition we find trackways (*Diplichnites*) and grazings (*Dimorphichnus*), some of which were presumably made by the same kinds of trilobites.

a) *Biologic and biostratinomic background*

In order to understand trilobite trace fossils, we must first discuss details of the trilobite *feeding process*. They can be derived from general features of trilobite construction (pleural duplicatures spreading-out, rather than contouring the body; outward-ratcheted terrace lines on the ventral surface; backwardly directed mouth; lack of mouth parts) and from the fact that the legs did not dig the sediment away from underneath the body (as would an animal that buries itself for protection). Rather the endopodites scraped the sediment together towards the midline. This makes sense only if the loosened sediment became subsequently suspended, processed and eventually flushed-out from under the hood by the beat of the exites – feathery appendages that were well-suited to combine the gill function with that of a food filter. The strained food could then be handed to the coxae and along the midline to the mouth, which sucked it in without needing the help of special mouth appendages.

The second important detail is one of *preservation*. Intuitively one would assume that positive relieves found on the sandstone soles are casts of impressions that had been made when the underlying mud was still exposed. But the switch from mud to sand sedimentation usually marks a high energy event, such as a storm or a flood, during which deposition is always preceded by an erosional phase. This phase is recorded by impact casts on the soles of storm-sand; but associated trilobite traces are unaffected, preserving minute details of scratch patterns, on which our taxonomic distinction largely depends. Instead we are faced with another kind of imperfection, namely that only the most deeply impressed parts of a burrow, or trackway, are preserved in the cast. Most of the

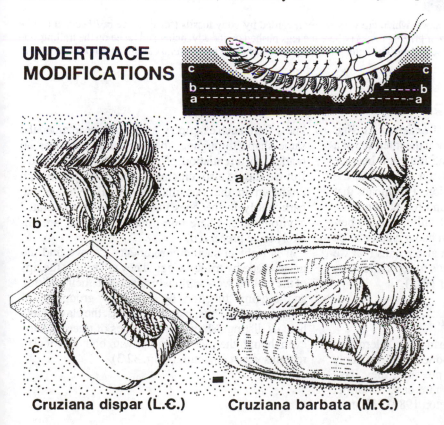

UNDERTRACE MODIFICATIONS

Cruziana dispar (L.Є.)   **Cruziana barbata (M.Є.)**

Figure 32.2 Preserved trilobite burrows did not originate at the sediment/water interface, but when the mud surface was already covered with a thin sheet of sand. This accounts for a lot of shape modifications and deficiencies depending on the undertrack level (a, b, c) (from Seilacher, in press).

ENDOPODIAL CLAW MARKS

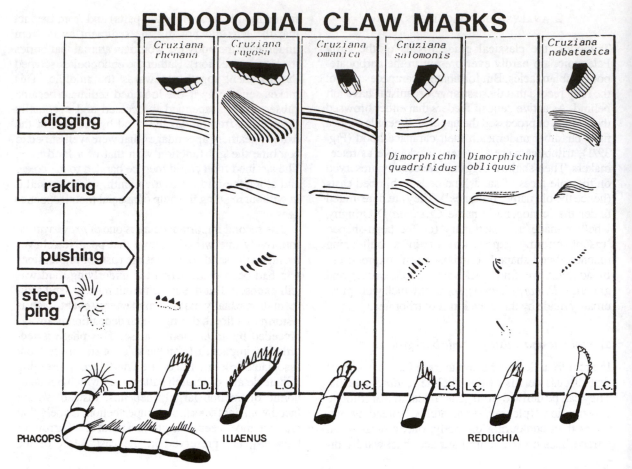

Figure 32.3 Shapes of endopodite claws, which are very rarely recorded by body fossils (for instance in *Phacops*) can be derived from impressions in walking or grazing tracks (where the legs pushed on the advancing and raked on the trailing side of the body) as well as from burrow scratches. The latter are an important criterion in *Cruziana* classification (modified from Seilacher 1962).

palichnological record consists of *undertraces* (Fig. 32/2), which originated internally at the interface of a sandy veneer with the underlying mud and became sand-cast without having ever been exposed to the destructive force of surface erosion. This means that the trilobites responsible lived and fed not on mud, but on *sandy* substrates and that only a minute percentage of their trace production had a chance to be preserved in the fossil record. It also means that we need a large number of specimens to reconstruct the complete trace pattern (Seilacher 1985).

b) *Endopodial claw shapes* (Fig. 32/3)

Since trilobite appendages had no sclerotized cuticles, but were largely supported (and probably also extended) by hydrostatic pressure, they have become fossilized only in very rare cases (Størmer 1939, Bergstrøm 1969, Whittington 1982). What we

know from such examples is that the legs retained the same shape throughout the body; i.e. there was no marked differentiation of mouth parts or pleopods. But we should also remember that leg shapes remain unknown in the vast majority of trilobite genera, particularly in the ones that inhabited well aerated environments, where non-mineralized parts are unlikely to be preserved.

It is right in such environments that we find the traces made by trilobite legs. They tell us that the tips of the endites were variously adapted to their walking and digging function by particular groupings of claws or setae. Thus, like fingerprints, the claw marks help us distinguish different species of trace makers, even though their identification with body fossil taxa remains very conjectural (Fig. 32/3).

EXOPODIAL BRUSHINGS

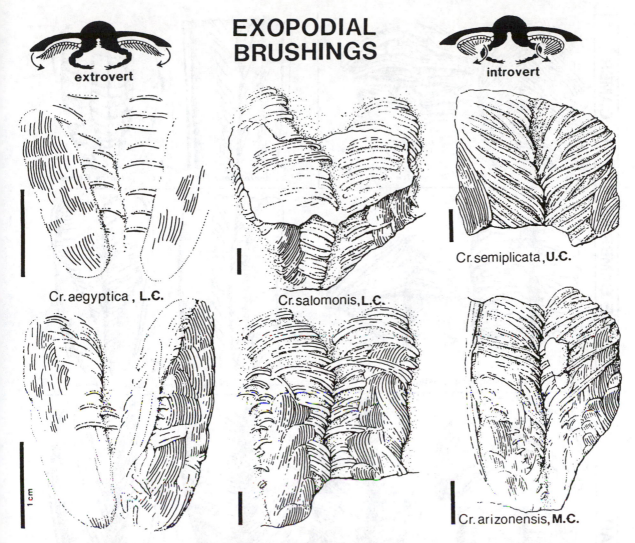

extrovert

introvert

Cr. aegyptica , **L.C.**

Cr. salomonis, **L.C.**

Cr. semiplicata , **U.C.**

Cr. arizonensis, **M.C.**

1 cm

Figure 32.4 Well-preserved *Cruziana* specimens (high undertrack levels) show the longitudinal brushings of the feathery exopodites in addition to the endopodial scratches. This proves that the outer branches acted not only as gills but participated also in sediment processing. The curvature of the brushings also allows us to distinguish species in which the flushing and straining exopodites were outstretched (extrovert) from those in which they worked in a bandy-legged fashion (introvert). All specimens in the Tübingen collection. *Cr. aegyptica* from loc. F27/86 (upper Wadi Qena); lower specimen Holotype GPIT 1660/1; upper: Berlin collection. *Cr. salomonis* from Umm Bogma (upper specimen, from loc. 13/86 in Berlin; lower loc. F8/86 GPIT 1660/2). *Cr. semiplicata* from Tremadocian (Molinos Beds), Portiella, N. Spain (GPIT 1660/3). *Cr. arizonensis* from mid-Cambrian (Bright Angel Shales) of the Grand Canyon (GPIT loc. 1086/5).

c) *Exopodial brushings* (Fig. 32/4)

In addition to the coarse endopodial scratches, trilobite burrows occasionally show much finer markings that are here referred to the activity of the feather-like exopodites. In rusophycoid burrows they are usually restricted to the posterolateral parts, as exemplified by *Cr. arizonensis* (Fig. 32/5; but note other position in *Cr. balsa*, Fig. 32/6). The exopodial brushings may also form separate lobes flanking the endopodial ones in cruzianaeform versions (*Cr. semiplicata*, Fig. 32/5).

The lineation in these brushings is so delicate, that it can be seen only in perfectly preserved specimens and only in tangent light. It has mostly a longitudinal direction. In some cases, however, we can distinguish individual sets of lineation, in which the parallel lines are slightly curved. The Egyptian material shows that the direction of the curvature basically differs between ichnospecies and can be used as an additional distinctive feature. In *Cr. aegyptica* the brush curves are convex to the outside (extrovert), as should be expected in an appendage that swings around a me-

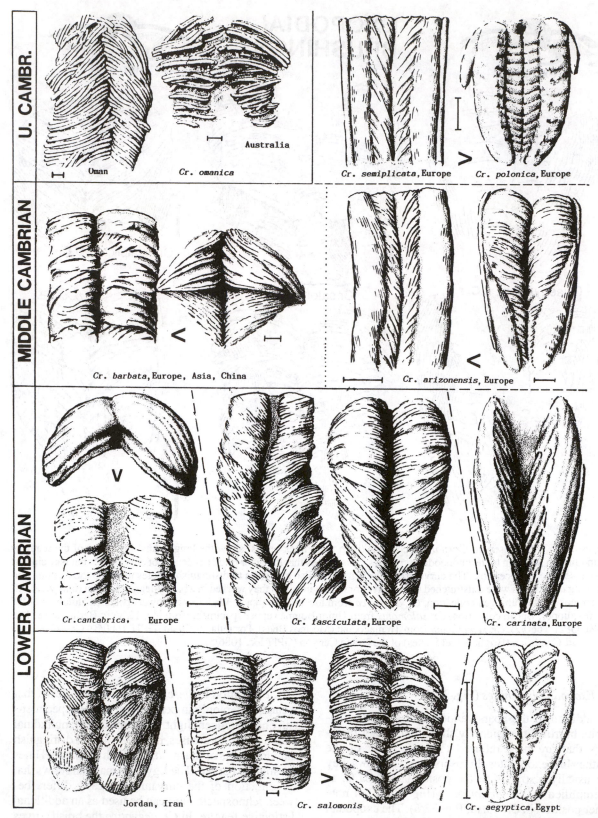

Figure 32.5 Cambrian *Cruziana* stratigraphy (modified from Seilacher 1970). *Cr. nabataeica*, Petra, Jordan (holotype GPIT 1660/4; see Plate 1a). *Cr. salomonis*; cruzianaeform version (left) from Zerka Main, Jordan (GPIT 1660/5; see Plate 1c), rusophyciform (right, loc. F5/86, GPIT 1660/16) from Umm Bogma. *Cr. aegyptica* from upper Wadi Qena, loc. F29/86 (GPIT 1660/7).

dian point of attachment (Fig. 32/4). In other species, such as *Cr. salomonis*, however, the curvature is reversed (introvert), as if these appendages had been attached at the lateral margin. The only plausible explanation is that in this case the exopodites were bandy-legged, like the endopodites.

d) Trilobite burrows (Cruziana)

Trilobite diggings are most telling, because they provide claw marks plus other distinguishing features. In their stationary (rusophyciform) expression we see that the scratches meet the midline at different angles in different species and that there is a general trend to form a larger angle in the broader and steeper front part of the burrow. In some species (Fig. 32/4), we also see the fine longitudinal brushings of the exopodites form separate lobes toward the rear end. But when stationary digging becomes combined with a forward bulldozing movement (cruzianaeform expression), the antero-posterior differences become wiped out and we see only one type of scratches, corresponding either to the front part (in prosocline attitude) or to the rear part (in opisthocline position) of the stationary burrow.

With these variables in mind, it has been possible to establish a standard sequence of *Cruziana* ichnospecies (Fig. 32/8), into which the Egyptian occurrences must be tentatively fitted.

e) Trilobite grazings (Dimorphichnus)

The digging mode of nutrition, in which right and left legs work symmetrically, could be modified into grazing over the surface. In order to forage a maximum surface without double coverage, grazing trilobites moved obliquely, or even at right angles, to their body axes. This implies a division of labor between the pushing legs of the leading flank and the raking endopodites on the trailing side of the body. Otherwise the legs worked similarly to the digging mode, namely by medio-posterior flexing during the active stroke.

The typical trackway of a grazing trilobite (*Dimorphichnus*) consists of alternating series, or sets, of pusher and raker impressions. In the clearest case, the pusher impressions are short and blunt, contrasting sharply with the long scratches of the raking sets (Fig. 32/10a). In other cases, the pusher impressions are elongate as well, although their elongation reflects the relaxation, rather than the active, stroke (Fig. 32/9a).

Since for diagnostic purposes it is important to distinguish between anterior and posterior claws within one footprint, we must also try to orient the trackway. In *Dimorphichnus*, the following criteria can be used as *directional* and *orientational clues*:

1. Because the raker or pusher impressions within one set were not made simultaneously, but in a metachronal sequence starting from the rear end, the axis of the set is always *oblique* to the body axis. This criterion does not by itself indicate the direction of movement or the head side, but in conjunction with the others it constrains our interpretation.

2. *Pusher impressions* tend to be drawn out in the direction of movement (relaxation stroke) and blunt at the other end (i.e. opposite to what we know from inorganic impact casts).

3. *Raker impressions* tend to be sigmoidal. This is because the mediocaudal activity stroke is always at an angle to the body movement (in order to make it parallel, the animal would have to move obliquely *backwards*). But since the actual course of the claws over the sediment is a combination of body movement *and* leg stroke, the ends of the raking marks bend sigmoidally into the marching direction, because they correspond to the slower initial and terminal phases of the raking stroke. This sigmoidal bend may become exaggerated if the leg touches ground before the relaxation stroke had terminated. Again, this feature by itself tells us only the strike, not the directionality of movement. The same is true for the retarded onset and lift-up of the anterior claws within individual rake marks, which results from a slight tilting of the leg during the active stroke.

Sigmoidal scratches can be read either way. There is, however, a directional component in the divergence of the claw marks, because the raking claws tend to open during the active stroke. Also the rakings (as well as the pusher impressions) should become larger towards the anterior end of the sets.

In contrast to the simple walking tracks (*Diplichnites*; Pl. 32/2, c), *Dimorphichnus* is almost exclusively found in the form of hypichnial undertracks on the soles of very thin sandstone layers. This means that the impressions are very sharp, but also tend to be incomplete, with regard to claw numbers as well as track pattern. The pusher impressions, in particular, are commonly missing (monomorphichnoid preservation). It also means that what we pick up in the scree are usually only small fragments. But even if we manage to salvage larger slabs by systematic excavation, the track patterns may be confused not only by undertrace incompleteness, but also by superposition of many trackways. In fact, it might have been impossible to decipher this complex trace fossil, had not the original specimens been a kind of Rosetta Stone, showing well-individualized and fairly continuous trackways over large surfaces (Seilacher 1955). In spite of these shortcomings, trilobite grazings have a good potential to be useful as substitute index fossils (Fig. 32/9).

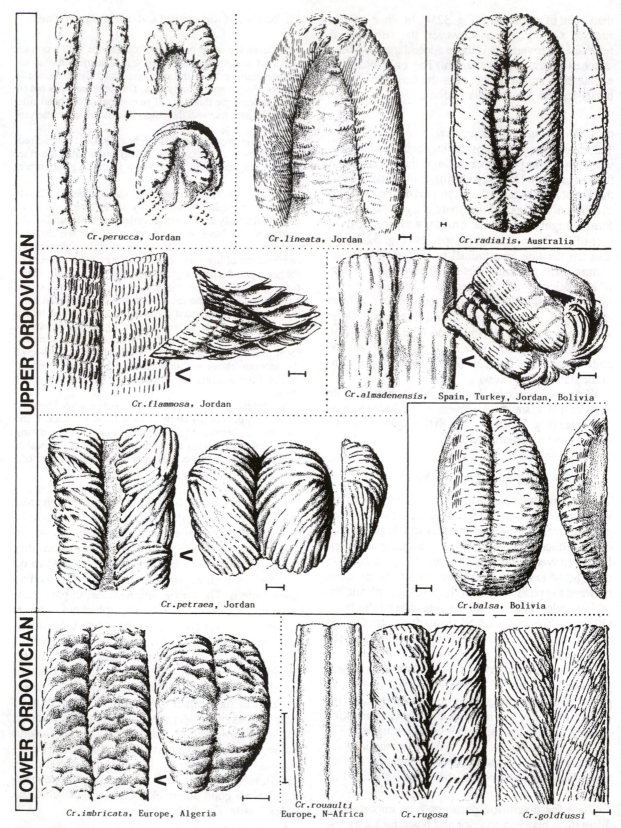

UPPER ORDOVICIAN

Cr.perucca, Jordan

Cr.lineata, Jordan

Cr.radialis, Australia

Cr.flammosa, Jordan

Cr.almadenensis, Spain, Turkey, Jordan, Bolivia

Cr.petraea, Jordan

Cr.balsa, Bolivia

LOWER ORDOVICIAN

Cr.imbricata, Europe, Algeria

Cr.rouaulti Europe, N–Africa

Cr.rugosa

Cr.goldfussi

Figure 32.6 Ordovician *Cruziana* stratigraphy (modified from Seilacher 1970). New forms (formally described in Seilacher, in print) are *Cr. balsa* from the Caradocian near Cochabamba and *Cr. radialis* from the upper Ordovician of the Georgina basin.

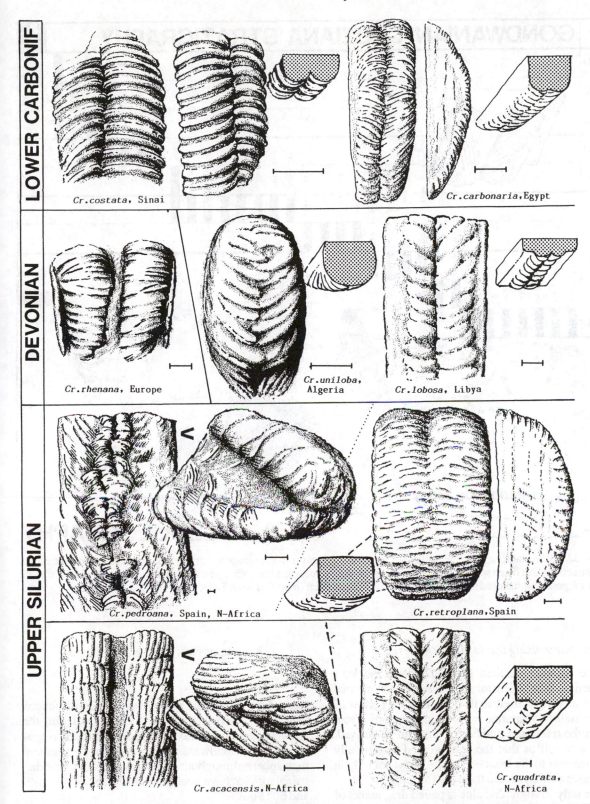

Figure 32.7 *Cruziana* stratigraphy of the Silurian through Carboniferous (modified from Seilacher 1970). New in the Silurian are part of a large cruzianaeform *Cr. pedroana* and *Cr. retroplana* (formally described in Seilacher, in print; both from Bonar and now in Tübingen museum) and in the Carboniferous the here-described Egyptian forms *Cr. costata* from loc. F9/86 near Umm Bogma (left GPIT 1660/8; right holotype GPIT 1660/9, see Plate 3d) and *Cr. carbonaria* from loc. F19/86 in south Wadi Feiran (holotype, GPIT 1660/21).

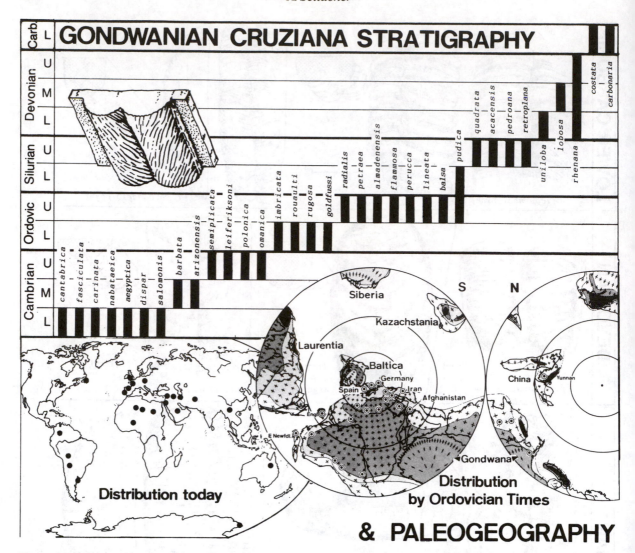

Figure 32.8 This table shows the value of *Cruziana* not only for stratigraphic correlation, but also for paleotectonic reconstructions. All listed species are Gondwanian and not yet recorded from other paleocontinents. Their occurrence in south China suggests that the used reconstruction is wrong in this point (modified from Seilacher 1983b).

f) *Trilobite walking tracks (Diplichnites)*

Simple walking resulted (Pl. 32/2, d) in the least distinctive type of trilobite traces. If the animal moved at a slight angle to its body axis, they allow us to distinguish disparate sets of impressions, from which the minimum number of legs can be derived. They also tell us that the legs worked sequentially from the rear to the front end of the body; but only in rare cases do they reveal the claw formula.

The only stratigraphic clue is paired drag marks of the caudal cerci, which are known only from early Cambrian *Diplichnites*. Such drag marks have not been found, however, in the Egyptian material.

3 DESCRIPTION OF ICHNOCOENOSES

a) *Cambrian*

According to current paleotectonic reconstructions (Fig. 32/8), the Nubian shield (then still including the Arabian Peninsula) was incorporated into the megacontinent of Gondwana and remained in the southern hemisphere throughout the Paleozoic. Its Cambrian paleolatitudes have been estimated to be between 40 and 60° South.

Within the Gondwana province, Cambrian trace fossil successions can be best gauged in north Spain (Seilacher 1970). Here the higher mobility of the continental crust allowed the accumulation of contin-

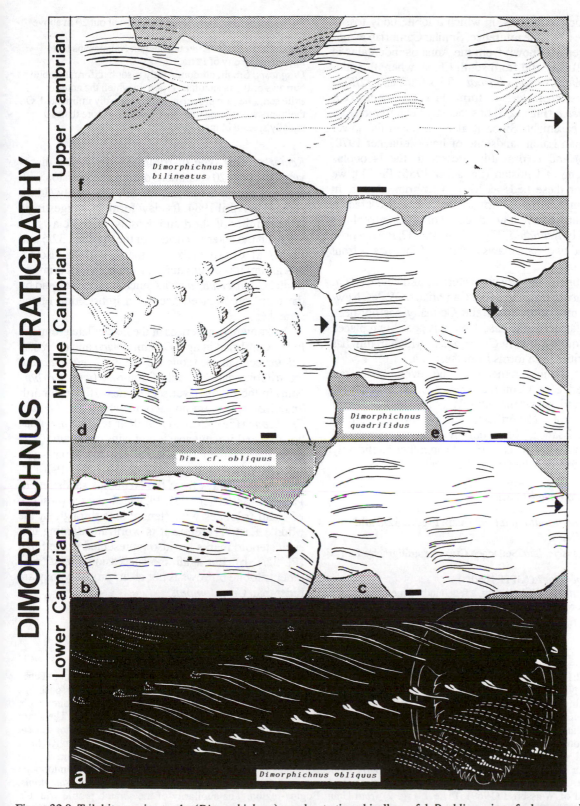

DIMORPHICHNUS STRATIGRAPHY

Upper Cambrian

Dimorphichnus bilineatus

f

Middle Cambrian

Dimorphichnus quadrifidus

d

e

Lower Cambrian

Dim. cf. obliquus

b

c

Dimorphichnus obliquus

a

Figure 32.9 Trilobite grazing tracks (*Dimorphichnus*) may be stratigraphically useful. *D. obliquus* is so far known only from the lower Cambrian in Pakistan (a, from Seilacher 1970) and in Lugnas, Sweden (b and c, GPIT 1660/10 and 11). *D. quadrifidus* with four equal claws and broad, pectinate pusher impressions, occurs in Spain in the middle Cambrian (d and e, Cerezedo, GPIT 1660/22-23). In *D. bilineatus* (Crimes) from the upper Cambrian of Wales, pusher as well as raker impressions are present, but the claw formula is indistinct (from cast of holotype).

uous rock sequences, in which trilobite body fossils provide a reliable reference. Similar Cambrian ichnocoenoses are known from the Amanos mountains in south Turkey and from south China, where the mid-Cambrian *Cruziana barbata* (Figs 32/2 and 32/5) is the most characteristic form. In Petra, Jordan, *Cr. nabataeica* (Fig. 32/3) has recently been found close to the basement. Since it also occurs in the lower Cambrian Lalun sandstone of Iran (Seilacher 1970: Pl. 1h) and is probably present in the Neobolus sandstone of Pakistan (Seilacher 1955: Pl. 21), we consider these beds as lower Cambrian in age. In Pakistan, *Cruziana* burrows are not well characterized, but we have a good record of grazing trilobites (possibly *Redlichia*, Figs 32/9a and 32/10a), whose legs had two main claws and two minor ones in front of them.

Strangely, the very characteristic and abundant *Cr. semiplicata* from the upper Cambrian of Wales, Newfoundland, north Spain and Germany has not been found farther to the east. Here it is possibly replaced by the quite different *Cr. omanica*, which in turn has similarities with forms from Australia.

We should remember, however, that our knowledge of Cambrian trace fossils is still very spotty and that we may also deal with a high degree of provinciality. Nevertheless, at the present state of reconnaissance, the occurrences in the Sinai and Eastern Desert areas of Egypt can rather confidently be attributed to the lower Cambrian.

Cambrian trace fossils

Cruziana nabataeica n.ichnosp. Figure 32/5 and Plate 32/1, a.

1970 *C. fasciculata*, Seilacher, Cruziana stratigraphy. Pl. 1, Fig. h
Holotype: Fig. 7a (GPIT 1660/4)
Locus typicus: Wadi Siyagh, a few hundred meters west of Petra, Jordan

Stratum typicum: Sand/shale sequence about 20 m above basement
Derivatio nominis: After the ancient Nabataeans that inhabited the city of Petra.
Diagnosis: Small, predominantly rusophyciform trilobite burrows of the fasciculata-group, in which the anterior exite scratches are proverse and backwardly imbricated. On their anterior surface they are lineated by more than 10 equally spaced blunt ridges.

Remarks. This type of trilobite burrow was discovered in about 20 specimens during an excursion with staff and students of the Yarmouk University (Irbid/Jordan) in April 1984. Its claw formula, suggesting a large number of short and densely spaced claws, or setae, on the anterior surface of the leg (Fig. 32/5 and Pl. 32/1a), clearly makes it a member of the fasciculata group. It is distinguished from both *Cr. cantabrica* and *Cr. fasciculata* by its much smaller size and by the more pronounced backward imbrication of the scratches.

All known occurrences of the fasciculata group are lower Cambrian in age: *Cr. cantabrica* and *Cr. fasciculata* occur, in this sequence, below *Cr. carinata* and the trilobite-dated middle Cambrian (with *Cr. barbata*) in the Porma river section near Bonar in north Spain. But there are no sections which also contain *Cr. dispar* (Fig. 32/2) so that their relative age relationship to this ichnospecies remains uncertain.

Similarly small burrows of the fasciculata-type from the Lalun sandstone in Iran (Seilacher 1970: Pl. 1h) and trackways from the Neobolus sandstone of Pakistan (Seilacher 1955: Pl. 21) signal the presence of small trilobites with ctenoid legs throughout the Middle East. Therefore it is worth mentioning that such ctenoid marks are also represented in the Sinai (Pl. 32/1, b). But again the successional relation to other trilobite burrows is not clear; probably they are older than *Cr. salomonis*.

Plate 32.1 Lower Cambrian Trilobite burrows (*Cruziana*): a. *Cr. nabataeica*; holotype: about 20 m above basement in Petra, Jordan. Note the ctenoid and strongly imbricated exopodial scratches (GPIT 1660/4); b. Similarly ctenoid markings (pushings and rakings) in Sinai (loc. F28/85, Berlin collection) may be made by a similar or identical early Cambrian trilobite; c. *Cr. salomonis*; holotype; higher part of lower Cambrian (dated by trilobites), Zerka Main, Jordan (GPIT 1660/16). The sharp exopodial scratches of this cruzianaeform burrow show the typical claw formula with secondary claws on the anterior side set apart from the principal one; d. *Cr. salomonis*; loc. F5/86 near Umm Bogma, Sinai (GPIT 1660/17). In this rusophyciform expression one recognizes the different angles at which the endopodial scratches meet in the median furrow in the anterior and posterior part; e. *Cr. salomonis*; loc. F21/85 (Berlin collection). In this procline cruzianaeform version the endopodial scratches are widely enough separated to show the complete claw formula with four secondary claws; f. *Cr. salomonis*. Original of Weissbrod (1969) (Plate 4, Fig. 2) from Umm Bogma region. The similarity of the scratch profile with *Cr. omanica* (Fig. 3) is a preservational artifact, caused by bending-over of the narrow scratch filling during compaction; g. *Cr.* cf. *salomonis*. Series of individual scratches with the complete claw formula (Berlin collection); h. *Cr. aegyptica*; loc. F28/85 (like b!), Berlin collection. In these unusually large rusophyciform versions the oblique ridge separating areas of endopodal scratches and exopodal brushings is clearly expressed. Note obliquely dug cruzianaeform version on right side. Diameter of coin 1,7 cm.

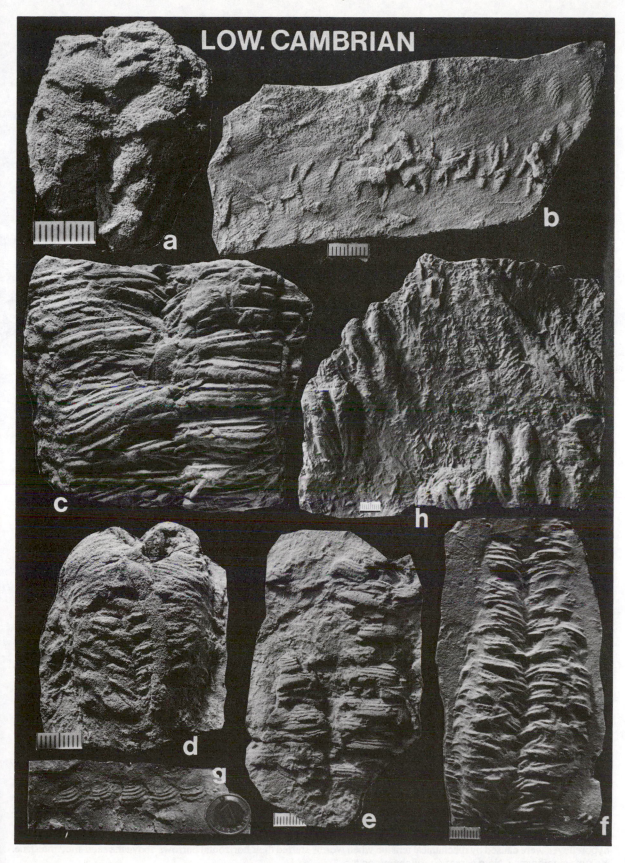

LOW. CAMBRIAN

Cruziana salomonis n.ichnosp. Figures 32/4-5 and Plate 32/1, c-g
1912 Meduses, Algues – Couyat & Fritel
1916 Meduses – Ball
1942 *Cruziana* – Picard, p. 9; Pl. I; Figs 3-5
1969 Bilobites – Weissbrod, Pl. 1, Fig. 2
Holotype: Pl. 1, c (GPIT 1660/16)
Derivat. nominis: Zerka Main, Jordan
Stratum typicum: From King Salomon, who ran his copper mines in this area.
Diagnosis: Rather large *Cruziana* from the barbata group, that was predominantly ploughed in a prosocline attitude. Scratches of the front legs run in a forwardly convex arc towards the midline and bear parallel ridges made by one major claw and up to four secondary claws in front of it. Rarely preserved exopodite brushings.

Discussion. The *barbata* group extends from the higher part of the lower Cambrian (*Cr. dispar* Linnarson from the Mickwitzia sandstone of Lugnas, Sweden, dated by the trilobite *Holmia*) to the middle Cambrian (*Cr. barbata* Seil. from trilobite-dated beds in north Spain; also found in Poland, Amanos mountains and south China). It is characterized by a functional differentiation between the strong front legs which work in a medio-anterior direction (proverse), and the weaker rear legs digging medio-posteriorly (retroverse). In stationary digging (rusophyciform expression), this may lead to a divergence, or discordance, of the scratch pattern in the deepest part of the burrow. Brushings are introvert. Another characteristic is the claw formula, recorded by parallel ridges on the front slopes of anterior scratches, which show one or two primary claws widely separated from up to four secondary claws, or setae, in front of it. Cruzianaeform versions tend to show only front leg scratches, indicating that the trilobites responsible ploughed in a tail-up (procline) attitude.

Distinction at the ichnospecific level, which may allow us to make stratigraphic subdivisions, uses a combination of these criteria. *Cr. dispar* (Fig. 32/2) tends to be deeply rusophyciform, sometimes with pleural edges impressed right on the steep lateral flanks of the burrow. Also, no more than three secondary claws can be counted. In *Cr. barbata*, rusophyciform burrows also predominate. They have a similar scratch pattern; but since the front scratches are much deeper than the rear ones, they constitute the only elements in deep undertrack preservation

and contrast as a 'moustache' against the 'goat beard' of the posterior scratches, if these show up at higher under-track levels (Fig. 32/2). Pleural impressions occur only in the very rare uppermost undertrack level (c in Fig. 32/2), and are separated from the endite diggings by a zone of exite brushings (Fig. 32/2), quite different from the situation in *Cr. dispar* and *Cr. salomonis*.

The specimens from the Dead Sea area of Jordan and from the Umm Bogma area of the Sinai differ from the other two ichnospecies by being predominantly cruzianaeform ploughings. With regard to the scratch pattern and claw formula they are intermediate between the two. Front scratches are as yet less different from the rear ones than in *Cr. barbata*. On the other hand, one can count up to four secondary claw impressions, i.e. more than in *Cr. dispar*, to which the general form of complete rusophyciform burrows would be most closely related (Pl. 32/1d). There are also two specimens that show exopodial brushings (Fig. 32/4). In contrast to *Cr. aegyptica* they diverge, rather than converge, in their backward strike and describe an arc that is convex toward the midline (introvert).

Range. We prefer to chose, for holo- and stratotypes, material that can be stratigraphically correlated by trilobite body fossils. This is the case in the Dead Sea area, where *Cruziana salomonis* is associated with trilobite body fossils of late lower Cambrian age (Richter & Richter 1941).

Material. Many specimens found by Professor E. Klitzsch and his group, as well as the author; Berlin TU collection.

Cruziana aegyptica n.ichnosp. Figures 32/4-5 and Plate 32/1, h
1969 Trilobite tracks – Weissbrod, Pl. 4, Fig. 1
Holotype: Locality F 29/86, upper Wadi Qena, Eastern Desert. (GPIT 1660/1); Fig. 32/4, lower specimen.
Diagnosis: Small, mostly rusophyciform *Cruziana* with indistinct endopodial scratches in the front and extrovert exopodial brushings in the rear parts of the lobes. In deeper rusophyciform casts the two areas are commonly separated by an oblique, retroverse ridge.

Remarks. Since this is a small to middle-sized ichnospecies, it is difficult to determine the claw formula

Plate 32.2 *Bergaueria sucta.* a. Holotype from trilobite – dated late lower Cambrian (Mickwitzia Sdst.) of Lugnås, Sweden (GPIT 1660/18); b. Specimen from loc. F5/86, Umm Bogma area, Sinai. Field photograph. Note the oblique radial markings near the rim of some disks! c. *Dimorphichnus obliquus*; loc. F5/86 (see above). Some of the sigmoidal rakings made in a right-hand gait (as this is a cast, it was actually left-hand) show the faint impressions of two smaller claws on the anterior side; e. *Dimorphichnus* cf. *quadrifidus.* Same specimen as Figure 32.11b; f. *Fucusopsis.* Loc. F5/86. The longitudinal cracks in the hyporelieves were caused by mechanical tension when worms stuffed their horizontal burrows above the sand-mud interface. No time significance. Note part of *Cr. salomonis* on top.

LOWER CAMBRIAN

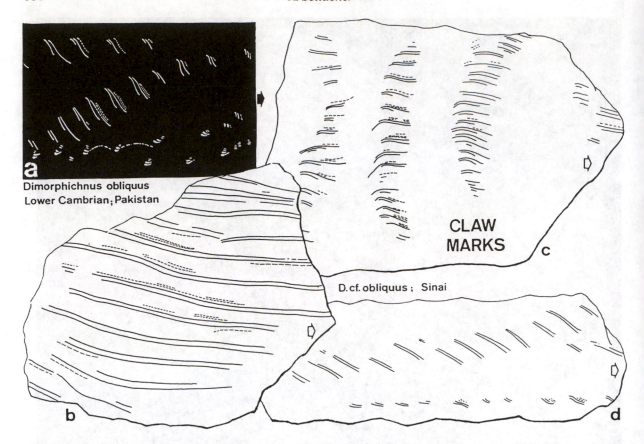

Figure 32.10 Specimens from the Sinai generally fail to show pusher impressions (probably due to undertrack deficiency), but the claw formula with two principal claws and two smaller ones in front of them agree well with *D. obliquus* topotype material (a, from Seilacher 1955; b, loc. F24/85; c and d, loc. F21/86, all Berlin collection).

of the endite scratches, which in contrast to the previously discussed ichnospecies are all retroverse (running in medio-posterior direction); i.e. they form an acute angle even in the anterior part of the burrow. There is a certain similarity with the mid-Cambrian *Cr. arizonensis* and the upper Cambrian *Cr. semiplicata* in the regular presence of longitudinally 'brushed' exite lobes in the posterior part of the burrow; but in contrast to these and all members of the barbata group, these brushings are extrovert (Fig. 32/4).

The Egyptian forms (from the Sinai as well as from the Eastern Desert) are generally small. They tend to occur in large numbers of 'coffee beans' and then show a conspicuous alignment with their broader front ends facing the current (Pl. 32/1, h). They do not co-occur with *Cr. salomonis*, neither on slabs nor in the same localities. This suggests that they represent a slightly lower level in all sections.

Dimorphichnus

Trilobite grazing tracks appear to be very common in

the Cambrian of the Sinai, but unfortunately their taxonomy and stratigraphic distribution in general has not yet been worked out in detail. As a first attempt to fill this gap, we present here three examples from Europe, all of them dated by trilobite body fossils and all associated with typical forms of *Cruziana*. Their succession will have to suffice as a preliminary standard for the new Egyptian occurrences.

Dimorphichnus cf. *obliquus* Seilacher (Figure 32/10 and Plate 32/2, c).

None of the European material is as well preserved as in the Pakistan generotype (Figs 32/9 and 32/10). Nevertheless some specimens from the lower Cambrian of Sweden, associated with *Cruziana dispar*, share with *Dimorphichnus obliquus* the presence of no more than two principal claws (Fig. 32/9, b-c). Most Sinai specimens (Fig. 32/10; Pl. 32/2, c) are even closer to the generotype in that they show in addition the faint scratches of two minute secondary claws, which are not preserved in the smaller tracks

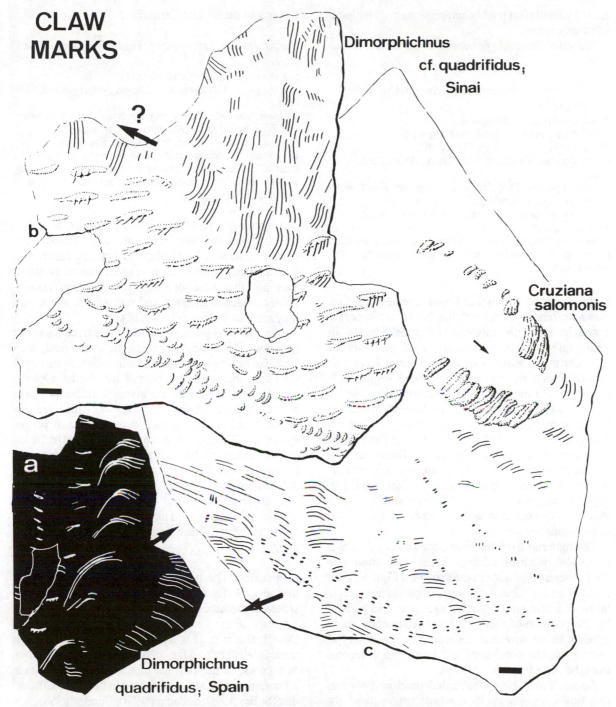

CLAW MARKS

Dimorphichnus cf. quadrifidus, Sinai

Cruziana salomonis

b

a

Dimorphichnus quadrifidus, Spain

c

Figure 32.11 Other specimens from Sinai appear to be more closely related to the middle Cambrian *D. quadrifidus* (see Fig. 32.9) by having four subequal scratches in the rakings and pectinate pusher impressions: a. Cerezedo near Bonar (GPIT 1660/12); b. loc. F24/85; c. loc. F21/86, both in Berlin collection.

from Sweden. But pusher impressions are lacking in most specimens.

Material. Several specimens in the Berlin collection.

Dimorphichnus quadrifidus n.ichnosp. Figure 32/11 and Plate 32/2, 3

?1922 *Eophyton* – Barthoux
?1942 *Protichnites* – Picard, Pl. 1, Figs. 1-2.
Holotype: Fig. 31/9, d (GPIT 1660/22)
Locus typicus: Cerecedo near Bonar (Prov. Leon. north Spain)
Stratum typicum: Mid-Cambrian sandstone, dated by trilobites in subjacent limestone.
Derivatio nominis: from the four subequal scratches in complete rakings.
Diagnosis: Dimorphichnus showing up to four subequal claw marks in the rakings and sometimes more in the pusher impressions.

In the holotype, which was found together with *Cruziana barbata*, several grazing tracks are superimposed. In our figure only one of them is depicted. Its rakings are in the majority bifid or trifid, probably due to undertrack incompleteness; but some have four scratches, as in the cotype (Fig. 32/9, e). Figure 32/9, d also shows three sets of pusher impressions, which bear claw impressions on what is here interpreted as the anterior flank. On their rear flanks we see the sediment interface deflected, including the rakings that were already inscribed on it. Therefore it is uncertain whether these push-impressions actually belong to the same trackway as the associated rakings. We also see a curved 'tail' of scratches merge into the (?) posterior ends in some of the pusher impressions.

The material from the Sinai, comprising six slabs, also show quadrifid rakings. Their association with the pusher impressions in the specimen Figure 32/11, c and Plate 32/2, 3 is, however, difficult to interpret in terms of a continuous grazing movement. In particular, the position of scooping and trailing claw marks appears to be reversed compared to the holotype. More complete specimens will be needed to reconstruct the complete track pattern.

Range. From the available information it appears that this ichnospecies as a whole ranges from the lower into the middle Cambrian.

Bergaueria sucta n.ichnosp. Figure 32/12 and Plate 32/2, a-b
Holotype: Pl. 32/2, a (GPIT 1660/18)
Locus typicus: Old mines near Lugnas (Vestergötland, Sweden).
Stratum typicum: Mickwitzia sandstone, dated as lower Cambrian by trilobites (*Holmia*) in overlying beds.
Derivatio nominis: Refers to the sucker disc shape.
Diagnosis: An ichnospecies of *Bergaueria*, in which the base is flat like a sucker disc and is commonly repeated by active lateral displacement. Marginal ring commonly with spiraling radial markings.

The name *Bergaueria* is usually applied to cylindrical hypichnial casts that tend to be deeper than wide and have a hemispherical bottom, sometimes with a central dimple. There is general agreement that they are the work of burrowing actinian coelenterates, or sea anemones, and it should be added that they are undertraces (Fig. 32/12).

The same origin is assumed for the trace fossils in question. But instead of being hemispherical, they look like the impressions of a sucker disk. Such might well have been the function of the flat base of the animal – the same way the inverted umbrella works in the modern jellyfish *Cassiopeia*, which has returned to a benthic life as a consequence of photosymbiosis. This would also explain the kind of creeping expressed by the multiple repetition of the same disc impression that is characteristic for *B. sucta.*

The numerous specimens from Sinai correspond to this description, except that they are somewhat smaller (Pl. 32/2, a-b). The oblique radial markings in some specimens also indicate a rotational movement.

Conclusion. The trace fossils so far studied suggest that several stages of the lower Cambrian are represented in the sandstones underlying the lower Carboniferous of the Umm Bogma area and the Eastern Desert. But it is also significant that no Ordovician trace fossils have so far been reported from this area, in contrast to the rich lower and upper Ordovician ichnofaunas that are represented in nearby south Jordan (Selley 1970, Seilacher 1970, Bender 1963).

Plate 32.3 a. *Cruziana rouaulti*; Jebel Ouweinat, southern Egypt (GPIT 1660/19; courtesy of Prof. Monod). This smooth form is well-known from the lower Ordovician of western Europe and Algiers; b, c. *Cruziana carbonaria*; lower Carboniferous, loc. F19/85, Wadi Feiran, Sinai (GPIT 1660/20); d. ?*Cruziana costata*; holotype (GPIT 1660/9), loc. F9/86, Umm Bogma, Sinai; e. *Planolites*, co-occurring with the previous form at loc. F9/86, but distinguished by lack of costae, backfill structure and convolute course; f. *Margaritichnus reptilis*; loc. F25/85. Compare Figure 12!; g-h. *Asteriacites gugelhupf* from loc. F19/86 (Berlin collection). In spite of the poor preservation the five-rayed symmetry can still be recognized.

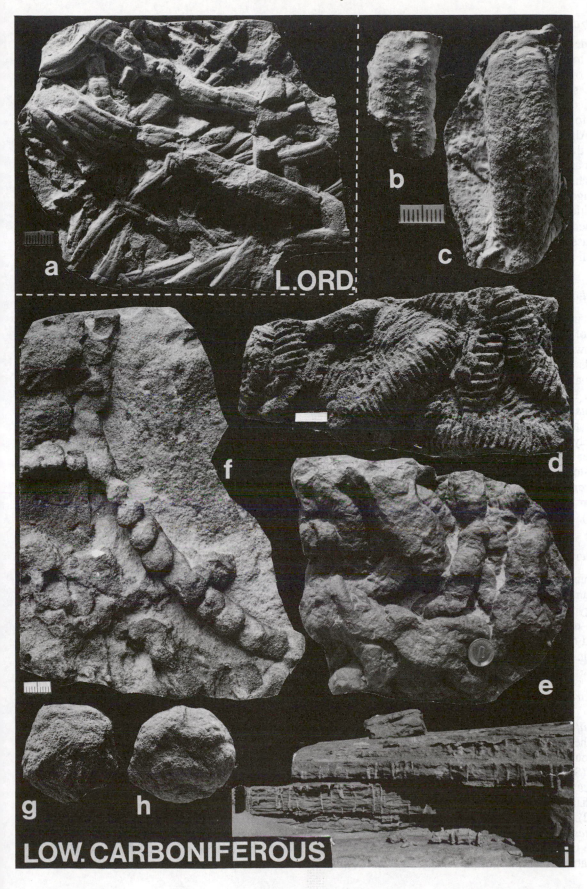

L.ORD.

LOW. CARBONIFEROUS

b) *Ordovician*

In order to meet typical Ordovician trace fossils, we have to turn to Jebel Uweinat in the extreme southwest of Egypt. From this area we know, through Professor Monod (Paris), about several slabs with the small *Cruziana rouaulti* Lebesconte (Pl. 32/3, a). This form, also known from the Arenigian of France and Algiers, is characterized by long and even furrows that show pleural ridges all along; but the exite lobes are completely smooth with not a single claw scratch. Nevertheless, we consider it as a member of the rugosa group, with which it co-occurs in the other localities. It is likely that the marine transgression reached the Uweinat area from the northwest, because similar ichnocoenoses are known from the Tassili and Tibesti areas.

c) *Silurian*

Silurian sandstone interbedded with clayey siltbeds west of Gilf Kebir and south of the Um Ras passage in southwest Egypt contains locally abundant trace fossils very similar to *Cruziana acacensis*, known originally from the Silurian Acacus sandstone of south Libya. The same strata also contain *Arthrophycus* (a worm burrow, also called *Harlania* by some authors), which in Libya is only found frequently in Silurian strata. South of the Egyptian border in northwest Sudan, transitional marine sandstone characterized by *Cruziana acacensis* and *Arthrophycus* covers large areas (K. Klitzsch pers. comm.) as they do in Benin (Seilacher & Alidou 1988).

d) *Devonian*

No trace fossils of Devonian age, as are known in the Fezzan, have come to my knowledge from localities in Egypt. Whether this means that marine transgressions during this period did not reach the Nubian shield or whether the deposits were simply eroded, remains to be seen.

e) *Carboniferous*

1. *Gilf Kebir*

A rather rich trace fossil assemblage of lower Carboniferous age has been reported from the Gilf Kebir area (Seilacher 1983). By Carboniferous times trilobites had declined. Instead, faunas are dominated by characteristic traces of other invertebrates (Fig. 32/12). Among these the burrows of starfishes (*Asteriacites gugelhupf*) indicate a suspension-feeding mode of life, which is rare among living asteroids. The attitude of the burrowed animal with steeply

bent-up arms, which was then inferred from the trace fossils, has since been observed in fossil asteroids from the German Muschelkalk (middle Triassic) and from the Eocene of Antarctica (D. Blake pers. comm.).

Another trace fossil that needs to be mentioned here is *Bifungites*, because it has created some confusion in the past. In the typical forms of Devonian and lower Carboniferous age, *Bifungites* resembles an arrow that has a bulky triangular arrowhead on either end. In older forms (Cambrian to Ordovician) the arrowheads are globular as in a dumbbell. This makes them less distinctive, because such forms can also result from the erosion of protrusive U-burrows with a connecting backfill structure (*Diplocraterion*), which are common throughout the Phanerozoic. This must be the origin of the structures that Issawi & Jux (1982) reported from the Nubian sandstones of Wadi Kirkaboub near Assuan, taking them to be an indication of Paleozoic age. In the meantime, *Inoceramus* shells have been found below that horizon, corroborating that we are dealing with Cretaceous deposits.

Among the other trace fossils found in the Gilf Kebir area are typically Paleozoic ichnogenera, such as *Trichophycus* with a teichichnoid backfill structure, the star-shaped *Asterichnus* and the cone-in-cone structures of *Conostichus* (Fig. 32/12). Other forms (*Scolicia*, *Neonereites*, *Zoophycos*, *Phycosiphon*) provide no time signature, while they may be used as indicating marine conditions.

2. *Sinai*

The lower Carboniferous of the Sinai, of which some horizons can be well dated by body fossils (Kora & Jux 1986), has recently yielded a broad variety of trace fossils. In contrast to the Cambrian, different species occur at different localities. This indicates that by this time ecological differentiation had progressed sufficiently to allow the distinction of small scale ichnofacies. Pending a more detailed mapping of the facies patterns, however, our present treatment has to be rather eclectic.

(a) In a high energy beach zone we find only the vertical tubes of *Skolithos*. (b) A more distal zone of thin sand/shale alternation (locality F 19/86) contains the richest assemblage of well-preserved forms. Among them are forms like *Asteriacites gugelhupf*, *Conostichus*, *Trichophycus* (Fig. 32/12) and *Cruziana*, which we know already from Gilf Kebir, while *Phycosiphon* is missing. But in addition we find the well-known *Curvolithus*, whose tongue-like epireliefves suggest the bulldozing of a flatworm (Fig. 32/12). (c) Less clean sands contain *Neonereites* when thin-bedded (locality F 20/86), while thicker beds (locality F 9/86) have their soles densely covered with the traces of sediment feeders, such as *Astersoma*, *Zoo-*

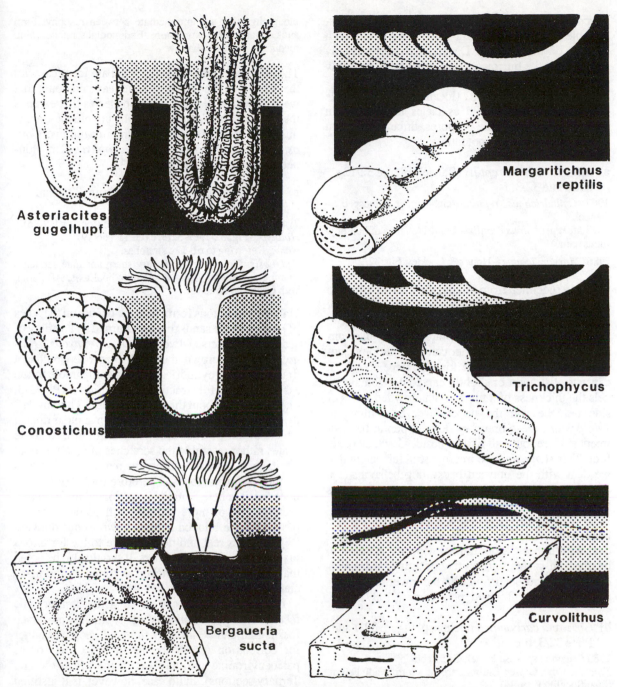

Figure 32.12 Characteristic marine trace fossils from the lower Cambrian and the lower Carboniferous. *Bergaueria sucta* is a typical actinian resting trace; but in contrast to other species of the same ichnogenus it reflects a sucker-like basal disc and is known only from the lower Cambrian. Figured specimen from Lugnas, Sweden (GPIT 1660/13) resembles those from Sinai (compare Plate 2a,b). *Conostichus* from the lower Carboniferous of the USA (specimen GPIT 1660/14 from Arkoma basin, Oklahoma) has a similar origin, but a conical shape with radial markings and an internal cone-in-cone structure. Specimens from Gilf Kebir and Sinai (loc. F19/86) are identical. *Asteriacites gugelhupf* is characteristically deeper than normal asteroid resting traces and known only from the lower Carboniferous (USA; Gilf Kebir; loc. F19/86, Wadi Feiran, Sinai, see Plate 3g, h). *Curvolithus* is a characteristic ichnogenus known from the Ordovician and from the lower Carboniferous through the Tertiary. So far it is not possible, however, to distinguish stratigraphically meaningful ichnospecies. The mode of backfilling and the geometry suggest a flatworm origin. Idealized diagram after specimens from loc. F19/86 (lower Carb., west Feiran) and elsewhere. *Trichophycus* is a teichichnoid backfill burrow known from the Ordovician (figured specimen from Seilacher 1983a) to the Carboniferous. As long as the characteristic scratch patterns have not been systematically evaluated, the ichnogenus is stratigraphically useless. *Margaritichnus* is a very distinctive backfill burrow that was so far known only from the upper Carboniferous of the USA (figured specimen GPIT 1660/15 from Minturn Formation, McCoy, Colorado). The occurrence of one identical specimen from the Sinai (Plate 3f) thus poses an interesting stratigraphic problem: longer range or younger rock?

phycos, *Fucusopsis* and *Planolites* as well as a questionable *Cuziana*. Dolomitic sandstones in the Wadi Dakhel of the Eastern Desert also show steeply inclined spreite burrows (*Rhizocorallium*), probably of crustacean origin, together with *Curvolithus*, *Neonereites* and *Trichophycus* (locality F 24/86). While all these forms are valid as facies indicators, but useless for stratigraphic purposes, three forms from the Sinai do have such potential.

a) *Margaritichnus reptilis* (Bandel) Figure 32/12
   and Plate 32/3, f

1967 *Cylindrichnus reptilis* n.ichnog., n.ichnospec. – Bandel

1973 *Margaritichnus reptilis* – Bandel (change of homonymous name)

1986? *Margaritichnus* – Houck & Lockley, Fig. 32/9

The only specimen from Sinai is identical in form and size with material from the upper Carboniferous of North America, which I was recently introduced to by Martin Lockley and collatorators during an excursion at the fourth North American Paleontology Convention. Unlike any other kind of trace fossils, it consists of straight or slightly curved horizontal tunnels the thickness of a finger, with crowded series of side branches on the upper side. In contrast to *Teichichnus*, these branches do not open to the sediment surface, but end like an upside-down elephant foot. This shape and the internal structure show that we deal with the unusual backfilling behavior of a sediment feeder, which so far has no counterparts outside the Carboniferous. Whether the Sinai occurrence extends the range of this distinctive ichnogenus back to the lower Carboniferous or whether we might deal with upper Carboniferous deposits, remains to be checked in the field. In the meantime *Margaritichinus* has also been found in the lower Carboniferous of Morocco (Tourani, pers. comm., 1987).

b) *Cruziana carbonaria* n.ichnosp. Figure 32/7 and
   Plate 32/3, b-c

1983 *Cruziana* sp. – Seilacher, Gilf Kebir

Type locality: Lower Carboniferous; south Wadi Feiran, Sinai (locality F 19/86).

Holotype: Fig. 32/7 (GPIT 1660/21).

Diagnosis: Small *Cruziana* in the shape of trough-like

ploughings that are intermediate between rusophyciform and cruzianaeform expressions. Endopodial scratches faint; median sulcus very shallow.

The new specimens from the Sinai are identical with the ones from the Gilf Kebir and leave no doubt that we deal with a distinctive ichnospecies that extends the range of *Cruziana* stratigraphy up into the Carboniferous. It occurs togther with *Curvolithus*, *Conostichus*, *Asteriacites* and *Trichophycus* in thin-bedded sand/shale alternations.

c) ?*Cruziana costata* n.ichnosp. Figure 32/7 and
   Plate 3, e

Type locality: Umm Bogma area of Sinai (locality F 9/86).

Holotype: Fig. 32/7; Pl. 32/3, e (GPIT 1660/9).

Name: Referring to rib-like stratches.

Diagnosis: Deep lobes with very sharp annulate scratches running at right angles to the longitudinal axis. Affiliation with *Cruziana* doubtful.

Unlike the previous form, this one occurs on the soles of rather impure sands that are churned by a variety of sediment feeders (*Asterosoma*, *Zoophycos*, *Planolites*). In the crowded assemblages, one would at a first glance see cylindrical bodies comparable to the associated backfill structures of *Planolites* (Pl. 32/3, e), which have about the same diameter. This similarity, however, is disproven by the sharp annular ridges, which would be unusual for a worm burrow. These ridges rather resemble the scratches of an arthropod, particularly since they show the marks of side claws on their anterior slopes. Therefore this very distinct trace fossil is here tentatively attributed to *Cruziana*. Examination of more samples will be necessary to prove that the bilobed nature shown in our drawing (Fig. 32/7) is real and not only due to the fortuitous juxtaposition of two cylinders. It should also be noted that these burrows are never as torted as the associated *Planolites*.

In conclusion, we can say that the stratigraphic record of Egypt contains many interesting trace fossils that can be used by the field geologist either for correlation (as in the lower Paleozoic) or as paleoenvironmental clues (as in the Mesozoic and Tertiary sections). In no case, however, is it justified to simply neglect this very telling part of the paleontological record.

PART 7

Annexes

References

ABBAS, H. L. 1961. *A monograph of Egyptian Cretaceous pelecypods.* Geol. Surv. Egypt, Monograph 1, 208pp.

ABBAS, H. L. 1963. *A monograph of Egyptian Cretaceous gastropods.* Geol. Surv. Egypt, Monograph 2, 145pp.

ABBAS, H. L. 1967. *A monograph of Egyptian Paleocene and Eocene gastropods.* Geol. Surv. Egypt, Monograph 4, 154pp.

ABBAS, H. L. 1972. A monograph of Egyptian Pliocene and Eocene pelecypods. *Egypt. J. Geol.* 16: 69-200.

ABBAS, H. L. & M. M. HABIB 1971. Hydrogeological studies on west Mawhoob area, south Western Desert, Egypt. *Bull. Inst. Desert Egypte* 21(2): 265-282.

ABDALLAH, A. M. 1964. *New Bathonian (middle Jurassic) occurrence at the western side of the Gulf of Suez.* Geol. Surv. Egypt, Paper 30, 8pp.

ABDALLAH, A. M. 1966. *Stratigraphy and structure of a portion of the north Western Desert of Egypt (El Alamein, Dabaa, Qattara, Moghra areas) with reference to economic potentialities.* Geol. Surv. Egypt, Paper 45, 19pp.

ABDALLAH, A. M. 1967. *Geology of some gypsum deposits in the northern Western Desert of Egypt.* Geol. Surv. Egypt, Paper 41, 11pp.

ABDALLAH, A. M. & F. M. ABDEL HADY 1968. Geology of the Sadat area, Gulf of Suez. *Egypt. J. Geol.* 10: 1-24.

ABDALLAH, A. M. & A. ADINDANI 1963. *Stratigraphy of upper Paleozoic rocks, western side of Gulf of Suez.* Geol. Surv. Egypt, Paper 25, 18pp.

ABDALLAH, A. M., A. ADINDANI & N. FAHMY 1963. *Stratigraphy of lower Mesozoic rocks, western side of Gulf of Suez, Egypt.* Egypt. Geol. Surv., Paper 27, 23pp.

ABDALLAH, A. M. & R. A. EISSA 1971. The Campanian rocks of the southern Galala. *Bull. Fac. Sci., Cairo Univ.* 44: 259-270.

ABDALLAH, A. M. & M. EL KADI 1974. Gypsum deposits in the Fayum province, Egypt. *Bull. Fac. Sci., Cairo Univ.* 47: 335-346.

ABDALLAH, A. M., M. A. EL SHARKAWI & M. MARZOUK 1971. Geology of Mersa Thelemt area, southern Galala, Eastern Desert, ARE. *Bull. Fac. Sci., Cairo Univ.* 44: 271-280.

ABDALLAH, A. M., M. Y. MENEISY & M. M. SHAABAN 1973. A geological study on the volcanic rocks of Wadi Araba and Abu Darag areas, Gulf of Suez region, Egypt. *Egypt. J. Geol.* 17: 111-123.

ABDEL AAL, A. Y. 1981. *Comparative petrological and geochemical studies of post-Cambrian basaltic rocks in Egypt.* Ph.D. Thesis, Minia Univ.

ABDEL AZIZ, A. M., S. GABRA & S. AHMED 1971. El Omayid gypsum deposit. *Ann. Geol. Surv. Egypt* 1: 111-116.

ABDEL AZIZ, M. G. 1968. *The geology of Wadi Kareim area.* Ph.D. Thesis, Cairo Univ.

ABDEL GAWAD, M. 1970. The Gulf of Suez; a brief review of stratigraphy and structure. *Phil. Trans. Roy. Soc., London* 267(A): 41-48.

ABDEL GAWAD, M. 1970. Interpretation of satellite photographs of the Red Sea and Gulf of Aden. *Phil. Trans. Roy. Soc., London* 267(A): 23-40.

ABDEL HADI, Y. 1969. Marine Pliocene planktonic foraminiferal zonation southeast of Salum. *Rev. Ital. Paleontol.* 75: 833-846.

ABDEL KHALEK, M. L. 1980. Tectonic evolution of basement rocks in the southern and central Eastern Desert of Egypt. *Bull. Inst. Applied Geol., King Abdul Aziz Univ., Jeddah* 3(1): 53-62.

ABDEL KIREEM, M. R. 1985. Planktontic foraminifera of Mokattam Formation (Eocene) of Gebel Mokattam, Cairo, Egypt. *Rev. Micropaleontol.* 28: 77-96.

ABDEL KIREEM, M. R. & H. F. ABDOU 1979. Upper Cretaceous-lower Tertiary foraminifera from south Galala Plateau, Eastern Desert, Egypt. *Rev. Espan. Micropaleontol.* 11(2): 175-222.

ABDEL KIREEM, M. R., A. BLONDEAU & K. H. SHAMAH 1985. Les nummulites de la localité-type de la formation de Gehannam, Le Fayoum, Egypte. *Bull. Sci. Hist. Toulouse* 121: 65-71.

ABDELMALIK, W. M. 1981. Biostratigraphy and history of sedimentation of the Jurassic rocks in the north Western Desert, Egypt. *Newsl. Stratigr.* 10: 148-155.

ABDELMALIK, W. M. 1982. Calcareous nannoplankton from the sequence between Dakhla and Esna Shale formations (upper Cretaceous-lower Eocene) in Quseir area, Egypt. *Rev. Espan. Micropaleontol.* 14: 73-84.

ABDEL MONEM, A. A. & M. A. HEIKAL 1981. Major element composition, magma type and tectonic environment of the Mesozoic to Recent basalts, Egypt. *Bull. Fac. Earth Sci., King Abdul Aziz Univ., Jeddah* 4: 121-148.

ABDEL MONEM, A. A. & P. M. HURLEY 1979. U-Pb dating of zircons from psammitic gneisses, Wadi Abu Rosheid-Wadi Sikait area, Egypt. *Bull. Inst. Applied Geol., King Abdul Aziz Univ., Jeddah* 3(2): 165-170.

ABDEL MONEM, A. A. & P. M. HURLEY 1980. U-Pb dating of detrital zircons from black sands, Nile Delta and Nubian Sandstones, Eastern Desert, Egypt. *Ann. Geol. Surv. Egypt* 10: 651-663.

ABDEL MONEM A. A. & P. M. HURLEY 1980. Age of Aswan monumental granite, Egypt by U-Pb dating of zircons. *Bull. Inst. Applied Geol., King Abdul Aziz Univ., Jeddah* 3(3): 141-144.

ABDEL RAHMAN, M. A. & H. A. EL ETR 1980. The orientational characteristics of the structural grain of the Eastern Desert of Egypt. *Bull. Inst. Applied Geol., King Abdul Aziz Univ., Jeddah* 3(3): 45-55.

ABDEL RAHMAN, M. A., A. A. HUSSEIN & H. A. EL ETR 1981. A statistical study of the relations between basement fractures, aeromagnetic lineaments and tectonic patterns, south Eastern Desert, Egypt. In: D.W. O'Leary & J.J. Earle (eds), *Proc. Intl. Conf. Basement Tectonics*: 103-118.

ABDEL RAZIK, T. M. 1967. Stratigraphy of the sedimentary cover of Anz-Atshan, south Duwi district. *Bull. Fac. Sci., Cairo Univ.* 41: 153-179.

ABDEL RAZIK, T. M. 1972. Comparative studies on the upper Cretaceous-early Paleocene sediments of the Red Sea coast, Nile Valley and Western Desert, Egypt. *8th Arab Petrol. Congr., Algiers*, Paper (B-3), 23pp.

ABDEL SATTAR, G. 1983. *Biostratigraphy of Paleozoic rocks in north Egypt*. Ph.D. Thesis, Ain Shams Univ.

ABDEL SHAFY, E. 1984. Stratigraphy of the subsurface Jurassic in Qattara Depression and its relation with similar sections in Egypt and in the Tethyan region. In: J. Klerkx & J. Michot (eds), *African Geology*, Tervuren: 247-259.

ABDEL TAWAB, M. M. 1961. *The ilmenite deposit of Abu Ghalaga, Eastern Desert, Egypt. Internal Report, Geol. Surv. Egypt.*

ABDEL TAWAB, M. M. 1978. A comparative study of the metallogenic provinces in the ophiolite belts and ring complexes in both Egypt and Saudi Arabia between Lat. 22-24N. In: H. Zapfe (ed.), *Scientific results of the Arabian projects of the IGCP until 1976, Oesterr. Akad. Wiss. Schriftr. Erdwiss. Komm.* 3: 45-55.

ABDEL WAHAB, S. & M. A. G. KHALIFA 1984. The lower Eocene dolomitized El Nashfa Formation in west Beni Mazar area, Western Desert, Egypt. *J. Afr. Earth Sci.* 2: 333-340.

ABDEL WAHAB, S. & M. A. G. KHALIFA 1984. Sedimentology of the middle Eocene Minia and Samalut Formations, west Beni Mazar, Western Desert, Egypt. *J. Afr. Earth Sci.* 2: 341-350.

ABDINE, A. S. 1981. Egypt's petroleum geology: good ground for optimism. *World Oil* December: 99-112

ABDINE, A.S.1974. *Oil and gas discoveries in the Western Desert of Egypt.* Internal Report 27, Western Desert Operating Oil Co. (WEPCO).

ABDINE, A. S. 1981. The Gulf of Suez has excellent potential. *World Oil* July: 147-149.

ABDINE, A. S. & S. DEIBIS 1972. Lower Cretaceous Aptian sediments and their oil prospects in the northern Western Desert, Egypt. *8th Arab Petrol. Congr.*, Algiers.

ABDOU, H. F. & M. R. ABDEL KIREEM 1975. Planktonic foraminiferal zonation of the middle and upper Eocene rocks of Fayoum Province, Egypt. *Rev. Espan. Micropaleontol.* 7(3): 15-46.

ABDOU, M. I. 1983. *Geological studies of some volcanic rocks in northern Egypt.* M.Sc. Thesis, Cairo Univ.

ABED, M. M., S. F. ANDRAWIS & A. E. EL BASSIOUNI 1978. Shebin El Kom, a new formation in the Nile Delta of Egypt. *Bull Fac. Sci., Mansoura Univ.* 6: 475-483.

ABOU KHADRAH & S. A. WAHAB 1984. Petrography and diagenesis of the Samh Formation and younger sediments, north Mersa Alam area, Red Sea coast, Egypt. *J. Afr. Earth Sci.* 2: 277-286.

ABRAMS, M. J. & B. S. SIEGAL 1980. Lithologic mapping. In: B. S. Siegal & A. R. Gillespie (eds), *Remote sensing in geology.* Wiley, New York: 381-418.

ABU EL ELA, F. F. 1985. *Ophiolitic melange of the Abu Mireiwa district, central Eastern Desert, Egypt.* Ph.D. Thesis, Assiut Univ.

ABU EL NAGA, M. 1984. Paleozoic and Mesozoic depocenters and hydrocarbon generating areas, northern Western Desert. *Proc. 7th Petrol. Explor. Seminar,* EGPC, Cairo: 269-287.

ABUL GADAYEL, A. 1974. *Contribution to geology and geochemistry of Wadi Natash lava flows.* M.Sc. Thesis, Cairo Univ.

ADAMS, R. D. 1977. Survey of practice in determining magnitudes of near earthquakes; part 2: Europe, Asia, Africa, Australasia, the Pacific. *World Data Center for Solid Earth Geophysics*, Report SE-8 (Boulder), 65pp.

ADAMS, R. D. 1983. Incident at the Aswan dam. *Nature* 301: 14.

ADINDANI, A., R. RABJEZEK, H. A. YOUSSEF & S. M. AWAD 1975. Evaluation of El-Gharbaniyat gypsum deposits. *Ann. Geol. Surv. Egypt* 5: 123-136.

ADINDANI, A. & SHAKHOV 1970. The occurrence of coal and some geological features of the Mesozoic and Paleozoic sediments of Egypt. In: O. Moharram et al. (eds), *Studies on some mineral deposits of Egypt.* Geol. Surv. Egypt: 67-86.

ADLAND, A. & A. A. HASSAN 1972. Hydrocarbon potential of the Abu Gharadig basin in the Western Desert, ARE. *8th Arab Petrol. Congr.*, Algiers.

AFFLECK, J. 1963. Magnetic anomaly trend and spacing patterns. *Geophysics* 28: 379-395.

AFIA, M. S. & I. IMAM 1979. *Mineral map of Egypt.* Geol. Surv. Egypt.

AGAH, A. 1981. *Structural map and plate reconstruction of the Gulf of Suez-Sinai area.* Internal report, Conoco Oil Co., Houston, Texas, USA.

AHMED, A. A. 1985. *Geology of uranium mineralization in El Missikat area, Qena-Safaga road, Eastern Desert, Egypt.* M.Sc. Thesis, Azhar Univ., Cairo.

AHMED, S. S. 1972. Geology and petroleum prospects in eastern Red Sea. *Bull. Am. Assoc. Petrol. Geol.* 56: 707-719.

AHWANY, M. M. 1982. Geological and sedimentological studies on Gebel Shabrawet area, Suez Canal district. *Ann. Geol. Surv. Egypt* 12: 305-381.

AIGNER, T. 1982. Event stratification in nummulite accumulations and in shell beds from the Eocene of Egypt. In: G. Einsele & A. Seilacher (eds), *Cyclic and event stratification.* Springer, Berlin: 248-262.

AKAAD, M. K. 1957. Petrography, structure and time relations of the Igla Formation. *Egypt. J. Geol.* 1: 93-102.

AKAAD, M. K. 1959. The migmatitic gneisses of Wadi Feiran, Sinai, Egypt. *Bull. Sci. Techn., Assiut Univ.* 2: 211-237.

AKAAD, M. K. 1972. A contemplation and assessment of the 1960-1961 classification of rocks of the central Eastern Desert. *Ann. Geol. Surv. Egypt* 2: 19-45.

AKAAD, M. K. & A. A. DARDIR 1983. Geology of Wadi El Dabbah iron ore deposits, Eastern Desert, Egypt. *Bull.*

Fac. Earth Sci., King Abdul Aziz Univ., Jeddah 6: 611-622.

AKAAD, M. K., S. EL GABY & A. A. ABBAS 1967. On the evolution of Feiran migmatites, Sinai. *Egypt. J. Geol.* 11: 49-58.

AKAAD, M. K., S. EL GABY & M. S. HABIB 1973. The Barud gneisses and the origin of the grey granite. *Bull. Fac. Sci., Assiut Univ.* 2: 55-69.

AKAAD, M. K. & M. F. EL RAMLY 1958. *Seven new occurrences of the Igla Formation in the Eastern Desert of Egypt.* Geol. Surv. Egypt, Paper 3, 37pp.

AKAAD, M. K. & M. F. EL RAMLY 1960. *Geological history and classification of the basement rocks of the central Eastern Desert of Egypt.* Geol. Surv. Egypt, Paper 9, 24pp.

AKAAD, M. K. & M. F. EL RAMLY 1961. *The nepheline-syenite ring complex of Gebel Abu Khrug (south Eastern Desert of Egypt).* Geol. Surv. Egypt, Paper 24, 22pp.

AKAAD, M. K. & M. F. EL RAMLY 1963. *Geology and structure of the Umm Lassaf-Umm Nar iron belt (Eastern Desert of Egypt).* Geol. Surv. Egypt, Paper 17, 23pp.

AKAAD, M. K. & M. F. EL RAMLY 1963. *The cataclastic-mylonitic gneisses north of Gabal El Maiyit and the origin of the granite of Shaitian type.* Geol. Surv. Egypt, Paper 26, 14pp.

AKAAD, M. K. & M. A. ESSAWY 1964. The meta-gabbro-diorite complex NE of Gabal Atud, Eastern Desert, and the term 'epidiorite'. *Bull. Sci. Techn., Assiut Univ.* 7: 83-108.

AKAAD, M. K. & M. A. ESSAWY 1965. Petrology, origin and sedimentation of the Atud Formation and its bearing on the early part of the geological history of the basement complex of the Eastern Desert of Egypt. *Bull. Sci. Techn., Assiut Univ.* 8: 75-102.

AKAAD, M. K. & M. A. ESSAWY 1965. Apinites and apinitic diorites from the Idfu-Mersa Alam road, Eastern Desert. *Bull. Sci. Techn., Assiut Univ.* 8: 55-74.

AKAAD, M. K. & G. A. MOUSTAFA 1963. *The Shaitian granite, cataclastic mylonitic granodiorite.* Geol. Surv. Egypt, Paper 19, 18pp.

AKAAD, M. K. & M. H. NAGGAR 1963. The deposit of Egyptian alabaster at Wadi El-Assiuty. *Bull. Soc. Geograph. Egypte* 36: 29-37.

AKAAD, M. K. & M. H. NAGGAR 1963. Geology of Wadi Sannur alabaster. *Bull. Inst. Desert Egypte* 13(2): 35-63.

AKAAD, M. K. & A. NOWEIR 1969. Lithostratigraphy of the Hammamat-Um Seleimat district, Eastern Desert, Egypt. *Nature* 223: 284-285.

AKAAD, M. K. & A. NOWEIR 1980. Geology and lithostratigraphy of the Arabian Desert orogenic belt of Egypt between Lat. 25° 35' and 26° 30' N. *Bull. Inst. Applied Geol., King Abdul Aziz Univ., Jeddah* 3(4): 127-135.

AKAAD, M. K. & A. G. SHAZLY 1972. Description and petrography of the Meatiq Group, Eastern Desert. *Ann. Geol. Surv. Egypt* 2: 215-238.

AKKAD, S. E. & A. M. ABDALLAH 1971. Contribution to the geology of Gebel Ataqa area. *Ann. Geol. Surv. Egypt* 1: 21-42.

AKKAD, S. E. & A. DARDIR 1966. *Geology of the Red Sea coast between Ras Shagra and Mersa Alam with short note on exploratory work at Gebel El Rusas lead-zinc deposits.* Geol. Surv. Egypt, Paper 35, 67pp.

AKKAD, S. E. & A. DARDIR 1966. *Geology and phosphate deposits of Wassif-Safaga area.* Geol. Surv. Egypt, Paper 36, 35pp.

AKKAD, S. E. & B. ISSAWI 1963. *Geology and iron ore deposits of Bahariya Oasis.* Geol. Surv. Egypt, Paper 18, 300pp.

ALBRITTON, C. C., J. E. BROOKS, B. ISSAWI & A. SWEDAN 1985. Origin of the Qattara depression, Egypt. Southern Methodist Univ., Dalas, Texas, USA (unpublished).

AL FAR, D. M. 1964. Some new formational names in central Sinai. *12th Annual Meeting, Geol. Soc. Egypt* (Abstr.).

AL FAR, D. M. 1966. *Geology and coal deposits of Gebel Maghara, north Sinai.* Geol. Surv. Egypt. Paper 37, 59pp.

AL FAR, D. M., H. W. HAGEMANN & S. OMARA 1965. Beiträge zur Geologie zur Kohlfürenden Gebietes von El Moghara, Nord Sinai (Ägypten). *Geol. Mitt.* 4: 397-429.

ALI, M. & A. FOUDA 1984. The petrophysics and reservoir geology of the Nubia Sandstone of the October oilfield, Gulf of Suez, Egypt. *Proc. 7th Petrol. Explor. Seminar, EGPC*, Cairo: 177 (Abstr.).

ALMAGOR, G. et al. 1976. The geology of the southeastern Mediterranean Sea. *Bull. Geol. Surv. Isr.*, 68.

ALLAM, A. 1985. Abu Madi field case history: evaluation of reserves and productivity. *8th Petrol. Explor. Seminar, EGPC*, Cairo.

ALLAM, A. M. 1986. A regional and paleoenvironmental study on the upper Cretaceous deposits of the Bahariya Oasis, Libyan Desert, Egypt. *J. Afr. Earth Sci.* 5: 407-412.

ALLAN, T. D. 1970. Magnetic and gravity fields over the Red Sea. *Phil. Trans. Roy. Soc., London* 267(A): 153-180.

ALLEN, G., A. AYYAD, G. DESFORGES, M. HADDADI & J. PIZON 1984. Subsurface sedimentological study of the Rudeis Formation in Kareem, Ayun, Yusr and Shukheir fields. *Proc. 7th Petrol. Explor. Seminar, EGPC*, Cairo: 164-176.

ALLEMANN, F. & T. PETERS 1972. The ophiolite-radiolarite belt of the north Oman mountains. *Eclog. Geol. Hevl.* 63: 657-697.

ALLUM, J. A. E. 1966. *Photogeology and regional mapping.* Pergamon Press, Oxford, 107pp.

ALMOND, D. C. 1979. Younger granite complexes of Sudan. *Bull. Inst. Applied Geol., King Abdul Aziz Univ., Jeddah* 3(1): 151-64.

AL SHANTI, A. M. S. & I. G. GASS 1983. The upper Proterozoic ophiolite melange zones of the easternmost Arabian-Nubian Shield. *J. Geol. Soc. London* 140: 867-876.

AL SHANTI, A. M. S. & A. H. G. MITCHELL 1976. Late Precambrian subduction and collision in the Al-Amar-Idsas region, Arabian Shield, Kingdom of Saudi Arabia. *Tectonophysics* 30: 41-47.

AL SHANTI, A. M. S. & M. J. ROOBOL 1979. Some thoughts on metallogenesis and evolution of the Arabian-Nubian Shield. *Bull. Inst. Applied Geol., King Abdul Aziz Univ., Jeddah.* 3(1): 87-96.

ALY, M. M. & M. E. MOUSTAFA 1984. Major chemistry statistics characterizing common igneous rocks of Egypt. *9th Intl. Congr. Statistics, Ain Shams Univ., Cairo* 4: 271-313.

ALY, M. M., M. F. EL RAMLY & M. L. KABESH 1983.

Analysis of Egyptian rocks. Geol. Surv. Egypt, Paper 6, 119pp.

AMBRASEYS, N. N. 1961. On the seismicity of southwest Asia (from a XV Century Arabic manuscript). *Rev. pour l'Etude des Calamites* (Geneva), 37.

AMBROGGI, R. P. 1966. Water under the Sahara. *Scientific American* 214: 21-29.

AMER, A., S. NOUR & M. MESHRIKI 1981. A finite element model for the Nubian aquifer system in Egypt. *Proc. Intl. Conf. on Water Resources Management in Egypt, Cairo*: 327-361.

AMER, A. F. & M. M. ABDEL TAWAB 1958. *Report on the geology of the ilmenite deposit at Abu Ghalaqa, Hamata, south Eastern Desert.* Internal Report, Geol. Surv. Egypt.

AMER, A. F. & A. O. MANSOUR 1955. *Geology of El Daghbag-El Gindi district.* Geol. Surv. Egypt, 69pp.

AMER, K., M. D. SAMUEL & G. S. SALEEB-ROUFAIEL 1982. Petrography and petrochemistry of subsurface basalts in Mit Ghamr and Abu Hammad wells. *Egypt. J. Geol.* 26: 83-94.

AMIN, M. S. 1947. A tin-tungsten deposit in Egypt. *Econ. Geol.* 43: 133-153.

AMIN, M. S. 1948. Origin and alteration of chromites from Egypt. *Econ. Geol.* 43: 133-153.

AMIN, M. S. 1954. The ilmenite deposit of Abu Ghalaga, Egypt. *Econ. Geol.* 49: 77-87.

AMIN, M. S. 1955. *Geology and mineral deposits of Umm Rus Sheet.* Geol. Surv. Egypt, 51pp.

AMIN, M. S. 1955. Some regional features of the Precambrian in the central Eastern Desert, Egypt. *Bull. Inst. Desert Egypte* 5(1): 193-207.

AMIN, M. S. 1955. Geological features of some mineral deposits in Egypt. *Bull. Inst. Desert Egypte* 5(1): 208-239.

AMIN, M. S. 1961. Subsurface features and oil prospects of the Western Desert, Egypt. *3rd Arab Petrol. Congr.*, Alexandria, 8pp.

AMIN, M. S. & M. S. AFIA 1965. Anthophyllite-vermiculite deposit of Hafafit, Eastern Desert, Egypt. *Econ. Geol.* 49: 317-327.

AMIN, M. S., M. S. MANSOUR, M. L. A. KABESH & D. M. AL FAR 1953. *Geology of the Naba district.* Geol. Surv. Egypt, 41pp.

AMIN, M. S. & I. H. MOHAMED 1954. *Geology of Umm Lassaf district.* Geol. Surv. Egypt, 13pp.

AMIN, M. S., G. A. MOUSTAFA & M. A. ZA'ATOUT 1954. *Geology of Abu Diab district.* Geol. Surv. Egypt, 42pp.

AMIN, M. S., A. H. SABET & A. O. S. MANSOUR 1953. *Geology of Atud district.* Geol. Surv. Egypt, 71pp.

ANDRAWIS, S. F. 1967. Biostraitgraphic study of the subsurface Paleocene sections of Ezz El Orban-Ras El Behar region in the western coast of the Gulf of Suez. *6th Arab Petrol. Congr.*, Baghdad, Paper 42.

ANDRAWIS, S. F. 1969. Bulimina and related genera from the Miocene of some wells in the Gulf of Suez region. *Proc. 3rd Afr. Micropaleontol. Colloq.* Cairo: 1-9.

ANDRAWIS, S. F. 1970. Planktonic zonation of the subsurface sections of Ezz El Orban-Ras El Behar region in the Gulf of Suez, UAR. *7th Arab Petrol. Congr.*, Kuwait, Paper 7, 12pp.

ANDRAWIS, S. F. 1972. New biostratigraphic contribution for the upper part of the Paleozoic rocks of Gibb Afia

well No. 2, Western Desert, Egypt. *8th Arab Petrol. Congr.* Algiers, Paper (B)76.

ANDRAWIS, S. F. & W. M. ABDEMALIK 1981. Lower/middle Miocene boundary in the Gulf of Suez region. *Newsl. Stratigr.* 10: 156-163.

ANDREW, G. 1934. Note on the 'Cephren diorite'. *Bull. Inst. Egypte* 16: 105-109.

ANDREW, G. 1934. The structure of Esh-Mellaha range (Eastern Desert of Egypt 27 30-28N). *Bull. Inst. Egypte* 16: 47-59.

ANDREW, G. 1935. On rocks from the south Eastern Desert and west central Sinai. *Bull. Inst. Egypte* 17: 205-221.

ANDREW, G. 1937. The late Tertiary igneous rocks of Egypt (field relations). *Bull. Fac. Sci., Cairo Univ.* 10: 61pp.

ANDREW, G. 1938. On Imperial Porphyry. *Bull. Inst. Egypte* 20: 63-81.

ANDREW, G. 1939. The greywackes of the Eastern Desert of Egypt. *Bull. Inst. Egypte* 21: 153-190.

ANDREWS, C. W. 1900. On a new species of Chelonia from the lower Miocene of Egypt. *Geol. Mag.* 7 (decade 4): 427-431.

ANDREWS, C. W. 1901. Preliminary note on some recently discovered extinct vertebrates from Egypt. *Geol. Mag.* 8 (decade 4): 400-409, 436-44, 528.

ANDREWS, C. W. 1902. Note on a Pliocene vertebrate fauna from the Wadi Natrun, Egypt. *Geol. Mag.* 9 (decade 4): 433-439.

ANDREWS, C. W. 1906. *The extinct mammals of Egypt. A descriptive catalogue of the Tertiary vertebrata of the Fayum, Egypt, based on the collection of the Egyptian Government in the Geological Museum, Cairo, and on the collection of the British Museum (Natural History).* Trustees of the British Museum, London, 324pp.

ANDREWS, C. W. 1916. Note on the sternum of a large carinate bird from the (?) Eocene of southern Nigeria. *Proc. Zool. Soc. London*: 519-524.

ANDREWS, C. W. & H. J. L. BEADNELL 1902. *A preliminary note on some new mammals from the Fayum (Phiomia and Saghatherium).* Egypt Surv. Dept., 9pp.

ANGELIER, J. 1985. Extension and rifting: the Zeit region, Gulf of Suez. *J. Struct. Geol.* 7: 605-612.

ANGELIER, J. & B. COLETTA 1983. Tensional fractures and extensional tectonics. *Nature* 301: 49-51.

ANONYMOUS 1938. Aerial photography used extensively in New Guinea oil search. *World Petrol.* 8: 44-47.

ANONYMOUS 1961. *Teknichsky instruktsia po gravimetrichiskoy razvedke* (in Russian). Moscow, 69pp.

ANONYMOUS 1972. Penrose field conference on ophiolites. *Geotimes* 17: 24-25.

ANONYMOUS 1974-1981. *Tidal Gravity corrections:* Tables prepared by Service hydrographique de la Marine et Compagnie Generale de Geophysique.

ANSARY, S. E. 1955. Report on the foraminiferal fauna from the upper Eocene of Egypt. *Bull. Inst. Desert Egypte* 6: 1-160.

ANSARY, S. E., S. F. ANDRAWIS & S. E. FAHMY 1963. Biostratigraphic studies of the upper Cretaceous sections in GPC wells, Eastern Desert and Sinai. *4th Arab Petrol Congr.*, Beirut, Paper 24.

ANSARY, S. E. & M. M. ISMAIL 1963. Middle Eocene foraminifera from Gebel Hof, north of Helwan, Egypt. *Bull. Fac. Sci., Alexandria Univ.* 5: 115-135.

ANSARY, S. E. & N. TEWFIK 1966. Planktonic foraminifera

and some new benthonic species from the subsurface upper Cretaceous of Ezz El Orban area, Gulf of Suez. *Egypt. J. Geol.* 10: 37-76.

ANWAR, Y. M. 1964. Note on the occurrence of copper minerals at Ras Benas, Eastern Desert. *Egypt J. Geol.* 8: 89-94.

ANWAR, Y., H. KOTB & N. ZOHNY 1969. Geochemistry of Egyptian chromites. *Bull. Fac. Sci., Alexandria Univ.* 9: 135-146.

ARAFA, A. A. 1982. Calcareous nannofossils from the Kareem Formation (middle Miocene), Gulf of Suez area, Egypt. *N. Jb. Geol. Paleontol., Mh.* 8(1982): 449-455.

ARGYRIADIS, I., P. C. DE GRACIANSKY, J. MARCOUX & L. E. RICO 1980. The opening of the Mesozoic Tethys between Eurasia and Arabia-Africa. In: *Geology of the Alpine chains born of the Tethys*. Mem. Bureau recherches geol. et min., 115: 199-214.

ARKELL, W. J. 1952. Jurassic ammonites from Jebel Tuwaiq, central Arabia. *Phil. Trans. Roy. Soc. London* 236(B):241-313.

ARKELL, W. J. 1956. *Jurassic geology of the world*. Oliver & Boyd, London, 806pp.

ASHMAWY, M. H. 1987. The ophiolte melange of the south Eastern Desert of Egypt: Remote sensing, field work and petrographic investigations. *Berl. Geowiss. Abh.* 84(A): 1-135.

ASHRI, A. H. 1973. The movement of sand dunes at Kharga Oasis. *Egypt. J. Geol.* 17: 37-46.

ASRAN, M. H. A. 1985. *Geology, petrography and geochemistry of the apogranites of the Nuweibei and Abu Dabbab areas. Eastern Desert. Egypt.* M.cs. Thesis Assiut Univ.

ASSAF, H. S. A. 1973. *Structure and radio-active mineralization of Wadi Arak area, Eastern Desert, Egypt.* Ph.D. Thesis, Ain Shams Univ., Cairo.

ATALLA, R. F. 1978. *Geology and petrography of the Hammamat conglomerate of Wadi Kareim* M.Sc. Thesis, Assiut Univ.

ATTAWIYA, M. Y. 1983. Geochemistry and genesis of the uranium mineralization of El Misikat area, Egypt. *Ann. Geol. Surv. Egypt* 13: 67-74.

ATTAWIYA, M. Y. 1983. On the geochemistry and genesis of the uranium mineralization of Egyptian chromites. *Bull. Fac. Sci., Alexandria Univ.* 9: 135-146.

ATTIA, M. I. 1948. *Geology of the Barramiya mining district.* Geol. Surv. Egypt, 65pp.

ATTIA, M. I. 1949. A new mode of occurrence of iron-ore deposits in the Eastern Desert of Egypt. *Bull. Inst. Egypte* 31: 49-68.

ATTIA, M. I. 1950. *The geology of the iron-ore deposits of Egypt.* Geol. Surv. Egypt, 34pp.

ATTIA, M. I. 1954. *Deposits in the Nile Valley and the Delta.* Geol. Surv. Egypt, 858pp.

ATTIA, M. I. 1955. *Topography, geology and iron-ore deposits of the district east of Aswan.* Geol. Surv. Egypt, 262pp.

ATTIA, M.I. 1956. Manganese deposits of Egypt. *20th Intl. Geol. Congr., Mexico* 2: 143-171.

ATTIA, M. I. & G. W. MURRAY 1952. Lower Cretaceous ammonites in marine intercalations in the 'Nubian Sandstone' of the Eastern Desert of Egypt. *Quart. J. Geol. Soc., London* 107: 442-443.

AUBOIN, J. 1965. *Geosynclines.* Elsevier, Amsterdam, 1335pp.

AWAD. G. H. 1946. On the occurrence of marine Triassic (Muschelkalk) deposits in Sinai. *Bull. Inst. Egypte* 27: 397-429.

AWAD, G. H. & A. M. ABDALLAH 1966. Upper Cretaceous in southern Galala, Eastern Desert, with emphasis on neighbouring areas. *Egypt. J. Geol.* 10: 125-142.

AWAD, G. H., M. I. FARIS & H. L. ABBAS 1953. Contributions to the stratigraphy of the Mokattam area east of Cairo. *Bull. Inst. Desert Egypte* 3(2): 106-107.

AWAD, G. H. & I. M. A. FAWZI 1956. The Cenomanian transgression of Egypt. *Bull. Inst. Desert Egypte* 6(1): 168-184.

AWAD, G. H. & M. G. GHOBRIAL 1965. *Zonal stratigraphy of the Kharga Oasis.* Geol. Surv. Egypt, Paper 34, 77pp.

AWAD, G. H. & B. ISSAWI 1974. Biostratigraphic zonation of upper Cretaceous Paleocene in Egypt. *Egypt. J. Geol.* 18: 61-76.

AWAD, G. M. 1984. Habitat and oil in Abu Gharadig and Fayum basins, Western Desert, Egypt. *Bull. Am. Assoc. Petrol. Geol.* 68: 564-573.

AWAD, G. M. 1985. A geophysical study on the Abu Gharadig basin. *Geophysics* 50: 5-15.

AWAD, H. 1951. *La montagne du Sinai central.* Soc. Geograph. Egypte, Cairo, 247pp.

AWAD, H. 1952. Presentation d'une carte morphologique du Sinai. *Bull. Inst. Desert Egypte* 2(1): 132-138.

AYOUTY, M. K. 1961. Geology of Belayim oilfield. *3rd Arab Petrol. Congr.*, Alexandria, 12pp.

AYOUTY, M. K. 1975. Exploration for oil in the Western Desert. *Ann. Geol. Surv. Egypt* 5: 99-104.

BABIKER, I. M. 1982. The chromiteferous ultrabasic rocks of Jebel El Ingessana, eastern Sudan. *Precambrian Res.* 16A: A44-A45 (Abstr.).

BA-BTTAT, M. A. & A. M. HUSSEIN 1983. Geology and mineralization of the Jabal Samran-Jabal Abu Mushut area. *Bull. Fac. Earth Sci., King Abdul Aziz Univ., Jeddah* 6: 571-578.

BAGNOLD, R. A. 1931. Journeys in the Libyan Desert, 1929 and 1930. *Geograph. J.* 78: 13-39, 524-535.

BAGNOLD, R. A. 1933. A further journey through the Libyan Desert. *Geograph. J.* 82: 103-129, 211-235.

BAGNOLD, R. A. 1941. *The physics of blown sand and desert dunes.* Methuen & Co., London, 265pp.

BAGNOLD, R. A. 1954. Physical aspects of dry deserts. In: J. L. Clovosley-Thompson (ed.), *Biology of Deserts*: 7-12.

BAGNOLD, R. A., R. F. PEEL, O. H. MYERS & H. A. WINKLER 1939. An expedition to the Gilf Kebir and Uweinat, 1938. *Geograph. J.* 93: 281-313.

BAEHR, H. P. & W. SCHUHR 1974. Versuche zur Ermittlung der geometrischen Genauigkeit von ERTS-Multispektral-Bildern. *Bildmess. u. Luftbildwiss.* 42: 22-25.

BAKER, B. H. & P. MORGAN 1981. Continental rifting: progress and outlook. *EOS, Trans. Am. Geophys. Union* 62: 585-586.

BAKER, R. N. 1975. LANDSAT data: a new perspective for geology. *Photogramm. Eng. Rem. Sens.* 41: 1223-1239.

BAKOR, A. R., I. G. GASS & C. R. NEARY 1976. Jabal Al Wask. northwest Saudi Arabia, an Eocambrian back-arc ophiolite. *Earth Palnet. Sci. Lett.* 30: 1-9.

BALL, J. 1900. *Kharga Oasis, its topography and geology.* Egypt Surv. Dept., 116pp.

BALL, J. 1902. *On the topographical and geological results of a reconnaissance Survey of Jebel Garra and the Oasis of Kurkur.* Egypt Srv. Dept., 40pp.

BALL, J. 1907. *A description of the first or Aswan Cataract of the Nile.* Egypt. Surv. Dept., 121pp.

BALL, J. 1912. *The geography and geology of south Eastern Desert.* Egypt. Surv. Dept., 394pp.

BALL, J. 1913. *A brief note on the phosphate deposits of Egypt.* Egypt. Surv. Dept., 6pp.

BALL, J. 1916. *The geography and geology of west central Sinai.* Egypt. Surv. Dept., 219pp.

BALL, J. 1927. Problems of the Libyan Desert. *Geograph. J.* 70: 21-83, 105-128, 209-224, 512.

BALL, J. 1933. The Qattara depression of the Libyan Desert and the possibility of its utilization for power production. *Geograph. J.* 82: 289-314.

BALL, J. 1939. *Contributions to the geography of Egypt.* Egypt. Surv. Dept., 300pp.

BALL, J. & H. J. L. BEADNELL 1903. *Baharia Oasis: Its topography and geology.* Egypt. Surv. Dept., 84pp.

BANDEL, K. 1967. Trace fossils from two upper Pensylvanian sandstones in Kansas. *Paleontol. Contr. Univ. Kansas,* Papoer 18.

BANDEL, K. 1973. A new name for the ichnogenus *Cylindrichnus* Bandel. *J. Paleontol.* 47: 1002.

BANDEL, K. & J. KUSS 1987. Depositional environment of the pre-rift sediments – Galala Heights (Gulf of Suez, Egypt).*Berl. Geowiss. Abh.,* 78 (A): 1-48.

BANDEL, K., J. KUSS & N. MALCHUS 1987. The sediments of Wadi Qena (Eastern Desert, Egypt). *J. Afr. Earth Sci.* 6: 427-456.

BARAKAT, H. 1982. Geochemical criteria for source rocks, Gulf of Suez. *6th Petrol. Explor. Seminar, EGPC, Cairo.*

BARAKAT, H. & P. MILLER 1984. Geology and petroleum exploration, Safaga concession, northern Red Sea, Egypt. *Proc. 7th Petrol. Explor. Seminar, EGPC, Cairo*: 191-214.

BARAKAT, M. G. 1970. A stratigraphic review of the Jurassic formations in Egypt and their oil potentialities. *7th Arab Petrol. Congr.,* Kuwait.

BARAKAT, M. G. & M. L. ABDEL HAMID 1973. Subsurface geology of Farafra oasis, Western Desert, Egypt. *Egypt. J. Geol.* 17: 97-110.

BARAKAT, M. G. & A. M. ABOU KHADRAH 1971. The geology of Gebel Abou Treifiya area, Cairo-Suez district. *Bull. Inst. Desert Egypte* 20(1): 3-19.

BARAKAT, M. G. & A. M. ABOU KHADRAH 1970. On the occurrence of lower Lutetian in Gebel Abou Treifiya area, Cairo-Suez district. *Egypt. J. Geol.* 15: 75-81.

BARAKAT, M. G. & N. M. ABUL ELA 1971. The geology of Gabel Geneifa-Gabal Gharra area, Cairo-Suez district. *Bull. Inst. Desert Egypte* 21(1): 31-48.

BARAKAT, M. G. & F. ASAAD 1965. Geological results of the Assiut-Kharga well. *J. Geol. UAR* 9: 81-87.

BARAKAT, M. G. & A. H. ASHRI 1972. Air photo interpretation of some structural features in the area southwest of Aswan. *Egypt. J. Geol.* 16: 247-254.

BARAKAT, M. G. & S. FAHMY 1969. Basinal evolution of Ez El Orban area during the Paleogene time. *Bull. Fac. Sci., Cairo Univ.* 42: 325-329.

BARAKAT, M. G. & G. S. MILAD 1966. Subsurface geology of Dakhla oasis. *Egypt. J. Geol.* 10: 145-154.

BARAKAT, N. & E. M. EL SHAZLY 1956. Spectograpic distribution of chemical elements in Egyptian minerals from lead, zinc, copper and gold deposists. *Bull. Inst. Egypte* 37: 31-46.

BARAZI, N. 1985. Sedimentologie und Stratigraphie des Abyad Beckens (NW Sudan). *Berl. Geowiss. Abh.* 64(A): 1-86.

BARBER, P. M. 1980. Paleogeographic evolution of the Proto-Nile Delta during the Messinian crisis. *Geol. Medit.* 7: 13-18.

BARBER, P. M. 1981. Messinian subaerial erosion of the proto-Nile Delta. *Marine Geol.* 44: 253-272.

BARBER W. M. & D. P. CARR 1976. *Groundwater pilot scheme in the New Valley Egypt: groundwater model of the Kharga-Dakhla area.* AGON/EGY, working document no. 7, UNDP/FAO, Rome, 108pp.

BARKER, F. 1979. Trondhjemite-definition, environment and hypothesis of origin. In: F. Barker (ed.), *Trondhjemites, dacites and related rocks.* Elsevier: 1-12.

BARNES, V. E. & J. R. UNDERWOOD 1976. New investigations of the strewn field of Libyan Desert glass and its petrography. *Earth Planet. Sci. Lett.* 30: 112-117.

BARR, F. T. & B. R. WALKER 1973. Late Tertiary channel system in northern Libya and its implications on Mediterranean sea level changes. In: W.B.F. Ryan & K.J. Hsu (eds), *Initial Reports of the Deep Sea Drilling Project, Leg 13,* US Government Printing Office 2: 1244-1251.

BARRON, T. 1907. *The topography and geology of the district between Cairo and Suez.* Egypt. Surv. Dept., 133p.

BARRON, T. 1907. *The topography and geology of the Peninsula of Sinia (Western Portion).* Egypt. Surv. Dept., 241pp.

BARRON, T. & W. F. HUME 1902. *Topography and goelogy of the Eastern Desert of Egypt (Central Portion).* Egypt. Surv. Dept., 331pp.

BARTHEL, K. W. & BOETTCHER 1978. Abu Ballas Formation: a significant lithostratigraphic unit of the former 'Nubian Series'. *Mitt. Bayer. Staats, Paleontol. Hist. Geol.* 18: 155-166.

BARTHEL, K. W. & W. HERRMANN-DEGEN 1981. Late Cretaceous and early Tertiary stratigraphy in the great Sand Sea and its SE margins (Farafra and Dakhla Oasis), SW Desert, Egypt. *Mitt. Bayer. Staats. Paleontol. Hist. Geol.* 21: 141-182.

BARTHOUX, J. C. 1919. Succession des roches eruptives anciennes dans le Désert Arabique. *C.R. Soc. Geol. France* 169: 660-663.

BARTHOUX, J. C. 1922. *Chronologie et description des roches ignées du Desert Arabique.* Mem. Inst. Egypte, 262pp.

BARTHOUX, J. 1924. Le Crétacé de l'Isthme de Suez et ses soulèvements diapyr. *Bull. Soc. Geol. France* 24(ser.A): 30-32.

BARTHOUX, J. C. & H. DOUVILLE 1913. Le Jurassique dans le désert a l'est de l'isthme de Suez. *C.R. Soc. Geol. France* 157: 265-268.

BARTHOUX, J. C. & P. H. FRITEL 1925. Flore cretacée du grès de Nubie. *Mém. Inst. Egypte* 7(2): 65-119.

BARTOV, Y. et al. 1972. Late Cretaceous and Tertiary stratigraphy and paleogeography of S. Israel. *Isr. J. Earth Sci.* 21: 69-97.

BARTOV, Y., Z. LEWY, G. STEINITZ & I. ZAK 1980. Mesozoic and Tertiary stratigraphy, paleogeography and structural history of Gebel Areif en Naqa area, eastern Sinai. *Isr. J. Earth Sci.* 29: 114-139.

BARTOV, Y. & G. STEINITZ 1977. The Judea and Mount Scopus Groups in the Negev and Sinai with trend surface analysis of the thickness data. *Isr. J. earth Sci.* 26: 119-148.

BARTOV, Y., G. STEINITZ, M. EYAL & U. EYAL 1980. Sinistral movement along the Gulf of Aqaba – its age and relation to the opening of the Red Sea. *Nature* 285: 220-221.

BAR-YOSEF, O. & J. L. PHILLIPS 1977. Prehistoric investigations in Gebel Maghara, northern Sinai. *Monograph. Inst. Archeol., Hebrew Univ.*, Jerusalem, 269pp.

BARZEL, A. & G. M. FRIEDMAN 1970. The Zohar Formation (Jurassic) in southern Israel: a model of shallow water marine carbonate sedimentation. *Isr. J. Earth Sci.* 1: 183-207.

BASAHEL, A. N. 1980. Tentative correlation of the Precambrian rock units of Saudi Arabia, Sudan and Egypt. *Bull. Inst. Applied Geol., King Abdul Aziz Univ., Jeddah* 4(3): 51-59.

BASAHEL, A. N., A. BAHAFAZALLA, U. JUX & S. OMARA 1982. Age and structural setting of a Proto-Red Sea embayment. *N. Jb. Geol. Palaeontol. Mh.* 8(1982): 456-468.

BASSIOUNI, M. A. 1965. Ostracoden aus dem 'Pliozaen' von Kom el Shelul, Pyramiden-Plateau, Gizeh (Aegypten). *Geol. Jb.* 82: 631-654.

BASSIOUNI, M. A. 1969. Ostracoden aus dem Eozaen von Aegypten. 1. Trachyleberidae. *Geol. Jb.* 87: 383-426.

BASSIOUNI, M. A. 1970. Ostracoden aus dem Eozaen von Aegypten. 2. Die Untrfamilien Hemicytherinae, Thacocytherinae und Compylocytherinae. *Geol. Jb.* 88: 203-234.

BASSIOUNI, M. A. 1972. Ostracoden aus dem Eozaen von Ägypten. 3. Die Unterfamilien Brachycytherinae und Buntonilinae. *Geol. Jb.* 89: 169-192.

BASSIOUNI, M. A., W. M. ABDELMALIK & M. BOUKHARY 1974. Litho- and biostratigraphy of middle and upper Eocene rocks in the Minia-Beni Suef reach of the Nile Valley, Egypt. *6th Afr. Micropaleontol. Colloq.*, Tunis: 101-113.

BASSIOUNI, M. A., M. BOUKHARY & H. S. ANNAN 1977. Ostracodes from Gebel Gurnah, Nile Valley, Egypt. *Proc. Egypt. Acad. Sci.* 30: 1-9.

BASSIOUNI, M. A., M. BOUKHARY & H. EL SHEIKH 1975. Biostratigraphy of lower and middle subsurface Oligocene sediments, Dabaa, west of Alexandria, Egypt. *9th Arab Petrol. Congr.*, Dubai.

BASSIOUNI, M. A., I. YOUSSEF, M. BOUKHARY & H. S. ANAN 1982. Stratigraphie des terrains d'age Paleogène de la region sud du desert au nord-ouest Egypte. *8th Afr. Micropleontol. Colloq., Paris, Cah. Micropaleontol.* 1: 41-52.

BASSYOUNI, F. A. 1960. *Geology of Abu Swayel copper-nickel mine area, southern province, UAR*. M.Sc. Thesis, Cairo Univ.

BASTA, E. Z. 1970. A note on an ilmenite occurrence in Wadi El Miyah, Eastern Desert. *Egypt. J. Geol.* 14: 37-41.

BASTA, E. Z. & Z. ABDEL KADER 1969. The mineralogy of Egyptian serpentinites and talc carbonates. *Mineral. Mag.* 37: 394-408.

BASTA, E. Z. & H. I. AMER 1969. El Gedida iron ores and their origin, Bahariya oasis, Western Desert, Egypt. *Econ. Geol.* 64: 424-444.

BASTA, E. Z. & M. H. GIRGIS 1968. The titaniferous ore of Umm Effin area. south Eastern Desert, Egypt. *Proc. Egypt. Acad. Sci.* 21: 127-144.

BASTA, E. Z. & M. H. GIRGIS 1969. Petrological, geological and geochemical studies of the magnetite-ilmenite-apatite (nelsonite) from Kolomnab, south Eastern Desert, Egypt. *Proc. Egypt. Acad. Sci.* 22: 145-156.

BASTA, E. Z., H. KOTB & M. F. AWADALLAH 1980. Petrochemical and geochemical characteristics of the Dokhan Formation at the type locality, Jabal Dokhan, Eastern Desert, Egypt. *Bull. Inst. Applied Geol., King Abdul Aziz Univ., Jeddah* 3(3): 121-140.

BASTA, E. Z., E. REFAI & N. WASSIF 1981. *Correlation between magnetic and mineralogical properties of some Tertiary basaltic rocks from Qatrani-Fayoum area*. Birkhauser Verlag, 119pp.

BASTA, E. Z. & W. S. SALEEB 1971. Elba manganese ores and their origin, south Eastern Desert, UAR. *Mineral. Mag.* 38: 235-244.

BASTA, E. Z. & M. A. TAKLA 1968. Petrological studies on Abu Ghalaga ilmenite occurrence, Eastern Desert. *Egypt. J. Geol.* 12: 43-71.

BASTA, E. Z. & M. A. TALKA 1974. Distribution of opaque minerals and the origin of the gabbroic rocks of Egypt. *Bull. Fac. Sci., Cairo Univ.* 47: 347-364.

BASTA, E. Z. & M. A. TAKLA 1974. Distribution of trace elements in the gabbroic rocks of Egypt. *Chem. Erde* 33: 282-299.

BASTA, E. Z. & G. M. ZAKI 1961. Geology and mineralization of Wadi Sikeit area, south Eastern Desert. *Egypt. J. Geol.* 5: 1-37.

BASTA, F. F. 1983. *Geology and geochemistry of the ophiolite melange and other rock units in the area around and west of Wadi Ghamis, Eastern Desert*, Egypt. Ph.D. Thesis, Cairo Univ.

BAUMGART, J. & F. QUIEL 1985. Erstellung einer Landnutzungskarte für Baden-Wuerttemberg mit Landsat-Daten. *Bildmess. u. Luftbildwiss.* 53: 85-89.

BAYOUMI, A. I. 1983. Tectonic origin of the Gulf of Suez, Egypt as deduced from gravity data. In: *Handbook Geophysical exploration at Sea*, CRC Press, Florida, USA.

BAYOUMI, A. I. & S. EL DASHLOUTY 1967. A preliminary study of Um Gheig zinc-lead deposits for gravity prospecting. *Egypt. J. Geol.* 11: 39-46.

BAYOUMI, A. I. & M. M. EL GAMILI 1970. A geophysical study on the Fayum-Rayan area, with reference to subsurface structures. *7th Arab Petrol. Congr.*, Kuwait.

BAYOUMI, A. I. & A. SABRI 1971. A contribution to magnetic anomalies in the Quatrani-El Natrun area. *Bull. Fac. Sci., Cairo Univ.* 40: 165-173.

BEADNELL, H. J. L. 1901. *Dakhla Oasis: Its topography and geology*. Egypt. Surv. Dept., 107pp.

BEADNELL, H. J. L. 1901. *Farafra Oasis: Its topography and geology*. Egypt. Surv. Dept., 39pp.

BEADNELL, H. J. L. 1902. *The Cretaceous region of Abu Roash, near the Pyramids of Giza*. Egypt. Surv. Dept., 48pp.

BEADNELL, H. J. L. 1905. The relations of the Eocene and Cretaceous systems in the Esna-Aswan reach of the Nile Valley. *Quart. J. Geol. Soc., London* 61: 667-678.

BEADNELL, H. J. L. 1905. *The topography and geology of the Fayum province of Egypt*. Egypt. Surv. Dept., 101pp.

BEADNELL, H. J. L. 1909. *An Egyptian Oasis: An account of the oasis of Kharga in the Libyan Desert*. Murray, London, 248pp.

BEADNELL, H. J. L. 1924. *Report on the geology of the Red Sea coast between Quseir and Wadi Ranga*. Petrol. Res.

Bull. 13, Government Press, Cairo.

BEADNELL, H. J. L. 1927. *The wilderness of Sinai*. Arnold, London, 180pp.

BEALL, A. O. & C. H. SQUYRES 1980. Modern frontier exploration strategy, a case history from upper Egypt. *Oil and Gas J.* 7(4): 106-110.

BEARD, J. H. et al. 1976. Fission track age of Pliocene volcanic glass from the Gulf of Mexico. *Trans. Gulf Coast Assoc. Geol. Soc.* 26: 156-163.

BEARD, J. H., J. B. SAGREE & L. A. SMITH 1982. Quaternary chronology, paleoclimate, depositional sequences and eustatic cycles. *Bull. Am. Assoc. Petrol. Geol.* 66: 158-169.

BECK, A. E. 1965. Techniques of measuring heat flow on land. In: W. H. K. Lee (ed.), *Terrestrial heat flow, Am. Geophys. Union*: 24-57.

BECKMAN, J. P., I. EL HEINY, M. T. KERDANY, R. SAID & C. VIOTTI 1969. Standard planktonic zones in Egypt. *Proc. 1st Intl. Conf. Plankt. Microfoss.*, Geneva, 1967, E. J. Brill, Leiden 1: 92-103.

BEETS, C. B. 1948. *Correlation of the Paleo-Mesozoic rocks in Egypt*. Internal Report 679, Anglo-Egyptian Oilfields Co. (Files of the General Petroleum Co., Cairo).

BEIN, A. & G. GVIRTZMAN 1977. A Mesozoic fossil edge of the Arabian plate along the Levant coast line and its bearing on the evolution of the eastern Mediterranean. In: B. Biju-Duval & L. Montadert (eds), *Structural history of the Mediterranean basins*, editions Technip, Paris: 95-110.

BELEITY, A., M. EL HINNAWI, M. GHONEIM, M. KAMEL, H. GEBALI & M. FATHI 1986. Paleozoic stratigraphy, paleogeography and paleotectonics in the Gulf of Suez. *8th Petrol. Explor. Seminar, EGPC*, Cairo.

BEN AVRAHAM, Z. 1978. The structure and tectonic setting of the Levant continental margin, eastern Mediterranean. *Tectonophysics* 46: 313-331.

BEN AVRAHAM, Z., G. ALMAGOR & Z. GARFUNKEL 1979. Sediments and structure of the Gulf of Elat (Aqaba), northern Red Sea. *Sed. Geol.* 23: 239-267.

BEN AVRAHAM, Z. & Y. MART 1981. Late Tertiary structure and stratigraphy of north Sinai continental margin. *Bull. Am. Assoc. Petrol. Geol.* 65: 1135-1145.

BENDER, F. 1963. Stratigraphie der 'Nubischen' Sandsteine in Süd-Jordanien. *Geol. Jb.* 81: 237.

BENFIELD, A. E. 1949. The effect of uplift and denudation on underground temperatures. *J. Applied Phys.* 20: 66-70.

BEN MENAHEM, A. 1981. A seismicity cycle of 1500 years on the Red Sea rift. *Boll. Geofis. Teorica e Applicata* 23(29): 349-354.

BEN MENAHEM, A. & E. ABOODI 1971. Tectonic patterns in the northern Red Sea region. *J. Geophys. Res.* 76: 2674-2689.

BEN MENAHEM, A., A. NUR & M. VERED 1976. Tectonics, seismicity and structure of the Afro-Eurasian junction – the breaking of an incoherent plate. *Phys. Earth Planet. Int.* 12: 1-50.

BENTOR, Y. K. 1985. The crustal evolution of the Arabo-Nubian massif with special reference to the Sinai Peninsula. *Precambrian Res.* 28: 1-74.

BENTZ, F. P. & S. I. GUTMAN 1977. Landsat data contribution of hydrocarbon exploration in foreign regions. *US Geol. Surv. Prof. Paper* 1015: 83-92.

BENTZ, F. P. & J. B. HUGHES 1981. New reflection seismic

evidence of a late Miocene Nile canyon. In: R. Said, *The Geological Evolution of the River Nile*. Springer, New York: 131-138.

BERG, R. R. 1962. Mountain flank thrusting in Rocky mountain foreland, Wyoming and Colorado. *Bull. Am. Assoc. Petrol. Geol.* 46: 2019-2032.

BERGGREN, W. A. 1964. Paleocene-lower Eocene biostratigraphy of Luxor and nearby Western Desert. *6th Annual Field Conf., Petroleum Explor. Soc. Libya*: 49-76.

BERGGREN, W. A. 1969. Biostratigraphy and planktonic foraminiferal zonation of the Tertiary system of the Sirte basin of Libya. *Proc. 1st Intl. Conf. Plankt. microfoss.* Geneva, 1967, E. J. Brill, Leiden 1: 104-120.

BERGGREN, W. A. & J. A. VAN COUVERING 1974. *The late Neogene*. Elsevier.

BERGSTROM, J. 1969. Remarks on the appendages of trilobites. *Lethaia* 2: 395-414.

BERNAU, R., D. P. F. DARBYSHIRE, G. FRANZ, U. HARMS, A. HUTH, N. MANSOUR, P. PASTEELS & H. SCHANDELMEIER 1987. Petrology, geochemistry and structural development of the Bir Safsaf-Aswan uplift, southern Egypt. *J. Afr. Earth Sci.* 6: 79-90.

BERNSTEIN, R. (ed.) 1983. Image geometry and rectification. In: R. N. Colwell (ed.), *Manual of remote sensing*: 873-922.

BERTRAND, J. M. L. & M. CABY 1978. Geodynamic evolution of the Pan-African orogenic belt; a new interpretation of the Hoggar Shield (Algerian Sahara). *Geol. Rundschau* 67: 337-338.

BEUS, A. A., E. A. SEVEROV, A. A. SITNIN & K. A. SUBBOTIN 1962. *Albitized and greissenized granites (Apogranites)*. Acad. Sci. Publishing House, Moscow, 196pp.

BEYTH, M. 1981. Paleozoic vertical movements in Um Bogma area, southwestern Sinai. *Bull. Am. Assoc. Petrol. Geol.* 65: 160-165.

BEYTH, M., H. GRUNHAGEN & A. SILBERFARB 1978. An ultramafic rock in the Precambrian of eastern Sinai. *Geol. Mag.* 115: 373-378.

(BGR) BUNDESANSTALT FÜR GEOWISSENSCHAFTEN UND ROHSTOFE GEOGRAPHISCHES INSTITUT UNIVERSITÄT WUERZBURG UND SALZGITTER CONSULT GmbH 1976. *Water resources and soil potential development in the New Valley, Egypt*. Mission Report (unpublished), v. II.

BHATTACHARYYA, D. P. & L. G. DUNN 1986. Sedimentologic evidence for repeated pre-Cenozoic vertical movements along the northern margin of the Nubian craton. *J. Afr. Earth Sci.* 5: 147-153.

BHATTACHARYYA, D. P. & J. C. LORENZ 1983. Different depositional settings of the Nubian lithofacies in Libya and southern Egypt. In: J. D. Collinson & J. Lewin (eds), *Modern and ancient fluvial systems, Intl. Assoc., Sedimentol.*, Special Publ. 6: 435-448.

BIBERSON, P., R. COQUE & F. DEBONO 1970. Découverte d'industries preacheuliennes *in situ* dans les formations du piedmont de la montagne de Thebes. *C.R. Acad. Sci., Paris* 285: 303-305.

BIELSKI, M. 1982. *Stages in the evolution of the Arabian-Nubian massif in Sinai*. Ph.D. Thesis, Hebrew Univ., Jerusalem.

BIELSKI, M. E. JAEGER & G. STEINITZ 1979. The geochronology of Igna Granite (Wadi Kid pluton), southern Sinai. *Contr. Mineral. Petrology* 70: 159-165.

BIJU-DUVAL, B. & J. DERCOURT 1980. Les bassins de la

Mediterranée orientale représentent-ils les restes d'un domain océanique, la Mesogée ouvert au Mesozoique et distance de la Tethys? *Bull. Soc. Geol. France* 22: 43-60.

BIJU-DUVAL , B. & J. DERCOURT & X. LE PICHON 1977. From Tethys Ocean to Mediterranean seas. In: B. Biju-Duval & L. Montadert (eds), *Structural History of the Mediterranean Basins*, Editions Technip, Paris: 143-164.

BIJU-DUVAL , B. J. LETOUZEY & L. MONTADERT 1979. Variety of margins and deep basins in the Mediterranean. *Bull. Am. Assoc. Petrol. Geol.* Mem. 29: 293-317.

BILIBIN, Y. A. 1968. *Metallogenic provinces and metallogenic epochs*. Queens College Press, Flushing, N.Y., USA.

BIRCH, F. 1950. Flow of heat in Front Range, Colorado. *Bull. Geol. Soc. Am.* 61: 567-630.

BISEWSKI, H. 1982. Zur Geologie des Dakhla Beckens (Südwest Ägyptens). *Berl. Geowiss. Abh.* 40(A): 1-85.

BISHADY, A. M. & M. F. EL RAMLY 1982. Petrographical and petrochemical studies on some alkaline volcanics from Uweinat area, south Western Desert, Egypt. *Ann. Geol. Surv. Egypt* 12: 29-45.

BISHARA, W. W. & M. E. HABIB 1973. The Precambrian banded iron-ore of Semna, Eastern Desert, Egypt. *N. Jb. Mineral, Abh. 120: 108-118.*

BISHAY, Y. 1961. Biostratigraphic study of the Eocene in the Eastern Desert between Samalut and Assiut by the large foraminifera. *3rd Arab Petrol. Congr.*, Alexandria.

BISHAY, Y. 1966. *Studies on the larger foraminifera of the Eocene (the Niley Valley between Assiut and Cairo and SW Sinai)*. Ph.D. Thesis, Alexandria University.

BLACK, C. C. 1978. Anthracotheriidae. In: V. J. Maglio & H. B. S. Cook (eds), *Evolution of African mammals*. Harvard Univ. Press, Cambridge: 423-434.

BLACKWELL, D. D. & D. S. CHAPMAN 1977. Interpretation of geothermal gradient and heat flow data for Basin and Range geothermal systems. *Geothermal resources Council Trans.* 1: 19-20.

BLANCKENHORN, M. 1900. Neues zur Geologie und Palaeontologie Aegyptens. II: Das Palaeogen (Eozaen und Oligozaen). *Z. Deutsch. Geol. Ges.* 52: 403-479.

BLANCKENHORN, M. 1901. Neues zur Geologie und Palaeontologie Aegyptens. III: Das Miozaen. *Z. Deutsch. Geol. Ges.* 53: 52-132.

BLANCKENHORN, M. 1901. Neues zur Geologie und Palaeontologie Aegyptens. IV: Das Pliozaen und Quartarzeitlater in Aegypten ausschliesslich des Rothen-Meergebietes. *Z. Deutsch. Geol. Ges.* 53: 307-502.

BLANCKENHORN, M. 1921. *Aegypten, Handbuch der regionalen Geologie*, Bd VII, Abt. 9, Heft 23. Carl Winters Universitätsbuchhandlung, Heidelberg, 244pp.

BLANK, H. R., J. H. HEALY, J. ROLLER, R. LAMSON, R. MCLEARN & S. ALLEN 1979. *Seismic refraction profile, Kingdom of Saudi Arabia: field operations instrumentation and initial results*. US Geol. Surv., Arabian project Report, 259: 49pp.

BLODGET, W. H. & G. F. BROWN 1982. *Geological mapping by making use of computer-enhanced imagery of western Saudi Arabia*. US Geol. Surv., Prof. Paper 1153, 10pp.

BLONDEAU, A., W. ABDELMALIK & M. BOUKHARY 1982. Les nummulites de la formation de Thebes, Vallée du Nil, Egypte. *Rev. Micropaleontol.* 25: 90-93.

BLONDEAU, A., M. BOUKHARY & K. SHAMAH 1984. L'evo-lution et la dispersion géographique de Nummulites gizehensis (Forskal). *Geol. Medit.* 11: 173-179.

BLOW, W. H. 1969. Late middle Eocene to Recent planktonic foraminiferal biostratigraphy. *Proc. 1st Intl. Conf. Plankt. Microfoss.*, Geneva, 1967, E. J. Brill, Leiden 1: 199-422.

BLOW, W. H. 1970. Validity of biostratigraphic correlation based on Globigerinacea. *Micropaleontology* 16: 257-268.

BLOW, W. H. 1979. *The Cainozoic Globigerinidae; a study of the morphology, evolutionary relationships and stratigraphical distribution of some Globigerinidae.* E. J. Brill, Leiden, 3 vols.

BOBITT, J. E. & J. D. GALLACHER 1978. The petroleum geology of the Gulf of Suez. *10th Annual Offshore Tech. Conf.*, Houston, Texas, USA.

BOETTCHER, R. 1982. Die Abu Ballas Formation (*Lingula* Shale) (Apt?) der Nubischen Gruppe Südwest Ägyptens. *Berl. Geowiss. Abh.* 39(A): 1-145.

BOETTCHER, R. 1985. Environmental model of the shallow marine Abu Ballas Formation (Aptian, Nubia Group) in southwestern Egypt. *N. Jb. Geol. Palaeontol. Abh.* 169: 261-283.

BOGOLEPOV, M. 1930. Die Dehnung der Lithosphaere. *Z. Deutsch. Geol. Ges.* 82: 206-228.

BOHANNON, R. G. 1986. How much divergence has occurred between Africa and Arabia as a result of opening of the Red Sea? *Geology* 14: 510-513.

BOLLI, H. 1959. Planktonic foraminifera as index fossils in Trinidad, British West Indies, and their value for world wide stratigraphic correlations. *Eclogae Geol. Hevl.* 55: 282-284.

BONATTI, E., P. HAMLYN & G. OTTONELLO 1981. Upper mantle beneath a young oceanic rift: peridotites from the island of Zabargad (Red Sea). *Geology* 9: 474-479.

BONATTI, E. R. CLOCCIATTI, P. COLANTONI, R. GELMINI, G. MARINELLI, G. OTTONELLO, R. SANTACROSE, M. TAVIANI, A. A. ABDEL MEGUID, H. S. ASSAF & M. S. EL TAHER 1983. Zabargad (St John's) island, an uplifted fragment of sub-Red Sea lithosphere. *J. Geol. Soc. London* 140: 677-690.

BONNET, R. 1904. Sur un *Nipadites* de l'Eocene d'Egypte. *Bull. Mus. Hist. Nat.* 10: 499-502.

BOUKHARY, M. & A. ABDALLAH 1982. Stratigraphy and microfacies of the Eocene limestones at Beni Hassan, Nile Valley, Egypt. *N. Jb. Geol. Palaeontol.*, Mh., 3(1982): 151-155.

BOUKHARY, M. & W. M. ABDELMALIK 1983. Revision of the stratigraphy of the Eocene deposits of Egypt. *N. Jb. Geol. Palaeontol.*, Mh., 6(1983): 321-337.

BOUKHARY, M., A. BLONDEAU & D. AMBROISE 1982. Etude sur les nummulites de la region de Minia-Samalut, Vallée du Nil, Egypte; 1. Biometrie et biostratigraphie. *8th Afr. Micropaleontol. Colloq., Paris, Cah. Micropaleontol.* 1: 65-78.

BOUKHARY, M., A. BLOUDEAU & D. AMBROISE 1982. Etude sur les nummulites de la region de Minia-Samalut, Vallée du Nil, Egypte; 2. Multivariate Analysis. *8th Afr. Micropaleontol. Colloq., Paris, Cah. Micropaleontol.* 1: 78-89.

BOUKHARY, M., C. L. GUERNET & H. MANSOUR 1982. Octracodes du Tertiare infèrieur de l'Egypte. *8th Afr. Micropaleontol. Colloq., Paris, Cah. Micropaleontol.* 1: 13-20.

BOUKHARY, M., M. TOUMARKINE, H. KHALIFA & M. ARIF 1982. Etude biostratigraphique à l'aide de foraminifères planctoniques et des ostracodes de l'Eocène de Beni Mazar, Vallée du Nil, Egypte. *8th Afr. Micropaleontol. Colloq., Paris, Cah. Micropaleontol.* 1: 53-64.

BOULOS, F. K., P. MORGAN & T. R. TOPPOZADA 1986. Microearthquake studies in Egypt carried out by the Geological Survey of Egypt. *J. Geodynamics 6*.

BOUSSAC, J. 1911. Sur la presence du Priabonien en Egypte. *Bull. Soc. Geol. France* 4: 485-486.

BOWDEN, P. 1974. Oversaturated alkaline rocks: granites, pantellerites and comendites. In: H. Sorensen (ed.), *The alkaline rocks*, Wiley: 109-123.

BOWEN, B. E. & C. F. VONDRA 1974. Paleoenvironmental interpretation of the Oligocene Gebel Qatrani Formation, Fayum depression, Egypt. *Ann. Geol. Surv. Egypt* 4: 115-138.

BOWLES, E. O. 1945. *Geological report on the southwest Gebel Yelleq anticline, central Sinai*. Internal report, Standard Oil Co., Egypt.

BOWMAN, T. S. 1925. *Report on boring for oil in Egypt, sec. I: Government petroleum research operations*. Mines & Quarries Dept., Cairo, 67pp.

BOWMAN, T. S. 1926. *Report on boring for oil in Egypt, sec. II: Sinai*. Mines & Quarries Dept., Cairo, 91pp.

BOWMAN, T. S. 1931. *Report on boring for oil in Egypt, sec. III: Eastern Desert and adjoining islands*. Mines & Quarries Dept., Cairo, 353pp.

BOWN, T. M. 1982. Ichnofossils and Rhizoliths of the near-shore fluvial Jebel Qatrani Formation (Oligocene), Fayum Province. *Paleogeograph., Paleoclimat., Paleoecol.* 40: 255-309.

BOWN, T. M. & M. J. KRAUS 1983. Trace fossils of the alluvial Willwood Formation (lower Eocene, NW Wyoming, USA). *Paleogeograph., Paleoclimat., Paleoecol.* 41: 89-125.

BOWN, T. M. & M. J. KRAUS 1988. Geology and paleoenvironments of the Oligocene Jebel Qatrani Formation and adjacent rocks, Fayum Province, Egypt. *US Geol. Surv., Prof. Paper* 1452: 1-60.

BOWN, T. M., M. K. KRAUS, S. L. WING, J. G. FLEAGLE, E. L. SIMONS & C. F. VONDRA 1982. The Fayum primate forest revisited. *J. Human Evol.* 11: 603-632.

BOWN, T. M. & E. L. SIMONS 1984. *Peratherium africanus* (Didelphidae: Polyprotodonta): the first African marsupial. *J. Mammol.* 65: 539-548.

BOWN, T. M. & E. L. SIMONS (in prep.). New fossil insectivores from the Oligocene Jebel Qatrani Formation, Fayum, Egypt.

BREED, C. S. & T. GROW 1979. Morphology and distribution of dunes in sand seas observed by remote sensing. In: E. D. McKee (ed.), *A study of global sand seas*. US Geol. Surv., Prof. Paper 1052: 253-302.

BREED, C. S., J. McCAULEY & M. J. GROLIER 1982. Relict drainage, conical hills and the eolian veneer in southwest Egypt – Applications to Mars. *J. Geophys. Res.* 87(B12): 9929-9950.

BROOKES, I. A. 1983. Dakhla Oasis – a geoarcheological reconnaissance. *J. Soc. Study Egypt. Antiquities* (SSEA), Toronto, Canada 13: 167-177.

BROWN, C. & R. H. GIRDLER 1982. Structure of the Red Sea at 20° N from gravity data and its implications for continental margins. *Nature* 298: 51-53.

BROWN, G. C. 1979. The changing pattern of batholith emplacement during earth history. In: M. P. Atherton & J. Tarney (eds), *Origin of granite batholiths: geochemical evidence*. Nantwich, Shiva: 106-115.

BROWN, G. F. & R. G. COLEMAN 1972. The tectonic framework of the Arabian Peninsula. *24th Intl. Geol. Congr.* 24 (sec. 3): 300-305.

BROWN, R. N. 1978. History of exploration and discovery of Morgan, Ramadan and July oilfields, Gulf of Suez, Egypt. *OAPEC Explor. Seminar*, Kuwait: 733-764.

BUGROV, V. 1972. *Technical report on follow-up geochemical operations in the period 1968-1972*. Aswan Mineral Survey Project, Geol. Surv. Egypt.

BUGROV, V. & M. KRS 1972. *Assessment of the mineral potential of the Aswan region*. Interim Review Technical Report, Aswan Mineral Survey Project, Geol. Surv. Egypt.

BUGROV, V., ABOU EL GADAYEL & M. M. SOLIMAN 1973. Rare metallic albitites; a new type of ore mineralization in Egypt. *Ann. Geol. Surv. Egypt* 3: 177-184.

BUGROV, V. & I. M. SHALABY 1973. First discovery of Cu-Ni sulphide mineralization in gabbro-peridotitic rocks in Eastern Desert of Egypt. *Ann. Geol. Surv. Egypt* 3: 177-184.

BULLARD, E. C. 1963. Gravity measurements in East Africa. *Phil. Trans. Roy. Soc. London* 235(A): 445-531.

BURGER, H. 1981. Untersuchungen zur Klassifizierung von Gesteinsoberflächen auf Landsat-Aufnahmen mit Hilfe von Signature-Texturparametern. *Berl. Geowiss. Abh.* 35(A): 1-102.

BURKE, K. & J. F. DEWEY 1974. Two plates in Africa during the Cretaceous? *Nature* 249: 313-316.

BURKE, K. & C. SENGOR 1986. Tectonic escape in the evolution of the continental crust. In: M. Barazangi & L. Brown (eds), *Reflection seismology, the Continental Crust*, Geodynamics Ser., 14, Am. Geophys. Union: 41-53.

BUROLLET, P. F. 1963. Reconnaissance géologique dans le sud-est du bassin de Kufra. *Rev. Inst. Franc. Petrole* 18: 219-227.

BUROLLET, P.F. & G. BUSSON 1983. Platforme saharienne et mesogée au cours du Cretace. *Comapgnie Franc. Petrole*, Notes et Mem. 18: 17-26.

BUTZER, K. W. 1959. Environment and human ecology in Egypt during pre-dynastic and early dynastic times. *Bull. Soc. Geograph. Egypte* 32: 42-88.

BUTZER, K. W. 1959. On the Pleistocene shore lines of Arabs' Gulf. *J. Geol.* 68: 626-637.

BUTZER, K. W. 1964. Pleistocene paleoclimates of the Kurkur Oasis. *Canadian Geographer* 8: 125-140.

BUTZER, K. W. & C. L. HANSEN 1968. *Desert and river in Nubia*. Univ. Wisconsin Press, Madison, 562pp.

CAHEN, L., N. J. SNELLING, J. DELHAL & J. R. VAIL 1984. *The geochronology and evolution of Africa*. Clarendon Press, Oxford, 512pp.

CAREY, S. W. 1958. *A tectonic approach to continental drift*. A symposium on continental drift, Dept. of Geology, Univ. of Tasmania: 177-355.

CARLE, C. 1982. Satellite mapping. Geometric correction of remote sensing images. *Inst. Landmal. Fotogramm. Medd.* (Lyngby), 12, 128pp.

CARPENTIER, A. & I. A. M. FARAG 1948. Sur la flore probablement Rhetienne à El Galala El Bahariya (rive occidentale du Golfe de Suez). *C.R. Acad. Sci., Paris* 226: 686-688.

CARTER, G. S. 1975. *Final report on the investigation of copper-nickel sulphide mineralization at Gabbro Akarem*. Internal Report, Aswan Mineral Survey Project, Geol. Surv. Egypt.

CARTER, G. S., M. S. GARSON, A. RASMY & P. R. SIMPSON 1978. A Proterozoic basic/ultrabasic intrusion in a transverse zone at Gabbro Akarem in the Eastern Desert of Egypt. *Precambrian Res.* 6: A10 (abstr.).

CATON-THOMPSON, G. 1952. *The Kharga Oasis in prehistory*. The Univ. of London, Athlone Press, 213pp.

CATON-THOMPSON, G. & E. W. GARDNER 1934. *The desert Fayum*. Roy. Anthropol. Inst., London, 167pp.

CHANDLER, M. E. J. 1954. Some upper Cretaceous and Eocene fruits from Egypt. *Bull. Brit. Mus. Nat. Hist.* 2(4): 147-187.

CHAPPEL, B. W. & A. J. R. WHITE 1974. Two contrasting granite types. *Pac. Geol.* (Canberra) 8: 173-174.

CHENET, P. Y. & J. LETOUZEY 1983. Tectonique de la zone comprise entre Abu Durba et Gebel Nezzazat (Sinai, Egypt) dans le contexte de l'evolution du rift de Suez. *Bull. Centre Rech. Explor. Elf Aquitaine* 7: 201-215.

CHENET, P. Y., J. LETOUZEY & E. ZAGHLOUL 1984. Some observations on the rift tectonics in the eastern part of the Suez rift. *Proc. Petrol. Explor. Seminar, EGPC*, Cairo: 18-36.

CHERIF, O. H. 1972. Some aspects of the paleoecology of the lower Miocene of the northern part of the Gulf of Suez, Egypt. *8th Arab Petrol. Congr.*, Algiers, Paper 82, 1pp.

CHERIF, O. H. 1972. Tertiary fauna from the Sadat area, southwest of Suez. *Bull. Inst. Egypte* 52: 91-123.

CHERIF, O. H. 1974. Remarques sur l'utilisation des Miogypsines et des Operuclines pour la chronostratigraphie du Miocene du nord du desert Arabique (Egypte). *6th Afr. Micropaleontol. Colloq.*, Tunis: 325-335.

CHERIF, O. H. 1976. Remarks on the tectonic evolution of the Gulf of Suez sedimentary basin during the upper Paleozoic. *9th Arab Petrol. Congr.*, Dubai.

CHERIF, O. H., M. A. BASSIOUNI & S. A. GHANIMA 1975. Stratigraphy, paleoecology and climates of the Neogene and Quaternary of the Mersa Matruh area, north Western Desert, Egypt. *Ann. Geol. Surv. Egypt* 5: 137-147.

CHERIF, O. H. & M. A. YHEIA 1977. Stratigraphy of the area between Wadi Gemal and Wadi Hommath, Gulf of Suez, Egypt. *Egypt. J. Geol.* 21: 185-203.

CHEVALIER, J. P. 1961. Recherches sur les Madreporaires et formations recifals Miocenes de la Mediterranée occidentale. *Mem. Soc. Geol. France* n.s. 60(93): 558pp.

CHOW, J. S. & J. O. WILSON 1981. A qualitative review of Nubian Sandstone regional aquifer behaviour. *Proc. Intl. Conf. on Water Resources Management in Egypt*, Cairo: 363-382.

CHURCH, W. R. 1976. Late Proterozoic ophiolites. *Intl. Colloq. C.N.R.S. 272, Associations mafiques-ultramafiques des les orogenes:* 105-117.

CHURCH, W. R. 1979. Granite and metamorphic rocks of the Taif area, western Saudi Arabia: discussion and reply. *Bull. Geol. Soc. Am.* 90: 893-896.

CHURCH, W. R. 1982. The northern Appalachians and the Eastern Desert of Egypt. *IGCP Syp. on Pan-African crustal evolution in the Arabian-Nubian Shield*, Jeddah (Proc.): 27-30.

CHURCH, W. R. 1983. Precambrian evolution of Afro-Arabian crust from ocean arc to craton: discussion. *Bull.*

Geol. Soc. Am. 94: 679-681.

CHURCH, W. R. 1986. Comment on the paper: 'Pan African (late Precambrian) tectonic terrains and the reconstruction of the Arabian-Nubian shield' by J. R. Vail. *Geology* 13: 839-842.

CHUMAKOV, I. S. 1967. Pliocene and Pleistocene deposits of the Nile Valley in Nubia and upper Egypt (in Russian). *Trans. Geol. Inst. Acad. Sci.* (USSR) 170: 1-110.

CITA, M. B. 1982. The Mediterranean salinity crisis. In: H. Berckheimer & K. Hsu (eds), Alpine Mediterranean geodynamics, Geodynamics Ser., 7, *Am. Geophys. Union*: 113-140.

CITA, M. B., M. A. CHIERICCI, M. MANCHARMONT, S. D'ONOFRIS, W. B. F. RYAN & R. SCORZIELLO 1973. The Quaternary record in the Ionian and Tyrhennian basins of the Mediterranean Sea. In: W. B. F. Ryan & K. J. Hsu (eds), *Initial reports of the Deep Sea Drilling Project*, US Government Printing Office, 13(1).

CLAESSON, S., J. S. PALLISTER & M. TATSUMOTO 1984. Samarium-neodymium data on two late Proterozoic ophiolites of Saudi Arabia and implications of crustal and mantle evolution. *Contr. Mineral. Petrology* 85: 244-252.

CLARK, S. P. JR 1966. Handbook of physical constants. *Mem. Geol. Soc. Am.* 97, 587pp.

CLAYTON, P. A. 1933. The western side of the Gilf Kebir. *Geograph. J.* 81: 254-259.

CLAYTON, P. A. & M. S. SPENCER 1932. Silica glass from the Libyan Desert. *Mineral. Soc.* 23: 501-508.

CLIFFORD, T. N. 1970. The structural framework of Africa. In: T. N. Clifford & I. G. Gass (eds), *African magmatism and tectonics*, Oliver & Boyd, Edinburgh: 1-26.

COCHRAN, J. R. 1981. The Gulf of Aden: structure and evolution of a young ocean basin and continental margin. *J. Geophys. Res.* 86: 263-288.

COCHRAN, J. R. 1982. The magnetic quiet zone in the eastern Gulf of Aden: implications for the early development of the continental margin. *Geophys. J. Roy. Astronom. Soc.* 68: 171-201.

COCHRAN, J. R. 1982. Reply. *J. Geophys. Res.* 87: 6765-6770.

COCHRAN, J. R. 1983. A model for development of Red Sea. *Bull. Am. Assoc. Petrol. Geol.* 67: 41-69.

COCHRAN, J. R., F. MARTINEZ, M. S. STECKLER & M. A. HOBART 1985. The northern Red Sea I: Pre-sea floor spreading tectonics. *Trans. Am. Geophus. Union* 66: 365.

COFER, C., K. LEE & J. WRAY 1984. Miocene carbonate microfacies, Esh El Mellaha range, Gulf of Suez. *Proc. 7th Petrol. Explor. Seminar, EGPC*, Cairo: 97-115.

COHEN, A. J. 1959. Origin of Libyan Desert silica glass. *Nature* 183: 1548-1549.

COHEN, A. J. 1961. The terrestrial origin of the Libyan Desert silica glass. *J. Glass. Technol.* 2(B): 83-86.

COIFFAIT, P. E., B. COIFFAIT, J. J. JAEGER & M. MAHBOUBY 1984. Un nouveau gisement à mammiferes fossiles d'age Eocène superieur sur le varsant sud des Nementcha (Algerie orientale): decouverte des plus anciens rongeurs d'Afrique. *C.R. Acad. Sci.* ser. II, 13: 893-898.

COLEMAN, R. G. 1974. Geologic background of the Red Sea. In: C. A. Burke & C. L. Drake (eds), *The geology of continental margins*. Springer, Berlin: 743-751.

COLEMAN, R. G. 1981. Tectonic setting for ophiolite Obduction in Oman. *J. Geophy. Res.* 86: 2497-2508.

COLEMAN, R. G. ET AL. 1975. The volcanic rocks of south-west Saudi Arabia and the opening of the Red Sea. *Bull. Saudi Arabia Dir. Gen. Mineral resources* 22: D1-D30.

COLLINS, R. J., F. P. McCOWN, L. P. STONIS, J. PETZEL & J. F. EVERETT 1973. An evaluation of the suitability of ERTS data for the purposes of petroleum exploration. *ERTS 1 Symp., 1(A) NASA* SP-351: 809-821.

COLYOCRESSES, A. P. 1972. Image resolution for ERTS, SKYLAB and GEMINI/APOLLO. *Photogramm.* 38: 33-35.

COLWELL, R. N. 1975. The 'multi' concept as applied to the acquisition and analysis of remote sensing data. In: R. G. Reeves (ed.), *Manual of remote sensing.* Falls Church, Virginia, USA 1: 5-15.

CONANT, L. C. & G. H. GOUDARZI 1964. *Geological map of the kingdom of Libya 1:2 000 000.* US Geol. Surv.

CONOCO INC. 1977. *Egypt's Depth of Basement Map* (unpiblished).

CONOCO INC. 1983. Abu Roda no. 1 well. Final geological well report (unpublished).

CONOCO INC. 1983. Kabrit no. 1 well. Final geological well report (unpublished).

CONOCO INC. 1983. Boughaz no. 1 well. Final geological well report (unpublished).

CONROY, G. C. 1976. Primate postcranial remains from the Oligocene of Egypt. *Contr. Primatol.* 8: 1-134.

COSTER, H. P. 1947. Terrestrial heat flow in Persia. *Mon. Not. Roy. Astronom. Soc., Geophys. suppl.* 5: 131-145.

COUYAT, J. 1910. Sur un nouveau gisement de feuilles fossiles en Egypte. *Bull. Soc. Geol. France* 4(10): 29.

COUYAT, J. & F. H. FRITEL 1910. Sur la presence d'empreintes vegetale dans les gres Nubien des environs d'Assouan. *C.R. Acad. Sci.* 151: 961-964.

COX, L. R. 1929. Notes on the post Miocene Osteridae and Pectenidae of the Red Sea region, with remarks on the geologic significance of their distribution. *Proc. Malacol. Soc. London* 18: 165-209.

COX, L. R. 1956. Lamellibranchiata from the Nubian Sandstone series of Egypt. *Bull. Inst. Egypte* 37: 456-480.

CRAWFORD, W. A., D. H. COULTER & H. B. HUBBARD 1984. The areal distribution, stratigraphy and major element chemistry of the Wadi Natash volcanic series, Eastern Desert, Egypt. *J. Afr. Earth Sci.* 2: 119-128.

CRONIN, T. M. & H. KHALIFA 1979. Middle and late Eocene ostracoda from Gebel El Mreir, Nile Valley, Egypt. *Micropaleontology* 25: 397-411.

CUVILLIER, J. 1924. Contribution à l'étude géologique du Mokattam. *Bull. Inst. Egypte* 6 (n. ser.): 93-102.

CUVILLIER, J. 1928. Les végétaux fossiles d'Egypte. *Bull. Soc. Geograph.* Egypte, 15.

CUVILLIER, J. 1930. Revision du nummulitique Egyptien. *Mem. Inst. Egypte* 16: 371pp.

DAGGETT, P. H. & P. MORGAN 1977. Egypt and the northern Red Sea: New microearthquake data. *Trans. Am. Geophys. Union* 58: 1189.

DAGGETT, P. H., P. MORGAN, F. K. BOULOS, S. F. HENNIN, A. A. EL SHERIF & Y. S. MELEK 1980. Microearthquake studies of the northeastern margin of the African plate. *Ann. Geol. Surv. Egypt* 10: 989-996.

DAGGETT, P. H., P. MORGAN, F. K. BOULOS, S. F. HENNIN, A. A. EL SHERIF, A. A. EL SAYED, N. Z. BASTA & Y. S. MELEK 1986. Seismicity and active tectonics of the Egyptian Red Sea margin and the northern Red Sea. *Tectonophysics* 125: 313-324.

DAMES W. B. 1894. Ueber Zeuglodonten aus Aegypten und die Beziehungen der archaeouten zu den ubrigen Cetacean. *Palaeontol. Abh.* 5: 1-36.

DARDIR, A. A. & K. M. ABU ZEID 1972. Geology of the basement rocks between latitudes 27 00 and 27 30 N, Eastern Desert. *Ann. Geol. Surv. Egypt* 2: 129-159.

DARDIR, A. A., M. F. AWADALLAH & K. M. ABU ZEID 1982. A new contribution to the geology of Gebel Dokhan, Eastern Desert, Egypt. *Ann. Geol. Surv. Egypt* 12: 19-27.

DARDIR, A. A., A. GAD ALLAH 1969. *Report on expedition 7/68, Eastern Desert.* Internal report, Geological Survey Egypt.

DAVIDSON, G. E. 1986. Ground control pointing and geometric transformation of satellite imagery. *Intl. J. Rem. Sens.* 7: 65-74.

DEGENS, E. T. & D. A. ROSS (eds) 1969. *Hot brines and recent heavy metal deposits in the Red Sea.* Springer, 600pp.

DE GRUYTER, P. & T. A. VOGEL 1981. A model for the origin of the alkaline complexes of Egypt. *Nature* 291: 571-574.

DEIBIS, S. 1976. Oil potential of the upper Cretaceous sediments in northern Western Desert, Egypt. *5th Petrol Explor. Seminr, EGPC*, Cairo.

DEIBIS, S. 1982. Abu Qir bay, a potential gas province area offshore Mediterranean, Egypt. *6th Petrol Explor. Seminar, EGPC*, Cairo.

DE HEINZELIN, J. 1968. Geological history of the Nile Valley in Nubia. In: F. Wendorf (ed.), *The prehistory of Nubia.* Southern Methodist University Press, Dallas, Texas, USA.

DE LA ROCHE, H., J. LETTERIER, P. GRANDCLAUDE & M. MARCHAL 1980. A classification of volcanic and plutonic rocks using R1, R2-diagram and major element analyses; its relationship with current nomenclature. *Chem. Geol.* 19: 183-210.

DEMAG 1960. Zwischenbericht über die Untersuchung von Kupfer-Vorkommen in der ostlichen Wüste. Unpublished report prepared for the Geol. Surv. Egypt.

DEUTSCH, E. R. 1960. First order tectonics of North America: A review. *Bull. Canadian Petrol. Geol.* 8: 228-232.

DEWEY, J. F., W. C. PITMAN, W. B. F. RYAN & J. BONNIN 1973. Plate tectonics and the evolution of the Arabian plate system. *Bull. Geol. Soc. Am.* 84: 3137-3180.

DEWEY, J. F. & A. M. C. SENGOR 1979. Aegean and surrounding regions: complex multi-plate and continuous tectonics in a convergent zone. *Bull. Geol. Soc. Am.* 90: 82-91.

DIA EL DIN, M. 1974. Stratigraphic and structural studies on Abu Gharadig oil and gas field. *4th Petrol. Explor. Seminar, EGPC*, Cairo.

DICKINSON, W. R. 1974. Plate tectonics and sedimentation. *Soc. Econ. Paleontol. & Mineral.*, Special Publ. 22: 1-17.

DIETZ, R. S. & J. C. HODEN 19770. Reconstruction of Pangaea, break-up and dispersion of continents, Permian to present. *J. Geophys. Res.* 75: 4939-4955.

DIXON, J. E. & A. H. F. ROBERTSON (eds) 1984. *The geological evolution of the eastern Mediterranean.* Blackwell Scientific Publishers, Oxford.

DIXON, T. H. 1979. *The evolution of continental crust in the late Precambrian Egyptian Shield*. Ph.D. Thesis, Univ. California, San Diego.

DIXON, T. H. 1981. Age and chemical characteristics of some pre-Pan African rocks in the Egyptian shield. *Precambrian Res.* 14: 113-119.

DIXON, T. H. 1981. Gabal Dahanib, Egypt: a late Precambrian layered sill of komatiitic composition. *Contr. Mineral. Petrology* 76: 42-52.

DOMINIK, W. 1985. Stratigraphie und Sedimentologie (Geochemie, Schwermineralanayse) der Oberkreide von Bahariya und ihre Korrelation zum Dakhla Becken (Western Desert, Ägypten). *Berl. Geowiss. Abh.* 62(A): 1-173.

DOMINIK, W. & S. SCHAAL 1984. Notes on the stratigraphy of the upper Cretaceous phosphate (Campanian) of the Western Desert (Egypt). *Berl. Geowiss. Abh.* 50(A): 153-176.

DOUVILLE, H. 1916. Les terrains secondaires dans le massif du Moghara, à l'est de l'Isthme de Suez, d'après les explorations de M. Couyat Barthoux. *Mém. Acad. Sci., Paris* 54 (ser. 2): 1-184.

DOYLE, F. J. (ed.) 1975. Cartographic presentation of remote sensing data. In: R. G. Reeves (ed.), *Manual of remote sensing*. Falls Church (ASP), 2: 1077-1106.

DRAKE, C. L. & R. W. GIRDLER 1964. A geophysical study of the Red Sea. *Geophys. J. Roy. Astronom. Soc.* 8: 473-495.

DRUCKMAN, Y. 1969. The petrography and environment of deposition of the Triassic Saharonim Formation and the dolomite member of the Mohilla Formation in Maktesh Ramon central Negev (southern Israel). *Bull. Geol. Surv. Isr.* 49: 1-35.

DRUCKMAN, Y. 1974. The stratigraphy of the Triassic sequence in southern Israel. *Bull. Geol. Surv. Isr.* 64: 1-92.

DRUCKMAN, Y., G. GVIRTZMAN & E. KASHAI 1975. Triassic oil prospects in Israel and northern Sinai in the light of the Ramallah no. 1 deep test and the new Triassic production in northern Syria and Iraq. *Geol. Surv. Isr., Report OD/1/75*, 57pp.

DUBERTRET, L. 1932. Les formes structurales de la Syrie et de la Palestine: leur origine. *C.R. Seances Acad. Sci.* 195: 65-66.

DUNCAN, R. A. 1981. Hotspots in the southern oceans – an absolute frame of reference for motion of the Gondwana continents. *Tectonophysics* 74: 29-42.

DUPERON-LAUDOUENEIX & A. LEJAL-NICOL 1981. Sur des bois homoxyles du sud-ouest de l'Egypte. *C.R. 106 Congr. Nat. Soc. Sav. Sci.* 1: 29-40.

DUYVERMAN, H. J. & N. B. W. HARRIS 1982. Late Precambrian evolution of Afro-Arabian crust from ocean arc to craton, discussion. *Bull. Geol. Soc. Am.* 93: 174-178.

EDWARDS, W. N. 1926. Fossil plants from the Nubian Sandstone of eastern Darfur. *Quart. J. Geol. Soc. London* 82: 94-100.

EGYPTIAN GENERAL PETROLEUM CORPORATION STRATIGRAPHIC COMMITTEE 1964. *Oligocene and Miocene rock stratigraphy of the Gulf of Suez region*, Cairo, 142pp.

EGYPTIAN GENERAL PETROLEUM CORPORATION (EGPC) 1982. *Petroleum potential evaluation, Western Desert, A.R. Egypt*. Special report prepared by Robertson Research International Ltd. and Associated Resources Consultants, Inc. in association with Scott Pickford & Associates Ltd. and Energy Resources Consultants Ltd., 8 volumes.

EGYPTIAN GENERAL PETROLEUM CORPORATION 1986. *Activity of oil exploration in Egypt: 1886-1986*, Cairo, 175pp.

EGYPTIAN GENERAL PETROLEUM CORPORATION 1986. *History of oil exploration: 1886-1986*, Cairo, 77pp.

EICHER, D. B. 1946. Conodonts from the Triassic of Sinai, Egypt. *Bull. Am. Assoc. Petrol. Geol.* 30: 613-616.

EICHER, D. B. 1947. Micropaleontology of the Triassic of north Sinai. *Bull. Inst. Egypte* 28: 87-92.

EICHER, D. B. & L. C. MOSHER 1974. Triassic conodonts from Sinai and Palestine. *J. Paleontology* 48: 727-739.

EISSA, R. H. 1974. Biostratigraphy and correlation of sections of the Dakhla Oasis, Western Desert, Egypt. *6th Afr. Micropaleontol. Colloq.*, Tunis: 259-275.

EL ALFY, E. M. 1940. *The tin tungsten deposits of Egypt*. Internal report, Mines & Quarries Dept., Cairo.

EL ALFY, E. M. 1946. The mineral resources of Egypt. *Trans. Mining Petrol. Assoc., Egypt* 1(3): 9-32.

EL AREF, M. M. 1984. Strata-bound and stratification iron sulphides, sulfur and galena in the Miocene evaporites in Ranga, Red Sea, Egypt (with special emphasis on their diagenetic crystallisation rhythmites). In: A. Wauschkuhn et al (eds), *Syngenesis and epigenesis in the formation of mineral deposists,* Springer Verlag: 457-467.

EL AREF, M. M., F. AWADALLAH & S. AHMED 1986. Karst landform development and related sediments in the Miocene rocks of the Red Sea coastal zone, Egypt. *Geol. Runschau* 75: 781-790.

EL AREF, M. M. & G. C. AMSTUTZ 1983. *Lead-zinc deposits along the Red Sea coast of Egypt: New observations and genetic models on the occurrences of Um Gheig, Wizr, Esel and Zug El Behar*. Monograph series on mineral deposits, Gebrueder Borntraeger, Stuttgart: 103pp.

EL AREF, M. M., A. KHUDEIR & G. HAMED 1985. On the geometry of strata-bound Fe, Cu, Zn and Pb sulphides in the metapyroclastics at Um Samiuki area, Eastern Desert, Egypt. *2nd Jordanian Geol. Conference, Amman (Abstr.)*.

EL AREF, M. M. & E. REFAI 1987. Paleokarst processes in the Eocene limestones of the Pyramids plateau, Giza, Egypt. *J. Afr. Earth Sci.* 6: 367-378.

EL ASHRY, M. T. 1985. Egypt. In: E. C. F. Bird & M. L. Schwartz (eds), *The world's coastlines*. Van Nostrand Reinhold, New York: 513-517.

EL AWADY, M. M., M. H. ELZARKA & A. A. HASSAN 1980. Subsurface goelogical studies on oil prospects of the coastal area between Fuka and Sidi Barrani, Western Desert, Egypt. *Delta J. Sci.* 4: 21-246.

EL AWADY, M. M., M. H. ELZARKA & T. A. ABDEL FATTAH 1981. Seismic geophysical investigation in the area west of the Nile delta, Egypt. *Delta J. Sci.* 5: 212-233.

EL BARKOUKY, A. N. 1979. *Preliminary investigations of ground water and soil resources in east Uweinat area, Western Desert, Egypt*. Internal report, General Petroleum Co., Cairo.

EL BASSYONY, A. A. 1978. Structure of the northwestern plateau of the Bahariya Oasis, Western Desert, Egypt. *Geol. en Mijnbouw* 57: 165-184.

ELBASSYONY, A. A. 1982. *Stratigraphical studies on Mio-*

cene and younger exposures btween Quseir and Be-
renice, Red Sea coast, Egypt. Ph.D. Thesis, Ain Shams
University, Cairo.

ELBASSYONY, A. A., A. F. KALMIKOV, S. AWAD, G. SIDAU-
GAS & V. A. TULYANKIN 1970. El Mahamid phosphorite
deposit. In: O. Moharram et al. (eds), *Studies on some
mineral deposits of Egypt.* Geol. Surv. Egypt: 135-152.

EL BAYOUMI, R. M. A. 1980. *Ophiolites and associated
rocks of Wadi Ghadir, east of Gebel Zabara, Eastern
Desert, Egypt.* Ph.D. Thesis, Cairo Univ.

EL BAYOUMI, R. M. A. 1984. Ophiolites and melange
complex of Wadi Ghadir, Eastern Desert, Egypt. *Bull.
Fac. Earth Sci., King Abdul Aziz Univ., Jeddah* 6: 324-
329.,

EL BAYOUMI, R. M. A. & R. GREILING 1984. Tectonic
evolution of a Pan-African plate margin in southeastern
Egypt – a suture zone overprinted by low angled thrusting. In: J. Klerkx & J. Michot (eds), *Geologie africaine.*
Tervuren: 47-65.

EL BAZ, F. 1978. The meaning of desert color in earth orbital
photographs. *Photogramm. Eng. Rem. Sens.* 44: 69-75.

EL BAZ, F. 1980. Space age developments in desert research.
Episodes 4: 12-17.

EL BAZ, F. 1981. Circular feature among dunes of the great
sand sea, Egypt. *Science* 213: 439-440.

EL BAZ, F. 1984. *The Geology of Egypt: An annotated
bibliography.* Brill, Leiden, 778pp.

EL BAZ, F., C. S. BREED, M. J. GROLIER & J. F. MCCAULEY
1979. Eolian features in the Western Desert of Egypt and
some applications to Mars. *J. Geophys. Res.* 84: 8205-
8221.

EL BAZ, F. & R. W. WOLFE 1982. Wind patterns in the
Western Desert. In: F. El Baz & T. A. Maxwell, *Desert
Landforms of southwest Egypt: A basis for comparison
with Mars.* NASA: 119-140.

EL BOUSEILY, A. M. & EL SOKKARY 1975. The relation
between Rb, Ba and Sr in granitic rocks. *Chem. Geol.* 16:
207-219.

EL DAWOODY, A. S. & M. G. BARAKAT 1972. Nannobios-
tratigraphy of the upper Paleocene and upper Eocene in
Duwi reigon, Quseir district, Egypt. *8th Arab Petrol.
Congr.*, Algiers, Paper 70(B-3), 43pp.

EL DEFTAR, T., B. ISSAWI & A. M. ABDALLAH 1978. Contributions to the geology of Abu Tartur and adjacent areas,
Western Desert, Egypt. *Ann. Geol. Surv. Egypt* 8: 51-90.

EL ETR, H. A. 1971. Analysis of air photo lineations of
Darhib district, south Eastern Desert, UAR. *Ann. Geol.
Surv. Egypt* 1: 93-108.

EL ETR, H. A., A. M. ABDALLAH & M. A. YEHIA 1973.
Structural analysis of the area southwest of Sidi Barrani,
northwestern Mediterranean coast. *Egypt. J. Geol.* 17:
125-145.

EL ETR, H. A. & M. S. M. YOUSSIF 1974. Air photo
lineations of Wadi El Assiuti, Wadi Mahariq and Sikket
El Agel, central Eastern Desert, Egypt. *Proc. Egypt.
Acad. Sci.* 28: 79-87.

EL ETR, H. A., M. S. M. YOUSSIF & A. A. DARDIR 1979.
Utilization of 'Landsat' images and conventional areal
photographs in the delineation of some aspects of the
geology of the central Eastern Desert, Egypt. *Ann. Geol.
Surv. Egypt* 9: 136-162.

EL ETR, H. A. & A. R. MOUSTAFA 1980. Utilization of
orbital imagery and conventional areal photography in

the delineation of the regional lineation pattern of the
central Western Desert, Egypt with particular emphasis
on the Bahriya region. In: M. J. Salem & M. T. Busrewil
(eds), *The Geology of Libya.* Academic Press 3: 933-
953.

ELF AQUITAINE 1982. North Alexandria area, geological
and geophysical study. *6th Petrol. Explor. Seminar,*
EGPC, Cairo.

EL GABY, S. 1975. Petrochemistry and geochemistry of
some granites from Egypt. *N. Jb. Mineral. Abh.* 125:
147-189.

EL GABY, S. 1983. Architecture of the Egyptian basement
complex. *Proc. 5th Intl. Conf. Basement Tectonics*, Cairo (Abstr.).

EL GABY, S. 1985. On the relation between tectonics and ore
mineral occurrences in the basement complex of the
Eastern Desert of Egypt. *Proc. 6th Intl. Conf. Basement
Tectonics*, Santa Fe, N.M., USA.

EL GABY, S. & A. A. AHMED 1980. The Feiran-Solaf gneiss
belt, SW Sinai, Egypt – evolution and mineralization of
Arabian Nubian shield. *Bull. Inst. Applied Geol., King
Abdul Aziz Univ., Jeddah* 3(4): 95-105.

EL GABY, S. & M. M. EL AREF 1977. Geological, petroche-
mical and geochemical studies on the Shait granite at
Wadi Shait, Eastern Desert, Egypt. *Bull. Fac. Sci., Assiut
Univ.* 6: 307-329.

EL GABY, S., O. M. EL NADY & A. M. KHUDEIR 1984.
Tectonic evolution of the basement complex in the
central Eastern Desert of Egypt. *Geol. Rundschau* 73:
1019-1036.

EL GABY, S. & M. E. HABIB 1980. The eugeosynclinal
filling of Abu Ziran Group in the area SW of Port Safaga,
Eastern Desert, Egypt. *Bull. Inst. Applied Geol., King
Abdul Aziz Univ., Jeddah* 3(4): 137-142.

EL GABY, S. & M. E. HABIB 1982. Geology of the area SW
of Port Safaga, with special emphasis on the granitic
rocks, Eastern Desert, Egypt. *Ann. Geol. Surv. Egypt* 12:
47-71.

EL GABY, S., F. K. LIST & R. TEHRANI 1988. Geology,
evolution and metallogenesis of the Pan African belt in
Egypt. In: El Gaby, S. & R.O. Greiling (eds). *The Pan
African belt of northeast Africa and adjacent areas.*
Braun-Schweig (Vieweg): 17-68.

EL-GAMILI, M. M. 1982. A geological interpretation of a
part of the Nile Valley based on gravity data. *Egypt. J.
Geol.*, special vol. pt. 2: 101-120.

EL GEMMIZI, M. A. 1984. On the occurrence and genesis of
mud zircon in the radioactive psammitic gneiss of Wadi
Nugrus, Eastern Desert, Egypt. *J. Univ. Kuwait Sci.* 11:
285-293.

EL GEMMIZI, M. A. 1985. Note on the occurrence of gold
and cassiterite in the Egyptian beach placer deposits.
Econ. Geol. 80: 769-772.

EL GEMMIZI, M. A. & M. A. HASSAN 1984. Morphology
and distribution of zircon in metasomatic rocks as a
guide to their origin. *27th Intl. Geol. Congr., Moscow* 5:
37-38.

EL GEZEERY, M. N. & I. M. MARZOUK 1974. Miocene rock
stratigraphy of Egypt. *Egypt. J. Geol.* 18: 1-59.

EL GEZEERY, M. N., S. M. MOHSEN & M. FARID 1972.
Sedimentary basins of Egypt and their petroleum prospects. *8th Arab Petrol. Congr.*, Algiers.

EL GHAWABY, M. A. 1973. *Structural geology and rad-*

ioactive mineralization of Wadi Zeidun area, Eastern Desert of Egypt. Ph.D. Thesis, Ain Shams Univ., Cairo.

EL GORESY, A. 1964. Neue Beobachtungen an der Nickel-Magnetkies-Lagerstätten von Abu Swayel, Aegypten. *N. Jb. Mineral. Abh.* 102: 107-113.

EL HEINY, I. 1981. Neogene stratigraphy and paleogeography of the southern Mediterranean-Red Sea area. *Ann. Geol. Helv., hors serie* 4: 291-304.

EL HEINY, I. 1982. Neogene stratigraphy of Egypt. *Newsl. Stratigr.* 11(2): 41-54.

EL HEINY, I. & E. MARTINI 1981. Miocene foraminiferal and nannoplankton assemblages from the Gulf of Suez region and correlations. *Geol. Medit.* 8: 101-108.

EL HEINY, I. & S. I. MORSI 1986. Review of the upper Eocene deposits in the Gulf of Suez, Egypt. *8th Petrol. Explor. Seminar, EGPC,* Cairo.

EL HINNAWI, E. 1965. Contributions to the study of Egyptian (UAR) iron ores. *Econ. Geol.* 60: 1497-1509.

EL HINNAWI, E. 1965. Petrographical and geochemical studies of Egyptian basalts. *Bull. Volc.* 28(3): 80.

EL HINNAWI, E. & M. A. ABDEL MAKSOUD 1968. Petrography of Cenozoic volcanic rocks of Egypt (UAR). *Geol. Rundschau* 57: 879-890.

EL HINNAWI, E. & M. A. ABDEL MAKSOUD 1972. Geochemistry of Egyptian Cenozoic volcanic rocks. *Chem. Erde* 31: 93-112.

EL HINNAWI, E. & S. M. LOUKINA 1972. A contribution to the geochemistry of 'Egyptian alabaster'. *Tschermacks Min. Petr. Mitt.* 17: 215-221.

EL HINNAWI, M., A. ABDALLAH & B. ISSAWI 1978. Geology of Abu Bayan, Bolaq stretch, Western Desert, Egypt. *Ann. Geol. Surv. Egypt* 8: 19-50.

EL KALIOUBI, B. A. 1974. *Petrological studies on the volcanic rocks of Bahriya Oasis.* M.Sc. Thesis, Ain Shams Univ., Cairo.

EL KAMMAR, A. M. 1977. Mineralogical and geochemical characteristics of Gebel El Hefuf phosphate-bearing rocks, Bahariya Oasis, Western Desert, Egypt. *6th Colloq. Geol. Aegean Region,* Athens.

EL KASSAS, I. A. 1974. *Radioactivity and geology of Wadi Atalla area, Eastern Desert, Egypt.* Ph.D. Thesis, Ain Shams University, Cairo.

EL KHASHAB, B. 1977. *A review of the reptile fauna of Egypt.* Geol. Surv. Egypt, Paper 62, 10pp.

EL KHASHAB, B., E. L. SIMONS & J. G. FLEAGLE 1983. *Annotated bibliography of Egyptian vertebrate fossils up to the end of 1980.* Geol. Surv. Egypt, Paper 65, 111pp.

EL KHAWAGA, M. L. 1979. A contribution to the fracture pattern of Abu Tartur plateau, Western Desert, Egypt. *Ann. Geol. Surv. Egypt* 9: 163-171.

EL MAHDY, A. 1985. Belayim land case history: reservoir management increases recoverable reserves. *8th Petrol. Production Seminar, EGPC,* Cairo.

EL MANHARAWY, M. S. 1977. *Geochronological investigation of some basement rocks in the central Eastern Desert between Lat. 25° and 26° N.* Ph.D. Thesis, Cairo University.

EL MEZAYEN, A. 1984. *Mineralogical and geochemical study of ophiolitic rocks along Qift-Quseir road, central Eastern Desert, Egypt.* Ph.D. Thesis, Azhar University, Cairo.

EL NAGGAR, Z. R. M. 1966. Stratigraphy and classification of type Esna Group of Egypt. *Bull. Am. Assoc. Petrol. Geol.* 50: 1455-1477.

EL NAGGAR, Z. R. M. 1966. Stratigraphy and planktonic foraminifera of the upper Cretaceous-lower Tertiary succession in the Esna-Idfu region, Nile Valley, Egypt. *Bull. Brit. Mus. Nat. Hist.,* suppl. 2, 279pp.

EL NAGGAR, Z. R. M. 1970. On a proposed lithostratigraphic subdivision for the late Cretaceous-early Paleogene succession in the Nile Valley, Egypt. *7th Arab Petrol. Congr.,* Kuwait.

ELOUI, M. & S. ABDINE 1972. *Rock units correlation chart of the northern Western Desert, Egypt.* Internal Report, Western Desert Petroleum Co., Alexandria.

EL RAMLY, I. M. 1964. The use of fissured limestone in locating ground water resources, and its application to Farafra Oasis, New Valley Area, Western Desert, UAR. *Trans. Mining Petrol. Assoc., Egypt* 19.

EL RAMLY, I. M. 1967. Contribution to the hydrological study of limestone terrains in the UAR. In: *Hydrology of fractured rocks, Proc. Dubrovnik Symp. (1965), Intl. Assoc. Sci. Hydrology,* Publ. 73.

EL RAMLY, I. M. 1969. Recent review of investigations on the thermal and mineral springs in the UAR. *Proc. 23rd Intl. Geol. Congr.* 19: 201-213.

EL RAMLY, M. F. 1962. *The absolute ages of some basement rocks from Egypt.* Egypt. Geol. Surv., Paper 15, 13pp.

EL RAMLY, M. F. 1972. A new geological map for the basement rocks in the Eastern and southwestern Deserts of Egypt, scale 1: 1 000 000. *Ann. Geol. Surv. Egypt* 2: 1-18.

EL RAMLY, M. F. & M. K. AKAAD 1960. *The basement complex in the central Eastern Desert of Egypt between Lat. 24° 30' and 25° 40' N.* Geol. Surv. Egypt, Paper 8, 35pp.

EL RAMLY, M. F., M. K. AKAAD & D. M. AL FAR 1959. *Cassiterite-wolframite mineralization near Gebel El Mueilha, Eastern Desert of Egypt.* Geol. Surv. Egypt, Paper 6, 19pp.

EL RAMLY, M. F., M. K. AKAAD & A. H. RASMY 1963. *Geology and structure of the Um Nar iron ore deposit (Eastern Desert, Egypt).* Geol. Surv. Egypt, Paper 28, 29pp.

EL RAMLY, M. F., M. K. AKAAD & A. H. SHAABAN 1973. *Geology and structure of the iron deposits of Gabal El Hadid (Eastern Desert, Egypt).* Geol. Surv. Egypt, Paper 16, 31pp.

EL RAMLY, M. F., L. K. ARMANIOUS & A. M. HUSSEIN 1979. The two ring complexes of Hadayib and Um Risha, south Eastern Desert. *Ann. Geol. Surv. Egypt* 9: 61-69.

EL RAMLY, M. F., V. I. BUDANOV & A. A. HUSSEIN 1971. *The alkaline rocks of south Eastern Desert.* Geol. Surv. Egypt, Paper 53, 111pp.

EL RAMLY, M. F., V. I. BUDANOV, A. A. HUSSEIN & N. E. DERENIUK 1970. Ring complexes in the south Eastern Desert of Egypt. In: O. Moharram et al. (eds), *Studies on some mineral deposits of Egypt.* Geol. Surv. Egypt: 181-194.

EL RAMLY, M. F., R. GREILING, A. KROENER & A. A. RASHWAN 1984. On the tectonic evolution of the Wadi Hafafit area and environs, Eastern Desert of Egypt. *Bull. Fac. Earth Sci., King Abdul Aziz Univ., Jeddah* 6: 113-126.

EL RAMLY, M. F., A. M. HASHAD, M. Y. ATTAWIYA & M. M. MANSOUR 1982. Geochemistry of Kolet Umm-Kabrit bimodal metavolcanics, south Eastern Desert,

Egypt. *Ann. Geol. Surv. Egypt* 12: 103-120.

EL RAMLY, M. F., M. E. HILMY, M. M. ALI & O. A. HASSAN 1982. Petrological and petrochemical studies on some young granite masses in the south Eastern Desert of Egypt. *Ann. Geol. Surv. Egypt* 12: 73-102.

EL RAMLY, M. F. & A. A. HUSSEIN 1982. *The alkaline ring complexes of Egypt.* Geol. Surv. Egypt, Paper 63, 16pp.

EL RAMLY, M. F. & A. A. HUSSEIN 1985. The ring complexes of the Eastern Desert of Egypt. *J. Afr. Earth Sci.* 3: 77-82.

EL RAMLY, M. F., A. A. HUSSEIN & M. H. FRANCIS 1982. The ring complexes of Wadi Dib. *Ann. Geol. Surv. Egypt* 12: 1-6.

EL RAMLY, M. F., S. S. IVANOV & G. G. KOCHIN 1970. General review of the mineral potential of Egypt. In: O. Moharram et al. (eds), *Studies on some mineral deposits of Egypt.* Geol. Surv. Egypt: 3-27.

EL RAMLY, M. F., S. S. IVANOV & G. G. KOCHIN 1970. The occurrence of gold in the Eastern Desert of Egypt. In: O. Moharram et al. (eds), *Studies on some mineral deposits of Egypt.* Geol. Surv. Egypt: 53-64.

EL SHARKAWI, M. A. 1983. Preservation and destruction of silicified wood in the Tertiary record of Egypt as revealed from petrographic and electron-scan studies. *21st Annual Meeting, Geol. Soc. Egypt* (Abstr.).

EL SHARKAWI, M. A. & A. M. ABOU KHADRAH 1968. The dolerite-limestone contact of Gebel Abu Treifiya, Cairo-Suez district. *Egypt. J. Geol.* 12: 11-19.

EL SHARKAWI, M. A. & R. M. EL BAYOUMI 1979. The ophiolites of Wadi Ghadir area, Eastern Desert, Egypt. *Ann. Geol. Surv. Egypt* 9: 125-135.

EL SHARKAWI, M. A., M. M. HIGAZI & N. A. KHALIL 1983. Three genetic iron ore dikes of iron ores at El Gedida mine, Western Desert, Egypt. *21st Annual Meeting, Geol. Soc. Egypt.* (Abstr.).

EL SHARKAWI, M. A. & M. A. KHALIL 1977. Glauconite, a possible source of iron for El Gedida iron ore deposit, Bahariya Oasis, Egypt. *Egypt. J. Geol.* 21: 109-116.

EL SHATOURY, H. M., M. E. MOSTAFA & E. F. NASR 1984. Granites and granitoid rocks in Egypt, a statistical approach of classification. *Chem. Erde* 43: 83-111.

EL SHATOURY, H. M., S. TAKENOUCHI & H. IMAI 1974. Fluid inclusion studies of some beryliferous pegmatites and a tin-tungsten lode from Egypt. *Mining Geol. (Soc. Mining Geol. Jap.)* 24: 307-314.

EL SHAZLY, E. M. 1957. Classification of Egyptian mineral deposits. *Egypt. J. Geol.* 1: 1-20.

EL SHAZLY, E. M. 1962. *Report on the results of drilling in the iron ore deposit of Gebel Ghorabi, Bahariya Oasis, Western Desert.* Geol. Surv. Egypt, 25pp.

EL SHAZLY, E. M. 1964. On the classification of the Precambrian and other rocks of magmatic affiliation in Egypt. *Proc. 22nd Intl. Geol. Congr., New Delhi* 10: 88-101.

EL SHAZLY, E. M. 1966. Structural development of Egypt. *14th Annual Meeting, Geol. Soc. Egypt:* 31-38 (Abstr.).

EL SHAZLY, E. M. 1974. Origin of uranium in Oligocene Qatrani sediments, Western Desert, Egypt. In: *Formation of uranium ore deposists; sedimentary basins and sandstone type deposits; sedimentary deposits in other areas.* IAEA, Vienna, Proc. Ser. STI/PUB/374: 467-478.

EL SHAZLY, E. M. 1977. The geology of the Egyptian region. In: A. E. M. Nairn, W. H. Kanes & F. G. Stehli (eds), *The ocean basins and margins.* Plenum Press 4(A): 379-444.

EL SHAZLY, E. M. 1982. The Red Sea region. In: A. E. M. Nairn, W. H. Kanes & F. G. Stehli (eds), *The ocean basins and margins.* Plenum Press 6: 205-252.

EL SHAZLY, E. M. & A. M. ABDALLAH 1964. *Geology of the sulphur occurrence in Ranga, Eastern Desert.* Geol. Surv. Egypt, Paper 31, 10pp.

EL SHAZLY, E. M., M. A. ABDEL HADY, M. A. EL GHA-WABY & I. A. EL KASSAS 1973. *Geologic interpretation of ERTS-1 satellite images of east Aswan area.* Interim Report 4, Rem. Sens. Center, Cairo, 19pp.

EL SHAZLY, E. M., M. A. ABDEL HADY & I. A. EL KASSAS 1977. Delineation of land features in Egypt by Landsat satellite images. In: Shahroki, F. (ed.). *Remote sensing of earth resources* 6: 277-294, Oklahoma.

EL SHAZLY, E. M., M. A. ABDEL HADY, I. A. EL KASSAS & M. M. EL SHAZLY 1974. *Geology of Sinai Peninsula from ERTS-2 satellite images.* Rem. Sens. Center, Cairo, 10pp.

EL SHAZLY, E. M., S. ABDEL NASSER & B. SHUKRI 1955. *Contributions to the mineralogy of the copper deposits om Sinai.* Geol. Surv. Egypt, Paper 1, 13pp.

EL SHAZLY, E. M. & M. S. AFIA 1958. Geology of Samiuki deposit, Eastern Desert. *Egypt. J. Geol.* 2: 25-42.

EL SHAZLY, E. M., F. A. BASSYOUNI & M. L. ABDEL KHALEK 1975. Geology of the greater Abu Swayel area, Eastern Desert, Egypt. *Egypt. J. Geol.* 19: 1-85.

EL SHAZLY, E. M., I. A. EL KASSAS, H. EL AMIN, M. A. ABDEL HADY, A. B. SALMAN, M. M. EL SHAZLY & A. A. ABDEL MEGUID 1976. *Geology of the Kharga/Dakla Oases area, Western Desert, Egypt, from Landsat-1 satellite images.* Rem. Sens. Center, Cairo, 41pp.

EL SHAZLY, E. M., I. A. EL KASSAS & M. A. EL TAHER 1982. Distribution and orientation of mafic dikes in Wadi Abu Zawal area, Eastern Desert. *Egypt. J. Geol.* 16: 95-105.

EL SHAZLY, E. M., I. A. M. FARAG & F. A. BASSYOUNI 1965. Contribution to the geology and mineralization of Abu Swayel area. *Egypt. J. Geol.* 9: 45-67.

EL SHAZLY, E. M., A. H. HASHAD, T. A. SAYYAH & F. A. BASSYOUNI 1973. Geochronology of Abu Swayel area, south Eastern Desert. *Egypt. J. Geol.* 17: 1-18.

EL SHAZLY, E. M. & A. K. HASSAN 1962. *Report on the results of drilling at Um Gheig mine, Eastern Desert.* Geol. Surv. Egypt, 34pp.

EL SHAZLY, E. M. & A. O. MANSOUR 1962. Note on the occurrence of native sulphur at Umm Reigha, Red Sea coast. *Egypt. J. Geol.* 6: 133-137.

EL SHAZLY, E. M., A. O. MANSOUR, M. S. AFIA & M. G. GHOBRIAL 1956. Miocene lead and zinc deposits in Egypt. *20th Intl. Geol. Congr., Mexico* 13: 19-34.

EL SHAZLY, E. M. & A. H. SABET 1955. *Preliminary report on El Atawi copper deposits, Eastern Desert.* Geol. Surv. Egypt, Paper 2, 5pp.

EL SHAZLY, E. M. & G. S. SALEEB-ROUFAIEL 1959. Contribution to the mineralogy of Egyptian manganese deposits. *Econ. Geol.* 54: 873-888.

EL SHAZLY, E. M. & G. S. SALEEB-ROUFAIEL 1972. Scapolite-cancrinite mineral association in St John's island, Egypt. *20th Intl. Geol. Congr., Mexico* 14: 192-195.

EL SHAZLY, E. M. & G. S. SALEEB-ROUFAIEL 1977. Metasomatism of Miocene sediments in St John's island and its bearing on the history of the Red Sea. *Egypt. J. Geol.* 21: 103-108.

EL SHAZLY , E. M. & G. S. SALEEB-ROUFAIEL 1978. Genesis of peridot in St John's island, Red Sea, and its relation to the metasomatism of the ultrabasic rocks. *Egypt. J. Geol.* 22: 103.

EL SHAZLY, E. M., G. S. SALEEB-ROUFAIEL & N. ZAKI 1974. Quaternary basalt in St John's island, Red Sea, Egypt. *Egypt. J. Geol.* 18: 137-148.

EL SHAZLY, E. M., A. B. SALMAN, I. E. EL ASSY, M. H. SHALABY & L. M. NOSIER 1982. Discovery of new Uranium-Thorium and Niobium-bearing pegmatite pink granite occurrences in the northern part of the Eastern Desert. *20th Annual Meeting, Geol. Soc. Egypt* (Abstr.).

EL SHAZLY, E. M., N. M. SHUKRI & G. S. SALEEB-ROUFAIEL 1963. Geological studies of Oleikat-Marahil and the Sakran manganese iron deposits, west central Sinai. *Egypt. J. Geol.* 7: 1-26.

EL SHAZLY, M. M. & A. SHATA 1960. Contribution to the study of the stratigraphy of El Kharga oasis. *Bull. Inst. Desert Egypte* 10(2): 1-10.

EL SHESHTAWI, Y. A. 1979. *Petrographical and petrochemical studies of the Qatrani basaltic ricks.* M.Sc. Thesis, Azhar Univ., Cairo.

EL SWEIFY, A. 1975. Subsurface Paleozoic stratigraphy of Siwa-Faghur area, Western Desert, Egypt. *9th Arab Petrol. Congr., Dubai*, Paper 119(B-3).

EL TAHIR, M. A. 1985. *Radioactivity and mineralization of granitic rocks of the Ereidiya occurrence and comparison to El Missikat-Rei El Garra occurrence, Eastern Desert, Egypt.* Ph.D. Thesis, Azhar Univ., Cairo.

ELZARKA, M. H. 1983. Mode of hydrocarbon generation and prospects of the northern part of the Western Desert, Egypt. *J. Afr. Earth Sci.* 1: 295-304.

ELZARKA, M. H. 1984. The application of facies parameters and mapping techniques to exploration of the subsurface lower Cretaceous of the Qattara depression, Western Desert, Egypt. *J. Petrol. Geol.* 7: 277-302.

ELZARKA, M. H. 1986. Subsurface geology of the Tertiary rocks of the northeastern district of the Western Desert of Egypt. *J. Afr. Earth Sci.* 5: 285-319.

ELZARKA, M. H., M. M. EL AWADI & T. A. ABDEL FATTAH 1980. Tectonic history of the area north of Wadi El Natrun, Western Desert, Egypt during the Tertiary period. *Delta J. Sci.* 4: 191-220.

ELZARKA, M. H. & I. A. RADWAN 1985. Tectonic history of the northeastern sector of the Western Desert during the Mesozoic era. *N. Jb. Geol. Palaeontol., Mh* (1985): 257-276.

ELZARKA, M. H. & M. Y. ZEIN EL-DIN 1985. Lithostratigraphy of some subsurface Eocene sections in the Gulf of Suez and Western Desert of Egypt. *Geoscience J., Mansoura* 4: 19-30.

EMBABI, N. S. 1982. Barchans of the Kharga depression. In: F. El Baz & T. A. Maxwell (eds), *Desert landforms of southwest Egypt, a basis for comparison with Mars*, NASA: 141-156.

EMBABI, N. S. & M. A. EL KAYALI 1979. A morphotectonic map of the Bhariya depression. *Ann. Geol. Surv. Egypt* 9: 179-183.

ENGEL, A. E., T. H. DIXON & R. J. STERN 1980. Late Precambrian evolution of Afro-Arabian crust from ocean arc to craton. *Bull. Geol. Soc. Am.* 91: 699-706.

EMSLIE, R. F. 1978. Anorthosite massives, rapakivi granites and late Proterozoic rifting of North America. *Precambrian Res.* 7: 61-98.

ENGELHARDT, H. 1907. Tertiäre Pflanzenreste aus dem Fajum. *Beitr. Palaeontol. u. Geol. Oesterreich. Ungarn. Orients* 20: 206-216.

ESSAWY, M. A. 1967. *Geology of the area around Bir Umm Khariga. Eastern Desert.* Ph.D. Thesis, Assiut Univ.

ESSAWY, M. A. & K. M. ABU ZEID 1972. Atalla felsite intrusion and its neighbouring flows and tuffs, Eastern Desert. *Ann. Geol. Surv. Egypt* 2: 271-280.

ESTES, J. E., J. R. JENSEN & D. S. SIMONETTE 1980. Impacts of remote sensing on US geography. *Rem. Sens. Environment* 10: 43-80.

ESTES, J. E. & D. S. SIMONETTE 1975. Fundamentals of image interpretation. In: R. G. Reeves (ed.), *Manual of remote sensing*, Falls Church (ASAP), 2: 869-1076.

EVANS, T. R. & N. C. COLEMAN 1974. North Sea geothermal gradients. *Nature* 247: 28-30.

EVANS, T. R. & H. U. TAMMEMAGI 1974. Heat flow and heat production in northeast Africa. *Earth Planet. Sci. Lett.* 23: 349-356.

EWER, R. F. 1973. *The carnivores.* Ithaca, Cornell Univ. Press.

EYAL, M., Y. BARTOV, A. E. SHIMRON & Y. K. BENTOR 1980. *Geological map of Sinai, scale 1 : 500 000.* Geol. Surv. Isr.

EYAL, M. & T. HEZKIYAHO 1980. Katherina pluton; outlines of petrographic framework. *Isr. J. Earth Sci.* 29: 41-52.

EYAL, Y. & Z. RECHES 1983. Tectonic analysis of the Dead Sea rift region since the late Cretaceous based on mesostructures. *Tectonics* 2: 167-185.

EYSINGA, F. W. B. VAN 1978. *Geological time table.* Elsevier.

EZZAT, M. A. 1959. *Origin of the underground water of the Libyan Desert and preliminary evaluation of its amount.* Internal Report, General Desert Development Organization, Cairo.

EZZAT, M. A. 1974. *Groundwater series in the Arab Republic of Egypt; exploitation of groundwater in El Wadi El Gedid Project area; Parts I-IV.* General Desert Development Authority, Ministry of Irrigation, Cairo.

EZZAT, M. A., S. NOUR & M. MISHRIKI 1977. *Groundwater model of south Qattara area, Western Desert of Egypt.* Internal Report, General Desert Development Authority, Cairo.

EZZAT, M. R. & M. DIA EL DIN 1974. Oil and gas discoveries in the Western Desert, Egypt (Abu Gharadig and Razzak fields). *4th Petrol. Explor. Seminar, EGPC*, Cairo.

FAHIM, M., E. Z. BASTA & Y. EL RASHIDI 1971. A paleomagnetic study on some Tertiary basalts from Egypt. *IAGA, 15th IUGS Meeting*, Moscow.

FAHMY, S. E., M. FAKHRY, V. KRASHENINNIKOV, D. MELNIKOV & V. SAMODUROV 1969. Biostratigraphic and correlation scheme of the Miocene deposits in the Gulf of Suez region. *Proc. 3rd Afr. Micropaleontol. Colloq.*, Cairo: 493-500.

FAHMY, S. E., V. KRASHENINNIKOV, I. MIKHAILOV & V. SAMODUROV 1969. Biostratigraphy of Paleogene deposits in Egypt. *Proc. 3rd Afr. Micropaleontol. Colloq.*, Cairo: 477-484.

FAIRHEAD, J. D. & R. W. GIRDLER 1970. The seismicity of the Red Sea, Gulf of Aden and Afar triangle. *Phil. Trans. Roy. Soc. London* 267(A): 49-74.

FANSHAW, J. R. 1939. Structural geology of the wind River

Canyon area, Wyoming. Bull. Am. Assoc. Petrol. Geol. 23: 1439-1492.

FARAG, I. A. M. 1948. Deux nouveaux gisements de Bathonien fossilifere sur la rive occidentale du Golf de Suez. C.R. sceances Soc. Geol. France 1948: 109-110.

FARAG, I. A. M. 1957. On the occurrence of Lias in Egypt. Egypt. J. Geol. 1: 49-63.

FARAG, I. A. M. & M. M. ISMAIL 1955. On the structure of Wadi Hof area (northeast of Helwan). Bull. Inst. Desert Egypte 5(1): 179-192.

FARAG, I. A. M. & M. M. ISMAIL 1959. A contribution to the stratigraphy of the wadi Hof area (northeast of Helwan). Bull. Fac. Sci., Cairo Univ. 34: 147-168.

FARAG, I. A. M. & M. M. ISMAIL 1959. A contribution to the structure of the area east of Helwan. Egypt. J. Geol. 3: 71-86.

FARAG, I. A. M. & A. SHATA 1954. Detailed goelogical survey of El Minsherah area. Bull. Inst. Desert Egypte 4(2): 5-82.

FARIS, M. I. & H. L. ABBAS 1963. The geology of Shabrawet area. Bull. Fac. Sci., Ain Shams Univ. 7: 37-59.

FARIS, M. I. & M. Y. HASSAN 1959. Report on the stratigraphy and fauna of the upper Cretaceous rocks of Um El Hueitat, Safaga area. Bull. Fac. Sci., Ain Shams Univ. 4: 191-207.

FARIS, M. I. & A. SHATA 1955. Stratigraphy of the Bir El Haleifiya-Gebel El Zeita area (west Sinai foreshore province, Egypt). 2nd Arab Sci. Congr., Cairo: 27-89.

FAUL, H. 1960. Geologic time scale. Bull. Geol. Soc. Am. 71: 637-644.

FAURE, G. 1977. Principles of isotope geology. Wiley, New York, 464pp.

FAWZI, M. A. 1959. Etude stratigraphique et paléontologique de la region du Gebel Shabrauet. Egypt. J. Geol. 3: 149-158.

FAWZI, M. I. 1960. Contribution to the study of the Cenomanian of Egypt and correlation of the Egyptian facies with those of neighbouring countries. Bull. Fac. Sci., Alexandria Univ. 4: 107-150.

FAWZI, M. I. 1963. La faune cénomanienne d'Egypte. Geol. Surv. Egypt, Monograph 2: 131pp.

FAWZY, H. & A. ABDEL AAL 1984. Regional study of Miocene evaporites and Pliocene-Recent sediments in the Gulf of Suez. Proc. 7th Petrol. Explor. Seminar, EGPC, Cairo: 49-74.

FENNINE. M. A. 1931. La formation géologique des gisements des mineraux de manganese au Sinai. Bull. Inst. Egypte 13: 15-26.

FELIX, J. P. 1904. Studien uber tertiäre und quartare Korallen aus Aegypten und der Sinai Halbinsel. Z. Deutsch. Geol. Ges. 56: 168-206.

FITCHES, W. R., R. H. GRAHAM, I. M. HUSSEIN, A. G. RIES, R. M. SCHACKLETON & R. C. PRICE 1983. The late Proterozoic ophiolite of Sol Hamed, NE Sudan. Precambrian Res. 19: 385-411.

FLEAGLE, J. G. 1980. Locomotor behavior of the earliest anthropoids; a review of the current evidence. Z. Morphol. Anthropol. 71: 149-156.

FLEAGLE, J. G. 1983. Locomotor adaptations of Oligocene and Miocene hominids and their phyletic implications. In: R. L. Ciochon & R. S. Corruccini (eds), New implications of ape and human ancestry. Plenum Press, New York: 301-324.

FLEAGLE, J. G., T. M. BROWN, J. D. OBRADOVICH & E. L.

SIMONS 1986. Age of the earliest African anthropoids. Science 234: 1247-1249.

FLEAGLE, J. G. & R. F. KAY 1985. The paleobiology of catarrhines. In: E. Delson (ed.), Ancestors, the hard evidence. Alan R. Liss Inc., New York: 23-36.

FLEAGLE, J. G., R. F. KAY & E. L. SIMONS 1980. Sexual dimorphism in early anthropoids. Nature 278: 328-330.

FLEAGLE, J. G. & E. L. SIMONS 1978. Humeral morphology of the earliest apes. Nature 276: 705-707.

FLEAGLE, J. G. & E. L. SIMONS 1979. Anatomy of the bony pelvis of parapithecid primates. Folia Primatol. 31: 176-186.

FLEAGLE, J. G. & E. L. SIMONS 1982. Skeletal remains of Propliopithecus chirobates from the Egyptian Oligocene. Folia primatol. 39: 161-177.

FLEAGLE, J. G. & E. L. SIMONS 1983. The tibio-fibular articulation in Apidium phiomense, an Oligocene anthropoid. Nature 301: 328-329.

FLECK, R. J., R. G. COLEMAN, H. R. CORNWALL, W. R. GREENWOOD, D. C. HADLEY, W. C. PRINZ, J. S. RATTE & D. L. SCHMIDT 1976. Potassium-argon geochronology of the Arabian shield. Bull. Geol. Soc. Am. 87: 9-21.

FLECK, R. J., W. R. GREENWOOD, D. C. HADLEY, R. E. ANDERSON & D. L. SCHMIDT 1980. Rubidium-strontium geochronology and plate tectonic evolution of the southern part of the Arabian shield. US Geol. Surv., Prof. Paper 1131: 38pp.

FLEXER, A. 1968. Stratigraphy and facies development of Mount Scopus Group (Senonian-Paleocene) in Israel and adjacent countries. Isr. J. Earth. Sci. 17: 85-114.

FLEXER, A. 1971. Late Cretaceous paleogeography of northern Israel and its significance for the Levant geology. Paleogeogr., Paleoclimat., Paleoecol. 10: 293-316.

FOURNIER, R. O. & R. W. POTTER 1979. Magnesium correction to the Na-K-Ca chemical geothermometer. Geochimica et Cosmochimica Acta 43: 1543-1550.

FOURNIER, R. O. & J. J. ROWE 1966. Estimation of underground temperatures from the silica content of water from hot springs and steam wells. Am.J.Sci. 264: 685-697.

FOURNIER, R. O. & A. H. TRUESDELL 1973. An empirical Na-K-Ca geothermometer for natural waters. Geochimica et Cosmochimica Acta 37: 1255-1275.

FOURTAU, R. 1898. Sur l'age des forets petrifiées des deserts d'Egypte. Bull. Soc. Geograph. Egypte 1898(2): 8pp.

FOURTAU, R. 1913. Catalogue des invertebres fossiles de l'Egypte. Terrains Tertiares. Geol. Surv. Egypt, 93pp.

FOURTAU, R. 1914. Catalogue des invertebres fossiles de l'Egypte. Terrains Cretacé. 1. Ecinodermes. Geol. Surv. Egypt, 109pp.

FOURTAU, R. 1917. Catalogue des invertebres fossiles de l'Egypte. Terrains Cretacé. 2. Mollusques lamellibraches. Geol. Surv. Egypt, 108pp.

FOURTAU, R. 1918. Contribution à l'étude des vertébrés miocènes de l'Egypte. Geol. Surv. Egypt, 109pp.

FOURTAU, R. 1920. Catalogue des invertébrés fossiles de l'Egypte. Terrains Tertiaire. 2. Echinodermes neogènes. Geol. Surv. Egypt, 101pp.

FOX, P. J. & N. D. OPDYKE 1973. Geology of the oceanic crust. Magnetic properties of oceanic rocks. J. Geophys. Res. 78: 5139-5154.

FRAAS, E. 1898. Streifzüge in der ägyptisch-arabischen Wüste. Ver. Naturkde. Württemburg (Stuttgart) 54: 5-7.

FRAAS, O. 1867. *Aus dem Orient: geologische Beobach-tungen am Nil, auf der Sinai Halbinsel und in Syrien.* Ender u. Suebert, Stuttgart, 222pp.

FRANCIS, M. H. 1972. Geology of the basement rocks in the north Eastern Desert between Lat. 27° 30' and 28° 00' N. *Ann. Geol. Surv. Egypt* 2: 161-180.

FRANZ, G., H. PUCHELT & P. PASTEELS 1987. Petrology, geochemistry and age relations of Triassic and Tertiary rocks from SW Egypt and NW Sudan. *J. Afr. Earth Sci.* 6: 335-352.

FREDRIKSON, G. 1974. *On the geology and mineralogy of El Genena El Gharbia, Eastern Desert, Egypt.* Internal report, Aswan Mineral Survey Project, Geol. Surv. Egypt.

FREUND, R. 1965. A model for the structural development of Israel and adjacent areas since upper Cretaceous times. *Geol. Mag.* 102: 190-205.

FREUND, R. 1970. Plate tectonics of the Red Sea and east Africa. *Nature* 228: 453.

FREUND, R., Z. GARFUNKEL, I. ZAK, M. GOLDBERG, T. WEISSBROD & B. DERIN 1970. The shear along the Dead Sea rift. *Phil. Trans. Roy. Soc. London* 267(A): 107-130.

FREUND, R. & M. RAAB 1969. Lower Turonian ammonites from Israel. *Isr. Geol. Surv., Spec. Papers Paleontology* 4: 83pp.

FRIEDMAN, G. M. ET AL. 1971. Paleoenvironments of the Jurassic in the coastal belt of northern and central Israel and their significance in the search for petroleum reser-voirs. *Isr. Geol. Surv., Oil Div. Report* no. OD/1/71.

FRISCH, W. 1982. The Wadi Dib ring complex, Nubian Desert, Egypt and its importance for the upper limit of the Pan African orogeny. *Precambrian Res.* 16: A20 (Abstr.).

FRISCH, W. & A. M. S. AL SHANTI 1977. Ophiolite belts and the collision of island arcs in the Arabian shield. *Tecto-nophysics* 43: 293-306.

FRISCHAT, G. H., C. KLOEPFER, W. BEYER & R. A. WEEKS 1982. Glastechnische Untersuchungen an libyschen Wüstenglas. *Glas. Technol. Ber.* 55: 228-234.

FRITEL, P. H. 1926. Remarques additionelles sur la flore fossile de grés de Nubie. *Bull. Mus. Hist. Nat.* 32: 315-319.

FRITZE, G., J. JNASA & K. KRAUS 1985. Orthophotos und stereopartner aus metrischen weltraumbildern. *Osterr. Z. Vermess. Photogramm.* 73: 159-174.

FULLAGER, P. D. 1980. Pan-African age granites of north-eastern Africa: new or reworked sialic materials? In: M. J. Salem & M. R. Busrewil (eds), *The Geology of Libya.* Academic Press 3: 1051-1058.

FURNES, H., A. E. SHIMRON & D. ROBERTS 1985. Geoche-mistry of Pan-African volcanic arc sequences in south-eastern Sinai Peninsula and plate tectonic implications. *Precambrian Res.* 29: 359-382.

GABRIEL, B. & S. KROEPELIN 1985. Parabeldünen am Wadi Howar. *Geowiss. unserer Zeit* 3: 105-112.

GAD, M. A., F. SOLIMAN, R. FAKHRY, E. A. HALIM, V. D. GAD & K. A. SOLIMAN 1978. *Geological and geoche-mical exploration of Hamata area.* Internal Report, Ex-pedition 8/76, Geol. Surv. Egypt.

GARFUNKEL, Z. & Y. BARTOV 1977. The tectonics of the Suez rift. *Bull. Geol. Surv. Isr.* 71: 1-48.

GARFUNKEL, Z. & B. DERIN 1984. Permian-early Mesozoic tectonism and continental margin formation of Israel and its implications for the history of the eastern Mediterra-nean. In: J. E. Dixon & A. H. F. Robertson (eds), *The geological evolution of the eastern Mediterranean.* Blackwell Scientific Publishers, Oxford: 187-201.

GARRISON, R. E., C. R. GLENN, P. D. SNAVELY & S. A. E. MANSOUR 1979. Sedimentology and origin of upper Cretaceous phosphorite deposits at Abu Tartur, Western Desert, Egypt. *Ann. Geol. Surv. Egypt* 9: 261-281.

GARSON, M. S. 1972. *Report on a visit at the UNDP Aswan Mineral Project.* Internal Report, Aswan Mineral Survey Project, Geol. Surv. Egypt.

GARSON, M. S. & G. FREDRICKSON 1975. *Geology and mineralization of El Genena area, south Eastern Desert.* Internal Report, Aswan Mineral Survey Project, Geol. Surv. Egypt.

GARSON, M. S. & M. KRS 1976. Geophysical and geologic-al evidence of the relationship of Red Sea transverse tectonics to ancient fractures. *Bull. Geol. Soc. Am.* 87: 169-181.

GARSON, M. S., A. A. HUSSEIN & G. FREDERIKSON 1974. *Report on the detailed study and results of drilling on Um Doweila dike.* Internal Report, Aswan Mineral Sur-vey Project, Geol. Surv. Egypt.

GARSON, M. S. & A. H. G. MITCHELL 1981. Precambrian ore deposits and plate tectonics. In: A. Kroener (ed.), *Precambrian plate tectonics* Elsevier: 689-731.

GARSON, M. S. & M. I. SHALABY 1976. Precambrian-lower Paleozoic plate tectonics and metallogenesis in the Red Sea region. *Geol. Assoc. Canada* Special paper 14: 573-596.

GASS, I. G. 1977. The evolution of the Pan-African crystal-line basement in NE Africa and south Arabia. *J. Geol. Soc. London* 134: 129-138.

GASS, I. G. 1979. Evolutionary model for the Pan-African crystalline basement. *Bull. Inst. Applied Geol., King Abdul Aziz Univ., Jeddah* 3(1): 11-20.

GASS, I. G. 1981. Pan-African (upper Proterozoic) plate tectonics of the Arabian Nubian shield. In: A. Kroener, *Precambrian plate tectonics.* Elsevier: 387-405.

GASS, I. G. 1982. Upper Proterozoic (Pan-African) calc-alkaline magmatism in northeastern Africa and Arabia. In: R. S. Thorpe (ed.), *Andesites.* Wiley, New York: 591-609.

(GDK) GROUNDWATER DEPARTMENT, KHARGA 1971. *Hy-drogeology and geochemistry of Bahariya Oasis.* In-ternal Report to General Desert Development Authority, Cairo.

GEBO, D. L. & D. T. RASMUSSEN 1985. The earliest fossil pangolin (Pholidota: Manidae) from Africa. *J. Mammol.* 66: 538-541.

GEBO, D. L. & E. L. SIMONS 1984. Puncture marks on early African anthropoids. *Am. J. Phys. Anthropol.* 65: 31-35.

GEBO, D. L. & E. L. SIMONS 1987. Morphology and lo-comotor adaptation in early Oligocene anthropoids. *Am. J. Phys. Anthropol.* 74: 83-101.

GEOLOGICAL MAP OF EGYPT (scale 1:500 000) 1987-1988. Egypt. Gen. Petrol. Corp. and Conoco, 20 sheets.

GEOLOGICAL SURVEY OF CANADA 1984. Canadian mineral deposit types: a geological synopsis. *Econ. Geol. Report* 36.

GOLOGICAL SURVEY OF EGYPT 1977. Mineral activities in Egypt (in Arabic). Country report, 3rd Arab. Mineral Conf., Rabat, Morocco.

GEOLOGICAL SURVEY OF EGYPT 1968-1983. *Series of geo-logical maps:* Idfu-Qena 1: 200 000 (1968); Aswan and Qena quadrangles 1:500 000 (1978); Egypt 1: 2 000 000 (1981); Wadi Qena, Gebel El Urf and Dakhla

1:250 000 (1983); Cairo 1: 100 000 (1983).

GEOLOGICAL SURVEY OF EGYPT 1984. *Potential mineral resources of Egypt.* 144pp.

GEOLOGICAL SURVEY OF EGYPT 1986. *Gold in Egypt: a commodity report.*

GEOLOGICAL SURVEY OF ISRAEL 1980-1984. *Research in Sinai, collected reprints,* 3 vols.

GERARDS, J. F. & H. E. LANDMIRANT 1962. Rapport sur l'application de la photogéologie 'pre-controllée' en Afrique centrale. *Intl. Arch. Photogramm.* 14: 115-119.

GERGAWI, A. & H. M. A. EL KHASHAB 1968. Seismicity of the UAR. *Bull. Helwan Observ.* 76: 1-27.

GERMANN, K., W. D. BOCK & T. SCHROEDER 1984. Facies development of upper Cretaceous phosphorites in Egypt: sedimentological and geochemical aspects. *Berl. Geowiss. Abh.* 50(A): 345-362.

GETTINGS, M. E. 1982. Heat flow measurements at shot points along the 1978 Saudi Arabian seismic deep-refraction line; discussion and interpretation. *US Geol. Surv., open file report* 82/794, 40pp.

GETTINGS, M. E. & A. SHOWAIL 1982. Heat flow measurements at shot points along the 1978 Saudi Arabian seismic deep refraction line, part A: results of the measurements. *US Geol. Surv., open file report* 82-793.

GHANEM, M. 1972. Geology of the basement rocks north of Lat 28° OON, Eastern Desert, Ras Gharib area. *Ann. Geol. Surv. Egypt* 2: 181-197.

GHANEM, M. 1972. Geology of Wadi Hodein area. *Ann. Geol. Surv. Egypt* 2: 199-214.

GHANEM, M., A. A. DARDIR, M. H. FRANCIS, A. A. ZALATA & K. M. ABU ZEID 1973. Basement rocks in the Eastern Desert of Egypt north of Lat. 26° 40' N. *Ann. Geol. Surv. Egypt* 3: 33-38.

GHANEM, M., I. A. MIKHAILOV, A. A. ZALATA, A. V. RAZVALIEV, T. M. ABDEL RAZIK, Y. V. MITROV & M. S. ABDEL GHANI 1970. Stratigraphy of the phosphate-bering Cretaceous and Paleocene sediments of the Nile Valley between Idfu and Qena: In O. Moharram et al. (eds), *Studies on some mineral deposits of Egypt.* Geol. Surv. Egypt: 109-134.

GHEITH, M. A. 1955. Classification and review of Egyptian iron ore deposits. *Symp. Applied Geol. Near East, UNESCO, Ankara:* 106-113.

GHEITH, M. A. 1959. Mineralogy, thermal analysis and origin of Bahariya iron ores of Egypt. *20th Intl. Geol. Congr., Mexico* 13: 135-151.

GHOBRIAL, M. G. 1963. *Some occurences of manganese deposits in Egypt.* Geol. Surv. Egypt, Paper 20, 25pp.

GHOBRIAL, M. G. 1967. *The structural geology of the Kharga oasis.* Geol. Surv. Egypt, paper 43, Cairo.

GHOBRIAL, M. G. & M. LOTFI 1967. *The geology of Gebel Gattar and Gebel Dokhan areas.* Geol. Surv. Egypt, Paper 40, 26pp.

GHORAB, M. A. 1961. Abnormal stratigraphic features in Ras Gharib oilfield. *3rd Arab Petrol. Congr.,* Alexandria, 10pp.

GHORAB, M. A. & M. S. AMIN 1959. Exploration efforts and their bearing on the oil prospecting of the UAR, part 1: the Egyptian region. *1st Arab Petrol. Congr.,* Cairo, 11p..

GHORAB, M. A. & M. M. ISMAIL 1957. A microfacies study of the Eocene and Pliocene east of Helwan. *Egypt. J. Geol.* 1: 105-125.

GHORAB, M. A. & I. M. MARZOUK 1967. *A summary report on the rock stratigraphic classification of the Miocene non-marine and coastal facies in the Gulf of Suez and Red Sea coast.* Internal Report 601, General Petrol. Co., Cairo.

GIBBS, A. D. 1984. Structural evolution of extensional basin margins. *J. Geol. Soc. London* 141: 609-620.

GIBOWICZ, S. J., Z. DROSTE, R. M. KEBEASY, E. M. IBRAHIM & R. N. H. ALBERT 1982. A microearthquake survey in the Abu Simbel area in Egypt. *Eng. Geol.* 19: 95-109.

GIEGENGACK, R. F. 1968. *Late Pleistocene history of the Nile Valley in Egypt and Nubia.* Ph.D. Thesis, Yale Univ., USA.

GIEGENGACK, R. F. 1970. Uranium series corals from Red Sea. *Nature* 126: 155-156.

GIEGENGACK, R. F. & B. ISSAWI 1975. Libyan desert silical glass, a summary of the problem of its origins. *Ann. Geol. Surv. Egypt* 5: 105-118.

GIESE, P., K. J. REUTTER, V. J. JACOBSHAGEN & R. NICILICH 1982. Explosion seismic crustal series in the Alpine Mediterranean region and their implications to tectonic processes. In: H. Berckhemer & K. Hsu (eds), *Alpine-Mediterranean Geodynamics, Am. Geophys. Union and Geol. Soc. Am.,* Geodynamics ser. 7: 39-73.

GILBOA, Y. 1980. Post Eocene clastics distribution along the El Qa' plain, southern Sinai. *Isr. J. Earth Sci.* 29: 197-206.

GILBOA, Y. & A. COHEN 1979. Oil trap patterns in the Gulf of Suez. *Isr. J. Earth Sci.* 28: 13-26.

GILL, D. & S. O. FORD 1956. Manganiferous iron ore deposits of the Um Bogma district, Sinai, Egypt. *20th Intl. Geol. Congr., Mexico* 2: 173-177.

GILLESPIE, J. G. & T. H. DIXON 1983. Lead isotope systematics of some igneous rocks from the Egyptian shield. *Precambrian Res.* 20: 63-77.

GINDY, A. R. 1954. The plutonic history of Aswan Area, Egypt. *Geol. Mag.* 91: 484-497.

GINDY, A. R. 1961. Radioactivity and Tertiary volcanic activity in Egypt. *Econ. Geol.* 56: 557-568.

GINDY, A. R. 1963. Stratigraphic sequence and structures of a post-lower Eocene section in Wadi Gasus El Bahari, Safaga district, Eastern Desert, UAR. *Bull. Fac. Sci., Alexandria Univ.* 5: 71-87.

GINDY, A. R. 1965. The Nubian Sandstone around Telal El Zouhour, its barite veins and sand crystals. *Bull. Inst. Egypte* 38: 1-70.

GINDY, A. R. 1966. The origin of copper mineralization in central Sinai, UAR. *5th Arab Sci. Congr., Baghdad* 3: 607-634.

GINDY, A. R. 1972. Toward a kinematic classification of the Egyptian basement. *Ann. Geol. Surv. Egypt* 2: 47-77.

GINDY, A. R. & M. A. EL ASKARI 1969. Stratigraphy, structure and origin of the Siwa depression, Western Desert of Egypt. *Bull. Am. Assoc. Petrol. Geol.* 53: 603-625.

GINDY, A. R., W. W. GHOBRIAL & E. E. BADRA 1973. Studies on Egyptian phosphates; I: thickness, phosphorous content, alpha radioactivity of the exploited phosphorite bed in Hamrwein area, Red Sea province, Egypt. *Proc. Egypt. Acad. Sci.* 26: 1-17.

GINDY, A. R., W. W. GHOBRIAL & E. E. BADRA 1976. Studies on Egyptian phosphates; II: intraphosphatic succession, mechanical analyses and radioactivity of the exploited bed units in Hamrawein area, Red Sea province, Egypt. *Proc. Egypt. Acad. Sci.* 29: 129-144.

GINDY, A. R. & M. R. MOHAMMED 1971. Precambrian laminated ignimbrite and its lithophysae in Wadi Nugara, southwest of Port Safaga, Red Sea province. *Bull. Fac. Sci., Alexandria Univ.* 1: 145-177.

GINDY, A. R. & M. O. TAMISH 1985. Some major and trace constituents of Phanerozoic Egyptian mud rocks and marls. *J. Afr. Earth Sci.* 3: 303-320.

GINGERICH, P. D. 1978. The Stuttgart collection of Oligocene primates from the Fayum province of Egypt. *Palaeontol. Z.* 52: 82-92.

GINZBURG, A. ET AL. 1975. Geology of Mediterranean shelf of Israel. *Bull. Am. Assoc. Petrol. Geol.* 59: 2142-2160.

GINZBURG, A. & G. GVIRTZMAN 1979. Changes in the crust and sedimentary cover across the transition from the Arabian platform to the Mediterranean basin: evidence from seismic refraction and sedimentary studies in Israel and N. Sinai. *J. Sedimentary Geol.* 23: 19-36.

GINZBURG, A., J. MAKRIS, K. FUCHS, B. PERATHONER & C. PRODEHL 1979. Detailed structure of the crust and upper mantle along the Jordan-Dead Sea rift. *J. Geophys. Res.* 84: 5605-5612.

GIRDLER, R. W. 1970. A review of Red Sea heat flow. *Phil. Trans. Roy. Soc. London* 267(A): 191-203.

GIRDLER, R. W. 1985. Problems concerning the oceanic lithosphere in northern Red Sea. *Tectonophysics* 116: 109-122.

GIRDLER, R. W. & B. W. DARCOTT 1972. African poles of rotation, comments on earth sciences. *Geophysics* 2: 131-138.

GIRDLER, R. W. & T. R. EVANS 1977. Red Sea heat flow. *Geophys. J. Roy. Astronom. Soc.* 51: 245-251.

GIRDLER, R. W. & P. STYLES 1974. Two stages in the Red Sea floor spreading. *Nature* 247: 7-11.

GIRDLER, R. W. & P. STYLES 1976. Opening of the Red Sea with two poles of rotation – some comments. *Earth and Planet. Sci. Lett.* 33: 169-172.

GIRDLER, R. W. & P. STYLES 1982. Comments on 'The Gulf of Aden: structure and evolution of a young ocean basin and continental margin' by J. R. Cochran. *J. Geophys. Res.* 87: 6761-6763.

GISCHLER, C. E. 1976. *Present and future trends in water resources development in the Arab countries.* UNESCO, Cairo.

GLENN, C. F & G. M. DENMAN 1980. Geologic literature of Egypt, 1933-1978. *US Geol. Surv., open file report* 0/930, 218pp.

GLENN, C. R. & S. E. A. MANSOUR 1979. Reconstruction of the depositional and diagenetic history of phosphorites and associated rocks of the Duwi Formation (late Cretaceous), Eastern Desert, Egypt.*Ann. Geol. Surv. Egypt* 9: 388-407.

GOETZ, A. F. H. 1980. Geological remote sensing in the 1980s. In: B. S. Siegal & A. R. Gillespie (eds), *Remote sensing in geology.* Wiley: 679-685.

GOLD, D. P., R. R. PARIZEK & S. A. ALEXANDER 1973. Analysis and application of ERTS-1 data for regional geologic mapping. *Symp. Sign. Res., ERTS-1, NASA* 1(A): 231-245.

GOLDBERG, M. & G. M. FRIEDMAN 1974. Paleoenvironments and paleogeographic evolution of the Jurassic system in northern Israel. *Bull. Geol. Surv. Isr.* 61: 1-44.

GONFIANTINI, R., G. CONRAD, J. C. FONTES, G. SAUZAY & B. R. PAYNE 1974. Etude isotopique de la nappe du continental intercalaire et de ses relations avec les autres nappes du Sahara. *Isotope Technique in Groundwater Hydrology, IAEA*, Vienna 1: 227-240.

GOUDARZI, G. H. 1980. Structure – Libya. In: M. J. Salem & M. T. Busrewil (eds), *The Geology of Libya.* Academic Press 3: 879-892.

GREENBERG, J. K. 1981. Characteristics and origin of Egyptian younger granites: summary. *Bull. Geol. Soc. Am.* 92: 224-232.

GREENWOOD, H. R. & R. E. ANDERSON 1975. Palinspastic map of the Red Sea prior to Miocene Red Sea floor spreading. *US Geol. Surv., Saudi Arabian Project Report* 197, 23pp.

GREENWOOD, H. R., R. E. ANDERSON, R. J. FLECK & R. J. ROBERTS 1980. Precambrian geologic history and plate tectonic evolution of the Arabian shield. *Bull. Saudi Arabian Dir. General Mineral Resources* 24: 35pp.

GREENWOOD, H. R., D. G. HADLEY, R. E. ANDERSON, R. J. FLECK & D. L. SCHMIDT 1976. Late Proterozoic cratonization in southwestern Saudi Arabia. *Phil. Trans. Roy. Soc. London* 280(A): 517-527.

GREENWOOD, J. E. G. W. 1962. Rock weathering in relation to the interpretation of igneous and metamorphic rocks in arid regions. *Intl. Arch. Photgramm., 14, Trans. Symp. Photo Interpret.*: 93-99.

GREENWOOD, R. J. 1969. *Radiometric dating of five basalt samples submitted by Gulf of Suez Petroleum Co.* Internal Report 297, Gulf of Suez Petrol. Co.

GREENWOOD, R. J. 1970. *K/Ar dating of three samples from SW Mubarak well-1.* Robertson Res., Internal Report 337 Gulf of Suez Petrol. Co.

GREGOR, H. J. & H. HAGN 1982. Fossil fructifications from the Cretaceous-Paleocene boundary of SW Egypt (Danian, BirAbu Mungar). *Tert. Res.* 4: 121-147.

GREGORY, J. W. 1921. *The rift valleys and geology of east Africa.* Seeley Service and Co., London, 479pp.

GROOTENBOER, J., K. ERIKSSON & J. TRUSWELL 1973. Stratigraphic subdivision of the Transvaal dolomite from ERTS imagery. *3rd ERTS-1 Symp. 1, A, NASA* SP-351: 657-664.

GROSS, G. A. 1965. Geology of iron ore deposits in Canada; I: general geology and evaluation of iron deposits. *Econ. Geol. Report 22, Geol. Surv. Canada.*

GROTHAUS, B., D. EPPLER & R. EHRLICH 1979. Depositional environment and structural implications of the Hamammat Formation, Egypt. *Ann. Geol. Surv. Egypt* 9: 564-590.

GUTENBERG, G. & C. F. RICHTER 1954. *Seismicity of the earth and associated phenomena.* Princeton Univ. Press, 310pp.

GVIRTZMAN, G. & B. BUCHBINDER 1977. The desiccation events in the eastern Mediterranean during Messinian times, as compared with other desiccation events in basins around the Mediterranean. In: B. Biju-Duval & L. Montadert (eds), *Structural history of the Mediterranean basins*, editions Technip: 411-420.

HABIB, M. E., A. A. AHMED & U. M. EL NADY 1985. Two orogenies in the Meatiq area of the central Eastern Desert, Egypt. *Precambrian Res.* 30: 83-111.

HABIB, M. E., S. EL GABY & O. M. EL NADY 1978. Geology of the area west of Gabel El Rubshi, Eastern Desert, Egypt. *Bull. Fac. Sci., Assiut Univ.* 7(1): 217-239.

HABIB, M. E., S. EL GABY & O. M. EL NADY 1978. Structure and deformational history of the area west of Gabal El Rubshi, Eastern Desert, Egypt. *Bull. Fac. Sci., Assiut Univ.* 7(2): 99-114.

HABIB, M. E., S. EL GABY & A. KHUDEIR 1978. Geology of the Precambrian basement of the Saqi district, Eastern Desert, Egypt. *Bull. Fac. Sci., Assiut Univ.* 7(1): 241-273.

HABIB, M. E. & O. M. EL NADY 1978. The first record of deep-water flysch sequence in the Precambrian basement of Egypt. *Bull. Fac. Sci., Assiut Univ.* 7(2): 53-70.

HABIB, M. E. & O. M. EL NADY 1985. Two orogenics in the Meatiq area of the central Eastern Desert, Egypt. *Precambrian Res.* 30: 83-111.

HADIDY, T. A. 1960. Bakr Oilfield. *2nd Arab Petrol. Congr.*, Beirut, 12pp.

HAENEL, R. 1972. Heat flow measurements in the Red Sea and the Gulf of Aden. *Z. für Geophyik* 38: 1035-1047.

HAFEZ, A. M. A. 1970. *Geology of Wadi Dib area, Esh Mellaha range, Eastern Desert*. M.Sc. Thesis, Cairo Univ.

HAFEZ, A. M. A. & H. EL AMIN 1983. Structure and metamorphism of a Precambrian sequence in Wadi El Miyah, Barramiya area, Eastern Desert, Egypt. *Proc. 5th Intl. Conf. Basement Tectonics*, Cairo (Abstr.).

HAFEZ, A. M. A. & I. M. SHALABY 1983. On the geochemical characteristics of the volcanic rocks at Umm Samiuki, Eastern Desert, Egypt. *Egypt. J. Geol.* 27: 73-92.

HAGEDORN, H. 1971. Untersuchungen über Relieftypen arider Räume aus Beispielen aus dem Tibesti-Gebirge und seiner Umgebung. *Z. Geomorph.* N.F., suppl. 11.

HAGRAS, M. 1976. The distribution and nature of the Miocene sediments in the Gulf of Suez. *5th Petrol. Explor. Seminar, EGPC*, Cairo.

HAGRAS, M. & S. SLOCKI 1982. Sand distribution of the Miocene clastics in the Gulf of Suez. *6th Petrol. Explor. Seminar, EGPC*, Cairo.

HALBOUTY, M. T. 1976. Application of Landsat imagery to petroleum and mineral exploration. *Bull. Am. Assoc. Petrol. Geol.* 60: 745-793.

HALBOUTY, M. T. 1980. Geologic significance of Landsat data for 15 giant oil and gas fields. *Bull. Am. Assoc. Petrol. Geol.* 64: 8-36.

HALL, M. J. K. 1980. *Bathymetric chart of the southeastern Mediterranean Sea*. Geol. Surv. Israel.

HALL, M. J. K. & Z. BENAVRAHAM, 1978. Bathymetric chart of the Gulf of Elat. *Geol. Survey Israel.*

HALL, S. A. 1979. A total intensity aeromagnetic map of the Red Sea and its interpretation. *US Geol. Surv. Saudi Arabian Project Report* 275: 260pp.

HALL, S. A., G. E. ANDREASEN & R. W. GIRDLER 1976. Red Sea total intensity magnetic anomaly map: compilation and primary interpretation. *US Geol. Surv. Interagency Report*, 197pp.

HAMAM, K. A. 1975. Larger foraminifera from the lower Eocene of Gebel Gurnah, Luxor, Egypt. *Paleontology* 18: 161-168.

HAMED, A. R. A. 1972. Environmental interpretation of the Aptian carbonates of the Western Desert, Egypt. *8th Arab Petrol. Congr.*, Algiers.

HAMROUSH, H. 1986. Geoarcheology: Egyptian predynastic ceramics and geochemistry. *Episodes* 9: 160-165.

HAMMOUDA, H. 1986. *Study and interpretation of basement structural configuration in the southern part of the Gulf of Suez using aeromagnetic and gravity data*. Ph.D. Thesis, Cairo Univ.

HANTAR, G. 1967. Remarks on the oil accumulations in the Gulf of Suez and Red Sea districts, Egypt. *6th Arab Petrol. Congr.*, Baghdad.

HANTAR, G. 1975. Contribution to the origin of the Nile delta. *9th Arab Petrol. Congr.*, Dubai.

HAQ, B. U., W. A. BERGGREN & J. A. VAN COUVERING 1977. Corrected age of the Pliocene/Pleistocene boundary. *Nature* 169: 483-488.

HAQ, B. U., J. HARDENBOL & P. R. VAIL 1987. Chronology of fluctuating sea levels since the Triassic. *Science* 235: 1156-1166.

HARDING, T. P. 1973. Newport-Inglewood trend, California – an example of wrenching style of deformation. *Bull. Am. Assoc. Petrol. Geol.* 57: 97-116.

HARDING, T. P. 1984. Graben hydrocarbon occurrences and structural styles. *Bull. Am. Assoc. Petrol. Geol.* 68: 333-362.

HARDY, M. 1982. *Report on the Halal well-1, NE Sinai*. Internal Report, Conoco, Egypt.

HARLAND, W. B. 1971. Tectonic transpression in Caledonian Spitzberg. *Geol. Mag.* 108: 27-42.

HARLAND, W. B., A. V. COX, P. C. LLEWELLYN, C. A. A. PICKTON, A. G. SMITH & R. WALTERS 1982. *A geologic time scale*. Cambridge Univ. Press, 131pp.

HARMS, J. C. 1979. Alluvial plain sediments of Nubia, southwestern Egypt. *Bull. Am. Assoc. Petrol. Geol.* 63: 829.

HARMS, J. C. 1985. *An interpretation of the stratigraphy of rock units within the Nubia Group*. Internal Report, Conoco, Egypt.

HARMS, J. C. SOUTHARD & R. G. WALKER 1982. Fluvial deposits and facies models. In: *Structures and sequences in clastic rocks*. Soc. Econ. Paleontol. Min., short course 95: 1-25.

HARPER, M. L. 1971. Approximate geothermal gradients in the North Sea. *Nature* 230: 235-236.

HARRIS, N. B. W. 1981. The role of fluorine and chlorine in the petrogenesis of a peralkaline complex from Saudi Arabia. *Chem. Geol.* 31: 303-310.

HARRIS, N. B. W., C. J. HAWKESWORTH & A. C. RIES 1984. Crustal evolution in northeast and east Africa from model Nd ages. *Nature* 309: 773-776.

HARSCH, W., T. KUPER, B. RAST & R. SAGRESSER 1981. Seismotectonic considerations of the Turkish-African plate boundary. *Geol. Rundschau* 70: 368-384.

HASHAD, A. H. 1980. Present status of geochronological data on the Egyptian basement complex. *Bull. Inst. Applied Geol., King Abdul Aziz Univ., Jeddah* 3(3): 31-46.

HASHAD, A. H. & M. W. EL REEDY 1979. Geochronology of the anorogenic alkalic rocks, south Eastern Desert, Egypt. *Ann. Geol. Surv. Egypt.* 9: 81-101.

HASHAD, A. H. & M. A. HASSAN 1979. On the validity of an ensimatic island arc cratonization model to the evolution of the Egyptian shield. *Ann. Geol. Surv. Egypt* 9: 70-80.

HASHAD, A. H., M. A. HASSAN & A. A. ABUL GADAYEL 1978. Trace element variations of Wadi Natash, Egypt. *Bull. Inst. Applied Geol., King Abdul Aziz Univ., Jeddah* 2: 195-204.

HASHAD, A. H., M. A. HASSAN & A. A. ABUL GADAYEL 1982. Geological and petrological study of Wadi Natash late Cretaceous volcanics. *Egypt. J. Geol.* 26: 19-38.

HASHAD, A. H., T. A. SAYYAH, S. B. EL KHOLY & A. YOUSSEF 1972. Rb-Sr isotopic age determination of some basement Egyptian granites. *Egypt. J. Geol.* 16: 169-181.

HASHAD, A. H., T. A. SAYYAH & M. W. EL REEDY 1981. Geochronological and strontium isotope study of the psammitic gneiss of Wadi Nugrus, Eastern Desert. *Egypt. J. Geol.* 25: 149-158.

HASHAD, A. H., T. A. SAYYAH & M. S. EL MANHARAWY 1981. Isotopic composition of stronium and origin of Wadi Kareim volcanics, Eastern Desert. *Egypt. J. Geol.* 25: 141-147.

HASSAN, A. A. 1967. A new Carboniferous occurrence in Abu Durba, Sinai, Egypt. *6th Arab Petrol. Congr.*, Baghdad.

HASSAN, A. A. 1974. Morgan Oilfield. *4th Petrol. Explor. Seminar, EGPC*, Cairo.

HASSAN, E. M. & M. I. EL SABH 1974. Changes in the current regime in the Suez Canal after construction of Aswan High Dam. *Nature* 248: 217-218.

HASSAN, F. & A. M. EL KAMMAR 1975. Environmental conditions affecting the distribution of uranium and rare earth elements in Egyptian phosphorites. *Egypt. J. Geol.* 19: 169-178.

HASSAN, F. A. 1978. Archeological explortion of the Siwa Oasis region, Egypt. *Current Anthropology* 19: 146-148.

HASSAN, F. A. 1986. Desert environment and origins of agriculture in Egypt. *Norwegian Archeol. Rev.* 19: 63-76.

HASSAN, M. A. 1972. On the occurrence of columbite mineralization in Wadi Abu Rasheid area, Eastern Desert, Egypt. *Egypt. J. Geol.* 16: 283-291.

HASSAN, M. A. 1973. Geology and geochemistry of radioactive columbite-bearing psammitic gneiss of Wadi Abu Rusheid, south Eastern Desert, Egypt. *Ann. Geol. Surv. Egypt* 3: 307-225.

HASSAN, M. A. & H. M. EL SHATOURY 1976. Beryl occurrences of Egypt. *Min. Geol.* 26: 253-262.

HASSAN, M. A. & M. A. ESSAWY 1977. Petrography of the metagabbro-diorite complex of Wadi Mubarak, Gabal Atud area, Eastern Desert, Egypt. *J. Univ. Kuwait (Sci.)* 4: 203-214.

HASSAN, M. A., A. H. HASHAD & R. M. EL BAYOUMI 1984. The recognition of ophiolitic melanges with associated ophiolites and their significance in the evolution of the Pan African orogenic belt. *27th Intl. Geol. Congr., Moscow* 2: 350-351.

HASSAN, M. A., A. H. HASHAD & A. EL MEZAYEN 1983. A proposed mode of emplacement of El Fawakhir ophiolites. Pan African crustal evolution the Arabian Nubian Shield. *IGCP Project 164 Newsletter* 5: 14-16.

HASSAN, M. M. 1977. Lead-Zinc mineralization on the western shore of the Red Sea and the mechanism of its distribution in space and time. *Vyssh. Uchebon. Zaved. Izu., Geol. Razved.* 1977(10): 96-105.

HASSAN, M. M. & A. H. SABET 1979. Geochemical and mineralogical studies on Esh El Mellaha manganese deposits, Eastern Desert, Egypt. *1st. Mid. East Geol. Congr., Mineral. Res. & Explor. Inst. Ankara, Turkey*: 143-178.

HASSAN, M. Y., M. BOUKHARY, G. SALLOUM & H. EL SHEIKH 1984. Biostratigraphy of the subsurface Oligocene sediments in the north Western Desert, Egypt. *Qatar Univ. Sci. Bull.* 4: 235-262.

HASSAN, M. Y., O. H. CHERIF & N. S. ZAHRAN 1981. Gelogy and stratigraphy of the Giza Pyramids plateau: *J. UAE Univ., Al' Ain* 2.

HASSANEIN, A. M. 1924. Through Kufra to Darfur. *Geograph. J.* 64: 273-291, 353-363.

HASSOUBA, A. B., A. SHAFY, A. A. NASHAAL & G. AZAZI 1984. Depositional history of Shoab Ali field; a model for hydrocarbon exploration. *Proc. 7th Petrol. Explor. Seminar, EGPC*, Cairo: 152-163.

HATCHER, R. D., I. ZIETZ, R. D. REGAN & M. ABU AJAMEIH 1981. Sinistral strike-slip motion on the Dead Sea rift: confirmation from new magnetic data. *Geology* 9: 458-462.

HAWARY, M. 1939. *Report on the concessions of talc, manganese, ilmenite, lead and tungsten held by the Hamata Mining Co.* Internal Report, Mines and Quarries Dept., Cairo.

HAYNES, C. V. 1980. Geological evidence of pluvial climates in the Nabta area of the Western Desert, Egypt. In: F. Wendorf & R. Schild, *Prehistory of the Eastern Sahara*, Academic Press: 353-371.

HAYNES, C. V. 1980. Geochronology of Wadi Tushka, lost tributary of the Nile. *Science* 210: 68-71.

HAYNES, C. V. 1982. The Darb El Arba'in desert: a product of Quaternary climatic change. In: F. El Baz & T. A. Maxwell (eds), *Desert landforms of southwest Egypt*, NASA: 91-118.

HAYNES, C. V. & H. HAAS 1980. Radiocarbon evidence for Holocene recharge of groundwater, Western Desert, Egypt. *Radiocarbon* 22: 705-717.

HECHT, F., M. FÜRST & E. KLITZSCH 1963. Zur Geologie von Libyen. *Geol. Rundschau* 53: 413-470.

HEDGE, C. E. 1984. Precambrian geochronology of part of southwestern Saudi Arabia. *Open File Report USGS-OF-04-31.*

HEIKAL, M. A., M. A. HASSAN & Y. EL SHESHTAWI 1983. The Cenozoic basalt of Gebel Qatrani, Western Desert, Egypt as an example of continental tholeiitic basalt. *Ann. Geol. Surv. Egypt* 13: 193-209.

HEIKAL, M. A., M. H. HEGAZY & M. M. EL RAHMANY 1980. Ignimbritic rhyolites in the Wassif area, Eastern Desert, Egypt. *Bull. Inst. Applied Geol., King Abdul Aziz Univ., Jeddah* 4(3): 107-114.

HEINL, M. & R. HOLLANDER 1984. Some aspects of a new groundwater model for the Nubian aquifer system. *Berl. Geowiss. Abh.* 50(A): 221-231.

HELAL, A. H. 1965. Jurassic spores and pollen grains from the Kharga Oasis, Western Desert, Egypt. *N. Jb. Geol. Palaeontol. Abh.* 123: 160-166.

HELAL, A. H. 1966. Jurassic plant microfossils from the subsurface of Kharga Oasis, Western Desert, Egypt. *Paleontographica* 117(A): 83-98.

HELBLING, R. 1938. Die Anwendung der Photogrammetrie bei geologischen Kartierungen. *Beitr. Geol. Karte Schweiz*, N.F. 76: 1-68.

HELLER-KALLAI, L., Y. NATHAN & T. WEISSBROD 1973. The clay mineralogy of Triassic sediments in southern Israel and Sinai. *Sedimentology* 20: 513-521.

HELLSTROM, B. 1939. The subterranean water in the Libyan Desert. *Geografika Annaler* 21: 206-239.

HENDRIKS, F. 1985. Upper Cretaceous to lower Tertiary

sedimentary environments and clay mineral associations in the Kharga Oasis area (Egypt). *N. Jb. Geol. Palaeontol., Mh* (1985) 10: 579-591.

HENDRIKS, F. 1986. The Maghrabi Formation of the El Kharga area (SW Egypt): deposits from a mixed estuarine and tidal flat environment of Cenomanian age. *J. Afr. Earth Sci.* 5: 481-489.

HENDRIKS, F. & P. LUGER 1987. The Rakhiyat Formation of the Gebel Qreiya area: evidence of middle Campanian to early Maastrichtian synsedimentary tectonism. *Berl. Geowiss. Abh.* 75(A): 83-96.

HENDRIKS, F., P. LUGER, J. BOVITZ & H. KALLENBACH 1987. Evolution of the depositional environments of SE Egypt during the Cretaceous and lower Tertiary. *Berl. Geowiss. Abh.* 75(A): 49-82.

HENDRIKS, F., P. LUGER, H. KALLENBACH & J. H. SCHROEDER 1984. Stratigraphical and sedimentological framework of the Kharga-Sin El Kaddab stretch (western and southern part of the upper Nile basin), Western Desert, Egypt. *Berl. Geowiss. Abh.* 50(A): 117-151.

HENDRIKS, F., P. LUGER, H. KALLENBACH & J. H. SCHROEDER 1985. Faziesmuster und Stratigraphie der Kampanen bis Palaeozänen Schichtenfolgen in südlichen Ober-Nil Becken (Western Desert, Aegypten). *Z. Deutsch. Ges.* 136: 207-233.

HERMINA, M. H. 1967. *Geology of the northwest approaches of Kharga.* Geol. Surv. Egypt, Paper 44, 87pp.

HERMINA, M. H. 1972. Review of the phosphate deposits in Egypt. *2nd Arab Conf. Mineral Resources, Jeddah*: 109-149.

HERMINA, M. H. 1973. Preliminary evaluation of Maghrabi-Liffiya phosphorites, Abu Tartur, Western Desert, Egypt. *Ann. Geol. Surv. Egypt* 3: 39-74.

HERMINA, M. H., M. G. GHOBRIAL & B. ISSAWI 1961. *The geology of the Dakhla area.* Geol. Surv. Egypt, 33pp.

HERMINA, M. H. & B. ISSAWI 1971. Rock stratigraphic classification of the upper Cretaceous-lower Tertiary exposures in southern Egypt. In: C. Gray (ed.), *Symp. on the Geology of Libya*, Fac. Sci. Univ. Libya: 147-154.

HERMINA, M. H. & A. WASSEF 1975. Geology and exploration of the large phosphate deposit in Abu Tartur plateau, the Libyan (Western) Desert. *Ann. Geol. Surv. Egypt* 5: 87-93.

HERMINA, M. H., A. WASSEF, A. EL TAHLAWY & A. KAMEL 1983. The subsurface Paleozoic section of Wadi Araba, Egypt. *Ann. Geol. Surv. Egypt* 13: 257-269.

HESTER, J. J. & P. M. HOBLER 1969. Prehistoric settlement patterns in the Libyan Desert. *Univ. Utah Anthropol. Papers* 92.

HEYBROEK, F. 1965. The Red Sea Miocene evaporite basin. In: *Salt Basins around Africa. Inst. Petrol. London*: 17-40.

HIGAZY, H. A. M. 1984. *Geology of Wadi El Gemal area, Eastern Desert, Egypt.* Ph.D. Thesis, Assiut Univ.

HIGAZY, R. A. & M. F. EL RAMLY 1960. *Potassium-argon ages of some rocks from the Eastern Desert of Egypt.* Geol. Surv. Egypt, Paper 7, 18pp.

HIGAZY, R. A. & H. A. HUSSEIN 1955. Remarks on the uranium content of some black shales and phosphates from Quseir and Safaga. *Proc. Egypt. Acad. Sci.* 11: 63-66.

HILMY, M. E. & A. A. HUSSEIN 1978. A proposed classification for the mineral deposits and occurrences in Egypt. *Precambrian Res.* 6 (abstr.).

HILMY, M. E. & L. A. MOHSEN 1965. Secondary copper minerals from west central Sinai. *Egypt. J. Geol.* 9: 1-11.

HILMY, M. E. & L. A. MOHSEN 1966. Geology of some occurrences of copper minerals in Sinai. *Bull. Fac. Sci., Ain Shams Univ.* 9: 512-539.

HILMY, M. E., F. M. NAKHLA & A. RASMY 1972. Contribution to the mineralogy, geochemistry and genesis of Miocene Pb/Zn deposits in Egypt. *Chem. Erde* 31: 373-390.

HILMY, M. E., A. STROUGO & S. A. HUSSEIN 1983. Natroalunite occurrence in middle Eocene beds of Darb El Fayum, Giza Pyramids area, Egypt. *Egypt. J. Geol.* 27: 1-10.

HIRMER, M. 1925. Die fossilen Pflanzen Aegyptens (Filicales). *Ergebn. Forsch. Reise E. Stromers. Abh. Bayer. Akad. Wiss., Math.-Naturw.* Kl., 30(3): 1-18.

HIRSCH, F. 1976. Sur l'origine des particularismes de la faune du Trias et du Jurassique de la plate-forme africano-arabe. *Bull. Soc. Geol. France* 18(7): 543-522.

HOLMAN, B. W. The ore deposits of Egypt: 1. ilmenite as an ore of titanium, 2. iron ores. *Trans. Min. Petrol. Assoc. Egypt* 9: 22-46.

HOROWITZ, A. 1970. Palynostratigraphy of the upper Paleozoic-lower Mesozoic sequence in Zohar-8 borehole (southern Israel). *Geol. Surv. Isr., Report Pal.* 70: 1-9.

HOROWITZ, A. 1973. *Noeggerathia dickeri* n.sp. from the Carboniferous of Sinai. *Rev. Paleobot. and Palyn.* 15: 51-56.

HOSNY, W., I. GAAFAR & A. SABOUR 1986. Miocene stratigraphic nomenclature in the Gulf of Suez region. *8th Petrol. Explor. Seminar*, EGPC, Cairo.

HOTTINGER, L. 1960. Über Palaeozän und Eozän Alveolinen. *Eclog. Geol. Helv.* 53: 265-284.

HSU, K. J. 1977. Tectonic evolution of the Mediterranean basins. In: A.E.M. Nairn, W.H. Kanes & F.G. Stehli (eds.), *The ocean basins and margins* 4(A): 29-75, Plenum Press, New York.

HSU, K. J. & D. BERNOULLI 1978. Genesis of the Tethys and Mediterranian. *Initial Report of the Deep Sea Drilling Project 42*, US Government Printing Office, Washington DC.

HSU, K. J., M. B. CITA & W. B. F. RYAN 1973. The origin of the Mediterranean evaporites. In: W. B. F. Ryan & K. J. Hsu (eds), *Initial Reports of the Deep Sea Drilling Project*, US Government Printing Office, 13/(2): 1011-1099.

HSU, K. J., L. MONTADERT, D. BERNOULLI, M. B. CITA, A. ERICKSON, R. E. GARRISON, R. B. KIDD, F. MELIERES, C. MULLER & R. WRIGHT 1978. History of the Messinian salinity crisis. In: K. J. Hsu & L. Montadert (eds), *Initial Reports of the Deep Sea Drilling Project*, US Government Printing Office, 42(1): 1053-1078.

HSU, K. J., W. B. F. RYAN & M. B. CITA 1973. Late Miocene desiccation of the Mediterranean. *Nature* 242: 239-243.

HUME, W. F. 1906. *The topography and geology of the Peninsula of Sinai (South Eastern portion).* Geol. Surv. Egypt, 280pp.

HUME, W. F. 1907. *A preliminary report on the geology of the Eastern Desert of Egypt between latitudes 22 and 25 N.* Egypt Surv. Dept., Cairo.

HUME, W. F. 1909. *The distribution of iron ores in Egypt.*

Egypt. Surv. Dept., Paper 20, 16pp.

HUME, W. F. 1910. The effects of secular oscillation in Egypt during the Cretaceous and Eocene periods. *Quart. J. Geol. Soc. London* 67: 118-148.

HUME, W. F. 1912. *Explanatory notes to accompany the goelogical map of Egypt*. Egypt. Surv. Dept., 50pp.

HUME, W. F. 1925. *Geology of Egypt. I. The surface features of Egypt, their determining causes and relation to geological structure*. Egypt. Surv. Dept., 408pp.

HUME, W. F. 1934, 1935, 1937. *Geology of Egypt. II. The fundamental Precambrian rocks of Egypt and the Sudan*: Part I, *The metamorphic rocks*: 1-300; Part II, *The later plutonic and minor intrusive rocks*: 301-688; Part III, *The minerals of economic value*: 689-900, Geol. Surv. Egypt.

HUME, W. F. 1962, 1965. *Geology of Egypt. III. The stratigraphical history of Egypt*: Part I, *From the close of the Precambrian episodes to the end of the Cretaceous Period*: 712pp.; Part II, *From the close of the Cretaceous Period to the end of the Oligocene*: 734pp. Geol. Surv. Egypt.

HUME, W. F. & F. HUGHES 1921. *The soils and water supply of the Maryut district, west of Alexandria*. Egypt Surv. Dept., Publ. 37.

HUME, W. F., T. G. MADGWICK, F. W. MOON & H. SADEK 1920. Preliminary general report of the occurrences of petroleum in western Sinai. *Petrol. Res. Bull.*, Government Press, Cairo, 15pp.

HUME, W. F., T. G. MADGWICK, F. W. MOON & H. SADEK 1920. Preliminary geological report on Gebel Tanka area. *Petrol. Res. Bull. 4*, Government Press, Cairo, 16pp.

HUME, W. F., T. G. MADGWICK, F. W. MOON & H. SADEK 1921. Preliminary geological report on south Zeit area. *Petrol. Res. Bull. 7*, Government Press, Cairo, 24pp.

HUME, W. F., T. G. MADGWICK, F. W. MOON, & H. SADEK 1921. Preliminary geological report on Abu Durba (western Sinai). *Petrol. Res. Bull. 1*, Government Press, Cairo, 20pp.

HUNTING GEOLOGY AND GEOPHYSICS LTD 1967. *Photogeological survey: The assessment of the mineral potential of the Aswan region, UAR*. UNDP/UAR regional planning of Aswan, 138pp.

HUNTING GEOLOGY AND GEOPHYSICS LTD 1974. *Geology of Jebel Al-Uwaynat area, Libyan Arab Republic*. Industrial Research Center, Tripoli.

HURLEY, P. M. 1972. Can the subduction process of mountain building be extended to Pan African and similar orogenic belts? *Earth and Planet. Sci. Lett.* 15: 305-314.

HUSSEIN, A. A. 1973. Results of mineral exploration program in south Eastern Desert, Egypt. *Ann. Geol. Surv. Egypt* 3: 109-124.

HUSSEIN, A. A. & K. M. ABU ZEID 1974. *On the geology and geochemistry of the radioactive dike of Um Doweila, south Eastern Desert*. Internal Report, 74/53, Geol. Surv. Egypt.

HUSSEIN, A. A., M. M. ALI & M. F. EL RAMLY 1982. A proposed new classification of the granites of Egypt. *J. Volc. Geoth. Res.* 14: 187-198.

HUSSEIN, A. A., G. S. CARTER & D. L. SEARLE 1978. Is the basement complex in Egypt complex? *Precambrian Res.* 6: A27-A28 (Abstr.).

HUSSEIN, A. A. & M. A. HASSAN 1973. Further contribution to geology and geochemistry of some Egyptian ring complexes. *Ann. Geol. Surv. Egypt* 3: 167-175.

HUSSEIN, A. A. & A. H. RASMY 1976. Kyanite in a contact aureole at Nasb Aluba, southern Eastern Desert. *14th Annual Meeting Geol. Soc. Egypt* (Abstr.).

HUSSEIN, A. A., I. M. SHALABY, M. A. GAD & A. H. RASMY 1977. On the origin of Zn/Cu/Pb deposits at Um Samiuki, Eastern Desert. *15th Annual Meeting, Geol. Soc. Egypt* (Abstr.).

HUSSEIN, H. A., Y. M. ANWAR & A. EL SOKKARY 1971. Radiogeologic studies of some Carboniferous rocks of west central Sinai. *Egypt. J. Geol.* 15: 119-127.

HUSSEIN, H. A., M. I. FARIS & H. S. ASSAF 1970. Some radiometric investigations of Wadi Kareim, Wadi Dabbah area, Eastern Desert. *Egypt. J. Geol.* 14: 13-20.

HUSSEIN, H. A., M. A. HASSAN, M. A. EL TAHIR & A. ABU DEIF 1986. Uranium-bearing siliceous veins in younger granites, Eastern Desert, Egypt. In: *Vein type uranium deposits*, Intl. Atomic Energy Agency: 143-158.

HUSSEIN, H. A. & I. A. EL KASSAS 1972. Occurrence of some primary uranium mineralization at El Atshan locality, central Eastern Desert. *Egypt. J. Geol.* 14: 97-110.

HUSSEIN, H. A. & I. A. EL KASSAS 1980. Some favorable host rocks for uranium and thorium mineralization in central Eastern Desert. *Ann. Geol. Surv. Egypt* 10: 897-908.

HUSSEIN, I. M., A. KROENER & S. DURR 1984. Wadi Oneib, a dismembered African ophiolite in the Red Sea hills of Sudan, Pan African crustal evolution the Arabian Nubian shield. *Bull. Fac. Earth Sci., King Abdul Aziz Univ., Jeddah* 6: 319-327.

HUTCHINSON, R. W. & G. G. EAGLES 1970. Tectonic significance of regional geology and evaporite lithofacies in northeastern Ethiopia. *Phil. Trans. Roy. Soc. London* 267(A): 313-329.

HUTH, A., G. FRANZ & H. SCHANDELMEIER 1984. Magmatic metamorphic rocks of NW Sudan; a reconnaissance survey. *Berl. Geowiss. Abh.* 50(A): 7-21.

IBRAHIM, E. M. & I. MARZOUK 1979. Seismotectonic study of Egypt. *Bull. Helwan Inst. Astronom. Geophys.* 191pp.

IBRAHIM, M. M. 1949. Ornamental stones in Egypt. *Trans. Mining Petrol. Assoc. Egypt* 4(1): 9-20.

INTERNATIONAL SEISMOLOGICAL CENTER. *International seismological summary* 1959-1977, Edinburgh.

IPCO 1963. *The copper/nickel of Abu Sweil, Eastern Desert (Egypt)*. Intl. Planning and Consulting GmbH, Diusburg/ DEMAG Aktiengesellschaft.

ISACHSEN, Y. W., R. H. FAKUNDINY & S. W. FORSTER 1973. Evaluation of ERTS imagery for spectral geological mapping in diverse terrains of New York state. *3rd ERTS-I Symp., NASA, 1, A, SP-351*: 691-717.

ISHIHARA, S. 1981. The granitoid series and mineralization. *Econ. Geol.* 75: 458-484.

ISMAIL, A. 1960. Near and local earthquakes of Helwan (1903-1950). *Bull. Helwan Observ.* 49: 33pp.

ISMAIL, M. M. & A. M. ABDALLAH 1966. Contribution to the stratigraphy of St Paul Monastery area by microfacies. *Bull. Fac. Sci., Alexandria Univ.* 7: 325-334.

ISMAIL, M. M. & I. A. M. FARAG 1957. Contribution to the stratigraphy of the area to the east of Helwan (Egypt). *Bull. Inst. Desert Egypte* 7(1): 95-134.

ISMAIL, M. M. & A. A. SELIM 1966. A contribution to the stratigraphy of Gebel Ataqa scarps, Eastern Desert, Egypt. *Bull. Inst. Desert Egypte* 16(1): 1-20.

ISSAR, A. & Y. ECKSTEIN 1969. The lacustrine beds of Wadi Feiran, Sinai: Their origin and significance. *Isr. J. Earth. Sci.* 18: 21-27.

ISSAR, A., E. ROSENTHAL, Y. ECKSTEIN & R. BOGOCH 1971. Formation waters, hot springs and mineralization phenomena along the eastern shore of the Gulf of Suez. *Bull. Intl. Assoc. Sci. Hydrol.* 16: 25-44.

ISSAWI, B. 1969. *The geology of Kurkur-Dungul area.* Geol. Surv. Egypt, Paper 46, 102pp.

ISSAWI, B. 1971. Geology of Darb El Arba'in, Western Desert. *Ann. Geol. Surv. Egypt* 1: 53-92.

ISSAWI, B. 1972. Review of upper Cretaceous-lower Tertiary stratigraphy in central and southern Egypt. *Bull. Am. Assoc. Petrol. Geol.* 56: 1448-1463.

ISSAWI, B., M. FRANCIS, M. EL HINNAWI & T. EL DEFTAR 1971. Geology of Safaga-Quseir coastal plain and of Mohamed Rabah area. *Ann. Geol. Surv. Egypt* 1: 1-19.

ISSAWI, B., M. FRANCIS, M. EL HINNAWI & A. MEHANNA 1969. *Contribution to the structure and phosphate deposits of Quseir area.* Geol. Surv. Egypt, Paper 50, 35pp.

ISSAWI, B., M. Y. HASSAN & S. A. N. ATTIA 1978. Geology of Abu Tartur plateau, Western Desert, Egypt. *Ann. Geol. Surv. Egypt* 8: 91-127.

ISSAWI, B. & U. JUX 1982. *Contributions to the stratigraphy of the Paleozoic rocks in Egypt.* Geol. Surv. Egypt, Paper 64, 28pp.

ISSEL, A. 1870. Malacologia del Mar Rosso. *Bibliotheca Malac.* (Pisa).

IVANOV, T. 1975. *Report for study of hydrothermal alterations in localities Um Garayat and Um Tundub, Eastern Desert.* Internal Report, Aswan Mineral Surv. Project, Geol. Surv. Egypt.

IVANOV, T. & A. A. HUSSEIN 1972. *Report on geological operations in the assessment of the mineral potential of the Aswan region project from July 1968 to June 1972.* Internal report, Aswan Mineral Survey project, Geol. Surv. Egypt.

IVANOV, T., I. M. SHALABY & A. A. HUSSEIN 1973. Metallogenic characteristics of south Eastern Desert, Egypt. *Ann. Geol. Surv. Egypt* 3: 139-166.

JACKSON, N. J. 1986. Petrogenesis and evolution of Arabian felsic plutonic rocks. *J. Afr. Earth Sci.* 4: 49-59.

JACKSON, N. J., C. J. DOUCH, J. ODELL, H. BEDWAI, H. AL HAZMI, E. PEGRAM & J. N. WALSH 1983. Late Precambrian granitoid plutonism and associated mineralization in the central Higaz. *Bull. Fac. Earth Sci., King Abdul Aziz Univ., Jeddah* 6: 511-537.

JACOBBERGER, P. A., R. E. ARVIDSON & D. L. RASHKA 1983. Application of Landsat multispectral scanner data and sediment spectral reflectance measurements to mapping the Meatiq dome, Egypt. *Geology* 11: 587-591.

JAEGER, E. 1977. The evolution of the central and west European continent. In: La chaine varsitique d'Europe moyenne et occidentale. *Colloq. Intl. CNRS* 234: 227-239.

JAEKEL, D. 1971. Erosion and Akkumulation im Ennedi bardageu-Araye des Tibesti Gebirges (Zentrale Zahara) wärhned des Pleistozäns und Holozäns. *Berl. Geograph. Abh.* 10.

JAKES, R. & A. J. R. WHITE 1972. Major and trace element abundance in volcanic rocks of orogenic areas. *Bull. Geol. Soc. Am.* 83: 29-40.

JAMES, C. T. & B. H. SLAUGHTER 1974. A primitive new middle Pliocene murid from Wadi El Natrun, Egypt. *Ann. Geol. Surv. Egypt* 4: 332-362.

JAMISON, W. R. 1983. *Compendium of structural styles*, Part 4, *Wrench tectonics*. Amoco Prod. Co. Res. Dept., Rep. F83-G-1D, June 30, 1983, 50pp.

JENKINS, D. 1980. *Evaluation of the Paleozoic, north Sinai, ARE.* Internal Report, Conoco, Egypt.

JENKINS, D. 1980. *Evaluation of the Triassic north Sinai.* Internal Report, Conoco, Egypt.

JENKINS, D., J. C. HARMS & T. W. OESLEBY 1982. Mesozoic sediments of Gebel Maghara, north Sinai, ARE. *6th Petrol. Explor. Seminar, EGPC*, Cairo.

JENSEN, M. L. & A. M. BATEMAN 1981. *Economic mineral deposits.* Wiley, 3rd ed., 539pp.

JOHNSON, D. W. 1921. Aerial observations of physiographic features (Reply to B. Willis). *Science*, N.S., 54: 435-436.

JONGMANS, W. J. 1955. Flore et faune du Carbonifère inférieur de l'Egypte. *Nether. Geol. Stichting Meded., n.ser.* 8: 59-75.

JONGMANS, W. J. & S. KOOPMANS 1940. Contribution to the flora of the Carboniferous of Egypt. *Meded. Geol. Bureau Heerlen*: 223-229.

JONGMANS, W. J. & S. VAN DER HEIDE 1953. Contribution à l'étude de la faune et de la flore Carbonifère infereure de l'Egypte. *C.R. 19th Intl. Geol. Congr.*, Algiers, sec. 2, fasc. 2: 5-70.

JONGMANS, W. J. & S. VAN DER HEIDE 1955. Flore et faune du Carbonifère inferieur de l'Egypte. *Meded. Geol. Stich.*, n.ser. 8: 59-75.

JORDAN, W. 1876. Uber die Verwerthung der Photographie zu geometrischen Aufnahmen (Photogrammetrie), mit einer photogrammetrischen Aufnahme der Oasenstadt Gassar-Dachel in der Libyschen Wuste. *Z. Vermessungswes.*: 1-17.

JORDI, H. A. 1984. A metatectonic concept of the Gulf of Suez. *Proc. 7th Petrol. Explor. Seminar, EGPC*, Cairo: 1-4.

(JVQ) JOINT VENTURE QATTARA 1978. Study of Qattara depression, special volume: Regional geology and hydrogeology. Unpublished Report, Lahmeyer International GmbH, Saalzgitter Consult GmbH and Deutsche Projekt Union GmbH.

JUX, U. 1954. Zur Geologie des Kreidegebietes von Abu Roasch bei Kairo. *N. Jb. Geol. Palaeontol. Abh.* 100: 159-207.

JUX, U. 1983. Diagenetic silica glass (formerly related to astroblems) from the Western Desert, Egypt. *Ann. Geol. Surv. Egypt* 13: 99-108.

JUX, U. 1983. Zusammensetzung und Ursprung von Wüstengläsern aus der Grossen Sand-See Ägyptens. *Z. Deutsch. Geol. Ges.* 134: 521-553.

JUX, U. & B. ISSAWI 1983. Cratonic sedimentation in Egypt during the Paleozoic. *Ann. Geol. Surv. Egypt* 13: 223-245.

KABESH, M. L., M. E. HILMY & A. M. BISHADY 1967. Geology of the basement rocks in the area around Um Rus gold mines, Eastern Desert. *Egypt. J. Geol.* 11: 59-85.

KABESH, M. L., E. E. EL HINNAWI & M. A. MANSOUR 1970. A note on the geochemistry of the iron ochre deposit of Um Greifat, Red Sea Coast. *Egypt. J. Geol.* 14: 43-46.

KABESH, M. L. & A. M. SHAHIN 1968. Preliminary studies of the dike rocks in Esh El Mellaha range, Eastern Desert. *Egypt. J. Geol.* 12: 21-32.

KALLENBACH, H. & F. HENDRIKS 1986. Transgressive and regressive sedimentary environments of Cretaceous to lower Tertiary age in upper Egypt, a case study from central Wadi Qena. *12th Intl. Sedimentol. Congr. Canberra*: 158 (Abstr.).

KAMEL EL-DIN, G. M. 1986. *Geology of Wadi Um Had area, Eastern Desert, Egypt.* M.Sc. Thesis, Assiut Univ.

KAMEL, H. 1974. *The international gravity base stations.* Internal Report 1281, General Petrol. Co., Cairo, 21pp.

KAMEL, H. & A. NAKHLA 1985. *Gravity map of Egypt.* A report prepared for the Egyptian Academy of Scientific Research and Technology: 43pp.

KAMEL, O. A. 1971. The Bhariya iron ores, their mineralogy and origin. *Ann. Geol. Surv. Egypt* 1: 117-134.

KAMEL, O. A., M. E. HILMY & R. BAKIR 1977. Mineralogy of Abu Tartur phosphorites. *Egypt. J. Geol.* 21: 133-158.

KAMEL, O. A., M. Y. MENEISY & A. Y. ABDEL AAL 1981. Petrography of Pahnerozoic basaltic rocks in Egypt. *Bull. Fac. Sci., Ain Shams Univ.* 23B: 93-125.

KAMEL, O. A., E. A. NIAZI & A. Y. ABDEL AAL 1985. Geochemical affinity of the main Fe-Mn ores of Egypt with the late Tertiary basaltic activity. *8th RCMNS Symp. European Cenozoic Mineral Resources, Hungarian Geol. Surv.*, Budapest.

KAMEL, O. A., A. H. RASMY, A. KHALIL & R. BAKIR 1973. Mineralogical analysis and evaluation of black sands at eastern part of east side Nile section, Abu Khashaba area, Rosetta, Egypt. *Ann. Geol. Surv. Egypt* 3: 227-247.

KAMEL, O. A., I. M. SHALABY & M. M. EL MAHALLAWY 1980. Petrological study of the basic-ultrabasic suite at El Genena El Gharbiya, south Eastern Desert. *Ann. Geol. Surv. Egypt* 10: 725-749.

KAPPELMEYER, O. 1957. The use of near surface temperature measurements for discovering anomalies due to causes at depth. *Geophys. Prosp.* 3: 239-258.

KARCZ, I. & M. BRAUN 1964. Sedimentary structures and paleocurrents in the Triassic sandstones of Arayif-En-Naga, Sinai. *Bull. Geol. Surv. Isr.* 38: 1-21.

KARCZ, I., U. KAFRI & Z. MESHEL 1977. Archeological evidence for subrecent seismic activity along the Dead Sea-Jordan rift. *Nature* 269: 234-235.

KARCZ, I. & I. ZAK 1968. Paleocurrents in the Triassic sandstone of Arayif-En-Naga, Sinai. *Isr. J. Earth Sci.* 17: 9-15.

KARNIK, V. 1969. *Seismicity of Europe, parts I and II.* Academic Publishing House, Czechoslovakia Acad. Sci., Prague.

KAY, R. F., J. G. FLEAGLE & E. L. SIMONS 1981. A revision of the Oligocene ages of the Fayum province, Egypt. *Am. J. Phys. Anthropol.* 55: 293-322.

KAY, R. F. & E. L. SIMONS 1980. The ecology of Oligocene African anthropoides. *Intl. J. Primatol.* 1: 21-38.

KAY, R. F. & E. L. SIMONS 1983. Dental formulae and dental eruption patterns in Parapithecidae. *Am. J. Phys. Anthropol.* 62: 363-375.

KEBEASY, R. M. 1971. The P-wave travel time anomaly and seimsotectonics and seismic activity. *Bull. Intl. Inst. Seismol. Earthquake Eng. (IISEE), Tokyo* 4: 1-16.

KEBEASY, R. M., M. MAAMOUN, R. N. H. ALBERT & M. MEGAHED 1981. Earthquake activity and earthquake risk around Alexandria. *Bull. Intl. Inst. Seismol. Earthquake Eng. (IISEE), Tokyo* 19: 93-113.

KEBEASY, R. M., M. MAAMOUN & E. M. IBRAHIM 1982. Aswan Lake induced earthquake. *Bull. Inst. Seismol. Earthquake Eng. (IISEE), Tokyo* 19: 155-160.

KEBEASY, R. M. & A. M. MEGAHED 1984. Earthquake activity and earthquake risk around Abu Madi, Gulf of Suez. *Rep. Bull. Helwan Inst. Astronom. & Geophys.*

KEBEASY, R. M., M. MAAMOUN, E. IBRAHIM, A. MEGAHED, D. W. SIMPSON & W. S. LEITH 1987. Earthquake studies at Aswan reservoir. In: A. M. Wassef, A. Boud & P. Vyskocil (eds), *Recent crustal movements in Africa, J. Geodynamics* 7: 173-193.

KELDANI, E. H. 1939. *A bibliography of geology and related sciences concerning Egypt up to the end of 1939.* Survey and Mines Dept., Cairo, 428pp.

KEMP, J., C., PELLATON & J. Y. CALVEZ 1980. *Geochronological investigations and geological history in the Precambrian of northwestern Saudi Arabia.* BRGM, Jeddah, BRGM-OF-01-1, 120pp.

KEMP, J. C., PELLATON & J. Y CALVEZ 1982. Cycles in the chemogenic evolution of Africa in the Pan-African (+500 Ma) tectonic episode. *8th Annual Report, Res. Inst. Afr. Geol.* Leeds Univ.: 24-27.

KENAWY, A. I. 1972. Large foraminifera from the Thebes Formation, Tramsa section, Qena, Egypt. *Magyar Allami Foeldr. Intez. Evi Jel.*: 271-321.

KENAWY, A. I. & N. M. EL BARADI 1977. Early and middle Eocene larger foraminifera in the environs of Assiut, Egypt. *Bull. Fac. Sci., Assiut Univ.* 6(1): 247-271.

KENAWY, A. I., H. KHALIFA & H. H. MANSOUR 1977. Biostratigraphic zonation of the middle Eocene in the Nile Valley based on larger foraminifera. *Bull. Fac. Sci., Assiut Univ.* 6(2): 237-259.

KENNEDY, W. O. 1964. The structural differentiation of Africa in the Pan-African tectonic episode. *8th Annual Report, Res. Inst. Afr. Geol.*, Leeds Univ.: 48-49.

KERDANY, M. T. 1968. Notes on the planktonic zonation of the Miocene of the Gulf of Suez region, UAR. *Girornale Geol.* 2(3): 157-166.

KERDANY, M. T. 1970. Lower Tertiary nannoplankton zones in Egypt. *Newsl. Stratigr.* 1: 35-48.

KERDANY, M. T. 1974. Jurassic prospects in the Western Desert, Egypt. *4th Petrol. Explor. Seminar, EGPC*, Cairo.

KERDANY, M. T. & A. A. ABDEL SALAM 1970. Bio- and lithostratigraphic studies of the pre-Miocene of some offshore exploration wells in the Gulf of Suez. *7th Arab Petrol. Congr.*, Kuwait, 22pp.

KHAIRY, E. H., M. K. HUSSEIN, F. M. NAKHLA & S. Z. EL TAWIL 1964. Analysis and composition of Egyptian ilmenite ores from Abu Ghalaga and Rosetta. *Egypt. J. Geol.* 8: 1-9.

KHALED, D. 1974. Jurassic prospects in the Western Desert, Egypt. *4th Petrol. Explor. Seminar, EGPC*, Cairo.

KHALIFA, H. 1974. *Late Eocene of the Nile Valley.* Ph.D. Thesis, Assiut Univ.

KHALIFA, H. & G. EL SAYED 1984. Biostratigraphic zonation of the late Cretaceous-early Paleogene succession along El Sheikh Fadl-Ras Gharib road, Eastern Desert. *Bull. Fac. Sci., Assiut Univ.* 13(1)C: 175-190.

KHALIL, N. A. & S. EL MOFTY 1972. Some geophysical anomalies in the Western Desert, Egypt; their geological significance and oil prospects. *8th Arab Petrol. Congr.*, Algiers.

KHATTAB, M. M. 1985. Geophysical investigation of an upper Tertiary subbasin in the southern Egyptian Red Sea shelf and its bearing on oil exploration. *J. Afr. Earth Sci.* 3: 321-330.

KHEDR, E. 1984. Sedimentological evolution of the Red Sea continental margin of Egypt and its relationship to sea level changes. *Sed. Geol.* 39: 71-86.

KHUDEIR, A. A. 1983. *Geology of the ophiolite suite of El Rubshi area. Eastern Desert, Egypt.* Ph.D. Thesis, Assiut Univ.

KIJKO, A., M. DESSOUKY, R. M. KEBEASY & G. H. HASEAB 1985. Preliminary estimation of seismic hazard in Aswan Lake area in Egypt. *Acta Geophys. Polonica* 33: 269-277

KLEBELSBERG, R. v. 1911. Ein Beitrag zur Kenntnis des Sinai Karbons. *Z. Deutsch. Geol. Ges.* 63: 594-603.

KLEINMANN, B. 1969. The breakdown of zircon observed in the Libyan Desert glass as evidence of its impact origin. *Earth Planet. Sci. Lett.* 5: 497-501.

KLEINMANN, B. 1969. The breakdown of zircon observed in the Libyan Desert glass as evidence of its own origin. *Earth Planet. Sci. Lett.* 5: 497-501.

KLERKX, J. 1980. Age and metamorphic evolution of the basement complex around Jabal Al Alwaynat. In: M. J. Salem & M. T. Busrewil (eds), *The geology of Libya,* Academic Press 3: 901-906.

KLERKX, J. & S. DEUTSCH 1977. Resultats préliminaires obtenus par la méthode Rb/Sr sur l'âge des formations precambriennes de la region d'Uweinat (Libye). *Rapp. Ann.* (1976), *Mus. Roy. Afr. Centrale, Dep. Geol. Min.,* Tervuren: 83-94.

KLERKX, J. & C. RUNDLE 1976. Preliminary K/Ar ages of different igneous rock formations from Gebel Uweinat region (SE Libya). *Rapp. Ann.* (1975), *Mus. Roy. Afr. Centrale, Dep. Geol. Min.,* Tervuren: 105-111.

KLITZSCH, E. 1970. Die Strukturgeschichte der Zentrasahara. *Geol. Rundschau* 59: 49-520.

KLITZSCH, E. 1978. Geologische Bearbeitung Südwest Ägyptens. *Geol. Rundschau* 67: 509-520.

KLITZSCH, E. 1979. Zur Geologie des Gilf Kebir Gebietes in der Ostsahara. *Clausthaler Geol. Abh.* 30: 113-132.

KLITZSCH, E. 1980. Neue stratigraphische und paläogeographische Ergebnisse aus dem NW Sudan. *Berl. Geowiss. Abh.* 20(A): 217-222.

KLITZSCH, E. 1981. Lower Paleozoic rocks of Libya, Egypt and Sudan. In: C. H. Holland (ed.), *Lower Paleozoic of the Middle East, eastern and southern Africa and Antarctica,* Wiley: 131-163.

KLITZSCH, E. 1983. Geological research in and around Nubia. *Episodes* 3: 15-19.

KLITZSCH, E. 1983. Paleozoic formations and a Carboniferous glaciation from the Gilf Kebir-Abu Ras area in southwestern Egypt. *J. Afr. Earth Sci.* 1: 17-19.

KLITZSCH, E. 1984. Northwestern Sudan and bordering areas; geological developments since Cambrian time. *Berl. Geowiss. Abh.* 50(A): 23-45.

KLITZSCH, E. 1986. Plate tectonics and cratonal geology in northeast Africa (Egypt/Sudan). *Geol. Runschau* 75: 755-768.

KLITZSCH, E., J. C. HARMS, A. LEJAL-NICOL & F. F. LIST 1979. Major subdivisions and depositional environments of Nubia strata, southwestern Egypt. *Bull. Am. Assoc. Petrol. Geol.* 63: 974-976.

KLITZSCH, E. & A. LEJAL-NICOL 1984. Flora and fauna from a strata in southern Egypt and northern Sudan (Nubia and surrounding areas). *Berl. Geowiss. Abh.* 50(A): 47-79.

KLITZSCH, E. & H. LINKE 1983. *Geological interpretation maps, Gulf of Suez 1:100 000, prepared for Conoco Egypt. Inst. Applied Geosci., Klaus Volger & TU, Berlin.*

KLITZSCH, E. & F. k. LIST (Eds.) 1978. Southwest Egypt 1:500 000 – Geological interpretation map, preliminary edition: Sheet 2523 Gilf Kebir, Sheet 2525 Ammonite Hills, Sheet 2823 Baris, Sheet 2825 El Kharga. Berlin.

KLITZSCH, E. & F. k. LIST (Eds.) 1979. Southwest Egypt 1:500 000 – Geological interpretation map, preliminary edition: Sheet 2521 Gebel Uweinat, Sheet 2527 Farafra, Sheet 2827 El Minya. Berlin.

KLITZSCH, E. & F. k. LIST (Eds.) 1980. Southwest Egypt 1:500 000 – Geological interpretation map, preliminary edition: Sheet 3123 Aswan, Sheet 3125 Luxor, Sheet 3127 Asyut. Berlin.

KLITZSCH, E., C. SONNTAG, K. WEISTROFFER & E. M. EL SHAZLY 1976. Grundwasser der Zentralsahara: Fossile Vorräte. *Geol. Rundschau* 65: 264-287.

KLITZSCH, E. & P. WYCISK 1987. Geology of sedimentary basins of northern Sudan and bordering areas. *Berl. Geowiss. Abh.* 75(A), 1: 97-136.

KNETSCH, G. & M. YALLOUZE 1955. Remarks on the origin of the Egyptian oases depressions. *Bull. Soc. Geograph. Egypte* 28: 21-33.

KNORRING, O. VON 1971. *Mineralogical, geochemical and economic aspects of columbite-cassiterite-bearing alkali granitic rocks, special reference to Wadi Nugrus mineralization in Eastern Desert.* Internal Report Aswan Mineral Survey Projectl., Geol. Surv. Egypt.

KNOTT, S. T., E. T. BUNCE & R. L. CHASE 1966. Red Sea seismic reflection studies: The world rift system. *Geol. Surv. Canada,* Paper 66/14: 31-61.

KOCHIN, G. G. & F. A. BASSIUNI 1968. *The mineral resources of the UAR. Report on generalization of geological data on mineral resources of the UAR carried out under Contract 1247 (1966 to 1968), part I: Metallic minerals.* Internal Report 18/68, Geol. Surv. Egypt.

KOHN, B. P. & M. EYAL 1981. History of uplift of the crystalline basement of Sinai and its relation to opening of Red Sea as revealed by fission track dating of apatites. *Earth Planet. Sci. Lett.* 52: 129-141.

KORA, M. 1984. *The Paleozoic outcrops of Um Bogma area, Sinai.* Ph.D. Thesis, Mansoura Univ., Mansoura, 253pp.

KORA, M. & U. JUX 1986. On the early Carboniferous macrofauna from the Um Bogma Formation, Sinai. *N. Jb. Geol. Palaeontol., Mh* (1986), 2: 85-98.

KORTLAND, A. 1980. The Fayum primate forest: Did it exist? *J. Human Evol.* 9: 277-297.

KOSTANDI, A. B. 1959. Facies maps for the study of the Paleozoic and Mesozoic sedimentary basins of the Egyptian region. *1st Arab Petrol. Congr., Cairo* 2: 44-62.

KOSTANDI, A. B. 1961. Hurgada Oilfield. *3rd Arab Petrol. Congr.,* Alexandria, 8pp.

KOSTANDI, A. B. 1963. Eocene facies maps and tectonic interpretation in the Western Desert, Egypt. *Rev. Inst. Franc. Petrole* 18: 1331-1343.

KOTB, H., A. M. EL KAMMAR & N. ZOHNY 1978. Mineralogical and geochemical studies on Abu Sabouna phosphorites (Mahamid, Sharawna), upper Egypt. *Tscher-*

maks Min. Petr. Mitt. 25: 171-183.

KOTB, H., M. B. KHAFFAGY & B. M. SEWIFI 1980. Geochemistry and petrochemistry of the Egyptian titaniferous gabbros. *Ann. Geol. Surv. Egypt* 10: 627-650.

KOTB, N. H. 1983. *Geochemistry of old metavolcanics at Sheikh El Shadli area, Eastern Desert, Egypt.* M.Sc. Thesis, Azhar Univ., Cairo.

KOVACIK, J. 1961. *Report on geological prospecting of Cu/Zn/Pb at Umm Samiuki, Eastern Desert, Egypt.* Internal Report prepared by Polytechna Foreign Trade Corporation, Prague, Czechoslovakia, for the Geol. Survey of Egypt.

KRASHENINNIKOV, V. A. & T. M. ABDEL RAZIK 1969. Zonal stratigraphy of the Paleocene in Quseir (Red Sea coast). *Proc. 3rd Afr. Micropaleontol. Colloq.*, Cairo: 299-310.

KRASHENINNIKOV, V. A. & B. P. PONIKAROV 1964. *Zonal stratigraphy of Paleogene in the Nile Valley.* Geol. Surv. Egypt, Paper 32, 26pp.

KRATKY, V. 1974. Cartographic accuracy of ERTS. *Photogramm. Eng.* 40(10): 203-212.

KRAUS, K. 1975. Die Entzerung von Multispektralbildern. *Bildmess. u. Luftbildmess.* 43: 129-134.

KRAÜSEL, R. 1939. Die fossilen Floren Ägyptens. *Abh. Bayer. Akad. Wiss. Math.-Nat. Abt.* 47: 1-140.

KRAÜSEL, R. & E. STROMER 1925. Ergebnisse d. Forschungsreisen Prof. E. Stromer in Wüsten Ägyptens IV: Die fossilen Floren Ägyptens. *Abh. Bayer. Akad. Wiss. Math. Nat.-Abt.* 30: 1-48.

KROEPELIN, S. 1987. Palaeoclimatic evidence from early to mid-Holocene playas in the Gilf Kebir (southwest Egypt). In: J. A. Coetzee (ed.), *Palaeoecology of Africa and surrounding islands*, Balkema 18: 189-208.

KRÖNER, A. 1977. Precambrian mobile belts of southern and eastern Africa: ancient sutures or sites of ensialic mobility? A case of crustal evolution towards plate tectonics. *Tectonphysics* 40: 101-135.

KRÖNER, A. 1979. Pan-African plate tectonics and its repercussions on the crust of northeast Africa. *Geol. Rundschau* 68: 565-583.

KRÖNER, A. (ed.) 1981. *Precambrian plate tectonics.* Elsevier, 781pp.

KRÖNER, A. 1984. Late Precambrian plate tectonics and orogeny: A need to redefine the term Pan-African. In: J. Klerkx & J. Michot (eds), *Geologie africaine*, Tervuren: 23-28.

KRÖNER, A. 1985. Ophiolites and the evolution of tectonic boundaries in late Proterozoic Arabian-Nubian shelf of NE Africa and Arabia. *Precambrian Res.* 27: 277-300.

KRONFELD, J., G. GVIRTZMAN & B. BUCHBINDER 1982. Geological evolution and Thorium/uranium ages of Quaternary coral reefs in southern Sinai. *Annual Meeting, Geol. Soc. Isr.*, Elat: 45-46.

KRS, M. 1972. *Technical report on follow-up geophysical survey in the period 1968-1972.* Internal Report, Aswan Mineral Survey Project, Geol. Surv. Egypt.

KRS, M. 1977. Rift tectonics development in the light of geophysical data, Red Sea region. *Studia Geophys. et Geol.* 21: 342-350.

KRS, M., A. A. SOLIMAN & H. A. AMIN 1973. Geophysical phenomena over deep-seated tectonic zones in southern part of Eastern Desert of Egypt. *Ann. Geol. Surv. Egypt* 3: 125-138.

KRUPP ROHSTUFFE 1963. *Iron ore project, Quseir (upper Egypt) main survey, 1st phase 1960-1962.* Internal Report prepared for the Geol. Surv. Egypt.

KULKE, H. 1982. A Miocene carbonate-anhydrite sequence in the Gulf of Suez as a complex oil reservoir. *6th Petrol. Explor. Seminar, EGPC*, Cairo.

KUMMEL, B. 1960. Middle Triassic nautiloids from Sinai, Egypt and Israel. *Bull. Mus. Comp. Zool., Harvard Univ.* 123(7): 285-302.

KUSS, J. 1986. Facies development of upper Cretaceous-lower Tertiary sediments from the Monastery of St Anthony, Eastern Desert, Egypt. *Facies* 15: 177-194.

KUSS, J. 1986. Upper Cretaceous calcareous algae from the Eastern Desert of Egypt. *N. Jb. Geol. Palaeontol., Mh* (1986), 4: 223-238.

LABRECQUE, J. L. & ZITTELLINI 1985. Continuous sea floor spreading in Red Sea: an alternative interpretation of magnetic anomaly pattern. *Bull. Am. Assoc. Petrol. Geol.* 69: 513-524.

LACAZE, M. J. & M. K. DUFRESNE 1982. North Alexandria area, geological and geophysical study. *6th Petrol. Explor. Seminar, EGPC*, Cairo.

LACHENBRUCH, A. H. & J. H. SASS 1978. Models of an extending lithosphere and heat flow in the Basin and Range province. In: R. B. Smith & G. P. Eaton (eds), *Cenozoic tectonics and regional geophysics of the western Cordillera. Mem. Geol. Soc. Am.* 152: 209-250.

LAHR, J. C. 1979. Hypoellipse, a computer program for determining local earthquake hypocentral parameters, magnitude and first motion pattern. *US Geol. Surv., open file report* 79/431, 233pp.

LAMBRECHT, K. 1933. *Handbuch der Palaeornithologie.* Gebrüder Bornträger, Berlin.

LARTET, L. 1869. Essay on the geology of Palestine. *Ann. Sci. Geol., Soc. Geol. France 1: 17-18.*

LAUBSCHER, H. & D. BERNOULLI 1977. Mediterranean and Tethys. In: A. E. M. Nairn, W. H. Kanes & F. G. Stehli (eds), *The ocean basins and margins*, Plenum Press 4(A): 1-28.

LAUGHTON, A. S. 1970. A new bathymetric chart of the Red Sea. *Phil. Trans. Roy. Soc. London* 267(A): 21-22.

LAWSON, A. C. 1927. The valley of the Nile. *Univ. California (Berkeley), Publ. Geol. Sci.* 29: 235-259.

LEBLING, C. 1919. Ergebnisse der Forschungsreisen Prof. E. Stromers in den Wüsten Agyptens. III. Forschuungen in der Bahariya Oase und anderen Gegenden Agyptens. *Abh. Bayer. Akad. Wiss., Math.-Phys. Kl.* 29, 44pp.

LEE, W. 1922. The face of the earth as seen from the air. *Am. Geograph. Soc.*, sp. publ. 4, 110pp.

LEE, W. H. K. 1970. On the global variations of terrestrial heat flow. *Phys. Earth Planet. Int.* 2: 332-341.

LEE, W. H. K., R. E. BENNETT & K. L. MEAGHER 1972. A method of investigating magnitude of local earthquakes from signal duration. *US Geol. Surv., open file report* 72/431.

LEE, W. H. K. & J. C. LAHR 1972. HYPO71, a computer program for determining hypocenter, magnitude and first motion of local earthquakes. *US Geol. Surv., open file report.*

LEJAL-NICOL, A. 1981. Nouvelles empreintes de la (Lingula Shale unit) dans la region d'Abu Ballas (Egypte). *C.R. 106ème Congr. Nat. Soc. Sav., Sci.* (1): 15-27.

LEJAL-NICOL, A. 1981. A propos de nouvelles flores plaéozoiques et mésozoiques de l'Egypte du sud-ouest. *C.R. Acad. Sci.* 292(II): 1337-1340.

LEJAL-NICOL, A. 1984. Sur une nouvelle espèce du genre *Pterophyllum* Brongniart du Nubien du Soudan. *C.R. 109ème Congr. Nat. Soc. Sav., Sci.* (2): 57-69.

LEJAL-NICOL, A. 1986. Découverte d'une flore a *Callipteris* dans la région du Suez (Egypte). *C.R. 111ème Congr. Nat. Soc. Sav., Sci.* (2): 9-22.

LIJAL-NICOL, A. 1987. Flores nouvelles du Paléozoique et du Mésozoique de l'Egypte et du Soudan septentrional. *Berl. Geowiss. Abh.* 75(A): 151-248.

LEJAL-NICOL, A. & W. DOMINIK 1987. Plant cover and paleoenvironment during the Cenomanian in Bahariya basin (Egypt). *14th Intl. Bot. Congr.*, Berlin: 402 (Abstr.).

LE PICHON, X., J. ANGELIER, & J. C. SIBUET 1982. Subsidence and stretching. In: J. S. Watkins & C. L. Drake (eds), *Studies in continental margin geology. Mem. 34, Am. Assoc. Petrol. Geol.*

LE PICHON, X. & J. FRANCHETEAU 1978. A plate tectonic analysis of the Red Sea-Gulf of Aden area. *Tectonophysics* 46: 369-406.

LE PICHON, X., J. FRANCHETEAU & J. BONNIN 1973. *Plate tectonics.* Elsevier, 330pp.

LERMAN, A. 1960. Triassic pelecypods from southern Israel and Sinai. *Bull. Res. Council Isr.* 9(G): 1-51.

LEROY, L. W. 1953. Biostratigraphy of the Maqfi secton, Egypt. *Mem. Geol. Soc. Am.* 54: 1-73.

LEWY, Z. 1975. The geological history of southern Israel and Sinai during the Coniacian. *Isr. J. Earth Sci.* 24: 19-43.

LEWY, Z. & M. RAAB 1977. Mid-Cretaceous stratigraphy in the Middle East, mid-Creatceous events *Mus. d'Hist. Nat. Nice* 4: xxxii.1-xxxii.19.

LIST, F. K. 1976. *Experiences of a geological remote sensing project in Africa (Tibesti mountains, Chad).* UN/PAO Int. reg. Train. Seminar Rem. Sens. Aplic. Lengries., Cologne: 69-76.

LIST, F. K., H. BURGER, E. KLITZSCH, B. MEISSNER, G. PÖHLMANN & H. SCHMITZ 1978. Geological interpretation of Landsat imagery of southwestern Egypt. *Proc. Intl. Syp. Rem. Sens., Int. Arch. Photogramm.* 22(7), 3: 2195-2208.

LIST, F. K., H. BURGER, E. KLITZSCH, B. MEISSNER, G. PÖHLMANN & H. SCHMITZ 1982. Application of visual interpretation and digital processing of Lndsat data for the preparation of a geological interpretation map of southwestern Egypt at a scale of 1:500 000. *Proc. Intl. Symp. rem. sens Environment, them. Conf.: Rem. Sens. arid semi-arid lands,* Cairo 2: 849-858.

LIST, F. K., D. HELMCKE & N. W. ROLAND 1975. Geologische Information im Satellitenbild und Luftbild-Erfahrungen aus dem Forschungsprojeckt Tibesti-Gebirge (NASA Landsat-1 SR-349). *DFVLR-DGP Symp. Erderkundung:* 139-145 Cologne.

LIST, F. K., D. HELMCKE, B. MEISSNER, G. PÖHLMANN & N. ROLAND 1978. Geologische Interpretation des Tibesti nach Aufnahmen von Landsat-1 (Republic Chad). Erläuterungen zur Karte Tibesti 1:1 000 000. *Bildmess. u. Luftbilmess.* 46(4): 139-145.

LIST, F. K., B. MEISSNER & M. ENDRISZEWITZ 1987. Operational remote sensing for thematic mapping in Egypt and Sudan. *Berl. Geowiss. Abh.* 75(A), 3: 907-926.

LIST, F. K., B. MEISSNER, & G. PÖHLMANN 1986. Landsat-MSS remote sensing and satellite cartography – an integrated aproach to the preparation of a new geological map of Egypt at a scale of 1:500 000. *Proc. IGARSS '86 Symp., Zurich, ESA SP-254:* 1503-1510, Paris.

LIST, F. K., B. MEISSNER, G. PÖHLMANN & U. RIPKE 1984. Medium to small scale geological maps based on Landsat MSS and RBV data – case histories of projects in north Africa. In: P. Teleki & C. Weber (eds), *Remote sensing for geological mapping.* BRGM 82. Proc. IUGS-UNESCO Seminar Orleans: 143-159.

LIST, F. K. & PÖHLMANN 1976. Geologische Interpretations – Karte des Tibesti 1:1 000 000 – Beispiel für eine thematische Auswertung von Landsat-1 Bildern. *Z. Deutsch. Geol. Ges.* 127: 435-498.

LIST, F. K., N. W. RICHTER & R. SCHOELE 1987. Digital image processing and ground reflectance measurements; applications in arid areas. *Berl. Geowiss. Abh.* 75(A), 3: 873-906.

LIST, F. K., N. W. ROLAND & D. HELMCKE 1974. Comparison of geological information from satellite imagery, areal photography, and ground truth investigations in the Tibesti Mountains, Chad. ISP Comm. VII Proc. *Symp. Rem. Sens. Photointerpret.*, Banff, Alberta 2: 543-553.

LIST, F. K. & P. STOCK 1969. Photogeologische Untersuchungen über Bruchtektonik und Entwässerungsnetz im Präkambrium des nordlichen Tibesti-Gebirges, Zentralsahara, Tschad. *Geol. Rundschau* 59: 228-256.

LITTLE, O. H. 1936. Recent geological work in the Fayium and in the adjoining portion of the Nile Valley. *Bull. Inst. Egypte* 18: 201-240.

LITTLE, O. H. & M. I. ATTIA 1943. *The development of the Aswan district, with notes on the minerals of southeastern Egypt.* Geol. Surv. Egypt, 107pp.

LOCKWOOD SURVEY CORPORATION LTD, TORONTO, CANADA 1968. *Airborne magnetometer, scintillation counter, duel frequency electromagnetometer survey of a part of the Aswan region, UAR.* Internal Report prepared for UNDP and Aswan Mineral Survey Project, Geol. Surv. Egypt.

LORT, J. M. 1971. The tectonics of the eastern Mediterranean, a geophysical review. *Rev. Geophys.* 9: 189-216.

LOUBIERE, H. 1935. Etude anatomique d'un bois minéralisé trouvé à Ouadi Halfa (Nubie). *Rev. Gen. Bot.* 47: 480.

LOUVET, P. 1973. Sur les affinités des flores tropicales ligneuses africaines tertiares et actuelles. *Bull. Soc. Bot. France* 120: 385-396.

LOVERING, T. S. 1948. Geothermal gradients, recent climatic changes, and rate of sulphide oxidation in the San Manuel district, Arizona. *Econ. Geol.* 43: 1-20.

LOWELL, J. D. 1972. Spitzbergen Tertiary orogenic belt and the Spitzbergen fracture zone. *Bull. Geol. Soc. Am.* 83: 3091-3101.

LOWELL, J. D. & G. J. GENIK 1972. Sea floor spreading and structural evolution of the southern Red Sea. *Bull. Am. Assoc. Petrol. Geol.* 56: 247-250.

LOWELL, J. D. & J. M. GILBERT 1970. Lateral and vertical alteration-mineralization zoning in porphyry deposits. *Econ. Geol.* 65: 373-408.

LOWMAN, P. D. 1969. Apollo 9 multispectral photography: Geologic analysis. *NASA X-644-69-423, Greenbelt (GSFC),* 24pp.

LOWMAN, P. D. 1985. Mechanical obstacles to the movement of continent-bearing plates. *Geophys. Res. Lett.* 12: 223-225.

LOWMAN, P. D., J. A. McDIVITT & E. H. WHITE 196 . *Terrain photography on the Gemini IV mission.* Preliminary report. NASA-X-644-69-423, Greenbelt (GSFC), 24pp.

LOWMAN, P. D. & H. A. TIEDEMANN 1971. *Terrain photography from Gemini spacecraft: Final geological report.* NASA-X-644-71-15, Greenbelt (GSFC), 75pp.

LOWRIE, W., R. LOVILE & N. D. OPDYKE 1973. Magnetic properties of Deep Sea Drilling Project basalts from the north Pacific Ocean. *J. Geophys. Res.* 78: 7647-7660.

LUCAS, A. 1927. Copper in ancient Egypt. *J. Egypt. Archeol.* 13: 169.

LUGER, P. 1985. Stratigraphie der marinen Oberkreide und des Alttertiärs im sudwestlichen Obernil-Becken (SW Ägypten) unter besonderer Berücksichtigung der Mikropaläontologie, Paläoekologie, Paläogeographie. *Berl. Geowiss. Abh.* 63(A), 151pp.

LUGER, P. & E. SCHRANK 1987. Mesozoic to Paleogene transgressions in middle and southern Egypt – summary of paleontological evidence. In: G. Matheis & H. Schandelmeier (eds), *Current research in African earth sciences*, Balkema: 199-202.

LYONS, H. G. 1906. *The physiography of the River Nile and its basin.* Egypt Surv. Dept., Publ. 58, 193pp.

MAAMOUN, K. 1979. Macroseismic observations of principal earthquakes in Egypt. *Bull. Helwan Inst. Astronom. and Geophys.* 183: 12pp.

MAAMOUN, K., A. ALLAM & A. MEGAHED 1984. Seismicity of Egypt. *Bull. Helwan Observ.* 4(B), 19pp.

MAAMOUN, K. & H. M. EL KHASHAB 1978. Seismic studies of the Shadwan (Red Sea) earthquake. *Bull. Helwan Inst. Astronom. and Geophys.* 184, 12pp.

MADGWICK, T. G., F. W. MOON & H. SADEK 1920. Preliminary geological report on the Abu Shaar El Qibli (Black Hill) district. *Petrol. Res. Bull.* 6, Government Press, Cairo, 11pp.

MADGWICK, T. G., F. W. MOON & H. SADEK 1920. Preliminary geological report on Ras Dib area. *Petrol. Res. Bull.* 8, Government Press, Cairo, 13pp.

MAGARITZ, M. & I. B. BRENNER 1979. Geochemistry of a lenticular manganese ore deposit, Um Bogma, Sinai. *Mineral. Deposita* 14: 1.

MAGNAVOX 1982. *Operation and service manual. Satellite surveyor MX 1502.* Torrence, California.

MAHBOUBI, M., R. AMEUR, J. Y. CROCHET & J. J. JAEGER 1986. El Kohol (Saharan Atlas, Algeria): A new Eocene mammal locality in northwestern Africa: Stratigraphic, phylogenetic and paleobiogeographic data. *Palaeontographica* 192(A): 15-49.

MAHGOUB, I. S. & E. E. MOHAMED 1985. A high closure reservoir with heterogeneous fluid properties, a field case history. *8th Petrol. Production Seminar*, EGPC, Cairo.

MAHMOUD, I. G. 1955. Etudes paleontologiques sur la faune cretacique du massif du Maghara, Sinai, Egypte. *Publ. Inst. Desert Egypte* 8: 1-192.

MAHMOUD, K. 1962. *Report on the logging done on well no. 15, Ayun Musa, Sinai.* Internal Report 3, Coal Project, Geol. Surv. Egypt.

MAMET, B. & S. OMARA 1969. Microfacies of the lower Carboniferous dolomitic limestone formation of the Um Bogma terrain (Sinai, Egypt). *Contr. Cushman Found. Foram. Res.* 20: 106-109.

MANN, P., M. R. HEMPTON, D. C. BRADLEY & K. BURKE 1983. Development of pull-apart basins. *J. Geol.* 91: 529-554.

MARRIOTT, R. P. 1981. *Gulf of Suez: Field structure study.* Internal Report, Conoco, Egypt.

MANSOUR, A. T., M. G. BARAKET & Y. ABDEL HADY 1969. Marine Pliocene planktonic foraminiferal zonation, southeast of Salum, Egypt. *Riv. Ital. Paeontol.* 75: 833-846.

MANSOUR, H. H., B. ISSAWI & M. M. ASKALANY 1982. Contribution to the geology of west Dakhla oasis area, Western Desert, Egypt. *Ann. Geol. Surv. Egypt* 12: 255-281.

MANSOUR, H. H. & E. R. PHILOBBOS 1983. Lithostratigraphic classification of the surface Eocene carbonates of the Nile Valley, Egypt: a review. *Bull. Fac. Sci., Assiut Univ.* 12(2)C: 129-153.

MANSOUR, H. H., E. R. PHILOBBOS, H. KHALIFA & S. GALIL 1983. Contribution to the stratigraphy and micropaleontology of the middle and upper Eocene exposures southeast of Cairo, Egypt. *Bull. Fac. Sci., Assiut Univ.* 12(2)C: 153-173.

MANSOUR, H. H., M. M. YOUSSEF & A. R. EL YOUNIS 1979. Petrology and sedimentology of the upper Cretaceous, Paleocene succession northwest of Kharga oasis, Egypt. *Ann. Geol. Surv. Egypt* 9: 471-497.

MANSOUR, M. S. & F. A. BASSYOUNI 1954. *Geology of Wadi Garf district (Barramiya east Sheet).* Geol. Surv. Egypt, 34pp.

MANSOUR, M. S., F. A. BASSYOUNI & D. M. AL FAR 1956. *Geology of Umm Salatit-El Hisinat (Barramiya east sheet).* Geol. Surv. Cairo, 28pp.

MANSOUR, M. S., S. EL MASRY, M. A. EL DANAF & N. B. EL SAYED 1962. *Report on the prospecting activities for copper-zinc deposits in Samiuki and Darhib regions.* Internal Report, General Mines and Quarries Co., Cairo.

MARATHON OIL COMPANY 1980. *Nile delta study. Technical review.* Internal Report.

MARATHON OIL COMPANY 1981. *Report on the subsurface geology of northern Sinai.* Internal Report.

MARHOLZ, W. W. 1968. *Geological exploration of the Kufra region, April-May 1965.* Ministry of Industry, Kingdom of Libya, Bull. Geol. Sec., 8.

MARINER, R. H. & L. W. WILLEY 1976. Geochemistry of thermal waters in Long Valley, Mono County, California. *J. Geophys. Res.* 81: 792-800.

MART, J. & E. SASS 1972. Geology and origin of the manganese ore of Um Bogma, Sinai. *Econ. Geol.* 67: 145-155.

MARTINI, E. 1971. Standard Tertiary and Quaternary calcareous nannoplankton zonation. *Proc. 2nd Intl. Conf. Plankt. Microfossils*, Rome 2: 739-785.

MARZOUK, I. 1969. Rock stratigraphy and oil potentialities of the Oligocene and Miocene in the Western Desert, UAR. *7th Arab Petrol. Congr.*, Kuwait, 54 (B-3).

MASTERS, B. A. 1984. Comparison of planktonic foraminifers at the Cretaceous-Tertiary boundary from the Haria Shale (Tunisia) and the Esna Shale (Egypt). *Proc. 7th Explor. Seminar, EGPC*, Cairo: 310-324.

MATHEIS & M. CAEN-VACHETTE 1983. Rb-Sr isotopic study of rare-metal bearing and barren pegmatites in the Pan-African reactivation zone of Nigeria. *J. Afr. Earth Sci.* 1: 35-40.

MATSUMOTO, H. 1922. Revision of *Palaeomastodon* and *Moeritherium: Paleomastodon intermedius* and *Phiomia osborni*, new species. *Am. Mus. Nov.* 51: 1-6.

MATSUMOTO, H. 1923. A contribution to the knowledge of *Moeritherium. Bull. Am. Mus. Nat. Hist.* 48: 97-140.

MATSUMOTO, H. 1924. A revision of *Palaeomastodon* dividing it into two genera, and with descriptions of two new species. *Bull. Am. Mus. Nat. Hist.* 50: 1-58.

MATSUMOTO, H. 1926. Contributions to the knowledge of the fossil Hydracoidea of the Fayum, Egypt, with a description of several new species. *Bull. Am. Mus. Nat. Hist.* 56: 253-350.

MAXWELL, T. A. 1982. Sand sheet and lag deposits in the south Western Desert. In: F. El Baz & T. A. Maxwell (eds), *Desert landforms of southwest Egypt: a basis for comparison with Mars*, NASA: 157-174.

MAZHAR, A., N. ENANY & A. K. YEHIA 1979. Contribution to the Cretaceous-early Tertiary stratigraphy of El Galala El Qibliya plateau. *Ann. Geol. Surv. Egypt* 9: 377-387.

MCCAULEY, J. F., C. S. BREED & M. J. GROLIER 1982. The interplay of fluvial, mass-wasting and eolian processes in the eastern Gilf Kebir region. In: F. El Baz & T. A. Maxwell (eds), *Desert landforms in southwest Egypt: a basis for comparison with Mars*, NASA: 297-340.

MCCAULEY, J. F. ET AL. 1982. Subsurface valleys and geoarcheology of the eastern Sahara as revealed by shuttle radar. *Science* 218: 1004-1021.

MCCAULEY, J. F., C. S. BREED, G. G. SCHABER, W. P. MCHUGH, C. V. HAYNES, M. J. GROLIER & A. EL KILANI 1986. Paleodrainages of the eastern Sahara, the radar rivers revisited (SIR-A/B implications for a mid-Tertiary trans-Africa drainage system). *IEEE Trans. Geosci. and Rem. Sens.* GE-24: 624-648.

MCCLURE, H. A. 1978. Early Paleozoic glaciation in Arabia. *Paleogeogr., Paleoclimat., Paleoecol.* 25: 315-326.

MCCLURE, H. A. 1980. Permian-Carboniferous glaciation in the Arabian Peninsula. *Bull. Geol. Soc. Am.* 91: 707-712.

MCHUGH, W. P. 1975. Some archeological results of the Bagnold Mond expeditions in the Gilf Kebir and Gebel Uweinat, southern Libyan Desert. *J. Near Eastern Studies* 34: 64-68.

MCHUGH, W. P. 1982. Archeological investigations in the Gilf Kebir and Abu Hussein dune field. In: F. El Baz & T. A. Maxwell (eds), *Desert landforms of southwestern Egypt: a basis for comparison with Mars*, NASA: 301-334.

MCKEE, E. D. (ed.) 1979. *A study of global sand seas*. US Geol. Surv., Prof. Paper 1052, 429pp.

MCKENZIE, D. P. 1970. Plate tectonics of the Mediterranean region. *Nature* 226: 239-248.

MCKENZIE, D. P. 1972. Active tectonics of the Mediterranean region. *Geophys. J. Roy. Astronom. Soc.* 30: 109-185.

MCKENZIE, D. P. 1977. Can plate tectonics describe continental deformation. In: B. Biju-Duval & L. Montadert (eds), *Structural history of the Mediterranean basins*, Editions Technip, Paris: 189-196.

MCKENZIE, D. P., D. DAVIES & P. MOLNAR 1970. Plate tectonics of the Red Sea and east Africa. *Nature* 226: 243-248.

MCWILLIAMS, M. O. 1981. Paleomagnetism and Precambrian evolution of Gondwana. In: A. Kröner (ed.), *Precambrian plate tectonics*, Elsevier 4: 649-687.

MEHNERT, K. R. 1968. *Migmatites and the origin of granitic rocks*. Elsevier, 393pp.

MELVILLE, C. P. 1984. The seismicity of the Gulf of Suez according to data provided by N. N. Ambraseys. Unpublished Report, Woodward-Clyde Consultants.

MENCHIKOFF, N. 1926. Observations géologiques faites au cours de l'exploration de S.A.A. le Prince Kamal el Dine Hussein dans le désèrt de Libyie (1925-1926). *C.R. Acad. Sci. Paris* 183: 1047-1049.

MENCHIKOFF, N. 1927. Etude petrographique des roches cristallines et volcaniques de la région d'Ouenat (Désèrt de Libyie). *Bull. Soc. Geol. France* 27: 337-354.

MENEISY, M. Y. 1986. Mesozoic igneous activity in Egypt. *Qatar Univ. Sci. Bull.* 6.

MENEISY, M. Y. & A. Y. ABDEL AAL 1984. Geochronology of Phanerozoic volcanic rocks in Egypt. *Bull. Fac. Sci., Ain Shams Univ.* 25.

MENEISY, M. Y. & B. EL KALIOUBI 1975. Isotopic ages of the volcanic rocks of the Bahariya Oasis. *Ann. Geol. Surv. Egypt* 5: 119-122.

MENEISY, M. Y. & H. KREUZER 1974. Potassium-argon ages of Egyptian basaltic rocks. *Geol. Jb.* D-9: 21-31.

MENEISY, M. Y. & H. KREUZER 1974. Potassium-argon ages of mepheline syenite ring complexes in Egypt. *Geol. Jb.* D-9: 33-39.

MENEISY, M. Y., A. A. OMAR & M. M. SHAABAN 1976. Petrology of volcanic rocks of Wadi Abu Darag, Eastern Desert. *Bull. Volc.* 37(3): 1-12.

MESHAL, A. H., 1975. Brine at the bottom of the Great Bitter Lake as a result of closing the Suez Canal. *Nature* 256: 297-298.

MESHREF, W. M., S. H. ABDEL BAKI, H. M. ABDEL HADY & S. A. SOLIMAN 1980. Magnetic trend analysis in northern part of the Arabian-Nubian shield and its tectonic implications. *Ann. Geol. Surv. Egypt* 10: 939-953.

MESHREF, W. M., A. BELEITY, H. HAMMOUDA & M. KAMEL 1988. Tectonic evaluation of the Abu Gharadig basin. *Am. Assoc. Petrol. Geol. Mediterranean basins Conference*, Nice. France (abstr.).

MESHREF, W. M. & M. M. EL SHEIKH 1973. Magnetic tectonic trend analysis in northern Egypt. *Egypt. J. Geol.* 17: 179-184.

MESHREF, W. M., E. M. REFAI & S. H. ABDEL BAKI 1976. Structural interpretation of the Gulf of Suez and its oil potentialities. *3rd Petrol. Explor. Seminar*, EGPC, Cairo.

METWALLI, M. H. 1963. *The study of some Miocene sediments in the Cairo-Suez district. M.Sc.* Thesis, Cairo Univ.

METWALLI, M. H. & Y. E. ABD EL-HADY 1975. Petrographic characteristics of oil bearing rocks in Alamein oilfield. Significance in source-reservoir relations in northern Western Desert. *Bull. Am. Assoc. Petrol. Geol.* 59: 510-523.

METWALLI, M. H., G. PHILIP & E. YOUSEF 1981. El Morgan crude oil and cycles of oil generation migration and accumulation in the Gulf of Suez petroleum province, Egypt. *Acta Geol. Acad. Sci. Hungaricae* 24(2-4): 369-387.

MEYER, G. E. 1973. A new Oligocene hyrax from the Jebel El Qatrani Formation, Fayum, Egypt. *Postilla* 63: 1-11.

MEYER, G. E. 1978. Hyracoidea. In: V. J. Maglio & H. B. S. Cooke (eds), *Evolution of African mammals*. Harvard Univ. Press: 284-314.

MILKEREIT, B. & E. R. FLUH 1985. Saudi Arabian refraction profiles: crustal structure of the Red Sea-Arabian

Shield transition. *Tectonophysics* 111: 283-298.

MITCHELL, A. H. G. & M. S. GARSON 1981. *Mineral deposits and global tectonic setting*. Academic Press, 405pp.

MIYASHIRO, A. 1973. The Troodos ophiolite complex was probably formed in an island arc. *Earth Planet. Sci. Lett.* 19: 218-224.

MIYASHIRO, A. 1975. Classification, characteristics and origin of ophiolites. *J. Geol.* 8: 249-281.

MOHAMED, A. F. 1940. The Egyptian exploration of the Red Sea. *Proc. Roy. Soc. London* 128: 306-316.

MOHAMED, M. M. 1979. The recent bottom sediments of the Gulf of Suez, Red Sea. *J. Univ. Kuwait (Sci.)* 6: 209-228.

MOHARRAM, O., D. Z. GACHECHILADZE, M. F. EL RAMLY, S. S. IVANOV & A. F. AMER 1970. *Studies on some mineral deposits of Egypt (summary of the results of the work carried out by a team of Egyptian and Soviet geologists under Conract 1247)*. Egypt. Geol. Surv., 269pp.

MOHR, P. A. 1973. Structural geology of the African rift system: summary of new data from ERTS-1 imagery. *ERTS-1 Symp., 1(A), NASA* SP-351: 767-782.

MONOD, O., J. MARCOUX, A. POISSON & J. F. DUMONT 1974. Le domaine d'Antalya, témoin de la fracturation de la plateforme africaine au cours du Trias. *Bull. Soc. Geol. France* 16: 116-127.

MONTASIR, A. H. 1937. Ecology of lake Manzala. *Bull. Fac. Sci., Cairo Univ.* 12: 50pp.

MOODY, J. D. & M. J. HILL 1956. Wrench-fault tectonics. *Bull. Geol. Soc. Am.* 67: 1207-1246.

MOON, F. W. 1923. *Preliminary geological report on St John's (Zeberged) Island in the Red Sea*. Geol. Surv. Egypt, 36pp.

MOON, F. W. & H. SADEK 1921. Topography and geology of northern Sinai. *Petrol. Res. Bull.* 10, Government Press, Cairo, 154pp.

MOON, F. W. & H. SADEK 1923. Preliminary geological report on Wadi Gharandel area. *Petrol. Res. Bull.* 12, Government Press, Cairo, 42pp.

MOON, F. W. & H. SADEK 1925. Preliminary geological report on Gebel Khoshera area (western Sinai). *Petrol. Res. Bull.* 9, Government Press, Cairo, 40pp.

MOONEY, M. D., M. E. GETTINGS, H. R. BLANK & J. H. HEALY 1985. Saudi Arabian seismic refraction profile: a traveltime interpretation of crustal and upper mantle structure. *Tectonophysics* 111: 173-246.

MOORE, J. M. 1979. Primary and secondary faulting in the Najd fault system, Kingdom of Saudi Arabia. *US Geol. Surv. Saudi Arabian Mission Project report 262, US open file report* 79/1661, 22pp.

MOORE, J. M. 1979. Tectonics of the Najd transcurrent fault system, Saudi Arabia. *J. Geol. Soc. London* 136: 441-454.

MORCOS, S. A. & S. N. MESSIEH, 1973. Change in the current regime in the Suez Canal after construction of Aswan High Dam. *Nature* 242: 38-39.

MORELLI, C. 1974. *The international gravity standardization net 1971*. Assoc. Intl. Geodesie, Paris, 194pp.

MORELLI, C. 1978. Eastern Mediterranean geophysical results and implications. *Tectonophysics* 46: 333-346.

MORETTI, I. & P. Y. CHENET 1986. The evolution of the Suez rift – a combination of stretching and secondary convection. *8th Petrol. Explor. Seminar*, EGPC, Cairo.

MORGAN, D. E. & A. N. EL BARKOUKY 1956. Geophysical

history of Ras Gharib field. *Soc. Explor. Geophys., Case Hist.* 2: 237-247.

MORGAN, H. J. 1982. Hotspot tracks and the opening of the Atlantic and Indian oceans. In: C. Emiliani (ed.), *The sea*, Wiley, 7: 443-487.

MORGAN, H. J. 1983. Hotspot tracks and the early rifting of the Atlantic. *Tectonophysics* 94: 123-139.

MORGAN, P. 1979. Cyprus heat flow with comments on the thermal regime of the eastern Mediterranean. In: V. Cermak & L. Ryback (eds), *Terrestrial heat flow in Europe*, Springer: 141-151.

MORGAN, P. 1984. The thermal structure and the thermal evolution of the continental lithosphere. *Phys. Chem. Earth* 15: 107-193.

MORGAN, P. & B. H. BAKER 1983. Introduction – processes of continental rifting. *Tectonophysics* 94: 1-10.

MORGAN, P., D. P. BLACKWELL, J. C. FARRIS, F. K. BOULOS & P. G. SALIB 1977. Preliminary geothermal gradient and heat flow values for northern Egypt and Gulf of Suez from oil well data. *Proc. Intl. Congr. Thermal Waters, Geothermal Energy and Volcanism of the Mediterranean area (1976)*, Nat. Tech. Athens: 424-438.

MORGAN, P., F. K. BOULOS, S. F. HENNIN, A. A. EL SHERIF, A. A. EL SAYED, N. Z. BASTA & Y. S. MELEK 1981. Geophysical investigations of a thermal anomaly at Wadi Ghadir, Eastern Desert, Egypt. *Proc. 1st Annual meeting Egypt. Geophys. Soc.*, Cairo.

MORGAN, P., F. K. BOULOS, S. F. HENNIN, A. A. EL SHERIF, A. A. EL SAYED, N. Z. BASTA & Y. S. MELEK 1985. Heat flow in eastern Egypt, the thermal signature of a continental break up. *J. Geodynamics* 4: 107-131.

MORGAN, P., F. K. BOULOS & C. A. SWANBERG 1983. Regional geothermal exploration in Egypt. *Geophys. Prosp.* 31: 361-376.

MORGAN, P. & K. BURKE 1985. Collisional plateaus. *Tectonophysics* 119: 137-151.

MORGAN, P., G. R. KELLER & F. K. BOULOS 1981. Earthquake cannons in the Egyptian Eastern Desert. *Bull. Seism. Soc. Am.* 71: 551-554.

MORGAN, P., C. A. SWANBERG, F. K. BOULOS, S. F. HENNIN, A. A. EL SAYED & N. Z. BASTA 1980. Geothermal studies in northeast Africa. *Ann. Geol. Surv. Egypt* 10: 971-987.

MOUSTAFA, A. R., A. YEHIA & S. ABDEL TAWAB 1985. Structural setting in the area east of Cairo, Maadi and Helwan. *Middle East Res. Center, Ain Shams Univ., Sci. Res. Series* 5: 40-64.

MOUSTAFA, F. H. 1974. July oilfield, Gulf of Suez. *4th Petrol. Explor. Seminar*, EGPC, Cairo.

MOUSTAFA, G. A. & A. M. ABDALLAH 1954. *Geology of the Abu Mireiwa district*. Geol. Surv. Egypt, 32pp.

MOUSTAFA, G. A. & M. K. AKAAD 1962. *Geology of the Hammash-Sufra district*. Geol. Surv. Egypt, Paper 12, 31pp.

MOUSTAFA, G. A. & G. M. AUSLENDER 1961. *Ayun Musa coal deposit, Sinai – geological report and estimation of reserves*. Internal Report, Geol. Surv. Egypt.

MOUSTAFA, G. A. & M. E. HILMY 1959. *Contribution to the geology and mineralogy of the Hammash copper deposit, south Eastern Desert of Egypt*. Geol. Surv. Egypt, Paper 5, 16pp.

MOUSTAFA, G. A., M. L. KABESH & A. M. ABDALLAH 1954. *Geology of Gebel El Ineigi district*. Geol. Surv. Egypt, 40pp.

MOUSTAFA, G. A. & M. YAROVOI 1961. *Geological report on calculation of coal reserves, Wadi Thora and Wadi Beda, west central Sinai*. Internal Report, Geol. Surv. Egypt.

MOUSTAFA, Y. S. 1974. Critical observations on the occurrence of Fayum fossil vertebrates. *Ann. Geol. Surv. Egypt* 4: 41-78.

MUNIER, C. H. 1983. Verarbeitung von Filmen für das Optronics Colorwrite System. *Berl. Geowiss. Abh.* 47(A): 33-37.

MURRAY, G. W. 1914. Ancient workings of amethysts. *Cairo Sci. J.* 8: 179.

MURRAY, G. W. 1933. *Sons of Ismael*. Routledge, London, 344pp.

MURRAY, G. W. 1951. The Egyptian climate. An historical outline. *Geograph. J.* 117: 422-434.

MURRAY, G. W. 1952. The water beneath the Egyptian Western Desert. *Geograph. J.* 118: 443-452.

MURRAY, G. W. 1967. Dare me to the desert. George Allen & Unwin, London, 214pp.

MURRIS, R. J. 1980. Middle East: stratigraphic evolution and oil habitat. *Bull. Am. Assoc. Petrol. Geol.* 64: 597-618.

MUSSA, M. A. 1979. The use of space imagery in mineral exploration. *Ann. Geol. Surv. Egypt* 9: 172-178.

MYERS, O. H. 1939. The Sir Robert Monod Expedition of the Egyptian Exploration Society. *Geograph. J.* 93: 278-291.

NAGA, M. A. 1984. Paleozoic and Mesozoic depocenters and hydrocarbon generating areas, northern Western Desert. *Proc. 7th Petrol. Explor. Seminar, EGPC*, Cairo: 269-287.

NAGATY, M. 1982. The seven reservoirs of the Shoab Ali field. *6th Petrol. Explor. Seminar, EGPC*, Cairo.

NAGY, R. M. 1978. *Geochemistry of Raba El Garrah granite pluton*. Ph.D. Thesis, Rice Univ., Houston, Texas, USA.

NAIM, G. M. 1983. Abu Ghalaga titanium, iron and vanadium ore reserves. *Trans. Mining Petrol. Assoc. Egypt* 39: 2-16.

NAIRN, A. E. M., R. RESSETAR & J. R. DAVIES 1980. Paleomagnetic results from Pan-African rocks of the Egyptian Eastern Desert. *Ann. Geol. Surv. Egypt* 19: 1013-1026.

NAKHLA, F. M. 1954. Notes on the mineralography of some Egyptian ore minerals. *Trans. Mining Petrol. Assoc. Egypt* 9: 129-165.

NAKHLA, F. M. 1958. Mineralogy of the Egyptian black sands and its applications. *Egypt. J. Geol.* 2: 1-22.

NAKHLA, F. M. 1961. The iron ore deposits of El-Bahariya Oasis, Egypt. *Econ. Geol.* 56: 1103-1111.

NAKHLA, F. M. & E. E. EL HINNAWI 1960. Contribution to the study of Egyptian barites. *Proc. Egypt. Acad. Sci.* 15: 11-16.

NAKHLA, F. M., M. K. HUSSEIN & S. A. TOMA 1972. Microstructures of Egyptian black sands ilmenite. *Chem. Erde* 31: 350-363.

NAKHLA, F. M., M. RAMSY & A. B. BASILI 1973. Contribution to the mineralogy and geochemistry of the sulphide mineralization at Gebel Derhib talc mine, Egypt. *N. Jb. Mineral. Abh.* 118: 149-158.

NAKHLA, F. M. & M. R. N. SHEHATA 1967. Contribution to the mineralogy and geochemistry of some iron ore deposits in Egypt. *Mineral. Deposita* 2: 357-371.

NAKKADY, S. E. 1950. A new foraminiferal fauna from the Esna Shales of Egypt. *J. Paleontology* 24: 675-692.

NAKKADY, S. E. 1955. The stratigraphy and geology of the district between the northern and southern Galala plateaux. *Bull. Inst. Egypte* 36: 254-268.

NAKKADY, S. E. 1959. Biostratigraphy of the Um Elghanayem section, Egypt. *Micropaleontology* 5: 453-472.

NASSEEF, A. O., A. R. BAKOR & A. H. HAHSAD 1980. Petrography of possible ophiolite rocks along the Qift-Quseir road, Eastern Desert, Egypt. *Bull. Inst. Applied Geol., King Abdul Aziz Univ., Jeddah* 3(4): 157-168.

NASSEEF, A. O. & I. G. GASS 1977. Granite and metamorphic rocks of the Taif area, western Saudi Arabia. *Bull. Geol. Soc. Am.* 88: 1721-1730.

NASSIM, G. L. 1949. The discovery of nickel in Egypt. *Econ. Geol.* 44: 143-150.

NEARY, C. R., I. G. GASS & B. J. CAVANAGH 1976. Granitic association of northeastern Sudan. *Bull. Geol. Soc. Am.* 87: 1501-1512.

NEEV, D. 1975. Tectonic evolution of the Middle East and the levantine basin (easternmost Mediterranean). *Geology* 3: 683-686.

NEEV, D. 1977. The Pelusium line – a major transcontinental shear. *Tectonophysics* 38: T1-T8.

NEEV, D. & G. M. FRIEDMAN 1978. Late Holocene tectonic activity along the margins of the Sinai subplate. *Science* 202: 427-429.

NEEV, D., J. K. HALL & J. M. SAUL 1982. The Pelusium megashear system across Africa and associated lineament swarms. *J. Geophys. Res.* 87: 1015-1030.

NELSON, R. A. 1986. *Oblique extension and its effect on structure in the Western Desert of Egypt*. Internal Report 27, Africa-Middle East region, AMOCO Oil Co.

NEUMANN, E. R. & I. B. RAMBERG 1978. Paleorifts – concluding remarks. In: E. R. Neumann & I. B. Ramberg (eds), *Tectonics and geophysics of continental rifts*. D. Reidel Publishing Co.: 409-424.

NEWTON, R. B. 1898. Notes on some lower Tertiary shells from Egypt. *Geol. Mag.* 6 (decade 3): 352-359.

NEWTON, R. B. 1909. On some fossils from the Nubian Sandstone of Egypt. *Geol. Mag.* 5 (Decade 4): 531-541.

NIAZI, M. 1968. Crustal thickness in the central Saudi Arabian Peninsula. *Geophys. J. Roy. Astronom. Soc.* 15: 545-547.

NIELSEN, E. (co-ordinator) 1974. *Proceedings of Seminar on Nile delta sedimentology, Alexandria*. UNESCO/UNDP.

NIR, Y. 1982. Asia, Middle East, coastal morphology: Israel and Sinai. In: M. L. Schwartz (ed.), *The encyclopedia of beaches and coastal environments*, Hutchinson Ross, Stroudsburg, Pennsylvania, USA: 86-98.

NORTON, P. 1967. *Rock stratigraphic nomenclature of the Western Desert*. Internal Report, Pan-American Oil Co., Cairo.

NOUR, S. 1984. *Groundwater resources in the east Uweinat region, hydrogeological conditions*. Internal Report, General Petrol. Co., Cairo.

NOWEIR, A. M. 1968. *Geology of El Hammamat-Wadi Um Seleimat area, Eastern Desert*. Ph.D. Thesis, Assiut Univ.

NOWEIR, A. M. & M. A. TAKLA 1975. Studies on the synorogenic plutonites of the central Eastern Desert between Qena-Safaga and Idfu-Mersa Alam roads. *Bull. Inst. Desert Egypte* 25(1-2): 77-99.

NUR, A. & Z. BEN AVRAHAM 1978. The eastern Mediterranean and the Levant, tectonics and continental collision. *Tectonophysics* 46: 297-311.

O'CONNER, J. T. 1965. A classification for quartz-rich igneous rocks based on feldspar ratios. *US Geol. Surv., Prof. Paper* 525-B: 79-84.

OESLEBY, T. W. 1981. *Report on Mesozoic carbonate sections, Egypt and Sinai*. Internal Report, Denver Research Center, Marathon Oil Co.

OESLEBY, T. W. 1981. *Gebel Yelleq field studies*. Internal Report, Conoco-Marathon, Cairo.

OLSON, S. L. & D. T. RASMUSSEN 1986. Paleoenvironment of the earliest hominoids: new evidence from the Oligocene avifauna of Egypt. *Science* 233: 1202-1204.

OMARA, S. 1956. New foraminifera from the Cenomanian of Sinai, Egypt. *J. Paleontology* 30: 883-890.

OMARA, S. 1965. A micropaleontological approach to the stratigraphy of the Carboniferous exposures of the Gulf of Suez region. *N. Jb. Geol. Palaeontol.*, Mh (1965), 6: 409-419.

OMARA, S. 1971. Early Carboniferous tabulate corals from Um Bogma area, southwestern Sinai, Egypt. *Riv. Ital. Palaeontol.* 77: 141-154.

OMARA, S. 1972. An early Cambrian outcrop in southwestern Sinai, Egypt. *N. Jb. Geol. Palaeontol.*, Mh (1972), 5: 306-314.

OMARA, S. & R. CONIL 1965. Lower Carboniferous foraminifers from southwestern Sinai, Egypt. *Ann. Soc. Geol. Belgique* 5: B221-242.

OMARA, S., M. R. EL TAHLAWI & H. ABDEL KIREEM 1973. Detailed geological mapping of the area between latitudes of Sohag and Girga, east of the Nile Valley. *Bull. Fac. Sci., Assiut Univ.* 1: 149-166.

OMARA, S., I. HEMIDA & S. SANAD 1970. Structure and hydrogeology of Farafra Oasis, Western Desert, UAR. *7th Arab Petrol. Congr., Kuwait*, Paper 65(B-3), 15pp.

OMARA, S. & A. KENAWY 1966. Upper Carboniferous microfossils from Wadi Araba, Eastern Desert, Egypt. *N. Jb. Geol. Paleontol. Abh.* 124: 56-83.

OMARA, S. & A. KENAWY 1979. *Craterocamerina*, a new nummulite genus from the Nile Valley and El Fayoum, Egypt. *N. Jb. Geol. Palaeontol. Abh.* 158: 123-138.

OMARA, S. & A. KENAWY 1984. *Nummulites rholfsi*, une nouvelle espèce de *Nummulites* du groupe *gizehensis* rencontrée dans la base de l'Eocéne superieur de la vallée du Nil. *Rev. Micropaleontol.* 27: 54-60.

OMARA, S., H. H. MANSOUR, M. M. YOUSSEF & H. KHALIFA 1977. Stratigraphy, paleoenvironment and structural features of the area to the east of Beni Mazar, Egypt. *Bull. Fac. Sci., Assiut Univ.* 6: 171-198.

OMARA, S. & K. OUDA 1969. Pliocene foraminifera from the subsurface rocks of Burg El Arab well no. 2, Western Desert, Egypt. *Proc. 3rd Afr. Micropaleontol. Colloq.*, Cairo: 581-601.

OMARA, S. & K. OUDA 1972. Review of the lithostratigraphy of the Oligocene and Miocene in the northern Western Desert. *8th Arab Petrol. Congr.*, Algiers.

OMARA, S., E. R. PHILOBBOS & S. S. HANNA 1972. Contribution to the geomorphology and goelogy of the area east of Minia. *Bull. Soc. Geograph. Egypte* 45/46: 125-147.

OMARA, S., E. R. PHILOBBOS & H. H. MANSOUR 1976. Contribution to the geology of the Dakhla oasis area, Western Desert, Egypt. *Bull. Fac. Sci., Assiut Univ.* 5: 319-339.

OMARA, S. & S. SANAD 1975. Rock stratigraphy and structural features of the area between wadi El Natrun and the Moghra depression (Western Desert, Egypt). *Geol. Jb.* 16: 45-73.

OMARA, S. & G. SCHULTZ 1965. A lower Carboniferous microflora from southwestern Sinai. *Palaeontographica* 117(A): 47-58.

OMARA, S. & E. F. VANGEROW 1965. Carboniferous (Westphalian) foraminifera from Abu El Darag, Eastern Desert, Egypt. *Geol. Mijnbouw* 44: 87-93.

OMARA, S., E. F. VANGEROW & A. KENAWY 1966. Neue Funde von Foraminiferen in Ober-Karbon, von Abu Darag, Aegypten. *Palaeontol. Z.* 40: 244-256.

ORWIG, E. R. 1982. Tectonic framework of northern Egypt and the eastern Mediterranean region. *6th Petrol. Explor. seminar, EGPC*, Cairo.

OSBORN, H. F. 1908. New fossil mammals from the Fayum Oligocene, Egypt. *Bull. Am. Mus. Nat. Hist.* 24: 265-272.

OSMAN, A. & A. DARDIR 1986. The wall rock alterations of El Barramiya gold mine, Eastern Desert, Egypt. *5th Symp. Precambrian and Development, IGCP*, Cairo (Abstr.).

OSMAN, A. & A. DARDIR 1986. On the gold-bearing rocks and alteration zones at Hanglia gold mine, Eastern Desert, Egypt. *5th Symp. Precambrian and Development, EGCP*, Cairo (Abstr.).

OWEN, H. G. 1983. *Atlas of continental displacement*. Cambridge Univ. Press, 159pp.

PACHUR, H. J. 1974. Geomorphologische Untersuchungen im Raum der Serir Tibesti (Zentralsahara). *Berl. Geograph. Abh.* 17: 1-62.

PACHUR, H. J. 1982. Das Abflussystem des Djebel Dalmar – eine Singularität? *Würzb. Geograph. Abh.* 56: 93-110.

PACHUR, H. J. 1987. Vergessene Flüsse und Seen der Ostsahara. *Geowiss. Unserer Zeit* 5(2): 55-64.

PACHUR, H. J. & G. BRAUN 1980. The paleoclimate of the central Sahara, Libya and the Libyan Desert. In: E.M. van Zinderen Bakker Sr. & J.A. Coetzee (eds), *Palaeoecology of Africa and the surroung islands*, Balkema 12: 351-364.

PACHUR, H. J. & G. BRAUN 1986. Drainage systems, lakes and ergs in the eastern Sahara as indicators of Quaternary climatic dynamics. *Berl. Geowiss. Abh.* 72(A): 3-16.

PACHUR, H. J. & S. KROEPLIN 1987. Wadi Howar, Paleoclimatic evidence from an extinct river system in the southwestern Sahara. *Science* 237: 298-300.

PACHUR, H. J. & H. P. ROEPPER 1984. Die Bedeutung paläoklimatischer Befunde aus den Flachbereichen der östlichen Sahara und des nördlichen Sudan. *Z. Geomorph. N.F.*, 50: 59-78.

PARKER, J. R. 1982. Hydrocarbon habitat of the Western Desert, Egypt. *6th Petrol. Explor. Seminar, EGPC*, Cairo.

PARNES, A. 1964. Coniacian ammonites from the Negev (southern Israel). *Bull. Geol. Surv. Isr.* 39: 1-42.

PARNES, A. 1974. Biostratigraphic synchronization of the middle Jurassic in Makhtash Ramon, Gebel Maghara and Morocco. *Annual Meeting, Geol. Soc. Isr.*: 14-15 (Abstr.).

PATTERSON, B. 1965. The fossil elephant shrews (Family

Macroscelididae). *Bull. Mus. Comp. Zool., Harvard Univ.* 133.

PAULISSEN, E. & P. M. VERMEERSCH 1987. Earth, man and climate in the Egyptian Nile Valley during the Pleistocene. In: A.E. Close (ed.), *Prehistory of arid north Africa*, Southern Methodist Univ. Press: 29-67.

PEARCE, J. A. 1976. Statistical analysis of major element patterns in basalts. *J. Petrology* 17: 15-43.

PEARCE, J. A., N. B. W. HARRIS & A. G. TINDLE 1984. Trace element discrimination diagrams for the tectonic interpretation of granitic rocks. *J. Petrology* 25: 956-983.

PEEL, R. F. 1941. Denudational landforms of the central Libyan Desert. *J. Geomorph.* 4: 3-23.

PERCH-NIELSEN, K., A. SADEK, M. G. BARAKAT & F. TELEB 1978. Late Cretaceous and early Tertiary calcareous nannofossil and planktonic foraminiferal zones from Egypt. *6th Afr. Micropaleontol. Colloq., Tunis, Ann. des Mines et Geol.* 28(II): 337-403.

PERRY, S. K. & S. SCHAMEL 1985. *Structural model of the southwestern Gulf of Suez and Gemsa plain, Egypt.* Technical Report, Earth Sci. and Resources Inst., Univ. South Carolina.

PERRY, S. K. & S. SCHAMEL 1986. *The role of low angle normal faulting and isostatic response in the evaluation of the Suez rift, Egypt.* Technical Report, Earth Sci. and Resources Inst., Univ. South Carolina.

PETRASCHECK, W. E. & F. M. NAKHLA 1961. Contribution to the study of Um Bogma coal, west central Sinai. *Egypt. J. Geol.* 5: 81-87.

PETRO, W. L., T. A. VOGEL & J. T. WILBAND 1979. Major element chemistry of plutonic rock suites from compressional and extensional plate boundaries. *Chem. Geol.* 26: 217-235.

PFANNENSTIEL, M. 1953. Das Quartär der Levante II. Die Entstehung der ägyptischen Oasendepressionen. *Abh. Akad. Wissen. u. Lit. Math.-Naturw. Kl.* 7: 337-411.

PHILIP, G. & F. A. ASAAD 1972. Contributions to the subsurface geology of the Nubia Sandstone: Lithology and zonation of some wells, south of Beris oasis, Kharga oasis, Western Desert, Egypt. *Bull. Fac. Sci., Cairo Univ.* 45: 287-307.

PHILLIPS, J. & D. A. ROSS 1970. Continuous seismic reflection profiles in the Red Sea. *Phil. Trans. Roy. Soc. London* 267(A): 153-180.

PHILOBBOS, E. R. & A. A. EL HADDAD 1983. Contribution to Miocene and Pliocene lithostratigraphy of the Red Sea coastal zone. *21st Annual Meeting, Geol. Soc. Egypt*: 5-6 (Abstr.).

PHILOBOS, E. R. & K. E. K. HASSAN 1975. The contribution of palaeosoil to Egyptian lithostratigraphy. *Nature* 253:33.

PHILOBBOS, E. R. & E. A. KEHEILA 1979. Depositional environments of the middle Eocene in the area southeast of Minia, Egypt. *Ann. Geol. Surv. Egypt* 9: 523-550.

PHILOBBOS, E. R., H. H. MANSOUR & H. A. MOSTAFA 1986. Sedimentology of lower Eocene carbonate algal buildup northeast of Sohag, Nile Valley. *Bull. Sohag Univ. Pure and Applied Sci.* 2: 41-85.

PICARD, L. 1942. New Cambrian fossils and Paleozoic problematica from the Dead Sea and Arabia. *Bull. Geol. Dept. Hebrew Univ.*, 3.

PICKFORD, M. 1986. Première decouverte d'une faune mammalienne terrestre paleogène d'Afrique subsaharienne. *C.R. Acad. Sci.*, Ser. II, 302: 1205-1210.

PITCHER, W. S. 1983. Granite: topology, geological environment and melting relationships. In: M. P. Atherton & C. G. Gribble (eds), *Migmatites, melting and metamorphism*, Shiva Publ., Cheshire: 227-285.

POHL, W. 1979. Metallogenic/minerogenic analysis – contribution to the differentiation between Mozambiquian basement and Pan-African superstructure in the Red Sea region. *Ann. Geol. Surv. Egypt* 9: 32-44.

POHL, W. 1984. Large scale metallogenic features of the Pan African in east Africa, Nubia and Arabia. *Bull. Fac. Earth Sci., King Abdul Aziz Univ., Jeddah* 6: 592-601.

PÖHLMANN, G. & B. MEISSNER (eds) 1984. Basiskarten für die thematische Kartierung arider Gebiete. *Berl. Geowiss. Abh.* 2(C): 106pp.

PÖHLMANN, G., B. MEISSNER & F. K. LIST (Eds.), 1981. Egypt 1:250 000 – Working sheets. NG 35-B Farafra: Sheets Qasr Farafra; Bir Karawein; Abu Minqar; Naqb el Khashabi; NG 35-C Mut: Sheets Qur el Malik; Abu Ballas; Six Hills. Berlin.

PÖHLMANN, G., B. MEISSNER & F. K. LIST (Eds.), 1982. Egypt 1:250 000 – Working sheets. NF 35-NE Bir Tarfawi: Sheets foot Hills; Bir Misaha; Nusab al Balgum; Bir Tarfawi; NG 36-NW Asyut: Sheets Asyut; Gebel el Urf; Tahta; Sohag; NG 36-SW Aswan: Sheets El Kharga; Baris; Luxor; Aswan; NH-35 NE Alexandria: Sheets Matruh; Alexandria; Bir Khalda; El Rammak; Qur el Hamra; Bahartiya Oasis. Berlin.

PÖHLMANN, G., B. MEISSNER & F. K. LIST (Eds.), 1983. Egypt 1:250 000 – Working sheets. NF 36-NW Al Saad al Ali: Sheets Bir Murr; Adindan; The High Dam; Abu Simbil; NF 36-NW NE Berenice: Sheets Gebel Hadaiyib; Wadi Gabgaba; Ras Banas; Marsa Sha'ab; NG 36-NE Gebel Hamata: Sheets Wadi el Barramiya; Wadi Shait; Marsa el Alam; Gebel Hamata; NH 35-NW Al Salum: Sheets Salum; Sidi Barrani, Bir Bayly, Bir Fuad; NH 35-SW Siwa: Sheets Siwa; El Bahrein; NH 36-SW Beni Suef: Sheets El Fayum; Suez; El Minya; Gebel Gharib. Berlin.

PÖHLMANN, G., B. MEISSNER & F. K. LIST (Eds.), 1984. Egypt 1:250 000 – Working sheets. NF 35-NW Al Gilf al Kebir Plateau: Sheets Wadi Sura, Wadi el-Dayik; Gebel Arkenu; Wadi el-Firaq; NG 35-NW Sakhret el Amud: Sheets El Tallein; Sakhret el Amud; Wadi el Blata, west el-Mingar; NG-35-SW Wadi el Gubba: Sheets Wadi el Gubba; Qaret el Hanash; Wadi Talh; NG 35C Al Dakhla: Sheet Mut; NH 35-SW Siwa: Sheets west el Bahrein; Girba Oasis; NH 36 NW Cairo: Sheets El Mansura; Port Said; Ciaro; Ismailiya; NH 36-NE N. Sinai: Sheets El Arish; El Qusaima; Maan; NH 36-SE S. Sinai: Sheets El Tor; Qal'at el Nakhl, El Aqaba; Gebel el Loz. Berlin.

POIRIER, J. P. & M. A. TAHER 1980. Historical seismicity in the Near and Middle East, North Africa and Spain from Arabic documents (7th-8th centuries). *Bull. Seism. Soc. Am.* 70: 2185-2201.

POIRIER, J. P., B. A. ROMANOWICZ & M. A. TAHER 1980. Large historical earthquakes and seismic risk in northwest Syria. *Nature* 285: 217-220.

POUMOT, C. & J. BOUROULLEC 1984. Palynoplanktological contribution to the Miocene-Pliocene environmental and paleoclimatological conditions of the Nile Delta area. *Proc. 7th Petrol. Explor. Seminar, EGPC*, Cairo: 325-334.

POWELL DUFFRYN TECHNICAL SERVICES LTD (London)

1963. *Mining economic report on the Ayun Musa coal field, UAR.* Technical Report to the General Organization for Executing the Five Year Industrial Plan (unpublished).

PRIOR, S. W. 1976. Matruh basin; possible failed arm of Mesozoic crustal rift. *5th Petrol. Explor. Seminar, EGPC*, Cairo.

PRODEHL, C. 1985. Interpretation of a seismic-refraction survey across the Arabian shield in western Saudi Arabia. *Tectonophysics* 111: 247-282.

PRUCHA, J. J., J. A. GRAHAM & R. P. NICKELSEN 1965. Basement controlled deformation in Wyoming province of the Rocky Mountains foreland. *Bull. Am. Assoc. Petrol. Geol.* 49: 966-992.

QUAAS, A. 1902. Beitrag zur Kenntnis der Fauna der obersten Kreidebildungen in der Libyschen Wüste (Overwegischichten und Blaetterthone). *Palaeontographica* 30: 153-334.

QUENNELL, A. M. 1958. The structure and evolution of the Dead Sea rift. *Quart. J. Geol. Soc. London* 64: 1-24.

QUENNELL, A. M. 1984. The western Arabian rift system. In: J. E. Dixon & A. H. F. Robertson, *The goelogical evolution of the eastern Mediterranean*, Blackwell Scientific. Publishers Oxford: 775-788.

RADINSKY, L. B. 1973. *Aegyptopithecus* endocast: oldest record of a pongid brain. *Am. J. Phys. Anthropol.* 39: 239-247.

RAGAB, A. I., M. Y. MENEISY & R. M. TAHER 1978. Contributions to the petrogenesis and age of Aswan granitic rocks, Egypt. *N. Jb. Mineral. Abh.* 133: 71-87.

RAMBERG, H. 1981. The role of gravity in orogenic belts. In: K. R. McClay & N. J. Rice (eds), *Thrust and nappe tectonics.* Geol. Soc. London, Special Publ. 9: 15-140.

RAMBERG, I. B. & P. MORGAN 1984. Physical characteristics and evolutionary trends of continental rift. *Proc. 27th Intl. Geol. Congr.* 7, Tectonics: 165-218.

RAMSAY, C. R., D. B. STOESER & A. R. DRYSDAL 1986. Guidelines to classification and nomenclature of Arabian felsic plutonic rocks. *J. Afr. Earth Sci.* 4: 13-20.

RAMSAY, C. R., A. R. DRYSDAL & M. D. CLARK 1986. Felsic plutonic rocks of the Medyan region, Kingdom of Saudi Arabia: I. Distribution, classification and resource potential. *J. Afr. Earth Sci.* 4: 63-77.

RAMZY, E. 1964. *Geology and petrography of the phosphates of the Quseir-Safaga region.* M. Sc. Thesis, Assiut Univ.

RANZ, E. & S. SCHNEIDER 1970. Der Äquidensitenfilm als Hilfsmittel bei der Photointerpretation. *Bildmess. u. Luftbildwes.* 38: 123-134.

RANZ, E. & S. SCHNEIDER 1972. Rasteräquidensiten in der Luftbild-Interpretation. *Bildmess. u. Luftbildwes.* 40: 189-193.

RASA, O. A. E. 1977. The ethnology and sociology of the dwarf mongoose (*Helogale undulata rufula*). *Z. Tierpsychol.* 43: 337-406.

RASMUSSEN, D. T., S. L. OLSON & E. L. SIMONS 1987. Fossil birds from the Oligocene Jebel Qatrani Formation, Fayum Province, Egypt. *Smith. Contr. Paleobiol.* 62: 1-21.

RASMUSSEN, D. T. & E. L. SIMONS 1988. New Oligocene hydracoids from Egypt. *J. Vert. Paleontol.* 8: 67-83.

RASMUSSEN, D. T. & E. L. SIMONS (in press.). New specimens of *Oligopithecus savagei*, early Oigocene primate from the Fayum Egypt. *Folia Primat.*

RASMY, A. H. 1974. *Geological and mineralogical study of corundum, anthophyllite, phlogopite and vermiculite from Hafafit, Egypt.* Ph.D. Thesis, Ain Shams Univ., Cairo.

RASMY, A. H. 1982. Mineralogy of copper-nickel mineralization at Akarem area. *Ann. Geol. Surv. Egypt* 12: 141-162.

RASMY, M. & A. A. MONTASSER 1982. Some geological indications of mineralization in four areas near Red Sea coast of Qoseir. *Bull. Nat. Res. Council*, Cairo 7: 119.

RAZVALYAYEV, A. & G. SHAKHOV 1978. Tectonic significance of location of some ring complexes in the Red Sea rift zone. *Bull. Moscow Soc., Trans. Geol. Sec.* 53(2): 56-66.

READ, H. H. 1955. *The granite problem.* Murby, London, 430pp.

REFAAT, A. M., M. L. KABESH & Z. M. ABDALLAH 1978. Petrographic studies of Ras Barud granitic rocks, Safaga district, Eastern Desert, Egypt. *J. Univ. Kuwait (Sci.)* 5: 181-197.

REFAAT, A. M., M. L. KABESH, M. E. HILMY & Z. M. ABDALLAH 1978. Geochemistry of granitic rocks in Ras Barud area, Eastern Desert, Egypt. *J. Univ. Kuwait (Sci.)* 5: 129-150.

REINACH, A. VON 1903. Vorläufige Mitteilungen über neue Schildkröten aus dem ägyptischen Tertiär. *Zool. Anz.* 26: 700.

RENNER, O. 1907. *Teichosperma*, eine Monokotylenfrucht aus dem Tertiar Aegyptens. *Beitr. z. Palaeontol. Geol. Oesterr. Ung. u.d. Orients* 20: 217-220.

RENOLDS, M. L. 1979. Geology of the northern Gulf of Suez. *Ann. Geol. Surv. Egypt* 9: 323-343.

RESSETAR, R. & J. R. MONRAD 1983. Chemical composition and tectonic setting of the Dokhan volcanic formation, Eastern Desert, Egypt. *J. Afr. Earth Sci.* 1: 103-111.

RESSETAR, R. & A. E. M. NAIRN 1980. Two phases of Cretaceous-Tertiary magmatism in the Egyptian Eastern Desert: Paleomagnetic and K-Ar evidence. *Ann. Geol. Surv. Egypt* 10: 997-1011.

REYMER, A. P. S. 1983. Metamorphism and tectonics of a Pan-African terrain in southeastern Sinai. *Precambrian Res.* 19: 225-238.

REYMER, A. P. S., A. MATHEWS & O. NAVON 1984. Pressure-temperature conditions in the Wadi Kid metamorphic complex: Implications for the Pan-African event in SE Sinai. *Contr. Mineral. Petrology* 85: 336-345.

REYMER, A. P. S. & G. SCHUBERT 1984. Phanerozoic addition rates to the continental crust and crustal growth. *Tectonics* 3: 63-77.

RHYS-DAVIS , P. 1984. *Oligocene interpretation, N. Sinai.* Internal Report, Conoco, Egypt.

RIAD, S. 1977. Shear zones in north Egypt interpreted from gravity data. *Geophysics* 42: 1207-1214.

RIAD, S., H. A. EL ETR & A. MOKHLES 1981. Basement tectonics of northern Egypt as interpreted from gravity data. *Proc. 4th Intl. Conf. Basement Tectonics* 4: 209-220.

RIAD, S., E. REFAI & M. GHALIB 1981. Bouguer anomalies and crustal structure in the eastern Mediterranean. *Tectonophysics* 71: 253-266.

RIAD, S. & H. MEYERS 1985. *Earthquake catalog for the Middle East countries 1900-1983.* World Data Center for solid earth Geophysics, Rep. SE-40, National Oceanic and Atmospheric Administration (NOAA), US Dept. of Commerce, Boulder, Colorado, USA.

RIAD, S., H. MEYERS & C. KISSLINGER 1985. On the seismicity of the Middle East. *Proc. 14th Symp. Hist. Seismograms and Earthquakes. Intl. Assoc. Seismol. and Physics of the Earth's Interior* (IASPEI)/UNESCO: 36(1)-36(22).

RICHARDSON, S. S. & C. G. A. HARRISON 1976. Opening of the Red Sea with two poles of rotation. *Earth Planet. Sci. Lett.* 30: 135-142.

RICHARDSON, E. S. & C. G. A. HARRISON 1976. Opening of the Red Sea with two poles of rotation – reply. *Earth Planet. Sci. Lett.* 33: 173-175.

RICHTER, A. 1986. Geologie der metamorphen und magmatischen Gesteine im Gebiet zwischen Gebel Uweinat und Gebel Kamil, SW Ägypten/NW-Sudan. *Berl. Geowiss. Abh.* 73(A): 1-201.

RICHTER, R. & E. RICHTER 1941. Das Kambrium am Toten Meer und die älteste Tethys. *Abh. Senckenberg. Naturf. Ges.* 460: 1-50.

RIES, A. C., R. M. SCHACKELTON, R. H. GRAHAM & W. R. FITCHES 1983. Pan-African structures, ophiolites and melange in the Eastern Desert of Egypt, a traverse at 26° N. *J. Geol. Soc. London* 140: 75-95.

RITTMANN, A. 1953. Some remarks on the geology of Aswan. *Bull. Inst. Desert Egypte* 3(2): 53-64.

RITTMANN, A. 1954. Remarks on eruption mechanism of the Tertiary volcanoes of Egypt. *Bull. Volcanol.* (ser. 2), 15: 109-117.

RITTMANN, A. 1958. Geosynclinal volcanism, ophiolites and Barramiya rocks. *Egypt. J. Geol.* 2: 61-65.

RITTMANN, A. & F. M. NAKHLA 1958. Contributions to the study of the Egyptian black sands. *Egypt. J. Chem.* 1: 127-135.

RIZZINI, A., F. VEZZANI, V. COCCETTA & G. MILAD 1978. Stratigraphy and sedimentation of a Neogene-Quaternary section in the Nile delta area. *Marine Geol.* 27: 327-348.

ROBERTSON, A. H. F. & J. E. DIXON 1984. Introduction: Aspects of the geological evolution of the eastern Mediterranean. In: J. E. Dixon & A. H. F. Richardson (eds), *The geological evolution of the eastern Mediterranean*, Blackwell Scientific Publishers, Oxford: 1-74.

ROBERTSON RESEARCH INTERNATIONAL (RRI) & ASSOCIATED RESEARCH CONSULTANTS in association with SCOTT PICKFORD & ASSOCIATES LIMITED and ERC ENERGY RESOURCE CONSULTANTS LIMITED 1982. *Petroleum potential evaluation, Western Desert, ARE.* Report prepared for EGPC, 8 vol.

ROBSON, D. A. 1971. The structure of the Gulf of Suez (Clysmic) rift, with special reference to the eastern side. *J. Geol. Soc. London* 127: 247-276.

ROCCI, G. 1965. Essai d'interpretation des mésures géochronologiques. La structure de l'ouest africain. *Sci. Terre* 10: 461-479.

ROE, D. A., J. W. OLSEN, J. R. UNDERWOOD & R. T. GIEGENGACK 1982. A handaxe of Libyan Desert glass. *Antiquity* 56: 406-410.

ROGERS, J., GHUMA, R. NAGY, J. GREENBERG & P. FULLAGER 1978. Plutonism in Pan-African belts and the geological evolution of northeast Africa. *Earth Planet. Sci. Lett.* 39: 109-17.

ROHRBACK, B. G. 1980. *Geochemistry of crude oil from Gulf of Suez and Western Desert oil provinces, Egypt.* Internal Report no. 69, Cities Service Co., Tulsa, Oklahoma, USA.

ROHRBACK, B. G. 1980. *Organic geochemistry of cores and cuttings from the Gulf of Suez and Western Desert oil provinces, Egypt.* Internal Report no. 75, Cities Service Co., Tulsa, Oklahoma, USA.

ROHRBACK, B. G. 1982. Crude oil geochemistry of the Gulf of Suez. *6th Petrol. Explor. Seminar, EGPC*, Cairo.

ROHRBACK, B. G. 1983. Crude oil geochemistry of the Gulf of Suez. In: *Advances of organic geochemistry,* John Wiley: 39-48.

ROOD, J. P. 1975. Population dynamics and food habits of the banded mongoose. *East Afr. Wildlife J.* 13: 89-111.

ROOD, J. P. 1978. Dwarf mongooses, helpers at the den. *Z. Tierpsychol.* 48: 277-287.

ROSE, J., S. ASKARY & M. ANDRIANI 1983. Ras Budran field, petrographical study and core calibration. *8th Petrol. Production Seminar, EGPC*, Cairo.

ROSENBERGER, A. L. 1985. In favour of the necrolemur-tarsier hypothesis. *Folia Primatol.* 45: 179-194.

ROSS, D. & J. SCHLEE 1973. Shallow structure and geologic development of the southern Red Sea. *Bull. Geol. Soc. Am.* 84: 3827-3848.

ROSS, D. & E. UCHUPI 1977. Structure and sedimentary history of northeastern Mediterranean Sea-Nile cone area. *Bull. Am. Assoc. Petrol. Geol.* 61: 872-902.

ROSSER, H. A. 1975. A detailed magnetic survey of the southern Red Sea. *Geol. Jb.* 13: 131-153.

ROTHE, J. P. 1970. *Moyen Orient et Afrique du nord.* UNESCO Mission d'Information Seismologique 1759/BMS, RD/SCE.

ROWAN, L. C. & E. H. LATHRAM 1980. Mineral exploration. In: B. S. Siegal & A. R. Gillespie (eds), *Remote sensing in geology*, Wiley: 553-605.

ROYDS, J. S., J. F. MASON & D. B. EICHER 1975. A history of exploration for petroleum. In: E. W. Owen (ed.), Trek of the oil finders: *A history of exploration for petroleum,* Am. Assoc. Petrol. Geol.: 1419-1436.

RRI (see under Robertson Research International).

RÜSSEGGER, J. R. 1837. Kreide und Sandstein: Einfluss von Granit auf letztern: *N. Jb. Mineral.* 1837: 665-669.

RÜSSEGGER, J. R. 1838. Geognostische Beobachtungen in Ägypten. *N. Jb. Mineral.* 1838: 623-637.

RYAN, W. B. F. 1978. Messinian badlands on the southeastern margin of the Mediterranean Sea. *Marine Geol.* 27: 349-363.

RYAN, W. B. F., D. J. STANLEY, J. B. HERSEY, D. A. FAHLQUIST & T. D. ALLAN 1970. The tectonics and geology of the Mediterranean Sea. In: A. E. Maxwell (ed.), *The sea*, Wiley, 4(2): 387-492.

RYAN, W. B. F. & K. J. HSU 1973. *Initial reports of the Deep Sea Drilling Project, Leg 13.* US Government Printing Office, Washington, DC.

SAAD, S. I. & G. GHAZALY 1976. Palynological studies in Nubia Sandstone from Kharga Oasis. *Pollen et Spores* 18: 407-470.

SABET, A. H. 1958. Geology of some dolerite flows, south of El Qoseir. *Egypt. J. Geol.* 2: 45-58.

SABET, A. H. 1962. *An example of photo interpretation of crystalline rocks.* ITC Publ., B, 14/15: 34pp, Delft.

SABET, A. H. 1969. *Basis of application of aerial photographs in geology.* Geol. Surv. Egypt, Paper 48, 76pp.

SABET, A. H. 1972. On the stratigraphy of basement rocks

of Egypt. *Ann. Geol. Surv. Egypt* 2: 79-102.

SABET, A. H., V. V. BESSONENKO & B. A. BYKOV 1976. The intrusive complexes of the central Eastern Desert of Egypt. *Ann. Geol. Surv. Egypt* 6: 53-73.

SABET, A. H., V. CHABANENCO & V. B. TSOGOEV 1973. Tin-tungsten and rare metal mineralization in central Eastern Desert of Egypt. *Ann. Geol. Surv. Egypt* 3: 75-86.

SABET, A. H., S. EL GABY & A. A. ZALATA 1972. Geology of basement rocks in the northern parts of El Shayib and Safaga sheets, Eastern Desert. *Ann. Geol. Surv. Egypt* 2: 111-128.

SABET, A. H., H. KOTB, M. M. ALY & A. H. HATHOUT 1980. Geochemistry of Abu Dabbab apogranite, central Eastern Desert, Egypt. *Bull. Inst. Applied Geol., King Abdul Aziz Univ., Jeddah* 3(4): 115-125.

SABET, A. H. & V. B. TSOGOEV 1973. Problems of geological and economic evaluation of tantalum deposits in apogranites during stages of prospection and exploration. *Ann. Geol. Surv. Egypt* 3: 87-107.

SABET, A. H., V. B. TSOGOEV, L. M. BABURIN & N. ZHORKOV 1976. Manifestation of rare metal mineralization of apogranite type in the central Eastern Desert of Egypt. *Ann. Geol. Surv. Egypt* 6: 75-93.

SABET, A. H., V. B. TSOGOEV, V. V. BESSONENKO, L. M. BABURIN & V. I. POKRYSHKIN 1976. Some geological and tectonic peculiarities of the central Eastern Desert. *Ann. Geol. Surv. Egypt* 6: 33-52.

SABET, A. H., V. B. TSOGOEV, V. A. BORDONOSOV, V. A. BELOSHITSKY, D. N. KUZENTSOV & H. A. HAKIM 1976. On some geological and structural peculiarities of localization of polymetal mineralization of the middle Miocene sediments of the Red Sea coast. *Ann. Geol. Surv. Egypt* 6: 223-236.

SABET, A. H. & G. S. SALEEB-ROUFAIEL 1968. On the geology and paragenesis of lead mineralization at Seweigat El Soda and Talet Id, Eastern Desert. *Egypt. J. Geol.* 12: 1-9.

SABET, A. H., V. B. TSOGOEV, L. P. SARIN, S. A. AZAZI, M. A. EL BEDEIWI & G. A. GHOBRIAL 1976. Tin-tantalum deposit of Abu Dabbab. *Ann. Geol. Surv. Egypt* 6: 93-118.

SADEK, A. 1968. Contribution to the Miocene stratigraphy of Egypt by means of miogypsinids. *Proc. 3rd Afr. Micropaleontol. Colloq.*, Cairo: 509-514.

SADEK, A. 1971. Determination of the upper Paleocene-lower Eocene boundary by means of calcareous nannofossils. *Rev. Espan. Micropaleontol.* 3: 277-282.

SADEK, A. & T. M. ABDEL RAZIK 1970. Zonal stratigraphy of the lower Tertiary of Gebel Um El Huetat, Red Sea by means of nannofossils. *7th Arab. Petrol. Congr.*, Kuwait, Paper 51, 16pp.

SADEK, H. 1926. *The geography and geology of the district between Gebel Ataqa and El Galala El Bahariya (Gulf of Suez).* Geol. Surv. Egypt, 120pp.

SADEK, H. 1928. The principal structural features of the Peninsula of Sinai. *14th Intl. Geol. Congr.*, Madrid (1926), fasc. 3: 895-900.

SADEK, H. 1959. *The Miocene in the Gulf of Suez region (Egypt).* Geol. Surv. Egypt, 118pp.

SADEK, M. 1944. *Nuweiba tin deposit; preliminary report.* Internal Report, Mines and Quarries Dept., Cairo.

SAFWAT, H. 1985. An engineering review of the October field. *8th Petrol. Production Seminar, EGPC*, Cairo.

SAID, R. 1950. The distribution of foraminifera in the northern Red Sea. *Contr. Cushman Found. Foram. Res.* 1: 9-29.

SAID, R. 1954. Remarks on the geomorphology of the area to the east of Helwan. *Bull. Soc. Geograph. Egypte* 27: 93-104.

SAID, R. 1955. Foraminifera from some Pliocene rocks of Egypt. *J. Washington Acad. Sci.* 45: 8-13.

SAID, R. 1958. Remarks on the geomorphology of the deltaic coastal plain between Rosetta and Port Said. *Bull. Soc. Geograph. Egypte* 31: 115-125.

SAID, R. 1960. Planktonic foraminifera from the Thebes Formation, Luxor. *Micropaleontology* 6: 277-286.

SAID, R. 1960. New light on the origin of the Qattara depression. *Bull. Soc. Geograph. Egypte* 33: 37-44.

SAID, R. 1961. Tectonic framework of Egypt and its influence on distribution of foraminifera. *Bull. Am. Assoc. Petrol. Geol.* 45: 198-218.

SAID, R. 1961. *Egitto.* Encyclopedia of Petroleum, Rome 4: 8-75.

SAID, R. 1962. Über das Miozän in der westlichen Wüste Ägyptens. *Geol. Jb.* 80: 349-366.

SAID, R. 1962. *The Geology of Egypt.* Elsevier, 377pp.

SAID, R. 1964. Trip to Gulf of Suez. *6th Annual Field Conf., Petrol. Explor. Soc. Libya*: 141-144.

SAID, R. 1969. General stratigraphy of the adjacent areas of the Red Sea. In: E. T. Degens & D. A. Ross (eds), *Hot brines and recent heavy metal deposits in the Red Sea*, Springer: 71-78.

SAID, R. 1969. Pleistocene geology of the Dungul region, southern Libyan Desert. In: J. J. Hester & P. M. Hobler, *Prehistoric settlement patterns in the Libyan Desert.* Univ. Utah Anthropol. Papers, 92: 7-18.

SAID, R. 1971. *The Geological Survey of Egypt: History and organization.* Geol. Surv. Egypt, Paper 55, 26pp.

SAID, R. 1971. *Explanatory notes to accompany the geological map of Egypt.* Geol. Surv. Egypt, Paper 56, 123pp.

SAID, R. 1971. Discovery of a large phosphate deposit in Abu Tartur Plateau, Western Desert, Egypt. *Bull. BRGM (2), Geol. applique chronique des mines*, II(39): 137-145.

SAID, R. 1975. Some observations on the geomorphology of the south Western Desert of Egypt and its relation to the origin of groundwater. *Ann. Geol. Surv. Egypt* 5: 61-70.

SAID, R. 1979. The Messinian in Egypt. *Intl. Congr. Mediterranean Neogene*, Athens. *Ann. Geol. Pays Helen.*, Hor. Ser., fasc. III: 1083-1090.

SAID, R. 1980. The Quaternary sediments of the southern Western Desert of Egypt. In: F. Wendorf & R. Schild, *Prehistory of eastern Sahara*, Academic Press: 283-289.

SAID, R. 1980. *The Paleozoic of the Gulf of Suez.* Internal Report, Conoco, Egypt.

SAID, R. 1981. *The geological evolution of the River Nile.* Springer, 151pp.

SAID, R. 1983. Proposed classification of the Quaternary of Egypt. *J. Afr. Earth Sci.* 1: 41-45.

SAID, R. 1983. Remarks on the origin of the landscape of the eastern Sahara. *J. Afr. Earth Sci.* 1: 153-158.

SAID, R. & S. F. ANDRAWIS 1961. Lower Carboniferous microfossils from the subsurface rocks of the Western Desert of Egypt. *Contr. Cushman Found. Foram. Res.* 12: 22-25.

SAID, R. & M. G. BARAKAT 1957. Lower Cretaceous foraminifera from Khashm El Mistan, northern Sinai, Egypt. *Micropaleontology* 3: 39-47.

SAID, R. & M. G. BARAKAT 1958. Jurassic microfossils from Gebel Maghara, Sinai, Egypt. *Micropaleontology* 4: 231-272.

SAID, R. & M. G. BARAKAT 1959. Foraminifera from the subsurface Jurassic rocks of Wadi El Natrun, Egypt. *Proc. Egypt. Acad. Sci.* 13: 3-8.

SAID, R. & M. A. BASSIOUNI 1958. Miocene foraminifera of Gulf of Suez region. *Bull. Am. Assoc. Petrol. Geol.* 42: 1958-1977.

SAID, R. & E. M. EL SHAZLY, 1957. Review of Egyptian geology. *Science Council.* Cairo.

SAID, R. & R. A. EISSA 1969. Some microfossils from upper Paleozoic rocks of western coastal plain of Gulf of Suez region. *Proc. 3rd Afr. Micropaleontol. Colloq.*, Cairo: 337-384.

SAID, R. & I. EL HEINY 1967. Planktonic foraminifera from the Miocene rocks of the Gulf of Suez region. *Contr. Cushman Found. Foram. Res.* 18: 14-26.

SAID, R. & B. ISSAWI 1963. *Geology of northern plateau, Bahariya oasis, Egypt.* Geol. Surv. Egypt, 41pp.

SAID, R. & B. ISSAWI 1964. Preliminary results of a geological expedition to lower Nubia and to Kurkur and Dungul. In: F. Wendorf (ed.), *Contribution to the prehistory of Nubia*, Southern Methodist Univ. Press, Dallas, Texas: 1-20.

SAID, R. & A. KENAWY 1956. Upper Cretaceous and lower Tertiary foraminifera from northern Sinai, Egypt. *Micropaleontology* 2: 105-173.

SAID, R. & I. KENAWY 1958. Foraminifera from the Turonian rocks of Abu Roash, Egypt. *Contr. Cushman Found. Foram. Res.* 8: 77-86.

SAID, R. & M. T. KERDANY 1961. The geology and micropaleontology of Farafra Oasis. Micropaleontology 7: 317-336.

SAID, R. & A. O. MANSOUR (eds) 1971. *The discovery of new kaolin deposits in Wadi Kalabsha, Western Desert.* Geol. Surv. Egypt, Paper 54: 138pp.

SAID, R., A. O. MANSOUR, M. F. MIKAIL & A. N. ELIAS, 1975. *Bibliography of geology and related sciences concerning Egypt for the period 1960-1973.* Geol. Surv. Egypt, 192pp.

SAID, R. & L. MARTIN 1964. Cairo area, geological excursion notes. *6th Annual Field Conf., Petrol. Explor. Soc. Libya*: 107-121.

SAID, R., G. PHILIP & N. M. SHUKRI 1956. Post-Tyrrhenian climatic fluctuations in northern Egypt. *Quaternaria* 3: 167-172.

SAID, R., A. H. SABET, A. A. ZALATA, V. A. TENIAKOV & V. I. POKRYSHKIN 1976. A review of theories on the geological distribution of bauxite and their application for bauxite prospecting in Egypt. *Ann. Geol. Surv. Egypt* 6: 6-32.

SAID, R. & H. SABRY 1964. Planktonic foraminifera from the type locality of the Esna Shale in Egypt. *Micropaleontology* 10: 375-395.

SAID, R. & M. YALLOUZE 1955. Miocene fauna from Gebel Oweibid, Egypt. *Bull. Fac. Sci., Cairo Univ.* 33: 61-81.

SAID, R. & S. ZAKI 1967. The distribution of the Miocene rock units of the East Belyim Oilfield, East Coast, Gulf of Suez, UAR. *6th Arab Petrol. Congr., Baghdad* 29pp.

SALAHCHOURIAN, M. H. 1986. Lithologische und tektonische Auswertung von Landsat-MSS-Daten und Luft-

bildern aus dem Tibesti-Gebirge/Zentralsahara mit Hilfe von visueller und digitaler Klassifizierung. *Berl. Geowiss. Abh.* 69(A): 1-82.

SALAMA, R. B. 1987. The evolution of the River Nile. The buried saline rift lakes in Sudan – I. Bahr El Arab rift, the Sudd buried saline lake. *J. Afr. Earth Sci.* 6: 899-914.

SALEEB-ROUFAIEL, G. S. & B. R. B. FASFOUS 1970. A note on the gold-bearing quartz-wolframite vein at Atud, Eastern Desert. *Egypt. J. Geol.* 14: 47-51.

SALEEB-ROUFAIEL, G. S., M. E. HILMY & M. T. AWAD 1976. Mineralization and geochemical features of the Hudi barite deposit, Egypt. *Bull. Nat. Res. Council, Cairo*, 1.

SALEEB-ROUFAIEL, G. S., M. D. SAMUEL, M. E. HILMY & H. E. MOUSSA 1982. Fluorite mineralization at El Bakriya, Eastern Desert, Egypt. *Egypt. J. Geol.* 26: 9-18.

SALEM, R. 1976. Evolution of Eocene-Miocene sedimentation patterns in parts of northern Egypt. *Bull. Am. Assoc. Petrol. Geol.* 60: 34-64.

SALES, J. K. 1968. Crustal mechanics of Cordilleran foreland: A regional and scale model approach. *Bull. Am. Assoc. Petrol. Geol.* 52: 2016-2044.

SAMUEL, M. D. 1977. Lithological and sedimentological studies on the redbeds of Wadi Igla, Eastern Desert. *Bull. Nat. Res. Council*, Cairo, 2: 287-279.

SAMUEL, M. D. & O. H. CHERIF 1978. An approach to the study of paleoclimatic and tectonic control of sedimentation of some Neogene rocks along the Red Sea coast, Egypt. *Bull. Nat. Res. Council, Cairo* 3: 209-215.

SAMUEL, M. D. & G. S. SALEEB-ROUFAIEL 1977. Lithostratigraphy and petrographic analysis of the Neogene sediments at Abu Ghusun, Um Mahara area, Red Sea coast, Egypt. *Beitr. zur Lithologie, Freiburg Forsch.* 323(c): 47-56.

SAMUEL, M. D., G. S. SALEEB-ROUFAIEL, M. E. HILMY & H. MOUSSA 1983. Study of Cretaceous foreland granite and associated rocks at Bakriya, Eastern Desert. *Egypt. J. Geol.* 27: 1-11.

SANDER, B. 1948. *Einführung in die Gefügekunde der geologischen Körper. 1: Allgemeine Gefügekunde und Arbeiten im Bereich Handstück bis Profil.* Springer, 215pp.

SANDER, B. 1950. *Einführung in die Gefügekunde der geologischen Körper. 2: Die Korngefüge.* Springer, 409pp.

SANDFORD, K. S. 1929. The Pliocene and Pleistocene deposits of wadi Qena and of the Nile Valley between Luxor and Assiut. *Quart. J. Geol. Soc. London* 75: 493-548.

SANDFORD, K. S. 1933. Geology and geomorphology of the southern Libyan Desert. *Geograph. J.* 82: 213-219.

SANDFORD, K. S. 1934. *Paleolithic man and the Nile Valley in upper and middle Egypt.* Chicago Univ. Press, Oriental Inst. publ. 3: 131pp.

SANDFORD, K. S. 1935. Geological observations on the southwestern frontiers of the Anglo-Egyptian Sudan and the adjoining part of the southern Libyan Desert. *Quart. J. Geol. Soc. London* 80: 323-381.

SANDFORD, K. S. 1935. Sources of water in the northwestern Sudan. *Geograph. J.* 85: 412-431.

SANDFORD, K. S. & K. J. ARKELL 1929. *Paleolithic man and the Nile-Fayum divide.* Chicago Univ. Press, Oriental Inst. Publ. 1: 77p.

SANDFORD, K. S. & W. J. ARKELL 1933. *Paleolithic man and the Nile Valley in Nubia and upper Egypt.* Chicago

Univ. Press, Oriental Inst. Publ. 17, 92pp.

SANDFORD, K. S. & W. J. ARKELL 1939. *Paleolithic man and the Nile Valley in lower Egypt.* Chicago Univ. Press, Oriental Inst. Publ. 36, 105pp.

SANGSTER, D. 1972. Precambrian volcanogenic massive sulphide deposits in Canada. *Geol. Surv. Canada*, Paper 72/22.

SAUODI, A. & B. KHALIL 1984. Distribution and hydrocarbon potential of Nukhul sediments in the Gulf of Suez. *Proc. 7th Petrol. Explor. Seminar*, EGPC, Cairo: 75-96.

SASS, J. H., A. H. LACHENBRUCH & R. J. MUNROE 1971. Thermal conductivity of rocks from measurements on fragments and its application at heat flow determinations. *J. Geophys. Res.* 76: 3391-3401.

SAVAGE, R. J. G. 1971. Review of the fossil mammals of Libya. In: C. Gray (ed.), *Symp. on the Geology of Libya, Fac. Sci. Univ. Libya*: 215-226.

SAVAGE, R. J. G. 1978. Carnivora. In: V. J. Maglio & H. B. S. Cooke (eds), *Evolution of African mammals*, Harvard Univ. Press: 249-267.

SAVAGE, W. 1984. Evaluation of regional seismicity. Woodward and Clyde Consultants, Internal Report to Aswan High Dam Authority (unpublished).

SAWKINS, F. J. 1984. *Mineral deposits in relation to plate tectonics.* Springer, 325pp.

SAYYAH, T. A. & H. M. EL SHATOURY 1973. Some geochemical features of Wadi Natash volcanics, south Eastern Desert. *Egypt. J. Geol.* 17: 85-95.

SAYYAH, T. A., A. H. HASHAD, A. M. IBRAHIM & M. Y ATTAWIYA 1973. Contribution to the geochemistry of some pink granites, central Eastern Desert. *Egypt. J. Geol.* 17: 57-69.

SAYYAH, T. A., A. H. HASHAD & M. EL MANHARAWY 1978. Radiometric Rb/Sr isochron ages for Wadi Kareim volcanics. *Arab. J. Nuclear Sci. and applications* 11: 1-9.

SAZHINA, N. & N. GRUSHINSKY 1966. *Gravity prospecting* (translated from Russian). NEDRA, Moscow, 491pp.

SCHABER, G. G., J. F. McCAULEY, C. S. BREED & G. R. OLHOEFT 1986. Shuttle imaging radar: Physical controls on signal penetrations and subsurface scattering in eastern Sahara. *IEEE Trans. Geosci. and Rem. Sens.* GE-24: 603-623.

SCHANDELMEIER, H. & F. DARBYSHIRE 1984. Metamorphic and magmatic events in the Uweinat-Safsaf uplift (Western Desert, Egypt). *Geol. Rundschau* 73: 819-831.

SCHANDELMEIER, H., E. KLITZSCH, F. HENDRIKS & P. WYCISK 1987. Structural development of northeast Africa since Precambrian times. *Berl. Geowiss. Abh.* 75(A), 1: 5-14.

SCHANDELMEIER, H., A. RICHTER & G. FRANZ 1983. Outline of the geology of magmatic and metamorphic units from Gebel Uweinat to Bir Safsaf (SW-Egypt/NW-Sudan). *J. Afr. Earth Sci.* 1: 275-283.

SCHANDELMEIER, H., A. RICHTER & U. HARMS 1987. Proterozoic deformation of the east Saharan craton in SE-Libya, S-Egypt and N-Sudan. *Tectonophysics*, 140:233-246.

SCHAUB, H. 1951. Stratigraphie und Palaeontologie des Schlieren Flysches mit besonderer Berücksichtigung der Palaozänen und unter Eozänen Nummuliten und Assilinen. *Schweizer. Palaeontol. Abh.* 68: 1-222.

SCHILD, R. & F. WENDORF 1981. *The prehistory of an Egyptian oasis.* Worclaw. Ossolineum, Poland.

SCHLOSSER, M. 1910. Über einige Säugetiere aus dem Oligozän von Aegypten. *Zool. Anz.* 34: 500-508.

SCHLOSSER, M. 1911. Beiträge zur Kenntnis der Oligozänen Landsäugetiere aus dem Fayum (Aegypten). *Beitr. Palaeontol. Oesterr.-Ungarns Orients* 6: 1-227.

SCHLUMBERGER MIDDLE EAST S.A. 1984. *Well evaluation Conference, Egypt.*

SCHLÜTER, T. & M. HARTUNG 1982. *Aegyptidium aburasiensis* gen. nov. spec. nov. (Gomphidea) aus mutmaasslicher Unterkreide Sudwest-Ägyptens (Anisoptera). *Odonatologica* (Utrecht), 11: 297-307.

SCHMIDT, D. L., D. G. HADLEY & D. B. STOESER 1979. Late Proterozoic crustal history of the Arabian shield, southern Najd Province, Kingdom of Saudi Arabia. *Bull. Inst. Applied Geol., King Abdul Aziz Univ., Jeddah* 3(2): 41-58.

SCHMIDT, M. 1913. Über Paarhufer des fluviomarinen Schichten des Fajum, Odontographisches und osteologisches Material. *Geol. Palaeontol. Abh.*: 153-264.

SCHNEIDER, N. M. & C. SONNTAG 1985. Hydrogeology of the Gebel Uweinat Aswan uplift system, eastern Sahara. *Proc. Intl. Congr. Hydrogeol. of Rocks of low Permeability*, Tucson, Arizona 17(2): 781-791.

SCHNELLMANN, D. A. 1959. Formation of sulphur by reduction of anhydrite at Ras Gemsa, Egypt. *Econ. Geol.* 54: 889-894.

SCHOELE, R. 1983. Das 'Geowissenschaftliche Multibild Auswerte-und Prozessor-System GEOMAPS'. *Berl. Geowiss. Abh.* 47(A): 39-47.

SCHRANK, E. 1982. Kretazische Pollen und Sporen aus dem 'Nubischen Sandstein' des Dakhla Beckens (Ägypten). *Berl. Geowiss. Abh.* 40(A): 87-109.

SCHRANK, E. 1983. Scanning electron and light microscopic investigations of Angiosperm pollen from the lower Cretaceous of Egypt. *Pollen et Spores* 25: 213-242.

SCHRANK, E. 1984. Organic-geochemical and palynological studies of a Dakhla Shale profile (late Cretaceous) in southeast Egypt, Part A: Succession of microfloras and depositional environment. *Berl. Geowiss. Abh.* 50(A): 189-207.

SCHRANK, E. 1984. Paleozoic and Mesozoic palynomorphs from the Foram-1 well (Western Desert, Egypt). *N. Jb. Geol. Palaeontol.*, Mh (1984), 2: 95-112.

SCHRANK, E. 1987. Paleozoic and Mesozoic palynomorphs from northeast Africa (Egypt and Sudan) with special reference to late Cretaceous pollen and dinoflagelates. *Berl. Geowiss. Abh.* 75(A), 1: 249-310.

SCHÜRMANN, H. M. E. 1937-1961. Massengesteine aus Aegypten, I-XX. *N. Jb. Mineral.*

SCHÜRMANN, H. M. E. 1966. *The Precambrian along the Gulf of Suez and the northern part of the Red Sea.* G. J. Brill, Leiden, 404pp.

SCHÜRMANN, H. M. E. 1974. *The Precambrian in North Africa.* Brill, Leiden, 351pp.

SCHÜRMANN, H. M. E., D. BURGER & J. DIJKSTRA 1963. Permian near Wadi Araba, Eastern Desert, Egypt. *Geol. Mijnbouw* 42: 329-336.

SCHWAGER, C. 1883. Die Foraminiferen aus den Eocänablagerungen der Libyschen Wüste Aegyptens. *Palaeontographica* 30(1): 79-154.

SCHWEINFURTH, G. 1864. Reise in die Gebirge der Ababdeh und Bisharin am Roten Meer. *Peterm. Mitt.* (Gotha), 23: 331-336.

714 References

SCHWEINFURTH, G. 1877. Reise durch die Arabische Wüste von Helouan bis Qeneh. *Peterm. Mitt.* (Gotha), 23: 387-389.

SCHWEINFURTH, G. 1885. Sur la dècouverte d'une faune palèozoique dans les grès d'Egypte. *Bull. Inst. Egypte* (ser. 2), 6: 239-255.

SCHWEINFURTH, G. 1885. Sur une ancienne digue un pierre aux environs de Helouan. *Bull. Inst. Egypte* (ser. 2), 6: 139-145.

SCHWEINFURTH, G. 1886. Reise in das Depressions Gebiet im Umkreise des Fajum im Januar 1886. *Z. Ges. Erdk.* 21: 96-149.

SCHWEINFURTH, G. 1899-1910. *Aufnahmen in der östlischen Wüste von Aegypten.* 1. Lieferung. Blatt 1-3 (1899); 2. Lieferung. Blatt 4-5 (1901); 3. Lieferung, Blatt 6, 10a, 10b (1902); 4. Lieferung, Blatt 7-8 (1910). Dietrich Reimer, Berlin.

SCOTT, R. & F. GOVEAN 1984. Early depositional history of a rift basin: Miocene in the western Sinai. *Proc. 7th Petrol. Explor. Seminar, EGPC,* Cairo: 37-48.

SEARLE, D. L. 1974. *Final report on the geology and assessment of the Mo prospect of Homr Akarem.* Internal Report, Aswan Mineral Survey Project, Geol. Surv. Egypt.

SEARLE, D. L. 1975. *Final report on the geology and mineralization at the Zn-Cu deposits of Um Samiuki, Eastern Desert.* Internal Report, Aswan Mineral Survey Project, Geol. Surv. Egypt.

SEARLE, D. L., G. S. CARTER, I. M. SHALABY & A. A. HUSSEIN 1976. Ancient volcanism of island arc type in the Eastern Desert of Egypt. *25th Intl. Geol. Congr.,* Sydney, Sec. 1: 62-63 (Abstr.).

SEILACHER, A. 1955. Spuren und Lebensweise der Trilobiten. In: Schindewolf & Seilacher, Beiträge zur Kenntnis des Kambriums in der Salt Range (Pakistan). *Akad. Wiss. Lit. Mainz. Abh. Math.-Naturw. Kl.*: 342-372.

SEILACHER, A. 1962. Form und Funktion des Trilobiten-Daktylus. *Palaeontol. Z.*: 218-227.

SEILACHER, A. 1970. *Cruziana* stratigraphy of 'non-fossiliferous' Paleozoic sandstones. In: T. P. Crimes & J. C. Harper (eds), *Trace fossils,* Geol. J., Special Issue 3: 447-476.

SEILACHER, A. 1983. Upper Paleozoic trace fossils from the Gilf Kebir-Abu Ras area in southwestern Egypt. *J. Afr. Earth Sci.* 1: 21-44.

SEILACHER, A. 1983. Paleozoic sandstones in southern Jordan: Trace fossils, depositional environments and biogeography. *Proc. 1st Jordan. Geol. Conf., Jordan Geol. Assoc.*

SEILACHER, A. 1985. Trilobite paleobiology and substrate relationships. *Trans. Roy. Soc. Edinburgh* 76: 231-237.

SEILACHER, A. (in press). An updated *Cruziana* stratigraphy of Gondwanian Paleozoic sandstones. *3rd Symp. on the Geology of Libya, Tripoli* (re-scheduled to 1987).

SELIM, E. T. H. 1966. *Thermal emission mass spectrometric analysis and its application to age determination.* M.Sc. Thesis, Cairo Univ.

SELLEY, R. C. 1970. Ichnology of Paleozoic sandstones in the southern desert of Jordan: a study of trace fossils in their sedimentologic context. In: T. P. Crimes & J. C. Harper (eds), *Trace fossils,* Geol. J., Special Issue 3: 477-488.

SELLWOOD, B. W. & R. E. NETHERWOOD 1984. Facies

evolution in the Gulf of Suez area: Sedimentation history as an indicator of rift initiation and development. *Modern Geol.* 9: 43-69.

SENGOR, A. M. C. 1979. Mid-Mesozoic closure of Permo-Triassic Tethys and its implications. *Nature* 279: 590-593.

SENGOR, A. M. C. 1985. The story of Tethys: how many wives did Okeanos have. *Episodes* 8: 3-12.

SENGOR, A. M. C. & K. BURKE 1978. Relative timing of rifting and volcanism on earth and its tectonic implications. *Geophys. Res. Lett.* 5: 419-421.

SENGOR, A. M. C. & Y. YILMAZ 1981. Tethyan evolution of Turkey; a plate tectonic approach. *Tectonophysics* 75: 181-241.

SENGOR, A. M. C., Y. YILMAZ & O. SUNGURLU 1984. Tectonics of the Mediterranean Cimmerides: Nature and evolution of the western termination of the Paleotethys. In: J. E. Dixon & A. H. F. Robertson (eds), *The geological evolution of the eastern Mediterranean,* Blackwell Scientific. Publishers, Oxford: 77-112.

SERENCSITS, C. McC., H. FAUL, K. A. ROLAND, A. A. HUSSEIN & T. M. LUTZ 1981. Alkaline ring complexes in Egypt: their ages and relationships in time. *J. Geophys. Res.* 86(B4): 3009-3013.

SERENCSITS, C. McC., H. FAUL, K. A. ROLAND, M. F. EL RAMLY & A. A. HUSSEIN 1979. Alkaline ring complexes in Egypt: their ages and relationship to tectonic development of the Red Sea. *Ann. Geol. Surv. Egypt* 9: 102-116.

SESTINI, G. 1976. Geomorphology of the Nile delta. *Proc. Seminar Nile delta Sedimentology,* Alexandria, UNESCO/UNDP: 12-14.

SESTINI, G. 1984. Tectonic and sedimentary history of NE African margin (Egypt/Libya). In: J. E. Dixon & A. H. F. Robertson (eds), *The geological evolution of the eastern Mediterranean,* Blackwell Scientific Publishers, Oxford: 161-175.

SEWARD, A. C. 1907. Fossil plants from Egypt. *Geol. Mag.* n.ser. 5: 253-257.

SEWARD, A. C. 1935. *Leaves of dicotyledons from the Nubian Sandstone of Egypt.* Geol. Surv. Egypt, 21pp.

SHAALAN, M. B. M. 1980. Mineralogical and geochemical studies on barites and barite-bearing rocks in Bahariya Oasis, Western Desert, Egypt. *Chem. Erde* 30: 63-73.

SHACKLETON, R. M. 1977. Possible late Precambrian ophiolites from Africa and Brazil. *20th Annual Report Res. Inst. Afr. Geol.,* Univ. Leeds: 3-7.

SHACKLETON, R. M. 1980. Precambrian tectonics of NE Africa. *Bull. Inst. Applied Geol., King Abdul Aziz Univ., Jeddah* 3(2): 1-6.

SHACKLETON, R. M., A. C. RIES, R. H. GRAHAM & W. R. FITCHES 1980. Late Precambrian ophiolite melange in the Eastern Desert of Egypt. *Nature* 285: 472-474.

SHAFIK, S. 1970. The nannoplankton assemblages of the Maastrichtian of Red Sea coast, Egypt. *Verh. Geol.*: A1103-A1104.

SHAHIN, A. N. & M. FATHI 1984. *Regional geotheral gradient map, Gulf of Suez.* Internal Rep., Gulf of Suez Petrol. Co.

SHAHIN, A. N. & M. M. SHEHAB 1984. Petroleum generation, migration and occurrence in the Gulf of Suez offshore of south Sinai. *Proc. 7th Petrol. Explor. Seminar, EGPC,* Cairo: 126-151.

SHAHIN, A. N., M. M. SHEHAB & H. F. MANSOUR 1986. Quantitative evaluation and timing of petroleum generation in Abu Gharadig basin, Western Desert, Egypt. *8th Petrol. Explor. Seminar*, EGPC, Cairo.

SHALEM, N. 1954. The Red Sea and the Erythrean disturbances. *C.R. 19th Intl. Geol. Congr.*, Algiers (1952), sec. 15(17): 223-231.

SHAMAH, K. & A. BLONDEAU 1979. Présence de *Nummulites striatus* (Bruguiere) dans le Lutetien superieur d'Egypte (biozone biarizzienne). *Rev. Micropaleontol.* 22: 191-194.

SHAMAH, K., A. BLONDEAU, Y. LE CALVEZ, K. PERCH-NIELSN & M. TOUMARKINE 1982. Biostratigraphie de l'Eocène de la Formation El Midawarah, region de Wadi El Rayan, Province de Fayoum, Egypte. *8th Afr. Micropaleontol. Colloq.* (1980), Paris, *Cah. Micropaleontol.* 1: 91-104.

SHATA, A. 1951. The Jurassic of Egypt. *Bull. Inst. Desert Egypte* 1(2): 68-73.

SHATA, A. 1953. New light on the structural development of the Western Desert of Egypt. *Bull. Inst. Desert Egypte* 3(1): 101-106.

SHATA, A. 1956. Structural development of the Sinai Peninsula, Egypt. *Bull. Inst. Desert Egypte* 6(2): 117-157.

SHATA, A. 1959. Ground water and geomorphology of the northern sector of Wadi El Arish basin. *Bull. Soc. Geograph. Egypte* 32: 247-262.

SHATA, A. 1960. The geology and geomorphology of El Qusaima area. *Bull. Soc. Geograph. Egypte* 33: 95-146.

SHAZLY, A. G. 1971. *Geology of Abu Ziran area, Eastern Desert*. Ph.D. Thesis, Assiut Univ.

SHEIKH, H. A. & M. FARIS 1985. The Eocene-Oligocene boundary in some wells of the western Desert, Egypt. *N. Jb. Geol. Palaeontol. Mh*(1985), 1: 23-28.

SHERBORN, C. D. 1910. *Bibliography of scientific and technical literature relating to Egypt*. Nat. Printing Dept., Cairo, 155pp.

SHIMRON, A. E. 1975. Petrogenesis of the Tarr albitite-carbonitite complex, Sinai Peninsula. *Min. Mag.* 40: 13-15.

SHIMRON, A. E. 1980. Proterozoic island arc volcanism and sedimentation in Sinai. *Precambrian Res.* 12: 437-458.

SHIMRON, A. E. 1984. Evolution of the Kid Group, southeastern Sinai Peninsula: thrusts, melanges and implications for accretionary tectonics during the late Proterozoic of the Arabian-Nubian shield. *Geology* 12: 242-247.

SHIMRON, A. E. & H. J. ZWART 1970. The occurrence of low pressure metamorphism in the Precambrian of the Middle East and northeast Africa. *Geol. Mijnbouw* 49: 369-374.

SHORT, N. & N. H. MACLEOD 1972. Analysis of multispectral images simulating ERTS observations. *NASA* X-430-72-118, 22pp, Greenbelt (GSFC).

SHOTTON, F. W. 1946. The main water table of the Miocene limestone in the coastal desert of Egypt. *Water and Water Eng.*: 1-16.

SHUKRI, N. M. 1950. The mineralogy of some Nile sediments. *Quart. J. Geol. Soc. London* 105: 511-534.

SHUKRI, N. M. 1953. The geology of the desert east of Cairo. *Bull. Inst. Desert Egypte* 3(2): 89-105.

SHUKRI, N. M. 1954. On cylindrial structures and colouration of Gebel Ahmar near Cairo, Egypt. *Bull. Fac. Sci., Cairo Univ.*, 32: 1-23.

SHUKRI, N. M. & G. AKMAL 1953. The geology of Gebel El Nasuri and Gebel El Anqabiya district. *Bull. Soc. Geograph. Egypte* 16: 243-276.

SHUKRI, N. M. & M. K. AYOUTY 1956. The geology of Gebel Iweibid-Gafra area, Cairo-Suez district. *Bull. Soc. Geograph. Egypte* 29: 67-109.

SHUKRI, N. M. & N. AZER 1952. The mineralogy of Pliocene and more recent sediments in the Faiyum. *Bull. Inst. Desert Egypte* 2(1): 10-53.

SHUKRI, N. M. & E. Z. BASTA 1959. A note on the occurence of a polysulphide deposit at Gabal Derhib, south Eastern Desert, Egypt. *Egypt. J. Geol.* 3: 167-173.

SHUKRI, N. M. & R. A. HIGAZY 1944. Mechanical analysis of some bottom deposits of the northern Red Sea. *J. Sed. Petrology* 14: 43-69.

SHUKRI, N. M. & R. A. HIGAZY 1944. The mineralogy of some bottom deposits of the northern Red Sea. *J. Sed. Petrology* 14: 70-85.

SHUKRI, N. M. & M. S. MANSOUR 1980. Lithostratigraphy of Um Samiuki district, Eastern Desert, Egypt. *Bull. Inst. Applied Geol., King Abdul Aziz Univ., Jeddah* 3(4): 83-94.

SHUKRI, N. M. & F. M. NAKHLA 1955. The sulphur deposits of Ras Gemsa coast, Egypt. *Symp. Applied Geol. Near East*, UNESCO, Ankara: 114-123.

SHUKRI, N. M., G. PHILIP & R. SAID 1956. The geology of the Mediterranean coast between Rosetta and Bardia. Part II: Pleistocene sediments; geomorphology and microfacies. *Bull. Inst. Egypte* 37(2): 395-427.

SIEBERG, A. 1932. *Erdbebengeographie*. Handbuch der Geophysik, Bd IV, Berlin.

SIEDNER, G. & A. HOROWITZ 1974. Radiometric ages of late Cenozoic basalts from northern Israel. *Nature* 250: 23-28.

SIEGAL, B. S. & A. R. GILLESPIE (eds) 1980. *Remote sensing in geology*. Wiley, 702pp.

SIGAEV, N. A. 1967. *The main tectonic features of Egypt; an explanatory note to the tectonic map of Egypt, scale 1:2 000 000*. Geol. Surv. Egypt, Paper 39, 26pp.

SILLITOE, R. H. 1978. Metallic mineralization affiliated to subaerial volcanism, a review. In: *Volcanic processes in ore genesis*, I.M.M. Geol. Soc. London: 99-116.

SIMONS, E. L. 1959. An anthropoid frontal bone from the Fayum Oligocene of Egypt: the oldest skull fragment of a higher primate. *Am. Mus. Novitates* 1976: 1-16.

SIMONS, E. L. 1961. Notes on Eocene tarsioids and a revision of some Necrolemurinae. *Bull. Brit. Mus. Nat. Hist. (Geol.)* 5: 45-69.

SIMONS, E. L. 1962. Two new primate species from the African Oligocene. *Postilla* 64: 1-12.

SIMONS, E. L. 1965. New fossil apes from Egypt and the initial identification of Hominoidea. *Nature* 205: 135-139.

SIMONS, E. L. 1967. The earliest apes. *Scientific American* 217: 28-35.

SIMONS, E. L. 1968. Early Cenozoic mammalian faunas, Fayum Province, Egypt. Part I: African Oligocene mammals: Introduction, History of study and faunal succession. *Bull. Peabody Mus. Nat. Hist.* (Yale Univ.) 28: 1-21.

SIMONS, E. L. 1971. Relationships of *Amphipithecus* and *Oligopithecus*. *Nature* 232: 489-491.

SIMONS, E. L. 1972. *Primate evolution: An introduction to man's place in nature*. Macmillan.

SIMONS, E. L. 1974. *Parapithecus grangeri* (Parapithecidae, old world higher primates); new species from the Oligocene of Egypt and the initial differentiation of cercopithecoidea. *Postilla* 166: 1-12.

SIMONS, E. L. 1986. *Parapithecus grangeri* of the African Oligocene: An archaic catarrhine without lower incisors. *J. Human Evol.* 15: 205-213.

SIMONS, E. L. 1987. New faces of *Aegyptopithecus* from the Oligocene of Egypt. *J. Human Evol.* 16:273-289.

SIMONS, E. L. & T. M. BOWN 1985. *Afrotarsius chatrathi*, first tarsiiform primate (Tarsiidae) from Africa. *Nature* 313: 475-477.

SIMONS, E. L. & T. M. BOWN 1987. New Oligocene Ptolemaiidae (Mammalia; ?Pantolesa) from the Jebel Qatrani Formation, Fayum depression, Egypt. *J. Vert. Paleontol.* 7(3): 311-324.

SIMONS, E. L., T. M. BOWN & D. T. RASMUSSEN 1987. Discovery of two additional prosimian primate families (Omomyidae, Lorisidae) in the African Oligocene. *J. Human Evol.* 15: 431-437.

SIMONS, E. L. & E. DELSON 1978. Circopithecidae and Parapithecidae. In: V. J. Maglio & H. B. S. Cooke (eds), *Evolution of African mammals*. Harvard Univ. Press: 100-119.

SIMONS, E. L. & P. D. GINGERICH 1974. New carnivorous mammals from the Oligocene of Egypt. *Ann. Geol. Surv. Egypt* 4: 157-166.

SIMONS, E. L. & R. F. KAY 1983. *Qatrania* new basal anthropoid primate from the Fayum Oligocene of Egypt. *Nature* 304: 624-626.

SIMONS, E. L., D. T. RASMUSSEN & D. L. GEBO 1987. A new species of *Propliopithecus* from the Fayum, Egypt. *Am. J. Phys. Anthropol.* 730: 139-147.

SIMONS, E. L. & D. E. RUSSELL 1960. Notes on the cranial anatomy of *Necrolemur*. *Breviora* 127: 1-14.

SIMPSON, D. W., R. M. KEBEASY, C. NICHOLSON, M. MAAMOUN, R. N. H. ALBERT, E. M. IBRAHIM, A. MEGAHED, A. GHARIB & A. HUSSAIN 1984. Aswan telemetered seismograph network. In: A. M. Wassef, A. Boud & P. Vyskocil (eds), *Recent crustal movements in Africa, J. Geodynamics* 7: 195-203.

SIMPSON, D. W., C. NICHOLSON, R. M. KEBEASY, M. MAAMOUN, R. ALBERT, E. IBRAHIM & S. MEGAHED 1982. Induced seismicity at Aswan reservoir, Egypt July-September 1982. *Trans. Am. Geophys. Union* 63: 1024.

SIMS, P. K. & H. L. JAMES 1984. Banded iron ore formation of late Proterozoic age in the central Eastern Desert, Egypt; Geological and tectonic setting. *Econ. Geol.* 79: 1777-1784.

SLAUGHTER, B. H. & J. T. GIDEON 1979. *Saidomys natrunensis* an arvicanthine rodent from the Pliocene of Egypt. *J. Mammology* 60: 421-425.

SMITH, A. G. 1971. Alpine deformation and the oceanic areas of the Tethys, Mediterranean and Atlantic. *Bull. Geol. Soc. Am.* 82: 2039-2070.

SMITH, A. G. & J. C. BRIDEN 1977. *Mesozoic and Cenozoic paleocontinental maps*, Cambridge Univ. Press.

SMITH, A. G., A. M. HURLEY & J. C. BRIDEN 1981. *Phanerozoic paleocontinental world maps*, Cambridge Univ. Press, 102pp.

SMITH, A. G. & N. H. WOODCOCK 1982. Tectonic synthesis of the alpine Mediterranean region; A review. In: H. Berckhemer & K. Hsu (eds), Alpine-Mediterranean geo-

dynamics. *Geodynamics* ser. 7, *Am. Geophys. Union and Geol. Soc. Am.*: 15-38.

SMITH, J. G. 1965. Fundamental transcurrent faulting in northern Rocky Mountains. *Bull. Am. Assoc. Petrol. Geol.* 49: 1398-1409.

SMITH, W. L. (ed.) 1977. *Remote sensing applications for mineral exploration*. Dowden, Hutchinson & Ross, Stroudsberg, Pennsylvania, 391pp.

SNAVELY, P. D., R. E. GARRISON & A. A. MEGUID 1979. Stratigraphy and regional depositional history of the Thebes Formation (Lower Eocene), Egypt. *Ann. Geol. Surv. Egypt* 9: 344-362.

SNEH, A. 1982. Drainage systems of the Quaternary in northern Sinai with emphasis on Wadi El Arish. *Z. Geomorph.* 26: 179-185.

SNEH, A., T. WEISSBROD, E. EHRLICH, A. HOROWITZ & S. MOSHKOVITZ 1986. Holocene evolution of the northeastern corner of the Nile delta. *Quaternary Res.* 26: 194-206.

SNELGROVE, A. K. 1972. *Report on a visit to Aswan Mineral Survey Project area, south Eastern Desert of ARE*. Internal Report, Aswan Mineral Survey Project, Geol. Surv. Egypt.

SNELGROVE, A. K. & M. S. GARSON 1972. *Possible application of plate tectonics in Aswan Mineral Survey Project area, south Eastern Desert of Egypt*. Internal Report, Aswan Mineral Survey Project, Geol. Surv. Egypt.

SOLIMAN, H. A. 1977. Foraminifères et microfossiles vegetaux provenant du 'Nubia Sandstone' de subsurface de l'oase El Kharga, désèrt de l'ouest, Egypte. *Rev. Micropaleontol.* 90: 114-124.

SOLIMAN, M. A., M. E. HABIB & E. A. AHMED 1986. Mineralogy and geochemistry of Wadi Qena phosphorites, Egypt. *N. Jb. Geol. Palaeontol.* Mh (1986), 2: 105-119.

SOLIMAN, M. A., M. E. HABIB & E. A. AHMED 1986. Sedimentologic and tectonic evolution of the upper Cretaceous-lower Tertiary succession at Wadi Qena, Egypt. *Sed. Geol.* 46: 111-133.

SOLIMAN, S. M. 1961. Geology of manganese deposits of Um Bogma, Sinai and its position in the African manganese production. *1st Iron and Steel Congr.*, Cairo: 1-21.

SOLIMAN, S. M. & O. EL BADRY 1970. Nature of Cretaceous sedimentation in Western Desert, Egypt. *Bull. Am. Assoc. Petrol. Geol.* 54: 2349-2370.

SOLIMAN, S. M. & M. A. EL. FETOUH 1969. Petrology of the Carboniferous sandstones in west central Sinai. *Egypt. J. Geol.* 13: 61-143.

SOLIMAN, S. M. & M. A. EL. FETOUH 1970. Carboniferous of Egypt, isopach and lithofacies maps. *Bull. Am. Assoc. Petrol. Geol.* 54: 1918-1930.

SOLIMAN, S. M. & M. M. HASSAN 1969. Contribution to the geology and geochemistry of lead-zinc and sulfur deposits of Gebel El Rousas, Anz and Ranga localities, Eastern Desert, Egypt. *6th Arab Sci. Congr.*, Damascus, 4B: 591-660.

SOLIMAN, S. M. & M. M. HASSAN 1971. Geochemical prospection for lead and zinc in Gebel El Rousas, Eastern Desert, Egypt. *Geol. Rundschau* 60: 1285-1301.

SONNENFELD, P. 1981. The Phanerozoic Tethys Sea. In: P. Sonnenfeld (ed.), *Tethys, the ancestral Mediterranean*, Hutchinson Ross, Stroudsburg, Pennsylvania: 18-55.

SONNTAG, C., E. KLITZSCH, E. M. EL SHAZLY, C. KALINKE & K. O. MUENNICH 1978. Paläoklimatische Information im Isotopengehalt C14-datierter Grundwässer: Kontinentaleffekt in D und O18. *Geol. Rundschau* 67: 413-423.

SONNTAG, C., E. KLITZSCH, E. P. LOEHNERT, K. O. MUENNICH, C. JUNGHANS, U. THORWEIHE, K. WEISTROFFER & F. M. SWAILEM 1978. Paleoclimatic information from Deuterium and Oxygen 18 in C14-dated north Saharan Groundwaters; Groundwater Formation in the past. *Proc. Intl. Symp. Isotope Hydrology*, IAEA, Vienna, II: 569-580.

SONNTAG, C., U. THORWEIHE, J. RUDOLPH, E. P. LOEHNERT, K. O. MUENNICH, E. KLITZSCH, E. M. EL SHAZLY & F. M. SWAILEM 1979. Isotopic identification of Saharian Groundwaters, Ground water formation in the past. *Ber. Zentralinst. f. Isotopen- u. Strahlenforschung, Akad. Wiss. Deutschen Demokratischen Republik* 30: 239-248.

SONNTAG, C., U.THORWEIHE & J.RUDOLPH 1982. Isotopenuntersuchungen zur Bildungsgeschichte Saharischer Paläowasser. *Geomethodica* (Bern), 7: 55-78.

SOUAYA, F. 1961. Contribution to the study of *Miogypsina* s.l. from Egypt. *Proc. Konikl. Ned. Akad. Wetenshap.* 64(B): 665-705.

SOUAYA, F. 1963. Micropaleontology of four sections south of Quseir, Egypt. *Micropaleontology* 9: 233-266.

SOUAYA, F. 1963. On the foraminifera of Gebel Gharra (Cairo-Suez road) and some other Miocene samples. *J. Paleontology* 37: 433-457.

SOUAYA, F. 1965. Miocene foraminifera of the Gulf of Suez region, UAR, Part 1. Systematics (Astrorhizoidea-Buliminoidea). *Micropaleontology* 11: 301-334.

SOUAYA, F. 1966. Miocene foraminifera of the Gulf of Suez region, UAR, Part 2. Systematics (Rotaloidea). *Micropaleontology* 12: 43-64.

SOUAYA, F. 1966. Miocene foraminifera of the Gulf of Suez region, UAR, Part 3. Biostratigraphy. *Micropaleontology* 12: 183-202.

SPANDERASHVILLI, G. I. & M. MANSOUR 1970. The Egyptian phosphates. In: O. Moharram et al. (eds), *Studies on some mineral deposits of Egypt*. Geol. Surv. Egypt: 89-106.

SPATH, L. F. 1946. The middle Triassic cephalopoda from Sinai. *Bull. Inst. Egypte* 27: 425-426.

STACEY, J. S. & C. E. HEDGE 1984. Geochronologic and isotopic evidence for early Proterozoic crust in the eastern Arabian shield. *Geology* 12: 310-313.

STACEY, J. S. & D. B. STOESER 1983. Distribution of oceanic and continental leads in the Arabian-Nubian shield. *Contr. Mineral. Petrology* 84: 91-105.

STAINFORTH, R. M. 1949. Foraminifera in the upper Tertiary of Egypt. *J. Paleontology* 23: 419-422.

STAINFORTH, R. M., J. L. LAMB, H. LUTHERBACHER, J. H. BEARD & R. M. JEFFORDS 1975. Cenozoic planktonic foraminiferal zonation and characteristics of index forms. *Univ. Kansas Paleontol. Contr.* 63: 1-425.

STANLEY, D. J., G. L. FREELAND & H. SHENG 1982. Dispersal of Mediterranean and Suez bay sediments in the Suez Canal. *Marine Geol.* 49: 61-79.

STANLEY, D. J. & H. SCHENG 1986. Volcanic shards from Santorini (upper Minoan) in the Nile delta, Egypt. *Nature* 320: 733-735.

STANTON, R. L. 1972. *Ore petrology*. McGraw Hill, 713pp.

STEARNS, D. W. 1971. Mechanics of drape folding in the Wyoming province. *23rd Annual Field conf., Wyoming Geol. Assoc.*: 125-143.

STECKLER, M. S. 1985. Uplift and extension at the Gulf of Suez: Indications of induced mantle convection. *Nature* 317: 135-139.

STEEN, G. 1982. Radiometric age dating and tectonic significance of some Gulf of Suez igneous rocks. *6th Petrol. Explor. Seminar, EGPC*, Cairo.

STEFFEN, E. M. 1983. Untersuchung zur Morphologie und Genese der aeolischen Akkumulationsformen der Ostsahara mit Hilfe der fernerkundung. *Berl. Geowiss. Abh.* 45(A): 1-137.

STEFFAN, E. M. 1983. The limitations and possibilities of the use of computer-enhanced imagery and digital image processing in geological mapping, as demonstrated on an image of the Dakhla Oasis, Egypt. *Berl. Geowiss. Abh.* 47(A): 109-116.

STEIGER, R. H. & E. JÄGER 1977. Subcommission in geochronology: convention on the use of decay constants in geo- and cosmochronology. *Earth Planet. Sci. Lett.* 36: 359-362.

STEINITZ, G., Y. BARTOV & J. C. HUNZIKER 1978. K/Ar age determinations of some Miocene-Pliocene basalts in Israel: Their significance to the tectonics of the rift valley. *Geol. Mag.* 115: 329-340.

STERN, R. J. 1979. *Late Precambrian ensimatic volcanism in the central Eastern Desert of Egypt*. Ph.D. Thesis, Univ. California at San Diego, 210pp.

STERN, R. J. 1979. Late Precambrian crustal environments as reconstructed from relict igneous minerals, central Eastern Desert of Egypt. *Ann. Geol. Surv. Egypt* 9: 9-31.

STERN, R. J. 1981. Petrogenesis and tectonic setting of late Precambrian ensimatic volcanic rocks, central Eastern Desert of Egypt. *Precambrian Res.* 16: 195-230.

STERN, R. J. & D. GOTTFRIED 1986. Petrogenesis of a late Precambrian (575-600 Ma) bimodal suite in northern Africa. *Contr. Mineral. Petrology* 92: 492-501.

STERN, R. J., D. GOTTFRIED & C. E. HEDGE 1984. Late Precambrian rifting and crustal evolution in the north Eastern Desert of Egypt. *Geology* 12: 168-172.

STERN, R. J. & C. E. HEDGE 1985. Geochronologic and isotopic constraints on late Precambrian crustal evolution in the Eastern Desert of Egypt. *Am. J. Sci.* 285: 97-172.

STIETZEL, H. J. 1987. Geologie und Petrographie im Gebiet des Wadi Hodein, SE Aegyptens. Gelandeuntersuchungen und Fernerkundung. *Berl. Geowiss. Abh.*

STOERMER, L. 1939. Studies on trilobite morphology. I. The thorasic appendages and their phylogenetic significance. *Norsk. Geol. Tidsskr.* 19: 143-273.

STOESER, D. B. & V. E. CAMP 1985. Pan African microplate accretion of the Arabian shield. *Bull. Geol. Soc. Am.* 96: 817-826.

STOFFERS, P. & D. A. ROSS 1974. *Initial reports of the Deep Sea Drilling Project, 23, Sedimentary history of the Red Sea*. US Government Printing Office, Washington, DC.

STORZER, D. & G. A. WAGNER 1971. Fission track ages of North American tektites. *Earth & Planet. Sci. lett.* 10: 435-440.

STRECKEISEN, A. 1976. To each plutonic rock its proper

name. *Earth Sci. Rev.* 12: 1-33.

STRINGFIELD, V. T., P. E. LAMOREAUX & H. E. LEGRAND 1974. Karst and paleohydrology of carbonate rock terrains in semi-arid regions with a comparison to a humid karst of Alabama. *Bull. Geol. Surv. Alabama* 105: 1-106.

STRØMER, L. 1939. Studies on trilobite morphology, I. The thorasic appendages and their phylogonetic significance. *Norsk. Geol. Tidsskr.* 19: 143-273.

STROMER VON REICHENBACH, E. 1905. Fossile Wirbeltier Reste aus dem Ouadi Faregh und Ouadi Natroun in Aegypten. *Abh. Senckenberg. Naturforsch. Ges.* 29: 99-132.

STROMER VON REICHENBACH, E. 1913. Mitteilung über die Wirbeltier Reste aus der mittel Pliozän in Natron Tales. *Z. Deutsch. Geol. Ges.*: 1-350.

STROMER VON REICHENBACH, E. & W. WEILER 1930. Beschreibung von Wirbeltier Resten aus dem nubischen Sandsteine Oberaegyptens und aus aegyptischen Phosphaten nebst Bemerkungen über die Geologie der Umgegen von Mahamid in Oberaegypten. *Abh. Bayer. Akad. Wiss., Math.-Naturw. Kl.* 11: 1-78.

STROUGO, A. 1974. L'horizon a *Cossmannella fajumensis*, niveau-repère dans la partie inférieure de l'Eocene superieure d'Egypte. *C.R. Acad. Sci., Paris* 279 (ser. D): 1841-1844.

STROUGO, A. 1976. Découverte d'une discontinuité de sedimentation dans l'Eocène supérieure du Gebel Mokattam (Egypte). *C. R. Somm., Soc. Geol. France*, fasc. 5: 213-215.

STROUGO, A. 1976. Le membere Ain Musa (Gebel Mokattam: Eocène supérieure) et ses équivalents chronostratigraphiques à l'ouest du Nil. *C.R. Acad. Sci., Paris* 283 (ser. D): 1137-1140.

STROUGO, A. 1977. *Le Biarrizien et le Priabonien en Egypte et leur faunes de bivalves*. Ph.D. Thesis, Univ. Paris sud, 247pp.

STROUGO, A. 1979. The middle Eocene-upper Eocene boundary in Egypt. *Ann. Geol. Surv. Egypt* 9: 454-470.

STROUGO, A. 1983. The genus *Carolia* (Bivalvia, Anomiidae) in the Egyptian Eocene. *Bull. Soc. Paleontol. Ital.* 22: 119-126.

STROUGO, A. 1985. Eocene stratigraphy of the eastern greater Cairo (Gebel Mokattam-Helwan) area. *Middle East Res. Center, Ain Shams Univ., Sci. Res. Series* 5: 1-39.

STROUGO, A. 1985. Eocene stratigraphy of the Giza Pyramids plateau. *Middle East Res. Center, Ain Shams Univ., Sci. Res. Series* 5: 79-99.

STROUGO, A. 1986. The *velascoensis* event: A significant episode of tectonic activity in the Egyptian Paleogene. *N. Jb. Geol. Palaeontol. Abh.* 173: 253-269.

STROUGO, A. 1986. Mokattam stratigraphy of eastern Maghagha-El Fashn district. *Middle East Res. Center, Ain Shams Univ., Sci. Res. Series* 6: 33-58.

STROUGO, A. & R. A. ABUL-NASR 1981. The age of the Thebes formation of Gebel Duwi, Quseir area, Egypt. *N. Jb. Geol. Palaeontol. Mh* (1981) 1: 49-53.

STROUGO, A., R. A. ABUL-NASR & M. A. Y. HAGGAG 1982. Contribution to the age of the middle Mokattam beds of Egypt. *N. Jb. Geol. Palaeontol. Mh* (1982) 4: 240-243.

STROUGO, A. & M. M. AZAB 1982. Middle Eocene mollusca from the basal beds of Gebel Qarara (upper Egypt) with remarks on the depositional environment of these beds. *N. Jb. Geol. Palaeontol. Mh* (1982) 11: 667-678.

STROUGO, A. & M. BOUKHARY 1987. The middle Eocene-upper Eocene boundary in Egypt: Present state of the problem. *Rev. Micropaleontol.* 30.

STROUGO, A. & M. A. Y. HAGGAG 1984. Contribution to the age determination of the Gehannam Formation in the Fayum province. *N. Jb. Geol. Palaeontol. Mh* (1984) 1: 46-52.

STROUGO, A., M. A. Y. HAGGAG, M. FARIS & M. M. AZAB 1984. Eocene stratigraphy of the Beni Suef area. *Bull. Fac. Sci., Ain Shams Univ.* (1982/83), (B), 24: 177-192.

STURCHIO, N. C., M. SULTAN & R. BATIZA 1983. Geology and origin of Meatiq dome, Egypt. A Precambrian metamorphic core complex? *Geology* 11: 72-76.

STURCHIO, N. C., M. SULTAN, P. SYLVESTER, R. BATIZA, C. E. HEDGE, E. M. EL SHAZLY & A. ABDEL MAGUID 1983. Geology, age and origin of the Meatiq dome: Implications for the Precambrian stratigraphy and tectonic evolution of the Eastern Desert of Egypt. *Bull. Fac. Earth Sci., King Abdul Aziz Univ., Jeddah* 6: 127-143.

STYLES, P. & S. A. HALL 1980. A comparison of sea floor spreading histories of the western Gulf of Aden and the central Red Sea. In: *Geodynamic evolution of the Afro-Arabian rift systems*. Accad. Naz. Licei, Rome: 587-606.

STYLES, P. & K. D. GERDES 1983. St John's island (Red Sea): a new geophysical model and its implications for the emplacement of ultramafic rocks in fracture zones and at continental margins. *Earth Planet. Sci. Lett.* 65: 353-368.

SUDRE, J. 1979. Nouveaux mammifères éocènes du Sahara occidental. *Palaeovertebrata* 9: 83-115.

SULTAN, F., H. ANTON, I. MOFTAH & S. HAFEZ 1985. Zeit Bay field – geological and reservoir engineering considerations. *8th Petrol. Production Seminar*, EGPC, Cairo.

SULTAN, F. & I. MOFTAH 1985. Ras Fanar field, a geological and engineering approach to field development. *8th Petrol. production Seminar*, EGPC, Cairo.

SULTAN, I. Z. 1976. Carboniferous microflora from a black shale unit in the Gulf of Suez. *4th Intl. Palynol. Conf.*, Lucknow: 175-176 (Abstr.).

SULTAN, I. Z. 1978. Palynostratigraphie du Bathonien-Callovien du puits No. 3 de Barga, Sinai nord, Egypte. *Rev. Micropaleontol.* 20: 222-229.

SULTAN, I. Z. 1985. Maastrichtian plant microfossils from the El Mahamid area, Nile Valley, southern Egypt. *Rev. Micropaleontol.* 28: 213-222.

SULTAN, I. Z. 1985. Palynological studies in the Nubia Sandstone Formation, east of Aswan, southern Egypt. *N. Jb. Geol. Palaeontol. Mh* (1985), 10: 605-617.

SULTAN, N. & K. SCHUTZ 1984. Cross faults in the Gulf of Suez area. *Proc. 7th Petrol. Explor. Seminar*, EGPC, Cairo: 5-17.

SWANBERG, C. A. & S. ALEXANDER 1979. Use of water quality file Watstore in geothermal exploration, an example from the Imperial valley, California. *Geology* 7: 108-111.

SWANBERG, C. A. & P. MORGAN 1979. The linear relation between temperatures based on the silica content of groundwater and regional heat flow, a new heat flow map of the United States. *Pure and Applied Geophys.* 117: 227-241.

SWANBERG, C. A. & P. MORGAN 1980. The silica heat flow interpretation technique, assumptions and applications.

J. Geophys. Res. 85: 7206-7214.

SWANBERG, C. A., P. MORGAN & F. K. BOULOS 1983. Geothermal potential of Egypt. *Tectonophysics* 96: 77-94.

SWANBERG, C. A., P. MORGAN & F. K. BOULOS 1984. Geochemistry of the groundwaters of Egypt. *Ann. Geol. Surv. Egypt* 14: 127-150.

SWANBERG, C. A., P. MORGAN, S. F. HENNIN, P. H. DAGGETT, Y. S. MELEK & A. A. EL SHERIF 1977. Preliminary report on the thermal springs of Egypt. *Proc. Intl. Congr. Thermal Waters, Geothermal Energy and Volcanism of the Mediterranean Area (1976)*, Nat. Tech. Univ., Athens 2: 540-554.

SWARTZ, D. H. & D. D. ARDEN 1960. Geologic history of Red Sea area. *Bull. Am. Assoc. Petrol. Geol.* 44: 1621-1637.

SYKES, L. R. 7 M. LANDSMAN 1964. The seismicity of east Africa, the Gulf of Aden and the Arabian and Red Seas. *Bull. Seismol. Soc. Am.* 54: 1927-1940.

SYLVESTER, A. G. & R. R. SMITH 1976. Tectonic transpression and basement-controlled deformation in San Andreas fault zone, Salton Trough, California. *Bull. Am. Assoc. Petrol. Geol.* 60: 1625-1640.

TAHER, M. 1976. *Matruh basin: a fossil submarine channel.* Internal Report, ARCO Oil Co. (unpublished).

TAKLA, M. A., A. M. NOWEIR & M. K. AKAAD 1977. Petrogenetic significance of opaque minerals contained in gabbroic rocks. *Proc. Egypt. Acad. Sci.* 30: 183-189.

TAKLA, M. A., M. A. EL SHARKAWI & F. F. BASTA 1982. Petrology of the basement rocks of Gebel Mohagara-Ghadir area, Eastern Desert, Egypt. *Ann. Geol. Surv. Egypt* 12: 121-140.

TANNER, L. 1978. Embrithropoda. In: V. J. Maglio & H. B. S. Cooke (eds), *Evolution of African mammals.* Harvard Univ. Press: 279-283.

TARABILI, E. E. 1966. General outline of epeirogenesis and sedimentation in region between Safaga and Quseir and southern Wadi Qena area, Eastern Desert, Egypt. *Bull. Am. Assoc. Petrol. Geol.* 50: 1890-1898.

TARABILI, E. E. 1969. Paleogeography, paleoecology and genesis of the phosphate sediments in Qusseir-Safaga area, UAR. *Econ. Geol.* 64: 172-182.

TARABILI, E. E. & N. ADAWY 1972. Geologic history of Nukhul-Baba area, Gulf of Suez, Sinai, Egypt. *Bull. Am. Assoc. Petrol. Geol.* 56: 882-902.

TATOR, B. A. 1960. Photointerpretation in geology. In: *Am. Soc. Photogramm.: Manual of photographic interpretation*: 78-117.

TAYLOR, P. & B. L. JONES 1982. Northwest delta region, Tertiary studies of the Beheira area. *6th Petrol. Explor. Seminar*, EGPC, Cairo.

TAYLOR, P. 1984. The key to prospect evaluation in Medoil's Matruh concession. *Proc. 7th Petrol. Explor. Seminar*, EGPC, Cairo: 288 (Abstr.).

TAYLOR, S. R. & S. M. MCLENNAN 1981. The rare earth element evidence in Precambrian sedimentary rocks: implications for crustal evolution. In: A. Kröner (ed.), *Precambrian plate tectonics*, Elsevier: 527-548.

TCHALENKO, J. S. 1970. Similarities between shear zones of different magnitudes. *Bull. Geol. Soc. Am.* 81: 1625-1640.

TEWFIC, R. 1975. *Geothermal gradients in the Gulf of Suez.* Internal Report, Gulf of Suez Oil Co., Cairo.

TEWFIK, N. & M. AYYAD 1982. Petroleum exploration in the Red Sea shelf of Egypt. *6th Petrol. Explor. Seminar, EGPC*, Cairo.

TEWFIK, N. & H. BURROUGH 1976. Overview of exploration techniques, Ras Banas area of Red Sea, Egypt. *3rd Petrol. Explor. Seminar, EGPC*, Cairo.

THIELE, J., F. GRAMANN & H. KLEINSORGE 1970. Zur Geologie zwischen dem Nordland der östlichen Kattara senke und der Mittelmeerkuste (Ägypten), westliche Wüste). *Geol. Jb.* 88: 321-354.

THISSE, Y., P. GUENNOC, G. POUIT & A. NAWAB 1983. The Red Sea: a natural geodynamic and metallogenic laboratory. *Episodes* 3: 3-9.

THORWEIHE, U. 1982. Hydrogeologie des Dakhla-Beckens, Ägypten. *Berl. Geowiss. Abh.* 38(A): 1-53.

THORWEIHE, U., M. SCHNEIDER & C. SONNTAG 1984. Aspects of hydrology in southern Egypt. *Berl. Geowiss. Abh.* 50(A): 209-216.

THUNNELL, R. C., D. F. WILLIAMS & J. P. KENNETT 1977. Late Quaternary paleoclimatology, stratigraphy and sapropel history in eastern Mediterranean deep-sea sediments. *Marine Micropaleontology* 2: 371-388.

TOPPOZADA, T. R., F. K. BOULOS, S. F. HENNIN, A. A. EL SHERIF, A. A. EL SAYED, N. Z. BASTA, F. A. SHATIYA, Y. S. MELEK, C. H. CRAMER & D. L. PARK 1984. Seismicity near Aswan High Dam, Egypt, following the November 1981 earthquake. *Ann. Geol. Surv. Egypt* 14: 107-126.

TOSSON, S. 1954. The Rennebaum volcano in Egypt. *Bull. Volcnol.*, ser. 2, 15: 99-108.

TOSSON, S. 1963. Note on El Ghorbaniat gypsum deposit near Alexandria. *Egypt. J. Geol.* 7: 71-72.

TOSSON, S. & N. A. SAAD 1972. Origin of the iron ore of Umm Greifat area, Egypt. *Chem. Erde* 9: 195-202.

TRÖGER, U. 1984. The oil shale potential of Egypt. *Berl. Geowiss. Abh.* 50(A): 375-380.

TRUESDELL, A. H. 1975. Geochemical techniques in exploration: summary of section III. *Proc. 2nd UN Symp. Development and Use of Geothermal Resources*, San Francisco, California (1975), US Government Printing Office, Washington DC: LIII-LXXIX.

TURNER, F. J. 1981. *Metamorphic petrology.* McGraw Hill, New York, 2nd ed., 524pp.

UNDERWOOD, J. R. & E. P. FISK 1980. Meteorite impact structures, southeast Libya. In: M. Salem & M. T. Busrewil (eds), *The Geology of Libya*, Academic Press 3: 893-900.

UNITED NATIONS 1973. *Assessment of the mineral potential of the Aswan region – Technical report, follow-up geophysical survey 1968-1972 for the Government of Egypt.* DP/SF/UN/114, New York.

UNITED NATIONS 1974. *Assessment of the mineral potential of the Aswan region – Technical report, geochemical operations 1968-1972 for the Government of Egypt.* DP/SF/UN/114, New York.

URBAN, L. L., L. V. MOORE & M. L. ALLEN 1976. Palynology, thermal alteration and source rock potential of three wells from Alamein area, Western Desert, Egypt. *5th Petrol. Explor. Seminar, EGPC.*

VACHETTE, M. 1974. Repartition des ages de biotites au strontium en Afrique. *C.R. 2ème Reun. Ann. Sci. Terre* (Nancy), 378pp.

VAIL, J. R. 1976. Outline of the geochronology and tectonic units of the basement complex of northeast Africa. *Proc. Roy. Soc. London* 350(A): 127-141.

VAIL, J. R. 1976. Location and geochronology of igneous ring complexes and related rocks in northeast Africa. *Geol. Jb.* 20(B): 97-114.

VAIL, J. R. 1978. Outline of the geology and mineral deposits of the Democratic Republic of the Sudan and adjacent areas. *Overseas Geol. and Mineral Resources* 49: 1-66.

VAIL, J. R. 1983. Pan African crustal accretion in northeast Africa. *J. Afr. Earth Sci.* 1: 285-294.

VAIL, J. R. 1985. Pan African (late Precambrian) tectonic terrains and the reconstruction of the Arabian shield. *Geology* 13: 839-842.

VAIL, P. R., R. M. MITCHUM, JR. & S. THOMPSON 1977. Seismic stratigraphy and global changes of sea level, Part 4: Global cycles of relative changes of sea level. In: *Seismic stratigraphy – application to hydrocarbon exploration, Am. Assoc. Petrol. Geol.*, Mem. 26: 83-97.

VAMI (The All Union Scientific and Design Institute of Aluminium, Magnesium and Electrode Industry, Leningrad, USSR) 1967. *Report on the technological testing of the nepheline syenite rock of the UAR for the purpose of determining the possibility of its complex processing for alumina and other products*. Report prepared for the Geol. Surv. Egypt.

VAN DER PLOEG, P. 1953. Egypt. In: V. C. Illing (ed.), *The World's oilfields: The Eastern Hemisphere*, The Science of Petroleum, Oxford Univ. Press 6(1): 151-157.

VAN HOUTEN, F. B. 1980. Latest Jurassic-Cretaceous regressive facies, northeast African craton. *Bull. Am. Assoc. Petrol. Geol.* 64: 857-868.

VAN HOUTEN, F. B. 1983. Sirte basin, north central Libya: Cretaceous rifting above fixed mantle hotspot? *Geology* 11: 115-118.

VAN HOUTEN, F. B. & D. P. BHATTACHARYYA 1979. Late Cretaceous Nubia Formation at Aswan, Southeastern Desert, Egypt. *Ann. Geol. Surv. Egypt* 9: 408-419.

VAN HOUTEN, F. B., D. P. BHATTACHARYYA & S. E. I. MANSOUR 1984. Cretaceous Nubia Formation and correlative deposits, eastern Egypt: Major regressive-transgressive complex. *Bull. Geol. Soc. Am.* 95: 397-405.

VAUDOIS-MIEJA, N. & A. LEJAL-NICOL 1987. Paleocarpologie africaine: apparition dès l'Aptien d'un palmier (*Hyphaeneocarpon aegyptiacum*, n.sp.). *C.R. Acad. Sci., Paris* 304, ser. II(6): 233-238.

VERMEERSCH, P. M. 1970. L'Elkabien, une nouvelle industrie epipalolithique d'Egypte. *Chronique d'Egypte* 45: 45-67.

VILJOEN, R. P., M. J. VILJOEN, J. GROOTENBOER & T. G. LONGSHAW 1975. ERTS-1 imagery: applications in geology and mineral exploration. *Mineral Sci. Eng.* 7(2): 132-168 (Johannesburg).

VINK, G. E., W. J. MORGAN & P. R. VOGT 1985. The earth's hot spots. *Scientific American*.

VIOTTI, C. & G. EL DEMERDASH 1969. Studies in Eocene sediments of Wadi Nukhul area, eastern coast of Gulf of Suez. *Proc. 3rd Afr. Micropaleontol. Colloq.*, Cairo: 403-423.

VIOTTI, C. & A. MANSOUR 1968. Tertiary planktonic foraminiferal zonation from the Nile delta, Egypt, Part 1: Miocene planktonic foraminiferal zonation. *Proc. 3rd Afr. Micropaleontol. Colloq.*, Cairo: 425-432.

VONDRA, C. F. 1974. Upper Eocene transitional and near-shore marine Qasr El Sagha Formation, Fayum depression, Egypt. *Ann. Geol. Surv. Egypt* 4: 79-94.

VON KNORRING, O. & J. M. ROOKE 1973. Trace element content of some alkali granites and gneisses from Egypt and Uganda. *17th Annual Report, Res. Inst. Afr. Geol.* Leeds Univ.: 34-35.

WALTHER, J. K., 1888. Die Korallenriffe der Sinaihalbinsel. *Abh. Sachs. Akad. Wiss. Leipzig. Math.-Naturwiss. Kl.* 14: 439-505.

WALTHER, J. K., 1890. Über eine Kohlenkalk Fauna aus der ägyptischen-Arabischen Wüste. *Z. deutsch. Geol. Ges.* 42:419-449.

WALTHER, J. K. 1900. *Das Gesetz der Wüstenbildung in Gegenwart und Vorzeit*. Reimer, Berlin, 157pp.

WANNER, J. 1902. Die Fauna der obersten Kreide der Libyschen Wüste. *Palaeontographica* 30: 91-152.

WARD, W. C. & K. C. McDONALD 1979. Nubia Formation of central Eastern Desert, Egypt – major subdivisions and depositional setting. *Bull. Am. Assoc. Petrol. Geol.* 63: 975-983.

WARD, W. C., K. C. McDONALD & S. E. I. MANSOUR 1979. The Nubia Formation of the Quseir-Safaga area, Egypt. *Ann. Geol. Surv. Egypt* 9: 420-431.

WARING, G. A. 1976. Thermal springs of the US and other countries of the world, a summary. *US Geol. Surv.*, Prof. Paper 492, 383pp.

WASFI, S. 1969. Miocene planktonic foraminiferal zones from the Gulf of Suez, Egypt. *Proc. 3rd Afr. Micropaleontol. Colloq.*, Cairo: 461-474.

WASFI, S. & G. AZAZI 1979. Stratigraphy of the northern Gulf of Suez. *Ann. Geol. Surv. Egypt* 9: 308-332.

WASFI, S. & H. HATABA 1981. Observations on the genus *Nezzazata* Omara and its significance to the Cenomanian-Turonian boundary in the Gulf of Suez. *2nd Intl. Symp. Benthonic Foraminifera* (Pau): 597-603.

WASFI, S., A. EL SWEIFY & W. ABDELMALIK 1982. Carboniferous-Jurassic microfauna from the northern part of the Gulf of Suez, Egypt. *Proc. 8th Afr. Micropaleontol. Colloq., Paris, Cah. Micropaleontol.* 1: 89-121.

WASSEF, A. S. 1977. On the results of the geological investigations and ore reserve calculations of Abu Tartur phosphate deposit. *Ann. Geol. Surv. Egypt* 7: 1-60.

WEBBER, P. J. 1961. *Phlebopteris branneri* from the Western Desert of Egypt. *Ann. Mag. Nat. Hist.* ser. 13, 4(37): 7-9.

WEBSTER, D. J. & N. RITSON 1982. Post-Eocene stratigraphy of the Suez rift in SW Sinai. *6th Petrol. Explor. Seminar, EGPC*, Cairo.

WEEKS, L. G. 1952. Factors of sedimentary basin development that control oil occurrence. *Bull. Am. Assoc. Petrol. Geol.* 36: 2071-2124.

WEISSBROD, T. 1969. The Paleozoic of Israel and adjacent countries: Part I, The subsurface Paleozoic stratigraphy of S. Israel. *Bull. Geol. Surv. Isr.* 47: 35pp.

WEISSBROD, T. 1969. The Paleozoic of Israel and adjacent countries: Part II, The Paleozoic outcrops in southwestern Sinai and their correlation with those of southern Israel. *Bull. Geol. Surv. Isr.* 48: 1-32.

WEISSBROD, T. 1970. 'Nubian Sandstone', discussion. *Bull. Am. Assoc. Petrol. Geol.* 54: 526-529.

WEISSBROD, T. 1976. The Permian in the Near East. In: H. Flake (ed.), *The continental Permian in central, west and south Europe*, D. Reidel Publishing Co.: 200-214.

WENDORF, F. & R. SCHILD 1976. *Prehistory of the Nile Valley*. Academic Press, 404pp.

WENDORF, F. & R. SCHILD 1980. *Prehistory of the eastern*

Sahara. Academic Press, 414pp.

WENDORF, F. & R. SCHILD 1986. *The prehistory of Wadi Kubbaniya*. Southern Methodist Univ. Press, 85pp.

WENDORF, F., R. SCHILD, R. SAID, V. C. HAYNES, A. GAUTIER & M. KOBUSIEWICZ 1976. The prehistory of the Egyptian Sahara. *Science* 193: 103-114.

WENDORF, F. ET AL. 1977. Late Pleistocene and recent climatic changes in the eastern Sahara. *Geograph. J.* 143: 211-234.

WERNICKE, B. & B. C. BURCHFIEL 1982. Models of extensional tectonics. *J. Struct. Geol.* 4: 105-115.

WHITTINGTON, H. B. 1982. Exoskeleton, moult stage, appendage with habits of the middle Cambrian tribolite *Oleonoides serratus*. *Palaeontology* 23: 171-204.

WILCOX, R. E., T. P. HARDING & D. R. SEELY 1973. Basic wrench tectonics. *Bull. Am. Assoc. Petrol. Geol.* 57: 74-96.

WIGHT, A. W. R. 1980. Paleogene vertebrate fauna and regressive sediments of Dur at Talhah, southern Sirt basin, Libya. In: M. J. Salem & M. T. Busrewil (eds), *The geology of Libya*, Academic Press 1: 309-325.

WILLIAMS, G. A. & J. O. SMALL 1984. A study of the Oligo-Miocene basalts in the Western Desert. *Proc. 7th Petrol. Explor. Seminar, EGPC*, Cairo: 252-268.

WILLIAMS, R. S. (ed.) 1983. Geological applications. In: R. N. Colwell (ed.), *Manual of remote sensing*, 2nd ed., Falls Church (ASP): 1667.

WILLIS, B. 1921. Aerial observation of earthquake rifts. *Science* n.ser., 54: 266.

WINDLEY, B. F. 1984. *The evolving continents*. Wiley, 2nd ed., 399pp.

WING, S. L. & B. H. TIFFNEY 1982. A paleotropical flora from the Oligocene Jebel Qatrani Formation of northern Egypt: a preliminary report. *Bot. Soc. Am., Misc. Ser. Publ.* 162: 67.

WINKLER, H. G. F. 1979. *Petrogenesis of metamorphic rocks*. Springer, 348pp.

WOENIG, F. 1897. *Die Pflanzen in alten Aegypten*. A. Aufl., Leipzig.

WOOD, A. E. 1968. Early Cenozoic mammalian faunas, Fayum Province, Egypt, Part II: The African Oligocene Rodentia. *Bull. Peabody Mus. Nat. Hist.* 28: 23-105.

WOODWARD CLYDE CONSULTANTS 1985. *Earthquake activity and dam stability for the Aswan High Dam, Egypt*. High and Aswan Dams Ministry of Irrigation, Cairo, 2.

WOOLARD, G. 1951. *Gravity base points in Egypt*. Internal Report no. 1012. Gen. Petrol. Co., Cairo, 21pp.

WRIGHT, J. B. 1969. A simple alkalinity ratio and its application to question of non-orogenic granite gneiss. *Geol. Mag.* 106: 370-394.

WYCISK, P. 1984. Depositional environment of Mesozoic strata from northwestern Sudan. *Berl. Geowiss. Abh.* 50(A): 81-97.

WYCISK, P. 1987. Contributions to the subsurface geology of the Misaha trough and the southern Dakhla basin (S. Egypt/N. Sudan). *Berl. Geowiss. Abh.* 75(A), 1: 137-150.

WYLLIE, P. J. 1983. Experimental studies on biotite and muscovite granite and some crustal magmatic sources. In: M. P. Atherton & C. D. Gribble (eds), *Migmatites, melting and metamorphism*, Shiva Publ., Cheshire: 12-26.

WYLLIE, P. J. 1983. Experimental and thermal constraints on the deep-seated parentage of some granitoid magmas in

subduction zones. In: M. P. Atherton & C. D. Gribble (eds), *Migmatites, melting and metamorphism*, Shiva Publ., Cheshier: 37-51.

YALLOUZE, M. & G. KNETSCH 1954. Linear structures in and around the Nile basin. *Bull. Soc. Geograph. Egypte* 27: 153-207.

YEHIA, M. A. 1985. Geologic structures of the Giza pyramids plateau. *Middle East Res. Center, Ain Shams Univ., Sci. Res. Ser.* 5: 100-120.

YORK, D. 1969. Least squares fitting of a straight line with correlated errors. *Earth Planet. Sci. Lett.* 5: 320-324.

YOUSSEF, E. A. A. 1986. Depositional diagenetic models of some Miocene evaporites in the Red Sea coast, Egypt. *Sedimentary Geol.* 48: 17-36.

YOUSSEF, M.I. 1954. Stratigraphy of Gebel Oweina section. *Bull. Inst. Desert Eqypte* 7(2): 35-54.

YOUSSEF, M. I. 1957. Upper Cretaceous rocks in Kosseir area. *Bull. Inst. Desert Egypte* 7(2): 35-54.

YOUSSEF, M. I. 1968. Structural pattern of Egypt and its interpretation. *Bull. Am. Assoc. Petrol. Geol.* 52: 601-614.

YOUSSEF, M. I. & W. ABDEL AZIZ 1971. Biostratigraphy of the upper Cretaceous-lower Tertiary in Farafra Oasis, Libyan Desert, Egypt. In L. C. Gray (ed.), *Symp. on the Geology of Libya, Fac. Sci. Univ. Libya*: 227-249.

YOUSSEF, M. I., M. A. BASSIOUNI & O. H. CHERIF 1971. Some stratigraphic and tectonic aspects of the Miocene in the northeastern part of the Eastern Desert, Egypt. *Bull. Inst. Egypte* 52: 119-158.

YOUSSEF, M. I., O. H. CHERIF, M. BOUKHARY & A. MOHAMED 1984. Geological studies on the Sakkara area, Egypt. *N. Jb. Geol. Palaeontol. Abh.* 168: 125-144.

YOUSSEF, M. S. & A. S. EL KHAWAGA 1966. *Geophysical and geochemical investigations at Umm Gheig area, Eastern Desert*. Geol. Surv. Egypt, Paper 33, 12pp.

ZA'ATOUT, M. A. 1956. *The dolomite and dolomitic rocks of Gebel Ataqa (eastern cliff, Suez sheet)*. Geol. Surv. Egypt, 27pp.

ZAGHLOUL, Z. M., S. F. ANDRAWIS & S. N. AYYAD 1979. New contribution to the stratigraphy of the Tertiary sediments of the Kafr El Dawar well-1, northern Nile delta, Egypt. *Ann. Geol. Surv. Egypt* 9: 292-307.

ZAGHLOUL, Z. M., A. EL SHAHAT & A. IBRAHIM 1983. On the discovery of Paleozoic trace fossil *Bifungites* in the Nubia Sandstone facies of Aswan. *Egypt. J. Geol.* 27: 65-72.

ZAGHLOUL, Z. M., M. A. ESSAWY & M. M. SOLIMAN 1976. Geochemistry of some younger granite masses, south Eastern Desert, Egypt. *J. Univ. Kuwait (Sci.)* 3: 231-242.

ZAGHLOUL, Z. M. & B. MABROUK 1964. On uranium in Dakhla and Mahamid phosphate deposits. *Egypt. J. Geol.* 8: 70-86.

ZAK, I. 1964. *The Triassic of Arayif en Naqa, Sinai*. Geol. Surv. Isr., Geochem. Div., Report 1/64, 28pp.

ZAK, I. 1968. *The geological map of Israel, Makhtesh Ramon, Har Gavanim* 1:50,000. Geol. Surv. Isr.

ZAKI, M. 1969. *A course in the theory of errors*. Fac. Eng., Cairo Univ., 74pp.

ZDANSKY, O. 1934. The occurrence of *Mosasaurus* in Egypt and in Africa in general. *Bull. Inst. Egypte* 22: 83-94.

ZIKO, A. 1985. Eocene bryozoa from Egypt: a paleontological and paleoecological study. *Tubinger Mikropalaeontol. Mitt.* 4: 1-183.

ZITTEL, A. K. 1883. Beiträge zur Geologie und Paläontologie der Libyschen Wüste und der angrenzenden Gebiete von Aegypten. *Palaeontographica* 30, 3.F., 1, 147pp, 2, 237pp.

ZOBRIST, A. L., N. A. BRYANT & R. McLEOD 1983. Technology for large digital mosaics of Landsat data. *Photogramm. Eng. & Rem. Sens.* 49: 1325-1335.

Index of subjects

Index of formations

For additional information concerning formation names consult M. Hermina, E. Klitzsch & F. List (eds.) 1989. *Stratigraphic Lexicon & Explanatory Notes to the Geological Map of Egypt 1:500 000*. Conoco Inc., Cairo, Egypt.

Index of oilfields, gasfields and oil wells

* Oilfield
+ Gasfield

List of contributors

ANDRAWIS, SAMIR F. *Consultant, 6 Ibn Ias Street, Heliopolis, Cairo, Egypt.*

AYOUTY, M.K. *Consultant, 8 Rustum Street, Garden City, Cairo, Egypt.*

BOULOS, FOUAD K. *Geological Survey of Egypt, 3 Salah Salem Street, Cairo, Egypt.*

CHERIF, O.H., *Geology Department, Faculty of Science, Ain Shams University, Abbassia, Cairo, Egypt.*

EL GABY, SAMIR, *Geology Department, Faculty of Science, Assiut University, Assiut, Egypt.*

EL SHARKAWI, MOHAMED A. *Geology Department, Faculty of Science, Cairo University, Gizeh, Egypt.*

GROESCHKE, M. *Institut für Geologie und Palaeontologie, Technische Universität Berlin, 1 Ernst Reuter Platz, D 1000 Berlin 12, Germany.*

HANTAR, GAMAL, *Consultant, 109 Haroun El Rashid Street, Heliopolis, Cairo, Egypt.*

HARMS, J.C. *Harms and Brady Consultants, P.O. Box 406, Littleton, Colorado 90160, USA.*

HASHAD, AHMED H. *Nuclear Materials Authority, Cairo; currently: Faculty of Earth Sciences, King Abdul Aziz University, P.O. Box 1744, Jeddah, Saudi Arabia.*

HASSAN, MAMDOUH A. *Nuclear Material Authority, Cairo; currently: Faculty of Earth Sciences, King Abdul Aziz University, P.O. Box 1744, Jeddah, Saudi Arabia.*

HERRMANN-DEGEN, W. *Gesellschaft für geologische Baugrunderkundung und Grundwasserhygiene mbH, Saatwinkler Damm 24-26, D-1000 Berlin 13, Germany.*

HERMINA, MAURICE H. *Conoco Egypt, P.O. Box 16, Maadi, Cairo, Egypt.*

HUSSEIN, ABDEL AZIZ A. *Geological Survey of Egypt, 3 Salah Salem Street, Cairo, Egypt.*

JENKINS, DAVID A. *Khalda Oil Co., P.O. Box 16, Maadi, Cairo, Egypt.*

KAMEL, HUSSEIN, *General Petroleum Co., Nasr City, Cairo, Egypt.*

KEBEASY, RASHAD M. *Helwan Institute of Astronomy and Geophysics, Helwan, Egypt.*

KERDANY, M.T. *Conoco Egypt, P.O. Box 16, Maadi, Cairo, Egypt.*

KHALIL, MOSBAH, *Gulf of Suez Petroleum Co., Maadi, Cairo, Egypt.*

KLITZSCH, E. *Institut für Geologie und Palaeontologie, Technische Universität Berlin, 1 Ernst Reuter Platz, D-1000 Berlin 12, Germany.*

LEJAL-NICOL, ANNIE, *Université Pierre et Marie Curie, Paleobotanique fondamentale et appliquée, 12 Rue Cuvier, 7005 Paris, France.*

LIST, FRANZ K. *Free University of Berlin, Malteserstrasse 74-100, D-1000 Berlin 46, Germany.*

MEISSNER, BERND, *College of Engineering (TFH), Luxemburgerstrasse 10, D-1000 Berlin 65, Germany.*

MENEISY, MOHAMED YOUSRI, *Geology Department, Faculty of Science, Ain Shams University, Abbassia, Cairo, Egypt.*

MESHREF, WAFIK M. *Gulf of Suez Petroleum Co., Maadi, Cairo, Egypt.*

MORGAN, PAUL, *Geology Department, P.O. Box 6030, Northern Arizona University, Flagstaff, Arizona 86011, USA.*

MOUSTAFA, ADEL R. *Geology Department, Faculty of Science, Ain Shams University, Abbassia, Cairo, Egypt.*

PÖHLMANN, GERHARD, *Consultant, Cimbernstrasse 11i, D-1000 Berlin 38, Germany.*

RASMUSSEN, D. TAB, *Department of Anthropology, UCLA, 405 Hilgard Ave., Los Angeles, California 90024, USA.*

RICHTER, AXEL, *Free University of Berlin, Malteserstrasse 74-100, D-1000 Berlin 46, Germany.*

SAID, RUSHDI, *Intergeosearch, Inc., 3801 Mill Creek Drive, Annandale, Virginia 22003, USA.*

SCHANDELMEIER, HEINZ, *Technische Universität Berlin, SFB 69, Ackerstrasse 71-76, D-1000 Berlin 65, Germany.*

SEILACHER, A. *Institut für Geologie, 74 Tubingen*

Universität, Tubingen, Germany.

SIMONS, ELWYN L. *Duke University, Center for the Study of Primate Biology and History, 3705 Erwin Road, Durham, North Carolina 27705, USA.*

TEHRANI, RESA, *Free University of Berlin, Malteserstrasse 74-100, D-1000 Berlin 46, Germany.*

THORWEIHE, ULF, *Technische Universität Berlin, SFB 69, Ackerstrasse 71-76, D-1000 Berlin 65, Germany.*

WRAY, J.L. *Consultant, 1537 West Briarwood Ave., Littleton, Colorado 90160, USA.*